FoodPix/Karl Petzke

Introduction to General, Organic, and Biochemistry

SEVENTH EDITION

Frederick A. Bettelheim
Adelphi University

William H. Brown
Beloit College

Jerry March
Adelphi University

THOMSON
★
BROOKS/COLE

Australia • Canada • Mexico • Singapore
Spain • United Kingdom • United States

DEDICATION: *This book is dedicated to the memory of John Vondeling, visionary editor and friend.*

THOMSON BROOKS/COLE

CHEMISTRY EDITOR: John Holdcroft
DEVELOPMENT EDITOR: Sandra Kiselica
ASSISTANT EDITOR: Alyssa White
EDITORIAL ASSISTANT: Lauren Raike
TECHNOLOGY PROJECT MANAGER: Ericka Yeoman-Saler
MARKETING MANAGER: Jeffrey L. Ward
MARKETING ASSISTANT: Mona Weltmer
ADVERTISING PROJECT MANAGER: Stacey Purviance
PROJECT MANAGER, EDITORIAL PRODUCTION: Lisa Weber
PRINT/MEDIA BUYER: Barbara Britton
PERMISSIONS EDITOR: Joohee Lee

PRODUCTION SERVICE: Thompson Steele, Inc.
TEXT DESIGNER: Baugher Design Inc.
PHOTO RESEARCHER: Jane Sanders
COPY EDITOR: Thompson Steele, Inc.
ILLUSTRATOR: J/B Woolsey and Thompson Steele, Inc.
COVER DESIGNER: Baugher Design Inc.
COVER IMAGE: ©FoodPix/Karl Petzke
COVER PRINTER: Transcontinental Printing/Interglobe
COMPOSITOR: Thompson Steele, Inc.
PRINTER: Transcontinental Printing/Interglobe

COPYRIGHT © 2004 Brooks/Cole, a division of Thomson Learning, Inc. Thomson Learning™ is a trademark used herein under license.

ALL RIGHTS RESERVED. No part of this work covered by the copyright hereon may be reproduced or used in any form or by any means—graphic, electronic, or mechanical, including but not limited to photocopying, recording, taping, Web distribution, information networks, or information storage and retrieval systems—without the written permission of the publisher.

Printed in Canada
1 2 3 4 5 6 7 07 06 05 04 03

For more information about our products, contact us at:
Thomson Learning Academic Resource Center
1-800-423-0563
For permission to use material from this text, contact us by:
Phone: 1-800-730-2214
Fax: 1-800-730-2215
Web: http://www.thomsonrights.com

ExamView® and ExamView Pro® are registered trademarks of FSCreations, Inc. Windows is a registered trademark of the Microsoft Corporation used herein under license. Macintosh and Power Macintosh are registered trademarks of Apple Computer, Inc. Used herein under license.

COPYRIGHT 2004 Thomson Learning, Inc. All Rights Reserved. Thomson Learning WebTutor™ is a trademark of Thomson Learning, Inc.

Library of Congress Control Number: 2003100305

Student Edition with InfoTrac College Edition: ISBN 0-534-40176-7
Student Edition without InfoTrac College Edition: ISBN-0-534-40189-9
Instructor's Edition: ISBN 0-534-40177-5
International Student Edition: ISBN 0-534-40211-9 (not for sale in the United States)

Brooks/Cole—Thomson Learning
10 Davis Drive
Belmont, CA 94002
USA

Asia
Thomson Learning
5 Shenton Way #01-01
UIC Building
Singapore 068808

Australia/New Zealand
Thomson Learning
102 Dodds Street
Southbank, Victoria 3006
Australia

Canada
Nelson
1120 Birchmount Road
Toronto, Ontario M1K 5G4
Canada

Europe/Middle East/Africa
Thomson Learning
High Holborn House
50/51 Bedford Row
London WC1R 4LR
United Kingdom

Latin America
Thomson Learning
Seneca, 53
Colonia Polanco
11560 Mexico D.F.
Mexico

Spain/Portugal
Paraninfo
Calle/Magallanes, 25
28015 Madrid, Spain

ABOUT THE AUTHORS

FREDERICK A. BETTELHEIM is a Distinguished University Research Professor at Adelphi University, and a Visiting Scientist at the National Eye Institute NIH. He has coauthored every edition of *Introduction to General, Organic & Biochemistry* and several laboratory manuals, including *Laboratory Experiments for General, Organic & Biochemistry* and *Experiments for Introduction to Organic Chemistry*. He is the author of *Experimental Physical Chemistry* and coauthor of numerous monographs and research articles. Professor Bettelheim received his Ph.D. from the University of California, Davis, and his areas of specialization have included the biochemistry of proteins and carbohydrates and the physical chemistry of polymers. He was Fulbright Professor at the Weizman Institute, Israel, and Visiting Professor at the University of Uppsala, Sweden, at Technion, Israel, and at the University of Florida. He was keynote lecturer at the 16th International Conference on Chemical Education, August 2000, in Budapest, Hungary, where his lecture topic was "Modern Trends in Teaching Chemistry to Nurses and Other Health-Related Professionals."

WILLIAM H. BROWN, a new coauthor for the sixth edition of *Introduction to General, Organic and Biochemistry,* is Professor of Chemistry at Beloit College, where he has twice been named Teacher of the Year. He is also the author of two best-selling undergraduate texts, *Introduction to Organic Chemistry,* sixth edition, and *Organic Chemistry,* third edition. His regular teaching responsibilities include organic chemistry, advanced organic chemistry, and special topics in pharmacology and drug synthesis. He was the leader of a team revising the organic chemistry material appearing in the upcoming *Encyclopaedia Britannica CD-ROM.* Professor Brown received his Ph.D. from Columbia University under the direction of Gilbert Stork and did postdoctoral work at the California Institute of Technology and the University of Arizona.

JERRY MARCH (d. 1997) was the coauthor of *Introduction to General, Organic & Biochemistry* for its first five editions. He was the sole author of the best-selling *Advanced Organic Chemistry* text, now in its fifth edition, which has been translated into many languages, including Russian and Japanese. He was a member of the Physical Organic Chemistry Commission of the International Union of Pure and Applied Chemistry, and as such was instrumental in the development of several nomenclature systems. Professor March received his Ph.D. at Pennsylvania State University and was Professor of Chemistry at Adelphi University, specializing in organic chemistry.

BRIEF CONTENTS

PART 1 GENERAL CHEMISTRY 1

CHAPTER
- 1 Matter, Energy, and Measurement 2
- 2 Atoms 27
- 3 Chemical Bonds 58
- 4 Chemical Reactions 89
- 5 Gases, Liquids, and Solids 119
- 6 Solutions and Colloids 153
- 7 Reaction Rates and Chemical Equilibrium 184
- 8 Acids and Bases 210
- 9 Nuclear Chemistry 242

PART 2 ORGANIC CHEMISTRY 269

CHAPTER
- 10 Organic Chemistry 270
- 11 Alkanes 286
- 12 Alkenes 314
- 13 Benzene and Its Derivatives 343
- 14 Alcohols, Ethers, and Thiols 359
- 15 Chirality: The Handedness of Molecules 382
- 16 Amines 403
- 17 Aldehydes and Ketones 419
- 18 Carboxylic Acids, Anhydrides, Esters, and Amides 438

PART 3 BIOCHEMISTRY 465

CHAPTER
- 19 Carbohydrates 466
- 20 Lipids 494
- 21 Proteins 525
- 22 Enzymes 553
- 23 Chemical Communications: Neurotransmitters and Hormones 572
- 24 Nucleotides, Nucleic Acids, and Heredity 596
- 25 Gene Expression and Protein Synthesis 620
- 26 Bioenergetics: How the Body Converts Food to Energy 645
- 27 Specific Catabolic Pathways: Carbohydrate, Lipid, and Protein Metabolism 666
- 28 Biosynthetic Pathways 691
- 29 Nutrition 704
- 30 Immunochemistry 722
- 31 Body Fluids 744

Appendix I Exponential Notation A-1
Appendix II Significant Figures A-5
Answers to In-Text and Odd-Numbered End-of-Chapter Problems A-8

CONTENTS

PART 1
GENERAL CHEMISTRY 1

CHAPTER 1
Matter, Energy, and Measurement 2

1.1 Introduction 2
1.2 The Scientific Method 4
1.3 Exponential Notation and Significant Figures 6
1.4 Measurements 6
1.5 Unit Conversions: The Factor-Label Method 11
1.6 The States of Matter 14
1.7 Density and Specific Gravity 15
1.8 Energy 18
1.9 Heat 19

Summary 22
Problems 23

CHEMICAL CONNECTIONS
1A Drug Dosage and Body Mass 10
1B Hypothermia and Hyperthermia 20
1C Cold Compresses 22

CHAPTER 2
Atoms 27

2.1 Introduction 27
2.2 Classification of Matter 28
2.3 Dalton's Atomic Theory 31
2.4 Inside an Atom 33
2.5 The Periodic Table 39
2.6 The Electron Configuration of Atoms 44
2.7 Electron Configuration and the Periodic Table 50
2.8 Ionization Energy of the Elements: A Periodic Property 52

Summary 53
Problems 54

CHEMICAL CONNECTIONS
2A Elements Necessary for Human Life 29
2B Abundance of Elements on Earth 32
2C Isotopic Abundance and Astrochemistry 38
2D Strontium-90 40
2E The Use of Metals as Historical Landmarks 41

CHAPTER 3
Chemical Bonds 58

3.1 Introduction 58
3.2 Atoms and Their Ions 58
3.3 Naming Ions 61
3.4 Formation of Chemical Bonds 63
3.5 Ionic Compounds 65
3.6 Naming Ionic Compounds 67
3.7 Molecular Compounds 70
3.8 Naming Binary Molecular Compounds 75
3.9 Bond Angles and Shapes of Molecules 77
3.10 Polar and Nonpolar Molecules 81

Summary 82
Problems 83

CHEMICAL CONNECTIONS
3A Coral Chemistry and Broken Bones 63
3B Ionic Compounds in Medicine 69
3C Nitric Oxide: Air Pollutant and Biological Messenger 76

CHAPTER 4
Chemical Reactions 89

4.1 Introduction 89
4.2 Formula Weight 89
4.3 The Mole 90
4.4 Chemical Equations 94
4.5 Mass Relationships in Chemical Reactions 98
4.6 Reactions Between Ions in Aqueous Solution 104
4.7 Oxidation–Reduction 107
4.8 Heat of Reaction 112

Summary 113
Problems 114

CHEMICAL CONNECTIONS
4A Solubility and Tooth Decay 107
4B Voltaic Cells 109
4C Oxidizing Antiseptics 111

CHAPTER 5
Gases, Liquids, and Solids 119

5.1 Introduction 119
5.2 Gas Pressure 121
5.3 Gas Laws 122

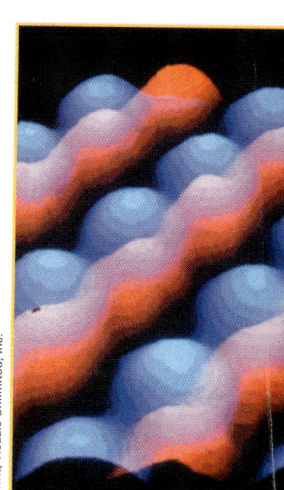

5.4 Avogadro's Law and the Ideal Gas Law   127
5.5 Dalton's Law of Partial Pressures   130
5.6 The Kinetic Molecular Theory   130
5.7 Forces of Attraction Between Molecules   131
5.8 Liquids   135
5.9 Vapor Pressure and Boiling Point   135
5.10 Solids   141
5.11 Phase Changes   143

Summary   147
Problems   148

CHEMICAL CONNECTIONS

5A Entropy: a Measure of Disorder   120
5B Breathing and Boyle's Law   123
5C Hyperbaric Medicine   129
5D Blood Pressure Measurement   136
5E Spray and Stretch—A New Use for Freons   137
5F The Densities of Ice and Water   140
5G Supercritical Carbon Dioxide   147

CHAPTER 6
Solutions and Colloids   153

6.1 Introduction   153
6.2 Types of Solutions   154
6.3 Characteristics of Solutions   154
6.4 Solubility   156
6.5 Concentration Units   159
6.6 Water as a Solvent   165
6.7 Colloids   170
6.8 Colligative Properties   172

Summary   178
Problems   179

CHEMICAL CONNECTIONS

6A Acid Rain   155
6B The Bends   158
6C Hydrates and Air Pollution: The Decay of Buildings and Monuments   167
6D Emulsions and Emulsifying Agents   172
6E Reverse Osmosis and Desalinization   176
6F Hemodialysis   178

CHAPTER 7
Reaction Rates and Chemical Equilibria   184

7.1 Introduction   184
7.2 Molecular Collisions   186
7.3 Activation Energy and Energy Diagrams   188
7.4 Factors Affecting Rates of Reaction   190
7.5 Reversible Reactions and Equilibrium   194
7.6 Equilibrium Constants   197
7.7 Equilibrium and Reaction Rates   200
7.8 Le Chatelier's Principle   200

Summary   206
Problems   206

CHEMICAL CONNECTIONS

7A Why High Fever Is Dangerous   192
7B The Effects of Lowering Body Temperature   193
7C Timed-Release Medication   194
7D Sunglasses and Le Chatelier's Principle   204
7E The Haber Process   205

CHAPTER 8
Acids and Bases   210

8.1 Introduction   210
8.2 Acid and Base Strength   212
8.3 Brønsted-Lowry Acids and Bases   214
8.4 The Position of Equilibrium in Acid–Base Reactions   216
8.5 Acid Ionization Constants   218
8.6 Some Properties of Acids and Bases   220
8.7 Self-Ionization of Water   223
8.8 pH and pOH   224
8.9 The pH of Aqueous Salt Solutions   226
8.10 Titration   228
8.11 Buffers   230
8.12 The Henderson-Hasselbalch Equation   234

Summary   236
Problems   237

CHEMICAL CONNECTIONS

8A Some Important Acids and Bases   213
8B Acid and Base Burns of the Cornea   217
8C Drugstore Antacids   221
8D Acidosis and Alkalosis   235

CHAPTER 9
Nuclear Chemistry   242

9.1 Introduction   242
9.2 What Happens When a Nucleus Emits Radioactivity?   245
9.3 Half-life   248
9.4 Characteristics of Radiation   251
9.5 Radiation Dosimetry and Human Health   253
9.6 Medical Uses of Radioactive Materials   256
9.7 Nuclear Fusion   261
9.8 Nuclear Fission and Atomic Energy   262

Summary   264
Problems   265

CHEMICAL CONNECTIONS

9A Three Kinds of Radiation   243
9B Radioactive Dating   249

9C The Indoor Radon Problem   255
9D How Radiation Damages Tissues—
    Free Radicals   255
9E CT, MRI, Ultrasound, and PET Scans   258
9F Energy from Mass   261
9G Radioactive Fallout from Nuclear Accidents   263

PART 2
ORGANIC CHEMISTRY   269

CHAPTER 10
Organic Chemistry   270

10.1 Introduction   270
10.2 Sources of Organic Compounds   272
10.3 Structure of Organic Compounds   274
10.4 Functional Groups   276

Summary   282
Problems   282

CHEMICAL CONNECTIONS
10A Taxol—A Story of Search and Discovery   273
10B Combinatorial Chemistry   274

CHAPTER 11
Alkanes   286

11.1 Introduction   286
11.2 Structure of Alkanes   287
11.3 Constitutional Isomerism in Alkanes   289
11.4 Nomenclature of Alkanes   291
11.5 Cycloalkanes   295
11.6 Shapes of Alkanes and Cycloalkanes   296
11.7 Cis-Trans Isomerism in Cycloalkanes   300
11.8 Physical Properties   302
11.9 Reactions of Alkanes   305
11.10 Important Haloalkanes   306
11.11 Sources of Alkanes   307

Summary   309
Key Reactions   309
Problems   309

CHEMICAL CONNECTIONS
11A The Poisonous Puffer Fish   299
11B The Environmental Impact of Freons   307
11C Octane Rating—What Those Numbers
     at the Pump Mean   308

CHAPTER 12
Alkenes   314

12.1 Introduction   314
12.2 Structure   315
12.3 Nomenclature   316
12.4 Physical Properties   323
12.5 Naturally Occurring Alkenes:
      The Terpenes   323
12.6 Addition Reactions of Alkenes   325
12.7 Polymerization of Ethylene and Substituted
      Ethylenes   332

Summary   337
Key Reactions   337
Problems   338

CHEMICAL CONNECTIONS
12A Ethylene, a Plant Growth Regulator   315
12B The Case of the Iowa and New York Strains of the
     European Corn Borer   320
12C Cis-Trans Isomerism in Vision   322
12D Recycling Plastics   336

CHAPTER 13
Benzene and Its Derivatives   343

13.1 Introduction   343
13.2 The Structure of Benzene   344
13.3 Nomenclature   345
13.4 Reactions of Benzene and Its Derivatives   349
13.5 Phenols   351

Summary   355
Key Reactions   355
Problems   356

CHEMICAL CONNECTIONS
13A DDT—A Boon and a Curse   347
13B Carcinogenic Polynuclear Aromatics
     and Smoking   348
13C Iodide Ion and Goiter   349
13D The Nitro Group in Explosives   350
13E FD & C No. 6 (a.k.a. Sunset Yellow)   351
13F Capsaicin, for Those Who Like It Hot   353

CHAPTER 14
Alcohols, Ethers, and Thiols   359

14.1 Introduction   359
14.2 Alcohols   360
14.3 Reactions of Alcohols   364
14.4 Ethers   370
14.5 Thiols   372
14.6 Important Alcohols   375

Summary   376
Key Reactions   377
Problems   377

CHEMICAL CONNECTIONS
14A Nitroglycerin, an Explosive and a Drug   362
14B Breath-Alcohol Screening   369

14C Ethylene Oxide: A Chemical Sterilant 371
14D Ethers and Anesthesia 373

CHAPTER 15
Chirality: The Handedness of Molecules 382

15.1 Introduction 382
15.2 Enantiomerism: A New Kind of Stereoisomerism 382
15.3 Naming Stereocenters: The R,S System 387
15.4 Molecules with Two or More Stereocenters 391
15.5 Optical Activity: How Chirality Is Detected in the Laboratory 395
15.6 The Significance of Chirality in the Biological World 397

Summary 398
Problems 399

CHEMICAL CONNECTIONS
15A Chiral Drugs 394

CHAPTER 16
Amines 403

16.1 Introduction 403
16.2 Structure and Classification 403
16.3 Nomenclature 406
16.4 Physical Properties 408
16.5 Basicity 409
16.6 Reaction with Acids 411
16.7 Epinephrine: A Prototype for the Development of New Bronchodilators 413

Summary 414
Key Reactions 414
Problems 415

CHEMICAL CONNECTIONS
16A Amphetamines (Pep Pills) 404
16B Alkaloids 405
16C Tranquilizers 408
16D The Solubility of Drugs in Body Fluids 412

CHAPTER 17
Aldehydes and Ketones 419

17.1 Introduction 419
17.2 Structure and Bonding 419
17.3 Nomenclature 420
17.4 Physical Properties 423
17.5 Oxidation 424
17.6 Reduction 425
17.7 Addition of Alcohols 428
17.8 Keto-Enol Tautomerism 430

Summary 431
Key Reactions 431
Problems 432

CHEMICAL CONNECTIONS
17A Some Naturally Occurring Aldehydes and Ketones 422

CHAPTER 18
Carboxylic Acids, Anhydrides, Esters, and Amides 438

18.1 Introduction 438
18.2 Carboxylic Acids 439
18.3 Anhydrides, Esters, and Amides 444
18.4 Preparation of Esters—Fischer Esterification 448
18.5 Preparation of Amides 449
18.6 Hydrolysis of Anhydrides, Esters, and Amides 451
18.7 Phosphoric Anhydrides and Esters 455
18.8 Step-Growth Polymerizations 455

Summary 458
Key Reactions 459
Problems 459

CHEMICAL CONNECTIONS
18A The Pyrethrins: Natural Insecticides of Plant Origin 446
18B The Penicillins and Cephalosporins: β-Lactam Antibiotics 447
18C From Willow Bark to Aspirin—and Beyond 450
18D Ultraviolet Sunscreens and Sunblocks 452
18E Barbiturates 453
18F Stitches that Dissolve 457

PART 3
BIOCHEMISTRY 465

CHAPTER 19
Carbohydrates 466

19.1 Introduction 466
19.2 Monosaccharides 467
19.3 The Cyclic Structure of Monosaccharides 471
19.4 Reactions of Monosaccharides 476
19.5 Disaccharides and Oligosaccharides 481
19.6 Polysaccharides 485
19.7 Acidic Polysaccharides 486

Summary 488
Key Reactions 489
Problems 490

CHEMICAL CONNECTIONS
19A Galactosemia 471
19B L-Ascorbic Acid (Vitamin C) 473
19C Testing for Glucose 479

19D A, B, AB, and O Blood Types 482
19E High-Fructose Corn Syrup 485

CHAPTER 20
Lipids 494

20.1 Introduction 494
20.2 The Structure of Triglycerides 495
20.3 Properties of Triglycerides 497
20.4 Complex Lipids 500
20.5 Membranes 501
20.6 Glycerophospholipids 503
20.7 Sphingolipids 505
20.8 Glycolipids 506
20.9 Steroids 509
20.10 Steroid Hormones 513
20.11 Bile Salts 517
20.12 Prostaglandins, Thromboxanes, and Leukotrienes 518

Summary 521
Problems 521

CHEMICAL CONNECTIONS

20A Rancidity 497
20B Waxes 498
20C Soaps and Detergents 500
20D Transport Across Cell Membranes 502
20E The Myelin Sheath and Multiple Sclerosis 506
20F Lipid Storage Diseases 508
20G Anabolic Steroids 514
20H Oral Contraception 517
20I Action of Anti-inflammatory Drugs 518

CHAPTER 21
Proteins 525

21.1 Introduction 525
21.2 Amino Acids 526
21.3 Zwitterions 528
21.4 Cysteine: A Special Amino Acid 530
21.5 Peptides and Proteins 530
21.6 Some Properties of Peptides and Proteins 532
21.7 The Primary Structure of Proteins 534
21.8 The Secondary Structure of Proteins 537
21.9 The Tertiary and Quaternary Structures of Proteins 540
21.10 Glycoproteins 546
21.11 Denaturation 546

Summary 549
Problems 549

CHEMICAL CONNECTIONS

21A Glutathione 532
21B AGE and Aging 534
21C The Use of Human Insulin 536
21D Sickle Cell Anemia 537
21E Protein/Peptide Conformation–Dependent Diseases 539
21F Proteomics, Ahoy! 543
21G The Role of Quaternary Structures in Mechanical Stress and Strain 544
21H Laser Surgery and Protein Denaturation 548

CHAPTER 22
Enzymes 553

22.1 Introduction 553
22.2 Naming and Classifying Enzymes 555
22.3 Common Terms in Enzyme Chemistry 556
22.4 Factors Affecting Enzyme Activity 557
22.5 Mechanism of Enzyme Action 559
22.6 Enzyme Regulation 563
22.7 Enzymes in Medical Diagnosis and Treatment 566

Summary 567
Problems 568

CHEMICAL CONNECTIONS

22A Muscle Relaxants and Enzyme Specificity 555
22B Acidic Environment and *Helicobacter* 559
22C Active Sites 561
22D Sulfa Drugs as Competitive Inhibitors 562
22E Noncompetitive Inhibition and Heavy Metal Poisoning 563
22F A Zymogen and Hypertension 564
22G One Enzyme, Two Functions 566

CHAPTER 23
Chemical Communications: Neurotransmitters and Hormones 572

23.1 Introduction 572
23.2 Chemical Messengers: Neurotransmitters and Hormones 574
23.3 Cholinergic Messenger: Acetylcholine 577
23.4 Amino Acids as Messengers: Neurotransmitters 582
23.5 Adrenergic Messengers: Neurotransmitters/Hormones 583
23.6 Peptidergic Messengers: Neurotransmitters/Hormones 589
23.7 Steroid Messengers: Hormones 592

Summary 593
Problems 593

CHEMICAL CONNECTIONS

23A Calcium as a Signaling Agent (Secondary Messsenger) 578
23B Nerve Gases and Antidotes 579

23C   Botulism and Acetylcholine Release   580
23D   Alzheimer's Disease and Acetylcholine Transferase   581
23E   Parkinson's Disease: Depletion of Dopamine   586
23F   Nitric Oxide as a Secondary Messenger   588
23G   Diabetes   590
23H   Tamoxifen and Breast Cancer   592

CHAPTER 24
Nucleotides, Nucleic Acids, and Heredity   596

24.1   Introduction   596
24.2   Components of Nucleic Acids   597
24.3   Structure of DNA and RNA   601
24.4   RNA   607
24.5   Genes, Exons, and Introns   608
24.6   DNA Replication   609
24.7   DNA Repair   613
24.8   Cloning   616
Summary   617
Problems   618

CHEMICAL CONNECTIONS
24A   Anticancer Drugs   601
24B   Telomers, Telomerase, and Immortality   610
24C   DNA Fingerprinting   611
24D   Pharmacogenomics: Tailoring Medication to an Individual's Predisposition   613
24E   Apoptosis: Programmed Cell Death   615

CHAPTER 25
Gene Expression and Protein Synthesis   620

25.1   Introduction   620
25.2   Transcription   621
25.3   The Role of RNA in Translation   624
25.4   The Genetic Code   625
25.5   Translation and Protein Synthesis   626
25.6   Gene Regulation   633
25.7   Mutations, Mutagens, and Genetic Diseases   636
25.8   Recombinant DNA   639
Summary   642
Problems   643

CHEMICAL CONNECTIONS
25A   "Antisense" Makes Sense   623
25B   Viruses   630
25C   AIDS   631
25D   The Promoter: A Case of Targeted Expression   634
25E   Mutations and Biochemical Evolution   637
25F   Oncogenes   638
25G   p53: A Central Tumor Suppressor Protein   638

CHAPTER 26
Bioenergetics: How the Body Converts Food to Energy   645

26.1   Introduction   645
26.2   Cells and Mitochondria   647
26.3   The Principal Compounds of the Common Catabolic Pathway   649
26.4   The Citric Acid Cycle   653
26.5   Electron and H^+ Transport   656
26.6   Phosphorylation and the Chemiosmotic Pump   658
26.7   The Energy Yield   660
26.8   Conversion of Chemical Energy to Other Forms of Energy   661
Summary   662
Problems   663

CHEMICAL CONNECTIONS
26A   The Importance of Iron–Sulfur Cluster Assembly   657
26B   2,4-Dinitrophenol as an Uncoupling Agent   658
26C   Protection Against Oxidative Damage   659

CHAPTER 27
Specific Catabolic Pathways: Carbohydrate, Lipid, and Protein Metabolism   666

27.1   Introduction   666
27.2   Glycolysis   668
27.3   The Energy Yield from Glucose   673
27.4   Glycerol Catabolism   674
27.5   β-Oxidation of Fatty Acids   675
27.6   The Energy Yield from Stearic Acid   677
27.7   Ketone Bodies   678
27.8   Catabolism of the Nitrogen of Amino Acids   680
27.9   Catabolism of the Carbon Skeleton of Amino Acids   684
27.10   Catabolism of Heme   686
Summary   687
Problems   688

CHEMICAL CONNECTIONS
27A   Lactate Accumulation   671
27B   Effects of Signal Transduction on Metabolism   674
27C   Ketoacidosis in Diabetes   679
27D   Lipomics   680
27E   Ubiquitin and Protein Targeting   683
27F   Hereditary Defects in Amino Acid Catabolism: PKU   686
27G   Jaundice   687

CHAPTER 28
Biosynthetic Pathways   691

28.1 Introduction   691
28.2 Biosynthesis of Carbohydrates   692
28.3 Biosynthesis of Fatty Acids   696
28.4 Biosynthesis of Membrane Lipids   697
28.5 Biosynthesis of Amino Acids   700

Summary   701
Problems   702

CHEMICAL CONNECTIONS
28A Photosynthesis   694
28B Prenylation of Ras Protein and Cancer   700
28C Amino Acid Transport and Blue Diaper Syndrome   701

CHAPTER 29
Nutrition and Digestion   704

29.1 Introduction   704
29.2 Nutrition   704
29.3 Calories   707
29.4 Carbohydrates in the Diet and Their Digestion   708
29.5 Fats in the Diet and Their Digestion   709
29.6 Proteins in the Diet and Their Digestion   710
29.7 Vitamins, Minerals, and Water   711

Summary   719
Problems   719

CHEMICAL CONNECTIONS
29A Parenteral Nutrition   705
29B Too Much Vitamin Intake?   716
29C Dieting and Artificial Sweetners   716
29D Food for Performance Enhancement   718

CHAPTER 30
Immunochemistry   722

30.1 Introduction   722
30.2 Organs and Cells of the Immune System   725
30.3 Antigens and Their Presentation by Major Histocompatibility Complex   727
30.4 Immunoglobulin   729
30.5 T Cells and Their Receptors   734
30.6 Control of Immune Response: Cytokines   736
30.7 How to Recognize "Self"   737

Summary   740
Problems   741

CHEMICAL CONNECTIONS
30A Antibodies and Cancer Therapy   733
30B Immunization   735
30C Mobilization of Leukocytes: Tight Squeeze   736
30D Myasthenia Gravis: An Autoimmune Disease   739
30E Glycomics   740

CHAPTER 31
Body Fluids   744

31.1 Introduction   744
31.2 Functions and Composition of Blood   745
31.3 Blood as a Carrier of Oxygen   749
31.4 Transport of Carbon Dioxide in the Blood   751
31.5 Blood Cleansing: A Kidney Function   751
31.6 Buffer Production: Another Kidney Function   754
31.7 Water and Salt Balance in Blood and Kidneys   754
31.8 Blood Pressure   755

Summary   756
Problems   757

CHEMICAL CONNECTIONS
31A Using the Blood–Brain Barrier to Eliminate Undesirable Side Effects of Drugs   745
31B Blood Clotting   748
31C Breathing and Dalton's Law   750
31D Sex Hormones and Old Age   755
31E Hypertension and Its Control   756

Appendix I
   Exponential Notation   A-1
Appendix II
   Significant Figures   A-5
Answers to In-Text and Odd-Numbered End-of-Chapter Problems   A-8
Glossary   G-1
Credits   C-1
Index   I-1

PREFACE

To see the world in a grain of sand
And heaven in a wild flower
Hold infinity in the palm of your hand
And eternity in an hour.

WILLIAM BLAKE,
AUGURIES OF INNOCENCE

PERCEIVING ORDER in the nature of the world is an ontological—not just pedagogical—need. It is our primary aim to convey the relationship among facts, thereby presenting a totality of the scientific edifice built over the centuries. In this process we marvel at the unity of laws that govern everything in the ever-exploding dimensions: from photons to protons, from hydrogen to water, from carbon to DNA, from genome to intelligence, from our planet to the galaxy and to the known universe. Unity in all diversity.

In writing the preface of the seventh edition of our textbook, we cannot help but ponder the paradigm shift that has occurred during the last 25 years. From the seventies slogan of "Better living through chemistry" to today's epitaph of "Life by chemistry," one is able to discern the change in focus. Chemistry not only helps to provide the amenities of good life but also lies at the heart of our concept and is the preoccupation of life itself. This shift in emphasis demands that our textbook, which is designed primarily for the education of future practitioners of health sciences, should attempt to provide both the basics and the scope of the horizon within which chemistry touches our life.

The increasing use of our textbook made this new edition possible, and we wish to thank our colleagues who adopted the previous editions for their courses. Testimony from colleagues and students indicates that we managed to convey our enthusiasm for the subject to students, who find this book great help in studying difficult concepts.

In the new edition, we strive further to present an easily readable and understandable text. At the same time, we emphasize the inclusion of new relevant concepts and examples in this fast-growing discipline, especially in the Biochemistry part. We maintain an integrated view of chemistry. From the very beginning in the General Chemistry part, we include organic compounds and biochemical substances to illustrate the principles. The progress is an ascension from the simple to the complex. We urge our colleagues to advance to the chapters of biochemistry as fast as possible, because there lies most of the material that is relevant to the future professions of our students.

Dealing with such a giant field in one course—and possibly the only course in which students get an exposure to chemistry—makes the selection of the material an overarching enterprise. Even though we tried to keep the book to a manageable size and proportion, we inevitably included more topics than could possibly be covered in a two-semester course. Our aim was to provide a variety of material, from which the instructor can select the topics he or she deems most important. We organized the sections so that each can stand independently; consequently, leaving out sections or even entire chapters will not cause fundamental cracks in the total edifice. We have increased the number of topics covered and provided 25 percent more new problems, many of them challenging and thought-provoking.

Audience

As with the previous editions, we developed this book for nonchemistry majors, mainly those entering health sciences and related fields, such as nursing, medical technology, physical therapy, and nutrition. Students in environmental studies can also benefit from it. In its entirety, the seventh edition can be used for a one-year (two-semester or three-quarter) course in chemistry, or parts of the book can be used for a one-term chemistry course.

We assume that the students using this book have little or no background in chemistry. Therefore, we introduce the basic concepts slowly at the beginning and increase the tempo and the level of sophistication as we go on. We progress from the basic tenets of general chemistry to organic and finally to biochemistry. We consider this progress as an ascent in terms of both practical importance and sophistication. Throughout the text, we integrate the three parts by keeping a unified view of chemistry. We do not consider the general chemistry sections to be the exclusive domain of inorganic compounds, so we frequently use organic and biological substances to illustrate general principles.

While teaching the chemistry of the human body is our ultimate goal, we try to show that each subsection of chemistry is important in its own right, besides being required for future understanding.

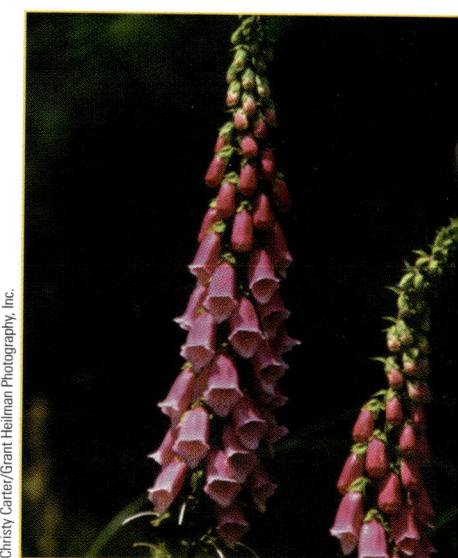

Chemical Connections (Medical and Other Applications of Chemical Principles)

The Chemical Connections features contain applications of the principles discussed in the text. Comments from users of the earlier editions indicate that these boxes have been especially well received and provide a much-requested relevance to the text. In the seventh edition, we have updated some of these features and added a number of new ones.

The majority of Chemical Connections boxes deal with health-related applications. Ones new to this edition focus on topics such as COX-2 inhibitors and their use in chemoprevention, amyloid structures in prion diseases including systematic amyloidosis and Alzheimer's disease, proteomics, zymogen and hypertension, botulinum toxin (Botox), phamacogenomics, "antisense drugs," four modes of combating AIDS, fusion inhibitors, iron–sulfur cluster assemblies, lipomics, structures of photosystems I and II, prenylation of Ras protein, potential anticancer vaccines, and glycomics.

The presence of these features allows for a considerable degree of flexibility. If an instructor wants to assign only the main text, the Chemical Connections do not interrupt continuity, and the essential material will be covered. However, because they enhance the core material, most instructors will probably wish to assign at least some of the Chemical Connections. In our experience, students are eager to read the relevant Chemical Connections, even without assignments, and they do so with discrimination. From such a large number of boxes, an instructor can judiciously select those that best fit the particular needs of the course. To enhance the material in the Chemical Connections, we provide problems at the end of each chapter covering all of the boxes.

Metabolism: Color Code

The biological functions of chemical compounds are explained in each of the biochemistry chapters and in many of the organic chapters. Throughout the book, emphasis is placed in chemistry rather than physiology. In the past, we have received much positive feedback regarding the way in which we have organized the topic of metabolism (Chapters 26, 27, and 28). We have maintained this organization in the current edition.

First we introduce the common metabolic pathway through which all food will be utilized (the citric acid cycle and oxidative phosphorylation). Only then do we discuss the specific pathways leading to the common pathway. We find this organization to be a useful pedagogic device, and it enables us to sum the caloric value of each type of food because its utilization through the common pathway has already been learned. Finally, we separate the catabolic pathways from the anabolic pathways by treating them in different chapters, emphasizing the different ways in which the body breaks down and builds up different molecules.

The topic of metabolism is a difficult one for most students, and we have tried to explain it as clearly as possible. As in the previous edition, we enhance the clarity of presentation by the use of a color code for the most important biological compounds discussed in Chapters 26, 27, and 28. Each type of compound is screened in a specific color, which remains the same throughout the three chapters. These colors are as follows:

- ATP and other nucleoside triphosphates
- ADP and other nucleoside diphosphates
- The oxidized coenzymes NAD^+ and FAD
- The reduced coenzymes NADH and $FADH_2$
- Acetyl coenzyme A

In figures showing metabolic pathways, we display the numbers of the various steps in yellow.

In addition to this major use of a color code, other figures in various parts of the book are color coded so that the same color is used for the same entity throughout. For example, in all Chapter 22 figures that show enzyme–substrate interactions, enzymes are always shown in blue and substrates in orange.

Features

Chemical Connections are a series of 150 in-depth and intriguing essays that describe applications of chemical concepts presented in the text. Examples include "Timed-Release Medications" (Chapter 7), "The Solubility of Drugs in Body Fluids" (Chapter 16), and "AGE and Aging" (Chapter 21).

Worked examples and corresponding problems are included throughout each chapter to help students develop sound problem-solving skills. Each of the 160 examples includes a detailed solution and is accompanied by a similar problem for students to try on their own.

CD-ROM icons in the text refer to **Interactive General, Organic and Biochemistry CD-ROM, version 2.0,** which accompanies this book. Students are encouraged to use the CD for additional help or information whenever they see this icon.

Margin notes offer additional bits of information, such as historical notes, and reminders, which complement nearby text.

Molecular models, including ball-and-stick models, space-filling models, and electron density maps, appear throughout the text and serve as appropriate aids to visualizing molecular properties and interactions.

A dynamic art program enhances the book's clarity and makes the text visually exciting. A number of new annotated art and macro/micro art pieces have been added. Many new photographs have been added to the book as well, illustrating reactions, chemical procedures, and applications.

Margin definitions of many new terms reinforce key terminology.

Key Reactions, which appear at the ends of many organic chapters, are an annotated list of new reactions introduced in that chapter (Chapters 10–18), along with an example of the reaction and the section in which it is introduced. These lists help students to summarize the organic reactions they have learned.

End-of-chapter problems are paired to give students two tries at solving a particular problem type. Problem numbers that appear in color indicate difficult problems; ones marked with this tetrahedral icon are application-type problems. Problems are broken up by section topics, followed by an additional set of unclassified problems. Problems based on the Chemical Connections help to reinforce the applications of the chemical principles.

Chapter Summaries are section-by-section synopses of the key terms and concepts introduced in each chapter.

The Glossary at the end of the book gives a definition of each new term along with the number of the section in which it is introduced.

Answers to all in-text and odd-numbered end-of-chapter problems are provided at the end of the book. Detailed, worked-out solutions to these same problems are provided in the *Student Solutions Manual*.

Style

Feedback from colleagues and students alike indicates that one of the book's major assets is its style, which addresses students directly using simple and clear phrasing. In the seventh edition, we continue to make special efforts to provide clear and concise writing. Our hope is that this style will facilitate the understanding and absorption of difficult concepts.

Problems

Approximately 25 percent of the problems in this edition are new. The number of more challenging, thought-provoking questions has also been increased. The end-of-chapter problems are grouped and given subheads in order of topic coverage. Problem numbers in color specify more challenging ideas; those marked with a tetrahedral icon are application-type problems; and a group of problems are based on the Chemical Connections boxes. In the last group, called "Additional Problems," problems are not arranged in any specific order.

The answers to all in-text and odd-numbered end-of-chapter problems appear at the end of the book. Complete worked-out solutions to all in-text and end-of-chapter problems are supplied in the *Instructor's Manual*.

Ancillaries

The following ancillaries are available to qualified adopters. Please contact your local Thomson • Brooks/Cole sales representative for details.

- **Interactive General, Organic and Biochemistry CD-ROM, version 2.0,** is an unparalleled multimedia presentation of GOB topics. Created by William Vining of the University of Massachusetts and Susan Young of Hartwick College, it features extensive use of videos and animations to create a strong link between observable macroscopic properties and molecular-scale chemical interpretations. It is supported by an interactive database of three-dimensional molecular models.

- **Student Study Guide,** by William Scovell of Bowling Green State University, includes reviews of chapter objectives, important terms and comparisons, focused reviews of concepts, and self-tests.

- **Student Solutions Manual,** by Mark Erickson of Hartwick College, William Brown of Beloit College, and Rodney Boyer of Hope College, contains complete worked-out solutions to all in-text and odd-numbered end-of-chapter problems.

- **Instructor's Manual,** by Mark Erickson of Hartwick College, William Brown of Beloit College, and Rodney Boyer of Hope College, contains worked-out solutions to all in-text and even-numbered end-of-chapter problems.

- **InfoTrac® College Edition** Every new copy of *General, Organic and Biochemistry,* seventh edition, comes packaged with four months of free access to **InfoTrac College Edition.** This online resource features a comprehensive database of reliable, full-length articles (not abstracts) from thousands of top academic journals and popular sources—updated daily and spanning 22 years! Just some of the journals available 24 hours a day, seven days a week: *American Scientist, Science, Science News, Science Weekly,* and thousands more.

- **MyCourse 2.0** is our new, free, online course builder. Whether you want only the easy-to-use tools to build it or the content to furnish it, MyCourse 2.0 offers you a simple solution for creating a custom course Web site that allows you to assign, track, and report on student progress. Contact your Thomson•Brooks/Cole sales representative for details or visit **http://mycourse.thomsonlearning.com** for a free demo.

Charles D. Winters

- **Multimedia Manager for Organic Chemistry: A Microsoft® PowerPoint® Link Tool** is a digital library and presentation tool that is available on one convenient multiplatform CD-ROM. Thanks to its easy-to-use interface, you can readily take advantage of Brooks/Cole's presentations, which consist of art and tables from the text, as well as more than 700 PowerPoint presentation lecture slides especially created for this book by William E. Brown. The presentations are available in a variety of e-formats that are easily exported into presentation software or used on Web-based course support materials. You can even customize your own presentation by importing your personal lecture slides or any other material you choose. The result is an interactive and fluid lecture that truly engages your students.

- **ExamView® Computerized Testing** allows you to enhance your range of assessment and tutorial activities—and save yourself time in the process. You can create, deliver, and customize tests and study guides (both print and online) in minutes with this user-friendly assessment and tutorial system. **ExamView** offers both a *Quick Test Wizard* and an *Online Test Wizard* that guide you step-by-step through the process of creating tests. Using **ExamView**'s complete word processing capabilities, you can enter an unlimited number of new questions or edit existing questions. And **ExamView** is the only test generator that offers a "what-you-see-is-what-you-get" feature that allows you to see the test you are creating on the screen exactly as it will print! You can explore **ExamView** online at http://www.examview.com/product_info/tour.htm.

- **Printed Test Bank,** contains approximately 50 multiple-choice questions per chapter for each of the 31 chapters in this text.

- **WebTutor™ Advantage on WebCT and Blackboard** Students have access to study tools that correspond chapter by chapter with the book, while professors can use **WebTutor Advantage** to provide virtual office hours, post syllabi, track student progress on the practice quizzes, and more. Visit http://webtutor.thomsonlearning.com for more information and to try WebTutor Advantage.

- **Overhead Transparencies** The 150 full-color overhead transparencies include figures and tables taken directly from the text.

- **Brooks/Cole Chemistry Book Companion Web Site at** http://www.brookscole.com/chemistry includes a rich array of teaching and learning tools for students and instructors. It includes online quizzes, tutorials, and more.

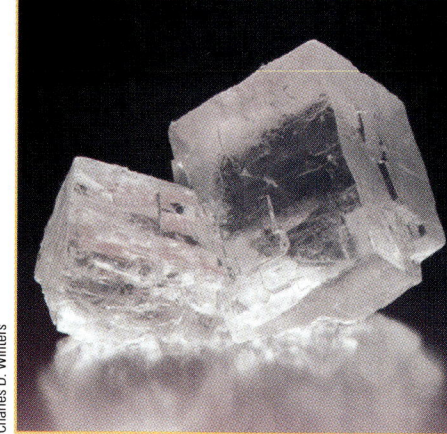
Charles D. Winters

- **Laboratory Experiments for General, Organic & Biochemistry, fifth edition,** by Frederick A. Bettelheim and Joseph M. Landesberg, includes 52 experiments that illustrate important concepts and principles in general, organic, and biochemistry. Three new experiments are featured: (1) fermentation of a carbohydrate: ethanol from sucrose; (2) isolation of DNA from onions; and (3) neurotransmission: an example of enzyme specificity. In addition, many experiments have been revised. All experiments have new pre- and post-lab questions. The large number of experiments allows sufficient flexibility for the instructor.

- **Instructor's Manual to Accompany Laboratory Experiments** will help instructors in grading the answers to questions and in assessing the range of experimental results obtained by students. The *Instructor's Manual* also contains reminders for professors to pass on to students and details on how to handle disposal of waste chemicals.

Acknowledgments

The publication of a book such as this one requires the efforts of many more people than merely the authors. We would like to thank the following professors who offered many valuable suggestions for this new edition:

JOHN T. BARBAS, Valdosta State University
MARK BENVENUTO, University of Detroit Mercy
VICKY BEVILACQUA, Kennesaw State University
ERWIN BOSCHMANN, Indiana University
SETH ELSHEIMER, University of Central Florida
MARK ERICKSON, Hartwick College
JOHN FOLKROD, University of Minnesota, Duluth
DONALD HARRISS, University of Minnesota, Duluth
JACK HEFLEY, Blinn College
JESSE W. JONES, Baylor University
LAURA KIBLER-HERZOG, Georgia State University
PAMELA MARKS, Arizona State University

Charles D. Winters

THOMAS NALLI, Winona State University
ELVA MAE NICHOLSON, Eastern Michigan University
KIMBERLY PACHECO, University of Northern Colorado
DAVID REINHOLD, Western Michigan University
KAREN SCHUSTER, Florida Community College at Jacksonville
ERIC SIMANEK, Texas A&M University
BOBBY STANTON, University of Georgia
RICHARD TREPTOW, Chicago State University
BURTON TROPP, Queens College, CUNY

We thank John Holdcroft, Chemistry Editor, for promoting and facilitating the production of this edition. We have special thanks for Sandi Kiselica, our Senior Development Editor, who has been a rock of support through the entire revision process. We greatly appreciate her ability to set challenging but manageable schedules for us and her constant encouragement as we worked to meet those deadlines. Sandi has also been a valuable resource person with whom we could discuss everything from pedagogy to details of the art program. We also thank Nicole Barone of Thompson Steele, Inc. for her determination and iron-hand coordination of the production processes.

REVISION SUMMARY

General Chemistry

In **Chapter 2, Atoms,** we introduce four of the five methods we use throughout the text to represent molecules: we show water as a molecular formula, a structural formula, a ball-and-stick model, and a space-filling model. We introduce electron density maps, a fifth form of representation, in Chapter 3. To draw the correlation between electron configuration and the Periodic Table, we have added a miniature version of the Periodic Table as a new graphic to show the locations of groups of elements as they are discussed. Figure 2.12 shows that the pairing of electron spins is a consequence of the alignment of the magnetic poles of spinning electrons. We have revised the discussion of electron configuration and the Periodic Table in Section 2.7 to focus on the underlying organization by s, p, d, and f blocks.

The material in **Chapter 3, Chemical Bonds,** has been reorganized to group the discussion of ionic compounds first, followed the discussion of molecular compounds. The major revision in **Chapter 4, Chemical Reactions,** is the deletion of the discussion and significance of the terms *exergonic* and *endergonic*. We now discuss the energetics of chemical reactions only in terms of how exothermic and endothermic they are.

In **Chapter 5, Gases, Liquids, and Solids,** we have rearranged the presentation of intermolecular forces of attraction to be in order of increasing energy—namely, London dispersion forces, dipole–dipole interactions, and hydrogen bonding. We have deleted Graham's law of diffusion, and added a Chemical Connections box on supercritical CO_2.

Chapter 8, Acids and Bases, now introduces the use of curved arrows to show the flow of electrons in organic reactions. Specifically, we use them here to show the flow of electrons in proton-transfer reactions. We have dropped the discussion of normality and equivalents from the discussion of the stoichiometry of acid–base reactions. The major revision in this chapter is an expansion of the discussion of acid–base buffers and the Henderson-Hasselbalch equation.

Chapter 9, Nuclear Chemistry, has new information on the creation of superheavy elements.

Organic Chemistry

Chapter 10, Organic Chemistry, is little changed except for an added Chemical Connections box on combinatorial chemistry.

In **Chapter 11, Alkanes and Cycloalkanes,** we introduce the concept of line-angle formulas and continue using these formulas throughout the organic chapters (Chapters 11–18). They are easier to draw than the usual condensed structural formulas as well as easier to visualize. In previous editions, line-angle formulas were used only to represent cycloalkane and aromatic rings. A new Section 11.9B describes the halogenation of alkanes. The following section treats chlorofluorocarbons (the Freons) and their replacements, plus haloalkane solvents. A new Chemical Connections box describes the environmental impact of Freons. Section 12.5, covering the terpenes, has been shortened considerably. A new Chemical Connections box in **Chapter 12, Alkenes,** treats ethylene, a plant growth regulator.

The positions of Chapters 13 and 14 have been reversed from the sixth edition. **Chapter 13, Benzene and Its Derivatives,** now follows immediately after the treatment of alkenes and alkynes. We have expanded the discussion of phenols to include phenols and antioxidants. In this subsection, we treat auto-oxidation of unsaturated fatty acid hydrocarbon chains as a radical-chain mechanism leading to hydroperoxidation. We also discuss how substituted phenols such as BHT and BHA act as antioxidants. **Chapter 14, Alcohols, Ethers, and Thiols,** has been reorganized so that the material now discusses alcohols first (Sections 14.2 and 14.3), then ethers (Section 14.4), and finally thiols (Section 14.5).

In **Chapter 15, Chirality: The Handedness of Molecules,** the introduction of the concept of a stereocenter in Section 15.2 is done much more slowly and carefully using 2-butanol as a prototype. Section 15.4, which treats molecules with two or more stereocenters, has been shortened to concentrate on recognizing a stereocenter by the placement of four different groups on a carbon. The relationship in cyclic compounds between cis-trans isomers and enantiomers has been reduced to a minimum.

In **Chapter 16, Amines,** a new Section 16.7, "Epinephrine: A Prototype for the Development of New Bronchodilators," traces the development of new asthma medications such as albuterol (Proventil) from epinephrine as a lead drug. To **Chapter 17, Aldehydes and Ketones,** we have added a discussion of $NaBH_4$ as a carbonyl-reducing agent with emphasis on it as a hydride transfer reagent. We then make the parallel to NADH as a carbonyl-reducing agent and as a hydride transfer agent.

To **Chapter 18, Carboxylic Acids, Anhydrides, Esters, and Amides,** we have added a brief discussion of the preparation of amides by treating a carboxylic acid with an amine (which gives a salt) and then heating this salt to eliminate water. A more useful preparation of amides is the treatment of an amine with an anhydride. We have added two new Chemical Connections: "Pyrethrins—Natural Insecticides of Plant Origin," and "Ultraviolet Sunscreens and Sunblocks."

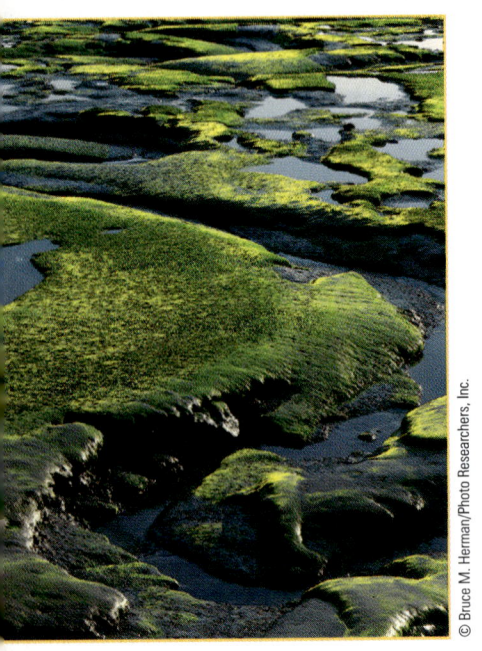

Biochemistry

The most significant revision in **Chapter 19, Carbohydrates,** is a new Chemical Connections box on high-fructose corn syrup.

In **Chapter 20, Lipids,** new topics include a description of potassium ion channels and the mode of transportation of K^+. The discussion of steroids in Section 20.9 has been reorganized and expanded. Detailed discussions on the structure, transport, and concentrations of the different lipoproteins have been added. The functions of steroidal and nonsteroidal anti-inflammatory agents are discussed, including the COX-2 inhibitors and their use in chemoprevention.

In **Chapter 21, Proteins,** we present an expanded Chemical Connections on protein conformation-dependent diseases that summarizes amyloid structures in prion-related diseases such as mad cow disease, Creutzfeld-Jacob disease, systematic amyloidosis, and Alzheimer's disease. New discussions focus on the quaternary structures of hemoglobin, collagen, and integral membrane proteins. A new Chemical Connections highlights the new discipline of proteomics and its function.

The discussion of enzymes now includes details of the chemistry of active sites in **Chapter 22, Enzymes.** In 65 percent of enzyme-catalyzed reactions, the active site operates with acid–base reactions. A new Chemical Connections, "Zymogen and Hypertension," describes the specificity of different ACE inhibitors.

New features in **Chapter 23, Chemical Communications: Neurotransmistters and Hormones,** are the updating of the list of drugs affecting neurotransmission, inclusion of the role of the phosphatidylinositol phosphates as secondary messengers, and the recognition of neurosteroids synthesized in the brain. New Chemical Connections describe the use of the botulinum toxin (Botox) in the cosmetics industry and the neurodegenerative role played by beta amyloid versus tau protein in Alzheimer's disease.

The ever-increasing developments in molecular biology necessitated rewriting of **Chapter 24, Nucleotides, Nucleic Acids, and Heredity.** It now inludes a new section on quaternary and higher structures of DNA leading to the superstructures of chromosomes. The section on DNA replication now includes material on the opening up of superstructures and the components of replisomes. A new section describes the different mechanisms of DNA repair. A new Chemical Connections feature deals with phamacogenomics' ability to tailor medication to an individual's predisposition.

Chapter 25, Gene Expression and Protein Synthesis, underwent major reorganization and expansion. We now concentrate on gene expression and protein synthesis as they operate in eukaryotes. Gene regulation has been expanded to cover different entries on the transcriptional, translational, and post-translational levels. New Chemical Connections cover antisense drugs, four modes of combating AIDS including the new "fusion inhibitors," and the role of the tumor suppression protein p53 in aging.

Chapter 26, Bioenergetics: How the Body Converts Food to Energy, has a new Chemical Connections on iron–sulfur cluster assemblies.

The only new feature in **Chapter 27, Specific Catabolic Pathways: Carbohydrate, Lipid, and Protein Metabolism,** is a Chemical Connections on lipomics, the collective lipid components of a cell.

In **Chapter 28, Biosynthetic Pathways,** there is extended coverage of photosynthesis, including the newly discovered structures of photosystems I and II. Stepwise synthesis of cholesterol up to the C10 and C15 units are detailed. A new Chemical Connections features prenylation of the Ras protein and its effect on signal transduction.

Chapter 29, Nutrition, has been reorganized to present separate sections on carbohydrates in diet and their digestion and similarly on fats and proteins. References to the latest Dietary Reference Intakes are included as well as discussion of current fad diets such as the Atkins diet and raw food diet.

Chapter 30, Immunochemistry, has been expanded to include a new section on how to recognize "self," which includes coverage of autoimmune diseases and their treatment. The existing sections were subdivided to facilitate the comprehension of difficult concepts. A new subsection was written to explain how the body acquires the diversity to react to different antigens. Among the Chemical Connections is a new feature on potential anticancer vaccines and on glycomics (a distant cousin of genomics and proteomics).

In **Chapter 31, Body Fluids,** we moved the Chemical Connections "Breathing and Dalton's Law" here from Chapter 5 to provide a better survey of gas transports in the blood.

HEALTH-RELATED TOPICS

Key:

ChemConn = Chemical Connections Box number
Sect. = Section number
Prob. = Problem number

Abundance of elements in the human body	ChemConn 2B
Acid and base burns to the cornea	ChemConn 8B
Acidic polysaccharides in the body	Sect. 19.8
Acidosis and alkalosis	ChemConn 8D
Active sites	ChemConn 22C
AGE and aging	ChemConn 21B
AIDS	ChemConn 25C
Albuterol (Proventil)	Sect. 16.7
Alkaloids	ChemConn 16B
Alzheimer's disease	ChemConn 23D
Amphetamines	ChemConn 16A
Anabolic steroids	ChemConn 20G
Angiotensin	ChemConn 22F
Antacids	ChemConn 8C
Antibodies and cancer therapy	ChemConn 30A
Anticancer drugs	ChemConn 24A
Antidepressants	Sect. 23.5F, Prob. 15.33
Antigens	Sect. 30.3
Antihistamines	Sect. 23.5G
Anti-inflammatory drugs	ChemConn 20I
Antioxidants	Sect. 13.5C
Antisense drugs	ChemConn 25A
Apoptosis, programmed cell death	ChemConn 24E
Artificial sweeteners	ChemConn 29C
Ascorbic acid (vitamin C)	ChemConn 19B
Aspartame	Prob. 18.56
Aspirin and other NSAIDs	ChemConn 18C
Asthma	Sect. 20.12
Atherosclerosis, levels of LDL and HDL	Sect. 20.9E
Atomic energy	Sect. 9.8
Atropine	Prob. 16.50
Attention deficit disorder (ADD)	ChemConn 23F
Autoimmune diseases	Sect. 30.7
Autoxidation	Sect. 13.5C
B cells	Sect. 30.2C
Barbiturates	ChemConn 18E
Basal caloric requirement	Sect. 29.3
Basic excision repair (BER) of DNA	Sect. 24.7
Bends	ChemConn 6B
BHT, an antioxidant in foods	Sect. 13.5C
Bile salts	Sect. 20.11
Blood buffers	Sect. 8.11D
Blood cleansing	Sect. 31.5
Blood clotting	ChemConn 31B
Blood pressure	Sect. 31.8
Blood pressure measurement	ChemConn 5D
Blood types	ChemConn 19D
Blue diaper syndrome	ChemConn 28C
Botox	ChemConn 23C
Botulism	ChemConn 23C
Breath-alcohol screening	ChemConn 14B
Breathing and Boyle's Law	ChemConn 5B
Breathing and Dalton's Law	ChemConn 31C
Bronchodilators and asthma	Sect. 16.7
Brown fat and hibernation	ChemConn 26B
Buffers	Sect. 8.11
Calcium, as a signaling agent	ChemConn 23A
Capsaicin	ChemConn 13F
Carbohydrates, as signals of virulence and malignancy	ChemConn 30E
Carcinogenic PAHs and smoking	Sect. 13.3B
Carcinogens	ChemConn 25E
β-Carotene	Prob. 12.60
Cephalosporins	ChemConn 18B
Chiral drugs	ChemConn 15A
Chirality, in biomolecules	Sect. 15.6
Cholera	Sect. 23.5D
Cis-trans isomerism, in vision	ChemConn 12C
Cocaine	ChemConn 16B
Cocaine addiction	ChemConn 23E
Cold compresses	ChemConn 1C
Coniine	ChemConn 16B
Coral chemistry and broken bones	ChemConn 3A
COX-2 inhibitor drugs	ChemConn 20I
Creatine, performance enhancement	ChemConn 29D
Crenation	Sect. 6.8B
CT, MRI, ultrasound, and PET scans	ChemConn 9E
Cystic fibrosis	ChemConn 27E
Cytochrome P-450, in detoxification	ChemConn 26C
Cytokines	Sect. 30.6
DDT	ChemConn 13A
Diabetes	ChemConn 23G
Dichloroacetic acid	Sect. 18.2D
Dietary Reference Intakes (DRI)	Sect. 29.2

Diets	Sect. 29.2	Innate immunity	Sect. 30.1A
2,4-Dinitrophenol, as an uncoupling agent	ChemConn 26B	Insulin	Sect. 21.7
DNA fingerprinting	ChemConn 24C	Iodide ion and goiter	ChemConn 13C
Drug dosage and body mass	ChemConn 1A	Ionic compounds in medicine	ChemConn 3B
		Ionizing radiation	Sect. 9.4
Elements necessary for human life	ChemConn 2A	Iron–sulfur cluster assembly	ChemConn 26A
Emulsions and emulsifying agents	ChemConn 6D		
Enzymes, in diagnosis	Sect. 22.7	Jaundice	ChemConn 27G
Enzymes, in therapy	Sect. 22.7		
Ephedrine	Prob. 15.26	Ketoacidosis, in diabetes	ChemConn 27C
Epibatadine	Prob. 16.51	Ketone bodies	Sect. 27.7
Epinephrine	Sect. 16.7		
Esters, as flavoring agents	Sect. 18.3	Lactate accumulation	ChemConn 27A
Ethers and anesthesia	ChemConn 14D	Laser in situ keratomileusis (LASIK)	ChemConn 21H
Ethylene, a plant growth regulator	ChemConn 12A	Laser surgery and protein denaturation	ChemConn 21H
Ethylene oxide, a chemical sterilant	ChemConn 14C	Lipid storage diseases	ChemConn 20F
		Lowering body temperature	ChemConn 7B
Fluid mosaic model of membranes	Sect. 20.5	Lycopene	Prob. 12.61
Free radicals	ChemConn 9D, Section 13.5C		
Freons	ChemConn 11B	Mad cow disease	ChemConn 21E
		Menstrual cycle	Sect. 20.10BB
Galactosemia	ChemConn 19A	Mobilization of leukocytes	ChemConn 30C
Gallstones	Sect. 20.9A	Monoclonal antibodies	Sect. 30.4C
Gene therapy	Sect. 25.8	Morphine and enkephalins	Sect. 23.6
Genetic code	Sect. 25.4	Mucins	Sect. 21.10
Genetic engineering	Sect. 25.8	Multiple sclerosis	ChemConn 20E
Glutathione	ChemConn 21A	Muscle relaxants	ChemConn 22A, Sect. 23.3E
Gout	Sect. 6.6B	Mutagens	ChemConn 25E
G-protein/cAMP cascade	Sect. 23.5C	Mutations and biochemical evolution	ChemConn 25E
		Myasthenia gravis	ChemConn 30D
Heart enzymes	Sect. 22.7		
Heavy metal poisoning	ChemConn 22E, Sect. 21.10	Naproxen	Sect. 15.3
Helicobacter	ChemConn 22B	Nerve gases and antidotes	ChemConn 23B
Heme products, in bruises	Sect. 27.10	Neurotransmitters	Sect. 23.2
Hemodialysis	ChemConn 6F	Nicotine	ChemConn 16B
Hemolysis	Sect. 6.8B	Nitric oxide	ChemConn 3C, 23F
Heparin	Sect. 19.8B	Nitroglycerin, an explosive and a drug	ChemConn 14A
High fever	ChemConn 7A		
High-fructose corn syrup	ChemConn 19E	Oncogenes	ChemConn 25F
Hormones	Sect. 23.2	Oral contraception	ChemConn 20H
Human insulin	ChemConn 21C	Osmotic pressure	Sect. 6.8B
Hyaluronic acid	Sect. 19.8	Oxidative damage	ChemConn 26C
Hyperbaric medicine	ChemConn 5C	Oxidizing antiseptics	ChemConn 4C
Hypertension and its control	ChemConn 31E		
Hypoglycemic awareness	Sect. 21C	p53, a central tumor suppressor protein	ChemConn 25G
Hypothermia and hyperthermia	ChemConn 1B	Parenteral nutrition	ChemConn 29A
		Parkinson's disease	ChemConn 23E
Ibuprofen	Sect. 15.3	Penicillins	ChemConn 18B
Immune system	Sect. 30.1B	pH of some common materials	Sect. 8.8
Immunization	ChemConn 30B	Phenols, as antioxidants	Sect. 13.5C
Immunoglobulin	Sect. 30.4	Phenylcyclidine (PCP)	Sect. 23.4B

HEALTH RELATED TOPICS

Topic	Reference
Phenylketonurea (PKU)	ChemConn 27F
Photorefractive keratectomy (PRK)	ChemConn 21H
Photosynthesis	ChemConn 28A
Poison ivy	Sect. 13.5A
Poisonous puffer fish	ChemConn 11A
Polynuclear aromatic hydrocarbons (PAHs)	Sect. 13.3D
Prenylation of Ras protein and cancer	ChemConn 28B
Promoter, a case of targeted expression	ChemConn 25D
Propofol	Prob. 14.36
Prostaglandin endoperoxide synthase, two functions	ChemConn 22G
Protein/peptide conformation-dependent diseases	ChemConn 21E
Proteoglycans of extracellular matrix	Sect. 21.10
Proteomics	ChemConn 21G
Proventil	Sect. 16.7
Pyrethrins	ChemConn 18A
Quaternary structures, in mechanical stress	ChemConn 21F
Radiation	ChemConn 9A
Radiation dosimetry	Sect. 9.5
Radiation sickness	Sect. 9.5
Radioactive fallout from nuclear accidents	ChemConn 9G
Radioactive isotopes, in medical diagnosis	Sect. 9.6
Radioactive isotopes, in medical therapy	Sect. 9.6
Radon	ChemConn 9C
Rancidity	ChemConn 20A
Recommended Daily Allowances (RDA)	Sect. 29.2
Relative sweetness	Sect. 19.5D
Reverse osmosis and desalination	ChemConn 6E
Roentgens, rads, and rems	Sect. 9.5
Senile systematic amyloidosis	ChemConn 21E
Sex hormones and old age	ChemConn 31D
Sickle cell anemia	ChemConn 21D
Side effects of drugs	ChemConn 31A
Sideroblastic anemia	ChemConn 26A
Signal transduction	ChemConn 27B
Smelling salts	Prob. 4.37
Smoking and carcinogenesis	ChemConn 13B
Solubility and tooth decay	ChemConn 4A
Solubility of drugs in body fluids	ChemConn 16D
Spray and stretch—a new use for freons	ChemConn 5E
Statins	Sect. 28.4
Stitches that dissolve	ChemConn 18F
Strontium-90	ChemConn 2D
Sulfa drugs	ChemConn 22D
Sunglasses and Le Chatelier's principle	ChemConn 7D
Sunscreens and sunblocks	ChemConn 18D
Synthetic food dyes	ChemConn 13E
T cells	Sect. 30.2C
Tailoring medication to an individual's predisposition	ChemConn 24D
Tamoxifen and breast cancer	ChemConn 23H
Taxol	ChemConn 10A
Telomers and immortality	ChemConn 24B
Testing for glucose	ChemConn 19C
Tetrodotoxin	ChemConn 11A
Timed-released medication	ChemConn 7C
Tranquilizers	ChemConn 16C
Transport across cell membranes	ChemConn 20D
Ubiquitin and protein targeting	ChemConn 27E
Urinometer	Sect. 1.7
Viagra and blood vessel dilation	ChemConn 23F
Viruses	ChemConn 25B
Vitamin A and vision	ChemConn 12C
Vitamin excess	ChemConn 29B
Vitamins in diet	Sect. 29.7
Xeroderma pigmentosa	Sect. 24.7
Zymogen and hypertension	ChemConn 22F

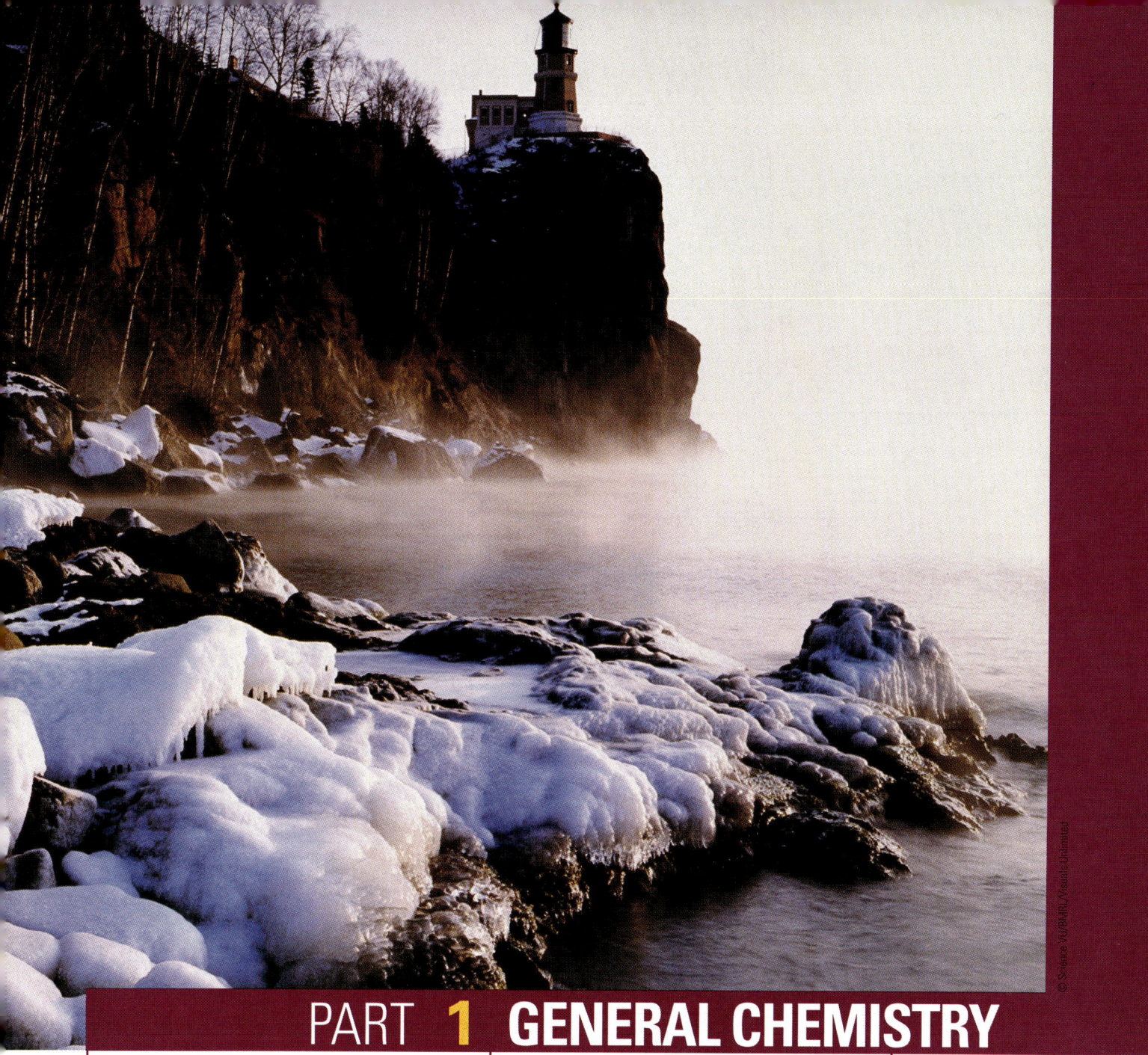

PART 1   GENERAL CHEMISTRY

Chapter 1
Matter, Energy, and Measurement

Chapter 2
Atoms

Chapter 3
Chemical Bonds

Chapter 4
Chemical Reactions

Chapter 5
Gases, Liquids, and Solids

Chapter 6
Solutions and Colloids

Chapter 7
Reaction Rates
and Chemical Equilibrium

Chapter 8
Acids and Bases

Chapter 9
Nuclear Chemistry

CHAPTER 1

1.1 Introduction
1.2 The Scientific Method
1.3 Exponential Notation and Significant Figures
1.4 Measurements
1.5 Unit Conversions: The Factor-Label Method
1.6 The States of Matter
1.7 Density and Specific Gravity
1.8 Energy
1.9 Heat

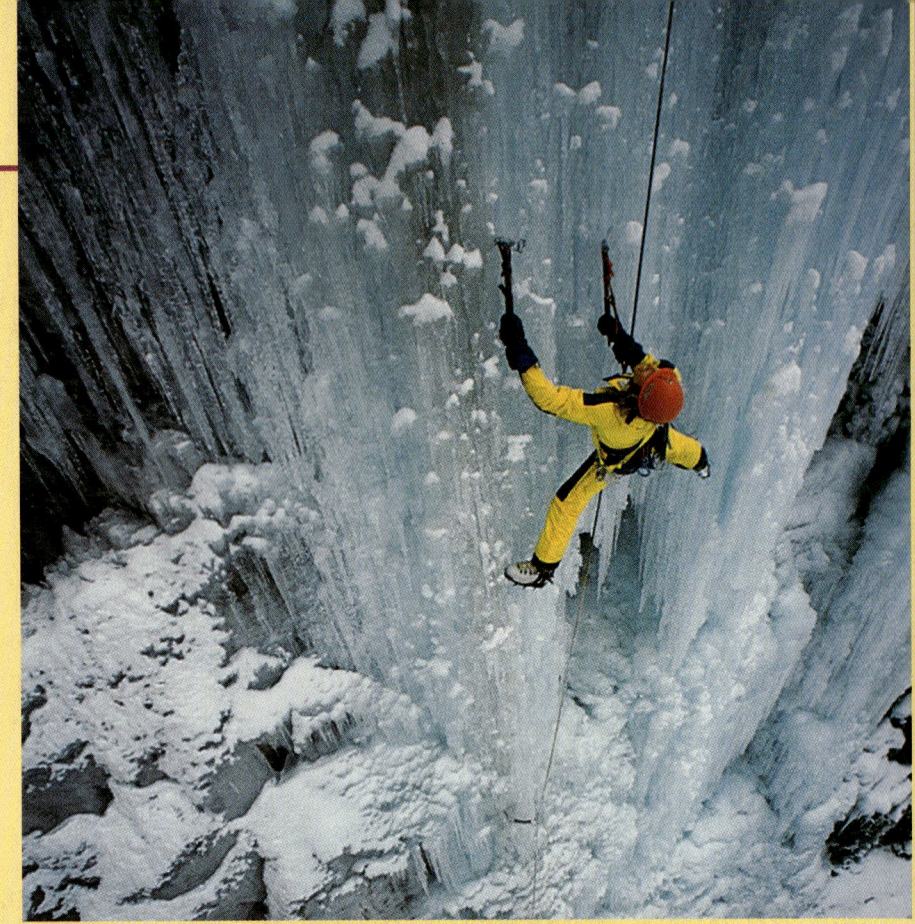

A woman climbing a frozen waterfall in British Columbia.

Matter, Energy, and Measurement

1.1 Introduction

The world around us is made of chemicals. Our food, our clothing, the buildings in which we live are all made of chemicals. Our bodies are made of chemicals, too. To understand the human body, its diseases and its cures, we must know all we can about those chemicals. There was a time—only a few hundred years ago—when physicians were powerless to treat many diseases. Cancer, tuberculosis, smallpox, typhus, plague, and many other sicknesses struck people seemingly at random; doctors, who had no idea what caused any of these diseases, could do little or nothing about them. Doctors treated them with magic as well as by such measures as bleeding (Figure 1.1), laxatives, hot plasters, and pills made from powdered stag horn, saffron, or gold. None of these treatments was effective, and the doctors, because they came into direct contact with highly contagious diseases, died at a much higher rate than the general public.

Medicine has made great progress since those times. We live much longer, and many once-feared diseases have been essentially eliminated or are curable. Smallpox has been completely eradicated, and polio, typhus, bubonic plague, diphtheria, and other diseases that once killed millions no longer pose a serious problem, at least not in the developed countries.

Figure 1.1 A woman being bled by a leech on her left forearm; a bottle containing more leeches is on the table. From a 1639 woodcut. *(Courtesy of the National Library of Medicine)*

How has this medical progress come about? The answer is that diseases could not be cured until they were understood, and this understanding has emerged through greater knowledge of how the body functions. It is progress in our understanding of the principles of biology, chemistry, and physics that has led to these advances in medicine.

Because so much of modern medicine depends on chemistry, it is essential that students who intend to enter the health professions have some understanding of basic chemistry. This book was written to help you achieve that goal. Even if you choose a different profession, you will find that the chemistry you learn in this course will greatly enrich your life.

Chemistry as the Study of Matter

The universe consists of matter, energy, and empty space. **Matter** is anything that has mass and takes up space. **Chemistry** is the science that deals with matter: the structure and properties of matter and the transformations from one form of matter to another. We will discuss energy in Section 1.8.

It has long been known that matter can change, or be made to change, from one form to another. In a **chemical change,** more commonly called a **chemical reaction,** substances are used up (disappear) and others are formed to take their places. An example is the burning of propane ("bottled gas"). When this chemical change takes place, propane and oxygen from the air are converted to carbon dioxide and water. Figure 1.2 shows another chemical change.

Matter also undergoes other kinds of changes, called **physical changes.** These changes differ from chemical reactions in that the identities of the substances do not change. Most physical changes involve changes of state—for example, the melting of solids and the boiling of liquids. Water remains water whether in the liquid state or in the form of ice or steam. Conversion from one state to another is a physical, not a chemical, change. Another important type of physical change involves making or separating mixtures. Dissolving sugar in water is a physical change.

When we talk of the **chemical properties** of a substance, we mean the chemical reactions that it undergoes. **Physical properties** are all properties that do not involve chemical reactions; for example, density, color, melting point, and physical state (liquid, solid, gas) are all physical properties.

Digestion of food also involves chemical changes.

The mixing of vinegar, salad oil, and eggs to make mayonnaise is a physical change.

(a) (b) (c)

Figure 1.2 A chemical reaction. (*a*) Bromine, an orange-brown liquid, and aluminum metal. (*b*) These two substances react so vigorously that the aluminum becomes molten and glows white hot on the bottom of the beaker. The yellow vapor consists of vaporized bromine and some of the product of the reaction, white aluminum bromide. (*c*) Once the reaction is complete, the beaker is coated with aluminum bromide and the products of its reaction with atmospheric moisture. (Note: This reaction is dangerous! Under no circumstances should it be done except under properly supervised conditions.)

1.2 The Scientific Method

Scientists learn what they know by using a tool called the **scientific method.** The heart of the scientific method is the testing of theories. It was not always so, however. Before about 1600, philosophers believed statements because they sounded right. For example, the great philosopher Aristotle (384–322 BCE) believed that if you took the gold out of a mine it would grow back. He believed this idea because it fitted in with a more general picture that he had about the workings of nature. In ancient times, most thinkers behaved in this way. If a statement sounded right, they believed it without testing it.

About 1600 the scientific method came into use. Let us look at an example to see how the scientific method operates. The Greek physician Galen (130–200) recognized that the blood on the left side of the heart somehow gets to the right side. This is a fact. A **fact** is a statement based on direct experience. It is a consistent and reproducible observation. Having observed this fact, Galen then proposed a hypothesis to explain it. A **hypothesis** is a statement that is proposed, without actual proof, to explain the facts and their relationship. Because Galen could not actually see how the blood got from the left side to the right side of the heart, he came up with the hypothesis that tiny holes must be present in the muscular wall that separates the two halves.

Up to this point, modern scientists and ancient philosophers would behave the same way. Each would offer a hypothesis to explain the facts. From this point on, however, their methods would differ. To Galen, his explanation sounded right and that was enough to make him believe it, even though he couldn't see any holes. His hypothesis was, in fact, believed by virtually all physicians for more than 1000 years. When we use the scientific method,

> Before about 1600, science as such did not exist except as a branch of philosophy.
>
> For example, it is a fact that humans must breathe to live.

> **Hypothesis** A statement that is proposed, without actual proof, to explain the facts and their relationship

however, we do not believe a hypothesis just because it sounds right. We test it, using the most rigorous testing we can imagine.

William Harvey (1578–1657) tested Galen's hypothesis by dissecting human and animal hearts and blood vessels. He discovered that there are one-way valves separating the upper chambers of the heart from the lower chambers. He also discovered that the heart is a pump that, by contracting and expanding, pushes the blood out. Harvey's teacher, Fabricius (1537–1619), had previously observed that one-way valves exist in the veins, so that blood in the veins can travel only toward the heart and not the other way.

Harvey put all these facts together to come up with a new hypothesis: Blood is pumped by the heart and circulates throughout the body. This was a better hypothesis than Galen's because it fitted the facts more closely. Even so, it was still a hypothesis and, according to the scientific method, had to be tested further. One important test took place in 1661, four years after Harvey died. Harvey had predicted that because there had to be a way for the blood to get from the arteries to the veins, tiny blood vessels must connect them. In 1661 the Italian anatomist Malpighi (1628–1694), using the newly invented microscope, found these tiny vessels, which are now called capillaries.

Malpighi's discovery supported the blood circulation hypothesis by fulfilling Harvey's prediction. When a hypothesis passes the tests, we have more confidence in it and call it a theory. A **theory** is the formulation of an apparent relationship of certain observed phenomena, which has been verified to some extent. In this sense, theory is the same as a hypothesis except that we have a stronger belief in it because more evidence supports it. No matter how much confidence we have in a theory, however, if we discover new facts that conflict with it or if it does not pass newly devised tests, the theory must be altered or rejected. In the history of science, many firmly established theories have eventually been thrown out because they could not pass new tests.

The scientific method is thus very simple. We don't accept a hypothesis or a theory just because it sounds right. We devise tests, and only if the hypothesis or theory passes the tests do we accept it. The enormous progress made since 1600 in chemistry, biology, and the other sciences is a testimony to the value of the scientific method.

You may get the impression from the preceding discussion that science progresses in one direction: facts first, hypothesis second, the theory last. Nothing is so simple, however. Hypotheses and theories call the attention of scientists to discover new facts. An example of this scenario is the discovery of the element germanium. In 1871, Mendeleev's Periodic Table, a graphic description of elements organized by properties, predicted the existence of a new element whose properties would be similar to those of silicon. Mendeleev called this element eka-silicon. In 1886, it was discovered in Germany (hence the name), and its properties were truly similar to those predicted by theory.

On the other hand, many scientific discoveries result from **serendipity**, or chance observation. An example of serendipity occurred in 1926 when James Sumner of Cornell University left an enzyme preparation of jack bean urease in a refrigerator over the weekend. Upon his return, he found that his solution contained crystals that turned out to be a protein. This chance discovery led to the hypothesis that all enzymes are proteins. But serendipity is not enough. Scientists must have the creativity and insight to recognize the significance of their observations. Sumner fought for more than 15 years for his hypothesis to gain acceptance because people believed that only small molecules can form crystals. Eventually his view won out, and he was awarded a Nobel Prize in Chemistry in 1946.

■ **Galen.**

Theory The formulation of an apparent relationship of certain observed phenomena, which has been verified. A theory explains many interrelated facts and can be used to make predictions about natural phenomena. Examples are Newton's theory of gravitation and the kinetic–molecular theory of gases, which we will encounter in Section 5.6. This type of theory is also subject to testing and will be discarded or modified if it is contradicted by new facts.

1.3 Exponential Notation and Significant Figures

Scientists often have to deal with numbers that are very large or very small. For example, an ordinary copper penny (dating from before 1982, when pennies in the United States were still made of copper) contains approximately

$$29{,}500{,}000{,}000{,}000{,}000{,}000{,}000 \text{ atoms of copper}$$

and a single copper atom weighs

$$0.0000000000000000000000023 \text{ pound}$$

> Exponential notation is also called scientific notation.
>
> For example, 10^6 means a one followed by six zeros, or 1,000,000, and 10^2 means 100.

Many years ago, an easy way to handle such large and small numbers was devised. This method, which is called **exponential notation,** is based on powers of 10. In exponential notation, the number of copper atoms in a penny is written

$$2.95 \times 10^{22} \text{ atoms}$$

and the weight of a single copper atom is written

$$2.3 \times 10^{-25} \text{ pound}$$

Problems 1.16 through 1.23 deal with exponential notation, and calculations using exponential notation can be found in Appendix I.

When it comes to measurements, not all the numbers you can generate in your calculator or computer are of equal importance. Only the number of digits that are known with certainty are significant. Suppose that you measured the weight of an object as 3.4 g on a balance that you can read to the nearest 0.1 g. You can report the weight as 3.4 g but not as 3.40 or 3.400 g because you do not know the added zeros with certainty. A detailed account of using **significant figures** is presented in Appendix II. Discussion on accuracy, precision and significant figures can be found in Laboratory Manuals [see Bettelheim and Landesberg, *Laboratory Experiments,* fifth edition (Experiment 2)].

1.4 Measurements

In our daily lives we are constantly making measurements. We measure ingredients for recipes, driving distances, gallons of gasoline, weights of fruits and vegetables, and the timing of TV programs. Doctors and nurses measure pulse rates, blood pressures, temperatures, and drug dosages. Chemistry, like other sciences, is based on measurements.

A measurement consists of two parts: a number and a unit. A number without a unit is usually meaningless. If you were told that a person's weight is 57, the information would be of very little use. Is it 57 pounds, which would indicate that the person is very likely a child or a midget, or 57 kilograms, which is the weight of an average woman or a small man? Or is it perhaps some other unit? Because so many units exist, a number by itself is not enough; the unit must also be stated.

In the United States, most measurements are made with the English system of units: pounds, miles, gallons, and so on. In most parts of the world, however, few people could tell you what a pound or an inch is. Most countries use the **metric system,** a system that originated in France about

> **Metric system** A system of units of measurement in which the divisions to subunits are made by a power of 10

Figure 1.3 A highway sign in Missouri.

1800 and that has since spread throughout the world. Even in the United States, metric measurements are slowly being introduced (Figure 1.3). For example, many soft drinks and most alcoholic beverages now come in metric sizes (Figure 1.4). Scientists in the United States have been using metric units all along.

Around 1960, international scientific organizations adopted another system, called the **International System of Units** (abbreviated **SI**). The SI is based on the metric system and uses some of the metric units. The main difference is that the SI is more restrictive: It discourages the use of certain metric units and favors others. Although the SI has advantages over the older metric system, it also has significant disadvantages. For this reason U.S. chemists have been very slow to adopt it. At this time, approximately 40 years after its introduction, not many U.S. chemists use the entire SI, although some of its preferred units are gaining ground.

In this book we will use the metric system (Table 1.1). Occasionally we will mention the preferred SI unit.

A Length

The key to the metric system (and the SI) is that there is one base unit for each kind of measurement and that other units are related to the base unit only by powers of 10. As an example, let us look at measurements of length. In the English system we have the inch, the foot, the yard, and the mile (not to mention such older units as the league, furlong, ell, and rod). If you want to convert one unit to another unit, you must memorize or look up these conversion factors:

$$12 \text{ inches} = 1 \text{ foot}$$
$$3 \text{ feet} = 1 \text{ yard}$$
$$1760 \text{ yards} = 1 \text{ mile}$$

All this is unnecessary in the metric system (and the SI). In both systems the base unit of length is the **meter (m).** To convert to larger or smaller units we do not use arbitrary numbers like 12, 3, and 1760, but only 10, 100, 1/10, 1/100, or other powers of 10. This means that *to convert from one met-*

Figure 1.4 Soft drinks in metric sizes.

"SI" is actually an abbreviation of the French title *Système Internationale*.

The official definition of the meter is the distance that light travels in 1/299,792,458 of a second.

Table 1.1	Base Units in the Metric System
Length	meter
Volume	liter
Mass	gram
Time	second
Temperature	°Celsius
Energy	calorie
Amount of substance	mole

Screen 1.3 Tutorial

Table 1.2 The Most Common Metric Prefixes

Prefix	Symbol	Value
giga	G	10^9 = 1,000,000,000 (one billion)
mega	M	10^6 = 1,000,000 (one million)
kilo	k	10^3 = 1000 (one thousand)
deci	d	10^{-1} = 0.1 (one-tenth)
centi	c	10^{-2} = 0.01 (one-hundredth)
milli	m	10^{-3} = 0.001 (one-thousandth)
micro	μ	10^{-6} = 0.000001 (one-millionth)
nano	n	10^{-9} = 0.000000001 (one-billionth)

ric or SI unit to another, we only have to move the decimal point. Furthermore, the other units are named by putting prefixes in front of "meter," and *these prefixes are the same throughout the metric system and the SI*. Table 1.2 lists the most important of these prefixes. If we put some of these prefixes in front of "meter," we have

$$1 \text{ kilometer (km)} = 1000 \text{ meters (m)}$$

$$1 \text{ centimeter (cm)} = 0.01 \text{ meter}$$

$$1 \text{ nanometer (nm)} = 10^{-9} \text{ meter}$$

For people who have grown up using English units, it is helpful to have some idea of the size of metric units. Table 1.3 shows some conversion factors.

B Volume

Volume is space. The volume of a liquid, solid, or gas is the space occupied by that substance. The base unit of volume in the metric system is the **liter (L)**. This unit is a little larger than a quart (Table 1.3). The only other common metric unit for volume is the milliliter (mL), which is equal to 10^{-3} L.

$$1000 \text{ mL} = 1 \text{ L}$$

One milliliter is exactly equal to one cubic centimeter (cc or cm^3):

$$1 \text{ mL} = 1 \text{ cc}$$

Thus there are 1000 cc in 1 L.

■ Hypodermic syringe. The numbers represent volumes in μL.

Volume measurements are important in chemistry and medicine. In a hypodermic syringe, the device for injecting the drug (the needle) is combined with the means for measuring the volume injected (the barrel), usually in cubic centimeters.

Table 1.3 Some Conversion Factors Between the English and Metric Systems

Length	Mass	Volume
1 in. = 2.54 cm	1 oz = 28.35 g	1 qt = 0.946 L
1 m = 39.37 in.	1 lb = 453.6 g	1 gal = 3.785 L
1 mile = 1.609 km	1 kg = 2.205 lb	1 L = 33.81 fl oz
	1 g = 15.43 grains	1 fl oz = 29.57 mL
		1 L = 1.057 qt

C Mass

Mass is the quantity of matter in an object. The base unit of mass in the metric system is the **gram (g)**. As always in the metric system, larger and smaller units are indicated by prefixes. The ones in common use are

$$1 \text{ kilogram (kg)} = 1000 \text{ g}$$

$$1 \text{ milligram (mg)} = 0.001 \text{ g}$$

The gram is a small unit; there are 453.6 g in one pound (Table 1.3).

We use a device called a balance to measure mass. Figure 1.5 shows two types of laboratory balances.

There is a fundamental difference between mass and weight. Mass is independent of location. The mass of a stone, for example, is the same whether we measure it at sea level, on top of a mountain, or in the depths of a mine. In contrast, weight is not independent of location. **Weight** is the force a mass experiences under the pull of gravity. This point was dramatically demonstrated when the astronauts walked on the surface of the Moon. The Moon, being a smaller body than the Earth, exerts a weaker gravitational pull. Consequently, even though the astronauts wore space suits and equipment that would be heavy on Earth, they felt lighter on the Moon and could execute great leaps and bounces during their walks.

Although mass and weight are different concepts, they are related to each other by the force of gravity. We frequently use the words interchangeably because we weigh objects by comparing their masses to standard reference masses (weights) on a balance, and the gravitational pull is the same on the unknown object and on the standard masses. Because the force of gravity is essentially constant, mass is always directly proportional to weight.

The official kilogram is defined by the mass of a particular platinum-iridium bar kept at the International Bureau of Standards in Paris.

Both mass and weight are independent of temperature.

The Moon's gravity is six times smaller than that of the Earth. Consequently, an astronaut weighing 240 lb on Earth will weigh only 40 lb on the Moon.

(a)

(b)

Figure 1.5 Two laboratory balances.

CHEMICAL CONNECTIONS 1A

Drug Dosage and Body Mass

In many cases, drug dosages are prescribed on the basis of body mass. For example, the recommended dosage of a drug may be 3 mg of drug for each kilogram of body weight. In this case, a 50 kg (110 lb) woman would receive 150 mg and an 82 kg (180 lb) man would get 246 mg. This adjustment is especially important for children, because a dose suitable for an adult will generally be too much for a child, who has much less body mass. For this reason, manufacturers package and sell smaller doses of certain drugs, such as aspirin, for children.

Drug dosage may also vary with age. Occasionally, when an elderly patient has an impaired kidney or liver function, the clearance of a drug from the body is delayed, and the drug may stay in the body longer than is normal. This persistence can cause dizziness, vertigo, and migraine-like headaches, resulting in falls and broken bones. Such delayed clearance must be monitored, and the drug dosage adjusted accordingly.

■ These packages of Advil and Tylenol have charts showing the proper doses for children of a given weight.

D Time

Time is the one quantity for which the units are the same in all systems: English, metric, and SI. The base unit is the **second (s)**:

$$60 \text{ s} = 1 \text{ min}$$
$$60 \text{ min} = 1 \text{ h}$$

E Temperature

Most people in the United States are familiar with the Fahrenheit scale of temperature. The metric system uses the centigrade, or **Celsius,** scale. In this scale, the boiling point of water is set at 100°C and the freezing point at 0°C. We can convert from one scale to the other by using the following formulas:

$$°F = \frac{9}{5}°C + 32$$

$$°C = \frac{5}{9}(°F - 32)$$

The 32 in these equations is a defined number and is, therefore, treated as if it had an infinite number of zeros following the decimal point. (See Appendix II.)

Screen 1.10 Tutorial

EXAMPLE 1.1

Normal body temperature is 98.6°F. Convert this temperature to Celsius.

Solution

$$°C = \frac{5}{9}(98.6 - 32) = \frac{5}{9}(66.6) = 37.0°C$$

> **Problem 1.1**
> Convert:
> (a) 64.0°C to Fahrenheit (b) 47°F to Celsius.

Figure 1.6 shows the relationship between the Fahrenheit and Celsius scales.

A third temperature scale is the **Kelvin (K)** scale, also called the absolute scale. The size of the Kelvin degree is the same as that of the Celsius degree; the only difference is the zero point. The temperature −273°C is taken as the zero point on the Kelvin scale. This makes conversions between Kelvin and Celsius very easy. To go from Celsius to Kelvin, just *add* 273; to go from Kelvin to Celsius, *subtract* 273:

$$K = °C + 273$$

$$°C = K - 273$$

Figure 1.6 also shows the relationship between the Kelvin and Celsius scales. Note that we don't use the degree symbol in the Kelvin scale: 100°C equals 373 K, not 373°K.

Why was −273°C chosen as the zero point on the Kelvin scale? The reason is that −273°C, or 0 K, is the lowest possible temperature. Because of this, 0 K is called **absolute zero**. Temperature reflects how fast molecules move. The more slowly they move, the colder it gets. At absolute zero, molecules stop moving altogether. Therefore, the temperature cannot get any lower. For some purposes it is convenient to have a scale that begins at the lowest possible temperature; the Kelvin scale fulfills this need. The Kelvin is the SI unit.

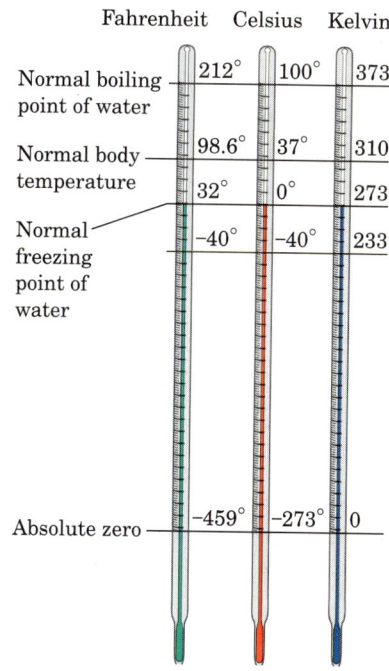

Figure 1.6 Three temperature scales.

Scientists in laboratories have been able to attain temperatures as low as 0.000001 K, but a temperature of 0 K is impossible.

1.5 Unit Conversions: The Factor-Label Method

We frequently need to convert a measurement from one unit to another. The best and most foolproof way to do this is the **factor-label method**. In this method we follow the rule that *when multiplying numbers we also multiply units, and when dividing numbers we also divide units.*

For conversions between one unit and another, it is always possible to set up two fractions, called **conversion factors.** Suppose we wish to convert 381 g to pounds. In Table 1.3, we see that there are 453.6 grams in 1 pound. The conversion factors between grams and pounds therefore are

$$\frac{1 \text{ lb}}{453.6 \text{ g}} \quad \text{and} \quad \frac{453.6 \text{ g}}{1 \text{ lb}}$$

To convert 381 grams to pounds, we must simply multiply by the proper conversion factor. But which one? Let us try both and see what happens. First let us multiply by 1 lb/453.6 g:

$$381 \text{ g} \times \frac{1 \text{ lb}}{453.6 \text{ g}} = 0.840 \text{ lb}$$

Following the procedure of multiplying and dividing units when we multiply and divide numbers, we find that dividing grams by grams cancels out the grams. We are left with pounds, which is the answer we want. Thus

> **Factor-label method** A procedure in which the equations are set up so that all the unwanted units cancel and only the desired units remain

> **Conversion factor** A ratio of two different units

Screen 1.4 Tutorial

1 lb/453.6 g is the correct conversion factor because it converts grams to pounds. Suppose we had done it the other way, multiplying by 453.6 g/1 lb:

$$381 \text{ g} \times \frac{453.6 \text{ g}}{1 \text{ lb}} = 173,000 \frac{\text{g}^2}{\text{lb}}$$

When we multiply grams by grams, we get g² (grams squared). Dividing by pounds gives g²/lb. This is not the unit we want, so we used the incorrect conversion factor.

The advantage of the factor-label method is that it lets us know when we have made a wrong calculation. *If the units of the answer are not the ones we are looking for, the calculation must be wrong.* Incidentally, this principle works not only in unit conversions but in all problems where we make calculations using measured numbers.

The factor-label method gives the correct mathematical solution for a problem. However, it is a mechanical technique and does not require you to think through the problem. Thus it may not provide a deeper understanding. For this reason and also to check your answer (because it is easy to make mistakes in arithmetic—for example, by punching the wrong numbers into a calculator), you should always ask yourself if the answer you have obtained is reasonable. For example, the question might ask the actual mass of a single oxygen atom. If your answer comes out 8.5×10^6 g, it is not reasonable. A single atom cannot weigh more than you do! In such a case, you have obviously made a mistake and should go back and take another look to see where you went wrong. Everyone makes mistakes at times, but if you check you can at least determine whether your answer is reasonable. If it is not, you will immediately know that you have made a mistake and can then correct it.

Checking whether an answer is reasonable gives you a deeper understanding of the problem because it forces you to think through the relationship between the question and the answer. We will now give a few examples of unit conversions and then test the answers to see if they are reasonable. To save space, we will practice this technique mostly in this chapter, but you should use a similar approach in all later chapters.

In unit conversion problems, you should always check two things. First, the numeric factor by which you multiply tells you whether the answer will be larger or smaller than the number being converted. Second, the factor tells you how much greater or smaller than the number you start with your answer should be. For example, if 100 kg is converted to lb and there are 2.205 lbs in every kg then an answer of about 200 is reasonable—but an answer of 0.2 or 2000 is not.

EXAMPLE 1.2

The distance between Rome and Milan (the largest cities in Italy) is 358 miles. How many kilometers separate the two?

Solution
We want to convert miles to kilometers. Table 1.3 shows that 1 mile = 1.609 km. From this we get two conversion factors:

$$\frac{1 \text{ mile}}{1.609 \text{ km}} \quad \text{and} \quad \frac{1.609 \text{ km}}{1 \text{ mile}}$$

Which should we use? We use the one that gives the answer in kilometers:

$$358 \text{ miles} \times \text{conversion factor} = ? \text{km}$$

This means that the miles must cancel, so the conversion factor 1.609 km/1 mile is appropriate.

$$358 \text{ miles} \times \frac{1.609 \text{ km}}{1 \text{ mile}} = 576 \text{ km}$$

Is this answer reasonable? We want to convert miles to kilometers. The conversion factor in Table 1.3 tells us that in a given distance the number of kilometers is larger than the number of miles. How much larger? The exact number is 1.609, which is approximately 1.5 times larger. Thus we expect that the answer in kilometers will be about 1.5 times greater than the number given in miles. The number given in miles is 358, which, *for the purpose of checking whether our answer is reasonable,* we can round off to, say, 400. Multiplying this number by 1.5 gives an approximate answer of 600 km. Our actual answer, 576 km, was of the same order of magnitude as the estimated answer, so we can say that it is reasonable. If the estimated answer had been 6 km, or 60 km, or 6000 km, we would suspect that we had made a mistake in calculating the actual answer.

Problem 1.2

How many kilograms are in 241 lb? Check your answer to see if it is reasonable.

EXAMPLE 1.3

The label on a container of olive oil says 1.844 gal. How many mL does the container hold?

Solution
Table 1.3 shows no factor for converting gallons to milliliters, but it does show that 1 gal = 3.785 L. Because we know that 1000 mL = 1 L, we can solve this problem by multiplying by two conversion factors, making certain that all units cancel except milliliters:

$$1.844 \text{ gal} \times \frac{3.785 \text{ L}}{1 \text{ gal}} \times \frac{1000 \text{ mL}}{1 \text{ L}} = 6980 \text{ mL}$$

Is this answer reasonable? The conversion factor in Table 1.3 tells us that there are more liters in a given volume than gallons. How much more? Approximately four times more. We also know that any volume in milliliters is 1000 times larger than the same volume in liters. Thus we expect that the volume expressed in milliliters will be 4 × 1000 or 4000 times more than the volume given in gallons. The estimated volume in milliliters will be approximately 1.8 × 4000, or 7000 mL. But we also expect that the actual answer should be somewhat less than the estimated figure because we overestimated the conversion factor (4 rather than 3.785). Thus the answer, 6980 mL, is quite reasonable.

Problem 1.3
Calculate the number of kilometers in 8.55 miles. Check your answer to see if it is reasonable.

EXAMPLE 1.4

The maximum speed limit on many roads in the United States is 65 mph. How many meters per second is this speed?

Solution
Here we have essentially a double conversion problem: We must convert miles to meters and hours to seconds. We use as many conversion factors as necessary, always making sure that we use them in such a way that the proper units cancel:

$$65 \frac{\text{miles}}{\text{h}} \times \frac{1.609 \text{ km}}{1 \text{ mile}} \times \frac{1000 \text{ m}}{1 \text{ km}} \times \frac{1 \text{ h}}{60 \text{ min}} \times \frac{1 \text{ min}}{60 \text{ s}} = 29 \frac{\text{m}}{\text{s}}$$

Is this answer reasonable? To estimate the 65 mph speed in meters per second, we must first establish the relationship between miles and meters. As in Example 1.2, we know that there are more kilometers than miles in a given distance. How much more? As there are approximately 1.5 km in a mile, there must be approximately 1500 times more meters. We also know that in 1 hour there are 60 × 60 = 3600 seconds. There are approximately half as many more meters in a mile than seconds in an hour: 1500/3600. Therefore, we estimate that the speed in meters per second will be about one-half of that in miles per hour or 32 m/s. Once again, the actual answer, 29 m/s, is not far from the estimate of 32 m/s, so the answer is reasonable.

As shown in these examples, when canceling units we do not cancel the numbers. The numbers are multiplied and divided in the ordinary way.

Problem 1.4
Convert the speed of sound, 332 m/s, to m/h. Check your answer to see if it is reasonable.

> Estimating the answer is a good thing to do when working any mathematical problem, not just unit conversions.

1.6 The States of Matter

Matter can exist in three states: gas, liquid, and solid. **Gases** have no definite shape or volume. They expand to fill whatever container they are put into. On the other hand, they are highly compressible and can be forced into small containers. **Liquids** also have no definite shape, but they do

> The most familiar gas is the air around us, which we depend on for life itself.

(a)　　　　　　　　(b)　　　　　　　　(c)

Figure 1.7 The three states of matter for bromine: (a) bromine as a solid, (b) bromine as a liquid, and (c) bromine as a gas.

have a definite volume that remains the same when they are poured from one container to another. Liquids are only slightly compressible. **Solids** have definite shapes and definite volumes. They are essentially incompressible.

Whether a substance is a gas, a liquid, or a solid depends on its temperature. On a cold winter day, a puddle of liquid water turns to ice; it becomes a solid. If we boil water, the liquid becomes a gas—we call it steam. Most substances can exist in the three states: They are gases at high temperature, liquids at a lower temperature, and solids when their temperature becomes low enough. Figure 1.7 shows a single substance in the three different states.

The chemical identity of a substance does not change when it is converted from one state to another. Water is still water whether it is in the form of ice, steam, or liquid water. We discuss the three states of matter, and the changes between one state and another, at greater length in Chapter 5.

1.7 Density and Specific Gravity

A Density

One of the many pollution problems that the world faces is the spillage of petroleum into the oceans from oil tankers or from offshore drilling. When oil spills into the ocean, it floats on top of the water. The oil doesn't sink because water has a higher density than oil. When two liquids are mixed (assuming that one does not dissolve in the other), the one of lower density floats on top (Figure 1.8).

The **density** of any substance is defined as its *mass per unit volume*. Not only do all liquids have a density, but so do all solids and gases. Density is calculated by dividing the mass of a substance by its volume:

$$d = \frac{m}{V} \qquad d = \text{density}, \quad m = \text{mass}, \quad V = \text{volume}$$

Figure 1.8 Two separatory funnels containing water and another liquid. The density of carbon tetrachloride is 2.961 g/mL that of water is 1.00 g/mL, and that of diethyl ether is 0.713 g/mL. In each case the liquid with the lower density is on top.

16 / CHAPTER 1 MATTER, ENERGY, AND MEASUREMENT

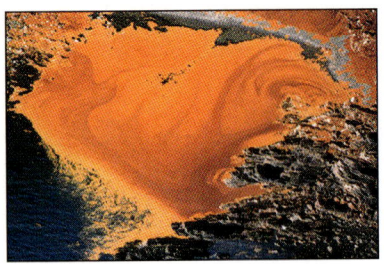

■ The spillage of more than 10 million gallons of petroleum in Prince William Sound, Alaska, in March 1989 caused a great deal of environmental damage.

EXAMPLE 1.5

If 73.2 mL of a liquid has a mass of 61.5 g, what is its density in g/mL?

Solution

$$d = \frac{m}{V} = \frac{61.5 \text{ g}}{73.2 \text{ mL}} = 0.840 \frac{\text{g}}{\text{mL}}$$

EXAMPLE 1.6

The density of iron is 7.86 g/mL. What is the volume in mL of an irregularly shaped piece of iron that has a mass of 524 g?

Solution
Here we are given the mass and the density. In this type of problem, it is useful to derive a conversion factor from the density. We know that the density is 7.86 g/mL. This means that 1 mL has a mass of 7.86 g. From this we can get two conversion factors:

$$\frac{1 \text{ mL}}{7.86 \text{ g}} \quad \text{and} \quad \frac{7.86 \text{ g}}{1 \text{ mL}}$$

As usual, we multiply the mass by whichever conversion factor results in the cancellation of all but the correct unit:

$$524 \text{ g} \times \frac{1 \text{ mL}}{7.86 \text{ g}} = 66.7 \text{ mL}$$

Is this answer reasonable? The density of 7.86 g/mL tells us that the volume in mL of any piece of iron is always less than its mass in grams. How much less? Approximately eight times less. Thus we expect the volume to be approximately 500/8 = 63 mL. As the actual answer is 66.7 mL, it is reasonable.

Screen 1.6 Tutorial

Problem 1.5
The density of titanium is 4.54 g/mL. What is the mass, in grams, of 17.3 mL of titanium? Check your answer to see if it is reasonable.

Problem 1.6
An unknown substance has a mass of 56.8 g and occupies a volume of 23.4 mL. What is its density, in g/mL? Check your answer to see if it is reasonable.

In old movies, villains frequently used a chloroform-wetted handkerchief to subdue their victims.

The density of any liquid or solid is a physical property that is constant, which means that it always has the same value at a given temperature. We use physical properties to help identify a substance. For example, the density of chloroform (a liquid formerly used as an anesthetic) is 1.483 g/mL at 20°C. If we want to find out if an unknown liquid is chloroform, one thing we might do is measure its density at 20°C. If the density is, say, 1.355 g/mL, we know the liquid isn't chloroform. If the density does come out to be 1.483 g/mL, we cannot be sure the liquid is chloroform, because other liquids might also have this density, but we can then measure other physical properties (the boiling point, for example). If all the physical properties we measure match those of chloroform, we can be reasonably sure the liquid is chloroform.

We have said that the density of a liquid or solid is a constant at a given temperature. Density does change when the temperature changes. Almost always, density decreases with increasing temperature. This is true because mass does not change when a substance is heated, but volume almost always increases because atoms and molecules tend to get farther apart as the temperature increases. Since $d = m/V$, if m stays the same and V gets larger, d must get smaller.

The most common liquid, water, provides a partial exception to this rule. As the temperature increases from 4°C to 100°C, the density of water does decrease, but from 0°C to 4°C the density increases. That is, water has its maximum density at 4°C. This anomaly and its consequences are due to the unique structure of water and will be discussed in Chemical Connections 5F.

B Specific Gravity

Because density is equal to mass divided by volume, it always has units, most commonly g/mL or g/cc (g/L for gases). **Specific gravity** is numerically the same as density, but it has no units (it is dimensionless). The reason is that specific gravity is defined as a comparison of the density of a substance with the density of water, which is taken as a standard. For example, the density of copper at 20°C is 8.92 g/mL. The density of water at the same temperature is 1.00 g/mL. Therefore, copper is 8.92 times as dense as water, and its specific gravity at 20°C is 8.92. Because water is taken as the standard and because the density of water is 1.00 g/mL at 20°C, the specific gravity of any substance is always numerically equal to its density, provided that the density is measured in g/mL or g/cc.

Specific gravity is often measured by a hydrometer. This simple device consists of a weighted glass bulb that is inserted into a liquid and allowed to float. The stem of the hydrometer has markings, and the specific gravity is read where the meniscus of the liquid hits the marking. The specific gravity of the acid in your car battery and that of a urine sample in a clinical laboratory are measured by hydrometers. A hydrometer measuring a urine sample is also called urinometer (Figure 1.9). Normal urine can vary in specific

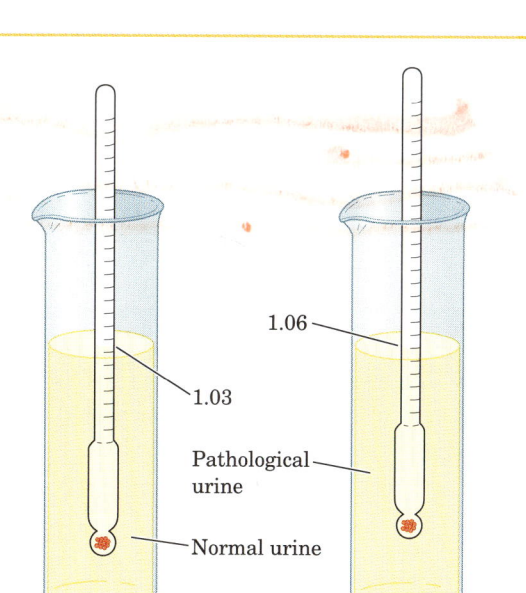

Figure 1.9 Urinometer.

gravity from about 1.010 to 1.030. Patients with diabetes mellitus have abnormally high, while those with Bright's disease have abnormally low specific gravity of their urine samples.

EXAMPLE 1.7

The density of ethanol at 20°C is 0.789 g/mL. What is its specific gravity?

Solution

$$\text{Specific gravity} = \frac{0.789 \text{ g/mL}}{1.00 \text{ g/mL}} = 0.789$$

Problem 1.7

The specific gravity of a urine sample is 1.016. What is its density, in g/mL?

1.8 Energy

Energy is defined as the capacity to do work. It can be described as being either kinetic energy or potential energy.

Kinetic energy (KE) is the energy of motion. Any object that is moving possesses kinetic energy. We can calculate how much energy by the formula $KE = 1/2 mv^2$, where m is the mass of the object and v is its velocity. This means that kinetic energy increases (1) when an object moves faster and (2) when a heavier object is moving. When a truck and a bicycle are moving at the same velocity, the truck has more kinetic energy.

Potential energy is stored energy. The potential energy possessed by an object arises from its capacity to move or to cause motion. For example, a book balanced at the edge of a table contains potential energy—it is capable of doing work. If given a slight push, it will fall. The potential energy the book has while it is on the table is converted to the kinetic energy it has while it is falling. Figure 1.10 shows another way in which potential energy is converted to kinetic energy.

An important principle in nature is that things have a tendency to seek their lowest possible potential energy. We all know that water always flows downhill and not uphill.

Several forms of energy exist, of which the most important are (1) mechanical energy, light, heat, and electrical energy, which are examples of kinetic energy possessed by all moving objects, whether elephants or molecules or electrons; and (2) chemical energy and nuclear energy, which are

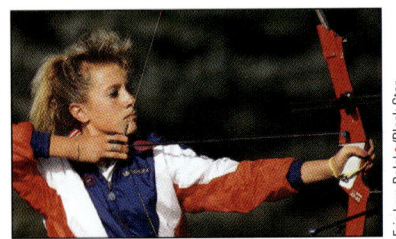

■ Potential energy is stored in this drawn bow and becomes kinetic energy in the arrow when released.

Hydroelectric plants use the kinetic energy of falling water.

Figure 1.10 The water held back by the dam possesses potential energy, which is converted to kinetic energy when the water is released.

Potential energy ready to do work

Kinetic energy doing work

examples of potential energy or stored energy. In chemistry the most important form of potential energy is chemical energy—the energy stored within chemical substances and given off when they take part in a chemical reaction. For example, a piece of paper possesses chemical energy. When the paper is ignited, the chemical energy (potential) is turned into energy in the form of heat and light.

The various forms of energy can be converted from one to another. In fact, we make such conversions all the time. A power plant operates either on the chemical energy derived from burning fuel or on nuclear energy. This energy is converted to heat, which is converted to the electricity that is sent over transmission wires into houses and factories. Here we convert the electricity to light, heat (in an electrical heater, for example), or mechanical energy (in the motors of refrigerators, vacuum cleaners, and other devices).

Although one form of energy can be converted to another, the *total amount* of energy in any system does not change. *Energy can neither be created nor destroyed.* This statement is called the **law of conservation of energy.***

■ An example of energy conversion: Light energy from the sun is converted to electrical energy by solar cells. The electricity runs the refrigerator on the back of the camel, keeping the vaccines inside cool so that they can be delivered to remote locations.

1.9 Heat

A Heat and Temperature

One form of energy that is particularly important in chemistry is **heat.** This is the form of energy that most frequently accompanies chemical reactions. Heat is not the same as temperature, however. Heat is a form of energy, but temperature is not.

The difference between heat and temperature can be seen in the following example. If we have two beakers, one containing 100 mL of water and the other 1 L of water at the same temperature, the heat content of the water in the larger beaker is ten times that of the water in the smaller beaker, even though the temperature is the same in both. If you were to dip your hand accidentally into a liter of boiling water, you would be much more severely burned than if only one drop fell on your hand. Even though the water is at the same temperature in both cases, the liter of boiling water has much more heat.

As we saw in Section 1.4, temperature is measured in degrees. Heat can be measured in various units, the most common of which is the **calorie,** which is defined as the amount of heat necessary to raise the temperature of 1 g of liquid water by 1°C. This is a small unit, and chemists more often use the kilocalorie (kcal):

$$1 \text{ kcal} = 1000 \text{ cal}$$

Nutritionists use the word "Calorie" (with a capital "C") to mean the same thing as "kilocalorie"; that is, 1 Cal = 1000 cal = 1 kcal. The calorie is not part of the SI. The official SI unit for heat is the **joule (J),** which is about one-fourth as big as the calorie:

$$1 \text{ cal} = 4.184 \text{ J}$$

Calories and joules are units not only of heat but of other forms of energy as well.

---

*This statement is not completely true. As discussed in Chemical Connections 9F, it is possible to convert matter to energy, and vice versa. Therefore, a more correct statement would be *matter-energy can be neither created nor destroyed.* However, the law of conservation of energy is valid for most purposes and is highly useful.

CHEMICAL CONNECTIONS 1B

Hypothermia and Hyperthermia

The human body cannot tolerate temperatures that are too low. A person outside in very cold weather (say, −20°F [−29°C]) who is not protected by heavy clothing will eventually freeze to death because the body loses heat. Normal body temperature is 37°C. When the outside temperature is lower than that, heat flows out of the body. When the air temperature is moderate (10°C to 25°C), this poses no problem and is, in fact, necessary because the body produces more heat than it needs and must lose some. At extremely low temperatures, however, too much heat is lost and body temperature drops, a condition called **hypothermia**. A drop in body temperature of 1 or 2°C causes shivering, which is the body's attempt to increase its temperature by the heat generated through muscular action. An even greater drop results in unconsciousness and eventually death.

The opposite condition is **hyperthermia**. It can be caused either by high outside temperatures or by the body itself when an individual develops a high fever. A sustained body temperature as high as 41.7°C (107°F) is usually fatal.

B Specific Heat

Specific heat is sometimes called specific heat capacity.

As we noted, it takes 1 cal to raise the temperature of 1 g of liquid water by 1°C. **Specific heat (SH) is the amount of heat necessary to raise the temperature of 1 g of any substance by 1°C.** Each substance has its own specific heat, which is a physical property of that substance, like density or melting point. Table 1.4 lists specific heats for a few common substances. For example, the specific heat of iron is 0.11 cal/g·°C. Therefore, if we had 1 g of iron at 20°C, it would require only 0.11 cal to increase the temperature to 21°C. Under the same conditions, aluminum would require twice as much heat. Thus cooking in an aluminum pan of the same weight as an iron pan would require more heat than cooking in the iron pan. Note from Table 1.4 that ice and steam do not have the same specific heat as liquid water.

Water has the highest specific heat of any common substance.

It is easy to make calculations involving specific heats. The equation is

$$\text{Amount of heat} = \text{specific heat} \times \text{mass} \times \text{change in temperature}$$
$$= SH \times m \times (T_2 - T_1)$$

where T_2 is the final temperature and T_1 is the initial temperature in °C.

Screen 1.12 Simulation

Table 1.4 Specific Heats for Some Common Substances

Substance	Specific Heat (cal/g · °C)	Substance	Specific Heat (cal/g · °C)
Water	1.00	Wood (typical)	0.42
Ice	0.48	Glass (typical)	0.22
Steam	0.48	Rock (typical)	0.20
Iron	0.11	Ethanol	0.59
Aluminum	0.22	Methanol	0.61
Copper	0.092	Ether	0.56
Lead	0.038	Carbon tetrachloride	0.21

EXAMPLE 1.8

How many calories are required to heat 352 g of water from 23°C to 95°C?

Solution

$$\text{Amount of heat} = \text{SH} \times m \times (T_2 - T_1)$$

$$= \frac{1.0 \text{ cal}}{g \cdot °C} \times 352 \text{ g} \times (95 - 23)°C$$

$$= 2.5 \times 10^4 \text{ cal} = 25 \text{ kcal}$$

Is this answer reasonable? Each gram of water requires one calorie to raise its temperature by one degree. We have approximately 350 g of water. To raise its temperature by one degree would therefore require approximately 350 calories. But we are raising the temperature not by one degree but by approximately 70 degrees (from 23 to 95). Thus the total number of calories will be approximately 70 × 350 = 24,500 cal, or 25 kcal. In this case the estimate comes out exactly the same as the calculated answer.

Problem 1.8

How many calories are required to heat 731 g of water from 8°C to 74°C? Check your answer to see if it is reasonable.

EXAMPLE 1.9

If we add 450 cal of heat to 37 g of ethanol at 20°C, what is the final temperature?

Solution

The specific heat of ethanol is 0.59 cal/g·°C (see Table 1.4).

$$\text{Amount of heat} = \text{SH} \times m \times (T_2 - T_1)$$

$$450 \text{ cal} = 0.59 \text{ cal/g} \cdot °C \times 37 \text{ g} \times (T_2 - T_1)$$

$$(T_2 - T_1) = 450 \text{ cal}/(0.59 \text{ cal/g} \cdot °C \times 37 \text{ g}) = 21°C$$

Since the starting temperature is 20°C, the final temperature is 41°C.

Is this answer reasonable? The specific heat of ethanol is 0.59 cal/g·°C. This value is close to 0.5, meaning that about half a calorie will raise the temperature of 1 g by 1°C. However, 37 g of ethanol needs approximately 40 times as many calories for a 1°C rise, and 40 × 1/2 = 20 calories. We are adding 450 calories, which is about 20 times as much. Thus we expect the temperature to rise by about 20°C, from 20°C to 40°C. The actual answer, 41°C, is quite reasonable.

Problem 1.9

A 100 g piece of iron at 25°C is heated by adding 230 cal. What will be the final temperature? Check your answer to see if it is reasonable.

CHEMICAL CONNECTIONS 1C

Cold Compresses

The high specific heat of water is useful in cold compresses and makes them last a long time. For example, consider two patients with cold compresses: one compress made by soaking a towel with water and the other made with a towel soaked in ethanol. Both are at 0°C. Each gram of water in the water compress requires 25 cal to make the temperature of the compress rise to 25°C (after which it must be changed). Because the specific heat of ethanol is 0.59 cal/g·°C (see Table 1.4), each gram of the ethanol requires only 15 cal to reach 25°C. If the two patients give off heat at the same rate, the ethanol compress is less effective because it will reach 25°C a good deal sooner than the water compress and will need to be changed sooner.

EXAMPLE 1.10

We heat 50.0 g of an unknown substance by adding 205 cal, and its temperature rises by 7.0°C. What is its specific heat? Using Table 1.4, identify the substance.

Solution

$$SH = \frac{\text{Amount of heat}}{m \times (T_2 - T_1)}$$

$$SH = \frac{205 \text{ cal}}{50.0 \text{ g} \times 7.0°C} = 0.59 \text{ cal/g} \cdot °C$$

The substance in Table 1.4 having a specific heat of 0.59 cal/g·°C is ethanol.

Is this answer reasonable? If we had water instead of an unknown substance with SH = 1 cal/g·°C, raising the temperature of 50.0 g by 7.0°C would require 50 × 7.0 = 350 cal. But we added only approximately 200 cal. Therefore, the SH of the unknown substance must be less than 1.0. How much less? Approximately 200/350 = 0.6. The actual answer, 0.59 cal/g·°C, is quite reasonable.

Problem 1.10

It required 88.2 cal to heat 13.4 g of an unknown substance from 23°C to 176°C. What is the specific heat of the unknown substance? Check your answer to see if it is reasonable.

SUMMARY

Chemistry is the science that deals with the structure of matter and the changes it can undergo. In a **chemical change** or **chemical reaction**, substances are used up and others are formed. Chemistry is also the study of energy changes during chemical reactions. In **physical changes** substances do not change their identity. There are three states of matter: **solid, liquid, and gas** (Section 1.1).

The **scientific method** is a tool used in science and medicine. The heart of the scientific method is the testing of **hypotheses** and **theories** by collecting facts (Section 1.2).

In chemistry we use the **metric system** for measurements. The base units are the meter for length, the liter for volume, the gram for mass, the second for time, and the calorie for heat. Other units are indicated by prefixes that represent powers of 10. Temperature is measured in degrees Celsius or in kelvins (Section 1.4). Conversions from one unit to another are best done by the **factor-label method,** in which units are multiplied and divided (Section 1.5).

Density is mass per unit volume. **Specific gravity** is density relative to water and thus has no units. Density usually decreases with increasing temperature (Section 1.7).

Kinetic energy is energy of motion; **potential energy** is stored energy. Energy can neither be created nor destroyed, but it can be converted from one form to another (Section 1.8).

Heat is a form of energy and is measured in calories. A calorie is the amount of heat necessary to raise the temperature of 1 g of liquid water by 1°C. Every substance has a **specific heat,** which is a physical constant. The specific heat is the number of calories required to raise the temperature of 1 g of a substance by 1°C (Section 1.9).

PROBLEMS

Numbers that appear in color indicate difficult problems.
◢ designates problems requiring application of principles.

Chemistry and the Scientific Method

1.11 The life expectancy of a citizen in the United States is 76 years. Eighty years ago it was 56 years. In your opinion, what was the major contributor to this spectacular increase in life expectancy? Explain your answer.

1.12 Define the following terms:
(a) Matter (b) Chemistry

1.13 In Table 1.4 you find four metals (iron, aluminum, copper, and lead) and three organic compounds (ethanol, methanol, and ether). What kind of hypothesis would you suggest about the specific heats of these chemicals?

1.14 In a newspaper, you read that Dr. X claimed that he has found a new remedy to cure diabetes. The remedy is an extract of carrots. How would you classify this claim: (a) fact, (b) theory, (c) hypothesis, or (d) hoax? Explain your choice of answer.

1.15 Classify each of the following as a chemical or physical change:
(a) Burning gasoline
(b) Making ice cubes
(c) Boiling oil
(d) Melting lead
(e) Rusting iron
(f) Making ammonia from nitrogen and hydrogen
(g) Digesting food

Exponential Notation

1.16 Write in exponential notation:
(a) 0.351 (b) 602.1
(c) 0.000128 (d) 628122

1.17 Write out in full:
(a) 4.03×10^5 (b) 3.2×10^3
(c) 7.13×10^{-5} (d) 5.55×10^{-10}

1.18 Multiply:
(a) $(4.73 \times 10^5)(1.37 \times 10^2)$
(b) $(2.7 \times 10^{-4})(5.9 \times 10^8)$
(c) $(6.49 \times 10^7)(7.22 \times 10^{-3})$
(d) $(3.4 \times 10^{-5})(8.2 \times 10^{-11})$

1.19 Divide:
(a) $\dfrac{6.02 \times 10^{23}}{3.01 \times 10^5}$ (b) $\dfrac{3.14}{2.30 \times 10^{-5}}$
(c) $\dfrac{7.08 \times 10^{-8}}{300}$ (d) $\dfrac{5.8 \times 10^{-6}}{6.6 \times 10^{-8}}$
(e) $\dfrac{7.05 \times 10^{-3}}{4.51 \times 10^5}$

1.20 Add:
(a) $(7.9 \times 10^4) + (5.2 \times 10^4)$
(b) $(8.73 \times 10^4) + (6.7 \times 10^3)$
(c) $(3.61 \times 10^{-4}) + (4.776 \times 10^{-3})$

1.21 Subtract:
(a) $(8.50 \times 10^3) - (7.61 \times 10^2)$
(b) $(9.120 \times 10^{-2}) - (3.12 \times 10^{-3})$
(c) $(1.3045 \times 10^2) - (2.3 \times 10^{-1})$

1.22 Solve:
$$\dfrac{(3.14 \times 10^3) \times (7.80 \times 10^5)}{(5.50 \times 10^2)}$$

1.23 Solve:
$$\dfrac{(9.52 \times 10^4) \times (2.77 \times 10^{-5})}{(1.39 \times 10^7) \times (5.83 \times 10^{-3})}$$

Significant Figures

1.24 How many significant figures are in the following:
(a) 0.012 (b) 0.10203
(c) 36.042 (d) 8401.0
(e) 32100 (f) 0.0402
(g) 0.000012

1.25 How many significant figures are in the following:
(a) 5.71×10^{13}
(b) 4.4×10^5
(c) 3×10^{-6}
(d) 4.000×10^{-11}
(e) 5.5550×10^{-3}

1.26 Round off to two significant figures:
(a) 91.621 (b) 7.329
(c) 0.677 (d) 0.003249
(e) 5.88

1.27 Multiply these numbers, using the correct number of significant figures in your answer:
(a) 3630.15×6.8
(b) 512×0.0081
(c) $5.79 \times 1.85825 \times 1.4381$

1.28 Divide these numbers, using the correct number of significant figures in your answer:
(a) $\dfrac{3.185}{2.08}$
(b) $\dfrac{6.5}{3.0012}$
(c) $\dfrac{0.0035}{7.348}$

1.29 Add these groups of measured numbers using the correct number of significant figures in your answer:
(a) 37.4083
5.404
10916.3
3.94
0.0006

(b) 84
8.215
0.01
151.7

(c) 51.51
100.27
16.878
3.6817

Measurements

1.30 In the SI system, the second is the base unit of time. We talk about atomic events that occur in picoseconds (10^{-12} s) or even in femtoseconds (10^{-15} s). But we don't talk about megaseconds or kiloseconds; the old standards of minutes, hours, and days prevail. How many minutes and hours are 20 kiloseconds?

1.31 How many g are in the following: (a) 1 kg (b) 1 mg

1.32 Estimate without actually calculating which one is the shorter distance:
(a) 20 mm or 0.3 m
(b) 1 inch or 30 mm
(c) 2000 m or 1 mile

1.33 For each of these, tell which answer is closest:
(a) A baseball bat has a length of 100 mm or 100 cm or 100 m
(b) A glass of milk holds 23 cc or 230 mL or 23 L
(c) A man weighs 75 mg or 75 g or 75 kg
(d) A tablespoon contains 15 mL or 150 mL or 1.5 L
(e) A paper clip weighs 50 mg or 50 g or 50 kg
(f) Your hand has a width of 100 mm or 100 cm or 100 m
(g) An audiocassette weighs 40 mg or 40 g or 40 kg

1.34 You are taken for a helicopter ride in Hawaii from Kona (sea level) to the top of the volcano Mauna Kea. Which property of your body would change during the helicopter ride: (a) height (b) weight (c) volume (d) mass

1.35 Convert to Celsius and to Kelvin:
(a) 320°F (b) 212°F
(c) 0°F (d) −250°F

1.36 Convert to Fahrenheit and to Kelvin:
(a) 25°C (b) 40°C
(c) 250°C (d) −273°C

Unit Conversions

1.37 Make the following conversions (conversion factors are given in Table 1.3):
(a) 42.6 kg to lb (b) 1.62 lb to g
(c) 34 in. to cm (d) 37.2 km to miles
(e) 2.73 gal to L (f) 62 g to oz
(g) 33.61 qt to L (h) 43.7 L to gal
(i) 1.1 miles to km (j) 34.9 mL to fl oz

1.38 Make the following metric conversions:
(a) 96.4 mL to L (b) 275 mm to cm
(c) 45.7 kg to g (d) 475 cm to m
(e) 21.64 cc to mL (f) 3.29 L to cc
(g) 0.044 L to mL (h) 711 g to kg
(i) 63.7 mL to cc (j) 0.073 kg to mg
(k) 83.4 m to mm (l) 361 mg to g

1.39 You drive in Canada where the distances are marked in km. The sign says you are 80 km from Ottawa. You are traveling at a speed of 75 mph. Would you reach Ottawa within one hour, after one hour, or later than that?

1.40 The speed limit in some European cities is 80 km/h. How many miles per hour is this?

1.41 Your car gets 25.00 miles on a gallon of gas. What would be your car efficiency in km/L?

Density and Specific Gravity

1.42 The volume of a rock weighing 1.075 kg is 334.5 mL. What is the density of the rock in g/mL? Express it to three significant figures.

1.43 The density of manganese is 7.21 g/mL, that of calcium chloride is 2.15 g/mL, and that of sodium acetate is 1.528 g/mL. You place these three solids in a liquid, in which they are not soluble. The liquid has a density of 2.15 g/mL. Which will sink to the bottom, which will stay on the top, and which will stay in the middle of the liquid?

1.44 The density of titanium is 4.54 g/mL. What is the volume, in mL, of 163 g of titanium?

1.45 A 335.0 cc sample of urine has a mass of 342.6 g. What is the density, in g/mL, to three decimal places?

1.46 The density of methanol at 20°C is 0.791 g/mL. What is the mass, in g, of a 280 mL sample?

1.47 The density of dichloromethane, a liquid insoluble in water, is 1.33 g/cc. If dichloromethane and water are placed in a separatory funnel, which will be the upper layer?

1.48 A sample of 10.00 g of oxygen has a volume of 6702 mL. The same weight of carbon dioxide occupies 5058 mL.
(a) What is the density of each gas in g/L?
(b) Carbon dioxide is used as a fire extinguisher to cut off the fire's supply of oxygen. Do the densities of these two gases explain the fire-extinguishing ability of carbon dioxide?

1.49 Crystals of a material are suspended in the middle of a cup of water at 2°C. This means that the densities of the crystal and that of the water are the same. How might you enable the crystals to rise to the surface of the water so that you can harvest them?

Energy

1.50 On many country roads you see telephones powered by a solar panel. What principle is at work in these devices?

1.51 While you drive your car, your battery is charged. How would you describe this process in terms of kinetic and potential energy?

Heat

1.52 How many calories are required to heat the following (specific heats are given in Table 1.4)?
(a) 52.7 g of aluminum from 100°C to 285°C
(b) 93.6 g of methanol from −35°C to 55°C
(c) 3.4 kg of lead from −33°C to 730°C
(d) 71.4 g of ice from −77°C to −5°C

1.53 If 168 g of an unknown liquid requires 2750 cal of heat to raise its temperature from 26° to 74°C, what is the specific heat of the liquid?

1.54 The specific heat of steam is 0.48 cal/g·°C. How many kcal are needed to raise the temperature of 10.5 kg steam from 120° to 150°C?

Chemical Connections

1.55 (Chemical Connections 1A) If the recommended dose of a drug is 445 mg for a 180 lb man, what would be a suitable dose for a 135 lb man?

1.56 (Chemical Connections 1A) The average lethal dose of heroin is 1.52 mg/kg of body weight. Estimate how many grams of heroin would be lethal for a 200 lb man.

1.57 (Chemical Connections 1B) How does the body react to hypothermia?

1.58 (Chemical Connections 1B) Low temperatures often cause people to shiver. What is the function of this involuntary body action?

1.59 (Chemical Connections 1C) Which would make a more efficient cold compress, ethanol or methanol? (Refer to Table 1.4.)

Additional Problems

1.60 The meter is a measure of length. Tell what each of the following units measures:
(a) cm^3 (b) mL   (c) kg
(d) cal   (e) g/cc   (f) joule
(g) °C   (h) cm/s

1.61 A brain weighing 1 lb occupied a volume of 620 mL. What is the specific gravity of the brain?

1.62 If the density of air is 1.25×10^{-3} g/cc, what is the mass in kg of the air in a room that is 5.3 m long, 4.2 m wide, and 2.0 m high?

1.63 Classify these as kinetic or potential energy:
(a) Water held by a dam
(b) A speeding train
(c) A book on its edge before falling
(d) A falling book
(e) Electric current in a light bulb

1.64 The kinetic energy possessed by an object with a mass of 1 g moving with a velocity of 1 cm/s is called 1 erg. What is the kinetic energy, in ergs, of an athlete with a mass of 127 lb running at a velocity of 14.7 miles/h?

1.65 A European car advertises an efficiency of 22 km/L, while an American car claims an economy of 30 miles/gal. Which car is more efficient?

1.66 In Potsdam, New York, you can buy gas for US$1.30/gal. In Montreal, Canada, you pay US$0.41/L. Which is the better buy? Is your calculation reasonable?

1.67 Shivering is the body's response to increase the body temperature. What kind of energy is generated by shivering?

1.68 When the astronauts walked on the Moon, they could make giant leaps in spite of their heavy gear.
(a) Why were their weights on the Moon so small?
(b) Were their masses different on the Moon than on the Earth?

1.69 Which of the following is the largest mass and which is the smallest?
(a) 41 g
(b) 3×10^3 mg
(c) 8.2×10^6 µg
(d) 4.1310^{-8} kg

1.70 Which quantity is bigger in each of the following pairs:
(a) 1 gigaton : 10 megaton
(b) 10 micrometer : 1 millimeter
(c) 10 centigram : 200 milligram

1.71 In Japan, high-speed "bullet trains" move with an average speed of 220 km/h. If Dallas and Los Angeles were connected by such a train, how long would it take to travel nonstop between these cities (a distance of 1490 miles)?

1.72 The specific heats of some elements at 25°C are as follows: aluminum = 0.215 cal/g·°C; carbon (graphite) = 0.170 cal/g·°C; iron = 0.107 cal/g·°C; mercury = 0.0331 cal/g·°C.
(a) Which element would require the smallest amount of heat to raise the temperature of 100 g of the element by 10°C?
(b) If the same amount of heat needed to raise the temperature of 1 g of aluminum by 25°C were applied to 1 g of mercury, by how many degrees would its temperature be raised?
(c) If a certain amount of heat is used to raise the temperature of 1.6 g of iron by 10°C, the temperature of 1 g of which element would also be raised by 10°C, using the same amount of heat?

1.73 The specific heat of heavy water at 25°C is 4.217 J/g·°C. Which requires more energy to raise the temperature of 10.0 g by 10°C, water or heavy water?

1.74 One quart of milk costs 80 cents and one liter costs 86 cents. Which is the better buy?

1.75 Consider butter, density 0.860 g/mL, and sand, density 2.28 g/mL.
(a) If 1.00 mL of butter is thoroughly mixed with 1.00 mL of sand, what is the density of the mixture?
(b) What would be the density of the mixture if 1.00 g of the same butter were mixed with 1.00 g of the same sand?

1.76 Which speed is the fastest?
(a) 70 miles/h
(b) 140 km/h
(c) 4.5 km/s
(d) 48 miles/min

1.77 In calculating the specific heat of a substance, the following data are used: mass = 92.15 g; heat = 3.200 kcal; rise in temperature = 45°C. How many significant figures should you report in calculating the specific heat?

1.78 A solar cell generates 500 kilojoules of energy per hour. To keep a refrigerator at 4°C, one needs 250 kcal/h. Can the solar cell supply sufficient energy per hour to maintain the temperature of the refrigerator?

1.79 The specific heat of urea is 1.339 J/g·°C. If one adds 60.0 J of heat to 10.0 g of urea at 20°C, what would be the final temperature?

InfoTrac College Edition

For additional readings, go to InfoTrac College Edition, your online research library, at

http://infotrac.thomsonlearning.com

Career Connections

Plan for your future! Find information about qualifications, salaries, and more at

http://chemistry.brookscole.com

CHAPTER 2

2.1 *Introduction*

2.2 *Classification of Matter*

2.3 *Dalton's Atomic Theory*

2.4 *Inside an Atom*

2.5 *The Periodic Table*

2.6 *The Electron Configuration of Atoms*

2.7 *Electron Configuration and the Periodic Table*

2.8 *Ionization Energy of the Elements: A Periodic Property*

SEM of Gallium arsenide.

Atoms

2.1 Introduction

What is matter made of? This question was discussed for thousands of years, long before humans had any reasonable way of getting an answer. In ancient Greece, two schools of thought tried to answer this question. One group, led by a scholar named Democritus (about 460–370 BCE), believed that all matter is made of very small particles—much too small to see. Democritus called these particles atoms (Greek *atomos,* meaning "not to cut"). Some of his followers developed the idea that there were different kinds of atoms, with different properties, and that the properties of the atoms caused ordinary matter to have the properties we all know.

These ideas, however, were not accepted by all ancient thinkers. A second group, led by Zeno of Elea (born about 450 BCE), did not believe in atoms at all. They insisted that matter is infinitely divisible. If you took any object, such as a piece of wood or a crystal of table salt, you could cut it or otherwise divide it into two parts, divide each of these parts into two more parts, and continue the process forever. According to Zeno and his followers, you would never reach a particle of matter that could no longer be divided.

27

Today we know that Democritus was right and Zeno was wrong. Atoms *are* the basic units of matter. Of course, there is a great difference in the way we now look at this question. Today our ideas are based on evidence. Democritus had no evidence to prove that matter cannot be divided an infinite number of times, just as Zeno had no evidence to support his claim that matter can be divided infinitely. Both claims were based not on evidence but on visionary belief: one in unity, the other in diversity. In Section 2.3 we will discuss the evidence for the existence of atoms, but first we need to look at the diverse forms of matter.

2.2 Classification of Matter

Matter can be divided into two classes: pure substances and mixtures. Each class is then subdivided as shown in Figure 2.1.

A Elements

An **element** is a substance (for example, carbon, hydrogen, and iron) that consists of identical atoms. At this time, 114 elements are known. Of these elements, 88 occur in nature; the others have been made by chemists and physicists. A list of the known elements appears on the inside back cover of this book, along with their symbols. Their symbols consist of one or two letters. Many symbols correspond directly to the name in English (for example, C for carbon, H for hydrogen, and Li for lithium), but a few come from the Latin or German names. Others are named for people who played significant roles in the development of science—in particular, atomic science. Still other elements are named for geographic locations.

B Compounds

A **compound** is a pure substance made up of two or more elements in a fixed ratio by mass. For example, water is a compound made up of hydrogen and oxygen, and table salt is a compound made up of sodium and chlorine. There are an estimated 20 million known compounds, only a few of which we will meet in this book.

Figure 2.1 Classification of matter. Matter is divided into pure substances and mixtures. A pure substance may be either an element or a compound. A mixture may be either homogeneous or heterogeneous.

Screen 2.2 Exercise

CHEMICAL CONNECTIONS 2A

Elements Necessary for Human Life

Only a few elements are necessary for human life. The six most important of these—carbon, hydrogen, nitrogen, oxygen, phosphorus, and sulfur—are the subjects of organic chemistry and biochemistry (Chapters 10–31). Seven other elements are also quite important, and our bodies use at least nine additional ones (trace elements) in very small quantities. The table lists these 20 elements and their functions in the human organism. Many of these elements are more fully discussed later in the book. For the average daily requirements of these elements, their sources in foods, and symptoms of their deficiencies, see Chapter 29.

Table 2A Elements and Their Functions in the Human Body

Element	Function	Element	Function
The Big Four		**The Trace Elements**	
Carbon (C)		Chromium (Cr)	Increases effectiveness of insulin
Hydrogen (H)	The subject of Chapters 10–18	Cobalt (Co)	Part of vitamin B_{12}
Nitrogen (N)	(organic chemistry) and 19–31	Copper (Cu)	Strengthens bones; assists in enzyme activity
Oxygen (O)	(biochemistry)		
		Fluorine (F)	Reduces the incidence of dental cavities
The Next Seven		Iodine (I)	An essential part of thyroid hormones
Calcium (Ca)	Strengthens bones and teeth; aids in blood clotting	Iron (Fe)	An essential part of some proteins, such as hemoglobin, myoglobin, cytochromes, and FeS proteins
Phosphorus (P)	Present in phosphates of bone, in nucleic acids (DNA and RNA); and involved in energy transfer		
Potassium (K)	Helps regulate electrical balance of body fluids; essential for nerve conduction	Manganese (Mn)	Present in bone-forming enzymes; aids in fat and carbohydrate metabolism
Sulfur (S)	An essential component of proteins	Molybdenum (Mo)	Helps regulate electrical balance in body fluids
Chlorine (Cl)	Necessary for normal growth and development	Zinc (Zn)	Necessary for the action of certain enzymes
Sodium (Na)	Helps regulate electrical balance in body fluids		
Magnesium (Mg)	Helps nerve and muscle action; present in bones		

A compound is characterized by its formula. The formula gives the ratios of the compound's constituent elements and identifies each element by its atomic symbol. For example, in table salt the ratio of sodium atoms to chlorine atoms is 1:1. Given that Na is the symbol for sodium and Cl is the symbol for chlorine, the formula of table salt is NaCl. In water, the combining ratio is two hydrogen atoms to one oxygen atom. The symbol for hydrogen is H, the symbol for oxygen is O, and the formula of water is H_2O. The subscripts following the atomic symbols indicate the ratio of the combining elements. The number 1 in these ratios is omitted from the subscript. It is understood that NaCl means a ratio of 1:1 and that H_2O represents a ratio of 2:1. You will find out more about the nature of combining elements in a compound and their names and formulas in Chapter 3.

Figure 2.2 shows four different representations for a water molecule. We will have more and more to say about molecular models as we move through this book.

Figure 2.2 A water molecule.

- Shows there are two H atoms and one O atom → Molecular formula: H_2O
- Lines represent connections between atoms → Structural formula: H—O—H
- Each element is represented by a different colored ball → Ball-and-stick model
- Space-filling model shows the relative sizes of H and O atoms in a water molecule → Space-filling model

---

EXAMPLE 2.1

(a) In a compound magnesium fluoride, magnesium (atomic symbol Mg) and fluorine (atomic symbol F) combine in a ratio of 1:2. What is the formula of magnesium fluoride?

(b) The formula of perchloric acid is $HClO_4$. What are the combining ratios of the elements in perchloric acid?

Solution

(a) The formula is MgF_2. We do not write a subscript of 1 after Mg.

(b) Both H and Cl have no subscripts. This means that hydrogen and chlorine have a combining ratio of 1:1. The subscript on oxygen is 4. Therefore, the combining ratios in $HClO_4$ are 1:1:4.

Problem 2.1

Write the formulas of compounds in which the combining ratios are as follows:

(a) Sodium : chlorine : oxygen, 1 : 1 : 3
(b) Aluminum (atomic symbol Al) : fluorine (atomic symbol F), 1 : 3

C Mixtures

A **mixture** is a combination of two or more pure substances. Most of the matter we encounter in our daily lives (including our own bodies) consists of mixtures rather than pure substances. For example, blood, butter, gasoline, soap, the metal in a wedding ring, the air we breathe, and the earth we walk on are all mixtures of pure substances. An important difference between a compound and a mixture is that the ratios by mass of the elements in a compound are fixed, whereas in a mixture the substances can be present in any mass ratio.

For some mixtures—blood, for example (Figure 2.3)—the texture of the mixture is even throughout. If you examine such a mixture under magnification, however, you can see that it is composed of different substances.

Other mixtures are homogeneous throughout, and no amount of magnification will reveal the presence of different substances. The air we breathe, for example, is a mixture of gases, primarily nitrogen (78%) and oxygen (21%).

An important characteristic of a mixture is that it consists of two or more substances, each having different physical properties. If we know the physical properties of the individual substances, we can use appropriate physical means to separate the mixture into its component parts. Figure 2.4 shows one example of how a mixture can be separated.

Figure 2.3 Blood cells. Blood appears to be homogeneous throughout to the unaided eye but magnification shows that it is heterogeneous; it contains a number of particles, including red and white blood cells. *(Pricked finger: Martin Dohrn/Science Photo Library/Photoresearchers; blood cells: Ken Edward/Science Photo Library/Photoresearchers)*

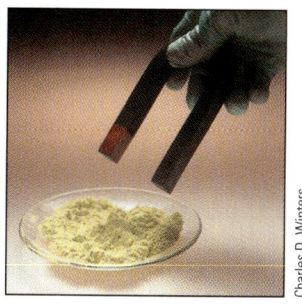

(a)　　　　　　　(b)　　　　　　　(c)

Figure 2.4 Separating a mixture of iron and sulfur. (*a*) The iron–sulfur mixture is stirred with a magnet, which attracts the iron filings. (*b*) Much of the iron is removed after the first stirring. (*c*) Stirring continues until no more iron filings can be removed.

2.3 Dalton's Atomic Theory

Around 1805 an English chemist, John Dalton (1766–1844), put forth a scientific atomic theory. The major difference between Dalton's theory and that of Democritus (Section 2.1) is that Dalton based his theory on evidence rather than a belief. First, let us state his theory. We will then see what kind of evidence supported it.

1. All matter is made up of very tiny, indivisible particles, which Dalton called **atoms.** The interaction among atoms accounts for the properties of matter.
2. All atoms of the same element have the same chemical properties. Conversely, atoms of different elements have different chemical properties.
3. Compounds are formed by the chemical combination of two or more different kinds of atoms.
4. A **molecule** is a tightly bound combination of two or more atoms that acts as a single unit.

> **Atom** The smallest particle of an element that retains all of the chemical properties of the element.

A Evidence for Dalton's Atomic Theory

The Law of Conservation of Mass

The great French chemist Antoine Laurent Lavoisier (1743–1794) discovered the **law of conservation of mass,** which states that matter can be neither created nor destroyed. Lavoisier proved this law by conducting many experiments in which he showed that the total mass of matter at the end of the experiment was exactly the same as that at the beginning. Dalton's theory explained this fact in the following way: If all matter consists of indestructible atoms, then any chemical reaction simply changes the attachments between atoms but does not destroy the atoms themselves.

In the following illustration, a carbon monoxide molecule reacts with a lead oxide molecule to give a carbon dioxide molecule and a lead atom. All of the original atoms are still present at the end; they have merely changed partners. Thus the total mass after this chemical change remains the same as it was before the change took place.

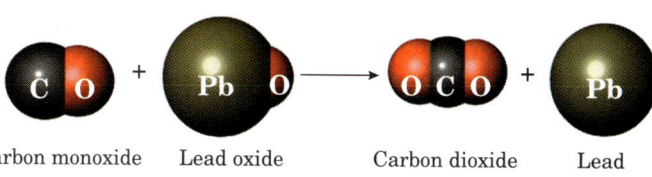

Carbon monoxide　　Lead oxide　　　Carbon dioxide　　Lead

CHEMICAL CONNECTIONS 2B

Abundance of Elements on Earth

The table shows the abundance of the elements present in the human body. As you can see, oxygen is the most abundant element by weight, followed by carbon, hydrogen, and nitrogen. If we go by number of atoms, however, hydrogen is even more abundant in the human body than oxygen is.

The table also shows the abundance of elements in the Earth's crust. Although 88 elements are found in the Earth's crust (we know very little about the interior of the Earth because we have not been able to penetrate into it very far), they are not present in anything close to equal amounts. In the Earth's crust as well as in the human body, the most abundant element by weight is oxygen, but there the similarity ends. Silicon, aluminum, and iron, which are not major elements in the body, are the second, third, and fourth most abundant elements in the Earth's crust, respectively. The table also shows that while carbon is the second most abundant element by weight in the human body, it is present to the extent of only 0.08% in the Earth's crust.

Table 2B Relative Abundance of Elements in the Human Body and in the Earth's Crust, Including the Atmosphere and Oceans

Element	Percentage in Human Body		Percentage in Earth's Crust by Weight
	By Number of Atoms	By Weight	
H	63.0	10.0	0.9
O	25.4	64.8	49.3
C	9.4	18.0	0.08
N	1.4	3.1	0.03
Ca	0.31	1.8	3.4
P	0.22	1.4	0.12
K	0.06	0.4	2.4
S	0.05	0.3	0.06
Cl	0.03	0.2	0.2
Na	0.03	0.1	2.7
Mg	0.01	0.04	1.9
Si	—	—	25.8
Al	—	—	7.6
Fe	—	—	4.7
Others	0.01	—	—

The Law of Constant Composition

Another French chemist, Joseph Proust (1754–1826), demonstrated the **law of constant composition,** which states that any compound is always made up of elements in the same proportion by mass. For example, if you decompose water, you will always get 8.0 g of oxygen for each 1.0 g of hydrogen. The mass ratio of oxygen to hydrogen in pure water is always 8.0 to 1.0, whether the water comes from the Atlantic Ocean or the Seine River or is collected as rain, squeezed out of a watermelon, or distilled from urine. This fact was also evidence for Dalton's theory. If a water molecule consists of one atom of oxygen and two atoms of hydrogen, and if an oxygen atom has a mass 16 times that of a hydrogen atom, then the mass ratio of these two elements in water must always be 8.0 to 1.0. The two elements can never be found in water in any other mass ratio.

Figure 2.5 Hydrogen, nitrogen, oxygen, and chlorine are diatomic elements. Ozone is a triatomic element. One form of sulfur, S_8, is a polyatomic element.

B Monatomic, Diatomic, and Polyatomic Elements

Some elements—for example, helium and neon—consist of single atoms that are not connected to each other. That is, they are **monatomic elements.** In contrast, oxygen, in its most common form, contains two atoms in each molecule, connected to each other by a chemical bond. We write the formula for an oxygen molecule as O_2, with the subscript showing the number of atoms in the molecule. Six other elements also occur as diatomic molecules (that is, they contain two atoms of the same element per molecule): hydrogen (H_2), nitrogen (N_2), fluorine (F_2), chlorine (Cl_2), bromine (Br_2), and iodine (I_2). It is important to understand that, under normal conditions, free atoms of O, H, N, F, Cl, Br, and I do not exist. Rather, these seven elements occur only as **diatomic elements** (Figure 2.5).

Some elements have even more atoms in each molecule. Ozone, O_3, has three oxygen atoms in each molecule. In one form of phosphorus, P_4, each molecule has four atoms. One form of sulfur, S_8, has eight atoms per molecule. Some elements have molecules that are much larger. For example, diamond has millions of carbon atoms all bonded together in a gigantic cluster. Diamond and S_8 are referred to as **polyatomic elements.**

2.4 Inside an Atom

A Three Subatomic Particles

Today we know that matter is more complex than Dalton believed. A wealth of experimental evidence obtained over the last 100 years or so has convinced us that atoms are not indivisible, but rather consist of even smaller particles, called subatomic particles. Three subatomic particles make up all atoms: protons, electrons, and neutrons. Table 2.1 shows the charge, mass, and location of these particles in an atom.

A **proton** has a positive charge. By convention, we say that the magnitude of the charge is +1. Thus one proton has a charge of +1, two protons have a charge of +2, and so forth. The mass of a proton is 1.6726×10^{-24} g,

There are many other subatomic particles, but we will not deal with them in this book.

Proton A subatomic particle with a charge of +1 and a mass of approximately 1 amu; it is found in a nucleus.

Table 2.1 Properties and Location of Three Subatomic Particles

Subatomic Particle	Charge	Mass (g)	Mass (amu)	Mass (amu); Rounded to One Significant Figure	Location in an Atom
Proton	+1	1.6726×10^{-24}	1.0073	1	In the nucleus
Electron	−1	9.1094×10^{-28}	5.4859×10^{-4}	0.0005	Outside the nucleus
Neutron	0	1.6749×10^{-24}	1.0087	1	In the nucleus

Atomic mass unit (amu) A unit of the scale of relative masses of atoms; 1 amu = 1.6605×10^{-24} g. By definition, 1 amu is 1/12 the mass of a carbon atom containing 6 protons and 6 neutrons.

but this number is so small that it is more convenient to use another unit, called the **atomic mass unit (amu),** to describe its mass.

$$1 \text{ amu} = 1.6605 \times 10^{-24} \text{ g}$$

Thus a proton has a mass of 1.0073 amu. For most purposes in this book, it is sufficient to round this number to one significant figure and, therefore, to say that the mass of a proton is 1 amu.

An **electron** has a charge of −1, equal in magnitude to the charge on a proton but opposite in sign. The mass of an electron is approximately 5.4859×10^{-4} amu or 1/1837 that of the proton. Thus it takes approximately 1837 electrons to equal the mass of one proton.

Electron A subatomic particle with a charge of −1 and a mass of approximately 0.0005 amu; it is found in the space surrounding a nucleus.

Like charges repel, and unlike charges attract. Two protons repel each other, just as two electrons repel each other. A proton and an electron, however, attract each other.

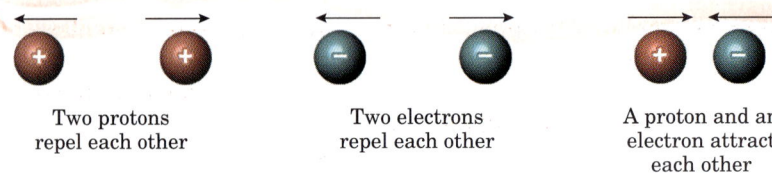

Two protons repel each other

Two electrons repel each other

A proton and an electron attract each other

Neutron A subatomic particle with a mass of approximately 1 amu and a zero charge; it is found in the nucleus.

A **neutron** has no charge. Therefore, neutrons do not attract or repel each other or any other particle. The mass of a neutron is slightly greater than that of a proton: 1.6749×10^{-24} g or 1.0087 amu. Again, for our purposes, we can round this number to 1 amu.

These three particles make up atoms, but where are they found? Protons and neutrons are found in a tight cluster in the center of an atom (Figure 2.6). This part of an atom is called the **nucleus.** We will discuss the nucleus in greater detail in Chapter 9. Electrons are found as a diffuse cloud outside the nucleus. One final point about an atom: Notice how small the nucleus is compared to the size of an atom.

B Mass Number

Each atom has a fixed number of protons, electrons, and neutrons. One way to describe an atom is by its **mass number,** which is the sum of the number of protons plus the number of neutrons in the nucleus of the atom. Note that an atom also contains electrons but, because the mass of an electron is so small compared to that of protons and neutrons (Table 2.1), electrons are not counted in determining mass number.

Mass number = number of protons
+ number of neutrons in the nucleus of an atom

For example, an atom with 5 protons, 5 electrons, and 6 neutrons has a mass number of 11.

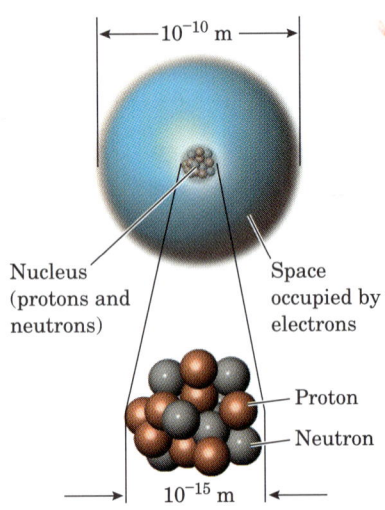

Figure 2.6 A typical atom. Protons and neutrons are found in the nucleus, and electrons are found as a cloud outside the nucleus.

EXAMPLE 2.2

What is the mass number of an atom containing:
(a) 58 protons, 58 electrons, and 78 neutrons?
(b) 17 protons, 17 electrons, and 20 neutrons?

Solution

The mass number of an atom is the sum of the number of its protons and neutrons.
(a) The mass number is 58 + 78 = 136.
(b) The mass number is 17 + 20 = 37.

Problem 2.2

What is the mass number of an atom containing:
(a) 15 protons, 15 electrons, and 16 neutrons?
(b) 86 protons, 86 electrons, and 136 neutrons?

C Atomic Number

The **atomic number** of an element is the number of protons in its nucleus.

Atomic number = number of protons in the nucleus of an element

At the present time, 114 elements are known. These elements have atomic numbers ranging from 1 to 112 plus 114 and 116. A few atoms with 114 protons and a few with 116 protons were created, but their existence still awaits confirmation. The smallest atomic number belongs to the element hydrogen, which has only one proton, and the largest (so far) to the yet unnamed heaviest known element, which contains 116 protons.

If you know the atomic number and the mass number of an element, you can properly identify it. For example, the element with 6 protons, 6 electrons, and 6 neutrons has an atomic number of 6 and a mass number of 12. The element with atomic number 6 is carbon, C. Because its mass number is 12, we call this atomic nucleus carbon-12. Alternatively, we can write the symbol for this atomic nucleus as $^{12}_{6}C$. In this symbol, the mass number of the element is always written in the upper-left corner (as a superscript) of the symbol of the element, and the atomic number in the lower-left corner (as a subscript) of it.

Atomic numbers for all the elements are given in the atomic weight table on the inside back cover. They are also given in the Periodic Table on the inside front cover.

Mass number (number of protons + neutrons) $\rightarrow$ $^{12}_{6}C$
Atomic number (number of protons) $\nearrow$

EXAMPLE 2.3

Name the elements given in Example 2.2 and write the symbols of their atomic nuclei.

Solution

(a) This element has 58 protons. We see from the Periodic Table on the inside front cover of the text that the element of atomic number 58 is cerium, and its symbol is Ce. This element has 58 protons and 78 neutrons and, therefore, its mass number is 136. We call it cerium-136. Its symbol is $^{136}_{58}Ce$.
(b) This atom has 17 protons, making it a chlorine (Cl) atom. Because its atomic mass number is 37, we call it chlorine-37. Its symbol is $^{37}_{17}Cl$.

If you know the name of an element, you can look up its atomic number and atomic weight from the atomic weight table on the inside back cover. Conversely, if you know the atomic number of an element, you can look up its symbol from the Periodic Table on the inside front cover. The atomic number of each element is shown in red.

Screen 2.9
Tutorial

Problem 2.3

Name the elements given in Problem 2.2. Write the symbols of their atomic nuclei.

EXAMPLE 2.4

A number of elements have an equal number of protons and neutrons in their nuclei. Among these are oxygen, nitrogen, and neon. What are the atomic numbers of these elements? How many protons and neutrons does an atom of each have? Write the name and the symbol of each of these atomic nuclei.

Solution

Atomic numbers for these elements are found in the list of elements on the inside back cover. This table shows that oxygen (O) has atomic number 8, nitrogen (N) has atomic number 7, and neon (Ne) has atomic number 10. This means that the oxygen has 8 protons and 8 neutrons. Its name is oxygen-16, and its symbol is $^{16}_{8}O$. Nitrogen has 7 protons and 7 neutrons. Its name is nitrogen-14, and its symbol is $^{14}_{7}N$. Neon has 10 protons and 10 neutrons. Its name is neon-20, and its symbol is $^{20}_{10}Ne$.

Problem 2.4
(a) What are the atomic numbers of mercury (Hg) and lead (Pb)?
(b) How many protons does an atom of each have?
(c) If both Hg and Pb have 120 neutrons in their nuclei, what is the mass number of each?
(d) Write the name and the symbol of each.

D Isotopes

Although we can say that a carbon atom always has 6 protons and 6 electrons, we cannot say that a carbon atom must have any particular number of neutrons. Some of the carbon atoms found in nature have 6 neutrons; the mass number of these atoms is 12, they are written as carbon-12, and their symbol is $^{12}_{6}C$. Other carbon atoms have 6 protons and 7 neutrons and, therefore a mass number of 13; they are written as carbon-13, and their symbol is $^{13}_{6}C$. Still other carbon atoms have 6 protons and 8 neutrons; they are written as carbon-14 or $^{14}_{6}C$. Atoms with the same number of protons but different numbers of neutrons are called **isotopes**. All isotopes of carbon contain 6 protons and 6 electrons (or they wouldn't be carbon). Each isotope, however, contains a different number of neutrons and, therefore, has a different mass number.

The properties of isotopes of the same element are almost identical, and for most purposes we regard them as identical. They differ, however, in radioactivity properties, which we discuss in Chapter 9.

The fact that isotopes exist means that the second statement of Dalton's theory (Section 2.3) is not correct.

EXAMPLE 2.5

How many neutrons are in each isotope of oxygen? Write the symbol of each isotope.
(a) Oxygen-16   (b) Oxygen-17   (c) Oxygen-18

Solution
Each oxygen atom has 8 protons. The difference between the mass number and the number of protons gives the number of neutrons.
(a) Oxygen-16 has $16 - 8 = 8$ neutrons. Its symbol is $^{16}_{8}O$.
(b) Oxygen-17 has $17 - 8 = 9$ neutrons. Its symbol is $^{17}_{8}O$.
(c) Oxygen-18 has $18 - 8 = 10$ neutrons. Its symbol is $^{18}_{8}O$.

Problem 2.5
Two iodine isotopes are used in medical treatments: iodine-125 and iodine-131. How many neutrons are in each isotope? Write the symbol for each isotope.

Most elements are found on Earth as mixtures of isotopes, in a more or less constant ratio. For example, all naturally occurring samples of the element chlorine contain 75.77% chlorine-35 (18 neutrons) and 24.23% chlorine-37 (20 neutrons). Silicon exists in nature in a fixed ratio of three isotopes, with 14, 15, and 16 neutrons, respectively. For some elements, the ratio of isotopes may vary slightly from place to place but, for most purposes, we can ignore these slight variations. The atomic masses and isotopic abundances are determined using an instrument called a mass spectrometer.

E Atomic Weight

The **atomic weight** of an element given in the Periodic Table is a weighted average of the masses (in amu) of its isotopes found on Earth. As an example of the calculation of atomic weight, let us examine chlorine. As we have just seen, two isotopes of chlorine exist in nature: chlorine-35 and chlorine-37. The mass of a chlorine-35 atom is 34.97 amu, and the mass of a chlorine-37 atom is 36.97 amu. Note that the atomic weight of each chlorine isotope (its mass in amu) is very close to its atomic number (the number of protons and neutrons in its nucleus). This statement holds true for the isotopes of chlorine and those of all elements, because protons and neutrons have a mass of approximately (but not exactly) 1 amu.

The atomic weight of chlorine is a weighted average of the masses of the two naturally occurring chlorine isotopes:

Atomic weight The weighted average of the masses of the naturally occurring isotopes of an element. The units of atomic weight are atomic mass units (amu).

$$\left(\frac{75.77}{100} \times 34.97 \text{ amu}\right) + \left(\frac{24.23}{100} \times 36.97 \text{ amu}\right) = 35.45 \text{ amu}$$

Chlorine-35 · Chlorine-37

14
Cl
35.4527

Atomic weight in the Periodic Table is given to four decimal places

Some elements—for example, gold, fluorine, and aluminum—occur naturally as only one isotope. The atomic weights of these elements are close to whole numbers (gold, 196.97 amu; fluorine, 18.998 amu; aluminum, 26.98 amu). A table of atomic weights is found on the inside back cover of this book.

EXAMPLE 2.6

The natural abundance of the three stable isotopes of magnesium are 78.99% magnesium-24 (23.99 amu), 10.00% magnesium-25 (24.99 amu), and 11.01% magnesium-26 (25.98 amu). Calculate the atomic weight of magnesium.

Figure 2.7 The surface of graphite is revealed with a scanning tunneling microscope. The contours represent the arrangement of individual carbon atoms on a crystal surface. *(Courtesy of Roger Freedman and Paul K. Hansma, University of California–Santa Barbara)*

Solution
To calculate the weighted average of the masses of the isotopes, multiply each atomic mass by its abundance and then add.

$$\left(\frac{78.99}{100} \times 23.99 \text{ amu}\right) + \left(\frac{10.00}{100} \times 24.99 \text{ amu}\right)$$
$$+ \left(\frac{11.01}{100} \times 25.98 \text{ amu}\right) = 24.31 \text{ amu}$$

The atomic weight of magnesium given in the Periodic Table to four decimal places is 24.3050.

Problem 2.6
The atomic weight of lithium is 6.941 amu. Lithium has only two naturally occurring isotopes: lithium-6 and lithium-7. Estimate which isotope is in greater natural abundance.

F The Mass and Size of an Atom

A typical heavy atom (although not the heaviest) is lead-208, a lead atom with 82 protons, 82 electrons, and 126 neutrons. It has a mass of 3.5×10^{-22} g. You would need 1.3×10^{24} atoms of lead-208 to make 1 lb of lead—a very large number. There are approximately six billion people on Earth right now. If you divided 1 lb of these atoms among all the people on Earth, each person would get about 2.2×10^{14} atoms.

An atom of lead-208 has a diameter of about 3.5×10^{-10} m. If you could line them up with the atoms just touching, it would take 73 million lead atoms to make a line 1 inch long. Despite their tiny size, we can actually see atoms, in certain cases, by using a special instrument called a scanning tunneling microscope (Figure 2.7).

CHEMICAL CONNECTIONS 2C

Isotopic Abundance and Astrochemistry

The term **isotopic abundance** refers to the relative amounts of the isotopes of an element present in a sample of that element. When we speak of isotopic abundances and the atomic weights derived from them, we refer to isotopic abundances on Earth. For example, there are 11 known isotopes of chlorine, Cl, ranging from chlorine-31 to chlorine-41. Only two of these occur in nature: Chlorine-35 with an abundance of 75.77% and chlorine-37 with an abundance of 24.23%.

With the advance of space exploration, it has become evident that other parts of the solar system, the Sun, the Moon, planets, asteroids, comets, and stars as well as intergalactic gases may have different iso-

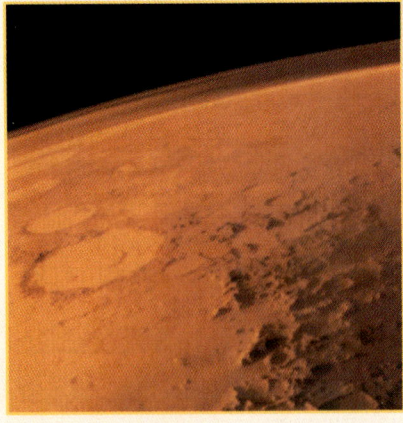

■ Mars.

topic abundances. For example, the deuterium/hydrogen ratio (^{2}H/^{1}H) on Mars is five times larger than that on Earth. The ^{17}O/^{18}O ratio is also higher on Mars than on Earth. Such differences in isotopic abundances are used to establish theories about the origins and history of the solar system.

Furthermore, a comparison of isotopic abundances in certain meteorites found on Earth enables us to guess their origin. For example, it is now believed that certain meteorites that landed on Earth came from Mars. Scientists speculate that they were ejected from its surface when the Red Planet collided with some other large body—perhaps an asteroid.

Virtually all of the mass of an atom is concentrated in its nucleus (because the nucleus contains the protons and neutrons of the atom). The nucleus of a lead-208 atom, for example, has a diameter of about 1.6×10^{-14} m. When we compare this with the diameter of a lead-208 atom, which is about 3.5×10^{-10} m, we see that the nucleus occupies only a tiny fraction of the total volume of the atom. If the nucleus of a lead-208 atom were the size of a baseball, then the entire atom would be much larger than a baseball stadium. In fact, it would be a sphere about one mile in diameter. Because a nucleus has such a relatively large mass concentrated in such a relatively small volume, a nucleus has a very high density. The density of a lead-208 nucleus, for example, is 1.8×10^{14} g/cm^3. Nothing in our daily life has a density anywhere near as high. If a paper clip had this density, it would weigh about 10 million (10^7) tons.

2.5 The Periodic Table

A Origin of the Periodic Table

In the 1860s, a number of scientists, most notably the Russian Dmitri Mendeleev (1834–1907), then professor at the University of St. Petersburg, produced one of the first periodic tables, the form of which we still use today. Mendeleev started by arranging the known elements in order of increasing atomic weight beginning with hydrogen. He soon discovered that when the elements are arranged in this manner, certain sets of properties recur periodically. Mendeleev then arranged those elements with recurring properties into **periods** (horizontal rows) by starting a new row each time he came to an element with properties similar to hydrogen. In this way, he discovered that lithium, sodium, potassium, and so forth each start new rows. All are metallic solids, all form ions with a charge of +1 (Li$^+$, Na$^+$, K$^+$, and so on), and all react with water to form metal hydroxides (LiOH, NaOH, KOH, and so on). Mendeleev also discovered that elements in other groups (vertical columns) have similar properties. For example, the elements fluorine, chlorine, bromine, and iodine all fall in the same column in the table. These elements, which are called **halogens,** are all colored substances, with the colors deepening as we go down the table (Figure 2.8). All form compounds with sodium that have the formula NaX (for example, NaCl and NaBr), but not NaX$_2$, Na$_2$X, Na$_3$X, or anything else. Only the elements in this column share this property.

■ Dmitri Mendeleev.

Screen 2.12
Exercise

Period, of the Periodic Table
A horizontal row of the Periodic Table.

"X" is a commonly used symbol for the halogens.

Figure 2.8 Three halogens. Chlorine is a gas, bromine is a liquid, and iodine is a solid.

CHEMICAL CONNECTIONS 2D

Strontium-90

Elements in the same column of the Periodic Table show similar properties. One important example is the similarity of strontium and calcium (strontium is just below calcium in Group 2A of the Periodic Table). Calcium is an important element for humans because our bones and teeth consist largely of calcium compounds. We need some of this mineral in our diet every day, and we get it mostly from milk, cheese, and other dairy products.

One of the products released by test nuclear explosions in the 1950s and 1960s was the isotope strontium-90. This isotope is radioactive, with a half-life of 28.1 years. (Half-life is discussed in Section 9.3.) Strontium-90 was present in the fallout from above-ground nuclear test explosions. It was carried all over Earth and slowly settled to the ground, where it was eaten by cows and other animals. Strontium-90 got into milk and eventually into human bodies as well. If it were not so similar to calcium, our bodies would eliminate it within a few days. Because it is similar, however, some of the strontium-90 became deposited in bones and teeth (especially in children), subjecting all of us to a small amount of radioactivity for long periods of time.

Above-ground nuclear testing was banned by a 1963 treaty between the United States and the former Soviet Union. Although a few other countries still conduct occasional above-ground tests, there is reason to hope that such tests will be completely halted in the near future.

At this point, we must say a word about the numbering of the columns of the Periodic Table. Mendeleev gave them numerals and added the letter A for some columns and B for others. This numbering pattern remains in common use in the United States today. An alternative international pattern uses the numbers 1 to 18, without added letters, going from left to right. Thus, in Mendeleev's numbering system, the halogens are in Group 7A; in the new international numbering system, they are in Group 17. Although this book uses the traditional numbering, both patterns are shown on the Periodic Table on the inside front cover.

If we look at the Periodic Table on the inside front cover of this book, we see that columns 1A and 2A extend above the rest, and so do columns 3A to 8A. The elements in these eight columns are called **main-group elements**. The elements in the B columns (Groups 3 to 12 in the new numbering system) are called **transition elements**. Notice that elements 58 to 71 and 90 to 103 are not included in the main body of the table but rather are shown separately at the bottom. These sets of elements, called **inner transition elements**, actually belong in the main body of the Periodic Table, between columns 3B and 4B (between La and Hf). As is customary, we put them outside the main body solely to make a more compact presentation. So, if you like, you may mentally take a pair of scissors, cut through the heavy line between columns 3B and 4B, move them apart, and insert the inner transition elements. You will now have a table with 32 columns.

B Classification of Elements

There are three classes of elements: metals, nonmetals, and metalloids. The majority of elements are **metals**—only 24 are not. Metals are solids (except for mercury, which is a liquid), shiny, conductors of electricity, ductile (they

> **Main-group element** An element in any of the A groups of the Periodic Table.

> **Metal** An element that is solid (except for mercury, which is a liquid), shiny, conducts electricity, ductile, and malleable, and that forms alloys.

■ Some metals. *(a)* Sodium, *(b)* magnesium, *(c)* copper, *(d)* calcium, and *(e)* gold.

(a)

(b)

(c)

(d)

(e)

CHEMICAL CONNECTIONS 2E

The Use of Metals as Historical Landmarks

The malleability of metals played an important role in the development of human society. In the Stone Age, tools were made from stone, which has no malleability. Then, about 11,000 BCE, it was discovered that the pure copper found on the surface of Earth could be hammered into sheets, which made it suitable for use in vessels, utensils, and religious or artistic objects. This period became known as the Copper Age. Pure copper on the surface of Earth, however, is scarce. Around 5000 BCE, humans found that copper could be obtained by putting malachite, $Cu_2CO_3(OH)_2$, a green copper-containing stone, into a fire. Malachite yielded pure copper at the relatively low temperature of 200°C.

Copper is a soft metal made of layers of large copper crystals. It can easily be drawn into wires because the layers of crystals can slip past each other. When hammered, the large crystals break into smaller ones with rough edges and the layers can no longer slip past each other. Therefore, hammered copper sheets are harder than drawn copper. Using this knowledge, the ancient profession of coppersmith was born, and beautiful plates, pots, and ornaments were produced.

■ Bronze Age artifact.

Around 4000 BCE, it was discovered that an even greater hardness could be achieved by mixing molten copper with tin. The resulting alloy is called bronze. The Bronze Age was born somewhere in the Middle East and quickly spread to China and all over the world. Because hammered copper takes an edge, knives and swords could be manufactured using it.

An even harder metal was soon to come. The first raw iron was found in meteorites. (The ancient Sumerian name of iron is "metal from heaven.") Later, around 2500 BCE, it was discovered that iron could be recovered from its ore by smelting (heating the ore to a very high temperature). Thus began the Iron Age. The process of recovering a metal from its ore is called smelting. More advanced technology was needed for smelting iron ores because iron melts at a high temperature (about 1500°C). For this reason, it took a longer time to perfect this smelting process and to learn how to manufacture steel, which is about 90 to 95% iron and 5 to 10% carbon. Steel objects appeared first in India around 100 BCE.

Modern historians look back at ancient cultures and use the discovery of a new metal as a landmark for that age.

can be drawn into wires), and malleable (they can be hammered and rolled into sheets). They also form alloys, which are solutions of one or more metals dissolved in another metal. Brass, for example, is an alloy of copper and zinc. Bronze is an alloy of copper and tin, and pewter is an alloy of tin, antimony, and lead. In their chemical reactions, metals tend to give up electrons (Section 3.2). Figure 2.9 shows a form of the Periodic Table in which the elements are classified by type.

Figure 2.9 Classification of the elements.

≠ Not yet named

Nonmetal An element that does not have the characteristic properties of a metal and tends to accept electrons.

Metalloid An element that displays some of the properties of metals and some of the properties of nonmetals.

Although hydrogen (H) appears in Group 1A, it is not an alkali metal; it is a nonmetal. It is placed here because its *s* shell is the one being filled (Section 2.7).

Halogen An element in Group 7A of the Periodic Table.

Alkali metal An element in Group 1A of the Periodic Table.

The second class of elements is the **nonmetals**. With the exception of hydrogen, the 18 nonmetals appear on the right side of the Periodic Table. With the exception of graphite, which is one form of carbon, they do not conduct electricity. At room temperature, nonmetals such as phosphorus and iodine are solids. Bromine is a liquid, and the noble gases helium through radon are gases. In their chemical reactions, nonmetals tend to accept electrons (Section 3.2). Virtually all of the compounds we will encounter in our study of organic and biochemistry are built from just six nonmetals: H, C, N, O, P, and S.

Six elements are classified **metalloids**: boron, silicon, germanium, arsenic, antimony, and tellurium. These elements have some of the properties of metals and some of the properties of nonmetals. For example, some metalloids are shiny like metals, but do not conduct electricity. One of these metalloids, silicon, is a semiconductor; that is, it does not conduct electricity under a certain applied voltage, but becomes a conductor at higher applied voltages. This semiconductor property of silicon makes it a vital element for Silicon Valley and the entire electronics industry.

C Examples of Periodicity in the Periodic Table

Not only do the elements in any particular column of the Periodic Table share similar properties, but the properties also vary in some fairly regular ways as we go up or down a column. For instance, Table 2.2 shows that the melting and boiling points of the **halogens** regularly increase as we go down a column.

Another example involves the Group 1A elements, also called the **alkali metals**. All alkali metals are soft enough to be cut with a knife, and their softness increases in going down the column. They all have relatively low melting and boiling points, which decrease in going down the column (Table 2.3).

All alkali metals react with water to form hydrogen gas, H_2, and a metal hydroxide with the formula MOH, where "M" stands for the alkali metal. The violence of their reaction with water increases in going down the column.

$$Na + H_2O \longrightarrow NaOH + H_2O$$
Sodium　　Water　　Sodium　　Hydrogen
　　　　　　　　　hydroxide　　gas

Table 2.2 Melting and Boiling Points of the Halogens (Group 7A)

Element	Melting Point (°C)	Boiling Point (°C)
Fluorine	−220	−188
Chlorine	−101	−35
Bromine	−7	59
Iodine	114	184
Astatine	302	337

Table 2.3 Melting and Boiling Points of the Alkali Metals (Group 1A)

Element	Melting Point (°C)	Boiling Point (°C)
Lithium	180	1342
Sodium	98	883
Potassium	63	760
Rubidium	39	686
Cesium	28	669

Table 2.4 Melting and Boiling Points of the Noble Gases (Group 8A)

Element	Melting Point (°C)	Boiling Point (°C)
Helium	−272	−269
Neon	−249	−246
Argon	−189	−186
Krypton	−157	−152
Xenon	−112	−107
Radon	−71	−62

They also form compounds with the halogens with the formula MX, where "X" stands for the halogen.

The elements in Group 8A, often called the **noble gases,** provide still another example of how properties of elements change gradually within a column. Group 8A elements are gases under normal conditions of temperature and pressure, and they form either no compounds or very few compounds. Notice how close the melting and boiling points for the elements in this series are to one another (Table 2.4).

The Periodic Table is so useful that it hangs in nearly all chemistry classrooms and most chemical laboratories throughout the world. What makes it so useful is that it correlates a vast amount of data about the elements and their compounds and allows us to make many predictions about both chemical and physical properties. For example, if you were told that the boiling point of germane (GeH_4) is −88°C and that of methane (CH_4) is −164°C, could you predict the boiling point of silane (SiH_4)? The position of silicon in the table, between germanium and carbon, might lead you to a prediction of about −126°C. The actual boiling point of silane is −112°C, not far from this prediction.

> **Noble gas** An element in Group 8A of the Periodic Table.

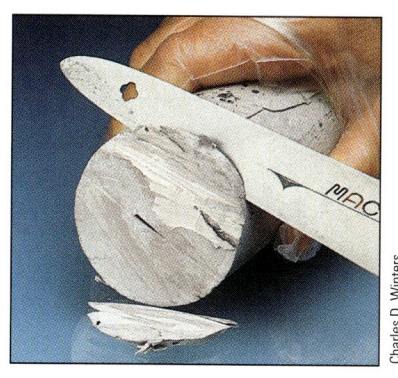

■ Sodium metal being cut with a knife.

Charles D. Winters

■ Niels Bohr of Denmark was awarded the Nobel Prize for Physics in 1922.

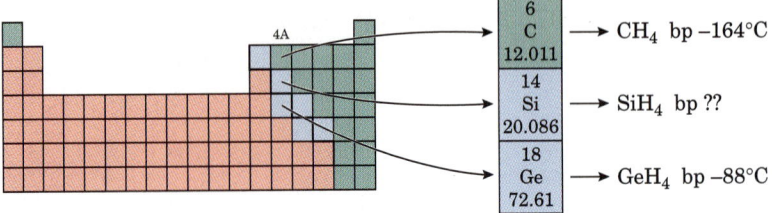

2.6 The Electron Configuration of Atoms

We have seen that the protons and neutrons of an atom are concentrated in the atom's very small nucleus, and that the electrons of an atom are located in the considerably larger space outside the nucleus. We can now ask how the electrons of an atom are arranged in this extranuclear space. Are they arranged randomly like raisins in a pudding, or are they organized into layers like the layers of an onion?

Let us begin with hydrogen because it has only one electron and is the simplest atom. Before we do so, however, it is necessary to describe a discovery made in 1913 by the Danish physicist Niels Bohr. At the time, it was known that an electron is always moving around the nucleus and so possesses kinetic energy. Bohr discovered that only certain values are possible for this energy. This was a very surprising discovery. If you were told that you could drive your car at 23.4 mph or 28.9 mph or 34.2 mph, but never at any speed in between these values, you wouldn't believe it. Yet that is just what Bohr discovered about electrons in atoms. The lowest possible energy level is the **ground state.** If an electron is to have more energy than it has in the ground state, only certain values are allowed; values in between are not permitted. Bohr was unable to explain why these energy levels of electrons exist in atoms, but the accumulated evidence forced him to the conclusion that they do. We say that the energy of electrons in atoms is quantized. We can liken quantization to walking up a flight of stairs (Figure 2.10). You can put your foot on any step, but you cannot stand any place between two steps.

A Electrons Are Distributed in Shells, Subshells, and Orbitals

One conclusion reached by Bohr is that electrons in atoms do not move freely in the space around the nucleus, but rather remain confined to specific regions of space called **principal energy levels,** or more simply, **shells.** These shells are numbered 1, 2, 3, 4, and so on, from the inside out. Table 2.5 gives the number of electrons that each shell can hold.

> **Ground-state electron configuration** The electron configuration of an atom of lowest energy.
>
> **Principal energy level** An energy level containing orbitals of all the same number (1, 2, 3, 4, and so forth).
>
> **Shell** All of the orbitals of a principal energy level of an atom.

Table 2.5 Distribution of Electrons in Shells

Shell	Number of Electrons Shell Can Hold	Relative Energies of Electrons in Each Shell
4	32	Higher
3	18	↑
2	8	
1	2	Lower

- Fourth shell (32 electrons)
- Third shell (18 electrons)
- Second shell (8 electrons)
- First shell (2 electrons)
- Nucleus

2.6 THE ELECTRON CONFIGURATION OF ATOMS | 45

Figure 2.10 An energy stairway. Unlike our stairways, the spaces between energy levels in an atom are not equal; they become smaller as they go up.

Electrons in the first shell are closest to the positively charged nucleus and are held most strongly by it; these electrons are said to be the lowest in energy. Electrons in higher-numbered shells are farther from the nucleus and are held less strongly to it; these electrons are said to be higher in energy.

Shells are divided into **subshells** designated by the letters s, p, d, and f. Within these subshells, electrons are grouped into **orbitals**. An orbital is a region of space that can hold two electrons. The first shell contains a single orbital called a $1s$ orbital and can hold two electrons (Table 2.6). The second shell contains one s orbital and three p orbitals. All p orbitals come in sets of three and can hold six electrons. The third shell contains one s orbital, three p orbitals, and five d orbitals. All d orbitals come in sets of five and can hold ten electrons. All f orbitals come in sets of seven and can hold 14 electrons.

Subshell All of the orbitals of an atom having the same principal energy level and the same letter designation (either s, p, d, or f).

Orbital A region of space around a nucleus that can hold up to two electrons.

B Orbitals Have Definite Shapes and Orientations in Space

All s orbitals have the shape of a sphere with the nucleus at the center of the sphere. Shown in Figure 2.11 are the shapes of the $1s$ and $2s$ orbitals. Of all the s orbitals, the $1s$ is the smallest; the $2s$ is a larger sphere, and the $3s$ is a still larger sphere. Also shown in Figure 2.11 are the three-dimensional shapes of the three $2p$ orbitals. Each $2p$ orbital has the shape of a dumbbell with the nucleus at the midpoint of the dumbbell. The three $2p$ orbitals are at right angles to each other, with one orbital on the x axis, the second on the y axis, and the third on the z axis. The shapes of $3p$ orbitals are similar, but larger. Because the vast majority of organic compounds and biomolecules consist of the elements H, C, N, O, P, and S, which use only $1s$, $2s$, $2p$, $3s$, and $3p$ orbitals for bonding, we concentrate on just these and other elements of the first, second, and third periods of the Periodic Table.

Table 2.6 Distribution of Orbitals within Shells

Shell	Orbitals Contained in Each Shell	Maximum Number of Electrons Shell Can Hold
4	One $4s$, three $4p$, five $4d$, and seven $4f$ orbitals	2 + 6 + 10 + 14 = 32
3	One $3s$, three $3p$, and five $3d$ orbitals	2 + 6 + 10 = 18
2	One $2s$ and three $2p$ orbitals	2 + 6 = 8
1	One $1s$ orbital	2

The d and f orbitals are less important for us, and we will not discuss their shapes.

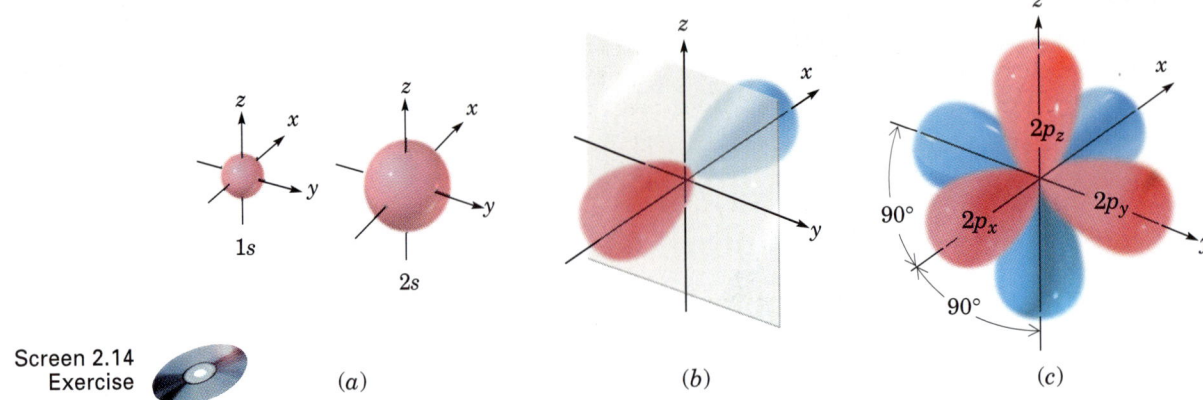

(a) (b) (c)

Figure 2.11 The 1s, 2s, and 2p orbitals. (a) An s orbital has the shape of a sphere with the nucleus at the center of the sphere. A 2s orbital is a larger sphere than a 1s orbital, and a 3s orbital is larger still. (b) A 2p orbital has the shape of a dumbbell with the nucleus at the midpoint of the dumbbell. (c) Each 2p orbital is perpendicular to the other two. The 3p orbitals are similar in shape but larger. To make it easier for you to see the two lobes of each 2p orbital, one lobe is colored red and the other is colored blue.

Screen 2.14 Exercise

Electron configuration A description of the orbitals of an atom or ion occupied by electrons.

C Electron Configurations of Atoms Are Governed by Three Rules

The **electron configuration** of an atom is a description of the orbitals that its electrons occupy. The orbitals available to all atoms are the same—namely, 1s, 2s, 2p, 3s, 3p, and so on. In the ground state of an atom, only the lowest energy orbitals are occupied; all other orbitals are empty. We determine the ground-state electron configuration of an atom using the following rules:

RULE 1. Orbitals fill in the order of increasing energy from lowest to highest.

Example: In this course we are concerned primarily with elements of the first, second, and third periods of the Periodic Table. Orbitals in these elements fill in the order 1s, 2s, 2p, 3s, and 3p. Figure 2.12 shows the order of filling through the 4f orbitals.

RULE 2. Each orbital can hold up to two electrons with spins paired.

Example: With four electrons, the 1s and 2s orbitals are filled and are written $1s^2 2s^2$. With an additional six electrons, the three 2p orbitals are filled and written either in an expanded form of $2p_x^2 2p_y^2 2p_z^2$, or in a condensed form of $2p^6$. Spin pairing means that the electrons spin in opposite directions (Figure 2.13).

RULE 3. When there is a set of orbitals of equal energy, each orbital becomes half-filled before any of them becomes completely filled.

Example: After the 1s and 2s orbitals are filled, a fifth electron is put into the $2p_x$ orbital, a sixth into the $2p_y$ orbital, and a seventh into the $2p_z$ orbital. Only after each 2p orbital has one electron is a second electron added to any 2p orbital.

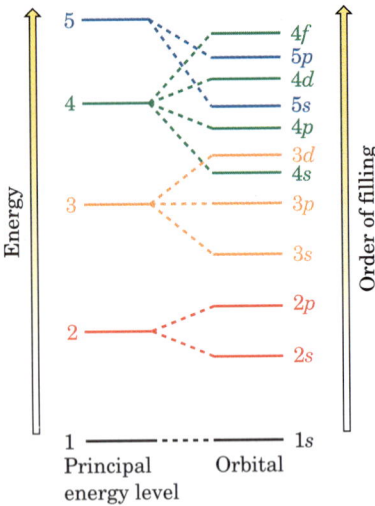

Figure 2.12 Energy levels for orbitals through the 4f orbitals.

D Electron Configurations and Orbital Box Diagrams

To illustrate how these rules are used, let us write ground-state electron configurations for several of the elements in periods 1, 2, and 3. In the following

Screen 2.18 Tutorial

2.6 THE ELECTRON CONFIGURATION OF ATOMS | 47

orbital box diagrams, we use a box to represent an orbital, an arrow with its head up to represent a single electron, and a pair of arrows with heads in opposite directions to represent two electrons with paired spins. In addition, we show both expanded and condensed electron configurations. Table 2.7 gives the complete condensed electron configurations for elements 1–18.

Hydrogen (H) The atomic number of hydrogen is 1, which means that it has a single electron. In the ground state, this electron is placed in the $1s$ orbital. Shown first is its orbital box diagram and then its electron configuration. Hydrogen has one unpaired electron.

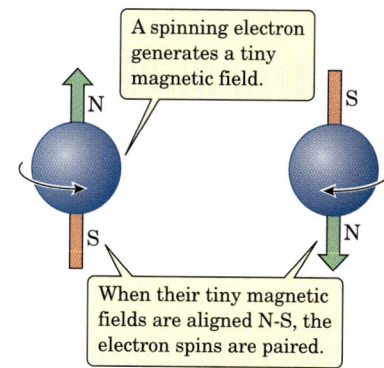

Figure 2.13 The pairing of electron spins.

There is one electron in this orbital

Table 2.7 Ground-State Electron Configurations of the First 18 Elements

	1s	2s	2p$_x$ 2p$_y$ 2p$_z$	3s	3p$_x$ 3p$_y$ 3p$_z$	Electron Configuration (Condensed)	Noble Gas Notation
H (1)	↑					$1s^1$	
He (2)	↑↓					$1s^2$	
Li (3)	↑↓	↑				$1s^2\, 2s^1$	[He] $2s^1$
Be (4)	↑↓	↑↓				$1s^2\, 2s^2$	[He] $2s^2$
B (5)	↑↓	↑↓	↑			$1s^2\, 2s^2\, 2p^1$	[He] $2s^2\, 2p^1$
C (6)	↑↓	↑↓	↑ ↑			$1s^2\, 2s^2\, 2p^2$	[He] $2s^2\, 2p^2$
N (7)	↑↓	↑↓	↑ ↑ ↑			$1s^2\, 2s^2\, 2p^3$	[He] $2s^2\, 2p^3$
O (8)	↑↓	↑↓	↑↓ ↑ ↑			$1s^2\, 2s^2\, 2p^4$	[He] $2s^2\, 2p^4$
F (9)	↑↓	↑↓	↑↓ ↑↓ ↑			$1s^2\, 2s^2\, 2p^5$	[He] $2s^2\, 2p^5$
Ne (10)	↑↓	↑↓	↑↓ ↑↓ ↑↓			$1s^2\, 2s^2\, 2p^6$	[He] $2s^2\, 2p^6$
Na (11)	↑↓	↑↓	↑↓ ↑↓ ↑↓	↑		$1s^2\, 2s^2\, 2p^6\, 3s^1$	[Ne] $3s^1$
Mg (12)	↑↓	↑↓	↑↓ ↑↓ ↑↓	↑↓		$1s^2\, 2s^2\, 2p^6\, 3s^2$	[Ne] $3s^2$
Al (13)	↑↓	↑↓	↑↓ ↑↓ ↑↓	↑↓	↑	$1s^2\, 2s^2\, 2p^6\, 3s^2\, 3p^1$	[Ne] $3s^2\, 3p^1$
Si (14)	↑↓	↑↓	↑↓ ↑↓ ↑↓	↑↓	↑ ↑	$1s^2\, 2s^2\, 2p^6\, 3s^2\, 3p^2$	[Ne] $3s^2\, 3p^2$
P (15)	↑↓	↑↓	↑↓ ↑↓ ↑↓	↑↓	↑ ↑ ↑	$1s^2\, 2s^2\, 2p^6\, 3s^2\, 3p^3$	[Ne] $3s^2\, 3p^3$
S (16)	↑↓	↑↓	↑↓ ↑↓ ↑↓	↑↓	↑↓ ↑ ↑	$1s^2\, 2s^2\, 2p^6\, 3s^2\, 3p^4$	[Ne] $3s^2\, 3p^4$
Cl (17)	↑↓	↑↓	↑↓ ↑↓ ↑↓	↑↓	↑↓ ↑↓ ↑	$1s^2\, 2s^2\, 2p^6\, 3s^2\, 3p^5$	[Ne] $3s^2\, 3p^5$
Ar (18)	↑↓	↑↓	↑↓ ↑↓ ↑↓	↑↓	↑↓ ↑↓ ↑↓	$1s^2\, 2s^2\, 2p^6\, 3s^2\, 3p^6$	[Ne] $3s^2\, 3p^6$

Helium (He) The atomic number of helium is 2, which means that it has two electrons. In the ground state, both electrons are placed in the 1s orbital with paired spins, which fills the 1s orbital. All electrons in helium are paired.

Lithium (Li) Lithium, atomic number 3, has three electrons. In the ground state, two electrons are placed in the 1s orbital with paired spins, and the third electron is placed in the 2s orbital. Lithium has one unpaired electron.

Carbon (C) Carbon, atomic number 6, has six electrons. Two are placed in the 1s orbital with paired spins and two are placed in the 2s orbital with paired spins. The fifth and sixth electrons are placed one each in the $2p_x$ and $2p_y$ orbitals. Carbon has two unpaired electrons.

Oxygen (O) Oxygen, atomic number 8, has eight electrons. The first four electrons fill the 1s and 2s orbitals. The next three electrons are placed in the $2p_x$, $2p_y$, and $2p_z$ orbitals so that each 2p orbital has one electron. The remaining electron now fills the $2p_x$ orbital. Oxygen has two unpaired electrons.

Neon (Ne) Neon, atomic number 10, has ten electrons, which completely fill all orbitals of the first and second shells. Neon has no unpaired electrons.

Sodium (Na) Sodium, atomic number 11, has 11 electrons. The first ten fill the $1s$, $2s$, and $2p$ orbitals. The eleventh electron is placed in the $3s$ orbital. Sodium has one unpaired electron.

Na (11) [↑↓] [↑↓] [↑↓][↑↓][↑↓] [↑]
          $1s$ $2s$ $2p_x\ 2p_y\ 2p_z$ $3s$

Electron configuration
Expanded: $1s^2 2s^2 2p_x^2 2p_y^2 2p_z^2 3s^1$
Condensed: $1s^2 2s^2 2p^6 3s^1$

Phosphorus (P) Phosphorus, atomic number 15, has 15 valence electrons. The first 12 fill the $1s$, $2s$, $2p$, and $3s$ orbitals. Electrons 13, 14, and 15 are placed one each in the $3p_x$, $3p_y$, and $3p_z$ orbitals. Phosphorus has three unpaired electrons.

P (15) [↑↓] [↑↓] [↑↓][↑↓][↑↓] [↑↓] [↑][↑][↑]
          $1s$ $2s$ $2p_x\ 2p_y\ 2p_z$ $3s$ $3p_x\ 3p_y\ 3p_z$

Electron configuration
Expanded: $1s^2 2s^2 2p_x^2 2p_y^2 2p_z^2 3s^2 3p_x^1 3p_y^1 3p_z^1$
Condensed: $1s^2 2s^2 2p^6 3s^2 3p^3$

E Showing Electron Configurations—Noble Gas Notations

In an alternative way of writing electron configurations, we use the symbol of the noble gas immediately preceding the particular atom to indicate the electron configuration of all filled shells. The first shell of lithium, for example, is abbreviated [He] and the single electron in its $2s$ shell is indicated by $2s^1$. Thus the electron configuration of lithium is [He]$2s^1$.

F Showing Electron Configurations—Lewis Dot Structures

When discussing the physical and chemical properties of an element, chemists often focus on the outermost shell of its atoms because electrons in this shell are the ones involved in the formation of chemical bonds (Chapter 3) and in chemical reactions (Chapter 4). Outer-shell electrons are called **valence electrons,** and the energy level in which they are found is called the **valence shell.** Carbon, for example, with a ground-state electron configuration of $1s^2 2s^2 2p^2$, has four valence (outer-shell) electrons.

To show the outermost electrons of an atom, we commonly use a representation called a **Lewis dot structure,** named after the American chemist Gilbert N. Lewis (1875–1946), who devised this notation. A Lewis structure shows the symbol of the element surrounded by a number of dots equal to the number of electrons in the outer shell of an atom of that element. In Lewis structures, the atomic symbol represents the nucleus and all filled inner shells. Table 2.8 shows Lewis structures for the first 18 elements of the Periodic Table.

The noble gases helium and neon have filled valence shells. The valence shell of helium is filled with two electrons; that of neon is filled with eight electrons. Neon and argon have in common an electron configuration in which the s and p orbitals of their valence shells are filled with eight electrons. The valence shells of all other elements shown in Table 2.8 contain fewer than eight electrons.

> **Valence electron** An electron in the outermost occupied (valence) shell of an atom.

> **Valence shell** The outermost occupied shell of an atom.

> **Lewis dot structure** The symbol of an element surrounded by a number of dots equal to the number of electrons in the valence shell of the atom.

Table 2.8 Lewis Dot Structures for Elements 1–18 of the Periodic Table

1A	2A	3A	4A	5A	6A	7A	8A
H·							He:
Li·	Be:	Ḃ:	·Ċ:	·N̈:	:Ö:	:F̈·	:N̈e:
Na·	Mg:	Ȧl:	·Si:	·P̈:	:S̈:	:C̈l:	:Är:

Each dot represents one valence electron.

Compare the Lewis structures given in Table 2.8 with the ground-state electron configurations given in Table 2.7. The Lewis structure of boron (B), for example, is shown in Table 2.8 with three valence electrons; these are the paired 2s electrons and the single $2p_x$ electron shown in Table 2.7. The Lewis structure of carbon (C) is shown in Table 2.8 with four valence electrons; these are the two paired 2s electrons and the unpaired $2p_x$ and $2p_y$ electrons shown in Table 2.7.

EXAMPLE 2.7

The Lewis dot structure for nitrogen shows five valence electrons. Write the expanded electron configuration for nitrogen and show to which orbitals its five valence electrons are assigned.

Solution
Nitrogen, atomic number 7, has the following ground-state electron configuration:

$$1s^2 2s^2 2p_x^1 2p_y^1 2p_z^1$$

The five valence electrons of the Lewis dot structure are the two paired electrons in the 2s orbital and the three unpaired electrons in the $2p_x$, $2p_y$, and $2p_z$ orbitals.

Problem 2.7
Write the Lewis dot structure for the element that has the following ground-state electron configuration. What is the name of this element?

$$1s^2 2s^2 2p_x^2 2p_y^2 2p_z^2 3s^2 3p_x^1$$

2.7 Electron Configuration and the Periodic Table

When Mendeleev published his first periodic table in 1869, he could not explain why it worked—that is, why elements with similar properties became aligned in the same column. Indeed, no one else had a good explanation for this phenomenon, either. It was not until the discovery of electron configurations that chemists finally understood why the table works. The answer is very simple. The Periodic Table works because elements in the same column have the same configuration of electrons in their outer shells. Figure 2.14 shows the relationship between shells (principal energy levels) and orbitals being filled.

All main-group elements have in common the fact that either their s or p orbitals are being filled. Notice that the 1s shell is filled with two elec-

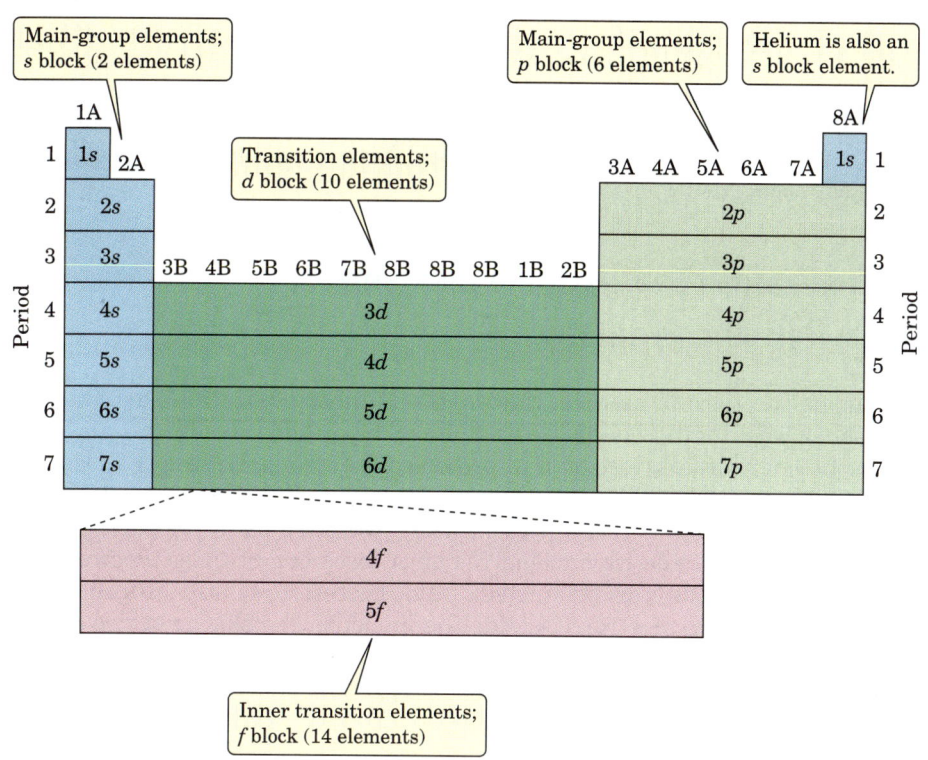

Figure 2.14 Electron configuration and the Periodic Table.

trons; there are only two elements in the first period. The 2s and 2p orbitals are filled with eight electrons; there are eight elements in period 2. Similarly, the 3s and 3p orbitals are filled with eight electrons; there are eight elements in Period 3.

To create the elements of period 4, one 4s, three 4p, and five 3d orbitals are available. These orbitals can hold a total of 18 electrons; there are 18 elements in period 4. Similarly, there are 18 elements in period 5. Inner transition elements are created by filling f orbitals, which come in sets of seven and can hold a total of 14 electrons; there are 14 inner transition elements in the lanthanide series and 14 in the actinide series.

To see the similarities in electron configurations within the Periodic Table, let us look at the elements in column 1A. We already know the configurations for lithium, sodium, and potassium (Table 2.7). To this list we can add rubidium and cesium. All elements in column 1A have one electron in their valence shells (Table 2.9).

Table 2.9 Lewis Dot Structures for the Alkali Metals (Group 1A)

Element	Noble Gas Notation	Lewis Dot Structure
Li	[He]$2s^1$	Li·
Na	[Ne]$3s^1$	Na·
K	[Ar]$4s^1$	K·
Rb	[Kr]$5s^1$	Rb·
Cs	[Xe]$6s^1$	Cs·

We will have more to say about atoms and their ions in Section 3.2.

All Group 1A elements are metals, with the exception of hydrogen, which is a nonmetal. The properties of elements largely depend on the electron configuration of their outer shell. As a consequence it is not surprising that Group 1A elements, all of which have the same outer-shell configuration, are metals (except for hydrogen) and have such similar physical and chemical properties.

2.8 Ionization Energy of the Elements: A Periodic Property

Atoms are electrically neutral; that is, the number of electrons outside the nucleus of an atom is equal to the number of protons inside the nucleus. Atoms do not normally lose or gain protons or neutrons, but they can lose or gain electrons. When a lithium atom, for example, loses one electron, it becomes a lithium **ion.** A lithium atom has three protons in its nucleus and three electrons outside its nucleus. When it loses one of these electrons, it still has three protons in its nucleus (and, therefore, is still lithium), but now it has only two electrons outside the nucleus. The charge of two of the protons is cancelled by the two remaining electrons, but no electron is present that can cancel the charge of the third proton. Therefore, a lithium ion has a charge of +1. We write it as Li^+.

Screen 2.20 Simulation

Ion An atom with an unequal number of protons and electrons.

$$Li + energy \longrightarrow Li^+ + e^-$$
Lithium   Ionization   Lithium   Electron
              energy   ion

Ionization energy The energy required to remove the most loosely held electron from an atom.

Ionization energy is a measure of how difficult it is to remove an electron from an atom in the gaseous state. The more difficult it is to remove the electron, the higher the ionization energy required. Figure 2.15 shows the ionization energies for atoms of the main-group elements 1–37 (hydrogen through rubidium).

Figure 2.15 Ionization energy versus atomic number for the first 37 elements.

In the figure, ionization energy generally increases as we go up a column of the Periodic Table and, with a few exceptions, increases as we go from left to right across a row. For example, within the Group 1A metals, rubidium gives up its 5s electron most easily and lithium gives up its 2s electron least easily.

We explain this trend by saying that the 5s electron of rubidium is farther from the positively charged nucleus than is the 4s electron in potassium, which in turn is farther from the positively charged nucleus than is the 3s electron of sodium, and so forth. Furthermore, the 5s electron of rubidium is more "shielded" by inner-shell electrons from the attractive force of the positive nucleus than is the 4s electron of potassium, and so forth. The greater the shielding, the lower the ionization energy. Thus, as atoms get bigger going down a column of the Periodic Table, their ionization energies decrease.

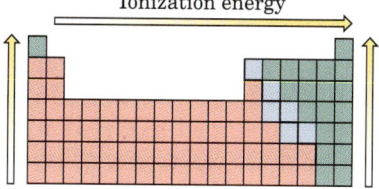

We explain the increase in ionization energy across a row by the fact that the valence electrons across a row are in the same shell (principal energy level) and, therefore, approximately the same distance from the nucleus. The number of protons in the nucleus, however, increases regularly across a row, which means that valence electrons experience an ever-stronger pull by the nucleus, which makes them more difficult to remove. Thus ionization energy increases from left to right across a row of the periodic table.

SUMMARY

Matter can be classified as an **element** (a pure substance composed of one type of atom), a **compound** (a pure substance made up of two or more types of elements), or a **mixture** (Section 2.2). Dalton's atomic theory proposes that all atoms of a given element are identical and that the atoms of any one element are different from those of any other element (Section 2.3A). His theory is based on the **law of conservation of mass** (matter can be neither created nor destroyed) and the **law of constant composition** (any compound is always made up of elements in the same proportion by mass). An **atom** is the smallest unit of an element that shows the chemical behavior of the element (Section 2.3A). A **molecule** is a cluster of two or more atoms that acts as a unit (Section 2.3B).

Atoms consist of **protons** and **neutrons** found inside the nucleus and **electrons** located outside it (Section 2.4A). An electron has a mass of approximately 0.0005 amu and a charge of −1. A proton has a mass of 1 amu and a charge of +1. A neutron has a mass of 1 amu and no charge. The **mass number** of an atom is the sum of the number of its protons and neutrons (Section 2.4B). The **atomic number** of an element is the number of protons in its nucleus; it determines the identity of the element (Section 2.3C). Neutral atoms have an equal number of electrons and protons. **Isotopes** are atoms with the same atomic number but different mass numbers; that is, they have the same number of protons in their nuclei but different numbers of neutrons (Section 2.3D). The **atomic weight** of an element is a weighted average of the masses (in amu) of the isotopes as they occur in nature (Section 2.3E). Atoms are very tiny, with a very small mass, almost all of which is concentrated in the nucleus (Section 2.4). The nucleus is extremely tiny, with a high density.

The **Periodic Table** arranges elements into columns of elements with similar properties, which gradually change as we move down a column (Section 2.5A). **Metals** are solids (except for mercury, which is a liquid), shiny, conductors of electricity, ductile, and malleable (Section 2.5B). They also form alloys, which are solutions of one or more metals dissolved in another metal. In their chemical reactions, metals tend to give up electrons. With the exception of hydrogen, the 18 **nonmetals** appear to the right side of the Periodic Table. With the exception of graphite, they do not conduct electricity. In their chemical reactions, nonmetals tend to accept electrons.

Six elements are classified **metalloids**: boron, silicon, germanium, arsenic, antimony, and tellurium. These elements have some of the properties of metals and some of the properties of nonmetals.

Electrons in atoms can exist in a number of **principal energy levels** or **shells** (Section 2.6A). All principal energy levels except the first are divided into **subshells** designated by the letters *s*, *p*, *d*, and *f*. Within each subshell, electrons are grouped in **orbitals**. An orbital is a region of space that can hold two electrons with paired spins. All *s* orbitals are spherical and can hold two electrons (Section 2.6B). All *p* orbitals come in sets of three, and each orbital is shaped like a dumbbell. A set of *p* orbitals can hold six electrons. All *d* orbitals come in sets of five and can hold ten electrons. All *f* orbitals come in sets of seven and can hold 14 electrons.

Electrons in the **ground state** (lowest energy state) of an atom fill up the available orbitals of lowest energy. The electron configuration of an atom is governed by three rules (Section 2.6C): (1) Orbitals fill in order of increasing energy; (2) each orbital can hold up to two electrons with paired spins; and (3) when filling orbitals of equivalent energy, each orbital adds one electron before any of them adds a second electron. Electron configurations may be shown by orbital notation (Section 2.6C), orbital box diagrams (Section 2.6D), or noble gas notation (Section 2.6E).

Electrons in the outermost or **valence shell** of an atom are called **valence electrons.** The valence shell contains all of the valence electrons (Section 2.6F). A **Lewis dot structure** is a representation of an atom in which the symbol of the element is surrounded by a number of dots equal to the number of its valence electrons (Section 2.6F).

The Periodic Table works because elements in the same column have the same outer-shell electron configuration (Section 2.7). **Ionization energy,** which is the energy necessary to remove an electron from an atom to form an **ion,** increases from bottom to top within a column of the Periodic Table and from left to right within a row (Section 2.8).

PROBLEMS

Numbers that appear in color indicate difficult problems.
▸ designates problems requiring application of principles.

Classification of Matter

2.8 Classify each of the following as an element, a compound, or a mixture.
(a) Oxygen
(b) Table salt
(c) Sea water
(d) Wine
(e) Air
(f) Silver
(g) Diamond
(h) A pebble
(i) Gasoline
(j) Milk
(k) Carbon dioxide
(l) Bronze

2.9 A jeweler can make a ring in which gold and silver are combined in any proportion. Is such a ring made of a compound or of a mixture?

2.10 Name these elements (try not to look at a table).
(a) O
(b) Pb
(c) Ca
(d) Na
(e) C
(f) Ti

2.11 Name these elements (try not to look at a table).
(a) S
(b) Fe
(c) H
(d) K
(e) Ag
(f) Au

2.12 The elements game, Part I. Name and give the symbol of the element that is named for each person.
(a) Niels Bohr (1885–1962), Nobel Prize for Physics in 1922
(b) Pierre and Marie Curie, Nobel Prize for Chemistry in 1903
(c) Albert Einstein (1879–1955), Nobel Prize for Physics in 1921
(d) Enrico Fermi (1901–1954), Nobel Prize for Physics in 1938
(e) Ernest Lawrence (1901–1958), Nobel Prize for Physics in 1939
(f) Lisa Meitner (1868–1968), codiscoverer of nuclear fission
(g) Dmitri Mendeleev (1834–1907), the first to formulate a workable periodic table
(h) Alfred Nobel (1833–1896), discoverer of dynamite
(i) Ernest Rutherford (1871–1937), Nobel Prize for Chemistry in 1908
(j) Glen Seaborg (1912–1999), Nobel Prize for Chemistry in 1951

2.13 The elements game, Part II. Name and give the symbol of the element that is named for each geographic location.
(a) The Americas
(b) Berkeley, California
(c) The state and University of California
(d) Dubna, location in Russia of the Joint Institute of Nuclear Research
(e) Europe
(f) France
(g) Gallia, Latin name for ancient France
(h) Germany
(i) Hafnia, Latin name for ancient Copenhagen
(j) Hesse, a German state
(k) Holmia, Latin name for ancient Stockholm
(l) Lutetia, Latin name for ancient Paris
(m) Magnesia, a district in Thessaly
(n) Poland, the native country of Marie Curie

(o) Rhenus, Latin name for the river Rhine
(p) Ruthenia, Latin name for ancient Russia
(q) Scandia, Latin name for ancient Scandinavia
(r) Strontian, a town in Scotland
(s) Ytterby, a village in Sweden (three elements)
(t) Thule, the earliest name for Scandinavia

2.14 The elements game, Part III. Give the names and symbols for the four elements named for planets.

2.15 Write the formulas of compounds in which the combining ratios are as follows:
(a) Potassium : oxygen, 2 : 1
(b) Sodium : phosphorus : oxygen, 3 : 1 : 4
(c) Lithium : nitrogen : oxygen, 1 : 1 : 3

2.16 Write the formulas of compounds in which the combining ratios are as follows:
(a) Sodium : hydrogen : carbon : oxygen, 1 : 1 : 1 : 3
(b) Carbon : hydrogen : oxygen, 2 : 6 : 1
(c) Potassium : manganese : oxygen, 1 : 1 : 4

Dalton's Atomic Theory

2.17 How does Dalton's atomic theory explain the following:
(a) The law of conservation of mass?
(b) The law of constant composition?

2.18 When 2.16 g of mercury oxide is decomposed to yield 2.00 g of mercury and 0.16 g of oxygen, which law is supported by the experiment?

2.19 The compound carbon monoxide contains 42.9% carbon and 57.1% oxygen. The compound carbon dioxide contains 27.2% carbon and 72.7% oxygen. Does this disprove Proust's law of constant composition?

Inside the Atom

2.20 Where in the atom are these particles located?
(a) Protons
(b) Electrons
(c) Neutrons

2.21 It has been said that "protons determine the atom." Do you agree or disagree with this statement? Explain.

2.22 What is the mass number of an atom with
(a) 22 protons, 22 electrons, and 26 neutrons?
(b) 76 protons, 76 electrons, and 114 neutrons?
(c) 34 protons, 34 electrons, and 45 neutrons?
(d) 94 protons, 94 electrons, and 150 neutrons?

2.23 Name and give the symbol for each element in Problem 2.22.

2.24 Given the mass number and the number of neutrons, what is the name and symbol of each element?

(a) mass number 45; 24 neutrons
(b) mass number 48; 26 neutrons
(c) mass number 107; 60 neutrons
(d) mass number 246; 156 neutrons
(e) mass number 36; 18 neutrons

2.25 If each of the atoms in Problem 2.24 acquired two more neutrons, what element would each then be?

2.26 How many neutrons are there in:
(a) A carbon atom of mass number 13?
(b) A germanium atom of mass number 73?
(c) An osmium atom of mass number 188?
(d) A platinum atom of mass number 195?

2.27 How many protons and how many neutrons does each of these isotopes of radon contain?
(a) Rn-210   (b) Rn-218   (c) Rn-222

2.28 How many neutrons and protons are there in each isotope?
(a) ^{22}Ne   (b) ^{104}Pd
(c) ^{35}Cl   (d) Tellurium-128
(e) Lithium-7   (f) Uranium-238

2.29 Tin-118 is one of the isotopes of tin. Name the isotopes of tin that contain two, three, and six more neutrons than tin-118.

2.30 What is the difference between atomic number and mass number?

2.31 Define:
(a) ion   (b) isotope

2.32 There are only two naturally occurring isotopes of antimony: ^{121}Sb (120.90 amu) and ^{123}Sb (122.90 amu). The atomic weight of antimony given in the Periodic Table is 121.75. Which of the two isotopes has the greater natural abundance?

2.33 The two most abundant naturally occurring isotopes of carbon are carbon-12 (98.90%, 12.000 amu) and carbon-13 (1.10%, 13.003 amu). From these abundances, calculate the atomic weight of carbon and compare your calculated value with that given in the Periodic Table.

2.34 Another isotope of carbon, carbon-14, occurs in nature but in such small amounts relative to carbon-12 and carbon-13 that it does not contribute to the atomic weight of carbon as recorded in the Periodic Table. Carbon-14 is invaluable in the science of radiocarbon dating (see Chemical Connections 9B). Give the number of protons, neutrons, and electrons in an atom of carbon-14.

2.35 The isotope carbon-11 does not occur in nature but has been made in the laboratory. This isotope is used in a medical imaging technique called positron emission tomography (PET; see Section 9.6). Give the number of protons, neutrons, and electrons in an atom of carbon-11.

2.36 Other isotopes used in PET therapy are fluorine-18, nitrogen-13, and oxygen-15. None of these isotopes occurs in nature; all must be produced in the laboratory. Give the number of protons, neutrons, and electrons in an atom of each of these artificial isotopes.

2.37 Americium-241 is used in household smoke detectors. This element has 11 known isotopes, none of which occurs in nature, but must be made in the laboratory. Give the number of protons, neutrons, and electrons in an atom of americium-241.

The Periodic Table

2.38 How many metals, metalloids, and nonmetals are there in the third period of the Periodic Table?

2.39 Which group(s) of the Periodic Table contains
  (a) Only metals?
  (b) Only metalloids?
  (c) Only nonmetals?

2.40 Which period(s) in the Periodic Table contain more nonmetals than metals? Which contain more metals than nonmetals?

2.41 Which of the following elements would you expect to have fairly similar properties (look at the Periodic Table): As, I, Ne, F, Mg, K, Ca, Ba, Li, He, N, P?

2.42 Which are transition elements?
  (a) Pd (b) K (c) Co
  (d) Ce (e) Br (f) Cr

2.43 Which element in each pair is more metallic?
  (a) Silicon or aluminum
  (b) Arsenic or phosphorus
  (c) Gallium or germanium
  (d) Gallium or aluminum

2.44 Classify these elements as metals, nonmetals, or metalloids:
  (a) Argon
  (b) Boron
  (c) Lead
  (d) Arsenic
  (e) Potassium
  (f) Silicon
  (g) Iodine
  (h) Antimony
  (i) Vanadium
  (j) Sulfur
  (k) Nitrogen

Electron Structure of Atoms

2.45 According to Table 2.5, the number of electrons that a shell can hold is 2 in the first shell, 8 in the second shell, 18 in the third shell, and 32 in the fourth shell. How do these numbers arise?

2.46 How many periods of the Periodic Table have two elements? How many have eight elements? How many have 18 elements? How many have 32 elements?

2.47 What is the correlation between group number of the main-group elements (1A, 2A, . . . , 8A) and the number of valence electrons in an element in the group?

2.48 Given your answer to Problem 2.47, write the Lewis dot structures for each of the following elements using no other information than the number of the group in the Periodic Table in which the element falls.
  (a) Carbon (4A) (b) Silicon (4A)
  (c) Oxygen (6A) (d) Sulfur (6A)
  (e) Aluminum (3A) (f) Bromine (7A)

2.49 Write the condensed ground-state electron configuration for each of the following elements. The element's atomic number is given in parentheses.
  (a) Li (3) (b) Ne (10)
  (c) Be (4) (d) C (6)
  (e) Mg (12)

2.50 Write the Lewis dot structure for each element in Problem 2.49.

2.51 Write the condensed ground-state electron configuration for each of the following elements. The element's atomic number is given in parentheses.
  (a) He (4) (b) Na (11) (c) Cl (17)
  (d) P (15) (e) H (1)

2.52 Write the Lewis dot structure for each element in Problem 2.51.

2.53 What is the same and what is different in the electron configurations of
  (a) Na and Cs (b) O and Te
  (c) C and Ge

2.54 Silicon, atomic number 14, is in Group 4A. How many orbitals are occupied by the valence electrons of Si in its ground state?

2.55 You are presented with a Lewis dot structure of element X as X:. To which two groups in the Periodic Table might this element belong?

2.56 The electron configurations for the elements with atomic numbers higher than 36 follow the same rules as given in the text for the first 36 elements. In fact, you can arrive at the correct order of filling of orbitals from Figure 2.14 by starting with H and reading the orbitals from left to right across the first row, then the second row, and so on. Write the condensed ground-state electron configuration for
  (a) Rb (b) Sr (c) Br

2.57 What is the total number of electrons that can fit into the 6s, 6p, 6d, and 6f orbitals?

Electron Configuration and the Periodic Table

2.58 Assume that a new element has been discovered with atomic number 117. Its chemical properties should be similar to those of At. Predict whether the new element's ionization energy will be greater than, the same as, or smaller than that of (a) At and (b) Ra.

2.59 Why do the elements in Group 1A of the Periodic Table (the alkali metals) have similar but not identical properties?

2.60 List these elements in order of increasing ionization energy:
(a) Li, Na, K
(b) C, N, Ne
(c) O, C, F
(d) Br, Cl, F

Chemical Connections

2.61 (Chemical Connections 2A) Why does the body need sulfur, calcium, and iron?

2.62 (Chemical Connections 2B) Which are the two most abundant elements, by weight, in
(a) The Earth's crust?
(b) The human body?

2.63 (Chemical Connections 2C) Consider the isotopic abundance of hydrogen on Mars. Would the atomic weight of hydrogen on Mars be greater than, the same as, or smaller than that on Earth?

2.64 (Chemical Connections 2D) Why is strontium-90 more dangerous to humans than most other radioactive isotopes that were present in the Chernobyl fallout?

2.65 (Chemical Connections 2E) Bronze is an alloy of which two metals?

2.66 (Chemical Connections 2E) Copper is a soft metal. How can one make it harder?

Additional Problems

2.67 Give the designation of all subshells in the:
(a) 1 shell
(b) 2 shell
(c) 3 shell
(d) 4 shell

2.68 Tell whether metals or nonmetals are more likely to have each characteristic:
(a) Conduct electricity and heat
(b) Accept electrons
(c) Be malleable
(d) Be a gas at room temperature
(e) Be a transition element
(f) Give up electrons

2.69 What is the outer-shell electron configuration of the elements in
(a) Group 3A?
(b) Group 7A?
(c) Group 5A?

2.70 Suppose that you face a problem similar to Mendeleev: You must predict the properties of an element not yet discovered. What will element 118 be like if and when enough of it is made for chemists to study its physical and chemical properties?

2.71 Determine the number of protons, electrons, and neutrons present in:
(a) ^{32}P
(b) ^{98}Mo
(c) ^{44}Ca
(d) ^{127}I
(e) ^{158}Gd
(f) ^{212}Bi

2.72 What percentage of the mass of each element is contributed by neutrons?
(a) Carbon-12
(b) Calcium-40
(c) Iron-55
(d) Bromine-79
(e) Platinum-195
(f) Uranium-238

2.73 Do isotopes of the heavy elements (for example, those from atomic number 37 to 53) contain more, the same, or fewer neutrons than protons?

2.74 What is the symbol for each of the following elements? (Try not to look at a table.)
(a) Phosphorus
(b) Potassium
(c) Sodium
(d) Nitrogen
(e) Bromine
(f) Silver
(g) Calcium
(h) Carbon
(i) Tin
(j) Zinc

2.75 The natural abundance of boron isotopes is as follows: 19.9% boron-10 (10.013 amu) and 80.1% boron-11 (11.009 amu). Calculate the atomic weight of boron (watch the significant figures) and compare your calculated value with that given in the Periodic Table. 10.8108

2.76 How many electrons are in the outer shell of each of the following elements?
(a) Si
(b) Br
(c) P
(d) K
(e) He
(f) Ca
(g) Kr
(h) Pb
(i) Se
(j) O

2.77 The mass of a proton is 1.6731^{-24} g. The mass of a grain of salt is 1.0310^{-2} g. How many protons would it take to have the same mass as a grain of salt?

2.78 (a) What are the charges of an electron, a proton, and a neutron?
(b) What are the masses (in amu, to one significant figure) of an electron, a proton, and a neutron?

2.79 Rubidium has two natural isotopes: rubidium-85 and rubidium-87. What is the natural abundance of each isotope if the atomic weight of rubidium is 85.47?

2.80 What is the name of this element and how many protons and neutrons does it have in its nucleus: $^{131}_{54}X$?

2.81 Based on the data presented in Figure 2.15, which atom would have the highest ionization energy: I, Cs, Sn, or Xe?

InfoTrac College Edition

For additional readings, go to InfoTrac College Edition, your online research library, at

http://infotrac.thomsonlearning.com

CHAPTER 3

3.1 Introduction

3.2 Atoms and Their Ions

3.3 Naming Ions

3.4 Formation of Chemical Bonds

3.5 Ionic Compounds

3.6 Naming Ionic Compounds

3.7 Molecular Compounds

3.8 Naming Binary Molecular Compounds

3.9 Bond Angles and Shapes of Molecules

3.10 Polar and Nonpolar Molecules

Sodium chloride crystal.

Chemical Bonds

3.1 Introduction

In Chapter 2, we stated that compounds are tightly bound groups of atoms. In this chapter, we will see that the atoms in compounds are held together by powerful forces of attraction called chemical bonds. There are two main types: ionic bonds and covalent bonds. We begin by examining ionic bonds. To talk about ionic bonds, however, we must first discuss why atoms form the ions they do.

3.2 Atoms and Their Ions

Noble Gas	Noble Gas Notation
He	$1s^2$
Ne	$[He]2s^22p^6$
Ar	$[Ne]3s^23p^6$
Kr	$[Ar]4s^24p^6$
Xe	$[Kr]5s^25p^6$

In 1916, Gilbert N. Lewis (Section 2.6) devised a beautifully simple model that unified many of the observations about chemical bonding and chemical reactions. He pointed out that the lack of chemical reactivity of the noble gases (Group 8A) indicates a high degree of stability of their electron configurations: helium with a valence shell of two electrons ($1s^2$), neon with a valence shell of eight electrons ($2s^22p^6$), argon with a valence shell of eight electrons ($3s^23p^6$), and so forth.

The tendency to react in ways that achieve an outer shell of eight valence electrons is particularly common among Group 1A–7A elements and is given the special name of the **octet rule.** An atom with almost eight valence electrons tends to gain the needed electrons to have eight electrons in its valence shell and an electron configuration like that of the noble gas nearest to it in atomic number. In gaining electrons, the atom becomes a negatively charged ion called an **anion.** An atom with only one or two valence electrons tends to lose the number of electrons required to have an electron configuration like the noble gas nearest to it in atomic number. In losing electrons, the atom becomes a positively charged ion called a **cation.** Note that when an ion forms, the number of protons and neutrons in the nucleus of the atom remains unchanged; only the number of electrons in the valence shell of the atom changes.

Octet rule When undergoing chemical reactions, atoms of Group 1A–7A elements tend to gain, lose, or share electrons to achieve an election configuration having eight valence electrons

Anion An ion with a negative electric charge.

Cation An ion with a positive electric charge.

EXAMPLE 3.1

Show how the following chemical changes obey the octet rule:
(a) A sodium atom loses an electron to form a sodium ion, Na^+.

$$Na \longrightarrow Na^+ + e^-$$
A sodium atom   A sodium ion   An electron

(b) A chlorine atom gains an electron to form a chloride ion, Cl^-.

$$Cl + e^- \longrightarrow Cl^-$$
A chlorine atom   An electron   A chloride ion

Screen 3.2
Tutorial

Solution

To see how each chemical change follows the octet rule, first write the condensed ground-state electron configuration (Section 2.6C) of the atom involved in the chemical change and the ion it forms, and then compare them.

(a) The condensed ground-state electron configurations for Na and Na^+ are as follows:

Na (11 electrons): $1s^2 2s^2 2p^6\ 3s^1$

Na^+ (10 electrons): $1s^2 2s^2 2p^6$

A sodium atom has one electron in its valence shell. The loss of this one valence electron changes the sodium atom to a sodium ion, Na^+, which has a complete octet of electrons in its valence shell and the same electron configuration as neon, the noble gas nearest to it in atomic number. We can write this chemical change using Lewis dot structures (Section 2.6F):

$$Na\cdot \longrightarrow Na^+ + e^-$$

(b) The condensed ground-state electron configurations for Cl and Cl^- are as follows:

Cl (17 electrons): $1s^2 2s^2 2p^6\ 3s^2 3p^5$

Cl^- (18 electrons): $1s^2 2s^2 2p^6\ 3s^2 3p^6$

A chlorine atom has seven electrons in its valence shell. The gain of one electron changes the chlorine atom to a chloride ion, Cl^-, which has a

■ G. N. Lewis.

complete octet of electrons in its valence shell and the same electron configuration as argon, the noble gas nearest to it in atomic number. We can write this chemical change using Lewis dot structures:

$$:\ddot{\underset{..}{Cl}}\cdot + e^- \longrightarrow :\ddot{\underset{..}{Cl}}:^-$$

Problem 3.1

Show how the following chemical changes obey the octet rule:
(a) A magnesium atom forms a magnesium ion, Mg^{2+}
(b) A sulfur atom forms a sulfide ion, S^{2-}

The octet rule gives us a good way to understand why Group 1A–7A elements form the ions that they do. It is not perfect, however, for two reasons:

1. Ions of period 1 and 2 elements with charges greater than +2 are unstable. Boron, for example, has three valence electrons. If it lost these three electrons, it would become B^{3+} and have a complete outer shell like that of helium. It seems, however, that this is far too large a charge for an ion of this period 2 element; consequently, this ion is not found in stable compounds. By the same reasoning, carbon does not lose its four valence electrons to become C^{4+}, nor does it gain four valence electrons to become C^{4-}. Either of these changes would place too great a charge on this period 2 element.

2. The octet rule does not apply to Group 1B–7B elements (the transition elements), most of which form ions with two or more different positive charges. Copper, for example, can lose one valence electron to form Cu^+; alternatively, it can lose two valence electrons to form Cu^{2+}.

It is important to understand that there are enormous differences between the properties of an atom and those of its ion(s). Atoms and their ions are completely different chemical species and have completely different chemical and physical properties. Consider, for example, sodium and chlorine. Sodium is a soft metal made of sodium atoms that react violently with water. Chlorine atoms are very unstable and even more reactive than sodium atoms. Both sodium and chlorine are poisonous. NaCl, common table salt, is made up of sodium ions and chloride ions. These two ions are quite stable and unreactive. Neither sodium ions nor chloride ions react with water at all.

■ Salt, sodium chloride. This chemical compound (right) is composed of the elements sodium (left) and chlorine (center) in chemical combination. Salt is very different in appearance and properties from the elements that constitute it.

Because atoms and their ions are different chemical species, we must be careful to distinguish one from the other. Consider the drug commonly known as "lithium," which is used to treat manic depression. The element lithium, like sodium, is a soft metal that reacts with water. The drug used to treat manic depression is not lithium atoms, Li, but rather lithium ions, Li^+, usually given in the form of lithium carbonate, Li_2CO_3. Another example comes from the fluoridation of drinking water, and of toothpastes and dental gels. The element fluorine, F_2, is an extremely poisonous and corrosive gas, and is not what is used for this purpose. Instead, fluoridation uses fluoride ions, F^-, in the form of sodium fluoride, NaF, a compound that is unreactive and nonpoisonous in the concentrations used.

Lithium carbonate is prescribed for manic depression under at least 18 different proprietary names.

3.3 Naming Ions

Names for anions and cations are formed by a system developed by the International Union of Pure and Applied Chemistry. We will refer to these names as "systematic" names. Note, however, that many ions have "common" names that were in use long before chemists undertook an effort to systematize their naming. In this and the following chapters, we will make every effort to use systematic names for ions, but where a long-standing common name remain in use, we will give it as well.

A Naming Cations

A cation forms when a metal loses one or more electrons. Elements of Groups 1A, 2A, and 3A form only one type of cation. For ions from these metals, the name of the cation is the name of the metal followed by the word "ion" (Table 3.1). There is no need to specify the charge on these cations, because only one charge is possible. For example, Na^+ is sodium ion and Ca^{2+} is calcium ion.

Most transition and inner transition elements form more than one type of cation and, therefore, the name of the cation must show its charge. To show the charge, we write a Roman numeral immediately following (with no space) the name of the metal (Table 3.2). For example, Cu^+ is copper(I) ion and Cu^{2+} is copper(II) ion. Note that even though silver is a transition metal, it forms only Ag^+; therefore, there is no need to use a Roman numeral to show this ion's charge.

In the older, common system for naming metal cations with two different charges, the suffix "-ous" is used to show the smaller charge and "-ic" is used to show the larger charge (Table 3.2).

■ Copper(I) oxide (left) and copper(II) oxide. The different copper ion charges result in different colors.

Table 3.1 Names for Cations from Some Metals That Form Only One Type of Positive Ion					
Group 1A		**Group 2A**		**Group 3A**	
Ion	Name	Ion	Name	Ion	Name
H^+	Hydrogen ion	Mg^{2+}	Magnesium ion	Al^{3+}	Aluminum ion
Li^+	Lithium ion	Ca^{2+}	Calcium ion		
Na^+	Sodium ion	Sr^{2+}	Strontium ion		
K^+	Potassium ion	Ba^{2+}	Barium ion		

Table 3.2 Systematic and Common Names for Four Metals That Form Two Different Cations

Ion	Systematic Name	Common Name	Origin of the Symbol of the Element or the Common Name of the Ion
Cu^+	Copper(I) ion	Cuprous ion	*Cupr-* from *cuprum*, the Latin name for copper
Cu^{2+}	Copper(II) ion	Cupric ion	
Fe^{2+}	Iron(II) ion	Ferrous ion	*Ferr-* from *ferrum*, the Latin name for iron
Fe^{3+}	Iron(III) ion	Ferric ion	
Hg^+	Mercury(I) ion	Mercurous ion	*Hg* from *hydrargyrum*, the Latin name for mercury
Hg^{2+}	Mercury(II) ion	Mercuric ion	
Sn^{2+}	Tin(II) ion	Stannous ion	*Sn* from *stannum*, the Latin name for tin
Sn^{4+}	Tin(IV) ion	Stannic ion	

Table 3.3 Names of the Most Common Monatomic Anions

Anion	Stem Name	Anion Name
F^-	fluor	fluoride
Cl^-	chlor	chloride
Br^-	brom	bromide
I^-	iod	iodide
O^{2-}	ox	oxide
S^{2-}	sulf	sulfide

B Naming Monatomic Anions

A monatomic (containing only one atom) anion is named by adding "-ide" to the stem part of the name. Table 3.3 gives the origin of the monatomic anions we deal with most often.

C Naming Polyatomic Ions

Screen 3.5 Tutorial

A **polyatomic ion** contains more than one atom. Examples are the hydroxide ion, OH^-, and the phosphate ion, PO_4^{3-}. We will not be concerned with how these ions are formed—only that they exist and are present in the materials around us. While rules for naming polyatomic ions have been developed, the simplest thing to do is just to memorize them.

Table 3.4 Names of Common Polyatomic Ions

Polyatomic Ion	Name	Polyatomic Ion	Name
NH_4^+	Ammonium	HCO_3^-	Hydrogen carbonate (bicarbonate)
OH^-	Hydroxide	SO_3^{2-}	Sulfite
NO_2^-	Nitrite	HSO_3^-	Hydrogen sulfite (bisulfite)
NO_3^-	Nitrate	SO_4^{2-}	Sulfate
CH_3COO^-	Acetate	HSO_4^-	Hydrogen sulfate (bisulfate)
CN^-	Cyanide	PO_4^{3-}	Phosphate
MnO_4^-	Permanganate	HPO_3^{2-}	Hydrogen phosphate
CrO_4^{2-}	Chromate	$H_2PO_4^-$	Dihydrogen phosphate
CO_3^{2-}	Carbonate		

Common names, where still widely used, are given in parentheses.

CHEMICAL CONNECTIONS 3A

Coral Chemistry and Broken Bones

Bone is a highly structured matrix consisting of both inorganic and organic materials. The inorganic material is chiefly hydroxyapatite, $Ca_5(PO_4)_3OH$, which makes up about 70% of bone by dry weight. By comparison, the enamel of teeth consists almost entirely of hydroxyapatite. Chief among the organic components of bone are collagen fibers (proteins; see Chemical Connections 21F) which thread through the inorganic matrix, providing extra strength and allowing bone to flex under stress. Also weaving through the hydroxyapatite-collagen framework are blood vessels that supply nutrients.

A problem faced by orthopedic surgeons is how to repair bone damage. For a minor fracture, usually a few weeks in a cast suffices for the normal process of bone growth to repair the damaged area. For severe fractures, especially those involving bone loss, a bone graft may be needed. An alternative to a bone graft is an implant of synthetic bone material. One such material, called Pro Osteon, is derived by heating coral with ammonium hydrogen phosphate to form a hydroxyapatite similar to that of bone. Throughout the heating process, the porous structure of the coral, which resembles that of bone, is retained.

■ A wrist fracture repaired with bone cement (white area).

The surgeon can then shape a piece of this material to match the bone void, implant it, stabilize the area by inserting metal plates and/or screws, and let new bone tissue grow into the pores of the implant.

In an alternative process, a dry mixture of calcium dihydrogen phosphate monohydrate $[Ca(H_2PO_4)_2 \cdot H_2O]$, calcium phosphate $[Ca_3(PO_4)_2]$, and calcium carbonate ($CaCO_3$), is prepared. Just before the surgical implant occurs, these chemicals are mixed with a solution of sodium phosphate to form a paste that is then injected into the bony area to be repaired. In this way, the fractured bony area is held in the desired position by the synthetic material while the natural process of bone rebuilding replaces the implant with living bone tissue.

$$\underset{\text{Coral}}{5CaCO_3} + 3(NH_4)_2HPO_4 \xrightarrow[\text{24–60 hours}]{200°C} \underset{\text{Hydroxyapatite}}{Ca_5(PO_4)_3OH} + 3(NH_4)_2CO_3 + 2H_2CO_3$$

The preferred system for naming polyatomic ions that differ in the number of hydrogen atoms is to show the number of hydrogens by the prefixes *di-, tri-,* and so forth, to show the presence of more than one hydrogen. For example, HPO_4^{2-} is hydrogen phosphate ion, and $H_2PO_4^-$ is dihydrogen phosphate ion. Table 3.4 lists some of the important polyatomic ions. Because several hydrogen-containing polyatomic anions have common names that are still widely used, you should memorize them as well. In these common names, the prefix "bi" is used to show the presence of one hydrogen.

3.4 Formation of Chemical Bonds

A Ionic and Covalent Bonds

According to the Lewis model of chemical bonding, atoms bond together in such a way that each atom participating in a bond acquires a valence-shell electron configuration the same as that of the noble gas nearest to it in atomic number. Atoms acquire completed valence shells in two ways:

1. An atom may lose or gain enough electrons to acquire a filled valence shell, becoming an ion as it does so (Section 3.2). An **ionic bond** results from the force of attraction between a cation and an anion.

> **Ionic bond** A chemical bond resulting from the attraction between a positive ion and a negative ion

Covalent bond A chemical bond resulting from the sharing of electrons between two atoms

2. An atom may share electrons with one or more other atoms to acquire a filled valence shell. A **covalent bond** results from the force of attraction between two atoms that share one or more pairs of electrons.

We can now ask how to determine whether two atoms in a compound are bonded by an ionic bond or a covalent bond. One way to do so is to consider the relative positions of the two atoms in the Periodic Table. Ionic bonds usually form between a metal and a nonmetal. An example of an ionic bond is that formed between the metal sodium and the nonmetal chlorine in the compound sodium chloride, Na^+Cl^-. When two nonmetals or a metalloid and a nonmetal combine, the bond between them is usually covalent. Examples of compounds containing covalent bonds between nonmetals include Cl_2, H_2O, CH_4, and NH_3. Examples of compounds containing covalent bonds between a metalloid and a nonmetal include BF_3, $SiCl_4$, and AsH_4.

Another way to determine the bond type is to compare the electronegativities of the atoms involved, which is the subject of the next subsection.

B Electronegativity and Chemical Bonds

Electronegativity is a measure of an atom's attraction for the electrons it shares in a chemical bond with another atom. The most widely used scale of electronegativities (Table 3.5) was devised by Linus Pauling in the 1930s. On the Pauling scale, fluorine, the most electronegative element, is assigned an electronegativity of 4.0, and all other elements are assigned values in relation to fluorine.

As you study the electronegativity values in Table 3.5, note that they generally increase from left to right across a row of the Periodic Table and increase from bottom to top within a column. Values increase from left to right because of the increasing positive charge on the nucleus, which leads to a stronger attraction for electrons in the valence shell. Values increase going up a column because of the decreasing distance of the valence electrons from the nucleus, which leads to stronger attraction between a nucleus and its valence electrons.

You might compare these trends in electronegativity with the trends in ionization energy (Section 2.8). Each illustrates the periodic nature of elements within the Periodic Table. Ionization energy reflects the energy necessary to remove an electron from an atom. Electronegativity measures the

Linus Pauling (1901–1994) was awarded the 1954 Nobel Prize in Chemistry for his pioneering studies of chemical bonding. Shortly after World War II, Pauling and his wife began a crusade to limit nuclear testing, a crusade that came to fruition in the form of the Limited Test Ban Treaty of 1963. For this effort, Pauling was awarded the 1963 Nobel Prize for Peace. Never before had any person received two unshared Nobel Prizes.

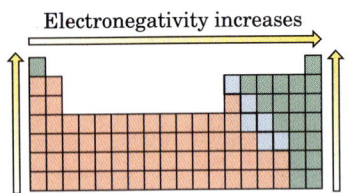
Electronegativity increases

Table 3.5 Electronegativity Values for the Elements (Pauling Scale)

1A	2A											3A	4A	5A	6A	7A
							H 2.1									
Li 1.0	Be 1.5											B 2.0	C 2.5	N 3.0	O 3.5	F 4.0
Na 0.9	Mg 1.2	3B	4B	5B	6B	7B	8B			1B	2B	Al 1.5	Si 1.8	P 2.1	S 2.5	Cl 3.0
K 0.8	Ca 1.0	Sc 1.3	Ti 1.5	V 1.6	Cr 1.6	Mn 1.5	Fe 1.8	Co 1.8	Ni 1.8	Cu 1.9	Zn 1.6	Ga 1.6	Ge 1.8	As 2.0	Se 2.4	Br 2.8
Rb 0.8	Sr 1.0	Y 1.2	Zr 1.4	Nb 1.6	Mo 1.8	Tc 1.9	Ru 2.2	Rh 2.2	Pd 2.2	Ag 1.9	Cd 1.7	In 1.7	Sn 1.8	Sb 1.9	Te 2.1	I 2.5
Cs 0.7	Ba 0.9	La 1.1	Hf 1.3	Ta 1.5	W 1.7	Re 1.9	Os 2.2	Ir 2.2	Pt 2.2	Au 2.4	Hg 1.9	Tl 1.8	Pb 1.8	Bi 1.9	Po 2.0	At 2.2

□ <1.0   □ 1.5–1.9   □ 2.5–2.9
□ 1.0–1.4 □ 2.0–2.4   □ 3.0–4.0

tightness with which an atom holds the electrons that it shares with another atom. Notice that for periods within the Periodic Table, both electronegativity and ionization potential generally increase from left to right across a row of the Periodic Table from columns 1A to 7A. In addition, both electronegativity and ionization potential increase in going up a column.

EXAMPLE 3.2

Judging from their relative positions in the Periodic Table, which element in each pair has the larger electronegativity?
(a) Lithium or carbon
(b) Nitrogen or oxygen
(c) Carbon or oxygen

Solution
The elements in each pair are in the second period of the Periodic Table. Within a period, electronegativity increases from left to right.
(a) C > Li
(b) O > N
(c) O > C

Problem 3.2

Judging from their relative positions in the Periodic Table, which element in each pair has the larger electronegativity?
(a) Lithium or potassium
(b) Nitrogen or phosphorus
(c) Carbon or silicon

3.5 Ionic Compounds

A Forming Ionic Bonds

According to the Lewis model of bonding, an ionic bond forms by the transfer of one or more valence-shell electrons from an atom of lower electronegativity to the valence shell of an atom of higher electronegativity. The more electronegative atom gains one or more valence electrons and becomes an anion; the less electronegative atom loses one or more valence electrons and becomes a cation. The compound formed by the combination of positive and negative ions is called an **ionic compound.**

As a guideline, we say that this type of electron transfer to form an ionic compound is most likely to occur if the difference in electronegativity between two atoms is approximately 1.9 or greater. A bond is more likely to be covalent if this difference is less than 1.9. You should be aware that the value of 1.9 for the formation of an ionic bond is somewhat arbitrary. Some chemists prefer a slightly larger value, others a slightly smaller value. The essential point is that the value of 1.9 gives us a guidepost against which to decide if a bond is more likely to be ionic or more likely to be covalent. Section 3.7 discusses covalent bonding.

An example of an ionic compound is that formed between the metal sodium (electronegativity 0.9) and the nonmetal chlorine (electronegativity 3.0). The difference in electronegativity between these two elements is 2.1. In forming the ionic compound NaCl, the single 3s valence electron of a

Figure 3.1 The structure of a sodium chloride crystal. (*a*) Ball-and-stick models show the relative positions of the ions. (*b*) Space-filling models show the relative sizes of the ions.

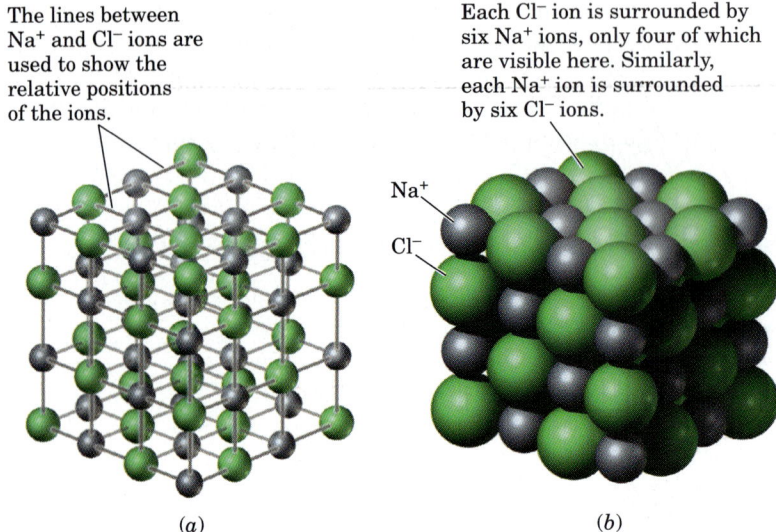

The lines between Na$^+$ and Cl$^-$ ions are used to show the relative positions of the ions.

Each Cl$^-$ ion is surrounded by six Na$^+$ ions, only four of which are visible here. Similarly, each Na$^+$ ion is surrounded by six Cl$^-$ ions.

(*a*) (*b*)

sodium atom is transferred to the partially filled valence shell of a chlorine atom.

$$\text{Na } (1s^22s^22p^63s^1) + \text{Cl } (1s^22s^22p^63s^23p^5) \longrightarrow \text{Na}^+ (1s^22s^22p^6) + \text{Cl}^- (1s^22s^22p^63s^23p^6)$$

Sodium atom   Chlorine atom   Sodium ion   Chloride ion

In the following equation, we use a single-headed curved arrow to show this transfer of one electron.

$$\text{Na}\cdot + \cdot\ddot{\underset{..}{\text{Cl}}}\!: \longrightarrow \text{Na}^+ \;\; :\!\ddot{\underset{..}{\text{Cl}}}\!:^-$$

The ionic bond in solid sodium chloride results from the force of attraction between positive sodium ions and negative chloride ions. In its solid (crystalline) form, sodium chloride consists of a three-dimensional array of Na$^+$ and Cl$^-$ ions arranged as shown in Figure 3.1.

Although ionic compounds do not consist of molecules, they do have a definite ratio of one kind of ion to another; their formulas give this ratio. For example, NaCl represents the simplest ratio of sodium ions to chloride ions—namely, 1 : 1.

B Predicting Formulas of Ionic Compounds

Ions are charged particles, but the matter we see all around us and deal with every day is electrically neutral (uncharged). If ions are present in any sample of matter, the total number of positive charges must equal the total number of negative charges. Therefore, we cannot have a sample containing only Na$^+$ ions. Any sample that contains Na$^+$ ions must also contain negative ions, such as Cl$^-$, Br$^-$, or S^{2-}, and the sum of the positive charges must equal the sum of the negative charges.

EXAMPLE 3.3

Write the formulas for the ionic compounds formed from the following ions:
(a) Lithium ion and bromide ion
(b) Barium ion and iodide ion
(c) Aluminum ion and sulfide ion

Solution

(a) Table 3.1 shows that the charge on a lithium ion is +1, and Table 3.3 shows that the charge on a bromide ion is −1. The formula for lithium bromide is LiBr.

(b) The charge on a barium ion is +2 and the charge on an iodide ion is −1. Two I⁻ ions are required to balance the charge of one Ba^{2+} ion. The formula for barium iodide is BaI_2.

(c) The charge on an aluminum ion is +3, and the charge on a sulfide ion is −2. For the compound to have an overall charge of zero, the ions must combine in the ratio of two aluminum ions to three sulfur ions. The formula is aluminum sulfide is Al_2S_3.

Problem 3.3

Write the formulas for the ionic compounds formed from the following ions:
(a) Potassium ion and chloride ion   KCl
(b) Calcium ion and fluoride ion   CaF₂
(c) Iron(III) ion and oxide ion   Fe₂O₃

Recall that the subscripts in the formulas for ionic compounds represent the ratio of the ions. Thus a crystal of BaI_2 has twice as many iodide ions as barium ions. For ionic compounds, when both charges are 2, as in the compound formed from Ba^{2+} and O^{2-}, we must "reduce to lowest terms." That is, barium oxide is BaO, and not Ba_2O_2. The reason is that we are looking at ratios only, and the ratio of ions in barium oxide is 1 : 1.

3.6 Naming Ionic Compounds

To name an ionic compound, we give the name of the cation first followed by the name of the anion.

Screen 3.6
Simulation

A Binary Ionic Compounds

A **binary compound** contains only two elements. In a **binary ionic compound,** both of these elements are present as ions. The name of the compound consists of the name of the metal from which the positive ion was formed followed by the name of the negative ion. Subscripts are ignored in naming binary ionic compounds. For example, $AlCl_3$ is named aluminum chloride. You know this compound must contain three chloride ions because the positive and negative charges must match; that is, one Al^{3+} ion must combine with three Cl^- ions to balance the charges.

EXAMPLE 3.4

Name these binary ionic compounds:
(a) LiBr
(b) Ag_2S
(c) NaBr

Solution
(a) Lithium bromide
(b) Silver sulfide
(c) Sodium bromide

68 / CHAPTER 3 CHEMICAL BONDS

Problem 3.4
Name these binary ionic compounds:
(a) MgO
(b) BaI$_2$
(c) KCl

EXAMPLE 3.5

Write the formulas for these binary ionic compounds:
(a) Barium hydride   BaH$_2$
(b) Sodium fluoride NaF
(c) Calcium oxide   CaO

Solution
(a) BaH$_2$
(b) NaF
(c) CaO

Problem 3.5
Write the formulas for these binary ionic compounds:
(a) Magnesium chloride   MgCl$_2$
(b) Aluminum oxide   Al$_2$O$_3$
(c) Lithium iodide   LiI

B Binary Ionic Compounds of Metals That Form Different Positive Ions

Table 3.2 shows that many transition metals form different positive ions. For example, copper can form both Cu$^+$ and Cu^{2+}. For systematic names, we use Roman numerals in the name to show the charge. For common names we use the *-ous/-ic* system.

EXAMPLE 3.6

Give each binary compound a systematic name and then a common name.
(a) CuO
(b) Cu$_2$O

Solution
Remember in answering part (b) that we ignore subscripts in naming binary ionic compounds. Therefore, the 2 in Cu$_2$O is not indicated. You know that two copper(I) ions are present because two positive charges are needed to balance the two negative charges on an O^{2-} ion.
(a) Systematic name: copper(II) oxide. Common name: cupric oxide.
(b) Systematic name: copper(I) oxide. Common name: cuprous oxide.

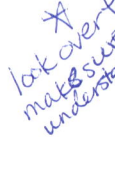
look over, make sure understand

Problem 3.6
Give each binary compound a systematic name and then a common name.
(a) FeO
(b) Fe$_2$O$_3$

CHEMICAL CONNECTIONS 3B

Ionic Compounds in Medicine

Many ionic compounds have medical uses, some of which are shown here.

Formula	Name	Medical Use
1. $AgNO_3$	Silver nitrate	Astringent (external)
2. $BaSO_4$	Barium sulfate	Radiopaque medium for X-ray work
3. $CaSO_4$	Calcium sulfate	Plaster casts
4. $FeSO_4$	Iron(II) sulfate	To treat iron deficiency
5. $KMnO_4$	Potassium permanganate	Anti-infective (external)
6. KNO_3	Potassium nitrate (saltpeter)	Diuretic
7. Li_2CO_3	Lithium carbonate	To treat manic depression
8. $MgSO_4$	Magnesium sulfate (epsom salts)	Cathartic
9. $NaHCO_3$	Sodium bicarbonate (baking soda)	Antacid
10. NaI	Sodium iodide	Iodine for thyroid hormones
11. NH_4Cl	Ammonium chloride	To acidify the digestive system
12. $(NH_4)_2CO_3$	Ammonium carbonate	Expectorant
13. SnF_2	Tin(II) fluoride	To strengthen teeth (external)
14. ZnO	Zinc oxide	Astringent (external)

■ Drinking a "barium cocktail" containing barium sulfate makes the intestinal tract visible on an X ray.

C Ionic Compounds That Contain Polyatomic Ions

To name ionic compounds containing polyatomic ions, name the positive ion first and then the negative ion, each as a separate word.

EXAMPLE 3.7

Name these ionic compounds, each of which contains a polyatomic ion:
(a) $NaNO_3$ nitrate
(b) $CaCO_3$
(c) $(NH_4)_2SO_3$
(d) NaH_2PO_4

Solution
Recall that the name of the $H_2PO_4^-$ ion is dihydrogen phosphate.
(a) Sodium nitrate
(b) Calcium carbonate
(c) Ammonium sulfite
(d) Sodium dihydrogen phosphate

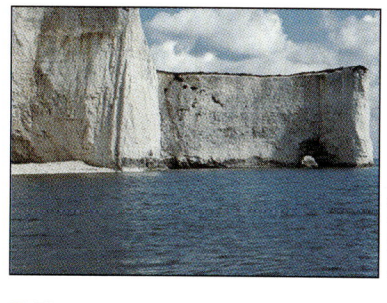

■ The White Cliffs of Dover, England, are composed mainly of calcium carbonate, $CaCO_3$.

Problem 3.7

Name these ionic compounds, each of which contains a polyatomic ion:
(a) K_2HPO_4 Potassium hydrogen phosphate
(b) $Al_2(SO_4)_3$ Aluminum sulfate
(c) $FeCO_3$ Iron(II) Carbonate

3.7 Molecular Compounds

A Formation of a Covalent Bond

A covalent bond forms when electron pairs are shared between two atoms whose difference in electronegativity is less than 1.9. As we have already mentioned, the most common covalent bonds occur between two nonmetals or between a nonmetal and a metalloid.

According to the Lewis model, a pair of electrons in a covalent bond functions in two ways simultaneously: It is shared by two atoms and at the same time it fills the valence shell of each atom. The simplest example of a covalent bond is that in a hydrogen molecule, H_2. When two hydrogen atoms bond, the single electrons from each atom combine to form an electron pair. A bond formed by sharing a pair of electrons is called a **single bond** and is represented by a single line between the two atoms. The electron pair shared between the two hydrogen atoms in H_2 completes the valence shell of each hydrogen. Thus, in H_2, each hydrogen has two electrons in its valence shell and an electron configuration like that of helium, the noble gas nearest to it in atomic number.

$$H\cdot + \cdot H \longrightarrow H\!-\!H$$

The single line represents a shared pair of electrons

B Nonpolar and Polar Covalent Bonds

Nonpolar covalent bond A covalent bond between two atoms whose difference in electronegativity is less than 0.5

Polar covalent bond A covalent bond between two atoms whose difference in electronegativity is between 0.5 and 1.9

Although all covalent bonds involve the sharing of electrons, they differ widely in the degree of sharing. We classify covalent bonds into two categories, **nonpolar covalent** and **polar covalent,** depending on the difference in electronegativity between the bonded atoms. In a nonpolar covalent bond, electrons are shared equally. In a polar covalent bond, they are shared unequally. It is important to realize that no sharp line divides these two categories, nor, for that matter, does a sharp line divide polar covalent bonds and ionic bonds. Nonetheless, the rule-of-thumb guidelines given in Table 3.6 will help you decide whether a given bond is more likely to be nonpolar covalent, polar covalent, or ionic.

An example of a polar covalent bond is that in H—Cl, the difference in electronegativity between the bonded atoms is $3.0 - 2.1 = 0.9$. A covalent bond between carbon and hydrogen, for example, is classified as nonpolar covalent because the difference in electronegativity between these two atoms is only $2.5 - 2.1 = 0.4$. You should be aware, however, that there is some slight polarity to a C—H bond but, because it is quite small, we arbitrarily say that it is nonpolar.

Table 3.6 Classification of Chemical Bonds

Electronegativity Difference Between Bonded Atoms	Type of Bond	Most Likely to Form Between
Less than 0.5	Nonpolar covalent	Two nonmetals or a nonmetal and a metalloid
0.5 to 1.9	Polar covalent	Two nonmetals or a nonmetal and a metalloid
Greater than 1.9	Ionic	A metal and a nonmetal

EXAMPLE 3.8

Classify each bond as nonpolar covalent, polar covalent, or ionic.
(a) O—H
(b) N—H
(c) Na—F
(d) C—Mg
(e) C—S

Solution

Bond	Difference in Electronegativity	Type of Bond
(a) O—H	3.5 − 2.1 = 1.4	Polar covalent
(b) N—H	3.0 − 2.1 = 0.9	Polar covalent
(c) Na—F	4.0 − 0.9 = 3.1	Ionic
(d) C—Mg	2.5 − 1.2 = 1.3	Polar covalent
(e) C—S	2.5 − 2.5 = 0	Nonpolar covalent

Problem 3.8

Classify each bond as nonpolar covalent, polar covalent, or ionic.
(a) S—H   (b) P—H
(c) C—F   (d) C—Cl

An important consequence of the unequal sharing of electrons in a polar covalent bond is that the more electronegative atom gains a greater fraction of the shared electrons and acquires a partial negative charge, indicated by the symbol δ− (read "delta minus"). The less electronegative atom has a lesser fraction of the shared electrons and acquires a partial positive charge, indicated by the symbol δ+ (read "delta plus"). This separation of charge produces a **dipole** (two poles). The presence of a bond dipole is most commonly shown by an arrow, with the head of the arrow near the negative end of the dipole and a cross on the tail of the arrow near the positive end.

We can show the polarity of a covalent bond by a type of molecular model called an electron density model. In this kind of model, a blue color shows the presence of a δ+ charge and a red color shows the presence of a δ− charge. Figure 3.2 shows an electron density model of HCl. The ball-and-stick model in the center shows the orientation of atoms in space. The transparent surface surrounding the ball-and-stick model shows the relative sizes of the atoms (equivalent to the size shown by a space-filling model). Colors on the surface show the distribution of electron density. We see by the blue color that hydrogen bears a δ+ charge and by the red color that chlorine bears a δ− charge.

Dipole A chemical species in which there is a separation of charge; there is a positive pole in one part of the species and a negative pole in another part

Figure 3.2 An electron density model of HCl. Red indicates a region of high electron density, and blue indicates a region of low electron density.

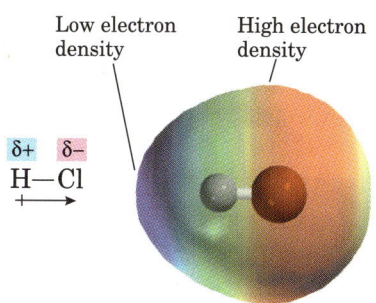

EXAMPLE 3.9

Using the symbols δ− and δ+, indicate the polarity in each polar covalent bond.
(a) C—O
(b) N—H
(c) C—Mg

Solution

For (a), carbon and oxygen are both in period 2 of the Periodic Table. Because oxygen is farther to the right than carbon, it is more electronegative than carbon. For (c), magnesium is a metal located to the far left in the Periodic Table and carbon is a nonmetal located to the right. All nonmetals, including hydrogen, have a greater electronegativity than do the metals in columns 1A and 2A. The electronegativity of each element is given below the symbol of the element.

(a) $\overset{\delta+}{C}—\overset{\delta-}{O}$ (b) $\overset{\delta-}{N}—\overset{\delta+}{H}$ (c) $\overset{\delta-}{C}—\overset{\delta+}{Mg}$
    2.5 3.5   3.0 2.1   2.5 1.2

Problem 3.9

Using the symbols $\delta-$ and $\delta+$, indicate the polarity in each polar covalent bond.
(a) C—N
(b) N—O
(c) C—Cl

C Drawing Lewis Structures

The ability to write Lewis structures for molecules is a fundamental skill for the study of chemistry. The following guidelines will help you with this task. As you study these guidelines, look at the examples in Table 3.7.

1. Determine the number of valence electrons in the molecule. To do so, add the number of valence electrons contributed by each atom (Section 2.6). The Lewis structure of a water molecule, H_2O, for example, must show eight valence electrons: one from each hydrogen and six from oxygen.

2. Decide on the arrangement of atoms in the molecule. For some molecules given as examples in the text, we ask you to propose an arrangement of atoms. For most, however, we give you the experimentally determined arrangement. Except for the simplest molecules, the arrangement of their atoms must be determined experimentally.

3. Connect the atoms with single bonds, and then arrange the remaining electrons in Lewis dot pairs so that each atom in the molecule has a complete outer shell. Each hydrogen atom must be surrounded by two electrons. Each carbon, oxygen, nitrogen, and halogen atom must be surrounded by eight electrons (per the octet rule).

4. A pair of electrons involved in a covalent bond (**bonding electrons**) is shown as a single line; an unshared pair of electrons (**nonbonding electrons**) is shown as a pair of Lewis dots.

5. In a **single bond,** two atoms share one pair of electrons. In a **double bond,** they share two pairs of electrons; a double bond is shown by two lines between the bonded atoms. In a **triple bond,** two atoms share three pairs of electrons; a triple bond is shown by three lines between the bonded atoms.

Table 3.7 gives the Lewis structures and names for several small molecules. The number of valence electrons that each molecule contains is shown in parentheses. Notice that, in these molecules, each hydrogen is surrounded by two valence electrons, and each carbon, nitrogen, oxygen, and chlorine is

---

Bonding electrons Valence electrons involved in forming a covalent bond; that is, shared electrons

Nonbonding electrons Valence electrons not involved in forming covalent bonds; that is, unshared electrons

Single bond A bond formed by sharing one pair of electrons; represented by a single line between two bonded atoms

Double bond A bond formed by sharing two pairs of electrons; represented by two lines between the two bonded atoms

Triple bond A bond formed by sharing three pairs of electrons; represented by three lines between the two bonded atoms

Table 3.7 Lewis Structures for Several Small Molecules

H$_2$O (8) Water   NH$_3$ (8) Ammonia   CH$_4$ (8) Methane   HCl (8) Hydrogen chloride

C$_2$H$_4$ (12) Ethylene   C$_2$H$_2$ (10) Acetylene   CH$_2$O (12) Formaldehyde   H$_2$CO$_3$ (24) Carbonic acid

The number of valence electrons in each molecule is given in parentheses after the molecular formula of the compound.

Lewis structure A formula for a molecule or ion showing all pairs of bonding electrons as single, double or triple lines, and all nonbonding electrons as pairs of Lewis dots.

Structural formula A formula showing how atoms in a molecule or ion are bonded to each other. Similar to a Lewis structure except that a structural formula shows only bonding pairs of electrons.

Screen 3.8 Tutorial

surrounded by eight valence electrons. Furthermore, each carbon has four bonds, each nitrogen has three bonds and one unshared pair of electrons, each oxygen has two bonds and two unshared pairs of electrons, and chlorine (as well as other halogens) has one bond and three unshared pairs of electrons.

EXAMPLE 3.10

State the number of valence electrons in each molecule, and draw a Lewis structure for each:
(a) Hydrogen peroxide, H$_2$O$_2$
(b) Methanol, CH$_3$OH
(c) Acetic acid, CH$_3$COOH

Solution

(a) A Lewis structure for hydrogen peroxide, H$_2$O$_2$, must contain six valence electrons from each oxygen and one from each hydrogen for a total of 12 + 2 = 14 valence electrons. We know that hydrogen forms only one covalent bond, so the order of attachment of atoms must be as follows:

$$H-O-O-H$$

The three single bonds account for six valence electrons. The remaining eight valence electrons must be placed on the oxygen atoms to give each a complete octet:

Ball-and-stick models show only nuclei and covalent bonds; they do not show unshared pairs of electrons

(b) A Lewis structure for methanol, CH$_3$OH, must contain four valence electrons from carbon, one from each hydrogen, and six from oxygen for a total of 4 + 4 + 6 = 14 valence electrons. The order of attachment of atoms in methanol is given on the left. The five single bonds in this partial structure account for ten valence electrons. The remaining four valence electrons must be placed on oxygen as two Lewis dot pairs to give it a complete octet.

The order of attachment of atoms   Lewis dot structure

(c) A molecule of acetic acid, CH_3COOH, must contain four valence electrons from each carbon, six from each oxygen, and one from each hydrogen for a total of $8 + 12 + 4 = 24$ valence electrons. The order of attachment of atoms, shown on the left, contains seven single bonds, which accounts for 14 valence electrons. The remaining ten electrons must be added in such a way that each carbon and oxygen atom has a complete outer shell of eight electrons. This can be done in only one way, which creates a double bond between carbon and one of the oxygens:

The order of attachment of atoms   Lewis dot structure   (Unshared electron pairs not shown)

In this Lewis dot structure, each carbon has four bonds: One carbon has four single bonds, and the other carbon has two single bonds and one double bond. Furthermore, each oxygen has two bonds and two unshared pairs of electrons: One oxygen has one double bond and two unshared pairs of electrons, and the other oxygen has two single bonds and two unshared pairs of electrons.

Problem 3.10

Draw a Lewis structure for each molecule. Each has only one possible order of attachment of its atoms, which is left for you to determine.
(a) Ethane, C_2H_6
(b) Chloromethane, CH_3Cl
(c) Hydrogen cyanide, HCN

D Exceptions to the Octet Rule

The Lewis model of covalent bonding focuses on valence electrons and the necessity for each atom other than H to have a completed valence shell containing eight electrons. Although most molecules formed by main-group elements (Groups 1A–7A) have structures that satisfy the octet rule, there are important exceptions. One exception involves molecules that contain an atom with more than eight electrons in its valence shell. Atoms of period

2 elements use one 2s and three 2p orbitals for bonding; these four orbitals can contain only eight valence electrons—hence the octet rule. Atoms of period 3 elements, however, have one 3s orbital, three 3p orbitals, and five 3d orbitals; they can accommodate more than eight electrons in their valence shells (Section 2.6A). In phosphine, PH_3, phosphorus has eight electrons in its valence shell and obeys the octet rule. The phosphorus atoms in phosphorus pentachloride, PCl_5, and phosphoric acid, H_3PO_4, have ten electrons in their valence shells and, therefore, are exceptions to the octet rule.

Sulfur, another period 3 element, forms compounds in which it contains 8, 10, and even 12 electrons in its valence shell. The sulfur atom in H_2S has 8 electrons in its valence shell and obeys the octet rule. The sulfur atoms in SO_2 and H_2SO_4 have 10 and 12 electrons in their valence shells and are exceptions to the octet rule.

3.8 Naming Binary Molecular Compounds

A **binary molecular compound** is a binary (two-element) compound that does not contain ions; that is, it is one in which all bonds are covalent. In naming a binary molecular compound:

1. Name the less electronegative element (see Table 3.5) first. Note that the less electronegative element is also generally written first in the formula.

2. Use the prefixes "di-," "tri-," "tetra-," and so on to show the number of atoms of each element. The prefix "mono-" is omitted when it refers to the first atom named, and it is rarely used with the second atom. An exception to this rule is CO, which is named carbon monoxide.

The name is then written as two words.

Screen 3.10 Tutorial

Name of the first element in the formula; use prefixes *di-* and so forth if necessary

Name of the second element; use prefixes *mono-* and so forth if necessary

CHEMICAL CONNECTIONS 3C

Nitric Oxide: Air Pollutant and Biological Messenger

Nitric oxide, NO, is a colorless gas whose importance in the environment has been known for several decades, but whose biological importance is only now being fully recognized. This molecule has 11 valence electrons. Because its number of electrons is odd, it is not possible to draw a structure for NO that obeys the octet rule; there must be one unpaired electron, here shown on the less electronegative nitrogen atom:

An unpaired electron → $:\!\overset{\cdot}{N}\!=\!\overset{\cdot\cdot}{O}\!:$

Nitric oxide

The importance of NO in the environment arises from the fact that it forms as a by-product of the combustion of fossil fuels. Under the temperature conditions of the internal combustion engine and other combustion sources, nitrogen and oxygen of the air react to form small quantities of NO:

$$N_2 + O_2 \xrightarrow{heat} 2NO$$
Nitric oxide

When inhaled, NO passes from the lungs into the bloodstream. There it interacts with the iron in hemoglobin, decreasing its ability to carry oxygen.

What makes nitric oxide so hazardous in the environment is that it reacts almost immediately with oxygen to form NO_2. When dissolved in water, NO_2 forms nitric acid, which is a major acidifying component of acid rain.

$$2NO + O_2 \longrightarrow 2NO_2$$
Nitric oxide Nitrogen dioxide

$$NO_2 + H_2O \longrightarrow HNO_3$$
Nitrogen dioxide Nitric acid

■ Colorless nitric oxide gas, coming from the tank, bubbles through the water. When it reaches the air, it is oxidized to brown NO_2.

Imagine the surprise when it was discovered within the last two decades that this highly reactive, seemingly hazardous compound is synthesized in humans and plays a vital role as a signaling molecule in the cardiovascular system (Chemical Connections 23F).

EXAMPLE 3.11

Name these binary molecular compounds:
(a) NO   (b) SF_2 (c) N_2O

Solution
(a) Nitrogen oxide (more commonly called nitric oxide)
(b) Sulfur difluoride
(c) Dinitrogen oxide (more commonly called laughing gas)

Problem 3.11

Name these binary molecular compounds:
(a) NO_2 (b) PBr_3 (c) SCl_2

Linear
(a)

Trigonal planar
(b)

Tetrahedral
(c)

Figure 3.3 Balloon models to predict bond angles. (a) Two balloons assume a linear shape with a bond angle of 180° about the tie point. (b) Three balloons assume a trigonal planar shape with bond angles of 120° about the tie point. (c) Four balloons assume a tetrahedral shape with bond angles of 109.5° about the tie point.

3.9 Bond Angles and Shapes of Molecules

In Section 3.7, we used a shared pair of electrons as the fundamental unit of a covalent bond and drew Lewis structures for several small molecules containing various combinations of single, double, and triple bonds (see, for example, Table 3.7). We can predict **bond angles** in these and other molecules using the **valence-shell electron-pair repulsion (VSEPR) model.** According to this model, the valence electrons of an atom may be involved in the formation of single, double, or triple bonds, or they may be unshared. Each combination creates a negatively charged region of electron density around the nucleus. Because like charges repel each other, the various regions of electron density around an atom spread out so that each is as far away from the others as possible.

You can demonstrate the bond angles predicted by this model in a very simple way. Imagine that a balloon represents a region of electron density. If two balloons are tied together by their ends, they assume the shapes shown in Figure 3.3. The point where they are tied together represents the atom about which you want to predict a bond angle, and the balloons represent regions of electron density about that atom.

We use the VSEPR model and the balloon model analogy in the following way to predict the shape of a molecule of methane, CH_4. The Lewis structure for CH_4 shows a carbon atom surrounded by four regions of electron density. Each region contains a pair of electrons forming a covalent bond to a hydrogen atom. According to the VSEPR model, the four regions radiate from carbon so that they are as far away from one another as possible. The maximum separation occurs when the angle between any two regions of electron density is 109.5°. Therefore, we predict all H—C—H bond angles to be 109.5°, and the shape of the molecule to be **tetrahedral** (Figure 3.4). The H—C—H bond angles in methane have been measured experimentally and found to be 109.5°. Thus the bond angles and shape of methane predicted by the VSEPR model are identical to those observed experimentally.

We can predict the shape of an ammonia molecule, NH_3, in the same way. The Lewis structure of NH_3 shows nitrogen surrounded by four regions of electron density. Three regions contain single pairs of electrons that form covalent bonds with hydrogen atoms. The fourth region contains an unshared pair of electrons [Figure 3.5(a)]. Using the VSEPR model, we predict

Bond angle The angle between two atoms bonded to a central atom

A regular tetrahedron is the geometric figure made up of four equilateral triangles. The central atom is inside the tetrahedron, and the four atoms or electron pairs point to the four corners. Another way to think of it is as a triangle-based pyramid.

Screen 3.12
Tutorial

Figure 3.4 The shape of a methane molecule, CH_4. (a) Lewis structure and (b) ball-and-stick model. The hydrogens occupy the four corners of a regular tetrahedron, and all H—C—H bond angles are 109.5°.

(a)

(b)

Figure 3.5 The shape of an ammonia molecule, NH$_3$. (a) Lewis structure and (b) ball-and-stick model. The H—N—H bond angles are 107.3°, slightly smaller than the H—C—H bond angles of methane.

Figure 3.6 The shape of a water molecule, H$_2$O. (a) Lewis structure, and (b) ball-and-stick model.

The geometry about an atom surrounded by three regions of electron density is said to be trigonal planar or, alternatively, triangular planar.

that the four regions are arranged in a tetrahedral manner and that the three H—N—H bond angles in this molecule are 109.5°. The observed bond angles are 107.3°. We can explain this small difference between the predicted and observed angles by proposing that the unshared pair of electrons on nitrogen repels adjacent electron pairs more strongly than the bonding pairs repel one another.

The geometry of an ammonia molecule is described as **pyramidal**; that is, the molecule is shaped like a triangular pyramid with the three hydrogens located at the base and the nitrogen located at the apex.

Figure 3.6 shows a Lewis structure and a ball-and-stick model of a water molecule. In H$_2$O, oxygen is surrounded by four regions of electron density. Two of these regions contain pairs of electrons used to form covalent bonds to hydrogens; the remaining two regions contain unshared electron pairs. Using the VSEPR model, we predict that the four regions of electron density around oxygen are arranged in a tetrahedral manner and that the H—O—H bond angle is 109.5°. Experimental measurements show that the actual H—O—H bond angle in a water molecule is 104.5°, a value smaller than that predicted. We can explain this difference between the predicted and observed bond angle by proposing, as we did for NH$_3$, that unshared pairs of electrons repel adjacent pairs more strongly than bonding pairs do. Note that the distortion from 109.5° is greater in H$_2$O, which has two unshared pairs of electrons, than it is in NH$_3$, which has only one unshared pair.

A general prediction emerges from this discussion. If a Lewis structure shows four regions of electron density around an atom, the VSEPR model predicts a tetrahedral distribution of electron density and bond angles of approximately 109.5°.

In many of the molecules we encounter, an atom is surrounded by three regions of electron density. Figure 3.7 shows Lewis structures and ball-and-stick models for molecules of formaldehyde, CH$_2$O, and ethylene, C$_2$H$_4$.

In the VSEPR model, we treat a double bond as a single region of electron density. In formaldehyde, carbon is surrounded by three regions of electron density. Two regions contain single pairs of electrons, each of which forms a single bond to a hydrogen; the third region contains two pairs of electrons, which form a double bond to oxygen. In ethylene, each carbon atom is also surrounded by three regions of electron density; two contain single pairs of electrons, and the third contains two pairs of electrons.

Three regions of electron density about an atom are farthest apart when they lie in a plane and make angles of 120° with one another. Thus the predicted H—C—H and H—C—O bond angles in formaldehyde and the H—C—H and H—C—C bond angles in ethylene are all 120°. Furthermore, all atoms in each molecule lie in a plane. Both formaldehyde and ethylene are planar molecules.

In still other types of molecules, a central atom is surrounded by two regions of electron density. Figure 3.8 shows Lewis structures and ball-and-stick models of molecules of carbon dioxide, CO$_2$, and acetylene, C$_2$H$_2$.

In carbon dioxide, carbon is surrounded by two regions of electron density; each contains two pairs of electrons and forms a double bond to an oxygen atom. In acetylene, each carbon is also surrounded by two regions of electron density; one contains a single pair of electrons and forms a single bond to a hydrogen atom, and the other contains three pairs of electrons and forms a triple bond to a carbon atom. In each case, the two regions of electron density are farthest apart if they form a straight line through the cen-

Figure 3.7 Shapes of formaldehyde, CH$_2$O, and ethylene, C$_2$H$_4$.

tral atom and create an angle of 180°. Both carbon dioxide and acetylene are linear molecules.

Table 3.8 summarizes the predictions of the VSEPR model. In this table, three-dimensional shapes are shown using a solid wedge to represent a bond coming toward you, out of the plane of the paper. A broken wedge represents a bond going away from you, behind the plane of the paper. A single, solid line represents a bond in the plane of the paper.

EXAMPLE 3.12

Predict all bond angles and the shape of each molecule:
(a) CH$_3$Cl
(b) CH$_2$=CHCl

■ Ammonia gas is "drilled" into the soil of a farm field. Most of the ammonia manufactured in the world is used as fertilizer because ammonia supplies the nitrogen needed by green plants.

Figure 3.8 Shapes of carbon dioxide, CO$_2$, and acetylene, C$_2$H$_2$.

Solution

(a) The Lewis structure for CH₃Cl shows carbon surrounded by four regions of electron density. Therefore, we predict that the distribution of electron pairs about carbon is tetrahedral, all bond angles are 109.5°, and the shape of CH₃Cl is tetrahedral.

(b) In the Lewis structure for CH₂=CHCl, each carbon is surrounded by three regions of electron density. Therefore, we predict that all bond angles are 120° and that the molecule is planar. The bonding about each carbon is trigonal planar.

Problem 3.12

Predict all bond angles for these molecules:
(a) CH₃OH
(b) CH₂Cl₂
(c) H₂CO₃ (carbonic acid)

Screen 3.13 Exercise Tutorial

Table 3.8 Predicted Molecular Shapes (VSEPR Model)

Regions of Electron Density Around Central Atom	Predicted Distribution of Electron Density	Predicted Bond Angles	Examples (Shape of the Molecule)
4	Tetrahedral	109.5°	Methane (tetrahedral), Ammonia (pyramidal), Water (bent)
3	Trigonal planar	120°	Ethylene (planar), Formaldehyde (planar)
2	Linear	180°	Carbon dioxide (linear), Acetylene (linear)

3.10 Polar and Nonpolar Molecules

In Section 3.7B, we used the terms "polar" and "dipole" to describe a covalent bond in which one atom bears a partial positive charge and the other bears a partial negative charge. We also saw that we can use the difference in electronegativity between bonded atoms to determine the polarity of a covalent bond and the direction of its dipole. We can now combine our understanding of bond polarity and molecular geometry (Section 3.9) to predict the polarity of molecules.

A molecule will be polar if (1) it has polar bonds and (2) its centers of partial positive charge and partial negative charge lie at different places within the molecule. Consider first carbon dioxide, CO_2, a molecule with two polar carbon–oxygen double bonds. The oxygen on the left pulls electrons of the O=C bond toward it, giving it a partial negative charge. Similarly, the oxygen on the right pulls electrons of the C=O bond toward it by the same amount, giving it the same partial negative charge as the oxygen on the left. Carbon bears a partial positive charge. We can show the polarity of these bonds by using the symbols δ^+ and δ^-. Alternatively, we can show that each carbon–oxygen bond has a dipole by using an arrow with the head of the arrow pointing to the negative end of the dipole and the crossed tail at the positive end of the dipole. Because carbon dioxide is a linear molecule, its centers of negative and positive partial charge coincide. Therefore, it is a nonpolar molecule; that is, it has no dipole.

$$\overset{\delta-}{\ddot{O}}=\overset{\delta+}{C}=\overset{\delta-}{\ddot{O}}$$

Carbon dioxide
(a nonpolar molecule)

In a water molecule, each O—H bond is polar, with oxygen, the more electronegative atom, bearing a partial negative charge and each hydrogen bearing a partial positive charge. The center of partial positive charge in a water molecule is located halfway between the two hydrogen atoms, and the center of partial negative charge is on the oxygen atom. Thus a water molecule has polar bonds and, because of its geometry, is a polar molecule.

Water
(a polar molecule)

The center of partial positive charge is midway between the two hydrogen atoms

Ammonia has three polar N—H bonds and, because of its geometry, the centers of partial positive and partial negative charges are found at different places within the molecule. Thus ammonia has polar bonds and, because of its geometry, is a polar molecule.

82 / CHAPTER 3 CHEMICAL BONDS

Ammonia
(a polar molecule)

The center of partial positive charge is midway between the three hydrogen atoms

Screen 3.15
Exercise

EXAMPLE 3.13

Which of these molecules are polar? Show the direction of the molecular dipole by using an arrow with a crossed tail.
(a) CH_2Cl_2 (b) CH_2O (c) C_2H_2

Solution
Both dichloromethane, CH_2Cl_2, and formaldehyde, CH_2O, have polar bonds and, because of their geometry, are polar molecules. Because acetylene, C_2H_2, contains no polar bonds, it is a nonpolar molecule.

This model shows the double bond as only a single cylinder

(a) Dichloromethane (b) Formaldehyde (c) Acetylene

Problem 3.13
Which of these molecules are polar? Show the direction of the molecular dipole by using an arrow with a crossed tail.
(a) CH_3Cl (b) HCN (c) C_2H_6

SUMMARY

The lack of chemical reactivity of the noble gases (Group 8A) indicates a high degree of stability of their electron configurations (Section 3.2). The tendency of elements of Groups 1A–7A to gain or lose electrons to achieve an outer shell containing eight valence electrons is called the **octet rule.** An atom with almost eight valence electrons tends to gain the needed electrons to have eight electrons in its valence shell—that is, to achieve the same electron configuration as the noble gas nearest to it in atomic number. In gaining electrons, the atom becomes a negatively charged ion called an **anion.** An atom with only one or two valence electrons tends to lose the number of electrons required to have eight valence electrons in its next lower shell—that is, to achieve the same electron configuration as the noble gas nearest to it in atomic number. In losing electrons, the atom becomes a positively charged ion called a **cation.**

For metals that form only one type of cation, the name of the cation is the name of the metal followed by the word "ion" (Section 3.3A). For metals that form more than one type of cation, we show the charge on the ion by placing a Roman numeral in parentheses immediately following the name of the metal. A monatomic anion is named by adding "ide" to the stem part of the name (Section 3.3B). A **polyatomic ion** contains more than one type of atom (Section 3.3C).

According to the Lewis model of chemical bonding (Section 3.4A), atoms bond together in such a way that each atom participating in a bond acquires a valence-shell electron configuration the same as that of the noble gas nearest to it in atomic number. **Electronegativity** is a measure of the force of attraction that an atom exerts on electrons it shares in a chemical bond (Section 3.4B). It increases from left to right across a row and from bottom to top in a column of the Periodic Table. An **ionic bond** forms between two atoms if the difference in electronegativity between them is greater than 1.9 (Section 3.4B). A **covalent bond** forms if the difference in electronegativity between the bonded atoms is 1.9 or less.

An **ionic bond** forms by the transfer of valence-shell electrons from an atom of lower electronegativity to the valence shell of an atom of higher electronegativity (Section 3.5A). In an ionic compound, the total number of positive charges must equal the total number of negative charges (Section 3.5B).

To name a **binary ionic compound,** we first name the cation, followed by the name of the anion (Section 3.6A). Where a metal ion may form different cations, we use a Roman numeral to show its positive charge (Section 3.6B). Alternatively, we can use the suffix "-ous" to show the lower positive charge, and "-ic" to show the higher positive charge. To name an ionic compound that contains polyatomic ions, we name the cation first, followed by the name of the anion (Section 3.6C).

According to the Lewis model, a **covalent bond** forms when pairs of electrons are shared between two atoms whose difference in electronegativity is 1.9 or less (Section 3.7A). A pair of electrons in a covalent bond is shared by two atoms and at the same time fills the valence shell of each atom. A **nonpolar covalent bond** is a covalent bond in which the difference in electronegativity between bonded atoms is less than 0.5 (Section 3.7). A **polar covalent bond** is a covalent bond in which the difference in electronegativity between bonded atoms is between 0.5 and 1.9. In a polar covalent bond, the more electronegative atom bears a partial negative charge ($\delta-$) and the less electronegative atom bears a partial positive charge ($\delta+$). This separation of charge produces a **dipole.**

A **Lewis structure** for a molecule (Section 3.7B) must show (1) the correct arrangement of atoms, (2) the correct number of valence electrons, (3) no more than two electrons in the outer shell of hydrogen, and (4) no more than eight electrons in the outer shell of any period 2 element. Exceptions to the octet rule include compounds of period 3 elements, such as phosphorus and sulfur, which may have as many as 10 and 12 electrons, respectively, in their valence shells.

To name a **binary molecular compound,** we use the prefixes "di-," "tri-," "tetra-," and so on to show the presence two or more atoms of the same kind (Section 3.8).

We can predict the bond angles of molecules and polyatomic ions by using Lewis structures and the **valence-shell electron-pair repulsion (VSEPR) model** (Section 3.9). For atoms surrounded by four regions of electron density, we predict bond angles of 109.5°; by three regions, we predict bond angles of 120°; and by two regions, we predict bond angles of 180°.

A molecule has a dipole if (1) it has polar bonds and (2) its centers of partial positive and partial negative charge do not coincide (Section 3.10). If it has polar bonds but the centers of partial positive and negative charges coincide, the molecule has no dipole.

PROBLEMS

Numbers that appear in color indicate difficult problems. ▶ designates problems requiring application of principles.

Ions and Ionic Bonding

3.14 How many electrons must each atom gain or lose to acquire an electron configuration identical to the noble gas nearest to it in atomic number?
   (a) Li   (b) Cl   (c) P
   (d) Al   (e) Sr   (f) S
   (g) Si   (h) O

3.15 Show how each chemical change obeys the octet rule.
   (a) Lithium forms Li^+
   (b) Oxygen forms O^{2-}

3.16 Show how each chemical change obeys the octet rule.
   (a) Hydrogen forms H^- (hydride ion)
   (b) Aluminum forms Al^{3+}

3.17 Write the formula for the most stable ion formed by each element.
   (a) Mg   (b) F   (c) Al
   (d) S   (e) K   (f) Br

3.18 Predict which ions are stable.
(a) I^-
(b) Se^{2+}
(c) Na^+
(d) S^{2-}
(e) Li^{2+}
(f) Ba^{3+}
(g) Br^{2-}
(h) C^{4-}
(i) Ca^+
(j) Ar^+
(k) Na^-
(l) Cs^+

3.19 Why is Li^- not a stable ion?

3.20 Why are carbon and silicon reluctant to form ionic bonds?

3.21 Table 3.2 shows the following ions of copper: Cu^+ and Cu^{2+}. Do these violate the octet rule? Explain.

3.22 Although not a transition metal, lead can form Pb^{2+} and Pb^{4+} ions. Write the formula for the compound formed between each of these lead ions and the following anions:
(a) Chloride ion
(b) Hydroxide ion
(c) Oxide ion

3.23 Describe the structure of sodium chloride in the solid state.

3.24 Because sodium sulfate, Na_2SO_4, is an ionic compound, no discrete Na_2SO_4 molecules exist. What does the 2 in Na_2SO_4 mean?

3.25 Complete the chart by writing formulas for the compounds formed:

	Br^-	ClO_4^-	O^{2-}	NO_3^-	SO_4^{2-}	PO_4^{3-}	OH^-
Li^+							
Ca^{2+}							
Co^{3+}							
K^+							
Cu^{2+}							

3.26 Write a formula for the ionic compound formed from each pair of elements.
(a) Sodium and bromine
(b) Sodium and oxygen
(c) Aluminum and chlorine
(d) Barium and chlorine
(e) Magnesium and oxygen

Polyatomic Ions

3.27 Name each polyatomic ion.
(a) HCO_3^-
(b) NO_2^-
(c) SO_4^{2-}
(d) HSO_4^-
(e) $H_2PO_4^-$

Naming Ionic Compounds

3.28 Potassium chloride and potassium bicarbonate are used as potassium supplements. Write the formula of each compound.

3.29 Potassium nitrite has been used as a vasodilator and as an antidote for cyanide poisoning. Write the formula of this compound.

3.30 Name the polyatomic ion in each compound.
(a) Na_2SO_3
(b) KNO_3
(c) Cs_2CO_3
(d) NH_4OH
(e) K_2HPO_4

3.31 Write the formulas for the ions present in each compound.
(a) NaBr
(b) $FeSO_3$
(c) $Mg_3(PO_4)_2$
(d) KH_2PO_4
(e) $NaHCO_3$
(f) $Ba(NO_3)_2$

3.32 What is the charge on each ion in these compounds?
(a) CaS
(b) MgF_2
(c) Cs_2O
(d) $ScCl_3$
(e) Al_2S_3

3.33 Write the formula for the compound formed from the following pairs of ions:
(a) Iron(III) ion and hydroxide ion
(b) Barium ion and chloride ion
(c) Calcium ion and phosphate ion
(d) Sodium ion and permanganate ion

3.34 Write the formula for the ionic compound formed from the following pairs of ions:
(a) Iron(II) ion and chloride ion
(b) Calcium ion and hydroxide ion
(c) Ammonium ion and phosphate ion
(d) Tin(II) ion and fluoride ion

3.35 Which formulas are not correct? For each that is not correct, write the correct formula.
(a) Ammonium phosphate, $(NH_4)_2PO_4$
(b) Barium carbonate, Ba_2CO_3
(c) Aluminum sulfide, Al_2S_3
(d) Magnesium sulfide, MgS

3.36 Which formulas are not correct? For each that is not correct, write the correct formula.
(a) Calcium oxide, CaO_2
(b) Lithium oxide, LiO
(c) Sodium hydrogen phosphate, $NaHPO_4$
(d) Ammonium nitrate, NH_4NO_3

Naming Ionic Compounds

3.37 Write formulas for the following ionic compounds:
(a) Potassium bromide
(b) Calcium oxide
(c) Mercury(II) oxide
(d) Copper(II) phosphate
(e) Lithium sulfate
(f) Iron(III) sulfide

3.38 Write formulas for the following ionic compounds:
(a) Ammonium hydrogen sulfite
(b) Magnesium acetate
(c) Strontium dihydrogen phosphate
(d) Silver carbonate
(e) Strontium chloride
(f) Barium permanganate

3.39 Name these ionic compounds:
(a) NaF
(b) MgS
(c) Al_2O_3
(d) $BaCl_2$
(e) $Ca(HSO_3)_2$
(f) KI
(g) $Sr_2(PO_4)_2$
(h) $Fe(OH)_2$
(i) NaH_2PO_4
(j) $Pb(CH_3COO)_2$
(k) BaH_2
(l) $(NH_4)_2HPO_4$

Covalent Bonding

3.40 Explain how a covalent bond is formed.

3.41 How many covalent bonds are normally formed by each element?
(a) N
(b) F
(c) C
(d) Br
(e) O

3.42 What is (a) a single bond? (b) a double bond? (c) a triple bond?

3.43 In Section 2.3B, we saw that there are seven diatomic elements.
(a) Draw Lewis structures for each of these diatomic elements.
(b) Which diatomic elements are gases at room temperature? Which are liquids? Which are solids?

3.44 Draw a Lewis structure formula for each covalent compound.
(a) CH_4
(b) C_2H_2
(c) C_2H_4
(d) BF_3
(e) CH_2O
(f) C_2Cl_6

Lewis Structures

3.45 What is the difference between a molecular formula, a structural formula, and a Lewis structure?

3.46 Draw a Lewis structure for each atom.
(a) K
(b) Se
(c) N
(d) I
(e) Ar
(f) Be
(g) Cl

3.47 Write the total number of valence electrons in each molecule.
(a) NH_3
(b) C_3H_6
(c) $C_2H_4O_2$
(d) C_2H_6O
(e) CCl_4
(f) HNO_2
(g) CCl_2F_2
(h) O_2

3.48 Draw a Lewis structure for each of these molecules. In each case, the atoms can be connected in only one way.
(a) Br_2
(b) H_2S
(c) N_2H_4
(d) N_2H_2
(e) CN^-
(f) NH_4^+
(g) N_2
(h) O_2

3.49 What is the difference between (a) a bromine atom, (b) a bromine molecule, and (c) a bromide ion? Draw the Lewis structure for each.

3.50 Acetylene (C_2H_2), hydrogen cyanide (HCN), and nitrogen (N_2) each contain a triple bond. Draw a Lewis structure for each molecule.

3.51 Some of the following structural formulas are incorrect because they contain one or more atoms that do not have their normal number of covalent bonds. Which structural formulas are incorrect, and which atom or atoms in each have the incorrect number of bonds?

(a) Cl—C=C—H with H, H, H substituents

(b) H—O—C—C—N—C—H with H substituents

(c) H—C—N—C—C—O with H substituents

(d) F=C—C—O—C—H with H, H—O, H substituents

(e) H—C=C=C—O—C—H with Br, H, H substituents

(f) H—C≡C—C=C—H with H, H, H substituents

The Shapes of Molecules

3.52 State the shape of a molecule whose central atom is surrounded by:
(a) Two regions of electron density
(b) Three regions of electron density
(c) Four regions of electron density

3.53 Hydrogen and oxygen combine in different ratios to form H_2O (water) and H_2O_2 (hydrogen peroxide).
(a) How many valence electrons are found in H_2O? in H_2O_2?
(b) Draw Lewis structures for each molecule in part (a). Be certain to show all valence electrons.
(c) Using the VSEPR model, predict the bond angles about the oxygen atom in water and about each oxygen atom in hydrogen peroxide.

3.54 If hydrogen and oxygen were to combine to form the compound H_2O_3, how many valence electrons would this compound have? What might its Lewis structure be? (There is no experimental evidence that this compound exists as a stable molecule.)

3.55 Hydrogen and nitrogen combine in different ratios to form three compounds: NH_3 (ammonia), N_2H_4 (hydrazine), and N_2H_2 (diimide).
(a) How many valence electrons must the Lewis structure of each molecule show?
(b) Draw a Lewis structure for each molecule.
(c) Predict the bond angles about the nitrogen atom(s) in each molecule.

3.56 Oxygen atoms combine in different ratios to form O_2 (oxygen) and O_3 (ozone).
(a) How many valence electrons must the Lewis structure of each molecule show?
(b) Draw a Lewis structure for each molecule. (Hint: The oxygens in O_3 are joined O—O—O.)
(c) Predict the bond angles about the central oxygen atom in ozone.

3.57 Predict the shape of each molecule or ion.
(a) CH_4 (b) PH_3 (c) CHF_3
(d) SO_2 (e) SO_3 (f) CCl_2F_2
(g) NO_2^- (h) NH_3 (i) NH_4^+
(j) PCl_3

Electronegativity and Dipoles

3.58 Why does electronegativity generally increase in going up a column of the Periodic Table?

3.59 Why does electronegativity generally increase in going from left to right across a row of the Periodic Table?

3.60 Judging from their relative positions in the Periodic Table, which element in each pair has the larger electronegativity?
(a) F or Cl (b) O or S
(c) C or N (d) C or F

3.61 Toward which atom are the electrons shifted in a covalent bond between each of the following pairs?
(a) H and Cl (b) N and O
(c) C and O (d) Cl and Br
(e) C and S (f) P and S
(g) H and O

3.62 Which of these bonds is the most polar? The least polar?
(a) C—N (b) C—C (c) C—O

3.63 Both CO_2 and SO_2 have polar bonds. Account for the fact that CO_2 is nonpolar and that SO_2 is polar.

3.64 Is it possible for a molecule to have polar bonds and yet have no dipole? Explain.

3.65 Is it possible for a molecule to have no polar bonds and yet have a dipole? Explain.

Kinds of Bonds

3.66 In each case, tell whether the bond is ionic, polar covalent, or nonpolar covalent.
(a) Br_2 (b) BrCl (c) HCl
(d) SrF_2 (e) SiH_4 (f) CO
(g) N_2 (h) CsCl

3.67 Predict whether a bond will form between the elements in each pair and, if so, whether it will be ionic or covalent.
(a) Cl and I (b) Li and K
(c) O and K (d) Al and N
(e) Na and Cr (f) Ne and S
(g) Na and Ca (h) Li and He

Reading Labels

3.68 Name and write the formula for the fluorine-containing compound present in fluoridated toothpastes and dental gels.

Charles D. Winters

3.69 If you read the labels of sun-blocking lotions, you will find that a common UV-blocking agent is a compound containing titanium. Name and write the formula of this titanium-containing compound.

3.70 On packaged table salt, it is common to see a label stating that the salt "supplies iodide, a necessary nutrient." Name and write the formula of the iodine-containing compound found in iodized salt.

3.71 We are constantly warned about the dangers of "lead-based" paints. Name and write the formula for a lead-containing compound found in lead-based paints.

3.72 If you read the labels of several liquid and tablet antacid preparations, you will find that in many of them, the active ingredients are compounds containing hydroxide ions. Name and write formulas for these hydroxide ion-containing compounds.

3.73 Iron forms Fe^{2+} and Fe^{3+} ions. Which ion is found in the over-the-counter preparations intended to treat "iron-poor blood"?

3.74 Read the labels of several multivitamin/multimineral formulations. Among their components you will find a number of so-called trace minerals—minerals required in the diet of a healthy adult in amounts less than 100 mg per day, or present in the body in amounts less than 0.01% of total body weight. Following are 18 trace minerals. Name at least one form of each trace mineral present in multivitamin formulations.

(a) Phosphorus
(b) Magnesium
(c) Potassium
(d) Iron
(e) Calcium
(f) Zinc
(g) Manganese
(h) Titanium
(i) Silicon
(j) Copper
(k) Boron
(l) Molybdenum
(m) Chromium
(n) Iodine
(o) Selenium
(p) Vanadium
(q) Nickel
(r) Tin

3.75 Write formulas for these compounds.
(a) Calcium sulfite, which is used in preserving cider and other fruit juices
(b) Calcium hydrogen sulfite, which is used in dilute aqueous solutions for washing casks in brewing to prevent souring and cloudiness of beer and to prevent secondary fermentation
(c) Calcium hydroxide, which is used in mortar, plaster, cement, and other building and paving materials
(d) Calcium hydrogen phosphate, which is used in animal feeds and as a mineral supplement in cereals and other foods

3.76 Many paint pigments contain transition metal compounds. Name the compounds in these pigments using a Roman numeral to show the charge on the metal ion.

Charles D. Winters

(a) Yellow, CdS
(b) Green, Cr_2O_3
(c) White, TiO_2
(d) Purple, $Mn_3(PO_4)_2$
(e) Blue, Co_2O_3
(f) Ochre, Fe_2O_3

Chemical Connections

3.77 (Chemical Connections 3A) What are the three main inorganic components of one dry mixture currently used to create synthetic bone?

3.78 (Chemical Connections 3B) Why is sodium iodide often present in the table salt we buy at the grocery store?

3.79 (Chemical Connections 3B) What is a medical use of barium sulfate?

3.80 (Chemical Connections 3B) What is a medical use of potassium permanganate?

3.81 (Chemical Connections 3A) What is the main metal ion present in bone and tooth enamel?

3.82 (Chemical Connections 3C) In what way does the gas nitric oxide, NO, contribute to the acidity of acid rain?

Additional Problems

3.83 Why can't hydrogen have more than two electrons in its valence shell?

3.84 Why can't second-row elements have more than eight electrons in their valence shells? That is, why does the octet rule work for second-row elements?

3.85 Why does carbon have four bonds and no unshared pairs of electrons in molecular compounds?

3.86 Draw a Lewis structure of a molecular compound in which carbon has:
(a) Four single bonds.
(b) Two single bonds and one double bond.
(c) One single bond and one triple bond.
(d) Two double bonds.

3.87 Why does nitrogen have three bonds and one unshared pair of electrons in molecular compounds?

3.88 Draw a Lewis structure of a molecular compound in which nitrogen has:
(a) Three single bonds and one pair of electrons.
(b) One single bond, one double bond, and one unshared pair of electrons.
(c) One triple bond and one unshared pair of electrons.

3.89 Why does oxygen have two bonds and two unshared pairs of electrons in molecular compounds?

3.90 Draw a Lewis structure of a molecular compound in which oxygen has:
(a) Two single bonds and two unshared pairs of electrons.
(b) One double bond and two unshared pairs of electrons.

3.91 The ion O^{6+} has a complete outer shell. Why is this ion not stable?

3.92 Draw a Lewis structure for a molecule in which a carbon atom is bonded by a double bond to (a) another carbon atom, (b) an oxygen atom, and (c) a nitrogen atom.

3.93 Classify each bond as nonpolar covalent, polar covalent, or ionic.
(a) C—Cl (b) C—Li (c) C—N

3.94 Classify each bond as nonpolar covalent, polar covalent, or ionic.
(a) C—Br (b) S—Cl (c) C—P

3.95 Why can't a carbon atom be bonded to a chlorine atom by a double bond?

3.96 Explain why argon does not form (a) ionic bonds or (b) covalent bonds.

3.97 Which of the following molecules have an atom that does not obey the octet rule (not all these are stable molecules)?
(a) BF_3 (b) CF_2 (c) BeF_2
(d) C_2H_4 (e) CH_3 (f) N_2
(g) NO

3.98 Show how to use the valence-shell electron-pair repulsion model to predict the shape of a molecule in which a central atom is surrounded by five regions of electron density—for example, phosphorus pentafluoride, PF_5. (Hint: Use molecular models or, if you do not have a set handy, use a marshmallow or gum drop and toothpicks.)

3.99 Show how to use the valence-shell electron-pair repulsion model to predict the shape of a molecule in which a central atom is surrounded by six regions of electron density—for example, sulfur hexafluoride, SF_6.

3.100 Account for the fact that chloromethane, CH_3Cl, which has only one polar C—Cl bond, is a polar molecule, but carbon tetrachloride, CCl_4, which has four polar C—Cl bonds, is a nonpolar molecule.

3.101 Knowing what you do about covalent bonding in compounds of carbon, nitrogen, and oxygen, and given the fact that silicon is just below carbon in the Periodic Table, phosphorus is just below nitrogen, and oxygen is just below sulfur, predict the molecular formula for the compound formed by (a) silicon and chlorine, (b) phosphorus and hydrogen, and (c) sulfur and hydrogen.

3.102 Perchloroethylene, which is a liquid at room temperature, is one of the most widely used solvents for commercial dry cleaning. It is sold for this purpose under several trade names, including Perclene. Does this molecule have polar bonds? Is it a polar molecule? Does it have a dipole?

Perchloroethylene

3.103 Vinyl chloride is the starting material for the production of poly(vinyl chloride), abbreviated PVC. Its recycling code is "V." The major use of PVC is for tubing in residential and commercial construction (Section 12.7).

Vinyl chloride

(a) Complete the Lewis structure for vinyl chloride by showing all unshared pairs of electrons.
(b) Predict the H—C—H, H—C—C, Cl—C—H, and Cl—C—C bond angles in this molecule.
(c) Does vinyl chloride have polar bonds? Is it a polar molecule? Does it have a dipole?

3.104 Tetrafluoroethylene is the starting material for the production of poly(tetrafluoroethylene), a polymer that is widely used for the preparation of nonstick coatings on kitchenware (Section 12.7). The most widely known trade name for this product is Teflon.

Tetrafluoroethylene

(a) Complete the Lewis structure for tetrafluoroethylene by showing all unshared pairs of electrons.
(b) Predict the F—C—F and F—C—C bond angles in this molecule.
(c) Does tetrafluoroethylene have polar bonds? Is it a polar molecule? Does it have a dipole?

3.105 Chlorine dioxide, ClO_2, is a yellow to reddish yellow gas at room temperature. This strong oxidizing agent is used for bleaching cellulose, paper pulp, and textiles, and for water purification. It is the gas that was used to kill anthrax spores in the anthrax-contaminated Hart Senate Office Building.

(a) How many valence electrons are present in ClO_2?
(b) Draw a Lewis structure for this molecule. (Hint: The order of attachment of atoms in this molecule is O—Cl—O. Chlorine is a period 3 element, and its valence shell may contain more than eight electrons.)

InfoTrac College Edition

For additional readings, go to InfoTrac College Edition, your online research library, at

http://infotrac.thomsonlearning.com

CHAPTER 4

4.1 Introduction
4.2 Formula Weight
4.3 The Mole
4.4 Chemical Equations
4.5 Mass Relationships in Chemical Reactions
4.6 Reactions Between Ions in Aqueous Solution
4.7 Oxidation–Reduction
4.8 Heat of Reaction

Fireworks are a spectacular display of chemical reactions.

Chemical Reactions

4.1 Introduction

In Chapter 1, we learned that chemistry is mainly concerned with two things: the structure of matter and the transformations of one form of matter to another. In Chapters 2 and 3, we discussed the first of these topics, and now we are ready to turn our attention to the second. In a chemical change, also called a **chemical reaction,** one or more **reactants** (starting materials) are converted into one or more **products.**

Chemical reactions occur all around us. They fuel and keep alive the cells of living tissues; they occur when we light a match, cook dinner, start a car, listen to a portable radio, or watch television. Most of the world's manufacturing processes involve chemical reactions; they include petroleum refining and food processing as well as the manufacture of drugs, plastics, synthetic fibers, explosives, and many other materials. In this chapter we discuss three aspects of chemical reactions: (1) mass relationships, (2) types of reactions, and (3) heat gains and losses.

4.2 Formula Weight

A table of atomic weights is given on the inside back cover. Atomic weights can also be found in the Periodic Table on the inside front cover.

We begin our study of mass relationships with a discussion of formula weight. The **formula weight** of any substance is the sum of the atomic weights in atomic mass units (amu) of all the atoms in the compound's formula. The

Table 4.1 Formula Weights for Two Ionic and Two Molecular Compounds

Ionic Compounds

Sodium chloride (NaCl)	23.0 amu Na + 35.5 amu Cl = 58.5 amu
Nickel(II) chloride hydrate ($NiCl_2 \cdot 6H_2O$)*	58.7 amu Ni + 2(35.5 amu Cl) + 12(1.0 amu H) + 6(16.0 amu O) = 237.7 amu

Molecular Compounds

Water (H_2O)	2(1.0 amu H) + 16.0 amu O = 18.0 amu
Aspirin ($C_9H_8O_4$)	9(12.0 amu C) + 8(1.0 amu H) + 4(16.0 amu O) = 180.0 amu

*Nickel(II) chloride crystallizes from an aqueous solution with six moles of water per mole of $NiCl_2$. The presence of water molecules in the crystal is indicated by the name "hydrate."

term "formula weight" can be used for both ionic and molecular compounds and tells nothing about whether the compound is ionic or molecular. Table 4.1 gives formula weights for two ionic compounds and two molecular compounds.

Another term, **molecular weight (MW)**, is strictly correct only when used for molecular compounds. In this book, we use *formula weight* for ionic and covalent compounds and *molecular weight* for covalent compounds only.

> **Molecular weight (MW)** The sum of the atomic weights of all atoms in a molecular compound expressed in atomic mass units (amu)

EXAMPLE 4.1

What is the molecular weight of (a) glucose, $C_6H_{12}O_6$, and (b) urea, $(NH_2)_2CO$?

Solution

(a) Glucose, $C_6H_{12}O_6$

$$
\begin{aligned}
&\text{C } 6 \times 12.0 = 72.0 \\
&\text{H } 12 \times 1.0 = 12.0 \\
&\text{O } 6 \times 16.0 = 96.0 \\
&\hline
&C_6H_{12}O_6 = 180.0 \text{ amu}
\end{aligned}
$$

(b) Urea, $(NH_2)_2CO$

$$
\begin{aligned}
&\text{N } 2 \times 14.0 = 28.0 \\
&\text{H } 4 \times 1.0 = 4.0 \\
&\text{C } 1 \times 12.0 = 12.0 \\
&\text{O } 1 \times 16.0 = 16.0 \\
&\hline
&(NH_2)_2CO = 60.0 \text{ amu}
\end{aligned}
$$

Problem 4.1

What is (a) the molecular weight of aspirin, $C_9H_8O_4$, and (b) the formula weight of barium phosphate, $Ba_3(PO_4)_2$?

4.3 The Mole

Atoms and molecules are so tiny (Section 2.4F) that chemists are seldom able to deal with them one at a time. When we weigh even a very small quantity of a compound, huge numbers of formula units (perhaps 10^{19}) are present. The formula unit may be atoms, molecules, or ions. To overcome this problem, chemists long ago defined a unit called the **mole (mol).** A mole is the amount of substance that contains as many atoms, molecules, or ions as there are atoms in exactly 12 g of carbon-12. The important point here is that whether we are dealing with a mole of iron atoms, a mole of methane molecules, or a mole of sodium ions, a mole always contains the same number of formula units.

The number of formula units in a mole is called **Avogadro's number** after the Italian physicist, Amadeo Avogadro (1776–1856), who first proposed the concept of a mole, but was not able to determine the number experimentally. Note that Avogadro's number is not a defined value, but rather a value that must be determined experimentally. Its value is now known to nine significant figures.

Avogadro's number = $6.02214199 \times 10^{23}$ formula units per mole

For most calculations in this text, we round this number to three significant figures to 6.02×10^{23} formula units per mole.

A mole of hydrogen atoms is 6.02×10^{23} hydrogen atoms, a mole of sucrose (table sugar) molecules is 6.02×10^{23} sugar molecules, a mole of apples is 6.02×10^{23} apples, and a mole of sodium ions is 6.02×10^{23} sodium ions. Just as we call 12 of anything a dozen, 20 a score, and 144 a gross, we call 6.02×10^{23} of anything a mole.

The **molar mass** of any substance (the mass of one mole of any substance) is the formula weight of the substance expressed in grams. For instance, the formula weight of glucose, $C_6H_{12}O_6$ (Example 4.1) is 180 amu; therefore, 180 g of glucose is one mol of glucose. Likewise, the formula weight of urea, $(NH_2)_2CO$, is 60.0 amu and, therefore, 60.0 g of $(NH_2)_2CO$ is one mol of urea. For atoms, one mole is the atomic weight expressed in grams: 12.0 g of carbon is one mol of carbon atoms; 32.1 g of sulfur is one mol of sulfur atoms; and so on. As you see, to talk about the mass of a mole, we need to know the chemical formula of the substance we are considering. Figure 4.1 shows one-mole quantities of several compounds.

Now that we know the relationship between moles and molar mass (g/mol), we can use molar mass as a conversion factor to convert from grams to moles, and from moles to grams. For this calculation, we use molar mass as a conversion factor.

Mole (mol) The formula weight of a substance expressed in grams

Avogadro's number 6.02×10^{23} formula units per mole; the amount of any substance that contains as many formula units as the number of atoms in 12 g of carbon-12

One mole of pennies placed side by side would stretch for more than one million light years, a distance far outside our solar system and even outside our own galaxy. Six moles of this text would weigh as much as the Earth.

Molar mass The mass of one mole of a substance expressed in grams; the formula weight of a compound expressed in grams

Screen 4.2
Tutorial

(a)

(b)

Figure 4.1 One-mole quantities of (a) six elements and (b) four compounds. (a) Top row (left to right): Cu beads (63.5 g), Al foil (27.0 g), and Pb shot (207.2 g). Bottom row (left to right): S powder (32.1 g), Cr chunks (52.0 g), and Mg shavings (24.4 g). (b) H_2O (18.0 g); small beaker NaCl (58.4 g); large beaker aspirin, $C_9H_8O_4$, (180.2 g); green $NiCl_2 \cdot 6H_2O$ (237.7 g).

92 | CHAPTER 4 CHEMICAL REACTIONS

Screen 4.4 Tutorial

Screen 4.5 Tutorial

Suppose we want to know the number of moles of water, H_2O, in a graduated cylinder that contains 36.0 g of water. We know that the molar mass of water is 18.0 g/mol. If 18.0 g of water is one mole of water, then 36.0 g must be two moles of water.

$$36.0 \text{ g } H_2O \times \frac{1 \text{ mol } H_2O}{18.0 \text{ g } H_2O} = 2.00 \text{ mol } H_2O$$

Molar mass can also be used to convert from moles to grams. Suppose you have a beaker that contains 0.753 mol of sodium chloride, NaCl, and you are asked to calculate the number of grams of sodium chloride in the beaker. As a conversion factor, use the fact that the molar mass of NaCl is 58.5 g/mol.

$$0.753 \text{ mol NaCl} \times \frac{58.5 \text{ g NaCl}}{1 \text{ mol NaCl}} = 44.1 \text{ g NaCl}$$

EXAMPLE 4.2

We have 27.5 g of sodium fluoride, NaF, the form of fluoride ions most commonly used in fluoride toothpastes. How many moles is this?

Solution
The formula weight of NaF = 23.0 + 19.0 = 42.0 amu. Thus each mole of NaF has a mass of 42.0 g, allowing us to use the conversion factor 1 mol NaF = 42.0 g NaF:

$$27.5 \text{ g NaF} \times \frac{1 \text{ mol NaF}}{42.0 \text{ g NaF}} = 0.655 \text{ mol NaF}$$

Note that such calculations can be performed for ionic compounds, such as NaF, as well as for molecular compounds, such as CO_2 and urea.

Problem 4.2
A person drinks 1500 g of water per day. How many moles is this?

EXAMPLE 4.3

We wish to weigh 3.41 mol of ethanol, C_2H_5OH. How many grams is this?

Solution
The formula weight of C_2H_5OH is $2(12.0) + 6(1.0) + 16.0 = 46.0$ amu, so the conversion factor is 1 mol C_2H_5OH = 46.0 g C_2H_5OH.

$$3.41 \text{ mol C}_2\text{H}_5\text{OH} \times \frac{46.0 \text{ g C}_2\text{H}_5\text{OH}}{1 \text{ mol C}_2\text{H}_5\text{OH}} = 157 \text{ g C}_2\text{H}_5\text{OH}$$

Problem 4.3
We wish to weigh 2.84 mol of sodium sulfide, Na_2S. How many grams is this?

EXAMPLE 4.4

How many moles of nitrogen atoms and oxygen atoms are there in 21.4 mol of the explosive trinitrotoluene (TNT), $C_7H_5N_3O_6$?

Solution
The formula $C_7H_5N_3O_6$ tells us that each molecule of TNT contains three nitrogen atoms and three oxygen atoms. It also tells us that each mole of TNT contains three moles of N atoms and six moles of O atoms. Therefore, we have the following conversion factors: 1 mol TNT = 3 mol N atoms, and 1 mol TNT = 6 mol O atoms. The number of moles of N atoms in 21.4 moles of TNT is

$$21.4 \text{ mol TNT} \times \frac{3 \text{ mol N atoms}}{1 \text{ mol TNT}} = 64.2 \text{ mol N atoms}$$

The number of moles of O atoms in 21.4 moles of TNT is

$$21.4 \text{ mol TNT} \times \frac{6 \text{ mol N atoms}}{1 \text{ mol TNT}} = 128 \text{ mol O atoms}$$

Problem 4.4
How many moles of C atoms, H atoms, and O atoms are there in 2.5 mol of glucose, $C_6H_{12}O_6$?

EXAMPLE 4.5

How many moles of sodium ions, Na^+, are there in 5.63 of sodium sulfate, Na_2SO_4?

Solution
First, we need to find out how many moles of Na_2SO_4 we have. The formula weight of Na_2SO_4 is $2(23.0) + 32.1 + 4(16.0) = 142.1$ amu. In the conversion of grams of Na_2SO_4, we can use the conversion factor 1 mol Na_2SO_4 = 142.1 g Na_2SO_4.

$$5.63 \text{ g Na}_2\text{SO}_4 \times \frac{1 \text{ mol Na}_2\text{SO}_4}{142.1 \text{ g Na}_2\text{SO}_4} = 0.0396 \text{ mol Na}_2\text{SO}_4$$

The formula Na_2SO_4 tells us that each mole of Na_2SO_4 contains two moles of Na^+ ions. Therefore, we use the conversion factor 1 mol

Na_2SO_4 = 2 mol Na^+. The number of moles of Na^+ ions in 5.63 g of Na_2SO_4 is

$$0.0396 \text{ mol } Na_2SO_4 \times \frac{2 \text{ mol } Na^+}{1 \text{ mol } Na_2SO_4} = 0.0792 \text{ mol } Na^+$$

Problem 4.5
How many moles of copper(I) ions, Cu^+, are there in 0.062 g of copper(I) nitrate, $CuNO_3$?

EXAMPLE 4.6

An aspirin tablet, $C_9H_8O_4$, contains 0.360 g of aspirin. (The rest of it is starch or other fillers.) How many molecules of aspirin are present in this tablet?

Solution
The formula weight of aspirin is 9(12.0) + 8(1.0) + 4(16.0) = 180.0 amu. First we find out how many mol of aspirin there are in 0.360 g:

$$0.360 \text{ g aspirin} \times \frac{1 \text{ mol aspirin}}{180.0 \text{ g aspirin}} = 0.00200 \text{ mol aspirin}$$

Each mole of aspirin contains 6.02×10^{23} molecules. Therefore, the number of molecules of aspirin in the tablet is

$$0.00200 \text{ mol} \times 6.02 \times 10^{23} \frac{\text{molecules}}{\text{mol}} = 1.20 \times 10^{21} \text{ molecules}$$

Problem 4.6
How many molecules of water, H_2O, are in a glass of water (235 g)?

4.4 Chemical Equations

A Writing a Chemical Equation

When propane, which is the major component in bottled gas or LP (liquefied petroleum) gas, burns in air, it reacts with the oxygen in the air. These reactants are converted to the products carbon dioxide and water in a chemical reaction called **combustion.** We can write this chemical reaction in the form of a **chemical equation,** using chemical formulas for the reactants and products, and an arrow to indicate the direction in which the reaction proceeds. In addition, it is important to show the state of each reactant and product—that is, whether it is a gas, liquid, or solid. We use the symbol (g) for gas, (ℓ) for liquid, (s) for solid, and (aq) for a substance dissolved in water (aqueous). We place the appropriate symbol immediately following each reactant and product. In our combustion equation, propane, oxygen, and car-

Combustion Burning in air

Chemical equation A representation using chemical formulas of the process that occurs when reactants are converted to products bon dioxide are gases, and the flame produced when propane burns is hot enough that the water forms as a gas (steam):

$$C_3H_8(g) + O_2(g) \longrightarrow CO_2(g) + H_2O(g)$$
Propane   Oxygen   Carbon   Water
                      dioxide

The equation we have written is incomplete, however. While it tells us the formulas of the starting materials and products (which every chemical equation must do) and the physical state of each reactant and product, it does not give the amounts correctly. It is not *balanced,* which means that the number of atoms on the left side of the equation is not the same as the number of atoms on the right side. From the law of conservation of mass (Section 2.3A), we know that, in chemical reactions, atoms are never destroyed or created; they merely shift from one substance to another. Thus all the atoms present at the start of the reaction (on the left side of the equation) must still be present at the end (on the right side of the equation). In the equation we have just written, there are three carbon atoms on the left but only one on the right.

B Balancing Chemical Equations

To balance an equation, we place numbers in front of the formulas until the number of each kind of atom in the products is the same as that in the starting materials. These numbers are called **coefficients.** As an example, let us balance our propane equation:

$$C_3H_8(g) + O_2(g) \longrightarrow CO_2(g) + H_2O(g)$$
Propane   Oxygen   Carbon   Water
                      dioxide

The following guidelines will help you balance an equation:

1. Begin with atoms that appear in only one compound on the left and only one compound on the right. In the equation for the reaction of propane and oxygen, begin with either carbon or hydrogen

2. If an atom occurs as a free element—as for example, as Mg, Na, O_2, or H_2—balance this element last.

3. You can change only coefficients in balancing an equation; you cannot change chemical formulas. For example, if you have H_2O on the left side of an equation but need two oxygens, you can add the coefficient "2" to read $2H_2O$. You cannot, however, get two oxygens by changing the formula to H_2O_2. The reactant is water, H_2O, and not hydrogen peroxide, H_2O_2.

For the equation for the reaction of propane with oxygen, we can begin with carbon. Three carbon atoms appear on the left and one on the right. If we put a 3 in front of the CO_2 (indicating that three CO_2 molecules are formed), there will be three carbons on each side, and the carbons will be balanced:

$$C_3H_8(g) + O_2(g) \longrightarrow 3\,CO_2(g) + H_2O(g)$$

Next we look at the hydrogens. There are eight on the left and two on the right. If we put a 4 in front of the H_2O, there will be eight hydrogens on each side and the hydrogens will be balanced:

$$C_3H_8(g) + O_2(g) \longrightarrow 3\,CO_2(g) + 4\,H_2O(g)$$

■ **Propane burning in air.**

Screen 4.8
Tutorial

That is why equations MUST be balanced.

The only atom still unbalanced is oxygen. Notice that we saved this reactant for last (rule 2). There are two oxygen atoms on the left and ten on the right. If we put a 5 in front of the O_2 on the left, we both balance the oxygen atoms and arrive at the balanced equation:

$$C_3H_8(g) + \boxed{5}\,O_2(g) \longrightarrow \boxed{3}\,CO_2(g) + \boxed{4}\,H_2O(g)$$

At this point, the equation ought to be balanced, but we should always check, just to make sure. In a balanced equation, there must be the same number of atoms of each element on both sides. A check of our work shows three C, ten O, and eight H atoms on each side. The equation is indeed balanced.

EXAMPLE 4.7

Balance this equation:

$$\underset{\text{Calcium hydroxide}}{Ca(OH)_2(s)} + HCl(g) \longrightarrow \underset{\text{Calcium chloride}}{CaCl_2(s)} + H_2O(\ell)$$

Solution
The calcium is already balanced—there is one on each side. There is one Cl on the left and two on the right. To balance them, we add the coefficient 2 in front of HCl:

$$Ca(OH)_2(s) + 2HCl(g) \longrightarrow CaCl_2(s) + H_2O(\ell)$$

Looking at hydrogens, we see that there are four hydrogens on the left but only two on the right. Placing the coefficient 2 in front of H_2O balances the hydrogens. It also balances the oxygens and completes the balancing of the equation:

$$Ca(OH)_2(s) + 2HCl(g) \longrightarrow CaCl_2(s) + 2H_2O(\ell)$$

Problem 4.7
Following is an unbalanced equation for photosynthesis, the process by which green plants convert carbon dioxide and water to glucose and oxygen. Balance this equation:

$$CO_2(g) + H_2O(\ell) \xrightarrow{\text{Photosynthesis}} \underset{\text{Glucose}}{C_6H_{12}O_6(aq)} + O_2(g)$$

EXAMPLE 4.8

Balance this equation for the combustion of butane, the fluid most commonly used in pocket lighters:

$$\underset{\text{Butane}}{C_4H_{10}(\ell)} + O_2(g) \longrightarrow CO_2(g) + H_2O(g)$$

Solution
The equation for the combustion of butane is very similar to the one we examined at the beginning of this section for the combustion of propane. Here

■ A pocket lighter contains butane in both the liquid and the gaseous state.

we put a 4 in front of the CO_2 (because there are four carbons on the left) and a 5 in front of the H_2O (because there are ten hydrogens on the left):

$$C_4H_{10}(g) + O_2(g) \longrightarrow 4CO_2(g) + 5H_2O(g)$$

When we count the oxygens, we find 2 on the left and 13 on the right. We can balance their numbers by putting 13/2 in front of the O_2.

$$C_4H_{10}(g) + \frac{13}{2}O_2(g) \longrightarrow 4CO_2(g) + 5H_2O(g)$$

Although chemists sometimes have good reason to write equations with fractional coefficients, it is common practice to use only whole-number coefficients. We accomplish that by multiplying everything by 2, which gives the balanced equation:

$$2C_4H_{10}(g) + 13O_2(g) \longrightarrow 8CO_2(g) + 10H_2O(g)$$

Problem 4.8
Balance this equation:

$$C_6H_{14}(g) + O_2(g) \longrightarrow CO_2(g) + H_2O(g)$$

EXAMPLE 4.9

Balance this equation:

$$Na_2SO_3(aq) + H_3PO_4(aq) \longrightarrow H_2SO_3(aq) + Na_3PO_4(aq)$$

Solution
The key to balancing equations like this one is to realize that polyatomic ions such as SO_3^{2-} and PO_4^{3-} usually remain intact on both sides of the equation. We can begin by balancing the Na^+ ions. We put a 3 in front of Na_2SO_3 and a 2 in front of Na_3PO_4, giving us six Na^+ ions on each side:

$$3Na_2SO_3(aq) + H_3PO_4(aq) \longrightarrow H_2SO_3(aq) + 2Na_3PO_4(aq)$$

There are three SO_3^{2-} units on the left and only one on the right, so we put a 3 in front of H_2SO_3:

$$3Na_2SO_3(aq) + H_3PO_4(aq) \longrightarrow 3H_2SO_3(aq) + 2Na_3PO_4(aq)$$

Now let's look at the PO_4^{3-} units. There are two PO_4^{3-} units on the right, but only one on the left. To balance them, we put a 2 in front of H_3PO_4. In doing so, we balance not only the PO_4^{3-} units but also the hydrogens, and arrive at the balanced equation:

$$3Na_2SO_3(aq) + 2H_3PO_4(aq) \longrightarrow 3H_2SO_3(aq) + 2Na_3PO_4(aq)$$

Problem 4.9
Balance this equation:

$$K_2C_2O_4(aq) + Ca_3(AsO_4)_2(s) \longrightarrow K_3AsO_4(aq) + CaC_2O_4(s)$$

One final point about balancing chemical equations. If

$$C_3H_8(g) + 5O_2(g) \longrightarrow 3CO_2(g) + 4H_2O(g)$$

is a correctly balanced equation (and indeed it is), would it be correct if we doubled all the coefficients?

$$2C_3H_8(g) + 10O_2(g) \longrightarrow 6CO_2(g) + 8H_2O(g)$$

This equation is mathematically and scientifically correct, but chemists do not normally write equations with coefficients that are all divisible by a common number. A correctly balanced equation is almost always written with the coefficients expressed in the lowest set of whole numbers.

4.5 Mass Relationships in Chemical Reactions

A Stoichiometry

As we saw in Section 4.4, a balanced chemical equation tells us not only which substances react and which are formed, but also the molar ratios in which they react. For example, using the molar ratios in a balanced chemical equation, we can calculate the mass of starting materials we need to produce a particular mass of a product. The study of mass relationships in chemical reactions is called **stoichiometry.**

Let us look once again at the balanced equation for the burning of propane:

$$C_3H_8(g) + 5O_2(g) \longrightarrow 3CO_2(g) + 4H_2O(g)$$
Propane

This equation tells us not only that propane and oxygen are converted to carbon dioxide and water, but also that one mole of propane combines with five moles of oxygen to produce three moles of carbon dioxide and four moles of water; that is, we know the mole ratios involved. The same is true for any other balanced equation. This fact allows us to answer questions such as the following:

1. How many moles of any particular product are formed if we start with a given mass of a starting material?

2. How many grams (or moles) of one starting material are necessary to react completely with a given number of grams (or moles) of another starting material?

3. How many grams (or moles) of starting material are needed if we want to form a certain number of grams (or moles) of a certain product?

4. How many grams (or moles) of another product are obtained when a certain amount of main product is produced?

It might seem as if we have four different types of problems here. In fact, we can solve them all by the same very simple procedure summarized in the following diagram:

Stoichiometry The mass relationships in a chemical reaction

"Stoichiometry" comes from the Greek *stoicheion,* meaning "element," and *metron,* meaning "measure."

In Section 4.4, we saw that the coefficients in an equation represent numbers of molecules. Because moles are proportional to molecules (Section 4.3), the coefficients in an equation also represent numbers of moles.

Screen 4.9
Exercise

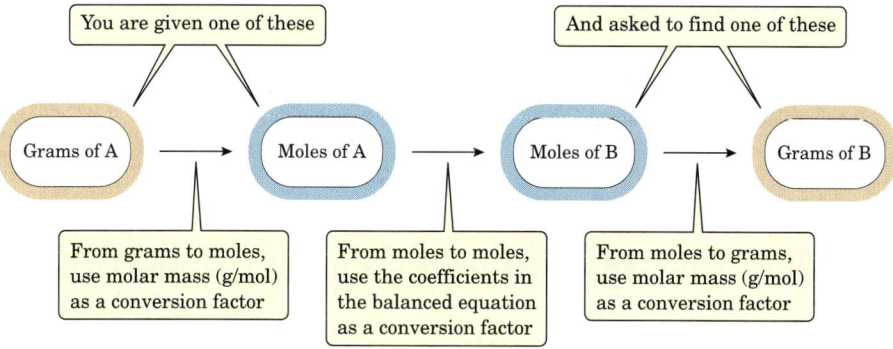

EXAMPLE 4.10

Ammonia is produced on an industrial scale by the reaction of nitrogen gas with hydrogen gas according to this balanced equation:

$$N_2(g) + 3H_2(g) \longrightarrow 2NH_3(g)$$
Ammonia

How many grams of N_2 are necessary to produce 7.50 g of NH_3?

Solution

Step 1 The coefficients in an equation refer to the relative numbers of moles, not grams. Therefore, we must first find out how many moles of NH_3 are in 7.50 g of NH_3. To convert grams of NH_3 to moles of NH_3, we use the conversion factor 1 mol NH_3 = 17.0 g NH_3. At this stage, we don't do the multiplication, but just leave the relationship set up:

$$7.50 \text{ g NH}_3 \times \frac{1 \text{ mol NH}_3}{17.0 \text{ g NH}_3} = \text{mol NH}_3$$

Step 2 Next we turn to the balanced equation, which tells us that 2 mol of NH_3 are produced from 1 mol of N_2. We multiply the number of moles of NH_3 by this fraction factor:

$$7.50 \text{ g NH}_3 \times \frac{1 \text{ mol NH}_3}{17.0 \text{ g NH}_3} \times \frac{1 \text{ mol N}_2}{2 \text{ mol NH}_3} = \text{mol N}_2$$

Step 3 To find the number of grams of N_2, we multiply by the conversion factor 28.0 g N_2 = 1 mol N_2. At this point, the conversion factors are set up properly and we have converted from grams of NH_3 to grams of N_2. Now we do the arithmetic and get an answer of 6.18 g N_2:

$$7.50 \text{ g NH}_3 \times \frac{1 \text{ mol NH}_3}{17.0 \text{ g NH}_3} \times \frac{1 \text{ mol N}_2}{2 \text{ mol NH}_3} \times \frac{28.0 \text{ g N}_2}{1 \text{ mol N}_2} = 6.18 \text{ g N}_2$$

In all such problems, we are given a mass (or number of moles) of one compound and asked to find the mass (or number of moles) of another compound. The two compounds can be on the same side of the equation, or on opposite sides. All such problems can be done by the three steps we have just used.

Screen 4.10
Tutorial

Problem 4.10

Pure aluminum is prepared by the electrolysis of aluminum oxide according to this equation:

$$Al_2O_3(s) \xrightarrow{\text{Electrolysis}} Al(s) + O_2(g)$$

Aluminum oxide

(a) Balance this equation.
(b) What mass of aluminum oxide is required to prepare 27 g (1 mol) of aluminum?

EXAMPLE 4.11

Silicon to be used in computer chips is manufactured by a process represented by the following reaction:

$$SiCl_4(s) + 2Mg(s) \longrightarrow Si(s) + 2MgCl_2(s)$$

Silicon tetrachloride   Magnesium chloride

A sample of 225 g of silicon tetrachloride, $SiCl_4$, is reacted with an excess more than necessary amount of Mg. How many moles of Si are produced?

Solution

We are given 225 g of $SiCl_4$ and asked for moles of Mg required to react completely with it. We must first convert grams of $SiCl_4$ to moles of $SiCl_4$. For this calculation we use the conversion factor 1 mol $SiCl_4$ = 170 g $SiCl_4$:

$$225 \text{ g SiCl}_4 \times \frac{1 \text{ mol SiCl}_4}{170 \text{ g SiCl}_4} = \text{mol SiCl}_4$$

Next we use the conversion factor 1 mol $SiCl_4$ = 1 mol Si, which we obtain from the balanced chemical equation. Now we do the arithmetic and obtain an answer of 1.32 mol Si:

$$225 \text{ g SiCl}_4 \times \frac{1 \text{ mol SiCl}_4}{170 \text{ g SiCl}_4} \times \frac{1 \text{ mol Si}}{1 \text{ mol SiCl}_4} = 1.32 \text{ mol Si}$$

Problem 4.11

In the industrial synthesis of acetic acid, methanol is reacted with carbon monoxide. How many moles of CO are required to produce 16.6 mol of acetic acid?

$$CH_3OH(g) + CO(g) \longrightarrow CH_3COOH(\ell)$$

Methanol   Carbon monoxide   Acetic acid

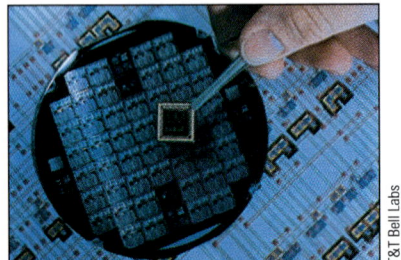

■ As microprocessor chips become smaller and smaller, the purity of the silicon becomes more important, because impurities can prevent a circuit from working properly.

EXAMPLE 4.12

When urea, $(NH_2)_2CO$, is acted on by the enzyme urease in the presence of water, ammonia and carbon dioxide are produced:

$$(NH_2)_2CO(aq) + H_2O(\ell) \xrightarrow{\text{Urease}} 2NH_3(aq) + CO_2(g)$$
$$\text{Urea} \qquad\qquad\qquad\qquad\qquad \text{Ammonia}$$

If excess water is present (more than necessary for the reaction), how many grams each of CO_2 and NH_3 are produced from 0.83 mol of urea?

Solution
For CO_2:
We are given moles of urea and asked for grams of CO_2. We first use the conversion factor 1 mol urea = 1 mol CO_2 to get the number of moles of CO_2 that will be formed:

$$0.83 \text{ mol urea} \times \frac{1 \text{ mol } CO_2}{1 \text{ mol urea}} = \text{mol } CO_2$$

Next we use the conversion factor 1 mol CO_2 = 44.0 g CO_2 to obtain the answer:

$$0.83 \text{ mol urea} \times \frac{1 \text{ mol } CO_2}{1 \text{ mol urea}} \times \frac{44 \text{ g } CO_2}{1 \text{ mol } CO_2} = 37 \text{ g } CO_2$$

For NH_3:
We follow the same procedure as for CO_2 but use different conversion factors:

$$0.83 \text{ mol urea} \times \frac{2 \text{ mol } NH_3}{1 \text{ mol urea}} \times \frac{17 \text{ g } NH_3}{1 \text{ mol } NH_3} = 28 \text{ g } NH_3$$

Problem 4.12
Ethanol is produced industrially by the reaction of ethylene with water in the presence of an acid catalyst. How many grams of ethanol are produced from 7.24 mol of ethylene? Assume that excess water is present.

$$C_2H_4(g) + H_2O(\ell) \xrightarrow{\text{Acid catalyst}} CH_3CH_2OH(\ell)$$
$$\text{Ethylene} \qquad\qquad\qquad\qquad \text{Ethanol}$$

B Limiting Reagents

Frequently reactants are mixed in molar proportions that differ from those that appear in a balanced equation. As an example, consider an experiment in which NO is prepared by mixing five moles of N_2 with one mole of O_2. Only one mole of N_2 will react, consuming the one mole of O_2. The oxygen is used up completely and four moles of nitrogen are left over. These mole relationships are summarized under the balanced equation:

	$N_2(g)$	+ $O_2(g)$	$\longrightarrow$ 2NO(g)
Before reaction (moles)	5.0	1.0	0
After reaction (moles)	4.0	0	2.0

Limiting reagent The reactant that is consumed, leaving an excess of another reagent or reagents unreacted

The **limiting reagent** is the reactant that is used up first. In this example, O_2 is the limiting reagent, because it governs how much NO can form. The other reagent, N_2, is in excess.

EXAMPLE 4.13

Suppose 12 g of C is mixed with 64 g of O_2, and the following reaction takes place:

$$C(s) + O_2(g) \longrightarrow CO_2(g)$$

(a) Which reactant is the limiting reagent, and which reactant is in excess?
(b) How many grams of CO_2 will be formed?

Solution

(a) We use the molar mass of each reactant to calculate the number of moles of each present before reaction.

$$12 \text{ g C} \times \frac{1 \text{ mol C}}{12 \text{ g C}} = 1 \text{ mol C}$$

$$64 \text{ g O}_2 \times \frac{1 \text{ mol O}_2}{32 \text{ g O}_2} = 2 \text{ mol O}_2$$

According to the balanced equation, reaction of one mole of C requires one mole of O_2. But two moles of O_2 are present at the start of the reaction. Therefore, C is the limiting reagent, and O_2 is in excess.

(b) To calculate the number of grams of CO_2 formed, we use the conversion factor 1 mol CO_2 = 44 g CO_2.

$$12 \text{ g C} \times \frac{1 \text{ mol C}}{12 \text{ g C}} \times \frac{1 \text{ mol CO}_2}{1 \text{ mol C}} \times \frac{44 \text{ g CO}_2}{1 \text{ mol CO}_2} = 44 \text{ g CO}_2$$

We can summarize these numbers in the following table. Note that, as required by the law of conservation of mass, the sum of the masses of the material present after reaction is the same as the amount present before the reaction took place—namely, 76 g of material.

	C	+	O_2	→	CO_2
Before reaction	12 g		64 g		0
Before reaction	1.0 mol		2.0 mol		0
After reaction	0		1.0 mol		1.0 mol
After reaction	0		32.0 g		44.0 g

Problem 4.13

Assume that 6.0 g of C and 2.1 g of H_2 are mixed and react according to the following balanced equation:

$$C(s) + 2H_2(g) \longrightarrow CH_4(g)$$

(a) Which reactant is the limiting reagent, and which reactant is in excess?
(b) How many grams of CH_4 are produced in the reaction?

Screen 4.11 Simulation

C Percent Yield

When carrying out a chemical reaction, we often get less of a product than we expect from the type of calculation discussed in Section 4.5A. For example, suppose we react 32.0 g (1 mol) of CH_3OH with excess CO to form acetic acid:

$$CH_3OH + CO \longrightarrow CH_3COOH$$
Methanol   Carbon monoxide   Acetic acid

If we calculate the expected yield based on the stoichiometry of the balanced equation, we find that we should get 1 mol (60.0 g) of acetic acid. Suppose, however, we get only 57.8 g of acetic acid. Does this result mean that the law of conservation of mass is being violated? No, it does not. We get less than 60.0 g of acetic acid either because some of the CH_3OH does not react or because some of it reacts in some other way, or perhaps because our laboratory technique is not perfect and we lose a little in transferring from one container to another.

At this point, we need to define three terms, all of which relate to yield of product in a chemical reaction:

Actual yield The mass of product formed in a chemical reaction
Theoretical yield The mass of product that should form in a chemical reaction according to the stoichiometry of the balanced equation
Percent yield The actual yield divided by the theoretical yield times 100

$$\text{Percent yield} = \frac{\text{actual yield}}{\text{theoretical yield}} \times 100$$

We summarize the data in the preceding preparation of acetic acid in the following table:

	CH_3OH	+	CO	→	CH_3COOH
Before reaction	32.0 g		Excess		0
Before reaction	1.00 mol		Excess		0
Theoretical yield					1.00 mol
Theoretical yield					60.0 g
Actual yield					57.8 g

We calculate the percent yield in this experiment as follows:

$$\text{Percent yield} = \frac{57.8 \text{ g acetic acid}}{60.0 \text{ g acetic acid}} \times 100 = 96.3\%$$

Why is it important to know the percent yield of a chemical reaction or a series of reactions? The most important reason often relates to cost. If the yield of commercial product is, say, only 10%, the chemists will most probably be sent back to the lab to vary experimental conditions in an attempt to improve the yield. As an example, consider a reaction in which starting material A is converted first to compound B, then to compound C, and finally to compound D.

$$A \longrightarrow B \longrightarrow C \longrightarrow D$$

Screen 4.12
Tutorial

A reaction that does not give the main product is called a side reaction.

Occasionally, the percent yield is larger than 100%. For example, if a chemist fails to dry a product completely before weighing it, the product weighs more than it should because it also contains water. In such cases, the actual yield may be larger than expected and the percent yield may be greater than 100%.

Suppose the yield is 50% at each step. In this case, the yield of compound D is 13% based on the mass of compound A. If, however, the yield at each step is 90%, the yield of compound D increases to 73%; and if the yield at each step is 99%, the yield of compound D is 97%:

If the Percent Yield per Step Is	The Percent Yield of Compound D Is
50%	$0.50 \times 0.50 \times 0.50 = 13\%$
90%	$0.90 \times 0.90 \times 0.90 = 73\%$
99%	$0.99 \times 0.99 \times 0.99 = 97\%$

EXAMPLE 4.14

In an experiment forming ethanol, the theoretical yield is 50.5 g. The actual yield is 46.8 g. What is the percent yield?

Solution

$$\% \text{ Yield} = \frac{46.8 \text{ g}}{50.5 \text{ g}} \times 100 = 92.7\%$$

Problem 4.14

In an experiment to prepare aspirin, the theoretical yield is 153.7 g. If the actual yield is 124 g, what is the percent yield?

4.6 Reactions Between Ions in Aqueous Solution

Many ionic compounds are soluble in water. As we saw in Section 3.5, ionic compounds always consist of both positive and negative ions. When they dissolve in water, the positive and negative ions become separated from each other by water molecules. We call such a separation a **dissociation.** For example,

$$NaCl(s) \xrightarrow{H_2O} Na^+(aq) + Cl^-(aq)$$

H_2O above the arrow means that the reaction takes place in water.

Aqueous solution A solution in which the solvent is water

What happens when we mix **aqueous solutions** of two different ionic compounds? Does a reaction take place between the ions? The answer depends on the ions. If any of the negative and positive ions come together to form a compound, then a reaction takes place; otherwise, no reaction occurs.

As an example, suppose we prepare one solution by dissolving sodium chloride, NaCl, in water and a second solution by dissolving silver nitrate, $AgNO_3$, in water.

Solution 1 $NaCl(s) \xrightarrow{H_2O} Na^+(aq) + Cl^-(aq)$
Solution 2 $AgNO_3(s) \xrightarrow{H_2O} Ag^+(aq) + NO_3^-(aq)$

If we now mix the two solutions, four ions are present in the solution: Ag^+, Na^+, Cl^-, and NO_3^-. Two of these ions, Ag^+ and Cl^-, react to form the compound AgCl (silver chloride), which is insoluble in water. A reaction therefore takes place, forming a white precipitate of AgCl that slowly sinks to the bottom of the container (Figure 4.2). We write this reaction as follows:

$$\underset{\text{Silver ion}}{Ag^+(aq)} + \underset{\text{Nitrate ion}}{NO_3^-(aq)} + \underset{\text{Sodium ion}}{Na^+(aq)} + \underset{\text{Chloride ion}}{Cl^-(aq)} \longrightarrow \underset{\text{Silver chloride}}{AgCl(s)} + Na^+(aq) + NO_3^-(aq)$$

Notice that the Na^+ and NO_3^- ions do not participate in the reaction, but merely remain dissolved in the water. Ions that do not participate in a reaction are called **spectator ions,** certainly an appropriate name.

We can simplify the equation for the formation of silver chloride by omitting all spectator ions:

$$\underset{\text{Silver ion}}{Ag^+(aq)} + \underset{\text{Chloride ion}}{Cl^-(aq)} \longrightarrow \underset{\text{Silver chloride}}{AgCl(s)}$$

This kind of equation that we write for ions in solution is called a **net ionic equation.** Like all other chemical equations, net ionic equations must be balanced. We balance them in the same way as we do other equations, except now we must make sure that charges balance as well as atoms.

Net ionic equations show only the ions that react—no spectator ions are shown. For example, the net ionic equation for the precipitation of arsenic(III) sulfide from aqueous solution is

$$2As^{3+}(aq) + 3S^{2-}(aq) \longrightarrow As_2S_3(s)$$

Not only are there two arsenic and three sulfur atoms on each side, but the total charge on the left side is the same as the total charge on the right side; they are both zero.

In general, ions in solution react with each other only when one of these four things can happen:

1. Two ions form a solid that is insoluble in water. AgCl is one example, as shown in Figure 4.2.
2. Two ions form a gas that escapes from the reaction mixture as bubbles. An example is the reaction of sodium bicarbonate, $NaHCO_3$, with HCl to form the gas carbon dioxide, CO_2 (Figure 4.3). The net ionic equation for this reaction is

$$\underset{\text{Bicarbonate ion}}{HCO_3^-(aq)} + H_3O^+(aq) \longrightarrow \underset{\text{Carbon dioxide}}{CO_2(g)} + 2H_2O(\ell)$$

3. An acid neutralizes a base. Acid–base reactions are so important that we devote Chapter 8 to them.
4. One of the ions can oxidize another. We discuss this type of reaction in Section 4.7.

In many cases, no reaction takes place when we mix solutions of ionic compounds because none of these situations holds. For example, if we mix solutions of copper(II) nitrate, $Cu(NO_3)_2$, and potassium sulfate, K_2SO_4, we merely have a mixture containing Cu^{2+}, K^+, NO_3^-, and SO_4^{2-} ions dissolved in water. None of these react with each other; therefore, we see nothing happening (Figure 4.4).

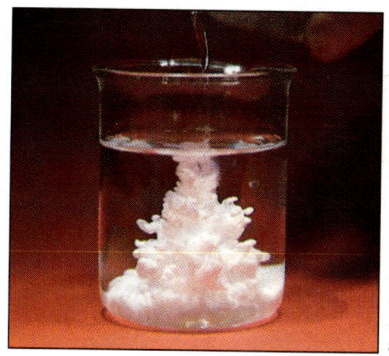

Figure 4.2 Adding Cl^- ions to a solution of Ag^+ ions produces a white precipitate of silver chloride, AgCl.

Spectator ion An ion that appears unchanged on both sides of a chemical equation

Net ionic equation A chemical equation that does not contain spectator ions

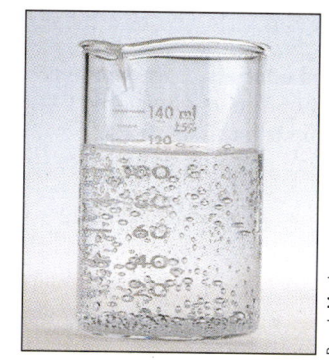

Figure 4.3 When aqueous solutions of $NaHCO_3$ and HCl are mixed, a reaction between HCO_3^- and H_3O^+ ions produces CO_2 gas, which can be seen as bubbles.

Figure 4.4 (a) The beaker on the left contains a solution of potassium sulfate (colorless), and the beaker on the right contains a solution of copper(II) nitrate (blue). (b) When the two solutions are mixed, the blue color becomes lighter because the copper(II) nitrate is less concentrated, but nothing else happens.

(a)   (b)

Screen 4.13 Tutorial

■ The mixing of aqueous solutions of barium chloride ($BaCl_2$) and sodium sulfate (Na_2SO_4) forms a precipitate of barium sulfate ($BaSO_4$).

■ Sea animals of the mollusk family often use insoluble $CaCO_3$ to construct their shells.

EXAMPLE 4.15

When a solution of barium chloride, $BaCl_2$, is added to a solution of sodium sulfate, Na_2SO_4, a white precipitate of barium sulfate, $BaSO_4$, forms. Write the net ionic equation for this reaction.

Solution
Because both barium chloride and sodium sulfate are ionic compounds, each exists in water as its dissociated ions:

$$Ba^{2+}(aq) + 2Cl^-(aq) + 2Na^+(aq) + SO_4^{2-}(aq)$$

We are told that a precipitate of barium sulfate forms:

$$Ba^{2+}(aq) + 2Cl^-(aq) + 2Na^+(aq) + SO_4^{2-}(aq) \longrightarrow$$
$$BaSO_4(s) + 2Na^+(aq) + 2Cl^-(aq)$$
Barium sulfate

Because Na^+ and Cl^- ions appear on both sides of the equation (they are spectator ions), we cancel them and are left with the following net ionic equation:

Net ionic equation $Ba^{2+}(aq) + SO_4^{2-}(aq) \longrightarrow BaSO_4(s)$

Problem 4.15

When a solution of copper(II) chloride, $CuCl_2$, is added to a solution of potassium sulfide, K_2S, a black precipitate of copper(II) sulfide, CuS, forms. Write the net ionic equation for the reaction.

Of the four ways for ions to react in water, one of the most common is the formation of an insoluble compound. We can predict when this result will happen if we know the solubilities of the ionic compounds. There is no simple way to remember which ionic solids are soluble in water and which are not, but we can make some useful generalizations:

1. All compounds containing Na^+, K^+, or NH_4^+ ions are soluble in water.
2. All nitrates (NO_3^-) and acetates (CH_3COO^-) are soluble in water.

CHEMICAL CONNECTIONS 4A

Solubility and Tooth Decay

The outermost protective layer of a tooth is the enamel, which is composed of approximately 95% hydroxyapatite, $Ca_{10}(PO_4)_6(OH)_2$, and 5% collagen (Chemical Connections 21G). Like most other phosphates and hydroxides, hydroxyapatite is insoluble in water. In acidic media, however, it dissolves to a slight extent, yielding Ca^{2+}, PO_4^{3-}, and OH^- ions. This loss of enamel creates pits and cavities in the tooth.

Acidity in the mouth is produced by bacterial fermentation of remnants of food, especially carbohydrates. Once pits and cavities form in the enamel, bacteria can hide there and cause further damage in the underlying softer material called dentin. The fluoridation of water brings F^- ions to the hydroxyapatite. There, F^- ions take the place of OH^- ions, forming the considerably less acid-soluble fluoroapatite, $Ca_{10}(PO_4)_6F_2$. Fluoride-containing toothpastes enhance this exchange process and provide further protection against tooth decay.

3. Most chlorides (Cl^-) and sulfates (SO_4^{2-}) are soluble in water. Important exceptions include silver chloride (AgCl), barium sulfate ($BaSO_4$), and lead sulfate ($PbSO_4$), which are insoluble in water.
4. Most carbonates (CO_3^{2-}), phosphates (PO_4^{3-}), sulfides (S^{2-}), and hydroxides (OH^-) are insoluble in water. Important exceptions include LiOH, NaOH, KOH, and NH_4OH, which are soluble in water.

4.7 Oxidation–Reduction

One of the most important and most common types of reaction is oxidation–reduction. **Oxidation** is the loss of electrons. **Reduction** is the gain of electrons. An **oxidation–reduction reaction** (or a **redox reaction**) involves the transfer of electrons from one species to another. An example is the oxidation of zinc by copper ions, the net ionic equation for which is

$$Zn(s) + Cu^{2+}(aq) \longrightarrow Zn^{2+}(aq) + Cu(s)$$

Oxidation The loss of electrons; the gain of oxygen or the loss of hydrogen

Reduction The gain of electrons; the loss of oxygen or the gain of hydrogen

Redox reaction An oxidation–reduction reaction

Negative ions such as Cl^- or NO_3^- must also be present to balance charges, but we do not show them because they are spectator ions.

(a)   (b)

■ Some insoluble compounds: (*a*) ferric hydroxide, and (*b*) copper carbonate.

Figure 4.5 When a piece of zinc is added to a solution containing Cu^{2+} ions, the Zn is oxidized by Cu^{2+} ions, and the Cu^{2+} ions are reduced by the Zn.

Blue solution of Cu^{2+} ions

Zinc bar

Coating of copper plating out, and zinc bar partially dissolved

Colorless solution of Zn^{2+} ions

If we put a piece of zinc metal into a beaker containing copper ions in aqueous solution, three things will happen (Figure 4.5):

1. Some of the zinc metal will dissolve and go into solution as Zn^{2+}.
2. Copper metal will deposit on the zinc metal.
3. The blue color of the Cu^{2+} ions will gradually disappear.

The zinc metal loses electrons to the copper ions:

$$Zn(s) \longrightarrow Zn^{2+}(aq) + 2e^- \qquad \text{Zn is oxidized}$$

At the same time, the Cu^{2+} ions gain electrons from the zinc. The copper ions are reduced:

$$Cu^{2+}(aq) + 2e^- \longrightarrow Cu(s) \qquad Cu^{2+} \text{ is reduced}$$

Oxidation and reduction are not independent reactions. That is, species cannot gain electrons from nowhere, nor can a species lose electrons to nothing. In other words, no oxidation can occur without an accompanying reduction, and vice versa. In the preceding reaction, Cu^{2+} is oxidizing Zn. We call Cu^{2+} an **oxidizing agent**. Similarly, Zn is reducing Cu^{2+}, and we call Zn a **reducing agent**.

Oxidizing agent An entity that accepts electrons in an oxidation–reduction reaction

Reducing agent An entity that donates electrons in an oxidation–reduction reaction

We summarize these oxidation–reduction relationships for the reaction of zinc metal with Cu(II) ion in the following way:

Electrons flow from Zn to Cu^{2+}

$$Zn(s) + Cu^{2+}(aq) \longrightarrow Zn^{2+}(aq) + Cu(s)$$

Loses electrons; is oxidized
Gives electrons to Cu^{2+}; is the reducing agent

Gains electrons; is reduced
Takes electrons from Zn; is the oxidizing agent

Although the definitions we have given for oxidation (loss of electrons) and reduction (gain of electrons) are easy to apply in many redox reactions, they are not so easy to apply in other cases. For example, another redox re-

CHEMICAL CONNECTIONS 4B

Voltaic Cells

In Figure 4.5, we see that when a piece of zinc metal is put in a solution containing Cu^{2+} ions, zinc atoms give electrons to Cu^{2+} ions. We can change the experiment by putting the zinc metal in one beaker and the Cu^{2+} ions in another, and then connecting the two beakers by a length of wire and a salt bridge (see the accompanying figure). A reaction still takes place; that is, zinc atoms still give electrons to Cu^{2+} ions, but now the electrons must flow through the wire to get from the Zn to the Cu^{2+}. This flow of electrons produces an electric current, and the electrons keep flowing until either the Zn or the Cu^{2+} is used up. In this way, the apparatus generates an electric current by using a redox reaction. We call this device a **voltaic cell** or, more commonly, a *battery*.

The electrons produced at the zinc end carry negative charges. This end of the battery is a negative electrode (called the **anode**). The electrons released at the anode as zinc is oxidized go through an outside circuit and, in doing so, produce the battery's electric current. At the other end of the battery, at the positively charged electrode (called the **cathode**), electrons are consumed as Cu^{2+} ions are reduced to copper metal.

To see why a salt bridge is necessary, we must look at the Cu^{2+} solution. Because we cannot have positive charges in any place without an equivalent number of negative charges, negative ions must be in the beaker as well—perhaps sulfate, nitrate, or some other anion. When electrons come over the wire, the Cu^{2+} is converted to Cu:

$$Cu^{2+}(aq) + 2e^- \longrightarrow Cu(s)$$

This reaction diminishes the number of Cu^{2+} ions, but the number of negative ions remains unchanged. The salt bridge is necessary to carry some of these negative ions to the other beaker, where they are needed to balance the Zn^{2+} ions being produced by the following reaction:

$$Zn(s) \longrightarrow Zn^{2+}(aq) + 2e^-$$

action is the combustion (burning) of methane, CH_4, in which CH_4 is oxidized to CO_2 while O_2 is reduced to CO_2 and H_2O:

$$CH_4(g) + O_2(g) \longrightarrow CO_2(g) + 2H_2O(g)$$
Methane

It is not easy to see the electron loss and gain in such a reaction, so chemists have developed another definition of oxidation and reduction, one that is easier to apply in many cases, especially where organic compounds are involved:

Oxidation is the gain of oxygen or the loss of hydrogen.
Reduction is the loss of oxygen or the gain of hydrogen.

Applying these alternative definitions to the reaction of methane with oxygen, we find the following:

$$CH_4(g) + O_2(g) \longrightarrow CO_2(g) + H_2O(g)$$

CH_4: Gains O and loses H; is oxidized — Is the reducing agent

O_2: Gains H; is reduced — Is the oxidizing agent

In fact, this second definition is much older than the one involving electron transfer; it is the definition given by Lavoisier when he first discovered oxidation and reduction more than 200 years ago. Note that we could not apply this definition to our zinc–copper example.

EXAMPLE 4.16

In each equation, identify the substance that is oxidized, the substance that is reduced, the oxidizing agent, and the reducing agent.

(a) $Al(s) + Fe^{3+}(aq) \longrightarrow Al^{3+}(aq) + Fe(s)$

(b) $CH_3OH(\ell) + O_2(g) \longrightarrow HCOOH(\ell) + H_2O(g)$
    Methanol   Formic Acid

Solution

(a) $Al(s)$ loses three electrons in going to Al^{3+}; therefore, aluminum is oxidized. In the process of being oxidized, $Al(s)$ gives its electrons to Fe^{3+}, so that $Al(s)$ is the reducing agent. Fe^{3+} gains three electrons in going to $Fe(s)$ and is reduced. In the process of being reduced, Fe^{3+} accepts three electrons from $Al(s)$, so Fe^{3+} is the oxidizing agent. To summarize:

Electrons flow from Al to Fe^{3+}

$$Al(s) + Fe^{3+}(aq) \longrightarrow Al^{3+}(aq) + Fe(s)$$

$Al(s)$: Loses electrons; is oxidized — Gives electrons to Fe^{3+}; is the reducing agent

$Fe^{3+}(aq)$: Gains electrons; is reduced — Takes electrons from Al; is the oxidizing agent

(b) Because it is not easy to see the loss or gain of electrons in this example, we apply the second set of definitions. In being converted to HCOOH, CH_3OH both gains oxygens and loses hydrogens; it is oxidized. In being converted to H_2O, O_2 gains hydrogens; it is reduced. The compound that is oxidized is the reducing agent; CH_3OH is the reducing agent. The compound reduced is the oxidizing agent; O_2 is the oxidizing agent. To summarize:

$$CH_3OH(\ell) + O_2(g) \longrightarrow HCOOH(\ell) + H_2O(g)$$

CH_3OH: Is oxidized; is the reducing agent

O_2: Is reduced; is the oxidizing agent

Screen 4.17 Exercise

CHEMICAL CONNECTIONS 4C

Oxidizing Antiseptics

An antiseptic is a compound that kills bacteria. Antiseptics are used to treat wounds—not to heal them any faster but to prevent them from becoming infected by bacteria. Some antiseptics operate by oxidizing (and hence destroying) compounds essential to the normal functioning of the bacteria. One example is iodine, I_2, which was used as a household antiseptic for minor cuts and bruises for many years, not in the pure form but as a dilute solution in ethanol, called a tincture. Pure I_2, is a steel-gray solid that gives off a purple vapor when heated. In contrast, the tincture is a brown liquid. Another example of an oxidizing antiseptic is a dilute solution (commonly 3%) of hydrogen peroxide, H_2O_2, which is used as a mouth rinse to heal gum infections. Oxidizing antiseptics, however, are often regarded as too harsh. They not only kill bacteria but also harm skin and other normal tissues. For this reason, oxidizing antiseptics have largely been replaced by phenolic antiseptics (Section 13.5).

Disinfectants are also employed to kill bacteria, but they are used on inanimate objects rather than on living tissues. Many disinfectants are oxidizing agents. Two important examples are chlorine, Cl_2, a pale green gas, and ozone, O_3, a colorless gas.

■ These tablets, which produce chlorine when mixed with water, are added to a swimming pool filter.

Both gases are added in small quantities to municipal water supplies to kill any harmful bacteria that may be present. Both gases must be handled carefully because they are very poisonous.

Problem 4.16

In each equation, identify the substance that is oxidized, the substance that is reduced, the oxidizing agent, and the reducing agent:

(a) $Ni^{2+}(aq) + Cr(s) \longrightarrow Ni(s) + Cr^{2+}(aq)$

(b) $\underset{\text{Formaldehyde}}{CH_2O(g)} + H_2(g) \longrightarrow \underset{\text{Methanol}}{CH_3OH(\ell)}$

We have said that redox reactions are extremely common. Here are some important categories:

1. **Combustion** All combustion (burning) reactions are redox reactions in which the compounds or mixtures that are burned are oxidized by oxygen, O_2. They include the burning of gasoline, diesel oil, fuel oil, natural gas, coal, wood, and paper. All these materials contain carbon, and all except coal also contain hydrogen. If the combustion is complete, carbon is oxidized to CO_2 and hydrogen is oxidized to H_2O. In an incomplete combustion, these elements are oxidized to other compounds, many of which cause air pollution. Unfortunately, much of today's combustion that takes place in gasoline and diesel engines and in furnaces is incomplete and so contributes to air pollution. In the incomplete combustion of carbon, for example, carbon is oxidized to carbon monoxide, CO, due to the lack of sufficient oxygen to complete the oxidation.

■ Air pollution is caused by incomplete fuel combustion.

Figure 4.6 (a) Dry cell batteries. (b) A lead storage battery.

■ The rusting of iron and steel can be a serious problem in an industrial society.

■ These household bleaches are oxidizing agents.

Heat of reaction The heat given off or absorbed in a chemical reaction

2. **Respiration** Humans and animals get their energy by respiration. The oxygen in the air we breathe oxidizes carbon-containing compounds in our cells to produce CO_2 and H_2O. In fact, this oxidation reaction is really the same as combustion, although it takes place more slowly and at a much lower temperature. We discuss these reactions more fully in Chapter 26. The important product of these reactions is not the CO_2 (which the body eliminates) or H_2O, but energy.

3. **Rusting** We all know that when iron or steel objects are left out in the open air, they eventually rust (steel is mostly iron but contains certain other elements, too). In rusting, iron is oxidized to a mixture of iron oxides. We can represent the main reaction by the following equation:

$$4Fe(s) + 3O_2(g) \longrightarrow 2Fe_2O_3(s)$$

4. **Bleaching** Most bleaching involves oxidation, and common bleaches are oxidizing agents. The colored compounds being bleached are usually organic compounds; oxidation converts them to colorless compounds.

5. **Batteries** A voltaic cell (Chemical Connections 4B) is a device in which electricity is generated from a chemical reaction. Such cells are often called batteries (Figure 4.6). We are all familiar with batteries in our cars and in such portable devices as radios, flashlights, cell phones, and computers. In all cases, the reaction taking place in the battery is a redox reaction.

4.8 Heat of Reaction

In almost all chemical reactions, not only are starting materials converted to products, but heat is also either given off or absorbed. For example, when one mole of carbon is oxidized by oxygen to produce one mole of CO_2, 94.0 kilocalories of heat is given off:

$$C(s) + O_2(g) \longrightarrow CO_2(g) + 94.0 \text{ kcal}$$

The heat given off or gained in a reaction is called the **heat of reaction.** A reaction that gives off heat is **exothermic;** a reaction that absorbs heat is **endothermic.** The amount of heat given off or absorbed is proportional to the amount of material. For example, when 2 mol of carbon is oxidized by oxygen to give carbon dioxide, $2 \times 94.0 = 188$ kcal of heat is given off.

The energy changes accompanying a chemical reaction are not limited to heat. In some such reactions, as in voltaic cells (Chemical Connections 4B), the energy given off takes the form of electricity. In other reactions, such as photosynthesis (the reaction whereby plants convert water and carbon dioxide to sugar and oxygen), the energy absorbed is in the form of light.

An example of an endothermic reaction is the decomposition of mercury(II) oxide:

$$2\text{HgO(s)} + 43.4 \text{ kcal} \longrightarrow 2\text{Hg}(\ell) + \text{O}_2(g)$$
Mercury(II) oxide
(mercuric oxide)

Exothermic reaction A chemical reaction that gives off heat

Endothermic reaction A chemical reaction that absorbs heat

This equation tells us that, if we want to decompose 2 mol of mercury(II) oxide into the elements Hg and O_2, we must add 43.4 kcal of energy to HgO. Incidentally, the law of conservation of energy tells us that the reverse reaction—the oxidation of mercury—must give off exactly the same amount of heat:

$$2\text{Hg}(\ell) + \text{O}_2(g) \longrightarrow 2\text{HgO(s)} + 43.4 \text{ kcal}$$

Especially important are the heats of reaction for combustion reactions. As we saw in Section 4.7, combustion reactions are the most important heat-producing reactions, because most of the energy required for modern society to function is derived from them. All combustions are exothermic. The heat given off in a combustion reaction is called the **heat of combustion.**

■ Mercury(II) oxide, a red compound, decomposes into two elements when heated, mercury (a metal) and oxygen (a nonmetal). Mercury vapor condenses on the cooler upper portion of the test tube.

SUMMARY

In a chemical reaction, the starting materials are called **reactants,** and the new materials formed are called **products** (Section 4.1) The **formula weight** (Section 4.2) of a compound is the sum of the atomic weights of all atoms in the compound expressed in atomic mass units (amu). Formula weight applies to both ionic and molecular compounds. The term **molecular weight,** also expressed in amu, applies to only molecular compounds. A **mole (mol)** of any substance is defined as Avogadro's number (6.02×10^{23}) of formula units of the substance (Section 4.3). The **molar mass** of a substance is its formula weight expressed in grams (Section 4.3).

A **chemical equation** is an expression showing which reactants are converted to which products (Section 4.4A). A balanced chemical equation shows how many moles of each starting material are converted to how many moles of each product (Section 4.4B).

Stoichiometry is the study of the mass relationships in chemical reactions (Section 4.5A). The reagent that is used up first in a reaction is called the **limiting reagent** (Section 4.5B) A reactant is in excess when the amount of it present is more than necessary for the reaction. The **percent yield** for a reaction equals the **actual yield** divided by **theoretical yield** multiplied by 100 (Section 4.5C).

When ions are mixed in aqueous solution, they react with one another only if (1) a precipitate forms, (2) a gas forms, (3) an acid neutralizes a base, or (4) oxidation–reduction takes place (Section 4.6). Ions that do not react are called **spectator ions.** A **net ionic equation** shows only those ions that react. In a net ionic equation, both the charges and the number of atoms must be balanced.

In **oxidation,** electrons are lost; in **reduction,** electrons are gained (Section 4.7). These two processes must take place together. The joint process is often called a **redox reaction.** Oxidation can also be defined as the gain of oxygen or the loss of hydrogen; reduction is the reverse process.

Either a gain or a loss of heat accompanies almost all reactions. This heat is called the **heat of reaction.** Reactions that give off heat are **exothermic;** those that absorb heat are **endothermic.** The heat given off in a combustion reaction is called the **heat of combustion.**

PROBLEMS

Numbers that appear in color indicate difficult problems. ▲ designates problems requiring application of principles.

Formula Weight, Moles, and Molar Mass

4.17 Calculate the formula weight of:
(a) Cl_2 (b) Ar (c) P_4
(d) N_2 (e) He

4.18 Calculate the formula weight of:
(a) KCl (b) Na_3PO_4
(c) $Fe(OH)_2$ (d) Sucrose, $C_{12}H_{22}O_{11}$
(e) $NaAl(SO_3)_2$ (f) Glycine, $C_2H_5NO_2$
(g) $Al_2(SO_4)_3$ (h) $(NH_4)_2CO_3$
(i) DDT, $C_{14}H_9Cl_5$

4.19 Calculate the number of moles in:
(a) 32 g of methane, CH_4
(b) 345.6 g of nitric oxide, NO
(c) 184.4 g of chlorine dioxide, ClO_2
(d) 720 g of glycerin, $C_3H_8O_3$

4.20 Calculate the number of grams in:
(a) 1.77 mol of nitrogen dioxide, NO_2
(b) 0.84 mol of 2-propanol, C_3H_8O (rubbing alcohol)
(c) 3.69 mol of uranium hexafluoride, UF_6
(d) 0.348 mol of galactose, $C_6H_{12}O_6$
(e) 4.9×10^{-2} mol of vitamin C, $C_6H_8O_6$

4.21 Calculate the number of moles of:
(a) O atoms in 18.1 mol of formaldehyde, CH_2O
(b) Br atoms in 0.41 mol of bromoform, $CHBr_3$
(c) O atoms in 3.5×10^3 mol of $Al_2(SO_4)_3$
(d) Hg atoms in 87 g of HgO

4.22 Calculate the number of moles of:
(a) S^{2-} ions in 6.56 mol of Na_2S
(b) Mg^{2+} ions in 8.320 mol of $Mg_3(PO_4)_2$
(c) acetate ion, CH_3COO^-, in 0.43 mol of $Ca(CH_3COO)_2$

4.23 A single atom of cerium weighs just about twice as much as a single atom of gallium. What is the weight ratio of 25 atoms of cerium to 25 atoms of gallium?

4.24 What is the mass in grams of each number of molecules of formaldehyde, CH_2O?
(a) 100 molecules
(b) 3000 molecules
(c) 5.0×10^6 molecules
(d) 2.0×10^{24} molecules

4.25 How many molecules are in 3.1×10^{-1} g of aspirin, $C_9H_8O_4$?

4.26 How many molecules are in
(a) 2.9 mol of TNT, $C_7H_5N_3O_6$?
(b) one drop (0.0500 g) of water?

4.27 A typical deposit of cholesterol, $C_{27}H_{46}O$, in an artery might have a mass of 3.9 mg. How many molecules of cholesterol are in this mass?

4.28 The molecular weight of hemoglobin is about 68,000 amu. What is the mass in grams of a single molecule of hemoglobin?

Balancing Equations

4.29 Balance each equation.
(a) $HI + NaOH \longrightarrow NaI + H_2O$
(b) $Ba(NO_3)_2 + H_2S \longrightarrow BaS + HNO_3$
(c) $CH_4 + O_2 \longrightarrow CO_2 + H_2O$
(d) $C_4H_{10} + O_2 \longrightarrow CO_2 + H_2O$
(e) $Fe + CO_2 \longrightarrow Fe_2O_3 + CO$

4.30 Balance each equation.
(a) $H_2 + I_2 \longrightarrow HI$
(b) $Al + O_2 \longrightarrow Al_2O_3$
(c) $Na + Cl_2 \longrightarrow NaCl$
(d) $Al + HBr \longrightarrow AlBr_3 + H_2$
(e) $P + O_2 \longrightarrow P_2O_5$

4.31 ▲ If you blow carbon dioxide gas into a solution of calcium hydroxide, a milky-white precipitate of calcium carbonate forms. Write a balanced equation for the formation of calcium carbonate in this reaction.

4.32 ▲ Calcium oxide is prepared by heating limestone (calcium carbonate, $CaCO_3$) to a high temperature, at which point it decomposes to calcium oxide and carbon dioxide. Write a balanced equation for this preparation of calcium oxide.

4.33 ▲ The brilliant white light in some fireworks displays is produced by burning magnesium in air. The magnesium reacts with oxygen in the air to form magnesium oxide, MgO. Write a balanced equation for this reaction.

4.34 ▲ The rusting of iron is a chemical reaction of iron with oxygen in air to form iron(III) oxide, Fe_2O_3. Write a balanced equation for this reaction.

4.35 When gaseous dinitrogen pentoxide, N_2O_5, is bubbled into water, nitric acid, HNO_3, forms. Write a balanced equation for this reaction.

4.36 ▲ When solid carbon burns in a limited supply of oxygen gas, carbon monoxide, CO, forms. This gas is deadly to humans because it combines with hemoglobin in the blood, making it impossible for the blood to transport oxygen. Write a balanced equation for the formation of carbon monoxide.

4.37 Solid ammonium carbonate, $(NH_4)_2CO_3$, decomposes at room temperature to ammonia, carbon dioxide, and water. Because of the ease of decomposition and the penetrating odor of ammonia, ammonium carbonate can be used as smelling salts. Write a balanced equation for this decomposition.

4.38 In the chemical test for arsenic, the gas arsine, AsH_3, is prepared. When arsine is decomposed by heating, arsenic metal deposits as a mirrorlike coating on the surface of a glass container, and hydrogen gas, H_2, is given off. Write a balanced equation for the decomposition of arsine.

4.39 When a piece of aluminum metal is dropped into hydrochloric acid, HCl, hydrogen is released as a gas, and a solution of aluminum chloride forms. Write a balanced equation for the reaction.

4.40 In the industrial chemical preparation of chlorine, Cl_2, electric current is passed through an aqueous solution of sodium chloride to give Cl_2 and H_2. The other product of this reaction is sodium hydroxide. Write a balanced equation for this reaction.

Mass Relationships in Chemical Reactions

4.41 For the reaction

$$2N_2(g) + 3O_2(g) \longrightarrow 2N_2O_3(g)$$

(a) How many moles of N_2 are required to react completely with one mole of O_2?
(b) How many moles of N_2O_3 are produced from the complete reaction of one mole of O_2?
(c) How many moles of O_2 are required to produce 8 moles of N_2O_3?

4.42 Magnesium reacts with sulfuric acid according to the following equation. How many moles of H_2 are produced by the complete reaction of 230 mg of Mg with sulfuric acid?

$$Mg + H_2SO_4 \longrightarrow MgSO_4 + H_2$$

4.43 Chloroform is prepared industrially by the reaction of methane with chlorine. How many grams of Cl_2 are needed to produce 1.50 moles of chloroform?

$$\underset{\text{Methane}}{CH_4(g)} + 3Cl_2(g) \longrightarrow \underset{\text{Chloroform}}{CHCl_3(\ell)} + 3HCl(g)$$

4.44 At one time, acetaldehyde was prepared industrially by the reaction of ethylene with air in the presence of a copper catalyst. How many grams of acetaldehyde can be prepared from 81.7 g of ethylene?

$$\underset{\text{Ethylene}}{C_2H_4(g)} + O_2(g) \longrightarrow \underset{\text{Acetaldehyde}}{C_2H_4O(g)}$$

4.45 Chlorine dioxide, ClO_2, is used for bleaching paper. It is also the gas used to kill the anthrax spores that contaminated the Hart Senate Office Building in fall 2001. Chlorine dioxide is prepared by treating sodium chlorite with chlorine gas.

$$\underset{\substack{\text{Sodium} \\ \text{chlorite}}}{NaClO_2(aq)} + Cl_2(g) \longrightarrow \underset{\substack{\text{Chlorine} \\ \text{dioxide}}}{ClO_2(g)} + NaCl(aq)$$

(a) Balance the equation for the preparation of chlorine dioxide.
(b) Calculate the weight of chlorine dioxide that can be prepared from 5.50 kg of sodium chlorite.

4.46 Ethanol, C_2H_5OH, is added to gasoline to produce "gasohol," a fuel for automobile engines. How many grams of O_2 are required for complete combustion of 421 g of C_2H_5OH?

$$\underset{\text{Ethanol}}{C_2H_5OH(\ell)} + 3O_2(g) \longrightarrow 2CO_2(g) + 3H_2O$$

4.47 In photosynthesis, green plants convert CO_2 and H_2O to glucose, $C_6H_{12}O_6$. How many grams of CO_2 are required to produce 5.1 g of glucose?

$$6CO_2(g) + 6H_2O(\ell) \xrightarrow{\text{Photosynthesis}} \underset{\text{Glucose}}{C_6H_{12}O_6(aq)} + 6O_2(g)$$

4.48 Iron ore is converted to iron by heating it with coal, C, and oxygen according to the following equation:

$$2Fe_2O_3(s) + 6C(s) + 3O_2(s) \longrightarrow 4Fe(s) + 6CO_2(s)$$

If the process is run until 3940 g of Fe is produced, how many grams of CO_2 will also be produced?

4.49 Given the reaction in Problem 4.48, how many grams of C are necessary to react completely with 0.58 g of Fe_2O_3?

4.50 Aspirin is made by the reaction of salicylic acid with acetic anhydride. How many grams of aspirin are produced if 85.0 g of salicylic acid is treated with excess acetic anhydride?

$$\underset{\substack{\text{Salicylic} \\ \text{acid}}}{C_7H_6O_3(s)} + \underset{\substack{\text{Acetic} \\ \text{anhydride}}}{C_4H_6O_3(\ell)} \longrightarrow \underset{\text{Aspirin}}{C_9H_8O_4(s)} + \underset{\substack{\text{Acetic} \\ \text{acid}}}{C_2H_4O_2(\ell)}$$

4.51 Benzene reacts with bromine to produce bromobenzene according to the following equation:

$$C_6H_6(\ell) + Br_2(\ell) \longrightarrow C_6H_5Br(\ell) + HBr(g)$$

Benzene Bromine Bromobenzene Hydrogen bromide

If 60.0 g of benzene is mixed with 135 g of bromine,
(a) Which is the limiting reagent?
(b) How many grams of bromobenzene are formed in the reaction?

4.52 Ammonia is prepared industrially by the reaction of nitrogen and hydrogen according to the following equation:

$$N_2(g) + 3H_2(g) \longrightarrow 2NH_3(g)$$

Ammonia

If 29.7 kg of N_2 is added to 3.31 kg of H_2,
(a) Which reactant is the limiting reagent?
(b) How many grams of the other reactant are left over?
(c) How many grams of NH_3 are formed if the reaction goes to completion?

Percent Yield

4.53 Ethyl chloride is prepared by the reaction of chlorine with ethane according to the following balanced equation. When 5.6 g of ethane is reacted with excess chlorine, 8.2 g of ethyl chloride forms. Calculate the percent yield of ethyl chloride.

$$C_6H_6(g) + Cl_2(g) \longrightarrow C_2H_5Cl(\ell) + HCl(g)$$

Ethane   Ethyl chloride

4.54 Diethyl ether is made from ethanol according to the following reaction:

$$2C_2H_5OH(\ell) \longrightarrow (C_2H_5)_2O(\ell) + H_2O(\ell)$$

Ethanol   Diethyl ether

In an experiment, 517 g of ethanol gave 391 g of diethyl ether. What was the percent yield in this experiment?

Reactions Between Ions in Aqueous Solution

4.55 Define:
(a) spectator ion
(b) net ionic equation
(c) aqueous solution

4.56 Balance these net ionic equations.
(a) $Ag^+(aq) + Br^-(aq) \longrightarrow AgBr(s)$
(b) $Cd^{2+}(aq) + S^{2-}(aq) \longrightarrow CdS(s)$
(c) $Sc^{3+}(aq) + SO_4^{2-}(aq) \longrightarrow Sc_2(SO_4)_3(s)$
(d) $Sn^{2+}(aq) + Fe^{2+}(aq) \longrightarrow Sn(s) + Fe^{3+}(aq)$
(e) $K(s) + H_2O(\ell) \longrightarrow K^+(aq) + OH^-(aq) + H_2(g)$

4.57 In the equation

$$Na^+(aq) + CO_3^{2-}(aq) + Sr^{2+}(aq) + Cl^-(aq) \longrightarrow SrCO_3(s) + Na^+(aq) + Cl^-(aq)$$

(a) Identify the spectator ions.
(b) Write the balanced net ionic equation.

4.58 Predict whether a precipitate will form when aqueous solutions of the following compounds are mixed. If a precipitate will form, write its formula. To make your predictions, use the solubility generalizations in Section 4.6.
(a) $CaCl_2(aq) + K_3PO_4(aq) \longrightarrow$
(b) $KCl(aq) + Na_2SO_4(aq) \longrightarrow$
(c) $(NH_4)_2CO_3(aq) + Ba(NO_3)_2(aq) \longrightarrow$
(d) $FeCl_2(aq) + KOH(aq) \longrightarrow$
(e) $Ba(NO_3)_2(aq) + NaOH(aq) \longrightarrow$
(f) $Na_2S(aq) + SbCl_3(aq) \longrightarrow$
(g) $Pb(NO_3)_2(aq) + K_2SO_4(aq) \longrightarrow$

4.59 When a solution of ammonium chloride, NH_4Cl, is added to a solution of lead(II) nitrate, $Pb(NO_3)_2$, a white precipitate lead(II) chloride, $PbCl_2$, forms. Write a balanced net ionic equation for this reaction. Both ammonium chloride and lead nitrate exist as dissociated ions in aqueous solution.

4.60 When a solution of hydrochloric acid, HCl, is added to a solution of sodium sulfite, Na_2SO_3, sulfur dioxide gas, SO_2, is released from the solution. Write a net ionic equation for this reaction. An aqueous solution of HCl contains H^+ and Cl^- ions, and Na_2SO_3 exists as dissociated ions in aqueous solution.

4.61 When a solution of sodium hydroxide, NaOH, is added to a solution of ammonium carbonate, $(NH_4)_2CO_3$, H_2O is formed and ammonia gas, NH_3, is released when the solution is heated. Write a net ionic equation for this reaction. Both NaOH and $(NH_4)_2CO_3$ exist as dissociated ions in aqueous solution.

4.62 Using the solubility generalizations given in Section 4.6, predict which of these ionic compounds are soluble in water.
(a) KCl (b) NaOH (c) $BaSO_4$
(d) Na_2SO_4 (e) Na_2CO_3 (f) $Fe(OH)_2$

4.63 Using the solubility generalizations given in Section 4.6, predict which of these ionic compounds are soluble in water.
(a) $MgCl_2$ (b) $CaCO_3$ (c) Na_2SO_3
(d) NH_4NO_3 (e) $Pb(OH)_2$

Oxidation–Reduction

4.64 Give two definitions of oxidation and two definitions of reduction.

4.65 Can a reduction take place without an oxidation? Explain.

4.66 In the reaction

$$Pb(s) + 2Ag^+(aq) \longrightarrow Pb^{2+}(aq) + 2Ag(s)$$

(a) Which species is oxidized and which is reduced?
(b) Which species is the oxidizing agent and which is the reducing agent?

4.67 In the reaction

$$C_7H_{12}(\ell) + 10O_2(g) \longrightarrow 7CO_2(g) + 6H_2O(\ell)$$

(a) Which species is oxidized and which is reduced?
(b) Which species is the oxidizing agent and which is the reducing agent?

4.68 When a piece of sodium metal is added to water, hydrogen is evolved as a gas and a solution of sodium hydroxide is formed.

(a) Write a balanced equation for this reaction.
(b) What is oxidized in this reaction? What is reduced?

Heat of Reaction

4.69 What is the difference between exothermic and endothermic?

4.70 Which of these reactions are exothermic and which are endothermic?

(a) $2NH_3(g) + 22.0 \text{ kcal} \longrightarrow N_2(g) + 3H_2(g)$
(b) $H_2(g) + F_2(g) \longrightarrow 2HF(g) + 124 \text{ kcal}$
(c) $C(s) + O_2(g) \longrightarrow CO_2(g) + 94.0 \text{ kcal}$
(d) $H_2(g) + CO_2(g) + 9.80 \text{ kcal} \longrightarrow H_2O(g) + CO(g)$
(e) $C_3H_8(g) + 5O_2(g) \longrightarrow$
$\quad\quad 3CO_2(g) + 4H_2O(g) + 531 \text{ kcal}$

4.71 In the following reaction, 9.80 kcal is absorbed per mole of CO_2 undergoing reaction. How much heat is given off if two moles of water is reacted with two moles of carbon monoxide?

$$H_2(g) + CO_2(g) + 9.80 \text{ kcal} \longrightarrow H_2O(g) + CO(g)$$

4.72 Following is the equation for the combustion of acetone:

$$2C_3H_6O(\ell) + 8O_2(g) \longrightarrow 6CO_2(g) + 6H_2O(g) + 853.6 \text{ kcal}$$
$\quad\text{Acetone}$

How much heat will be given off if 0.37 mol of acetone is burned completely?

4.73 The oxidation of glucose, $C_6H_{12}O_6$, to carbon dioxide and water is exothermic: The heat liberated is the same whether glucose is metabolized in the body or burned in air.

$$C_6H_{12}O_6 + 6O_2 \longrightarrow 6CO_2 + 6H_2O + 670 \text{ kcal/mol}$$
$\quad\text{Glucose}$

Calculate the heat liberated when 15.0 g of glucose is metabolized to carbon dioxide and water in the body.

4.74 The heat of combustion of glucose, $C_6H_{12}O_6$, is 670 kcal/mol. The heat of combustion of ethanol, C_2H_5OH, is 327 kcal/mol. The heat liberated by oxidation of each compound is the same whether it is burned in air or metabolized in the body. On a kcal/g basis, metabolism of which compound liberates more heat?

4.75 A plant requires approximately 4178 kcal for the production of 1.00 kg of starch (Section 19.6A) from carbon dioxide and water.

(a) Is the production of starch in a plant an exothermic process or an endothermic process?
(b) Calculate the energy in kilocalories required by a plant for the production of 6.32 g of starch.

4.76 The heat of combustion of methane, CH_4, the major component of natural gas, is 213 kcal/mol. The heat of combustion of propane, C_3H_8, the major component of LPG or bottled gas, is 530 kcal/mol.

(a) Write a balanced equation for the complete combustion of each to CO_2 and H_2O.
(b) On a kcal/mol basis, which of these two fuels is the better source of heat energy?
(c) On a kcal/g basis, which of these two fuels is the better source of heat energy?

4.77 To convert 1 mol of iron(III) oxide to its elements requires 196.5 kcal:

$$Fe_2O_3(s) + 196.5 \text{ kcal} \longrightarrow 2Fe(s) + \tfrac{3}{2}O_2(g)$$

How many grams of iron can be produced if 156.0 kcal of heat is absorbed by a large-enough sample of iron(III) oxide?

Chemical Connections

4.78 (Chemical Connections 4A) How does fluoride ion protect the enamel against tooth decay?

4.79 (Chemical Connections 4A) What kind of ions is hydroxyapatite made of?

4.80 (Chemical Connections 4B) A voltaic cell is represented by the following equation:

$$Fe(s) + Zn^{2+}(aq) \longrightarrow Fe^{2+}(aq) + Zn(s)$$

Which electrode is the anode and which is the cathode?

4.81 (Chemical Connections 4C) Hydrogen peroxide is not only an antiseptic but also an oxidizing agent. The following equation shows the reaction of hydrogen peroxide with acetaldehyde to give acetic acid:

$$C_2H_4O(\ell) + H_2O_2(\ell) \longrightarrow C_2H_4O_2(\ell) + H_2O(\ell)$$
$\quad\text{Acetaldehyde}\quad\text{Hydrogen}\quad\quad\text{Acetic}\quad\quad\text{Water}$
$\quad\quad\quad\quad\quad\quad\quad\text{peroxide}\quad\quad\quad\text{acid}$

In this reaction, which species is oxidized and which is reduced? Which species is the oxidizing agent and which is the reducing agent?

Additional Problems

4.82 In a certain reaction, Cu^+ is converted to Cu^{2+}. Is Cu^+ ion oxidized or reduced in this reaction? Is Cu^+ ion an oxidizing agent or a reducing agent in this reaction?

4.83 Using the equation

$$Fe_2O_3(s) + 3CO(g) \longrightarrow 2Fe(s) + 3CO_2(g)$$

(a) Show that this is a redox reaction. Which species is oxidized and which is reduced?
(a) How many moles of Fe_2O_3 are required to produce 38.4 mol of Fe?
(b) How many grams of CO are required to produce 38.4 mol of Fe?

4.84 Explain this statement: Our modern civilization depends on the heat obtained from chemical reactions.

4.85 When an aqueous solution of Na_3PO_4 is added to an aqueous solution of $Cd(NO_3)_2$, a precipitate forms. Write a net ionic equation for this reaction, and identify the spectator ions.

4.86 The active ingredient in an analgesic tablet is 488 mg of aspirin, $C_9H_8O_4$. How many moles of aspirin does the tablet contain?

4.87 Chlorophyll, the compound responsible for the green color of leaves and grasses, contains one atom of magnesium in each molecule. If the percentage by weight of magnesium in chlorophyll is 2.72%, what is the molecular weight of chlorophyll?

4.88 If 7.0 kg of N_2 is added to 11.0 kg of H_2 to form NH_3, which reactant is in excess?

$$N_2(g) + 3H_2(g) \longrightarrow 2NH_3(g)$$

4.89 Lead(II) nitrate and aluminum chloride react according to the following equation:

$$3Pb(NO_3)_2 + 2AlCl_3 \longrightarrow 3PbCl_2 + 2Al(NO_3)_3$$

In an experiment, 8.00 g of lead nitrate was reacted with 2.67 g of aluminum chloride, and 5.55 g of lead chloride was formed.

(a) Which reactant was the limiting reagent?
(b) What was the percent yield?

4.90 Assume that the average red blood cell has a mass of 2×10^{-8} g and that 20% of its mass is hemoglobin (a protein whose molar mass is 68,000). How many molecules of hemoglobin are present in one red blood cell?

4.91 Reaction of pentane, C_5H_{12}, with oxygen, O_2, gives carbon dioxide and water.

(a) Write a balanced equation for this reaction.
(b) In this reaction, what is oxidized and what is reduced?
(c) What is the oxidizing agent and what is the reducing agent?

4.92 The two major sources of energy in our diets are fats and carbohydrates. Palmitic acid, one of the major components of both animal fats and vegetable oils, belongs to a group of compounds called fatty acids. The metabolism of these fatty acids is responsible for the energy from fats. The major carbohydrates in our diets are sucrose (table sugar; Section 19.5A) and starch (Section 19.6A). Both starch and sucrose are first converted in the body to glucose, and then glucose is metabolized to produce energy. The heat of combustion of palmitic acid is 2385 kcal/mol and that of glucose is 670 kcal/mol. Following are equations for the metabolism of each body fuel:

$$C_{16}H_{32}O_2(s) + O_2(g) \longrightarrow CO_2(g) + H_2O(\ell) + 2385 \text{ kcal/mol}$$
Palmitic acid
(256 g/mol)

$$C_6H_{12}O_6(s) + O_2(g) \longrightarrow CO_2(g) + H_2O(\ell) + 670 \text{ kcal/mol}$$
Glucose
(180 g/mol)

(a) Balance the equation for the metabolism of each fuel.
(b) Calculate the heat of combustion of each in kcal/g.
(c) In terms of kcal/mol, which of the two is the better source of energy for the body?
(d) In terms of kcal/g, which of the two is the better source of energy for the body?

InfoTrac College Edition

For additional readings, go to InfoTrac College Edition, your online research library, at

http://infotrac.thomsonlearning.com

CHAPTER 5

- 5.1 Introduction
- 5.2 Gas Pressure
- 5.3 Gas Laws
- 5.4 Avogadro's Law and the Ideal Gas Law
- 5.5 Dalton's Law of Partial Pressures
- 5.6 The Kinetic Molecular Theory
- 5.7 Forces of Attraction Between Molecules
- 5.8 Liquids
- 5.9 Vapor Pressure and Boiling Point
- 5.10 Solids
- 5.11 Phase Changes

Hot air balloon, Utah.

Gases, Liquids, and Solids

5.1 Introduction

Various forces hold matter together and cause it to take different forms. In an atomic nucleus, very strong forces of attraction keep the protons and neutrons together (Chapter 9). In an atom itself, there are attractions between the positive nucleus and the negative electrons that surround it. Within molecules, atoms are attracted to each other by covalent bonds, the arrangement of which causes the molecules to assume a particular shape. Within an ionic crystal, three-dimensional shapes arise because of electrostatic attractions between ions.

In addition to all these forces, there are also attractive forces between molecules. These forces, which are the subject of this chapter, are weaker than any of the forces already mentioned; nevertheless, they help to determine whether a particular compound is a solid, a liquid, or gas at room temperature.

These attractive forces hold matter together; in effect, they counteract another form of energy—kinetic energy—which tends to disorganize matter. In the absence of attractive forces, the kinetic energy that particles possess keeps them constantly moving, mostly in random, disorganized ways. This

119

CHEMICAL CONNECTION 5A

Entropy: a Measure of Disorder

As noted earlier, the molecules of a gas move randomly. The higher the temperature, the faster they move. In general, random motion means disorder. When the temperature decreases, the molecules slow down and more order becomes apparent. When all the molecules of a system become motionless and line up perfectly, we achieve the greatest possible order. At this point the substance is a solid. The measure of such order is called **entropy**. When the order is perfect, the entropy of the system is zero. When molecules rotate or move from one place to another, the disorder increases and so does the entropy. Thus when a crystal melts, entropy increases; when a liquid vaporizes, it likewise increases. When the temperature of a gas, a liquid, a solid, or a mixture of these states increases, the entropy of the system increases because an increase in temperature always increases molecular motions. When we combine two pure substances and they mix together, disorder increases—and so does the entropy.

We learned about the absolute, or kelvin, temperature scale in Section 1.4E. Absolute zero (0 K or −273°C) is the lowest possible temperature. Although scientists have not been able to reach absolute zero, they have been able to produce temperatures within a few billionths of a degree of it. At this low temperature, virtually all molecular motion ceases, almost perfect order reigns, and the

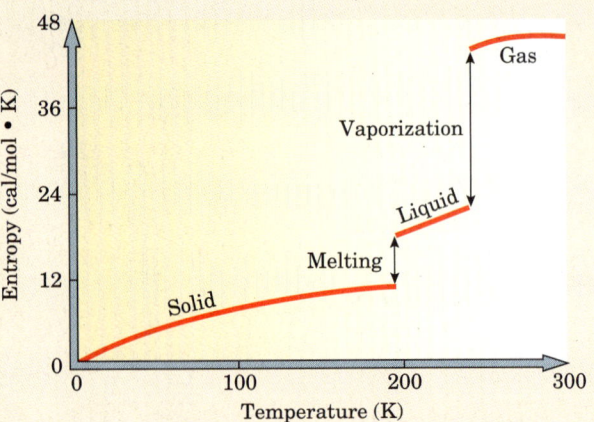

The entropy of ammonia, NH_3, as a function of absolute temperature. Note the large increase in entropy upon melting (change from a solid to a liquid) and upon vaporization (change from a liquid to a gas).

entropy of the substance is nearly zero. In fact, a completely ordered pure crystalline solid at 0 K has an entropy of zero.

---

kinetic energy increases with increasing temperature. Therefore, the higher the temperature, the greater the tendency of particles to fly randomly in every direction.

The physical state of matter thus depends on a balance between the kinetic energy of particles, which tend to keep them apart, and the attractive forces between them, which tend to bring them together (Figure 5.1).

Figure 5.1 The three states of matter. A gas has no definite shape, and its volume is the volume of the container. A liquid has a definite volume but no definite shape. A solid has a definite shape and a definite volume.

Gas
- Molecules far apart and disordered
- Negligible interactions between molecules

Liquid
- Intermediate situation

Solid
- Molecules close together and ordered
- Strong interactions between molecules

At high temperatures, molecules possess a high kinetic energy and move so fast that the attractive forces between them are too weak to hold the molecules together. This situation is called the **gaseous state.** At lower temperatures, molecules move more slowly, to the point where the forces of attraction between them become important. When the temperature is low enough, a gas condenses to form a **liquid state.** Molecules in the liquid state still move past each other, but they travel much more slowly than they do in the gaseous state. When the temperature is even lower, molecules no longer have enough velocity to move past each other. In this **solid state,** each molecule has a certain number of nearest neighbors, and these neighbors do not change.

The attractive forces between molecules are the same in all three states. The difference is that, in the gaseous state (and to a lesser degree in the liquid state), the kinetic energy of the molecules is great enough to overcome the attractive forces.

Most substances can exist in any of the three states. Typically a solid, when heated to a sufficiently high temperature, melts and becomes a liquid. The temperature at which this change takes place is called the melting point. Further heating causes the temperature to rise to the point at which the liquid boils and becomes a gas. This temperature is called the boiling point. Not all substances, however, can exist in all three states. For example, wood and paper cannot be melted. On heating, they either decompose or burn (depending on whether air is present), but they do not melt. Another example is sugar, which does not melt when heated but rather forms a dark substance called caramel.

5.2 Gas Pressure

On Earth we live under a blanket of air that presses down on us and on everything else around us. As we all know from weather reports, the **pressure** of the atmosphere varies from day to day. We use an instrument called a **barometer** (Figure 5.2) to measure atmospheric pressure.

Pressure is most commonly measured in **millimeters of mercury (mm Hg).** Pressure is also measured in **torr,** a unit named in honor of the Italian physicist and mathematician Evangelista Torricelli (1608–1647), who invented the barometer. At sea level the average pressure of the atmosphere is

> **Pressure** The force per unit area exerted against a surface

Figure 5.2 A mercury barometer. A long glass tube completely filled with mercury is inverted into a pool of mercury in a dish. Because there is no air at the top of the mercury column in the tube (there is no way air could get in), no gas pressure is exerted on the mercury column. The entire atmosphere, however, exerts its pressure on the mercury in the open dish. The difference in the heights of the two mercury levels is the atmospheric pressure.

Figure 5.3 A manometer consists of a U-shaped tube containing mercury. Arm A has been evacuated, sealed, and has zero pressure. Arm B is connected to the container in which the gas sample is enclosed. The pressure of the gas depresses the level of the mercury in arm B. The difference between the two mercury levels gives the pressure directly in mm Hg. If more gas is added to the sample container, the mercury level in B will be pushed down and that in A will rise as the pressure in the bulb increases.

760 mm Hg. We use this number to define still another unit of pressure, the **atmosphere (atm)**.

There are several other units to measure pressure. The SI unit is the pascal, and meteorologists report pressure in inches of mercury. In this book we use only mm Hg and atm.

1 atm = 760 mm Hg
= 760 torr
= 101,325 pascals
= 28.96 in. Hg

A barometer is adequate for measuring the pressure of the atmosphere, but to measure the pressure of a gas in a container, we use a simpler instrument called a **manometer.** One form of manometer consists of a U-shaped tube containing mercury (Figure 5.3).

5.3 Gas Laws

These gas laws hold not only for pure gases but also for mixtures of gases.

By observing the behavior of gases under different sets of temperatures and pressures, scientists have established a number of relationships. In this section, we study three of the most important.

A Boyle's Law and the Pressure–Volume Relationship

Robert Boyle (1627–1691), author of *The Sceptical Chymist* (1661), studied the expansion and compression of gases.

Boyle's law states that for a fixed mass of gas at a constant temperature, the volume of the gas is inversely proportional to the pressure. If the pressure doubles, for example, the volume decreases by one-half. This law can be stated mathematically in the following equation, where P_1 and V_1 are the initial pressure and volume, and P_2 and V_2 are the final pressure and volume:

$$PV = \text{constant} \quad \text{or} \quad P_1V_1 = P_2V_2$$

This relationship between pressure and volume accounts for what happens in the cylinder of a car engine (Figure 5.4).

CHEMICAL CONNECTIONS 5B

Breathing and Boyle's Law

Under normal resting conditions, we breathe about 12 times per minute, each time inhaling and exhaling about 500 mL of air. When we inhale, we lower the diaphragm or raise the rib cage, either of which increases the volume of the chest cavity. In accord with Boyle's law, as the volume of the chest cavity increases, the pressure within it decreases and becomes lower than the outside pressure. As a result, air flows from the higher-pressure area outside the body into the lungs. While the difference in these two pressures is only about 3 mm Hg, it is enough to cause air to flow into the lungs. In exhaling, we reverse the process: We raise the diaphragm or lower the rib cage. The resulting decrease in volume increases the pressure inside the chest cavity, causing air to flow out of the lungs.

In certain diseases the chest becomes paralyzed, and the affected person cannot move either the diaphragm or the rib cage. In such a case, a respirator is used to help the person breathe. The respirator first pushes down on the chest cavity and forces air out of the lungs. The pressure of the respirator is then lowered below atmospheric pressure, causing the rib cage to expand and draw air into the lungs.

Schematic drawings of the chest cavity. (*a*) The lungs fill with air. (*b*) Air empties from the lungs.

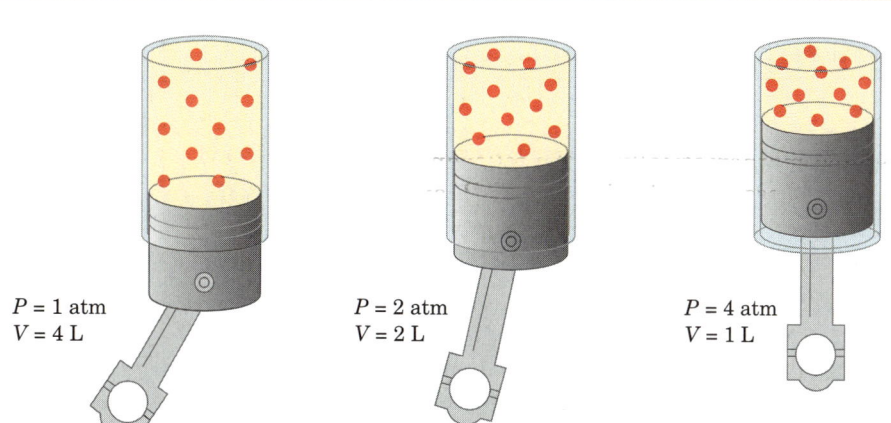

Figure 5.4 Boyle's law illustrated in an automobile cylinder. When the piston moves up, the volume occupied by the gas (a mixture of air and gasoline vapor) decreases and the pressure increases. The spark plug then ignites the compressed gasoline vapor.

Figure 5.5 Charles's law illustrated in a hot-air balloon. Because the balloon can stretch, the pressure inside it remains constant. When the air in the balloon is heated, its volume increases, expanding the balloon. As the air in the balloon expands, it becomes less dense than the surrounding air, providing the lift for the balloon. (Charles was one of the first balloonists.)

Although Jacques Charles (1746–1823) had earlier formulated a crude version of his law, it was first stated mathematically by Joseph Gay-Lussac (1778–1850) in 1802.

When using the gas laws, temperature must be expressed in kelvins (K). The zero in this scale is the lowest possible temperature.

■ **An autoclave used to sterilize hospital equipment.**

B Charles's Law and the Temperature–Volume Relationship

Charles's law states that the volume of a fixed mass of gas at a constant pressure is directly proportional to the temperature in kelvins (K). In other words, as long as the pressure on a gas remains constant, increasing the temperature of the gas causes an increase in the volume occupied by the gas. Charles's law can be stated mathematically this way:

$$\frac{V}{T} = \text{a constant} \quad \text{or} \quad \frac{V_1}{T_1} = \frac{V_2}{T_2}$$

This relationship between volume and temperature is the basis of the hot-air balloon's operation (Figure 5.5).

C Gay-Lussac's Law and the Temperature–Pressure Relationship

For a fixed mass of a gas at constant volume, the pressure is directly proportional to the temperature in kelvins:

$$\frac{P}{T} = \text{a constant} \quad \text{or} \quad \frac{P_1}{T_1} = \frac{P_2}{T_2}$$

As the temperature of the gas increases, the pressure increases proportionately. Consider, for example, what happens inside an autoclave. Steam generated inside an autoclave at 1 atm pressure has a temperature of 100°C. As the steam is heated further, the pressure within the autoclave increases as well. A valve controls the pressure inside the autoclave; if the pressure exceeds the designated maximum, the valve opens, releasing the steam. At

Table 5.1 Mathematical Expressions of the Three Gas Laws for a Fixed Mass of Gas

Name	Expression	Constant
Boyle's law	$P_1V_1 = P_2V_2$	T
Charles's law	$\dfrac{V_1}{T_1} = \dfrac{V_2}{T_2}$	P
Gay-Lussac's law	$\dfrac{P_1}{T_1} = \dfrac{P_2}{T_2}$	V

maximum pressure, the temperature may reach as high as 120° to 150°C. All microorganisms in the autoclave are destroyed at such temperatures.

Table 5.1 shows mathematical expressions of these three laws.

The three gas laws can be combined and expressed by a mathematical equation called the **combined gas law**:

$$\frac{PV}{T} = \text{a constant} \qquad \text{or} \qquad \frac{P_1V_1}{T_1} = \frac{P_2V_2}{T_2}$$

Combined gas law The pressure, volume, and temperature in kelvins of two samples of the same gas are related by the equation $P_1V_1/T_1 = P_2V_2/T_2$.

EXAMPLE 5.1

A gas occupies 3.00 L at 2.00 atm pressure. Calculate its volume when we increase the pressure to 10.15 atm at the same temperature.

Solution

First we identify the known quantities. Because T_1 and T_2 are the same in this example and consequently cancel each other, we don't need to know the temperature.

$$\text{Initial: } P_1 = 2.00 \text{ atm} \qquad V_1 = 3.00 \text{ L}$$
$$\text{Final: } P_2 = 10.15 \text{ atm} \qquad V_2 = ?$$

We want the final volume, so we solve the combined gas law equation for V_2:

$$V_2 = \frac{P_1V_1\cancel{T_2}}{\cancel{T_1}P_2} = \frac{(2.00 \; \cancel{\text{atm}})(3.00 \text{ L})}{10.15 \; \cancel{\text{atm}}} = 0.591 \text{ L}$$

Problem 5.1

A gas occupies 3.8 L at 0.70 atm pressure. If we expand the volume at constant temperature to 6.5 L, what is the final pressure?

EXAMPLE 5.2

In an autoclave, steam at 100°C is generated at 1.00 atm. After the autoclave is closed, the steam is heated at constant volume until the pressure gauge indicates 1.13 atm. What is the final temperature in the autoclave?

Solution

All temperatures in gas law calculations must be in kelvins; therefore, we must convert the Celsius temperature to kelvins:

$$100°C = 100 + 273 = 373 \text{ K}$$

We then identify the known quantities. Because V_1 and V_2 are the same in this example and consequently cancel each other, we don't need to know the volume of the autoclave.

$$\text{Initial: } P_1 = 1.00 \text{ atm} \qquad T_1 = 373 \text{ K}$$
$$\text{Final: } P_2 = 1.13 \text{ atm} \qquad T_2 = ?$$

Because we want a final temperature, we now solve the combined gas law equation for T_2:

$$T_2 = \frac{P_2 V_2 T_1}{P_1 V_1} = \frac{(1.13 \text{ atm})(373 \text{ K})}{1.00 \text{ atm}} = 421 \text{ K}$$

The final temperature is 421 K, or $421 - 273 = 148°C$.

Problem 5.2

A constant volume of oxygen gas, O_2, is heated from 120°C to 212°C. The final pressure is 20.3 atm. What was the initial pressure?

EXAMPLE 5.3

A gas in a flexible container has a volume of 0.50 L and a pressure of 1.0 atm at 393 K. When the gas is heated to 500 K, its volume expands to 3.0 L. What is the new pressure of the gas in the flexible container?

Solution
The known quantities are

$$\text{Initial: } P_1 = 1.0 \text{ atm} \qquad V_1 = 0.50 \text{ L} \qquad T_1 = 393 \text{ K}$$
$$\text{Final: } P_2 = ? \qquad V_2 = 3.0 \text{ L} \qquad T_2 = 500 \text{ K}$$

Solving the combined gas law for P_2, we find

$$P_2 = \frac{P_1 V_1 T_2}{T_1 V_2} = \frac{(1.0 \text{ atm})(0.50 \text{ L})(500 \text{ K})}{(3.0 \text{ L})(393 \text{ K})} = 0.21 \text{ atm}$$

Problem 5.3

A gas is expanded from an initial volume of 20.5 L at 0.92 atm at room temperature (23.0°C) to a final volume of 340.6 L. During the expansion the gas cools to 12.0°C. What is the new pressure?

5.4 Avogadro's Law and the Ideal Gas Law

The relationship between the amount of gas present and its volume is described by **Avogadro's law,** which states that equal volumes of gases at the same temperature and pressure contain equal numbers of molecules. Thus, if the temperature, pressure, and volumes of two gases are the same, then the two gases contain the same number of molecules, regardless of their identities (Figure 5.6). Avogadro's law is valid for all gases, no matter what they are.

The actual temperature and pressure at which we compare two or more gases don't matter. It is convenient, however, to select one temperature and one pressure as standard, and chemists have chosen 1 atm as the standard pressure and 0°C (273 K) as the standard temperature. These conditions are called **standard temperature and pressure (STP).**

All gases at STP or any other combination of temperature and pressure contain the same number of molecules in any given volume. But how many is that? In Chapter 4 we saw that a mole contains 6.02×10^{23} formula units. What volume of a gas at STP contains one mole of molecules? This quantity has been measured experimentally and found to be 22.4 L. Thus one mole of any gas at STP occupies a volume of 22.4 L.

Avogadro's law allows us to write a gas law that is valid not only for any pressure, volume, and temperature, but also for any quantity of gas. This law, called **the ideal gas law,** is

$$PV = nRT$$

where

P = pressure of the gas in atmospheres (atm)
V = volume of the gas in liters (L)
n = amount of the gas in moles (mol)
T = temperature of the gas in kelvins (K)
R = a constant for all gases, called the **ideal gas constant**

■ Molar volume. The cube has a volume of 22.4 L, which is the volume of one mole of a gas at STP.

Avogadro's law Equal volumes of gases at the same temperature and pressure contain the same number of molecules

Standard temperature and pressure (STP) One atmosphere pressure and 0°C (273 K)

Ideal gas law $PV = nRT$

Ideal gas constant (R) $0.0821 \; L \cdot atm \cdot mol^{-1} \cdot K^{-1}$

Figure 5.6 Avogadro's law. Two tanks of gas of equal volume at the same temperature and pressure contain the same number of molecules.

We can find the value of R by using the fact that one mole of any gas at STP occupies a volume of 22.4 L:

$$R = \frac{PV}{nT} = \frac{(1.00 \text{ atm})(22.4 \text{ L})}{(1.00 \text{ mol})(273 \text{ K})} = 0.0821 \frac{\text{L} \cdot \text{atm}}{\text{mol} \cdot \text{K}}$$

Real gases behave most like ideal gases at low pressures (1 atm or less) and high temperatures (300 K or higher).

The ideal gas law holds for all ideal gases at any temperature, pressure, and volume. But the only gases we have around us in the real world are real gases. How valid is it to apply the ideal gas law to real gases? The answer is that, under most experimental conditions, real gases behave sufficiently like ideal gases that we can use the ideal gas law for them with little trouble. Thus, using $PV = nRT$, we can calculate any one quantity—P, V, T, or n—if we know the other three quantities.

Screen 5.5
Simulation Tutorial

EXAMPLE 5.4

One mole of CH_4 gas occupies 20.0 L at 1.00 atm pressure. What is the temperature of the gas in kelvins?

Solution
Solve the ideal gas law for T and plug in the given values:

$$T = \frac{PV}{nR} = \frac{(1.00 \text{ atm})(20.0 \text{ L})}{(1.00 \text{ mol})(0.0821 \text{ L} \cdot \text{atm} \cdot \text{mol}^{-1} \cdot \text{K}^{-1})} = 244 \text{ K}$$

Note that we calculated the temperature for 1.00 mol of CH_4 gas under these conditions. The answer would be the same for 1.00 mol of CO_2, N_2, NH_3, or any other gas under these conditions.

Problem 5.4
If 2.00 mol of NO gas occupies 10.0 L at 295 K, what is the pressure of the gas in atm?

EXAMPLE 5.5

If there is 5.0 g of CO_2 gas in a 10-L cylinder at 350 K, what is the gas pressure within the cylinder?

Solution
We are given the quantity of CO_2 in grams but, to use the ideal gas law, we must express the quantity in moles. Therefore, we must first convert grams of CO_2 to moles, and then use this value in the ideal gas law. To convert from grams to moles, we use the conversion factor 1 mol CO_2 = 44 g.

$$5.0 \text{ g } CO_2 \times \frac{1 \text{ mol } CO_2}{44 \text{ g } CO_2} = 0.114 \text{ mol } CO_2$$

We now use this value in the ideal gas equation to solve for the pressure of the gas.

$$P = \frac{nRT}{V}$$

$$= \frac{(0.114 \text{ mol } CO_2)(0.0821 \text{ L} \cdot \text{atm} \cdot \text{mol}^{-1} \cdot \text{K}^{-1})(350 \text{ K})}{(10 \text{ L})} = 0.32 \text{ atm}$$

CHEMICAL CONNECTIONS 5C

Hyperbaric Medicine

Ordinary air contains 21% oxygen. Under certain conditions, the cells of tissues can become starved for oxygen (hypoxia), and quick oxygen delivery is needed. Increasing the percentage of oxygen in the air supplied to a patient is one way to remedy this situation, but sometimes even breathing pure (100%) oxygen may not be enough. For example, in carbon monoxide poisoning, hemoglobin, which normally carries most of the O_2 from the lungs to the tissues, binds CO and cannot take up any O_2 in the lungs. Without any help, the tissues would soon be starved for oxygen and the patient would die. When oxygen is administered under a pressure of 2 to 3 atm, it dissolves in the plasma to such a degree that the tissues receive enough of it to recover without the help of the poisoned hemoglobin molecules. Other conditions for which hyperbaric medicine is used are gas gangrene, smoke inhalation, cyanide poisoning, skin grafts, and thermal burns.

Breathing pure oxygen for prolonged periods, however, is toxic. For example, if O_2 is administered at 2 atm for more than 6 hours, both lung tissue and the central nervous system may be damaged. In addition, this treatment may cause nuclear cataract

■ Hyperbaric oxygen chamber at Medical City Dallas Hospital.

formation, necessitating postrecovery eye surgery. Therefore, recommended exposures of O_2 are 2 hours at 2 atm and 90 minutes at 3 atm. The benefits of hyperbaric medicine must be carefully weighed against these and other contraindications.

Problem 5.5
A certain quantity of neon gas is under 1.05 atm pressure at 303 K in a 10.0-L vessel. How many moles of neon are present?

EXAMPLE 5.6

If 3.3 g of a gas at 40°C and 1.15 atm pressure occupies a volume of 1.0 L, what is the mass of one mole of the gas?

Solution
This problem is more complicated than previous ones. We are given grams of gas and P, T, and V values. We are asked to calculate the mass of one mole of the gas (g/mol). We can solve this problem in two steps.

Step 1 Use the P, V, and T measurements and the ideal gas law to calculate the number of moles of gas present in the sample. To use the ideal gas law, we must first convert 40°C to kelvins: 40 + 273 = 313 K.

$$n = \frac{PV}{RT} = \frac{(1.15 \text{ atm})(1.0 \text{ L})}{(0.0821 \text{ L} \cdot \text{atm} \cdot \text{mol}^{-1} \cdot \text{K}^{-1})(313 \text{ K})} = 0.0448 \text{ mol}$$

Step 2 Calculate the mass of one mole of the gas by dividing grams by moles.

$$\text{Mass of one mole} = \frac{3.3 \text{ g}}{0.0448 \text{ mol}} = 74 \text{ g} \cdot \text{mol}^{-1}$$

Problem 5.6

An unknown amount of He gas occupies 30.5 L at 2.00 atm pressure and 300 K. What is the mass of the gas in the container?

5.5 Dalton's Law of Partial Pressures

In a mixture of gases, each molecule acts independently of all the others, assuming that the gases behave as ideal gases and that the gases do not react with each other. For this reason, the ideal gas law works for mixtures of gases as well as for pure gases. **Dalton's law of partial pressures** states that the total pressure, P_T, of a mixture of gases is the sum of the partial pressures of each individual gas:

$$P_T = P_1 + P_2 + P_3 + \cdots$$

Partial pressure The pressure that a gas in a mixture of gases would exert if it were alone in the container

The **partial pressure** of a gas in a mixture is the pressure that the gas would exert if it were alone in the container. The equation holds separately for each gas in the mixture as well as for the mixture as a whole.

EXAMPLE 5.7

To a tank containing N_2 at 2.0 atm and O_2 at 1.0 atm, we add an unknown quantity of CO_2 until the total pressure within the tank is 4.6 atm. What is the partial pressure of the CO_2?

Solution
Dalton's law tells us that the addition of CO_2 does not affect the partial pressures of the N_2 or O_2 already present in the tank. The partial pressures of N_2 and O_2 remain at 2.0 atm and 1.0 atm, respectively, and their sum is 3.0 atm. If the new pressure is 4.6 atm, the partial pressure of the added CO_2 must be 1.6 atm. Thus, when the final pressure is 4.6 atm, the partial pressures are

4.6 atm	=	2.0 atm	+	1.0 atm	+	1.6 atm
Total pressure		Partial pressure of N_2		Partial pressure of O_2		Partial pressure of CO_2

Problem 5.7

A vessel under 2.015 atm pressure contains nitrogen, N_2, and water vapor, H_2O. The partial pressure of N_2 is 1.908 atm. What is the partial pressure of the water vapor?

5.6 The Kinetic Molecular Theory

To this point, we have studied the macroscopic properties of gases—namely, the various laws dealing with the relationships among temperature, pressure, volume, mass, and number of molecules in a sample of gas. Now let us examine the behavior of gases at the molecular level and see how we can explain their behavior in terms of molecules and the interactions between them.

The relationship between the observed behavior of gases and the behavior of individual gas molecules within the gas can be explained by the **kinetic molecular theory,** which makes the following assumptions about the molecules of a gas:

1. Gases consist of particles, either atoms or molecules, constantly moving through space in straight lines, in random directions, and with various speeds. Because these particles move in random directions, different gases mix very readily.
2. The average kinetic energy of gas particles is proportional to the temperature in kelvins. The higher the temperature, the greater the kinetic energy of gas particles, and the faster they move through space.
3. Molecules collide with each other, much as billiard balls do, bouncing off each other and changing direction. Each time they collide, they may exchange kinetic energies (one moves faster than before, the other slower), but the total kinetic energy of the gas sample remains the same.
4. Gas particles have no volume. Most of the volume taken up by a gas is empty space, which explains why gases can be compressed so easily.
5. There are no attractive forces between gas particles. They do not stick together after a collision occurs.
6. Molecules collide with the walls of the container, and these collisions constitute the pressure of the gas. The greater the number of collisions per unit time, the greater the pressure.

These six assumptions of the kinetic molecular theory give us an idealized picture of the molecules of a gas and their interactions with one another (Figure 5.7). In real gases there are forces of attraction between molecules, and molecules really do occupy some volume. Because of these factors, a gas described by these six assumptions of the kinetic molecular theory is called an **ideal gas.** In reality, there is no ideal gas; all gases are real. At STP, however, most real gases behave in much the same way that an ideal gas would, so we can safely use these assumptions.

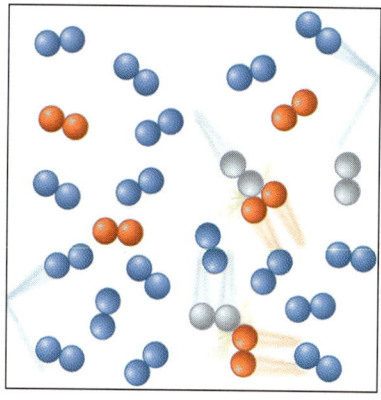

Figure 5.7 The kinetic molecular model of a gas. Molecules of nitrogen (blue) and oxygen (red) are in constant motion and collide with each other and with the walls of the container. Collisions with the walls of the container cause gas pressure. In air at STP, 6.02×10^{23}, molecules undergo approximately 10 billion collisions per second.

5.7 Forces of Attraction Between Molecules

As noted in Section 5.1, the strength of the intermolecular forces (forces between molecules) in any sample of matter determines whether the sample is a gas, liquid, or solid under normal conditions of temperature and pressure. In general, the closer the molecules are to each other, the greater the effect of the intermolecular forces. When the temperature of a gas is high (room temperature or higher) and the pressure is low (1 atm or less), molecules of the gas are so far apart that we can effectively ignore attractions between them and treat the gas as ideal. When the temperature decreases, the pressure increases, or both, the distances between molecules decrease so that we can no longer ignore intermolecular forces. As a matter of fact, these forces become so important that they cause **condensation** (change from a gas to a liquid) and **solidification** (change from a liquid to a solid). Therefore, before discussing the structures and properties of liquids and solids, we must look at the nature of these intermolecular forces of attraction.

In this section, we discuss three types of intermolecular forces: London dispersion force, dipole–dipole interactions, and hydrogen bonding.

Condensation The change of a substance from the vapor or gaseous state to the liquid state

Table 5.2	Three Types of Forces of Attraction between Molecules	
Intermolecular forces of attraction between molecules	London dispersion forces	0.01–2 kcal/mol
	Dipole–dipole interactions	1–6 kcal/mol
	Hydrogen bonding	2–10 kcal/mol
Covalent bonds	Single, double, and triple covalent bonds	70–200 kcal/mol

For purposes of comparison, the strengths of single, double, and triple covalent bonds are also given.

Table 5.2 shows the strengths of these forces. Also shown for comparison are the strengths of covalent bonds. Note that covalent bonds are considerably stronger than any of the three types of intermolecular forces. Although intermolecular forces are relatively weak compared to covalent bonds, it is the former that determine many of the physical properties of molecules, such as melting point, boiling point, and viscosity. As we will see in Chapters 20–31, these forces are also extremely important in determining the three-dimensional shapes of biomolecules such as proteins and nucleic acids, and how these types of biomolecules recognize and interact with one another.

A London Dispersion Forces

There are attractive forces between all molecules, whether they are polar or nonpolar. If the temperature falls far enough, even nonpolar molecules such as helium, He, and hydrogen, H_2, can be liquefied. The fact that these and other nonpolar gases can be liquefied means that some sort of interaction must occur between them to make them stick together in the liquid state. These weak attractive forces are called **London dispersion forces,** after the American chemist Fritz London (1900–1954), who studied them extensively.

London dispersion forces have their origin in electrostatic interactions. To visualize the origin of these forces, it is necessary to think in terms of instantaneous distributions of electrons within an atom or molecule. Consider, for example, a sample of neon atoms. Neon is a gas at room temperature and atmospheric pressure. It can be liquefied if cooled to −246°C. Over time, the distribution of electron density in a neon atom is symmetrical, and a neon atom has no dipole; that is, it has no separation of positive and negative charges. However, at any given instant, the electron density in a neon atom may be shifted more toward one part of the atom than another, thereby creating temporary dipoles (Figure 5.8). These temporary dipoles, which last for only tiny fractions of a second, induce temporary dipoles in adjacent neon atoms. The attractions between these temporary induced dipoles are called London dispersion forces, and they make nonpolar molecules stick together to form the liquid state.

London dispersion forces exist between all molecules, but they are the only forces of attraction between nonpolar molecules. They range in strength from 0.01 to 2.0 kcal/mol depending on the mass, size, and shape of the interacting molecules. In general, their strength increases as the mass and number of electrons in a molecule increases. Even though London dispersion forces are very weak, they contribute significantly to the attractive forces between large molecules because they act over large surface areas.

London dispersion forces
Extremely weak attractive forces between atoms or molecules caused by the electrostatic attraction between temporary induced positive and negative dipoles

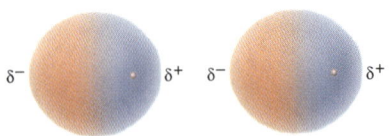

Figure 5.8 London dispersion forces. A temporary polarization of electron density in one neon atom creates positive and negative dipoles, which in turn induce temporary positive and negative dipoles in an adjacent atom. The intermolecular attractions between temporary dipoles of opposite charge are called London dispersion forces.

B Dipole–Dipole Interactions

As mentioned in Section 3.10, may molecules are dipoles. The attraction between the positive end of one dipole and the negative end of another dipole is called a **dipole–dipole interaction.** These interactions can exist between two identical polar molecules or between two different polar molecules. To see the importance of dipole–dipole interactions, we can look at the differences in boiling points between nonpolar and polar molecules of comparable molecular weight. Butane, C_4H_{10}, with a molecular weight of 58 amu, is a nonpolar molecule with a boiling point of 0.5°C. Acetone, C_3H_6O, with the same molecular weight, has a boiling point of 58°C. Acetone is a polar molecule, and its molecules are held together in the liquid state by dipole–dipole attractions between the negative oxygen dipole of one acetone molecule and the positive carbon dipole of another acetone molecule. Because it requires more energy to break the dipole-dipole interactions between acetone molecules than it does to break the considerably weaker London dispersion forces between butane molecules, acetone has a higher boiling point than butane.

> **Dipole–dipole attraction** The attraction between the dipole of one molecule and a dipole of opposite charge of another molecule

$$CH_3-CH_2-CH_2-CH_3 \qquad CH_3-\underset{\delta+}{\overset{\overset{\displaystyle O\;\delta-}{\|}}{C}}-CH_3$$

Butane
(bp 0.5°C)

Acetone
(bp 58°C)

C Hydrogen Bonds

As we have just seen, the attraction between the positive end of one dipole and the negative end of another results in dipole–dipole attraction. When the positive end of one dipole is a hydrogen atom bonded to an O or N (atoms of high electronegativity; see Table 3.5) and the negative end of the other dipole is an O or N atom, the attractive interaction between dipoles is particularly strong and is given the special name of **hydrogen bond.** An example is the hydrogen bonding that occurs between molecules of water in both the liquid and solid states (Figure 5.9).

The strength of hydrogen bonding ranges from 2 to 10 kcal/mol. The strength in liquid water, for example, is approximately 5 kcal/mol. By comparison, the strength of the O—H covalent bond in water is approximately 119 kcal/mol. As can be seen by comparing these numbers, an O----H

Screen 5.10
Exercise Tutorial

> **Hydrogen bond** A noncovalent force of attraction between the partial positive charge on a hydrogen atom bonded to an atom of high electronegativity, most commonly oxygen or nitrogen, and the partial negative charge on a nearby oxygen or nitrogen

Figure 5.9 Two water molecules joined by a hydrogen bond. (a) Structural formulas, (b) ball-and-stick models, and (c) electron density models.

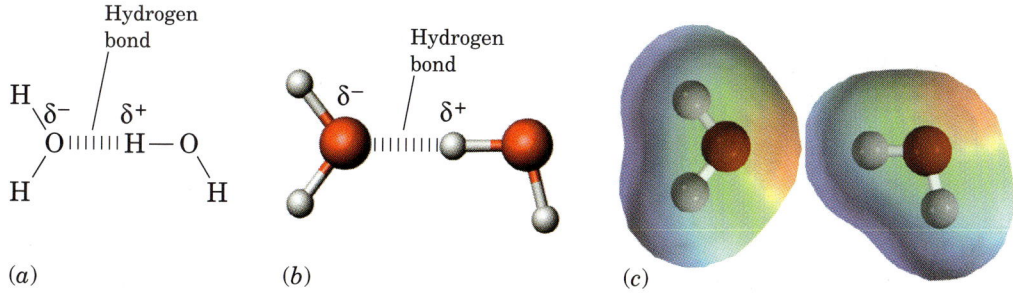

hydrogen bond is considerably weaker than an O—H covalent bond. Nonetheless, the presence of hydrogen bonds in liquid water has an important effect on the physical properties of water. ==Because of hydrogen bonding, extra energy is required to separate each water molecule from its neighbors—hence the relatively high boiling point of water.== As we will see in later chapters, hydrogen bonds play an especially important role in biological molecules.

Hydrogen bonds are not restricted to water, however. They form between two molecules whenever one molecule has a hydrogen atom covalently bonded to O or N, and the other molecule has an O or N atom.

EXAMPLE 5.8

Can a hydrogen bond form between
(a) Two molecules of methanol, CH_3OH?
(b) Two molecules of formaldehyde, CH_2O?
(c) One molecule of methanol, CH_3OH, and one of formaldehyde, CH_2O?

Solution

(a) Yes. Methanol is a polar molecule and has a hydrogen covalently bonded to an oxygen atom.

(b) No. Although formaldehyde is a polar molecule, it does not have a hydrogen covalently bonded to an oxygen or nitrogen atom.

(c) Yes. Methanol has a hydrogen atom bonded to an oxygen atom and formaldehyde has an oxygen atom.

> **Problem 5.8**
> Will the molecules in each of the following pairs form a hydrogen bond between them?
> (a) A molecule of water and a molecule of methanol, CH₃OH
> (b) Two molecules of methane, CH₄

5.8 Liquids

We have seen that we can describe the behavior of gases under most circumstances by the ideal gas law, which assumes that there are no attractive forces between molecules. As pressure increases in a real gas, however, the molecules of the gas become squeezed into a smaller space, with the result that attractions between molecules become more and more effective in causing molecules to stick together.

If the distances between molecules decrease so that almost all the molecules touch or almost touch each other, the gas condenses to a liquid. Unlike gases, liquids do not fill all the available space, but they do have a definite volume, irrespective of the container. Because gases have a lot of empty space between molecules, it is easy to compress them into a smaller volume. There is very little empty space in liquids; consequently, they are difficult to compress. A great increase in pressure is needed to cause even a very small decrease in the volume of a liquid. Thus liquids, for all practical purposes, are incompressible. In addition, the density of liquids is much greater than that of gases because the same mass occupies a much smaller volume in liquid form than it does in gaseous form.

The position of the molecules in the liquid state is random, and some irregular empty space is available into which molecules can slide. The molecules therefore constantly change their position with respect to neighboring molecules. This property causes liquids to be fluid and explains why they do not have a constant shape, only a constant volume.

Unlike gases, liquids have surface properties, one of which is **surface tension** (Figure 5.10). The surface tension of a liquid is directly related to the strength of the intermolecular attraction between its molecules. Water has a high surface tension because of strong hydrogen bonding among water molecules. As a result, a steel needle can easily be made to float on the surface of water. However, if the same needle is pushed below the elastic skin into the interior of the liquid, it sinks to the bottom. Similarly, water bugs gliding on the surface of a pond appear to be walking on an elastic skin of water.

5.9 Vapor Pressure and Boiling Point

A Vapor Pressure

An important property of liquids is their tendency to evaporate. A few hours after a heavy rain, for example, most of the puddles have dried up. The water has evaporated, gone into the air. The same thing occurs if we leave a container of water or any other liquid out in the open. Let us explore how this change occurs.

In any liquid, there is a distribution of velocities among its molecules. Some of the molecules have high kinetic energy and move rapidly. Others have low kinetic energy and move slowly. Whether fast or slow, a molecule in the interior of the liquid cannot go very far before it hits another molecule

Figure 5.10 Surface tension. Molecules in the interior of a liquid have equal intermolecular attractions in every direction. Molecules at the liquid–gas interface, however, experience greater attractions toward the interior of the liquid than toward the gaseous state above it. Therefore, molecules on the surface are preferentially pulled toward the center of the liquid. This pull crowds the molecules on the surface, thereby creating a layer, like an elastic skin, that is tough to penetrate.

■ **A water strider standing on water.**

Screen 5.12
Simulation

CHEMICAL CONNECTIONS 5D

Blood Pressure Measurement

Liquids, like gases, exert a pressure on the walls of their containers. Blood pressure, for example, results from pulsating blood pushing against the walls of the blood vessels. When the heart ventricles contract, pushing blood out into the arteries, the blood pressure is high (systolic pressure); when the ventricles relax, the blood pressure is low (diastolic pressure). Blood pressure is usually expressed as a fraction showing systolic over diastolic pressure—for instance, 120/80. The normal range in young adults is 100 to 120 mm Hg systolic and 60 to 80 mm Hg diastolic. In older adults, the corresponding normal ranges are 115 to 135 and 75 to 85 mm Hg, respectively.

A sphygmomanometer—the instrument used to measure blood pressure—consists of a bulb, a cuff, a manometer, and a stethoscope. The cuff is wrapped around the upper arm and inflated by squeezing the bulb. It exerts a pressure on the arm, which is read on the manometer. When the cuff is sufficiently inflated, its pressure collapses the brachial artery, preventing pulsating blood from flowing to the lower arm. At this pressure, no sound is heard in the stethoscope because the applied pressure in the cuff is greater than the blood pressure. Next, the cuff is slowly deflated, which decreases the pressure on the arm. The first faint tapping sound is heard when the pressure in the cuff just matches the systolic pressure as the ventricle contracts—that is, when the pressure in the cuff is low enough to allow pulsating blood to begin flowing into the lower arm. As the cuff pressure continues to decrease, the tapping first becomes louder and then begins to fade. At the point when the last faint tapping sound is heard, the cuff pressure matches the diastolic pressure when the ventricle is relaxed, allowing continuous blood flow into the lower arm.

Digital blood pressure monitors are now available for home or office use. In these instruments, the stethoscope and manometer are combined in a sensory device that records the systolic and diastolic blood pressures together with the pulse rate. The cuff and the inflation bulb are used the same way as in traditional sphygmomanometers.

(a)   (b) Systolic pressure   (c) Diastolic pressure

■ Measurement of blood pressure.

and has its speed and direction changed by the collision. A molecule at the surface, however, is in a different situation (Figure 5.11). If it is moving slowly, it cannot escape from the liquid because of the attractions of neighboring molecules. If it is moving rapidly (has a high kinetic energy) and upward, however, it can escape from the liquid and enter the gaseous space above it.

In an open container, this process continues until all molecules have escaped. If the liquid is in a closed container, as in Figure 5.12, the molecules

Figure 5.11 Evaporation. Some molecules at the surface of a liquid are moving fast enough to escape into the gaseous space.

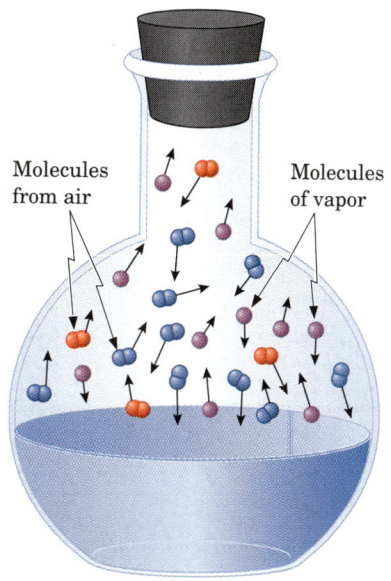

Figure 5.12 Evaporation and condensation. In a closed container, molecules of liquid escape into the vapor phase, and vapor molecules are recaptured by the liquid.

in the gaseous state cannot diffuse away (as they do if the container is open). Instead, they remain in the air space above the liquid, where they move rapidly in straight lines until they strike something. Some of these **vapor** molecules move downward, strike the surface of the liquid, and are recaptured by it.

At this point we have reached **equilibrium.** As long as the temperature does not change, the number of vapor molecules reentering the liquid equals the number escaping from it. At equilibrium, the space above the liquid shown in Figure 5.12 contains air and vapor molecules, and we can measure

> **Vapor** A gas

> **Equilibrium** A condition in which two opposing physical forces are equal

At equilibrium, the rate of vaporization is equal to the rate of liquefaction.

CHEMICAL CONNECTIONS 5E

Spray and Stretch—A New Use for Freons

Fluori-Methane, a liquid mixture of 85% Freon-11 (CCl_3F, bp 23.7°C) and 15% Freon-12 (CCl_2F_2, bp −29.8°C), is commonly used by physical therapists to treat restricted motion. Both compounds in Fluori-Methane are gases at STP, but can be liquefied under pressure. The liquid is sprayed on a trigger point and, because of the low boiling points of its two components, evaporates quickly, the skin cools, and the underlying muscles relax. As the affected muscles relax, they can then be stretched to increase their range of motion.

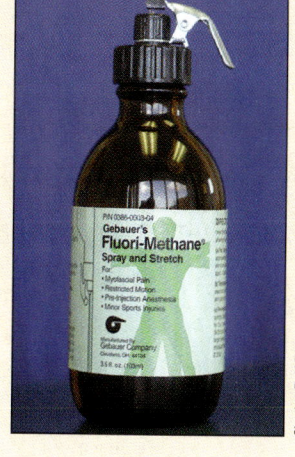

■ Fluori-Methane.

Figure 5.13 The change in vapor pressure with temperature for four liquids. The normal boiling point of a liquid is defined as the temperature at which the vapor pressure of the liquid equals 760 mm Hg.

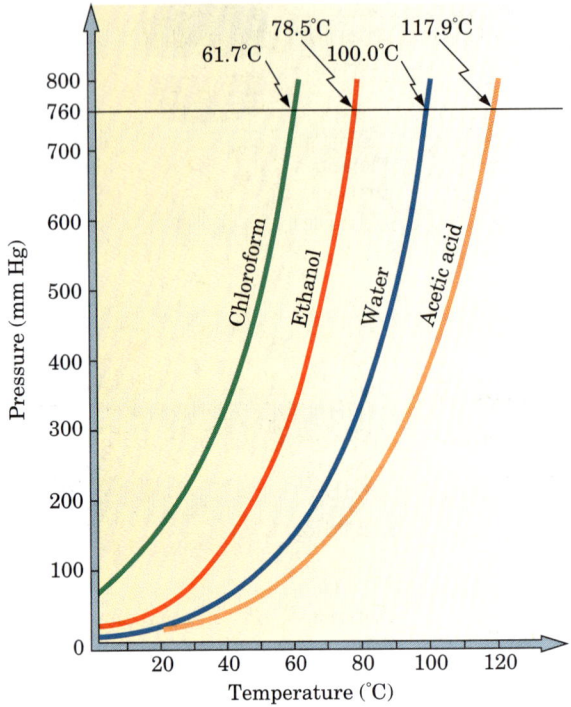

Vapor pressure The pressure of gas in equilibrium with its liquid form in a closed container

the partial pressure of the vapor. We call this measurement the **vapor pressure** of the liquid. Note that we measure the partial pressure of a gas but call it the vapor pressure of the liquid.

The vapor pressure of a liquid is a physical property of the liquid and a function of temperature (Figure 5.13). As the temperature of a liquid increases, the average kinetic energy of its molecules increases and the easier it becomes for molecules to escape from the liquid state to the gaseous state. As the temperature of the liquid increases, its vapor pressure continues to increase until it equals the atmospheric pressure. At this point, bubbles of vapor form under the surface of the liquid and then force their way upward through the surface of the liquid, and the liquid boils.

The molecules that evaporate from a liquid surface are those that have a higher kinetic energy. When they enter the gas phase, the molecules left behind are those with lower kinetic energy. Because the temperature of a sample is proportional to the average kinetic energy of its molecules, the temperature of the liquid drops as a result of evaporation. This evaporation of water from your skin produces the cooling effect you feel when you come out of a swimming pool and the layer of water evaporates from your skin.

B Boiling Point

Boiling point The temperature at which the vapor pressure of a liquid is equal to the atmospheric pressure

Normal boiling point The temperature at which a liquid boils under a pressure of 1 atm

The **boiling point** of a liquid is the temperature at which its vapor pressure is equal to the pressure of the atmosphere in contact with its surface. The boiling point when the atmospheric pressure is 1 atm is called the **normal boiling point.** For example, 100°C is the normal boiling point of water because it is the temperature at which water boils at 1 atm pressure.

The use of a pressure cooker is an example of boiling water at higher temperatures. In this type of pot, food is cooked at, say, 2 atm, where the boiling point of water is 121°C. Because the food has been raised to a higher

Table 5.3	Names, Molecular Formulas, Molecular Weights, and Normal Boiling Points for the Four Liquids in Figure 5.13		
Name	Molecular Formula	Molecular Weight (amu)	Boiling Point (°C)
Chloroform	$CHCl_3$	120	61.7
Ethanol	CH_3CH_2OH	46	78.5
Water	H_2O	18	100.0
Acetic acid	CH_3COOH	60	117.9

temperature, it cooks faster than it would in an open pot, in which boiling water cannot get hotter than 100°C. Conversely, at low pressures water boils at lower temperatures. For example, at the top of a mountain where the atmospheric pressure is less than 760 mm Hg, the boiling point of water might be 95°C.

C Factors That Affect Boiling Point

As Figure 5.13 shows, different liquids have different normal boiling points. Table 5.3 gives the molecular formulas, molecular weights, and normal boiling points for four liquids.

As you study the information in this table, note that chloroform, which has the largest molecular weight of the four compounds, has the lowest boiling point. Water, which has the lowest molecular weight, has the second highest boiling point. From a study of these and other compounds, chemists have determined that the boiling point of covalent compounds depends primarily on two factors: the nature and strength of their intermolecular forces, and their molecular shape.

1. **Intermolecular Forces** Water (H_2O, MW 18) and methane (CH_4, MW 16) have about the same molecular weight. The normal boiling point of water is 100°C, while that of methane is −164°C. The difference in boiling points reflects the fact that CH_4 molecules in the liquid state must overcome only the weak London dispersion forces to escape to the vapor state (low boiling point). In contrast, water molecules, being hydrogen-bonded to each other, need more kinetic energy (and a higher boiling temperature) to escape into the vapor phase. Thus the difference in boiling points between these two compounds is due to the greater strength of hydrogen bonding compared with the much weaker London dispersion forces.

   As another example, consider the boiling points of methane, CH_4 and hexane, C_6H_{14}. Both are nonpolar compounds with no possibility for hydrogen bonding or dipole–dipole interactions between their molecules. The only forces of attraction between molecules of either compound are London dispersion forces. The normal boiling point of hexane is 69°C, and that of methane is −164°C. The difference in their boiling points reflects the fact that hexane has more electrons and a larger surface area than methane. Because of its larger surface area, the London dispersion forces between hexane molecules are stronger than those between methane molecules and, therefore, hexane has the higher boiling point.

2. **Molecular Shape** When molecules are similar in every way except shape, the strength of their London dispersion forces determines their

CHEMICAL CONNECTIONS 5F

The Densities of Ice and Water

The hydrogen-bonded superstructure of ice has empty spaces in the middle of each hexagon because the H_2O molecules in ice are not as closely packed as those in liquid water. For this reason, ice has a lower density (0.917 g/cm^3) than does liquid water (1.00 g/cm^3). As ice melts, some of the hydrogen bonds are broken and the hexagonal superstructure of ice collapses to the more densely packed organization of water. This change explains why ice floats on top of water instead of sinking to the bottom. Such behavior is highly unusual—most substances are denser in the solid state than they are in the liquid state. The lower density of ice keeps fish and microorganisms alive in many rivers and lakes that would freeze solid each winter if the ice sank to the bottom. The presence of ice on top insulates the remaining water and keeps it from freezing.

The fact that ice has a lower density than water means that a given mass of ice takes up more space than the same mass of liquid water. This factor explains the damage done to biological tissues by freezing. When parts of the body (usually fingers, toes, nose, and ears) are subjected to extreme cold, they develop a condition called frostbite. The water in cells freezes despite the blood's attempt to keep the temperature at 37°C. As liquid water freezes, it expands and in doing so ruptures the cell walls, causing damage. In some cases, frostbitten fingers or toes must be amputated.

Cold weather can damage plants in a similar way. Many plants are killed when the air temperature drops below the freezing point of water for several hours. Trees can survive cold winters because they have a low water content inside their trunks and branches.

Slow freezing is often more damaging to plant and animal tissues than quick freezing is. In slow freezing, only a few crystals form and they can then grow to large sizes, rupturing the cells. In quick freezing, such as can be achieved in liquid nitrogen (at a temperature of $-196°C$), many tiny crystals form. Because they do not grow much, tissue damage may be minimal.

■ Snowflakes are six-sided structures, reflecting the underlying structure of ice. Ice consists of six-sided rings formed by water molecules, in which each side of a ring has two O atoms and an H atom.

relative boiling points. Consider pentane, bp 36.2°C, and 2,2-dimethylpropane, bp 9.5°C (Figure 5.14).

Both compounds have the same molecular formula, C_5H_{12}, and the same molecular weight, but the boiling point of pentane is approximately 26° higher than that of 2,2-dimethylpropane. These differences in boiling point are related to molecular shape in the following way. The only forces between these nonpolar molecules are London dispersion forces. Pentane is a roughly linear molecule, whereas 2,2-dimethylpropane has a spherical shape and a smaller surface area than pentane. As surface area decreases, contact between adjacent molecules, the strength of London dispersion forces, and boiling points all decrease.

$CH_3-CH_2-CH_2-CH_2-CH_3$ $CH_3-\underset{\underset{CH_3}{|}}{\overset{\overset{CH_3}{|}}{C}}-CH_3$

Pentane
(bp 36.2°C)

2,2-Dimethylpropane
(bp 9.5°C)

Figure 5.14 Pentane and 2,2-dimethylpropane have the same molecular formula, C_5H_{12}, but quite different shapes.

Consequently, London dispersion forces between molecules of 2,2-dimethylpropane are less than those between molecules of pentane and, therefore, 2,2-dimethylpropane has a lower boiling point.

5.10 Solids

When liquids are cooled, their molecules come so close together and attractive forces between them become so strong that random motion stops, and a solid forms. Formation of a solid from a liquid is called solidification or, alternatively, **crystallization.**

All crystals have a regular shape that, in many cases, is obvious to the eye (Figure 5.15). This regular shape often reflects the arrangement of the particles within the crystal. In table salt, for example, Na^+ and Cl^- ions are arranged in a cubic system (Figure 3.1). Metals also consist of particles arranged in a regular crystal lattice (generally not cubic), but here the particles are atoms rather than ions. Because the particles in a solid are almost always closer together than they are in the corresponding liquid, solids almost always have a higher density than liquids.

As can be seen in Figure 5.15, crystals have characteristic shapes and symmetries. We are familiar with the cubic nature of table salt and the hexagonal ice crystals in snowflakes. A less well-known fact is that some compounds have more than one type of solid state. The best-known example is the element carbon, which has five crystalline forms (Figure 5.16). Diamond occurs when solidification takes place under very high pressure (thousands of atmospheres). Another form of carbon is the graphite in a pencil. Carbon atoms are packed differently in high-density, hard diamonds than they are in low-density, soft graphite.

Even in the solid state, molecules and ions do not stop moving completely. They vibrate around fixed points.

Crystallization The formation of a solid from a liquid

Figure 5.15 Some crystals.

Garnet

Sulfur

Quartz

Pyrite

142 | CHAPTER 5 GASES, LIQUIDS, AND SOLIDS

(a)   (b)   (c)   (d)   (e)

Figure 5.16 Solid forms of carbon: (a) graphite, (b) diamond, (c) "buckyball," (d) nanotube, and (e) soot.

Richard E. Smalley (1943–), Robert F. Curl, Jr. (1933–), and Harold Kroto (1939–) were awarded the 1996 Nobel Prize in Chemistry for the discovery of these compounds.

In a third form of carbon, each molecule contains 60 carbon atoms arranged in a structure having 12 pentagons and 20 hexagons as faces, resembling a soccer ball [Figure 5.16(c)]. Because the famous architect Buckminster Fuller (1895–1983) invented domes of a similar structure (he called them geodesic domes), the C-60 substance was named buckminsterfullerene, or "buckyball" for short. The discovery of buckyball has generated a whole new area of carbon chemistry. Similar cage-like structures containing 72, 80, and even larger numbers of carbon have been synthesized. As a group, they are called fullerenes.

New variations on the fullerenes are nanotubes [Figure 5.16(d)]. The *nano-* part of the name comes from the fact that the cross section of each tube is only nanometers (10^{-9} m) across. Nanotubes come in a variety of forms. Single-walled carbon nanotubes can vary in diameter from 1 to 3 nm and are about 20 mm long. These compounds have generated great industrial interest because of their optical and electronic properties. They may play a role in miniaturization of instruments, giving rise to a new generation of nanoscale devices.

Soot is the fifth form of solid carbon. This substance solidifies directly out of carbon vapor and is an **amorphous solid;** that is, its atoms have no set pattern and are arranged randomly [Figure 5.16(e)]. Another example of an amorphous solid is glass. In essence, glass is a frozen liquid.

As we have now seen, some crystalline solids consist of orderly arrays of ions (ionic solids; Figure 3.1), and others consist of molecules (molecular solids). Ions are held in the crystal lattice by ionic bonds. Molecules are held only by intermolecular forces, which are much weaker than ionic bonds. Therefore, molecular solids generally have much lower melting points than do ionic solids.

Other types of solids exist as well. Some are extremely large molecules, with each molecule having as many as 10^{23} atoms, all connected by covalent

Table 5.4 Types of Solids			
Type	Made Up Of	Characteristics	Examples
Ionic	Ions in a crystal lattice	High melting point	NaCl, K_2SO_4
Molecular	Molecules in a crystal lattice	Low melting point	Ice, aspirin
Polymeric	Giant molecules; can be crystalline, semicrystalline, or amorphous	Low melting point or cannot be melted; soft or hard	Rubber, plastics, proteins
Network	A very large number of atoms connected by covalent bonds	Very hard; very high melting point or cannot be melted	Diamond, quartz
Amorphous	Randomly arranged atoms or molecules	Mostly soft, can be made to flow, but no melting point	Soot, tar, glass

bonds. In such a case, the entire crystal is one big molecule. We call such molecules **network solids** or network crystals. A good example is diamond [Figure 5.16(b)]. When you hold a diamond in your hand, you are holding a single gigantic molecule. Like ionic crystals, network solids have very high melting points—if they can be melted at all. In many cases they cannot be. Table 5.4 summarizes the types of solids.

5.11 Phase Changes

A The Heating Curve for $H_2O(s)$ to $H_2O(g)$

Imagine the following experiment: We heat a piece of ice that is initially at $-20°C$. At first we don't see any difference in its physical state. The temperature of the ice increases, but its appearance does not change. At $0°C$, the ice begins to melt and liquid water appears. As we continue heating, more and more of the ice melts, but the temperature stays constant at $0°C$ until all the ice has melted and only liquid water remains. After all the ice has melted, the temperature of the water again increases as heat is added. At $100°C$, the water boils. We continue heating as it continues to evaporate, but the temperature of the remaining liquid water does not change. Only after the liquid water has changed to gaseous water (steam) does the temperature of the sample rise above $100°C$.

These changes in state are called **phase changes**. A **phase** is any part of a system that looks uniform (homogeneous) throughout. Solid water (ice) is one phase, liquid water is another phase, and gaseous water is still another phase. Table 5.5 summarizes the energies for each step in the conversion of ice to steam.

Let us use the data in Table 5.5 to calculate the heat required to raise the temperature of 1.0 g of ice at $-20°C$ to water vapor at $120°C$. We begin with ice, whose specific heat is 0.48 cal/g·°C (Table 1.4). It requires $0.48 \times 20 = 9.6$ cal to raise the temperature of 1.0 g of ice from $-20°C$ to $0°C$.

$$0.48 \frac{\text{cal}}{\text{g} \cdot °C} \times 1.0 \text{ g} \times 20 °C = 9.6 \text{ cal}$$

After the ice reaches $0°C$, additional heat causes a phase change: Solid water melts and becomes liquid water. The heat necessary to melt 1.0 g of

Screen 5.17
Exercise

Phase change A change from one physical state (gas, liquid, or solid) to another

The criterion of uniformity is the way it appears to our eyes and not as it is on the molecular level.

144 | CHAPTER 5 GASES, LIQUIDS, AND SOLIDS

Table 5.5 Energy Required to Heat 1.0 g of Solid Water at −20°C to 120°C

Physical Change	Energy (cal/g)	Basis for Calculation of Energy Required
Warming ice from −20°C to 0°C	9.6	Specific heat of ice = 0.48 cal/g · °C
Melting ice; temperature = 0°C	80	Heat of fusion of ice = 80 cal/g
Warming water from 0°C to 100°C	100	Specific heat of liquid water = 1.0 cal/g · °C
Boiling water; temperature = 100°C	540	Heat of vaporization = 540 cal/g
Warming steam from 100°C to 120°C	9.6	Specific heat of steam = 0.48 cal/g · °C

any solid is called its **heat of fusion**. The heat of fusion of ice is 80 cal/g. Thus it requires 80 cal to melt 1.0 g of ice—that is, to change 1.0 g of ice at 0°C to liquid water at 0°C.

Only after the ice has completely melted does the temperature of the water rise again. The **specific heat** of liquid water is 1 cal/g · °C (Table 1.4). Thus it requires 100 cal to raise the temperature of 1.0 g of liquid water from 0° to 100°C. Contrast this with the 80 cal required to melt 1.0 g of ice.

When liquid water reaches 100°C, the normal boiling point of water, the temperature of the sample remains constant as another phase change takes place: Liquid water vaporizes to gaseous water. The amount of heat necessary to vaporize 1.0 g of a liquid at its normal boiling point is called its **heat of vaporization**. For water, this value is 540 cal/g. Once all of the liquid water has been vaporized, the temperature again rises as the water vapor (steam) is heated. The specific heat of steam is 0.48 cal/g (Table 1.4). Thus it requires 9.6 cal to heat 1.0 g of steam from 100°C to 120°C. The data for heating 1.0 g of water from −20°C to 120°C can be shown in a graph called a **heating curve** (Figure 5.17).

An important aspect of these phase changes is that each one of them is reversible. If we start with liquid water at room temperature and cool it by immersing the container in a dry ice bath, the reverse process is observed. The temperature drops until it reaches 0°C, and then ice begins to crystallize. During this phase change, the temperature of the sample stays constant but heat is given off. The amount of heat given off when 1.0 g of liquid water at 0°C freezes is exactly the same as the amount of heat absorbed when the 1.0 g of ice at 0°C melts.

Screen 5.18 Tutorial

Figure 5.17 The heating curve of ice. The graph shows the effect of adding heat to 1.0 g of ice initially at −20°C and raising its temperature to 120°C.

A transition from the solid state directly into the vapor state without going through the liquid state is called **sublimation.** Solids usually sublime only at reduced pressures (less than 1 atm). At high altitudes, where the atmospheric pressure is low, snow sublimes. Solid CO_2 (dry ice) sublimes at $-78.5°C$ under 1 atm pressure. At 1 atm pressure, CO_2 can exist only as a solid or as a gas, never as a liquid.

■ In this photo the cold vapors of CO_2 are causing moisture (seen as wispy white clouds) to condense. More dense than air at room temperature, the CO_2 vapors glide slowly toward the table top or floor.

EXAMPLE 5.9

The heat of fusion of ice is 80 cal/g. How many calories are required to melt 1.0 mol of ice?

Solution
One mole of H_2O has a mass of 18 g. Using the factor-label method:

$$\frac{80 \text{ cal}}{\text{g ice}} \times 18 \text{ g ice} = 1.4 \times 10^3 \text{ cal} = 1.4 \text{ kcal}$$

Problem 5.9
What mass of water at 100°C can be vaporized by the addition of 45.0 kcal of heat?

EXAMPLE 5.10

What will be the final temperature if we add 1000 cal of heat to 10.0 g of ice at 0°C?

Solution
The first thing the added heat will do is to melt the ice. This phase change will use up 10.0 g × 80 cal/g = 800 cal. The remaining 200 cal will be used to heat the liquid water. The specific heat (SH; Section 1.9) of liquid water is 1.00 cal/g·°C (Table 1.4). The amount of heat required to bring about this change (Section 1.9) is

$$\text{Amount of heat} = \text{SH} \times m \times (T_2 - T_1)$$
$$200 \, cal = (1.00 \text{ cal/g·°C}) \times 10.0 \text{ g} \times (T_2 - T_1)$$
$$T_2 - T_1 = \frac{200 \, cal}{(1.00 \, cal/\text{g·°C}) \times 10.0 \text{g}} = 20°C$$

Thus the temperature of the liquid water will rise by 20°C from 0°C, and it is now 20°C.

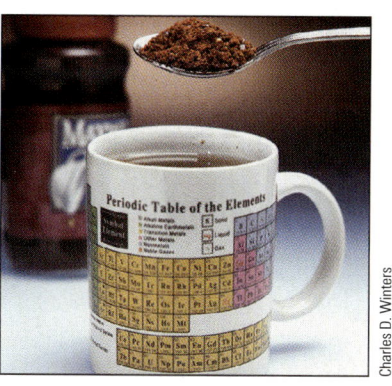

■ These freeze-dried coffee crystals were prepared by subliming water from frozen coffee.

Problem 5.10
The specific heat of iron is 0.11 cal/g·°C (Table 1.4). The heat of fusion of iron—that is, the heat required to convert iron from a solid to a liquid at its melting point—is 63.7 cal/g. Iron melts at 1530°C. How much heat must be added to 1.0 g of iron at 25°C to completely melt it?

B Phase Diagrams

We can show all the phase changes for any substance on a **phase diagram.** Figure 5.18 is a phase diagram for water. Temperature is plotted on the x-axis and pressure on the y-axis. Three areas with different colors are labeled solid, liquid, and vapor. Within these areas, water exists either as

Figure 5.18 Phase diagram of water. Temperature and pressure scales are greatly reduced.

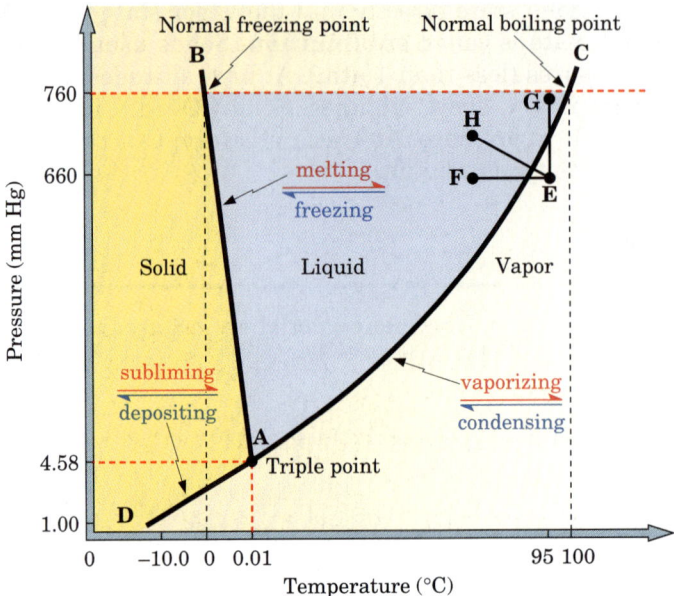

ice or liquid water or water vapor. The line (A-B) separating the solid phase from the liquid phase contains all the freezing (melting) points of water—for example, 0°C at 1 atm and 0.005°C at 400 mm Hg.

At the melting point, the solid and liquid phases coexist. The line separating the liquid phase from the gas phase (A-C) contains all the boiling points of water—for example, 100°C at 760 mm Hg and 84°C at 400 mm Hg. At the boiling points, the liquid and gas phases coexist.

Finally, the line separating the solid phase from the gas phase (A-D) contains all the sublimation points. At the sublimation points, the solid and gas phases coexist.

At a unique point (A) on the phase diagram, called the **triple point,** all three phases coexist. The triple point for water occurs at 0.01°C and 4.58 mm Hg pressure.

A phase diagram illustrates how one may go from one phase to another. For example, suppose we have water vapor at 95°C and 660 mm Hg (E). We want to condense it to liquid water. We can decrease the temperature to 70°C without changing the pressure (moving horizontally from E to F). Alternatively, we can increase the pressure to 760 mm Hg without changing the temperature (moving vertically from E to G). Or we can change both temperature and pressure (moving from E to H). Any of these processes will condense the water vapor to liquid water, although the resulting liquids will be at different pressures and temperatures. The phase diagram allows us to visualize what will happen to the phase of a substance when we change the experimental conditions from one set of temperature and pressure to another set.

EXAMPLE 5.11

What will happen to ice at 0°C if the pressure decreases from 1 atm to 0.03 atm?

Solution
According to Figure 5.18, when the pressure decreases while the temperature remains constant, we move vertically from 1 atm to 0.001 atm

CHEMICAL CONNECTION 5G

Supercritical Carbon Dioxide

We are conditioned to think that a compound may exist in three phases: solid, liquid, and gas. Under certain pressures and temperatures, however, more phases may exist. A case in point is the plentiful nonpolar substance carbon dioxide. At room temperature and one atmosphere pressure, CO_2 is a gas. Even when it is cooled to $-78°C$ it does not become a liquid but rather goes directly from a gas to a solid, which we call dry ice. At room temperature, a pressure of 60 atm is necessary to force molecules of CO_2 close enough together that they condense to a liquid.

Much more esoteric is the form of carbon dioxide called supercritical CO_2 which has some of the properties of a gas and some of a liquid. It has the density of a liquid but maintains its gaslike property of being able to flow with little viscosity or surface tension. What makes supercritical CO_2 particularly useful is that it is an excellent solvent for many organic materials. Supercritical CO_2, for example, can extract caffeine from ground coffee beans and, after extraction when the pressure is released, it simply evaporates leaving no traces behind. Similar processes can be performed with organic solvents, but traces of solvent may be left behind in the decaffeinated coffee, and may be detrimental to taste.

To understand the supercritical state, it is necessary to think about the interactions of molecules in the gas and liquid states. In the gaseous state, molecules are far apart, there is very little interaction between them, and most of the volume occupied by the gas is empty space. In the liquid state, molecules are held close together by the attractive forces between their molecules,

and there is very little empty space between them. The supercritical state is something between these two states. Molecules are close enough together to give the sample some of the properties of a liquid but at the same time far enough apart to give it some of the properties of a gas.

The critical temperature and pressure for carbon dioxide are 31°C and 73 atm. When supercritical CO_2 is cooled below the critical temperature and/or compressed, there is a phase transition and gas and liquid coexist. At a critical temperature and pressure the two phases merge. Above critical conditions only the gas phase exists.

(0.76 mm Hg). During this process we cross the boundary separating the solid phase from the vapor phase. Thus, when the pressure drops to 0.001 atm, ice sublimes and becomes vapor.

Problem 5.11
What will happen to water vapor if it is cooled from 100°C to $-30°C$ while the pressure stays at 1 atm?

SUMMARY

Matter can exist in three different states: gas, liquid, and solid (Section 5.1). Attractive forces between molecules tend to hold matter together, whereas the kinetic energy of the molecules tends to disorganize matter.

The pressure of the atmosphere is measured with a barometer. Three units of pressure are 1 mm Hg = 1 torr and 760 mm Hg = 1 atm (Section 5.2).

Boyle's law states that for a gas at constant temperature, the volume of the gas is inversely proportional to the pressure (Section 5.3A). **Charles's law** states that the volume of a gas at constant pressure is directly proportional to the temperature in kelvins (Section 5.3B). **Gay-Lussac's law** states that for gas at constant volume, the pressure is directly proportional to the temperature in kelvins (Section 5.3C).

These laws are combined and expressed as the **combined gas law** (Section 5.3).

$$\frac{P_1 V_1}{T_1} = \frac{P_2 V_2}{T_2}$$

Avogadro's law states that equal volumes of gases at the same temperature and pressure contain the same number of molecules (Section 5.4). The ideal gas law, $PV = nRT$, incorporates Avogadro's law into the combined gas law. **Dalton's law of partial pressures** states that the total pressure of a mixture of gases is the sum of the partial pressures of each individual gas (Section 5.5).

The **kinetic molecular theory** explains the behavior of gases (Section 5.6). Molecules in the gaseous state move randomly, allowing gases to fill all the available space of their container. Gas molecules, in their random motion, collide with the walls of the container and thereby exert pressure.

Intermolecular forces of attraction are responsible for the condensation of gases into the liquid state and for the solidification of liquids to the solid state (Section 5.7). In order of increasing strength, the intermolecular forces of attraction are **London dispersion forces, dipole–dipole attractions,** and **hydrogen bonds.**

Surface tension is the energy of intermolecular attractive forces at the surface of a liquid (Section 5.8). **Vapor pressure** is the pressure of a vapor (gas) above its liquid in a closed container (Section 5.9A).

The vapor pressure of a liquid increases with increasing temperature. The **boiling point** of a liquid is the temperature at which its vapor pressure equals the atmospheric pressure (Section 5.9B). The boiling point of a liquid is determined both by its shape and by the strength of the forces of attraction between its molecules in the liquid state (Section 5.9C).

Solids crystallize in well-formed geometrical shapes that often reflect the patterns in which the atoms are arranged within the crystals (Section 5.10). The **melting point** is the temperature at which a substance changes from the solid state to the liquid state. **Crystallization** is the formation of a solid from a liquid.

A **phase** is any part of a system that looks uniform throughout (Section 5.11A). A **phase change** involves a change of matter from one physical state to another—that is, from a solid, liquid, or gaseous state to any one of the other two states. **Sublimation** is a change from a solid state directly to a gaseous state. The heat necessary to convert 1.0 g of any solid to a liquid is its **heat of fusion** (Section 5.11). The heat necessary to convert 1.0 g of any liquid to the gaseous state is its **heat of vaporization.** A **phase diagram** allows for the visualization of what happens to the phase of a substance when experimental conditions in pressure and temperature change (Section 5.11B). It contains all melting points, boiling points, and sublimation points where any two phases coexist. It also contains a **triple point** where all three phases coexist.

PROBLEMS

Numbers that appear in color indicate difficult problems. ▶ designates problems requiring application of principles.

Boyle's Law and the Pressure–Volume Relationship

5.12 A sample of gas has a volume of 6.20 L at 20°C and 1.10 atm pressure. What is its volume at the same temperature and a pressure of 0.925 atm?

5.13 Methane gas is compressed from 20 L to 2.5 L at a constant temperature. The final pressure is 12.2 atm. What was the original pressure?

5.14 A gas syringe at 20°C contains 20 mL of CO_2 gas. The pressure of the gas in the syringe is 1.0 atm. What is the pressure in the syringe at 20°C if the plunger is depressed to 10 mL?

Charles's Law and the Temperature–Volume Relationship

5.15 Suppose that the pressure in an automobile tire is 2.30 atm at a temperature of 20°C. What will the pressure in the tire be if, after 10 miles of driving, the temperature of the tire increases to 47°C?

5.16 A sample of 23.0 L of NH_3 gas at 10.0°C is heated at constant pressure until it fills a volume of 50.0 L. What is the new temperature in °C?

5.17 If a sample of 4.17 L of ethane gas, C_2H_6, at 725°C is cooled to 175°C at constant pressure, what is the new volume?

5.18 A sample of SO_2 gas has a volume of 5.2 L. It is heated at constant pressure from 30°C to 90°C. What is its new volume?

Gay-Lussac's Law and the Temperature–Pressure Relationship

5.19 A sample of B_2H_6 gas in a 35-mL container is at a pressure of 450 mm Hg and a temperature of 625°C. If the gas is allowed to cool at constant volume until the pressure is 375 mm Hg, what is the new temperature in °C?

5.20 A gas in a bulb as in Figure 5.3 registers a pressure of 833 mm Hg in the manometer in which the reference arm of the U-shaped tube (A) is sealed and evacuated. What will the difference in the mercury level be if the reference arm of the U-shaped tube is open to atmospheric pressure (760 mm Hg)?

5.21 In an autoclave, a constant amount of steam is generated at a constant volume. Under 1.00 atm pressure, the steam temperature is 100°C. What pressure setting should be used to obtain a 165°C steam temperature for the sterilization of surgical instruments?

5.22 A sample of the inhalation anesthetic gas Halothane, $C_2HBrClF_3$, in a 500-mL cylinder has a pressure of 2.3 atm at 0°C. What will be the pressure of the gas if its temperature is warmed to 37°C (body temperature)?

Combined Gas Law

5.23 Complete this table:

V_1	T_1	P_1	V_2	T_2	P_2
546 L	43°C	6.5 atm	——	65°C	1.9 atm
43 mL	−56°C	865 torr	——	43°C	1.5 atm
4.2 L	234 K	0.87 atm	3.2 L	29°C	——
1.3 L	25°C	740 mm Hg	——	0°C	1.0 atm

5.24 Complete this table:

V_1	T_1	P_1	V_2	T_2	P_2
6.35 L	10°C	0.75 atm	——	0°C	1.0 atm
75.6 L	0°C	1.0 atm	——	35°C	735 torr
1.06 L	75°C	0.55 atm	3.2 L	0°C	——

5.25 A balloon filled with 1.2 L of helium at 25°C and 0.98 atm pressure is submerged in liquid nitrogen at −196°C. Calculate the final volume of the helium in the balloon at this pressure.

5.26 A balloon used for atmospheric research has a volume of 1×10^6 L. Assume that the balloon is filled with helium gas at STP and then allowed to ascend to an altitude of 10 km, where the pressure of the atmosphere is 243 mm Hg and the temperature is −33°C. What will the volume of the balloon be under these atmospheric conditions?

5.27 A gas occupies 56.44 L at 2.00 atm and 310 K. If the gas is compressed to 23.52 L and the temperature is lowered to 281 K, what is the new pressure?

5.28 A certain quantity of helium gas is at a temperature of 27°C and a pressure of 1.00 atm. What will the new temperature be if its volume is doubled at the same time that its pressure is decreased to one-half of its original value?

5.29 A sample of 30.0 mL of krypton gas, Kr, is at 756 mm Hg and 25.0°C. What is the new volume if the pressure is decreased to 325 mm Hg and the temperature is decreased to −12.5°C?

5.30 A 26.4-mL sample of ethylene gas, C_2H_4, has a pressure of 2.50 atm at 2.5°C. If the volume is increased to 36.2 mL and the temperature is raised to 10°C, what is the new pressure?

Avogadro's Law and the Ideal Gas Law

5.31 A sample of a gas at 77°C and 1.33 atm occupies a volume of 50.3 L.
(a) How many moles of the gas are present?
(b) Does your answer depend on knowing what gas it is?

5.32 What is the volume in liters occupied by 1.21 g of Freon-12 gas, CCl_2F_2, at 0.980 atm and 35°C?

5.33 An 8.00-g sample of a gas occupies 22.4 L at 2.00 atm and 273 K. What is the molecular weight of the gas?

5.34 What volume is occupied by 5.8 g of propane gas, C_3H_8, at 23°C and 1.15 atm pressure?

5.35 Does the density of a gas increase, decrease, or stay the same as the pressure increases at constant temperature? As the temperature increases at constant pressure?

5.36 What volume in mL does 0.275 g of uranium hexafluoride, UF_6, gas occupy at its boiling point of 56°C at 365 torr?

5.37 A hyperbaric chamber has a volume of 200 L.
(a) How many moles of oxygen are needed to fill the chamber at room temperature (23°C) and 3.00 atm pressure?
(b) How many grams of oxygen are needed?

5.38 One gulp of air has a volume of 2 L at STP. If air contains 20.9% oxygen, how many molecules of oxygen are in one gulp?

5.39 An average pair of lungs has a volume of 5.5 L. If the air they contain is 21% oxygen, how many molecules of O_2 do the lungs contain at 1.1 atm and 37°C?

5.40 Calculate the molecular weight of a gas if 3.30 g of the gas occupies 660 mL at 735 mm Hg and 27°C.

5.41 The three main components of dry air are N_2 (78.08%), O_2 (20.95%), and Ar (0.93%).
(a) Calculate the mass of one mole of air.
(b) Given the mass of one mole of air, calculate the density of air in g/L at STP.

5.42 Calculate the density in g/L of each of these gases at STP. Which gases are more dense than air? Which are less dense than air?
(a) SO_2 (b) CH_4 (c) H_2
(d) He (e) CO_2

5.43 The density of Freon-12, CCl_2F_2, at STP is 4.99 g/L, which means that it is approximately four times more dense than air. Show how the kinetic molecular theory of gases accounts for the fact that, although Freon-12 is more dense than air, it nevertheless finds its way to the stratosphere, where it is implicated in the destruction of the Earth's protective ozone layer.

5.44 The density of liquid octane, C_8H_{18}, is 0.7025 g/mL. If 1.00 mL of liquid octane is vaporized at 100°C and 725 torr, what volume does the vapor occupy?

5.45 How many molecules of CO are in 100 L of CO at STP?

5.46 The density of acetylene gas, C_2H_2, in a 4-L container at 0°C and 2 atm pressure is 0.02 g/mL. What would be the density of the gas under identical temperature and pressure if the container were partitioned into two 2-L compartments?

5.47 Automobile air bags are inflated by nitrogen gas. When a significant collision occurs, an electronic sensor triggers the decomposition of sodium azide to form nitrogen gas and sodium metal. The nitrogen gas then inflates nylon bags, which protect the driver and front-seat passenger from impact with the dashboard and windshield.

$$2NaN_3(s) \longrightarrow 2Na(s) + 3N_2(g)$$
Sodium azide

What volume of nitrogen gas measured at 1 atm and 27°C is formed by the decomposition of 100 g of sodium azide?

Dalton's Law of Partial Pressures

5.48 The three main components of dry air are nitrogen (78.08%), oxygen (20.95%), and argon (0.93%).
(a) Calculate the partial pressure of each gas in a sample of dry air at 760 mm Hg.
(b) Calculate the total pressure exerted by these three gases combined.

5.49 Air in the trachea contains 19.4% oxygen, 0.4% carbon dioxide, 6.2% water vapor, and 74.0% nitrogen. If the pressure in the trachea is assumed to be 1.0 atm, what are the partial pressures of these gases in this part of the body?

5.50 The partial pressures of a mixture of gases were as follows: oxygen, 210 mm Hg; nitrogen, 560 mm Hg; and carbon dioxide, 15 mm Hg. The total pressure of the gas mixture was 790 mm Hg. Was there another gas present in the mixture?

Intermolecular Forces

5.51 Which forces are stronger, intramolecular covalent bonds or intermolecular hydrogen bonds?

5.52 Under which condition does water vapor behave most ideally?
(a) 0.5 atm, 400 K
(b) 4 atm, 500 K
(c) 0.01 atm, 500 K

5.53 Can water and dimethyl sulfoxide, $(CH_3)_2S{=}O$, form hydrogen bonds between them?

5.54 What kind of intermolecular interactions take place (a) in liquid CCl_4 and (b) liquid CO? Which will have the highest surface tension?

5.55 Ethanol, C_2H_5OH, and carbon dioxide, CO_2, have approximately the same molecular weight, yet carbon dioxide is a gas at STP and ethanol is a liquid. How do you account for this difference in physical property?

5.56 Can dipole–dipole interactions ever be weaker than London dispersion forces? Explain.

5.57 Which compound has a higher boiling point: butane, C_4H_{10}, or hexane, C_6H_{14}?

Evaporation and Condensation

5.58 The melting point of ethyl chloride is −136°C and its boiling point is 12°C. Is ethyl chloride a gas, a liquid, or a solid at STP?

5.59 Calculate the specific heat (Section 1.9) of gaseous Freon-12, CCl_2F_2, if it requires 170.0 cal to change the temperature of 36.6 g of Freon-12 from 30°C to 50°C.

5.60 The heat of vaporization of liquid Freon-12, CCl_2F_2, is 4.71 kcal/mol. Calculate the energy required to vaporize 39.2 g of this compound. The molecular weight of Freon-12 is 120.9 amu.

5.61 The specific heat (Section 1.9) of mercury is 0.0332 cal/g·°C. Calculate the energy necessary to raise the temperature of one mole of liquid mercury by 36°C.

5.62 Using Figure 5.13, estimate the vapor pressure of ethanol at (a) 30°C, (b) 40°C, and (c) 60°C.

5.63 CH_4 and H_2O have about the same molecular weights. Which has the higher vapor pressure at room temperature? Explain.

5.64 The normal boiling point of a substance depends on both the mass of the molecule and the attractive forces between molecules. Arrange the compounds in each set in order of increasing boiling point and explain your answer.
(a) HCl, HBr, HI
(b) O_2, HCl, H_2O_2

Phase Changes

5.65 Refer to Figure 5.17. How many calories are required to bring one mole of ice at 0°C to a liquid state at room temperature (23°C)?

5.66 Compare the number of calories absorbed when 100 g of ice at 0°C is changed to liquid water at 37°C with the number absorbed when 100 g of liquid water is warmed from 0°C to 37°C.

5.67 (a) How much energy is released when 10 g of steam at 100°C is condensed and cooled to body temperature (37°C)? (b) How much energy is released when 10 g of liquid water at 100°C is cooled to body temperature (37°C)? (c) Why are steam burns more painful than hot-water burns?

5.68 When iodine vapor hits a cold surface, iodine crystals form. Name the phase change that is the reverse of this condensation.

5.69 If a 156-g block of dry ice, CO_2, is sublimed at 25°C and 740 mm Hg, what volume does the gas occupy?

5.70 Trichlorofluoromethane (Freon-11, CCl_3F), as a spray (Chemical Connections 5E), is used to temporarily numb the skin around minor scrapes and bruises. It accomplishes this by reducing the temperature of the treated area, thereby numbing the nerve endings that perceive pain. Calculate the heat in kilocalories that can be removed from the skin by 1.00 mL of Freon-11. The density of Freon-11 is 1.49 g/mL, and its heat of vaporization is 6.42 kcal/mol.

5.71 Explain what will happen when you heat a quantity of ice from −10°C at 1 atm to 20°C and at the same time reduce the pressure to 0.1 atm.

Chemical Connections

5.72 (Chemical Connections 5A) Which has lower entropy, a gas at 100°C or one at 200°C? Explain.

5.73 (Chemical Connections 5A) Which form of carbon presented in Figure 5.16 has the highest entropy?

5.74 (Chemical Connections 5B) What happens when a person lowers the diaphragm in their chest cavity?

5.75 (Chemical Connections 5C) In carbon monoxide poisoning, the hemoglobin is incapable of transporting oxygen to the tissues. How does the oxygen get delivered to the cells when a patient is put into a hyperbaric chamber?

5.76 (Chemical Connections 5D) In a sphygmomanometer one listens to the first tapping sound as the constrictive pressure of the arm cuff is slowly released. What is the significance of this tapping sound?

5.77 (Chemical Connections 5F) Why is the damage by severe frostbite irreversible?

5.78 (Chemical Connections 5F) If you fill a glass bottle with water, cap it, and cool to −10°C, the bottle will crack. Explain.

5.79 (Chemical Connections 5G) In what way does supercritical CO_2 have some of the properties of a gas, and some of the properties of a liquid?

Additional Problems

5.80 Why is it difficult to compress a liquid or a solid?

5.81 Explain in terms of the kinetic molecular theory what causes (a) the pressure of a gas and (b) the temperature of a gas.

5.82 The unit of pressure most commonly used for checking the inflation of automobile and bicycle tires is pounds per square inch (lb/in^2), abbreviated psi. The conversion factor between atm and psi is 1 atm = 14.7 psi. Suppose an automobile tire is filled to a pressure of 34 psi. What is the pressure in the tire in atm?

5.83 The gas in an aerosol can is at a pressure of 3.0 atm at 23°C. What will the pressure of the gas in the can be if the temperature is raised to 400°C?

5.84 Why do aerosol cans carry the warning "Do not incinerate"?

5.85 Under certain weather conditions (just before rain), the air becomes less dense. How does this change affect the barometric pressure reading?

5.86 An ideal gas occupies 387 mL at 275 mm Hg and 75°C. If the pressure changes to 1.36 atm and the temperature increases to 105°C, what is the new volume?

5.87 Which compound has greater intermolecular interactions, CO or CO_2?

5.88 On the basis of what you have learned about intermolecular forces, predict which liquid has the highest boiling point:
(a) Pentane, C_5H_{12}
(b) Chloroform, $CHCl_3$
(c) Water, H_2O

5.89 A 10-L gas cylinder is filled with N_2 to a pressure of 35 in. Hg. How many moles of N_2 do you have to add to your container to raise the pressure to 60 in. Hg? Assume a constant temperature of 27°C.

5.90 When filled, a typical tank for an outdoor grill contains 20 lb of LP (liquefied petroleum) gas, the major component of which is propane, C_3H_8. For this problem, assume that propane is the only substance present.
(a) How do you account for the fact that when propane is put under pressure, it can be liquefied?
(b) How many kilograms of propane does a full tank contain?
(c) How many moles of propane does a full tank contain?
(d) If the propane in a full tank were released into a flexible container, what volume would it occupy at STP?

5.91 Explain why gases are transparent.

5.92 The density of a gas is 0.00300 g/cm^3 at 100°C and 1.00 atm. What is the mass of one mole of the gas?

5.93 The normal boiling point of hexane, C_6H_{14}, is 69°C, and that of pentane, C_5H_{12}, is 36°C. Predict which of these compounds has a higher vapor pressure at 20°C.

5.94 If 60.0 g of NH_3 occupies 35.1 L under a pressure of 77.2 in. Hg, what is the temperature of the gas, in °C?

5.95 Water is a liquid at STP. Hydrogen sulfide, H_2S, a heavier molecule, is a gas under the same conditions. Explain.

5.96 Why does the temperature of a liquid drop as a result of evaporation?

5.97 What volume of air (21% oxygen) measured at 25°C and 0.975 atm is required to completely oxidize 3.42 g of aluminum to aluminum oxide, Al_2O_3?

5.98 Diving, particularly SCUBA (Self-Contained Underwater Breathing Apparatus) diving, subjects the body to increased pressure. Each 10 m (approximately 33 ft) of water exerts an additional pressure of 1 atm on the body.
(a) What is the pressure on the body at a depth of 100 ft?
(b) The partial pressure of nitrogen gas in air at 1 atm is 593 mm Hg. Assuming a SCUBA diver breathes compressed air, what is the partial pressure of nitrogen entering the lungs from a breathing tank at a depth of 100 ft?
(c) The partial pressure of oxygen gas in the air at 1 atm is 158 mm Hg. What is the partial pressure of oxygen in the air in the lungs at a depth of 100 ft?
(d) Why is it absolutely essential to exhale vigorously in a rapid ascent from a depth of 100 ft?

5.99 In what phase(s) will water be at 0.2 atm pressure and 120°C?

InfoTrac College Edition

For additional readings, go to InfoTrac College Edition, your online research library, at

http://infotrac.thomsonlearning.com

CHAPTER 6

6.1 Introduction
6.2 Types of Solutions
6.3 Characteristics of Solutions
6.4 Solubility
6.5 Concentration Units
6.6 Water as a Solvent
6.7 Colloids
6.8 Colligative Properties

Human red blood cells in an isotonic solution.

Solutions and Colloids

6.1 Introduction

In Chapter 5, we discussed pure substances—systems having only one component. Such systems are the easiest to study, so it was convenient to begin with them. In our daily lives, however, we more frequently encounter systems having more than one component. Such systems—for example, air, smoke, sea water, milk, blood, and rocks—are **mixtures** (Section 2.2C).

If a mixture is uniform throughout at the molecular level, we call it a **homogeneous mixture** or, more commonly, a **solution.** Air and sea water, for example, are both solutions. They are clear and transparent.

In contrast, in most rocks, we can see distinct regions separated from each other by well-defined boundaries. Such rocks are **heterogeneous mixtures.** Another example is a mixture of sand and sugar. We can easily distinguish between the two components; the mixing does not occur at the molecular level (Figure 2.4). Thus mixtures are classified on the basis of how they look to the unaided eye.

Some systems, however, fall between homogeneous and heterogeneous mixtures. Cigarette smoke, milk, and blood plasma may look homogeneous, but they do not have the transparency of air or sea water. These mixtures

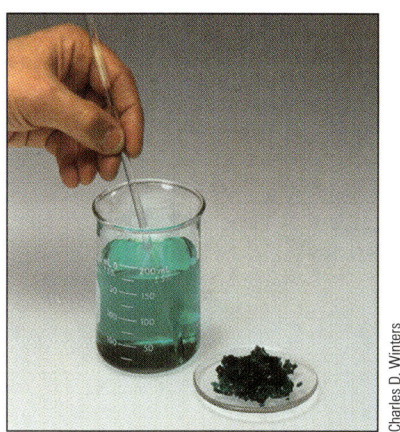

■ **Making a homogeneous solution. A green solid, nickel nitrate, is stirred into water, where it dissolves with stirring to form a homogeneous solution.**

154 / CHAPTER 6 SOLUTIONS AND COLLOIDS

■ Beer is a solution in which a liquid (alcohol), a solid (malt), and a gas (CO_2) are dissolved in the solvent water.

■ Mixtures can be homogeneous, as with brass, which is a solid solution of copper and zinc. Alternatively, they can be heterogeneous, as with granite, which contains discrete regions of different minerals (feldspar, mica, and quartz).

Alloys Homogeneous mixtures of metals

Many alloys are solid solutions. One example is stainless steel, which is mostly iron but also contains carbon, chromium, and other elements. (See also Chemical Connections 2E.)

We normally do not use the terms "solute" and "solvent" in talking about solutions of gases in gases or solids in solids.

Table 6.1 Types of Solutions

Solute		Solvent	Appearance of Solution	Example
Gas	in	Liquid	Liquid	Carbonated water
Liquid	in	Liquid	Liquid	Wine
Solid	in	Liquid	Liquid	Salt water (saline solution)
Gas	in	Gas	Gas	Air
Solid	in	Solid	Solid	14-Carat gold

are classified as **colloidal dispersions** (suspensions). We will deal with such systems in Section 6.7.

Although mixtures can contain many components, we will generally restrict our discussion here to two-component systems, with the understanding that everything we say can be extended to multicomponent systems.

6.2 Types of Solutions

When we think of a solution, we normally think of a liquid. Liquid solutions, such as sugar in water, are the most common kind, but there are also solutions that are gases or solids. In fact, all mixtures of gases are solutions. Because gas molecules are far apart from each other and much empty space separates them, two or more gases can mix with each other in any proportions. Because the mixing takes place at the molecular level, a true solution always forms; that is, there are no heterogeneous mixtures of gases.

With solids, we are at the other extreme. Whenever we mix solids, we get a heterogeneous mixture. Because even microscopic pieces of solid still contain many billions of particles (molecules, ions, or atoms), there is no way to achieve mixing at the molecular level. Homogeneous mixtures of solids (or **alloys**), such as brass, do exist, but we make them by first melting the solids then mixing the molten components, and allowing the mixture to solidify.

Table 6.1 lists the five most common types of solutions. Examples of other types (gas in solid, liquid in gas, and so on) are also known but are much less important.

When a solution consists of a solid or a gas dissolved in a liquid, the liquid is called the **solvent,** and the solid or gas is called the **solute.** A solvent may have several solutes dissolved in it, even of different types. A common example is spring water, in which gases (carbon dioxide and oxygen) and solids (salts) are dissolved in the solvent, water.

When one liquid is dissolved in another, a question may arise regarding which is the solvent and which is the solute. In most cases, the one present in the greater amount is called the solvent, but no rigid rule applies.

6.3 Characteristics of Solutions

The following are some properties of solutions:

1. *The distribution of particles in a solution is uniform.* Every part of the solution has exactly the same composition and properties as every other part. That, in fact, is the definition of "homogeneous." As a consequence, we cannot usually tell a solution from a pure solvent simply by

CHEMICAL CONNECTIONS 6A

Acid Rain

The water vapor evaporated by the sun from oceans, lakes, and rivers condenses and forms clouds of water vapor that eventually fall as rain. The raindrops contain small amounts of CO_2, O_2, and N_2. Table 6A shows that, of these gases, CO_2 is the most soluble in water. When CO_2 dissolves in water, it forms carbonic acid, H_2CO_3.

The acidity caused by the CO_2 is not harmful; however, contaminants that result from industrial pollution may create a serious acid rain problem. Burning coal or oil that contains sulfur generates sulfur dioxide, which has a high solubility in water. Sulfur dioxide in the air is oxidized to sulfur trioxide. The combination of water and sulfur dioxide results in sulfurous acid, and the reaction of sulfur trioxide and water gives sulfuric acid.

Smelting industries produce other soluble gases as well. In many parts of the world, especially those downwind from heavily

■ Trees killed by acid rain in Mt. Mitchell in NC.

industrialized areas, the result is acid rain that pours down on forests and lakes. It damages vegetation and kills fish. Such is the situation in the eastern United States, where North Carolina, in the Adirondacks and parts of New England, as well as in eastern Canada, acid rain has been observed with increasing frequency.

Table 6A	Solubility of Some Gases in Water
Gas	Solubility (g/kg H_2O at 20°C and 1 atm)
O_2	0.0434
N_2	0.0190
CO_2	1.688
H_2S	3.846
SO_2	112.80
NO_2	0.0617

looking at it. A glass of pure water looks the same as a glass of water containing dissolved salt or sugar. In some cases, we *can* tell by looking—for example, if the solution is colored and we know that the solvent is colorless.

2. *The components of a solution do not separate on standing.* A solution of vinegar (acetic acid in water), for example, would never separate.

3. *A solution cannot be separated into its components by filtration.* Both the solvent and the solute pass through the filter paper.

4. *For any given solute and solvent, it is possible to make solutions of many different compositions.* For example, we can easily make a solution of 1 g of glucose in 100 g of water, or 2 g, or 6 g, or 8.7 g, or any other amount up to the solubility limit (Section 6.4).

5. *Solutions are almost always transparent.* They may be colorless or colored, but you can usually see through them. Solid solutions are exceptions.

6. *Solutions can be separated into pure components.* Common separation methods include distillation and chromatography, which you may learn about in the laboratory portion of this course. The separation of a solution into its components is a physical change, not a chemical one.

We use the word "clear" to mean transparent. A solution of copper sulfate in water is blue, but clear.

Tables of solubility values are found in handbooks.

Screen 6.2 Exercise

We use the word "miscible" to refer to a liquid dissolving in a liquid.

Figure 6.1 Diethyl ether and water form two layers. A separatory funnel permits the bottom layer to be drawn off.

Polar compounds dissolve in polar compounds because the positive end of one molecule attracts the negative end of the other molecule.

6.4 Solubility

The **solubility** of a solid in a liquid is the maximum amount of that solid that will dissolve in a given amount of a particular liquid at a given temperature. Suppose we wish to make a solution of table salt (NaCl) in water. We take some water, add a few grams of salt, and stir. At first we see the particles of salt suspended in the water. Soon, however, all the salt dissolves. Now let us add more salt and continue to stir. Again, the salt dissolves. Can we repeat this process indefinitely? The answer is no—there is a limit. The solubility of table salt at 25°C is 36.2 g per 100 g of water. If we add more salt than that amount, the excess solid will not dissolve but rather will remain suspended as long as we keep stirring; it will sink to the bottom after we stop.

Solubility is a physical constant, like melting point or boiling point. Each solid has a different solubility in every liquid. Some solids have a very low solubility in a particular solvent; we often call these solids *insoluble*. Others have a much higher solubility; we call these *soluble*. Even for soluble solids however, there is always a solubility limit (see also Section 4.6).

The same is true for gases dissolved in liquids. Different gases have different solubilities in a solvent (see Table 6A). Some liquids are essentially insoluble in other liquids (gasoline in water), whereas others are soluble to a limit. For example, 100 g of water dissolves about 6 g of diethyl ether (another liquid). If we add more ether than that amount, we see two layers (Figure 6.1). Some liquids, however, are completely soluble in other liquids, no matter how much is present. An example is ethanol and water, which form a solution no matter what quantities of each are mixed. We say that the water and ethanol are *miscible* in all proportions.

When a solvent contains all the solute it can hold at a given temperature, we call the solution **saturated.** Any solution containing a lesser amount of solute is **unsaturated.**

A **supersaturated** solution holds more solute in the solvent than it can normally hold at a given temperature. A supersaturated solution is not stable; when disturbed in any way, such as by stirring or shaking, the excess solute precipitates, and the solution becomes saturated. If we add more solute to a saturated solution at constant temperature, it looks as if none of the additional solid dissolves, because the solution already holds all the solute that it can. Actually an equilibrium, similar to the one discussed in Section 5.9A, is at work in this situation. Some particles of the additional solute dissolve, whereas an equal quantity of dissolved solute comes out of solution. So, even though the concentration of dissolved solute does not change, the solute particles themselves are constantly going in and out of solution.

Whether a particular solute dissolves in a particular solvent depends on several factors, as discussed next.

A Nature of the Solvent and the Solute

The more similar two compounds are, the more likely that one will be soluble in the other. Here the rule is *"like dissolves like."* This is not an absolute rule, but it does apply in most cases.

When we say "like," we mostly mean alike in terms of *polarity*. In other words, polar compounds dissolve in polar compounds, and nonpolar compounds dissolve in nonpolar compounds. For example, the liquids benzene (C_6H_6) and carbon tetrachloride (CCl_4) are nonpolar compounds. They dissolve in each other, and other nonpolar materials, such as gasoline and camphor ($C_{10}H_{16}O$), dissolve in them. In contrast, ionic compounds (such as NaCl) and polar compounds (such as table sugar, $C_{12}H_{22}O_{11}$) are insoluble in all of these solvents.

The most important polar solvent is water. We have already seen that most ionic compounds are soluble in water, as are small covalent compounds that can form hydrogen bonds with water. It is worth noting that even polar molecules are usually insoluble in water if they cannot either react with the water or form hydrogen bonds with water molecules. Water as a solvent is discussed in Section 6.6.

B Temperature

For most solids and liquids that dissolve in liquids, the rule is *solubility increases with increasing temperature*. Sometimes the increase is great, sometimes it is moderate. For a few substances, solubility even decreases with increasing temperature (Figure 6.2).

Screen 6.4 Tutorial

For example, the solubility of glycine, H_2N-CH_2-COOH, a polar building block of proteins, is 52.8 g in 100 g of water at 80°C but only 33.2 g at 30°C. If, for instance, we prepare a saturated solution of glycine in 100 g of water at 80°C, it will hold 52.8 g of glycine. If we then allow it to cool to 30°C, where the solubility is 33.2 g, we might expect the excess glycine, 19.6 g, to precipitate out of solution and form crystals. It often does, but on many occasions it does not. The latter case is an example of a **supersaturated solution**. Even though the solution contains more glycine than the water can normally hold at 30°C, the excess glycine stays in solution because the molecules need a seed, a surface on which to begin crystallizing, and no such surface is present.

Supersaturated solutions are not indefinitely stable, however. If we shake or stir the solution, we may find that all the excess solid precipitates

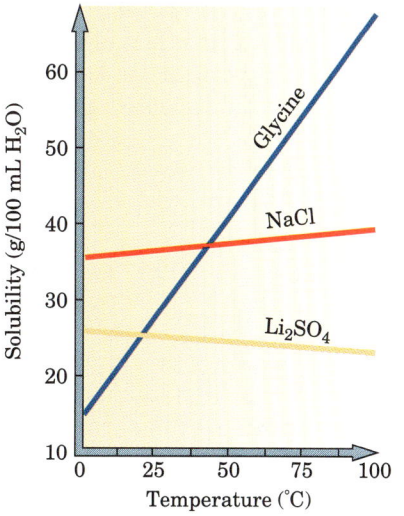

Figure 6.2 The solubilities of some solids in water as a function of temperature. The solubility of glycine increases rapidly, that of NaCl barely increases, and that of Li_2SO_4 decreases with temperature.

CHEMICAL CONNECTIONS 6B

The Bends

Deep-sea divers encounter high pressures. For them to breathe properly under such conditions, oxygen must be supplied under pressure. At one time, this goal was achieved with compressed air. As pressure increases, the solubility of the gases in the blood increases. This is especially true for nitrogen, which constitutes almost 80% of our air.

When divers come up and the pressure on their bodies decreases, the solubility of nitrogen in their blood decreases as well. As a consequence, the previously dissolved nitrogen in blood and tissues starts to form small bubbles, especially in the veins. The formation of gas bubbles, called the **bends,** can prevent blood circulation. If this condition is allowed to develop uncontrolled, the resulting pulmonary embolism can be fatal.

If the divers' ascent is gradual, the dissolved gases are removed by regular exhalation and diffusion through the skin. Divers use decompression chambers, where the high pressure is gradually reduced to normal pressure.

If decompression disease develops after a dive, patients are put into hyperbaric chambers (see Chemical Connections 5C), where they breathe pure oxygen at 2.8 atm pressure. In the standard form of treatment, the pressure is reduced to 1 atm over a period of 6 hours.

Nitrogen also has a narcotic effect in divers with compressed air below 40 m. This "rapture of the deep" is similar to alcohol-induced intoxication.

Because of the problem caused by nitrogen, divers' tanks often hold a helium–oxygen mixture instead of air. The solubility of helium in blood is affected less by pressure than is that of nitrogen.

■ **Ascending too rapidly will cause dissolved nitrogen bubbles to be released and form bubbles in the blood.**

Sudden decompression and ensuing bends are important not only in deep-sea diving but also in high-altitude flight, especially orbital flight.

---

at once (Figure 6.3). Another way to crystallize the excess solute is to add a grain of crystal, a process called **seeding.** The seed crystal provides the surface onto which the solute molecules can converge.

For gases, *solubility in liquids almost always decreases with increasing temperature.* The effect of temperature on the solubility of gases in water can have important consequences for fish, for example. Oxygen is only slightly soluble in water, but fish need that oxygen to live. When the temperature of the water increases, perhaps because of the output from a power plant, the solubility of the oxygen decreases and may become so low that fish die. This situation is called *thermal pollution.*

C Pressure

Pressure has little effect on the solubility of liquids or solids. For gases, however, **Henry's law** applies: The higher the pressure, the greater the solubility of a gas in a liquid. This concept is the basis of the hyperbaric medicine discussed in Chemical Connections 5C. When the pressure increases, more oxygen dissolves in the blood plasma and reaches tissues at higher-than-normal pressures (2 to 3 atm).

Henry's law also explains why a bottle of beer or a carbonated soft drink foams when it is opened. The bottle is sealed under greater than 1 atm of CO_2 pressure. When opened at 1 atm, the solubility of CO_2 in the liquid

Figure 6.3 When a supersaturated solution is disturbed, the dissolved salt (here sodium acetate, $CH_3COO^-Na^+$) rapidly crystallizes.

Henry's law The solubility of a gas in a liquid is directly proportional to the pressure decreases. The excess CO_2 is released, forming bubbles, and the gas pushes out some of the liquid.

Screen 6.5
Exercise

6.5 Concentration Units

We can express the amount of a solute dissolved in a given quantity of solvent, called the **concentration** of the solution, in a number of ways. Some concentration units are better suited than others for some purposes. Sometimes qualitative terms are good enough. For example, we may say that a solution is *dilute* or *concentrated*. These terms tell us little specific information about the concentration, but we know that a concentrated solution contains more solute than a dilute solution.

For most purposes, however, we need quantitative concentrations. For example, a nurse must know how much glucose to give to a patient. Many methods of expressing concentration exist, but in this chapter we deal with only the three most important: percent concentration, molarity, and parts per million (ppm).

A Percent Concentration

There are three ways of representing **percent concentration.** The most common is weight of solute per volume of solution (w/v):

% Concentration = (weight of solute/volume of solution) × 100

If 10 g of table sugar is dissolved in enough water so that the total volume is 100 mL, the concentration is 10% w/v. Note that we need to know the total volume of the solution, *not* the volume of the solvent. For this purpose, we use volumetric flasks, which come in different sizes (see Figures 6.4 and 6.5).

> **Percent concentration (% w/v)** The number of grams of solute in 100 mL of solution

Figure 6.4 (*a*) A 0.0100 *M* solution of $KMnO_4$ is made by adding enough water to 0.395 g of $KMnO_4$ to make 0.250 L of solution. (*b*) To ensure the correct solution volume, the $KMnO_4$ is placed in a volumetric flask and dissolved in a small amount of water. (*c*) After dissolving is complete, sufficient water is added to fill the flask to the mark. The flask now contains 0.250 L of solution.

(*a*)

(*b*)

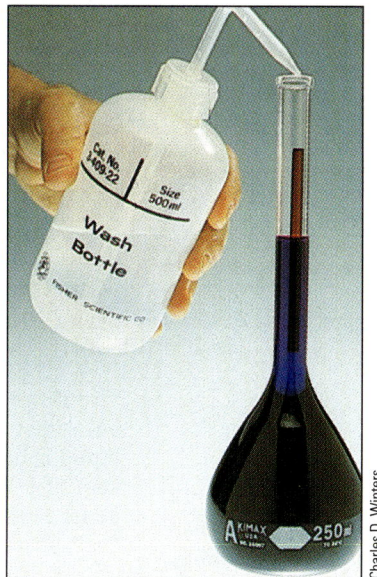
(*c*)

EXAMPLE 6.1

The label on a bottle of vinegar says it contains 5.0% acetic acid, CH_3COOH. The bottle contains 240 mL of vinegar. How many grams of acetic acid are in the bottle?

Solution
To find out how many grams of CH_3COOH are in this solution, we take the percentage and multiply it by the volume:

$$\frac{5.0 \text{ g } CH_3COOH}{100 \text{ mL solution}} \times 240 \text{ mL solution} = 12 \text{ g } CH_3COOH$$

Problem 6.1
How would you make 250 mL of a 4.4% w/v KBr solution in water? Assume that a 250-mL volumetric flask is available.

EXAMPLE 6.2

If 6.0 g of NaCl is dissolved in enough water to make a 300-mL solution, first estimate and then calculate the w/v percentage of NaCl.

Solution
To estimate, we note that the percentage is based on a 100-mL solution. We have 300 mL, which is three times larger than 100 mL. Therefore, in 100 mL, we will have only one-third as much NaCl, which is 2 g. Thus the solution is about 2% w/v.

To calculate, we divide the weight of the solute by the volume of the solution and multiply by 100:

$$(6.0 \text{ g NaCl}/300 \text{ mL solution}) \times 100 = 2.0\% \text{ w/v}$$

Problem 6.2
If 7.7 g of lithium iodide, LiI, is dissolved in enough water to make 400.0 mL of solution, what is the w/v percentage of LiI?

A second way to represent percent concentration is weight of solute per weight of solution (w/w). Calculations are essentially the same as the w/v calculations, except that we use the weight of the solution instead of its volume. A volumetric flask is not used for these solutions. (Why not?)

Finally, we can represent percent concentration with volume of solute per volume of solution (v/v). This unit is used only for solutions of liquids in liquids, most notably alcoholic beverages. For example, 40% v/v ethanol in water means that 40 mL of ethanol has been added to enough water to make 100 mL of solution.*

A 40% v/v solution of ethanol in water is 80 proof. Proof is twice the percent concentration (v/v) of ethanol in water.

---

*A 40% v/v ethanol solution contains not 60 mL of water for every 40 mL of alcohol, but somewhat less than that. *Volumes of liquids are not always additive.* When one liquid is dissolved in another liquid, the new volume may be more or less than the sum of the volumes of the individual liquids because the molecules move either closer to each other (because of greater attraction) or farther apart.

B Molarity

For many purposes, it is easiest to express concentration by using the weight or volume percentage methods just discussed. When we want to focus on the number of molecules present, however, we need another way. For example, a 5% solution of glucose in water does not contain the same number of solute molecules as a 5% solution of ethanol in water. That is why chemists often use molarity. The **molarity, M,** of a solution is defined as the number of moles of solute dissolved in 1 L of solution. The units of molarity are moles per liter.

Screen 1.6.6
Tutorial

$$\text{Molarity } (M) = \text{moles } (n)/\text{liter } (V)$$

Thus, in the same volume of solution, a 0.2-molar solution of glucose, $C_6H_{12}O_6$, in water contains the same number of molecules of solute as a 0.2-molar solution of ethanol in water. In fact, this holds true for equal volumes of any solution, as long as the molarities are the same.

We can prepare a solution of a given molarity in essentially the same way that we prepare a solution of given w/v concentration, except that we use moles instead of grams in our calculations. We can always find out how many moles of solute are in any volume of a solution of known molarity by the following formula:

$$\text{Molarity} \times \text{volume in liters} = \text{number of moles}$$

$$M \times V = \text{moles}$$

The solution is then prepared as shown in Figure 6.4.

EXAMPLE 6.3

How do we prepare 2.0 L of a 0.15 M aqueous solution of sodium hydroxide, NaOH?

Solution
First, we find out how many moles of NaOH there will be in this solution:

$$M \times V = \text{moles}$$

$$(0.15 \text{ mol NaOH}/1 \text{ L}) \times 2.0 \text{ L} = 0.30 \text{ mol NaOH}$$

We need 0.30 mol of NaOH to make up this solution. How many grams is this? To convert moles to grams, we multiply by the appropriate conversion factor (Section 4.3). The molar mass of NaOH = 40.0 g/mol, so

$$0.30 \text{ mol NaOH} \times (40.0 \text{ g NaOH}/1 \text{ mol NaOH}) = 12 \text{ g NaOH}$$

We therefore weigh 12 g of NaOH, put it into a 2-L volumetric flask, add some water, swirl until the solid has dissolved, and then fill the flask with water to the 2-L mark.

Problem 6.3

How would you prepare 2.0 L of a 1.06 M aqueous solution of KCl?

EXAMPLE 6.4

Screen 6.7
Exercise Tutorial

We dissolve 18.0 g Li_2O (molar mass = 30.0 g/mol) in sufficient water to make 500 mL of solution. Estimate and then calculate the molarity of the solution.

Solution

We are looking for molarity in units of mol/L. To estimate, we note that 18.0 g is about half of the formula weight, 30. Therefore, we have about 0.5 mol of Li_2O in 500 mL of solution. But molarity is based on 1 L, which is twice as much as 500 mL. Therefore, there will be 1.0 mol of Li_2O in 1 L; that is, the solution will be approximately 1.0 M.

To calculate, we use the conversion factor: 1 mol of Li_2O/30.0 g Li_2O:

$$18.0 \text{ g } Li_2O \times (1 \text{ mol } Li_2O / 30 \text{ g } Li_2O) = 0.60 \text{ mol } Li_2O$$

To find liters of solution, we divide mL by 1000:

$$500 \text{ mL} \times (1 \text{ L}/1000 \text{ mL}) = 0.500 \text{ L}$$

To find the moles of Li_2O per liter of solution, we divide:

$$(0.60 \text{ mol } Li_2O / 0.500 \text{ L}) = 1.2 \text{ } M$$

The estimate of 1.0 M is reasonably close to the calculated value of 1.2 M.

Problem 6.4

If 0.440 g of KSCN is dissolved in enough water to make 340 mL of solution, what is the molarity of the solution?

Blood serum is the liquid part of the blood that remains after removal of the cellular particulates and the fibrinogen.

EXAMPLE 6.5

The concentration of sodium chloride in blood serum is approximately 0.14 M. What volume of blood serum contains 2.0 g of NaCl?

Solution

As we know the concentration in moles per liter, we must first find out how many moles are in 2.0 g of NaCl (formula weight = 58.5g). We first multiply by the proper conversion factor:

$$2.0 \text{ g NaCl} \times (1 \text{ mol NaCl}/58.5 \text{ g NaCl}) = 0.034 \text{ mol NaCl}$$

We next find liters:

$$0.034 \text{ mol NaCl} \times (1 \text{ L}/0.14 \text{ mol NaCl}) = 0.24 \text{ L} = 240 \text{ mL}$$

Problem 6.5

If a 0.300 M glucose ($C_6H_{12}O_6$) solution is available for intravenous infusion, how many milliliters are needed to deliver 10.0 g of glucose?

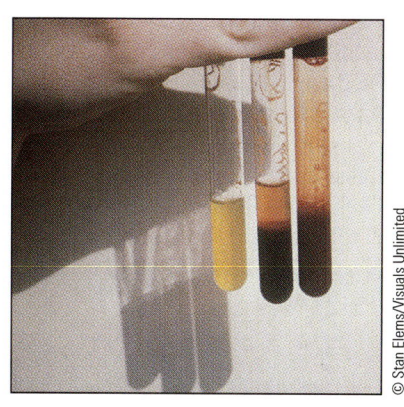

■ Blood (left to right): serum, coagulated, and whole blood.

EXAMPLE 6.6

How many grams of HCl are in 225 mL HCl, whose concentration is 6.00 M?

Solution
First, we find out how many moles of HCl are present:

$$M \times V = \text{moles}$$

$$(6.00 \text{ mol}/1 \text{ L}) \times 0.225 \text{ L} = 1.35 \text{ mol}$$

We get the number of grams by multiplying by the conversion factor 1 mol = 36.5 g:

$$1.35 \text{ mol HCl} \times (36.5 \text{ g HCl}/1 \text{ mol HCl}) = 49.3 \text{ g HCl}$$

Problem 6.6

A certain wine contains 0.010 M NaHSO$_3$ (sodium bisulfite) as a preservative. How many grams of sodium bisulfite must be added to a 100-gallon barrel containing wine to reach this concentration? Assume no change in volume of wine upon addition of the sodium bisulfite.

C Dilution

We frequently prepare solutions by diluting concentrated solutions rather than by weighing out pure solute (Figure 6.5). Because we add only solvent, the number of moles of solute remains unchanged. Before we dilute, the equation that applies is

$$M_1V_1 = \text{moles}$$

After we dilute, the volume and molarity have both changed and we have

$$M_2V_2 = \text{moles}$$

Screen 6.7
Exercise Tutorial

Figure 6.5 Making a solution by dilution. (a) A 100-mL volumetric flask is filled to the mark with 0.100 M $K_2Cr_2O_7$. (b) This is transferred to a 1.00-L volumetric flask. (c) The 1.00-L volumetric flask is filled to the mark with distilled water. The concentration of the now diluted $K_2Cr_2O_7$ is 0.0100 M.

(a) (b) (c)

The number of moles of solute is the same both before and after, however, so we can say that

$$M_1V_1 = M_2V_2$$

We can use this handy equation for dilution problems.

EXAMPLE 6.7

How do we prepare 200 mL of a 3.5 M solution of acetic acid if we have a bottle of concentrated acetic acid (6.0 M)?

Solution
We use the equation

$$M_1V_1 = M_2V_2$$
$$(6.0 \text{ mol}/1 \text{ L}) \times V_1 = (3.5 \text{ mol}/1 \text{ L}) \times 0.200 \text{ L}$$
$$V_1 = (3.5 \text{ mol/L})(0.200 \text{ L})/6.0 \text{ mol/L} = 0.12 \text{ L}$$

We put 0.12 L, or 120 mL, of the concentrated acetic acid in a 200-mL volumetric flask, add some water and mix, and then fill to the mark with water.

Problem 6.7

A concentrated HCl solution is 12.0 M. You have to make 300 mL of 0.600 M solution. How would you prepare it?

A similar equation can be used for dilution problems involving percent concentrations:

$$\%_1 V_1 = \%_2 V_2$$

EXAMPLE 6.8

How do we prepare 500 mL of a 0.5% w/v solution of NaOH if we have a solution of 50% w/v NaOH on hand?

Solution

We add 5 mL of the concentrated solution to a 500-mL volumetric flask, add some water and mix, and then fill to the mark with water.

$$(50\%)V_1 = (0.5\%)(500 \text{ mL})$$

$$V_1 = (0.5\%)(500 \text{ mL})/50\% = 5 \text{ mL}$$

Problem 6.8

A concentrated solution of 15% w/v KOH solution is available. How would you prepare 20.0 mL of a 0.10% w/v KOH solution?

D Parts per Million

Sometimes we need to deal with very dilute solutions—for example, 0.0001%. In such cases, it is more convenient to use **parts per million (ppm)** to express concentration. For example, if drinking water is polluted with lead ions to the extent of 1 ppm, it means that there is 1 mg of lead ions in 1 kg (1 L) of the water. Some solutions are so dilute that we use **parts per billion (ppb)** to express their concentrations.

Modern methods of analysis allow us to detect such miniscule concentrations. Some substances are harmful even at concentrations measured in ppb. One such substance is dioxin, an impurity in the 2,4,5-T herbicide sprayed by the United States as a defoliant in Vietnam.

ppm = [(g solute)/(g solution)] × 10^6

ppb = [(g solute)/(g solution)] × 10^9

6.6 Water as a Solvent

Water covers about 75% of the Earth's surface in the form of oceans, ice caps, glaciers, lakes, and rivers. Water vapor is always present in the atmosphere. Life evolved in water, and without it life as we know it could not exist. The human body is about 60% water. This water is found both inside the cells of the body (intracellular) and outside the cells (extracellular).

Most of the important chemical reactions in living tissue occur in aqueous solution; water serves as a solvent to transport reactants and products from one place in the body to another. Water is also itself a reactant or product in many biochemical reactions. The properties that make water such a good solvent are its polarity and its hydrogen-bonding capacity (Section 5.7C).

A How Water Dissolves Ionic Compounds

We learned in Section 3.5 that ionic compounds in the solid state are composed of a regular array of ions. The crystal is held together by ionic bonds, which are attractions between positive and negative ions. Water, of course, is a polar molecule. When a solid ionic compound is added to water, the ions at the surface of the crystal become surrounded by water molecules. The negative ions (anions) attract the positive ends of water molecules, and the positive ions (cations) attract the negative ends of water molecules (Figure 6.6). Each ion attracts two to four water molecules. When the combined force of attraction to water molecules is greater than the ionic bond that would keep the ion in the crystal, the ion will be completely dislodged. Once the ion is removed from the crystal, it becomes surrounded by water molecules (Figure 6.7). These ions are said to be **hydrated.** A more general

Figure 6.6 Water molecules remove anions and cations from the surface of an ionic solid.

Screen 6.8 Animation

term, covering all solvents, is **solvated.** The solvation layer—that is, the surrounding shell of solvent molecules—acts as a cushion. It prevents a solvated anion from colliding directly with a solvated cation, thereby keeping the solvated ions in solution.

Not all ionic solids are soluble in water. Some rules for predicting solubilities were given in Section 4.6.

B Solid Hydrates

The attraction between ions and water molecules is so strong in some cases that the water molecules are an integral part of the crystal structure of the solids. Water molecules in a crystal are called **water of hydration.** The substances that contain water in their crystals are themselves called **hy-**

The dot in the formula indicates that the H_2O is present in the crystal, but it is not covalently bonded to the ions of Ca^{2+} or SO_4^{2-}.

Figure 6.7 Anions and cations solvated by water.

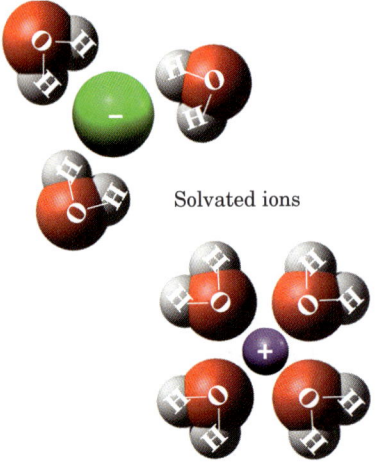

Solvated ions

CHEMICAL CONNECTIONS 6C

Hydrates and Air Pollution: The Decay of Buildings and Monuments

Many buildings and monuments in urban areas throughout the world are decaying, ruined by air pollution. The main culprit in this process is gypsum, an end product of air pollution. The most common stones used for buildings and monuments are limestone and marble, both of which are largely calcium carbonate. In the absence of polluted air, these stones can last for thousands of years, and many statues and buildings from ancient times (Babylonian, Egyptian, Greek, and others) survived until recently with little change. Indeed, they remain intact in many rural areas.

In urban areas, however, the air is polluted with SO_2 and SO_3 which mostly come from the combustion of coal and petroleum products containing small amounts of sulfur compounds as impurities (see Chemical Connections 6A). They react with the calcium carbonate at the surface of the stones to form calcium sulfate. When calcium sulfate interacts with water in the rain, it forms the dihydrate gypsum. The problem is that gypsum has a larger volume than the original marble or limestone, causing the surface of the stone to expand. This activity, in turn, results in

■ Acid rain damage to stonework on the walls of York Minster.

flaking. Eventually, statues such as those in the Parthenon (in Athens, Greece) become noseless and later faceless.

---

drates. For example, gypsum and plaster of Paris are hydrates of calcium sulfate: Gypsum is $CaSO_4 \cdot 2H_2O$, calcium sulfate dihydrate, and plaster of Paris is $(CaSO_4)_2 \cdot H_2O$, calcium sulfate monohydrate.

Some of these crystals hold on to their water tenaciously. To remove it, the crystals must be heated for some time at a high temperature. The crystal without its water is called **anhydrous.** In many cases, the anhydrous crystals are so strongly attracted to water that they take it from the water vapor in the air. That is, some anhydrous crystals become hydrated upon standing in air. Crystals that do so are called **hygroscopic.**

Hydrated crystals often look different from the anhydrous forms. For example, copper sulfate pentahydrate, $CuSO_4 \cdot 5H_2O$, is blue, but the anhydrous form is white (Figure 6.8).

The difference between hydrated and anhydrous crystals can sometimes have an effect in the body. For example, the compound sodium urate exists as spherical crystals in the anhydrous form, but the crystals are needle-shaped in the monohydrate form (Figure 6.9). The deposition of sodium urate monohydrate in the joints (mostly in the big toe) causes gout.

C Electrolytes

Ions in water migrate from one place to another, maintaining their charge along the way. As a consequence, solutions of ions conduct electricity. They can do so because the ions in the solution migrate independently of one another. As shown in Figure 6.10, cations migrate to the negative electrode, called the **cathode,** and anions migrate to the positive electrode, called the

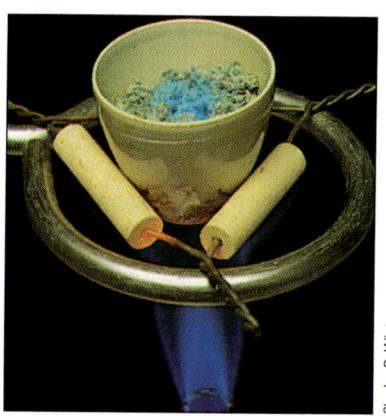

Figure 6.8 When blue hydrated copper(II) sulfate is heated, it changes to a white anhydrous salt.

Hygroscopic substances Compounds able to absorb water vapor from the air

If we want hygroscopic compounds to remain anhydrous, we must place them in sealed containers that contain no water vapor.

Figure 6.9 (a) The needle-shaped sodium urate monohydrate crystals that cause gout. (b) The pain of gout as depicted by a cartoonist.

anode. The movement of ions constitutes an electric current. The migration of ions completes the circuit initiated by the battery and can cause an electric bulb to light up (see also Chemical Connections 4B).

Substances that conduct an electric current when dissolved in water or when in the molten state are called **electrolytes.** Sodium chloride, for example, is an electrolyte. Hydrated Na^+ ions carry positive charges, and hydrated Cl^- ions carry negative charges; as a result, the bulb in Figure 6.10 lights brightly if these ions are present. Substances that do not conduct electricity are called **nonelectrolytes.** Distilled water, for example, is a nonelectrolyte. The light bulb shown in Figure 6.10 does not light up if only distilled water is placed in the beaker. However, when tap water is placed in

Figure 6.10 Conductance by an electrolyte.

the beaker, the bulb lights dimly. Tap water contains enough ions to carry electricity, but their concentration is so low that the solution conducts only a small amount of electricity.

As you see, electric conductance depends on the concentration of ions. The higher the ion concentration, the greater the electric conductance. Nevertheless, differences in electrolytes exist. If we take a 0.1 M aqueous solution of NaCl and compare it with a 0.1 M HF solution, we find that the NaCl solution lights a bulb brightly, but the HF solution lights it only dimly. We might have expected the two solutions to behave similarly. They have the same concentration, 0.1 M, and each "molecule" provides two ions, a cation and an anion (Na^+ and Cl^-; H^+ and F^-). The reason they behave differently is that, whereas Na^+Cl^- dissociates completely to two ions (each hydrated and each moving independently), in the case of HF only a few molecules are dissociated into ions. Most of the HF molecules do not dissociate, and undissociated molecules do not conduct electricity. Compounds that dissociate completely are called **strong electrolytes,** and those that dissociate into ions only partially are called **weak electrolytes.**

Electrolytes are important components of the body because they help to maintain the acid–base balance and the water balance. The most important cations in body tissues are Na^+, K^+, Ca^{2+}, and Mg^{2+}. The most abundant anions in the body are bicarbonate, HCO_3^-; chloride, Cl^-; hydrogen phosphate, HPO_4^{2-}; and dihydrogen phosphate, $H_2PO_4^-$.

■ Sports drinks help to maintain the body's electrolytic balance.

D How Water Dissolves Covalent Compounds

Water is a good solvent not only for ionic compounds but also for many covalent compounds. In a few cases, the covalent compounds dissolve because they react with water. An important example is the covalent compound HCl. HCl is a gas (with a penetrating, choking odor) that attacks the mucous membranes of eyes, nose, and throat. When dissolved in water, the HCl molecules react with water to give ions:

$$HCl(g) + H_2O(\ell) \longrightarrow Cl^-(aq) + H_3O^+(aq)$$
Hydronium ion

Another example is the gas sulfur trioxide, which reacts as follows:

$$SO_3(g) + 2H_2O(\ell) \longrightarrow HSO_4^-(aq) + H_3O^+(aq)$$

Because HCl and SO_3 are completely converted to ions in dilute aqueous solution, these solutions are ionic solutions and behave just as other electrolytes do (they conduct a current). Nevertheless, HCl and SO_3 are themselves covalent compounds, unlike salts such as NaCl.

Most covalent compounds that dissolve in water do not, in fact, react with water. They dissolve because the water molecules surround the entire covalent molecule and solvate it. For example, when methanol, CH_3OH, is dissolved in water, the methanol molecules are solvated by the water molecules (Figure 6.11).

There is a simple way to predict which covalent compounds will dissolve in water and which will not. Covalent compounds will dissolve in water if they can form hydrogen bonds with water, provided that the solute molecules are fairly small. Hydrogen bonding is possible between two molecules if one of them contains an O, N, or F atom and the other contains an O—H, N—H, or F—H bond. Every water molecule contains an O atom and O—H bonds. Therefore, water can form hydrogen bonds with any molecule that

In aqueous solutions, H^+ does not exist; it combines with water and forms hydronium ion, H_3O^+.

Figure 6.11 Solvation of a polar covalent compound by water. The dotted lines represent hydrogen bonds.

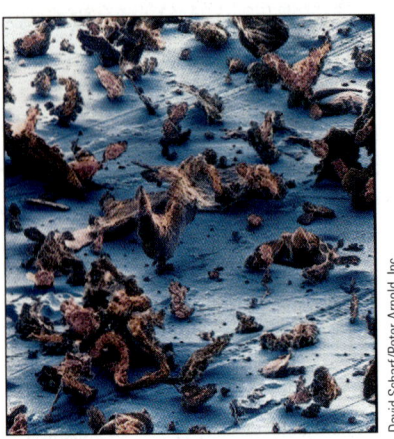

■ Ordinary household dust particles, magnified 2200 times.

contains an O, N, or F atom or an O—H, N—H, or F—H bond. If these molecules are small enough, they will be soluble in water. How small? In general, they should have no more than three C atoms for each O, N, or F atom.

For example, acetic acid, CH_3COOH, is soluble in water, but benzoic acid, C_6H_5COOH, is not. Similarly, ethanol, C_2H_5OH, is soluble, but dipropyl ether, $C_6H_{14}O$, is not. Table sugar, $C_{12}H_{22}O_{11}$ (Section 19.5A), is very soluble in water. Although it contains a large number of carbon atoms, it has so many oxygen atoms that it forms many hydrogen bonds with water molecules; thus this molecule is very well solvated.

Covalent molecules that do not contain O, N, or F atoms are almost always insoluble in water. For example, methanol, CH_3OH, is infinitely soluble in water, but chloromethane, CH_3Cl, is not. The exception is the rare case where a covalent compound reacts with water—for instance, HCl.

E Water in the Body

Water is important in the body not only because it dissolves ionic substances as well as some covalent compounds, but also because it hydrates all polar molecules in the body. In this way, it serves as a vehicle to carry most of the organic compounds, nutrients, and fuels used by the body as well as waste material. Blood and urine are two examples of aqueous body fluids.

In addition, the hydration of macromolecules such as proteins, nucleic acids, and polysaccharides allows the proper motions within these molecules, which are necessary for such functions as enzyme activity (see Chapter 22).

6.7 Colloids

$1 \text{ nm} = 1 \times 10^{-9}$ m

Colloid A homogeneous mixture containing particles ranging from 1 to 1000 nm in diameter

Colloids are neither true solutions nor heterogeneous mixtures. They constitute a third kind of mixture.

Up to now we have discussed only solutions. In a true solution, the maximum diameter of the solute particles is about 1 nm. If the diameter of the solute particles exceeds this size, then we no longer have a true solution—we have a **colloid**. In a colloid (also called a colloidal dispersion or colloidal system), the diameter of the solute particles ranges from about 1 to 1000 nm. The term *colloid* has acquired a new name recently. In Chapter 5, we encountered the term *nanotube*. The "nano" part refers to dimensions in the nanometer range—that is, the range of colloids. Thus, when you encounter terms such as "nanoparticle" or "nanoscience," they are equivalent to "colloidal particle" or "colloid science," although the former refer mostly to particles with well-defined geometrical shape (such as tubes), while the latter are more general.

Particles of this size usually have a very large surface area, which accounts for the two basic characteristics of colloidal systems:

1. They scatter light and therefore appear turbid, cloudy, or milky.
2. Although colloidal particles are large, they form stable dispersions; that is, they do not form separate phases that settle out. As with true solutions, colloids can exist in a variety of phases (Table 6.2).

Tyndall effect Light passing through and scattered by a colloid viewed at a right angle

All colloids exhibit the following characteristic effect. When we shine light through a colloid and look at the system from a 90° angle, we see the pathway of the light without seeing the individual particles (Figure 6.12). This **Tyndall effect** is due to light scattering of colloidal particles. Smoke, serum, and fog, to name a few examples, all exhibit the Tyndall effect. We are all familiar with the sunbeams that can be seen when sunlight passes through dusty air. This, too, is an example of the Tyndall effect. Again, we

Table 6.2 Types of Colloidal Systems

Type	Example
Gas in gas	None
Gas in liquid	Whipped cream
Gas in solid	Marshmallows
Liquid in gas	Clouds, fog
Liquid in liquid	Milk, mayonnaise
Liquid in solid	Cheese, butter
Solid in gas	Smoke
Solid in liquid	Jelly
Solid in solid	Dried paint

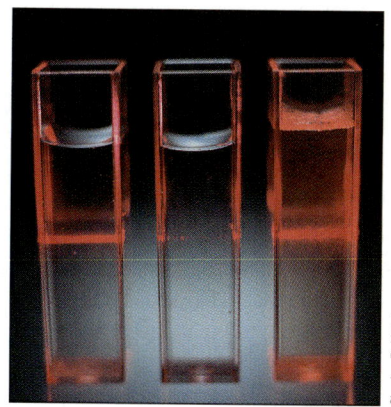

Figure 6.12 Light scattering by a colloidal suspension. A narrow beam of light passes through a colloidal mixture (left), then through a salt solution, and finally through another colloidal mixture. This illustrates the light scattering ability of the colloid-sized particles.

do not see the individual solid particles in dusty air, but only the light scattered by them. Ordinary solutions do not show this effect.

Colloidal systems are stable. Mayonnaise, for example, stays emulsified and does not separate into oil and water. When the size of colloidal particles is larger than about 1000 nm, however, the system is unstable and separates into phases. Such systems are called **suspensions.**

For example, if we take a lump of soil and disperse it in water, we get a muddy suspension. The soil particles are anywhere from 10^3 to 10^6 nm in diameter. The muddy mixture scatters light and, therefore, appears turbid. It is not a stable system, however. If left alone, the soil particles soon settle to the bottom, with clear water found above the sediment. Therefore, soil in water is a suspension, not a colloidal system.

Table 6.3 summarizes the properties of solutions, colloids, and suspensions.

What makes a colloidal dispersion stable? To answer this question, we must first realize that the colloidal particles are in constant motion. Just look at the dust particles dancing in a ray of sunlight that enters your room. Actually, you do not see the dust particles themselves; they are too small. Rather, you see flashes of scattered light. The motion of the dust particles dispersed in air is a random, chaotic motion. This motion of any colloidal particle suspended in a solvent is called **Brownian motion** (Figure 6.13).

The constant buffeting and collisions by solvent molecules cause the colloidal particles to move in random Brownian motion. (In the case of the dust particles, the solvent is air.) This ongoing motion creates favorable conditions for collisions between particles. When such large particles collide, they stick together, combine to give larger particles, and finally settle out of the solution. That is what happens in a suspension.

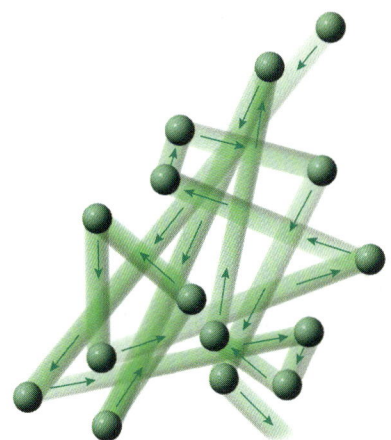

Figure 6.13 Brownian motion.

Brownian motion is named for Robert Brown (1773–1858), the Scottish botanist who first observed it in 1827.

Table 6.3 Properties of Three Types of Mixtures

Property	Solutions	Colloids	Suspensions
Particle size (nm)	0.1–1.0	1–1000	>1000
Filterable with ordinary paper	No	No	Yes
Homogeneous	Yes	Borderline	No
Settles on standing	No	No	Yes
Behavior to light	Transparent	Tyndall effect	Translucent or opaque

CHEMICAL CONNECTIONS 6D

Emulsions and Emulsifying Agents

Oil and water do not mix. Even when we stir them vigorously and the oil droplets become dispersed in the water, the two phases separate as soon as we stop stirring. There are, however, a number of stable colloidal systems made of oil and water; known as **emulsions**. For example, the oil droplets in milk are dispersed in an aqueous solution. This is possible because milk contains a protective colloid—the milk protein called casein. Casein molecules surround the oil droplets and, because they are polar and carry a charge, protect and stabilize the oil droplets. Casein is thus an emulsifying agent.

A similar emulsifying agent is egg yolk. This ingredient in mayonnaise coats the oil droplets and prevents them from separating.

■ Freshly made wines are often cloudy because of colloidal particles (left). Removing the particles clarifies the wine (right).

Particles with the dimensions of colloids have many new practical applications.

Nanotubes, nanowires, and nanopores in composite coatings have unusual electronic and optical properties because of their enormous surface areas. For example, titanium oxide particles smaller than 20 nm are used to coat surfaces of plastics, glass, and other materials. These thin coatings have self-cleaning, antifogging, antifouling, and sterilizing properties.

So why do colloidal particles remain in solution despite all the collisions due to their Brownian motion? Two reasons explain this phenomenon:

1. Most colloidal particles carry a large solvation layer. If the solvent is water, as in the case of protein molecules in the blood, the colloidal particles are surrounded by a large number of water molecules, which move together with the colloidal particles and cushion them. When two colloidal particles collide as a result of Brownian motion, they do not actually touch each other; only the solvent layers collide. As a consequence, the particles do not stick together and thus stay in solution.

2. The large surface area of colloidal particles acquires charges from the solution. All colloids in a particular solution acquire the same kind of charge—for example, a negative charge. This development leaves a net positive charge in the solvent. When a charged colloidal particle encounters another colloidal particle, the two repel each other because of their like charges.

Thus the combined effects of the solvation layer and the surface charge keep colloidal particles in a stable dispersion. By taking advantage of these effects, chemists can either increase or decrease the stability of a colloidal system. If we want to get rid of a colloidal dispersion, we can remove the solvation layer, the surface charge, or both. For example, proteins in the blood form a colloidal dispersion. If we want to isolate a protein from blood, we may want to precipitate it. We can accomplish this task in two ways: by removing the hydration layer or by removing the surface charges. If we add solvents such as ethanol or acetone, both of which have great affinity for water, the water will be removed from the solvation layer of the protein, and then the unprotected protein molecules will stick together when they collide and form a sediment. Similarly, by adding electrolytes such as NaCl to the solution, we can remove the charges from the surface of the proteins (by a mechanism too complicated to discuss here). Without their protective charges, two protein molecules will no longer repel each other. Instead, when they collide, they will stick together and precipitate from the solution.

6.8 Colligative Properties

Colligative property A property of a solution that depends only on the number of solute particles and not on the chemical identity of the solute

Any property of a solution that depends only on the *number* of solute particles dissolved in a given amount of solvent and not on the *nature* of these particles is called a **colligative property**.

A Freezing-Point Depression

One colligative property is **freezing-point depression.** One mole of any particle dissolved in 1000 g of water lowers the freezing point of the water by 1.86°C. The nature of the solute does not matter, only the number of particles.

$$\Delta T = (1.86°C/\text{mol}) \times \text{mole of particles}$$

This principle is used in a number of practical ways. In winter, we use salts (sodium chloride and calcium chloride) to melt snow and ice on our streets. The salts dissolve in the melting snow and ice, which lowers the freezing point of the water. Another application is the use of antifreeze in automobile radiators. Because water expands upon freezing (see Chemical Connections 5F), the ice formed in a car's cooling system when the outside temperature falls below 0°C can crack the engine block. The addition of antifreeze prevents this problem, because it makes the water freeze at a much lower temperature. The most common antifreeze is ethylene glycol.

Freezing-point depression The decrease in the freezing point of a liquid caused by adding a solute

Note that in preparing a solution for this purpose we do not use molarity. That is, we do not need to measure the total volume of the solution.

■ Salting lowers the freezing point of ice.

Screen 6.10
Exercise Tutorial

EXAMPLE 6.9

If we add 275 g of a nondissociating molecular compound, ethylene glycol, $C_2H_6O_2$, per 1000 g of water in a car radiator, what will the freezing point of the solution be?

Solution
Because $C_2H_6O_2$ has a molar mass of 62.0 g/mol, 275 g contains

$$275 \text{ g } C_2H_6O_2 \times (1 \text{ mol } C_2H_6O_2/62.0 \text{ g } C_2H_6O_2) = 4.44 \text{ mol } C_2H_6O_2$$

Each mole lowers the freezing point by 1.86°C, so the freezing point is lowered by

$$(1.86°C/1 \text{ mol}) \times 4.44 \text{ mol} = 8.26°C$$

The freezing point of the water will be lowered from 0°C to −8.26°C, and the radiator will not crack if the outside temperature remains above −8.26°C (17.18°F).

Problem 6.9

If we add 215 g of methanol, CH_3OH, to 1000 g of water, what will be the freezing point of the solution?

If a solute is ionic, then each mole of solute dissociates to more than 1 mole of particles. For example, if we dissolve one mole (58.5 g) of NaCl in 1000 g of water, the solution contains two moles of solute particles: one mole each of Na^+ and Cl^-. The freezing point of water will be lowered by twice 1.86°C, or 3.72°C.

EXAMPLE 6.10

What will be the freezing point of the solution if we dissolve one mole of potassium sulfate, K_2SO_4, in 1000 g of water?

Solution
One mole of K_2SO_4 dissociates to produce three moles of ions: two moles of K^+ and one mole of SO_4^{2-}. The freezing point will be lowered by $3 \times 1.86°C = 5.58°C$, and the solution will freeze at $-5.58°C$.

Problem 6.10
Which aqueous solution would have the lowest freezing point?
(a) 6.2 M NaCl (b) 2.1 M Al(NO$_3$)$_3$ (c) 4.3 M K$_2$SO$_3$

Besides freezing-point depression, several other colligative properties exist, including vapor-pressure lowering, boiling-point elevation, and osmotic pressure. We discuss only the last of these because biologically it is the most important.

B Osmotic Pressure

To understand osmotic pressure, let us consider the beaker shown in Figure 6.14(b). In this beaker, two compartments are separated by an osmotic semipermeable membrane. A **semipermeable membrane** is a thin slice of some material, such as cellophane, that contains very tiny holes (far too small for us to see) that are big enough to let small solvent molecules pass through but not big enough to let large solute molecules pass. In the right compartment of the beaker is pure solvent—in this example, water. In the left compartment is a dispersion of starch (Section 19.6A) in water.

The starch molecules are too big to pass through the membrane and remain in the left compartment. The water molecules, however, easily go back and forth; as far as they are concerned, no membrane is present. Therefore, water molecules go from the right side to the left side because the concentration of water in the right compartment is higher (100%) than in the left compartment. Molecules will always diffuse from an area of higher concentration to an area of lower concentration, which is what the water molecules do. If no membrane separated the two solutions, the starch molecules would also follow this rule, moving from the higher concentration area (in this case, the left side) to the lower one (the right side). The membrane, of course, prevents such movement. However, it does not hinder the water

An osmotic membrane is a very selective semipermeable membrane that allows only solvent molecules, and nothing else, to pass through.

Screen 6.11
Tutorial

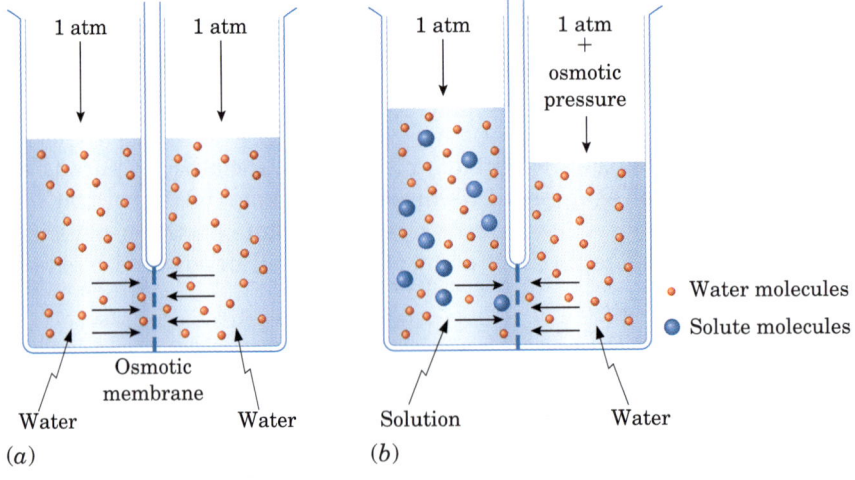

Figure 6.14 Osmotic pressure. (a) Two compartments separated by an osmotic semipermeable membrane both contain only solvent molecules that can pass through the membrane. The liquid level on each side is the same. (b) The compartment on the right contains only solvent; the one on the left contains both solute and solvent. Solute molecules cannot pass through the membrane. In an effort to dilute the solution, more solvent molecules move from the right to the left compartment than vice versa, raising the liquid level on that side.

molecules from moving into the left compartment. This passage of solvent molecules from the right (solvent or dilute solution) to the left (a more concentrated solution) across a semipermeable membrane is called **osmosis.** When a significant number of water molecules have moved from the right to the left side, the liquid level on the right side goes down and that on the left side goes up, as shown in Figure 6.14(b).

The process cannot continue indefinitely because gravity prevents the difference in levels from becoming too great. Eventually the process stops with the levels unequal. The levels can be made equal again if we apply an external pressure to the higher side. The amount of external pressure required to equalize the levels is called the **osmotic pressure.**

Although this discussion assumes that one compartment contains pure solvent, the same principle applies if both compartments contain solutions, as long as the concentrations are different. The solution of higher concentration always has a higher osmotic pressure than the one of lower concentration, which means that the flow of solvent always goes from the more dilute to the more concentrated side. Of course, the number of particles is what is important. We must remember that, in ionic solutions, each mole of solute gives rise to more than one mole of particles. For convenience in calculation, we define a new term, **osmolarity** (in osmol), as molarity (M) multiplied by the number of particles (i) produced by each molecule of solute.

$$\text{Osmolarity} = M \times i$$

Osmotic pressure The amount of external pressure applied to the more concentrated solution to stop the passage of solvent molecules across a semipermeable membrane

Some ions are small but still do not go through the membrane because they are solvated by a shell of water molecules (see Figure 6.7).

EXAMPLE 6.11

A 0.89% w/v NaCl solution is referred to as a physiological or isotonic saline solution because it has the same concentration of salts as normal human blood, although blood contains several salts and saline solution has only NaCl. What is the osmolarity of this solution?

Solution

$$0.89\% \text{ w/v NaCl} = 8.9 \text{ g NaCl in 1 L of solution}$$

$$(8.9 \text{ g NaCl}/1 \text{ L solution}) \times (1 \text{ mol NaCl}/58.5 \text{ g NaCl}) =$$

$$0.15 \text{ mol NaCl/ L solution} = 0.15 \, M$$

Each NaCl dissociates into two particles, Na^+ and Cl^- in aqueous solutions.

$$\text{Osmolarity} = 0.15 \times 2 = 0.30 \text{ osmol}$$

Problem 6.11
What is the osmolarity of a 3.3% w/v Na_3PO_4 solution?

As noted earlier, osmotic pressure is a colligative property. The osmotic pressure generated by a solution across a semipermeable membrane—the difference between the heights of the two columns in Figure 6.14(b)—depends on the osmolarity of the solution. If the osmolarity increases by a factor of 2, the osmotic pressure will also increase by a factor of 2.

Osmotic pressure is very important in biological organisms because cell membranes are semipermeable. Therefore, biological fluids must have the proper osmolarity. For example, red blood cells in the body are suspended in

■ The effect of osmosis on a carrot. The carrot on the left is in a solution that contains a high concentration of NaCl. Water flowing out of the carrot has caused it to shrink. The carrot on the right, in pure water, is slightly swollen.

CHEMICAL CONNECTIONS 6E

Reverse Osmosis and Desalinization

In osmosis, the solvent flows spontaneously from the dilute solution compartment into the concentrated solution compartment. In reverse osmosis, the opposite happens. One must apply pressures greater than the osmotic pressure for this flow to occur.

Reverse osmosis is used to make drinkable water from sea water or brackish water. In large plants in the Persian Gulf countries, for example, more than 100 atm pressure is applied to sea water containing 35,000 ppm salt. The water that passes through the semipermeable membrane under this pressure contains only 400 ppm salt—well within the limits set by the World Health Organization for drinkable water.

■ An emergency hand-operated water desalinator that works by reverse osmosis. It can produce 4.5 L of pure water per hour from sea water, which can save someone adrift at sea.

Iso-osmotic, hypo-osmotic, and hyper-osmotic are synonyms for isotonic, hypotonic, and hypertonic, respectively.

A 5.5% glucose solution is also isotonic and is used in intravenous feeding.

a medium called plasma, which must have the same osmolarity as the red blood cells. Two solutions with the same osmolarity are called **isotonic,** so plasma is said to be isotonic with red blood cells. As a consequence, no osmotic pressure is generated across the cell membrane.

What would happen if we suspended red blood cells in distilled water instead of in plasma? Inside the red blood cells the osmolarity is approximately the same as in a physiological saline solution, 0.30 osmol. Distilled water has zero osmolarity. As a consequence, water flows into the red blood cells. The volume of the cells increases, and the cells swell, as shown in Figure 6.15(c). The membrane cannot resist the osmotic pressure, and the red blood cells eventually burst, spilling their contents into the water. We call this process **hemolysis.**

Solutions with an osmolarity (and hence osmotic pressure) lower than that of suspended cells are called **hypotonic solutions.** Obviously, it is very important that we always use isotonic solutions and never hypotonic solutions in intravenous feeding and blood transfusion. Hypotonic solutions would simply kill the red blood cells by hemolysis.

Equally important, we should not use **hypertonic solutions.** A hypertonic solution has a greater osmolarity (and greater osmotic pressure) than

---

Figure 6.15 Red blood cells in solutions of different tonicity.

(a) Isotonic solution

(b) Hypertonic solution

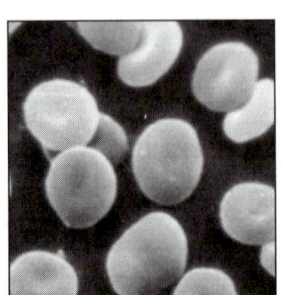
(c) Hypotonic solution

the red blood cells. If red blood cells are placed in a hypertonic solution (for example, 0.5 osmol glucose solution), water flows from the cells into the glucose solution through the semipermeable cell membrane. This process, called **crenation,** shrivels the cells, as shown in Figure 6.15(b).

As already mentioned, 0.89% NaCl (physiological saline) is isotonic with red blood cells and is used in intravenous injections.

EXAMPLE 6.12

Is a 0.50% w/v solution of KCl (a) hypertonic, (b) hypotonic, or (c) isotonic compared to red blood cells?

Solution
The 0.50% KCl contains 5.0 g KCl in 1.0 L of solution:

$$(5.0 \text{ g KCl}/1.0 \text{ L soln}) \times (1.0 \text{ mol KCl}/74.6 \text{ g KCl}) =$$
$$0.067 \text{ mol KCl/L} = 0.067 \, M$$

Because each KCl yields two particles, the osmolarity is $0.067 \times 2 = 0.13$ osmol; this is smaller than the osmolarity of the red blood cells, which is 0.30 osmol. Therefore, the KCl solution is hypotonic.

Problem 6.12
Which solution is isotonic compared to red blood cells: (a) $0.1 \, M$ Na_2SO_4, (b) $1.0 \, M$ Na_2SO_4, or (c) $0.2 \, M$ Na_2SO_4?

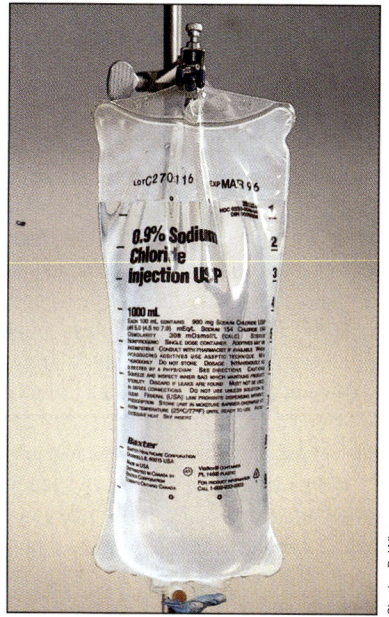

■ An isotonic saline solution.

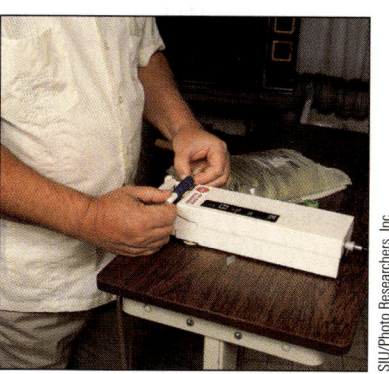

■ A portable dialysis unit.

C Dialysis

An osmotic semipermeable membrane allows only solvent and not solute molecules to pass. If, however, the openings in the membrane are somewhat larger, then small solute molecules can also get through, but large solute molecules, such as macromolecular and colloidal particles, cannot. This process is called **dialysis.**

For example, ribonucleic acids are important biological molecules that we will study in Chapter 24. When biochemists prepare ribonucleic acid solutions, they must remove small molecules, such as NaCl, from the solution to obtain a pure nucleic acid preparation. To do so, they place the nucleic acid solution in a dialysis bag (cellophane) of sufficient pore size to allow all the small molecules to diffuse and retain only the large nucleic acid molecules. If the dialysis bag is suspended in flowing distilled water, all the NaCl and small molecules will leave the bag. After a certain amount of time, the bag contains only the pure nucleic acids dissolved in water.

Our kidneys work in much the same way. The millions of nephrons, or kidney cells, have very large surface areas in which the capillaries of the blood vessels come in contact with the nephrons. The kidneys serve as a gigantic filtering machine. The waste products of the blood dialyse out through semipermeable membranes in the glomeruli (see Sections 31.5 and 31.6) and enter collecting tubes that carry the urine to the ureter. Meanwhile, large protein molecules and cells are retained in the blood.

Dialysis A process in which a solution containing particles of different sizes is placed in a bag made of a semipermeable membrane. The bag is placed into a solvent or solution containing only small molecules. The solution in the bag reaches equilibrium with the solvent outside, allowing the small molecules to diffuse across the membrane but retaining the large molecules.

The glomeruli of the kidney are fine capillary blood vessels in which the body's waste products are removed from the blood.

CHEMICAL CONNECTIONS 6F

Hemodialysis

The kidneys' main function is to remove toxic waste products from the blood. When the kidneys are not functioning properly, these waste products may threaten life. **Hemodialysis** is a process that performs the same filtration function (see Figure 31.4).

In hemodialysis, the patient's blood circulates through a long tube of cellophane membrane suspended in an isotonic solution and then returns to the patient's vein. The cellophane membrane retains all the large molecules (for example, proteins) but allows the small ones, including the toxic wastes, to pass through. In this way, dialysis removes the wastes from the blood.

If the cellophane tube were suspended in distilled water, other small molecules and ions, such as Na^+, Cl^-, and glucose, would also be removed from the blood. That is something you *don't* want to happen. The isotonic solution used in hemodialysis consists of 0.6% NaCl, 0.04% KCl, 0.2% $NaHCO_3$, and 0.72% glucose (all w/v). It ensures that no glucose or Na^+ is lost from the blood.

A patient usually remains on an artificial kidney machine for four to seven hours. During this time, the isotonic bath is changed every two hours. Kidney machines allow people with kidney failure to lead a fairly normal life, although they must take these hemodialysis treatments regularly.

■ A schematic diagram of the hollow-fiber (or capillary) dialyzer, the most commonly used artificial kidney. The blood flows through small tubes constructed of a semipermeable membrane; these tubes are bathed in the dialyzing solution.

SUMMARY

Systems containing more than one component are **mixtures** (Section 6.1). **Homogeneous mixtures** are uniform throughout. **Heterogeneous mixtures** exhibit well-defined boundaries between phases (Section 6.1). When a solution consists of a solid or gas dissolved in a liquid, the liquid acts as the **solvent**, and the solid or gas is the **solute** (Sections 6.2 and 6.3). When one liquid is dissolved in another, the liquid present in greater amount is considered to be the solvent. The **solubility** of a substance is the maximum amount of the substance that dissolves in a given amount of solvent at a given temperature (Sections 6.2 and 6.3).

"Like dissolves like" means that polar molecules are soluble in polar solvents and that nonpolar molecules are soluble in nonpolar solvents. The solubility of solids and liquids in liquids usually increases with increasing temperature; the solubility of gases in liquids usually decreases with increasing temperature (Section 6.4).

Percent concentration is given in either weight per unit volume of solution (w/v), weight per unit weight of solvent (w/w), or volume of solute per unit volume of solution (v/v). **Molarity** is the number of moles of solute in 1 L of solution (Section 6.5).

Water is the most important solvent because it dissolves polar compounds and ions through hydrogen bonding and dipole–dipole interactions. Hydrated ions are surrounded by water molecules (solvation layer) that move together with the ion, cushioning it from collisions with other ions. Aqueous solutions of ions and molten salts are **electrolytes** and conduct electricity (Section 6.6).

Colloids exhibit a scattering of light called the **Tyndall effect** (Section 6.7). They also exhibit a chaotic random motion, called **Brownian motion**. Colloids are stable mixtures despite the relatively large size of the colloidal particles (1 to 1000 nm). The stability results from the solvation layer that cushions the colloids from direct collisions and from the electric charge on the surface of colloidal particles (Section 6.7).

Colligative properties are properties that depend only on the number of solute particles present (Section 6.8). **Freezing-point depression** and **osmotic pressure** are two examples of colligative properties. Osmotic pressure operates across an osmotic semipermeable membrane that allows only solvent molecules to pass but screens out all other molecules. In osmotic pressure calculations, concentration is measured in **osmolarity,** which is the molarity of the solution multiplied by the number of particles produced by dissociation of the solute.

Red blood cells in **hypotonic solution** swell and burst, a process called **hemolysis.** Red blood cells in **hypertonic solution** shrink, a process called **crenation.** Some semipermeable membranes allow small solute molecules to pass through along with solvent molecules. In **dialysis,** such membranes are used to separate large molecules from smaller ones (Section 6.8).

PROBLEMS

Numbers that appear in color indicate difficult problems. ▶ designates problems requiring application of principles.

Solutions and Their Characteristics

6.13 Vinegar is a homogeneous aqueous solution containing 6% acetic acid. Which is the solvent?

6.14 A solution is made by dissolving glucose in water. Which is the solvent and which is the solute?

6.15 In each of the following, tell whether the solutes and solvents are gases, liquids, or solids.
(a) Bronze (see Chemical Connections 2E)
(b) Cup of coffee
(c) Car exhaust
(d) Champagne

6.16 Give a familiar example of solutions of each of these types:
(a) Liquid in liquid
(b) Solid in liquid
(c) Gas in liquid
(d) Gas in gas

6.17 Are mixtures of gases true solutions or heterogeneous mixtures? Explain.

Solubility

6.18 You dissolved 0.32 g of aspartic acid in 115.0 mL of water and obtained a clear solution. After two days standing at room temperature, you notice a white powder at the bottom of the beaker. What may have happened?

6.19 The solubility of a compound is 2.5 g in 100 mL of aqueous solution at 25°C. If you put 1.12 g of the compound in a 50-mL volumetric flask at 25°C and add sufficient water to fill it to the 50-mL mark, what kind of solution do you get—saturated or unsaturated? Explain.

6.20 To a separatory funnel with two layers—the nonpolar diethyl ether and the polar water—is added a small amount of solid. After shaking the separatory funnel, in which layer will you find each of the following solids?
(a) NaCl
(b) camphor, $(C_{10}H_{16}O)$
(c) KOH

6.21 On the basis of polarity and hydrogen bonding, which solute would be the most soluble in benzene, C_6H_6?
(a) CH_3OH
(b) H_2O
(c) $CH_3CH_2CH_2CH_3$
(d) H_2SO_4

6.22 ▶ Suppose that you have a stain on an oil painting that you want to remove without damaging the painting. The stain is not water soluble. Knowing the polarities of the following solvents, which one would you try first and why?
(a) Benzene, C_6H_6
(b) Isopropyl (rubbing) alcohol, C_3H_7OH
(c) Hexane, C_6H_{14}

6.23 Which pairs of liquids are likely to be miscible?
(a) H_2O and CH_3OH
(b) H_2O and C_6H_6
(c) C_6H_{14} and CCl_4
(d) CCl_4 and CH_3OH

6.24 The solubility of aspartic acid in water is 0.500 g in 100 mL at 25°C. If you dissolve 0.251 g of aspartic acid in 50.0 mL water at 50°C and let the solution cool to 25°C without stirring, shaking, or otherwise disturbing the solution, would you get a saturated, unsaturated, or supersaturated solution? Explain.

6.25 ▶ Near a power plant, warm water is discharged into a river. Sometimes dead fish are observed in the area. Why do fish die from the warm water?

180 | CHAPTER 6 SOLUTIONS AND COLLOIDS

6.26 If a bottle of beer is allowed to stand for several hours after being opened, it becomes "flat" (it loses CO_2). Explain.

6.27 Would you expect the solubility of ammonia gas in water at 2 atm pressure to be
(a) greater,
(b) the same, or
(c) smaller than at 0.5 atm pressure?

Concentration Units

6.28 Describe how you would make the following solutions:
(a) 500.0 g of 5.32% w/w H_2S solution in water
(b) 342.0 g of 0.443% w/w benzene in toluene
(c) 12.5 g of 34.2% w/w dimethyl sulfoxide in acetone

6.29 Describe how you would make the following solutions:
(a) 280 mL of a 27% v/v solution of ethanol, C_2H_5OH, in water
(b) 435 mL of a 1.8% v/v solution of ethyl acetate, $C_4H_8O_2$, in water
(c) 1.65 L of an 8.00% v/v solution of benzene, C_6H_6, in chloroform, $CHCl_3$

6.30 Describe how you would make the following solutions:
(a) 250 mL of a 3.6% w/v solution of NaCl in water
(b) 625 mL of a 4.9% w/v solution of glycine, $C_2H_5NO_2$, in water
(c) 43.5 mL of a 13.7% w/v solution of Na_2SO_4 in water
(d) 518 mL of a 2.1% w/v solution of acetone, C_3H_6O, in water

6.31 Calculate the w/v percentage of each of these solutes:
(a) 623 mg of casein in 15.0 mL of milk
(b) 74 mg vitamin C in 250 mL of orange juice
(c) 3.25 g of sucrose in 186 mL coffee

6.32 Describe how you would prepare 250 mL of 0.10 M NaOH from solid NaOH and water.

6.33 Assuming that the appropriate volumetric flasks are available, describe how you would make these solutions:
(a) 175 mL of a 1.14 M solution of NH_4Br in water
(b) 1.35 L of a 0.825 M solution of NaI in water
(c) 330 mL of a 0.16 M solution of ethanol, C_2H_5OH, in water

6.34 What is the molarity of each solution?
(a) 47 g of KCl dissolved in enough water to give 375 mL of solution
(b) 82.6 g of sucrose, $C_{12}H_{22}O_{11}$, dissolved in enough water to give 725 mL of solution
(c) 9.3 g of $(NH_4)_2SO_4$ dissolved in enough water to give 2.35 L of solution

6.35 A teardrop with a volume of 0.5 mL contains 5.0 mg NaCl. What is the molarity of the NaCl in the teardrop?

6.36 The concentration of stomach acid, HCl, is approximately 0.10 M. What volume of stomach acid contains 0.25 mg of HCl?

6.37 The label on a sparkling cider says it contains 22.0 g glucose ($C_6H_{12}O_6$), 190 mg K^+, and 4.00 mg Na^+ per serving of 240 mL cider. Calculate the molarities of these ingredients in the sparkling cider.

6.38 If 3.18 g of $BaCl_2$ is dissolved in enough solvent to make 500.0 mL of solution, what is the molarity?

6.39 The label on a jar of jam says it contains 13 g of sucrose, $C_{12}H_{22}O_{11}$, per tablespoon (15 mL). What is the molarity of sucrose in the jam?

6.40 A toothpaste contains 0.17 g NaF in 75 mL toothpaste. What is the percent w/v and the molarity of NaF in the toothpaste?

6.41 A student has a bottle labeled 0.750% albumin solution. The bottle contains exactly 5.00 mL. How much water must the student add to make the concentration of albumin 0.125%?

6.42 How many grams of solute are present in each of the following aqueous solutions?
(a) 575 mL of a 2.00 M solution of HNO_3
(b) 1.65 L of a 0.286 M solution of alanine, $C_3H_7NO_2$
(c) 320 mL of a 0.0081 M solution of $CaSO_4$

6.43 A student has a stock solution of 30.0% w/v H_2O_2 (hydrogen peroxide). Describe how the student should prepare 250 mL of a 0.25% H_2O_2 solution.

6.44 To make 5.0 L of a fruit punch that contains 10% v/v ethanol, how much 95% v/v ethanol must be mixed with how much fruit juice?

6.45 A pill weighing 325 mg contains the following. What is the concentration of each in ppm?
(a) 12.5 mg Captopril, a medication for high blood pressure
(b) 22 mg Mg^{2+}
(c) 0.27 mg Ca^{2+}

6.46 One slice of enriched bread weighing 80 g contains 70 µg folic acid. What is this concentration in ppm and ppb?

6.47 Dioxin is considered to be poisonous in concentrations above 2 ppb. If a lake containing 1×10^7 L has been contaminated by 0.1 g of dioxin, did the concentration reach a dangerous level?

6.48 An industrial waste water contains 3.60 ppb cadmium, Cd^{2+}. How many mg of Cd^{2+} could be recovered from a ton (1016 kg) of waste water?

6.49 According to the label on a piece of cheese, one serving of 28 g provides the following daily values: 2% of Fe, 6% of Ca, and 6% of vitamin A. The recommended daily allowance (RDA) of each of these nutrients are as follows: 15 mg Fe, 1200 mg Ca, and 0.800 mg vitamin A. Calculate the concentrations of each of these nutrients in the cheese in ppm.

Water as a Solvent

6.50 Considering polarities, electronegativities, and similar concepts learned in Chapter 3, classify the following as a strong electrolyte, a weak electrolyte, or a nonelectrolyte.
(a) KCl
(b) C_2H_5OH (ethanol)
(c) NaOH
(d) HF
(e) $C_6H_{12}O_6$ (glucose)

6.51 Which of the following would produce the brightest light in the conductance apparatus shown in Figure 6.10?
(a) 0.1 M KCl
(b) 0.1 M $(NH_4)_3PO_4$
(c) 0.5 M sucrose

6.52 Ethanol, C_2H_5OH, is very soluble in water. Describe how water dissolves ethanol.

6.53 Predict which of these covalent compounds is soluble in water.
(a) C_2H_6
(b) CH_3OH
(c) HF
(d) NH_3
(e) CCl_4

Colloids

6.54 Your car tire is made of synthetic rubber in which carbon black particles of the size of 200–500 nm are randomly dispersed. Because carbon black absorbs light, you do not see any turbidity (that is, a Tyndall effect). Do you consider a tire to be a colloidal system and, if so, what kind? Explain.

6.55 On the basis of Tables 6.1 and 6.2, classify the following systems as homogeneous, heterogeneous, or colloidal mixtures.
(a) Physiological saline solution
(b) Orange juice
(c) A cloud
(d) Wet sand
(e) Suds
(f) Milk

6.56 Table 6.2 shows no examples of a gas-in-gas colloidal system. Considering the definition of a colloid, explain why.

6.57 A solution of protein is transparent at room temperature. When it is cooled to 10°C, it becomes turbid. What causes this change in appearance?

6.58 What gives nanotubes their unique optical and electrical properties?

Colligative Properties

6.59 Calculate the freezing points of solutions made by dissolving one mole of each of the following ionic solutes in 1000 g of H_2O.
(a) NaCl
(b) $MgCl_2$
(c) $(NH_4)_2CO_3$
(d) $Al(HCO_3)_3$

6.60 If we add 175 g of ethylene glycol, $C_2H_6O_2$, per 1000 g of water to a car radiator, what will be the freezing point of the solution?

6.61 Methanol is used as an antifreeze. How many grams of methanol, CH_3OH, do you need per 1000 g of water for an aqueous solution to stay liquid at $-20°C$?

6.62 In winter, after a snowstorm, salt (NaCl) is spread to melt the ice on roads. How many grams of salt per 1000 g of ice is needed to make it liquid at $-5.0°C$?

6.63 A 4 M acetic acid (CH_3COOH) solution lowered the freezing point by 8°C; a 4 M KF solution yields a 15°C; freezing-point depression. What can account for this difference?

Osmosis

6.64 In an apparatus using a semipermeable membrane, a 0.005 M glucose (small molecule) solution yielded an osmotic pressure of 10 mm Hg. What kind of osmotic pressure change would you expect if instead of a semipermeable membrane you used a dialysis membrane?

6.65 In each case, tell which side (if either) rises and why. The solvent is water.

Osmotic membrane

	A	B
(a)	1% glucose	5% glucose
(b)	0.1 M glucose	0.5 M glucose
(c)	1 M NaCl	1 M glucose
(d)	1 M NaCl	1 M K_2SO_4
(e)	3% NaCl	3% NaI
(f)	1 M NaBr	1 M NaCl

6.66 An osmotic semipermeable membrane that allows only water to pass separates two compartments, A and B. Compartment A contains 0.9% NaCl, and compartment B contains 3% glycerol, $C_3H_8O_3$.
(a) In which compartment will the level of solution rise?
(b) Which compartment has the higher osmotic pressure (if either)?

6.67 Calculate the osmolarity of each of the following solutions.
(a) 0.39 M Na_2CO_3
(b) 0.62 M $Al(NO_3)_3$
(c) 4.2 M LiBr
(d) 0.009 M K_3PO_4

6.68 Two compartments are separated by a semipermeable osmotic membrane through which only water molecules can pass. Compartment A contains a 0.3 M KCl solution, and compartment B contains a 0.2 M Na_3PO_4 solution. Predict from which compartment the water will flow to the other compartment.

6.69 A 0.9% NaCl solution is isotonic with blood plasma. Which solution would crenate red blood cells?
(a) 0.3% NaCl
(b) 0.9 M glucose
(c) 0.9% glucose (MW = 180)

Chemical Connections

6.70 (Chemical Connections 6A) Oxides of nitrogen (NO, NO_2, N_2O_3) are also responsible for acid rain. Which acids do you suppose can be formed from these oxides?

6.71 (Chemical Connections 6A) What makes normal rain water slightly acidic?

6.72 (Chemical Connections 6B) Why do deep-sea divers use a helium–oxygen mixture in the tank instead of air?

6.73 (Chemical Connections 6B) What is nitrogen narcosis?

6.74 (Chemical Connections 6C) Write balanced equations (two steps) for the conversion of marble to gypsum dihydrate.

6.75 (Chemical Connections 6C) What is the chemical formula for the main component of limestone and marble?

6.76 (Chemical Connections 6D) What is the protective colloid in milk?

6.77 (Chemical Connections 6E) What is the minimum pressure on sea water that will force water to flow from the concentrated solution into the dilute solution?

6.78 (Chemical Connections 6E) The osmotic pressure generated across a semipermeable membrane by a solution is directly proportional to its osmolarity. Given the data in Chemical Connections 6E on the purification of sea water, estimate what pressure you need to apply to purify a brackish water containing 5000 ppm salt by reverse osmosis.

6.79 (Chemical Connections 6F) A manufacturing error occurred in the isotonic solution used in hemodialysis. Instead of 0.2% $NaHCO_3$, an amount of 0.2% of $KHCO_3$ was added. Did this error make a change in the labeled tonicity of the solution? If so, is the resulting solution hypotonic or hypertonic? Would such an error create an electrolyte imbalance in the patient's blood? Explain.

6.80 (Chemical Connections 6F) The artificial kidney machine uses a solution containing 0.6% NaCl, 0.04% KCl, 0.2% $NaHCO_3$, and 0.72% glucose (all w/v). Show that this is an isotonic solution.

Additional Problems

6.81 Synovial fluid that exists in joints is a colloidal solution of hyaluronic acid (Section 19.7A) in water. When a biochemist wants to isolate hyaluronic acid from synovial fluid, he or she adds ethanol to bring the solution to 65% ethanol. The hyaluronic acid precipitates upon standing. What makes the hyaluronic acid solution unstable and causes it to precipitate?

6.82 When a cucumber is put into a saline solution to pickle it, the cucumber shrinks; when a prune is put into the same solution, the prune swells. Explain what happens in each case.

6.83 A solution of As_2O_3 has a molarity of 2×10^{-5} M. What is this concentration in ppm? (Assume that the density of the solution is 1.00 g/mL.)

6.84 Two bottles of water are carbonated, with CO_2 gas being added, under 2 atm pressure and then capped. One bottle is stored at room temperature; the other is stored in the refrigerator. When the bottle stored at room temperature is opened, large bubbles escape, along with a third of the water. The bottle stored in the refrigerator is opened without frothing or bubbles escaping. Explain.

6.85 Both methanol, CH_3OH, and ethylene glycol, $C_2H_6O_2$, are used as antifreeze. Which is more efficient—that is, which produces a lower freezing point if equal weights of each are added to the same weight of water?

6.86 You know that a 0.89% saline (NaCl) solution is isotonic with blood. In a real-life emergency, you run out of physiological saline solution and have only KCl as a salt and distilled water. Would it be acceptable to make a 0.89% aqueous KCl solution and use it for intravenous infusion? Explain.

6.87 Carbon dioxide and sulfur dioxide are soluble in water because they react with water. Write possible equations for these reactions.

6.88 A reagent label shows that it contains 0.05 ppm lead as a contaminant. How many grams of lead are present in 5.0 g of the reagent?

6.89 A concentrated nitric acid solution contains 35% HNO_3. How would you prepare 300 mL of 4.5% solution?

6.90 Which will have greater osmotic pressure:
(a) A 0.9% w/v NaCl solution?
(b) A 25% w/v solution of a nondissociating dextran with a molecular weight of 15,000?

6.91 Government regulations permit a 6 ppb concentration of a certain pollutant. How many grams of pollutant are allowed in 1 ton (1016 kg) of water?

6.92 The average osmolarity of sea water is 1.18. How much pure water would have to be added to 1.0 mL of sea water for it to achieve the osmolarity of blood (0.30 osmol)?

6.93 A swimming pool containing 20,000 L of water is chlorinated to have a final Cl_2 concentration of 0.00500 M. What is the Cl_2 concentration in ppm? How many kilograms of Cl_2 were added to the swimming pool to reach this concentration?

InfoTrac College Edition

For additional readings, go to InfoTrac College Edition, your online research library, at

http://infotrac.thomsonlearning.com

CHAPTER 7

7.1   *Introduction*

7.2   *Molecular Collisions*

7.3   *Activation Energy and Energy Diagrams*

7.4   *Factors Affecting Rates of Reaction*

7.5   *Reversible Reactions and Equilibrium*

7.6   *Equilibrium Constants*

7.7   *Equilibrium and Reaction Rates*

7.8   *Le Chatelier's Principle*

When magnesium burns in air, it creates a brilliant white light.

Reaction Rates and Chemical Equilibrium

7.1 Introduction

Some chemical reactions take place rapidly; others are very slow. For example, glucose and oxygen gas react with each other to form water and carbon dioxide:

$$C_6H_{12}O_6(s) + 6O_2(g) \longrightarrow 6CO_2(g) + 6H_2O(\ell)$$
Glucose

In the course of several years, a few molecules of glucose and O_2 will react, but not enough for us to detect within a laboratory period.

This reaction is extremely slow, however. A box of glucose exposed to O_2 in the air shows no measurable change even after many years.

In contrast, consider what happens when you take one or two aspirin tablets for a slight headache. Very often, the pain disappears in half an hour or so. Thus the aspirin must have reacted with compounds in the body within that time.

Many reactions occur even faster. For example, if we add a solution of silver nitrate ($AgNO_3$) to a solution of sodium chloride (NaCl) (see Sec-

184

tion 4.6), a precipitate of silver chloride (AgCl) forms almost instantaneously. The net ionic equation is as follows:

$$Ag^+(aq) + Cl^-(aq) \longrightarrow AgCl(s)$$

This reaction is essentially complete in considerably less than 1 s.

The study of reaction rates is called **chemical kinetics.** The **rate of a reaction** is the change in concentration of a reactant (or product) per unit time. Every reaction has its own rate, which must be measured in the laboratory.

Consider the following reaction carried out in the solvent acetone:

$$CH_3-Cl + I^- \xrightarrow{\text{Acetone}} CH_3-I + Cl^-$$
$$\text{Chloromethane} \hspace{2cm} \text{Iodomethane}$$

This is a net ionic equation. Spectator ions are not shown.

To determine the reaction rate, we can measure the concentration of the product, iodomethane (CH_3I), in the acetone at periodic time intervals—say, every 10 min. For example, the concentration might increase from 0 to 0.12 mol/L over a period of 30 min. The rate of the reaction is the change in the concentration of iodomethane divided by the time interval:

The rate could also be determined by following the decrease in concentration of CH_3Cl, if that is more convenient.

$$\frac{(0.12 \text{ mol } CH_3I/L) - (0 \text{ mol } CH_3I/L)}{30 \text{ min}} = \frac{0.12 \text{ mol } CH_3I/L}{30 \text{ min}}$$

$$= \frac{0.0040 \text{ mol } CH_3I/L}{\text{min}}$$

Screen 7.2

This unit is read "0.0040 mole per liter per minute." During each minute of the reaction, an average of 0.0040 mol of chloromethane, CH_3Cl, is converted to iodomethane for each liter of solution.

The rate of a reaction is not constant over a long period of time. At the beginning, in most reactions, the change in concentration is directly proportional to time. This period is shown as the linear portion of the curve in Figure 7.1. The rate calculated during this period, called the **initial rate,** is constant during this time interval. Later, as the reactant is used up, the rate of reaction decreases.

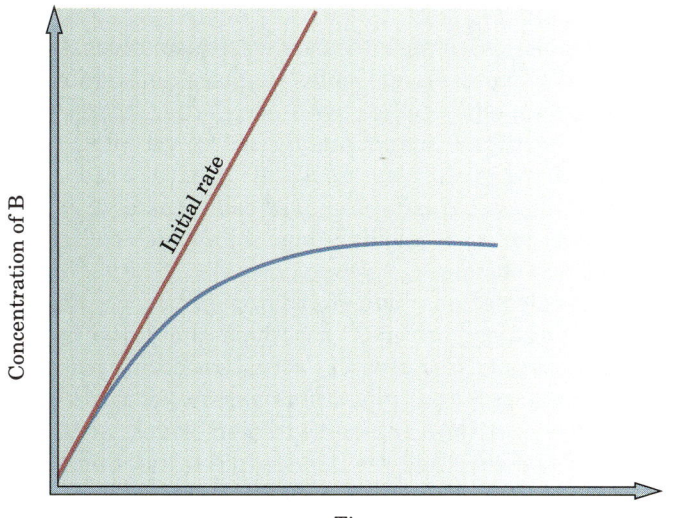

Figure 7.1 Changes in the concentration of B in the system A → B with respect to time. At the beginning the change is linear. This is the initial rate of reaction.

EXAMPLE 7.1

Another way of determining the rate of the preceding reaction is to measure the disappearance of the I^- from the solution. Suppose that the concentration of I^- was 0.24 mol I^-/L at the start of the reaction. At the end of 20 min, the concentration dropped to 0.16 mol I^-/L. This difference is equal to a change in concentration of 0.08 mol I^-/L. What is the rate of reaction?

Solution
The rate of the reaction is

$$(0.16 \text{ mol } I^-/L - 0.24 \text{ mol } I^-/L)/20 \text{ min} = (-0.0040 \text{ mol } I^-/L)/\text{min}$$

As the stoichiometry of the components is 1:1 in this reaction, we get the same numerical answer for the rate whether we monitor a reactant or a product. Note, however, that when we measure the concentration of a reactant that disappears with time, the rate of reaction is represented by a negative number.

Problem 7.1

In the reaction

$$2HgO(s) \longrightarrow 2Hg(l) + O_2(g)$$

we measure the evolution of O_2 gas to determine the rate of reaction. At the beginning of the reaction, at 0 min, 0.020 L of O_2 is present. After 15 min, the volume of O_2 gas is 0.35 L. What is the rate of reaction?

The rates of chemical reactions—both the ones that we carry out in the laboratory and the ones that take place inside our bodies—are very important. A reaction that goes more slowly than we need may be useless, whereas a reaction that goes too fast may be dangerous. Ideally, we would like to know what causes the enormous variety in reaction rates. In the next three sections, we examine this question.

7.2 Molecular Collisions

For two molecules or ions to react with each other, they must first collide. As we saw in Chapter 5, molecules in gases and liquids are constantly in motion and frequently collide with each other. If we want a reaction to take place between two compounds, A and B, we allow them to mix if they are gases or dissolve them in a solvent. In either case, the constant motion of the molecules will lead to frequent collisions between molecules of A and B. In fact, we can even calculate how many such collisions will take place in a given period of time.

Such calculations indicate that there are so many collisions between A and B molecules that most reactions should be over in considerably less than one second. Because the actual reactions are generally much slower, we must conclude that *most collisions do not result in a reaction*. Typically, when a molecule of A collides with a molecule of B, the two simply bounce apart without reacting. Every once in a while, though, a molecule of A and a molecule of B collide and react to form a new compound. A collision that results in a reaction is called an **effective collision.**

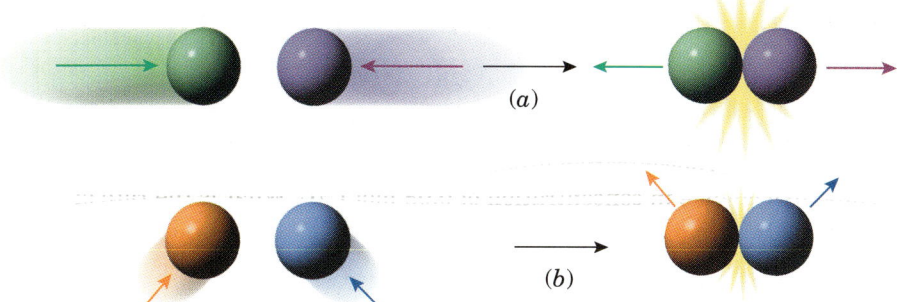

Figure 7.2 The energy of molecular collisions varies. Two fast-moving molecules colliding head-on (a) have a higher collision energy than two slower-moving molecules colliding at an angle (b).

Why are some collisions effective while others are not? There are two main reasons:

1. In most cases, for a reaction to take place between A and B, one or more covalent bonds must be broken in A or B or both, and energy is required for this to happen. The energy comes from the collision between A and B. If the energy of the collision is large enough, the bonds will break and the reaction will take place. If the collision energy is too low, the molecules will bounce apart without reacting. The minimum energy necessary for the reaction to happen is called the **activation energy.**

    The energy of any collision depends on the relative speed of the colliding objects and on the angle of approach. Much greater damage is done in a head-on collision of two cars both going 40 mph than in a collision in which a car going 20 mph sideswipes one going 10 mph. The same consideration applies with molecules, as Figure 7.2 shows.

2. Even if two molecules collide with an energy greater than the activation energy, a reaction still may not take place if the molecules are not oriented properly when they collide. Consider, for example, the reaction between H_2O and HCl:

$$H_2O(l) + HCl(g) \longrightarrow H_3O^+(aq) + Cl^-(aq)$$

For this reaction to take place, the molecules must collide in such a way that the H of the HCl hits the O of the water, as shown in Figure 7.3(a). A collision in which the Cl hits the O, as shown in Figure 7.3(b), cannot lead to a reaction, even if sufficient energy is available.

Screen 7.3
Animation

(a)   (b)

Figure 7.3 (a) The two molecules are properly oriented for a reaction to take place. (b) The orientation is such that a reaction cannot take place, because the Cl, not the H, collides with the O. The colored arrows show the path of the molecules.

*For an exothermic reaction, the products have less energy than the reactants.

Figure 7.4 Energy diagram for the exothermic reaction:

$$HCl(g) + H_2O(\ell) \longrightarrow H_3O^+(aq) + Cl^-(aq)$$

The energy of the reactants is greater than the energy of the products. The positions of two reacting molecules before, at, and after the transition state are indicated on the diagram.

Figure 7.5 The reaction

$$3H_2 + N_2 \rightarrow 2NH_3$$

In this reaction, six covalent bonds (three single bonds and one triple bond) are broken, and six new covalent bonds form.

Molecules tend to go spontaneously from higher to lower energy states.

Returning to the example given at the beginning of this chapter, we can now see why the reaction between $C_6H_{12}O_6$ and O_2 is so slow. The O_2 molecules are constantly colliding with glucose, but the percentage of effective collisions is extremely tiny at room temperature.

7.3 Activation Energy and Energy Diagrams

Figure 7.4 shows a typical energy diagram for an exothermic reaction. The products have a lower energy than the reactants; we might, therefore, expect the reaction to take place rapidly. As the curve shows, however, the reactants cannot be converted to products without the necessary activation energy. The activation energy is like a hill. If we are in a mountainous region, we may find that the only way to go from one point to another is to climb over a hill. It is the same in a chemical reaction. Even though the products may have a lower energy than the reactants [$H_3O^+(aq) + Cl^-(aq)$ have a lower energy than $HCl(g) + H_2O(l)$], the products cannot form unless the reactants "go over the hill"—that is, gain the necessary activation energy.

Let us look into this issue more closely. In a typical reaction, existing bonds are broken and new bonds form. For example, when H_2 reacts with N_2 to give NH_3, six covalent bonds (counting a triple bond as three bonds) must break, and six new covalent bonds must form (Figure 7.5). Breaking a bond requires an input of energy, but forming a bond releases energy. In a "downhill" reaction of the type shown in Figure 7.4, the amount of energy released in creating the new bonds is greater than that required to break the original bonds. In other words, the reaction is exothermic. Yet it may well have a substantial activation energy, or energy barrier, because, in most cases, at least one bond must break before any new bonds can form. Thus energy must be put into the system before we get any back. This is analogous to the following situation: Somebody offers to let you buy into a business from which, for an investment of $100,000, you could get an income of $40,000 per year, beginning in one year. In the long run, you would do very well. First, however, you need to put up the initial $100,000 (the activation energy).

"Downhill" reactions are exothermic.

Every reaction has a different energy diagram. Sometimes, the energy of the products is higher than that of the reactants ("uphill" reactions; see Figure 7.6). In almost all cases, however, there is an energy hill—the activation energy. The activation energy is inversely related to the rate of the reaction. *The lower the activation energy, the faster the reaction; the higher the activation energy, the slower the reaction.*

"Uphill" reactions are endothermic.

The top of the energy hill is called the **transition state.** When the reacting molecules reach this point, one or more original bonds are partially broken, and one or more new bonds may be in the process of formation. Figure 7.4 shows a typical situation.

Screen 17.4 Simulation

The speed of a reaction is proportional to the probability of effective collisions. In a single-step reaction, the probability that two particles will collide is greater than the probability of a simultaneous collision of five particles. If you consider the net ionic reaction

$$H_2O_2 + 3I^- + 2H^+ \longrightarrow I_3^- + 2H_2O$$

it is highly unlikely that six particles will collide simultaneously; thus this reaction should be slow. In reality, this reaction is very fast. This fact indicates that the reaction cannot occur in one step but rather takes place in

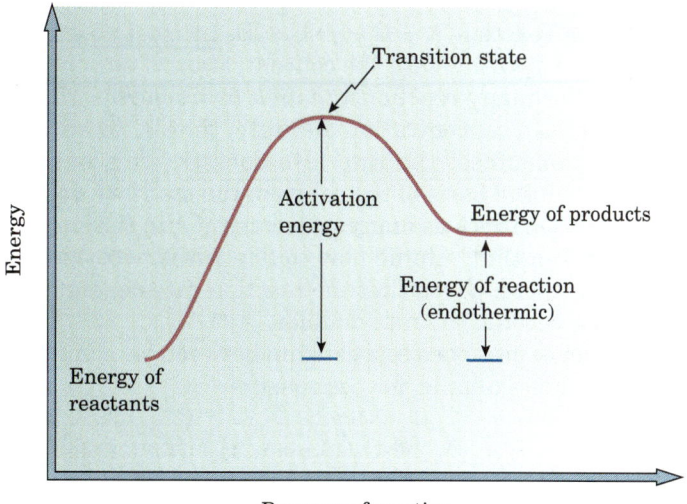

Figure 7.6 Energy diagram for an endothermic reaction. The energy of the products is greater than that of the reactants.

multiple steps. In each of those steps, the probability is high for collisions between two particles. Even a simple reaction such as

$$H_2(g) + Br_2(g) \longrightarrow 2HBr(g)$$

occurs in three steps:

1. $Br_2 \longrightarrow 2Br\cdot$
2. $H_2 + Br\cdot \longrightarrow HBr + H\cdot$
3. $H\cdot + Br_2 \longrightarrow HBr + Br\cdot$

The · indicates the single unpaired electron in the atom. The overall rate of the reaction will be controlled by the slowest of the three steps, just as the slowest-moving car controls the flow of traffic on a street. In the preceding reaction, step 1 is the slowest, because it has the highest activation energy.

7.4 Factors Affecting Rates of Reaction

In Section 7.2, we saw that reactions occur as a result of collisions between fast-moving molecules possessing a certain minimum energy (the activation energy). In this section, we examine some of the factors that affect activation energies and reaction rates.

A Nature of the Reactants

In general, reactions that take place between ions in aqueous solution (Section 4.6) are extremely rapid, occurring almost instantaneously. Activation energies for these reactions are very low because usually no covalent bonds must be broken. As we might expect, reactions between covalent molecules, whether in aqueous solution or not, are much slower. Many of these reactions require 15 min to 24 h or longer for most of the starting compounds to be converted to the products. Some reactions take a good deal longer, of course, but they are seldom useful.

B Concentration

Consider a reaction $A + B \longrightarrow C + D$. In most cases, the reaction rate increases when we increase the concentration of either or both reactants (Figure 7.7). For many reactions, *though by no means all,* a direct relationship exists between concentration and rate; that is, when the concentration of a reactant is doubled, the rate also doubles. This outcome is easily understandable on the basis of the collision theory. If we double the concentration of A, there are twice as many molecules of A in the same volume, so the molecules of B in that volume now collide with twice as many A molecules per second as before. Given that the reaction rate depends on the number of collisions per second, the rate doubles.

We can express the relationship between rate and concentration mathematically. For example, for the reaction

$$2H_2O_2(\ell) \longrightarrow 2H_2O(\ell) + O_2(g)$$

the rate was determined to be -0.01 mol H_2O_2/L/min at a constant temperature when the initial concentration of $[H_2O_2]$ was 1 mol/L. In other words, every minute 0.01 mol/L of hydrogen peroxide was used up. Researchers

---

For reactions in the gas phase, an increase in pressure usually increases the rate.

In the case where one of the reactants is a solid, the rate is affected by the surface area of the solid. For this reason, a substance in powder form reacts faster than the same substance in the form of large chunks.

The brackets [ ] stand for the concentration of the chemical species whose formula is given between the brackets.

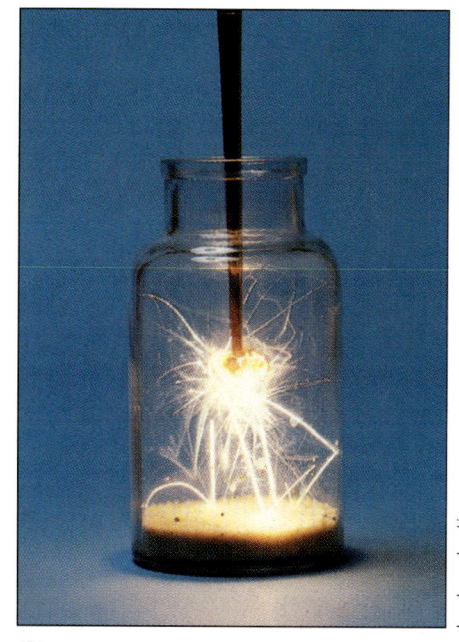

(a) (b)

Figure 7.7 (a) When heated in air, steel wool glows but does not burn rapidly because the O_2 concentration of the air is only about 20%. (b) When the glowing steel wool is put into 100% O_2, it burns vigorously.

also found that every time the concentration of H_2O_2 was doubled, the rate also doubled. Thus the rate is directly proportional to the concentration of H_2O_2. We can write this relationship as

$$\text{Rate} = k[H_2O_2]$$

where the brackets around the H_2O_2 indicate the concentration and k is a constant, called the **rate constant.** Rate constants are usually calculated from the **initial rates of reaction** (Figure 7.1).

Screen 17.5
Simulation

EXAMPLE 7.2

Calculate the rate constant for the reaction

$$2H_2O_2(\ell) \longrightarrow 2H_2O(\ell) + O_2(g)$$

using the rate and the initial H_2O_2 concentration mentioned in the preceding discussion:

$$\text{Rate} = -0.01 \text{ mol } H_2O_2/\text{L/min}; \ [H_2O_2] = 1 \text{ mol/L}$$

Solution
We solve the equation

$$\text{Rate} = k[H_2O_2]$$

for k.

$$k = \text{Rate}/[H_2O_2]$$
$$= (-0.01 \text{ mol } H_2O_2/\text{L/min})/(1 \text{ mol } H_2O_2/\text{L})$$
$$= -0.01/\text{min}$$

Problem 7.2
Calculate the rate of reaction for the reaction in Example 7.2 when the initial concentration of H_2O_2 is 0.36 mol/L.

C Temperature

In virtually all cases, reaction rates increase with increasing temperature. A rule of thumb for many reactions is that every time the temperature goes up by 10°C, the rate of reaction doubles. This rule is far from exact, but it is not far from the truth in many cases. As you can see, this effect is quite large. It says, for example, that if we run a reaction at 90°C instead of at room temperature (20°C), the reaction will go about 128 times faster. Put another way, if it takes 20 h to convert 100 g of reactant A to product C at 20°C, then it would take only 10 min at 90°C. Temperature, therefore, is a powerful tool that lets us increase the rates of reactions that are inconveniently slow. It also lets us decrease the rates of reactions that are inconveniently fast. For example, we might choose to run reactions at low temperatures because explosions would result or the reactions would otherwise be out of control at room temperature.

There are seven 10° increments between 20° and 90°, and $2^7 = 128$.

What causes reaction rates to increase with increasing temperature? Once again, we turn to collision theory. Here temperature has two effects:

1. In Section 5.6, we learned that temperature is related to the average kinetic energy of molecules. When the temperature increases, molecules move more rapidly, which means that they collide more frequently. More frequent collisions mean higher reaction rates. However, this factor is much less important than the second factor.

2. Recall from Section 7.2 that a reaction between two molecules takes place only if an effective collision occurs—a collision with an energy equal to or greater than the activation energy. When the temperature increases, not only is the average speed of the molecules greater, but there is also a different distribution of speeds. The number of very fast molecules increases much more than the average speed (Figure 7.8). As a consequence, the number of effective collisions rises even more than

CHEMICAL CONNECTIONS 7A

Why High Fever Is Dangerous

Chemical Connections 1B points out that a sustained body temperature of 41.7°C (107°F) is invariably fatal. We can now see why a high fever is dangerous. Normal body temperature is 37°C (98.6°F), and all the many reactions in the body—including respiration, digestion, and the synthesis of various compounds—take place at that temperature. If an increase of 10°C causes the rates of most reactions to approximately double, then an increase of even 1°C makes them go significantly faster than normal.

Fever is a protective mechanism, and a small increase in temperature allows the body to kill germs faster by mobilizing the immune defense mechanism. This increase must be small, however: A rise of 1°C brings the temperature to 38°C (100.4°F); an increase of 3°C brings it to 40°C (104°F). A temperature higher than 104°F increases reaction rates to the danger point.

One can easily detect the increase in reaction rates when a patient has a high fever. The pulse rate increases and breathing becomes faster as the body attempts to supply increased amounts of oxygen for the accelerated reactions. A marathon runner, for example, may become overheated on a hot and humid day. After a time, perspiration can no longer cool his or her body effectively, and the runner may suffer hyperthermia or heat stroke, which, if not treated properly, can cause brain damage.

CHEMICAL CONNECTIONS 7B

The Effects of Lowering Body Temperature

Like a significant increase in body temperature, a substantial decrease in body temperature below 37°C (98.6°F) can prove harmful because reaction rates are abnormally low. It is sometimes possible to take advantage of this effect. In some heart operations, for example, it is necessary to stop the flow of oxygen to the brain for a considerable time. At 37°C (98.6°F), the brain cannot survive without oxygen for longer than about 5 min without suffering permanent damage. When the patient's body temperature is deliberately lowered to about 28 to 30°C (82.4 to 86°F), however, the oxygen flow can be stopped for a considerable time without causing damage because reaction rates slow down. At 25.6°C (78°F), the body's oxygen consumption is reduced by 50%.

■ An operating table unit monitors a patient packed in ice.

the total number of collisions. Not only do more collisions take place, but the percentage of collisions that have an energy greater than the activation energy also rises. This factor is mainly responsible for the sharp increase in reaction rates with increasing temperature.

D Presence of a Catalyst

Any substance that increases the rate of a reaction without itself being used up is called a **catalyst**. Many catalysts are known—some that increase the rate of only one reaction and others that can affect several reactions. Although we have seen that we can speed up reactions by increasing the temperature, in some cases they remain too slow even at the highest temperatures we can conveniently reach. In other cases, it is not feasible to increase the temperature—perhaps because other, unwanted reactions would be speeded up, too. In such cases, a catalyst, if we can find the right one for a given reaction, can prove very valuable. Many important industrial processes rely on catalysts (see Chemical Connections 7E), and virtu-

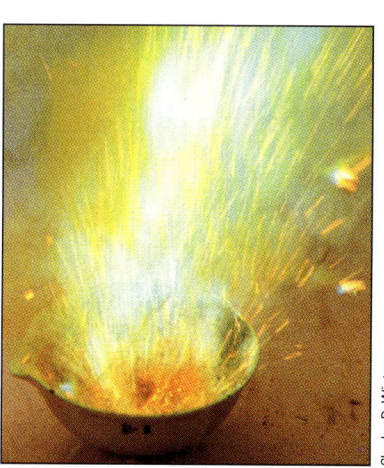

■ In this dish, Cl^- ions act as a catalyst in the decomposition of NH_4NO_3.

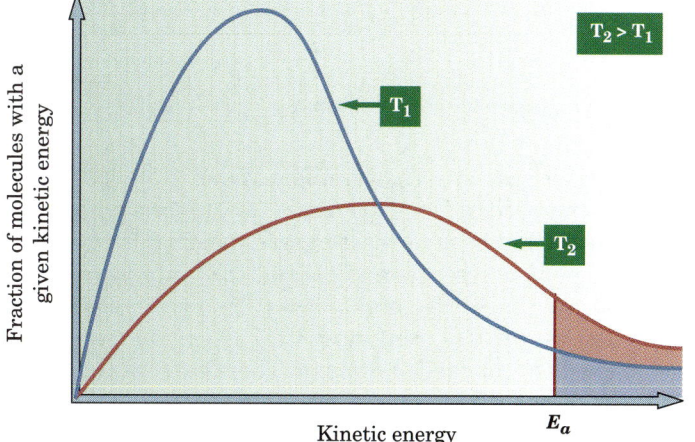

Figure 7.8 Distribution of kinetic energies (molecular velocities) at two temperatures. The kinetic energy on the x-axis designated by E_a indicates the energy (molecular velocity) necessary to pass through the activation energy barrier. The shaded areas represent the fraction of molecules that have kinetic energies (molecular velocities) greater than the activation energy.

CHEMICAL CONNECTIONS 7C

Timed-Release Medication

It is often desirable that a particular medicine act slowly and maintain its action evenly in the body for 24 h. We know that a solid in powder form reacts faster than the same weight in pill form because the powder has a greater surface area at which the reaction can take place. To slow the reaction and to have the drug be delivered evenly to the tissues, pharmaceutical companies coat beads of some of their drugs. The coating prevents the drug from reacting for a time. The thicker the coating, the longer it takes the drug to react. A drug with a smaller bead size has more surface area than a drug with a larger bead size; hence, drugs packaged in a smaller bead size will react more rapidly.

■ Two packages of timed-release medicines.

By combining the proper bead size with the proper amount of coating, the drug can be designed to deliver its effect over a 24-h period. In this way, the patient needs to take only one pill per day.

Coating can also prevent the problems related to stomach irritation. For example, aspirin can cause stomach ulceration or bleeding in some people. Enteric-coated aspirin tablets have a polymeric coat that is acid resistant. Such a drug does not dissolve until it reaches the intestines, where it causes no harm.

Screen 7.6 Exercise

Enteric (from the Greek *enteron*) means affecting the intestines.

Heterogeneous catalyst A catalyst in a separate phase from the reactants—for example, the solid platinum in the reaction between $C_6H_{10}O$ and H_2

Homogeneous catalyst A catalyst in the same phase as the reactants—for example, enzymes in body tissues

We often write the catalyst over or under the arrow.

ally all reactions that take place in living organisms are catalyzed by enzymes (Chapter 22).

Catalysts work by allowing the reaction to take a different pathway, one with a lower activation energy. Without the catalyst, the reactants would have to get over the higher energy hill shown in Figure 7.9. The catalyst provides a lower hill. As we have seen, a lower activation energy means a higher reaction rate.

Each catalyst has its own way of providing an alternative pathway. Many catalysts provide a surface on which the reactants can meet. For example, the reaction between cyclohexanone, $C_6H_{10}O$, and hydrogen gas, H_2, to give cyclohexanol, $C_6H_{12}O$, goes so slowly without a catalyst that it is not practical, even if we increase the temperature to a reasonable level. If the mixture of gases is shaken with finely divided solid platinum, however, the reaction takes place at a convenient rate (Section 12.6D). The $C_6H_{10}O$ molecules and the H_2 molecules meet each other on the surface of the platinum, where the proper bonds can be broken and the reaction can proceed:

$$C_6H_{10}O(\ell) + H_2(g) \xrightarrow{Pt} C_6H_{12}O(\ell)$$

7.5 Reversible Reactions and Equilibrium

Many reactions are irreversible. When a piece of paper is completely burned, the products are CO_2 and H_2O. Anyone who takes pure CO_2 and H_2O and tries to make them react to give paper and oxygen will not succeed.

A tree, of course, turns CO_2 and H_2O into wood and oxygen, and we, in sophisticated factories, make paper from the wood. These activities are not the same as directly combining CO_2, H_2O, and energy in a single process to get paper and oxygen, however. Therefore, we can certainly consider the burning of paper to be an irreversible reaction.

Other reactions are reversible. A **reversible reaction** is one that can be made to go in either direction. For example, if we mix carbon monoxide

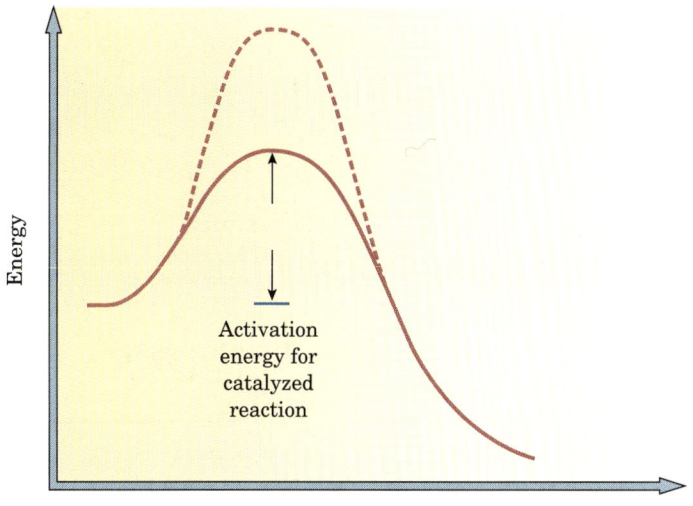

Figure 7.9 Energy diagram for a catalyzed reaction. The dashed line shows the energy curve for the uncatalyzed process. The catalyst provides an alternate pathway whose activation energy is lower.

with water in the gas phase at a high temperature, hydrogen and carbon dioxide are produced:

$$CO(g) + H_2O(g) \longrightarrow CO_2(g) + H_2(g)$$

If we desire, we can also make this reaction take place the other way. That is, we can mix hydrogen and carbon dioxide to get carbon monoxide and water vapor:

$$CO_2(g) + H_2(g) \longrightarrow CO(g) + H_2O(g)$$

Let us see what happens when we run a reversible reaction. We will add some carbon monoxide to water vapor in the gas phase. The two compounds begin to react at a certain rate (the forward reaction):

$$CO(g) + H_2O(g) \longrightarrow CO_2(g) + H_2(g)$$

As the reaction proceeds, the concentrations of CO and H_2O gradually decrease because both reactants are being used up. In turn, the rate of the reaction gradually decreases because the rate of a reaction depends on the concentrations of the reactants (Section 7.4).

But what is happening in the other direction? Before we added the carbon monoxide, no carbon dioxide or hydrogen was present. As soon as the forward reaction began, it produced small amounts of these substances, and we now have some CO_2 and H_2. These two compounds will now, of course, begin reacting with each other (the reverse reaction):

$$CO_2(g) + H_2(g) \longrightarrow CO(g) + H_2O(g)$$

At first, the reverse reaction is very slow. As the concentrations of H_2 and CO_2 (produced by the forward reaction) gradually increase, the rate of the reverse reaction also gradually increases.

We have a situation, then, in which the rate of the forward reaction gradually decreases, while the rate of the reverse reaction (which began at zero) gradually increases. Eventually the two rates become equal. At this point, we say that the process is in **dynamic equilibrium** (or just **equilibrium**).

The units of concentration in the gas phase are the same as in solution: moles per liter.

Dynamic equilibrium A state in which the rate of the forward reaction equals the rate of the reverse reaction

Figure 7.10 Changes in the concentrations of A, B, C, and D in the A + B ⇌ C + D system, as equilibrium is approached. Only A and B are present at the beginning of the reaction.

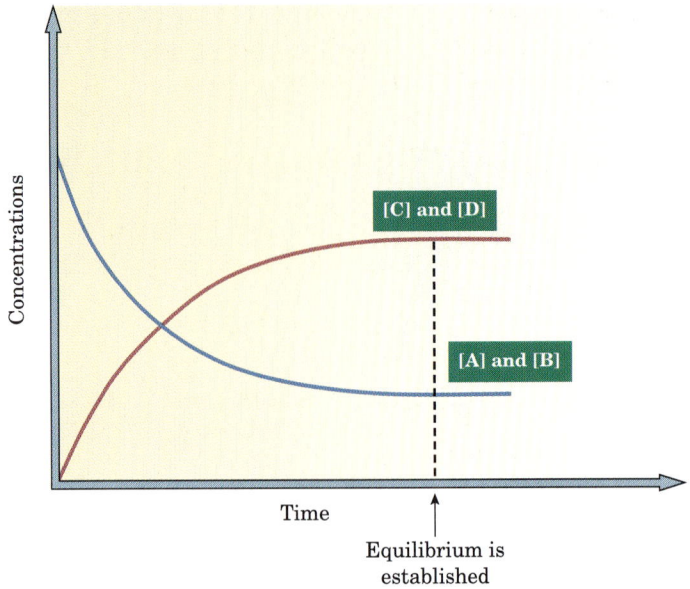

In the reaction

$$CO_2(g) + H_2(g) \rightleftharpoons CO(g) + H_2O(g)$$

Screen 7.8 Simulation

we use a double arrow to indicate that the reaction is reversible.

What happens in the reaction container once we reach equilibrium? If we measure the concentrations of the substances in the container, we find that *no change in concentration takes place after equilibrium is reached* (Figure 7.10). Whatever the concentrations of all the substances are at equilibrium, they remain the same forever unless something is done to disturb the equilibrium (as discussed in Section 7.8). This does not mean that all the concentrations must be the same—all of them can, in fact, be different and usually are—but it does mean that, whatever they are, they no longer change once equilibrium has been reached, no matter how long we wait.

Given that the concentrations of all the reactants and products no longer change, can we say that nothing is happening? No, we know that both reactions are going on; all the molecules are constantly reacting—the CO and H_2O are being changed to CO_2 and H_2, and the CO_2 and H_2 are being changed to CO and H_2O. Because the rates of the forward and reverse reactions are the same, however, none of the concentrations change.

In the example just discussed, we approached equilibrium by adding carbon monoxide to water vapor. We could also have added carbon dioxide to hydrogen. In either case, we eventually get an equilibrium mixture containing the same four compounds (Figure 7.11).

Another way to look at this situation is to say that the concentration of carbon monoxide (and of the other three compounds) does not change because it is being used up as fast as it is being formed.

Figure 7.11 An equilibrium can be approached from either direction.

It is not necessary to begin with equal amounts. We could, for example, take 10 moles of carbon monoxide and 0.2 mole of water vapor. We would still arrive at an equilibrium mixture of all four compounds.

7.6 Equilibrium Constants

We have already met two examples of dynamic equilibrium in this book: the equilibrium between a liquid and its vapor (Section 5.9) and the equilibrium involving solute particles in saturated solutions (Section 6.4).

Chemical equilibria can be treated by a simple mathematical expression. First, let us write

$$a\text{A} + b\text{B} \rightleftharpoons c\text{C} + d\text{D}$$

as the general equation for all reversible reactions. In this equation, the capital letters stand for substances—CO_2 and H_2O, for instance—and the small letters are the coefficients of the balanced equation. The double arrow shows that the reaction is reversible. In general, any number of substances can be present on either side.

Once equilibrium is reached, the following equation is valid:

$$K = \frac{[\text{C}]^c[\text{D}]^d}{[\text{A}]^a[\text{B}]^b} \quad \text{The \textbf{equilibrium expression}}$$

This expression was first derived in 1863 by Cato Guldberg (1836–1902) and Peter Waage (1833–1900), two Norwegian chemists.

Let us examine this equation. K is a constant called the **equilibrium constant.** The equilibrium expression tells us that, when we multiply the equilibrium concentrations of the substances on the right side of the chemical equation and divide this product by the equilibrium concentrations of the substances on the left side (after raising each number to the appropriate power), we get the equilibrium constant, a number that does not change.

Let us look at several examples of how to set up equilibrium expressions.

Equilibrium constant A value calculated from the equilibrium expression for a given reaction indicating in which direction the reaction goes

Screen 7.11
Tutorial

EXAMPLE 7.3

Write the equilibrium expression for the reaction

$$\text{CO(g)} + \text{H}_2\text{O(g)} \rightleftharpoons \text{CO}_2\text{(g)} + \text{H}_2\text{(g)}$$

Solution

$$K = \frac{[\text{CO}_2][\text{H}_2]}{[\text{CO}][\text{H}_2\text{O}]}$$

This expression tells us that, at equilibrium, the concentration of carbon dioxide multiplied by the concentration of hydrogen and divided by the concentrations of water and carbon monoxide is a constant, K. Note that no exponent is written in this equation because all the coefficients of the balanced chemical equation are one, and, by convention, an exponent of 1 is not written. Mathematically, it would be just as correct to write the left-side compounds on top, but the universal custom is to write them as shown here, with the products on top and the reactants on the bottom.

It is understood that the concentration of the species within the brackets is always expressed in moles per liter.

The concentration terms in the brackets for K are for species in solution, either liquid or gaseous.

Problem 7.3
Write the equilibrium expression for the reaction

$$\text{SO}_3\text{(g)} + \text{H}_2\text{O}(\ell) \rightleftharpoons \text{H}_2\text{SO}_4\text{(aq)}$$

EXAMPLE 7.4

Write the equilibrium expression for the reaction

$$O_2(g) + 4ClO_2(g) \rightleftharpoons 2Cl_2O_5(g)$$

Solution

$$K = \frac{[Cl_2O_5]^2}{[O_2][ClO_2]^4}$$

In this case, the chemical equation has coefficients other than unity, so the equilibrium expression contains exponents.

Problem 7.4

Write the equilibrium expression for the reaction

$$2NH_3(g) \rightleftharpoons N_2(g) + 3H_2(g)$$

Screen 7.9 Simulation

Now let us see how K is calculated.

EXAMPLE 7.5

Some H_2 is added to I_2 at 427°C, and the following reaction is allowed to come to equilibrium:

$$I_2(g) + H_2(g) \rightleftharpoons 2\,HI(g)$$

When equilibrium is reached, the concentrations are found to be $[I_2] = 0.42$ mol/L, $[H_2] = 0.025$ mol/L, and $[HI] = 0.76$ mol/L. Calculate K at 427°C.

Solution
The equilibrium expression is

$$K = \frac{[HI]^2}{[I_2][H_2]}$$

Substituting the concentrations, we get

$$K = \frac{[(0.76)^2 M^2]}{[0.42\,M][0.025\,M]} = 55$$

The equilibrium constant of this reaction is unitless because the molarities cancel. Depending on the stoichiometry of the reaction, equilibrium constants may have different units, such as M, M^2, M^{-1}, and so on.

Problem 7.5

What is the equilibrium constant for the following reaction if the equilibrium concentrations are $[PCl_3] = 1.66\,M$, $[Cl_2] = 1.66\,M$, and $[PCl_5] = 1.66\,M$?

$$PCl_3 + Cl_2 \rightleftharpoons PCl_5$$

The last example shows us that the reaction between I_2 and H_2 to give HI has an equilibrium constant of 55. What does this value mean? At constant temperature, equilibrium constants remain the same no matter what concentrations we have. That is, at 427°C, if we begin by adding, say, 5 moles of H_2 to 5 moles of I_2, the forward and backward reactions will take place, and equilibrium will eventually be reached. At that point, the value of $[HI]^2/[I_2][H_2]$ will equal 55. If we begin at 427°C with different numbers of moles of H_2 and I_2, perhaps 7 moles of H_2 and 2 moles of I_2, once equilibrium is reached, the value of $[HI]^2/[I_2][H_2]$ will again be 55. It makes no difference what the initial concentrations of the three substances are. At 427°C, as long as all three substances are present and equilibrium has been reached, the concentrations of the three substances will adjust themselves so that the value of the equilibrium expression equals 55.

The equilibrium constant is different for every reaction. Some reactions have a large K; others have a small K. A reaction with a very large K proceeds almost to completion (to the right). For example, K for the following reaction is about 100,000,000, or 10^8, at 25°C:

$$N_2(g) + 3H_2(g) \rightleftharpoons 2NH_3(g)$$

The symbol $\rightleftharpoons$ means that the equilibrium lies far to the right.

This value of K means that, at equilibrium, $[NH_3]$ must be very large and $[N_2]$ and $[H_2]$ must be very small so that $[NH_3]^2/[N_2][H_2]^3$ can be equal to 10^8. Thus if we add N_2 to H_2, we can be certain that, when equilibrium is reached, an essentially complete reaction has taken place.

On the other hand, a reaction such as

$$AgCl(s) \rightleftharpoons Ag^+(aq) + Cl^-(aq)$$

with a very small K, 1.77×10^{-10}, hardly goes forward at all. Equilibrium effects are most obvious in reactions with K values between 10^3 and 10^{-3}. In such cases, the reaction goes part of the way, and significant concentrations of all substances are present at equilibrium. An example is the reaction between carbon monoxide and water discussed in Section 7.5, for which K is equal to 10 at 600°C.

Screen 7.10
Simulation

In *dilute solutions* it is customary to omit the concentration of the solvent from the equilibrium expression. Consider

$$NH_3(aq) + H_2O(\ell) \rightleftharpoons NH_4^+(aq) + OH^-(aq)$$

If the concentration of ammonia is small, and consequently the concentrations of ammonium ion and hydroxide ion are small, *the concentration of water molecules remains essentially the same*. Because the molar concentration of water is effectively constant, we do not include it in the equilibrium expression:

$$K = \frac{[NH_4^+][OH^-]}{[NH_3]}$$

EXAMPLE 7.6

Calculate the equilibrium constant for the above reaction (a) with and (b) without the inclusion of water in the equilibrium expression. The equilibrium concentrations are as follows: $[NH_3] = 0.0100\ M$; $[NH_4^+] = 0.000400\ M$; $[OH^-] = 0.000400\ M$.

Solution

(a) Since molarity is based on 1 L,

$$[H_2O] = (1000.0 \text{ g } H_2O)/(18.00 \text{ g } H_2O/\text{mol})/L = 55.56 \, M$$

$$K = \frac{[0.000400 \, M][0.000400 \, M]}{[0.0100 \, M][55.56 \, M]} = 2.88 \times 10^{-7}$$

(b)

$$K = \frac{[0.000400 \, M][0.000400 \, M]}{[0.0100 \, M]} = 1.60 \times 10^{-5} \, M$$

The inclusion of the concentration of the solvent, water, gives an equilibrium constant two magnitudes smaller than without it. The equilibrium constant without the inclusion of the solvent is the accepted one.

Problem 7.6

Ethyl acetate is a common solvent in many industrial products, ranging from nail enamel and remover to liquid cement for plastics. Write the equilibrium expressions for the reaction

$$\underset{\text{Acetic acid}}{CH_3COOH(\ell)} + \underset{\text{Ethanol}}{C_2H_5OH(\ell)} \rightleftharpoons \underset{\text{Ethyl acetate}}{CH_3COOC_2H_5(\ell)} + H_2O(\ell)$$

(a) when we start with pure reactants and (b) when the reaction is run in dilute aqueous solution and the concentration of water is not included in the equilibrium expression.

The equilibrium constant for a given reaction remains the same no matter what happens to the concentrations, but the same is not true for changes in temperature. The value of K does change when the temperature changes.

7.7 Equilibrium and Reaction Rates

As pointed out in Section 7.6, the equilibrium expression is valid only after equilibrium has been reached. Before that point, there is no equilibrium, and the equilibrium expression is not valid. But how long does it take for a reaction to reach equilibrium? There is no easy answer to this question. Some reactions, if the reactants are well mixed, reach equilibrium in less than one second; others will not get there even after millions of years.

There is no relationship between the rate of a reaction (how long it takes to reach equilibrium) and the value of K. It is possible to have a large K and a low rate, as in the reaction between glucose and O_2 to give H_2O and CO_2, which does not reach equilibrium for many years (Section 7.1), or a small K and a high rate. In other reactions, both the rate and K are large or both are small.

7.8 Le Chatelier's Principle

When a reaction reaches equilibrium, the forward and reverse reactions take place at the same rate, and the concentrations of all the components do not change as long as we don't do anything to the system. But what happens if we do? In 1888 Henri Le Chatelier (1850–1936) put forth the statement known as **Le Chatelier's principle:** If an external stress is applied to a system in equilibrium, the system reacts in such a way as to partially relieve that stress. Let us look at three types of stress that can be put on chemical equilibria: adding and removing a reactant or product, and changing the temperature.

A Addition of a Reaction Component

Suppose that the reaction

$$CH_3COOH + C_2H_5OH \underset{}{\overset{HCl}{\rightleftarrows}} CH_3COOC_2H_5 + H_2O$$
Acetic acid · Ethanol · Ethyl acetate

HCl is a catalyst, but it does not affect the equilibrium.

has reached equilibrium. This means that the reaction flask contains all these substances and that the concentrations no longer change. We now disturb the system by adding some acetic acid. The result is that the concentration of acetic acid suddenly increases, which increases the rate of the forward reaction. As a consequence, the concentrations of the products (ethyl acetate and water) begin to increase. At the same time, the concentrations of reactants decrease. Now, an increase in the concentrations of products causes the rate of the reverse reaction to increase, but the rate of the forward reaction is decreasing, so eventually the two rates will be equal again and a new equilibrium will be established.

Screen 7.14
Simulation Tutorial

When that happens, the concentrations are once again constant, but they are not the same as they were before the addition of the acetic acid. The concentrations of ethyl acetate and water are higher now, and the concentration of ethanol is lower. The concentration of acetic acid is higher because we added some, but it is less than it was immediately after we made the addition.

The equilibrium constant K, of course, remains the same.

When we add more of any component to a system in equilibrium, that addition constitutes a stress. The system relieves this stress by increasing the concentrations of the components on the other side of the equation. We say that the equilibrium shifts in the opposite direction:

Adding CH_3COOH $CH_3COOH + C_2H_5OH \overset{HCl}{\rightleftarrows} CH_3COOC_2H_5 + H_2O$
Equilibrium shift →

The addition of acetic acid, on the left side of the equation, causes the rate of the forward reaction to increase and the reaction to move toward the right: More ethyl acetate and water form, and some of the acetic acid and ethanol are used up. The same thing happens if we add ethanol, C_2H_5OH.

On the other hand, if we add water or ethyl acetate, the rate of the reverse reaction increases, and the reaction shifts to the left:

$CH_3COOH + C_2H_5OH \overset{HCl}{\rightleftarrows} CH_3COOC_2H_5 + H_2O$ Adding $CH_3COOC_2H_5$
← Equilibrium shift

We can summarize by saying that the addition of any component causes the equilibrium to shift to the opposite side.

■ **Headstone damaged by acid rain.**

■ The tube at the left contains a saturated solution of silver acetate (Ag$^+$ ions and CH$_3$COO$^-$ ions) in equilibrium with solid silver acetate. When silver ions in the form of silver nitrate solution are added, the equilibrium

Ag$^+$(aq) + CH$_3$COO$^-$(aq) $\rightleftharpoons$
 CH$_3$COOAg(s)

shifts to the right, producing more solid silver acetate, as can be seen in the tube on the right.

EXAMPLE 7.7

When dinitrogen tetroxide, N$_2$O$_4$, a colorless gas, is enclosed in a vessel, a color indicating the formation of brown nitrogen dioxide, NO$_2$, soon appears (see Figure 7.12 later in this chapter). The intensity of the brown color indicates the amount of nitrogen dioxide formed. The equilibrium reaction is

$$N_2O_4(g) \rightleftharpoons 2NO_2(g)$$

When more N$_2$O$_4$ is added to the equilibrium mixture, the brownish color becomes darker. Explain what happened.

Solution
The darker color indicates that more nitrogen dioxide is formed. This happens because the addition of the reactant, N$_2$O$_4$, shifts the equilibrium to the right, forming more product, NO$_2$.

Problem 7.7
What happens to the equilibrium reaction

$$2NOBr(g) \rightleftharpoons 2NO(g) + Br_2(g)$$
Nitrosyl bromide

when Br$_2$ gas is added to the equilibrium mixture?

B Removal of a Reaction Component

It is not always as easy to remove a component from a reaction mixture as it is to add one, but there are often ways to do it. The removal of a component, or even a decrease in its concentration, lowers the corresponding reaction rate and changes the position of the equilibrium. If we remove a reactant, the reaction shifts to the left, toward the side from which the reactant was removed. If we remove a product, the reaction shifts to the right, toward the side from which the product was removed.

In the case of the acetic acid–ethanol equilibrium, ethyl acetate has the lowest boiling point of the four components and can be removed by distillation. The equilibrium then shifts to that side so that more ethyl acetate is produced to compensate for the removal:

$$CH_3COOH + C_2H_5OH \underset{}{\overset{HCl}{\rightleftharpoons}} H_2O + CH_3COOC_2H_5 \quad \text{Removing } CH_3COOC_2H_5$$
Equilibrium shift $\longrightarrow$

The concentrations of CH$_3$COOH and C$_2$H$_5$OH decrease, and the concentration of H$_2$O increases. The effect of removing a component is thus the opposite of adding one. The removal of a component causes the equilibrium to shift to the side from which the component was removed.

No matter what happens to the individual concentrations, the value of the equilibrium constant remains unchanged.

EXAMPLE 7.8

The beautiful stone we know as marble is mostly calcium carbonate. When acid rain containing sulfuric acid attacks marble, the following equilibrium reaction can be written:

$$CaCO_3(s) + H_2SO_4(aq) \rightleftharpoons CaSO_4(s) + CO_2(g) + H_2O(\ell)$$

How does the fact that CO_2 is a gas influence the equilibrium?

Solution
The gaseous CO_2 diffuses away from the reaction site, meaning that this product is removed from the equilibrium mixture. The equilibrium shifts to the right, so that the statue keeps eroding more and more.

Problem 7.8
In the reaction

$$2H_2O_2(aq) \rightleftharpoons 2H_2O(\ell) + O_2(g)$$

oxygen has limited solubility in water (see the table in Chemical Connections 6A). What happens to the equilibrium after the solution becomes saturated with oxygen?

C Change in Temperature

The effect of a change in temperature on a reaction that has reached equilibrium depends on whether the reaction is exothermic (gives off heat) or endothermic (requires heat). Let us look at an exothermic reaction:

Screen 7.13
Tutorial

$$2H_2(g) + O_2(g) \rightleftharpoons 2H_2O(\ell) + 137,000 \text{ cal}$$

If we consider the heat to be a product of this reaction, then we can use the same type of reasoning as we did before. An increase in temperature means that we are adding heat. Because heat is a product, its addition pushes the equilibrium to the opposite side. We can therefore say that, if the reaction is at equilibrium and we increase the temperature, the reaction goes to the left—the concentrations of H_2 and O_2 increase and that of H_2O decreases. This is true of all exothermic reactions.

- An increase in temperature drives an exothermic reaction toward the reactants (to the left).
- A decrease in temperature drives an exothermic reaction toward the products (to the right).

For an endothermic reaction, of course, the opposite is true.

- An increase in temperature drives an endothermic reaction toward the products.
- A decrease in temperature drives an endothermic reaction toward the reactants.

Recall from Section 7.4 that a change in temperature changes not only the position of equilibrium but also the value of K, the equilibrium constant.

CHEMICAL CONNECTIONS 7D

Sunglasses and Le Chatelier's Principle

Heat is not the only form of energy that affects equilibria. The statements made in the text regarding endothermic or exothermic reactions can be generalized to reactions involving also other forms of energy. A practical illustration of this is the use of sunglasses with adjustable shading. The compound silver chloride, AgCl, is incorporated in the glasses. This compound, on exposure to sunlight, produces metallic silver, Ag:

$$\text{Light} + 2Ag^+ + 2Cl^- \rightleftharpoons 2Ag + Cl_2$$

The more silver produced, the darker the glasses. At night, or when the wearer goes indoors, the reaction is reversed according to Le Chatelier's principle. In this case, the addition of sunlight drives the equilibrium to the right; its removal drives it to the left.

EXAMPLE 7.9

The conversion of nitrogen dioxide to dinitrogen tetroxide is an exothermic reaction:

$$2NO_2(g) \rightleftharpoons N_2O_4(g) + 13{,}700 \text{ cal}$$

NO_2 is a brown gas; N_2O_4 is a colorless gas. In Figure 7.12 we see that at 50°C the brown color is darker; hence, the NO_2 concentration is higher than at 0°C. Why?

Solution

To go from 0 to 50°C, heat must be added. But heat is a product of this equilibrium reaction, as written in the equation given here. The addition of heat, therefore, will shift the equilibrium to the left. This shift produces more NO_2, leading to the darker brown color.

Problem 7.9

In a reaction $A \rightleftharpoons B$, the amount of B increased when the temperature of the reaction rose. Was this an exothermic or an endothermic reaction? Explain.

Figure 7.12 Effect of temperature on the N_2O_4–NO_2 system at equilibrium. (*top*) At 50°C, the deep brown color indicates the predominance of NO_2. (*bottom*) At 0°C, N_2O_4, which is colorless, predominates.

D Change in Pressure

A change in pressure influences the equilibrium only if one or more of the components of the reaction is a gas. For example, in the reaction

$$N_2O_4(g) \rightleftharpoons 2NO_2(g)$$

we have one mole of gas as a reactant and two moles of gas as products. According to Le Chatelier's principle, an increase in pressure shifts the equilibrium in the direction that will decrease the moles in the gas phase and hence decreases the pressure. In the preceding reaction, the equilibrium will shift to the left.

- An increase in pressure shifts the reaction toward the side with fewer moles of gas.

CHEMICAL CONNECTIONS 7E

The Haber Process

Human beings and animals need proteins and other nitrogen-containing compounds to live. Ultimately, the nitrogen in these compounds comes from the plants that we eat. Although the atmosphere contains plenty of N_2, nature converts this N_2 to compounds usable by biological organisms in only one way: Certain bacteria have the ability to "fix" atmospheric N_2—that is, convert it to ammonia, NH_3. Most of these bacteria live in the roots of certain plants such as clover, alfalfa, peas, and beans. However, the amount of N_2 fixed by such bacteria each year is far less than the amount necessary to feed all the humans and animals in the world.

The world today can support its population only by using fertilizers made by artificial fixing, primarily the **Haber process,** which converts N_2 to NH_3 by the reaction

$$N_2(g) + 3H_2(g) \rightleftharpoons 2NH_3(g) + 22 \text{ kcal}$$

Early workers who focused on the problem of fixing nitrogen were troubled by a conflict between equilibrium and rate. As this is an exothermic reaction, an increase in temperature drives the equilibrium to the left, so the best results (largest possible yield of NH_3) should be obtained at low temperatures. At low temperatures, however, the rate is too slow to produce any meaningful

■ Ammonia is produced by bacteria living in these root nodules.

amounts of NH_3 In 1908 Fritz Haber (1868–1934) solved this problem when he discovered a catalyst that permits the reaction to take place at a convenient rate at 500°C.

The NH_3 produced by the Haber process is converted to fertilizers, which are used all over the world. Without these fertilizers, food production would diminish so much that widespread starvation would result.

- A decrease in pressure shifts the reaction toward the side with more moles of gas.

EXAMPLE 7.10

In the production of ammonia, both reactants and products are gases:

$$N_2(g) + 3H_2(g) \rightleftharpoons 2NH_3(g)$$

What kind of pressure change would increase the yield of ammonia?

Solution
There are four moles of gases on the left side and two moles on the right side. To increase the yield of ammonia, we must shift the equilibrium to the right. An increase in pressure shifts the equilibrium toward the side with fewer moles—that is, to the right. Thus an increase in pressure will increase the yield of ammonia.

Problem 7.10
What happens to the equilibrium reaction

$$O_2(g) + 4ClO_2(g) \rightleftharpoons 2Cl_2O_5(g)$$

when the pressure is increased?

E Effect of Catalyst

As we saw in Section 7.4, a catalyst increases the rate of a reaction without itself being changed. For a reversible reaction, catalysts always increase the rates of both the forward and reverse reactions to the same extent. Therefore, the addition of a catalyst has no effect on the position of equilibrium. However, adding a catalyst to a system not yet at equilibrium causes it to reach equilibrium faster than it would without the catalyst.

SUMMARY

Some reactions are fast; others are slow. The **rate of a reaction** is the change in concentration of a reactant or a product per unit time (Section 7.1). Reactions take place when molecules or ions collide. The rate of a reaction increases with an increasing number of **effective collisions**—that is, collisions that lead to a reaction (Section 7.2). The energy necessary for a reaction to begin is the **activation energy.** Effective collisions are those that have (1) more than the activation energy required for the reaction to proceed forward and (2) the proper orientation in space.

The lower the activation energy, the faster the reaction. An energy diagram shows the progress of a reaction. The position at the top of the curve is called the **transition state** (Section 7.3).

Reaction rates generally increase with increasing concentration and temperature; they also depend on the nature of the reactants (Section 7.4). The rate of some reactions can be increased by adding a **catalyst,** which provides a pathway with a lower activation energy.

The rate constant that gives the relationship between the rate of the reaction and the concentration of the reactants remains constant at a constant temperature.

Many reactions are reversible and eventually reach equilibrium—some slowly, others quickly. At **equilibrium,** the forward and reverse reactions take place at equal rates, and concentrations do not change (Section 7.5). Every equilibrium has an **equilibrium expression** and an **equilibrium constant,** K, which does not change when concentrations change but does change when temperature changes (Section 7.6). The equilibrium constant is large for some reactions and small for others. K and the rate of a reaction are not related (Section 7.7).

Le Chatelier's principle tells us what happens when we put stress on a system in equilibrium (Section 7.8). The addition of a component causes the equilibrium to shift to the opposite side. The removal of a component causes the equilibrium to shift to the side from which the component is taken. Increasing the temperature drives an exothermic equilibrium to the side of the reactants. Addition of a catalyst has no effect on the position of equilibrium.

PROBLEMS

Numbers that appear in color indicate difficult problems.
▶ designates problems requiring application of principles.

Reaction Rates

7.11 In the reaction

$$CH_3Br + I^- \longrightarrow CH_3I + Br^-$$

we start the reaction with an initial CH_3I concentration of 0.260 M. This concentration increases to 0.840 M over a period of 1 h 20 min. What is the rate of reaction?

Molecular Collisions and Activation Energy

7.12 Two kinds of gas molecules are reacted at a set temperature. The gases are blown into the reaction vessel from two tubes. In setup A, the two tubes are aligned parallel to each other; in B they are 90° to each other; and in C they are aligned against each other. Which setup would yield the most effective collisions?

7.13 Why are reactions between ions in aqueous solution generally much faster than reactions between covalent molecules?

7.14 A certain reaction is exothermic by 9 kcal/mol and has an activation energy of 14 kcal/mol. Draw an energy diagram for this reaction, and label the transition state.

7.15 What is the likelihood that the reaction

$$O_2(g) + 4ClO_2(g) \longrightarrow 2Cl_2O_5(g)$$

occurs in a single step? Explain.

Factors Affecting Rates of Reaction

7.16 A quart of milk quickly spoils if left at room temperature but keeps for several days in a refrigerator. Explain.

7.17 If a certain reaction takes 16 h to go to completion at 10°C, at what temperature should we run it if we want it to go to completion in 1 h?

7.18 In most cases, when we run a reaction by mixing a fixed quantity of A with a fixed quantity of B, the rate of the reaction begins at a maximum and then decreases as time goes by. Explain.

7.19 If you were running a reaction and wanted it to go faster, what three things might you try to accomplish this goal?

7.20 What factors determine whether a reaction run at a given temperature will be fast or slow?

7.21 Explain how a catalyst increases the rate of a reaction.

7.22 If you add a piece of marble, $CaCO_3$, to a 6 M HCl solution at room temperature, you will see some bubbles form around the marble as CO_2 gas slowly rises. If you crush another piece of marble and add it to the same solution at the same temperature, you will see vigorous gas formation, so much so that the solution appears to be boiling. Explain.

Reversible Reactions and Equilibrium

7.23 Burning a piece of paper is an irreversible reaction. Give some other examples of irreversible reactions.

7.24 If the reaction $PCl_3 + Cl_2 \rightleftharpoons PCl_5$ is at equilibrium, are the concentrations of PCl_3, Cl_2, and PCl_5 necessarily equal? Is the concentration of PCl_3 necessarily equal to that of Cl_2? Explain.

Equilibrium Constants

7.25 Write equilibrium expressions for these reactions.
(a) $2H_2O_2 \rightleftharpoons 2H_2O + O_2$
(b) $2N_2O_5 \rightleftharpoons 2N_2O_4 + O_2$
(c) $6H_2O + 6CO_2 \rightleftharpoons C_6H_{12}O_6 + 6O_2$

7.26 Write the chemical equations corresponding to the following equilibrium expressions.
(a) $K = \dfrac{[H_2CO_3]}{[CO_2][H_2O]}$
(b) $K = \dfrac{[P_4][O_2]^5}{[P_4O_{10}]}$
(c) $K = \dfrac{[F_2]^3[PH_3]}{[HF]^3[PF_3]}$

7.27 When a mixture containing H_2, CO_2, H_2O, and CO at 987°C reached equilibrium, the concentrations were [CO] = 0.933 M, [H_2O] = 0.72 M, [CO_2] = 0.133 M, and [H_2] = 3.37 M. What is the equilibrium constant at 987°C for the reaction

$$CO(g) + H_2O(g) \rightleftharpoons CO_2(g) + H_2(g)$$

7.28 For the reaction

$$2NO_2 \rightleftharpoons N_2O_4$$

the equilibrium constant at 325 K was found to be 172. When a sample was taken of the equilibrium mixture, it was found to contain 0.0714 M NO_2. What was the equilibrium concentration of N_2O_4?

7.29 When a mixture containing NOCl, NO, and Cl_2 at 25°C reached equilibrium, the concentrations were [Cl_2] = 0.34 M, [NOCl] = 2.6 M, and [NO] = 1.4 M. What is the equilibrium constant at 25°C for the reaction

$$2NOCl(g) \rightleftharpoons 2NO(g) + Cl_2(g)$$

in dilute solution.

7.30 Write the equilibrium expression for the reaction:

$$HNO_3(aq) + H_2O(\ell) \rightleftharpoons NO_3^-(aq) + H_3O^+(aq)$$

7.31 Here are equilibrium constants for several reactions. Which of them favor the formation of products and which favor the formation of reactants?
(a) 4.5×10^8 (b) 32
(c) 4.5   (d) 3×10^{-7}
(e) 0.0032

7.32 A particular reaction has an equilibrium constant of 1.13 under one set of conditions and an equilibrium constant of 1.72 under a different set of conditions. Which conditions would be more advantageous in an industrial process to obtain the maximum amount of products? Explain.

Equilibrium and Reaction Rates

7.33 (a) If a reaction is very exothermic—that is, if the products have a much lower energy than the reactants—can we be reasonably certain that it will take place rapidly?
(b) If a reaction is very endothermic—that is, if the products have a much higher energy than the reactants—can we be reasonably certain that it will take place extremely slowly or not at all? (Think carefully about this one.)

7.34 A reaction has a high rate constant but a small equilibrium constant. What does this mean in terms of producing an industrial product?

Le Chatelier's Principle

7.35 Assume that the following exothermic reaction is at equilibrium:

$$H_2(g) + I_2(g) \rightleftharpoons 2HI(g)$$

Tell whether the position of equilibrium will shift to the right or the left if we
(a) Remove some HI.
(b) Add some H_2.
(c) Remove some I_2.
(d) Increase the temperature.
(e) Add a catalyst.

7.36 The following reaction is endothermic:

$$3O_2 \rightleftharpoons 2O_3$$

If the reaction is at equilibrium, tell whether the equilibrium will shift to the right or the left if we

(a) Remove some O_3.
(b) Remove some O_2.
(c) Add some O_3.
(d) Decrease the temperature.
(e) Add a catalyst.
(f) Increase the pressure

7.37 The following reaction is exothermic:

$$2NO(g) + Br_2(g) \rightleftharpoons 2NOBr(g)$$

After it reaches equilibrium, we add a few drops of Br_2.

(a) What will happen to the equilibrium?
(b) What will happen to the equilibrium constant?

7.38 Is there any change in conditions that changes the equilibrium constant, K, of a given reaction?

7.39 The equilibrium constant at 1127°C for the following endothermic reaction is 571:

$$2H_2S(g) \rightleftharpoons 2H_2(g) + S_2(g)$$

If the mixture is at equilibrium, what happens to K if we

(a) Add some H_2S?
(b) Add some H_2?
(c) Lower the temperature to 1000°C?

Chemical Connections

7.40 (Chemical Connections 7A) In a bacterial infection, body temperature may rise to 101°F. Does this body defense kill the bacteria directly by heat or by another mechanism? If so, by which mechanism?

7.41 (Chemical Connections 7A and 7B) Why is a high fever dangerous? Why is a low body temperature dangerous?

7.42 (Chemical Connections 7B) Why do surgeons sometimes lower body temperatures during heart operations?

7.43 (Chemical Connections 7C) A pain killer—for example, Tylenol—can be purchased in two forms both containing the same amount of drug. One form is a solid coated pill, and the other is a capsule that contains tiny beads and has the same coat. Which medication will act faster? Explain.

7.44 (Chemical Connections 7D) What reaction takes place when sunlight hits the compound silver chloride?

7.45 (Chemical Connections 7E) If the equilibrium for the Haber process is unfavorable at high temperatures, why do factories nevertheless use high temperatures?

Additional Problems

7.46 In the reaction

$$H_2(g) + Cl_2(g) \rightleftharpoons 2HCl(g)$$

a 10°C increase in temperature doubles the rate of reaction. If the rate of reaction at 15°C is 2.8 moles of HCl per liter per second, what is the rate at −5°C and at 45°C?

7.47 Draw an energy diagram for an exothermic reaction that yields 75 kcal/mol. The activation energy is 30 kcal/mol.

7.48 Draw a diagram similar to Figure 7.4. Draw a second line of energy profile starting and ending at the same level as the first but having a smaller peak than the first line. Label them 1 and 2. What may have occurred to change the energy profile of a reaction from 1 to 2?

7.49 For the reaction of

$$2NOBr(g) \longrightarrow 2NO(g) + Br_2(g)$$

the rate of the reaction was −2.3 mol NOBr/L/h when the initial NOBr concentration was 6.2 mol NOBr/L. What is the rate constant of the reaction?

7.50 The equilibrium constant for

$$2NOBr(g) \rightleftharpoons 2NO(g) + Br_2(g)$$

is 25. A measurement made on the equilibrium mixture found that the concentrations of NO and Br_2 were each 0.80 M. What is the concentration of NOBr at equilibrium?

7.51 In the reaction

$$N_2O_4(g) \longrightarrow 2NO_2(g)$$

the concentration of N_2O_4 was measured in mol/L at the end of the times shown:

Time(s)	[N_2O_4]
0	0.200
10	0.180
20	0.160
30	0.140

What is the rate of reaction?

7.52 How could you increase the rate of a gaseous reaction without adding more reactants or a catalyst and without changing the temperature?

7.53 The equilibrium constant for the reaction

$$\alpha\text{-D-glucose(aq)} \rightleftharpoons \beta\text{-D-glucose(aq)}$$

is 1.5 at 30°C. This reaction takes place only in aqueous solution. If you begin with a fresh solution of 1 M α-D-glucose in water, what will be its concentration when equilibrium is reached?

7.54 In an endothermic reaction, A $\rightleftharpoons$ B, the activation energy was 100 kcal/mol. Would the activation energy of the reverse reaction, B $\rightleftharpoons$ A, also be 100 kcal/mol, or would it be more or less? Explain with the aid of a diagram.

7.55 A reaction has an equilibrium expression

$$K = \frac{[NO_2]^4[H_2O]^6}{[NH_3]^4[O_2]^7}$$

Write the reaction.

7.56 For the reaction

$$N_2O_4(g) \longrightarrow 2NO_2(g)$$

the rate of reaction at 300K was found to be 0.22 M NO_2/min. What would be the approximate rate of reaction at 320 K?

7.57 Assume that two different reactions are taking place at the same temperature. In reaction A, two different spherical molecules collide to yield a product. In reaction B, the shape of the colliding molecules is rod-like. Each reaction has the same number of collisions per second and the same activation energy. Which reaction goes faster?

7.58 Is it possible for an endothermic reaction to have zero activation energy?

7.59 In the reaction

$$2HI(g) \rightleftharpoons H_2(g) + I_2(g)$$

the rate of appearance of I_2 is measured. The following concentrations in mol/L were found at the end of the times shown:

Time(s)	[I_2]
0	0
10	0.30
20	0.60
30	0.90

What is the rate of reaction?

7.60 (Chemical Connections 7D) You have a recipe to manufacture sunglasses: 3.5 g AgCl/kg glass. A new order comes in to manufacture sunglasses to be used in deserts like the Sahara. How would you change the recipe?

7.61 A reaction A ⟶ D occurs in three steps:
1. A ⟶ B;
2. B ⟶ C;
3. C ⟶ D

The rate constants of the three steps are:

$k_1 = 0.3\ M;\ k_2 = 0.05\ M;\ k_3 = 4.2\ M$

(a) Which step is the rate-determining step?
(b) Which step has the lowest activation energy?

InfoTrac College Edition

For additional readings, go to InfoTrac College Edition, your online research library, at

http://infotrac.thomsonlearning.com

CHAPTER 8

8.1 Introduction

8.2 Acid and Base Strength

8.3 Brønsted-Lowry Acids and Bases

8.4 The Position of Equilibrium in Acid–Base Reactions

8.5 Acid Ionization Constants

8.6 Some Properties of Acids and Bases

8.7 Self-Ionization of Water

8.8 pH and pOH

8.9 The pH of Aqueous Salt Solutions

8.10 Titration

8.11 Buffers

8.12 The Henderson-Hasselbalch Equation

These U.S runners have just won the gold medal in the 4 × 400 m relay race at the 1996 Olympic games. The buildup of an organic acid called lactic acid has caused severe muscle pain. In a short time the pain will be gone.

Acids and Bases

8.1 Introduction

We frequently encounter acids and bases in our daily lives. Oranges, lemons, and vinegar are examples of acidic foods, and sulfuric acid is in our automobile batteries. As for bases, we take antacid tablets for heartburn and use household ammonia as a cleaning agent. What do these substances have in common? Why are acids and bases usually discussed together?

In 1884, a young Swedish chemist named Svante Arrhenius (1859–1927) answered the first question by proposing what was then a new definition of acids and bases. According to the Arrhenius definition, an **acid** is a substance that produces H_3O^+ ions in aqueous solution, and a **base** is a substance that produces OH^- ions in aqueous solution.

This definition of acid is a slight modification of the original Arrhenius definition, which was that an acid produces hydrogen ions, H^+. Today we know that H^+ ions cannot exist in water. A H^+ ion is a bare proton, and a charge of +1 is too concentrated to exist on such a tiny particle (Section 3.2). Therefore, a H^+ ion in water immediately combines with an H_2O molecule to give a **hydronium ion,** H_3O^+.

Hydronium ion The H_3O^+ ion

$$H^+(aq) + H_2O(\ell) \longrightarrow H_3O^+(aq)$$
Hydronium ion

Although we know that acidic aqueous solutions do not contain H^+ ions, we frequently use the terms "H^+" and "proton" when we really mean "H_3O^+." The three terms are generally interchangeable.

Apart from this modification, the Arrhenius definitions of acid and base are still valid and useful today, as long as we are talking about aqueous solutions.

When an acid dissolves in water, it reacts with the water to produce H_3O^+. For example, hydrogen chloride, HCl, in its pure state is a poisonous, choking gas. When HCl dissolves in water, it reacts with a water molecule to give hydronium ion and chloride ion:

$$H_2O(\ell) + HCl(aq) \longrightarrow H_3O^+(aq) + Cl^-(aq)$$

Thus a bottle labeled aqueous "HCl" is actually not HCl at all, but rather an aqueous solution of H_3O^+ and Cl^- ions in water.

We can show the transfer of a proton from an acid to a base by using a symbol called a **curved arrow.** First we write the Lewis structure of each reactant and product. Then we use curved arrows to show the change in position of electron pairs during the reaction. The tail of the curved arrow is located at the electron pair. The head of the curved arrow shows the new position of the electron pair.

Screen 8.2
Exercise Tutorial

$$H-\overset{..}{\underset{|}{O}}: + H-\overset{..}{\underset{..}{Cl}}: \longrightarrow H-\overset{..}{\underset{|}{O}}-H + :\overset{..}{\underset{..}{Cl}}:^-$$
$$H H$$

In this equation, the curved arrow on the left shows that an unshared pair of electrons on oxygen forms a new covalent bond with hydrogen. The curved arrow on the right shows that the pair of electrons of the H—Cl bond is given entirely to chlorine to form a chloride ion. Thus, in the reaction of HCl with H_2O, a proton is transferred from HCl to H_2O and, in the process, an O—H bond forms and an H—Cl bond is broken.

With bases, the situation is slightly different. Many bases are metal hydroxides, such as KOH, NaOH, $Mg(OH)_2$, and $Ca(OH)_2$. These compounds are ionic solids and, when they dissolve in water, their ions merely separate, and each ion is solvated by water molecules (Section 6.6). For example,

$$NaOH(s) \xrightarrow{H_2O} Na^+(aq) + OH^-(aq)$$

Other bases are not hydroxides. Instead, they produce OH^- ions in water by reacting with water molecules. The most important example of this kind of base is ammonia, NH_3, a poisonous, choking gas. When ammonia dissolves in water, it reacts with water to produce ammonium ions and hydroxide ions.

The boiling point of liquid ammonia is $-33°C$.

$$NH_3(aq) + H_2O(\ell) \rightleftharpoons NH_4^+(aq) + OH^-(aq)$$

We show the equation with equilibrium arrows because it is reversible.

As we will see in Section 8.2, ammonia is a weak base and the position of the equilibrium for its reaction with water lies considerably toward the left. In a 1.0 M solution of NH_3 in water, for example, only about 4 molecules of NH_3 out of every 1000 react with water to form NH_4^+ and OH^-. Thus, when ammonia is dissolved in water, it exists primarily as NH_3 molecules. Nevertheless, some OH^- ions are produced and, therefore, NH_3 is a base.

Bottles of NH_3 in water are sometimes labeled "ammonium hydroxide" or "NH_4OH," but this gives a false impression of what is really in the bottle. Most of the NH_3 molecules have not reacted with the water, so the bottle contains mostly NH_3 and H_2O and only a little NH_4^+ and OH^-.

We indicate how the reaction of ammonia with water takes place by using curved arrows to show the transfer of a proton from a water molecule to an ammonia molecule. Here, the curved arrow on the left shows that the unshared pair of electrons on nitrogen forms a new covalent bond with a hydrogen of a water molecule. At the same time as the new N—H bond forms, an O—H bond of a water molecule breaks and the pair of electrons forming the H—O bond moves entirely to oxygen forming OH^-.

$$H-\underset{\underset{H}{|}}{\overset{\overset{H}{|}}{N}}\colon + H-\overset{..}{\underset{..}{O}}-H \longrightarrow H-\underset{\underset{H}{|}}{\overset{\overset{H}{|}}{N}}\overset{+}{-}H + \colon\overset{..}{\underset{..}{O}}-H$$

Thus ammonia produces an OH^- ion by taking H^+ from a water molecule and leaving OH^- behind.

8.2 Acid and Base Strength

All acids are not equally strong. According to the Arrhenius definition, a **strong acid** is one that reacts completely or almost completely with water to form H_3O^+ ions. Table 8.1 gives the names and molecular formulas for six of the most common strong acids. These six acids are strong acids because, when they dissolve in water, they dissociate completely to give H_3O^+ ions.

Weak acids produce a much smaller concentration of H_3O^+ ions. Acetic acid, for example, is a weak acid. In water it exists primarily as acetic acid molecules; only a few acetic acid molecules (4 out of every 1000) are converted to acetate ions.

$$\underset{\text{Acetic acid}}{CH_3COOH(aq)} + H_2O(\ell) \rightleftharpoons \underset{\text{Acetate ion}}{CH_3COO^-(aq)} + H_3O^+(aq)$$

There are only four common **strong bases** (Table 8.1), all of which are metal hydroxides. They are strong bases because, when they dissolve in water, they ionize completely to give OH^- ions. As we saw in Section 8.1, ammonia is a **weak base** because the equilibrium for its reaction with water lies far to the left.

Strong acid An acid that ionizes completely in aqueous solution

Weak acid An acid that is only partially ionized in aqueous solution

Strong base A base that ionizes completely in aqueous solution

Weak base A base that is only partially ionized in aqueous solution

Table 8.1 Six Strong Acids and Four Strong Bases

Formula	Name	Formula	Name
HCl	Hydrochloric acid	LiOH	Lithium hydroxide
HBr	Hydrobromic acid	NaOH	Sodium hydroxide
HI	Hydroiodic acid	KOH	Potassium hydroxide
HNO_3	Nitric acid	$Ba(OH)_2$	Barium hydroxide
H_2SO_4	Sulfuric acid		
$HClO_4$	Perchloric acid		

CHEMICAL CONNECTIONS 8A

Some Important Acids and Bases

STRONG ACIDS Sulfuric acid, H_2SO_4, is used in many industrial processes. In fact, more tons of sulfuric acid are manufactured in the United States than any other chemical, organic or inorganic.

Hydrochloric acid, HCl, is an important acid in chemistry laboratories. Pure HCl is a gas, and the HCl in laboratories is an aqueous solution. HCl is the acid in the gastric fluid in your stomach, where it is secreted at a strength of about 5%.

Nitric acid, HNO_3, is a strong oxidizing agent. A drop of it causes the skin to turn yellow because the acid reacts with proteins on the skin. A yellow color upon contact with nitric acid has long been a test for proteins.

WEAK ACIDS Acetic acid, CH_3COOH, is present in vinegar (about 5%). Pure acetic acid is called glacial acetic acid because of its melting point of 17°C, which means that it freezes on a moderately cold day.

Boric acid, H_3BO_3, is a solid. Solutions of boric acid in water were once used as antiseptics, especially for eyes. Boric acid is toxic when swallowed.

Phosphoric acid, H_3PO_4, is one of the strongest of the weak acids. The ions produced from it—$H_2PO_4^-$, HPO_4^{2-}, and PO_4^{3-}—are important in biochemistry (see also Section 26.3).

STRONG BASES Sodium hydroxide, NaOH, also called lye, is the most important of the strong bases. It is a solid whose aqueous solutions are used in many industrial processes, including the manufacture of glass and soap. Potassium hydroxide, KOH, also a solid, is used for many of the same purposes as NaOH.

WEAK BASES Ammonia, NH_3, the most important weak base, is a gas with many industrial uses. One of its chief uses is for fertilizers. A 5% solution is sold in supermarkets as a cleaning agent, and weaker solutions are used as "spirits of ammonia" to revive people who have fainted.

Magnesium hydroxide, $Mg(OH)_2$, is a solid that is insoluble in water. A suspension of about 8% $Mg(OH)_2$ in water is called milk of magnesia and is used as a laxative. $Mg(OH)_2$ is also used to treat waste water in metal-processing plants and as a flame retardant in plastics.

■ Weak acids are found in many common materials. In the foreground are strips of litmus paper that have been dipped into solutions of these materials. Acids turn litmus paper red.

■ Weak bases are also common in many household products. These cleaning agents all contain weak bases. Bases turn litmus paper blue.

It is important to understand that the strength of an acid or base is not related to its concentration. HCl is a strong acid, whether it is concentrated or dilute, because it dissociates completely in water to chloride ions and hydronium ions. Acetic acid is a weak acid, whether it is concentrated or dilute, because the equilibrium for its reaction with water lies far to the left. When acetic acid dissolves in water, most of it is present as undissociated CH_3COOH molecules.

$$HCl(aq) + H_2O(\ell) \longrightarrow Cl^-(aq) + H_3O^+(aq)$$

$$\underset{\text{Acetic acid}}{CH_3COOH(aq)} + H_2O(\ell) \rightleftharpoons \underset{\text{Acetate ion}}{CH_3COO^-(aq)} + H_3O^+(aq)$$

In Section 6.6, we saw that electrolytes (substances that produce ions in aqueous solution) can be strong or weak. The strong acids and bases in Table 8.1 are strong electrolytes. Almost all other acids and bases are weak electrolytes.

8.3 Brønsted-Lowry Acids and Bases

The Arrhenius definitions of acid and base are very useful in aqueous solutions. But what if water is not involved? In 1923 the Danish chemist Johannes Brønsted and the English chemist Thomas Lowry independently proposed the following definitions: An **acid** is a proton donor, a **base** is a proton acceptor, and an **acid–base reaction** is a proton transfer reaction. Furthermore, according to the Brønsted-Lowry definitions, any pair of molecules or ions that can be interconverted by transfer of a proton is called a **conjugate acid–base pair.** When an acid transfers a proton to a base, the acid is converted to its **conjugate base.** When a base accepts a proton, it is converted to its **conjugate acid.**

We can illustrate these relationships by examining the reaction of hydrogen chloride with water to form chloride ion and hydronium ion. The acid HCl donates a proton and is converted to its conjugate base Cl^-. The base H_2O accepts a proton and is converted to its conjugate acid H_3O^+.

> **Conjugated acid–base pair** A pair of molecules or ions that are related to one another by the gain or loss of a proton

> **Conjugate base** In the Brønsted-Lowry theory, a substance formed when an acid donates a proton to another molecule or ion

> **Conjugate acid** In the Brønsted-Lowry theory, a substance formed when a base accepts a proton

$$HCl(aq) + H_2O(\ell) \longrightarrow Cl^-(aq) + H_3O^+(aq)$$

Hydrogen chloride (Acid) Water (Base) Chloride ion (Conjugate base of HCl) Hydronium ion (Conjugate acid of water)

We have illustrated the application of the Brønsted-Lowry definitions using water as a reactant. The Brønsted-Lowry definitions, however, are more general than the Arrhenius definitions and do not require water as a reactant. Consider the following reaction between acetic acid and ammonia:

$$CH_3COOH + NH_3 \rightleftharpoons CH_3COO^- + NH_4^+$$

Acetic acid (Acid) Ammonia (Base) Acetate ion (Conjugate base of acetic acid) Ammonium ion (Conjugate acid of ammonia)

We can use curved arrows to show how this reaction takes place. The curved arrow on the right shows that the unshared pair of electrons on nitrogen becomes shared to form a new H—N bond. At the same time that the H—N bond forms, the O—H bond breaks and the electron pair of the O—H bond moves entirely to oxygen to form —O⁻ of the acetate ion. The result of these two electron-pair shifts is the transfer of a proton from an acetic acid molecule to an ammonia molecule.

$$CH_3-\overset{\overset{\ddot{O}:}{\|}}{C}-\ddot{\underset{\ddot{}}{O}}-H + :\underset{H}{\overset{H}{N}}-H \rightleftharpoons CH_3-\overset{\overset{\ddot{O}:}{\|}}{C}-\ddot{\underset{\ddot{}}{O}}:^- + H-\underset{H}{\overset{H}{\overset{+}{N}}}-H$$

Acetic acid (Proton donor) Ammonia (Proton acceptor) Acetate ion Ammonium ion

Table 8.2 Some Acids and Their Conjugate Bases

	Acid	Name	Conjugate Base	Name	
Strong Acids ↑	HI	Hydroiodic acid	I$^-$	Iodide ion	**Weak Bases**
	HCl	Hydrochloric acid	Cl$^-$	Chloride ion	
	H$_2$SO$_4$	Sulfuric acid	HSO$_4^-$	Hydrogen sulfate ion	
	HNO$_3$	Nitric acid	NO$_3^-$	Nitrate ion	
	H$_3$O$^+$	Hydronium ion	H$_2$O	Water	
	HSO$_4^-$	Hydrogen sulfate ion	SO$_4^{2-}$	Sulfate ion	
	H$_3$PO$_4$	Phosphoric acid	H$_2$PO$_4^-$	Dihydrogen phosphate ion	
	CH$_3$COOH	Acetic acid	CH$_3$COO$^-$	Acetate ion	
	H$_2$CO$_3$	Carbonic acid	HCO$_3^-$	Bicarbonate ion	
	H$_2$S	Hydrogen sulfide	HS$^-$	Hydrogen sulfide ion	
	H$_2$PO$_4^-$	Dihydrogen phosphate ion	HPO$_4^{2-}$	Hydrogen phosphate ion	
	NH$_4^+$	Ammonium ion	NH$_3$	Ammonia	
	HCN	Hydrocyanic acid	CN$^-$	Cyanide ion	
	C$_6$H$_5$OH	Phenol	C$_6$H$_5$O$^-$	Phenoxide ion	
	HCO$_3^-$	Bicarbonate ion	CO$_3^{2-}$	Carbonate ion	
	HPO$_4^{2-}$	Hydrogen phosphate ion	PO$_4^{3-}$	Phosphate ion	
Weak Acids	H$_2$O	Water	OH$^-$	Hydroxide ion	↓ **Strong Bases**
	C$_2$H$_5$OH	Ethanol	C$_2$H$_5$O$^-$	Ethoxide ion	

Table 8.2 gives examples of common acids and their conjugate bases. As you study the examples of conjugate acid–base pairs in Table 8.2, note the following points:

1. An acid can be positively charged, neutral, or negatively charged. Examples of these charge types are H$_3$O$^+$, H$_2$CO$_3$, and H$_2$PO$_4^-$, respectively.

2. A base can be negatively charged or neutral. Examples of these charge types are Cl$^-$ and NH$_3$, respectively.

3. Acids are classified as monoprotic, diprotic, or triprotic depending on the number of protons each may give up. Examples of **monoprotic acids** include HCl, HNO$_3$, and CH$_3$COOH. Examples of **diprotic acids** include H$_2$SO$_4$ and H$_2$CO$_3$. An example of a **triprotic acid** is H$_3$PO$_4$. Carbonic acid, for example, loses one proton to become bicarbonate ion, and then a second proton to become carbonate ion.

$$H_2CO_3 + H_2O \rightleftharpoons HCO_3^- + H_3O^+$$
Carbonic acid   Bicarbonate ion

$$HCO_3^- + H_2O \rightleftharpoons CO_3^{2-} + H_3O^+$$
Bicarbonate ion   Carbonate ion

4. Several molecules and ions appear in both the acid and conjugate base columns; that is, each can function as either an acid or a base. The bicarbonate ion, HCO$_3^-$, for example, can give up a proton to become CO$_3^{2-}$ (in which case it is an acid) or it can accept a proton to become H$_2$CO$_3$ (in which case it is a base). A substance that can act as either an acid or a base is called **amphiprotic.** The most important amphiprotic substance in Table 8.2 is water, which can accept a proton to become H$_3$O$^+$ or lose a proton to become OH$^-$.

5. A substance cannot be a Brønsted-Lowry acid unless it contains a hydrogen atom, but not all hydrogen atoms can be given up. For example,

> **Monoprotic acid** An acid that can give up only one proton
>
> **Diprotic acid** An acid that can give up two protons
>
> **Triprotic acid** An acid that can give up three protons
>
> **Amphiprotic** A substance that can act as either an acid or a base

A box of Arm & Hammer baking soda (sodium bicarbonate). Sodium bicarbonate is composed of Na^+ and HCO_3^-, the amphiprotic bicarbonate ion.

acetic acid, CH_3COOH, has four hydrogens but is monoprotic; it gives up only one of them. Similarly, phenol, C_6H_5OH, gives up only one of its six hydrogens:

$$C_6H_5OH + H_2O \rightleftharpoons C_6H_5O^- + H_3O^+$$
Phenol   Phenoxide ion

6. There is an inverse relationship between the strength of an acid and the strength of its conjugate base: The stronger the acid, the weaker its conjugate base. HI, for example, is the strongest acid listed in Table 8.2 and I^-, its conjugate base, is the weakest base. As another example, CH_3COOH (acetic acid) is a stronger acid than H_2CO_3 (carbonic acid); conversely, CH_3COO^- (acetate ion) is a weaker base than HCO_3^- (bicarbonate ion).

8.4 The Position of Equilibrium in Acid–Base Reactions

We know that HCl reacts with H_2O according to the following equilibrium:

$$HCl + H_2O \longrightarrow Cl^- + H_3O^+$$

We also know that HCl is a strong acid, which means that the position of this equilibrium lies very far to the right. In fact, this equilibrium lies so far to the right that out of every 10,000 HCl molecules dissolved in water, all but one react with water molecules to give Cl^- and H_3O^+.

As we have also seen, acetic acid reacts with H_2O according to the following equilibrium:

$$CH_3COOH + H_2O \rightleftharpoons CH_3COO^- + H_3O^+$$
Acetic acid   Acetate ion

Acetic acid is a weak acid. Only a few acetic acid molecules react with water to give acetate ions and hydronium ions, and the major species present in equilibrium in aqueous solution is CH_3COOH. The position of this equilibrium, therefore, lies very far to the left.

In these two acid–base reactions, water is the base. But what if we have a base other than water as the proton acceptor? How can we determine which are the major species present at equilibrium? That is, how can we determine if the position of equilibrium lies toward the left or toward the right?

As an example, let us examine the acid–base reaction between acetic acid and ammonia to form acetate ion and ammonium ion. As indicated by the question mark over the equilibrium arrow, we want to determine if the position of this equilibrium lies toward the left or toward the right.

$$CH_3COOH + NH_3 \overset{?}{\rightleftharpoons} CH_3COO^- + NH_4^+$$
Acetic acid   Ammonia   Acetate ion   Ammonium ion
(Acid)   (Base)   (Conjugate base   (Conjugate acid
                            of CH_3COOH)   of NH_3)

In this equilibrium, there are two acids present: acetic acid and ammonium ion. There are also two bases present: ammonia and acetate ion. One way to analyze this equilibrium is to view it as a competition of the two bases, ammonia and acetate ion, for a proton. Which is the stronger base? The information we need to answer this question is found in Table 8.2. We first

CHEMICAL CONNECTIONS 8B

Acid and Base Burns of the Cornea

The cornea is the outermost part of the eye. This transparent tissue is very sensitive to chemical burns, whether caused by acids or by bases. Acids and bases in small amounts and in low concentrations that would not seriously damage other tissues, such as skin, may cause severe corneal burns. If not treated promptly, these burns can lead to permanent loss of vision.

For acid splashed in the eyes, immediately wash the eyes with a steady stream of cold water. Quickly removing the acid is of utmost importance. Time should not be wasted looking for a solution of some mild base to neutralize it because removing it is more helpful than neutralizing it. The extent of the burn can be assessed after all the acid has been washed out.

Concentrated ammonia or sodium hydroxide of the "liquid plumber" type can also cause severe damage when splattered in the eyes. Again, immediately washing the eyes is of utmost importance; however, burns by bases can later develop ulcerations in which the healing wound deposits scar tissue that is not transparent, thereby impairing vision. For this reason, first aid is not sufficient for burns by bases, and a physician should be consulted immediately.

■ Cleaning products like these contain bases. Care must be taken to make sure that they are not splashed into anyone's eyes.

Fortunately, the cornea is a tissue with no blood vessels and, therefore, can be transplanted without immunological rejection problems. Today, corneal grafts are common.

---

determine which conjugate acid is the stronger acid and then use this information along with the fact that the stronger the acid, the weaker its conjugate base. From Table 8.2, we see that CH_3COOH is the stronger acid, which means that CH_3COO^- is the weaker base. Conversely, NH_4^+ is the weaker acid, which means than NH_3 is the stronger base. We can now label the relative strengths of each acid and base in this equilibrium:

$$CH_3COOH + NH_3 \overset{?}{\rightleftharpoons} CH_3COO^- + NH_4^+$$

Acetic acid (Stronger acid) Ammonia (Stronger base) Acetate ion (Weaker base) Ammonium ion (Weaker acid)

In an acid–base reaction, the position of equilibrium always favors reaction of the stronger acid and stronger base to form the weaker acid and weaker base. Thus, at equilibrium, the major species present are the weaker acid and weaker base. In the reaction between acetic acid and ammonia, therefore, the equilibrium lies to the right and the major species present are acetate ion and ammonium ion:

$$CH_3COOH + NH_3 \rightleftharpoons CH_3COO^- + NH_4^+$$

Acetic acid (Stronger acid) Ammonia (Stronger base) Acetate ion (Weaker base) Ammonium ion (Weaker acid)

To summarize, the four steps we use in determining the position of an acid–base equilibrium are as follows:

1. Identify the two acids in the equilibrium; one is on the left side of the equilibrium, and the other on the right side.

2. Using the information in Table 8.2, determine which acid is the stronger acid and which is the weaker acid.
3. Identify the stronger base and weaker base in each equilibrium. Remember that the stronger acid gives the weaker conjugate base and the weaker acid gives the stronger conjugate base.
4. The stronger acid and stronger base react to give the weaker acid and weaker base. The position of equilibrium, therefore, lies on the side of the weaker acid and weaker base.

EXAMPLE 8.1

For each acid–base equilibrium, label the stronger acid, the stronger base, the weaker acid, and the weaker base. Then predict whether the position of equilibrium lies toward the right or toward the left.
(a) $H_2CO_3 + OH^- \rightleftharpoons HCO_3^- + H_2O$
(b) $HPO_4^{2-} + NH_3 \rightleftharpoons PO_4^{3-} + NH_4^+$

Solution
Arrows over each equilibrium show the conjugate acid–base pairs. The position of equilibrium in (a) lies toward the right. In (b) it lies toward the left.

(a) H_2CO_3 + OH^- $\rightleftharpoons$ HCO_3^- + H_2O
    Stronger   Stronger   Weaker   Weaker
    acid   base   base   acid

(b) HPO_4^{2-} + NH_3 $\rightleftharpoons$ PO_4^{3-} + NH_4^+
    Weaker   Weaker   Stronger   Stronger
    acid   base   base   acid

Problem 8.1
For each acid–base equilibrium, label the stronger acid, the stronger base, the weaker acid, and the weaker base. Then predict whether the position of equilibrium lies toward the right or toward the left.
(a) $H_3O^+ + I^- \rightleftharpoons H_2O + HI$
(b) $CH_3COO^- + H_2S \rightleftharpoons CH_3COOH + HS^-$

8.5 Acid Ionization Constants

In Section 8.2, we learned that acids vary in the extent to which they produce H_3O^+ when added to water. Because the ionizations of weak acids in water are all equilibria, we can use equilibrium constants to tell us quantitatively just how strong any weak acid is. The reaction that takes place when a weak acid, HA, is added to water is

HA is the general formula for an acid.

$$HA + H_2O \rightleftharpoons A^- + H_3O^+$$

The equilibrium expression (Section 7.6) for this ionization is

$$K_{eq} = \frac{[A^-][H_3O^+]}{[HA][H_2O]}$$

Table 8.3 K_a and pK_a Values for Some Weak Acids

Formula	Name	K_a	pK_a
H_3PO_4	Phosphoric acid	7.5×10^{-3}	2.12
HCOOH	Formic acid	1.8×10^{-4}	3.75
$CH_3CH(OH)COOH$	Lactic acid	8.4×10^{-4}	3.08
CH_3COOH	Acetic acid	1.8×10^{-5}	4.75
H_2CO_3	Carbonic acid	4.3×10^{-7}	6.37
$H_2PO_4^-$	Dihydrogen phosphate ion	6.2×10^{-8}	7.21
H_3BO_3	Boric acid	7.3×10^{-10}	9.14
NH_4^+	Ammonium ion	5.6×10^{-10}	9.25
HCN	Hydrocyanic acid	4.9×10^{-10}	9.31
C_6H_5OH	Phenol	1.3×10^{-10}	9.89
HCO_3^-	Bicarbonate ion	5.6×10^{-11}	10.25
HPO_4^{2-}	Hydrogen phosphate ion	2.2×10^{-13}	12.66

Increasing acid strength ↑

Notice that this expression contains the concentration of water. Because water is the solvent and its concentration changes very little when we add HA to it, we can treat the concentration of water, $[H_2O]$, as a constant equal to 1000 g/L or approximately 55.5 mol/L. We can then combine these two constants (K_{eq} and $[H_2O]$) to define a new constant called an **acid ionization constant**, K_a.

$$K_a = K_{eq}[H_2O] = \frac{[A^-][H_3O^+]}{[HA]}$$

> **Acid ionization constant (K_a)** An equilibrium constant for the ionization of an acid in aqueous solution to H_3O^+ and its conjugate base; also called an acid dissociation constant

The value of the acid ionization constant for acetic acid, for example, is 1.8×10^{-5}. Because acid ionization constants for weak acids are numbers with negative exponents, acid strengths are often expressed as pK_a, where p$K_a = -\log K_a$. The pK_a of acetic acid is 4.75. Table 8.3 gives names, molecular formulas, and values of K_a and pK_a for some weak acids. As you study the entries in this table, note the inverse relationship between the values of K_a and pK_a. The weaker the acid, the smaller its K_a, but the larger its pK_a.

One reason for the importance of K_a is that it immediately tells us how strong an acid is. For example, Table 8.3 shows us that although acetic acid, formic acid, and phenol are all weak acids, their strengths as acids are not the same. Formic acid, with a K_a of 1.8×10^{-4}, is stronger than acetic acid whereas phenol, with a K_a of 1.3×10^{-10}, is much weaker than acetic acid.

EXAMPLE 8.2

The K_a for benzoic acid is 6.5×10^{-5}. What is the pK_a of this acid?

Solution
Take the logarithm of 6.5×10^{-5} on your scientific calculator. The answer is -4.19. Because pK_a is equal to $-\log K_a$, you must multiply this value by -1 to get pK_a. The pK_a of benzoic acid is 4.19.

Problem 8.2
The K_a for hydrocyanic acid, HCN, is 4.9×10^{-10}. What is its pK_a?

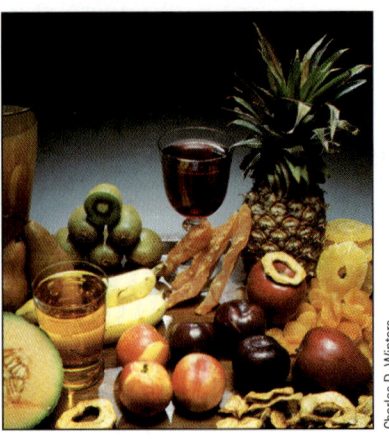

All these fruits and fruit drinks contain organic acids.

EXAMPLE 8.3

Which is the stronger acid:
(a) Benzoic acid with a K_a of 6.5×10^{-5}, or hydrocyanic acid with a K_a of 4.9×10^{-10}?
(b) Boric acid with a pK_a of 9.14 or carbonic acid with a pK_a of 6.37?

Solution
(a) Benzoic acid is the stronger acid; it has the greater K_a value.
(b) Carbonic acid is the stronger acid; it has the smaller pK_a value.

Problem 8.3

Which is the stronger acid:
(a) Carbonic acid, $pK_a = 6.37$, or ascorbic acid (vitamin C), $pK_a = 4.1$?
(b) Aspirin, $pK_a = 3.49$, or acetic acid, $pK_a = 4.75$?

8.6 Some Properties of Acids and Bases

Chemists of today do not taste the substances they work with, but 150 and 200 years ago they routinely did so. That is how we know that acids taste sour and bases taste bitter. The sour taste of lemons, vinegar, and many other foods, for example, is due to the acids they contain.

A Neutralization

The most important reaction of acids and bases is that they react with each other in a process called neutralization. This name is appropriate because, when a strong corrosive acid such as hydrochloric acid reacts with a strong corrosive base such as sodium hydroxide, the product (a solution of ordinary table salt in water) has neither acidic nor basic properties. We call such a solution *neutral*. Section 8.10 discusses neutralization reactions in detail.

B Reactions with Metals

Strong acids react with certain metals (called active metals) to produce hydrogen gas, H_2, and a salt. Hydrochloric acid, for example, reacts with magnesium metal to give the salt magnesium chloride and hydrogen gas (Figure 8.1):

$$Mg(s) + 2HCl(aq) \longrightarrow MgCl_2(aq) + H_2(g)$$
Magnesium   Hydrochloric   Magnesium   Hydrogen
                  acid   chloride

The reaction of an acid with an active metal to give a salt and hydrogen gas is a redox reaction (Section 4.7). The metal is oxidized to a metal ion and H^+ is reduced to H_2.

C Reaction with Metal Hydroxides

Acids react with metal hydroxides to give a salt and water.

$$HCl(aq) + KOH(aq) \longrightarrow H_2O(\ell) + KCl(aq)$$
Hydrochloric   Potassium   Water   Potassium
     acid   hydroxide   chloride

Figure 8.1 A ribbon of magnesium metal reacts with aqueous HCl to give H_2 gas and aqueous $MgCl_2$.

CHEMICAL CONNECTIONS 8C

Drugstore Antacids

Stomach fluid is normally quite acidic because of its HCl content. At some time, everyone has undoubtedly gotten "heartburn" caused by excess stomach acidity. At that time, you may have taken an antacid, which, as the name implies, is a substance that neutralizes acids—in other words, it is a base.

The word "antacid" is a medical term, not one used by chemists. It is, however, found on the labels of many medications available in drugstores and supermarkets. Almost all of them use bases such as $CaCO_3$, $Mg(OH)_2$, $Al(OH)_3$, and $NaHCO_3$ to decrease the acidity of the stomach. Simethicone is often added as an antiflatulant.

Also in drugstores and supermarkets are nonprescription drugs labeled "acid reducers." Among these are the brands Zantac, Tagamet, Pepcid, and Axid. Instead of neutralizing acidity, these compounds reduce the secretion of acid into the stomach. In larger doses (sold only with a prescription), some of these drugs are used in the treatment of stomach ulcers.

■ Commercial remedies for excess stomach acid.

Both the acid and metal hydroxide are ionized in aqueous solution. Furthermore, the salt formed is an ionic compound that is present in aqueous solution as anions and cations. Therefore, the actual equation for the reaction of HCl and KOH is more accurately written showing all of the ions present (Section 4.6):

$$H_3O^+ + Cl^- + K^+ + OH^- \longrightarrow 2H_2O + Cl^- + K^+$$

We can simplify this equation by omitting the spectator ions (Section 4.6), which gives the following equation for the net ionic reaction of any strong acid and strong base to give a salt and water:

$$H_3O^+ + OH^- \longrightarrow 2H_2O$$

D Reaction with Metal Oxides

Strong acids react with metal oxides to give water and a salt as shown in the following net ionic equation:

$$2H_3O^+(aq) + \underset{\text{Calcium oxide}}{CaO(s)} \longrightarrow 3H_2O(\ell) + Ca^{2+}(aq)$$

E Reactions with Carbonates and Bicarbonates

When a strong acid is added to a carbonate such as sodium carbonate, bubbles of carbon dioxide gas are rapidly given off. The overall reaction is a summation of two reactions. In the first reaction, carbonate ion reacts with H_3O^+ to give carbonic acid. Almost immediately, in the second reaction, carbonic acid decomposes to carbon dioxide and water. The following equations show the individual reactions and then the overall reaction:

$$2H_3O^+(aq) + CO_3^{2-}(aq) \longrightarrow H_2CO_3(aq) + 2H_2O(\ell)$$
$$\underline{H_2CO_3(aq) \longrightarrow CO_2(g) + H_2O(\ell)}$$
$$2H_3O^+(aq) + CO_3^{2-}(aq) \longrightarrow CO_2(g) + 3H_2O(\ell)$$

Strong acids also react with bicarbonates such as potassium bicarbonate to give carbon dioxide and water:

$$H_3O^+(aq) + HCO_3^-(aq) \longrightarrow H_2CO_3(aq) + H_2O(\ell)$$
$$\underline{H_2CO_3(aq) \longrightarrow CO_2(g) + H_2O(\ell)}$$
$$H_3O^+(aq) + HCO_3^-(aq) \longrightarrow CO_2(g) + 2H_2O(\ell)$$

To generalize, any acid stronger than carbonic acid will react with a carbonate or bicarbonate to give CO_2 gas.

The production of CO_2 is what makes bread and biscuit doughs and cake batters rise. The earliest method used to generate CO_2 for this purpose involved the addition of yeast, which catalyzes the fermentation of carbohydrates to produce carbon dioxide and ethanol (Chapter 27):

$$\underset{\text{Glucose}}{C_6H_{12}O_6} \xrightarrow{\text{Yeast}} 2CO_2 + \underset{\text{Ethanol}}{2C_2H_5OH}$$

■ Baking powder contains the weak acid dihydrogen phosphate ion and the weak base bicarbonate ion. When mixed together in water, the acid and base react to produce carbon dioxide gas. Bubbles of the gas are seen in the picture.

The production of CO_2 by fermentation, however, is slow. Sometimes it is desirable to have its production take place more rapidly, in which case bakers use the reaction of $NaHCO_3$ (sodium bicarbonate, also called **baking soda**) and a weak acid. But which weak acid? Vinegar (a 5% solution of acetic acid in water) would work, but it has a potential disadvantage—it imparts a particular flavor to foods. For a weak acid that imparts little or no flavor, bakers use either sodium dihydrogen phosphate, NaH_2PO_4, or potassium dihydrogen phosphate, KH_2PO_4. The two salts do not react with $NaHCO_3$ when they are dry but, when mixed with water in a dough or batter, they react quite rapidly to produce CO_2. The production of CO_2 is even more rapid in an oven!

$$H_2PO_4^-(aq) + H_2O(\ell) \rightleftharpoons HPO_4^{2-}(aq) + H_3O^+(aq)$$
$$\underline{HCO_3^-(aq) + H_3O^+(aq) \longrightarrow CO_2(g) + 2H_2O(\ell)}$$
$$H_2PO_4^-(aq) + HCO_3^-(aq) \longrightarrow HPO_4^{2-}(aq) + CO_2(g) + H_2O(\ell)$$

F Reactions with Ammonia and Amines

Any acid stronger than NH_4^+ (Table 8.2) is strong enough to react with NH_3 to form a salt. In the following reaction, the salt formed is ammonium chloride, NH_4Cl, which is shown as it would be ionized in aqueous solution:

$$HCl(aq) + NH_3(aq) \rightarrow NH_4^+(aq) + Cl^-(aq)$$

In Chapter 16 we will meet a family of compounds called amines, which are similar to ammonia except that one or more of the three hydrogen atoms of ammonia are replaced by carbon groups. A typical amine is methylamine, CH_3NH_2. The base strength of most amines is similar to that of NH_3, which means that amines also react with acids to form salts. The salt formed in the reaction of methylamine with HCl is methylammonium chloride, shown here as it would be ionized in aqueous solution:

$$HCl(aq) + \underset{\text{Methylamine}}{CH_3NH_2(aq)} \longrightarrow \underset{\text{Methylammonium ion}}{CH_3NH_3^+(aq)} + Cl^-(aq)$$

The reaction of ammonia and amines with acids to form salts is very important in the chemistry of the body, as we will see in later chapters.

8.7 Self-Ionization of Water

We have seen that an acid produces H_3O^+ ions in water and that a base produces OH^- ions. Suppose that we have absolutely pure water, with no added acid or base. Surprisingly enough, even pure water contains a very small number of H_3O^+ and OH^- ions. They are formed by the transfer of a proton from one molecule of water (the proton donor) to another (the proton acceptor).

$$\underset{\text{Acid}}{H_2O} + \underset{\text{Base}}{H_2O} \rightleftharpoons \underset{\substack{\text{Conjugate} \\ \text{base of } H_2O}}{OH^-} + \underset{\substack{\text{Conjugate} \\ \text{acid of } H_2O}}{H_3O^+}$$

What is the extent of this reaction? We know from the information in Table 8.2 that, in this equilibrium, H_3O^+ is the stronger acid and OH^- is the stronger base. Therefore, as shown by the arrows, the equilibrium for this reaction lies far to the left. We shall soon see exactly how far, but first let us write the equilibrium expression:

$$K_{eq} = \frac{[H_3O^+][HO^-]}{[H_2O]^2}$$

Because the degree of self-ionization of water is so slight, we can treat the concentration of water, $[H_2O]$, as a constant equal to 1000 g/L or approximately 55.5 mol/L, just as we did in Section 8.5 in developing the K_a for a weak acid. We can then combine these two constants (K_{eq} and $[H_2O]^2$) to define a new constant called the **ion product of water, K_w**. In pure water at room temperature, K_w has a value of 1.0×10^{-14}.

$$K_w = K_{eq}[H_2O]^2 = [H_3O^+][OH^-]$$

$$K_w = 1.0 \times 10^{-14}$$

In pure water H_3O^+ and OH^- form in equal amounts (see the balanced equation for the self-ionization of water), so their concentrations must be equal. That is, in pure water,

$$\left. \begin{array}{l} [H_3O^+] = 1.0 \times 10^{-7} \text{ mol/L} \\ [OH^-] = 1.0 \times 10^{-7} \text{ mol/L} \end{array} \right\} \text{In pure water}$$

These are very small concentrations, not enough to make pure water a conductor of electricity. Pure water is not an electrolyte (Section 6.6C).

The equation for the ionization of water is important because it applies not only to pure water but also to any water solution. The product of $[H_3O^+]$ and $[OH^-]$ in any aqueous solution is equal to 1.0×10^{-14}. If, for example, we add 0.010 mol of HCl to 1 L of pure water, it reacts completely to give H_3O^+ ions and Cl^- ions. The concentration of H_3O^+ will be 0.010 M, or 1.0×10^{-2} M. This means that the $[OH^-]$ must be $1.0 \times 10^{-14}/1.0 \times 10^{-2}$ = 1.0×10^{-12} M.

EXAMPLE 8.4

The $[OH^-]$ of an aqueous solution is 1.0×10^{-4} M. What is its $[H_3O^+]$?

Solution
We substitute into the equation:

$$[H_3O^+][OH^-] = 1.0 \times 10^{-14}$$

$$[H_3O^+] = \frac{1.0 \times 10^{-14}}{1.0 \times 10^{-4}} = 1.0 \times 10^{-10} M$$

Problem 8.4
The $[OH^-]$ of an aqueous solution is 1.0×10^{-12} M. What is its $[H_3O^+]$?

Aqueous solutions can have a very high $[H_3O^+]$ but the $[OH^-]$ must then be very low, and vice versa. Any solution with a $[H_3O^+]$ greater than 1.0×10^{-7} M is acidic. In such solutions, of necessity $[OH^-]$ must be less than 1.0×10^{-7} M. The higher the $[H_3O^+]$, the more acidic the solution. Similarly, any solution with an $[OH^-]$ greater than 1.0×10^{-7} M is basic. Pure water, in which $[H_3O^+]$ and $[OH^-]$ are equal (they are both 1.0×10^{-7} M), is neutral—that is, neither acidic nor basic.

8.8 pH and pOH

Screen 8.3 Simulation Tutorial

Because hydronium ion concentrations for most solutions are numbers with negative exponents, these concentrations are more commonly expressed as pH, where

$$pH = -\log[H_3O^+]$$

In Section 8.7, we saw that a solution is acidic if its $[H_3O^+]$ is greater than 1.0×10^{-7}, and that it is basic if its $[H_3O^+]$ is less than 1.0×10^{-7}. We can now state the definitions of acidic and basic solutions in terms of pH:

A solution is acidic if its pH is less than 7.0
A solution is basic if its pH is greater than 7.0
A solution is neutral if its pH is equal to 7.0

Although pure water has a pH of 7, tap water usually has a pH of approximately 6 because it contains dissolved CO_2 (from the atmosphere), which reacts with water to give dissolved carbonic acid, H_2CO_3.

EXAMPLE 8.5

(a) The $[H_3O^+]$ of a certain liquid detergent is 1.4×10^{-9} M. What is its pH?
(b) The pH of black coffee is 5.3. What is its $[H_3O^+]$?

Solution

(a) On your calculator, take the log of 1.4×10^{-9}. The answer is -8.85. Multiply this value by -1 to give the pH of 8.85.

(b) Enter 5.3 into your calculator and then press the $+/-$ key to change the sign to minus and give -5.3. Then take the antilog of this number. The $[H_3O^+]$ of black coffee is 5×10^{-6}.

Problem 8.5

(a) The $[H_3O^+]$ of an acidic solution is $3.5 \times 10^{-3}\,M$. What is its pH?
(b) The pH of tomato juice is 4.1. What is its $[H_3O^+]$?

Just as pH is a convenient way to designate the concentration of H_3O^+, pOH is a convenient way to designate the concentration of OH^-:

$$\text{pOH} = -\log\,[OH^-]$$

In aqueous solutions, the ion product of water, K_w, is 1×10^{-14} and thus $pK_w = -\log K_w = 14$. Therefore, we can write

$$14 = \text{pH} + \text{pOH}$$

Thus, once we know the pH of a solution, we can easily calculate the pOH.

EXAMPLE 8.6

The $[OH^-]$ of a strongly basic solution is 1.0×10^{-2}. What are the pOH and pH of this solution?

Solution
The pOH is 2, and the pH is $14 - 2 = 12$.

Problem 8.6

The $[OH^-]$ of a solution is 1.0×10^{-4}. What are the pOH and pH of this solution?

All fluids in the human body are aqueous; that is, the only solvent present is water. Consequently, all body fluids have a pH value. Some of them have a narrow pH range; others have a wide pH range. The pH of blood, for example, must be between 7.35 and 7.45 (slightly basic). If it goes outside these limits, illness and even death may result (see Chemical Connections 8D). In contrast, the pH of urine can vary from 5.5 to 7.5. Table 8.4 gives the pH values of some common materials.

One thing you must remember when you see a pH value is that, because pH is a logarithmic scale, an increase (or decrease) of one pH unit means a tenfold decrease (or increase) in the $[H_3O^+]$. For example, a pH of 3 does not sound very different from a pH of 4. The first, however, means a $[H_3O^+]$ of $10^{-3}\,M$, whereas the second means a $[H_3O^+]$ of $10^{-4}\,M$. The $[H_3O^+]$ of the pH 3 solution is ten times the $[H_3O^+]$ of the pH 4 solution.

There are two ways to measure the pH of an aqueous solution. One way is to use a pH paper, which is made by soaking plain paper with a mixture of pH indicators. A pH **indicator** is a substance that changes color at a certain pH. When we place a drop of solution on this paper, the paper turns a certain color. To determine the pH, we compare the color of the paper with the colors on a chart supplied with the paper.

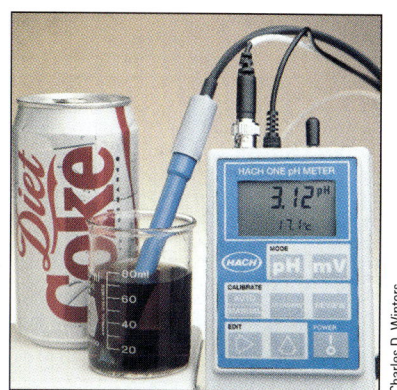

■ The pH of this soft drink is 3.12. Soft drinks are often quite acidic.

■ The pH of three household substances. The colors of the acid–base indicators in the flasks show that vinegar is more acidic than club soda, and the cleaner is basic.

Litmus paper turns red in acidic solution and blue in basic solution. It is often used as an approximate guide to acidity or basicity.

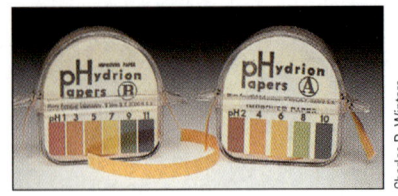

■ Strips of paper impregnated with indicator are used to find an approximate pH.

Table 8.4 pH Values of Some Common Materials

Material	pH	Material	pH
Battery acid	0.5	Saliva	6.5–7.5
Gastric juice	1.0–3.0	Pure water	7.0
Lemon juice	2.2–2.4	Blood	7.35–7.45
Vinegar	2.4–3.4	Bile	6.8–7.0
Tomato juice	4.0–4.4	Pancreatic fluid	7.8–8.0
Carbonated beverages	4.0–5.0	Sea water	8.0–9.0
Black coffee	5.0–5.1	Soap	8.0–10.0
Urine	5.5–7.5	Milk of magnesia	10.5
Rain (unpolluted)	6.2	Household ammonia	11.7
Milk	6.3–6.6	Lye (1.0 M NaOH)	14.0

Figure 8.2 Some acid–base indicators. Note that some indicators have two color changes.

One acid–base indicator is the compound methyl orange. When a drop of methyl orange is added to an aqueous solution with a pH of 3.2 or lower, this indicator turns red and the entire solution becomes red. When added to an aqueous solution with a pH of 4.4 or higher, this indicator turns yellow. These particular limits and colors apply only to methyl orange. Other indicators have other limits and colors (Figure 8.2).

The second way for determining pH is more accurate. In this method, we use a pH meter (Figure 8.3). We dip the electrode of the pH meter into the solution whose pH is to be measured, and then read the pH on a dial. The most commonly used pH meters read pH to the nearest hundredth of a unit.

8.9 The pH of Aqueous Salt Solutions

When an acid is added to water, the solution becomes acidic (pH less than 7); when a base is added to water, the solution becomes basic (pH greater than 7). Suppose we add a salt to water? What happens to the pH then? At first thought, you might suppose that all salts are neutral and that adding any salt to pure water leaves the pH at 7. This is, in fact, true for some salts. Many salts, however, are acidic or basic; as a result, they cause the pH of pure water to change. To understand how this change in pH occurs, let us look at sodium acetate, a typical basic salt.

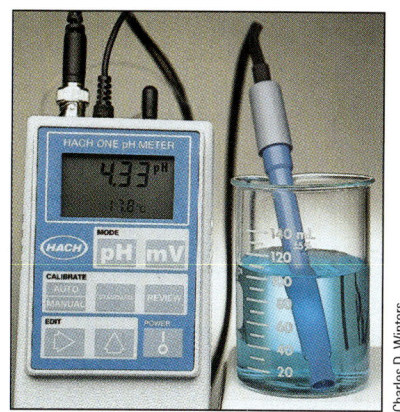

Figure 8.3 Copper(II) sulfate, CuSO₄, is an acidic salt. Because copper(II) sulfate is the salt of a strong acid and a weak base, aqueous solutions of it is acidic. The pH of the 0.1 M solution shown here is 4.33.

Like almost all salts, sodium acetate is 100% ionized in water so that, when we add CH₃COONa to water, we get a solution containing Na⁺ ions and CH₃COO⁻ ions. The Na⁺ ions do not react with the water, but the CH₃COO⁻ ions do according to the following equilibrium:

$$\underset{\text{Acetate ion}\atop\text{(Weaker acid)}}{CH_3COO^-} + \underset{\text{Water}\atop\text{(Weaker base)}}{H_2O} \rightleftharpoons \underset{\text{Acetic acid}\atop\text{(Stronger acid)}}{CH_3COOH} + \underset{\text{Hydroxide ion}\atop\text{(Stronger base)}}{OH^-}$$

Using the reasoning we developed in Section 8.4 to predict the position of equilibrium in an acid–base reaction, we see that this equilibrium lies to the left—that is, toward the weaker acid and the weaker base. It has been determined experimentally that at equilibrium in a 0.1 M solution of sodium acetate, only about 1 out of every 13,500 acetate ions is converted to acetic acid and OH⁻ ions. Nevertheless, enough OH⁻ ions are produced to increase the pH to about 8.88. Because this pH is higher than 7, the solution is basic, and sodium acetate is called a basic salt. Although acetic acid is also being produced, the pH of an aqueous solution depends only on the concentration of H₃O⁺ and OH⁻ and not on the concentration of any other un-ionized acid or base.

We can predict whether a salt will be acidic, neutral, or basic in water solution if we know something about the strength of the acid and base. We have four cases:

1. A salt of a strong base and a weak acid raises the pH of pure water and, therefore, is called a **basic salt.** An example is sodium acetate, which we just discussed.

2. A salt of a strong acid and a weak base lowers the pH of pure water and, therefore, is called an **acidic salt.** An example is ammonium chloride, NH₄Cl. The Cl⁻ ion does not react with H₂O, but the NH₄⁺ ion does to a slight extent to form NH₃ and H₃O⁺. The presence of H₃O⁺ lowers the pH.

$$\underset{\text{Ammonium ion}\atop\text{(Weaker acid)}}{NH_4^+} + \underset{\text{Water}\atop\text{(Weaker base)}}{H_2O} \rightleftharpoons \underset{\text{Ammonia}\atop\text{(Stronger base)}}{NH_3} + \underset{\text{Hydronium ion}\atop\text{(Stronger acid)}}{H_3O^+}$$

Copper(II) sulfate (Figure 8.3) is another example of an acidic salt.

3. A salt of a strong acid and a strong base does not change the pH of pure water and, therefore, is called a **neutral salt.** Examples of neutral salts are NaCl, KNO₃, and Na₂SO₄.

■ Red cabbage juice is a naturally occurring acid–base indicator. From left to right are solutions of pH 1, 4, 7, 10, and 13.

Basic salt The salt of a weak acid and a strong base; an aqueous solution of a basic salt is acidic

Acidic salt The salt of a strong acid and a weak base; an aqueous solution of an acidic salt is basic

Neutral salt A salt whose aqueous solution is neither acidic nor basic

■ Because CH₃COONa, sodium acetate, is the salt of a weak acid and a strong base, aqueous solutions of this salt are basic. The pH of the solution shown here is 8.88. An inert, insoluble solid has been added to the flask to enhance the visibility of the liquid in the flask.

Titration An analytical procedure whereby we react a known volume of a solution of known concentration with a known volume of a solution of unknown concentration

End point The point at which there is an equal amount of acid and base in a neutralization reaction

Screen 8.11 Simulation

4. A salt of a weak acid and a weak base may or may not change the pH of pure water. Both ions react with water. Sometimes the effects cancel each other, and sometimes they do not. We will not try to make predictions in such cases.

8.10 Titration

Laboratories, whether medical, academic, or industrial, are frequently asked to determine the exact concentration of a particular substance in solution—for example, the concentration of acetic acid in a given sample of vinegar, or the concentrations of iron, calcium, and magnesium ions in a sample of "hard" water. Determinations of solution concentrations can be made using an analytical technique called a **titration.**

In a titration, we react a known volume of a solution of known concentration with a known volume of a solution of unknown concentration. The solution of unknown concentration may contain an acid (such as stomach acid), a base (such as ammonia), an ion (such as Fe^{2+} ion), or any other substance whose concentration we are asked to determine. If we know the titration volumes and the mole ratio in which the solutes react, we can then calculate the concentration of the second solution.

Titrations must meet several requirements:

1. We must know the equation for the reaction so that we can determine the stoichiometric ratio of reactants to use in our calculations.
2. The reaction must be rapid and complete.
3. When the reactants have combined exactly, there must be a clear-cut change in some measurable property of the reaction mixture. We call the point at which the reactants combine exactly the **end point** of the titration.
4. We must have accurate measurements of the amount of each reactant.

Let us apply these requirements to the titration of a solution of sulfuric acid of known concentration with a solution of sodium hydroxide of unknown concentration. We know the balanced equation for this acid–base reaction, so requirement 1 is met.

$$2NaOH(aq) + H_2SO_4(aq) \longrightarrow Na_2SO_4(aq) + 2H_2O(\ell)$$

(Concentration not known)   (Concentration known)

Sodium hydroxide ionizes in water to form sodium ions and hydroxide ions; sulfuric acid ionizes to form hydronium ions and sulfate ions. The reaction between hydroxide and hydronium ions is rapid and complete, so requirement 2 is met.

To meet requirement 3, we must be able to observe a clear-cut change in some measurable property of the reaction mixture at the end point. For acid–base titrations, we use the sudden pH change that occurs at this point. Suppose we add the sodium hydroxide solution slowly. As it is added, it reacts with hydronium ions to form water. As long as any unreacted hydronium ions are present, the solution is acidic. When the number of hydroxide ions added exactly equals the original number of hydronium ions, the solution becomes neutral. Then, as soon as any extra hydroxide ions are added, the solution becomes basic. We can observe this sudden change in pH by reading a pH meter.

Another way to observe the change in pH at the end point is to use an acid–base indicator (Section 8.8). Such an indicator changes color when the

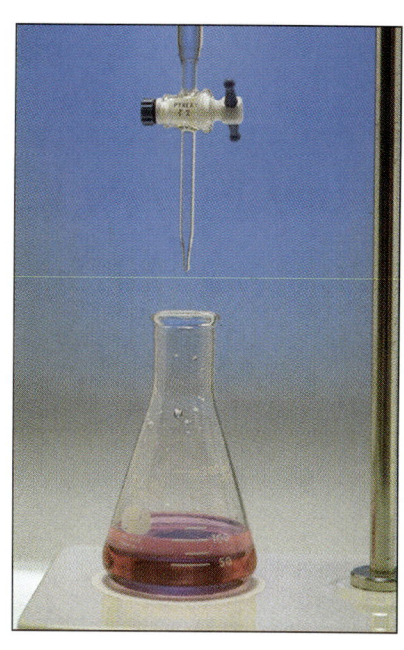

(a) (b) (c)

Figure 8.4 An acid–base titration. (a) An acid of known concentration is in the Erlenmeyer flask. (b) When a base is added from the buret, the acid is neutralized. (c) The end point is reached when the color of the indicator changes from colorless to pink.

solution changes from acidic to basic. Phenolphthalein, for example, is colorless in acid solution and pink in basic solution. If this indicator is added to the original sulfuric acid solution, the solution remains colorless as long as excess hydronium ions are present [Figure 8.4(a)]. After enough sodium hydroxide solution has been added to react with all of the hydronium ions, the next drop of base provides excess hydroxide ions, and the solution turns pink (Figure 8.4(c)). Thus we have a clear-cut indication of the end point.

To meet requirement 4, which is that the volume of each solution used must be known, we use volumetric glassware such as volumetric flasks, burets, and pipets.

Data for a typical acid–base titration are given in Example 8.7. Note that the experiment is run in triplicate, a standard procedure for checking the precision of a titration.

EXAMPLE 8.7

Following are data for the titration of 0.108 M H_2SO_4 with a solution of NaOH of unknown concentration. What is the concentration of the NaOH solution?

	Volume of 0.108 M H_2SO_4	Volume of NaOH
Trial I	25.0 mL	33.48 mL
Trial II	25.0 mL	33.46 mL
Trial III	25.0 mL	33.50 mL

Solution

From the balanced equation for this acid–base reaction, we know the stoichiometry: Two moles of NaOH react with one mole of H_2SO_4. From the experimental data, we calculate that the average volume of the NaOH required for complete reaction is 33.48 mL. Because the units of molarity are moles/liter, we must convert volumes of reactants from milliliters to liters. We can then use the factor-label method (Section 1.5) to calculate the molarity of the NaOH solution. What we wish to calculate is the number of moles of NaOH per liter of NaOH.

$$\frac{\text{mol NaOH}}{\text{L NaOH}} = \frac{0.108 \text{ mol } H_2SO_4}{1 \text{ L } H_2SO_4} \times \frac{0.0250 \text{ L } H_2SO_4}{0.03348 \text{ L NaOH}} \times \frac{2 \text{ mol NaOH}}{1 \text{ mol } H_2SO_4}$$

$$= \frac{0.161 \text{ mol NaOH}}{\text{L NaOH}} = 0.161 \, M$$

Problem 8.7

Calculate the concentration of an acetic acid solution using the following data. Three 25.0-mL samples of acetic acid were titrated to a phenolphthalein end point with 0.121 M NaOH. The volumes of NaOH were 19.96 mL, 19.73 mL, and 19.79 mL.

It is important to understand that titration is not a method of determining the acidity (or basicity) of a solution. If we want to do that, we measure the sample's pH, which is the only measurement of solution acidity or basicity. Rather, titration is a method for determining the total acid or base concentration of a solution, which is not the same as the acidity. For example, a 0.1 M solution of HCl in water has a pH of 1, but a 0.1 M solution of acetic acid has a pH of 2.9. These two solutions have the same concentration of acid and each neutralizes the same volume of NaOH solution, but they have very different acidities.

8.11 Buffers

A What Is a Buffer and How Does One Work?

As noted earlier, the body must keep the pH of blood between 7.35 and 7.45. Yet we frequently eat acidic foods such as oranges, lemons, sauerkraut, and tomatoes, and doing so eventually adds considerable quantities of H_3O^+ to the blood. Despite these additions of acidic or basic substances, the body manages to keep the pH of blood remarkably constant. The body does it with buffers. A **buffer** is a solution whose pH changes very little when H_3O^+ or OH^- ions are added to it. In a sense, a pH buffer is an acid or base "shock absorber."

The most common buffers consist of approximately equal molar amounts of a weak acid and a salt of the weak acid. Put another way, they consist of approximately equal amounts of a weak acid and its conjugate base. For example, if we dissolve 1.0 mol of acetic acid (a weak acid) and 1.0 mol of its conjugate base (in the form of CH_3COONa, sodium acetate) in 1.0 L of water, we have a good buffer solution. The equilibrium present in this buffer solution is

> **Buffer** A solution that resists change in pH when limited amounts of an acid or base are added to it; an aqueous solution containing a weak acid and its conjugate base

$$\underset{\substack{\text{Acetic acid}\\\text{(A weak acid)}}}{\overset{\overset{\text{Added as}}{\text{CH}_3\text{COOH}}}{\text{CH}_3\text{COOH}}} + \text{H}_2\text{O} \rightleftharpoons \underset{\substack{\text{Acetate ion}\\\text{(Conjugate base}\\\text{of a weak acid)}}}{\overset{\overset{\text{Added as}}{\text{CH}_3\text{COO}^-\text{Na}^+}}{\text{CH}_3\text{COO}^-}} + \text{H}_3\text{O}^+$$

A buffer resists any change in pH upon the addition of small quantities of acid or base. To see how, we will use an acetic acid–sodium acetate buffer as an example. If a strong acid such as HCl is added to this buffer solution, the added H_3O^+ ions react with CH_3COO^- ions and are removed from solution.

$$\underset{\substack{\text{Acetate ion}\\\text{(Conjugate base}\\\text{of a weak acid)}}}{\text{CH}_3\text{COO}^-} + \text{H}_3\text{O}^+ \longrightarrow \underset{\substack{\text{Acetic acid}\\\text{(A weak acid)}}}{\text{CH}_3\text{COOH}} + \text{H}_2\text{O}$$

There is a slight increase in the concentration of CH_3COOH as well as a slight decrease in the concentration of CH_3COO^-, but there is no appreciable change in pH. We say that this solution is buffered because it resists a change in pH on the addition of small quantities of a strong acid.

If NaOH or another strong base is added to the buffer solution, the added OH^- ions react with CH_3COOH molecules and are removed from solution.

Screen 8.9
Simulation Tutorial

$$\underset{\substack{\text{Acetic acid}\\\text{(A weak acid)}}}{\text{CH}_3\text{COOH}} + \text{OH}^- \longrightarrow \underset{\substack{\text{Acetate ion}\\\text{(Conjugate base}\\\text{of a weak acid)}}}{\text{CH}_3\text{COO}^-} + \text{H}_2\text{O}$$

Here there is a slight decrease in the concentration of CH_3COOH as well as a slight increase in the concentration of CH_3COO^-, but, again, there is no appreciable change in pH.

The important point about this or any other buffer solution is that when the conjugate base of the weak acid removes H_3O^+, it is converted to the undissociated weak acid. Because a substantial amount of weak acid is already present, there is no appreciable change in its concentration and, because H_3O^+ ions are removed from solution, there is no appreciable change in pH. By the same token, when the weak acid removes OH^- ions from solution, it is converted to its conjugate base. Because OH^- ions are removed from solution, there is no appreciable change in pH.

The effect of a buffer can be quite powerful. Addition of either dilute HCl or NaOH to pure water, for example, causes a dramatic change in pH (Figure 8.5).

When HCl or NaOH is added to a phosphate buffer, the results are quite different. Suppose we have a phosphate buffer solution of pH 7.21 prepared by dissolving 0.10 mol NaH_2PO_4 (a weak acid) and 0.10 mol Na_2HPO_4 (its conjugate base) in enough water to make 1.00 L of solution. If we add 0.010 mol of HCl to 1.0 L of this solution, the pH decreases to only 7.12. If we add 0.1 mol of NaOH, the pH increases to only 7.30.

Phosphate buffer (pH 7.21) + 0.010 mol HCl   pH 7.21 → 7.12

Phosphate buffer (pH 7.21) + 0.010 mol NaOH   pH 7.21 → 7.30

Figure 8.6 shows the effect of adding acid to a buffer solution.

(a) pH 7.00

(b) pH 2.00

(c) pH 12.00

Figure 8.5 The addition of HCl and NaOH to pure water. (a) The pH of pure water is 7.0. (b) The addition of 0.01 mol of HCl to 1 L of pure water causes the pH to decrease to 2. (c) The addition of 0.010 mol of NaOH to 1 L of pure water causes the pH to increase to 12.

B Buffer pH

In the previous example, the pH of the buffer containing equal molar amounts of $H_2PO_4^-$ and HPO_4^{2-} is 7.21. From Table 8.3, we see that 7.21 is the pK_a of the acid $H_2PO_4^-$. This is not a coincidence. If we make a buffer solution by mixing equimolar concentrations of any weak acid and its conjugate base, the pH of the solution will equal the pK_a of the weak acid.

Figure 8.6 Buffer solutions. The solution in the Erlenmeyer flask on the right in both (a) and (b) is a buffer of pH 7.40, the same pH as human blood. The buffer solution also contains bromocresol green, an acid–base indicator that is blue at pH 7.40 (see Figure 8.2). (a) The beaker contains some of the pH 7.40 buffer and the bromocresol green indicator to which has been added 5 mL of 0.1 M HCl. After the addition of the HCl, the pH of the buffer solution drops only 0.65 unit to 6.75. (b) The beaker contains pure water and bromocresol green indicator to which has been added 5 mL of 0.10 M HCl. After the addition of the HCl, the pH of the unbuffered solution drops by 4.38 units to pH 3.02.

(a)

(b)

This fact allows us to prepare buffer solutions to maintain almost any pH. For example, if we want to maintain a pH of 9.14, we could make a buffer solution from boric acid, H_3BO_3, and sodium dihydrogen borate, NaH_2BO_3, the sodium salt of its conjugate base (see Table 8.3).

EXAMPLE 8.8

What is the pH of a buffer solution containing equimolar quantities of:
(a) H_3PO_4 and NaH_2PO_4?
(b) H_2CO_3 and $NaHCO_3$?

Solution
The pH is equal to the pK_a of the weak acid, which we find in Table 8.3.
(a) pH = 2.12
(b) pH = 6.37

Screen 8.10
Simulation Tutorial

Problem 8.8
What is the pH of a buffer solution containing equimolar quantities of:
(a) NH_4Cl and NH_3?
(b) CH_3COOH and CH_3COONa?

C Buffer Capacity

Buffer capacity is the amount of hydrogen or hydroxide ions that a buffer can absorb without a significant change in its pH. We have already mentioned that a pH buffer is an acid–base "shock absorber." We now ask what makes one solution a better acid–base shock absorber than another solution. The capacity of a pH buffer depends on both its pH and its concentration:

> pH: The closer the pH of the buffer is to the pK_a of the weak acid, the greater the buffer capacity.
>
> Concentration: The greater the concentration of the weak acid and its conjugate base, the greater the buffer capacity.

Buffer capacity The extent to which a buffer solution can prevent a significant change in pH of a solution upon addition of an acid or a base

An effective buffer has a pH equal to the pK_a of the weak acid ±1. For acetic acid, for example, the pK_a is 4.75. Therefore, a solution of acetic acid and sodium acetate functions as an effective buffer within the pH range of approximately 3.75–5.75. The most effective buffer is one with equal concentrations of the weak acid and its salt—that is, one in which the pH of the buffer solution equals the pK_a of the weak acid.

Buffer capacity also depends on concentration. The greater the concentration of weak acid and its conjugate base, the greater the buffer capacity. We could make a buffer solution by dissolving 1.0 mol each of CH_3COONa and CH_3COOH in 1 L of H_2O, or we could use only 0.10 mol of each. Both solutions have the same pH of 4.75. However, the former has a buffer capacity ten times that of the latter. If we add 0.2 mol of HCl to the former solution, it performs the way we expect—the pH drops to 4.57. If we add 0.2 mol of HCl to the latter solution, however, the pH drops to 1.0 because the buffer has been "swamped out." That is, the amount of H_3O^+ added has exceeded the buffer capacity. The first 0.10 mol of HCl completely neutralizes essentially all the CH_3COO^- present. After that, the solution contains only CH_3COOH and is no longer a buffer, so that the second 0.10 mol of HCl decreases the pH to 1.0.

D Blood Buffers

The average pH of human blood is 7.4. Any change larger than 0.10 pH unit in either direction may cause illness. If the pH goes below 6.8 or above 7.8, death may result. To hold the pH of the blood close to 7.4, the body uses three buffer systems: carbonate, phosphate, and proteins (proteins are discussed in Chapter 21).

The most important of these systems is the carbonate buffer. The weak acid of this buffer is carbonic acid, H_2CO_3; the conjugate base is the bicarbonate ion, HCO_3^-. The pK_a of H_2CO_3 is 6.37 (from Table 8.3). Because the pH of an equal mixture of a weak acid and its salt is equal to the pK_a of the weak acid, a buffer with equal concentrations of H_2CO_3 and HCO_3^- has a pH of 6.37.

Blood, however, has a pH of 7.4. The carbonate buffer can maintain this pH only if $[H_2CO_3]$ and $[HCO_3^-]$ are not equal. In fact, the necessary $[HCO_3^-]/[H_2CO_3]$ ratio is about 10:1. The normal concentrations of these species in blood are about 0.025 M HCO_3^- and 0.0025 M H_2CO_3. This buffer works because any added H_3O^+ is neutralized by the HCO_3^- and any added OH^- is neutralized by the H_2CO_3.

The fact that the $[HCO_3^-]/[H_2CO_3]$ ratio is 10:1 means that this system is a better buffer for acids, which lower the ratio and thus improve buffer efficiency, than for bases, which raise the ratio and decrease buffer capacity. This result is in harmony with the actual functioning of the body—under normal conditions, larger amounts of acidic than basic substances enter the blood. The 10:1 ratio is easily maintained under normal conditions, because the body can very quickly increase or decrease the amount of CO_2 entering the blood.

The second most important buffering system of the blood is a phosphate buffer made up of hydrogen phosphate ion, HPO_4^{2-}, and dihydrogen phosphate ion, $H_2PO_4^-$. In this case, a 1.6:1 $[HPO_4^{2-}]/[H_2PO_4^-]$ ratio is necessary to maintain a pH of 7.4. This ratio is well within the limits of good buffering action.

8.12 The Henderson-Hasselbalch Equation

Suppose we want to make a phosphate buffer solution of pH 7.00. The weak acid with a pK_a closest to this desired pH is $H_2PO_4^-$; it has a pK_a of 7.21. If we use equal concentrations of NaH_2PO_4 and Na_2HPO_4, however, we will have a buffer of pH 7.21. We want a phosphate buffer that is slightly more acidic than 7.21, so it would seem reasonable to use more of the weak acid, $H_2PO_4^-$, and less of its conjugate base, HPO_4^{2-}. But what proportions of these two salts do we use? Fortunately, we can calculate these proportions using the **Henderson-Hasselbalch equation.**

The Henderson-Hasselbalch equation is a mathematical relationship between pH, the pK_a of a weak acid, and the concentrations of the weak acid and its conjugate base. The equation is derived in the following way. Assume that we are dealing with weak acid, HA, and its conjugate base, A^-.

$$HA + H_2O \rightleftharpoons A^- + H_3O^+$$

$$K_a = \frac{[A^-][H_3O^+]}{[HA]}$$

Taking the logarithm of this equation gives

$$\log K_a = \log [H_3O^+] + \log \frac{[A^-]}{[HA]}$$

CHEMICAL CONNECTIONS 8D

Acidosis and Alkalosis

The pH of blood is normally between 7.35 and 7.45. If the pH goes lower than this level, the condition is called **acidosis**. A blood pH higher than 7.45 produces **alkalosis**. Both are abnormal conditions. Acidosis leads to depression of the nervous system. Mild acidosis can result in fainting; a more severe case can cause coma. Alkalosis leads to overstimulation of the nervous system, muscle cramps, and convulsions. If the acidosis or alkalosis persists for a sufficient period of time, or if the pH gets too far away from 7.35 to 7.45, death may result.

Acidosis has several causes. One type, called respiratory acidosis, results from difficulty in breathing (hypoventilation). An obstruction in the windpipe or diseases such as pneumonia, emphysema, asthma, or congestive heart failure may diminish the amount of CO_2 that leaves the body through the lungs. (You can even produce mild acidosis by holding your breath.) The pH of the blood decreases because the CO_2, unable to escape fast enough, remains in the blood, where it lowers the $[HCO_3^-]/[H_2CO_3]$ ratio.

Respiratory alkalosis involves the reverse process. It arises from rapid or heavy breathing, called hyperventilation. This condition can come about from fever, infection, the action of certain drugs, or even hysteria. Here the excessive loss of CO_2 raises the $[HCO_3^-]/[H_2CO_3]$ ratio and the pH.

Acidosis caused by other factors is called metabolic acidosis. Two causes of this condition are starvation (or fasting) and heavy exercise. When the body doesn't get enough food, it burns its own fat, and the products of this reaction are acidic compounds that enter the blood. This problem sometimes arises in people on fad diets. Heavy exercise causes the muscles to produce excessive amounts of lactate (the anion of lactic acid), which makes muscles feel tired and sore. Metabolic acidosis is also caused by a number of metabolic irregularities. For example, the disease diabetes mellitus produces acidic compounds called ketone bodies (Section 27.7).

Metabolic alkalosis can result from various metabolic irregularities as well. In addition, it can be caused by excessive vomiting. The contents of the stomach are strongly acidic (pH 1.0–3.0), and loss of substantial amounts of this acidic material raises the pH of the blood.

■ **High blood pH brought on by severe anxiety can cause alkalosis. One remedy is to breathe into a paper bag.**

Rearranging terms gives us a new expression, in which $-\log K_a$ is by definition pK_a, and $-\log [H_3O^+]$ is by definition pH. Making these substitutions gives the Henderson-Hasselbalch equation:

$$-\log [H_3O^+] = -\log K_a + \log \frac{[A^-]}{[HA]}$$

$$pH = pK_a + \log \frac{[A^-]}{[HA]} \quad \text{Henderson-Hasselbalch equation}$$

The Henderson-Hasselbalch equation gives us a convenient way to calculate the pH of a buffer when the concentrations of the weak acid and its conjugate base are not equal.

EXAMPLE 8.9

What is the pH of a phosphate buffer solution containing 1.0 mol/L of sodium dihydrogen phosphate, NaH_2PO_4, and 0.50 mol/L of sodium hydrogen phosphate, Na_2HPO_4?

Solution
The weak acid in this problem is $H_2PO_4^-$; its ionization produces HPO_4^{2-}. The pK_a of this acid is 7.21 (from Table 8.3). Under the weak acid and its conjugate base are shown their concentrations.

$$H_2PO_4^- + H_2O \rightleftharpoons HPO_4^{2-} + H_3O^+ \quad pK_a = 7.21$$
$$\text{1.0 mol/L} \qquad\qquad \text{0.50 mol/L}$$

Substituting these values in the Henderson-Hasselbalch equation gives a pH of 6.91.

$$pH = 7.21 + \log \frac{0.50}{1.0}$$

$$= 7.21 - 0.30 = 6.91$$

Problem 8.9
What is the pH of a boric acid buffer solution containing 0.25 mol/L of boric acid, H_3BO_3, and 0.50 mol/L of sodium dihydrogen borate, NaH_2BO_3? See Table 8.3 for the pK_a of boric acid.

Returning to the problem posed at the beginning of this section, how do we calculate the proportions of NaH_2PO_4 and Na_2HPO_4 needed to make up a phosphate buffer of pH 7.00? We know that the pK_a of $H_2PO_4^-$ is 7.21 and that the buffer we wish to prepare has a pH of 7.00. We can substitute these two values in the Henderson-Hasselbalch equation as follows:

$$7.00 = 7.21 + \log \frac{[HPO_4^{2-}]}{[H_2PO_4^-]}$$

Rearranging and solving gives

$$\log \frac{[HPO_4^{2-}]}{[H_2PO_4^-]} = 7.00 - 7.21 = -0.21$$

Taking the antilog of both sides gives

$$\frac{[HPO_4^{2-}]}{[H_2PO_4^-]} = \frac{0.62}{1}$$

Thus, to prepare a phosphate buffer of pH 7.00, we can use 0.62 mol of Na_2HPO_4 and 1.0 mol of NaH_2PO_4. Alternatively, we can use any other amounts of these two salts, as long as their mole ratio is 0.62 : 1.0.

SUMMARY

By the **Arrhenius** definitions, acids are substances that produce H_3O^+ ions in aqueous solution, and bases are substances that produce OH^- ions in aqueous solution (Section 8.1). A strong acid reacts completely or almost completely with water to form H_3O^+ ions (Section 8.2). A strong base reacts completely or almost completely with water to form OH^- ions.

The **Brønsted-Lowry** definitions expand the definitions of acid and base to beyond water (Section 8.3): An acid is a proton donor, and a base is a proton acceptor (Section 8.3). Every acid has a **conjugate base,** and every base has a **conjugate acid.** The stronger the acid, the weaker its conjugate base. Conversely, the stronger the base, the weaker its conjugate acid. An **amphiprotic substance,** such as water, can act as either an acid or a base (Section 8.3).

In an acid–base reaction, the position of equilibrium favors the reaction of the stronger acid and the stronger base to form the weaker acid and the weaker base (Section 8.4).

The strength of a weak acid is expressed by its **ionization constant, K_a** (Section 8.5). The larger the value of K_a, the stronger the acid $pK_a = -\log [K_a]$.

Acids react with metals, metal hydroxides, and metal oxides to give **salts,** which are ionic compounds made up of cations from the base and anions from the acid (Sections 8.6B, 8.6C, and 8.6D). Acids also react with carbonates, bicarbonates, ammonia, and amines (Sections 8.6E and 8.6F) to give salts.

In pure water, a small percentage of molecules undergo self-ionization:

$$H_2O + H_2O \rightleftharpoons H_3O^+ + OH^-$$

As a result, pure water has a concentration of 10^{-7} M for H_3O^+ and 10^{-7} M for OH^- (Section 8.7). The **ion product of water, K_w,** is equal to 1.0×10^{-14}. $pK_w = 14$.

Hydronium ion concentrations are generally expressed in **pH** units, with pH = $-\log [H_3O^+]$ (Section 8.8). **pOH** = $-\log [OH^-]$. Solutions with pH less than 7 are acidic; those with pH greater than 7 are basic. A **neutral solution** has a pH of 7. The pH of an aqueous solution is measured with an acid–base indicator or with a pH meter.

Aqueous solutions of salts of strong acids and weak bases are acidic; those of weak acids and strong bases are basic; and those of strong acids and strong bases are neutral (Section 8.9).

We can measure the concentration of aqueous solutions of acids and bases by using a **titration,** (Section 8.10). In an acid–base titration, a base of known concentration is added to an acid of unknown concentration (or vice versa) until an end point is reached, at which point the acid or base being titrated is completely neutralized.

A **buffer** does not change its pH very much when either hydronium ions or hydroxide ions are added to it (Section 8.11A). Buffer solutions consist of approximately equal concentrations of a weak acid and its conjugate base. The **buffer capacity** depends on both its pH and its concentration (Section 8.11C). The most effective buffer solutions have a pH equal to the pK_a of the weak acid (Section 8.11C). The greater the concentration of the weak acid and its conjugate base, the greater the buffer capacity (Section 8.11C). The most important buffers for blood are carbonate and phosphate (Section 8.11D).

The Henderson-Hasselbalch equation is a mathematical relationship between pH, the pK_a of a weak acid, and the concentrations of the weak acid and its conjugate base (Section 8.12):

$$pH = pK_a + \log \frac{[A^-]}{[HA]}$$

PROBLEMS

Numbers that appear in color indicate difficult problems. ▶ designates problems requiring application of principles.

Arrhenius Acids and Bases

8.10 Define:
(a) An Arrhenius acid
(b) An Arrhenius base

8.11 Write an equation for the reaction that takes place when each acid is added to water. For a diprotic or triprotic acid, consider only its first ionization.
(a) HNO_3 (b) HBr (c) H_2SO_3
(d) H_2SO_4 (e) HCO_3^- (f) H_3BO_3

8.12 Which of these acids are monoprotic, which are diprotic, and which are triprotic? Which are amphiprotic?
(a) $H_2PO_4^-$ (b) HBO_3^{2-} (c) $HClO_4$
(d) C_2H_5OH (e) HSO_3^- (f) HS^-
(g) H_2CO_3

8.13 Write an equation for the reaction that takes place when each base is added to water.
(a) LiOH
(b) $(CH_3)_2NH$

Brønsted-Lowry Acids and Bases

8.14 Define (a) Brønsted-Lowry acid and (b) Brønsted-Lowry base.

8.15 Write the formula for the conjugate base of each acid.
(a) H_2SO_4 (b) H_3BO_3 (c) HI
(d) H_3O^+ (e) NH_4^+ (f) HPO_4^{2-}

8.16 Write the formula for the conjugate base of each acid.
(a) $H_2PO_4^-$ (b) H_2S (c) HCO_3^-
(d) CH_3CH_2OH (e) H_2O

CHAPTER 8 ACIDS AND BASES

8.17 Write the formula for the conjugate acid of each base.
(a) OH^- (b) HS^- (c) NH_3
(d) $C_6H_5O^-$ (e) CO_3^{2-} (f) HCO_3^-

8.18 Write the formula for the conjugate acid of each base.
(a) H_2O (b) HPO_4^{2-} (c) CH_3NH_2
(d) PO_4^{3-} (e) H_2O (f) HBO_3^{2-}

8.19 For each equilibrium, label the stronger acid, the stronger base, the weaker acid, and the weaker base. For which reaction(s) does the position of equilibrium lie toward the right? For which does it lie toward the left?
(a) $H_3PO_4 + OH^- \rightleftharpoons H_2PO_4^- + H_2O$
(b) $H_2O + Cl^- \rightleftharpoons HCl + OH^-$
(c) $HCO_3^- + OH^- \rightleftharpoons CO_3^{2-} + H_2O$

8.20 For each equilibrium, label the stronger acid, the stronger base, the weaker acid, and the weaker base. For which reaction(s) does the position of equilibrium lie toward the right? For which does it lie toward the left?
(a) $C_6H_5OH + C_2H_5O^- \rightleftharpoons C_6H_5O^- + C_2H_5OH$
(b) $HCO_3^- + H_2O \rightleftharpoons H_2CO_3 + OH^-$
(c) $CH_3COOH + H_2PO_4^- \rightleftharpoons CH_3COO^- + H_3PO_4$

8.21 Unless under pressure, carbonic acid in aqueous solution breaks down into carbon dioxide and water, and carbon dioxide is evolved as bubbles of gas. Write an equation for the conversion of carbonic acid to carbon dioxide and water.

8.22 Will carbon dioxide be evolved as a gas when sodium bicarbonate is added to an aqueous solution of each compound? Explain.
(a) Sulfuric acid
(b) Ethanol, C_2H_5OH
(c) Ammonium chloride, NH_4Cl

Acid Ionization Constants

8.23 Which has the larger numerical value?
(a) The pK_a of a strong acid or the pK_a of a weak acid
(b) The K_a of a strong acid or the K_a of a weak acid

8.24 In each pair, select the stronger acid.
(a) Pyruvic acid ($pK_a = 2.49$) or lactic acid ($pK_a = 3.08$)
(b) Citric acid ($pK_a = 3.08$) or phosphoric acid ($pK_a = 2.10$)
(c) Benzoic acid ($K_a = 6.5 \times 10^{-5}$) or lactic acid ($K_a = 8.4 \times 10^{-4}$)
(d) Carbonic acid ($K_a = 4.3 \times 10^{-7}$) or boric acid ($K_a = 7.3 \times 10^{-10}$)

8.25 Which solution will be more acidic—that is, which will have a lower pH?
(a) $0.10\ M\ CH_3COOH$ or $0.10\ M\ HCl$
(b) $0.10\ M\ CH_3COOH$ or $0.10\ M\ H_3PO_4$
(c) $0.010\ M\ H_2CO_3$ or $0.010\ M\ NaHCO_3$
(d) $0.10\ M\ NaH_2PO_4$ or $0.10\ M\ Na_2HPO_4$
(e) $0.10\ M$ aspirin ($pK_a = 3.47$) or $0.10\ M$ acetic acid

8.26 Which solution will be more acidic—that is, which will have a lower pH?
(a) $0.10\ M\ C_6H_5OH$ (phenol) or $0.10\ M\ C_2H_5OH$ (ethanol)
(b) $0.10\ M\ NH_3$ or $0.10\ M\ NH_4Cl$
(c) $0.10\ M\ NaCl$ or $0.10\ M\ NH_4Cl$
(d) $0.10\ M\ CH_3CH(OH)COOH$ (lactic acid) or $0.10\ M\ CH_3COOH$
(e) $0.10\ M$ ascorbic acid (vitamin C, $pK_a = 4.1$) or $0.10\ M$ acetic acid

Acid and Base Properties

8.27 Write an equation for the reaction of HCl with each compound. Which are acid–base reactions? Which are redox reactions?
(a) Na_2CO_3 (b) Mg (c) $NaOH$
(d) Fe_2O_3 (e) NH_3 (f) CH_3NH_2
(g) $NaHCO_3$

8.28 When a solution of sodium hydroxide is added to a solution of ammonium carbonate and the solution is heated, ammonia gas, NH_3, is released. Write a net ionic equation for this reaction. Both NaOH and $(NH_4)_2CO_3$ exist as dissociated ions in aqueous solution.

Self-Ionization of Water

8.29 Given the following values of $[H_3O^+]$, calculate the corresponding value of $[OH^-]$ for each solution.
(a) $10^{-11}\ M$ (b) $10^{-4}\ M$
(c) $10^{-7}\ M$ (d) $10\ M$

8.30 Given the following values of $[OH^-]$, calculate the corresponding value of $[H_3O^+]$ for each solution.
(a) $10^{-10}\ M$ (b) $10^{-2}\ M$
(c) $10^{-7}\ M$ (d) $10\ M$

pH and pK_a

8.31 What is the pH of each solution, given the following values of $[H_3O^+]$? Which solutions are acidic, which are basic, and which are neutral?
(a) $10^{-8}\ M$ (b) $10^{-10}\ M$ (c) $10^{-2}\ M$
(d) $10\ M$ (e) $10^{-7}\ M$

8.32 What is the pH and pOH of each solution given the following values of $[OH^-]$? Which solutions are acidic, which are basic, and which are neutral?
(a) $10^{-3}\ M$ (b) $10^{-1}\ M$
(c) $10^{-5}\ M$ (d) $10^{-7}\ M$

8.33 What is the pH of each solution, given the following values of [H$_3$O$^+$]? Which solutions are acidic, which are basic, and which are neutral?
(a) 3.0×10^{-9} M
(b) 6.0×10^{-2} M
(c) 8.0×10^{-12} M
(d) 5.0×10^{-7} M

8.34 Which is more acidic, a beer with [H$_3$O$^+$] = 3.16×10^{-5} or a wine with [H$_3$O$^+$] = 5.01×10^{-4}?

8.35 What is the [OH$^-$] and pOH of each solution?
(a) Potassium hydroxide (0.10 M), pH 13.0
(b) Sodium carbonate (0.10 M), pH 11.6
(c) Trisodium phosphate (0.10 M), pH 12.0
(d) Sodium bicarbonate (0.10 M), pH 8.4

8.36 Following are pH ranges for several human biological materials. From the pH at the midpoint of each range, calculate the corresponding [H$_3$O$^+$]. Which materials are acidic, which are basic, and which are neutral?
(a) Milk, pH 6.6–7.6
(b) Gastric contents, pH 1.0–3.0
(c) Spinal fluid, pH 7.3–7.5
(d) Saliva, pH 6.5–7.5
(e) Urine, pH 4.8–8.4
(f) Blood plasma, pH 7.35–7.45
(g) Feces, pH 4.6–8.4
(h) Bile, pH 6.8–7.0

The pH of Aqueous Salt Solutions

8.37 Write a balanced equation for the reaction of an acid and a base that gives each salt.
(a) BaCl$_2$
(b) NaNO$_3$
(c) HCOONH$_4$
(d) CaSO$_4$
(e) MgSO$_4$
(f) Na$_2$HPO$_4$
(g) NH$_4$Cl
(h) Li$_2$CO$_3$

8.38 Which salts in Problem 8.37 are acidic, which are basic, and which are neutral?

8.39 When sodium carbonate is dissolved in pure water, the resulting solution is basic; that is, its pH is greater than 7. Explain how this happens.

Titration

8.40 What is the purpose of an acid–base titration?

8.41 What is the molarity of a solution made by dissolving 12.7 g of HCl in enough water to make 1.00 L of solution?

8.42 What is the molarity of a solution made by dissolving 3.4 g of Ba(OH)$_2$ in enough water to make 450 mL of solution? Assume that Ba(OH)$_2$ ionizes completely in water to Ba^{2+} and OH$^-$ ions.

8.43 Describe how you would prepare each of the following solutions (in each case assume that you have the solid bases).
(a) 400.0 mL of 0.75 M NaOH
(b) 1.0 L of 0.071 M Ba(OH)$_2$

8.44 If 25.0 mL of an aqueous solution of H$_2$SO$_4$ requires 19.7 mL of 0.72 M NaOH to reach the end point, what is the molarity of the H$_2$SO$_4$ solution?

8.45 A sample of 27.0 mL of 0.310 M NaOH is titrated with 0.740 M H$_2$SO$_4$. How many milliliters of the H$_2$SO$_4$ solution are required to reach the end point?

8.46 A 0.300 M solution of H$_2$SO$_4$ was used to titrate 10.00 mL of an unknown base; 15.00 mL of acid was required to neutralize the basic solution. What was the molarity of the base?

8.47 A solution of an unknown base was titrated with 0.150 M HCl, and 22.0 mL of acid was needed to reach the end point of the titration. How many moles of the unknown base were in the solution?

8.48 The usual concentration of HCO$_3^-$ ions in blood plasma is approximately 24 millimoles per liter (mmol/L). How would you make up 1.00 L of a solution containing this concentration of HCO$_3^-$ ions?

Buffers

8.49 Write equations to show what happens when, to a buffer solution containing equimolar amounts of CH$_3$COOH and CH$_3$COO$^-$, we add (a) H$_3$O$^+$ (b) OH$^-$.

8.50 Write equations to show what happens when, to a buffer solution containing equimolar amounts of HPO$_4^{2-}$ and H$_2$PO$_4^-$, we add (a) H$_3$O$^+$ (b) OH$^-$.

8.51 What is the pH of a buffer solution made by dissolving 0.10 mol of formic acid, HCOOH, and 0.10 mol of sodium formate, HCOONa, in 1 L of water?

8.52 We commonly refer to a buffer as consisting of approximately equal molar amounts of a weak acid and its conjugate base—for example, CH$_3$COOH and CH$_3$COO$^-$. Is it also possible to have a buffer consisting of approximately equal molar amounts of a weak base and its conjugate acid? Explain.

8.53 What is meant by buffer capacity?

8.54 How can you change the pH of a buffer? How can you change the capacity of a buffer?

8.55 The pH of a solution made by dissolving 1.0 mol of propanoic acid and 1.0 mol of sodium propanoate in 1.0 L of water is 4.85.
(a) What would the pH be if we used 0.10 mol of each (in 1 L of water) instead of 1.0 mol?
(b) With respect to buffer capacity, how would the two solutions differ?

8.56 What is the connection between buffer action and Le Chatelier's principle?

The Henderson-Hasselbalch Equation

8.57 Show that when the concentration of the weak acid, [HA], in an acid–base buffer equals that of the conjugate base of the weak acid, [A$^-$], the pH of the buffer solution is equal to the pK_a of the weak acid.

240 | CHAPTER 8 ACIDS AND BASES

8.58 Calculate the pH of an aqueous solution containing the following:
(a) 0.80 M lactic acid and 0.40 M lactate ion
(b) 0.30 M NH_3 and 1.50 M NH_4^+

8.59 What is the ratio of $HPO_4^{2-}/H_2PO_4^-$ in a phosphate buffer of pH 7.40 (the average pH of human blood plasma)?

8.60 What is the ratio of $HPO_4^{2-}/H_2PO_4^-$ in a phosphate buffer of pH 7.9 (the pH of human pancreatic fluid)?

8.61 The pH of 0.10 M HCl is 1.0. When 0.10 mol of sodium acetate, CH_3COONa, is added to this solution, its pH changes to 2.9. Explain why the pH changes, and why it changes to this particular value.

Chemical Connections

8.62 (Chemical Connections 8A) Which weak base is used as a flame retardant in plastics?

8.63 (Chemical Connections 8B) What is the most important immediate first aid in any chemical burn of the eyes?

8.64 (Chemical Connections 8B) With respect to corneal burns, are strong acids or strong bases more dangerous?

8.65 (Chemical Connections 8C) Name the most common bases used in over-the-counter antacids.

8.66 (Chemical Connections 8D) What causes (a) hypoventilation, (b) respiratory alkalosis, and (c) metabolic alkalosis?

Additional Problems

8.67 4-Methylphenol, $CH_3C_6H_4OH$ (pK_a = 10.26), is only slightly soluble in water, but its sodium salt, $CH_3C_6H_4O^-Na^+$, is quite soluble in water. In which of the following solutions will 4-methylphenol dissolve more readily than in pure water?
(a) Aqueous NaOH
(b) Aqueous $NaHCO_3$
(c) Aqueous NH_3

8.68 Benzoic acid, C_6H_5COOH (pK_a = 4.19), is only slightly soluble in water, but its sodium salt, $C_6H_5COO^-Na^+$, is quite soluble in water. In which of the following solutions will benzoic acid dissolve more readily than in pure water?
(a) Aqueous NaOH
(b) Aqueous $NaHCO_3$
(c) Aqueous Na_2CO_3

8.69 Assume that you have a dilute solution of HCl (0.10 M) and a concentrated solution of acetic acid (5.0 M). Which solution is more acidic? Explain.

8.70 If the [OH^-] of a solution is 1×10^{-14},
(a) What is the pH of the solution?
(b) What is the [H_3O^+]?

8.71 What is the molarity of a solution made by dissolving 0.583 g of the diprotic acid oxalic acid, $H_2C_2O_4$, in enough water to make 1.75 L of solution?

8.72 Following are three organic acids and the pK_a of each: butanoic acid, 4.82; barbituric acid, 5.00; and lactic acid, 3.85.
(a) What is the K_a of each acid?
(b) Which of the three is the strongest acid and which is the weakest?

8.73 The pK_a value of barbituric acid is 5.0. If the H_3O^+ and barbiturate ion concentrations are each 0.0030 M, what is the concentration of the undissociated barbituric acid?

8.74 If pure water self-ionizes to give H_3O^+ and OH^- ions, why doesn't pure water conduct an electric current?

8.75 Can an aqueous solution have a pH of zero? Explain your answer using aqueous HCl as your example.

8.76 A scale of K_b values for bases could be set up in a manner similar to that for the K_a scale for acids. However, this step is generally considered unnecessary. Explain.

8.77 Do a 1.0 M CH_3COOH solution and a 1.0 M HCl solution have the same pH? Explain.

8.78 Suppose you wish to make a buffer whose pH is 8.21. You have available 1.00 L of 0.100 M NaH_2PO_4 and solid Na_2HPO_4. How many grams of the solid Na_2HPO_4 must be added to the solution to accomplish this task? (Assume that the volume remains 1.00 L.)

8.79 In the past, boric acid has been used to rinse an inflamed eye. How would you make up one liter of a $H_3BO_3/H_2BO_3^-$ buffer solution that has a pH of 8.40?

8.80 Suppose you want to make a CH_3COOH/CH_3COO^- buffer solution with a pH of 5.60. The acetic acid concentration is to be 0.10 M. What should the acetate ion concentration be?

8.81 For an acid–base reaction, one way to determine the position of equilibrium is to say that the larger of the equilibrium arrow pair points to the acid with the higher value of pK_a. For example,

$$CH_3COOH + HCO_3^- \rightleftharpoons CH_3COO^- + H_2CO_3$$
pK_a = 4.75 $\qquad\qquad\qquad\qquad$ pK_a = 6.37

Explain why this rule works.

8.82 When a solution prepared by dissolving 4.00 g of an unknown acid in 1.00 L of water is titrated with 0.600 M NaOH, 38.7 mL of the NaOH solution is needed to neutralize the acid. What was the molarity of the acid solution?

8.83 Write equations to show what happens when, to a buffer solution containing equal amounts of HCOOH and $HCOO^-$, we add (a) H_3O^+ (b) OH^-.

8.84 If we add 0.10 mol of NH_3 to 0.50 mol of HCl dissolved in enough water to make 1.0 L of solution, what happens to the NH_3? Will any NH_3 remain? Explain.

8.85 Suppose you have an aqueous solution prepared by dissolving 0.050 mol of NaH_2PO_4 in 1 L of water. This solution is not a buffer, but suppose you want to make it into one. How many moles of solid Na_2HPO_4 must you add to 1.0 L of this aqueous solution to make it into (a) a buffer of pH 7.21, (b) a buffer of pH 6.21, and (c) a buffer of pH 8.21?

8.86 The pH of a 0.10 M solution of acetic acid is 2.93. When 0.10 mol of sodium acetate, CH_3COONa, is added to this solution, its pH changes to 4.74. Explain why the pH changes, and why it changes to this particular value.

8.87 Suppose you have a phosphate buffer of pH 7.21. If you add more solid NaH_2PO_4 to this buffer, would you expect the pH of the buffer to increase, decrease, or remain unchanged? Explain.

8.88 Suppose you have a bicarbonate buffer containing carbonic acid, H_2CO_3, and sodium bicarbonate, $NaHCO_3$, and that the pH of the buffer is 6.37. If you add more solid $NaHCO_3$ to this buffer solution, would you expect its pH to increase, decrease, or remain unchanged? Explain.

InfoTrac College Edition

For additional readings, go to InfoTrac College Edition, your online research library, at

http://infotrac.thomsonlearning.com

CHAPTER 9

- *9.1* Introduction
- *9.2* What Happens When a Nucleus Emits Radioactivity?
- *9.3* Half-life
- *9.4* Characteristics of Radiation
- *9.5* Radiation Dosimetry and Human Health
- *9.6* Medical Uses of Radioactive Materials
- *9.7* Nuclear Fusion
- *9.8* Nuclear Fission and Atomic Energy

The sun's energy is the result of nuclear fusion.

Nuclear Chemistry

9.1 Introduction

Every so often, a scientist makes the kind of discovery that changes the future of the world in some significant way. In 1896 a French scientist, Henri Becquerel (1852–1908), made one of the most important of these discoveries. Completely by accident, he found that a certain rock was giving off mysterious radiation. The discovery was totally unexpected: No one had ever had the faintest idea that anything of the kind was happening, and yet that rock, and others like it, had been giving off radiation since long before there were humans on the earth. It was simply that, before 1896, nobody recognized that fact.

Becquerel's rock contained uranium. The radiation given off by the uranium caused photographic plates to "fog" even though they were wrapped in black paper. This meant that the radiation coming from the rock had to be more energetic than visible light because the former could pass through the paper but the latter couldn't.

Shortly after Becquerel's discovery, other scientists—most notably Pierre (1859–1906) and Marie (1867–1934) Curie and Ernest Rutherford (1871–1937)—began to study this phenomenon, which Marie Curie called

Becquerel and the Curies were awarded the Nobel Prize in physics in 1903 for this discovery. Rutherford won the Nobel Prize in chemistry in 1908.

CHEMICAL CONNECTIONS 9A

Three Kinds of Radiation

Early experiments identified three kinds of radiation. When a radioactive material was placed in a lead container that had a small opening, the emitted radiation passed through the opening and then between the two poles of a magnet. One ray was deflected to the left, as negatively charged particles would be, and another to the right, as expected from positively charged particles. A third ray passed through without deflection, indicating no charge. The three rays were detected when they darkened a photographic plate. Rutherford named the positively charged rays *alpha rays*, the negatively charged rays *beta rays*, and the undeflected rays *gamma rays*.

Figure 9A Types of radioactivity.

radioactivity. They soon learned that there were three types, which they named alpha, beta, and gamma, after the first three letters of the Greek alphabet.

Alpha particles are helium nuclei. Each contains two protons and two neutrons and has an atomic number of 2.

Beta particles are electrons. Beta "radiation" is just a stream of electrons.

Gamma rays are not particles; they are a high-energy form of electromagnetic radiation.

Gamma rays are only one form of electromagnetic radiation. There are many others, including visible light. All, however, consist of waves (Figure 9.1).

The difference between one form of electromagnetic radiation and another is the **wavelength** (λ), which is the distance from one wave crest to the next. All electromagnetic radiation travels at the same speed in a vacuum (the speed of light, which is 3.0×10^8 m/s). Because the wavelengths differ but the speed always remains constant, the different forms of electromagnetic radiation have different frequencies. The **frequency** (*f*) of a radiation is the number of crests that pass a given point in one second. The

Screen 9.2
Exercises

Figure 9.1 Two electromagnetic waves with different wavelengths.

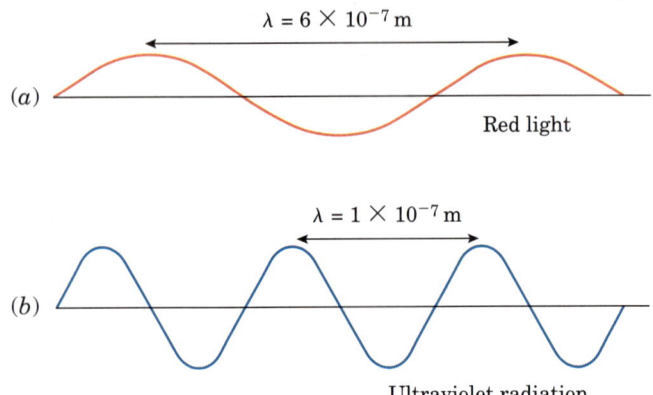

longer the wavelength, the lower the frequency. Mathematically they are related by the following equation, where c is the speed of light:

$$\lambda = \frac{c}{f}$$

A direct relationship also exists between the frequency of electromagnetic radiation and its energy: *The higher the frequency, the higher the energy.* Electromagnetic radiation comes in packets; the smallest units are called **photons**.

Figure 9.2 shows the wavelengths of various types of radiations of the electromagnetic spectrum. Gamma rays are electromagnetic radiation of very high frequency (very short wavelength). Humans cannot see them because our eyes are not sensitive to waves of this frequency, but they can be detected by instruments (Section 9.4). Another kind of radiation, called X rays, has frequencies (and energies) higher than those of visible light but less than those of gamma rays. Materials that emit radiation (alpha, beta, or gamma) are called **radioactive**. Radioactivity comes from the atomic nucleus and not from the

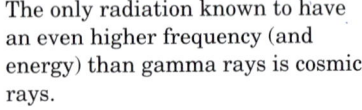

The only radiation known to have an even higher frequency (and energy) than gamma rays is cosmic rays.

UV stands for ultraviolet and IR for infrared.

Figure 9.2 The electromagnetic spectrum.

Table 9.1 Particles and Rays Frequently Encountered in Radiation

Particle or Ray	Common Name of Radiation	Symbol	Charge	Atomic Mass Units	Penetrating Power[a]	Energy Range[b]
Proton	Proton beam	^1_1H	+1	1	1–3 cm	60 MeV
Electron	Beta particle	$^{0}_{-1}e$ or β	−1	0.00055 $\left(\frac{1}{1835}\right)$	0–4 mm	1–3 MeV
Neutron	Neutron beam	^1_0n	0	1	—	—
Positron	—	$^0_{+1}e$ or β^+	+1	0.00055	—	—
Helium nucleus	Alpha particle	^4_2He or α	+2	4	0.02–0.04 mm	3–9 MeV
Energetic radiation	Gamma ray	γ	0	0	1–20 cm	0.1–10 MeV
	X ray		0	0	0.01–1 cm	0.1–10 keV

[a]Distance at which half the radiation has been stopped in water.
[b]MeV = 1.602×10^{-13} J = 3.829×10^{-14} cal.

electron cloud. Table 9.1 summarizes the properties of the particles and rays that come out of radioactive nuclei, along with the properties of some other particles and rays. Note that X rays are not considered to be a form of radioactivity, because they do not come out of the nucleus but are generated in other ways.

We have said that humans cannot see gamma rays. We cannot see alpha or beta particles either. Likewise, we cannot hear them, smell them, or feel them. They are undetectable by our senses. Radioactivity can be detected only by instruments, as discussed in Section 9.4.

The electronvolt (eV) is a non-SI energy unit used frequently in nuclear chemistry.

9.2 What Happens When a Nucleus Emits Radioactivity? SKIP

As mentioned in Section 2.4, different nuclei consist of different numbers of protons and neutrons. It is customary to indicate these numbers with subscripts and superscripts placed to the left of the atomic symbol. The number of protons (the atomic number of the element) is shown as a subscript and the total number of protons and neutrons (the mass number) as a superscript. For example, ^1_1H, ^2_1H, and ^3_1H are the symbols for the three isotopes of hydrogen. We call them hydrogen-1, hydrogen-2, and hydrogen-3, respectively.

Hydrogen-2 is also called deuterium, and hydrogen-3 is called tritium.

A Radioactive and Stable Nuclei

Some isotopes are radioactive, whereas others are stable. More than 300 naturally occurring isotopes have been identified. Of these, 264 are stable, meaning that the nuclei of these isotopes never give off any radioactivity. As far as we can tell, they will last forever. The remainder are **radioactive isotopes**—they do give off radioactivity. Furthermore, scientists have made more than 1000 artificial isotopes in laboratories. All artificial isotopes are unstable.

Isotopes in which the number of protons and neutrons are balanced seem to be stable. In the lighter elements, this balance occurs when the numbers of protons and neutrons are approximately equal. For example, $^{12}_6C$ is a stable nucleus (six protons and six neutrons) as are $^{16}_8O$ (eight protons and eight neutrons) and $^{20}_{10}Ne$ (ten protons and ten neutrons). Among the heavier elements, stability requires more neutrons than protons. Lead-206, one of the most stable isotopes of lead, contains 82 protons and 124 neutrons.

Radioactive isotope A radiation-emitting isotope of an element

The role of neutrons seems to be to provide binding energy to overcome the repulsion between protons.

B Beta Emission

If there is a serious imbalance in the proton-to-neutron ratio, a nucleus will emit an alpha or beta particle—in a **nuclear reaction**—and thereby make the ratio more favorable and the nucleus more stable. For example, if a nucleus has more neutrons than it needs for stability, it can stabilize itself by converting a neutron to a proton and an electron:

$$^{1}_{0}n \longrightarrow \, ^{1}_{1}H + \, ^{0}_{-1}e$$

> **Nuclear reaction** A reaction that changes atomic nuclei of elements (usually to atomic nuclei of other elements)

This process is called **beta emission** because the electrons (beta particles) that are produced are emitted from the nucleus. For example, phosphorus-32 is a beta emitter:

$$^{32}_{15}P \longrightarrow \, ^{32}_{16}S + \, ^{0}_{-1}e$$

For balancing purposes, the electron is given an atomic number of −1.

The electrons of the beta emission come from the nucleus, not from the surrounding electron cloud.

When the unstable phosphorus-32 is converted to sulfur-32, stability is achieved (16 protons and 16 neutrons). *Every time a nucleus emits a beta particle, it is transformed into another nucleus with the same mass number but an atomic number one unit greater.* The changing of one element into another is called **transmutation**. It happens naturally every time an element gives off beta particles.

In balancing nuclear equations, *the sum of the mass numbers (superscripts) must be equal on both sides of the equation. The same is true for the atomic numbers (subscripts).* To find the symbol for a nucleus corresponding to a certain atomic number, use the Periodic Table on the inside front cover of this book. In writing nuclear equations we consider the nucleus only, and disregard the surrounding electrons.

EXAMPLE 9.1

To which nucleus is $^{14}_{6}C$, a beta emitter, converted?

$$^{14}_{6}C \longrightarrow ? + \, ^{0}_{-1}e$$

Solution
The $^{14}_{6}C$ nucleus has six protons and eight neutrons. One of its neutrons is converted to a proton and an electron. The electron is emitted (it is the beta particle), but the proton remains. The nucleus now has seven protons (and therefore must be nitrogen) and seven neutrons:

$$^{14}_{6}C \longrightarrow \, ^{14}_{7}N + \, ^{0}_{-1}e$$

Note that this equation is balanced: The sum of the mass numbers is 14 on both sides, and the sum of the atomic numbers is 6 on both sides.

Problem 9.1
Iodine-139 is a beta emitter. What is the reaction product?

C Alpha Emission

Transmutation also takes place in alpha emission. For heavy elements, the loss of α particles is an especially important stabilization process. For example,

$$^{238}_{92}U \longrightarrow {}^{234}_{90}Th + {}^{4}_{2}He$$

$$^{210}_{84}Po \longrightarrow {}^{206}_{82}Pb + {}^{4}_{2}He + \gamma$$

The symbol γ, the Greek letter gamma, represents a gamma ray.

In the second example, stable lead-206 is obtained after alpha emission. Note that the polonium-210 emits both α particles and gamma rays.

Both alpha and beta emissions can be either "pure" or mixed with gamma rays.

Can you deduce, from these examples, the general rule for alpha emission? *The new nucleus always has a mass number four units lower and an atomic number two units lower than the original.*

EXAMPLE 9.2

Polonium-218 is an α emitter. Write an equation for this α decay and identify the product.

Solution
The atomic number of polonium is 84, so the equation is

$$^{218}_{84}Po \longrightarrow {}^{4}_{2}He + ?$$

Subtract 4 from the mass number of 218 to get the new mass number, 214. Subtract 2 from atomic number of 84 to get the new atomic number, 82. In the Periodic Table we find that the element with an atomic number of 82 is lead, Pb. Therefore, the product is $^{214}_{82}Pb$.

Problem 9.2
Thorium-223 is an α emitter. Identify the isotope that is formed.

$^{233}_{90}Th \longrightarrow {}^{229}_{88}Ra + {}^{4}_{2}H$

D Positron Emission

A positron is a particle that has the same mass as an electron, but a charge of +1 rather than −1. Its symbol is β^+ or $^{0}_{+1}e$. Positron emission is much rarer than beta emission, but some nuclei do emit positrons. Because a positron has no appreciable mass, the nucleus is transmuted into another nucleus with the same mass number but an atomic number that is one unit less. For example:

$$^{11}_{6}C \longrightarrow {}^{0}_{+1}e + {}^{11}_{5}B$$

EXAMPLE 9.3

Nitrogen-13 is a positron emitter. Write an equation for this nuclear reaction and identify the product.

$^{13}_{7}N \longrightarrow {}^{13}_{6}C + {}^{0}_{+1}e$

Solution
The mass number of the new isotope will still be 13, because a positron has no appreciable mass. To get the atomic number we subtract 1 from 7, the atomic number of nitrogen. The new atomic number is 6 and the element with an atomic number of 6 is carbon. Therefore, the equation is

$$^{13}_{7}N \longrightarrow {}^{0}_{+1}e + {}^{13}_{6}C$$

Problem 9.3
Arsenic-74 is a positron emitter used in locating brain tumors. What is the product nucleus formed?

E Gamma Emission

Besides alpha, beta, and positron emitters, one occasionally encounters pure gamma emitters:

$$^{11}_{5}B^* \longrightarrow {}^{11}_{5}B + \gamma$$

In this equation, the $^{11}_{5}B^*$ symbolizes the boron nucleus in a high-energy (excited) state. In this case, no transmutation takes place. The boron is still boron, but it is in a lower-energy form (more stable) after the emission of excess energy in the form of gamma rays. When all excess energy has been emitted (sometimes in a discrete, stepwise manner), the nucleus returns to its most stable, lowest energy state—the **ground state.**

9.3 Half-life

When a nucleus gives off radiation, it is said to decay.

Suppose we have 40 g of a radioactive isotope—say, $^{90}_{38}Sr$. It contains many atoms, of course (about 2.7×10^{23}). These atoms are unstable and decay. In this case they give off beta particles and thereby become $^{90}_{39}Y$ atoms. But do all of the Sr atoms decay at once? No, they do not. They decay one at a time, at a fixed rate. For strontium-90, the decay rate is such that one-half of our sample (about 1.35×10^{23} atoms) will have decayed by the end of 28.1 years. The time it takes for one-half of any sample of radioactive material to decay is called the **half-life,** $t_{1/2}$.

It does not matter how big or small a sample is. For example, in the case of our 40 g of strontium-90, 20 g will be left at the end of 28.1 years (the rest will be converted to yttrium-90). It will then take another 28.1 years for half of the remainder to decay, so that at the end of that time we will have 10 g. If we wait for a third span of 28.1 years, then 5 g will be left. If we had begun with 100 g, then 50 g would be left after the first 28.1-year period.

Figure 9.3 shows the radioactive decay curve of $^{131}_{53}I$. Inspection of this graph shows that, at the end of 8 days, half of the original $^{131}_{53}I$ has disappeared. Thus the half-life of iodine-131 is 8 days. It would take a total of 16 days, or two half-lives, for three-fourths of the original amount of iodine-131 to decay.

Figure 9.3 The decay curve of iodine-131.

CHEMICAL CONNECTIONS 9B

Radioactive Dating

Carbon-14, with a half-life of 5730 years, can be used to date archeological objects as old as 60,000 years. This dating technique relies on the principle that the carbon-12/carbon-14 ratio of an organism—whether plant or animal—remains constant during its lifetime. When the organism dies, the carbon-12 level remains constant but the carbon-14 decays through the following reaction:

$$^{14}_{6}C \longrightarrow ^{14}_{7}N + ^{0}_{-1}e$$

Using this fact, a scientist can calculate the changed carbon-12/carbon-14 ratio to determine the date of an artifact.

For example, in charcoal made from a tree that has recently died, the carbon-14 gives a radioactive count of 13.70 disintegrations per minute per gram of carbon. In a piece of charcoal found in a cave in France near some ancient cave paintings, the carbon-14 count was 1.71 disintegrations/min for each gram of carbon. From this information, the cave paintings can be dated. After one half-life, the number of disintegrations per minute per gram is 6.85; after two half-lives, it is 3.42, and after three half-lives, it is 1.71. Therefore, three half-lives have passed since the paintings were created. As carbon-14 has a half-life of 5730 years, the paintings are about $3 \times 5730 = 17{,}190$ years old.

The famous Shroud of Turin, a piece of linen cloth with the image of a man's head on it, was believed by many to be the original cloth that was wrapped around the body of Jesus Christ after his death. However, radioactive dating showed, with 95%

Figure 9B.1 Cro-Magnon cave painting.

certainty, that the plants from which the linen was obtained were alive sometime between AD 1260 and 1380, proving that it could not have been the shroud of Christ. Note that it was not necessary to destroy the shroud to perform the tests. In fact, scientists in different laboratories used just a total of a few square centimeters of cloth from its edge.

Rock samples can be dated on the basis of their lead-206 and uranium-238 content. The assumption is that all of the lead-206 has come from the decay of uranium-238,* which has a half-life of 4.5 billion years. One of the oldest rocks found on the earth is a granite outcrop in Greenland, dated at 3.7×10^9 years old.

On the basis of dating of meteorites, the estimated age of the solar system is 4.6×10^9 years.

---

*Uranium-238 does not decay directly to lead-206. It first goes to thorium-234, which then also decays to radium-230, and so on. Lead-206 is the final result of a series of such transmutations.

(a)

(b)

Figure 9B.2 The Shroud of Turin (*a*) as it appears to the eye and (*b*) as a photographic negative, which brings out the images more clearly.

EXAMPLE 9.4

If 10 mg of $^{131}_{53}$I is administered to a patient, how much is left in the body after 32 days?

Solution
We know that 32 days corresponds to four half-lives because $t_{1/2}$ for this isotope is eight days. After one half-life, 5 mg is left; after two half-lives, 2.5 mg; after three half-lives, 1.25 mg; and after four half-lives, 0.63 mg.

Problem 9.4
Barium-122 has a half-life of 2 minutes. You just obtained a sample weighing 10 g. It takes 10 minutes to set up the experiment in which barium-122 will be used. How many grams of barium-122 will be left when you begin the experiment?

It must be noted that, in theory, it would take an infinite amount of time for all of a radioactive sample to disappear. In reality, most of the radioactivity disappears after five half-lives; by that time, only 3% of the original is left. After ten half-lives less than 0.1% of the activity remains.

The half-life of an isotope is independent of temperature and pressure—and, indeed, of all other conditions—and is a property of the particular isotope only. It does not depend on what kind of other atoms surround the particular nucleus (that is, what kind of molecule the nucleus is part of). *We do not know any way to speed up radioactive decay or slow it down.*

Table 9.2 gives some half-lives. Even this brief sampling indicates that there are tremendous differences in half-lives. Some isotopes, such as technetium-99m, decay and disappear in a day; others, such as uranium-238, remain radioactive for billions of years. Very short-lived isotopes, especially the artificial heavy elements (Section 9.7) with atomic numbers greater than 100, have half-lives of the order of seconds.

The usefulness or inherent danger in radioactive isotopes is related to their half-lives. In assessing the long-range health effects of atomic-bomb damage or of nuclear power plant accidents, like those at Three Mile Island,

The "m" in technetium-99m means *metastable*.

Table 9.2 Half-lives of Some Radioactive Nuclei

Name	Symbol	Half-life	Radiation
Hydrogen-3 (tritium)	$^3_1$H	12.26 years	Beta
Carbon-14	$^{14}_6$C	5730 years	Beta
Phosphorus-28	$^{28}_{15}$P	0.28 second	Positrons
Phosphorus-32	$^{32}_{15}$P	14.3 days	Beta
Potassium-40	$^{40}_{19}$K	1.28×10^9 years	Beta + gamma
Scandium-42	$^{42}_{21}$Sc	0.68 second	Positrons
Cobalt-60	$^{60}_{27}$Co	5.2 years	Gamma
Strontium-90	$^{90}_{38}$Sr	28.1 years	Beta
Technetium-99m	$^{99m}_{43}$Tc	6.0 hours	Gamma
Indium-116	$^{116}_{49}$In	14 seconds	Beta
Iodine-131	$^{131}_{53}$I	8 days	Beta + gamma
Mercury-197	$^{197}_{80}$Hg	65 hours	Gamma
Polonium-210	$^{210}_{84}$Po	138 days	Alpha
Radon-205	$^{205}_{86}$Rn	2.8 minutes	Alpha
Radon-222	$^{222}_{86}$Rn	3.8 days	Alpha
Uranium-238	$^{238}_{92}$U	4×10^9 years	Alpha

Pennsylvania, in 1979 and Chernobyl (in the former Soviet Union) in 1986 (Chemical Connections 9G), we can see that radioactive isotopes with long half-lives, such as $^{85}_{36}Kr$ ($t_{1/2}$ = 10 years) or $^{60}_{27}Co$ ($t_{1/2}$ = 5.2 years), are more important than short-lived ones. On the other hand, when a radioactive isotope is used in medical diagnosis or therapy, short-lived isotopes are more useful because they disappear faster from the body—for example, $^{99m}_{43}Tc$, $^{32}_{15}P$, $^{131}_{53}I$, and $^{197}_{80}Hg$.

9.4 Characteristics of Radiation SKIP

As already noted, radioactivity is not detectable by our senses. We cannot see it, hear it, feel it, or smell it. Then how do we know it is there? Alpha, beta, gamma, and X rays, as well as proton beams, all have a property we can use to detect them: When they interact with matter, they usually knock electrons out of the electron cloud surrounding the atomic nucleus, thereby creating positively charged ions from neutral atoms. For this reason, we call all of them **ionizing radiation.**

Radiation is characterized by two physical measurements: (1) its **intensity,** which is the number of particles or photons emerging per unit time, and (2) the **energy** of each particle or photon.

A Intensity

To measure intensity, we take advantage of the ionizing property of radiation. Instruments such as the **Geiger-Müller counter** (Figure 9.4) and the **proportional counter** contain a gas such as helium or argon. When a radioactive nucleus emits beta particles (electrons), these particles ionize the gas, and the instrument registers this fact by indicating that an electric current is passing between two electrodes. In this way, the instrument counts particle after particle.

Other measuring devices, such as **scintillation counters,** have a material called a phosphor that emits a unit of light when an alpha or beta particle or gamma ray strikes it. Once again, the particles are counted one by one. The quantitative measure of radiation intensity can be reported in counts/minute or counts/second.

■ The Geiger-Müller counter.

A television picture tube works on a similar principle.

Figure 9.4 A schematic drawing of a Geiger-Müller counter. The counter is made of a glass cylinder, the inner surface of which is coated with a material that conducts electricity. In the center is the collector electrode, made of tungsten wire. It is positively charged to about 1200 volts. The space between is filled with helium or argon gas at low pressure. A thin Mylar plate (window) at the end keeps the gas in, yet allows the ionizing radiation to pass through. When the radiation enters the tube, it converts an argon or helium atom to a positive ion, which goes to the negative electrode and is counted.

The intensity of any radiation decreases with the square of the distance:

$$\frac{I_1}{I_2} = \frac{d_1^2}{d_2^2}$$

mCi and other units of radiation are defined in the next section.

EXAMPLE 9.5

If the intensity of radiation is 28 mCi at a distance of 1.0 m, what is the intensity at a distance of 2.0 m?

Solution
From the preceding equation we have

$$\frac{28\ \text{mCi}}{I_2} = \frac{2.0^2}{1.0^2}$$

$$I_2 = \frac{28\ \text{mCi}}{4.0} = 7.0\ \text{mCi}$$

Problem 9.5

If the intensity of radiation 1.0 cm from a source is 300 mCi, what is the intensity at 3.0 m?

B Energy

The energies of different particles or photons vary. As shown in Table 9.1, each particle has a certain range of energy. For example, beta particles have an energy range of 1 to 3 MeV. This range may overlap with the energy range of another particle—for example, gamma rays. The penetrating power of a radiation depends on its energy as well on the mass of the particle. Because the photons of gamma radiation have no mass, but the beta particles do have a mass, gamma rays have a greater penetrating power than do beta particles.

One easy way to protect against ionizing radiation is to wear lead aprons, covering sensitive organs. This practice is followed routinely when diagnostic X rays are taken. Another way to lessen the damage from ionizing radiation is to move farther away from the source.

Penetration of radioactive emissions. Alpha particles penetrate the least; they are stopped by a piece of paper. Beta particles easily penetrate paper but cannot go through a 0.5-cm sheet of lead. Gamma rays have the greatest penetrating power. It takes 10 cm of lead to stop them.

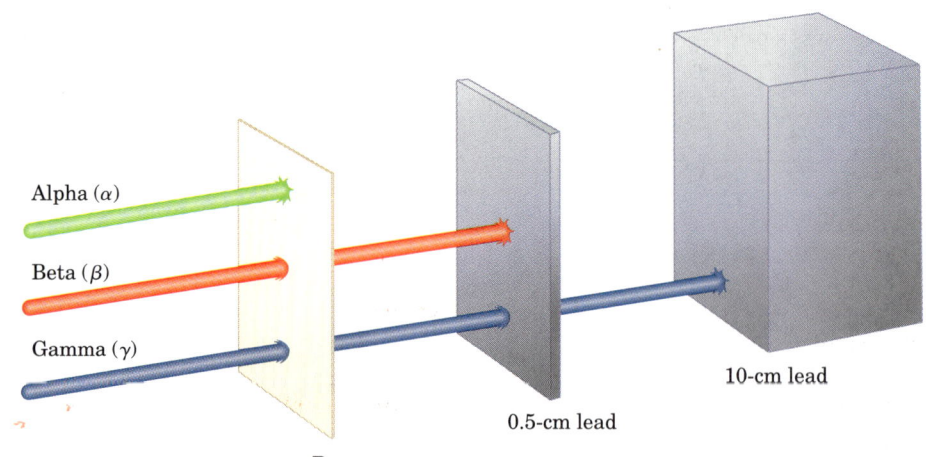

9.5 Radiation Dosimetry and Human Health

A common unit of radiation intensity is the **curie** (Ci), named in honor of Marie Curie, whose lifelong work with radioactive materials greatly helped our understanding of nuclear phenomena. One curie is defined as 3.7×10^{10} disintegrations per second (or counts per second). This is radiation of very high intensity, the amount a person would get from 1 g of pure $^{286}_{88}Ra$. This intensity is too high for regular medical use, and the most common units used in the health sciences are small fractions of it:

$$1 \text{ millicurie (mCi)} = 3.7 \times 10^7 \text{ counts/s}$$
$$1 \text{ microcurie } (\mu\text{Ci}) = 3.7 \times 10^4 \text{ counts/s}$$

The term *activity of a radiation* is the same as *intensity of a radiation*.

EXAMPLE 9.6

A radioactive isotope with an intensity (activity) of 100 mCi per vial is sold to a hospital. The vial contains 10 mL of liquid. The instruction is to administer 2.5 mCi intravenously. How much of the liquid should one use?

Solution
The intensity (activity) of a sample is directly proportional to the amount present, so

$$2.5 \text{ mCi} \times \frac{10 \text{ mL}}{100 \text{ mCi}} = 0.25 \text{ mL}$$

Problem 9.6
A radioactive isotope in a 9.0-mL vial has an intensity of 300 mCi. A patient is required to take 50 mCi intravenously. How much liquid should be used for the injection?

Another, albeit much smaller, unit of radiation activity (intensity) is the **becquerel** (Bq). One becquerel is one disintegration per second, so $1 \text{ Bq} = 2.8 \times 10^{-11}$ Ci. Becquerel is the SI unit.

In studying the effect of radiation on the body, neither the energy of the radiation (in kcal/mol) nor its intensity (in Ci) alone or in combination is of particular importance. Rather, the critical question is what kind of effects such radiation produces in the body. Three different units are used to describe the effects of radiation: roentgens, rads, and rems.

Roentgens Roentgens (R) measure the ability of the radiation to affect matter, in particular humans and animals. One roentgen is the amount of radiation that produces ions having 2.58×10^{-4} coulomb per kilogram (a coulomb is a unit of electrical charge). The number of roentgens is the *effective dose delivered* by the radiation and is therefore a measure of exposure to a particular form of radiation.

Wilhelm Roentgen (1845–1923) discovered X rays in 1895.

Rads Exposure, or delivered energy, does not take into account the effect on tissue. Radiation damages body tissue by causing ionization. For ionization to occur, the tissue must absorb the delivered energy. The common measure of the amount of **radiation absorbed** is the rad, which stands for *r*adiation *a*bsorbed *d*ose. The SI unit is the gray (Gy), where 1 Gy = 100 rad.

The relationship between the delivered dose in roentgens and the absorbed dose in rads can be summarized as follows: Exposure to 1 roentgen yields 0.97 rad of absorbed radiation in water, 0.96 rad in muscle, and 0.93 rad in bone. This relationship holds for high-energy photons. For lower-energy photons, such as "soft" X rays, each roentgen yields 3 rads of absorbed dose in bone. This principle underlies diagnostic X rays, wherein soft tissue lets the radiation through to strike a photographic plate but the bones absorb the radiation and cast a shadow on the plate.

Rems The most important biological measure of radiation is the rem, which stands for *r*oentgen *e*quivalent for *m*an. One rem is a measure of the effect of the radiation when a person absorbs 1 roentgen. Other units are the **millirem** (mrem; 1 mrem = 1×10^{-3} rem) and the **sievert** (Sv; 1 Sv = 100 rem). The sievert is the SI unit. The need for such units arises because 1 rad of absorbed energy from different sources can cause different damage to the tissues. One rad from alpha rays or neutrons, for example, causes ten times more damage than 1 rad from X rays or gamma rays. Table 9.3 summarizes the various radiation units.

Although they cause more damage than X rays or gamma rays because of their absorption by the tissues, alpha particles have a very low penetrating power (Table 9.1) and cannot pass through the skin. Consequently, they are not harmful to humans or animals as long as they do not enter the body some other way. If they do get in, however, they can prove quite harmful. They can get inside, for example, if a person swallows or inhales a small particle of a substance that emits alpha particles. Beta particles are less damaging to tissue than alpha particles but penetrate farther and so are generally more harmful. Gamma rays, which can completely penetrate the skin, are by far the most dangerous and harmful form of radiation.

Therefore, for comparison purposes and for determining exposure from all kinds of sources, the equivalent dose is an important measure. If an organ receives radiation from different sources, the total effect can be summed up in rems (or mrem or Sv). For example, 10 mrem of alpha particles and 15 mrem of gamma radiation give a total of 25 mrem absorbed equivalent dose. Table 9.4 shows the amount of radiation exposure that an average person obtains yearly from both natural and artificial sources.

The naturally occurring background radiation varies with the geological location. In some phosphate mines, for example, tenfold higher than the average radiation has been detected. People who work in nuclear medicine are, of course, exposed to greater amounts. To ensure that exposures do not get too high, they wear radiation badges (Figure 9.5). A single whole-body irradiation of 25 rem is noticeable in the blood count, and 100 rem causes the typical symptoms of radiation sickness. A dose of 400 rem causes death within one month in 50% of exposed persons, and 600 rem is almost invariably lethal within a short time. It should be noted that as much as 50,000 rem is needed to kill bacteria and as much as 10^6 rem to inactivate viruses.

Figure 9.5 Worker with a radiation badge.

Table 9.3 Radiation Dosimetry

Measurement	SI Unit	Conventional Unit
Intensity/activity	1 becquerel (Bq) = 1 disintegration/s	1 curie (Ci) = 3.7×10^{10} disintegrations/s
Production of ions	1 roentgen (R) = 2.58×10^{-4} coulomb/kg	1 roentgen (R) = 2.58×10^{-4} coulomb/kg
Absorbed dose	1 gray (Gy) = 1 J/kg	1 rad = 0.01 Gy
Dose absorbed by humans	1 sievert (Sv) = 1 J/kg	1 rem = 0.01 Sv

CHEMICAL CONNECTIONS 9C

The Indoor Radon Problem

Most of our exposure to ionizing radiation comes from natural sources (Table 9.4), with radon gas being the main cause. Radon has more than 20 isotopes, all of which are radioactive. The most important is radon-222, an alpha-particle emitter. Radon-222 is a natural decay product of uranium-238, which is widely distributed in the earth's crust.

Radon poses a particular health hazard among radioactive elements because it is a gas at normal temperatures and pressures. As a consequence, it can enter our lungs with the air we breathe and become trapped in the mucous lining of the lungs. Radon-222 has a half-life of 3.8 days. It decays naturally and produces, among other isotopes, two harmful alpha emitters: polonium-218 and polonium-214. The Po isotopes are solid and do not leave the lungs with exhalation. In the long run, they can cause lung cancer.

■ Testing devices are available to determine whether radon is building up in a home.

The U.S. Environmental Protection Agency has set a standard of 4 pCi/L (one picocurie, pCi, is 10^{-12} curie) as a safe exposure level. A survey of single-family homes in the United States showed that 7% exceeded this level. Most radon seeps into dwellings through cracks in cement foundations and around pipes, then accumulates in basements. The remedy is to ventilate both basements and houses enough to reduce the radiation levels. In a notorious case, in Grand Junction, Colorado, houses were built from bricks made from uranium tailings. Obviously, the radiation levels in these buildings were unacceptably high. Because they could not be controlled, the buildings had to be destroyed. In our modern radiation-conscious age, more and more home buyers choose to request a certification of radon levels before buying a house.

Fortunately, most of us never get a single dose of more than a few rem and so never suffer from any form of radiation sickness. This does not mean, however, that small doses are totally harmless. The harm may arise in two ways:

1. Small doses of radioactivity over a period of years can cause cancer, especially blood cancers such as leukemia. Repeated exposure to sunlight also carries a risk of tissue damage. Most of the high-energy UV radiation of

CHEMICAL CONNECTIONS 9D

How Radiation Damages Tissues—Free Radicals

As mentioned earlier, high-energy radiation damages tissue by causing ionization. That is, the radiation knocks electrons out of the molecules that make up the tissue (generally one electron per molecule), thereby forming unstable ions. For example, the interaction of high-energy radiation with water forms an unstable cation:

$$\text{Energy} + H_2O \longrightarrow H_2O^+ + e^-$$

The positive charge on the H_2O^+ means that one of the electrons normally present in the water molecule, either in a covalent bond or an unshared pair, is missing in this cation; it has been knocked out. Once formed, the cation is unstable and decomposes:

$$H_2O^+ \longrightarrow H^+ + \cdot OH$$

The $\cdot OH$ has an unpaired electron (indicated by the dot), which makes it extremely reactive. Compounds that have unpaired electrons are called **free radicals**. They rapidly interact with other molecules, causing chemical reactions that damage the tissues. These reactions have especially serious consequences if they occur inside the cell nucleus and damage genetic material. In addition, they affect rapidly dividing cells more than they do stationary cells. Thus the damage is greater to embryonic cells, cells of the bone marrow and intestines, and cells in the lymph. Symptoms of radiation sickness include nausea, vomiting, a decrease in the white blood cell count, and loss of hair.

Cosmic rays High-energy particles, mainly protons, from outer space bombarding the Earth

Table 9.4 Average Exposure to Radiation from Common Sources

Source	Dose (mrem/year)
Naturally Occurring Radiation	
Cosmic rays	27
Terrestrial radiation (rocks, buildings)	28
Inside human body (K-40 and Ra-226 in the bones)	39
Radon in the air	200
Total	294
Artificial Radiation	
Medical X rays[a]	39
Nuclear medicine	14
Consumer products	10
Nuclear power plants	0.5
All others	1.5
Total	65.0
Grand total	359[b]

[a] Individual medical procedures may expose certain parts of the body to much higher doses. For instance, one chest X ray gives 27 mrem and a diagnostic GI series gives 1970 mrem.

[b] The federal safety standard for allowable occupational exposure is about 5000 mrem/year. It has been suggested that this level be lowered to 4000 mrem/year or even lower to reduce the risk of cancer stemming from low levels of radiation.

SOURCE: *National Council on Radiation Protection and Measurements,* NCRP Report No. 93 (1993).

the sun is absorbed by the protective ozone layer in the stratosphere. In tanning, however, the repeated overexposure to UV radiation can cause skin cancer (See Chemical Connections 18D). No one knows how many cancers have resulted from this practice, because the doses are so small and continue for so many years that they cannot generally be measured. Also, because so many other causes of cancer exist, it is difficult or impossible to decide if any particular case is caused by radiation.

2. If any form of radiation strikes an egg or sperm cell, it can cause a change in the genes (see Chemical Connections 25E). Such changes are called mutations. If an affected egg or sperm cell mates, grows, and becomes a new individual, that individual may have mutated characteristics, which are usually harmful and frequently lethal.

Because radiation carries so much potential for harm, it would be nice if we could totally escape it. But can we? Table 9.4 shows that this is impossible. Naturally occurring radiation, called **background radiation,** is present everywhere on earth. As Table 9.4 shows, this background radiation vastly outstrips the average radiation level from artificial sources (mostly diagnostic X rays). If we eliminated all forms of artificial radiation, including medical uses, we would still be exposed to the background radiation.

9.6 Medical Uses of Radioactive Materials

Radioactive isotopes have two major uses in medicine: diagnosis and therapy (Table 9.5).

A Diagnosis

Diagnosis is the more widely used aspect of nuclear medicine. Chemically and metabolically, a radioactive isotope in the body behaves exactly the same way as do the nonradioactive isotopes of the same element. In the sim-

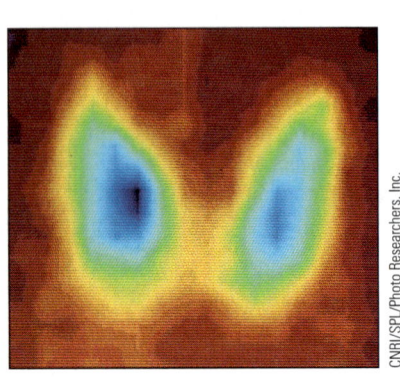

A scan of radiation released by radioactive iodine concentrated in thyroid tissue gives an image of the thyroid gland.

9.6 MEDICAL USES OF RADIOACTIVE MATERIALS

Table 9.5 Some Medically Useful Radioactive Isotopes

	Isotope	Use
$^{3}_{1}H$	Tritium	Measure water content of body
$^{11}_{6}C$	Carbon-11	Brain scan with positron emission transverse tomography (PET) to trace glucose pathway
$^{14}_{6}C$	Carbon-14	Radioimmunoassay
$^{32}_{15}P$	Phosphorus-32	Detection of eye tumors
$^{51}_{24}Cr$	Chromium-51	Albumism diagnosis, size and shape of spleen, gastrointestinal disorders
$^{59}_{26}Fe$	Iron-59	Bone marrow function, diagnosis of anemias
$^{60}_{27}Co$	Cobalt-60	Treatment of cancer
$^{67}_{31}Ga$	Gallium-67	Whole-body scan for tumors
$^{75}_{34}Se$	Selenium-75	Pancreas scan
$^{81m}_{36}Kr$	Krypton-81m	Lung ventilation scan
$^{85}_{38}Sr$	Strontium-85	Bone scan for bone diseases, including cancer
$^{99m}_{43}Tc$	Technetium-99m	Brain, liver, kidney, bone marrow scans; diagnosis of damaged heart muscles
$^{125}_{53}I$	Iodine-125	Prostate cancer treatment
$^{131}_{53}I$	Iodine-131	Diagnosis of thyroid malfunction; treatment of hyperthyroidism and thyroid cancer
$^{197}_{80}Hg$	Mercury-197	Kidney scan
$^{201}_{81}Tl$	Thallium-201	Heart scan and exercise stress test
$^{225}_{89}Ac$	Actinium-225	Treatment of prostate cancer

plest form of diagnosis, a radioactive isotope is injected intravenously and a technician uses the various types of detectors to monitor how the radiation is distributed in the body of the patient.

The use of iodine-131 to diagnose malfunctioning thyroid glands is a good example. The thyroid glands in the neck produce a hormone, thyroxine, that controls the overall rate of metabolism (use of food) in the body. Thyroxine contains iodine atoms. When radioactive iodine-131 is administered into the bloodstream, the thyroid glands take it up and incorporate it into the thyroxine (Chemical Connections 13C). A normally functioning thyroid absorbs about 12% of the administered iodine within a few hours. An overactive thyroid (hyperthyroidism) absorbs and localizes iodine-131 in the glands faster, and an underactive thyroid (hypothyroidism) does so much more slowly than normal. By counting the radioactive particles emitted from the neck, then, one can determine the rate of uptake of iodine-131 into the glands and diagnose hyperthyroidism or hypothyroidism.

Most organ scans are similarly based on the preferential uptake of some radioactive isotopes by a particular organ (Figure 9.6).

Another important use of radioactive isotopes is to learn what happens to ingested material. The foods and drugs swallowed or otherwise taken in by the body are transformed, decomposed, and excreted. To understand the pharmacology of a drug, it is important to know how these processes occur. For example, a certain drug may be effective in treating certain bacterial infections. Before beginning a clinical trial of the drug, its manufacturer must prove that the drug is not harmful to humans. In a typical case, the drug is first tested in animal studies. It is synthesized, and some radioactive isotope, such as hydrogen-3, carbon-14, or phosphorus-32, is incorporated into its structure. The drug is administered to the test animals, and after a certain period the animals are sacrificed. The fate of the drug is then determined by isolating from the body any radioactive compounds formed.

One typical pharmacological experiment studied the effects of tetracycline. This powerful antibiotic tends to accumulate in bones and is not given to pregnant women because it is transferred to the bones of the fetus. A

Normal

Meningioma (brain tumor)

"Brain death"

Scalp tumor

Figure 9.6 A comparison of dynamic scan patterns for normal and pathological brains. The studies were performed by injecting technetium-99m into blood vessels.

This process is called *tagging a drug*.

CHEMICAL CONNECTIONS 9E

CT, MRI, Ultrasound, and PET Scans

Computer-assisted tomography (CT or CAT scan), an X-ray diagnostic technique, has seen a phenomenal growth in its use in the more than 20 years since its introduction. Allan M. Cormack of Tufts University and Godfrey N. Hounsfield, an English scientist, shared the 1979 Nobel Prize in medicine for this invention.

CT scans can reveal small differences in the density of tissues because different tissues slow down X rays to different extents. A modern CT scan needs one rotation of the X ray source around the body. It uses a large number of detectors to measure simultaneously the X rays passing through the body. Using these data, a computer then clearly reconstructs the internal structure of the body. CT scans are used mainly to detect brain tumors because no other technique can locate tissues within the skull so accurately. Not only are internal structures displayed (Figure 9E.1), but one can even distinguish among the various fluid-filled cavities in the brain and between gray and white matter. CT scans are also extensively used to detect the damage caused by strokes. Their utility is not restricted to the brain—any part of the body can be examined with this tool.

Another diagnostic tool, magnetic resonance imaging (MRI), is especially good for examining the soft tissues that contain many hydrogen atoms, mostly in the form of water or fat. It probes the nuclear spin of atoms, mainly protons, which at a certain frequency invert their spin when subjected to radio-frequency waves. An advantage of MRI is that bones contain few hydrogen atoms, so they seem almost transparent with this imaging technique. Thus the soft tissues with their differing water contents stand out in the image (Figure 9E.2). Furthermore, MRI has an advantage in that it does not use ionizing radiation, as CT (and PET) scans do.

Figure 9E.2 An MRI scan.

Another diagnostic scan employing non-ionizing radiation is the ultrasound scan (sonograph). Ultrasound can image soft tissues with high water content—for example, kidney cysts, tumors, or stones. (See Figure 9E.4.)

The fourth important type of imaging is called positron emission tomography (PET scan). It is based on the property that certain isotopes (such as carbon-11) emit positrons (Section 9.2). Positrons have very short lives: When a positron collides with an electron, the two annihilate each other, resulting in a gamma ray. Because electrons are present in every atom, there are always lots of them around, so positrons generated in the body cannot live long. For example, if a person is fed glucose labeled with carbon-11, the glucose soon enters the blood and from there moves into the brain. Detectors for the gamma rays can pick up the signals that come from the areas where the glucose is used up. In this way, one can see which areas of the brain are involved when we process, for example, visual information (Figure 9E.5). PET scans can diagnose early stages of epilepsy and other

Figure 9E.1 A CT scan of the human brain.

Figure 9E.3 A medical MRI imager.

CHEMICAL CONNECTIONS 9E

CT, MRI, Ultrasound, and PET Scans, continued

Figure 9E.4 Ultrasound scan of a human kidney showing a tumor. The size of the tumor is indicated on the scan.

diseases that involve abnormal glucose metabolism, such as schizophrenia.

The favorite nuclide used in PET scans is fluorine-18. It is a positron emitter:

$$^{18}_{9}F \longrightarrow {}^{18}_{8}O + {}^{0}_{+1}e$$

When deoxyglucose is tagged with fluorine-18, one can follow the metabolic fate of glucose. Because tumors have high metabolic rates, PET scans using 18-fluorodeoxyglucose (FDG) are becoming more and more the diagnostic choice for their detection and localization. FDG/PET has been used in the diagnosis of malignant melanoma, malignant lymphoma, and ovarian cancer, among other conditions.

The latest imaging technique combines the CT and PET scan. The CT scan provides a high-resolution, three-dimensional image of tissues with different densities. The PET scan highlights the tissues with the highest metabolic rates, indicating the position of the possible malignancy.

Figure 9E.5 Positron emission tomography brain scans showing that glucose can cross the blood–brain barrier. The second scan in each case shows the result of visual stimulation. When a patient is visually stimulated, it increases the blood flow and the glucose to certain areas of the brain. These areas are shown in red.

particular tetracycline was tagged with radioactive hydrogen-3, and its uptake in rat bones was monitored in the presence and the absence of a sulfa drug. With the aid of a scintillation counter, researchers measured the radiation intensity of the maternal and fetal bones. They found that the sulfa drug helped to minimize this undesirable side effect of tetracycline—namely, its accumulation in the fetal bones.

The metabolic fate of essential chemicals in the body can also be followed with radioactive tracers. The use of radioactive isotopes has illuminated a number of normal and pathological body functions as well.

B Therapy

The main use of radioactive isotopes in therapy is the selective destruction of pathological cells and tissues. Recall that radiation, whether from gamma rays or other sources, is detrimental to cells. Ionizing radiation damages cells, especially those that divide rapidly. This damage may be great enough to destroy the cells or to sufficiently alter the genes so that multiplication of the cells slows down.

In therapy applications, cancerous cells are the main target for such ionizing radiation. Radiation is typically used when the cancer is not well localized and therefore surgical removal is not the complete answer; it may also be employed when the cancerous cells spread and are in a metastatic state. In addition, it is used for preventive purposes, to eliminate any possible remaining cancerous cells after surgery has been performed. The idea, of course, is to kill cancerous cells but not normal ones. Therefore, radiation such as X rays or gamma rays from cobalt-60 is highly *collimated*—that is, aimed at a small part of the body where cancerous cells are suspected to reside. Besides the gamma rays from cobalt-60, other ionizing radiation is used to treat inoperable tumors. Proton beams from cyclotrons, for instance, have been used to treat ocular melanoma and tumors of the skull base and spine.

Despite this pinpointing technique, the radiation inevitably kills normal cells along with the cancerous cells. Because the radiation is most effective against rapidly dividing cancer cells, rather than normal cells, and because the radiation is aimed at a specific location, the damage to healthy tissues is minimized. Figure 9.7 shows a patient receiving radiation treatment.

Another way to localize radiation damage in therapy is to use specific radioactive isotopes. In the case of thyroid cancer, large doses of iodine-131 are administered, which are taken up by the glands. The isotope, which has high radioactivity, kills all the cells of the gland (cancerous as well as normal ones), but does not appreciably damage other organs.

Another radioisotope, iodine-125, is used in the treatment of prostate cancer. Seeds of iodine-125, a gamma emitter, are implanted in the cancerous area of the prostate gland while being imaged with ultrasound. The seeds deliver 160 Gy (16,000 rad) over their lifetime.

A newer form of treatment of prostate cancer with great potential relies on actinium-225, an alpha emitter. As discussed in Section 9.5, alpha particles cause more damage to the tissues than any other form of radiation, but they have low penetrating power. Researchers have developed a very clever way to deliver actinium-225 to the cancer region of the prostate gland without damaging healthy tissues. The cancer has a high concentration of prostate-specific antigen (PSA) on its surface. A monoclonal antibody (Section 30.4) hones in on the PSA and interacts with it. A single actinium-225 atom attached to such a monoclonal antibody can deliver the desired radiation, thereby destroying the cancer. Actinium-225 is especially effective, because it has a half-life of ten days and it decays to three nuclides,

A metastatic state exists when the cancerous cells have broken off from their primary site(s) and begin moving to other parts of the body.

Radiation sickness is a side effect of radiation therapy for this reason.

Figure 9.7 Patient receiving radiation treatment.

themselves alpha emitters. In clinical trials, a single injection of antibody with an intensity in the kBq range (nanocuries) provided tumor regression without toxicity.

9.7 Nuclear Fusion

An estimated 98% of all matter in the universe is made up of hydrogen and helium. The "big bang" theory of the formation of the universe postulates that our universe started with an explosion (big bang) in which matter was formed out of energy and that, at the beginning, only the lightest element, hydrogen, was in existence. Later, as the universe expanded, stars were born when the hydrogen clouds collapsed under gravitational forces. In the cores of these stars, hydrogen nuclei fused together to form helium.

The transmutation of hydrogen nuclei into a helium nucleus liberates a very large amount of energy in the form of photons (Chemical Connections 9F), largely by the following reaction:

$$^{2}_{1}H + ^{3}_{1}H \longrightarrow ^{4}_{2}He + ^{1}_{0}n + 5.3 \times 10^{8} \text{ kcal/mol He}$$

This process, called **fusion**, is how the sun makes its energy. Uncontrolled fusion is employed in the "hydrogen bomb." If we can ever achieve a controlled version of this fusion reaction (which is unlikely to happen in the near term), we should be able to solve our energy problems.

All of the **transuranium elements** (elements with atomic numbers greater than 92) from 93 to 116 are artificial and have been prepared by a fusion process in which heavy nuclei are bombarded with light ones. Many, as their names indicate, were first prepared at the Lawrence Laboratory of the University of California, Berkeley, by Glenn Seaborg (1912–1999; Nobel laureate in chemistry, 1951) and his colleagues:

$$^{244}_{96}Cm + ^{4}_{2}He \longrightarrow ^{245}_{97}Bk + ^{1}_{1}H + 2\,^{1}_{0}n$$

$$^{238}_{92}U + ^{12}_{6}C \longrightarrow ^{246}_{98}Cf + 4\,^{1}_{0}n$$

$$^{252}_{98}Cf + ^{10}_{5}B \longrightarrow ^{257}_{103}Lr + 5\,^{1}_{0}n$$

> **Nuclear fusion** Joining together atomic nuclei to form nuclides heavier than the starting material

The reactions occurring in the sun are essentially the same as those in the hydrogen bomb.

CHEMICAL CONNECTIONS 9F

Energy from Mass

When a deuterium nucleus and a tritium nucleus are transmuted into a helium nucleus and a neutron, they give off a large amount of energy. What is the source of this energy? If we weighed the nuclei of the two hydrogen isotopes and compared it with the weight of the helium nucleus plus that of a neutron, we would find that the combined mass of the former is greater than the combined mass of the latter. When the deuterium and tritium nuclei are converted to helium and a neutron, the extra mass has to go somewhere. Where does it go? It is converted to energy. We even know, from the equation developed by Albert Einstein (1879–1955), how much energy we can get from the conversion of any amount of mass:

$$E = mc^2$$

This equation says that the mass (m) that was lost, in kilograms, multiplied by the square of the velocity of light (c^2), in meters squared per seconds squared (m^2/s^2), is equal to the amount of energy created (E), in joules. For example, 1 g of matter completely converted to energy would produce 8.8×10^{13} J, which is enough energy to boil 34,000,000 L of water initially at 20°C. This is equivalent to the amount of water you may find in an Olympic-size swimming pool. As you can see, we get a lot of energy from a little bit of mass.

$$c = 3.0 \times 10^8 \text{ m/s}$$

These transuranium elements are unstable, and most have very short half-lives. For example, the half-life of $^{257}_{103}Lr$ is 0.65 second. Many of the new superheavy elements have been obtained by bombarding lead isotopes with calcium-48 or nickel-64. So far, the creation of elements 110, 112, 114, and 116 have been reported, even though their detection was based on the observation of the decay of a single atom.

9.8 Nuclear Fission and Atomic Energy

In the 1930s, Enrico Fermi (1901–1954) and his colleagues in Rome, and Otto Hahn (1879–1968), Lisa Meitner (1878–1968), and Fritz Strassman (1902–1980) in Germany tried to produce new transuranium elements by bombarding uranium-235 with neutrons. To their surprise, Hahn and his co-workers found that, rather than fusion, they obtained **nuclear fission** (fragmentation of large nuclei into smaller pieces):

$$^{235}_{92}U + ^{1}_{0}n \longrightarrow ^{141}_{56}Ba + ^{92}_{36}Kr + 3\,^{1}_{0}n + \gamma + \text{energy}$$

In this reaction, the uranium-235 nucleus breaks into smaller particles, but the most important product is energy, which is produced because the products have less mass than the starting materials (Chemical Connections 9F). This form of energy, called **atomic energy**, has been used for both war (in the atomic bomb) and peace.

With uranium-235, each fission produces three neutrons, which in turn can generate more fissions by colliding with other uranium-235 nuclei. If even one of these neutrons produces a new fission, the process becomes a self-propagating **chain reaction** (Figure 9.8) that continues at a constant

Figure 9.8 A chain reaction is begun when a neutron collides with a nucleus of uranium-235.

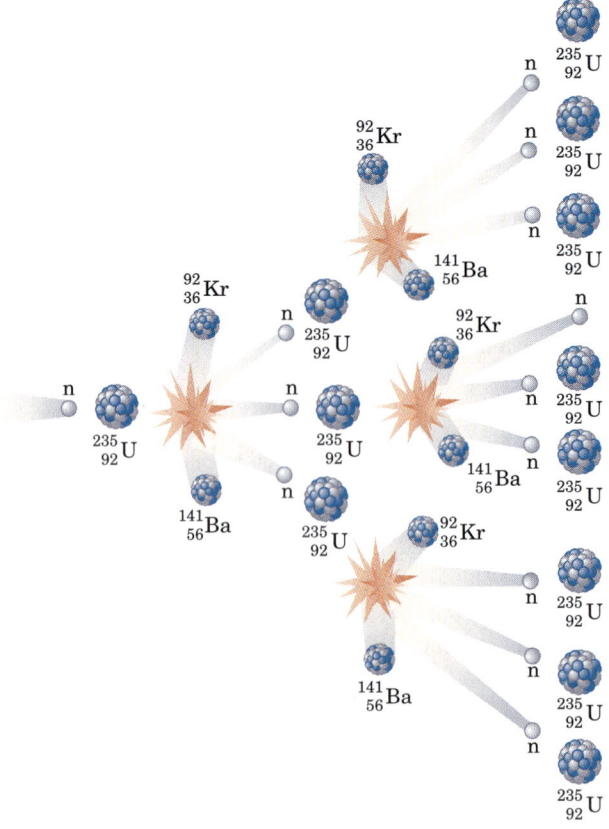

CHEMICAL CONNECTIONS 9G

Radioactive Fallout from Nuclear Accidents

On April 26, 1986, an accident occurred at the nuclear reactor in the town of Chernobyl in the former Soviet Union. It was a clear reminder of the dangers involved in this industry and of the far-reaching contamination that such accidents can produce. In Sweden, more than 500 miles away from the accident, the radioactive cloud increased the background radiation from 4 to 15 times the normal level. The radioactive cloud reached England, about 1300 miles away, one week later. There, it increased the natural background radiation by 15%. The radioactivity from iodine-131 was measured at 400 Bq/L in milk and 200 Bq/kg in leafy vegetables. Even some 4000 miles away in Spokane, Washington, an iodine-131 activity of 242 Bq/L was found in rainwater, and smaller activities—1.03 Bq/L of ruthenium-103 and 0.66 Bq/L of cesium-137—were recorded as well. These levels are not harmful.

Closer to the source of the nuclear accident, in neighboring Poland, potassium iodide pills were given to children. This step was taken to prevent radioactive iodine-131 (which might come from contaminated food) from concentrating in their thyroid glands, which could lead to cancer. In the wake of the September 11, 2001, terrorist attacks, Massachusetts became the first state to authorize the storage of KI pills in case of nuclear-related terrorist activity.

Figure 9.G Map showing those areas most affected by the Chernobyl accident.

---

rate. If all three neutrons are allowed to produce new fission, the rate of the reaction increases constantly and eventually culminates in a nuclear explosion. In nuclear plants, the rate of reaction can be controlled by inserting boron control rods to absorb neutrons.

In nuclear power plants, the energy produced by fission is sent to heat exchangers and used to generate steam, which drives a turbine to produce electricity (Figure 9.9). Today, more than 15% of the electrical energy in the United States is supplied by such plants. The opposition to nuclear plants is based on safety considerations and on the unsolved problems of waste disposal. Although nuclear plants in general have good safety records, accidents such as those at Chernobyl (Chemical Connections 9G) and Three Mile Island have caused concern.

Waste disposal is a long-range problem. The fission products of nuclear reactors are highly radioactive themselves, with long half-lives. Spent fuel contains these high-level fission products as nuclear wastes, together with

Figure 9.9 A nuclear power plant at Salem, New Jersey.

Figure 9.10 Storage of nuclear wastes in a storage room carved out of an underground salt bed.

uranium and plutonium that can be recovered and reused as mixed oxide (MOX) fuel. Reprocessing is costly: Although done routinely in Europe and Russia, it is not practiced by nuclear plants in the United States for economic reasons. However, this situation may change since cleaner extraction processes have been developed that use supercritical carbon dioxide (Chemical Connections 5G), thereby eliminating the need to dispose of solvent.

The United States has about 50,000 metric tons of spent fuel, stored under water and in dry casks at power plants. The additional nuclear wastes from the nuclear weapons program, research reactors, and other sources are stored by the Department of Energy in three major sites. After 40 years, the level of radioactivity that the wastes had immediately after their removal from the reactor is reduced a thousandfold. Such nuclear waste is a good candidate for underground burial (Figure 9.10). Recently, the U.S. federal government gave its final approval to a plan to store nuclear waste in the Yucca Mountain of Nevada.

Environmental concerns persist, however. The site cannot be guaranteed to stay dry for centuries. Moisture may corrode the steel cylinders and even the inner glass/ceramic cylinders surrounding the nuclear waste. Some fear that leaked materials from such storage tanks may escape if carbon-14 is oxidized to radioactive carbon dioxide or, less likely, that other radioactive nuclides may contaminate the ground water, which lies far below the desert rock of Yucca Mountain.

To keep these problems in perspective, one must remember that most other ways of generating large amounts of electrical power have their own environmental problems. For example, burning coal or oil contributes to the accumulation of CO_2 in the atmosphere and to acid rain (see Chemical Connections 6A).

SUMMARY

Certain isotopes are radioactive; others are stable. The three major types of radioactivity are **alpha particles** (helium nuclei), **beta particles** (electrons), and **gamma rays** (high-energy photons) (Section 9.1). **Radioactivity** comes from the nucleus, not from the electron cloud. When a nucleus emits a beta particle, it changes to a nucleus of another element that has the same mass number but an atomic number one unit higher (Section 9.2). When it emits an alpha particle, the new element has an atomic number two units lower and a mass number four units lower. **Positron** (positive electron) emission leads to a new element having the same mass number but an atomic number one unit lower. In gamma emission, no transmutation takes place; only the energy of the isotope is lowered.

Each radioactive isotope decays at a fixed rate described by its **half-life,** which is the time required for half of the sample to decay (Section 9.3).

Radiation is detected and counted by devices such as Geiger-Müller counters (Section 9.4). The main unit of intensity of radiation is the **curie,** which is equal to 3.7×10^{10} disintegrations per second. For medical purposes and to measure potential radiation damage, the absorbed dose, measured in **rads,** is an important factor (Section 9.5). Different particles damage body tissues differently; the **rem** is a measure of such damage.

Radioactive nuclei are used in medicine for both diagnosis and therapy (Section 9.6).

Helium is synthesized in the interiors of stars by **fusion** of hydrogen nuclei (Section 9.7). The energy released in this process is the energy of our sun. Researchers have created artificial elements by fusion, by bombarding atomic nuclei with high-energy nuclei. With certain nuclei, such bombardment causes the nucleus to split **(fission)** (Section 9.8). Nuclear fission releases large amounts of energy, which can be either controlled (nuclear reactors) or uncontrolled (nuclear weapons).

PROBLEMS

Numbers that appear in color indicate difficult problems. ▲ designates problems requiring application of principles.

Radioactivity and Electromagnetic Radiation

9.7 Do gamma rays travel with the speed of light in a vacuum?

9.8 What is the difference between alpha rays and protons?

9.9 Microwaves are a form of electromagnetic radiation that is used for the rapid heating of foods. What is the frequency of a microwave with a wavelength of 5.8 cm?

9.10 ▲ In each case, given the frequency, give the wavelength in centimeters or nanometers (10^{-9} m) and tell what kind of radiation it is.
(a) 7.5×10^{14}/s (b) 1.0×10^{10}/s
(c) 1.1×10^{15}/s (d) 1.5×10^{18}/s

9.11 A red light has a wavelength of 650 nm. What is its frequency?

9.12 Which has the longest wavelength: (a) infrared, (b) ultraviolet, or (c) X rays? Which has the highest energy?

Radioactive and Stable Isotopes

9.13 Write the symbol for a nucleus with the following components:
(a) 9 protons and 10 neutrons
(b) 15 protons and 17 neutrons
(c) 37 protons and 50 neutrons

9.14 In each pair, tell which isotope is more likely to be radioactive:
(a) nitrogen-14 and nitrogen-13
(b) phosphorus-31 and phosphorus-33
(c) lithium-7 and lithium-9
(d) calcium-39 and calcium-40

9.15 Which boron isotope is the most stable: boron-8, boron-10, or boron-12?

Natural Transmutation and Nuclear Reactions

9.16 Samarium-151 is a beta emitter. What is the transmutation product of the nuclear reaction?

9.17 The following nuclei turn into new nuclei by emitting beta particles. In each case, write the formula for the new nucleus.
(a) $^{159}_{63}$Eu (b) $^{141}_{56}$Ba (c) $^{242}_{95}$Am

9.18 ▲ Chromium-51 is used in diagnosing the pathology of the spleen. The nucleus of this isotope captures an electron according to the following equation. What is the transmutation product?

$$^{51}_{24}\text{Cr} + ^{0}_{-1}\text{e} \longrightarrow ?$$

9.19 The following nuclei turn into new nuclei by emitting alpha particles. In each case, write the formula for the new nucleus.
(a) $^{210}_{83}$Bi (b) $^{238}_{94}$Pu (c) $^{174}_{72}$Hf

9.20 Curium-248 was bombarded, yielding antimony-116 and cesium-160. What was the bombarding nucleus?

9.21 Phosphorus-29 is a positron emitter. Write the balanced nuclear equation for its activity.

9.22 For each of the following, write a balanced nuclear equation and identify the radiation emitted.
(a) beryllium-10 changes to boron-10
(b) europium-151* changes to europium-151
(c) thallium-195 changes to mercury-195
(d) plutonium-239 changes to uranium-235

9.23 In the first three steps in the decay of uranium-238, the following isotopic species appear: uranium-238 → thorium 234 → protactinium-234 → uranium 234. What kind of emission occurs in each step?

9.24 What kind of emission does *not* result in transmutation?

9.25 Complete the following nuclear reactions.
(a) $^{16}_{8}$O + $^{16}_{8}$O $\longrightarrow$? + $^{4}_{2}$He
(b) $^{235}_{92}$U + $^{1}_{0}$n $\longrightarrow$ $^{90}_{38}$Sr + ? + 3 $^{1}_{0}$n
(c) $^{13}_{6}$C + $^{4}_{2}$He $\longrightarrow$ $^{16}_{8}$O + ?
(d) $^{210}_{83}$Bi $\longrightarrow$ $^{0}_{-1}$e + ?
(e) $^{12}_{6}$C + $^{1}_{1}$H $\longrightarrow$? + γ

Half-life

9.26 Iodine-125 emits gamma rays and has a half-life of 60 days. If a 20-mg pellet of iodine-125 is implanted onto a prostate gland, how much iodine-125 remains there after one year?

9.27 Polonium-218, a decay product of radon-222 (see Chemical Connections 9C), has a half-life of 3 min. What percentage of the polonium-218 formed will remain in the lung 9 minutes after inhalation?

9.28 ▲ A rock containing 1 mg plutonium-239/kg rock is found in a glacier. The half-life of plutonium-239 is 25,000 years. If this rock was deposited 100,000 years ago during an ice age, how much plutonium-239, per kilogram of rock, was in the rock at that time?

9.29 The element radium is extremely radioactive. If you converted a piece of radium metal to radium chloride (with the weight of the radium remaining the same), would it become less radioactive?

9.30 In what ways can we increase the rate of radioactive decay? Decrease it?

9.31 Suppose 50.0 mg of potassium-45, a beta emitter, was isolated in pure form. After one hour, only 3.1 mg of the radioactive material was left. What is the half-life of potassium-45?

9.32 A patient receives 200 mCi of iodine-131, which has a half-life of eight days.
  (a) If 12% of this amount is taken up by the thyroid gland after two hours, what will be the activity of the thyroid after two hours, in millicuries and in counts per minute?
  (b) After 24 days, how much activity will remain in the thyroid gland?

Characteristics of Radiation

9.33 In scintillation counters, do the different radiations elicit different responses?

9.34 If you work in a lab containing radioisotopes emitting all kinds of radiation, from which emission should you seek most protection?

9.35 What do Geiger-Müller counters measure: (a) the intensity or (b) the energy of radiation?

9.36 It is known that radioactivity is being emitted with an intensity of 175 mCi at a distance of 1.0 m from the source. How far, in meters, from the source should you stand if you wish to be subjected to no more than 0.20 mCi?

Radiation Dosimetry

9.37 State clearly the properties that are measured in
  (a) rads   (b) rems
  (c) roentgens   (d) curies
  (e) grays   (f) becquerels
  (g) sieverts

9.38 A radioactive isotope with an activity (intensity) of 80.0 mCi per vial is delivered to a hospital. The vial contains 7.0 cc of liquid. The instruction is to administer 7.2 mCi intravenously. How many cubic centimeters of liquid should be used for one injection?

9.39 Why does exposure of a hand to alpha rays not cause serious damage to the person, whereas entry of an alpha emitter into the lung as an aerosol produces very serious damage to the person's health?

9.40 A certain radioisotope has an intensity of 1 million Bq at 1-cm distance from the source. What would be the intensity at 20 cm? Give your answer in both Bq and μCi units.

9.41 Assuming the same amount of effective radiation, in rads, from three sources, which would be the most damaging to the tissues: alpha particles, beta particles, or gamma rays?

9.42 In an accident involving radioactive exposure, person A received 3.0 Sv while person B received 0.50 mrem exposure. Who was hurt more seriously?

Medical Uses of Radioactive Materials

9.43 Which radioactive isotopes are used for the following purposes?
  (a) cancer treatment   (b) pancreas scans
  (c) bone scans   (d) brain scans
  (e) kidney scans

9.44 In 1986, the nuclear reactor in Chernobyl had an accident and spewed radioactive nuclei that were carried by the winds for hundreds of miles. Today, among the child survivors of the event, the most common damage is thyroid cancer. What radioactive nucleus do you expect to be responsible for these cancers?

9.45 Cobalt-60, with a half-life of 5.26 years, is used in cancer therapy. The energy of the radiation from cobalt-62 (half-life = 14 minutes) is even higher. Why isn't this isotope also used for cancer therapy?

9.46 Match the radioactive isotope with its proper use:
  _____ (a) cobalt-60   1. heart scan in exercise
  _____ (b) thallium-201   2. measure water content of body
  _____ (c) tritium   3. kidney scan
  _____ (d) mercury-197   4. cancer therapy

Nuclear Fusion and Fission

9.47 What is the most abundant nucleus in the universe?

9.48 What is the product of the fusion of hydrogen-2 and hydrogen-3 nuclei?

9.49 Assuming that one proton and two neutrons will be produced in an alpha-bombardment fusion reaction, what target nucleus would you use to obtain berkelium-249?

9.50 Element 109 was first prepared in 1982. A single atom of this element, with a mass number of 266 ($^{266}_{109}$Mt), was made by bombarding a bismuth-209 nucleus with an iron-58 nucleus. What other products, if any, must have been formed besides $^{266}_{109}$Mt?

9.51 A new element was formed when lead-208 was bombarded by krypton-86. One could detect four neutrons as the product of the fusion. Identify the new element.

9.52 Boron-10 is used as control rods in nuclear reactors. It absorbs a neutron while it emits an alpha particle. What is the resulting product in the control rods?

Chemical Connections

9.53 (Chemical Connections 9A) Explain why, in the experiment with radioactive material in a lead box, the gamma rays were not deflected at all.

9.54 (Chemical Connections 9B) In a recent archeological dig in the Amazon region of Brazil, charcoal paintings were found in a cave. The carbon-14 content of the charcoal was one-fourth of what is found in charcoal prepared from this year's tree harvest. How long ago was the cave settled?

9.55 (Chemical Connections 9B) Carbon-14 dating of the Shroud of Turin indicated that the plant from which the shroud was made was alive around AD 1350. To how many half-lives does this correspond?

9.56 (Chemical Connections 9B) The half-life of carbon-14 is 5730 years. The wrapping of an Egyptian mummy gave off 7.5 counts per minute per gram of carbon. A piece of linen purchased today would give an activity of 15 counts per minute per gram of carbon. How old is the mummy?

9.57 (Chemical Connections 9C) How does radon-222 produce polonium-218?

9.58 (Chemical Connections 9D) How does the sun's ultraviolet radiation damage the skin?

9.59 (Chemical Connections 9E) What are the advantages of an MRI scan over a CT scan?

9.60 (Chemical Connections 9E) Which radionuclide is used in FDG/PET scans?

9.61 (Chemical Connections 9E) Why do positrons have a very short lifetime?

9.62 (Chemical Connections 9F) Explain how the fusion of hydrogen nuclei to produce a helium nucleus gives off energy.

9.63 (Chemical Connections 9G) In a nuclear accident, one of the radioactive nuclei that concerns people is iodine-131. Iodine is easily vaporized, can be carried by the winds, and can cause radioactive fallout hundreds—even thousands—of miles away. Why is iodine-131 especially harmful?

Additional Problems

9.64 During the bombardment of argon-40 with protons for each proton absorbed one neutron is emitted. What new element is formed?

9.65 Neon-19 and sodium-20 are positron emitters. What products result in each case?

9.66 The half-life of nitrogen-16 is 7 seconds. How long does it take for 100 mg of nitrogen-16 to be reduced to 6.25 mg?

9.67 Do the curie and the becquerel measure the same or different properties of radiation?

9.68 Selenium-75 has a half-life of 120.4 days, so it would take 602 days (five half-lives) to diminish to 3% of the original quantity. Yet this isotope is used for pancreatic scans without any fear that the radioactivity will cause undue harm to the patient. Suggest a possible explanation.

9.69 Use Table 9.4 to determine the percentage of annual radiation we receive from the following sources:
(a) naturally occurring sources
(b) diagnostic medical sources
(c) nuclear power plants

9.70 $^{225}_{89}$Ac is an alpha emitter. In its decay process, it produces three more alpha emitters in succession. Identify each of the decay products.

9.71 Which radiation will cause more ionization, X rays or radar?

9.72 You have an old wristwatch that still has radium paint on its dial. Measurement of the radioactivity of the watch shows a beta-ray count of 0.50 count/s. If 1.0 microcurie radiation of this sort produces 1000 mrem/year, how much radiation, in mrem, do you expect from the wristwatch if you wear it for one year?

9.73 Americium-241, which is used in some smoke detectors, has a half-life of 475 years and is an alpha emitter. What is the decay product of americium-241, and approximately what percentage of the original americium-241 will be still around after 1000 years?

9.74 On rare occasions a nucleus, instead of emitting a beta particle, captures one. Berkelium-246 is such a nucleus. What is the product of this nuclear transmutation?

9.75 A patient is reported to have been irradiated by a dose of 1 sievert in a nuclear accident. Is he in mortal danger?

9.76 What is the ground state of a nucleus?

9.77 Explain the following:
(a) It is impossible to have a completely pure sample of any radioactive isotope.
(b) Beta emission of a radioactive isotope creates a new isotope with an atomic number one unit higher than that of the radioactive isotope.

9.78 Yttrium-90, which emits beta particles, is used in radiotherapy. What is the decay product of yttrium-90?

9.79 The half-lives of some oxygen isotopes are as follows: oxygen-14 = 71 s; oxygen-15 = 124 s; oxygen-19 = 29 s; and oxygen-20 = 14 s. Oxygen-16 is the stable, nonradioactive isotope. Do the half-lives indicate anything about the stability of the other oxygen isotopes?

9.80 $^{225}_{89}$Ac is effective in prostate cancer therapy when administered at kBq levels. If an antibody tagged with $^{225}_{89}$Ac has an intensity of 2 million Bq/mg and if a solution contains 5 mg/L tagged antibody, how many milliliters of the solution should you use for an injection to administer 1 kBq intensity?

9.81 When $^{208}_{82}$Pb is bombarded with $^{64}_{28}$Ni, a new element and six neutrons are produced. Identify the new element.

InfoTrac College Edition

For additional readings, go to InfoTrac College Edition, your online research library, at

http://infotrac.thomsonlearning.com

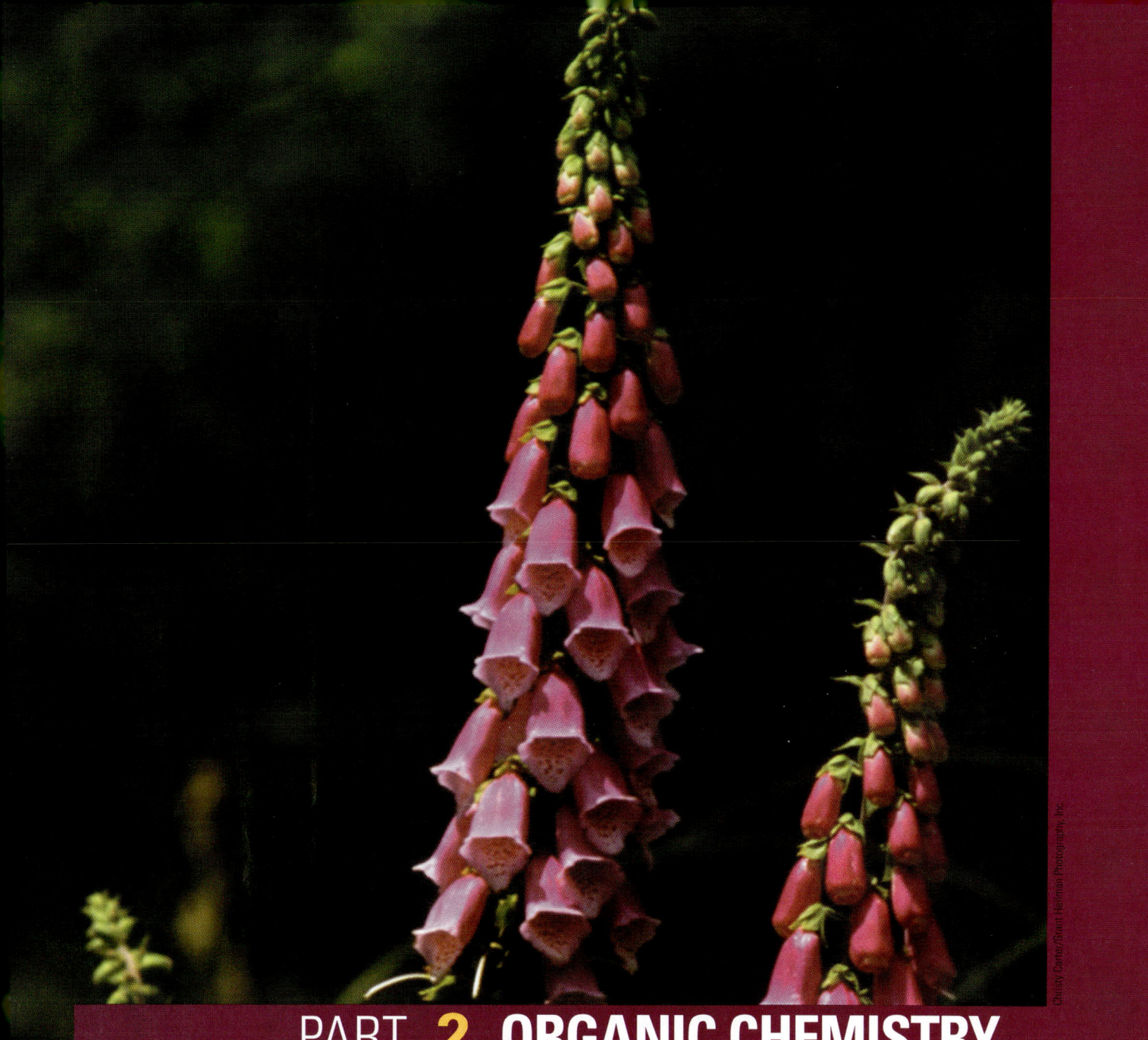

PART 2 ORGANIC CHEMISTRY

Chapter 10
Organic Chemistry

Chapter 11
Alkanes

Chapter 12
Alkenes

Chapter 13
Benzene and Its Derivatives

Chapter 14
Alcohols, Ethers, and Thiols

Chapter 15
Chirality: The Handedness of Molecules

Chapter 16
Amines

Chapter 17
Aldehydes and Ketones

Chapter 18
Carboxylic Acids, Anhydrides, Esters, and Amides

CHAPTER 10

10.1 Introduction

10.2 Sources of Organic Compounds

10.3 Structure of Organic Compounds

10.4 Functional Groups

The bark of the Pacific yew contains paclitaxel, a substance that has proved effective in treating certain types of ovarian and breast cancer.

Organic Chemistry

 ## 10.1 Introduction

Organic chemistry is the chemistry of the compounds of carbon. As you study Chapters 10–18 (organic chemistry) and 19–31 (biochemistry), you will see that organic compounds are everywhere around us. They are in our foods, flavors, and fragrances; in our medicines, toiletries, and cosmetics; in our plastics, films, fibers, and resins; in our paints, varnishes, and glues; and, of course, in our bodies and the bodies of all other living organisms.

Perhaps the most remarkable feature of organic compounds is that they involve the chemistry of carbon and only a few other elements—chiefly, hydrogen, oxygen, and nitrogen. While the majority of organic compounds contain carbon and just these three elements, many also contain sulfur, a halogen (fluorine, chlorine, bromine, or iodine), or phosphorus.

When this book was written, there were 114 known elements. Organic chemistry concentrates on carbon, just one of the 114. The chemistry of the other 113 elements comes under the field of inorganic chemistry. As we see in Figure 10.1, carbon is far from being among the most abundant elements in Earth's crust. In terms of elemental abundance, approximately 75% of

270

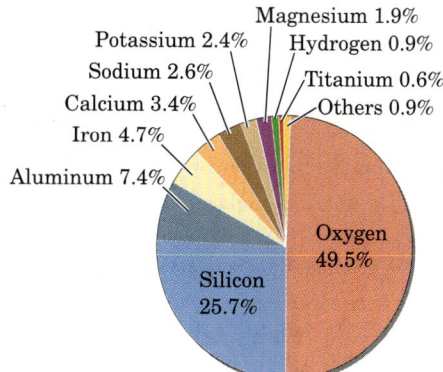

Figure 10.1 Abundance of the elements in the Earth's crust.

Earth's crust is composed of just two elements: oxygen and silicon. These two elements are the components of silicate minerals, clays, and sand. In fact, carbon is not even among the ten most abundant elements. Instead, it is merely one of the components making up the remaining 0.9% of Earth's crust. Why, then, do we pay this special attention to just one element from among 114?

The first reason is largely historical. In the early days of chemistry, scientists thought organic compounds were those produced by living organisms, and that inorganic compounds were those found in rocks and other nonliving matter. At the time, they believed that a "vital force," possessed only by living organisms, was necessary to produce organic compounds. In other words, chemists believed that they could not synthesize any organic compound starting with only inorganic compounds. This theory was very easy to disprove if, indeed, it was wrong: It required only one experiment in which an organic compound was made from inorganic compounds. In 1828, Friedrich Wöhler (1800–1882) carried out such an experiment. He heated an aqueous solution of ammonium chloride and silver cyanate, both inorganic compounds and, to his surprise, obtained urea, an "organic" compound found in urine.

Organic and inorganic compounds differ in properties because they differ in structure, not because they obey different natural laws. One set of natural laws applies to all compounds.

$$NH_4Cl + AgNCO \xrightarrow{heat} H_2N-\underset{\underset{O}{\|}}{C}-NH_2 + AgCl$$

Ammonium chloride   Silver cyanate   Urea   Silver chloride

Although this single experiment of Wöhler's sufficed to disprove the "doctrine of vital force," it took several years and a number of additional experiments for the entire scientific community to accept the fact that organic compounds could be produced in the laboratory. This discovery meant that the terms "organic" and "inorganic" no longer had their original meaning because, as Wöhler demonstrated, organic compounds could be obtained from inorganic materials. A few years later, August Kekulé (1829–1896) put forth a new definition—organic compounds are those containing carbon—and his definition has been accepted ever since.

A second reason for the study of carbon compounds as a separate discipline is the sheer number of organic compounds. Chemists have discovered or made more than 10 million of them, and an estimated 10,000 new ones are discovered or prepared in the laboratory each year. By comparison,

Table 10.1	A Comparison of Properties of Organic and Inorganic Compounds
Organic Compounds	**Inorganic Compounds**
Bonding is almost entirely covalent.	Most have ionic bonds.
May be gases, liquids, or solids with low melting points (less than 360°C).	Most are solids with high melting points.
Most are insoluble in water.	Many are soluble in water.
Most are soluble in organic solvents such as diethyl ether, toluene, and dichloromethane.	Almost all are insoluble in organic solvents.
Aqueous solutions do not conduct electricity.	Aqueous solutions conduct electricity.
Almost all burn.	Very few burn.
Reactions are usually slow.	Reactions are often very fast.

chemists have discovered or made an estimated 1.7 million inorganic compounds. Thus approximately 85% of all known compounds are organic compounds.

A third reason—and one particularly important for those of you who plan to study biochemistry—is that carbohydrates, lipids, proteins, enzymes, nucleic acids (DNA and RNA), hormones, vitamins, and almost all other important chemicals in living systems are organic compounds. Furthermore, their reactions are often strikingly similar to those observed occurring in test tubes. For this reason, knowledge of organic chemistry is essential for an understanding of biochemistry.

One final point about organic compounds. They generally differ from inorganic compounds in many of their properties, some of which are shown in Table 10.1. Most of these differences stem from the fact that the bonding in organic compounds is almost entirely covalent, while most inorganic compounds have ionic bonds.

Of course, there are many exceptions to the generalizations in Table 10.1. Some organic compounds behave like inorganic compounds, and vice versa, but the generalizations remain largely true for the vast majority of compounds of both types.

10.2 Sources of Organic Compounds

Chemists obtain organic compounds in two principal ways: isolation from nature and synthesis in the laboratory.

A Isolation from Nature

Living organisms are chemical "factories." Each plant and animal—even microorganisms such as bacteria—makes thousands of organic compounds by a process called biosynthesis. One way, then, to get organic compounds is to extract them from biological sources. In this book, we will encounter many compounds that are or have been obtained in this way. Some important examples include vitamin E, the penicillins, table sugar, insulin, quinine, and the anticancer drug Taxol (see Chemical Connections 10A). Nature also supplies us with three other important sources of organic compounds: natural gas, petroleum, and coal; we discuss them in Section 11.11.

■ Sugar cane, Hawaii.

CHEMICAL CONNECTIONS 10A

Taxol—A Story of Search and Discovery

In the early 1960s, the National Cancer Institute undertook a program to analyze samples of native plant materials in the hope of discovering substances that would prove effective in the fight against cancer. Among the materials tested was an extract of the bark of the Pacific yew, *Taxus brevifolia,* a slow-growing tree found in the old-growth forests of the Pacific Northwest. This extract proved to be remarkably effective in treating certain types of ovarian and breast cancer, even in cases where other forms of chemotherapy failed. The structure of the cancer-fighting component of yew bark was determined in 1962, and the compound was named paclitaxel (Taxol).

Paclitaxel (Taxol)

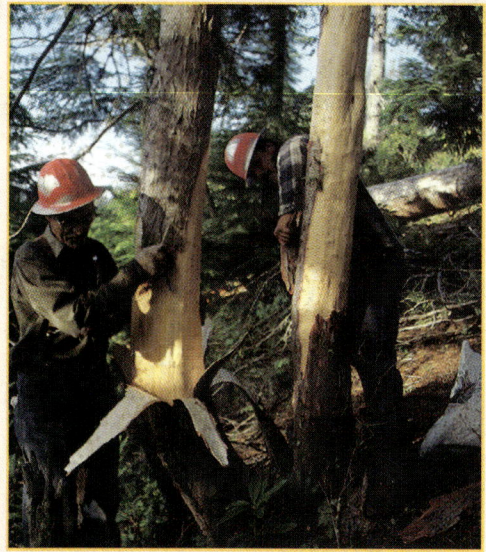

■ Pacific yew bark being stripped for Taxol extraction.

Unfortunately, the bark of a single 100-year-old tree yields only about 1 g of Taxol, not enough for effective treatment of even one cancer patient. Furthermore, getting Taxol means stripping the bark from trees, which kills them. In 1994, chemists succeeded in synthesizing Taxol in the laboratory, but the drug's cost was far too high to be economical. Fortunately, an alternative natural source of the drug was found. Researchers in France discovered that the needles of a related plant, *Taxus baccata,* contain a compound that can be converted to Taxol in the laboratory. Because the needles can be gathered without harming the plant, it is not necessary to kill trees to obtain the drug.

Taxol inhibits cell division by acting on microtubules, a key component of the scaffolding of cells. Before cell division can take place, the cell must disassemble these microtubule units, and Taxol prevents this disassembly. Because cancer cells are the fastest dividing cells, the drug effectively controls their spreading.

The remarkable success of Taxol in the treatment of breast and ovarian cancer has stimulated research efforts to discover and/or synthesize other substances that work the same way in the body and that may be even more effective anticancer agents than Taxol.

B Synthesis in the Laboratory

Ever since Wöhler synthesized urea, organic chemists have sought to develop more ways to make the same compounds found in nature. In recent years, the methods for doing so have become so sophisticated that there are few natural organic compounds, no matter how complicated, that chemists cannot synthesize in the laboratory.

Compounds made in the laboratory are identical to those found in nature, assuming, of course, that each is 100% pure. There is no way that anyone can tell whether a sample of a compound was made by chemists or obtained from nature. If, however, a compound is not 100% pure, we can usually tell from the identity of the impurities whether the compound was isolated from nature or made in a laboratory. As a consequence, pure ethanol made by chemists has exactly the same physical and chemical properties as the pure ethanol prepared by distilling wine. The same is true for

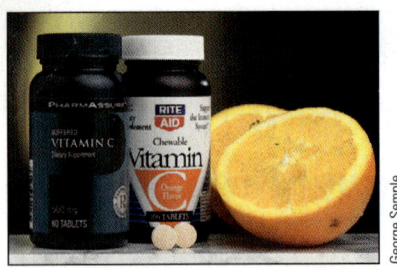

■ Vitamin C in an orange is identical to its synthetic tablet form.

ascorbic acid (vitamin C). There is no advantage, therefore, in paying more money for vitamin C obtained from a natural source than for synthetic vitamin C, because the two are identical in every way.

Organic chemists, however, have not been satisfied with just duplicating nature's compounds. They have also synthesized compounds not found in nature. In fact, the majority of the more than 10 million known organic compounds are purely synthetic and do not exist in living organisms. For example, many modern drugs—Valium, Vasotec, Prozac, Zantac, Cardizem, Lasix, Viagra, and Enovid—are synthetic organic compounds not found in nature. Even the over-the-counter drugs aspirin and ibuprofen are synthetic organic compounds not found in nature.

10.3 Structure of Organic Compounds

A structural formula shows all the atoms present in a molecule as well as the bonds that connect the atoms to each other. The structural formula for ethanol, for example, shows all nine atoms and the eight bonds that connect them:

$$\begin{array}{c c} \text{H} & \text{H} \\ | & | \\ \text{H}-\text{C}-\text{C}-\text{O}-\text{H} \\ | & | \\ \text{H} & \text{H} \end{array}$$

The Lewis model of bonding (Section 3.7) enables us to see how carbon forms four covalent bonds that may be various combinations of single, double, and triple bonds. Furthermore, the valence-shell electron-pair repulsion (VSEPR) model (Section 3.9) tells us that the most common bond angles about carbon atoms in covalent compounds are approximately 109.5°, 120°, and 180°.

CHEMICAL CONNECTIONS 10B

Combinatorial Chemistry

Since the beginning of synthetic organic chemistry about 200 years ago, the goal has been to produce pure organic compounds with high percentage yields. More often than not, a particular use for a compound was discovered accidentally. Such was the case, for example, with aspirin and the barbiturates. Later, when researchers discovered the mode of action of certain drugs, it became possible to design new compounds that would have a reasonable probability of being more effective or having fewer side effects than some existing drug. Such was the case with ibuprofen and albuterol (see Section 16.7). Even in such a "rational" approach (as compared to the "shotgun" approach, in which compounds were synthesized at random), the goal was still to obtain a single pure compound, no matter how many steps were involved in the synthesis or how difficult the steps were. In the last few years this strategy has changed, mostly in pharmaceutical research.

Combinatorial chemistry is a process in which chemists try to produce as many compounds from as few building blocks as possible. For example, if we allow 20 compounds (building blocks) to react at random, and could find a method to restrict the combination in such a way that only molecules containing three of these building blocks are produced, then a mixture of 8000 different compounds (20^3) would result. Perhaps this mixture might include two or three candidates with the potential to be useful drugs. The task, then, is to select from this synthetic soup those compounds that promise to be biologically active. Researchers accomplish this goal by screening the mixture for a desired biological activity and then separating promising candidates for further study. In essence, it is cheaper and faster to produce a large "library" of combinatorial synthetic compounds and then "fish out" the biologically active ones than to pursue the old way of randomly synthesizing new compounds and screening them one at a time.

10.3 STRUCTURE OF ORGANIC COMPOUNDS | 275

Table 10.2 Single, Double, and Triple Bonds in Compounds of Carbon (bond angles are predicted using the VSEPR model)

Compound	Bond angles
Ethane	109.5°
Ethylene	120°
Acetylene	180°
Chloroethane	109.5°
Methanol	109.5°
Formaldehyde	120°
Methylamine	109.5°
Methyleneimine	120°
Hydrogen cyanide	180°

Table 10.2 shows several covalent compounds containing carbon bonded to hydrogen, oxygen, nitrogen, and chlorine. From these examples, we see the following:

- Carbon normally forms four covalent bonds and has no unshared pairs of electrons.
- Nitrogen normally forms three covalent bonds and has one unshared pair of electrons.
- Oxygen normally forms two covalent bonds and has two unshared pairs of electrons.
- Hydrogen forms one covalent bond and has no unshared pairs of electrons.
- Chlorine (and fluorine, bromine, and iodine) normally forms one covalent bond and has three unshared pairs of electrons.

EXAMPLE 10.1

The structural formulas for acetic acid, CH_3COOH, and ethylamine, $CH_3CH_2NH_2$, are

Acetic acid   Ethylamine

(a) Complete the Lewis structure for each molecule by adding unshared pairs of electrons so that each atom of carbon, oxygen, and nitrogen has a complete octet.
(b) Using the VSEPR model (Section 3.9), predict all bond angles in each molecule.

Solution

(a) Each carbon atom is surrounded by eight valence electrons and therefore has a complete octet. To complete the octet of each oxygen, add two unshared pairs of electrons. To complete the octet of nitrogen, add one unshared pair of electrons.

(b) To predict bond angles about a carbon, nitrogen, or oxygen atom, count the number of regions of electron density about it. If it is surrounded by four regions of electron density, predict bond angles of 109.5°. If it is surrounded by three regions, predict bond angles of 120°. If it is surrounded by two regions, predict a bond angle of 180°.

Acetic acid

Ethylamine

Problem 10.1

The structural formulas for ethanol, CH_3CH_2OH, and propene, $CH_3CH=CH_2$, are

Ethanol   Propene

(a) Complete the Lewis structure for each molecule showing all valence electrons.
(b) Using the VSEPR model, predict all bond angles in each molecule.

10.4 Functional Groups

As noted in Section 10.1, more than 10 million organic compounds have been discovered or made by organic chemists. It might seem an almost impossible task to learn the physical and chemical properties of so many compounds. Fortunately, the study of organic compounds is not as formidable a task as you might think. While organic compounds can undergo a wide variety of chemical reactions, only certain portions of their structures are changed in any particular reaction. We call the part of an organic molecule that undergoes chemical reactions a **functional group.** As we will see, the

Functional group An atom or group of atoms within a molecule that shows a characteristic set of physical and chemical properties

Screen 10.4
Tutorial

Table 10.3 Five Common Functional Groups

Family	Functional Group	Example	Name
Alcohol	—OH	CH_3CH_2OH	Ethanol
Amine	—NH_2	$CH_3CH_2NH_2$	Ethanamine
Aldehyde	$-\overset{\overset{O}{\|\|}}{C}-H$	$CH_3\overset{\overset{O}{\|\|}}{C}H$	Ethanal
Ketone	$-\overset{\overset{O}{\|\|}}{C}-$	$CH_3\overset{\overset{O}{\|\|}}{C}CH_3$	Acetone
Carboxylic acid	$-\overset{\overset{O}{\|\|}}{C}-OH$	$CH_3\overset{\overset{O}{\|\|}}{C}OH$	Acetic acid

same functional group, in whatever organic molecule it occurs, undergoes the same types of chemical reactions. Therefore, we do not have to study the chemical reactions of even a fraction of the 10 million known organic compounds. Instead, we need to identify only a few characteristic functional groups and then study the chemical reactions that each undergoes.

Functional groups are also important because they are the units by which we divide organic compounds into families of compounds. For example, we group those compounds that contain an —OH (hydroxyl) group bonded to a tetrahedral carbon into a family called alcohols; compounds containing a —COOH group belong to a family called carboxylic acids. Table 10.3 introduces five of the most common functional groups. A complete list of all functional groups we will study appears on the inside back cover of the text.

At this point, our concern is simply pattern recognition—that is, how to recognize one of these five functional groups when you see it, and how to draw structural formulas of molecules containing them. We will have more to say about the physical and chemical properties of these and several other functional groups in Chapters 11–18.

Functional groups also serve as the basis for naming organic compounds. Ideally, each of the 10 million or more organic compounds must have a name different from the name of every other compound. We will show how these names are derived in Chapters 11–18 as we study individual functional groups in detail.

To summarize, functional groups

- Are sites of chemical reaction—a particular functional group, in whatever compound it is found, undergoes the same types of chemical reactions;
- Determine in large measure the physical properties of a compound;
- Are the units by which we divide organic compounds into families; and
- Serve as a basis for naming organic compounds.

A Alcohols

The functional group of an **alcohol** is an **—OH (hydroxyl) group** bonded to a tetrahedral carbon atom (a carbon having bonds to four atoms). In the general formula of an alcohol (shown on the next page on the left), the symbol R— indicates either a hydrogen or another carbon group. The im-

Alcohol A compound containing an —OH (hydroxyl) group bonded to a tetrahedral carbon atom

Hydroxyl group An —OH group bonded to a tetrahedral carbon atom

278 / CHAPTER 10 ORGANIC CHEMISTRY

■ Table wine contains about 10 to 13% ethanol.

portant point of the general structure is the —OH group bonded to a tetrahedral carbon atom.

$$\begin{array}{c} R \\ | \\ R-C-\ddot{O}-H \\ | \\ R \end{array} \qquad \begin{array}{c} H \quad H \\ | \quad | \\ H-C-C-O-H \\ | \quad | \\ H \quad H \end{array} \qquad CH_3CH_2OH$$

Functional group   Structural formula   Condensed structural formula

R = H or carbon group   An alcohol (Ethanol)

Here we represent the alcohol as a **condensed structural formula**, CH_3CH_2OH. In a condensed structural formula, CH_3 indicates a carbon bonded to three hydrogens, CH_2 indicates a carbon bonded to two hydrogens, and CH indicates a carbon bonded to one hydrogen. Unshared pairs of electrons are generally not shown in a condensed structural formula.

Alcohols are classified as **primary (1°), secondary (2°),** or **tertiary (3°),** depending on the number of carbon atoms bonded to the carbon bearing the —OH group.

$$\begin{array}{c} H \\ | \\ CH_3-C-OH \\ | \\ H \end{array} \qquad \begin{array}{c} H \\ | \\ CH_3-C-OH \\ | \\ CH_3 \end{array} \qquad \begin{array}{c} CH_3 \\ | \\ CH_3-C-OH \\ | \\ CH_3 \end{array}$$

A 1° alcohol   A 2° alcohol   A 3° alcohol

EXAMPLE 10.2

Draw Lewis structures and condensed structural formulas for the two alcohols with molecular formula C_3H_8O. Classify each as primary, secondary, or tertiary.

Solution
Begin by drawing the three carbon atoms in a chain. The oxygen atom of the hydroxyl group may be bonded to the carbon chain at two different positions on the chain: either to an end carbon or to the middle carbon.

$$\begin{array}{c} \\ C-C-C \\ \\ \text{Carbon chain} \end{array} \qquad \begin{array}{c} \\ C-C-C-OH \quad C-C-C \\ \\ \text{The two locations for the —OH group} \end{array} \qquad \begin{array}{c} OH \\ | \\ \end{array}$$

Finally, add seven more hydrogens, giving a total of eight as shown in the molecular formula. Show unshared electron pairs on the Lewis structures but not on the condensed structural formulas.

Lewis structures

$$\begin{array}{c} H \quad H \quad H \\ | \quad | \quad | \\ H-C-C-C-\ddot{O}-H \\ | \quad | \quad | \\ H \quad H \quad H \end{array}$$

Condensed structural formulas

$CH_3CH_2CH_2OH$

A primary alcohol

10.4 FUNCTIONAL GROUPS

H—C(H)(H)—C(:O:H)(H)—C(H)(H)—H   CH₃CHCH₃ (with OH)
A secondary alcohol

■ 2-Propanol (isopropyl alcohol) is used to disinfect cuts and scrapes.

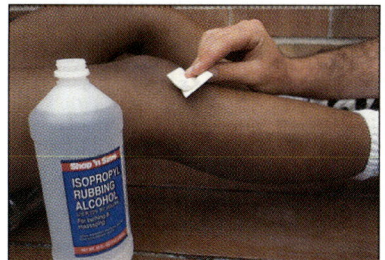

The secondary alcohol, whose common name is isopropyl alcohol, is the cooling, soothing component in rubbing alcohol.

Problem 10.2
Draw Lewis structures and condensed structural formulas for the four alcohols with molecular formula $C_4H_{10}O$. Classify each alcohol as primary, secondary, or tertiary. (*Hint:* First consider the order of attachment of the four carbon atoms; they can be bonded either four in a chain or three in a chain with the fourth carbon as a branch on the middle carbon. Then consider the points at which the —OH group can be bonded to each carbon chain.)

B Amines

The functional group of an **amine** is an **amino group**—a nitrogen atom bonded to one, two, or three carbon atoms. In a **primary (1°) amine**, nitrogen is bonded to one carbon group. In a **secondary (2°) amine**, it is bonded to two carbon groups, and in a **tertiary (3°) amine**, it is bonded to three carbon groups. The second and third structural formulas can be written in a more abbreviated form by collecting the CH₃ groups and writing them as $(CH_3)_2NH$ and $(CH_3)_3N$.

Amine An organic compound in which one, two, or three hydrogens of ammonia are replaced by carbon groups: RNH_2, R_2NH, or R_3N.

CH₃NH₂

Methylamine
(a 1° amine)

CH₃NH   or   (CH₃)₂NH
|
CH₃

Dimethylamine
(a 2° amine)

CH₃NCH₃   or   (CH₃)₃N
|
CH₃

Trimethylamine
(a 3° amine)

EXAMPLE 10.3
Draw condensed structural formulas for the two primary amines with molecular formula C_3H_9N.

Solution
For a primary amine, draw a nitrogen atom bonded to two hydrogens and one carbon.

280 | CHAPTER 10 ORGANIC CHEMISTRY

```
              NH₂
              |
C—C—C—NH₂   C—C—C
```
The three carbons may be bonded to nitrogen in two ways

CH₃CH₂CH₂NH₂   CH₃CHCH₃
                  |
                  NH₂

Add seven hydrogens to give each carbon four bonds and give the correct molecular formula

Problem 10.3
Draw structural formulas for the three secondary amines with molecular formula $C_4H_{11}N$.

Carbonyl group A C=O group

Aldehyde A compound containing a carbonyl group bonded to a hydrogen; a —CHO group

Ketone A compound containing a carbonyl group bonded to two carbon groups

C Aldehydes and Ketones

Both aldehydes and ketones contain a C=O (**carbonyl**) **group**. The **aldehyde** functional group contains a carbonyl group bonded to a hydrogen. In formaldehyde, CH₂O, the simplest aldehyde, two hydrogens are bonded to its carbonyl carbon. In a condensed structural formula, the aldehyde group may be written showing the carbon–oxygen double bond as CH=O, or, alternatively, it may be written —CHO. The functional group of a **ketone** is a carbonyl group bonded to two carbon atoms. In the general structural formula of each functional group, we use the symbol R to represent other groups bonded to carbon to complete the tetravalence of carbon.

Functional group | Acetaldehyde (an aldehyde) | | Functional group | Acetone (a ketone)

EXAMPLE 10.4

Draw condensed structural formulas for the two aldehydes with molecular formula C_4H_8O.

Solution
First draw the functional group of an aldehyde, and then add the remaining carbons. These may be bonded in two ways. Then, add seven hydrogens to complete the four bonds of each carbon.

$$\begin{array}{c}\text{O}\\\|\\\text{CH}_3\text{CH}_2\text{CH}_2\text{CH}\end{array}$$

or

$$\text{CH}_3\text{CH}_2\text{CH}_2\text{CHO}$$

$$\begin{array}{c}\text{O}\\\|\\\text{CH}_3\text{CHCH}\\|\\\text{CH}_3\end{array}$$

or

$$\begin{array}{c}\text{CH}_3\text{CHCHO}\\|\\\text{CH}_3\end{array}$$

■ Acetone is a ketone.

Problem 10.4
Draw condensed structural formulas for the three ketones with molecular formula $C_5H_{10}O$.

D Carboxylic Acids

The functional group of a **carboxylic acid** is a **—COOH (carboxyl: carbonyl + hydroxyl) group.** In a condensed structural formula, a carboxyl group may also be written —CO_2H.

Carboxylic acid A compound containing a —COOH group

Carboxyl group A —COOH group

$$\begin{array}{c}\text{O}\\\|\\\text{RCOH}\end{array}\qquad\begin{array}{c}\text{O}\\\|\\\text{CH}_3\text{COH}\end{array}$$

Functional   Acetic acid
group   (a carboxylic acid)

EXAMPLE 10.5

Draw a condensed structural formula for the single carboxylic acid with molecular formula $C_3H_6O_2$.

Solution
The only way the carbon atoms can be written is three in a chain, and the —COOH group must be on an end carbon of the chain.

$$\begin{array}{c}\text{O}\\\|\\\text{CH}_3\text{CH}_2\text{COH}\end{array}\quad\text{or}\quad\text{CH}_3\text{CH}_2\text{COOH}$$

Problem 10.5
Draw condensed structural formulas for the two carboxylic acids with molecular formula $C_4H_8O_2$.

SUMMARY

Organic chemistry (Section 10.1) is the study of compounds containing carbon. Chemists obtain organic compounds either by isolation from plant and animal sources or by synthesis in the laboratory (Section 10.2).

Carbon normally forms four bonds and has no unshared pairs of electrons (Section 10.3). Its four bonds may be four single bonds, two single bonds and one double bond, or one triple bond and one single bond. Nitrogen normally forms three bonds and has one unshared pair of electrons. Its bonds may be three single bonds, one single bond and one double bond, or one triple bond. Oxygen normally forms two bonds and has two unshared pairs of electrons. Its bonds may be two single bonds or one double bond.

Functional groups (Section 10.4) are sites of chemical reactivity; a particular functional group, in whatever compound it is found, always undergoes the same types of reactions. In addition, functional groups are the characteristic structural units by which we both classify and name organic compounds. Important functional groups include the **hydroxyl group** (Section 10.4A) of 1°, 2°, and 3° alcohols; the **amino group** of 1°, 2°, and 3° amines (Section 10.4B); the **carbonyl group** of aldehydes and ketones (Section 10.4C); and the **carboxyl group** of carboxylic acids (Section 10.4D).

PROBLEMS

Numbers that appear in color indicate difficult problems.
▶ designates problems requiring application of principles.

Review of Lewis Structures

10.6 How many electrons are in the valence shell of each of the following atoms? Write a Lewis dot structure for an atom of each element.
   (a) Carbon   (b) Oxygen
   (c) Nitrogen   (d) Fluorine

10.7 What is the relationship between the number of electrons in the valence shell of each of the following atoms and the number of covalent bonds it forms?
   (a) Carbon   (b) Oxygen
   (c) Nitrogen   (d) Hydrogen

10.8 Write Lewis structures for these compounds. Show all valence electrons. None of them contains a ring of atoms. (*Hint:* Remember that carbon has four bonds, nitrogen has three bonds and one unshared pair of electrons, oxygen has two bonds and two unshared pairs of electrons, and each halogen has one bond and three unshared pairs of electrons.)
   (a) H_2O_2, hydrogen peroxide
   (b) N_2H_4, hydrazine
   (c) CH_3OH, methanol
   (d) CH_3SH, methylethiol
   (e) CH_3NH_2, methanamine
   (f) CH_3Cl, chloromethane

10.9 Write Lewis structures for these compounds. Show all valence electrons. None of them contains a ring of atoms.
   (a) CH_3OCH_3, dimethyl ether
   (b) C_2H_6, ethane
   (c) C_2H_4, ethylene
   (d) C_2H_2, acetylene
   (e) CO_2, carbon dioxide
   (f) CH_2O, formaldehyde
   (g) H_2CO_3, carbonic acid
   (h) CH_3COOH, acetic acid

10.10 Write Lewis structures for these ions.
   (a) HCO_3^-, bicarbonate ion
   (b) CO_3^{2-}, carbonate ion
   (c) CH_3COO^-, acetate ion
   (d) Cl^-, chloride ion

10.11 Why are the following molecular formulas impossible?
   (a) CH_5 (b) C_2H_7

Review of the VSEPR Model

10.12 Explain how to use the valence-shell electron-pair repulsion (VSEPR) model to predict bond angles about atoms of carbon, oxygen, and nitrogen.

PROBLEMS | 283

10.13 Suppose you forget to take into account the presence of the unshared pair of electrons on nitrogen in the molecule NH₃. What would you predict for the H—N—H bond angles?

10.14 Suppose you forget to take into account the presence of the two unshared pairs of electrons on the oxygen atom of ethanol, CH₃CH₂OH. What would you predict for the C—O—H bond angle?

10.15 Use the VSEPR model to predict the bond angles about each highlighted atom. (*Hint:* Remember to take into account the presence of unshared pairs of electrons.)

(a) H—C—C—O—H (with H's on each C)

(b) H—C=C—Cl (with H's)

(c) H—C—C≡C—H (with H's on first C)

10.16 Use the VSEPR model to predict the bond angles about each highlighted atom.

(a) H—C(=O)—O—H

(b) H—C—N—H (with H's)

(c) H—O—N=O

10.17 Following is a structural formula and a ball-and-stick model of benzene, C₆H₆.

(a) Predict each H—C—C and C—C—C bond angle in benzene.
(b) Predict the shape of a benzene molecule.

Sources of Organic Compounds

10.18 Is there any difference between vanillin made synthetically and vanillin extracted from vanilla beans?

10.19 Suppose that you are told that only organic substances are produced by living organisms. How would you rebut this assertion?

10.20 What important experiment did Wöhler carry out in 1828?

Organic Compounds

10.21 List the four principal elements that make up organic compounds and give the number of bonds each typically forms.

10.22 Think about the types of substances in your immediate environment, and make a list of those that are organic—for example, textile fibers. We will ask you to return to this list later in the course and to refine, correct, and possibly expand it.

Functional Groups

10.23 What is meant by the term "functional group"?

10.24 List three reasons why functional groups are important in organic chemistry.

10.25 Draw Lewis structures for each of the following functional groups. Show all valence electrons.
(a) Carbonyl group
(b) Carboxyl group
(c) Hydroxyl group
(d) Primary amino group

10.26 Complete the following structural formulas by adding enough hydrogens to complete the tetravalence of each carbon. Then write the molecular formula of each compound.

(a) C—C=C—C—C (with C branch)

(b) C—C—C—C(=O)—OH

(c) C—C—C(=O)—C

(d) C—C—C(=O)—H (with C branch)

(e) C—C—C—C—NH₂ (with C branches)

(f)
$$\text{C}-\underset{\underset{\text{NH}_2}{|}}{\text{C}}-\overset{\overset{\text{O}}{\|}}{\text{C}}-\text{OH}$$

(g)
$$\text{C}-\underset{\underset{}{|}}{\overset{\overset{\text{OH}}{|}}{\text{C}}}-\text{C}-\text{C}-\text{C}$$

(h)
$$\text{C}-\underset{\underset{}{|}}{\overset{\overset{\text{OH}}{|}}{\text{C}}}-\text{C}-\overset{\overset{\text{O}}{\|}}{\text{C}}-\text{OH}$$

(i) C=C—C—OH

10.27 Some of the following structural formulas are incorrect (that is, they do not represent real compounds) because they have atoms with an incorrect number of bonds. Which structural formulas are incorrect, and which atoms in them have an incorrect number of bonds?

(a)
$$\text{H}-\text{C}\equiv\text{C}-\underset{\underset{\text{H}}{|}}{\overset{\overset{\text{H}}{|}}{\text{C}}}-\text{H}$$

(b)
$$\text{H}-\underset{\underset{\text{H}}{|}}{\text{C}}=\underset{\underset{\text{H}}{|}}{\overset{\overset{\text{Cl}}{|}}{\text{C}}}-\text{H}$$

(c)
$$\text{H}-\underset{\underset{\text{H}}{|}}{\overset{\overset{\text{H}}{|}}{\text{N}}}-\underset{\underset{\text{H}}{|}}{\overset{\overset{\text{H}}{|}}{\text{C}}}-\underset{\underset{\text{H}}{|}}{\overset{\overset{\text{H}}{|}}{\text{C}}}-\text{O}-\text{H}$$

(d)
$$\text{H}-\underset{\underset{\text{H}}{|}}{\overset{\overset{\text{H}}{|}}{\text{C}}}-\underset{\underset{\text{H}}{|}}{\overset{\overset{\text{H}}{|}}{\text{C}}}=\text{O}-\text{H}$$

(e)
$$\text{H}-\text{O}-\underset{\underset{\text{H}}{|}}{\overset{\overset{\text{H}}{|}}{\text{C}}}-\underset{\underset{\text{H}}{|}}{\overset{\overset{\text{H}}{|}}{\text{C}}}-\overset{\overset{\text{O}}{\|}}{\text{C}}-\text{O}-\text{H}$$

(f)
$$\text{H}-\underset{\underset{\text{H}}{|}}{\overset{\overset{\text{H}}{|}}{\text{C}}}-\underset{\underset{\text{H}}{|}}{\overset{\overset{\text{H}}{|}}{\text{C}}}-\overset{\overset{\text{O}}{\|}}{\text{C}}-\text{H}$$

10.28 What is the meaning of the term "tertiary (3°)" when it is used to classify alcohols?

10.29 Draw a structural formula for the one tertiary (3°) alcohol with molecular formula $C_4H_{10}O$.

10.30 What is the meaning of the term "tertiary (3°)" when it is used to classify amines?

10.31 Draw a structural formula for the one tertiary (3°) amine with molecular formula $C_4H_{11}N$.

10.32 Identify the functional group(s) in each compound.

(a) $\text{CH}_3-\underset{\underset{}{|}}{\overset{\overset{\text{OH}}{|}}{\text{CH}}}-\overset{\overset{\text{O}}{\|}}{\text{C}}-\text{OH}$

Lactic acid
(produced in actively respiring muscle; gives sourness to sour cream)

(b) $\text{HO}-\text{CH}_2-\text{CH}_2-\text{OH}$

Ethylene glycol
(a component of antifreeze)

(c) $\text{CH}_3-\underset{\underset{\text{NH}_2}{|}}{\text{CH}}-\overset{\overset{\text{O}}{\|}}{\text{C}}-\text{OH}$

Alanine
(one of the building blocks of proteins)

(d) $\text{HOCH}_2\overset{\overset{\text{HO}}{|}}{\text{CH}}\overset{\overset{\text{O}}{\|}}{\text{CH}}$

Glyceraldehyde
(the simplest carbohydrate)

(e)

1,6-Hexanediamine
(one component of nylon-66)

10.33 Identify the functional group(s) in each compound.

(a) $\text{CH}_3\text{CH}_2\overset{\overset{\text{O}}{\|}}{\text{C}}\text{CH}_3$

2-Butanone
(a solvent for paints and lacquers)

(b) $\overset{\overset{\text{O}}{\|}}{\text{HOC}}\text{CH}_2\text{CH}_2\text{CH}_2\overset{\overset{\text{O}}{\|}}{\text{COH}}$

Hexanedioic acid
(the second component of nylon-66)

(c) $\text{H}_2\text{NCH}_2\text{CH}_2\text{CH}_2\text{CH}_2\underset{\underset{\text{NH}_2}{|}}{\text{CH}}\overset{\overset{\text{O}}{\|}}{\text{COH}}$

Lysine
(one of the 20 amino acid building blocks of proteins)

(d) $\text{HOCH}_2\overset{\overset{\text{O}}{\|}}{\text{C}}\text{CH}_2\text{OH}$

Dihydroxyacetone
(a component of several artificial tanning lotions)

10.34 Draw condensed structural formulas for all compounds with molecular formula C_4H_8O that contain a carbonyl group (there are two aldehydes and one ketone).

10.35 Draw structural formulas for the following:
  (a) The four primary (1°) alcohols with molecular formula $C_5H_{12}O$
  (b) The three secondary (2°) alcohols with molecular formula $C_5H_{12}O$
  (c) The one tertiary (3°) alcohol with molecular formula $C_5H_{12}O$

10.36 Draw structural formulas for the six ketones with molecular formula $C_6H_{12}O$.

10.37 Draw structural formulas for the eight carboxylic acids with molecular formula $C_6H_{12}O_2$.

10.38 Draw structural formulas for the following:
  (a) The four primary (1°) amines with molecular formula $C_4H_{11}N$
  (b) The three secondary (2°) amines with molecular formula $C_4H_{11}N$
  (c) The one tertiary (3°) amine with molecular formula $C_4H_{11}N$

Chemical Connections

10.39 (Chemical Connections 10A) How was Taxol discovered?

10.40 (Chemical Connections 10A) In what way does Taxol interfere with cell division?

10.41 (Chemical Connections 10B) What is the goal of combinatorial chemistry?

Additional Problems

10.42 Use the VSEPR model to predict the bond angles about each atom of carbon, nitrogen, and oxygen in these molecules. (*Hint:* First add unshared pairs of electrons as necessary to complete the valence shell of each atom and then make your predictions of bond angles.)
  (a) $CH_3CH_2CH_2OH$
  (b) $CH_3CH_2\overset{\overset{O}{\|}}{C}H$
  (c) $CH_3CH=CH_2$
  (d) $CH_3C{\equiv}CCH_3$
  (e) $CH_3\overset{\overset{O}{\|}}{C}OCH_3$
  (f) $CH_3\overset{\overset{CH_3}{|}}{N}CH_3$

10.43 Silicon is immediately below carbon in the Periodic Table. Predict the C—Si—C bond angle in tetramethylsilane, $(CH_3)_4Si$.

10.44 Phosphorus is immediately below nitrogen in the Periodic Table. Predict the C—P—C bond angle in trimethylphosphine, $(CH_3)_3P$.

10.45 Draw the structure for a compound with molecular formula
  (a) C_2H_6O that is an alcohol
  (b) C_3H_6O that is an aldehyde
  (c) C_3H_6O that is a ketone
  (d) $C_3H_6O_2$ that is a carboxylic acid

10.46 Draw structural formulas for the eight aldehydes with molecular formula $C_6H_{12}O$.

10.47 Draw structural formulas for the three tertiary (3°) amines with molecular formula $C_5H_{13}N$.

10.48 Which of these covalent bonds are polar and which are nonpolar? (*Hint:* Review Section 3.7B.)
  (a) C—C   (b) C=C   (c) C—H
  (d) C—O   (e) O—H   (f) C—N
  (g) N—H   (h) N—O

10.49 Of the bonds in Problem 10.48, which is the most polar? Which is the least polar?

10.50 Using the symbol δ+ to indicate a partial positive charge and δ− to indicate a partial negative charge, indicate the polarity of the most polar bond (or bonds if two or more have the same polarity) in each of the following molecules.
  (a) CH_3OH
  (b) CH_3NH_2
  (c) $HSCH_2CH_2NH_2$
  (d) $CH_3\overset{\overset{O}{\|}}{C}CH_3$
  (e) $H\overset{\overset{O}{\|}}{C}H$
  (f) $CH_3\overset{\overset{O}{\|}}{C}OH$

InfoTrac College Edition

For additional readings, go to InfoTrac College Edition, your online research library, at

http://infotrac.thomsonlearning.com

CHAPTER 11

11.1 Introduction
11.2 Structure of Alkanes
11.3 Constitutional Isomerism in Alkanes
11.4 Nomenclature of Alkanes
11.5 Cycloalkanes
11.6 Shapes of Alkanes and Cycloalkanes
11.7 Cis-Trans Isomerism in Cycloalkanes
11.8 Physical Properties
11.9 Reactions of Alkanes
11.10 Important Haloalkanes
11.11 Sources of Alkanes

A petroleum refinery. Petroleum, along with natural gas, provides nearly 90 percent of the organic raw materials for the synthesis and manufacture of synthetic fibers, plastics, detergents, drugs, dyes, adhesives, and a multitude of other products.

Alkanes

11.1 Introduction

In this chapter, we examine the physical and chemical properties of **alkanes**, the simplest type of organic compounds. Actually, alkanes are members of a larger class of organic compounds called hydrocarbons. A **hydrocarbon** is a compound composed of only carbon and hydrogen. Figure 11.1 shows the four classes of hydrocarbons, along with the characteristic type of bonding between carbon atoms in each class. Alkanes are **saturated hydrocarbons;** that is, they contain only carbon–carbon single bonds. Saturated in this context means that each carbon in the hydrocarbon has the maximum number of hydrogens bonded to it. A hydrocarbon that contains one or more carbon–carbon double bonds, triple bonds, or benzene rings is classified as an **unsaturated hydrocarbon.** We study alkanes (saturated hydrocarbons) in this chapter and alkenes, alkynes, and arenes (unsaturated hydrocarbons) in Chapters 12 and 13.

We commonly refer to alkanes as **aliphatic hydrocarbons** because the physical properties of the higher members of this class resemble those of the long carbon-chain molecules we find in animal fats and plant oils (Greek: *aleiphar,* fat or oil).

Alkane A saturated hydrocarbon whose carbon atoms are arranged in an open chain.

Hydrocarbon A compound that contains only carbon and hydrogen atoms

Saturated hydrocarbon A hydrocarbon that contains only carbon–carbon single bonds

Aliphatic hydrocarbon An alkane

Figure 11.1 The four classes of hydrocarbons.

11.2 Structure of Alkanes

Methane, CH$_4$, and ethane, C$_2$H$_6$, are the first two members of the alkane family. Figure 11.2 shows Lewis structures and ball-and-stick models for these molecules. The shape of methane is tetrahedral, and all H—C—H bond angles are 109.5°. Each carbon atom in ethane is also tetrahedral, and the bond angles in it are all approximately 109.5°. Although the three-dimensional shapes of larger alkanes are more complex than those of methane and ethane, the four bonds about each carbon atom are still arranged in a tetrahedral manner, and all bond angles are still approximately 109.5°.

The next members of the alkane family are propane, butane, and pentane. In the representations on page 288, these hydrocarbons are drawn first as condensed structural formulas, which show all carbons and hydrogens. They can also be drawn in a more abbreviated form called a **line-angle formula**. In this type of representation, a line represents a carbon–carbon bond and an angle represents a carbon atom. A line ending in space represents a —CH$_3$ group. To count hydrogens from a line-angle formula, you simply add enough hydrogens in your mind to give each carbon its required four bonds.

Structural formulas for alkanes can be written in yet another condensed form. For example, the structural formula of pentane contains three CH$_2$ (**methylene**) groups in the middle of the chain. We can group them

> **Line-angle formula** An abbreviated way to draw structural formulas in which each angle and line terminus represents a carbon atom and each line represents a bond

Figure 11.2 Methane and ethane.

288 / CHAPTER 11 ALKANES

Ball-and-stick model			
Line-angle formula	∧	∧∕	∧∕∧
Condensed structural formula	CH₃CH₂CH₃ Propane	CH₃CH₂CH₂CH₃ Butane	CH₃CH₂CH₂CH₂CH₃ Pentane

■ Butane, CH₃CH₂CH₂CH₃, is the fuel in this lighter. Butane molecules are present in the liquid and gaseous states in the lighter.

together and write the structural formula CH₃(CH₂)₃CH₃. Table 11.1 gives the first ten alkanes with unbranched chains. Note that the names of all these alkanes end in "-ane." We will have more to say about naming alkanes in Section 11.4.

EXAMPLE 11.1

Table 11.1 gives the condensed structural formula for hexane. Draw a line-angle formula for this alkane, and number the carbons on the chain beginning at one end and proceeding to the other end.

Solution
Hexane contains six carbons in a chain. Its line-angle formula is

$$\underset{123456}{\diagup\!\!\diagdown\!\!\diagup\!\!\diagdown\!\!\diagup\!\!\diagdown}$$

Problem 11.1
Following is a line-angle formula for an alkane. What is the name and molecular formula of this alkane?

Table 11.1 The First Ten Alkanes with Unbranched Chains

Name	Molecular Formula	Condensed Structural Formula	Name	Molecular Formula	Condensed Structural Formula
methane	CH₄	CH₄	hexane	C₆H₁₄	CH₃(CH₂)₄CH₃
ethane	C₂H₆	CH₃CH₃	heptane	C₇H₁₆	CH₃(CH₂)₅CH₃
propane	C₃H₈	CH₃CH₂CH₃	octane	C₈H₁₈	CH₃(CH₂)₆CH₃
butane	C₄H₁₀	CH₃(CH₂)₂CH₃	nonane	C₉H₂₀	CH₃(CH₂)₇CH₃
pentane	C₅H₁₂	CH₃(CH₂)₃CH₃	decane	C₁₀H₂₂	CH₃(CH₂)₈CH₃

11.3 Constitutional Isomerism in Alkanes

Constitutional isomers are compounds that have the same molecular formula but different structural formulas. By "different structural formulas," we mean that they differ in the kinds of bonds (single, double, or triple) and/or in their connectivity (the order of attachment among their atoms). For the molecular formulas CH_4, C_2H_6, and C_3H_8, only one order of attachment of their atoms is possible so there are no isomers for these molecular formulas. For the molecular formula C_4H_{10}, two structural formulas are possible: In butane, the four carbons are bonded in a chain; in 2-methylpropane, three carbons are bonded in a chain with the fourth carbon as a branch on the chain. The two constitutional isomers with molecular formula C_4H_{10} are drawn here both as condensed structural formulas and as line-angle formulas. Also shown are ball-and-stick models of each.

> **Constitutional isomers** Compounds with the same molecular formula but a different order of attachment of their atoms

Constitutional isomers have also been called structural isomers, an older term that is still in use.

$CH_3CH_2CH_2CH_3$
Butane
(bp −0.5°C)

CH_3 |
CH_3CHCH_3
2-Methylpropane
(bp −11.6°C)

We refer to butane and 2-methylpropane as constitutional isomers; they are different compounds and have different physical and chemical properties. Their boiling points, for example, differ by approximately 11°C.

In Section 10.4, we encountered several examples of constitutional isomers, although we did not call them that at the time. We saw, for example, that there are two alcohols with molecular formula C_3H_8O, two primary amines with molecular formula C_3H_9N, two aldehydes with molecular formula C_4H_8O, and two carboxylic acids with molecular formula $C_4H_8O_2$.

To determine whether two or more structural formulas represent constitutional isomers, we write the molecular formula of each and then compare them. All compounds that have the same molecular formula but different structural formulas are constitutional isomers.

EXAMPLE 11.2

Do the structural formulas in each set represent the same compound or constitutional isomers? You will find it helpful to redraw each molecule as a line-angle formula, which will make it easier for you to see similarities and differences in molecular structure.

(a) $CH_3CH_2CH_2CH_2CH_2CH_3$ and $CH_3CH_2CH_2$
                                           |
                                      $CH_2CH_2CH_3$ (Each is C_6H_{14})

(b) CH_3 CH_3
    | |
CH_3CHCH_2CH and $CH_3CH_2CHCHCH_3$ (Each is C_7H_{16})
  | | |
CH_3 CH_3 CH_3

Solution
First, find the longest chain of carbon atoms. Note that it makes no difference whether the chain is drawn straight or bent. As structural formulas are drawn in this problem, there is no attempt to show three-dimensional shapes. Second, number the longest chain from the end nearest the first branch. Third, compare the lengths of the two chains and the sizes and locations of any branches. Structural formulas that have the same molecular formula and the same order of attachment of atoms represent the same compound; those that have the same molecular formula but different orders of attachment of their atoms represent constitutional isomers.

(a) Each structural formula has an unbranched chain of six carbons; they are identical and represent the same compound.

$$CH_3CH_2CH_2CH_2CH_2CH_3 \quad \text{and} \quad CH_3CH_2CH_2CH_2CH_2CH_3$$

(b) Each structural formula has the same molecular formula, C_7H_{16}. In addition, each has a chain of five carbons with two CH_3 branches. Although the branches are identical, they are found at different locations on the chains. Therefore, these structural formulas represent constitutional isomers.

Problem 11.2
Do the structural formulas in each set represent the same compound or constitutional isomers?

(a)

(b)

EXAMPLE 11.3

Draw structural formulas for the five constitutional isomers with molecular formula C_6H_{14}.

Solution

In solving problems of this type, you should devise a strategy and then follow it. Here is one possible strategy. First, draw a line-angle formula for the constitutional isomer with all six carbons in an unbranched chain. Then, draw line-angle formulas for all constitutional isomers with five carbons in a chain and one carbon as a branch on the chain. Finally, draw line-angle formulas for all constitutional isomers with four carbons in a chain and two carbons as branches. No constitutional isomers for C_6H_{14} having only three carbons in the longest chain are possible.

Six carbons in an unbranched chain

Five carbons in a chain; one carbon as a branch

Four carbons in a chain; two carbons as branches

Problem 11.3

Draw structural formulas for the three constitutional isomers with molecular formula C_5H_{12}.

The ability of carbon atoms to form strong, stable bonds with other carbon atoms results in a staggering number of constitutional isomers, as the following table shows:

Molecular Formula	Number of Constitutional Isomers
CH_4	1
C_5H_{12}	3
$C_{10}H_{22}$	75
$C_{15}H_{32}$	4,347
$C_{25}H_{52}$	36,797,588
$C_{30}H_{62}$	4,111,846,763

Thus, for even a small number of carbon and hydrogen atoms, a very large number of constitutional isomers is possible. In fact, the potential for structural and functional group individuality among organic molecules made from just the basic building blocks of carbon, hydrogen, nitrogen, and oxygen is practically limitless.

11.4 Nomenclature of Alkanes

A The IUPAC System

Ideally, every organic compound should have a name from which its structural formula can be drawn. For this purpose, chemists have adopted a set of rules established by the **International Union of Pure and Applied Chemistry (IUPAC)**.

The IUPAC name for an alkane with an unbranched chain of carbon atoms consists of two parts: (1) a prefix to show the number of carbon atoms in the chain and (2) the suffix -**ane** to show that the compound is a saturated

Screen 11.3
Tutorial

Table 11.2 Prefixes Used in the IUPAC System to Show the Presence of One to Ten Carbons in an Unbranched Chain

Prefix	Number of Carbon Atoms	Prefix	Number of Carbon Atoms
meth-	1	hex-	6
eth-	2	hept-	7
prop-	3	oct-	8
but-	4	non-	9
pent-	5	dec-	10

hydrocarbon. Table 11.2 gives the prefixes used to show the presence of one to ten carbon atoms.

The IUPAC chose the first four prefixes listed in Table 11.2 because they were well established long before the nomenclature was systematized. For example, the prefix *but-* appears in the name butyric acid, a compound of four carbon atoms formed by the air oxidation of butter fat (Latin: *butyrum*, butter). The prefixes to show five or more carbons are derived from Latin numbers. Table 11.1 gives the names, molecular formulas, and condensed structural formulas for the first ten alkanes with unbranched chains.

IUPAC names of alkanes with branched chains consist of a parent name that shows the longest chain of carbon atoms and substituent names that indicate the groups attached to the parent chain. For example,

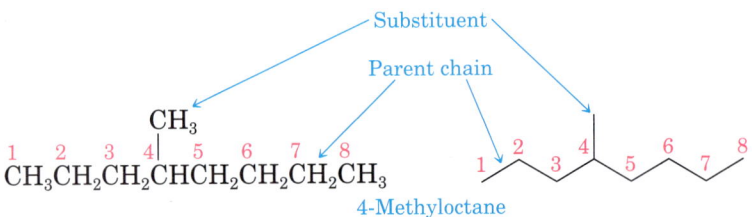

4-Methyloctane

A substituent group derived from an alkane by removal of a hydrogen atom is called an **alkyl group** and is commonly represented by the symbol **R—**. Alkyl groups are named by dropping the "-ane" from the name of the parent alkane and adding the suffix **-yl.** Tables 11.3 gives the names and condensed structural formulas for eight of the most common alkyl groups. The prefix *sec-* is an abbreviation for "secondary," meaning a carbon bonded to two other carbons. The prefix *tert-* is an abbreviation for "tertiary," meaning a carbon bonded to three other carbons.

The rules of the IUPAC system for naming alkanes are as follows:

1. The name for an alkane with an unbranched chain of carbon atoms consists of a prefix showing the number of carbon atoms in the chain and the ending *-ane*.

2. For branched-chain alkanes, take the longest chain of carbon atoms as the parent chain and its name becomes the root name.

3. Give each substituent on the parent chain a name and a number. The number shows the carbon atom of the parent chain to which the substituent is bonded. Use a hyphen to connect the number to the name.

> **Alkyl group** A group derived by removing a hydrogen from an alkane; given the symbol R—
>
> **R—** A symbol used to represent an alkyl group

2-Methylpropane

Table 11.3 Names of the Most Common Alkyl Groups

Name	Condensed Structural Formula	Name	Condensed Structural Formula
methyl	—CH$_3$	isobutyl	—CH$_2$CHCH$_3$ 　　　　| 　　　　CH$_3$
ethyl	—CH$_2$CH$_3$	sec-butyl	—CHCH$_2$CH$_3$ 　| 　CH$_3$
propyl	—CH$_2$CH$_2$CH$_3$	tert-butyl	CH$_3$ 　| —CCH$_3$ 　| 　CH$_3$
isopropyl	—CHCH$_3$ 　| 　CH$_3$		
butyl	—CH$_2$CH$_2$CH$_2$CH$_3$		

4. If there is one substituent, number the parent chain from the end that gives the substituent the lower number.

$$\begin{array}{c} CH_3 \\ | \\ CH_3CH_2CH_2CHCH_3 \end{array}$$

2-Methylpentane
(not 4-methylpentane)

5. If the same substituent occurs more than once, number the parent chain from the end that gives the lower number to the substituent encountered first. Indicate the number of times the substituent occurs by a prefix di-, tri-, tetra-, penta-, hexa-, and so on. Use a comma to separate position numbers.

$$\begin{array}{c} CH_3 \quad\; CH_3 \\ | \qquad\; | \\ CH_3CH_2CHCH_2CHCH_3 \end{array}$$

2,4-Dimethylhexane
(not 3,5-dimethylhexane)

6. If there are two or more different substituents, list them in alphabetical order and number the chain from the end that gives the lower number to the substituent encountered first. If there are different substituents in equivalent positions on opposite ends of the parent chain, give the substituent of lower alphabetical order the lower number.

$$\begin{array}{c} CH_3 \\ | \\ CH_3CH_2CHCH_2CHCH_2CH_3 \\ | \\ CH_2CH_3 \end{array}$$

3-Ethyl-5-methylheptane
(not 3-methyl-5-ethylheptane)

7. Do not include the prefixes di-, tri-, tetra-, and so on, or the hyphenated prefixes *sec-* and *tert-* in alphabetizing. Alphabetize the names of substituents first, and then insert these prefixes. In the following example, the alphabetizing parts are ethyl and methyl, not ethyl and dimethyl.

$$\underset{\underset{CH_3}{|}}{\overset{\overset{CH_3}{|}\;\;\overset{CH_2CH_3}{|}}{CH_3CCH_2CHCH_2CH_3}}$$

4-Ethyl-2,2-dimethylhexane
(not 2,2-dimethyl-4-ethylhexane)

EXAMPLE 11.4

Write the molecular formula and IUPAC name for each alkane.

(a)

(b)

Solution

The molecular formula of (a) is C_5H_{12}, and that of (b) is $C_{11}H_{24}$. In (a), number the longest chain from the end that gives the methyl substituent the lower number (rule 4). In (b), list the isopropyl and methyl substituents in alphabetical order (rule 6).

(a) 2-Methylbutane

(b) 4-Isopropyl-2-methylheptane

Problem 11.4

Write the molecular formula and IUPAC name for each alkane.

(a)

(b)

B Common Names

In the older system of **common nomenclature,** the total number of carbon atoms in an alkane, regardless of their arrangement, determines the name. The first three alkanes are methane, ethane, and propane. All alkanes of formula C_4H_{10} are called butanes, all those of formula C_5H_{12} are called pentanes, and all those of formula C_6H_{14} are called hexanes. For alkanes beyond propane, **iso** indicates that one end of an otherwise unbranched chain terminates in a $(CH_3)_2CH-$ group. Following are examples of common names:

$$CH_3CH_2CH_2CH_3 \qquad CH_3\overset{\overset{\displaystyle CH_3}{|}}{C}HCH_3 \qquad CH_3CH_2CH_2CH_2CH_3 \qquad CH_3CH_2\overset{\overset{\displaystyle CH_3}{|}}{C}HCH_3$$

Butane　　　　Isobutane　　　　　Pentane　　　　　　Isopentane

This system of common names has no way of handling other branching patterns; for more complex alkanes, we must use the more flexible IUPAC system.

In this book, we concentrate on IUPAC names. From time to time we also use common names, especially when chemists and biochemists use them almost exclusively in their everyday discussions. When the text uses both IUPAC and common names, we always give the IUPAC name first, followed by the common name in parentheses. In this way, you should have no doubt about which name is which.

11.5 Cycloalkanes

A hydrocarbon that contains carbon atoms joined to form a ring is called a **cyclic hydrocarbon.** When all carbons of the ring are saturated, the hydrocarbon is called a **cycloalkane.** Cycloalkanes of ring sizes ranging from 3 to more than 30 carbon atoms are found in nature, and in principle there is no limit to ring size. Five-membered (cyclopentane) and six-membered (cyclohexane) rings are especially abundant in nature; for this reason, we concentrate on them in this book.

> **Cycloalkane** A saturated hydrocarbon that contains carbon atoms bonded to form a ring

Organic chemists rarely show all carbons and hydrogens when writing structural formulas for cycloalkanes. Rather, we use line-angle formulas to represent cycloalkane rings. We represent each ring by a regular polygon having the same number of sides as there are carbon atoms in the ring. For example, we represent cyclobutane by a square, cyclopentane by a pentagon, and cyclohexane by a hexagon (Figure 11.3).

To name a cycloalkane, prefix the name of the corresponding open-chain alkane with *cyclo-*, and name each substituent on the ring. If there is only one substituent on the ring, there is no need to give it a location number. If there are two substituents, number the ring beginning with the substituent of lower alphabetical order.

Figure 11.3 Examples of cycloalkanes.

EXAMPLE 11.5

Write the molecular formula and IUPAC name for each cycloalkane.

(a)

(b)

Solution

(a) The molecular formula of this compound is C_8H_{16}. Because there is only one substituent on the ring, there is no need to number the atoms of the ring. The IUPAC name of this cycloalkane is isopropylcyclopentane.

(b) The molecular formula is $C_{11}H_{22}$. To name this compound, first number the atoms of the cyclohexane ring beginning with *tert*-butyl, the substituent of lower alphabetical order (remember, alphabetical order here is determined by the b of butyl, and not by the t of *tert*-). The name of this cycloalkane is 1-*tert*-butyl-4-methylcyclohexane.

Problem 11.5

Write the molecular formula and IUPAC name for each cycloalkane.

(a)

(b)

(c)

11.6 Shapes of Alkanes and Cycloalkanes

In this section, we concentrate on ways to visualize molecules as three-dimensional objects and to visualize bond angles and relative distances between various atoms and functional groups within a molecule. We urge you to build molecular models of these compounds and to study and manipulate the models we have prepared for you on the CD-ROM supplied with this text. Organic molecules are three-dimensional objects, and it is essential that you become comfortable in dealing with them as such.

A Alkanes

Although the VSEPR model (Section 3.9) gives us a way to predict the geometry about each carbon atom in an alkane, it provides no information about the three-dimensional shape of an entire molecule. The fact is that there is

At this point, you should review Section 3.9 on use of the valence-shell electron-pair (VSEPR) model to predict bond angles and shapes of molecules.

11.6 SHAPES OF ALKANES AND CYCLOALKANES | 297

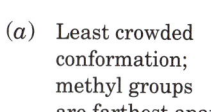

(a) Least crowded conformation; methyl groups are farthest apart.

rotate by 120°

(b) Intermediate crowding; methyl groups are closer to each other.

rotate by 60°

(c) Most crowded conformation; methyl groups are closest to each other.

Figure 11.4 Three conformations of a butane molecule.

free rotation about each carbon–carbon bond in an alkane and, as a result, even a molecule as simple as ethane has an infinite number of possible three-dimensional shapes, or **conformations.**

Figure 11.4 shows three conformations for a butane molecule. Conformation (a) is the most stable because the methyl groups at the ends of the four-carbon chain are farthest apart. Conformation (b) is formed by a rotation of 120° about the single bond joining carbons 2 and 3. In this conformation, there is some crowding of groups because the two methyl groups are closer together. Rotation about the C_2—C_3 single bond by another 60° gives conformation (c), which is the most crowded because the two methyl groups face each other.

Figure 11.4 shows only three of the possible conformations for a butane molecule. In fact, there are an infinite number of possible conformations that differ only in the angles of rotation about the various C—C bonds within the molecule. Within an actual sample of butane, the conformation of each molecule constantly changes as a result of collisions with other butane molecules and with the walls of the container. Even so, at any given time, a majority of butane molecules are in the most stable, fully extended conformation. There are the fewest butane molecules in the most crowded conformation.

To summarize, for any alkane (except, of course, for methane), there are an infinite number of conformations. The majority of molecules in any sample will be in the least crowded conformation; the fewest will be in the most crowded conformation.

Conformation Any three-dimensional arrangement of atoms in a molecule that results from rotation about a single bond

B Cycloalkanes

We limit our discussion to the conformations of cyclopentanes and cyclohexanes, because they are the most common carbon rings found in nature.

Cyclopentane

The most stable conformation of cyclopentane is the **envelope conformation** shown in Figure 11.5. In it, four carbon atoms are in a plane, and the fifth is bent out of the plane, like an envelope with its flap bent upward. All bond angles in cyclopentane are approximately 109.5°.

Cyclohexane

The most stable conformation of cyclohexane is the **chair conformation** (Figure 11.6), in which all bond angles are approximately 109.5°.

Figure 11.5 Cyclopentane. The most stable conformation is an envelope conformation.

Chair conformation The most stable conformation of a cyclohexane ring; all bond angles are approximately 109.5°

Figure 11.6 Cyclohexane. The most stable conformation is a chair conformation.

(a) Skeletal model

(b) Ball-and-stick model viewed from the side

(c) Ball-and-stick model viewed from above

Axial position A position on a chair conformation of a cyclohexane ring that extends from the ring parallel to the imaginary axis of the ring

Equatorial position A position on a chair conformation of a cyclohexane ring that extends from the ring roughly perpendicular to the imaginary axis of the ring

In a chair conformation, the twelve C—H bonds are arranged in two different orientations. Six of them are **axial bonds,** and the other six are **equatorial bonds.** One way to visualize the difference between these two types of bonds is to imagine an axis running through the center of the chair, (Figure 11.7). Axial bonds are oriented parallel to this axis. Three of the axial bonds point up; the other three point down. Notice also that axial bonds alternate, first up and then down, as you move from one carbon to the next.

Equatorial bonds are oriented approximately perpendicular to our imaginary axis and also alternate first slightly up and then slightly down as you move from one carbon to the next. Notice also that if the axial bond on a carbon points upward, the equatorial bond on that carbon points slightly downward. Conversely, if the axial bond on a particular carbon points downward, the equatorial bond on that carbon points slightly upward.

Finally, notice that each equatorial bond is oriented parallel to two ring bonds on opposite sides of the ring. A different pair of parallel C—H bonds is shown in each of the following structural formulas, along with the two ring bonds to which each pair is parallel:

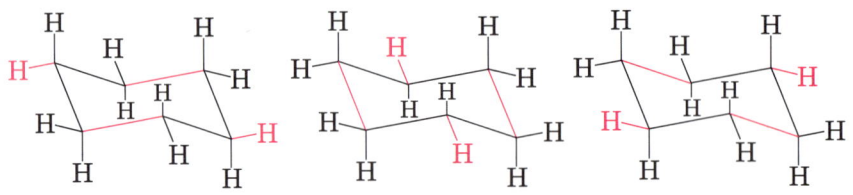

Figure 11.7 Chair conformation of cyclohexane showing equatorial and axial C—H bonds.

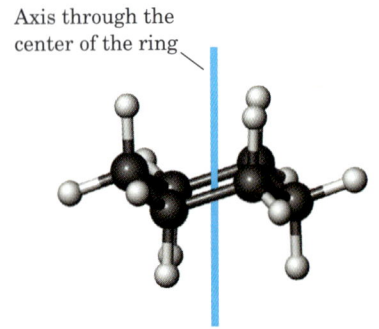

(a) Ball-and-stick model showing all 12 hydrogens

(b) The six equatorial C—H bonds shown in red

(c) The six axial C—H bonds shown in blue

CHEMICAL CONNECTIONS 11A

The Poisonous Puffer Fish

Nature is by no means limited to carbon in six-membered rings. Tetrodotoxin, one of the most potent toxins known, is composed of a set of interconnected six-membered rings, each in a chair conformation. All but one of these rings contain atoms other than carbon.

Tetrodotoxin is produced in the liver and ovaries of many species of *Tetraodontidae,* especially the puffer fish, so called because it inflates itself to an almost spherical spiny ball when alarmed. It is evidently a species highly preoccupied with defense, but the Japanese are not put off by its prickly appearance. They regard the puffer, called *fugu* in Japanese, as a delicacy. To serve it in a public restaurant, a chef must be registered as sufficiently skilled in removing the toxic organs so as to make the flesh safe to eat.

Tetrodotoxin blocks the sodium ion channels, which are essential for neurotransmission (Section 23.3). This blockage prevents communication between neurons and muscle cells and results in weakness, paralysis, and eventual death.

■ A puffer fish with its body inflated.

Tetrodotoxin

EXAMPLE 11.6

Following is a chair conformation of methylcyclohexane showing a methyl group and one hydrogen. Indicate by a label whether each is equatorial or axial.

Solution
The methyl group is axial, and the hydrogen is equatorial.

Problem 11.6

Following is a chair conformation of cyclohexane with carbon atoms numbered 1 through 6. Draw methyl groups that are equatorial on carbons 1, 2, and 4.

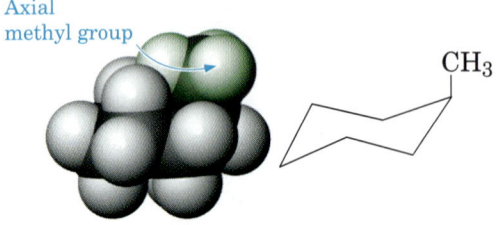

(a) Equatorial methylcyclohexane   (b) Axial methylcyclohexane

Figure 11.8 Methylcyclohexane. The three hydrogens of the methyl group are shown in green to make them stand out more clearly.

Suppose that —CH_3 or another group on a cyclohexane ring may occupy either an equatorial or axial position. Chemists have discovered that a six-membered ring is more stable when the maximum number of substituent groups are equatorial. Perhaps the simplest way to see that this is so is to examine molecular models. Figure 11.8(a) shows a space-filling model of methylcyclohexane with the methyl group equatorial. In this position, the methyl group is as far away as possible from other atoms of the ring. When methyl is axial [Figure 11.8(b)], it quite literally bangs into two hydrogen atoms on the top side of the ring. Thus the more stable conformation of a substituted cyclohexane ring has the substituent group(s) equatorial.

Screen 11.6 Tutorial

11.7 Cis-Trans Isomerism in Cycloalkanes

Cis-trans isomers have also been called geometric isomers.

Cis-trans isomers Isomers that have the same order of attachment of their atoms but a different arrangement of their atoms in space due to the presence of either a ring or a carbon–carbon double bond.

Cycloalkanes with substituents on two or more carbons of the ring show a type of isomerism called **cis-trans isomerism.** Cycloalkane cis-trans isomers have (1) the same molecular formula and (2) the same order of attachment of their atoms, but (3) a different arrangement of their atoms in space because of restricted rotation around the carbon–carbon single bonds of the ring. We study cis-trans isomerism in cycloalkanes in this chapter and in alkenes in Chapter 12.

We can illustrate cis-trans isomerism in cycloalkanes by using 1,2-dimethylcyclopentane as an example. In the following structural formulas, the cyclopentane ring is drawn as a planar pentagon viewed edge-on. (In determining the number of cis-trans isomers in a substituted cycloalkane, it is adequate to draw the cycloalkane ring as a planar polygon.) Carbon–carbon bonds of the ring that project forward are shown as heavy lines. When viewed from this perspective, substituents attached to the cyclopentane ring project above and below the plane of the ring. In one isomer of 1,2-dimethylcyclopentane, the methyl groups are on the same side of the ring (either both above or both below the plane of the ring); in the other isomer, they are on opposite sides of the ring (one above and one below the plane of the ring). The prefix **cis** (Latin, on the same side) indicates that the substituents are on the same side of the ring; the prefix **trans** (Latin, across) indicates that they are on opposite sides of the ring.

Planar representations of five- and six-membered rings are not spatially accurate because these rings normally exist as envelope and chair conformations. Planar representations, however, are adequate for showing cis-trans isomerism.

cis-1,2-Dimethyl-cyclopentane

trans-1,2-Dimethyl-cyclopentane

Occasionally hydrogen atoms are written before the carbon, H$_3$C—, to avoid crowding or to emphasize the C—C bond, as in H$_3$C—CH$_3$.

Alternatively, we can view the cyclopentane ring from above, with the ring in the plane of the paper. Substituents on the ring then either project toward you (that is, they project up above the page) and are shown by solid wedges, or they project away from you (project down below the page) and are shown by broken wedges. In the following structural formulas, we show only the two methyl groups; we do not show hydrogen atoms of the ring.

cis-1,2-Dimethyl-cyclopentane

trans-1,2-Dimethyl-cyclopentane

Similarly, we can draw a cyclohexane ring as a planar hexagon and view it through the plane of the ring. Alternatively, we can view it from above with substituent groups pointing toward you shown by solid wedges, and those pointing away from you shown by broken wedges.

trans-1,4-Dimethylcyclohexane

cis-1,4-Dimethylcyclohexane

Because cis-trans isomers differ in the orientation of their atoms in space, they are **stereoisomers**. Cis-trans isomerism is one type of stereoisomerism. We will study another type of stereoisomerism, called enantiomerism, in Chapter 15.

Stereoisomers Isomers that have the same connectivity (the same order of attachment of their atoms) but a different orientation of their atoms in space

EXAMPLE 11.7

Which of the following cycloalkanes show cis-trans isomerism? For each that does, draw both isomers.
(a) Methylcyclopentane
(b) 1,1-Dimethylcyclopentane
(c) 1,3-Dimethylcyclobutane

Solution

(a) Methylcyclopentane does not show cis-trans isomerism; it has only one substituent on the ring.
(b) 1,1-Dimethylcyclobutane does not show cis-trans isomerism because only one arrangement is possible for the two methyl groups. Because both groups are bonded to the same carbon, they must be trans to each other—one above the ring, the other below it.
(c) 1,3-Dimethylcyclobutane shows cis-trans isomerism. The two methyl groups may be cis or they may be trans.

cis-1,3-Dimethylcyclobutane   trans-1,3-Dimethylcyclobutane

Problem 11.7

Which of the following cycloalkanes show cis-trans isomerism? For each that does, draw both isomers.
(a) 1,3-Dimethylcyclopentane
(b) Ethylcyclopentane
(c) 1,3-Dimethylcyclohexane

11.8 Physical Properties

The most important property of alkanes and cycloalkanes is their almost complete lack of polarity. As we saw in Section 3.4B, the electronegativity difference between carbon and hydrogen is 2.5 − 2.1 = 0.4 on the Pauling scale. Given this small difference, we classify a C—H bond as nonpolar covalent. Therefore, alkanes are nonpolar compounds and the only interactions between their molecules are the very weak London dispersion forces (Section 5.7A).

A Melting and Boiling Points

The boiling points of alkanes are lower than those of almost any other type of compound of the same molecular weight. In general, both boiling and melting points of alkanes increase with increasing molecular weight (Table 11.4).

Alkanes containing 1 to 4 carbons are gases at room temperature. Alkanes containing 5 to 17 carbons are colorless liquids. High-molecular-weight alkanes (those containing 18 or more carbons) are white, waxy solids. Several plant waxes are high-molecular-weight alkanes. The wax found in apple skins, for example, is an unbranched alkane with molecular formula $C_{27}H_{56}$. Paraffin wax, a mixture of high-molecular-weight alkanes, is used for wax candles, in lubricants, and to seal home-canned jams, jellies, and other preserves. Petrolatum, so named because it is derived from petroleum refining, is a liquid mixture of high-molecular-weight alkanes. It is sold as mineral oil and Vaseline, used as an ointment base in pharmaceuticals and cosmetics, and as a lubricant and rust preventive.

Alkanes that are constitutional isomers are different compounds and have different physical and chemical properties. Table 11.5 lists the boiling

■ Paraffin wax and mineral oil are mixtures of alkanes.

Screen 11.7
Simulation

Table 11.4 Physical Properties of Some Unbranched Alkanes

Name	Condensed Structural Formula	Molecular Weight (amu)	Melting Point (°C)	Boiling Point (°C)	Density of Liquid (g/mL) at 0°C*
methane	CH_4	16.0	−182	−164	(a gas)
ethane	CH_3CH_3	30.1	−183	−88	(a gas)
propane	$CH_3CH_2CH_3$	44.1	−190	−42	(a gas)
butane	$CH_3(CH_2)_2CH_3$	58.1	−138	0	(a gas)
pentane	$CH_3(CH_2)_3CH_3$	72.2	−130	36	0.626
hexane	$CH_3(CH_2)_4CH_3$	86.2	−95	69	0.659
heptane	$CH_3(CH_2)_5CH_3$	100.2	−90	98	0.684
octane	$CH_3(CH_2)_6CH_3$	114.2	−57	126	0.703
nonane	$CH_3(CH_2)_7CH_3$	128.3	−51	151	0.718
decane	$CH_3(CH_2)_8CH_3$	142.3	−30	174	0.730

*For comparison, the density of H_2O is 1 g/mL at 4°C.

points of the five constitutional isomers with molecular formula C_6H_{14}. The boiling point of each branched-chain isomer is lower than that of hexane itself; the more branching there is, the lower the boiling point. These differences in boiling points are related to molecular shape in the following way. As branching increases, the alkane molecule becomes more compact, and its surface area decreases. As we learned in Section 5.7A, as surface area decreases, London dispersion forces act over a smaller surface area. Hence the attraction between molecules decreases, and boiling point decreases. Thus, for any group of alkane constitutional isomers, the least branched isomer generally has the highest boiling point, and the most branched isomer generally has the lowest boiling point.

B Solubility: A Case of "Like Dissolves Like"

Because alkanes are nonpolar compounds, they are not soluble in water, which dissolves only ionic and polar compounds. Recall from Section 5.7C that water is a polar substance, and that its molecules associate with one another through hydrogen bonding. Alkanes do not dissolve in water because they cannot form hydrogen bonds with water. Alkanes, however, are

Table 11.5 Boiling Points of the Five Isomeric Alkanes with Molecular Formula C_6H_{14}

Name	bp (°C)
hexane	68.7
3-methylpentane	63.3
2-methylpentane	60.3
2,3-dimethylbutane	58.0
2,2-dimethylbutane	49.7

Hexane

2,2-Dimethylbutane

soluble in each other, an example of "like dissolves like" (Section 6.4). Alkanes are also soluble in other nonpolar organic compounds, such as toluene and diethyl ether.

C Density

The average density of the liquid alkanes listed in Table 11.4 is about 0.7 g/mL; that of higher-molecular-weight alkanes is about 0.8 g/mL. All liquid and solid alkanes are less dense than water (1.0 g/mL) and, because they are insoluble in water, they float on water.

EXAMPLE 11.8

Arrange the alkanes in each set in order of increasing boiling point.
(a) Butane, decane, and hexane
(b) 2-Methylheptane, octane, and 2,2,4-trimethylpentane

Solution
(a) All three compounds are unbranched alkanes. As the number of carbon atoms in the chain increases, London dispersion forces between molecules increase, and boiling points increase (Section 5.9A). Decane has the highest boiling point, and butane has the lowest boiling point.

Butane
bp −0.5°C

Hexane
bp 69°C

Decane
bp 174°C

(b) These three alkanes are constitutional isomers with the molecular formula C_8H_{18}. Their relative boiling points depend on the degree of branching. 2,2,4-Trimethylpentane is the most highly branched isomer and, therefore, has the smallest surface area and the lowest boiling point. Octane, the unbranched isomer, has the largest surface area and the highest boiling point.

2,2,4-Trimethylpentane
(bp 99°C)

2-Methylheptane
(bp 118°C)

Octane
(bp 126°C)

Problem 11.8
Arrange the alkanes in each set in order of increasing boiling point.
(a) 2-Methylbutane, pentane, and 2,2-dimethylpropane
(b) 3,3-Dimethylheptane, nonane, and 2,2,4-trimethylhexane

11.9 Reactions of Alkanes

The most important chemical property of alkanes and cycloalkanes is their inertness. They are quite unreactive toward any of the normal ionic reaction conditions we studied in Chapters 4, 7, and 8. Under certain conditions, however, alkanes do react with oxygen, O_2. By far their most important reaction with oxygen is oxidation (combustion) to form carbon dioxide and water. They also react with bromine and chlorine to form halogenated hydrocarbons.

A Reaction with Oxygen: Combustion

Oxidation of hydrocarbons, including alkanes and cycloalkanes, is the basis for their use as energy sources for heat [natural gas, liquefied petroleum gas (LPG), and fuel oil] and power (gasoline, diesel fuel, and aviation fuel). Following are balanced equations for the complete combustion of methane, the major component of natural gas, and propane, the major component of LPG or bottled gas. The heat liberated when an alkane is oxidized to carbon dioxide and water is called its heat of combustion (Section 4.8).

$$\underset{\text{Methane}}{CH_4} + 2O_2 \longrightarrow CO_2 + 2H_2O + 212 \text{ kcal/mol}$$

$$\underset{\text{Propane}}{CH_3CH_2CH_3} + 5O_2 \longrightarrow 3CO_2 + 4H_2O + 530 \text{ kcal/mol}$$

B Reaction with Halogens: Halogenation

If we mix methane with chlorine or bromine in the dark at room temperature, nothing happens. If, however, we heat the mixture to 100°C or higher or expose it to light, a reaction begins at once. The products of the reaction between methane and chlorine are chloromethane and hydrogen chloride. What occurs is a substitution reaction—in this case, the substitution of a chlorine for a hydrogen in methane.

$$\underset{\text{Methane}}{CH_4} + Cl_2 \xrightarrow{\text{heat or light}} \underset{\substack{\text{Chloromethane} \\ \text{(Methyl chloride)}}}{CH_3Cl} + HCl$$

If chloromethane is allowed to react with more chlorine, further chlorination produces a mixture of dichloromethane, trichloromethane, and tetrachloromethane.

$$\underset{\substack{\text{Chloromethane} \\ \text{(Methyl chloride)}}}{CH_3Cl} + Cl_2 \xrightarrow{\text{heat}} \underset{\substack{\text{Dichloromethane} \\ \text{(Methylene chloride)}}}{CH_2Cl_2} + HCl$$

$$\underset{\substack{\text{Dichloromethane} \\ \text{(Methylene chloride)}}}{CH_2Cl_2} \xrightarrow[\text{heat}]{Cl_2} \underset{\substack{\text{Trichloromethane} \\ \text{(Chloroform)}}}{CHCl_3} \xrightarrow[\text{heat}]{Cl_2} \underset{\substack{\text{Tetrachloromethane} \\ \text{(Carbon tetrachloride)}}}{CCl_4}$$

In the last equation, the reagent Cl_2 is placed over the reaction arrow and the equivalent amount of HCl formed is not shown. Placing reagents over reaction arrows and omitting by-products is commonly done to save space.

We derive IUPAC names of haloalkanes by naming the halogen atom as a substituent (fluoro-, chloro-, bromo-, iodo-) and alphabetizing it along with other substituents. Common names consist of the common name of the alkyl group followed by the name of the halogen (chloride, bromide, and so forth) as a separate word. Dichloromethane (methylene chloride) is a widely used solvent for organic compounds.

EXAMPLE 11.9

Write a balanced equation for the reaction of ethane with chlorine to form chloroethane, C_2H_5Cl.

Solution
The reaction of ethane with chlorine results in the substitution of one of the hydrogen atoms of ethane by a chlorine atom.

$$CH_3CH_3 + Cl_2 \xrightarrow{\text{heat or light}} CH_3CH_2Cl + HCl$$

Ethane   Chloroethane (Ethyl chloride)

Problem 11.9
The reaction of propane with chlorine gives two products with the molecular formula C_3H_7Cl. Draw structural formulas for these two compounds, and give each an IUPAC and common name.

11.10 Important Haloalkanes

One of the major uses of haloalkanes is as intermediates in the synthesis of other organic compounds. Just as we can replace a hydrogen atom of an alkane by a halogen, we can replace the halogen atom, in turn, by a number of other functional groups. In this way, we can construct more complex molecules. In contrast, alkanes that contain several halogens are often quite unreactive, a fact that has proved especially useful in the design of several classes of consumer products.

A The Chlorofluorocarbons

Of all the haloalkanes, the **chlorofluorocarbons (CFCs)** manufactured under the trade name Freons are the most widely known. CFCs are nontoxic, nonflammable, odorless, and noncorrosive. Originally, they seemed to be ideal replacements for the hazardous compounds such as ammonia and sulfur dioxide formerly used as heat-transfer agents in refrigeration systems. Among the CFCs most widely used for this purpose were trichlorofluoromethane (CCl_3F, Freon-11) and dichlorodifluoromethane (CCl_2F_2, Freon-12). The CFCs also found wide use as industrial cleaning solvents to prepare surfaces for coatings, to remove cutting oils and waxes from millings, and to remove protective coatings. In addition, they were employed as propellants for aerosol sprays.

B Solvents

Several low-molecular-weight haloalkanes are excellent solvents in which to carry out organic reactions and to use as cleaners and degreasers. Carbon tetrachloride ("carbon tet") was the first of these compounds to find wide ap-

CHEMICAL CONNECTIONS 11B

The Environmental Impact of Freons

Concern about the environmental impact of CFCs arose in the 1970s when researchers found that more than 4.5×10^5 kg/yr of these compounds were being emitted into the atmosphere. In 1974, Sherwood Rowland and Mario Molina, both of the United States, announced their theory, which has since been amply confirmed, that these compounds destroy the stratospheric ozone layer. When released into the air, CFCs escape to the lower atmosphere. Because of their inertness, however, they do not decompose there. Slowly, they find their way to the stratosphere, where they absorb ultraviolet radiation from the sun and then decompose. As they do so, they set up a chemical reaction that leads to the destruction of the stratospheric ozone layer, which shields Earth against short-wavelength ultraviolet radiation from the sun. An increase in short-wavelength ultraviolet radiation reaching Earth is believed to promote the destruction of certain crops and agricultural species, and even to increase the incidence of skin cancer in light-skinned individuals.

This concern prompted two conventions, one in Vienna in 1985 and one in Montreal in 1987, held by the United Nations Environmental Program. The 1987 meeting produced the Montreal Protocol, which set limits on the production and use of ozone-depleting CFCs and urged the complete phase-out of their production by the year 1996. This phase-out resulted in enormous costs for manufacturers and is not yet complete in developing countries.

Rowland, Molina, and Paul Crutzen (a Dutch chemist at the Max Planck Institute for Chemistry in Germany) were awarded the 1995 Nobel Prize for chemistry. As noted in the award citation by the Royal Swedish Academy of Sciences, "By explaining the chemical mechanisms that affect the thickness of the ozone layer, these three researchers have contributed to our salvation from a global environmental problem that could have catastrophic consequences."

The chemical industry responded to the crisis by developing replacement refrigerants that have much lower ozone-depleting potential. The most prominent of these replacements are the hydrofluorocarbons (HFCs) and hydrochlorofluorocarbons (HCFCs). These compounds are much more chemically reactive in the atmosphere than the Freons and are destroyed before they reach the stratosphere. However, they cannot be used in air conditioners in 1994- and earlier-model cars.

HFC-134a   HCFC-141b

---

plication, but its use for this purpose has since been discontinued because it is now known that carbon tet is both toxic and carcinogenic. Today, the most widely used haloalkane solvent is dichloromethane, CH_2Cl_2.

11.11 Sources of Alkanes *go to end of chapter*

The two major sources of alkanes are natural gas and petroleum. **Natural gas** consists of approximately 90 to 95% methane, 5 to 10% ethane, and a mixture of other relatively low-boiling alkanes, chiefly propane, butane, and 2-methylpropane.

Petroleum is a thick, viscous, liquid mixture of thousands of compounds, most of them hydrocarbons, formed from the decomposition of marine plants and animals. Petroleum and petroleum-derived products fuel automobiles, aircraft, and trains. They provide most of the greases and lubricants required for the machinery of our highly industrialized society. Furthermore, petroleum, along with natural gas, provides nearly 90 percent of the organic raw materials for the synthesis and manufacture of synthetic fibers, plastics, detergents, drugs, dyes, and a multitude of other products.

The fundamental separation process in refining petroleum is fractional distillation (Figure 11.9). Practically all crude petroleum that enters a refinery goes to distillation units, where it is heated to temperatures as high as 370 to 425°C and separated into fractions. Each fraction contains a mixture of hydrocarbons that boils within a particular range.

■ A petroleum fractional distillation tower.

Figure 11.9 Fractional distillation of petroleum. The lighter, more volatile fractions are removed from higher up the column; the heavier, less volatile fractions are removed from lower down.

 CHEMICAL CONNECTIONS 11C

Octane Rating: What Those Numbers at the Pump Mean

Gasoline is a complex mixture of C_6 to C_{12} hydrocarbons. Its quality as a fuel for internal combustion engines depends on how much it makes an engine "knock." Engine knocking occurs when some of the air–fuel mixture explodes prematurely and independently of ignition by the spark plug.

Two compounds were selected for rating gasoline quality. 2,2,4-Trimethylpentane (isooctane) has very good antiknock properties (the fuel–air mixture burns smoothly in the combustion chamber) and was assigned an octane rating of 100. Heptane, in contrast, has poor antiknock properties and was assigned an octane rating of 0.

2,2,4-Trimethylpentane (octane rating 100)   Heptane (octane rating 0)

The **octane rating** of a particular gasoline is a number equal to the percentage of 2,2,4-trimethylpentane in a mixture of 2,2,4-

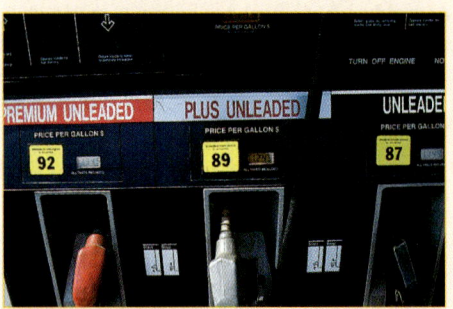

■ Typical octane ratings of commonly available gasolines.

trimethylpentane and heptane that has the same antiknock properties as the gasoline being rated. For example, the antiknock properties of 2-methylhexane are the same as those of a mixture of 42% 2,2,4-trimethylpentane and 58% heptane; therefore, the octane rating of 2-methylhexane is 42. Ethanol, which is added to gasoline to produce gasohol, has an octane rating of 105. Octane itself has an octane rating of −20.

SUMMARY

A **hydrocarbon** is a compound that contains only carbon and hydrogen (Section 11.1). A **saturated hydrocarbon** contains only single bonds. An **alkane** is a saturated hydrocarbon whose carbon atoms are arranged in an open chain (Section 11.2). **Constitutional isomers** have the same molecular formula but a different order of attachment of their atoms (Section 11.3).

Alkanes are named according to a set of rules developed by the **International Union of Pure and Applied Chemistry** (Section 11.4). The IUPAC name of an alkane consists of two parts: a prefix that tells the number of carbon atoms in the parent chain, and the ending **-ane**. Substituents derived from alkanes by removal of a hydrogen atom are called **alkyl groups** and are given the symbol **R—**.

An alkane that contains carbon atoms bonded to form a ring is called a **cycloalkane** (Section 11.5). To name a cycloalkane, we prefix the name of the open-chain alkane with **cyclo-**.

A **conformation** is any three-dimensional arrangement of the atoms of a molecule that results from rotation about a single bond (Section 11.6). The lowest-energy conformation of cyclopentane is an **envelope conformation**. The lowest-energy conformation of cyclohexane is a **chair conformation**. In a chair conformation, six bonds are **axial** and six bonds are **equatorial**. A substituent on a six-membered ring is more stable when it is equatorial than when it is axial.

Cis-trans isomers of cycloalkanes have (1) the same molecular formula and (2) the same order of attachment of their atoms, (3) but a different orientation of their atoms in space because of the restricted rotation around the C—C bonds of the ring (Section 11.7). With cis-trans isomers of cycloalkanes, **cis** means that substituents are on the same side of the ring; **trans** means that they are on opposite sides of the ring.

Alkanes are nonpolar compounds, and the only forces of attraction between their molecules are London dispersion forces (Section 11.8). At room temperature, low-molecular-weight alkanes are gases, higher-molecular-weight alkanes are liquids, and very-high-molecular-weight alkanes are solids. For any group of alkane constitutional isomers, the least branched isomer generally has the highest boiling point, and the most branched isomer generally has the lowest boiling point.

Alkanes are insoluble in water but soluble in each other and in other nonpolar organic solvents such as toluene (Section 11.8B). All liquid and solid alkanes are less dense than water (Section 11.8C).

Natural gas consists of 90 to 95 percent methane with lesser amounts of ethane and other lower-molecular-weight hydrocarbons (Section 11.11). **Petroleum** is a liquid mixture of thousands of different hydrocarbons.

KEY REACTIONS

1. **Oxidation of Alkanes (Section 11.9A)** Oxidation of alkanes to carbon dioxide and water is the basis for their use as sources of heat and power.

   $CH_3CH_2CH_3 + 5O_2 \longrightarrow 3CO_2 + 4H_2O + 530$ kcal/mol
   Propane

2. **Halogenation of Alkanes (Section 11.9B)** Reaction of an alkane with chlorine or bromine results in the substitution of a halogen atom for a hydrogen.

   $CH_3CH_3 + Cl_2 \xrightarrow{\text{heat or light}} CH_3CH_2Cl + HCl$
   Ethane   Chloroethane
   (Ethyl chloride)

PROBLEMS

Numbers that appear in color indicate difficult problems. ▶ designates problems requiring application of principles.

Alkanes

11.10 Define:
    (a) Hydrocarbon   (b) Alkane
    (c) Saturated hydrocarbon

11.11 Why is it not accurate to describe unbranched alkanes as "straight chain" alkanes?

11.12 What is meant by the term "line-angle formula" as applied to alkanes and cycloalkanes?

11.13 For each condensed structural formula, write a line-angle formula.

(a) CH₃CH₂CHCHCH₂CHCH₃ with substituents CH₂CH₃, CH₃, and CH(CH₃)₂

(b) CH₃CCH₃ with two CH₃ substituents (top and bottom)

(c) (CH₃)₂CHCH(CH₃)₂

(d) CH₃CH₂CCH₂CH₃ with two CH₂CH₃ substituents

(e) (CH₃)₃CH

(f) CH₃(CH₂)₃CH(CH₃)₂

11.14 Write the molecular formula for each alkane.

(a), (b), (c) [line-angle structures]

Constitutional Isomerism

11.15 Which statements are true about constitutional isomers?
(a) They have the same molecular formula.
(b) They have the same molecular weight.
(c) They have the same order of attachment of atoms.
(d) They have the same physical properties.

11.16 Each member of the following set of compounds is an alcohol; that is, each contains an —OH (hydroxyl group, Section 10.4A). Which structural formulas represent the same compound, and which represent constitutional isomers?

(a), (b), (c), (d), (e), (f), (g), (h) [structures]

11.17 Each member of the following set of compounds is an amine; that is, each contains a nitrogen bonded to one, two, or three carbon groups (Section 10.4B). Which structural formulas represent the same compound, and which represent constitutional isomers?

11.18 Each member of the following set of compounds is either an aldehyde or ketone (Section 10.4C). Which structural formulas represent the same compound, and which represent constitutional isomers?

11.19 For which parts do the pairs of structural formulas represent constitutional isomers?

(a) and (isobutane structure)

(b) (2-methyl-1,3-butadiene) and (3-methyl-1-butyne)

(c) (tetrahydrofuran) and (methyl ethyl ketone)

(d) (tetrahydrofuran) and (ethyl vinyl ether / ethoxyethene)

(e) (3-methylpyrrolidine) and (cyclopentylamine)

(f) (tetrahydrofuran) and (3-buten-1-ol)

11.20 Draw line-angle formulas for the nine constitutional isomers with molecular formula C_7H_{16}.

Nomenclature of Alkanes and Cycloalkanes

11.21 Name these alkyl groups:

(a) CH_3CH_2-

(b) $CH_3\overset{\underset{\mid}{CH_3}}{CH}-$

(c) $CH_3\overset{\underset{\mid}{CH_3}}{CH}CH_2-$

(d) $CH_3\overset{\underset{\mid}{CH_3}}{\underset{\underset{\mid}{CH_3}}{C}}-$

11.22 Write the IUPAC names for isobutane and isopentane.

11.23 Write the IUPAC names for these alkanes and cycloalkanes

(a) $CH_3CHCH_2CH_2CH_3$
     $\quad\quad|$
     $\quad\quad CH_3$

(b) $CH_3CHCH_2CH_2CHCH_3$
     $\quad\quad|\quad\quad\quad\quad|$
     $\quad\quad CH_3\quad\quad CH_3$

(c) $CH_3(CH_2)_4CHCH_2CH_3$
     $\quad\quad\quad\quad\quad|$
     $\quad\quad\quad\quad\quad CH_2CH_3$

(d) (methylisopropylcyclohexane structure)

(e) (isobutylcyclopentane structure)

(f) (ethyl, methyl, methylcyclohexane structure)

11.24 Write line-angle formulas for these alkanes and cycloalkanes.

(a) 2,2,4-Trimethylhexane
(b) 2,2-Dimethylpropane
(c) 3-Ethyl-2,4,5-trimethyloctane
(d) 5-Butyl-2,2-dimethylnonane
(e) 4-Isopropyloctane
(f) 3,3-Dimethylpentane
(g) *trans*-1,3-Dimethylcyclopentane
(h) *cis*-1,2-Diethylcyclobutane

Conformations of Alkanes and Cycloalkanes

11.25 Define conformation.

11.26 The condensed structural formula of butane is $CH_3CH_2CH_2CH_3$. Explain why this formula does not show the geometry of the real molecule.

11.27 Draw a conformation of ethane in which hydrogen atoms on adjacent carbons are as far apart as possible. Also draw a conformation in which they are as close together as possible. In a sample of ethane molecules at room temperature, which conformation is the more likely?

Cis-Trans Isomerism in Cycloalkanes

11.28 What structural feature of cycloalkanes makes cis-trans isomerism in them possible?

11.29 Is cis-trans isomerism possible in alkanes?

11.30 Name and draw structural formulas for the cis and trans isomers of 1,2-dimethylcyclopropane.

11.31 Name and draw structural formulas for the six cycloalkanes with molecular formula C_5H_{10}. Include cis-trans isomers as well as constitutional isomers.

11.32 Why is equatorial methylcyclohexane more stable than axial methylcyclohexane?

11.33 Following is a structural formula for 2-isopropyl-5-methylcyclohexanol.

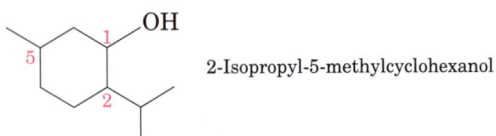

2-Isopropyl-5-methylcyclohexanol

Using a planar hexagon representation for the cyclohexane ring, draw a structural formula for the cis-trans isomer with isopropyl trans to —OH and methyl cis to —OH. If you answered this part correctly, you have drawn the isomer found in nature and given the name menthol.

11.34 On the left is a representation of the glucose molecule. Convert this representation to the alternative representation using the ring on the right. (We discuss the structure and chemistry of glucose in Chapter 19.)

11.35 On the left is a representation for 2-deoxy-D-ribose. This molecule is the "D" of DNA. Convert this representation to the alternative representation using the ring on the right. (We discuss the structure and chemistry of this compound in more detail in Chapter 19.)

2-Deoxy-D-ribose

Physical Properties of Alkanes and Cycloalkanes

11.36 In Problem 11.20, you drew structural formulas for the nine constitutional isomers with molecular formula C_7H_{16}. Predict which isomer has the lowest boiling point and which has the highest boiling point.

11.37 What unbranched alkane (Table 11.4) has about the same boiling point as water? Calculate the molecular weight of this alkane, and compare it with the molecular weight of water.

11.38 What generalizations can you make about the densities of alkanes relative to the density of water?

11.39 What generalization can you make about the solubility of alkanes in water?

11.40 Suppose that you have samples of hexane and octane. Could you tell the difference by looking at them? What color would they be? How could you tell which is which?

11.41 As you can see from Table 11.4, each CH_2 group added to the carbon chain of an alkane increases its boiling point. This increase is greater going from CH_4 to C_2H_6 and from C_2H_6 to C_3H_8 than it is going from C_8H_{18} to C_9H_{20} or from C_9H_{20} to $C_{10}H_{22}$. What do you think is the reason for this difference?

11.42 How are the boiling points of hydrocarbons during petroleum refining related to their molecular weight?

Reactions of Alkanes

11.43 Write balanced equations for the combustion of the following hydrocarbons. Assume that each is converted completely to carbon dioxide and water.
(a) Hexane
(b) Cyclohexane
(c) 2-Methylpentane

11.44 The heat of combustion of methane, a component of natural gas, is 212 kcal/mol. That of propane, a component of LP gas, is 530 kcal/mol. On a gram-for-gram basis, which hydrocarbon is the better source of heat energy?

11.45 Draw structural formulas for these haloalkanes.
(a) Bromomethane
(b) Chlorocyclohexane
(c) 1,2-Dibromoethane
(d) 2-Chloro-2-methylpropane
(e) Dichlorodifluoromethane (Freon-12)

11.46 The reaction of chlorine with pentane gives a mixture of three chloroalkanes with molecular formula $C_5H_{11}Cl$. Write a line-angle formula and IUPAC name for each.

Chemical Connections

11.47 (Chemical Connections 11A) How many rings in tetrodotoxin contain only carbon atoms? How many contain nitrogen atoms? How many contain two oxygen atoms?

11.48 (Chemical Connections 11B) What are Freons? Why were they considered ideal compounds to use as heat-transfer agents in refrigeration systems? Give structural formulas of two Freons used for this purpose.

11.49 (Chemical Connections 11B) In what way do Freons negatively affect the environment?

11.50 (Chemical Connections 11B) What are HFCs and HCFCs? How does their use in refrigeration systems avoid the environmental problems associated with the use of Freons?

11.51 (Chemical Connections 11C) What is an "octane rating"? What two reference hydrocarbons are used for setting the scale of octane ratings?

11.52 (Chemical Connections 11C) Octane has an octane rating of –20. Will it produce more or less engine knocking than heptane does?

11.53 (Chemical Connections 11C) When ethanol is added to gasoline to produce "gasohol," it promotes more complete combustion of the gasoline and is an octane booster. Compare the heats of combustion of 2,2,4-trimethylpentane (1304 kcal/mol) and ethanol (327 kcal/mol). Which has the higher heat of combustion in kcal/mol? In kcal/g?

Additional Problems

11.54 Tell whether the compounds in each set are constitutional isomers.

(a) CH_3CH_2OH and CH_3OCH_3

(b) $CH_3\overset{O}{\overset{\|}{C}}CH_3$ and $CH_3CH_2\overset{O}{\overset{\|}{C}}H$

(c) $CH_3\overset{O}{\overset{\|}{C}}OCH_3$ and $CH_3CH_2\overset{O}{\overset{\|}{C}}OH$

(d) $CH_3\overset{OH}{\underset{|}{C}}HCH_2CH_3$ and $CH_3\overset{O}{\overset{\|}{C}}CH_2CH_3$

(e) ⬠ and $CH_3CH_2CH_2CH_2CH_3$

(f) ⬠ and $CH_2{=}CHCH_2CH_2CH_3$

11.55 Explain why each of the following is an incorrect IUPAC name. Write the correct IUPAC name for the compound.

(a) 1,3-Dimethylbutane
(b) 4-Methylpentane
(c) 2,2-Diethylbutane
(d) 2-Ethyl-3-methylpentane
(e) 2-Propylpentane
(f) 2,2-Diethylheptane
(g) 2,2-Dimethylcyclopropane
(h) 1-Ethyl-5-methylcyclohexane

11.56 Which compounds can exist as cis-trans isomers? For each that does, draw both isomers using solid and dashed wedges to show the orientation in space of the —OH and —CH_3 groups.

(a) [cyclohexane with OH and CH_3]

(b) [cyclohexane with OH and CH_3 on same carbon]

(c) [cyclohexane with HO and CH_3]

11.57 Tetradecane, $C_{14}H_{30}$, is an unbranched alkane with a melting point of 5.9°C and a boiling point of 254°C. Is tetradecane a solid, liquid, or gas at room temperature?

11.58 Dodecane, $C_{12}H_{26}$, is an unbranched alkane. Predict the following:

(a) Will it dissolve in water?
(b) Will it dissolve in hexane?
(c) Will it burn when ignited?
(d) Is it a liquid, solid, or gas at room temperature and atmospheric pressure?
(e) Is it more or less dense than water?

11.59 As stated in Section 11.8, the wax found in apple skins is an unbranched alkane with molecular formula $C_{27}H_{56}$. Explain how the presence of this alkane prevents the loss of moisture from within an apple.

InfoTrac College Edition

For additional readings, go to InfoTrac College Edition, your online research library, at

http://infotrac.thomsonlearning.com

CHAPTER 12

- 12.1 Introduction
- 12.2 Structure
- 12.3 Nomenclature
- 12.4 Physical Properties
- 12.5 Naturally Occurring Alkenes: The Terpenes
- 12.6 Addition Reactions of Alkenes
- 12.7 Polymerization of Ethylene and Substituted Ethylenes

Carotene is a naturally occurring polyene in carrots and tomatoes (Problem 12.61).

Alkenes

12.1 Introduction

In this chapter, we begin our study of unsaturated hydrocarbons. Recall from Section 11.1 that these compounds contain one or more carbon–carbon double bonds, triple bonds, or benzene-like rings. In this chapter we study **alkenes.** The simplest alkene is ethylene.

Alkene An unsaturated hydrocarbon that contains a carbon–carbon double bond

Alkyne An unsaturated hydrocarbon that contains a carbon–carbon triple bond

Ethylene
(an alkene)

Acetylene
(an alkyne)

Alkynes are unsaturated hydrocarbons that contain one or more carbon–carbon triple bonds. The simplest alkyne is acetylene. Because alkynes are not widespread in nature and have little importance in biochemistry, we will not study their chemistry.

CHEMICAL CONNECTIONS 12A

Ethylene, a Plant Growth Regulator

As noted below, ethylene occurs only in trace amounts in nature. True enough, but scientists have discovered that this small molecule is a natural ripening agent for fruits. Thanks to this knowledge, fruit growers can now pick fruit while it is still green and less susceptible to bruising. When they are ready to pack the fruit for shipment, the growers can treat it with ethylene gas. Alternatively, fruit can be treated with ethephon (Ethrel), which slowly releases ethylene and initiates fruit ripening.

$$Cl-CH_2-CH_2-\overset{\overset{O}{\|}}{\underset{\underset{OH}{|}}{P}}-OH$$

Ethephon

The next time you see ripe bananas in the market, you might wonder when they were picked and whether their ripening was artificially induced.

Compounds containing carbon–carbon double bonds are especially widespread in nature. Furthermore, several low-molecular-weight alkenes, including ethylene and propene, have enormous commercial importance in our modern, industrialized society. The organic chemical industry worldwide produces more pounds of ethylene than any other chemical. Annual production in the United States alone exceeds 55 billion pounds.

What is unusual about ethylene is that it occurs only in trace amounts in nature. The enormous amounts of it required to meet the needs of the chemical industry are derived the world over by thermal cracking of hydrocarbons. In the United States and other areas of the world with vast reserves of natural gas, the major process for the production of ethylene is thermal cracking of the small quantities of ethane extracted from natural gas. In **cracking,** a saturated hydrocarbon is converted to an unsaturated hydrocarbon plus H_2. Ethane is cracked by heating it in a furnace to 800–900°C for a fraction of a second.

$$CH_3CH_3 \xrightarrow[\text{(thermal cracking)}]{800-900°C} CH_2{=}CH_2 + H_2$$

Ethane · Ethylene

Europe, Japan, and other parts of the world with limited supplies of natural gas depend almost entirely on thermal cracking of petroleum for their ethylene.

From the perspective of the chemical industry, the single most important reaction of ethylene and other low-molecular-weight alkenes is polymerization, which we discuss in Section 12.7. The crucial point to recognize is that ethylene and all of the commercial and industrial products made from it are derived from either natural gas or petroleum—both nonrenewable natural resources!

12.2 Structure

A Alkenes

Using the VSEPR model (Section 3.9), we predict bond angles of 120° about each carbon in a double bond. The observed H—C—C bond angle in ethylene, for example, is 121.7°, close to the predicted value. In other alkenes, deviations from the predicted angle of 120° may be somewhat larger because of interaction between alkyl groups bonded to the doubly bonded carbons. The C—C—C bond angle in propene, for example, is 124.7°.

Ethylene   Propene

If we look at a molecular model of ethylene, we see that the two carbons of the double bond and the four hydrogens bonded to them all lie in the same plane; that is, ethylene is a flat or planar molecule. Furthermore, chemists have discovered that, under normal conditions, no rotation is possible about the carbon–carbon double bond of ethylene or, for that matter, of any other alkene. Whereas free rotation occurs about each carbon–carbon single bond in an alkane (Section 11.6A), rotation about the carbon–carbon double bond in an alkene does not normally take place. For an important exception to this generalization about carbon–carbon double bonds, however, see Chemical Connections 12C: Cis-Trans Isomerism in Vision.

B Cis-Trans Isomerism in Alkenes

Cis-trans isomerism Isomers that have the same order of attachment of their atoms but a different arrangement of their atoms in space due to the presence of either a ring or a carbon–carbon double bond

Because of the restricted rotation about a carbon–carbon double bond, an alkene in which each carbon of the double bond has two different groups bonded to it shows **cis-trans isomerism.** For example, 2-butene has two cis-trans isomers. In *cis*-2-butene, the two methyl groups are located on one side of the double bond and the two hydrogens are on the other side. In *trans*-2-butene, the two methyl groups are on opposite sides of the double bond. *Cis*-2-butene and *trans*-2-butene are different compounds and have different physical and chemical properties.

cis-2-Butene
mp –139°C, bp 4°C

trans-2-Butene
mp –106°C, bp 1°C

12.3 Nomenclature

Alkenes and alkynes are named using the IUPAC system. As we will see, some are still referred to by their common names.

A IUPAC Names

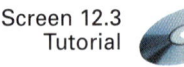
Screen 12.3 Tutorial

The key to the IUPAC names of alkenes is the ending **-ene.** Just as the ending "-ane" tells us that a hydrocarbon chain contains only carbon–carbon single bonds, so the ending "-ene" tells us that it contains a carbon–carbon double bond. To name an alkene:

1. Find the longest carbon chain that includes the double bond. Indicate the length of the parent chain by using a prefix that tells the number of carbon atoms in it (Table 11.2).

2. Number the chain from the end that gives the lower set of numbers to the carbon atoms of the double bond. Designate the position of the double bond by the number of its first carbon.

3. Branched alkenes are named in a manner similar to alkanes; substituent groups are located and named.

$CH_3CH_2CH_2CH_2CH=CH_2$
1-Hexene

$CH_3CH_2\overset{\underset{\displaystyle CH_3}{|}}{C}HCH_2CH=CH_2$
4-Methyl-1-hexene

$CH_3CH_2\overset{\underset{\displaystyle CH_2CH_3}{|}}{\underset{\underset{\displaystyle CH_2CH_3}{|}}{C}}HC=CH_2$
2,3-Diethyl-1-pentene

Note that, although 2,3-diethyl-1-pentene has a six-carbon chain, the longest chain that contains the double bond has only five carbons. The parent alkane is, therefore, a pentane rather than a hexane, and the molecule is named as a disubstituted 1-pentene.

The key to the IUPAC name of an alkyne is the ending **-yne,** which shows the presence of a carbon–carbon triple bond. Thus HC≡CH is ethyne and $CH_3C≡CH$ is propyne. In higher alkynes, number the longest carbon chain that contains the triple bond from the end that gives the lower set of numbers to the triply bonded carbons. Indicate the location of the triple bond by the number of its first carbon atom.

$CH_3\overset{\underset{\displaystyle CH_3}{|}}{C}HC≡CH$
3-Methyl-1-butyne

$CH_3CH_2C≡C\overset{\underset{\displaystyle CH_3}{|}}{C}H_2\overset{\underset{\displaystyle CH_3}{|}}{C}CH_3$
6,6-Dimethyl-3-heptyne

EXAMPLE 12.1

Write the IUPAC name of each unsaturated hydrocarbon.

(a) $CH_2=CH(CH_2)_5CH_3$

(b)
$$\begin{array}{c} H_3C \\ \diagdown \\ C=C \\ \diagup \diagdown \\ H_3C H \end{array}$$
with CH_3 on upper right

(c) $CH_3(CH_2)_2C≡CCH_3$

Solution

(a) The parent chain contains eight carbons; thus the parent alkane is octane. To show the presence of the carbon–carbon double bond, change -ane to -ene. Number the chain beginning with the first carbon of the double bond. This alkene is 1-octene.

(b) Because there are four carbon atoms in the chain containing the carbon–carbon double bond, the parent alkane is butane. The double bond is between carbons 2 and 3 of the chain, and there is a methyl group on carbon 2. This alkene is 2-methyl-2-butene.

(c) There are six carbons in the parent chain, with the triple bond between carbons 2 and 3. This alkyne is 2-hexyne.

Problem 12.1
Write the IUPAC name of each unsaturated hydrocarbon.

(a)

(b)

(c)

B Common Names

Despite the precision and universal acceptance of IUPAC nomenclature, some alkenes and alkynes—particularly those of low molecular weight—are known almost exclusively by their common names. Three examples are:

	$CH_2=CH_2$	$CH_3CH=CH_2$	$CH_3\underset{\underset{CH_3}{\mid}}{C}=CH_2$
IUPAC name:	Ethene	Propene	2-Methylpropene
Common name:	Ethylene	Propylene	Isobutylene

We derive common names for alkynes by prefixing the names of the substituents on the carbon–carbon triple bond to the name *acetylene*:

	$HC\equiv CH$	$CH_3C\equiv CH$	$CH_3C\equiv CCH_3$
IUPAC name:	Ethyne	Propyne	2-Butyne
Common name:	Acetylene	Methylacetylene	Dimethylacetylene

C Cis and Trans Configurations of Alkenes

The orientation of the carbon atoms of the parent chain determines whether an alkene is cis or trans. If the carbons of the parent chain are on the same side of the double bond, the alkene is cis; if they are on opposite sides, it is a trans alkene. In the first example on the next page, they are on opposite sides and the compound is a trans alkene. In the second example, they are on the same side and the compound is a cis alkene.

12.3 NOMENCLATURE | 319

trans-3-Hexene

cis-3,4-Dimethyl-2-pentene

EXAMPLE 12.2

Name each alkene and specify its configuration.

(a)
```
CH₃CH₂CH₂      H
         \    /
          C=C
         /    \
        H      CH₂CH₃
```

(b)
```
CH₃CH₂CH₂      CH₂CH₃
         \    /
          C=C
         /    \
       CH₃     H
```

Solution

(a) The chain contains seven carbon atoms and is numbered from the right to give the lower number to the first carbon of the double bond. The carbon atoms of the parent chain are on opposite sides of the double bond. This alkene is *trans*-3-heptene.

(b) The longest chain contains seven carbon atoms and is numbered from the right so that the first carbon of the double bond is carbon 3 of the chain. The carbon atoms of the parent chain are on the same side of the double bond. This alkene is *cis*-4-methyl-3-heptene.

Problem 12.2

Name each alkene and specify its configuration.

(a)

(b)

CHEMICAL CONNECTIONS 12B

The Case of the Iowa and New York Strains of the European Corn Borer

Although humans communicate largely by sight and sound, the vast majority of other species in the animal world communicate by chemical signals. Often, communication within a species is specific for one of two or more stereoisomers. For example, a member of a given species may respond to the cis isomer of a chemical but not to the trans isomer. Alternatively, it might respond to a precise blend of cis and trans isomers but not to other blends of these same isomers.

Several groups of scientists have studied the components of the sex **pheromones** of both the Iowa and New York strains of the European corn borer. Females of these closely related species secrete the sex attractant 11-tetradecenyl acetate. Males of the Iowa strain show maximum response to a mixture containing 96% of the cis isomer and 4% of the trans isomer. When the pure cis isomer is used alone, males are only weakly attracted. Males of the New York strain show an entirely different response pattern. They respond maximally to a mixture containing 3% of the cis isomer and 97% of the trans isomer.

Evidence suggests that an optimal response to a narrow range of stereoisomers as we see here is widespread in nature and that most insects maintain species isolation for mating and reproduction by the stereochemistry of their pheromones.

D Cycloalkenes

In naming **cycloalkenes,** we number the carbon atoms of the ring double bond 1 and 2 in the direction that gives the substituent encountered first the lower number. Therefore, it is not necessary to use a location number for the carbons of the double bond. Number substituents and list them in alphabetical order.

3-Methylcyclopentene
(not 5-methylcyclopentene)

4-Ethyl-1-methylcyclohexene
(not 5-ethyl-2-methylcyclohexene)

EXAMPLE 12.3

Write the IUPAC name for each cycloalkene.

(a) (b) (c)

Solution
(a) 3,3-Dimethylcyclohexene
(b) 1,2-Dimethylcyclopentene
(c) 4-Isopropyl-1-methylcyclohexene

Problem 12.3
Write the IUPAC name for each cycloalkene.

(a)   (b) [octagon with one double bond]

(c) [cyclohexene with tert-butyl substituent]

E Dienes, Trienes, and Polyenes

We name alkenes that contain more than one double bond as alkadienes, alkatrienes, and so on. We refer to those that contain several double bonds more generally as polyenes (Greek: *poly*, many). Following are three dienes:

$CH_2=CHCH_2CH=CH_2$
1,4-Pentadiene

$CH_2=\overset{\overset{\displaystyle CH_3}{|}}{C}CH=CH_2$
2-Methyl-1,3-butadiene
(Isoprene)

1,3-Cyclopentadiene

We saw earlier that for an alkene with one carbon–carbon double bond that can show cis-trans isomerism, two cis-trans isomers are possible. For an alkene with n carbon–carbon double bonds, each of which can show cis-trans isomerism, 2^n cis-trans isomers are possible.

EXAMPLE 12.4

How many cis-trans isomers are possible for 2,4-heptadiene?

$$CH_3-CH=CH-CH=CH-CH_2-CH_3$$
2,4-Heptadiene

Solution
This molecule has two carbon–carbon double bonds, each of which shows cis-trans isomerism. As shown in the following table, $2^2 = 4$ cis-trans isomers are possible. Line-angle formulas for two of these dienes are shown below.

Double-Bond	
C_2-C_3	C_4-C_5
trans	trans
trans	cis
cis	trans
cis	cis

trans,trans-2,4-Heptadiene

trans,cis-2,4-Heptadiene

Problem 12.4
Draw structural formulas for the other two cis-trans isomers of 2,4-heptadiene.

CHEMICAL CONNECTIONS 12C

Cis-Trans Isomerism in Vision

The retina, the light-detecting layer in the back of our eyes, contains reddish compounds called visual pigments. Their name, rhodopsin, is derived from the Greek word meaning "rose-colored." Each rhodopsin molecule is a combination of one molecule of a protein called opsin and one molecule of 11-*cis*-retinal, a derivative of vitamin A in which the CH_2OH group of carbon 15 is converted to an aldehyde group, —CH=O.

When rhodopsin absorbs light energy, the less stable 11-cis double bond, is converted to the more stable 11-trans double bond. This isomerization changes the shape of the rhodopsin molecule, which in turn causes the neurons of the optic nerve to fire and produce a visual image.

The retina of vertebrates contain two kinds of cells that contain rhodopsin: rods and cones. Cones function in bright light and are used for color vision; they are concentrated in the central portion of the retina, called the macula, and are responsible for the greatest visual acuity. The remaining area of the retina consists mostly of rods, which are used for peripheral and night vision. 11-*cis*-Retinal is present in both cones and rods. Rods have one kind of opsin, whereas cones have three kinds—one for blue, one for green, and one for red color vision.

EXAMPLE 12.5

Draw all cis-trans isomers possible for the following unsaturated alcohol:

$$CH_3\overset{\displaystyle CH_3}{C}=CHCH_2CH_2\overset{\displaystyle CH_3}{C}=CHCH_2OH$$

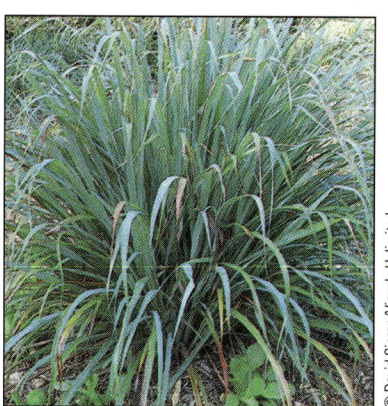

Lemon grass from Shelton Herb Farm, North Carolina.

Solution
Cis-trans isomerism is possible only about the double bond between carbons 2 and 3 of the chain. It is not possible for the other double bond because carbon 7 has two identical groups on it (review Section 12.2B). Thus, $2^1 = 2$ cis-trans isomers are possible. The trans isomer of this alcohol, named geraniol, is a major component of the oils of rose, citronella, and lemon grass.

The trans isomer

The cis isomer

Problem 12.5
How many cis-trans isomers are possible for the following unsaturated alcohol?

$$\underset{}{CH_3}\underset{}{\overset{CH_3}{|}}=CHCH_2CH_2\overset{CH_3}{\underset{|}{C}}=CHCH_2CH_2\overset{CH_3}{\underset{|}{C}}=CHCH_2OH$$

An example of a biologically important polyunsaturated alcohol for which a number of cis-trans isomers are possible is vitamin A. Each of the four carbon–carbon double bonds in the chain of carbon atoms bonded to the substituted cyclohexene ring has the potential for cis-trans isomerism. Therefore, $2^4 = 16$ cis-trans isomers are possible for this structural formula. Vitamin A, the isomer shown here, is the all-trans isomer.

Vitamin A (retinol)

12.4 Physical Properties

Alkenes and alkynes are nonpolar compounds, and the only attractive forces between their molecules are very weak London dispersion forces (Section 5.7A). Their physical properties, therefore, are similar to those of alkanes having the same carbon skeletons. Alkenes and alkynes that are liquid at room temperature have densities less than 1.0 g/mL (they float on water). They are insoluble in water but soluble in one another and in other nonpolar organic liquids.

12.5 Naturally Occurring Alkenes: The Terpenes

Among the compounds found in the essential oils of plants are a group of substances called **terpenes,** all of which have in common the fact that their carbon skeletons can be divided into two or more carbon units that are iden-

Terpene A compound whose carbon skeleton can be divided into two or more units identical to the carbon skeleton of isoprene tical with the carbon skeleton of isoprene. Carbon 1 of an **isoprene unit** is called the head, and carbon 4 is called the tail. A terpene is a compound in which the tail of one isoprene unit becomes bonded to the head of another isoprene unit.

$$CH_2=\overset{\overset{\displaystyle CH_3}{|}}{C}-CH=CH_2 \qquad \overset{Head}{\underset{1\ \ 2\ \ 3\ \ 4}{C-C-C-C}}\overset{Tail}{}$$

2-Methyl-1,3-butadiene
(Isoprene)

Isoprene unit

Terpenes are among the most widely distributed compounds in the biological world, and a study of their structure provides a glimpse into the wondrous diversity that nature can generate from a simple carbon skeleton. Terpenes also illustrate an important principle of the molecular logic of living systems: In building large molecules, small subunits are bonded together by a series of enzyme-catalyzed reactions and then chemically modified by additional enzyme-catalyzed reactions. Chemists use the same principles in the laboratory, but their methods do not have the precision and selectivity of the enzyme-catalyzed reactions of cellular systems.

Probably the terpenes most familiar to you, at least by odor, are components of the so-called essential oils obtained by steam distillation or ether extraction of various parts of plants. Essential oils contain the relatively low-molecular-weight substances that are largely responsible for characteristic plant fragrances. Many essential oils, particularly those from flowers, are used in perfumes.

One example of a terpene obtained from an essential oil is myrcene (Figure 12.1), a component of bayberry wax and oils of bay and verbena. Myrcene is a triene with a parent chain of eight carbon atoms and two one-carbon branches. The two isoprene units in myrcene are joined by bonding the tail of one unit to the head of the other unit. Figure 12.1 also shows three other terpenes, each of ten carbon atoms. In limonene and menthol, nature has formed an additional bond between two carbons to form a six-membered ring.

Farnesol, a terpene with molecular formula $C_{15}H_{26}O$, includes three isoprene units. Derivatives of both farnesol and geraniol are intermediates in the biosynthesis of cholesterol (Section 28.4).

Farnesol
(Lily-of-the-valley)

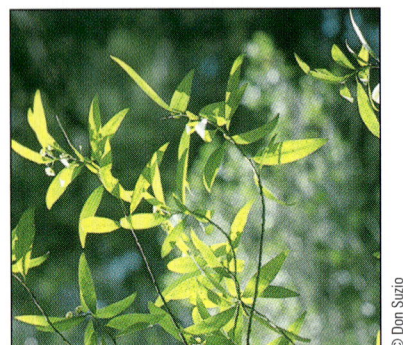

■ California laurel, *Umbelluria californica,* one source of myrcene.

Figure 12.1 Four terpenes, each derived from two isoprene units bonded from the tail of the first unit to the head of the second unit. In limonene and menthol, formation of an additional carbon–carbon bond creates a six-membered ring.

Myrcene
(Bay oil)

Geraniol
(Rose and other flowers)

Limonene
(Lemon and orange)

Menthol
(Peppermint)

Vitamin A (Section 12.3E), a terpene with molecular formula $C_{20}H_{30}O$, consists of four isoprene units bonded head-to-tail and cross-linked at one point to form a six-membered ring.

12.6 Addition Reactions of Alkenes

The most characteristic reaction of alkenes is addition to the carbon–carbon double bond: The double bond is broken and in its place single bonds form to two new atoms or groups of atoms. Table 12.1 shows several examples of alkene addition reactions, along with the descriptive name(s) associated with each reaction.

At this point, you might ask why a carbon–carbon double bond is a site of chemical reactivity, whereas carbon–carbon single bonds are quite unreactive under most experimental conditions. One way to answer this question is to focus on the changes in bonding that occur as a result of an alkene addition reaction. Consider, for example, the addition of hydrogen to ethylene. As a result of this addition, one double bond and one single bond (the H—H bond of H_2) are replaced by three single bonds, giving a net conversion of one bond of the double bond to two single bonds.

$$\underset{\text{one double bond}\atop\text{and one single bond}}{H_2C=CH_2} + H-H \xrightarrow{\text{are replaced by}} \underset{\text{three single bonds}}{H-CH_2-CH_2-H} + \text{heat}$$

This and almost all other addition reactions of alkenes are exothermic, which means that the products are more stable (have lower energy) than the reactants. Remember, though, that just because an alkene addition reaction is exothermic doesn't mean that it occurs rapidly. The rate of a chem-

Table 12.1 Characteristic Addition Reactions of Alkenes

Reaction	Descriptive Name(s)
$\text{C=C} + HCl \longrightarrow -\overset{H}{\underset{\vert}{C}}-\overset{Cl}{\underset{\vert}{C}}-$	hydrochlorination
$\text{C=C} + H_2O \longrightarrow -\overset{H}{\underset{\vert}{C}}-\overset{OH}{\underset{\vert}{C}}-$	hydration
$\text{C=C} + Br_2 \longrightarrow -\overset{Br}{\underset{\vert}{C}}-\overset{Br}{\underset{\vert}{C}}-$	bromination
$\text{C=C} + H_2 \longrightarrow -\overset{H}{\underset{\vert}{C}}-\overset{H}{\underset{\vert}{C}}-$	hydrogenation (reduction)

Screen 12.4 Simulation

ical reaction depends on its activation energy, not on how exothermic or endothermic it is (Section 7.3). In fact, the addition of hydrogen to an alkene is immeasurably slow room temperature but, as we will see soon, proceeds quite rapidly in the presence of a suitable catalyst.

A Addition of Hydrogen Halides

The hydrogen halides HCl, HBr, and HI add to alkenes to give haloalkanes (alkyl halides). Addition of HCl to ethylene, for example, gives chloroethane (ethyl chloride):

$$CH_2=CH_2 + HCl \longrightarrow \underset{\text{Chloroethane (Ethyl chloride)}}{CH_2-CH_2}$$
$$\underset{\text{Ethylene}}{}$$

with H and Cl on adjacent carbons.

Addition of HCl to propene gives 2-chloropropane (isopropyl chloride); hydrogen adds to carbon 1 of propene and chlorine adds to carbon 2. If the orientation of addition were reversed, 1-chloropropane (propyl chloride) would form. The observed result is that almost no 1-chloropropane forms. Because 2-chloropropane is the observed product, we say that addition of HCl to propene is **regioselective.**

Regioselective reaction A reaction in which one direction of bond forming or bond breaking occurs in preference to all other directions

$$\underset{\text{Propene}}{CH_3\overset{2}{CH}=\overset{1}{CH_2}} + HCl \longrightarrow \underset{\text{2-Chloropropane}}{CH_3\overset{Cl}{\underset{|}{CH}}-\overset{H}{\underset{|}{CH_2}}} \quad \underset{\text{1-Chloropropane (not formed)}}{\cancel{CH_3\overset{H}{\underset{|}{CH}}-\overset{Cl}{\underset{|}{CH_2}}}}$$

This regioselectivity was noted by Vladimir Markovnikov (1838–1904), who made the following generalization known as **Markovnikov's rule:** In the addition of HX (where X = halogen) to an alkene, hydrogen adds to the doubly bonded carbon that has the greater number of hydrogens already bonded to it; halogen adds to the other carbon.

Markovnikov's rule In the addition of HX or H_2O to an alkene, hydrogen adds to the carbon of the double bond having the greater number of hydrogens

Markovnikov's rule is often paraphrased "the rich get richer."

EXAMPLE 12.6

Draw a structural formula for the product of each alkene addition reaction.

(a) $CH_3\overset{\overset{CH_3}{|}}{C}=CH_2 + HI \longrightarrow$

(b) [cyclopentene with CH_3] $+ HCl \longrightarrow$

Solution
(a) Markovnikov's rule predicts that the hydrogen of HI adds to carbon 1 and the iodine adds to carbon 2 to give 2-iodo-2-methylpropane.
(b) H adds to carbon 2 of the ring and Cl adds to carbon 1 to give 1-chloro-1-methylcyclopentane.

$$\underset{\text{2-Iodo-2-methylpropane}}{CH_3\underset{\underset{I}{|}}{\overset{\overset{CH_3}{|}}{C}}CH_3} \qquad \underset{\text{1-Chloro-1-methylcyclopentane}}{[\text{cyclopentane with } \overset{1}{Cl}, CH_3]}$$

Screen 12.5
Simulation

Problem 12.6
Draw a structural formula for the product of each alkene addition reaction.

(a) $CH_3CH=CH_2 + HBr \longrightarrow$

(b) [cyclohexyl]$=CH_2 + HBr \longrightarrow$

Markovnikov's rule tells us what happens when we add HCl, HBr, or HI to a carbon–carbon double bond. We know that in the addition of HCl or other halogen acid, one bond of the double bond and the H—Cl bond are broken, and that new C—H and C—Cl bonds form. But chemists also want to know *how* this conversion happens. Are the C=C and H—X bonds broken and both new covalent bonds formed all at the same time? Or does this reaction take place in a series of steps? If the latter, what are these steps, and in what order do they take place?

Chemists account for the addition of HX to an alkene by a two-step **reaction mechanism,** which we illustrate for the reaction of 2-butene with hydrogen chloride to give 2-chlorobutane. Step 1 is the addition of H^+ to 2-butene. To show this addition, we use a curved arrow that shows the repositioning of an electron pair from its origin (the tail of the arrow) to its new location (the head of the arrow). Recall that we used curved arrows in Section 8.1 to show bond breaking and bond formation in proton-transfer reactions. We now use curved arrows in the same way to show bond breaking and bond formation in a reaction mechanism.

Step 1 results in the formation of an organic cation. One carbon atom in this cation has only six electrons in its valence shell and so carries a charge of +1. A species containing a positively charged carbon atom is called a **carbocation** (carbon + cation). Carbocations are classified as primary (1°), secondary (2°), or tertiary (3°), depending on the number of carbon groups bonded to the carbon bearing the positive charge.

> **Reaction mechanism** A step-by-step description of how a chemical reaction occurs

> **Carbocation** A species containing a carbon atom with only three bonds to it and bearing a positive charge

Mechanism: Addition of HCl to 2-Butene

Step 1: Reaction of the carbon–carbon double bond of the alkene with H^+ forms a 2° carbocation intermediate. In forming this intermediate, one bond of the double bond breaks and its pair of electrons forms a new covalent bond with H^+. One carbon of the double bond is left with only six electrons in its valence shell and, therefore, has a positive charge.

$$CH_3CH=CHCH_3 + H^+ \longrightarrow CH_3\overset{+}{C}H-\overset{\overset{H}{|}}{C}HCH_3$$

A 2° carbocation intermediate

Step 2: Reaction of the 2° carbocation intermediate with chloride ion completes the valence shell of carbon and gives 2-chlorobutane.

$$:\!\ddot{\underset{..}{Cl}}\!:^- + CH_3\overset{+}{C}HCH_2CH_3 \longrightarrow CH_3\overset{\overset{:\ddot{Cl}:}{|}}{C}HCH_2CH_3$$

Chloride ion   A 2° carbocation intermediate

EXAMPLE 12.7

Propose a two-step mechanism for the addition of HI to methylenecyclohexane to give 1-iodo-1-methylcyclohexane.

[Reaction: Methylenecyclohexane + HI → 1-Iodo-1-methylcyclohexane]

Solution
The mechanism is similar to that proposed for the addition of HCl to propene.

Step 1: Reaction of H^+ with the carbon–carbon double bond forms a new C—H bond to the carbon bearing the greater number of hydrogens and gives a 3° carbocation intermediate.

[Reaction showing methylenecyclohexane + H^+ → A 3° carbocation intermediate]

Step 2: Reaction of the 3° carbocation intermediate with iodide ion completes the valence shell of carbon and gives the product.

[Reaction showing carbocation + I^- → product]

Problem 12.7

Propose a two-step mechanism for the addition of HBr to 1-methylcyclohexene to give 1-bromo-1-methylcyclohexane.

B Addition of Water: Acid-Catalyzed Hydration

In the presence of an acid catalyst, most commonly concentrated sulfuric acid, water adds to the carbon–carbon double bond of an alkene to give an alcohol. Addition of water is called **hydration.** In the case of simple alkenes, hydration follows Markovnikov's rule: H adds to the carbon of the double bond with the greater number of hydrogens and OH adds to the carbon with the fewer hydrogens.

Hydration Addition of water

Most industrial ethanol is made by the acid-catalyzed hydration of ethylene.

$$CH_3CH{=}CH_2 + H_2O \xrightarrow{H_2SO_4} CH_3CH{-}CH_2$$
Propene   2-Propanol
(with OH and H added)

$$CH_3\underset{CH_3}{\overset{CH_3}{C}}{=}CH_2 + H_2O \xrightarrow{H_2SO_4} CH_3\underset{HO}{\overset{CH_3}{C}}{-}\underset{H}{CH_2}$$
2-Methylpropene   2-Methyl-2-propanol

EXAMPLE 12.8

Draw a structural formula for the alcohol formed by the acid-catalyzed hydration of 1-methylcyclohexene.

Solution
Markovnikov's rule predicts that H adds to the carbon with the greater number of hydrogens, which, in this case, is carbon 2 of the cyclohexene ring. OH then adds to carbon 1.

$$\text{1-Methylcyclohexene} + H_2O \xrightarrow{H_2SO_4} \text{1-Methylcyclohexanol}$$

Problem 12.8

Draw a structural formula for the alcohol formed from acid-catalyzed hydration of each alkene:
(a) 2-Methyl-2-butene
(b) 2-Methyl-1-butene

The mechanism for the acid-catalyzed hydration of an alkene is similar to what we proposed for the addition of HCl, HBr, and HI to an alkene and is illustrated by the hydration of propene. This mechanism is consistent with the fact that acid is a catalyst. One H^+ is consumed in Step 1, but another is generated in Step 3.

Mechanism: Acid-Catalyzed Hydration of Propene

Step 1: Addition of H^+ to the carbon of the double bond with the greater number of hydrogens gives a 2° carbocation intermediate.

$$CH_3CH=CH_2 + H^+ \longrightarrow CH_3\overset{+}{C}HCH_2H$$

A 2° carbocation intermediate

Step 2: The carbocation intermediate completes its valence shell by forming a new covalent bond with an unshared pair of electrons of the oxygen atom of H_2O to give an **oxonium ion.**

$$CH_3\overset{+}{C}HCH_3 + :\!\ddot{O}\!-\!H \longrightarrow CH_3CHCH_3$$
$$\qquad\qquad\qquad\qquad\;\; |\qquad\qquad\qquad |$$
$$\qquad\qquad\qquad\qquad\;\; H \qquad\qquad\;\; H\!-\!\overset{+}{\underset{..}{O}}\!-\!H$$

An oxonium ion

Oxonium ion An ion in which oxygen is bonded to three other atoms and bears a positive charge

Step 3: Loss of H^+ from the oxonium ion gives the alcohol and generates a new H^+ catalyst.

$$\underset{CH_3CHCH_3}{H\!-\!\overset{+}{\underset{..}{O}}\!-\!H} \longrightarrow \underset{CH_3CHCH_3}{:\!\ddot{O}H} + H^+$$

EXAMPLE 12.9

Propose a three-step reaction mechanism for the acid-catalyzed hydration of methylenecyclohexane to give 1-methylcyclohexanol.

Solution
The reaction mechanism is similar to that for the acid-catalyzed hydration of propene.

Step 1: Reaction of the carbon–carbon double bond with H^+ gives a 3° carbocation intermediate.

$$\text{cyclohexane}=CH_2 + H^+ \longrightarrow \text{cyclohexane}^+-CH_3$$

A 3° carbocation intermediate

Step 2: Reaction of the carbocation intermediate with water completes the valence shell of carbon and gives an oxonium ion.

$$\text{cyclohexane}^+-CH_3 + :\overset{H}{\underset{}{O}}-H \longrightarrow \text{cyclohexane}\begin{matrix}\overset{+}{O}H_2\\ CH_3\end{matrix}$$

An oxonium ion

Step 3: Loss of H^+ from the oxonium ion completes the reaction and generates a new H^+ catalyst.

$$\text{cyclohexane}\begin{matrix}\overset{+}{O}-H\\ CH_3\end{matrix} \longrightarrow \text{cyclohexane}\begin{matrix}OH\\ CH_3\end{matrix} + H^+$$

Problem 12.9
Propose a three-step reaction mechanism for the acid-catalyzed hydration of 1-methylcyclohexene to give 1-methylcyclohexanol.

C Addition of Bromine and Chlorine

Chlorine, Cl_2, and bromine, Br_2, react with alkenes at room temperature by addition of halogen atoms to the carbon atoms of the double bond. This reaction is generally carried out either by using the pure reagents or by mixing them in an inert solvent, such as dichloromethane, CH_2Cl_2.

$$CH_3CH=CHCH_3 + Br_2 \xrightarrow{CH_2Cl_2} CH_3\underset{Br}{\overset{Br}{\underset{|}{CH}}}-\underset{}{\overset{Br}{\underset{|}{CH}}}CH_3$$

2-Butene   2,3-Dibromobutane

$$\text{cyclohexene} + Br_2 \xrightarrow{CH_2Cl_2} \text{1,2-dibromocyclohexane}$$

Cyclohexene   1,2-Dibromocyclohexane

Addition of bromine is a useful qualitative test for the presence of an alkene. If we dissolve bromine in carbon tetrachloride, the solution is red. In

(a)

(b)
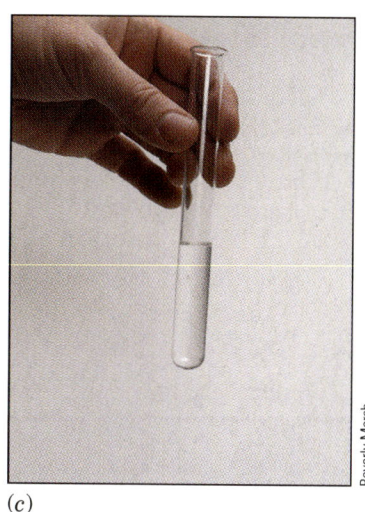
(c)

■ (a) A drop of Br_2 dissolved in CCl_4 is added to an unknown liquid. If the color remains after stirring (b), it indicates the absence of unsaturation. If the color disappears (c), the unknown is unsaturated.

contrast, alkenes and dibromoalkanes are colorless. If we mix a few drops of the red bromine solution with an unknown sample suspected of being an alkene, disappearance of the red color as bromine adds to the double bond tells us that an alkene is, indeed, present.

EXAMPLE 12.10

Complete these reactions.

(a) cyclopentene + $Br_2 \xrightarrow{CH_2Cl_2}$

(b) 1-methylcyclohexene + $Cl_2 \xrightarrow{CH_2Cl_2}$

Solution

In addition of Br_2 or Cl_2 to a cycloalkene, one halogen adds to each carbon of the double bond.

(a) cyclopentene + $Br_2 \xrightarrow{CH_2Cl_2}$ 1,2-dibromocyclopentane

(b) 1-methylcyclohexene + $Cl_2 \xrightarrow{CH_2Cl_2}$ 1,2-dichloro-1-methylcyclohexane

Problem 12.10

Complete these reactions.

(a) $CH_3\underset{\underset{CH_3}{|}}{\overset{\overset{CH_3}{|}}{C}}CH=CH_2 + Br_2 \xrightarrow{CH_2Cl_2}$

(b) methylenecyclohexane + $Cl_2 \xrightarrow{CH_2Cl_2}$

D Addition of Hydrogen: Reduction

Virtually all alkenes react quantitatively with molecular hydrogen, H_2, in the presence of a transition metal catalyst to give alkanes. Commonly used transition metal catalysts include platinum, palladium, ruthenium, and nickel. Because the conversion of an alkene to an alkane involves reduction by hydrogen in the presence of a catalyst, the process is called **catalytic reduction** or, alternatively, **catalytic hydrogenation.**

In Section 20.3 we will see how catalytic hydrogenation is used to solidify liquid vegetable oils to margarines and semisolid cooking fats.

$$\underset{\text{trans-2-Butene}}{\begin{array}{c}H_3C\\ \\ H\end{array}\!\!C\!\!=\!\!C\!\!\begin{array}{c}H\\ \\ CH_3\end{array}} + H_2 \xrightarrow[\text{25°C, 3 atm}]{Pd} \underset{\text{Butane}}{CH_3CH_2CH_2CH_3}$$

$$\underset{\text{Cyclohexene}}{\bigcirc\!\!=} + H_2 \xrightarrow[\text{25°C, 3 atm}]{Pd} \underset{\text{Cyclohexane}}{\bigcirc}$$

The metal catalyst is used in the form of a finely powdered solid. The reaction is carried out by dissolving the alkene in ethanol or another nonreacting organic solvent, adding the solid catalyst, and exposing the mixture to hydrogen gas at pressures ranging from 1 to 100 atm.

12.7 Polymerization of Ethylene and Substituted Ethylenes

A Structure of Polyethylenes

From the perspective of the chemical industry, the single most important reaction of alkenes is the formation of **chain-growth polymers** (Greek words: *poly*, many, and *meros*, part). In the presence of certain compounds called initiators, many alkenes form polymers made by the stepwise addition of **monomers** (Greek words: *mono*, one, and *meros*, part) to a growing polymer chain, as illustrated by the formation of polyethylene from ethylene. In alkene polymers of industrial and commercial importance, n is a large number, typically several thousand.

Polymer From the Greek *poly*, many, and *meros*, parts; any long-chain molecule synthesized by bonding together many single parts (monomers)

Monomer From the Greek *mono*, single, and *meros*, parts; the simplest nonredundant unit from which a polymer is synthesized

$$n\,CH_2\!\!=\!\!CH_2 \xrightarrow[\text{(polymerization)}]{\text{initiator}} -(CH_2CH_2)_n-$$

$$\underset{\text{Ethylene}}{} \qquad \underset{\text{Polyethylene}}{}$$

To show the structure of a polymer, we place parentheses around the repeating monomer unit. The structure of an entire polymer chain can be reproduced by repeating this enclosed structure in both directions. A subscript n is placed outside the parentheses to indicate that this unit is repeated n times.

12.7 POLYMERIZATION OF ETHYLENE AND SUBSTITUTED ETHYLENES

Monomer units shown in red

$$-CH_2\overset{\overset{CH_3}{|}}{CH}-CH_2\overset{\overset{CH_3}{|}}{CH}-CH_2\overset{\overset{CH_3}{|}}{CH}-CH_2\overset{\overset{CH_3}{|}}{CH}-$$

Part of an extended polymer chain

$$\left(CH_2\overset{\overset{CH_3}{|}}{CH}\right)_n$$

The repeating unit

The most common method of naming a polymer is to attach the prefix **poly-** to the name of the monomer from which the polymer is synthesized—for example, polyethylene and polystyrene. Where the name of the monomer consists of two words (for example, the monomer vinyl chloride), its name is enclosed in parentheses.

Vinyl chloride $\xrightarrow{\text{polymerization}}$ Poly(vinyl chloride) (PVC)

Table 12.2 lists several important polymers derived from ethylene and substituted ethylenes along with their common names and most important uses.

Know

Table 12.2 Polymers Derived from Ethylene and Substituted Ethylenes

Monomer Formula	Common Name	Polymer Name(s) and Common Uses
$CH_2=CH_2$	ethylene	polyethylene, Polythene; break-resistant containers and packaging materials
$CH_2=CHCH_3$	propylene	polypropylene, Herculon; textile and carpet fibers
$CH_2=CHCl$	vinyl chloride	poly(vinyl chloride), PVC; construction tubing
$CH_2=CCl_2$	1,1-dichloroethylene	poly(1,1-dichloroethylene); Saran Wrap is a copolymer with vinyl chloride
$CH_2=CHCN$	acrylonitrile	polyacrylonitrile, Orlon; acrylics and acrylates
$CF_2=CF_2$	tetrafluoroethylene	polytetrafluoroethylene, PTFE; Teflon, nonstick coatings
$CH_2=CHC_6H_5$	styrene	polystyrene, Styrofoam; insulating materials
$CH_2=CHCOOCH_2CH_3$	ethyl acrylate	poly(ethyl acrylate), latex paint
$CH_2=\underset{\underset{CH_3}{|}}{C}COOCH_3$	methyl methacrylate	poly(methyl methacrylate), Lucite; Plexiglas; glass substitutes

■ Some articles made from chain-growth polymers. (*a*) Saran Wrap, a copolymer of vinyl chloride and 1,1-dichloroethylene. (*b*) Plastic containers for various supermarket products, made mostly from polyethylene and polypropylene. (*c*) Teflon-coated kitchenware. (*d*) Articles made from polystyrene.

(*a*)

(*b*)

(*c*)

(*d*)

A peroxide is any compound that contains an —O—O— bond, for example, hydrogen peroxide, H—O—O—H.

■ Polyethylene films are produced by extruding molten plastic through a ringlike gap and inflating the film into a balloon.

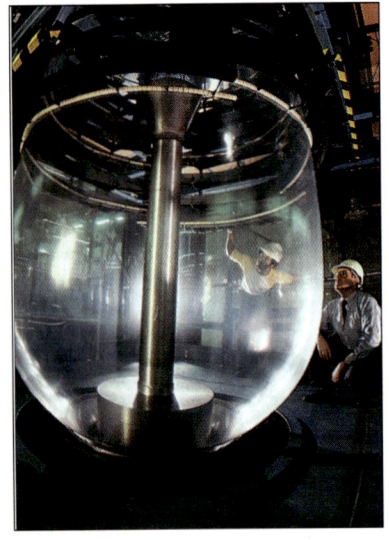

B Low-Density Polyethylene

The first commercial process for ethylene polymerization used peroxide initiators at 500°C and 1000 atm and yielded a tough, transparent polymer known as **low-density polyethylene (LDPE).** At the molecular level, LDPE chains are highly branched, with the result that they do not pack well together and London dispersion forces (Section 5.7A) between them are weak. LDPE softens and melts at about 115°C, which means that it cannot be used in products that are exposed to boiling water.

Today, approximately 65% of all LDPE is used for the manufacture of films by the blow-molding technique illustrated in Figure 12.2. LDPE film is inexpensive, which makes it ideal for packaging such consumer items as baked goods and vegetables and for the manufacture of trash bags.

C High-Density Polyethylene

An alternative method for polymerization of alkenes, and one that does not involve peroxide initiators, was developed by Karl Ziegler of Germany and Giulio Natta of Italy in the 1950s. Polyethylene from Ziegler-Natta systems, termed **high-density polyethylene (HDPE),** has little chain branching. Consequently, its chains pack together more closely than those of LDPE, with the result that London dispersion forces between chains of HDPE are

Figure 12.2 Fabrication of LDPE film. A tube of melted LDPE along with a jet of compressed air is forced through an opening and blown into a gigantic, thin-walled bubble. The film is then cooled and taken up onto a roller. This double-walled film can be slit down the side to give LDPE film or sealed at points along its length to make LDPE bags.

stronger than those in LDPE. HDPE has a higher melting point than LDPE and is three to ten times stronger.

Approximately 45% of all HDPE products are made by the blow-molding process shown in Figure 12.3. HDPE is used for consumer items such as milk and water jugs, grocery bags, and squeezable bottles.

Figure 12.3 Blow molding an HDPE container. (*a*) A short length of HDPE tubing is placed in an open die and the die is closed, sealing the bottom of the tube. (*b*) Compressed air is forced into the hot polyethylene/die assembly, and the tubing is literally blown up to take the shape of the mold. (*c*) After the assembly cools, the die is opened, and there is the container!

CHEMICAL CONNECTIONS 12D

Recycling Plastics

Plastics are polymers that can be molded when hot and that retain their shape when cooled. Because they are durable and lightweight, plastics are probably the most versatile synthetic materials in existence. In fact, the current production of plastics in the United States exceeds the U.S. production of steel. Plastics have come under criticism, however, for their role in the trash crisis. They account for approximately 21% of the volume and 8% of the weight of solid wastes, with most plastic waste consisting of disposable packaging and wrapping.

Six types of plastics are commonly used for packaging applications. In 1988, manufacturers adopted recycling code letters developed by the Society of the Plastics Industry as a means of identifying them.

Code	Polymer	Common Uses
1 PET	poly(ethylene terephthalate)	soft drink bottles, household chemical bottles, films, textile fibers
2 HDPE	high-density polyethylene	milk and water jugs, grocery bags, squeezable bottles
3 V	poly(vinyl chloride), PVC	shampoo bottles, pipes, shower curtains, vinyl siding, wire insulation, floor tiles
4 LDPE	low-density polyethylene	shrink wrap, trash and grocery bags, sandwich bags, squeeze bottles
5 PP	polypropylene	plastic lids, clothing fibers, bottle caps, toys, diaper linings
6 PS	polystyrene	Styrofoam cups, egg cartons, disposable utensils, packaging materials, appliances
7	all other plastics	various

Currently, only poly(ethylene terephthalate) (PET) and high-density polyethylene are recycled in large quantities. The synthesis and structure of PET, a polyester, is described in Section 18.8B.

The process for recycling most plastics is simple, with separation of the plastic from other contaminants being the most labor-intensive step. For example, PET soft drink bottles usually have a paper label and adhesive that must be removed before the PET can be reused. Recycling begins with hand or machine sorting, after which the bottles are shredded into small chips. An air cyclone then removes paper and other lightweight materials. After any remaining labels and adhesives are eliminated with a detergent wash, the PET chips are dried. The PET produced by this method is 99.9% free of contaminants and sells for about half the price of the virgin material.

■ These students are wearing jackets made from recycled PET soda bottles.

SUMMARY

An **alkene** is an unsaturated hydrocarbon that contains a carbon–carbon double bond (Section 12.1). An **alkyne** is an unsaturated hydrocarbon that contains a carbon–carbon triple bond.

The structural feature that makes **cis-trans isomerism** possible in alkenes is restricted rotation about the two carbons of the double bond (Section 12.2B). The cis or trans configuration of an alkene is determined by the orientation of the atoms of the parent chain about the double bond. If atoms of the parent chain are located on the same side of the double bond, the configuration of the alkene is cis; if they are located on opposite sides, the configuration is trans.

In IUPAC names, the presence of a carbon–carbon double bond is indicated by a prefix showing the number of carbons in the parent chain and the ending **-ene**. (Section 12.3A). Substituents are numbered and named in alphabetical order. The presence of a carbon–carbon triple bond is indicated by a prefix showing the number of carbons in the parent chain and the ending **-yne**. The carbon atoms of a cycloalkene are numbered 1 and 2 in the direction that gives the smaller number to the first substituent (Section 12.3D). Compounds containing two double bonds are called dienes, those with three double bonds are trienes, and those containing four or more double bonds are called polyenes (Section 12.3E).

Because alkenes and alkynes are nonpolar compounds and the only interactions between their molecules are London dispersion forces, their physical properties are similar to those of alkanes with similar carbon skeletons (Section 12.4).

The characteristic structural feature of a **terpene** is a carbon skeleton that can be divided into two or more **isoprene units** (Section 12.5).

A characteristic reaction of alkenes is addition to the double bond (Section 12.6). In addition, the double bond breaks and bonds to two new atoms or groups of atoms form in its place. A **reaction mechanism** is a step-by-step description of how a chemical reaction occurs, including the role of the catalyst (if one is present). A **carbocation** contains a carbon with only six electrons in its valence shell and bears a positive charge.

Polymerization is the process of bonding together many small **monomers** into large, high-molecular-weight **polymers** (Section 12.7).

KEY REACTIONS

1. **Addition of HX (Section 12.6A)** Addition of HX to the carbon–carbon double bond of an alkene follows Markovnikov's rule. The reaction occurs in two steps and involves formation of a carbocation intermediate.

2. **Acid-Catalyzed Hydration (Section 12.6B)** Addition of H_2O to the carbon–carbon double bond of an alkene follows Markovnikov's rule. The reaction occurs in three steps and involves formation of carbocation and oxonium ion intermediates.

3. **Addition of Bromine and Chlorine (Section 12.6C)** Addition to a cycloalkene gives a 1,2-dihalocycloalkane.

4. **Reduction—Formation of Alkanes (Section 12.6D)** Catalytic reduction involves addition of hydrogen to form two new C—H bonds.

5. **Polymerization of Ethylene and Substituted Ethylenes (Section 12.7)** In polymerization of alkenes, monomer units bond together without the loss of any atoms.

PROBLEMS

Numbers that appear in color indicate difficult problems.
▲ designates problems requiring application of principles.

Structure of Alkenes and Alkynes

12.11 What is the difference in structure between a saturated hydrocarbon and an unsaturated hydrocarbon?

12.12 Each carbon atom in ethane and in ethylene is surrounded by eight valence electrons and has four bonds to it. Explain how the VSEPR model (Section 3.9) predicts a bond angle of 109.5° about each carbon in ethane but an angle of 120° about each carbon in ethylene.

12.13 Predict all bond angles about each highlighted carbon atom.

(a)

(b)

(c) HC≡C—CH=CH$_2$

(d)

12.14 Predict all bond angles about each highlighted carbon atom.

(a), (b), (c), (d)

Nomenclature of Alkenes and Alkynes

12.15 Draw a structural formula for each compound.
(a) *trans*-2-Methyl-3-hexene
(b) 2-Methyl-3-hexyne
(c) 2-Methyl-1-butene
(d) 3-Ethyl-3-methyl-1-pentyne
(e) 2,3-Dimethyl-2-pentene
(f) *cis*-2-Hexene

12.16 Draw a structural formula for each compound.
(a) 3-Chloropropene
(b) 3-Methylcyclohexene
(c) 1,2-Dimethylcyclohexene
(d) *trans*-3,4-Dimethyl-3-heptene
(e) Cyclopropene
(f) 3-Hexyne

12.17 Write the IUPAC name for each unsaturated hydrocarbon.
(a) CH$_2$=CH(CH$_2$)$_4$CH$_3$
(b)
(c) (structure with CH$_3$ groups on cyclohexene)
(d) (CH$_3$)$_2$CHCH=C(CH$_3$)$_2$
(e) CH$_3$(CH$_2$)$_5$C≡CH
(f) CH$_3$CH$_2$C≡CC(CH$_3$)$_3$

12.18 Write the IUPAC name for each unsaturated hydrocarbon.

(a) (branched alkene structure) (b) (methylcyclopentene structure)

(c) CH$_3$CH$_2$CCH$_3$ with =CH$_2$

(d) CH$_3$CH$_2$CH$_2$ \ C=CH$_2$ / CH$_3$CH$_2$CH$_2$

12.19 Explain why each name is incorrect, and then write a correct name.
(a) 1-Methylpropene
(b) 3-Pentene
(c) 2-Methylcyclohexene
(d) 3,3-Dimethylpentene
(e) 4-Hexyne
(f) 2-Isopropyl-2-butene

12.20 Explain why each name is incorrect, and then write a correct name.
(a) 2-Ethyl-1-propene
(b) 5-Isopropylcyclohexene
(c) 4-Methyl-4-hexene
(d) 2-*sec*-Butyl-1-butene
(e) 6,6-Dimethylcyclohexene
(f) 2-Ethyl-2-hexene

Cis-Trans Isomerism in Alkenes and Cycloalkenes

12.21 Which of these alkenes show cis-trans isomerism? For each that does, draw structural formulas for both isomers.
(a) 1-Hexene
(b) 2-Hexene
(c) 3-Hexene
(d) 2-Methyl-2-hexene
(e) 3-Methyl-2-hexene
(f) 2,3-Dimethyl-2-hexene

12.22 Which of these alkenes shows cis-trans isomerism? For each that does, draw structural formulas for both isomers.
(a) 1-Pentene
(b) 2-Pentene
(c) 3-Ethyl-2-pentene
(d) 2,3-Dimethyl-2-pentene
(e) 2-Methyl-2-pentene
(f) 2,4-Dimethyl-2-pentene

12.23 Name and draw structural formulas for all alkenes with molecular formula C_5H_{10}. As you draw these alkenes, remember that cis and trans isomers are different compounds and must be counted separately.

12.24 In Chapter 20 on the biochemistry of lipids, we will study the three long-chain unsaturated carboxylic acids shown below. Each has 18 carbons and is a component of animal fats, vegetable oils, and biological membranes. Because of their presence in animal fats, they are called fatty acids. How many cis-trans isomers are possible for each fatty acid?

Oleic acid $CH_3(CH_2)_7CH{=}CH(CH_2)_7COOH$
Linoleic acid $CH_3(CH_2)_4(CH{=}CHCH_2)_2(CH_2)_6COOH$
Linolenic acid $CH_3CH_2(CH{=}CHCH_2)_3(CH_2)_6COOH$

12.25 The fatty acids in Problem 12.24 occur in animal fats, vegetable oils, and biological membranes almost exclusively as the all-cis isomers. Draw line-angle formulas for each fatty acid showing the cis configuration about each carbon–carbon double bond.

12.26 In Chapter 20, we also encounter this 20-carbon fatty acid. How many cis-trans isomers are possible for it?

$$CH_3(CH_2)_4(CH{=}CHCH_2)_4CH_2CH_2COOH$$
Arachidonic acid

12.27 Draw a line-angle formula for arachidonic acid showing the cis configuration about each double bond.

12.28 What structural feature in alkenes makes cis-trans isomerism in them possible? What structural feature in cycloalkanes makes cis-trans isomerism in them possible? What do these two structural features have in common?

12.29 For each molecule that shows cis-trans isomerism, draw the cis isomer.

12.30 Draw structural formulas for all compounds with molecular formula C_5H_{10} that are
(a) Alkenes that do not show cis-trans isomerism.
(b) Alkenes that do show cis-trans isomerism.
(c) Cycloalkanes that do not show cis-trans isomerism.
(d) Cycloalkanes that do show cis-trans isomerism.

12.31 β-Ocimene, a triene found in the fragrance of cotton blossoms and several essential oils, has the IUPAC name *cis*-3,7-dimethyl-1,3,6-octatriene. (Cis refers to the configuration of the double bond between carbons 3 and 4, the only double bond in this molecule about which cis-trans isomerism is possible). Draw a structural formula for β-ocimene.

Terpenes

12.32 Which of these terpenes (Section 12.5) show cis-trans isomerism?
(a) Myrcene
(b) Geraniol
(c) Limonene
(d) Farnesol

12.33 Show that the structural formula of vitamin A (Section 12.3E) can be divided into four isoprene units joined by head-to-tail linkages and cross-linked at one point to form the six-membered ring.

Addition Reactions of Alkenes

12.34 Define "alkene addition reaction." Write an equation for an addition reaction of propene.

12.35 What reagent and/or catalysts are necessary to bring about each conversion?

(a) $CH_3CH{=}CHCH_3 \longrightarrow CH_3CH_2\overset{Br}{\underset{|}{C}}HCH_3$

(b) $CH_3\overset{CH_3}{\underset{|}{C}}{=}CH_2 \longrightarrow CH_3\overset{CH_3}{\underset{|}{\underset{OH}{C}}}CH_3$

(c) cyclopentene → iodocyclopentane

(d) $CH_3\underset{CH_3}{\underset{|}{C}}=CH_2 \longrightarrow CH_3\underset{Br\;\;Br}{\underset{|\;\;\;|}{\overset{CH_3}{\overset{|}{C}}-CH_2}}$

12.36 Complete these equations.

(a) [cyclopentene]—CH_2CH_3 + HCl ⟶

(b) [cyclopentene]—CH_2CH_3 + H_2O $\xrightarrow{H_2SO_4}$

(c) $CH_3(CH_2)_5CH=CH_2$ + HI ⟶

(d) [cyclohexyl]—$\underset{CH_3}{\underset{|}{C}}=CH_2$ + HCl ⟶

(e) $CH_3CH=CHCH_2CH_3 + H_2O \xrightarrow{H_2SO_4}$

(f) $CH_2=CHCH_2CH_2CH_3 + H_2O \xrightarrow{H_2SO_4}$

12.37 Draw structural formulas for all possible carbocations formed by the reaction of each alkene with HCl. Label each carbocation as primary, secondary, or tertiary.

(a) $CH_3CH_2\underset{|}{\overset{CH_3}{\overset{|}{C}}}=CHCH_3$

(b) $CH_3CH_2CH=CHCH_3$

(c) [cyclopentene]—CH_3

(d) [cyclohexylidene]=CH_2

12.38 Draw a structural formula for the product formed by treatment of 2-methyl-2-pentene with each reagent.
(a) HCl
(b) H_2O in the presence of H_2SO_4

12.39 Draw a structural formula for the product of each reaction.
(a) 1-Methylcyclohexene + Br_2
(b) 1,2-Dimethylcyclopentene + Cl_2

12.40 Draw a structural formula for an alkene with the indicated molecular formula that gives the compound shown as the major product. Note that more than one alkene may give the same compound as the major product.

(a) $C_5H_{10} + H_2O \xrightarrow{H_2SO_4} CH_3\underset{OH}{\underset{|}{\overset{CH_3}{\overset{|}{C}}CH_2CH_3}}$

(b) $C_5H_{10} + Br_2 \longrightarrow CH_3\underset{Br\;\;Br}{\underset{|\;\;\;|}{\overset{CH_3}{\overset{|}{C}HCHCH_2}}}$

(c) C_7H_{12} + HCl ⟶ [cyclohexane with CH_3 and Cl on same carbon]

12.41 Draw a structural formula for an alkene with molecular formula C_5H_{10} that reacts with Br_2 to give each product.

(a) $CH_3\underset{Br\;\;Br}{\underset{|\;\;\;|}{\overset{CH_3}{\overset{|}{C}}-CHCH_3}}$

(b) $CH_2\underset{Br\;\;Br}{\underset{|\;\;\;|}{\overset{CH_3}{\overset{|}{C}CH_2CH_3}}}$

(c) $\underset{Br\;\;Br}{\underset{|\;\;\;|}{CH_2CHCH_2CH_3}}$

12.42 Draw a structural formula for an alkene with molecular formula C_5H_{10} that reacts with HCl to give the indicated chloroalkane as the major product. More than one alkene may give the same compound as the major product.

(a) $CH_3\underset{Cl}{\underset{|}{\overset{CH_3}{\overset{|}{C}CH_2CH_3}}}$

(b) $CH_3\underset{Cl}{\underset{|}{\overset{CH_3}{\overset{|}{C}HCHCH_3}}}$

(c) $CH_3\underset{Cl}{\underset{|}{CHCH_2CH_2CH_3}}$

12.43 Draw the structural formula of an alkene that undergoes acid-catalyzed hydration to give the indicated alcohol as the major product. More than one alkene may give each alcohol as the major product.
  (a) 3-Hexanol
  (b) 1-Methylcyclobutanol
  (c) 2-Methyl-2-butanol
  (d) 2-Propanol

12.44 Terpin, $C_{10}H_{20}O_2$, is prepared commercially by the acid-catalyzed hydration of limonene (Figure 12.1).
  (a) Propose a structural formula for terpin.
  (b) How many cis-trans isomers are possible for the structural formula you propose?
  (c) Terpin hydrate, the isomer of terpin in which the methyl and isopropyl groups are trans to each other, is used as an expectorant in cough medicines. Draw a structural formula for terpin hydrate showing the trans orientation of these groups.

12.45 Draw the product formed by treatment of each alkene with H_2/Ni.

(a) H_3C\\C=C//H — H\\ //CH_2CH_3

(b) H\\C=C//H — H_3C\\ //CH_2CH_3

(c) cyclopentene

(d) 1-methylcyclopentene

12.46 Hydrocarbon A, C_5H_8, reacts with 2 moles of Br_2 to give 1,2,3,4-tetrabromo-2-methylbutane. What is the structure of hydrocarbon A?

12.47 Show how to convert ethylene to these compounds.
  (a) Ethane
  (b) Ethanol
  (c) Bromoethane
  (d) 1,2-Dibromoethane
  (e) Chloroethane

12.48 Show how to convert 1-butene to these compounds.
  (a) Butane
  (b) 2-Butanol
  (c) 2-Bromobutane
  (d) 1,2-Dibromobutane

Chemical Connections

12.49 (Chemical Connections 12A) What is one function of ethylene as a plant growth regulator?

12.50 (Chemical Connections 12B) What is the meaning of the term "pheromone"?

12.51 (Chemical Connections 12B) What is the molecular formula of 11-tetradecenyl acetate? What is its molecular weight?

12.52 (Chemical Connections 12B) Assume that 1×10^{-12} g of 11-tetradecenyl acetate is secreted by a single corn borer. How many molecules is this?

12.53 (Chemical Connections 12C) What different functions are performed by the rods and cones in the eye?

12.54 (Chemical Connections 12C) In which isomer of retinal is the end-to-end distance longer, the all-trans isomer or the 11-cis isomer?

12.55 (Chemical Connections 12D) What types of consumer products are made of high-density polyethylene? What types of products are made of low-density polyethylene? One type of polyethylene is currently recyclable and the other is not. Which is which?

12.56 (Chemical Connections 12D) In recycling codes, what do these abbreviations stand for?
  (a) V   (b) PP   (c) PS

Additional Problems

12.57 Write line-angle formulas for all compounds with molecular formula C_4H_8. Which are sets of constitutional isomers? Which are sets of cis-trans isomers?

12.58 Knowing what you do about the meaning of the terms "saturated" and "unsaturated" as applied to alkanes and alkenes, what do you think these same terms mean when they are used to describe animal fats such as those found in butter and animal meats? What might the term "polyunsaturated" mean in this same context?

12.59 Name and draw structural formulas for all alkenes with molecular formula C_6H_{12} that have these carbon skeletons. Remember to consider cis and trans isomers.

(a) C—C—C—C—C with C branch on second C

(b) C—C—C—C with C branches on second and third C

(c) C—C—C—C with two C branches on second C

342 / CHAPTER 12 ALKENES

12.60 Following is the structural formula of lycopene, $C_{40}H_{56}$, a deep-red compound that is partially responsible for the red color of ripe fruits, especially tomatoes. Approximately 20 mg of lycopene can be isolated from 1 kg of fresh, ripe tomatoes.
(a) Show that lycopene is a terpene—that is, its carbon skeleton can be divided into two sets of four isoprene units with the units in each set joined head-to-tail.
(b) How many of the carbon–carbon double bonds in lycopene have the possibility for cis-trans isomerism? Lycopene is the all-trans isomer.

12.61 As you might suspect, β-carotene, $C_{40}H_{56}$, precursor to vitamin A, was first isolated from carrots. Dilute solutions of β-carotene are yellow—hence its use as a food coloring. In plants, this compound is almost always present in combination with chlorophyll to assist in the harvesting of the energy of sunlight. As tree leaves die in the fall, the green of their chlorophyll molecules is replaced by the yellows and reds of carotene and carotene-related molecules. Compare the carbon skeletons of β-carotene and lycopene. What are the similarities? What are the differences?

12.62 Draw the structural formula for a cycloalkene with molecular formula C_6H_{10} that reacts with Cl_2 to give each compound.

(a) [structure with two Cl]
(b) [structure with Cl, Cl, CH₃]
(c) [structure with H₃C, Cl, Cl]
(d) [structure with Cl, CH₂Cl]

12.63 Propose a structural formula for the product(s) when each of the following alkenes is treated with H_2O/H_2SO_4. Why are two products formed in part (b), but only one in parts (a) and (c)?
(a) 1-Hexene gives one alcohol with molecular formula $C_6H_{14}O$.
(b) 2-Hexene gives two alcohols, each with molecular formula $C_6H_{14}O$.
(c) 3-Hexene gives one alcohol with molecular formula $C_6H_{14}O$.

12.64 cis-3-Hexene and trans-3-hexene are different compounds and have different physical and chemical properties. Yet, when treated with H_2O/H_2SO_4, each gives the same alcohol. What is the alcohol and how do you account for the fact that each alkene gives the same one?

12.65 Draw the structural formula of an alkene that undergoes acid-catalyzed hydration to give each of the following alcohols as the major product. More than one alkene may give each compound as the major product.

(a) [cyclohexanol]
(b) [1-methylcyclohexanol]
(c) [1,2-dimethylcyclopentanol with CH₃, OH, CH₃]
(d) [2-methyl-2-butanol]

12.66 Show how to convert cyclopentene into these compounds.
(a) 1,2-Dibromocyclopentane
(b) Cyclopentanol
(c) Iodocyclopentane
(d) Cyclopentane

InfoTrac College Edition

For additional readings, go to InfoTrac College Edition, your online research library, at

http://infotrac.thomsonlearning.com

Lycopene

β-Carotene

CHAPTER 13

13.1 Introduction
13.2 The Structure of Benzene
13.3 Nomenclature
13.4 Reactions of Benzene and Its Derivatives
13.5 Phenols

Peppers of the capsicum family (see Chemical Connections 13F).

Benzene and Its Derivatives

13.1 Introduction

So far we have described the three classes of hydrocarbons—alkanes, alkenes, and alkynes—called aliphatic hydrocarbons. More than 150 years ago, organic chemists realized that yet another class of hydrocarbons existed—one whose properties are quite different from those of aliphatic hydrocarbons. Because some of these new hydrocarbons have pleasant odors, they were called **aromatic compounds.** Today we know that not all aromatic compounds share this characteristic. Some do have a pleasing odors, but some have no odor at all, and others have downright unpleasant odors. A more appropriate definition of an aromatic compound is any compound that has one or more benzene-like rings.

We use the term **arene** to describe aromatic hydrocarbons. Just as a group derived by removal of an H from an alkane is called an alkyl group and given the symbol R—, so a group derived by removal of an H from an arene is called an **aryl group** and given the symbol **Ar—**.

Aromatic compound Benzene or one of its derivatives

Arene A compound containing one or more benzene rings

13.2 The Structure of Benzene

Benzene is an important compound in both the chemical industry and the laboratory, but it must be handled carefully. Not only is it poisonous if ingested in liquid form, but its vapor is also toxic and can be absorbed either by breathing or through the skin. Long-term inhalation can cause liver damage and cancer.

Benzene, the simplest aromatic hydrocarbon, was discovered by Michael Faraday (1791–1867) in 1825. Its structure presented an immediate problem to chemists of the day. Benzene has the molecular formula C_6H_6, and a compound with so few hydrogens for its six carbons (compare hexane, C_6H_{14}), chemists argued, should be unsaturated. But benzene does not behave like an alkene (the only class of unsaturated hydrocarbons known at that time). Whereas 1-hexene, for example, reacts instantly with Br_2 (Section 12.6C), benzene does not react at all with this reagent. Nor does benzene react with HBr, H_2O/H_2SO_4, or H_2/Pd, reagents that normally add to carbon–carbon double bonds.

A Kekulé's Structure of Benzene

The first structure for benzene was proposed by Friedrich August Kekulé in 1872 and consisted of a six-membered ring with alternating single and double bonds, with one hydrogen attached to each carbon.

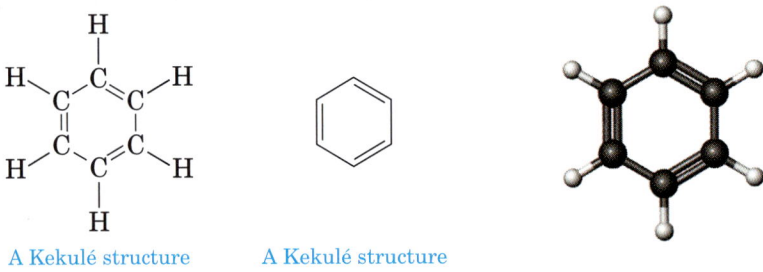

A Kekulé structure showing all atoms

A Kekulé structure as a line-angle formula

Although Kekulé's proposal was consistent with many of the chemical properties of benzene, it was contested for years. The major objection was its failure to account for the unusual chemical behavior of benzene. If benzene contains three double bonds, Kekulé's critics asked, why doesn't it undergo reactions typical of alkenes?

B Resonance Structure of Benzene

Resonance hybrid A molecule best described as a composite of two or more Lewis structures

The concept of resonance, developed by Linus Pauling in the 1930s, provided the first adequate description of the structure of benzene. According to the theory of resonance, certain molecules and ions are best described by writing two or more Lewis structures and considering the real molecule or ion to be a **resonance hybrid** of these structures. Each individual Lewis structure is called a **contributing structure.** To show that the real molecule is a resonance hybrid of the two Lewis structures, we position a double-headed arrow between them.

The two contributing structures for benzene are often called Kekulé structures.

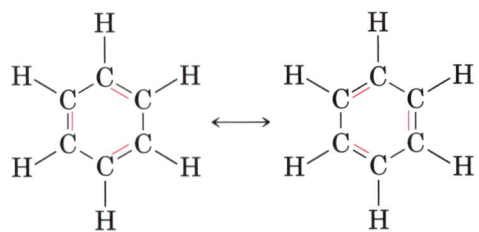

Alternative Lewis contributing structures for benzene

The resonance hybrid has some of the characteristics of each Lewis contributing structure. For example, the carbon–carbon bonds are neither single nor double but something intermediate. It has been determined experimentally that the length of the carbon–carbon bond in benzene is not as long as a carbon–carbon single bond nor as short as a carbon–carbon double bond, but rather midway between the two. The closed loop of six electrons (two from the second bond of each double bond) characteristic of a benzene ring is sometimes called an **aromatic sextet.**

Wherever we find resonance, we find stability. The real structure is generally more stable than any of the fictitious Lewis contributing structures. The benzene ring is greatly stabilized by resonance, which explains why it does not undergo the addition reactions typical of alkenes.

13.3 Nomenclature

■ Household products containing benzene derivatives.

A One Substituent

Monosubstituted alkylbenzenes are named as derivatives of benzene—as for example, ethylbenzene. The IUPAC system retains certain common names for several of the simpler monosubstituted alkylbenzenes, including **toluene** and **styrene.**

Ethylbenzene   Toluene   Styrene

The IUPAC system also retains the common names for the following compounds:

Phenol   Anisole   Aniline   Benzaldehyde   Benzoic acid

The substituent group derived by loss of an H from benzene is called a **phenyl group**, C_6H_5-, for which the common symbol is Ph—. In molecules containing other functional groups, phenyl groups are often named as substituents.

Phenyl group (C_6H_5-; Ph—)   1-Phenylcyclohexene   4-Phenyl-1-pentene

Phenyl group C_6H_5-, the aryl group derived by removing a hydrogen atom from benzene

B Two Substituents

When two substituents occur on a benzene ring, three isomers are possible. We locate the substituents either by numbering the atoms of the ring or by using the locators *ortho* (*o*), *meta* (*m*), and *para* (*p*). The numbers 1,2- are equivalent to *ortho* (Greek: straight); 1,3- to *meta* (Greek: after); and 1,4- to *para* (Greek: beyond).

1,2- or *ortho*-   1,3- or *meta*   1,4- or *para*

When one of the two substituents on the ring imparts a special name to the compound (for example, —CH$_3$, —OH, —NH$_2$, or —COOH), we name the compound as a derivative of that parent molecule, and assume that the substituent occupies ring position number 1. The IUPAC system retains the common name **xylene** for the three isomeric dimethylbenzenes. Where neither substituent imparts a special name, we locate the two substituents and list them in alphabetical order before the ending "benzene." The carbon of the benzene ring with the substituent of lower alphabetical ranking is numbered C-1.

p-Xylene is a starting material for the synthesis of poly(ethylene terephthalate). Consumer products derived from this polymer include Dacron polyester and Mylar films (Section 18.8).

4-Bromobenzoic acid (*p*-Bromobenzoic acid)   3-Chloroaniline (*m*-Chloroaniline)   1,3-Dimethylbenzene (*m*-Xylene)   1-Chloro-4-ethylbenzene (*p*-Chloroethylbenzene)

C Three or More Substituents

When three or more substituents are present on a benzene ring, we specify their locations by numbers. If one of the substituents imparts a special name, then we name the molecule as a derivative of that parent molecule. If none of the substituents imparts a special name, then we locate the substituents, number them to give the smallest set of numbers, and list them in alphabetical order before the ending "benzene." In the following examples, the first compound is a derivative of toluene, and the second is a derivative of phenol. Because no substituent in the third compound imparts a special name, we list its three substituents in alphabetical order followed by the word "benzene."

4-Chloro-2-nitrotoluene   2,4,6-Tribromophenol   2-Bromo-1-ethyl-4-nitrobenzene

CHEMICAL CONNECTIONS 13A

DDT: A Boon and a Curse

Probably the best-known insecticide worldwide is *d*ichloro*d*iphenyl*t*richloroethane (not an IUPAC name), commonly abbreviated DDT.

Dichlorodiphenyltrichloroethane
(DDT)

This compound was first prepared in 1874, but it was not until the late 1930s that its potential as an insecticide was recognized. First used for this purpose in 1939, it proved extremely effective in ridding large areas of the world of the insect hosts that transmit malaria and typhus. In addition, it was so effective in killing crop-destroying insect pests that crop yields in many areas of the world increased dramatically.

Widespread use of DDT, however, has been a double-edged sword. Despite DDT's well-known benefits, it has an enormous disadvantage. Because it resists biodegradation, it remains in the soil for years—and this persistence in the environment creates the problem. Scientists estimate that the tissues of adult humans contain, on average, five to ten parts per million of DDT.

The dangers associated with the persistence of DDT in the environment were dramatically portrayed by Rachel Carson in her 1962 book *The Silent Spring*, which documented the serious

■ DDT was sprayed on crops in the United States until its use was banned in the 1970s.

decline in the population of eagles and other raptors as well as many other kinds of birds. Scientists discovered soon thereafter that DDT inhibits the mechanism by which these birds incorporate calcium into their egg shells, with the result that shells become so thin and weak that they break during incubation.

Because of these problems, almost all nations have now banned the use of DDT for agricultural purposes. It is still used, however, in some areas to control the population of disease-spreading insects.

EXAMPLE 13.1

Write names for these compounds.

(a) 3-methyl-1-iodobenzene structure   (b) 3,5-dibromobenzoic acid structure   (c) 4-chloroaniline structure

Solution
(a) The parent is toluene, and the compound is 3-iodotoluene or *m*-iodotoluene.
(b) The parent is benzoic acid, and the compound is 3,5-dibromobenzoic acid.
(c) The parent is aniline, and the compound is 4-chloroaniline or *p*-chloroaniline.

Problem 13.1
Write names for these compounds.

(a) 2,4,6-tri-tert-butylphenol structure with OH
(b) 2,4-dichloroaniline structure with NH₂ and two Cl
(c) 3-nitrobenzoic acid structure with COOH and NO₂

D Polynuclear Aromatic Hydrocarbons

Polynuclear aromatic hydrocarbon A hydrocarbon containing two or more benzene rings, each of which shares two carbon atoms with another benzene ring

Polynuclear aromatic hydrocarbons (PAHs) contain two or more benzene rings, with each pair of rings sharing two adjacent carbon atoms. Naphthalene, anthracene, and phenanthrene, the most common PAHs, and substances derived from them are found in coal tar and high-boiling petroleum residues. At one time, naphthalene was used as moth balls and an insecticide in preserving woolens and furs, but its use decreased after the introduction of chlorinated hydrocarbons such as *p*-dichlorobenzene.

Naphthalene   Anthracene   Phenanthrene

CHEMICAL CONNECTIONS 13B

Carcinogenic Polynuclear Aromatics and Smoking

A **carcinogen** is a compound that causes cancer. The first carcinogens to be identified were a group of polynuclear aromatic hydrocarbons, all of which have at least four aromatic rings. Among them is benzo[a]pyrene, one of the most carcinogenic of the aromatic hydrocarbons. It is formed whenever there is incomplete combustion of organic compounds. Benzo[a]pyrene is found, for example, in cigarette smoke, automobile exhaust, and charcoal-broiled meats.

Benzo[a]pyrene causes cancer in the following way. Once it is absorbed or ingested, the body attempts to convert it into a more soluble compound that can be excreted easily. By a series of enzyme-catalyzed reactions, benzo[a]pyrene is transformed into a **diol** (two —OH groups) **epoxide** (a three-membered ring, one atom of which is oxygen). This compound can bind to DNA by reacting with one of its amino groups, thereby altering the structure of DNA and producing a cancer-causing mutation.

Benzo[a]pyrene →(enzyme-catalyzed oxidation)→ A diol epoxide

CHEMICAL CONNECTIONS 13C

Iodide Ion and Goiter

One hundred years ago, goiter, an enlargement of the thyroid gland caused by iodine deficiency, was common in the central United States and central Canada. This disease results from underproduction of thyroxine, a hormone synthesized in the thyroid gland. Young mammals require this hormone for normal growth and development. Its lack during fetal development results in mental retardation. Low levels of thyroxine in adults result in hypothyroidism, commonly called goiter, the symptoms of which are lethargy, obesity, and dry skin.

Iodine is an element that comes primarily from the sea. Rich sources of it, therefore, are fish and other seafoods. The iodine in our diets that doesn't come from the sea most commonly is derived from food additives. Most of the iodide ion in the North American diet comes from table salt fortified with sodium iodide, commonly referred to as iodized salt. Another source is dairy products, which accumulate iodide because of the iodine-containing additives used in cattle feeds and the iodine-containing disinfectants used on milking machines and milk storage tanks.

Thyroxine

13.4 Reactions of Benzene and Its Derivatives

By far the most characteristic reaction of aromatic compounds is substitution at a ring carbon, which we give the name **aromatic substitution.** Groups we can introduce directly on the ring include the halogens, the nitro ($-NO_2$) group, and the sulfonic acid ($-SO_3H$) group.

Screen 13.4
Simulation

A Halogenation

As noted in Section 13.2, chlorine and bromine do not react with benzene, in contrast to their instantaneous reaction with cyclohexene and other alkenes (Section 12.6C). In the presence of an iron catalyst, however, chlorine reacts rapidly with benzene to give chlorobenzene and HCl:

Benzene + Cl_2 $\xrightarrow{FeCl_3}$ Chlorobenzene + HCl

Treatment of benzene with bromine in the presence of $FeCl_3$ results in formation of bromobenzene and HBr.

B Nitration

When we heat benzene or one of its derivatives with a mixture of concentrated nitric and sulfuric acids, one of the hydrogen atoms bonded to the ring is replaced by a nitro ($-NO_2$) group.

Benzene + HNO_3 $\xrightarrow{H_2SO_4}$ Nitrobenzene + H_2O

CHEMICAL CONNECTIONS 13D

The Nitro Group in Explosives

Treatment of toluene with three moles of nitric acid in the presence of sulfuric acid as a catalyst results in nitration of toluene three times to form the explosive 2,4,6-trinitrotoluene, TNT. It is the presence of these three nitro groups that confers the explosive properties to TNT. Similarly, it is the presence of three nitro groups that confers the explosive properties to nitroglycerin.

In recent years, several new explosives have been discovered, all of which contain multiple nitro groups. Among them are RDX and PETN. The plastic explosive Semtex, for example, is a mixture of RDX and PETN. It was used in the destruction of Pan Am flight 103 over Lockerbie, Scotland, in December 1988.

2,4,6-Trinitrotoluene (TNT)

Trinitroglycerin (Nitroglycerin)

Cyclonite (RDX)

Pentaerythritol tetranitrate (PETN)

A particular value of nitration is that we can reduce the resulting —NO_2 group to a primary amino group, —NH_2, by catalytic reduction using hydrogen in the presence of a transition-metal catalyst. In the following example, neither the benzene ring nor the carboxyl group is affected by these experimental conditions:

O_2N—⟨benzene⟩—COOH + 3H_2 $\xrightarrow[\text{3 atm}]{\text{Ni}}$ H_2N—⟨benzene⟩—COOH + 2H_2O

4-Nitrobenzoic acid (*p*-Nitrobenzoic acid)

4-Aminobenzoic acid (*p*-Aminobenzoic acid, PABA)

Bacteria require *p*-aminobenzoic acid to synthesize folic acid (Section 29.5), which is in turn required for the synthesis of the heterocyclic aromatic amine bases of nucleic acids (Section 24.2). Whereas bacteria can synthesize folic acid from *p*-aminobenzoic acid, folic acid is a vitamin for humans and must be obtained in the diet. Chemical Connections 22D describes how sulfa drugs inhibit the synthesis of this vitamin.

C Sulfonation

Heating an aromatic compound with concentrated sulfuric acid results in formation of a sulfonic acid, all of which are strong acids, comparable in strength to sulfuric acid.

⟨benzene⟩—H + H_2SO_4 ⟶ ⟨benzene⟩—SO_3H + H_2O

Benzenesulfonic acid

A major use of sulfonation is in the preparation of synthetic detergents, an important example of which is sodium 4-dodecylbenzenesulfonate. To prepare this type of detergent, a linear alkylbenzene such as dodecylben-

CHEMICAL CONNECTIONS 13E

FD & C No. 6 (a.k.a. Sunset Yellow)

Did you ever wonder what gives gelatin desserts their red, green, orange, or yellow color? Or what gives margarines their yellow color? Or what gives maraschino cherries their red color? If you read the content labels, you will see code names such as FD & C Yellow No. 6 and FD & C Red No. 40.

At one time, the only colorants for foods were compounds obtained from plant or animal materials. Beginning as early as the 1890s, however, chemists discovered a series of synthetic food dyes that offered several advantages over natural dyes, such as greater brightness, better stability, and lower cost. Opinion remains divided on the safety of their use. No synthetic food colorings are allowed, for example, in Norway and Sweden. In the United States, the Food and Drug Administration (FDA) has certified seven synthetic dyes for use in foods, drugs, and cosmetics (FD & C). They include two yellows, two reds, two blues, and one green. When these dyes are used alone or in combinations, they can approximate the color of almost any natural food.

Following are structural formulas for Allura Red (Red No. 40) and Sunset Yellow (Yellow No. 6). These and the other five food dyes certified in the United States have in common three or more benzene rings and two or more ionic groups, either the sodium salt of a carboxylic acid group, $-COO^-Na^+$, or the sodium salt of sulfonic acid group, $-SO_3^-Na^+$. These ionic groups make the dyes soluble in water.

To return to our original questions, maraschino cherries are colored with FD & C Red No. 40, and margarines are colored with FD & C Yellow No 6. Gelatin desserts use either one or a combination of the seven certified food dyes to produce their color.

Allura Red
(FD & C Red No. 40)

Sunset Yellow
(FD & C Yellow No. 6)

---

zene is treated with concentrated sulfuric acid to give an alkylbenzenesulfonic acid. The sulfonic acid is then neutralized with sodium hydroxide.

Dodecylbenzene → Sodium 4-dodecylbenzenesulfonate, SDS (an anionic detergent)

Alkylbenzenesulfonate detergents were introduced in the late 1950s, and today they claim nearly 90% of the market once held by natural soaps. Chemical Connections 20C discusses the chemistry and cleansing action of soaps and detergents.

13.5 Phenols

A Structure and Nomenclature

The functional group of a **phenol** is a hydroxyl group bonded to a benzene ring. Substituted phenols are named either as derivatives of phenol or by common names.

> **Phenol** A compound that contains an —OH bonded to a benzene ring

Phenol in crystalline form.

Phenol — 3-Methylphenol (*m*-Cresol) — 1,2-Benzenediol (Catechol) — 1,3-Benzenediol (Resorcinol) — 1,4-Benzenediol (Hydroquinone)

Phenols are widely distributed in nature. Phenol itself and the isomeric cresols (*o*-, *m*-, and *p*-cresol) are found in coal tar. Thymol and vanillin are important constituents of thyme and vanilla beans, respectively. Urushiol is the main component of the irritating oil of poison ivy. It can cause severe contact dermatitis in sensitive individuals.

2-Isopropyl-5-methylphenol (Thymol) — 4-Hydroxy-3-methoxy-benzaldehyde (Vanillin) — Urushiol

Vanilla beans.

B Acidity of Phenols

Phenols are weak acids, with pK_a values of approximately 10 (Table 8.3). Most phenols are insoluble in water, but they react with strong bases, such as NaOH and KOH, to form water-soluble salts.

C₆H₅—OH + NaOH ⟶ C₆H₅—O⁻Na⁺ + H₂O

Phenol
pK_a = 9.95
(stronger acid)

Sodium hydroxide
(stonger base)

Sodium phenoxide
(weaker base)

Water
pK_a = 15.7
(weaker acid)

Most phenols are such weak acids that they do not react with weak bases such as sodium bicarbonate; that is, they do not dissolve in aqueous sodium bicarbonate.

C Phenols as Antioxidants

An important reaction for living systems, foods, and other materials that contain carbon–carbon double bonds is **autoxidation**—that is, oxidation requiring oxygen and no other reactant. If you open a bottle of cooking oil that has stood for a long time, you will notice a hiss of air entering the bottle. This sound occurs because the consumption of oxygen by autoxidation of the oil creates a negative pressure inside the bottle.

Cooking oils contain esters of polyunsaturated fatty acids. You need not worry now about what esters are; we will discuss them in Chapter 18. The important point here is that all vegetable oils contain fatty acids with long hydrocarbon chains, many of which have one or more carbon–carbon double bonds (see Problems 12.24 and 12.26 for the structures of four of these fatty acids). Autoxidation takes place adjacent to these double bonds.

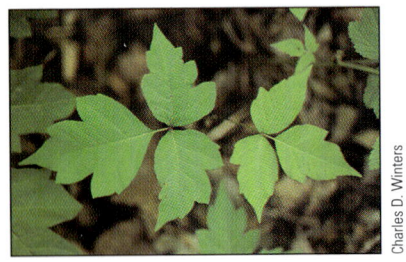

Poison ivy.

CHEMICAL CONNECTIONS 13F

Capsaicin, for Those Who Like It Hot

Capsaicin, the pungent principal from the fruit of various species of peppers (*Capsicum* and *Solanaceae*), was isolated in 1876, and its structure was determined in 1919. Capsaicin contains both a phenol and a phenol ether.

Capsaicin
(from various types of peppers)

The inflammatory properties of capsaicin are well known; the human tongue can detect as little as one drop in 5 L of water. We all know of the burning sensation in the mouth and sudden tearing in the eyes caused by a good dose of hot chili peppers. For this reason, capsaicin-containing extracts from these flaming foods are used in sprays to ward off dogs or other animals that might nip at your heels while you are running or cycling.

Ironically, capsaicin is able to both cause and relieve pain. Currently, two capsaicin-containing creams, Mioton and Zostrix, are prescribed to treat the burning pain associated with postherpetic neuralgia, a complication of the disease known as shingles. They are also prescribed for diabetics to relieve persistent foot and leg pain.

■ Red chilies being dried in the sun.

Autoxidation is a radical chain process that converts an R—H group to an R—O—O—H group, called a hydroperoxide. This process begins when a hydrogen atom with one of its electrons (H·) is removed from a carbon adjacent to one of the double bonds in a hydrocarbon chain. The carbon losing the H· has only seven electrons in its valence shell, one of which is unpaired. An atom or molecule with an unpaired electron is called a **radical**.

Step 1: Chain Initiation—Formation of a Radical from a Nonradical Compound

Removal of a hydrogen atom (H·) may be initiated by light or heat. The product formed is a carbon radical; that is, it contains a carbon atom with one unpaired electron.

$$-CH_2CH=CH-CH- \xrightarrow{\text{light or heat}} -CH_2CH=CH-\overset{\cdot}{CH}-$$

Section of a fatty acid hydrocarbon chain → A carbon radical

Step 2a: Chain Propagation—Reaction of a Radical to Form a New Radical

The carbon radical reacts with oxygen, itself a diradical, to form a hydroperoxy radical. The new covalent bond of the hydroperoxy radical forms by the combination of one electron from the carbon radical and one electron from the oxygen diradical.

$$-CH_2CH=CH-\overset{\bullet}{C}H- \;+\; \cdot O-O\cdot \;\longrightarrow\; -CH_2CH=CH-\underset{}{\overset{\overset{O-O\cdot}{|}}{C}H}-$$

Oxygen is a diradical A hydroperoxy radical

Step 2b: Chain Propagation—Reaction of a Radical to Form a New Radical

The hydroperoxy radical removes a hydrogen atom (H·) from a new fatty acid hydrocarbon chain to complete the formation of a hydroperoxide and at the same time produce a new carbon radical.

$$-CH_2CH=CH-\underset{}{\overset{\overset{O-O\cdot}{|}}{C}H}- \;+\; -CH_2CH=CH-\underset{}{\overset{\overset{H}{|}}{C}H}- \;\longrightarrow$$

Section of a new fatty acid hydrocarbon chain

$$-CH_2CH=CH-\underset{}{\overset{\overset{O-O-H}{|}}{C}H}- \;+\; -CH_2CH=CH-\overset{\bullet}{C}H-$$

A hydroperoxide A new carbon radical

The most important point about the pair of chain propagation steps is that they form a continuous cycle of reactions. The new radical formed in Step 2b next reacts with another molecule of O₂ by step 2a to give a new hydroperoxy radical. This new hydroperoxy radical then reacts with a new hydrocarbon chain to repeat Step 2b, and so forth. This cycle of propagation steps repeats over and over in a chain reaction. Thus, once a radical is generated in Step 1, the cycle of propagation steps may repeat many thousands of times, generating thousands and thousands of hydroperoxide molecules. The number of times the cycle of chain propagation steps repeats is called the **chain length.**

Hydroperoxides themselves are unstable and, under biological conditions, degrade to short-chain aldehydes and carboxylic acids with unpleasant "rancid" smells. These odors may be familiar to you if you have ever smelled old cooking oil or aged foods that contain polyunsaturated fats or oils. Similar formation of hydroperoxides in the low-density lipoproteins (Section 26.4) deposited on the walls of arteries leads to cardiovascular disease in humans. In addition, many effects of aging are thought to result from the formation and subsequent degradation of hydroperoxides.

Fortunately, nature has developed a series of defenses against the formation of these and other destructive hydroperoxides, including the phenol vitamin E (Section 29.5). This compound is one of nature's "scavengers." It inserts itself into either Step 2a or 2b, donates an H· from its —OH group to the carbon radical, and converts the carbon radical back to its original hydrocarbon chain. Because the vitamin E radical is stable, it breaks the cycle of chain propagation steps, thereby preventing further formation of destructive hydroperoxides. While some hydroperoxides may form, their numbers are very small and they are easily decomposed to harmless materials by one of several enzyme-catalyzed reactions.

Vitamin E

Butylated hydroxy-toluene (BHT)

Butylated hydroxy-anisole (BHA)

Unfortunately, vitamin E is removed in the processing of many foods and food products. To make up for this loss, phenols such as BHT and BHA are added to foods to "retard spoilage" (as they say on the packages) by autoxidation. Likewise, similar compounds are added to other materials, such as plastics and rubber, to protect them against autoxidation.

SUMMARY

Benzene and its alkyl derivatives are classified as **aromatic hydrocarbons** or **arenes** (Section 13.1). The first structure for benzene was proposed by Friedrich August Kekulé in 1872 (Section 13.2A). The theory of **resonance**, developed by Linus Pauling in the 1930s, provided the first adequate structure for benzene (Section 13.2B).

Aromatic compounds are named according to the IUPAC system (Section 13.3). The C_6H_5— group is named **phenyl**. Two substituents on a benzene ring may be located by either numbering the atoms of the ring or by using the locators **ortho (o), meta (m),** and **para (p)** (Section 13.3B). **Polynuclear aromatic hydrocarbons** contain two or more benzene rings, each sharing two adjacent carbon atoms with another ring (section 13.3D).

A characteristic reaction of aromatic compounds is **aromatic substitution,** in which another atom or group of atoms is substituted for a hydrogen atom of the aromatic ring (Section 13.4). Typical aromatic substitution reactions are halogenation, nitration, and sulfonation.

The functional group of a **phenol** is an —OH group bonded to a benzene ring (Section 13.5A). Phenol and its derivatives are weak acids, with pK_a equal to approximately 10.0 (Section 13.5B). Vitamin E, a phenolic compound, is an antioxidant (Section 13.5C). Phenolic compounds such as BHT and BHA are synthetic antioxidants.

KEY REACTIONS

1. **Halogenation (Section 13.4A)** Treatment of an aromatic compound with Cl_2 or Br_2 in the presence of an $FeCl_3$ catalyst substitutes a halogen for an H.

$$C_6H_6 + Cl_2 \xrightarrow{FeCl_3} C_6H_5-Cl + HCl$$

2. **Nitration (Section 13.4B)** Treatment of an aromatic compound with a mixture of concentrated nitric and sulfuric acids substitutes a nitro group for an H.

$$C_6H_6 + HNO_3 \xrightarrow{H_2SO_4} C_6H_5-NO_2 + H_2O$$

3. **Sulfonation (Section 13.4C)** Treatment of an aromatic compound with concentrated sulfuric acid substitutes a sulfonic acid group for an H.

$$C_6H_6 + H_2SO_4 \longrightarrow C_6H_5-SO_3H + H_2O$$

4. **Reaction of Phenols with Strong Bases (Section 13.5B)** Phenols are weak acids and react with strong bases to form water-soluble salts.

$$C_6H_5-OH + NaOH \longrightarrow C_6H_5-O^-Na^+ + H_2O$$

PROBLEMS

Numbers that appear in color indicate difficult problems. ▶ designates problems requiring application of principles.

Aromatic Compounds

13.2 What is the difference in structure between a saturated and an unsaturated compound?

13.3 Define *aromatic compound*.

13.4 Why are alkenes, alkynes, and aromatic compounds said to be unsaturated?

13.5 Do aromatic rings have double bonds? Are they unsaturated? Explain.

13.6 Can an aromatic compound be a saturated compound?

13.7 Draw at least two structural formulas for the following. (Several constitutional isomers are possible for each part.)
   (a) An alkene of six carbons
   (b) A cycloalkene of six carbons
   (c) An alkyne of six carbons
   (d) An aromatic hydrocarbon of eight carbons

13.8 Write a structural formula and the name for the simplest (a) alkane, (b) alkene, (c) alkyne, and (d) aromatic hydrocarbon.

13.9 Account for the fact that the six-membered ring in benzene is planar but the six-membered ring in cyclohexane is not.

13.10 The compound 1,4-dichlorobenzene (*p*-dichlorobenzene) has a rigid geometry that does not allow free rotation. Yet no cis-trans isomers exist for this structure. Explain why it does not show cis-trans isomerism.

13.11 One analogy often used to explain the concept of a resonance hybrid is to relate a rhinoceros to a unicorn and a dragon. Explain the reasoning in this analogy and how it might relate to a resonance hybrid.

Nomenclature and Structural Formulas

13.12 Name these compounds.

(a) *p-nitrochlorobenzene structure (NO₂ and Cl para)*

(b) *o-bromotoluene structure (CH₃ and Br ortho)*

(c) $C_6H_5CH_2CH_2CH_2Cl$

(d) $C_6H_5\overset{Br}{\underset{CH_3}{C}}CH_2CH_3$

(e) *benzene with NH₂ and NO₂ ortho*

(f) *benzene with OH and C₆H₅ ortho*

(g) $\underset{H}{\overset{C_6H_5}{>}}C=C\underset{C_6H_5}{\overset{H}{<}}$

(h) *benzene with CH₃, Cl (2), Cl (4)*

13.13 Draw structural formulas for these compounds.
   (a) 1-Bromo-2-chloro-4-ethylbenzene
   (b) 4-Bromo-1,2-dimethylbenzene
   (c) 2,4,6-Trinitrotoluene
   (d) 4-Phenyl-1-pentene
   (e) *p*-Cresol
   (f) 2,4-Dichlorophenol

13.14 We say that naphthalene, anthracene, phenanthrene, and benzo[a]pyrene are polynuclear aromatic hydrocarbons. In this context, what does "polynuclear" mean? What does "aromatic" mean? What does "hydrocarbon" mean?

Reactions of Aromatic Compounds

13.15 Suppose you have unlabeled bottles of benzene and cyclohexene. What chemical reaction could you use to tell which bottle contains which chemical? Explain what you would do, what you would expect to see, and how you would explain your observations.

13.16 Three possible products with molecular formula C_6H_4BrCl can form when bromobenzene is treated with chlorine, Cl_2, in the presence of $FeCl_3$ as a catalyst. Name and draw a structural formula for each product.

13.17 The reaction of bromine with toluene in the presence of FeCl₃ gives a mixture of three products, all with the molecular formula C_7H_7Br. Name and draw a structural formula for each product.

13.18 What reagents and/or catalysts are necessary to carry out each conversion? Each conversion requires only one step.
(a) Benzene to nitrobenzene
(b) 1,4-dichlorobenzene to 2-bromo-1,4-dichlorobenzene

13.19 What reagents and/or catalysts are necessary to carry out each conversion? Each conversion requires two steps.
(a) Benzene to 3-nitrobenzenesulfonic acid
(b) Benzene to 1-bromo-4-chlorobenzene

13.20 Aromatic substitution can be done on naphthalene. Treatment of naphthalene with concentrated H_2SO_4 gives two (and only two) different sulfonic acids. Draw a structural formula for each.

Phenols

13.21 Both phenol and cyclohexanol are only slightly soluble in water. Account for the fact that phenol dissolves in aqueous sodium hydroxide but cyclohexanol does not.

13.22 Define *autoxidation*.

13.23 Autoxidation is described as a radical chain reaction. What is meant by the term "radical" in this context? By the term "chain"? By the term "chain length"?

13.24 Show that, if you add Steps 2a and 2b of the radical chain mechanism for the autoxidation of a fatty acid hydrocarbon chain, you arrive at the following net equation:

$$—CH_2CH=CH—\overset{H}{\underset{|}{CH}}— + \cdot O—O\cdot$$
Section of a fatty acid hydrocarbon chain   Oxygen

$$\longrightarrow —CH_2CH=CH—\overset{O—O—H}{\underset{|}{CH}}—$$
A hydroperoxide

13.25 How does vitamin E function as an antioxidant?

13.26 What structural features are common to vitamin E, BHT, and BHA (the three antioxidants presented in Section 13.5C)?

Chemical Connections

13.27 (Chemical Connections 13A) From what parts of its common name are the letters *DDT* derived?

13.28 (Chemical Connections 13A) What are the advantages and disadvantages of using DDT as an insecticide?

13.29 (Chemical Connections 13A) Would you expect DDT to be soluble or insoluble in water? Explain.

13.30 (Chemical Connections 13A) One of the degradation products of DDT is dichlorodiphenyldichloroethylene (DDE), which is formed by the loss of HCl from DDT. DDE inhibits the enzyme responsible for the incorporation of calcium ion into bird egg shells. Draw a structural formula for DDE.

13.31 (Chemical Connections 13A) What is meant by the term "biodegradable"?

13.32 (Chemical Connections 13B) What is a carcinogen? What kind of carcinogens are found in cigarette smoke?

13.33 (Chemical Connections 13C) In the absence of iodine in the diet, goiter develops. Explain why goiter is a regional disease.

13.34 (Chemical Connections 13D) Calculate the molecular weight of each explosive shown in this Chemical Connection. In which explosive do the nitro groups contribute the largest percentage of molecular weight?

13.35 (Chemical Connections 13E) What are the differences in structure between Allura Red and Sunset Yellow?

13.36 (Chemical Connections 13E) Which feature of Allura Red and Sunset Yellow makes them water-soluble?

13.37 (Chemical Connections 13E) What color would you get if you mix Allura Red and Sunset Yellow? (*Hint:* Remember the color wheel.)

13.38 (Chemical Connections 13E) Problems 12.60 and 12.61 gave structural formulas for lycopene and β-carotene, two natural food-coloring agents. Compare the structural formulas for these compounds with those of Allura Red and Sunset Yellow.
(a) What structural feature or features do these four compounds have in common that might account for the fact that each is colored?
(b) Why are these two food and cosmetic dyes soluble in water, while lycopene and β-carotene are not?

13.39 (Chemical Connections 13F) From what types of plants is capsaicin isolated?

13.40 (Chemical Connections 13F) Identify the phenol group in capsaicin.

13.41 (Chemical Connections 13F) How many cis-trans isomers are possible for capsaicin? Is the structural formula shown the cis isomer or the trans isomer?

13.42 (Chemical Connections 13F) In what ways is capsaicin used in medicine?

Additional Problems

13.43 The structure for naphthalene given in Section 13.3D is only one of three possible resonance structures. Draw the other two.

13.44 Draw structural formulas for these compounds.
(a) 1-Phenylcyclopropanol
(b) Styrene
(c) *m*-Bromophenol
(d) 4-Nitrobenzoic acid
(e) Isobutylbenzene
(f) *m*-Xylene

13.45 2,6-Di-*tert*-butyl-4-methylphenol (BHT, Section 13.5C) is an antioxidant added to processed foods to "retard spoilage." How does BHT accomplish this goal?

13.46 Write the structural formula for the product of each reaction.

(a) [benzene] + HNO$_3$ $\xrightarrow{H_2SO_4}$

(b) [1,4-dimethylbenzene] + Br$_2$ $\xrightarrow{FeCl_3}$

(c) [1,4-dibromobenzene] + H$_2$SO$_4$ $\longrightarrow$

13.47 Styrene reacts with bromine to give a compound with molecular formula $C_8H_8Br_2$. Draw a structural formula for this compound.

InfoTrac College Edition

For additional readings, go to InfoTrac College Edition, your online research library, at

http://infotrac.thomsonlearning.com

CHAPTER 14

14.1 Introduction
14.2 Alcohols
14.3 Reactions of Alcohols
14.4 Ethers
14.5 Thiols
14.6 Important Alcohols

Fermentation vats of wine grapes at the Beaulieu Vineyards, California.

Alcohols, Ethers, and Thiols

14.1 Introduction

In this chapter, we study the physical and chemical properties of alcohols and ethers, two classes of oxygen-containing organic compounds. We also study thiols, a class of sulfur-containing organic compounds. Thiols are like alcohols in structure, except that they contain an —SH group rather than an —OH group.

$$CH_3CH_2OH \qquad CH_3CH_2OCH_2CH_3 \qquad CH_3CH_2SH$$
Ethanol   Diethyl ether   Ethanethiol
(An alcohol)   (An ether)   (A thiol)

These three compounds are certainly familiar to you. Ethanol is the fuel additive in gasohol, the alcohol in alcoholic beverages, and an important laboratory and industrial solvent. Diethyl ether was the first inhalation anesthetic used in general surgery. It too is an important laboratory and industrial solvent. Ethanethiol, like other low-molecular-weight thiols, has a stench. Traces of ethanethiol are added to natural gas so that gas leaks can be detected by the smell of the thiol.

Figure 14.1 Methanol, CH_3OH. (*a*) Lewis structure and (*b*) ball-and-stick model. The H—C—O bond angle is 108.6°, very close to the tetrahedral angle of 109.5°.

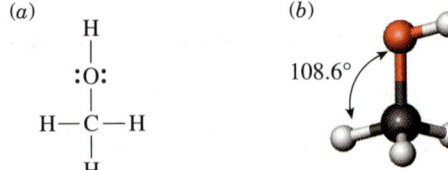

14.2 Alcohols

A Structure

The functional group of an **alcohol** is an **—OH (hydroxyl) group** bonded to a tetrahedral carbon atom (Section 10.4A). Figure 14.1 shows a Lewis structure and a ball-and-stick model of methanol, CH_3OH, the simplest alcohol.

B Nomenclature

IUPAC names of alcohols are derived in the same manner as those for alkenes and alkynes, with the exception that the ending of the parent alkane is changed from *-e* to *-ol*. The ending *-ol* tells us that the compound is an alcohol.

1. Select the longest carbon chain that contains the —OH group as the parent alkane and number it from the end that gives the —OH the lower number. In numbering the parent chain, the location of the —OH group takes precedence over alkyl and aryl groups, and halogens.

2. Change the ending of the parent alkane from **-e** to **-ol,** and use a number to show the location of the —OH group. For cyclic alcohols, numbering begins at the carbon bearing the —OH group; this carbon is automatically carbon-1.

3. Name and number substituents, and list them in alphabetical order.

To derive common names for alcohols, we name the alkyl group bonded to —OH and then add the word "alcohol." Following are IUPAC names and, in parentheses, common names for eight low-molecular-weight alcohols:

Solutions in which ethanol is the solvent are called tinctures.

EXAMPLE 14.1

Write the IUPAC name for each alcohol.

(a) [structure: 4-methyl-2-pentanol skeletal with OH]

(b) [structure: cyclohexane with OH wedge and methyl dash — trans-2-methylcyclohexanol]

Solution

(a) The parent alkane is pentane. We number the parent chain from the direction that gives the lower number to the carbon bearing the —OH group. This alcohol is 4-methyl-2-pentanol.
(b) The parent cycloalkane is cyclohexane. We number the atoms of the ring beginning with the carbon bearing the —OH group as carbon 1. This alcohol is *trans*-2-methylcyclohexanol.

Problem 14.1

Write the IUPAC name for each alcohol.

(a) [structure: hexane with OH on C2]

(b) [structure: neopentyl alcohol, (CH₃)₃C—CH₂OH]

(c) [structure: cyclohexane with OH wedge and isopropyl dash]

We classify alcohols as **primary (1°), secondary (2°),** or **tertiary (3°),** depending on the number of carbon groups bonded to the carbon bearing the —OH group (Section 10.4A).

EXAMPLE 14.2

Classify each alcohol as primary, secondary, or tertiary.

(a) [structure: cyclohexyl—C(OH)(H)(CH₃)]

(b) $CH_3\underset{\underset{CH_3}{|}}{\overset{\overset{CH_3}{|}}{C}}OH$

(c) [structure: cyclopentyl—CH₂OH]

Solution
(a) Secondary (2°) (b) Tertiary (3°)
(c) Primary (1°)

Problem 14.2
Classify each alcohol as primary, secondary, or tertiary.

(a)   (b) ▷—OH

(c) ⌒⌒—OH   (d) ⬠—OH

CHEMICAL CONNECTIONS 14A

Nitroglycerin, an Explosive and a Drug

In 1847, Ascanio Sobrero (1812–1888) discovered that treatment of 1,2,3-propanetriol, more commonly named glycerin, with nitric acid in the presence of sulfuric acid gives a pale yellow, oily liquid called nitroglycerin. Sobrero also discovered the explosive properties of this compound; when he heated a small quantity of it, it exploded!

$$\begin{array}{c}CH_2-OH\\|\\CH-OH\\|\\CH_2-OH\end{array} + 3HNO_3 \xrightarrow{H_2SO_4} \begin{array}{c}CH_2-ONO_2\\|\\CH-ONO_2\\|\\CH_2-ONO_2\end{array} + 3H_2O$$

1,2,3-Propanetriol (Glycerol, Glycerin) → 1,2,3-Propanetriol trinitrate (Nitroglycerin)

Nitroglycerin very soon became widely used for blasting in the construction of canals, tunnels, roads, and mines and, of course, for warfare.

One problem with the use of nitroglycerin was soon recognized: It was difficult to handle safely, and accidental explosions occurred frequently. This problem was solved by the Swedish chemist Alfred Nobel (1833–1896), who discovered that a clay-like substance called diatomaceous earth absorbs nitroglycerin so that it will not explode without a fuse. He gave the name *dynamite* to this mixture of nitroglycerin, diatomaceous earth, and sodium carbonate.

Surprising as it may seem, nitroglycerin is used in medicine to treat angina pectoris, the symptoms of which are sharp chest pains caused by reduced flow of blood in the coronary artery. Nitroglycerin, which is available in liquid form (diluted with alcohol to render it nonexplosive), tablet form, or paste form, relaxes the smooth muscles of blood vessels, causing dilation of the coronary artery. This dilation, in turn, allows more blood to reach the heart.

When Nobel became ill with heart disease, his physicians advised him to take nitroglycerin to relieve his chest pains. He refused, saying he could not understand how the explosive could relieve chest pains. It took science more than 100 years to find the answer. We now know that it is nitric oxide, NO, derived from the nitro groups of nitroglycerin, that relieves the pain. (See Chemical Connections 23 F.)

■ (*a*) Nitroglycerin is more stable if absorbed onto an inert solid, a combination called dynamite. (*b*) The fortune of Alfred Nobel, 1833–1896, built on the manufacture of dynamite, now funds the Nobel Prizes.

(*a*)   (*b*)

In the IUPAC system, we name a compound containing two hydroxyl groups as a **diol,** one containing three hydroxyl groups as a **triol,** and so on. We retain the final *-e* in the name of the parent alkane—for example, 1,2-ethanediol.

As with many other organic compounds, common names for certain diols and triols have persisted. We often refer to compounds containing two hydroxyl groups on adjacent carbons as **glycols.** Ethylene glycol and propylene glycol are synthesized from ethylene and propylene, respectively; hence their common names.

■ Ethylene glycol is a polar molecule and dissolves readily in the polar solvent water.

CH₂—CH₂   CH₃—CH—CH₂   CH₂—CH—CH₂
| | | | | | |
OH   OH   OH   OH   OH   OH OH

1,2-Ethanediol   1,2-Propanediol   1,2,3-Propanetriol
(Ethylene glycol)   (Propylene glycol)   (Glycerol, Glycerin)

C Physical Properties of Alcohols

The most important physical property of alcohols is the polarity of their —OH groups. Because of the large difference in electronegativity (Table 3.5) between oxygen and carbon (3.5–2.5 = 1.0) and between oxygen and hydrogen (3.5–2.1 = 1.4), both the C—O and O—H bonds of an alcohol are polar covalent, and alcohols are polar molecules, as illustrated in Figure 14.2 for methanol.

Alcohols have higher boiling points than do alkanes, alkenes, and alkynes of similar molecular weight (Table 14.1), because alcohol molecules associate with one another in the liquid state by **hydrogen bonding** (Section 5.7C). The strength of hydrogen bonding between alcohol molecules is approximately 2 to 5 kcal/mol, which means that extra energy is required to separate hydrogen-bonded alcohols from their neighbors (Figure 14.3).

Because of increased London dispersion forces (Section 5.7A) between larger molecules, the boiling points of all types of compounds, including alcohols, increase with increasing molecular weight.

Figure 14.2 Polarity of the C—O—H bonds in methanol. (*a*) There are partial positive charges on carbon and hydrogen and a partial negative charge on oxygen. (*b*) An electron density map showing a partial negative charge (in red) around oxygen and a partial positive charge (in blue) around hydrogen of the OH group.

Figure 14.3 The association of ethanol molecules in the liquid state. Each O—H can participate in up to three hydrogen bonds (one through hydrogen and two through oxygen). Only two of these three possible hydrogen bonds per molecule are shown here.

Table 14.1 Boiling Points and Solubilities in Water of Four Sets of Alcohols and Alkanes of Similar Molecular Weight

Structural Formula	Name	Molecular Weight (amu)	Boiling Point (°C)	Solubility in Water
CH_3OH	methanol	32	65	infinite
CH_3CH_3	ethane	30	−89	insoluble
CH_3CH_2OH	ethanol	46	78	infinite
$CH_3CH_2CH_3$	propane	44	−42	insoluble
$CH_3CH_2CH_2OH$	1-propanol	60	97	infinite
$CH_3CH_2CH_2CH_3$	butane	58	0	insoluble
$CH_3CH_2CH_2CH_2OH$	1-butanol	74	117	8 g/100 g
$CH_3CH_2CH_2CH_2CH_3$	pentane	72	36	insoluble

Alcohols are much more soluble in water than are hydrocarbons of similar molecular weight (Table 14.1), because alcohol molecules interact by hydrogen bonding with water molecules. Methanol, ethanol, and 1-propanol are soluble in water in all proportions. As molecular weight increases, the water solubility of alcohols becomes more like those of hydrocarbons of similar molecular weight. Higher-molecular-weight alcohols are much less soluble in water because the size of the hydrocarbon portion of their molecules (which decreases water solubility) becomes so large relative to the size of the —OH group (which increases water solubility).

14.3 Reactions of Alcohols

In this section, we study the acidity of alcohols, their dehydration to alkenes, and their oxidation to aldehydes, ketones, or carboxylic acids.

A Acidity of Alcohols

Alcohols have about the same pK_a values as water (Table 8.2) which means that aqueous solutions of alcohols have the same pH as that of pure water. In Section 13.5B, we studied phenols, another class of compounds that contains an —OH group. Phenols are weak acids and react with aqueous sodium hydroxide to form water-soluble salts.

C₆H₅—OH + NaOH $\xrightarrow{H_2O}$ C₆H₅—O⁻Na⁺ + H₂O
Phenol   Sodium phenoxide
                             (a water-soluble salt)

Alcohols are considerably weaker acids than phenols and do not react in this manner.

B Acid-Catalyzed Dehydration Alkenes

We can convert an alcohol to an alkene by eliminating a molecule of water from adjacent carbon atoms in a reaction called **dehydration**. In the laboratory, dehydration of an alcohol is most often brought about by heating it with either 85% phosphoric acid or concentrated sulfuric acid. Primary alcohols—the most difficult to dehydrate—generally require heating in concentrated sulfuric acid at temperatures as high as 180°C. Secondary alcohols undergo acid-catalyzed dehydration at somewhat lower temperatures. Tertiary alcohols generally undergo acid-catalyzed dehydration at temperatures only slightly above room temperature.

Dehydration Elimination of a molecule of water from an alcohol; an OH is removed from one carbon, and an H is removed from an adjacent carbon

$$CH_3CH_2OH \xrightarrow[180°C]{H_2SO_4} CH_2{=}CH_2 + H_2O$$
   Ethanol   Ethylene

Cyclohexanol $\xrightarrow[140°C]{H_2SO_4}$ Cyclohexene + H₂O

$$\underset{\underset{OH}{|}}{CH_3\overset{\overset{CH_3}{|}}{C}CH_3} \xrightarrow[50°C]{H_2SO_4} CH_3\overset{\overset{CH_3}{|}}{C}{=}CH_2 + H_2O$$
2-Methyl-2-propanol   2-Methylpropene
(*tert*-Butyl alcohol)   (Isobutylene)

Screen 14.5
Simulation

Thus the ease of acid-catalyzed dehydration of alcohols follows this order:

       1° alcohols   2° alcohols   3° alcohols
             → Ease of dehydration of alcohols →

When the acid-catalyzed dehydration of an alcohol yields isomeric alkenes, the alkene having the greater number of alkyl groups on the double bond generally predominates. In the acid-catalyzed dehydration of 2-butanol, for example, the major product is 2-butene, which has two alkyl groups (two methyl groups) on its double bond. The minor product is 1-butene, which has only one alkyl group (an ethyl group) on its double bond.

$$\underset{\underset{OH}{|}}{CH_3CH_2CHCH_3} \xrightarrow[heat]{H_3PO_4} CH_3CH{=}CHCH_3 + CH_3CH_2CH{=}CH_2 + H_2O$$
     2-Butanol   2-Butene   1-Butene
                                 (80%)   (20%)

EXAMPLE 14.3

Draw structural formulas for the alkenes formed by the acid-catalyzed dehydration of each alcohol. For each reaction, predict which alkene will be the major product.
(a) 3-Methyl-2-butanol
(b) 2-Methylcyclopentanol

Solution

(a) Elimination of H_2O from carbons 2–3 gives 2-methyl-2-butene; elimination of H_2O from carbons 1–2 gives 3-methyl-1-butene. 2-Methyl-2-butene has three alkyl groups (three methyl groups) on its double bond and is the major product. 3-Methyl-1-butene has only one alkyl group (an isopropyl group) on its double bond and is the minor product.

$$\underset{\text{3-Methyl-2-butanol}}{\overset{4\ 3\ 2\ 1}{CH_3\underset{\underset{OH}{|}}{\overset{\overset{CH_3}{|}}{CH}}CHCH_3}} \xrightarrow[\text{acid-catalyzed dehydration}]{H_2SO_4} \underset{\underset{\text{(major product)}}{\text{2-Methyl-2-butene}}}{CH_3\overset{\overset{CH_3}{|}}{C}=CHCH_3} + \underset{\text{3-Methyl-1-butene}}{CH_3\overset{\overset{CH_3}{|}}{CH}CH=CH_2} + H_2O$$

(b) The major product, 1-methylcyclopentene, has three alkyl groups on its double bond. The minor product, 3-methylcyclopentene, has only two alkyl groups on its double bond.

2-Methylcyclopentanol $\xrightarrow[\text{acid-catalyzed dehydration}]{H_2SO_4}$ 1-Methylcyclopentene (major product) + 3-Methylcyclopentene + H_2O

Problem 14.3

Draw structural formulas for the alkenes formed by the acid-catalyzed dehydration of each alcohol. For each reaction, predict which alkene will be the major product.
(a) 2-Methyl-2-butanol
(b) 1-Methylcyclopentanol

In Section 12.6B, we discussed the acid-catalyzed hydration of alkenes to give alcohols. In this section, we discuss the acid-catalyzed dehydration of alcohols to give alkenes. In fact, hydration–dehydration reactions are reversible. Alkene hydration and alcohol dehydration are competing reactions, and the following equilibrium exists:

$$\underset{\text{An alkene}}{\overset{}{C=C}} + H_2O \underset{\text{hydration}}{\overset{\text{dehydration}}{\rightleftharpoons}} \underset{\text{An alcohol}}{\overset{}{-\underset{\underset{H}{|}}{\overset{|}{C}}-\underset{\underset{OH}{|}}{\overset{|}{C}}-}}$$

In accordance with Le Chatelier's principle (Section 7.8), large amounts of water (in other words, using dilute aqueous acid) favor alcohol formation, whereas scarcity of water (using concentrated acid) or experimental conditions where water is removed (heating the reaction mixture above 100°C) favor alkene formation. Thus, depending on the experimental conditions, we can use the hydration–dehydration equilibrium to prepare either alcohols or alkenes, each in high yields.

EXAMPLE 14.4

In part (a), acid-catalyzed dehydration of 2-methyl-3-pentanol gives Compound A. Treatment of Compound A with water in the presence of sulfuric acid in part (b) gives Compound B. Propose structural formulas for Compounds A and B.

(a) $CH_3CHCHCH_2CH_3$ with CH_3 branch and OH $\xrightarrow[\text{acid-catalyzed dehydration}]{H_2SO_4}$ Compound A (C_6H_{12}) + H_2O

(b) Compound A (C_6H_{12}) + H_2O $\xrightarrow{H_2SO_4}$ Compound B ($C_6H_{14}O$)

Solution
(a) Acid-catalyzed dehydration of 2-methyl-3-pentanol gives 2-methyl-2-pentene, an alkene with three substituents on its double bond: two methyl groups and one ethyl group.
(b) Acid-catalyzed addition of water to this alkene gives 2-methyl-2-pentanol in accord with Markovnikov's rule (Section 12.6B).

$CH_3C(CH_3)=CHCH_2CH_3 + H_2O$ $\xrightarrow[\text{acid-catalyzed hydration}]{H_2SO_4}$ $CH_3C(CH_3)(OH)CH_2CH_2CH_3$

Compound A (C_6H_{12})   Compound B ($C_6H_{14}O$)

Problem 14.4
Acid-catalyzed dehydration of 2-methylcyclohexanol gives Compound C (C_7H_{12}). Treatment of Compound C with water in the presence of sulfuric acid gives Compound D ($C_7H_{14}O$). Propose structural formulas for Compounds C and D.

C Oxidation of Primary and Secondary Alcohols

Oxidation of a primary alcohol can give an aldehyde or a carboxylic acid, depending on the experimental conditions. Following is a series of transformations in which a primary alcohol is oxidized first to an aldehyde and then to a carboxylic acid. The fact that each transformation involves oxidation is indicated by the letter O in brackets over the reaction arrow.

$CH_3-CH(OH)-H$ $\xrightarrow{[O]}$ $CH_3-C(=O)-H$ $\xrightarrow{[O]}$ $CH_3-C(=O)-OH$

A primary alcohol   An aldehyde   A carboxylic acid

368 | CHAPTER 14 ALCOHOLS, ETHERS, AND THIOLS

Recall from Section 4.7 that, according to one definition, oxidation is either the loss of hydrogens or the gain of oxygens. Using this definition, conversion of a primary alcohol to an aldehyde is an oxidation because the alcohol loses hydrogens. Conversion of an aldehyde to a carboxylic acid is an oxidation because the aldehyde gains an oxygen.

The reagent most commonly used in the laboratory for the oxidation of a primary alcohol to a carboxylic acid is potassium dichromate, $K_2Cr_2O_7$, dissolved in aqueous sulfuric acid. Using this reagent, oxidation of 1-octanol, for example, gives octanoic acid. This experimental condition is more than sufficient to oxidize the intermediate aldehyde to a carboxylic acid.

$$CH_3(CH_2)_6CH_2OH \xrightarrow[H_2SO_4]{K_2Cr_2O_7} CH_3(CH_2)_6\overset{O}{\overset{\|}{C}}H \xrightarrow[H_2SO_4]{K_2Cr_2O_7} CH_3(CH_2)_6\overset{O}{\overset{\|}{C}}OH$$

1-Octanol → Octanal → Octanoic acid

Although the usual product of oxidation of a primary alcohol is a carboxylic acid, it is often possible to stop the oxidation at the aldehyde stage by distilling the mixture. We remove the aldehyde, which usually has a lower boiling point than either the primary alcohol or the carboxylic acid, from the reaction mixture before it can be oxidized further.

Oxidation of a secondary alcohol using potassium dichromate as the oxidizing agent gives a ketone. Menthol, a secondary alcohol that is present in peppermint and other mint oils, is used in liqueurs, cigarettes, cough drops, perfumery, and nasal inhalers. Its oxidation product, menthone, is also used in perfumes and artificial flavors.

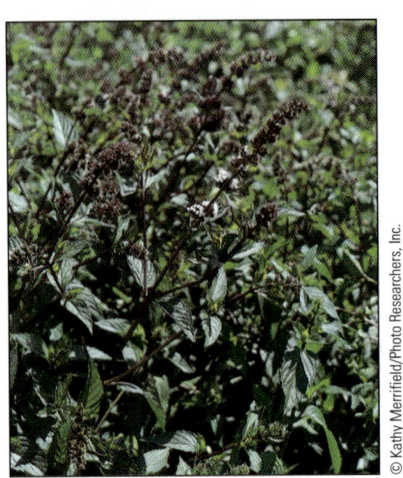

■ The peppermint plant, *Mentha piperita*, is a perennial herb with aromatic qualities used in candies, gums, hot and cold beverages, and garnishes for punch and fruit.

2-Isopropyl-5-methyl-cyclohexanol (Menthol) $\xrightarrow[H_2SO_4]{K_2Cr_2O_7}$ 2-Isopropyl-5-methyl-cyclohexanone (Menthone)

Tertiary alcohols are resistant to oxidation because the carbon bearing the —OH is bonded to three carbon atoms and, therefore, cannot form a carbon–oxygen double bond.

1-Methylcyclopentanol $\xrightarrow[H_2SO_4]{K_2Cr_2O_7}$ (no oxidation)

EXAMPLE 14.5

Draw a structural formula for the product formed by oxidation of each alcohol with potassium dichromate.
(a) 1-Hexanol
(b) 2-Hexanol

CHEMICAL CONNECTIONS 14B

Breath-Alcohol Screening

Potassium dichromate oxidation of ethanol to acetic acid is the basis for the original breath-alcohol screening test used by law enforcement agencies to determine a person's blood alcohol content (BAC). The test is based on the difference in color between the dichromate ion (reddish orange) in the reagent and the chromium(III) ion (green) in the product.

$$\underset{\text{Ethanol}}{CH_3CH_2OH} + \underset{\substack{\text{Dichromate ion}\\\text{(reddish orange)}}}{Cr_2O_7^{2-}}$$

$$\xrightarrow[H_2O]{H_2SO_4} \underset{\text{Acetic acid}}{CH_3\overset{\overset{O}{\|}}{C}OH} + \underset{\substack{\text{Chromium(III) ion}\\\text{(green)}}}{Cr^{3+}}$$

In its simplest form, breath-alcohol screening uses a sealed glass tube that contains a potassium dichromate–sulfuric acid reagent impregnated on silica gel. To administer the test, the ends of the tube are broken off, a mouthpiece is fitted to one end, and the other end is inserted into the neck of a plastic bag. The person being tested then blows into the mouthpiece to inflate the plastic bag.

As breath containing ethanol vapor passes through the tube, reddish orange dichromate ion is reduced to green chromium(III) ion. To estimate the concentration of ethanol in the breath, one measures how far the green color extends along the length of the tube. When it extends beyond the halfway point, the person is judged as having a sufficiently high blood alcohol content to warrant further, more precise testing.

The test described here measures the alcohol content of the breath. The legal definition of being under the influence of alcohol, however, is based on alcohol content in the blood, not in the breath. The correlation between these two measurements is based on the fact that air deep within the lungs is in equilibrium with blood passing through the pulmonary arteries, and an equilibrium is established between blood alcohol and breath alcohol. Based on tests in persons drinking alcohol, researchers have determined that 2100 mL of breath contains the same amount of ethanol as 1.00 mL of blood.

Person forces breath through mouthpiece into the tube.

Glass tube containing potassium dichromate–sulfuric acid coated on silica gel particles

As the person blows into the tube, the plastic bag becomes inflated.

Solution
Oxidation of 1-hexanol, a primary alcohol, gives either hexanal or hexanoic acid, depending on the experimental conditions. Oxidation of 2-hexanol, a secondary alcohol, gives 2-hexanone.

(a) Hexanal or Hexanoic acid   (b) 2-Hexanone

Problem 14.5
Draw the product formed by oxidation of each alcohol with potassium dichromate.
(a) Cyclohexanol
(b) 2-Pentanol

Figure 14.4 Dimethyl ether, CH_3OCH_3. (a) Lewis structure and (b) ball-and-stick model. The C—O—C bond angle is 110.3°, close to the tetrahedral angle of 109.5°.

Ether A compound containing an oxygen atom bonded to two carbon atoms

14.4 Ethers

A Structure

The functional group of an **ether** is an atom of oxygen bonded to two carbon atoms. Figure 14.4 shows a Lewis structure and a ball-and-stick model of dimethyl ether, CH_3OCH_3, the simplest ether.

B. Nomenclature

Although we can use the IUPAC system to name ethers, chemists almost invariably use common names for low-molecular-weight ethers. We derive common names by listing the alkyl groups attached to oxygen in alphabetical order and adding the word ether. Alternatively, we name one of the groups on oxygen as an alkoxy group. The —OCH_3 group, for example, is named "methoxy" to indicate a methyl group bonded to oxygen.

$CH_3CH_2OCH_2CH_3$ [cyclohexyl]—OCH_3

Diethyl ether   Cyclohexyl methyl ether
                      (Methoxycyclohexane)

EXAMPLE 14.6

Write the common name for each ether.

(a) $CH_3\underset{CH_3}{\overset{CH_3}{\underset{|}{\overset{|}{C}}}}OCH_2CH_3$ (b) [cyclohexyl]—O—[cyclohexyl]

Solution
(a) The groups bonded to the ether oxygen are *tert*-butyl and ethyl. The compound's common name is *tert*-butyl ethyl ether.
(b) Two cyclohexyl groups are bonded to the ether oxygen. The compound's common name is dicyclohexyl ether.

Problem 14.6

Write the common name for each ether.

(a) (isopropyl)—O—(ethyl)   (b) [cyclopentyl]—OCH_3

CHEMICAL CONNECTIONS 14C

Ethylene Oxide: A Chemical Sterilant

Ethylene oxide is a colorless, flammable gas, with a boiling point of 11°C. Because it is such a highly strained molecule (the normal tetrahedral bond angles of both C and O are compressed to approximately 60° in this molecule), ethylene oxide reacts with the amino ($-NH_2$) and sulfhydryl ($-SH$) groups present in biological materials.

At sufficiently high concentrations, it reacts with enough molecules in cells to cause the death of microorganisms. This toxic property is the basis for ethylene oxide's use as a fumigant in foodstuffs and textiles and its use in hospitals to sterilize surgical instruments.

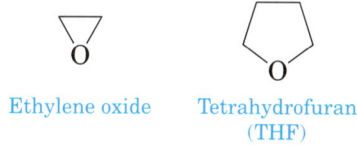

Cyclic ethers are ethers in which one of the atoms in a ring is oxygen. These ethers are also known by their common names. Ethylene oxide is an important building block for the organic chemical industry (Section 14.6). Tetrahydrofuran is a useful laboratory and industrial solvent.

> **Cyclic ether** An ether in which the ether oxygen is one of the atoms of a ring

Ethylene oxide   Tetrahydrofuran (THF)

C Physical Properties

Ethers are polar compounds in which oxygen bears a partial negative charge and each attached carbon bears a partial positive charge (Figure 14.5). However, only weak forces of attraction exist between ether molecules in the pure liquid; consequently, the boiling points of ethers are close to those of hydrocarbons of similar molecular weight.

We can illustrate the effect of hydrogen bonding on physical properties by comparing the boiling points of ethanol (78°C) and its constitutional isomer dimethyl ether (-24°C). The difference in boiling point between these

---

Figure 14.5 Ethers are polar molecules, but only weak attractive interactions exist between ether molecules in the liquid state.

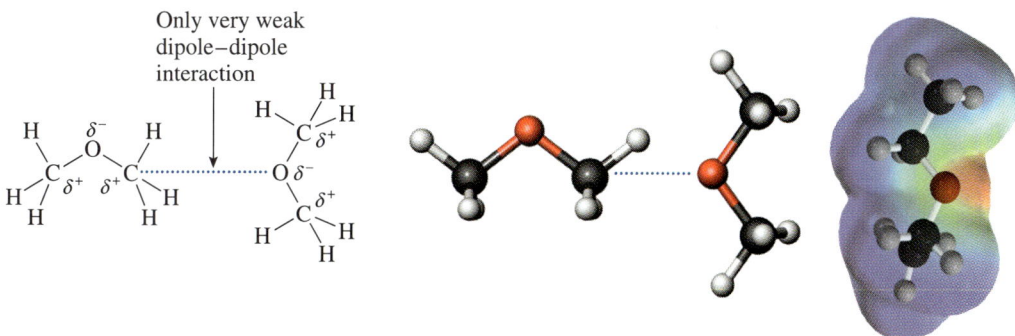

two compounds is due to the presence in ethanol of a polar O—H group, which is capable of forming hydrogen bonds. This hydrogen bonding increases intermolecular associations and thus gives ethanol a higher boiling point than dimethyl ether.

CH$_3$CH$_2$OH   CH$_3$OCH$_3$
Ethanol   Dimethyl ether
bp 78°C   bp −24°C

Ethers are more soluble in water than hydrocarbons of similar molecular weight and shape. Their greater solubility is due to the fact that the oxygen atom of an ether carries a partial negative charge and forms hydrogen bonds with water.

D Reactions of Ethers

Ethers resemble hydrocarbons in their resistance to chemical reaction. They do not react with oxidizing agents, such as potassium dichromate. Likewise, they do not react with reducing agents such as H$_2$ in the presence of a metal catalyst (Section 12.6D). Furthermore they are not affected by most acids or bases at moderate temperatures. Because of their general inertness to chemical reaction and their good solvent properties, ethers are excellent solvents in which to carry out many organic reactions. The most important ether solvents are diethyl ether and tetrahydrofuran.

■ The scent of the spotted skunk, native to the Sonoran Desert, is a mixture of two thiols, 2-butene-1-thiol and 3-methyl-1-butanethiol.

14.5 Thiols

The most outstanding property of low-molecular-weight **thiols** is their stench. They are responsible for unpleasant odors such as those from skunks, rotten eggs, and sewage. The scent of skunks is due primarily to two thiols:

$$\text{CH}_3\text{CH}=\text{CHCH}_2\text{SH} \qquad \text{CH}_3\overset{\overset{\displaystyle \text{CH}_3}{|}}{\text{CH}}\text{CH}_2\text{CH}_2\text{SH}$$

2-Butene-1-thiol   3-Methyl-1-butanethiol

> **Thiol** A compound containing an —SH (sulfhydryl) group bonded to a tetrahedral carbon atom
>
> **Mercaptan** A common name for any molecule containing an —SH group

A Structure

The functional group of a thiol is an **—SH (sulfhydryl) group** bonded to a tetrahedral carbon atom. Figure 14.6 shows a Lewis structure and a ball-and-stick model of methanethiol, CH$_3$SH, the simplest thiol.

B Nomenclature

The sulfur analog of an alcohol is called a thiol (*thi-* from the Greek: *theion*, sulfur) or, in the older literature, a **mercaptan,** which literally means "mercury capturing." Thiols react with Hg^{2+} in aqueous solution to give sulfide salts as insoluble precipitates. Thiophenol, C$_6$H$_5$SH, for example, gives (C$_6$H$_5$S)$_2$Hg.

In the IUPAC system, we name thiols by selecting the longest carbon chain that contains the —SH group as the parent alkane. To show that the

Figure 14.6 Methanethiol, CH$_3$SH. (*a*) Lewis structure and (*b*) ball-and-stick model. The H—S—C bond angle is 100.3°, somewhat smaller than the tetrahedral angle of 109.5°.

CHEMICAL CONNECTIONS 14D

Ethers and Anesthesia

■ This painting by Robert Hinckley shows the first use of ether as an anesthetic in 1846. Dr. Robert John Collins was removing a tumor from the patient's neck, and dentist W. T. G. Morton—who discovered its anesthetic properties—administered the ether.

Before the mid–1800s, surgery was performed only when absolutely necessary, because no truly effective general anesthetic was available. More often than not, patients were drugged, hypnotized, or simply tied down.

In 1772, Joseph Priestly isolated nitrous oxide, N_2O, a colorless gas. In 1799, Sir Humphry Davy demonstrated this compound's anesthetic effect, naming it "laughing gas." In 1844, an American dentist, Horace Wells, introduced nitrous oxide into general dental practice. One patient awakened prematurely, however, screaming with pain; another died during the procedure. Wells was forced to withdraw from practice, became embittered and depressed, and committed suicide at age 33. In the same period, a Boston chemist, Charles Jackson, anesthetized himself with diethyl ether; he also persuaded a dentist, William Morton, to use it. Subsequently, they persuaded a surgeon, John Warren, to give a public demonstration of surgery under anesthesia. The operation was completely successful, and soon general anesthesia by diethyl ether became routine practice for general surgery.

Diethyl ether was easy to use and caused excellent muscle relaxation. Blood pressure, pulse rate, and respiration were usually only slightly affected. The gas's chief drawbacks were its irritating effect on the respiratory passages and its after-effect of nausea.

Among the inhalation anesthetics used today are several halogenated ethers, the most important being enflurane and isoflurane.

Enflurane (Ethrane)   Isoflurane (Forane)

---

compound is a thiol, we add the suffix *-thiol* to the name of the parent alkane. We number the parent chain in the direction that gives the —SH group the lower number.

We derive common names for simple thiols by naming the alkyl group attached to —SH and adding the word mercaptan.

CH_3CH_2SH

Ethanethiol
(Ethyl mercaptan)

CH_3CHCH_2SH with CH_3 branch

2-Methyl-1-propanethiol
(Isobutyl mercaptan)

EXAMPLE 14.7

Write the IUPAC name for each thiol.

(a) ~~~SH (b) SH on isopropyl

■ Mushrooms, onions, garlic, and coffee all contain sulfur compounds. This thiol is present in the aroma of coffee.

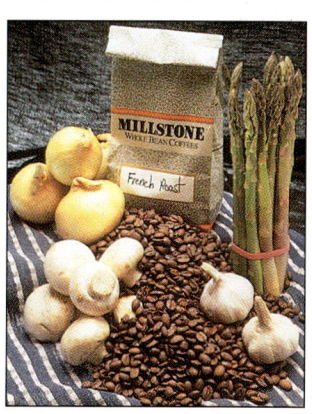

furan—CH_2SH

> **Solution**
> (a) The parent alkane is pentane. We show the presence of the —SH group by adding *thiol* to the name of the parent alkane. The IUPAC name of this thiol is 1-pentanethiol. Its common name is pentyl mercaptan.
> (b) The parent alkane is butane. The IUPAC name of this thiol is 2-butanethiol. Its common name is *sec*-butyl mercaptan.

> **Problem 14.7**
> Write the IUPAC name for each thiol.
>
> (a) [structure: CH₃CH₂CH(CH₃)CH₂SH]   (b) [structure: (CH₃)₂CHCH(SH)—]

C Physical Properties

Because of the small difference in electronegativity between sulfur and hydrogen (2.5 − 2.1 = 0.4), we classify the S—H bond as nonpolar covalent. Because of this lack of polarity, thiols show little association by hydrogen bonding. Consequently, they have lower boiling points and are less soluble in water and other polar solvents than alcohols of similar molecular weight. Table 14.2 gives boiling points for three low-molecular-weight thiols. Shown for comparison are boiling points of alcohols with the same number of carbon atoms.

Earlier we illustrated the importance of hydrogen bonding in alcohols by comparing the boiling points of ethanol (78°C) and its constitutional isomer dimethyl ether (−24°C). By comparison, the boiling point of ethanethiol is 35°C and that of its constitutional isomer dimethyl sulfide is 37°C. Because the boiling points of these constitutional isomers are almost identical, we know that little or no association by hydrogen bonding occurs between thiol molecules.

$$CH_3CH_2SH \qquad CH_3SCH_3$$
Ethanethiol Dimethyl sulfide
bp 35°C bp 37°C

D Reactions of Thiols

Thiols are weak acids (pK_a 10), and are comparable in strength to phenols (Section 13.5B). Thiols react with strong bases such as NaOH to form thiolate salts.

$$CH_3CH_2SH + NaOH \xrightarrow{H_2O} CH_3CH_2S^-Na^+ + H_2O$$
Ethanethiol Sodium
(pK_a 10) ethanethiolate

Disulfide A compound containing an —S—S— group

The most common reaction of thiols in biological systems is their oxidation to **disulfides,** the functional group of which is a disulfide (—S—S—) bond. Thiols are readily oxidized to disulfides by molecular oxygen. In fact, they are so susceptible to oxidation that they must be protected from con-

Table 14.2 Boiling Points of Three Thiols and Alcohols with the Same Number of Carbon Atoms

Thiol	Boiling Point (°C)	Alcohol	Boiling Point (°C)
methanethiol	6	methanol	65
ethanethiol	35	ethanol	78
1-butanethiol	98	1-butanol	117

tact with air during storage. Disulfides, in turn, are easily reduced to thiols by several reducing agents. This easy interconversion between thiols and disulfides is very important in protein chemistry, as we will see in Chapters 21 and 22.

$$2\text{HOCH}_2\text{CH}_2\text{SH} \underset{\text{reduction}}{\overset{\text{oxidation}}{\rightleftharpoons}} \text{HOCH}_2\text{CH}_2\text{S}-\text{SCH}_2\text{CH}_2\text{OH}$$

A thiol   A disulfide

We derive the common names of disulfides by listing the names of the groups attached to sulfur and adding the word "disulfide."

14.6 Important Alcohols

As you study the alcohols described in this section, you should pay particular attention to two key points. First, they are derived almost entirely from petroleum, natural gas, or coal—all nonrenewable resources. Second, many are themselves starting materials for the synthesis of valuable commercial products, without which our modern industrial society could not exist.

At one time **methanol** was derived by heating hard woods in a limited supply of air—hence the name "wood alcohol." Today methanol is derived entirely from the catalytic reduction of carbon monoxide. Methanol, in turn, is the starting material for the preparation of several important industrial and commercial chemicals, including acetic acid and formaldehyde. Treatment of methanol with carbon monoxide in the presence of a rhodium catalyst gives acetic acid. Partial oxidation of methanol gives formaldehyde. An important use of this one-carbon aldehyde is in the preparation of phenol-formaldehyde and urea-formaldehyde glues and resins, which are used as molding materials and as adhesives for plywood and particle board for the construction industry.

$$\text{Coal or methane} \xrightarrow{[O]} \underset{\substack{\text{Carbon}\\\text{monoxide}}}{\text{CO}} \xrightarrow{2\text{H}_2} \underset{\text{Methanol}}{\text{CH}_3\text{OH}} \begin{array}{c} \xrightarrow[\text{catalyst}]{\text{CO}} \underset{\text{Acetic acid}}{\text{CH}_3\text{COOH}} \\ \xrightarrow[\text{oxidation}]{\text{O}_2} \underset{\text{Fomaldehyde}}{\text{CH}_2\text{O}} \end{array}$$

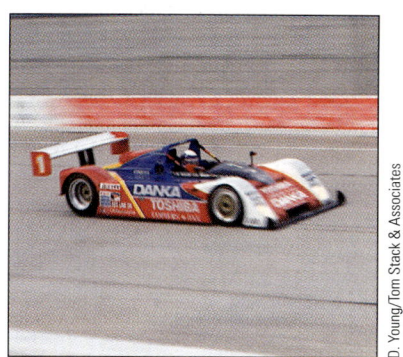

■ Methanol, CH_3OH, is the fuel used in cars of the type that race in Indianapolis.

The bulk of the **ethanol** produced worldwide is prepared by acid-catalyzed hydration of ethylene, itself derived from the cracking of the ethane separated from natural gas (Section 11.11). Ethanol is also produced by fermentation of the carbohydrates in plant materials, particularly corn and molasses. The bulk of the fermentation-derived ethanol is used as an "oxygenate" additive to produce gasohol, which is a blend containing as much as 10% ethanol in gasoline. Combustion of gasohol produces less air pollution than combustion of gasoline itself.

$$CH_2\!=\!CH_2 \xrightarrow{\begin{array}{c}H_2O, H_2SO_4\end{array}} CH_3CH_2OH \xrightarrow[180°C]{H_2SO_4} CH_3CH_2OCH_2CH_3$$

Ethylene → Ethanol → Diethyl ether

$$\xrightarrow[\text{catalyst}]{O_2} H_2C\!-\!CH_2 \xrightarrow{H_2O, H_2SO_4} HOCH_2CH_2OH$$

Ethylene oxide → Ethylene glycol

Acid-catalyzed dehydration of ethanol gives **diethyl ether,** an important laboratory and industrial solvent. Ethylene is also the starting material for the preparation of **ethylene oxide.** This compound has few direct uses. Rather, ethylene oxide's importance is as an intermediate for the production of **ethylene glycol,** a major component of automobile antifreeze. Ethylene glycol freezes at −12°C and boils at 199°C, which makes it ideal for this purpose. In addition, reaction of ethylene glycol with the methyl ester of terephthalic acid gives the polymer poly(ethylene terephthalate), abbreviated PET or PETE (Section 18.8B). Ethylene glycol is also used as a solvent in the paint and plastics industry, and in the formulation of printer's inks, ink pads, and inks for ball-point pens.

Isopropyl alcohol, the alcohol in rubbing alcohol, is made by acid-catalyzed hydration of propene. It is also used in hand lotions, after-shave lotions, and similar cosmetics. A multistep process also converts propene to epichlorohydrin, one of the key components in the production of epoxy glues and resins.

Glycerin is a by-product of the manufacture of soaps by saponification of animal fats and tropical oils (Section 20.3). The bulk of the glycerin used for industrial and commercial purposes, however, is prepared from propene. Perhaps the best-known use of glycerin is for the manufacture of nitroglycerin. Glycerin is also used as an emollient in skin care products and cosmetics, in liquid soaps, and in printing inks.

$$CH_2\!=\!CHCH_3 \xrightarrow{\text{several steps}} ClCH_2CH\!-\!CH_2 \xrightarrow{\text{several steps and other reagents}} \text{Epoxy glues and resins}$$

Epichlorohydrin

$$\xrightarrow{H_2O, H_2SO_4} CH_3\underset{|}{\overset{OH}{C}}HCH_3$$

Propene → Isopropyl alcohol

$$\xrightarrow{\text{several steps}} \underset{|}{\overset{HO}{C}}H_2\underset{|}{\overset{HO}{C}}H\underset{|}{\overset{OH}{C}}H_2$$

Glycerin, Glycerol

SUMMARY

The functional group of an **alcohol** is an —OH (**hydroxyl**) **group** bonded to a tetrahedral carbon atom (Section 14.2A). IUPAC names of alcohols are derived by changing the **-e** of the parent alkane to **-ol** (Section 14.2B). The parent chain is numbered from the end that gives the carbon bearing the —OH group the lower number. Common names for alcohols are derived by naming the alkyl group bonded to the —OH group and adding the word *alcohol*. Alcohols are classified as 1°, 2°, or 3°, depending on the number of carbon groups bonded to the carbon bearing the —OH group. Compounds containing two hydroxyl groups on adjacent carbons are called **glycols**.

Alcohols are polar compounds with oxygen bearing a partial negative charge and both the carbon and hydrogen bonded to it bearing partial positive charges (Section 14.2C). They associate in the liquid state by hydrogen bonding; as a consequence, their boiling points are higher than those of hydrocarbons of similar molecular weight. Because of increased London dispersion forces, the boiling points of alcohols increase with increasing molecular weight. Alcohols interact with water by hydrogen bonding and are more soluble in water than are hydrocarbons of similar molecular weight.

Alcohols have about the same pK_a value as pure water. For this reason, aqueous solutions of alcohols have the same pH as that of pure water (Section 14.3A).

The functional group of an **ether** is an atom of oxygen bonded to two carbon atoms (Section 14.4A). Common names for ethers are derived by naming the two groups attached to oxygen followed by the word *ether* (Section 14.4B). In **cyclic ethers,** oxygen is one of the atoms in a ring. Ethers are weakly polar compounds. Their boiling points are close to those of hydrocarbons of similar molecular weight (Section 14.4C). Because they form hydrogen bonds with water, however, ethers are more soluble in water than are hydrocarbons of similar molecular weight.

A **thiol** contains an **—SH (sulfhydryl) group** (Section 14.5A). Thiols are named in the same manner as alcohols, but the suffix **-e** of the parent alkane is retained, and **-thiol** is added (Section 14.5B). Common names for thiols are derived by naming the alkyl group bonded to —SH and adding the word *mercaptan*. The S—H bond is nonpolar, and the physical properties of thiols resemble those of hydrocarbons of similar molecular weight (Section 14.4C).

KEY REACTIONS

1. **Acid-Catalyzed Dehydration of an Alcohol (Section 14.3B)** When isomeric alkenes are possible, the major product is generally the more substituted alkene.

   $$CH_3CH_2CHCH_3(OH) \xrightarrow[\text{heat}]{H_3PO_4} CH_3CH=CHCH_3 + CH_3CH_2CH=CH_2 + H_2O$$

   Major product

2. **Oxidation of a Primary Alcohol (Section 14.3C)** Oxidation of a primary alcohol by potassium dichromate gives either an aldehyde or a carboxylic acid, depending on the experimental conditions.

   $$CH_3(CH_2)_6CH_2OH \xrightarrow[H_2SO_4]{K_2Cr_2O_7} CH_3(CH_2)_6CHO$$

   $$\xrightarrow[H_2SO_4]{K_2Cr_2O_7} CH_3(CH_2)_6COOH$$

3. **Oxidation of a Secondary Alcohol (Section 14.3C)** Oxidization of a secondary alcohol by potassium dichromate gives a ketone.

   $$CH_3(CH_2)_4CHCH_3(OH) \xrightarrow[H_2SO_4]{K_2Cr_2O_7} CH_3(CH_2)_4CCH_3(O)$$

4. **Acidity of Thiols (Section 14.5D)** Thiols are weak acids, with pK_a equal to approximately 10. They react with strong bases to form water-soluble thiolate salts.

   $$CH_3CH_2SH + NaOH \xrightarrow{H_2O} CH_3CH_2S^-Na^+ + H_2O$$

   Ethanethiol (pK_a 10) → Sodium ethanethiolate

5. **Oxidation of a Thiol to a Disulfide (Section 14.5D)** Oxidation of a thiol gives a disulfide. Reduction of a disulfide gives two thiols.

   $$2HOCH_2CH_2SH \underset{\text{reduction}}{\overset{\text{oxidation}}{\rightleftharpoons}} HOCH_2CH_2S-SCH_2CH_2OH$$

   A thiol → A disulfide

PROBLEMS

Numbers that appear in color indicate difficult problems.
▲ designates problems requiring applications of principles.

Structure and Nomenclature of Alcohols, Ethers, and Thiols

14.8 What is the difference in structure between a primary, secondary, and tertiary alcohol?

14.9 Which are secondary alcohols?

(a) cyclohexane with CH₃ and OH on same carbon

(b) $(CH_3)_3COH$

(c) CH₃CH=CHCH(OH)CH₂CH₃

(d) cyclopentane with OH

CHAPTER 14 ALCOHOLS, ETHERS, AND THIOLS

14.10 Which of the alcohols in Problem 14.9 are primary? Which are tertiary?

14.11 Write the IUPAC name of each compound.
(a) [structure: CH₃CH₂CH₂CH₂OH-like chain with OH]
(b) HO—CH₂CH₂CH₂—OH
(c) structure with OH on C2 and OH on C1
(d) HO— attached to isobutyl-like structure
(e) cyclohexane with two OH groups (cis/trans)
(f) cyclohexane with OH and two methyl substituents

14.12 Draw a structural formula for each alcohol.
(a) Isopropyl alcohol
(b) Propylene glycol
(c) 5-Methyl-2-hexanol
(d) 2-Methyl-2-propyl-1,3-propanediol
(e) 1-Octanol
(f) 3,3-Dimethylcyclohexanol

14.13 Draw a structural formula for each alcohol.
(a) Isobutyl alcohol
(b) 1,4-Butanediol
(c) 5-Methyl-1-hexanol
(d) 1,3-Pentanediol
(e) *trans*-1,4-Cyclohexanediol
(f) 1-Chloro-2-propanol

14.14 Write the common name for each ether.
(a) cyclopentyl—O—cyclopentyl
(b) [CH₃(CH₂)₄]₂O
(c) CH₃CHOCHCH₃ with CH₃ groups on each carbon

14.15 Write the common name for each ether.
(a) ethyl isopropyl ether structure
(b) dibutyl ether structure
(c) diphenyl ether structure

14.16 Write the IUPAC name for each thiol.
(a) CH₃CH₂CHCH₃ with SH
(b) CH₃CH₂CH₂CH₂SH
(c) cyclohexyl—SH

14.17 Write the common name for each thiol in Problem 14.16.

14.18 Both alcohols and phenols contain an —OH group. What structural feature distinguishes these two classes of compounds? Illustrate your answer by drawing the structural formulas of a phenol of six carbon atoms and an alcohol of six carbon atoms.

14.19 Name the functional groups in each compound.

(a) Prednisone (a synthetic anti-inflammatory steroid)

(b) Estradiol (a female sex hormone; Section 20.10)

Physical Properties of Alcohols, Ethers, and Thiols

14.20 Explain in terms of noncovalent interactions why the low-molecular-weight alcohols are soluble in water but the low-molecular-weight alkanes and alkenes are not.

14.21 Explain in terms of noncovalent interactions why the low-molecular-weight alcohols are more soluble in water than the low-molecular-weight ethers.

14.22 Why does the water solubility of alcohols decrease as molecular weight increases?

14.23 Show hydrogen bonding between methanol and water in the following ways:
(a) Between the oxygen of methanol and a hydrogen of water
(b) Between the hydrogen of methanol's OH group and the oxygen of water

14.24 Show hydrogen bonding between the oxygen of diethyl ether and a hydrogen of water.

14.25 Arrange these compounds in order of increasing boiling point. Values in °C are −42, 78, 117, and 198.
(a) $CH_3CH_2CH_2CH_2OH$
(b) CH_3CH_2OH
(c) $HOCH_2CH_2OH$
(d) $CH_3CH_2CH_3$

14.26 Arrange these compounds in order of increasing boiling point. Values in °C are 0, 35, and 97.
(a) $CH_3CH_2CH_2OH$
(b) $CH_3CH_2OCH_2CH_3$
(c) $CH_3CH_2CH_2CH_3$

14.27 Following are structural formulas for 1-butanol and 1-butanethiol. One of these compounds has a boiling point of 98°C, the other has a boiling point of 117°C. Which compound has which boiling point?

$CH_3CH_2CH_2CH_2OH$ $CH_3CH_2CH_2CH_2SH$
1-Butanol   1-Butanethiol

14.28 Explain why methanethiol, CH_3SH, has a lower boiling point (6°C) than methanol, CH_3OH (65°C), even though methanethiol has a higher molecular weight.

14.29 2-Propanol (isopropyl alcohol) is commonly used as rubbing alcohol to cool the skin. 2-Hexanol, also a liquid, is not suitable for this purpose. Why not?

14.30 Explain why glycerol is much thicker (more viscous) than ethylene glycol, which in turn is much thicker than ethanol.

14.31 From each pair, select the compound that is more soluble in water.
(a) CH_3OH or CH_3OCH_3
(b) $CH_3\underset{\underset{OH}{|}}{C}HCH_3$ or $CH_3\underset{\underset{CH_2}{\|}}{C}CH_3$
(c) $CH_3CH_2CH_2SH$ or $CH_3CH_2CH_2OH$

14.32 Arrange the compounds in each set in order of decreasing solubility in water.
(a) Ethanol, butane, and diethyl ether
(b) 1-Hexanol, 1,2-hexanediol, and hexane

Synthesis of Alcohols

14.33 Give the structural formula of an alkene or alkenes from which each alcohol can be prepared.
(a) 2-Butanol
(b) 1-Methylcyclohexanol
(c) 3-Hexanol
(d) 2-Methyl-2-pentanol
(e) Cyclopentanol

Reactions of Alcohols

14.34 Show how to distinguish between cyclohexanol and cyclohexene by a simple chemical test. Tell what you would do, what you would expect to see, and how you would interpret your observation.

14.35 Compare the acidity of alcohols and phenols, both classes of organic compounds that contain an —OH group.

14.36 Both 2,6-diisopropylcyclohexanol and the intravenous anesthetic Propofol are insoluble in water. Show how these two compounds can be distinguished by their reaction with aqueous sodium hydroxide.

2,6-Diisopropylcyclohexanol   2,6-Diisopropylphenol (Propofol)

14.37 Write equations for the reaction of 1-butanol, a primary alcohol, with these reagents.
(a) H_2SO_4, heat
(b) $K_2Cr_2O_7$, H_2SO_4

14.38 Write equations for the reaction of 2-butanol with these reagents.
(a) H_2SO_4, heat
(b) $K_2Cr_2O_7$, H_2SO_4

14.39 Write the formula for the product formed by treating each of the following compounds with $K_2Cr_2O_7 / H_2SO_4$.
(a) 1-Octanol
(b) 1,4-Butanediol

14.40 Show how to convert cyclohexanol to the following compounds.
(a) Cyclohexene
(b) Cyclohexane
(c) Cyclohexanone
(d) Bromocyclohexane

14.41 Show reagents and experimental conditions to synthesize each compound from 1-propanol. More than one step is required for some parts of this problem.

(a) $CH_3CH_2\overset{\overset{O}{\|}}{C}H$

(b) $CH_3CH_2\overset{\overset{O}{\|}}{C}OH$

(c) $CH_3CH=CH_2$

(d) $CH_3\overset{\overset{OH}{|}}{C}HCH_3$

(e) $CH_3\overset{\overset{O}{\|}}{C}CH_3$

(f) $CH_3CH_2CH_3$

(g) $CH_3\overset{\overset{Br}{|}}{C}HCH_3$

(h) $CH_3\overset{\overset{Br}{|}}{C}H\overset{\overset{Br}{|}}{C}H_2$

14.42 Show how to convert the given alcohol to compounds (a) and (b).

cyclopentyl—CH_2OH

(a) cyclopentyl—$\overset{\overset{O}{\|}}{C}H$

(b) cyclopentyl—$\overset{\overset{O}{\|}}{C}OH$

14.43 Name two important alcohols derived from ethylene and give two important uses of each.

14.44 Name two important alcohols derived from propene and give an important use of each.

Reaction of Thiols

14.45 Following is a structural formula for the amino acid cysteine:

$$HS-CH_2-\underset{\underset{NH_2}{|}}{C}H-\overset{\overset{O}{\|}}{C}-OH$$

(a) Name the three functional groups in cysteine.

(b) In the human body, cysteine is oxidized to a disulfide. Draw a structural formula for this disulfide.

14.46 Lipoic acid is a growth factor for many bacteria and protozoa and an essential component of several enzymes involved in human metabolism.

Lipoic acid

(a) Name the two functional groups in lipoic acid.

(b) At one stage in its function in human metabolism, the disulfide bond of lipoic acid is reduced to two thiol groups. Draw a structural formula for this reduced form of lipoic acid.

Chemical Connections

14.47 (Chemical Connections 14A) When was nitroglycerin discovered? Is this substance a solid, liquid, or gas?

14.48 (Chemical Connections 14A) What was Alfred Nobel's discovery that made nitroglycerin safer to handle?

14.49 (Chemical Connections 14A) What is the relationship between the medical use of nitroglycerin to relieve the sharp chest pains (angina) associated with heart disease and the gas nitric oxide, NO?

14.50 (Chemical Connections 14B) What is the color of dichromate ion? Of chromium(III) ion? Explain how the conversion of one to the other is used in breath-alcohol screening.

14.51 (Chemical Connections 14B) The legal definition of being under the influence of alcohol is based on blood alcohol content. What is the relationship between breath alcohol content and blood alcohol content?

14.52 (Chemical Connections 14C) What does it mean to say that ethylene oxide is a highly strained molecule?

14.53 (Chemical Connections 14D) What are the advantages and disadvantages of using diethyl ether as an anesthetic?

14.54 (Chemical Connections 14D) Show that enflurane and isoflurane are constitutional isomers.

14.55 (Chemical Connections 14D) Would you expect enflurane and isoflurane to be soluble in water? Would you expect them to be soluble in organic solvents such as hexane?

Additional Problems

14.56 Write a balanced equation for the complete combustion of ethanol, the alcohol added to gasoline to produce gasohol.

14.57 Write a balanced equation for the complete combustion of methanol, the alcohol used in the engines of formula racing cars.

14.58 Knowing what you do about electronegativity, the polarity of covalent bonds, and hydrogen bonding, would you expect an N—H----N hydrogen bond to be stronger than, the same strength as, or weaker than an O—H----O hydrogen bond?

14.59 Draw structural formulas and write IUPAC names for the eight isomeric alcohols with molecular formula $C_5H_{12}O$.

14.60 Draw structural formulas and write common names for the six isomeric ethers with molecular formula $C_5H_{12}O$.

14.61 Explain why the boiling point of ethylene glycol, (198°C) is so much higher than that of 1-propanol, (97°C), even though their molecular weights are about the same.

14.62 1,4-Butanediol, hexane, and 1-pentanol have similar molecular weights. Their boiling points, arranged from lowest to highest, are 69°C, 138°C, and 230°C. Which compound has which boiling point?

14.63 Of the three compounds given in Problem 14.62, one is insoluble in water, another has a solubility of 2.3 g/100 g water, and one is infinitely soluble in water. Which compound has which solubility?

14.64 Each of the following compounds is a common organic solvent. From each pair of compounds, select the solvent with the greater solubility in water.
(a) CH_2Cl_2 or CH_3CH_2OH
(b) $CH_3CH_2OCH_2CH_3$ or CH_3CH_2OH

14.65 Show how to prepare each compound from 2-methyl-1-propanol.
(a) 2-Methylpropene
(b) 2-Methyl-2-propanol
(c) 2-Methylpropanoic acid, $(CH_3)_2CHCOOH$

14.66 Show how to prepare each compound from 2-methylcyclohexanol.

Info Trac College Edition

For additional readings, go to InfoTrac College Edition, your online research library, at

http://infotrac.thomsonlearning.com

CHAPTER 15

15.1 Introduction

15.2 Enantiomerism: A New Kind of Stereoisomerism

15.3 Naming Stereocenters: The (R,S) System

15.4 Molecules with Two or More Stereocenters

15.5 Optical Activity: How Chirality Is Detected in the Laboratory

15.6 The Significance of Chirality in the Biological World

Median cross section through the shell of a chambered nautilus found in the deep waters of the Pacific Ocean. The shell shows handedness; this cross section is a left-handed spiral.

Chirality: The Handedness of Molecules

15.1 Introduction

In this chapter, we study the relationship between objects and their mirror images; that is, we study isomers called enantiomers and diastereomers. Figure 15.1 summarizes the relationship among these isomers and those we studied in Chapters 11 through 14.

The significance of enantiomers is that, except for inorganic and a few simple organic compounds, the vast majority of molecules in the biological world show this type of isomerism. Furthermore, approximately one-half of all medications used to treat humans show it as well.

15.2 Enantiomerism: A New Kind of Stereoisomerism

As an example of a molecule that shows enantiomerism, let us consider 2-butanol. As we discuss this molecule, we will focus on carbon 2, the carbon

15.2 ENANTIOMERISM: A NEW KIND OF STEREOISOMERISM

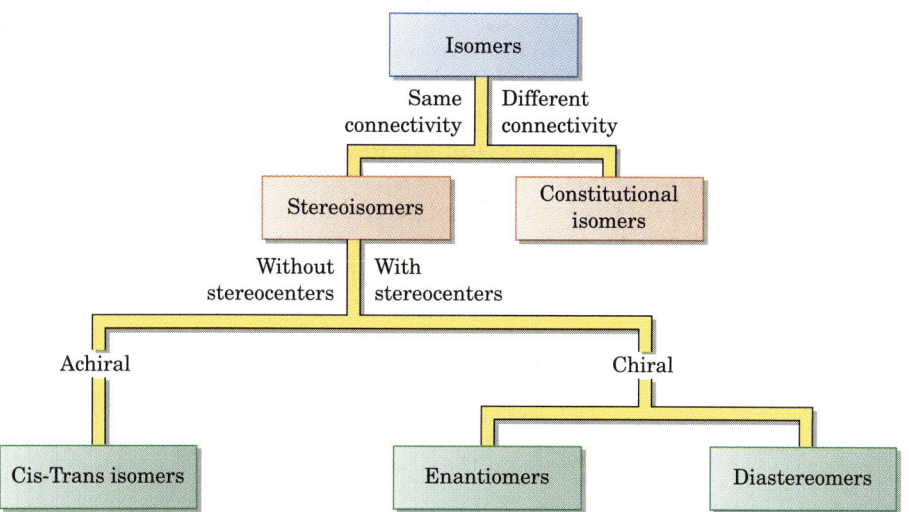

Figure 15.1 Relationships among isomers. In this chapter, we study enantiomers and diastereomers.

bearing the —OH group. What makes this carbon of interest is the fact that it has four different groups bonded to it.

$$\underset{\text{2-Butanol}}{CH_3\overset{\overset{\displaystyle OH}{|}}{C}HCH_2CH_3}$$

This structural formula does not show the shape of 2-butanol or the orientation of its atoms in space. To do so, we must consider the molecule as a three-dimensional object.

On the left is what we will call the "original molecule" and a ball-and-stick model of it. In this drawing, the —OH and —CH₃ groups are in the plane of the paper, the —H is behind it, and the —CH₂CH₃ group is in front of it. In the middle is a mirror. On the right is a **mirror image** of the original molecule along with a ball-and-stick model of the mirror image. Every molecule—and, in fact, every object in the world around us—has a mirror image.

The question we now need to ask is, "What is the relationship between the original molecule of 2-butanol and its mirror image?" To answer this question, you need to imagine that you can pick up the mirror image and move it in space in any way you wish. If you can move the mirror image in space and find that it fits over the original so that every bond, atom, and de-

Screen 15.3
Interactive Example

384 | *CHAPTER 15* CHIRALITY: THE HANDEDNESS OF MOLECULES

The terms "superposable" and "superimposable" mean the same thing, and both are used currently.

tail of the mirror image matches exactly the bonds, atoms, and details of the original molecule, then the two are **superposable.** In other words, the mirror image and the original represent the same molecule; they are merely oriented differently in space. If, no matter how you turn the mirror image in space, it will not fit exactly on the original with every detail matching, then the two are **nonsuperposable;** they are different molecules.

One way to see that the mirror image of 2-butanol is not superposable on the original molecule is illustrated in the following drawings. Imagine that you hold the mirror image by the C—OH bond and rotate the bottom part of molecule by 180° about this bond. The —OH group retains its position in space, but the —CH₃ group, which was to the right and in the plane of the paper, remains in the plane of the paper but is now to the left. Similarly, the —CH₂CH₃ group, which was in front of the plane of the paper and to the left, is now behind the plane and to the right.

Now move the mirror image in space and try to fit it on the original molecule so that all bonds and atoms match.

■ The threads of a drill or screw twist along the axis of the helix, and some plants climb by sending out tendrils that twist helically.

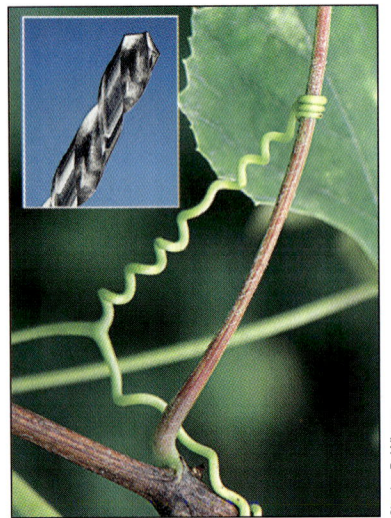

Charles D. Winters

By turning the mirror image as we did, its —OH and —CH₃ groups now fit exactly on top of the —OH and —CH₃ groups of the original molecule. However, the —H and —CH₂CH₃ groups of the two do not match. The —H is away from you in the original but toward you in the mirror image; the —CH₂CH₃ group is toward you in the original but away from you in the mirror image. We conclude that the original molecule of 2-butanol and its mirror image are nonsuperposable and, therefore, different compounds.

To summarize, we can turn and rotate the mirror image of 2-butanol in any direction in space, but as long as no bonds are broken and rearranged, we can make only two of the four groups bonded to carbon 2 of the mirror image coincide with the groups on the original molecule. Because 2-butanol

15.2 ENANTIOMERISM: A NEW KIND OF STEREOISOMERISM | 385

(a)

(b)

■ (a) Mirror images of two wood carvings. The mirror image cannot be superposed on the actual model. The man's right arm rests on the camera in the mirror image; in the actual statue, however, the man's left arm rests on the camera. (b) Left- and right-handed sea shells. If you cup a right-handed shell in your right hand with your thumb pointing from the narrow end to the wide end, the opening will be on your right.

and its mirror image are not superposable, they are isomers. Isomers such as these are called **enantiomers**. Enantiomers, like gloves, always occur in pairs.

Objects that are not superposable on their mirror images are said to be **chiral** (pronounced "ki-ral," rhymes with "spiral;" from the Greek: *cheir*, hand); that is, they show handedness. We encounter chirality in three-dimensional objects of all sorts. Your left hand is chiral, and so is your right hand. A spiral binding on a notebook is chiral. A machine screw with a right-handed twist is chiral. A ship's propeller is chiral. As you examine the objects in the world around you, will undoubtedly conclude that the vast majority of them are chiral.

As we said after we examined the original and mirror image of 2-butanol, the most common cause of enantiomerism in organic molecules is the presence of a carbon with four different groups bonded to it. Let us examine this statement further by considering 2-propanol, which has no such carbon atom. In this molecule, one carbon atom is bonded to three different groups, but none is bonded to four different groups.

On the left is a three-dimensional representation of 2-propanol; on the right is its mirror image. Also shown are ball-and-stick models of each molecule.

> **Enantiomers** Stereoisomers that are nonsuperposable mirror images; refers to a relationship between pairs of objects

> **Chiral** From the Greek *cheir*, hand; an object that is not superposable on its mirror image

Original molecule | Mirror image

The question we now ask is, "What is the relationship of the mirror image to the original?" This time, let us rotate the mirror image by 120° about the C—OH bond, and then compare it to the original. After performing this rotation, we see that all atoms and bonds of the mirror image fit exactly on the original. Thus the structures we first drew for the original

Screen 15.3
Interactive Example

Figure 15.2 Three-dimensional representations of lactic acid and its mirror image.

molecule and its mirror image are, in fact, the same molecule—just viewed from different perspectives.

Achiral Objects that lack chirality; an object that is superposable on its mirror image

Stereocenter A tetrahedral carbon atom that has four different groups bonded to it

Racemic mixture A mixture of equal amounts of two enantiomers

If an object and its mirror image are superposable, then they are identical and no enantiomerism is possible. We say that such an object is **achiral** (without chirality). Examples of achiral objects include an undecorated cup, an unmarked baseball bat, a regular tetrahedron, a cube, and a sphere.

To repeat, the most common cause of chirality in organic molecules is a tetrahedral carbon atom with four different groups bonded to it. We call such a carbon atom a **stereocenter.** 2-Butanol has one stereocenter; 2-propanol has none. As another example of a molecule with a stereocenter, let us consider 2-hydroxypropanoic acid, more commonly named lactic acid. Lactic acid is a product of anaerobic glycolysis. (See Section 27.2 and also Chemical Connections 27A.) It is also what gives sour cream its sour taste.

Figure 15.2 shows three-dimensional representations for lactic acid and its mirror image. In these representations, all bond angles about the central carbon atom are approximately 109.5°, and the four bonds from it are directed toward the corners of a regular tetrahedron. Lactic acid shows enantiomerism; that is, the original molecule and its mirror image are not superposable but rather are different molecules.

An equimolar mixture of two enantiomers is called a **racemic mixture,** a term derived from the name racemic acid (Latin: *racemus,* a cluster of grapes). Racemic acid is the name originally given to the equimolar mixture of the enantiomers of tartaric acid that forms as a by-product during the fermentation of grape juice.

EXAMPLE 15.1

Each of the following molecules has one stereocenter marked by an asterisk. Draw three-dimensional representations for its enantiomers.

(a) $CH_3\overset{*}{C}HCH_2CH_3$ with Cl on the stereocenter

(b) phenyl–$\overset{*}{C}HCH_3$ with NH_2 on the stereocenter

Solution

First, draw the carbon stereocenter showing the tetrahedral orientation of its four bonds. One way to do so is to draw two bonds in the plane of the paper, a third bond toward you in front of the plane, and the fourth bond away from you behind the plane. Now place the four groups bonded to the stereocenter on these positions. How you position them is purely arbitrary; there are 12 equivalent ways to do it. To draw an original of (a), for example, place the —Cl group in the vertical position in the plane of the paper, and the —H to the left also in the plane. Place the —CH_2CH_3 away from you and the —CH_3 toward you; this orientation gives the enantiomer of (a) on the left. Its mirror image is on the right.

Problem 15.1

Each of the following molecules has one stereocenter marked by an asterisk. Draw three-dimensional representations for its enantiomers.

15.3 Naming Stereocenters: The *R,S* System

Because enantiomers are two different compounds, each must have a different name. The over-the-counter drug ibuprofen, for example, shows enantiomerism and can exist as the pair of enantiomers shown here:

The inactive enantiomer of ibuprofen   *The active enantiomer*

Only one enantiomer of ibuprofen is biologically active. It reaches therapeutic concentrations in the human body in approximately 12 minutes, whereas the racemic mixture takes approximately 30 minutes. However, in this case, the inactive enantiomer is not wasted. The body converts it to the active enantiomer, but that process takes time.

What we need is a way to designate which enantiomer of ibuprofen (or one of any other pair of enantiomers, for that matter) is which without hav-

Screen 15.4
Tutorial

R,S system A set of rules for specifying configuration about a stereocenter

ing to draw and point to one or the other of the enantiomers. To do so, chemists have developed the **R,S system.** The first step in assigning an R or S configuration to a stereocenter is to arrange the groups bonded to it in order of priority. Priority is based on atomic number: The higher the atomic number, the higher the priority. If a priority cannot be assigned on the basis of the atoms bonded directly to the stereocenter, look at the next atom or set of atoms and continue to the first point of difference; that is, continue until you can assign a priority.

Table 15.1 shows the priorities of the most common groups encountered in organic and biochemistry. In the R,S system, a C=O is treated as if carbon were bonded to two oxygens by single bonds; thus CH=O has a higher priority than —CH_2OH, in which carbon is bonded to only one oxygen.

EXAMPLE 15.2

Assign priorities to the groups in each set.
(a) —CH_2OH and —CH_2CH_2OH
(b) —CH_2CH_2OH and —CH_2NH_2

Solution
(a) The first point of difference is O of the —OH group compared to C of the —CH_2OH group.

    First point of difference

    —CH_2OH —CH_2CH_2OH
    Higher priority   Lower priority

(b) The first point of difference is C of the —CH_2OH group compared to N of the —NH_2 group.

    First point of difference

    —CH_2CH_2OH —CH_2NH_2
    Lower priority   **Higher priority**

Problem 15.2
Assign priorities to the groups in each set.

(a) —CH_2OH and —$CH_2CH_2\overset{\overset{\displaystyle O}{\|}}{C}OH$

(b) —CH_2NH_2 and —$CH_2\overset{\overset{\displaystyle O}{\|}}{C}OH$

To assign an R or S configuration to a stereocenter:

1. Assign a priority from 1 (highest) to 4 (lowest) to each group bonded to the stereocenter.

Table 15.1 R,S Priorities of Some Common Groups

Atom or Group	Reason for Priority: First Point of Difference (Atomic Number)
—I	iodine (53)
—Br	bromine (35)
—Cl	chlorine (17)
—SH	sulfur (16)
—OH	oxygen (8)
—NH$_2$	nitrogen (7)
—COOH (—C(=O)OH)	carbon to oxygen, oxygen, then oxygen (6 ⟶ 8, 8, 8)
—C(=O)NH$_2$	carbon to oxygen, oxygen, then nitrogen (6 ⟶ 8, 8, 7)
—C(=O)H	carbon to oxygen, oxygen, then hydrogen (6 ⟶ 8, 8, 1)
—CH$_2$OH	carbon to oxygen (6 ⟶ 8)
—CH$_2$NH$_2$	carbon to nitrogen (6 ⟶ 7)
—CH$_2$CH$_3$	carbon to carbon (6 ⟶ 6)
—CH$_2$H	carbon to hydrogen (6 ⟶ 1)
—H	hydrogen (1)

2. Orient the molecule in space so that the lowest-priority group (4) is directed away from you as would be, for instance, the steering column of a car. The three higher-priority groups (1–3) then project toward you, as would the spokes of a steering wheel.

3. Read the three groups projecting toward you in order from highest (1) to lowest (3) priority.

Group of lowest priority points away from you

R Used in the *R,S* system to show that, when the lowest-priority group is oriented away from you, the order of priority of groups on a stereocenter is clockwise

4. If reading the groups 1-2-3 proceeds in a clockwise direction, the configuration is designated as **R** (Latin: *rectus,* straight); if reading the groups 1-2-3 proceeds in a counterclockwise direction, the configuration is **S** (Latin: *sinister,* left). You can also visualize this system as follows: Turning the steering wheel to the right equals R and turning it to the left equals S.

S Used in the *R,S* system to show that, when the lowest-priority group is oriented away from you, the order of priority of groups on a stereocenter is counterclockwise

EXAMPLE 15.3

Assign an R or S configuration to each stereocenter.

(a) 2-Butanol: H₃C—C(OH)(H)—CH₂CH₃

(b) Alanine: H₃C—C(NH₂)(H)—COOH

Solution

View each molecule through the stereocenter and along the bond from the stereocenter to the group of lowest priority.

(a) The order of decreasing priority is
—OH > —CH₂CH₃ > —CH₃ > —H. Therefore, view the molecule along the C—H bond with the H pointing away from you. Reading the other three groups in the order 1-2-3 occurs in the clockwise direction. Therefore, the configuration is R and this enantiomer is (R)-2-butanol.

With —H, the lowest-priority group, pointing away from you, this is what you see

(b) The order of decreasing priority is
—NH₂ > —COOH > —CH₃ > —H. View the molecule along the C—H bond with H pointing away from you. Reading the groups in the order 1-2-3 occurs in the clockwise direction; therefore, the configuration is R and this enantiomer is (R)-alanine.

With —H, the lowest-priority group, pointing away from you, this is what you see

Problem 15.3

Assign an R or S configuration to each stereocenter.

(a) 2-Methylbutanoic acid

(b) Glyceraldehyde

Now let us return to our three-dimensional drawing of the enantiomers of ibuprofen and assign each an R or S configuration. In order of decreasing priority, the groups bonded to the stereocenter are —COOH (1) >

—C₆H₅ (2) > —CH₃ (3) > —H (4). In the enantiomer on the left, reading the groups on the stereocenter in order of priority is clockwise and, therefore, this enantiomer is (R)-ibuprofen. Its mirror image is (S)-ibuprofen.

(R)-Ibuprofen
(the inactive enantiomer)

(S)-Ibuprofen
(the active enantiomer)

The R,S system can be used to specify the configuration of any stereocenter in any molecule. It is not, however, the only system used for this purpose. There is also a D,L system, which is used primarily to specify the configuration of carbohydrates (Chapter 19) and amino acids (Chapter 21).

15.4 Molecules with Two or More Stereocenters

For a molecule with n stereocenters, the maximum number of stereoisomers possible is 2^n. We have already verified that, for a molecule with one stereocenter, $2^1 = 2$ stereoisomers (one pair of enantiomers) are possible. For a molecule with two stereocenters, a maximum of $2^2 = 4$ stereoisomers (two pairs of enantiomers) is possible; for a molecule with three stereocenters, a maximum of $2^3 = 8$ stereoisomers (four pairs of enantiomers) is possible; and so forth.

A Molecules with Two Stereocenters

We begin our study of molecules with two stereocenters by considering 2,3,4-trihydroxybutanal, a molecule with two stereocenters.

CHO
|
*CHOH
|
*CHOH
|
CH₂OH

2,3,4-Trihydroxybutanal

The maximum number of stereoisomers possible for this molecule is $2^2 = 4$, each of which is drawn in Figure 15.3.

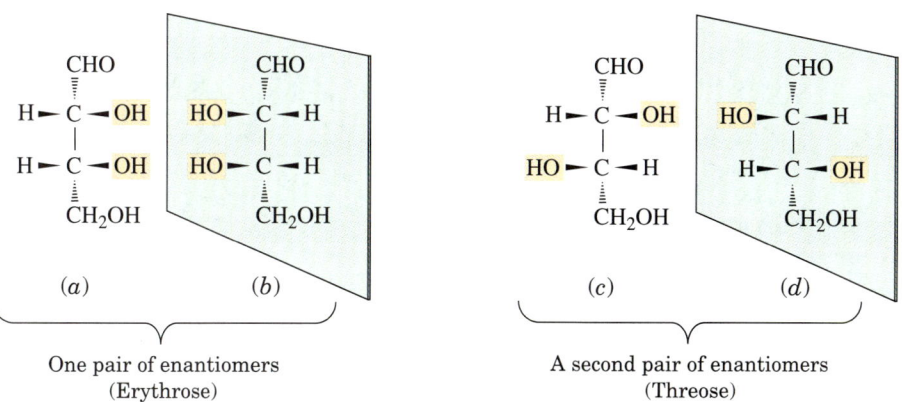

Figure 15.3 The four stereoisomers of 2,3,4-trihydroxybutanal.

Stereoisomers (a) and (b) are nonsuperposable mirror images and are, therefore, a pair of enantiomers. Stereoisomers (c) and (d) are also nonsuperposable mirror images and are a second pair of enantiomers. We describe the four stereoisomers of 2,3,4-trihydroxybutanal by saying that they consist of two pairs of enantiomers. Enantiomers (a) and (b) are named **erythrose**. Erythrose is synthesized in erythrocytes (red blood cells); hence the derivation of its name. Enantiomers (c) and (d) are named **threose**. Erythrose and threose belong to the class of compounds called carbohydrates, which we discuss in Chapter 19.

We have specified the relationship between (a) and (b) and that between (c) and (d). What is the relationship between (a) and (c), between (a) and (d), between (b) and (c), and between (b) and (d)? The answer is that they are **diastereomers**—stereoisomers that are not mirror images.

> **Diastereomers** Stereoisomers that are not mirror images of each other

Screen 15.4
Exercise

EXAMPLE 15.4

1,2,3-Butanetriol has two stereocenters (carbons 2 and 3); thus $2^2 = 4$ stereoisomers are possible for it. Following are three-dimensional representations for each.

```
       CH₂OH         CH₂OH         CH₂OH         CH₂OH
    H—C—OH        H—C—OH        HO—C—H        HO—C—H
   HO—C—H         H—C—OH        HO—C—H         H—C—OH
       CH₃           CH₃           CH₃           CH₃
        (1)           (2)           (3)           (4)
```

(a) Which stereoisomers are pairs of enantiomers?
(b) Which stereoisomers are diastereomers?

Solution
(a) Enantiomers are stereoisomers that are nonsuperposable mirror images. Compounds (1) and (4) are one pair of enantiomers, and compounds (2) and (3) are a second pair of enantiomers.
(b) Diastereomers are stereoisomers that are not mirror images. Compounds (1) and (2), (1) and (3), (2) and (4), and (3) and (4) are diastereomers.

Problem 15.4

3-Amino-2-butanol has two stereocenters (carbons 2 and 3); thus $2^2 = 4$ stereoisomers are possible for it.

```
         CH₃           CH₃           CH₃           CH₃
      H—C—OH        H—C—OH        HO—C—H         HO—C—H
     H₂N—C—H        H—C—NH₂       H—C—NH₂       H₂N—C—H
         CH₃           CH₃           CH₃           CH₃
          (1)           (2)           (3)           (4)
```

(a) Which stereoisomers are pairs of enantiomers?
(b) Which sets of stereoisomers are diastereomers?

We can analyze chirality in cyclic molecules with two stereocenters in the same way we analyzed it in acyclic compounds.

EXAMPLE 15.5

How many stereoisomers are possible for 3-methylcyclopentanol?

Solution
Carbons 1 and 3 of this compound are stereocenters. There are, therefore, $2^2 = 4$ stereoisomers possible for this molecule.

$$H_3C \overset{*}{} \triangle \overset{*}{} OH$$

Problem 15.5
How many stereoisomers are possible for 3-methylcyclohexanol?

EXAMPLE 15.6

Mark the stereocenters in each compound with an asterisk. How many stereoisomers are possible for each?

(a) [cyclopentane ring with OH and two CH₃ groups on adjacent carbon]

(b) [cyclohexane ring with OH and isopropyl group]

(c) $CH_3-\underset{\underset{NH_2}{|}}{\overset{\overset{OH}{|}}{CH}}-CH-\overset{\overset{O}{\|}}{C}OH$

Solution
Each stereocenter is marked with an asterisk, and under each compound is given the number of stereoisomers possible for it. In (a), the carbon bearing the two methyl groups is not a stereocenter; this carbon has only three different groups bonded to it.

(a) [cyclopentane with *OH and C(CH₃)₂]
$2^1 = 2$

(b) [cyclohexane with *OH and *isopropyl]
$2^2 = 4$

(c) $CH_3-\overset{*}{\underset{\underset{NH_2}{|}}{\overset{\overset{OH}{|}}{CH}}}-\overset{*}{CH}-\overset{\overset{O}{\|}}{C}OH$
$2^2 = 4$

Problem 15.6
Mark all stereocenters in each compound with an asterisk. How many stereoisomers are possible for each?

(a) HO—C₆H₃(OH)—CH₂CHCOOH with NH₂ (b) $CH_2=CHCHCH_2CH_3$ with OH (c) cyclohexane with OH and NH₂

B Molecules with Three or More Stereocenters

The 2^n rule applies equally well to molecules with three or more stereocenters. In the following disubstituted cyclohexanol with three stereocenters, each stereocenter is marked with an asterisk. A maximum of $2^3 = 8$ stereoisomers are possible for this molecule. Menthol, one of the eight, has the configuration shown on the right. The configuration at the carbon bearing the —OH group is R. Menthol is present in peppermint and other mint oils.

CHEMICAL CONNECTIONS 15A

Chiral Drugs

Some common drugs used in human medicine—for example, aspirin (Chemical Connections 18C)—are achiral. Others, such as the penicillin and erythromycin classes of antibiotics and the drug Captopril, are chiral and are sold as single enantiomers. Captopril is very effective for the treatment of high blood pressure and congestive heart failure (See Chemical Connections 22F). It is manufactured and sold as the (s,s)-stereoisomer.

Captopril

A large number of chiral drugs, however, are sold as racemic mixtures. The popular analgesic ibuprofen (the active ingredient in Motrin, Advil, and many other nonaspirin analgesics) is an example.

Recently, the U.S. Food and Drug Administration established new guidelines for the testing and marketing of chiral drugs. After reviewing these guidelines, many drug companies have decided to develop only single enantiomers of new chiral drugs.

In addition to regulatory pressure, pharmaceutical developers must deal with patent considerations. If a company has a patent on a racemic mixture of a drug, a new patent can often be taken out on one of its enantiomers.

■ Ibuprofen (for example, Advil) is sold as a racemic mixture, and naproxen (for example, Aleve) is sold as the s enantiomer.

2-Isopropyl-5-methyl-cyclohexanol

Menthol

The R configuration at this carbon

Cholesterol is a more complicated molecule with eight stereocenters. To identify them, remember to add an appropriate number of hydrogens to complete the tetravalence of each carbon you think might be a stereocenter.

Cholesterol
(has 8 stereocenters; 256 stereoisomers are possible)

Cholesterol
(this is the stereoisomer present in human metabolism)

15.5 Optical Activity: How Chirality Is Detected in the Laboratory

A Plane-Polarized Light

As we have already established, the two members of a pair of enantiomers are different compounds, and we must expect, therefore, that some of their properties differ. One such property relates to their effect on the plane of polarized light. Each member of a pair of enantiomers rotates the plane of polarized light; for this reason, each enantiomer is said to be **optically active**. To understand how optical activity is detected in the laboratory, we must first understand plane-polarized light and how a polarimeter, the instrument used to detect optical activity, works.

Optically active Showing that a compound rotates the plane of polarized light

Plane-polarized light Light with waves vibrating in only parallel planes

Ordinary light consists of waves vibrating in all planes perpendicular to its direction of propagation. Certain materials, such as a Polaroid sheet (a plastic film like that used in polarized sunglasses), selectively transmit light waves vibrating only in parallel planes. Electromagnetic radiation vibrating in only parallel planes is said to be **plane polarized.**

B A Polarimeter

A **polarimeter** consists of a light source emitting unpolarized light, a polarizer, an analyzer, and a sample tube (Figure 15.4). If the sample tube is empty, the intensity of light reaching the detector (in this case, your eye) is at its maximum when the axes of the polarizer and analyzer are parallel to each other. If the analyzer is turned either clockwise or counterclockwise, less light is transmitted. When the axis of the analyzer is at right angles to the axis of the polarizer, the field of view is dark.

When a solution of an optically active compound is placed in the sample tube, it rotates the plane of the polarized light. If it rotates the plane clockwise, we say it is **dextrorotatory;** if it rotates the plane counterclockwise, we say it is **levorotatory.** Each member of a pair of enantiomers rotates the plane of polarized light by the same number of degrees, but in opposite directions. If one enantiomer is dextrorotatory, the other is levorotatory.

Dextrorotatory Clockwise (to the right) rotation of the plane of polarized light in a polarimeter

Levorotatory Counterclockwise (to the left) rotation of the plane of polarized light in a polarimeter

The number of degrees by which an optically active compound rotates the plane of polarized light is called its **specific rotation** and is given the symbol $[\alpha]$. Specific rotation is defined as the observed rotation of an optically active substance at a concentration of 1 g/mL in a sample tube 10 cm long. A dextrorotatory compound is indicated by a plus sign in parentheses, $(+)$, and a levorotatory compound is indicated by a minus sign in parentheses, $(-)$. Common practice is to report the temperature (in °C) at which the measurement is made and the wavelength of light used. The most common wavelength of light used in polarimetry is the sodium D line, the same wavelength responsible for the yellow color of sodium-vapor lamps.

Figure 15.4 Schematic diagram of a polarimeter with its sample tube containing a solution of an optically active compound. The analyzer has been turned clockwise by α degrees to restore the light field.

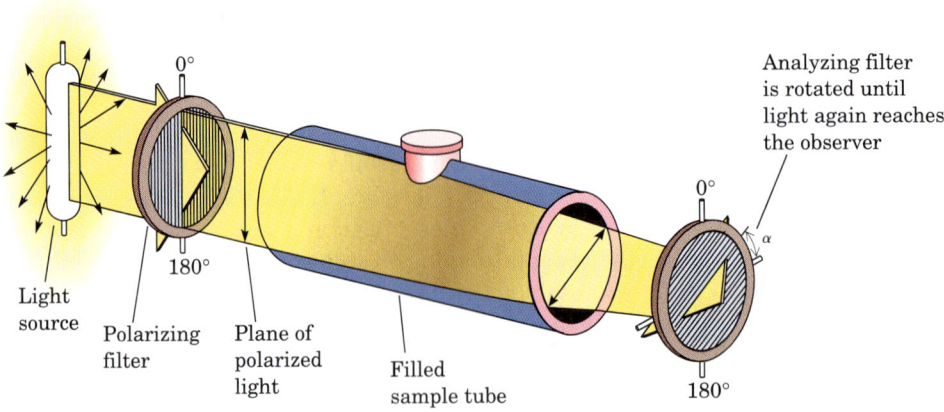

Following are specific rotations for the enantiomers of lactic acid measured at 21°C and using the D line of a sodium-vapor lamp as the light source:

$$\underset{\substack{(S)\text{-}(+)\text{-Lactic acid} \\ [\alpha]_D^{21} = +2.6°}}{\text{H}_3\text{C}-\overset{\text{COOH}}{\underset{\text{OH}}{\text{C}}}-\text{H}} \qquad \underset{\substack{(R)\text{-}(-)\text{-Lactic acid} \\ [\alpha]_D^{21} = -2.6°}}{\text{H}-\overset{\text{COOH}}{\underset{\text{CH}_3}{\text{C}}}-\text{OH (HO on left)}}$$

The (+) enantiomer of lactic acid is produced by muscle tissue in humans. The (−) enantiomer is found in sour milk.

15.6 The Significance of Chirality in the Biological World

Except for inorganic salts and a few low-molecular-weight organic substances, the molecules in living systems—both plant and animal—are chiral. Although these molecules can exist as a number of stereoisomers, almost invariably only one stereoisomer is found in nature. Of course, instances do occur in which more than one stereoisomer is found, but these isomers rarely exist together in the same biological system.

A Chirality in Biomolecules

Perhaps the most conspicuous examples of chirality among biological molecules are the enzymes, all of which have many stereocenters. Consider chymotrypsin, an enzyme in the intestines of animals that catalyzes the digestion of proteins (Chapter 22). Chymotrypsin has 251 stereocenters. The maximum number of stereoisomers possible is 2^{251}—a staggeringly large number, almost beyond comprehension. Fortunately, nature does not squander its precious energy and resources unnecessarily; any given organism produces and uses only one of these stereoisomers.

B How an Enzyme Distinguishes Between a Molecule and Its Enantiomer

An enzyme catalyzes a biological reaction of a molecule by first positioning it at a **binding site** on its surface. An enzyme with specific binding sites for three of the four groups on a stereocenter can distinguish between a molecule and its enantiomer or one of its diastereomers. Assume, for example, that an enzyme involved in catalyzing a reaction of glyceraldehyde has three binding sites: one specific for —H, a second specific for —OH, and a third specific for —CHO. Assume further that the three sites are arranged on the enzyme surface as shown in Figure 15.5. The enzyme can distinguish (R)-glyceraldehyde (the natural or biologically active form) from its enantiomer because the natural enantiomer can be absorbed with three groups interacting with their appropriate binding sites; for the S enantiomer, at best only two groups can interact with these binding sites.

■ The horns of this African gazelle show chirality; one horn is the mirror image of the other.

Figure 15.5 A schematic diagram of an enzyme surface that can interact with (R)-glyceraldehyde at three binding sites, but with (S)-glyceraldehyde at only two of these sites.

(R)-Glyceraldehyde fits the three binding sites on surface

(S)-Glyceraldehyde fits only two of the three binding sites

Because interactions between molecules in living systems take place in a chiral environment, it should come as no surprise that a molecule and its enantiomer or one of its diastereomers elicit different physiological responses. As we have already seen, (S)-ibuprofen is active as a pain and fever reliever while its R enantiomer is inactive. The S enantiomer of the closely related analgesic naproxen is also the active pain reliever of this compound, but its R enantiomer is a liver toxin!

(S)-Ibuprofen   (S)-Naproxen

SUMMARY

A **mirror image** is the reflection of an object in a mirror (Section 15.2). **Enantiomers** are a pair of stereoisomers that are nonsuperposable mirror images of each other. A **racemic mixture** contains equal amounts of two enantiomers. An object that is not superposable on its mirror image is said to be **chiral**. An **achiral** object is one that lacks chirality; that is, it has a superposable mirror image (Section 15.2). The most common cause of chirality in organic molecules is the presence of a tetrahedral carbon atom with four different groups bonded to it. Such a carbon is called a **stereocenter**. The configuration at a stereocenter can be specified using the **R,S system** (Section 15.3). For a molecule with n stereocenters, the maximum number of stereoisomers possible is 2^n.

Light with waves that vibrate in only parallel planes is said to be **plane polarized** (Section 15.5A).

To measure optical activity, we use a **polarimeter** (Section 15.5B). A compound is said to be **optically active** if it rotates the plane of polarized light. If it rotates the plane clockwise, it is **dextrorotatory**; if it rotates the plane counterclockwise, it is **levorotatory**. Each member of a pair of enantiomers rotates the plane of polarized light by an equal number of degrees, but in opposite directions.

An enzyme catalyzes biological reactions of molecules by first positioning them at binding sites on its surface (Section 15.6). An enzyme with binding sites specific for three of the four groups on a stereocenter can distinguish between a molecule and its enantiomer or one of its diastereomers.

PROBLEMS

Numbers that appear in color indicate difficult problems.
▲ designates problems requiring application of principles.

Chirality

15.7 What does the term "chiral" mean? Give an example of a chiral molecule.

15.8 What does the term "achiral" mean? Give an example of an achiral molecule.

15.9 Define the term "stereoisomer." Name three types of stereoisomers.

15.10 In what way do constitutional isomers differ from stereoisomers? In what way are they the same?

15.11 Which of the following objects are chiral (assume that there is no label or other identifying mark)?
(a) Pair of scissors
(b) Tennis ball
(c) Paper clip
(d) Beaker
(e) The swirl created in water as it drains out of a sink or bathtub

15.12 2-Pentanol is chiral but 3-pentanol is not. Explain.

15.13 2-Butene exists as a pair of cis-trans isomers. Is cis-2-butene chiral? Is trans-2-butene chiral? Explain.

15.14 Explain why the carbon of a carbonyl group cannot be a stereocenter.

Enantiomers

15.15 Which of the following compounds contain stereocenters?
(a) 2-Chloropentane
(b) 3-Chloropentane
(c) 3-Chloro-1-butene
(d) 1,2-Dichloropropane

15.16 Which of the following compounds contain stereocenters?
(a) Cyclopentanol
(b) 1-Chloro-2-propanol
(c) 2-Methylcyclopentanol
(d) 1-Phenyl-1-propanol

15.17 Using only C, H, and O, write structural formulas for the lowest-molecular-weight chiral molecule of each of the following.
(a) Alkane
(b) Alkene
(c) Alcohol
(d) Aldehyde
(e) Ketone
(f) Carboxylic acid

15.18 Draw the mirror image for each molecule.

(a) H_3C, H, OH, COOH on central C

(b) H—C—OH with CHO up and CH_2OH down

(c) tetrahydrofuran ring with OH and H

(d) cyclohexane with OH and CH_3

15.19 Draw the mirror image for each molecule.

(a) H_2N—C—H with COOH up and CH_3 down

(b) cyclohexene with H and OH

(c) sugar ring with CH_2OH, OH, H, HO, H, OH

(d) H_3C, NH_2 on cyclopentane

15.20 Mark each stereocenter in these molecules with an asterisk. Note that not all contain stereocenters.

(a)

(b) [structure: pentan-3-one with OH]

(c) [structure: valine-like with NH₂ and COOH]

(d) [structure: ketone]

15.21 Mark each stereocenter in these molecules with an asterisk. Note that not all contain stereocenters.

(a) HO–CH₂–CH(OH)–CH₂–OH

(b) HO–CH₂–CH(OH)–CH₃

(c) [structure with OH and alkene]

(d) [structure with OH]

Molecules with Two or More Stereocenters

15.22 Label all stereocenters in each molecule with an asterisk. How many stereoisomers are possible for each molecule?

(a) CH₃CHCHCOOH
        | |
        OH OH

(b) HO—CHCOOH
         |
         CHCOOH
         |
         CH₂COOH

(c) [menthol-like cyclohexane with OH and isopropyl]

(d) [limonene-like structure]

15.23 Label all stereocenters in each molecule with an asterisk. How many stereoisomers are possible for each molecule?

(a) [cyclopentane with OH and CH₃]

(b) [geraniol/linalool-like structure with OH]

(c) [tetrahydrofuran with CH₂OH]

(d) [decalone structure]

The R,S System

15.24 Assign priorities to the groups in each set.
(a) —H, —CH₃, —OH, —CH₂OH
(b) —CH₃, —H, —COOH, —NH₂
(c) —CH₃, —CH₂SH, —NH₂, —COOH

15.25 Which molecules have R configurations?

(a) Br—C(CH₃)(H)—CH₂OH

(b) HOCH₂—C(CH₃)(H)—Br

(c) H₃C—C(CH₂OH)(H)—Br

(d) Br—C(CH₂OH)(H)—CH₃

15.26 For centuries, Chinese herbal medicine has used extracts of *Ephedra sinica* to treat asthma. The asthma-relieving component of this plant is ephedrine, a very potent dilator of the air passages of the lungs. The naturally occurring stereoisomer is levorotatory and has the following structure. Assign an R or S configuration to each stereocenter.

Ephedrine $[\alpha]_D^{21}$ −41

15.27 The specific rotation of naturally occurring ephedrine, shown in Problem 15.26, is −41°. What is the specific rotation of its enantiomer?

15.28 What is a racemic mixture? Is a racemic mixture optically active—that is, will it rotate the plane of polarized light?

Chemical Connections

15.29 (Chemical Connection 15A) What does it mean to say that a drug is chiral? If a drug is chiral, will it be optically active—that is, will it rotate the plane of polarized light?

Additional Problems

15.30 Which of the eight alcohols with molecular formula $C_5H_{12}O$ are chiral?

15.31 Write the structural formula of an alcohol with molecular formula $C_6H_{14}O$ that contains two stereocenters.

15.32 Which carboxylic acids with molecular formula $C_6H_{12}O_2$ are chiral?

15.33 Following are structural formulas for three of the most widely prescribed drugs to treat depression. Label all stereocenters in each, and state the number of stereoisomers possible for each.

(a) Fluoxetine (Prozac)

(b) Sertraline (Zoloft)

(c) Paroxetine (Paxil)

15.34 Label the four stereocenters in amoxicillin, which belongs to the family of semisynthetic penicillins.

Amoxicillin

15.35 Following is a chair conformation of glucose:

(a) Identify the five stereocenters in this molecule.
(b) How many stereoisomers are possible?
(c) How many pairs of enantiomers are possible?

15.36 Consider a cyclohexane ring substituted with one hydroxyl group and one methyl group. Draw a structural formula for a compound of this composition that

(a) Does not show cis-trans isomerism and has no stereocenters.
(b) Shows cis-trans isomerism but has no stereocenters.
(c) Shows cis-trans isomerism and has two stereocenters.

15.37 Triamcinolone acetonide, the active ingredient in Azmacort Inhalation Aerosol, is a steroid used to treat bronchial asthma.

Triamcinolone acetonide

(a) Label the eight stereocenters in this molecule.
(b) How many stereoisomers are possible for it? (Of these, only one is the active ingredient in Azmacort.)

15.38 The next time you have the opportunity to view a collection of sea shells that have a helical twist, study the chirality (handedness) of their twists. For each kind of shell, do you find an equal number of left-handed and right-handed twists or, for example, do they all have the same handedness? What about the handedness among various kinds of shells with helical twists?

15.39 The next time you have an opportunity to examine any of the seemingly endless varieties of blond spiral pasta (rotini, fusilli, radiatori, tortiglione, and so forth), examine their twists. Do the twists of any one kind all have a right-handed twist or a left-handed twist, or are they a racemic mixture?

15.40 Think about the helical coil of a telephone cord or the spiral binding on a notebook. Suppose that you view the spiral from one end and find that it has a left-handed twist. If you view the same spiral from the other end, does it have a left-handed twist from that end as well or does it have a right-handed twist?

InfoTrac College Edition

For additional readings, go to InfoTrac College Edition, your online research library, at

http://infotrac.thomsonlearning.com

CHAPTER 16

16.1 Introduction

16.2 Structure and Classification

16.3 Nomenclature

16.4 Physical Properties

16.5 Basicity

16.6 Reaction with Acids

16.7 Epinephrine: A Prototype for the Development of New Bronchodilators

This inhaler delivers puffs of albuterol (Proventil), a potent synthetic bronchodilator whose structure is patterned after that of epinephrine (adrenaline). See Section 16.7.

Amines

16.1 Introduction

Carbon, hydrogen, and oxygen are the three most common elements in organic compounds. Because of the wide distribution of amines in the biological world, nitrogen is the fourth most common element of organic compounds. The most important chemical property of amines is their basicity.

16.2 Structure and Classification

Amines are classified as **primary (1°), secondary (2°),** or **tertiary (3°),** depending on the number of carbon groups bonded to nitrogen (Section 10.4B).

$$CH_3-NH_2 \qquad CH_3-\underset{\underset{H}{|}}{N}-CH_3 \qquad CH_3-\underset{\underset{CH_3}{|}}{N}-CH_3$$

Methylamine (a 1° amine)   Dimethylamine (a 2° amine)   Trimethylamine (a 3° amine)

Amines are further classified as aliphatic or aromatic. In an **aliphatic amine,** all the carbons bonded to nitrogen are derived from alkyl groups. In an **aromatic amine,** one or more of the groups bonded to nitrogen are aryl groups.

Aliphatic amine An amine in which nitrogen is bonded only to alkyl groups

Aromatic amine An amine in which nitrogen is bonded to one or more aromatic rings

403

CHEMICAL CONNECTIONS 16A

Amphetamines (Pep Pills)

Amphetamine, methamphetamine, and phentermine—all synthetic amines—are powerful stimulants of the central nervous system. Like most other amines, they are stored and administered as their salts. The sulfate salt of amphetamine is prescribed as Benzedrine, the hydrochloride salt of the s enantiomer of methamphetamine as Methedrine, and the hydrochloride salt of phentermine as Fastin.

These three amines all have similar physiological effects and are referred to by the general name *amphetamines*. Structurally, they have in common a benzene ring with a three-carbon side chain and an amine nitrogen on the second carbon of the chain. Physiologically, they have in common the ability to reduce fatigue and diminish hunger by raising the glucose level of the blood. Because of these properties, amphetamines are widely prescribed to counter mild depression, reduce hyperactivity in children, and suppress appetite in people who are trying to lose weight. They are also used illegally to reduce fatigue and elevate mood.

Abuse of amphetamines can have severe effects on both the body and the mind. They are addictive, concentrate in the brain and nervous system, and can lead to long periods of sleeplessness, loss of weight, and paranoia.

The action of amphetamines is similar to that of epinephrine (Section 16.7), the hydrochloride salt of which is named Adrenalin.

Amphetamine (Benzedrine)

(s)-Methamphetamine (Methedrine)

Phentermine (Fastin)

Aniline (a 1° aromatic amine)

N-Methylaniline (a 2° aromatic amine)

Benzyldimethylamine (a 3° aliphatic amine)

Heterocyclic amine An amine in which nitrogen is one of the atoms of a ring

Heterocyclic aliphatic amine A heterocyclic amine in which nitrogen is bonded only to alkyl groups

Heterocyclic aromatic amine An amine in which nitrogen is one of the atoms of an aromatic ring

An amine in which the nitrogen atom is part of a ring is classified as a **heterocyclic amine.** When the ring is saturated, the amine is classified as a **heterocyclic aliphatic amine.** When the nitrogen is part of an aromatic ring (Section 13.2), the amine is classified as a **heterocyclic aromatic amine.** The most important of these compounds are pyridine, pyrimidine, imidazole, and purine. Pyridine and pyrimidine are analogs of benzene in which first one and then two CH groups are replaced by nitrogen atoms. Pyrimidine and purine serve as the building blocks for the amine bases of DNA and RNA (Chapter 24).

Pyrrolidine Piperidine
(heterocyclic aliphatic amines)

Pyridine Pyrimidine Imidazole Purine
(heterocyclic aromatic amines)

CHAPTER 16

16.1 Introduction

16.2 Structure and Classification

16.3 Nomenclature

16.4 Physical Properties

16.5 Basicity

16.6 Reaction with Acids

16.7 Epinephrine: A Prototype for the Development of New Bronchodilators

This inhaler delivers puffs of albuterol (Proventil), a potent synthetic bronchodilator whose structure is patterned after that of epinephrine (adrenaline). See Section 16.7.

Amines

16.1 Introduction

Carbon, hydrogen, and oxygen are the three most common elements in organic compounds. Because of the wide distribution of amines in the biological world, nitrogen is the fourth most common element of organic compounds. The most important chemical property of amines is their basicity.

16.2 Structure and Classification

Amines are classified as **primary (1°)**, **secondary (2°)**, or **tertiary (3°)**, depending on the number of carbon groups bonded to nitrogen (Section 10.4B).

$$CH_3-NH_2 \qquad CH_3-\underset{\underset{H}{|}}{N}-CH_3 \qquad CH_3-\underset{\underset{CH_3}{|}}{N}-CH_3$$

Methylamine (a 1° amine)   Dimethylamine (a 2° amine)   Trimethylamine (a 3° amine)

Amines are further classified as aliphatic or aromatic. In an **aliphatic amine,** all the carbons bonded to nitrogen are derived from alkyl groups. In an **aromatic amine,** one or more of the groups bonded to nitrogen are aryl groups.

Aliphatic amine An amine in which nitrogen is bonded only to alkyl groups

Aromatic amine An amine in which nitrogen is bonded to one or more aromatic rings

CHEMICAL CONNECTIONS 16A

Amphetamines (Pep Pills)

Amphetamine, methamphetamine, and phentermine—all synthetic amines—are powerful stimulants of the central nervous system. Like most other amines, they are stored and administered as their salts. The sulfate salt of amphetamine is prescribed as Benzedrine, the hydrochloride salt of the s enantiomer of methamphetamine as Methedrine, and the hydrochloride salt of phentermine as Fastin.

These three amines all have similar physiological effects and are referred to by the general name *amphetamines.* Structurally, they have in common a benzene ring with a three-carbon side chain and an amine nitrogen on the second carbon of the chain. Physiologically, they have in common the ability to reduce fatigue and diminish hunger by raising the glucose level of the blood. Because of these properties, amphetamines are widely prescribed to counter mild depression, reduce hyperactivity in children, and suppress appetite in people who are trying to lose weight. They are also used illegally to reduce fatigue and elevate mood.

Abuse of amphetamines can have severe effects on both the body and the mind. They are addictive, concentrate in the brain and nervous system, and can lead to long periods of sleeplessness, loss of weight, and paranoia.

The action of amphetamines is similar to that of epinephrine (Section 16.7), the hydrochloride salt of which is named Adrenalin.

Amphetamine (Benzedrine)   (s)-Methamphetamine (Methedrine)   Phentermine (Fastin)

Aniline (a 1° aromatic amine)   N-Methylaniline (a 2° aromatic amine)   Benzyldimethylamine (a 3° aliphatic amine)

Heterocyclic amine An amine in which nitrogen is one of the atoms of a ring

Heterocyclic aliphatic amine A heterocyclic amine in which nitrogen is bonded only to alkyl groups

Heterocyclic aromatic amine An amine in which nitrogen is one of the atoms of an aromatic ring

An amine in which the nitrogen atom is part of a ring is classified as a **heterocyclic amine.** When the ring is saturated, the amine is classified as a **heterocyclic aliphatic amine.** When the nitrogen is part of an aromatic ring (Section 13.2), the amine is classified as a **heterocyclic aromatic amine.** The most important of these compounds are pyridine, pyrimidine, imidazole, and purine. Pyridine and pyrimidine are analogs of benzene in which first one and then two CH groups are replaced by nitrogen atoms. Pyrimidine and purine serve as the building blocks for the amine bases of DNA and RNA (Chapter 24).

Pyrrolidine   Piperidine
(heterocyclic aliphatic amines)

Pyridine   Pyrimidine   Imidazole   Purine
(heterocyclic aromatic amines)

CHEMICAL CONNECTIONS 16B

Alkaloids

Alkaloids are basic nitrogen-containing compounds found in the roots, bark, leaves, berries, or fruits of plants. In almost all alkaloids, the nitrogen atom is part of a ring. The name "alkaloid" was chosen because these compounds are alkali-like (*alkali* is an older term for a basic substance) and react with strong acids to give water-soluble salts. Thousands of different alkaloids have been extracted from plant sources, many of which are used in modern medicine.

When administered to animals, including humans, alkaloids have pronounced physiological effects. Whatever their individual effects, most alkaloids are toxic in large enough doses. For some, the toxic dose is very small!

■ **Tobacco plants.**

(*s*)-Coniine is the toxic principal of water hemlock (a member of the carrot family). Its ingestion can cause weakness, labored respiration, paralysis, and eventually death. It was the toxic substance in the "poison hemlock" used in the death of Socrates. Water hemlock is easily confused with Queen Anne's lace, a type of wild carrot—a mistake that has killed numerous people.

(*s*)-Nicotine occurs in the tobacco plant. In small doses, it is an addictive stimulant. In larger doses, this substance causes depression, nausea, and vomiting. In still larger doses, it is a deadly poison. Solutions of nicotine in water are used as insecticides.

Cocaine is a central nervous system stimulant obtained from the leaves of the coca plant. In small doses, it decreases fatigue and gives a sense of well-being. Prolonged use of cocaine leads to physical addiction and depression.

EXAMPLE 16.1

How many hydrogens does piperidine have? How many does pyridine have? Write the molecular formula of each amine.

Solution
Recall that hydrogen atoms are not shown in line-angle formulas. Piperidine has 11 hydrogens, and its molecular formula is $C_5H_{11}N$. Pyridine has 5 hydrogens, and its molecular formula is C_5H_5N.

Problem 16.1
How many hydrogens does pyrrolidine have? How many does purine have? Write the molecular formula of each amine.

16.3 Nomenclature

A IUPAC Names

IUPAC names for aliphatic amines are derived just as they are for alcohols. The final **-e** of the parent alkane is dropped and replaced by **-amine.** The location of the amino group on the parent chain is indicated by a number.

CH_3CHCH_3 with NH_2
2-Propanamine

Cyclohexanamine

$H_2N\text{~~~~}NH_2$
1,6-Hexanediamine

IUPAC nomenclature retains the common name **aniline** for $C_6H_5NH_2$, the simplest aromatic amine. Its simple derivatives are named using numbers to locate substituents or, alternatively, the prefixes ortho (o), meta (m), and para (p). Several derivatives of aniline have common names that remain in wide use. Among them is **toluidine** for a methyl-substituted aniline.

Aniline

4-Nitroaniline
(*p*-Nitroaniline)

3-Methylaniline
(*m*-Toluidine)

Unsymmetrical secondary and tertiary amines are commonly named as *N*-substituted primary amines. The largest group bonded to nitrogen is taken as the parent amine; the smaller groups bonded to nitrogen are named, and their locations are indicated by the prefix *N* (indicating that they are bonded to nitrogen).

N-Methylaniline

N,N-Dimethylcyclopentanamine

EXAMPLE 16.2

Write the IUPAC name for each amine. Be certain to specify the configuration of the stereocenter in (c).

(a) [structure: $CH_3CH(NH_2)CH_2CH_3$]

(b) $H_2N(CH_2)_5NH_2$

(c) [structure: benzyl group with stereocenter bearing H and NH_2]

Solution

(a) The parent alkane has four carbon atoms and is therefore butane. The amino group is on carbon 2, giving the IUPAC name 2-butanamine.

(b) The parent chain has five carbon atoms and is therefore pentane. There are amino groups on carbons 1 and 5, giving the IUPAC name 1,5-pentanediamine. The common name of this diamine is cadaverine, which should give you a hint of where it occurs in nature and of its odor. Cadaverine, one of the end products of decaying flesh, is quite poisonous.

(c) The parent chain has three carbon atoms and is therefore propane. To have the lowest numbers possible, we number from the end that places the phenyl group on carbon 1 and the amino group on carbon 2. The priorities for determining R or S configuration are $NH_2 > C_6H_5CH_2 > CH_3 > H$. Its systematic name is (S)-1-phenyl-2-propanamine. Its common name is amphetamine.

Problem 16.2
Write a structural formula for each amine.
(a) 2-Methyl-1-propanamine
(b) Cyclopentanamine
(c) 1,4-Butanediamine

B Common Names

Common names for most aliphatic amines list the groups bonded to nitrogen in alphabetical order in one word ending in the suffix **-amine.**

Propylamine   sec-Butylamine   Diethylmethylamine   Cyclohexylamine

EXAMPLE 16.3
Write a structural formula for each amine.
(a) Isopropylamine
(b) Cyclohexylmethylamine
(c) Triethylamine

Solution

(a) $(CH_3)_2CHNH_2$ (b) ⬡—$NHCH_3$ (c) $(CH_3CH_2)_3N$

Problem 16.3
Write a structural formula for each amine.
(a) Isobutylamine
(b) Diphenylamine
(c) Diisopropylamine

408 | CHAPTER 16 AMINES

■ Several over-the-counter mouthwashes contain an *N*-alkylpyridinium chloride as an antibacterial.

When four atoms or groups of atoms are bonded to a nitrogen atom—as, for example, in NH_4^+ and $CH_3NH_3^+$—nitrogen bears a positive charge and is associated with an anion as a salt. The compound is named as a salt of the corresponding amine. The ending **-amine** (or aniline or pyridine or the like) is replaced by **-ammonium** (or anilinium, pyridinium, or the like) and the name of the anion (chloride, acetate, and so on) is added.

$$(CH_3CH_2)_3NH^+Cl^-$$

Triethylammonium chloride

16.4 Physical Properties

Like ammonia, low-molecular-weight amines have very sharp, penetrating odors. Trimethylamine, for example, is the pungent principal in the smell of rotting fish. Two other particularly pungent amines are 1,4-butanediamine (putrescine) and 1,5-pentanediamine (cadaverine).

Amines are polar compounds because of the difference in electronegativity between nitrogen and hydrogen ($3.0 - 2.1 = 0.9$). Both primary and secondary amines have N—H bonds, and they can form hydrogen bonds with one another (Figure 16.1). Tertiary amines do not have a hydrogen bonded to nitrogen and, therefore, do not form hydrogen bonds with one another.

An N—H -----N hydrogen bond is weaker than an O—H -----O hydrogen bond because the difference in electronegativity between nitrogen and hydrogen ($3.0 - 2.1 = 0.9$) is less than that between oxygen and hydrogen ($3.5 - 2.1 = 1.4$). To see the effect of hydrogen bonding between molecules of comparable molecular weight, we can compare the boiling points of

Figure 16.1 Hydrogen bonding between two molecules of a secondary amine.

CHEMICAL CONNECTIONS 16C

Tranquilizers

Most people have experienced anxiety and stress at some time in their lives, and each person develops various ways to cope with these factors. Perhaps it is meditation, or exercise, or psychotherapy, or drugs. One modern way of coping is to use tranquilizers.

The first modern tranquilizers were derivatives of a compound called benzodiazepine. The first of these compounds, chlorodiazepoxide, better known as Librium, was introduced in 1960 and was soon followed by more than two dozen related compounds. Diazepam, better known as Valium, became one of the most widely used of these drugs.

Librium, Valium, and other benzodiazepines are central nervous system sedatives/hypnotics. As sedatives, they diminish activity and excitement, thereby exerting a calming effect. As hypnotics, they produce drowsiness and sleep. We will discuss their action on the central nervous system in Section 23.4.

Benzodiazepine

Chlorodiazepoxide (Librium)

Diazepam (Valium)

ethane, methanamine, and methanol. Ethane is a nonpolar hydrocarbon, and the only attractive forces between its molecules are weak London dispersion forces (Section 5.7A). Both methanamine and methanol have polar molecules that interact in the pure liquid by hydrogen bonding. Methanol has the higher boiling point because hydrogen bonding between its molecules is stronger than that between methanamine molecules.

Screen 16.4
Simulation

	CH_3CH_3	CH_3NH_2	CH_3OH
Molecular weight (amu)	30.1	31.1	32.0
Boiling point (°C)	−88.6	−6.3	65.0

All classes of amines form hydrogen bonds with water and are more soluble in water than are hydrocarbons of comparable molecular weight. Most low-molecular-weight amines are completely soluble in water, but higher-molecular-weight amines are only moderately soluble in water or are insoluble.

16.5 Basicity *skip*

Like ammonia, amines are weak bases, and aqueous solutions of amines are basic. The following acid–base reaction between an amine and water is written using curved arrows to emphasize that, in this proton-transfer reaction (Section 8.1), the unshared pair of electrons on nitrogen forms a new covalent bond with hydrogen and displaces a hydroxide ion.

$$CH_3-\underset{H}{\overset{H}{N}}: + H-\overset{..}{\underset{..}{O}}-H \rightleftharpoons CH_3-\underset{H}{\overset{H}{N^+}}-H \quad :\overset{..}{\underset{..}{O}}-H^-$$

Methylamine (a base)　　　Methylammonium hydroxide

The base dissociation constant, K_b, for the reaction of an amine with water has the following form, illustrated here for the reaction of methylamine with water to give methylammonium hydroxide. pK_b is defined as the negative logarithm of K_b.

$$K_b = \frac{[CH_3NH_3^+][OH^-]}{[CH_3NH_2]} = 4.37 \times 10^{-4}$$

$$pK_b = -\log 4.37 \times 10^{-4} = 3.360$$

All aliphatic amines have about the same base strength, pK_b 3.0–4.0, and are slightly stronger bases than ammonia (Table 16.1). Aromatic

Table 16.1 Approximate Base Strengths of Amines

Class	pK_a	Example	Name	
Aliphatic	3.0–4.0	$CH_3CH_2NH_2$	Ethanamine	Stronger base
Ammonia	4.74			↑
Aromatic	8.5–9.5	$C_6H_5NH_2$	Aniline	Weaker base

amines and heterocyclic aromatic amines (pK_b 8.5–9.5) are considerably weaker bases than aliphatic amines. One additional point about the basicities of amines: While aliphatic amines are weak bases by comparison with inorganic bases such as NaOH, they are strong bases among organic compounds.

Given the basicities of amines, we can determine which form of an amine exists in body fluids—say, blood. In a normal, healthy person, the pH of blood is approximately 7.40, which is slightly basic. If an aliphatic amine is dissolved in blood, it is present predominantly as its protonated or conjugate acid form.

Dopamine

Conjugate acid of dopamine
(the major form present
in blood plasma)

We can show that an aliphatic amine such as dopamine dissolved in blood is present largely as its protonated or conjugate acid form in the following way. Assume that the amine, RNH_2, has a pK_b of 3.50 and that it is dissolved in blood, pH 7.40. We first write the base dissociation constant for the amine and then solve for the ratio of RNH_3^+ to RNH_2.

$$RNH_2 + H_2O \rightleftharpoons RNH_3^+ + OH^-$$

$$K_b = \frac{[RNH_3^+][OH^-]}{[RNH_2]}$$

$$\frac{K_b}{[OH^-]} = \frac{[RNH_3^+]}{[RNH_2]}$$

We now substitute the appropriate values for K_b and $[OH^-]$ in this equation. Taking the antilog of 3.50 gives a K_b of 4.0×10^{-4}. Calculating the concentration of hydroxide requires two steps. First recall from Section 8.8 that pH + pOH = 14. If the pH of blood is 7.40, then its pOH is 6.60 and its $[OH^-]$ is 2.5×10^{-7}. Substituting these values in the appropriate equation gives a ratio of 1600 parts RNH_3^+ to 1 part RNH_2.

$$\frac{4.0 \times 10^{-4}}{2.5 \times 10^{-7}} = \frac{[RNH_3^+]}{[RNH_2]} = 1600$$

As this calculation demonstrates, an aliphatic amine present in blood is more than 99.9% in the protonated form. Thus, even though we may write the structural formula of dopamine as the free amine, it is present as the protonated form. It is important to realize, however, that the amine and ammonium ion forms are always in equilibrium, so some of the unprotonated form is nevertheless present in solution.

Aromatic amines, on the other hand, are considerably weaker bases than aliphatic amines and are present in blood largely in the unprotonated form. Performing the same type of calculation for an aromatic amine, $ArNH_2$, with pK_b of approximately 10, we find that the aromatic amine is more than 99.9% in its unprotonated ($ArNH_2$) form.

EXAMPLE 16.4

Select the stronger base in each pair of amines.

(a) [pyridine] (A) or [morpholine] (B)

(b) [2-methylaniline with NH₂ and CH₃] (C) or [benzylamine CH₂NH₂] (D)

Solution
(a) Morpholine (B), a 2° aliphatic amine, is the stronger base. Pyridine (A), a heterocyclic aromatic amine, is the weaker base.
(b) Benzylamine (D), a 1° aliphatic amine, is the stronger base. Even though it contains an aromatic ring, it is not an aromatic amine because the amine nitrogen is not bonded to the aromatic ring. o-Toluidine (C), a 1° aromatic amine, is the weaker base.

Problem 16.4
Select the stronger base from each pair of amines.

(a) [pyridine] (A) or [cyclohexylamine] (B)

(b) NH_3 (C) or [aniline] (D)

16.6 Reaction with Acids

The most important chemical property of amines is their basicity. Amines, whether soluble or insoluble in water, react quantitatively with strong acids to form water-soluble salts as illustrated by the reaction of (R)-norepinephrine (noradrenaline) with aqueous HCl to form a hydrochloride salt.

[(R)-Norepinephrine structure] + HCl $\xrightarrow{H_2O}$ [(R)-Norepinephrine hydrochloride structure with NH_3^+ Cl^-]

(R)-Norepinephrine
(only slightly soluble in water)

(R)-Norepinephrine hydrochloride
(a water-soluble salt)

CHEMICAL CONNECTIONS 16D

The Solubility of Drugs in Body Fluids

Many drugs have "·HCl" or some other acid as part of their chemical formulas and occasionally as part of their generic names. Invariably these drugs are amines that are insoluble in aqueous body fluids such as blood plasma and cerebrospinal fluid. For the administered drug to be absorbed and carried by body fluids, it must be treated with an acid to form a water-soluble ammonium salt. Methadone, a narcotic analgesic, is marketed as its water-soluble hydrochloride salt. Novocain, one of the first local anesthetics, is the hydrochloride salt of procaine.

Methadone ·HCl

Procaine ·HCl
(Novocain, a local anesthetic)

■ These two drugs are amine salts and are labeled as hydrochlorides.

There is another reason besides increased water solubility for preparing these and other amine drugs as salts. Amines are very susceptible to oxidation and decomposition by atmospheric oxygen, with a corresponding loss of biological activity. By comparison, their amine salts are far less susceptible to oxidation; they retain their effectiveness for a much longer time.

EXAMPLE 16.5

Complete the equation for each acid–base reaction, and name the salt formed.

(a) $(CH_3CH_2)_2NH + HCl \longrightarrow$

(b) pyridine $+ CH_3COOH \longrightarrow$

Solution

(a) $(CH_3CH_2)_2NH_2^+Cl^-$
Diethylammonium chloride

(b) pyridinium CH_3COO^-
Pyridinium acetate

Problem 16.5

Complete the equation for each acid–base reaction and name the salt formed.

(a) $(CH_3CH_2)_3N + HCl \longrightarrow$

(b) cyclohexyl-NH + $CH_3COOH \longrightarrow$

Screen 16.5 Simulation

16.7 Epinephrine: A Prototype for the Development of New Bronchodilators

Epinephrine was first isolated in pure form in 1897 and its structure determined in 1901. It occurs in the adrenal gland (hence the common name adrenalin) as a single enantiomer with the R configuration at its stereocenter. Epinephrine is commonly referred to as a catecholamine: The common name of 1,2-dihydroxybenzene is catechol (Section 13.5A), and amines containing a benzene ring with ortho hydroxyl groups are called catecholamines.

Early on, it was recognized that epinephrine is a vasoconstrictor, a bronchodilator, and a cardiac stimulant. The fact that it has these three major effects has stimulated much research, one line of which has sought to develop compounds that are even more effective bronchodilators than epinephrine but that lack epinephrine's cardiac-stimulating and vasoconstricting effects.

Soon after epinephrine became commercially available, it became an important treatment for asthma and hayfever. It has been marketed for the relief of bronchospasms under several trade names, including Bronkaid Mist and Primatine Mist.

Epinephrine

(R)-Isoproterenol

One of the most important of the synthetic catecholamines was isoproterenol, the levorotatory enantiomer of which retains the bronchodilating effects of epinephrine but is free from its cardiac-stimulating effects. (R)-Isoproterenol was introduced into clinical medicine in 1951 and for the next two decades was the drug of choice for the treatment of asthmatic attacks. Interestingly, the hydrochloride salt of (S)-isoproterenol is a nasal decongestant and was marketed under several trade names, including Sudafed.

A problem with the first synthetic catecholamines (and with epinephrine itself) is the fact that they are inactivated by an enzyme-catalyzed reaction that converts one of the —OH groups on the catechol unit to an OCH_3 group. A strategy to circumvent this enzyme-catalyzed inactivation was to replace the catechol unit with one that would allow the drug to bind to the catecholamine receptors in the bronchi but would not be inactivated by this enzyme.

In terbutaline (Brethaire), inactivation is prevented by placing the —OH groups meta to each other on the aromatic ring. In addition, the isopropyl group of isoproterenol is replaced by a *tert*-butyl group. In albuterol (Proventil), the commercially most successful of the antiasthma medications, one —OH group of the catechol unit is replaced by a —CH_2OH group and the isopropyl group is replaced by a *tert*-butyl group. When terbutaline and albuterol were introduced into clinical medicine in the 1960s, they almost immediately replaced isoproterenol as the drugs of choice for the treatment of asthmatic attacks.

Terbutaline

Albuterol

In their search for a longer-acting bronchodilator, scientists reasoned that extending the side chain on nitrogen might strengthen the binding of the drug to the adrenoreceptors in the lungs, thereby increasing the duration of the drug's action. A result of this line of reasoning is salmeterol (Serevent), a bronchodilator that is approximately ten times more potent than albuterol and much longer acting.

Salmeterol

SUMMARY

Amines are classified as **primary, secondary, or tertiary,** depending on the number of carbon atoms bonded to nitrogen (Section 16.2). In an **aliphatic amine,** all carbon atoms bonded to nitrogen are derived from alkyl groups. In an **aromatic amine,** one or more of the groups bonded to nitrogen are aryl groups. In a **heterocyclic amine,** the nitrogen atom is part of a ring.

In IUPAC nomenclature, aliphatic amines are named by changing the final -e of the parent alkane to -amine and using a number to indicate the location of the amino group on the parent chain (Section 16.3A). In the common system of nomenclature, aliphatic amines are named by listing carbon groups in alphabetical order in one word ending in the suffix -amine (Section 16.3B).

Amines are polar compounds, and primary and secondary amines associate by intermolecular hydrogen bonding (Section 16.4). All classes of amines form hydrogen bonds with water and are more soluble in water than are hydrocarbons of comparable molecular weight.

Amines are weak bases, and aqueous solutions of amines are basic. The base ionization constant for an amine in water is denoted by the symbol K_b (Section 16.5). Aliphatic amines are stronger bases than are aromatic amines.

All amines, whether soluble or insoluble in water, react with strong acids to form water-soluble salts (Section 16.6). We can use this chemical property to separate water-insoluble amines from water-insoluble neutral organic compounds.

KEY REACTIONS

1. **Basicity of Aliphatic Amines (Section 16.5)** Most aliphatic amines have about the same basicity (pK_b 3.0–4.0) and are slightly stronger bases than ammonia (pK_b 4.74).

   $CH_3NH_2 + H_2O \rightleftharpoons CH_3NH_3^+ + OH^-$
   $pK_b = 3.36$

2. **Basicity of Aromatic Amines (Section 16.5)** Most aromatic amines (pK_b 9.0–10.0) are considerably weaker bases than ammonia and aliphatic amines.

   $C_6H_5-NH_2 + H_2O \rightleftharpoons C_6H_5-NH_3^+ + OH^-$
   $pK_b = 9.37$

3. **Reaction with Acids (Section 16.5)** All amines, whether water-soluble or water-insoluble, react quantitatively with strong acids to form water-soluble salts.

PROBLEMS

Numbers that appear in color indicate difficult problems.
▶ designates problems requiring application of principles.

Structure and Nomenclature

16.6 We use the terms primary, secondary, and tertiary to classify both alcohols and amines. Cyclohexanol, for example, is classified as a secondary alcohol, and cyclohexanamine is classified as a primary amine. In each compound, the functional group is attached to a carbon of the cyclohexane ring. Explain why one compound is classified as secondary whereas the other is primary.

16.7 What is the difference in structure between an aliphatic amine and an aromatic amine?

16.8 In what way are pyridine and pyrimidine related to benzene?

16.9 Draw a structural formula for each amine.
(a) 2-Butanamine
(b) 1-Octanamine
(c) 2,2-Dimethyl-1-propanamine
(d) 1,5-Pentanediamine
(e) 2-Bromoaniline
(f) Tributylamine

16.10 Draw a structural formula for each amine.
(a) 4-Methyl-2-pentanamine
(b) *trans*-2-Aminocyclohexanol
(c) *N,N*-Dimethylaniline
(d) Dicyclohexylamine
(e) *sec*-Butylamine
(f) 2,4-Dimethylaniline

16.11 Classify each amino group as primary, secondary, or tertiary, and as aliphatic or aromatic.

(a) Serotonin (a neurotransmitter)

(b) Benzocaine (a topical anesthetic)

(c) Diphenhydramine (the hydrochloride salt is the antihistamine Benadryl)

16.12 Classify each amino group as primary, secondary, or tertiary, and as aliphatic or aromatic.

(a) Lysine (an amino acid)

(b) Chloroquine (an antimalaria drug)

(c) 4-Aminobutanoic acid (a neurotransmitter)

416 / CHAPTER 16 AMINES

16.13 There are eight constitutional isomers with molecular formula $C_4H_{11}N$.
   (a) Name and draw a structural formula for each amine.
   (b) Classify each amine as primary, secondary, or tertiary.
   (c) Which are chiral?

16.14 There are eight primary amines with molecular formula $C_5H_{13}N$.
   (a) Name and draw a structural formula for each amine.
   (b) Which are chiral?

Physical Properties

16.15 Propylamine (bp 48°C), ethylmethylamine (bp 37°C), and trimethylamine (bp 3°C) are constitutional isomers with molecular formula C_3H_9N. Account for the fact that trimethylamine has the lowest boiling point of the three.

16.16 Account for the fact that 1-butanamine (bp 78°C) has a lower boiling point than 1-butanol (bp 117°C).

16.17 2-Methylpropane (bp −12°C), 2-propanol (bp 82°C), and 2-propanamine (bp 32°C) all have approximately the same molecular weight, yet their boiling points are quite different. Explain the reason for these differences.

16.18 Account for the fact that most low-molecular-weight amines are very soluble in water whereas low-molecular-weight hydrocarbons are not.

Basicity of Amines

16.19 Compare the basicities of amines with those of alcohols.

16.20 Write structural formulas for these amine salts.
   (a) Ethyltrimethylammonium hydroxide
   (b) Dimethylammonium iodide
   (c) Tetramethylammonium chloride
   (d) Anilinium bromide

16.21 Name these amine salts.
   (a) $CH_3CH_2NH_3^+Cl^-$
   (b) $(CH_3CH_2)_2NH_2^+Cl^-$
   (c) $C_6H_5-NH_3^+HSO_4^-$

16.22 From each pair of compounds, select the stronger base.
   (a) piperidine or pyridine
   (b) cyclohexyl-$N(CH_3)_2$ or phenyl-$N(CH_3)_2$
   (c) phenyl-$NHCH_3$ or phenyl-CH_2NH_2

16.23 Following are two structural formulas for 4-aminobutanoic acid, a neurotransmitter. Is this compound better represented by structural formula (A) or (B)? Explain.

H_2N–CH$_2$CH$_2$CH$_2$–COOH (A)   or $H_3\overset{+}{N}$–CH$_2$CH$_2$CH$_2$–COO$^-$ (B)

16.24 Suppose you have two test tubes, one containing 2-methylcyclohexanol and the other containing 2-methylcyclohexanamine, both insoluble in water, but you do not know which test tube contains which compound. Describe a simple chemical test by which you could tell which compound is the alcohol and which is the amine.

16.25 Complete the equations for the following acid–base reactions.

(a) CH_3COH + pyridine ⟶
      Acetic acid   Pyridine

(b) phenyl-CH$_2$CH(NH$_2$)CH$_3$ + HCl ⟶
    1-Phenyl-2-propanamine
    (Amphetamine)

(c) phenyl-CH$_2$CH(NHCH$_3$)CH$_3$ + H_2SO_4 ⟶
    Methamphetamine

16.26 The pK_b of amphetamine [Problem 16.25(b)] is approximately 3.2.
   (a) Which form of amphetamine would you expect to be present at pH 1.0, the pH of stomach acid?
   (b) Which form of amphetamine would you expect to be present at pH 7.40, the pH of blood plasma?

16.27 Pyridoxamine is one form of vitamin B_6.

(a) Which nitrogen atom of pyridoxamine is the stronger base?
(b) Draw the structural formula of the salt formed when pyridoxamine is treated with one mole of HCl.

16.28 Many tumors of the breast are correlated with estrogen levels in the body. Drugs that interfere with estrogen binding have antitumor activity and may even help prevent tumor occurrence. A widely used antiestrogen drug is tamoxifen.

(a) Name the functional groups in tamoxifen.
(b) Classify the amino group in tamoxifen as primary, secondary, or tertiary.
(c) How many stereoisomers are possible for tamoxifen?

16.29 Classify each amino group in epinephrine and albuterol (Section 16.7) as primary, secondary, or tertiary. In addition, list the similarities and differences between the structural formulas of these two compounds.

Chemical Connections

16.30 (Chemical Connections 16A) What are the differences in structure between the natural hormone epinephrine (Section 16.7) and the synthetic pep pill amphetamine? Between amphetamine and methamphetamine?

16.31 (Chemical Connections 16A) What are the possible negative effects of illegal use of amphetamines such as methamphetamine?

16.32 (Chemical Connections 16B) What is an alkaloid? Are all alkaloids basic as measured by a litmus test?

16.33 (Chemical Connections 16B) Identify all stereocenters in coniine and nicotine. How many stereoisomers are possible for each?

16.34 (Chemical Connections 16B) Of the two nitrogen atoms in nicotine, which is converted to its salt by reaction with one mole of HCl? Draw the structural formula of this salt.

16.35 (Chemical Connections 16B) Cocaine has four stereocenters. Identify each. Draw the structural formula for the salt formed by treatment of cocaine with one mole of HCl.

16.36 (Chemical Connections 16C) What structural feature is common to all benzodiazepines?

16.37 (Chemical Connections 16C) Is Librium chiral? Is Valium chiral?

16.38 (Chemical Connections 16C) Benzodiazepines affect neural pathways in the central nervous system mediated by GABA, whose IUPAC name is 4-aminobutanoic acid. Draw a structural formula of GABA.

16.39 (Chemical Connections 16D) Suppose you saw this label on a decongestant: phenylephrine · HCl. Should you worry about being exposed to a strong acid such as HCl? Explain.

16.40 (Chemical Connections 16D) Give two reasons why amine-containing drugs are most commonly administered as their salts.

Additional Problems

16.41 Draw a structural formula for a compound with each molecular formula.
(a) A 2° aromatic amine, C_7H_9N
(b) A 3° aromatic amine, $C_8H_{11}N$
(c) A 1° aliphatic amine, C_7H_9N
(d) A chiral 1° amine, $C_4H_{11}N$
(e) A 3° heterocyclic amine, $C_5H_{11}N$
(f) A trisubstituted 1° aromatic amine, $C_9H_{13}N$
(g) A chiral quaternary ammonium salt, $C_9H_{22}NCl$

16.42 Arrange these three compounds in order of decreasing ability to form intermolecular hydrogen bonds: CH_3OH, CH_3SH, and $(CH_3)_2NH$.

16.43 Consider these three compounds: CH_3OH, CH_3SH, and $(CH_3)_2NH$.
(a) Which is the strongest acid?
(b) Which is the strongest base?
(c) Which has the highest boiling point?
(d) Which forms the strongest hydrogen bonds in the pure state?

16.44 Arrange these compounds in order of increasing boiling point: $CH_3CH_2CH_2CH_3$, $CH_3CH_2CH_2OH$, and $CH_3CH_2CH_2NH_2$.

16.45 Account for the fact that amines have about the same solubility in water as alcohols of similar molecular weight.

16.46 If you dissolved $CH_3CH_2CH_2OH$ and $CH_3CH_2CH_2NH_2$ in the same container of water and lowered the pH of the solution to 2 by adding HCl, would anything happen to the structures of these compounds? If so, write the formula of the species present in solution at pH 2.

16.47 The compound phenylpropanolamine hydrochloride is used as both a decongestant and an anorexic. The chemical name of this compound is 1-phenyl-2-amino-1-propanol.

(a) Draw the structural formula of 1-phenyl-2-amino-1-propanol.

(b) How many stereocenters are present in this molecule? How many stereoisomers are possible for it?

16.48 Procaine was one of the first local anesthetics. Its hydrochloride salt is marketed as Novocain.

(a) Is procaine chiral? Does it contain a stereocenter?

(b) Which nitrogen atom of procaine is the stronger base?

(c) Draw a structural formula of the salt formed by treating procaine with one mole of HCl, showing which nitrogen is protonated and bears the positive charge.

16.49 Following are two structural formulas for gabapentin, a drug used to treat epilepsy. Is gabapentin better represented by structure (A) or (B)? Explain.

16.50 Several poisonous plants, including *Atropa belladonna*, contain the alkaloid atropine. The name "belladonna" (which means beautiful lady) probably comes from the fact that Roman women used extracts from this plant to make themselves more attractive. Atropine is widely used by ophthalmologists and optometrists to dilate the pupils for eye examination.

(a) Classify the amino group in atropine as primary, secondary, or tertiary.

(b) Locate the single stereocenter in atropine.

(c) Account for the fact that atropine is almost insoluble in water (1 g in 455 mL cold water), but atropine hydrogen sulfate is very soluble (1 g in 5 mL of cold water).

(d) Account for the fact that a dilute aqueous solution of atropine is basic (pH approximately 10.0).

16.51 Epibatadine, a colorless oil isolated from the skin of the Equadorian poison frog *Epipedobates tricolor*, has several times the analgesic potency of morphine. It is the first chlorine-containing, non-opioid (nonmorphine-like in structure) analgesic ever isolated from a natural source.

Epibatadine

■ Poison arrow frog.

(a) Which of the two nitrogens in epibatadine is the stronger base?

(b) Mark the three stereocenters in this molecule.

InfoTrac College Edition

For additional readings, go to InfoTrac College Edition, your online research library, at

http://infotrac.thomsonlearning.com

CHAPTER 17

17.1 Introduction
17.2 Structure and Bonding
17.3 Nomenclature
17.4 Physical Properties
17.5 Oxidation
17.6 Reduction
17.7 Addition of Alcohols
17.8 Keto-Enol Tautomerism

Benzaldehyde is found in the kernels of bitter almonds, and cinnamaldehyde is found in Ceylonese and Chinese cinnamon oils.

Aldehydes and Ketones

17.1 Introduction

In this and the following chapter, we study the physical and chemical properties of compounds containing the **carbonyl group,** C=O. Because the carbonyl group is present in aldehydes, ketones, and carboxylic acids and their derivatives, it is one of the most important functional groups in organic chemistry. Its chemical properties are straightforward, and an understanding of its characteristic reaction patterns leads very quickly to an understanding of a wide variety of organic and biochemical reactions.

17.2 Structure and Bonding

The functional group of an **aldehyde** is a carbonyl group bonded to a hydrogen atom (Section 10.4C). In methanal, the simplest aldehyde, the carbonyl group is bonded to two hydrogen atoms. In other aldehydes, it is bonded to one hydrogen atom and one carbon atom. The functional group of a ketone is a carbonyl group bonded to two carbon atoms (Section 10.4C).

Methanal (Formaldehyde)   Ethanal (Acetaldehyde)   Propanone (Acetone)

Because aldehydes always contain at least one hydrogen bonded to the C=O group, they are often written RCH=O or RCHO. Similarly, ketones are often written RCOR'.

17.3 Nomenclature

Screen 17.2 Tutorial

A IUPAC Names

The IUPAC names for aldehydes and ketones follow the familiar pattern of selecting as the parent alkane the longest chain of carbon atoms that contains the functional group (Section 11.4A). To name an aldehyde, we change the suffix -e of the parent alkane to -al. Because the carbonyl group of an aldehyde can appear only at the end of a parent chain and numbering must start with it as carbon 1, there is no need to use a number to locate the aldehyde group.

For **unsaturated aldehydes,** we show the presence of a carbon–carbon double bond and an aldehyde by changing the ending of the parent alkane from -ane to -enal: "-en-" to show the carbon–carbon double bond, and "-al" to show the aldehyde. We number the carbon chain with the aldehyde as carbon 1. We show the location of the carbon–carbon double bond by the number of its first carbon.

Hexanal   3-Methylbutanal   2-Propenal (Acrolein)

In the IUPAC system, we name ketones by selecting as the parent alkane the longest chain that contains the carbonyl group and then indicating the presence of this group by changing the -e of the parent alkane -one. The parent chain is numbered from the direction that gives the smaller number to the carbonyl carbon. While the systematic name of the simplest ketone is 2-propanone, the IUPAC retains its common name, acetone.

Acetone   5-Methyl-3-hexanone   2-Methylcyclohexanone

EXAMPLE 17.1

Write the IUPAC name for each compound:

(a) [structure with O, H, methyl and ethyl branches] (b) [cyclohexanone with two methyls] (c) [benzene ring with CHO and ethyl]

Solution
(a) The longest chain has six carbons, but the longest chain that contains the carbonyl carbon has five carbons. Its IUPAC name is 2-ethyl-3-methylpentanal.
(b) We number the six-membered ring beginning with the carbonyl carbon. Its IUPAC name is 3,3-dimethylcyclohexanone.
(c) This molecule is derived from benzaldehyde. Its IUPAC name is 2-ethylbenzaldehyde.

Problem 17.1

Write the IUPAC name for each compound. Specify the R or S configuration of (c).

(a) [structure with tert-butyl CH2CHO] (b) [cyclopentanone] (c) C_6H_5-C with CHO, H, CH$_3$

EXAMPLE 17.2

Write structural formulas for all ketones with molecular formula $C_6H_{12}O$ and give the IUPAC name of each. Which of these ketones are chiral?

Solution
There are six ketones with this molecular formula: two with a six-carbon chain, three with a five-carbon chain and a methyl branch, and one with a four-carbon chain and two methyl branches. Only 3-methyl-2-pentanone has a stereocenter and is chiral.

2-Hexanone   3-Hexanone   4-Methyl-2-pentanone

3-Methyl-2-pentanone (Stereocenter)   2-Methyl-3-pentanone   3,3-Dimethyl-2-butanone

Problem 17.2

Write structural formulas for all aldehydes with molecular formula $C_6H_{12}O$ and give each its IUPAC name. Which of these aldehydes are chiral?

In naming aldehydes or ketones that also contain an —OH or —NH₂ group elsewhere in the molecule, the parent chain is numbered to give the lower number to the carbonyl group. An —OH substituent is indicated by *hydroxy*, and an —NH₂ substituent is indicated by *amino-*. Hydroxy and amino substituents are numbered and alphabetized along with any other substituents that might be present.

EXAMPLE 17.3

Write the IUPAC name for each compound.

(a) [structure: 3-hydroxy-4-methylpentanal]   (b) [structure: 3-amino-4-ethyl-2-hexanone]

Solution

(a) We number the parent chain beginning with the CHO as carbon 1. There is a hydroxyl group on carbon 3 and a methyl group on carbon 4. The IUPAC name of this compound is 3-hydroxy-4-methylpentanal.

(b) The IUPAC name of this compound is 3-amino-4-ethyl-2-hexanone.

Problem 17.3

Write the IUPAC name for each compound.

(a) CH₂CHCH
    | | ‖
   OH OH O

(b) [ortho-aminobenzaldehyde structure]

(c) H₂N—[CH₂CH₂CH₂—C(=O)—CH₃]

CHEMICAL CONNECTIONS 17A

Some Naturally Occurring Aldehydes and Ketones

Benzaldehyde (oil of almonds)

Cinnamaldehyde (oil of cinnamon)

Citronellal (citronella oils; also in lemon and lemon grass oils)

Muscone (from the musk deer; used in perfumes)

Vanillin (vanilla bean)

β-Ionone (from violets)

B Common Names

We derive the common name for an aldehyde from the common name of the corresponding carboxylic acid. The word "acid" is dropped, and the suffix "-ic" or "-oic" is changed to "-*aldehyde*." Because we have not yet studied common names for carboxylic acids, we are not in a position to discuss common names for aldehydes. We can, however, illustrate how they are derived by reference to two common names with which you are familiar. The name formaldehyde is derived from formic acid, and the name acetaldehyde is derived from acetic acid.

$$\underset{\text{Formaldehyde}}{\text{HCH}} \quad \underset{\text{Formic acid}}{\text{HCOH}} \quad \underset{\text{Acetaldehyde}}{\text{CH}_3\text{CH}} \quad \underset{\text{Acetic acid}}{\text{CH}_3\text{COH}}$$
(each with a C=O group)

We derive common names for ketones by naming each alkyl or aryl group bonded to the carbonyl group as a separate word, followed by the word "ketone." The alkyl or aryl groups are generally listed in order of increasing molecular weight.

Ethyl isopropyl ketone   Methyl ethyl ketone   Dicyclohexyl ketone

2-Butanone, more commonly called methyl ethyl ketone (MEK), is used as a solvent for paints and varnishes.

17.4 Physical Properties

Oxygen is more electronegative than carbon (3.5 compared with 2.5; Table 3.5): Therefore, a carbon–oxygen double bond is polar, with oxygen bearing a partial negative charge and carbon bearing a partial positive charge (Figure 17.1).

In liquid aldehydes and ketones, intermolecular attractions occur between the partial positive charge on the carbonyl carbon of one molecule and the partial negative charge on the carbonyl oxygen of another molecule. There is no possibility for hydrogen bonding between aldehyde or ketone molecules, which explains why these compounds have lower boiling points than alcohols (Section 14.2C) and carboxylic acids (Section 18.2), compounds in which there is hydrogen bonding between molecules.

Table 17.1 lists structural formulas and boiling points of six compounds of similar molecular weight. Of the six, pentane and diethyl ether have the lowest boiling points. The boiling point of 1-butanol, which can associate by

Polarity of a carbonyl group

Figure 17.1 The polarity of a carbonyl group. Oxygen bears a partial negative charge and carbon bears a partial positive charge.

Table 17.1	Boiling Points of Six Compounds of Comparable Molecular Weight		
Name	Structural Formula	Molecular Weight	Boiling Point (°C)
diethyl ether	$CH_3CH_2OCH_2CH_3$	74	34
pentane	$CH_3CH_2CH_2CH_3CH_3$	72	36
butanal	$CH_3CH_2CH_2CHO$	72	76
2-butanone	$CH_3CH_2COCH_3$	72	80
1-butanol	$CH_3CH_2CH_2CH_2OH$	74	117
propanoic acid	CH_3CH_2COOH	72	141

Screen 17.3 Simulation

intermolecular hydrogen bonding, is higher than that of either butanal or 2-butanone. Propanoic acid, in which intermolecular association by hydrogen bonding is the strongest, has the highest boiling point.

Because the oxygen atom of their carbonyl groups are hydrogen bond acceptors, the low-molecular-weight aldehydes and ketones are more soluble in water than are nonpolar compounds of comparable molecular weight. Formaldehyde, acetaldehyde, and acetone are infinitely soluble in water. As the hydrocarbon portion of the molecule increases in size, aldehydes and ketones become less soluble in water.

Most aldehydes and ketones have strong odors. The odors of ketones are generally pleasant, and many are used in perfumes and as flavoring agents. The odors of aldehydes vary. You may be familiar with the smell of formaldehyde; if so, you know that it is not pleasant. Many higher aldehydes, however, have pleasant odors and are used in perfumes.

17.5 Oxidation

Aldehydes are oxidized to carboxylic acids by a variety of oxidizing agents, including potassium dichromate (Section 14.3C).

The body uses nicotinamide adenine dinucleotide, NAD^+, for this type of oxidation (Section 26.3).

Hexanal $\xrightarrow{K_2Cr_2O_7, H_2SO_4}$ Hexanoic acid

They are also oxidized to carboxylic acids by the oxygen in the air. In fact, aldehydes that are liquid at room temperature are so sensitive to oxidation that they must be protected from contact with air during storage. Often this is done by sealing the aldehyde in a container under an atmosphere of nitrogen.

Benzaldehyde $+ O_2 \longrightarrow$ Benzoic acid

Ketones, in contrast, resist oxidation by most oxidizing agents, including potassium dichromate and molecular oxygen.

The fact that aldehydes are so easy to oxidize and ketones are not allows us to use simple chemical tests to distinguish between these types of compounds. Suppose that we have a compound we know is either an aldehyde or a ketone. To identify which it is, all we need to do is treat it with a mild oxidizing agent. If it can be oxidized, it is an aldehyde; otherwise, it is a ketone. One reagent that has been used for this purpose is Tollens' reagent.

Screen 17.4 Simulation

Tollens' reagent contains silver nitrate and ammonia in water. When these two compounds are mixed, silver ion combines with NH_3 to form the ion $Ag(NH_3)_2^+$. When this solution is added to an aldehyde, the aldehyde acts as a reducing agent and reduces the complexed silver ion to silver metal. If this reaction is carried out properly, the silver metal precipitates as a smooth, mirror-like deposit, leading to the name **silver-mirror test.** If the remaining solution is then acidified with HCl, the carboxylic anion, $RCOO^-$, is converted to the carboxylic acid, RCOOH.

$$\underset{\text{Aldehyde}}{R-\overset{\overset{O}{\|}}{C}-H} + \underset{\text{Tollens' reagent}}{2Ag(NH_3)_2^+} + 3OH^- \longrightarrow \underset{\text{Carboxylic anion}}{R-\overset{\overset{O}{\|}}{C}-O^-} + \underset{\text{Silver mirror}}{2Ag} + 4NH_3 + 2H_2O$$

Today, silver(I) is rarely used for the oxidation of aldehydes because of its high cost and because of the availability of other, more convenient methods for this oxidation. This reaction, however, is still used for making (silvering) mirrors.

■ A silver mirror has been deposited on the inside of this flask by the reaction between an aldehyhde and Tollens' reagent.

EXAMPLE 17.4

Draw a structural formula for the product formed by treating each compound with Tollens' reagent followed by acidification with aqueous HCl.
(a) Pentanal
(b) 4-Hydroxybenzaldehyde

Solution
The aldehyde group in each compound is oxidized to a carboxylic anion, $-COO^-$. Acidification with HCl converts the anion to a carboxylic acid, $-COOH$.

(a) Pentanoic acid

(b) 4-Hydroxybenzoic acid

Problem 17.4
Complete the equations for these oxidations.
(a) Hexanedial + $O_2 \longrightarrow$
(b) 3-Phenylpropanal + $Ag(NH_3)_2^+ \longrightarrow$

17.6 Reduction

In Section 12.6D, we saw that the C=C double bond of an alkene can be reduced by hydrogen in the presence of a transition-metal catalyst to a C—C single bond. The same is true of the C=O double bond of an aldehyde or ketone. Aldehydes are reduced to primary alcohols and ketones are reduced to secondary alcohols.

$$\text{Pentanal} + H_2 \xrightarrow{\text{Transition metal catalyst}} \text{1-Pentanol}$$

$$\text{Cyclopentanone} + H_2 \xrightarrow{\text{Transition metal catalyst}} \text{Cyclopentanol}$$

The reduction of a C=O double bond under these conditions is slower than the reduction of a C=C double bond. Thus, if the same molecule contains both C=C and C=O double bonds, the C=C double bond is reduced first.

The most common reagent used in the laboratory for the reduction of an aldehyde or ketone is sodium borohydride, $NaBH_4$. This reagent contains hydrogen in the form of hydride ion, $H{:}^-$. In the hydride ion, hydrogen has two valence electrons and bears a negative charge. In a reduction by sodium borohydride, hydride ion is attracted to and then adds to the partially positive carbonyl carbon, which leaves a negative charge on the carbonyl oxygen. Reaction of this intermediate with aqueous acid gives the alcohol.

$$H{:}^- + \text{C=O} \longrightarrow H-C-O{:}^- \xrightarrow{H_3O^+} H-C-O-H$$

Of the two hydrogens added to the carbonyl group in this reduction, one comes from the reducing agent and the other comes from aqueous acid. Reduction of cyclohexanone, for example, with this reagent gives cyclohexanol:

$$\text{cyclohexanone} \xrightarrow{NaBH_4} \text{intermediate} \xrightarrow{H_3O^+} \text{cyclohexanol}$$

An advantage of $NaBH_4$ over the H_2/metal catalyst reaction is that $NaBH_4$ does not reduce carbon–carbon double bonds. The reason for this selectivity is quite straightforward. There is no polarity (no partial positive or negative charges) on a carbon–carbon double bond. Therefore, it has no site to attract the hydride ion. In the following reaction, $NaBH_4$ selectively reduces the aldehyde to a primary alcohol:

$$\text{Cinnamaldehyde} \xrightarrow[\text{2. } H_2O]{\text{1. } NaBH_4} \text{Cinnamyl alcohol}$$

In biological systems, the agent for the reduction of aldehydes and ketones is the reduced form of the coenzyme nicotinamide adenine dinucleotide, abbreviated NADH (Section 26.3). This reducing agent, like NaBH$_4$, delivers a hydride ion to the carbonyl carbon of the aldehyde or ketone. Reduction of pyruvate by NADH, for example, gives lactate:

$$CH_3-\underset{\text{Pyruvate}}{\overset{O}{\underset{\|}{C}}}-COO^- \xrightarrow{\text{NADH}} CH_3-\underset{H}{\overset{O^-}{\underset{|}{C}}}-COO^- \xrightarrow{H_3O^+} CH_3-\underset{\underset{\text{Lactate}}{H}}{\overset{O-H}{\underset{|}{C}}}-COO^-$$

Pyruvate is the end product of glycolysis, a series of enzyme-catalyzed reactions that converts glucose to two molecules of this ketoacid (Section 27.1). Under anaerobic conditions, NADH reduces pyruvate to lactate. The buildup of lactate in the bloodstream leads to acidosis, and in muscle tissue is associated with muscle fatigue. When blood lactate reaches a concentration of about 0.4 mg/100 mL, muscle tissue becomes almost completely exhausted.

EXAMPLE 17.5

Complete the equations for these reductions.

(a) CH$_3$CH$_2$CH$_2$C(=O)H + H$_2$ $\xrightarrow{\text{Transition metal catalyst}}$

(b) CH$_3$O—C$_6$H$_4$—C(=O)CH$_3$ $\xrightarrow[\text{2. H}_3\text{O}^+]{\text{1. NaBH}_4}$

Solution
The carbonyl group of the aldehyde in (a) is reduced to a primary alcohol and that of the ketone in (b) is reduced to a secondary alcohol.

(a) CH$_3$CH$_2$CH$_2$CH$_2$OH

(b) CH$_3$O—C$_6$H$_4$—CH(OH)CH$_3$

Problem 17.5
Which aldehyde or ketone gives these alcohols upon reduction with H$_2$/ metal catalyst?

(a) cyclohexyl—OH

(b) CH$_3$O—C$_6$H$_4$—CH$_2$CH$_2$OH

(c) CH$_3$—CH(OH)—CH$_2$—CH(OH)—CH$_3$ (2,4-pentanediol type structure with OH groups)

17.7 Addition of Alcohols

Hemiacetal A molecule containing a carbon bonded to one —OH and one —OR group; the product of adding one molecule of alcohol to the carbonyl group of an aldehyde or ketone

Addition of a molecule of alcohol to the carbonyl group of an aldehyde or ketone forms a **hemiacetal** (a half-acetal). The functional group of a hemiacetal is a carbon bonded to one —OH group and one —OR group. In forming a hemiacetal, the H of the alcohol adds to the carbonyl oxygen and the OR adds to the carbonyl carbon. Shown here are the hemiacetals formed by addition of one molecule of ethanol to benzaldehyde and to cyclohexanone:

Benzaldehyde + Ethanol ⇌ A hemiacetal

Cyclohexanone + Ethanol ⇌ A hemiacetal

Hemiacetals are generally unstable and are only minor components of an equilibrium mixture, except in one very important type of molecule. When a hydroxyl group is part of the same molecule that contains the carbonyl group and a five- or six-membered ring can form, the compound exists almost entirely in a cyclic hemiacetal form. In this case, the —OH group adds to the C=O group of the same molecule. We will have much more to say about cyclic hemiacetals when we consider the chemistry of carbohydrates in Chapter 19.

4-Hydroxypentanal → (Redraw to show —OH and —CHO close to each other) ⇌ A cyclic hemiacetal

Acetal A molecule containing two —OR groups bonded to the same carbon

Hemiacetals can react further with alcohols to form **acetals** plus water. This reaction is acid-catalyzed. The functional group of an acetal is a carbon bonded to two —OR groups.

A hemiacetal (from benzaldehyde) + Ethanol ⇌ An acetal + H_2O

A hemiacetal (from cyclohexanone) + Ethanol ⇌ An acetal + H_2O

All steps in hemiacetal and acetal formation are reversible. As with any other equilibrium, we can make this one go in either direction by using Le Chatelier's principle (Section 7.8). If we want to drive it to the right (formation of the acetal), we either use a large excess of alcohol or remove water from the equilibrium mixture. If we want to drive it to the left (hydrolysis of the acetal to the original aldehyde or ketone and alcohol), we use a large excess of water.

EXAMPLE 17.6

Show the reaction of 2-butanone with one molecule of ethanol to form a hemiacetal and then with a second molecule of ethanol to form an acetal.

Solution
Given are structural formulas of the hemiacetal and then the acetal.

2-Butanone → A hemiacetal → An acetal + H_2O

Problem 17.6
Show the reaction of benzaldehyde with one molecule of methanol to form a hemiacetal and then with a second molecule of methanol to form an acetal.

EXAMPLE 17.7

Identify all hemiacetals and acetals in the following structures, and tell whether they are formed from aldehydes or ketones.

(a) $CH_3CH_2\underset{\underset{OCH_3}{|}}{\overset{\overset{OCH_3}{|}}{C}}CH_2CH_3$

(b) $CH_3CH_2OCH_2CH_2OH$

(c) [cyclic structure: tetrahydropyran ring with OH substituent]

Solution
Compound (a) is an acetal derived from a ketone. Compound (b) is neither a hemiacetal nor an acetal because it does not have a carbon bonded

430 | CHAPTER 17 ALDEHYDES AND KETONES

to two oxygens; its functional groups are an ether and a primary alcohol. Compound (c) is a hemiacetal derived from an aldehyde.

(a) $CH_3CH_2\underset{\underset{OCH_3}{|}}{\overset{\overset{OCH_3}{|}}{C}}CH_2CH_3 + H_2O \xrightarrow{H^+} CH_3CH_2\overset{\overset{O}{\|}}{C}CH_2CH_3 + 2CH_3OH$

An acetal　　　　　　　　　　　　2-Pentanone

(c) [cyclic hemiacetal] $\xrightarrow{H^+, H_2O}$ [open-chain aldehyde with OH] $\xrightarrow{\text{Redraw the carbon chain}}$ HO—(chain)—CHO

A hemiacetal　　　　　　　　　　　5-Hydroxypentanal

Problem 17.7
Identify all hemiacetals and acetals in the following structures, and tell whether they are formed from aldehydes or ketones.

(a) $CH_3CH_2\underset{\underset{OCH_2CH_3}{|}}{\overset{\overset{OH}{|}}{C}}CH_2CH_3$

(b) $CH_3OCH_2CH_2OCH_3$

(c) [tetrahydropyran ring with OCH₃ substituent on carbon adjacent to ring O]

17.8 Keto-Enol Tautomerism   *Skip*

A carbon atom adjacent to a carbonyl group is called an α-carbon, and a hydrogen atom bonded to it is called an α-hydrogen.

$CH_3-\overset{\overset{O}{\|}}{C}-CH_2-CH_3$

α-hydrogens (arrow pointing to CH₂)
α-carbons (arrow pointing to CH₃ and CH₂)

Enol A molecule containing an —OH group bonded to a carbon of a carbon–carbon double bond

A carbonyl compound that has a hydrogen on an α-carbon is in equilibrium with a constitutional isomer called an **enol**. The name *enol* is derived from the IUPAC designation of it as both an alkene (*-en-*) and an alcohol (*-ol*).

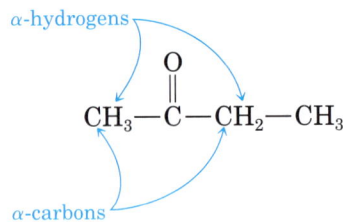

$CH_3-\overset{\overset{O}{\|}}{C}-CH_3 \rightleftharpoons CH_3-\underset{\underset{}{}}{\overset{\overset{OH}{|}}{C}}=CH_2$

Acetone　　　　　　Acetone
(keto form)　　　　(enol form)

Keto and enol forms are examples of **tautomers,** constitutional isomers in equilibrium with each other that differ in the location of a hydrogen atom relative to an oxygen atom. This type of isomerism is called **keto-enol tautomerism.** For any pair of keto-enol tautomers, the keto form generally predominates at equilibrium.

> **Tautomers** Constitutional isomers that differ in the location of a hydrogen atom and a double bond relative to an oxygen or nitrogen atom

EXAMPLE 17.8

Draw structural formulas for the two enol forms for each ketone.

(a) 2-methylcyclohexanone

(b) 2-hexanone

Solution

(a) Two enol forms shown (OH on ring with double bond on each side of methyl-bearing carbon)

(b) Two enol forms of 2-hexanone

Problem 17.8

Draw a structural formula for the keto form of each enol.

(a) cyclohexenol with OH

(b) cyclohexene-1,2-diol

(c) cyclohexane with =CHOH

SUMMARY

An **aldehyde** contains a carbonyl group bonded to at least one hydrogen atom (Section 17.2). A **ketone** contains a carbonyl group bonded to two carbon atoms. The IUPAC name of an aldehyde is derived by changing -e of the parent alkane to -al (Section 17.3A). The IUPAC name of a ketone is derived by changing -e of the parent alkane to -one and using a number to locate the carbonyl carbon.

Aldehydes and ketones are polar compounds. They have higher boiling points and are more soluble in water than nonpolar compounds of comparable molecular weight (Section 17.4).

KEY REACTIONS

1. **Oxidation of an Aldehyde to a Carboxylic Acid (Section 17.5)** The aldehyde group is among the most easily oxidized functional groups. Oxidizing agents include $K_2Cr_2O_7$, Tollens' reagent, and O_2.

$$PhCHO + Ag(NH_3)_2^+ \xrightarrow{\text{Tollens' reagent}} PhCO_2^- + Ag$$

432 | CHAPTER 17 ALDEHYDES AND KETONES

2. Reduction (Section 17.6) Aldehydes are reduced to primary alcohols and ketones to secondary alcohols by H_2 in the presence of a transition metal catalyst such as Pt or Ni. They are also reduced to alcohols by sodium borohydride, $NaBH_4$, followed by protonation.

$$\text{cyclohexanone} + H_2 \xrightarrow{\text{Transition metal catalyst}} \text{cyclohexanol-OH}$$

$$\text{C}_6\text{H}_5\text{CH=CHCHO} \xrightarrow[\text{2. H}_2\text{O}]{\text{1. NaBH}_4} \text{C}_6\text{H}_5\text{CH=CHCH}_2\text{OH}$$

3. Addition of Alcohols to Form Hemiacetals (Section 17.7) Hemiacetals are only minor components of an equilibrium mixture of aldehyde or ketone and alcohol, except where the —OH and C=O groups are parts of the same molecule and a five- or six-membered ring can form.

(structure: 4-hydroxypentanal ⇌ cyclic hemiacetal)

4. Addition of Alcohols to Form Acetals (Section 17.7) Formation of acetals is catalyzed by acid. Acetals are hydrolyzed in aqueous acid to an aldehyde or ketone and two molecules of an alcohol.

$$\underset{\text{Cyclohexanone}}{\text{C}_6\text{H}_{10}\text{=O}} + \underset{\text{Methanol}}{2\text{CH}_3\text{OH}} \xrightleftharpoons{H^+} \underset{\text{An acetal}}{\text{C}_6\text{H}_{10}(\text{OCH}_3)_2} + H_2O$$

5. Keto-Enol Tautomerism (Section 17.8) The keto form generally predominates at equilibrium.

$$\underset{\substack{\text{Keto form} \\ \text{(approximately 99.9\%)}}}{CH_3\overset{O}{\underset{\|}{C}}CH_3} \rightleftharpoons \underset{\text{Enol form}}{CH_3\overset{OH}{\underset{|}{C}}=CH_2}$$

PROBLEMS

Numbers that appear in color indicate difficult problems.
▶ designates problems requiring application of principles.

Preparation of Aldehydes and Ketones

17.9 Complete the equations for these reactions.

(a) cyclooctanol $\xrightarrow{K_2Cr_2O_7 / H_2SO_4}$

(b) cyclopentylmethanol $\xrightarrow{K_2Cr_2O_7 / H_2SO_4}$

17.10 Write an equation for each conversion.
(a) 1-Pentanol to pentanal
(b) 1-Pentanol to pentanoic acid
(c) 2-Pentanol to 2-pentanone
(d) 2-Propanol to acetone
(e) Cyclohexanol to cyclohexanone

Structure and Nomenclature

17.11 What is the difference in structure between an aldehyde and a ketone?

17.12 What is the difference in structure between an aromatic aldehyde and an aliphatic aldehyde?

17.13 Is it possible for the carbon atom of a carbonyl group to be a stereocenter? Explain.

17.14 Which compounds contain carbonyl groups?

(a) $CH_3\overset{OH}{\underset{|}{C}H}CH_3$

(b) $CH_3CH_2\overset{O}{\underset{\|}{C}}H$

(c) $C_6H_5\overset{O}{\underset{\|}{C}}OCH_2CH_3$

(d) cyclopentanone

(e) 2-hydroxytetrahydrofuran (cyclic with O and —OH)

(f) $CH_3CH_2CH_2\overset{O}{\underset{\|}{C}}OH$

17.15 Following are structural formulas for two steroid hormones. Name the functional groups in each.

(a) Cortisone

(b) Aldosterone

17.16 Draw structural formulas for the four aldehydes with molecular formula $C_5H_{10}O$. Which of these aldehydes are chiral?

17.17 Draw structural formulas for these aldehydes.
(a) Formaldehyde
(b) Propanal
(c) 3,7-Dimethyloctanal
(d) Decanal
(e) 4-Hydroxybenzaldehyde
(f) 2,3-Dihydroxypropanal

17.18 Draw structural formulas for these ketones.
(a) Ethyl isopropyl ketone
(b) 2-Chlorocyclohexanone
(c) 2,4-Dimethyl-3-pentanone
(d) Diisopropyl ketone
(e) Acetone
(f) 2,5-Dimethylcyclohexanone

17.19 Write IUPAC names for these compounds.
(a) $(CH_3CH_2CH_2)_2C=O$

(b)

(c)

(d)

(e)

(f) $HC(CH_2)_4CH$ with O groups

17.20 Write IUPAC names for these compounds.

(a) (b)

(c) CHO–CHOH–CHOH–CH$_2$OH   (d)

17.21 Explain why each name is incorrect. Write the correct IUPAC name for the intended compound.
(a) 3-Butanone
(b) 1-Butanone
(c) 4-Methylbutanal
(d) 2,2-Dimethyl-3-butanone

17.22 Explain why each name is incorrect. Write the correct IUPAC name for the intended compound.
(a) 2-Pentanal
(b) Cyclopentanal
(c) 3-Ethyl-2-butanone
(d) 5-Aminobenzaldehyde

Physical Properties

17.23 In each pair of compounds, select the one with the higher boiling point.
(a) Acetaldehyde or ethanol
(b) Acetone or 3-pentanone
(c) Butanal or butane
(d) Butanone or 2-butanol

17.24 Acetone is completely soluble in water, but 4-heptanone is completely insoluble. Explain.

17.25 Why does acetone have a higher boiling point (56°C) than ethyl methyl ether (11°C), even though their molecular weights are almost the same?

17.26 Pentane, 1-butanol, and butanal all have approximately the same molecular weights but different boiling points. Arrange them in order of increasing boiling point. Explain the basis for your ranking.

17.27 Show how acetaldehyde can form hydrogen bonds with water.

17.28 Why can't two molecules of acetone form a hydrogen bond with each other?

Oxidation–Reduction of Aldehydes and Ketones

17.29 Draw the structural formula for the principal organic product formed when each compound is treated with $K_2Cr_2O_7/H_2SO_4$. If no reaction takes place, say so.
(a) Butanal
(b) Benzaldehyde
(c) Cyclohexanone
(d) Cyclohexanol

17.30 Draw the structural formula for the principal organic product formed when each compound in Problem 17.29 is treated with Tollens' reagent. If no reaction takes place, say so.

17.31 What simple chemical test could you use to distinguish between the members of each pair of compounds? Tell what you would do, what you would expect to observe, and how you would interpret your experimental observation.
(a) Pentanal and 2-pentanone
(b) 2-Pentanone and 2-pentanol

17.32 Explain why liquid aldehydes are often stored under an atmosphere of nitrogen rather than air.

17.33 Suppose that you take a bottle of benzaldehyde (a liquid, bp 179°C) from a shelf and find a white solid in the bottom of the bottle. The solid turns litmus red; that is, it is acidic. Yet aldehydes are neutral compounds. How can you explain these observations?

17.34 Explain why the reduction of an aldehyde always gives a primary alcohol and the reduction of a ketone always gives a secondary alcohol.

17.35 Write the structural formula for the principal organic product formed by treating each compound with H_2/transition metal catalyst.

(a) $CH_3CCH_2CH_3$ (with C=O)
(b) $CH_3(CH_2)_4CH$ (with C=O)
(c) cyclopentanone with CH₃ substituent
(d) 2-hydroxybenzaldehyde

17.36 Write the structural formula for the principal organic product formed by treating each compound in Problem 17.35 with $NaBH_4$ followed by H_2O.

17.37 1,3-Dihydroxy-2-propanone, more commonly known as dihydroxyacetone, is the active ingredient in artificial tanning agents such as Man-Tan and Magic Tan.
(a) Write a structural formula for this compound.
(b) Would you expect it to be soluble or insoluble in water?
(c) Write the structural formula of the product formed by its reduction with H_2/metal catalyst.

17.38 Draw a structural formula for the product formed by treatment of butanal with each set of reagents.
(a) H_2/metal catalyst
(b) $NaBH_4$, then H_2O
(c) $Ag(NH_3)_2^+$ (Tollens' reagent)
(d) $K_2Cr_2O_7/H_2SO_4$

17.39 Draw a structural formula for the product formed by treatment of acetophenone, $C_6H_5COCH_3$, with each set of reagents given in Problem 17.38.

Addition of Alcohols

17.40 What is the characteristic structural feature of a hemiacetal? Of an acetal?

17.41 Which compounds are hemiacetals, which are acetals, and which are neither?

(a) $C_6H_5-CH(OCH_3)(OCH_3)$
(b) $CH_3CH_2CH(OH)OCH_3$
(c) $CH_3OCH_2OCH_3$
(d) tetrahydrofuran with CH_2CH_3 and O substituents
(e) tetrahydrofuran with OCH_2CH_3 substituent
(f) tetrahydrofuran with OH substituent

17.42 Which compounds are hemiacetals, which are acetals, and which are neither?

(a) [structure: HO, OCH₃ on carbon with two alkyl groups]

(b) [structure: CH₃O, OCH₃ on carbon attached to phenyl and ethyl]

(c) [γ-butyrolactone structure]

(d) [cyclopentane with OCH₃ and OH]

(e) [tetrahydrofuran with OCH₃]

(f) [cyclopentane with OCH₃ and OCH₃]

17.43 Draw the hemiacetal and then the acetal formed in each reaction. In each case, assume an excess of the alcohol.

(a) Propanal + methanol ⟶
(b) Cyclopentanone + methanol ⟶

17.44 Draw the structures of the aldehydes or ketones and alcohols formed when these acetals are treated with aqueous acid and hydrolyzed.

(a) [1,3-dioxolane with ethyl substituent]
(b) [benzene with CH(OCH₃)₂ and OCH₃ substituent]
(c) [cyclohexane with two OCH₃ groups]
(d) [tetrahydropyran with OCH₃]

17.45 What is the difference in meaning between the terms "hydration" and "hydrolysis"? Give an example of each.

17.46 What is the difference in meaning between the terms "hydration" and "dehydration"? Give an example of each.

Keto-Enol Tautomerism

17.47 Which of these compounds undergo keto-enol tautomerism?

(a) CH₃CHO
(b) CH₃CCH₃ (acetone)
(c) PhCHO (benzaldehyde)
(d) PhCOCH₃ (acetophenone)
(e) Ph-CO-Ph (benzophenone)
(f) cyclohexanone

17.48 Draw all enol forms of each aldehyde and ketone.

(a) CH₃CH₂CHO
(b) CH₃COCH₂CH₃
(c) 2-methylcyclopentanone

17.49 Draw the structural formula of the keto form of each enol.

(a) 1-hydroxycyclopentene
(b) CH₃C=CHCH₂CH₂CH₃ with OH on C2
(c) PhCH=C(OH)CH₃

17.50 Starting with cyclohexanone, show how to prepare these compounds. In addition to the given starting material, use any other organic or inorganic reagents as necessary.

(a) Cyclohexanol
(b) Cyclohexene
(c) 1,2-Dibromocyclohexane
(d) Cyclohexane
(e) Chlorocyclohexane

17.51 Draw the structural formula of an aldehyde or ketone that can be reduced to produce each alcohol. If none exists, say so.

(a) CH₃CHOHCH₃
(b) PhCH₂OH
(c) CH₃OH
(d) 1-methylcyclohexanol

436 | CHAPTER 17 ALDEHYDES AND KETONES

17.52 Draw the structural formula of an aldehyde or ketone that can be reduced to produce each alcohol. If none exists, say so.

(a) cyclopentanol (cyclopentane with —OH)

(b) cyclohexane—CH$_2$OH

(c) CH$_3$—C(CH$_3$)(CH$_3$)—OH

(d) HO—CH$_2$CH$_2$CH$_2$CH$_2$CH$_2$—OH

17.53 1-Propanol can be prepared by the reduction of an aldehyde, but it cannot be prepared by the acid-catalyzed hydration of an alkene. Explain why it can't be prepared from an alkene.

17.54 Show how to bring about these conversions. In addition to the given starting material, use any other organic or inorganic reagents as necessary.

(a) C$_6$H$_5$CCH$_2$CH$_3$ (with =O) → C$_6$H$_5$CHCH$_2$CH$_3$ (with OH) → C$_6$H$_5$CH=CHCH$_3$

(b) cyclopentanone (=O) → cyclopentanol (—OH) → cyclopentene → cyclopentyl chloride (—Cl)

17.55 Show how to bring about these conversions. In addition to the given starting material, use any other organic or inorganic reagents as necessary.

(a) 1-Pentene to 2-pentanone
(b) Cyclohexene to cyclohexanone

17.56 Describe a simple chemical test by which you could distinguish between the members of each pair of compounds.

(a) Cyclohexanone and aniline
(b) Cyclohexene and cyclohexanol
(c) Benzaldehyde and cinnamaldehyde

Additional Problems

17.57 Indicate the aldehyde or ketone group in these compounds.

(a) HCCH$_2$CH$_2$CH$_2$CCH$_3$ (with two C=O groups)

(b) cyclohexanone with —CH=O substituent

(c) HOCH$_2$CHCH (with HO and =O)

(d) tetralone-type bicyclic with =O

(e) C$_6$H$_5$CCH$_2$CH$_3$ (with =O)

(f) HO—(CH$_3$O substituted benzene)—CH=O

17.58 Draw the structural formula for the product formed by treatment of each compound in Problem 17.57 with sodium borohydride, NaBH$_4$.

17.59 Sodium borohydride is a laboratory reducing agent. NADH is a biological reducing agent. In what way is the chemistry by which they reduce aldehydes and ketones similar?

17.60 Name the functional groups in each steroid hormone. Methandrostenolone is a synthetic anabolic steroid; that is, it promotes tissue and muscle growth and development. What are the differences in structural formula between testosterone and methandrostenolone?

(a) Testosterone

(b) Methandrostenolone

17.61 Draw structural formulas for the (a) one ketone and (b) two aldehydes with molecular formula C_4H_8O.

17.62 Draw structural formulas for these compounds.
(a) 1-Chloro-2-propanone
(b) 3-Hydroxybutanal
(c) 4-Hydroxy-4-methyl-2-pentanone
(d) 3-Methyl-3-phenylbutanal
(e) 1,3-Cyclohexanedione
(f) 5-Hydroxyhexanal

17.63 Why does acetone have a lower boiling point (56°C) than 2-propanol (82°C), even though their molecular weights are almost the same?

17.64 Propanal (bp 49°C) and 1-propanol (bp 97°C) have about the same molecular weight, yet their boiling points differ by almost 50°C. Explain this fact.

17.65 What simple chemical test could you use to distinguish between the members of each pair of compounds? Tell what you would do, what you would expect to observe, and how you would interpret your experimental observation.
(a) Benzaldehyde and cyclohexanone
(b) Acetaldehyde and acetone

17.66 5-Hydroxyhexanal forms a six-membered cyclic hemiacetal, which predominates at equilibrium in aqueous solution.
(a) Draw a structural formula for this cyclic hemiacetal.
(b) How many stereoisomers are possible for 5-hydroxyhexanal?
(c) How many stereoisomers are possible for this cyclic hemiacetal?

17.67 The following molecule is an enediol; each carbon of the double bond carries an —OH group. Draw structural formulas for the α-hydroxyketone and the α-hydroxyaldehyde with which this enediol is in equilibrium.

α-hydroxyaldehyde ⇌
HC—OH
‖
C—OH
|
CH₃
An enediol
⇌ α-hydroxyketone

17.68 Alcohols can be prepared by the acid-catalyzed hydration of alkenes (Section 12.6B) and by the reduction of aldehydes and ketones (Section 17.6). Show how you might prepare each of the following alcohols by (1) acid-catalyzed hydration of an alkene and (2) reduction of an aldehyde or ketone.
(a) Ethanol
(b) Cyclohexanol
(c) 2-Propanol
(d) 1-Phenylethanol

17.69 Glucose, $C_6H_{12}O_6$, contains an aldehyde group but exists predominantly in the form of the cyclic hemiacetal shown here:

β-D-Glucose

(a) A cyclic hemiacetal is formed when the —OH group of one carbon bonds to the carbonyl group of another carbon. Which carbon in glucose provides the —OH group and which provides the CHO group?
(b) Write the structural formula of the free aldehyde form of glucose.

17.70 Ribose, $C_5H_{10}O_5$, contains an aldehyde group but exists predominantly in the form of the cyclic hemiacetal shown here:

β-D-Ribose

(a) Which carbon of ribose provides the —OH group and which provides the CHO group for formation of this cyclic hemiacetal?
(b) Write the structural formula of the free aldehyde form of ribose.

InfoTrac College Edition

For additional readings, go to InfoTrac College Edition, your online research library, at

http://infotrac.thomsonlearning.com

CHAPTER 18

18.1 Introduction
18.2 Carboxylic Acids
18.3 Anhydrides, Esters, and Amides
18.4 Preparation of Esters—Fischer Esterification
18.5 Preparation of Amides
18.6 Hydrolysis of Anhydrides, Esters, and Amides
18.7 Phosphoric Anhydrides and Esters
18.8 Step-Growth Polymerizations

Citrus fruits are sources of citric acid, a tricarboxylic acid.

Carboxylic Acids, Anhydrides, Esters, and Amides

18.1 Introduction

In this chapter, we study carboxylic acids, another class of organic compounds containing the carbonyl group. We also study three classes of organic compounds derived from carboxylic acids: anhydrides, esters, and amides. Below, under the general formula of each carboxylic acid derivative, is a drawing to help you see how the functional group of the derivative is formally related to a carboxyl group. The loss of —OH from a carboxyl group and H— from ammonia, for example, gives an amide.

Of these three carboxylic derivatives, anhydrides are so reactive that they are rarely found in nature. Esters and amides, however, are widespread in the biological world.

18.2 Carboxylic Acids

The functional group of a **carboxylic acid** is a **carboxyl group** (Section 10.4D), which can be represented in any one of three ways:

$$-\overset{\overset{\displaystyle O}{\|}}{C}-OH \quad -COOH \quad -CO_2H$$

A IUPAC Names

We derive the IUPAC name of an acyclic carboxylic acid from the name of the longest carbon chain that contains the carboxyl group. The final **-e** from the name of the parent alkane is dropped and replaced by **-oic acid.** The chain is numbered beginning with the carbon of the carboxyl group. Because the carboxyl carbon is understood to be carbon 1, there is no need to give it a number. In the following examples, the common name is given in parentheses.

Hexanoic acid
(Caproic acid)

3-Methylbutanoic acid
(Isovaleric acid)

When a carboxylic acid also contains an alcohol, we indicate the presence of the —OH group by adding the prefix **hydroxy-.** When it contains an amine, we indicate the presence of the —NH$_2$ group by **amino-.**

5-Hydroxyhexanoic acid

4-Aminobenzoic acid

To name dicarboxylic acids, we add the suffix **-dioic acid** to the name of the parent alkane that contains both carboxyl groups. The numbers of the carboxyl carbons are not indicated because they can be only at the ends of the parent chain.

Ethanedioic acid
(Oxalic acid)

Propanedioic acid
(Malonic acid)

■ Leaves of the rhubarb plant contain the poison oxalic acid as its potassium or sodium salt.

Butanedioic acid
(Succinic acid)

Pentanedioic acid
(Glutaric acid)

Hexanedioic acid
(Adipic acid)

440 | CHAPTER 18 CARBOXYLIC ACIDS, ANHYDRIDES, ESTERS, AND AMIDES

The unbranched carboxylic acids having between 12 and 20 carbon atoms are known as fatty acids. We study them in Chapter 20.

Table 18.1 Several Aliphatic Carboxylic Acids and Their Common Names

Structure	IUPAC Name	Common Name	Derivation
HCOOH	methanoic acid	formic acid	Latin: *formica*, ant
CH_3COOH	ethanoic acid	acetic acid	Latin: *acetum*, vinegar
CH_3CH_2COOH	propanoic acid	propionic acid	Greek: *propion*, first fat
$CH_3(CH_2)_2COOH$	butanoic acid	butyric acid	Latin: *butyrum*, butter
$CH_3(CH_2)_3COOH$	pentanoic acid	valeric acid	Latin: *valere*, to be strong
$CH_3(CH_2)_4COOH$	hexanoic acid	caproic acid	Latin: *caper*, goat
$CH_3(CH_2)_6COOH$	octanoic acid	caprylic acid	Latin: *caper*, goat
$CH_3(CH_2)_8COOH$	decanoic acid	capric acid	Latin: *caper*, goat
$CH_3(CH_2)_{10}COOH$	dodecanoic acid	lauric acid	Latin: *laurus*, laurel
$CH_3(CH_2)_{12}COOH$	tetradecanoic acid	myristic acid	Greek: *myristikos*, fragrant
$CH_3(CH_2)_{14}COOH$	hexadecanoic acid	palmitic acid	Latin: *palma*, palm tree
$CH_3(CH_2)_{16}COOH$	octadecanoic acid	stearic acid	Greek: *stear*, solid fat
$CH_3(CH_2)_{18}COOH$	eicosanoic acid	arachidic acid	Greek: *arachis*, peanut

The name *oxalic acid* is derived from one of its sources in the biological world—plants of the genus *Oxalis*, one of which is rhubarb. Oxalic acid also occurs in human and animal urine, and calcium oxalate is a major component of kidney stones. Succinic acid is an intermediate in the citric acid cycle (Section 26.4). Adipic acid is one of the two monomers required for the synthesis of the polymer nylon-66 (Section 18.8A).

B Common Names

Common names for aliphatic carboxylic acids, many of which were known long before the development of IUPAC nomenclature, are often derived from the name of a natural substance from which the acid can be isolated. Table 18.1 lists several of the unbranched aliphatic carboxylic acids found in the biological world along with the common name of each. Those with 16, 18, and 20 carbon atoms are particularly abundant in animal fats and vegetable oils (Section 20.2), and the phospholipid components of biological membranes (Section 20.5).

When common names are used, the Greek letters alpha (α), beta (β), gamma (γ), and so forth, are often added as a prefix to locate substituents.

■ Formic acid was first obtained in 1670 from the destructive distillation of ants, whose Latin genus is *Formica*. It is one of the components of the venom injected by stinging ants.

GABA is a neurotransmitter in the central nervous system.

$$\overset{4}{C}-\overset{3}{\underset{\gamma}{C}}-\overset{2}{\underset{\beta}{C}}-\overset{1}{\underset{\alpha}{C}}-OH \qquad H_2N-\overset{4}{\underset{\gamma}{\quad}}\underset{\quad}{\quad}\overset{1}{C}-OH \qquad \underset{\underset{OH}{\alpha}}{\overset{2}{\quad}}\overset{1}{C}-OH$$

4-Aminobutanoic acid   2-Hydroxypropanoic acid
(γ-Aminobutyric acid; GABA)   (Lactic acid)

EXAMPLE 18.1

Write the IUPAC name for each carboxylic acid:

(a) [structure: hexanoic acid with ethyl branch, ending in COOH]

(b) HO—⟨benzene ring⟩—COOH

(c) [structure: CH₃—CH(OH)—COOH with stereochemistry]

Solution

(a) The longest carbon chain that contains the carboxyl group has five carbons and, therefore, the parent alkane is pentane. The IUPAC name is 2-ethylpentanoic acid.
(b) 4-Hydroxybenzoic acid
(c) (R)-2-Hydroxypropanoic acid

Problem 18.1

Each of these compounds has a well-recognized common name. A derivative of glyceric acid is an intermediate in glycolysis (Section 27.2). β-Alanine is a building block of pantothenic acid (Section 26.5). Mevalonic acid is an intermediate in the biosynthesis of steroids (Section 28.4). Write the IUPAC name for each compound.

(a) COOH
     |
     CHOH
     |
     CH$_2$OH
     Glyceric acid

(b) H$_2$NCH$_2$CH$_2$COOH
    β-Alanine

(c) Mevalonic acid: HOCH$_2$CH$_2$C(OH)(CH$_3$)CH$_2$COOH
    Mevalonic acid

C Physical Properties

A major feature of carboxylic acids is the polarity of the carboxyl group. This group contains three polar covalent bonds: C=O, C—O, and O—H. It is the polarity of these bonds that determines the major physical properties of carboxylic acids.

Carboxylic acids have significantly higher boiling points than other types of organic compounds of comparable molecular weight (Table 18.2). Their higher boiling points result from their polarity and the fact that hydrogen bonding between two carboxyl groups creates a dimer that behaves as a higher-molecular-weight compound.

Hydrogen bonding between two molecules

$$H_3C-C\begin{array}{c}O\cdots H-O\\ \| \quad\quad\quad\;\;\|\\ O-H\cdots O\end{array}C-CH_3$$

Carboxylic acids are more soluble in water than are alcohols, ethers, aldehydes, and ketones of comparable molecular weight. This increased sol-

■ Acetic acid molecules can interact through hydrogen bonds. Shown here is partly solid glacial acetic acid. Like most substances (water being a notable exception), the solid is denser than the liquid.

Table 18.2 Boiling Points and Solubilities in Water of Two Groups of Carboxylic Acids, Alcohols, and Aldehydes of Comparable Molecular Weight

Structure	Name	Molecular Weight	Boiling Point (°C)	Solubility (g/100 mL H$_2$O)
CH$_3$COOH	acetic acid	60.5	118	infinite
CH$_3$CH$_2$CH$_2$OH	1-propanol	60.1	97	infinite
CH$_3$CH$_2$CHO	propanal	58.1	48	16
CH$_3$(CH$_2$)$_2$COOH	butanoic acid	88.1	163	infinite
CH$_3$(CH$_2$)$_3$CH$_2$OH	1-pentanol	88.1	137	2.3
CH$_3$(CH$_2$)$_3$CHO	pentanal	86.1	103	slight

ubility is due to their strong association with water molecules by hydrogen bonding through both their carbonyl and hydroxyl groups. The first four aliphatic carboxylic acids (formic, acetic, propanoic, and butanoic) are infinitely soluble in water. As the size of the hydrocarbon chain increases relative to that of the carboxyl group, however, water solubility decreases. The solubility of hexanoic acid (six carbons) in water is 1.0 g/100 mL water.

We must mention two other properties of carboxylic acids. First, the liquid carboxylic acids from propanoic acid to decanoic acid have sharp, often disagreeable odors. Butanoic acid is found in stale perspiration and is a major component of "locker room odor." Pentanoic acid smells even worse, and goats, which secrete C$_6$, C$_8$, and C$_{10}$ acids (Table 18.1), are not famous for their pleasant odors. Second, carboxylic acids have a characteristic sour taste. The sour taste of pickles and sauerkraut is due to the presence of lactic acid. The sour taste of limes (pH 1.9), lemons (pH 2.3), and grapefruit (pH 3.2) is due to the presence of citric and other acids.

D Acidity

Carboxylic acids are weak acids. Values of K_a for most unsubstituted aliphatic and aromatic carboxylic acids fall within the range of 10^{-4} to 10^{-5} (pK_a = 4.0–5.0). The value of K_a for acetic acid, for example, is 1.74×10^{-5}; its pK_a is 4.76.

$$CH_3\overset{O}{\overset{\|}{C}}OH + H_2O \rightleftharpoons CH_3\overset{O}{\overset{\|}{C}}O^- + H_3O^+ \quad K_a = \frac{[CH_3COO^-][H_3O^+]}{[CH_3COOH]} = 1.74 \times 10^{-5}$$

$$pK_a = 4.76$$

Substituents of high electronegativity (especially —OH, —Cl, and —NH$_3^+$) near the carboxyl group increase the acidity of carboxylic acids, often by several orders of magnitude. Compare, for example, the acidities of acetic acid and the chlorine-substituted acetic acids. Both dichloroacetic acid and trichloroacetic acid are stronger acids than H$_3$PO$_4$ (pK_a 2.1).

Formula:	CH$_3$COOH	ClCH$_2$COOH	Cl$_2$CHCOOH	Cl$_3$CCOOH
Name:	Acetic acid	Chloroacetic acid	Dichloroacetic acid	Trichloroacetic acid
pK_a:	4.76	2.86	1.48	0.70

Increasing acid strength →

Dichloroacetic acid is used as a topical astringent and as a treatment for genital warts in males.

Dentists use a 50% aqueous solution of trichloroacetic acid to cauterize gums. This strong acid kills the bleeding and diseased tissue and allows the growth of healthy gum tissue.

Electronegative atoms on the carbon adjacent to a carboxyl group increase acidity because they pull electron density away from the O—H bond, thereby facilitating ionization of the carboxyl group and making it a stronger acid.

E Reaction with Bases

All carboxylic acids, whether soluble or insoluble in water, react with NaOH, KOH, and other strong bases to form water-soluble salts.

$$C_6H_5\text{—COOH} + \text{NaOH} \xrightarrow{H_2O} C_6H_5\text{—COO}^-\text{Na}^+ + H_2O$$

Benzoic acid (slightly soluble in water) → Sodium benzoate (60 g/100 mL water)

■ Sodium benzoate and calcium propanoate are used as preservatives in baked goods.

Carboxylic acids also form water-soluble salts with ammonia and amines.

$$C_6H_5\text{—COOH} + NH_3 \xrightarrow{H_2O} C_6H_5\text{—COO}^-NH_4^+$$

Benzoic acid (slightly soluble in water) → Ammonium benzoate (20 g/100 mL water)

Carboxylic acids, like inorganic acids (Section 8.6E), react with sodium bicarbonate and sodium carbonate to form water-soluble sodium salts and carbonic acid. Carbonic acid then decomposes to give water and carbon dioxide, which evolves as a gas.

$$CH_3COOH + NaHCO_3 \xrightarrow{H_2O} CH_3COO^-Na^+ + CO_2 + H_2O$$

Acetic acid → Sodium acetate

Salts of carboxylic acids are named in the same manner as the salts of inorganic acids: the cation is named first and then the anion. We derive the name of the anion from the name of the carboxylic acid by dropping the suffix *-ic acid* and adding the suffix *-ate*. For example, $CH_3CH_2COO^-Na^+$ is named sodium propanoate.

EXAMPLE 18.2

Complete the equation for each acid–base reaction and name the salt formed.

(a) $CH_3(CH_2)_2COOH + NaOH \longrightarrow$

(b) $CH_3\overset{\overset{\displaystyle OH}{|}}{C}HCOOH + NaHCO_3 \longrightarrow$

Solution
Each carboxylic acid is converted to its sodium salt. In (b), carbonic acid forms, which then decomposes to carbon dioxide and water.

(a) $CH_3(CH_2)_2COOH + NaOH \longrightarrow CH_3(CH_2)_2COO^-Na^+ + H_2O$

Butanoic acid → Sodium butanoate

(b) $\underset{\text{2-Hydroxypropanoic acid}\atop\text{(Lactic acid)}}{\text{CH}_3\overset{\overset{\text{OH}}{|}}{\text{CH}}\text{COOH}} + \text{NaHCO}_3$

$\longrightarrow \underset{\text{Sodium 2-hydroxypropanoate}\atop\text{(Sodium lactate)}}{\text{CH}_3\overset{\overset{\text{OH}}{|}}{\text{CH}}\text{COO}^-\text{Na}^+} + \text{H}_2\text{O} + \text{CO}_2$

Problem 18.2
Write an equation for the reaction of each acid in Example 18.2 with ammonia and name the salt formed.

One last point about the acidity of carboxylic acids: The form in which they exist in an aqueous solution depends on the solution's pH. If we dissolve a carboxylic acid in water, the solution will be slightly acidic due to ionization of the carboxyl group. If we then add enough HCl to the solution to bring its pH to 2.0 or lower (more acidic), almost all carboxylic acid molecules are present in their un-ionized (—COOH) form. When the pH of the solution is equal to the pK_a of the acid, the un-ionized and ionized forms are present in approximately equal amounts. When the pH of the solution is brought to 8.0 or higher (more basic), almost all carboxylic molecules are present in their ionized (—COO$^-$) form.

$$\underset{\substack{\text{at pH 2.0}\\\text{or lower}\\\text{(more acidic)}}}{\text{R}-\overset{\overset{\text{O}}{\|}}{\text{C}}-\text{OH}} \underset{\text{H}^+}{\overset{\text{OH}^-}{\rightleftharpoons}} \underset{\substack{\text{at pH} = pK_a = 4.0-5.0\\\text{(both forms are present in}\\\text{approximately equal amounts)}}}{\text{R}-\overset{\overset{\text{O}}{\|}}{\text{C}}-\text{OH} + \text{R}-\overset{\overset{\text{O}}{\|}}{\text{C}}-\text{O}^-} \underset{\text{H}^+}{\overset{\text{OH}^-}{\rightleftharpoons}} \underset{\substack{\text{at pH 8.0}\\\text{or higher}\\\text{(more basic)}}}{\text{R}-\overset{\overset{\text{O}}{\|}}{\text{C}}-\text{O}^-}$$

18.3 Anhydrides, Esters, and Amides

A Anhydrides

Anhydride A compound in which two carbonyl groups are bonded to the same oxygen atom; RCO—O—COR'

The functional group of an **anhydride** consists of two carbonyl groups bonded to the same oxygen atom. The anhydride may be symmetrical (from two identical **acyl groups**), or mixed (from two different acyl groups). To name anhydrides, we drop the word "acid" from the name of the carboxylic acid from which the anhydride is derived and add the word "anhydride."

Acyl group An $\overset{\overset{\text{O}}{\|}}{\text{RC}}$ group

$$\underset{\text{Acetic anhydride}}{\text{CH}_3\overset{\overset{\text{O}}{\|}}{\text{C}}-\text{O}-\overset{\overset{\text{O}}{\|}}{\text{C}}\text{CH}_3}$$

B Esters

Ester A compound in which the H atom of a carboxyl group, RCOOH, is replaced by an —R' group; RCOOR'

The functional group of an **ester** is a carbonyl group bonded to an —OR group. Both IUPAC and common names of esters are derived from the names of the parent carboxylic acids. The alkyl group bonded to oxygen is named first, followed by the name of the acid in which the suffix **-ic acid** is replaced by the suffix **-ate.**

Ethyl ethanoate
(Ethyl acetate)

Diethyl pentanedioate
(Diethyl glutarate)

Cyclic esters are called **lactones.** Following are structural formulas for a five-membered and a seven-membered lactone. Note that the oxygen atom counts as one atom of the ring.

A five-membered lactone

A seven-membered lactone

> **Lactone** A cyclic ester

Esters are slightly polar compounds because of the presence of the C=O and C—O—C bonds. Because they have no available hydrogens for hydrogen bonding, however, they are generally insoluble in water. Almost all low-molecular-weight esters are liquids.

EXAMPLE 18.3

Write the IUPAC name of each ester.

(a)   (b)

Solution
The IUPAC name is given first and the common name is given in parentheses.
(a) Methyl 3-methylbutanoate (methyl isovalerate, from isovaleric acid)
(b) Ethyl butanoate (ethyl butyrate, from butyric acid)

Problem 18.3
Write the IUPAC name for each ester.

(a) —COCH$_2$CH$_3$ (b)

■ Some products that contain ethyl acetate, a common carboxylic ester.

Screen 18.7
Simulation Tutorial

C Amides

The functional group of an **amide** is a carbonyl group bonded to a nitrogen atom. Amides are named by dropping the suffix **-oic acid** from the IUPAC name of the parent acid, or **-ic acid** from its common name, and adding **-amide.** If the nitrogen atom of the amide is bonded to an alkyl group, the

> **Amide** A compound in which a carbonyl group is bonded to a nitrogen atom, RCONR'$_2$; one or both of the R' groups may be H or alkyl or aryl groups

CHEMICAL CONNECTIONS 18A

The Pyrethrins: Natural Insecticides of Plant Origin

Pyrethrym is a natural insecticide obtained from the powdered flower heads of several species of *Chrysanthemum*, particularly *C. cinerariaefolium*. The active substances in pyrethrum, principally pyrethrins I and II, are contact poisons for insects and cold-blooded vertebrates. Because their concentrations in the pyrethrum powder used in chrysanthemum-based insecticides are nontoxic to plants and higher animals, pyrethrum powder has found wide application in household and livestock sprays as well as in dusts for edible plants. Natural pyrethrins are esters of chrysanthemic acid.

While pyrethrum powders are effective insecticides, the active substances in them are destroyed rapidly in the environment. In an effort to develop synthetic compounds as effective as these natural insecticides but with greater biostability, chemists have prepared a series of esters related in structure to chrysanthemic acid. Permethrin is one of the most commonly used synthetic pyrethrin-like compounds in household and agricultural products.

Pyrethrin I

Permethrin

---

group is named and its location on nitrogen is indicated by *N*-. Two alkyl groups are indicated by *N,N*-di-.

$$CH_3\overset{O}{\underset{\|}{C}}NH_2$$

Acetamide
(a 1° amide)

$$CH_3\overset{O}{\underset{\|}{C}}NHCH_3$$

N-Methylacetamide
(a 2° amide)

$$H\overset{O}{\underset{\|}{C}}N(CH_3)_2$$

N, N-Dimethylformamide
(a 3° amide)

Lactam A cyclic amide

Cyclic amides are called **lactams**. Following are structural formulas for a four-membered and a seven-membered lactam. A four-membered lactam is essential to the function of the penicillin and cephalosporin antibiotics (Chemical Connections 18B).

A four-membered lactam

A seven-membered lactam

CHEMICAL CONNECTIONS 18B

The Penicillins and Cephalosporins: β-Lactam Antibiotics

The **penicillins** were discovered in 1928 by the Scottish bacteriologist Sir Alexander Fleming. Thanks to the brilliant experimental work of Sir Howard Florey, an Australian pathologist, and Ernst Chain, a German chemist who fled Nazi Germany, penicillin G was introduced into the practice of medicine in 1943. For their pioneering work in developing one of the most effective antibiotics of all time, Fleming, Florey, and Chain were awarded the Nobel Prize for medicine or physiology in 1945.

The mold from which Fleming discovered penicillin was *Penicillium notatum*, a strain that gives a relatively low yield of penicillin. It was replaced in commercial production of the antibiotic by *P. chrysogenum*, a strain cultured from a mold found growing on a grapefruit in a market in Peoria, Illinois. The structural feature common to all penicillins is the four-membered lactam ring, called a **β-lactam**, bonded to a five-membered, sulfur-containing ring. The penicillins owe their antibacterial activity to a common mechanism that inhibits the biosynthesis of a vital part of bacterial cell walls.

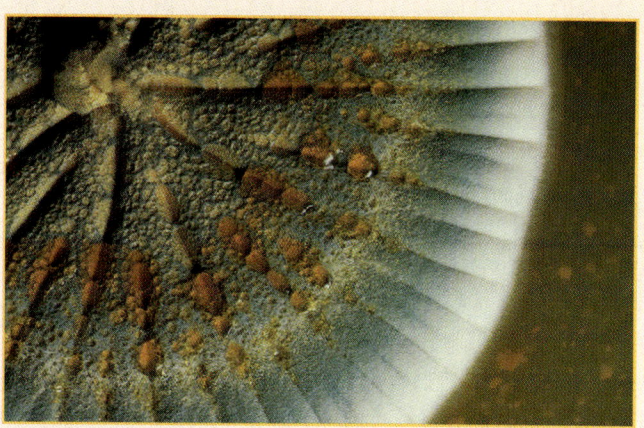

■ Macrophotograph of the fungus *Penicillium notatum* growing in a petri dish culture of Whickerham's agar. This fungus is the species that was used as an early source of the first penicillin antibiotic.

Another approach is to search for newer, more effective β-lactam antibiotics. The most effective of these agents discovered so far are the **cephalosporins**, the first of which was isolated from the fungus *Cephalosporium acremonium*.

Soon after the penicillins were introduced into medical practice, penicillin-resistant strains of bacteria began to appear. They have since proliferated dramatically. One approach to combating resistant strains is to synthesize newer, more effective penicillins, such as ampicillin, methicillin, and amoxicillin.

This class of β-lactam antibiotics has an even broader spectrum of antibacterial activity than the penicillins and is effective against many penicillin-resistant bacterial strains. Cephalexin (Keflex) is presently one of the most widely prescribed of the cephalosporin antibiotics.

EXAMPLE 18.4

Write the IUPAC name for each amide.

(a) CH₃CH₂CH₂C(O)NH₂

(b) H₂N–C(O)–(CH₂)₄–C(O)–NH₂

Solution
Given is the IUPAC name and then, in parentheses, the common name.
(a) Butanamide (butyramide, from butyric acid)
(b) Hexanediamide (adipamide, from adipic acid)

Problem 18.4
Draw a structural formula for each amide.
(a) *N*-Cyclohexylacetamide
(b) Benzamide

18.4 Preparation of Esters—Fischer Esterification

One of the most commonly used preparations of esters involves treating a carboxylic acid with an alcohol in the presence of an acid catalyst, such as concentrated sulfuric acid. In this reaction, two changes take place. First, OH is removed from the carboxyl group and H from the alcohol to form a molecule of H_2O. Second, the carbon of the C=O group bonds to the O of the alcohol.

$$\underset{\substack{\text{Ethanoic acid}\\\text{(Acetic acid)}}}{CH_3\overset{O}{\overset{\|}{C}}-OH} + \underset{\substack{\text{Ethanol}\\\text{(Ethyl alcohol)}}}{H-OCH_2CH_3} \underset{}{\overset{H_2SO_4}{\rightleftharpoons}} \underset{\substack{\text{Ethyl ethanoate}\\\text{(Ethyl acetate)}}}{CH_3\overset{O}{\overset{\|}{C}}OCH_2CH_3} + H_2O$$

(Removal of OH and H gives the ester)

Fischer esterification The process of forming an ester by refluxing a carboxylic acid and an alcohol in the presence of an acid catalyst, commonly sulfuric acid

Preparing an ester in this manner is given the special name **Fischer esterification**, after the German chemist Emil Fischer (1852–1919).

In the process of Fischer esterification, the alcohol adds to the carbonyl group of the carboxylic acid to form a tetrahedral carbonyl addition intermediate. Note how closely this step resembles the addition of an alcohol to the carbonyl group of an aldehyde or ketone to form a hemiacetal (Section 17.7). In the case of Fischer esterification, the intermediate then loses a molecule of water to give an ester.

$$\underset{\substack{}}{\overset{O}{\overset{\|}{CH_3C}}\underset{OH}{|}} + \overset{H}{\underset{}{\overset{|}{OCH_2CH_3}}} \overset{H_2SO_4}{\rightleftharpoons} \left[\underset{\substack{\text{A tetrahedral carbonyl}\\\text{addition intermediate}}}{\overset{O-H}{\underset{OH}{\overset{|}{CH_3C}-OCH_2CH_3}}}\right] \overset{H_2SO_4}{\rightleftharpoons} CH_3\overset{O}{\overset{\|}{C}}OCH_2CH_3 + H_2O$$

Fischer esterification is reversible, and generally, at equilibrium, the quantities of carboxylic acid and alcohol present are appreciable. By controlling reaction conditions, however, we can use Fischer esterification to prepare esters in high yields. If the alcohol is inexpensive, we can use a large excess of it to drive the equilibrium to the right and increase the yield of ester (recall Le Chatelier's principle; Section 7.8). Alternatively, we can remove water from the reaction mixture as it forms.

EXAMPLE 18.5

Complete these Fischer esterification reactions:

(a) C₆H₅—COH + CH₃OH ⇌ (H⁺)

Benzoic acid

(b) HO—C(=O)—CH₂CH₂—C(=O)—OH + CH₃CH₂OH ⇌ (H⁺)

Butanedioic acid (Succinic acid)   (excess)

Solution

Here is a structural formula for the ester produced in each reaction.

(a) C₆H₅—COCH₃

Methyl benzoate

(b) CH₃CH₂O—C(=O)—CH₂CH₂—C(=O)—OCH₂CH₃

Diethyl butanedioate (Diethyl succinate)

Problem 18.5

Complete these Fischer esterification reactions:

(a) HO—CH₂CH₂CH₂—C(=O)—OH ⇌ (H⁺) (a lactone)

(b) (CH₃)₂CH—C(=O)—OH + HO—C₆H₁₁ ⇌ (H⁺)

18.5 Preparation of Amides

In principle, we can form an amide by treating a carboxylic acid with an amine and removing —OH from the acid and —H from the amine. In practice, mixing the two leads to an acid–base reaction that forms an ammonium salt. If this salt is heated to a high enough temperature, water splits out and an amide forms.

Removal of OH and one H gives the amide

CH₃C(=O)—OH + H₂NCH₂CH₃ ⟶ CH₃C(=O)—O⁻ ⁺H₃NCH₂CH₃ —Heat→ CH₃C(=O)—NHCH₂CH₃ + H₂O

Acetic acid   Ethanamine (Ethylamine)   An ammonium salt   An amide

CHEMICAL CONNECTIONS 18C

From Willow Bark to Aspirin—and Beyond

The story of the development of this modern pain reliever goes back more than 2000 years. In 400 BCE, the Greek physician Hippocrates recommended chewing the bark of the willow tree to alleviate the pain of childbirth and to treat eye infections.

The active component of willow bark was found to be salicin, a compound composed of salicyl alcohol bonded to a unit of β-D-glucose (Section 19.3). Hydrolysis of salicin in aqueous acid followed by oxidation gave salicylic acid. Salicylic acid proved to be an even more effective reliever of pain, fever, and inflammation than salicin, and without the latter's extremely bitter taste. Unfortunately, patients quickly recognized salicylic acid's major side effect; it causes severe irritation of the mucous membrane lining of the stomach.

■ Two nonprescription pain relievers. The active ingredient in Advil is ibuprofen; that in Aleve is sodium naproxen.

Salicin

Salicylic acid

In the search for less irritating but still effective derivatives of salicylic acid, chemists at the Bayer division of I. G. Farben in Germany in 1883 treated salicylic acid with acetic anhydride and prepared acetylsalicylic acid. They gave this new compound the name *aspirin*.

Aleve. Both ibuprofen and naproxen have one stereocenter and can exist as a pair of enantiomers. For each drug, the active form is the s enantiomer. Naproxen is administered as its water-soluble sodium salt.

Acetylsalicylic acid (Aspirin)

(S)-Ibuprofen

(S)-Naproxen

Aspirin proved to be less irritating to the stomach than salicylic acid as well as more effective in relieving the pain and inflammation of rheumatoid arthritis. Aspirin, however, remains irritating to the stomach and frequent use of it can cause duodenal ulcers in susceptible persons.

In the 1960s, in a search for even more effective and less irritating analgesics and nonsteroidal anti-inflammatory drugs (NSAIDs), chemists at the Boots Pure Drug Company in England, who were studying compounds structurally related to salicylic acid, discovered an even more potent compound, which they named *ibuprofen*. Soon thereafter, Syntex Corporation in the United States developed naproxen, the active ingredient in

In the 1960s, researchers discovered that aspirin acts by inhibiting cyclooxygenase (COX), a key enzyme in the conversion of arachidonic acid to prostaglandins (Chemical Connections 20I). With this discovery, it became clear why only one enantiomer of ibuprofen and naproxen is active: Only the s enantiomer of each has the correct handedness to bind to COX and inhibit its activity.

It is much more common, however, to prepare amides by treating an amine with an anhydride.

$$\underset{\text{Acetic anhydride}}{CH_3\overset{O}{\overset{\|}{C}}-O-\overset{O}{\overset{\|}{C}}CH_3} + H_2NCH_2CH_3 \longrightarrow \underset{\text{An amide}}{CH_3\overset{O}{\overset{\|}{C}}-NHCH_2CH_3} + CH_3\overset{O}{\overset{\|}{C}}OH$$

18.6 Hydrolysis of Anhydrides, Esters, and Amides

A Anhydrides

Carboxylic anhydrides, particularly the low-molecular-weight ones, react readily with water to give two carboxylic acids.

Screen 18.6, 18.8 Simulations

B Esters

Esters are hydrolyzed only very slowly, even in boiling water. Hydrolysis becomes considerably more rapid, however, when the ester is heated in aqueous acid or base. When we discussed acid-catalyzed Fischer esterification in Section 18.4, we pointed out that it is an equilibrium reaction. Hydrolysis of esters in aqueous acid, also an equilibrium reaction, is the reverse of Fischer esterification. A large excess of water drives the equilibrium to the right to form the carboxylic acid and alcohol (Le Chatelier's principle; Section 7.8).

$$\underset{\text{Ethyl acetate}}{CH_3\overset{O}{\overset{\|}{C}}OCH_2CH_3} + H_2O \underset{}{\overset{H^+}{\rightleftharpoons}} \underset{\text{Acetic acid}}{CH_3\overset{O}{\overset{\|}{C}}OH} + \underset{\text{Ethanol}}{CH_3CH_2OH}$$

Hydrolysis of an ester can also be carried out using a hot aqueous base, such as aqueous NaOH. This reaction is often called **saponification,** a reference to its use in the manufacture of soaps (Chemical Connections 20C). The carboxylic acid formed in the hydrolysis reacts with hydroxide ion to form a carboxylic acid anion. Thus each mole of ester hydrolyzed requires one mole of base, as shown in the following balanced equation:

Saponification Hydrolysis of an ester in aqueous NaOH or KOH to an alcohol and the sodium or potassium salt of a carboxylic acid

$$\underset{\text{Ethyl acetate}}{CH_3\overset{O}{\overset{\|}{C}}OCH_2CH_3} + \underset{\substack{\text{Sodium}\\\text{hydroxide}}}{NaOH} \overset{H_2O}{\longrightarrow} \underset{\substack{\text{Sodium}\\\text{acetate}}}{CH_3\overset{O}{\overset{\|}{C}}O^-Na^+} + \underset{\text{Ethanol}}{CH_3CH_2OH}$$

EXAMPLE 18.6

Complete the equation for each hydrolysis reaction. Show the products as they are ionized under these experimental conditions.

CHEMICAL CONNECTIONS 18D

Ultraviolet Sunscreens and Sunblocks

Ultraviolet (UV) radiation (Section 9.1) penetrating the earth's ozone layer is arbitrarily divided into two regions: UVB (290–320 nm) and UVA (320–400 nm). UVB, a more energetic form of radiation than UVA, creates more radicals and hence does more oxidative damage (Section 24.7). UVB radiation interacts directly with biomolecules of the skin and eyes, causing skin cancer, skin aging, eye damage leading to cataracts, and delayed sunburn that appears 12–24 hours after exposure. UVA radiation, by contrast, causes tanning. It also damages skin, albeit much less efficiently than UVB. Its role in promoting skin cancer is less well understood.

Commercial sunscreen products are rated according to their sun protection factor (SPF), which is defined as the minimum effective dose of UV radiation that produces a delayed sunburn on protected skin compared to unprotected skin. Two types of active ingredients are found in commercial sunblocks and sunscreens. The most common sunblock agent is titanium dioxide, TiO_2, which reflects and scatters UV radiation. (TiO_2 is also widely used in paints because it covers better than any other inorganic white pigment.) Sunscreens, the second type of active ingredient, absorb UV radiation and then reradiate it as heat. These compounds are most effective in screening UVB radiation, but they do not screen UVA radiation. Thus they allow tanning but prevent the UVB-associated damage. Given here are structural formulas for three common esters used as UVB-screening agents, along with the name by which each is most commonly listed in the Active Ingredients label on commercial products:

Octyl *p*-methoxycinnamate

Homosalate

Padimate A

(a) [benzoic acid isopropyl ester] + NaOH $\xrightarrow{H_2O}$

(b) $CH_3COCH_2CH_2OCCH_3$ + 2NaOH $\xrightarrow{H_2O}$

Solution
The products of hydrolysis of compound (a) are benzoic acid and 2-propanol. In aqueous NaOH, benzoic acid is converted to its sodium salt. In this reaction, one mole of NaOH is required for hydrolysis of each mole of this ester. Compound (b) is a diester of ethylene glycol. Two moles of NaOH are required for its hydrolysis.

(a) [PhCOO⁻Na⁺] + HO—CH(CH₃)₂
Sodium benzoate   2-Propanol (Isopropyl alcohol)

(b) $2CH_3CO^-Na^+$ + $HOCH_2CH_2OH$
Sodium acetate   1,2-Ethanediol (Ethylene glycol)

Problem 18.6
Complete the equation for each hydrolysis reaction. Show all products as they are ionized under these experimental conditions.

(a) [structure: 1,2-benzenediyl bis(methyl ester)] + 2NaOH →(H₂O)

(b) [structure: ethyl 5-oxohexanoate] + H₂O →(HCl)

CHEMICAL CONNECTIONS 18E

Barbiturates

In 1864, Adolph von Baeyer (1835–1917) discovered that heating the diethyl ester of malonic acid with urea in the presence of sodium ethoxide (like sodium hydroxide, a strong base) gives a cyclic compound that he named *barbituric acid*.

Diethyl propanedioate (Diethyl malonate) + Urea

$\xrightarrow[\text{CH}_3\text{CH}_2\text{OH}]{\text{CH}_3\text{CH}_2\text{O}^-\text{Na}^+}$ Barbituric acid + 2CH₃CH₂OH

Pentobarbital

Sodium pentobarbital (Nembutal)

Phenobarbital

Some say that Baeyer named this compound after a friend named Barbara. Others claim that he named it after St. Barbara, the patron saint of artillerymen.

A number of derivatives of barbituric acid have powerful sedative and hypnotic effects. One such derivative is pentobarbital. Like the other derivatives of barbituric acid, pentobarbital is quite insoluble in water and body fluids. To increase its solubility in these fluids, pentobarbital is converted to its sodium salt, which is given the name Nembutal. Phenobarbital, also administered as its sodium salt, is an anticonvulsant, sedative, and hypnotic.

Technically speaking, only the sodium salts of these compounds should be called barbiturates. In practice, however, all derivatives of barbituric acid are called barbiturates, whether they are the un-ionized form or the ionized, water-soluble salt form.

Barbiturates have two principal effects. In small doses, they are sedatives (tranquilizers); in larger doses, they induce sleep. Barbituric acid, in contrast, has neither of these effects. Barbiturates are dangerous because they are addictive, which means that a regular user will suffer withdrawal symptoms when their use is stopped. They are especially dangerous when taken with alcohol because the combined effect (called a *synergistic effect*) is usually greater than the sum of the effects of either drug taken separately.

C Amides

Screen 18.10 Simulation

Amides require more vigorous conditions for hydrolysis in both acid and base than do esters. Hydrolysis in hot aqueous acid gives a carboxylic acid and an ammonium ion. This reaction is driven to completion by the acid–base reaction between ammonia or the amine and the acid to form an ammonium ion. For complete hydrolysis, one mole of acid is required per mole of amide.

$$CH_3CH_2CH_2CNH_2 + H_2O + HCl \xrightarrow[\text{heat}]{H_2O} CH_3CH_2CH_2COH + NH_4^+ \, Cl^-$$

Butanamide → Butanoic acid

In aqueous base, the products of amide hydrolysis are a carboxylic acid salt and ammonia or an amine. This hydrolysis is driven to completion by the acid–base reaction between the carboxylic acid and base to form a salt. To carry out this process, one mole of base is required per mole of amide.

$$CH_3CNH{-}\text{C}_6\text{H}_5 + NaOH \xrightarrow[\text{heat}]{H_2O} CH_3CO^-Na^+ + H_2N{-}\text{C}_6\text{H}_5$$

N-Phenylethanamide (N-Phenylacetamide, Acetanilide) → Sodium acetate + Aniline

EXAMPLE 18.7

Write a balanced equation for the hydrolysis of each amide in concentrated aqueous HCl. Show all products as they exist in aqueous HCl.

(a) $CH_3CN(CH_3)_2$ (b) [piperidinone: six-membered ring with NH and C=O]

Solution

(a) Hydrolysis of N,N-dimethylacetamide gives acetic acid and dimethylamine. Dimethylamine, a base, reacts with HCl to form dimethylammonium ion, shown here as dimethylammonium chloride.

$$CH_3CN(CH_3)_2 + H_2O + HCl \xrightarrow{\text{Heat}} CH_3COH + (CH_3)_2NH_2^+Cl^-$$

(b) Hydrolysis of this lactam gives the protonated form of 5-aminopentanoic acid.

$$\text{[lactam]} + H_2O + HCl \xrightarrow{\text{heat}} HOCCH_2CH_2CH_2CH_2NH_3^+Cl^-$$

> **Problem 18.7**
> Write a balanced equation for the hydrolysis of the amides in Example 18.7 in concentrated aqueous NaOH. Show all products as they exist in aqueous NaOH.

18.7 Phosphoric Anhydrides and Esters

A Phosphoric Anhydrides

Because of the special importance of phosphoric anhydrides in biochemical systems, we discuss them here to show the similarity between them and the anhydrides of carboxylic acids. The functional group of a **phosphoric anhydride** consists of two phosphoryl (P=O) groups bonded to the same oxygen atom. Shown here are structural formulas for two anhydrides of phosphoric acid and the ions derived by ionization of the acidic hydrogens of each:

At physiological pH (approximately 7.3–7.4), pyrophosphoric acid exists as the pyrophosphate ion and triphosphoric acid exists as the triphosphate ion.

$$\text{HO}-\overset{\overset{\text{O}}{\|}}{\underset{\underset{\text{OH}}{|}}{\text{P}}}-\text{O}-\overset{\overset{\text{O}}{\|}}{\underset{\underset{\text{OH}}{|}}{\text{P}}}-\text{OH} \qquad {}^-\text{O}-\overset{\overset{\text{O}}{\|}}{\underset{\underset{\text{O}^-}{|}}{\text{P}}}-\text{O}-\overset{\overset{\text{O}}{\|}}{\underset{\underset{\text{O}^-}{|}}{\text{P}}}-\text{O}^- \qquad \text{HO}-\overset{\overset{\text{O}}{\|}}{\underset{\underset{\text{OH}}{|}}{\text{P}}}-\text{O}-\overset{\overset{\text{O}}{\|}}{\underset{\underset{\text{OH}}{|}}{\text{P}}}-\text{O}-\overset{\overset{\text{O}}{\|}}{\underset{\underset{\text{OH}}{|}}{\text{P}}}-\text{OH} \qquad {}^-\text{O}-\overset{\overset{\text{O}}{\|}}{\underset{\underset{\text{O}^-}{|}}{\text{P}}}-\text{O}-\overset{\overset{\text{O}}{\|}}{\underset{\underset{\text{O}^-}{|}}{\text{P}}}-\text{O}-\overset{\overset{\text{O}}{\|}}{\underset{\underset{\text{O}^-}{|}}{\text{P}}}-\text{O}^-$$

Diphosphoric acid (Pyrophosphoric acid)   Diphosphate ion (Pyrophosphate ion)   Triphosphoric acid   Triphosphate ion

B Phosphoric Esters

Phosphoric acid has three —OH groups. It forms mono-, di-, and triphosphoric esters, which we name by giving the name(s) of the alkyl group(s) bonded to oxygen followed by the word "phosphate" (for example, dimethyl phosphate). In more complex **phosphoric esters**, it is common practice to name the organic molecule and then indicate the presence of the phosphoric ester by including either the word "phosphate" or the prefix *phospho-*. Dihydroxyacetone phosphate, for example, is an intermediate in glycolysis (Section 27.2). Pyridoxal phosphate is one of the metabolically active forms of vitamin B_6. The last two phosphoric esters are shown here as they are ionized at pH 7.4, the pH of blood plasma.

$$\text{CH}_3\text{O}-\overset{\overset{\text{O}}{\|}}{\underset{\underset{\text{OCH}_3}{|}}{\text{P}}}-\text{OH} \qquad \begin{array}{c}\text{CH}_2\text{OH}\\|\\\text{C}=\text{O}\\|\\\text{CH}_2-\text{O}-\overset{\overset{\text{O}}{\|}}{\underset{\underset{\text{O}^-}{|}}{\text{P}}}-\text{O}^-\end{array}$$

Dimethyl phosphate   Dihydroxyacetone phosphate   Pyridoxal phosphate

■ Vitamin B_6, pyridoxal.

18.8 Step-Growth Polymerizations

Step-growth polymers form from the reaction of molecules containing two functional groups, with each new bond being created in a separate step. In this section, we discuss three types of step-growth polymers: polyamides, polyesters, and polycarbonates.

> **Step-growth polymerization**
> A polymerization in which chain growth occurs in a stepwise manner between difunctional monomers, as, for example, between adipic acid and hexamethylenediamine to form nylon-66

A Polyamides

Polyamide A polymer in which each monomer unit is bonded to the next by an amide bond—for example, nylon-66

In the early 1930s, chemists at E. I. DuPont de Nemours & Company began fundamental research into the reactions between dicarboxylic acids and diamines to form **polyamides**. In 1934, they synthesized nylon-66, the first purely synthetic fiber. Nylon-66 is so named because it is synthesized from two different monomers, each containing six carbon atoms.

In the synthesis of nylon-66, hexanedioic acid and 1,6-hexanediamine are dissolved in aqueous ethanol and then heated in an autoclave to 250°C and an internal pressure of 15 atm. Under these conditions, —COOH and —NH_2 groups react by loss of H_2O to form a polyamide, similar to the formation of amides described in Section 18.5.

Hexanedioic acid
(Adipic acid)

1,6-Hexanediamine
(Hexamethylenediamine)

Nylon-66
(a polyamide)

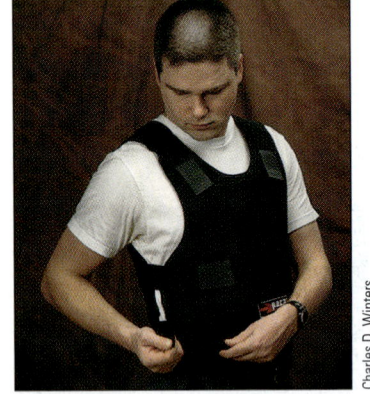

■ Bulletproof vests have a thick layer of Kevlar.

Based on extensive research into the relationships between molecular structure and bulk physical properties, scientists at DuPont reasoned that a polyamide containing benzene rings would be even stronger than nylon-66. Their line of reasoning eventually produced the following polyamide that DuPont named Kevlar.

1,4-Benzenedicarboxylic acid
(Terephthalic acid)

1,4-Benzenediamine
(*p*-Phenylenediamine)

Kevlar
(a polyaromatic amide)

One remarkable feature of Kevlar is that it weighs less than other materials of similar strength. For example, a cable woven of Kevlar has a strength equal to that of a similarly woven steel cable. Yet the Kevlar cable has only 20% of the weight of the steel cable! Kevlar now finds use in such articles as anchor cables for offshore drilling rigs and reinforcement fibers for automobile tires. It is also woven into a fabric that is so tough that it can be used for bulletproof vests, jackets, and raincoats.

B Polyesters

Polyester A polymer in which each monomer unit is bonded to the next by an ester bond—for example, poly(ethylene terephthalate)

The first **polyester**, developed in the 1940s, involved polymerization of benzene 1,4-dicarboxylic acid with 1,2-ethanediol to give poly(ethylene terephthalate), abbreviated PET. Virtually all PET is now made from the dimethyl ester of terephthalic acid by the following reaction:

18.8 STEP-GROWTH POLYMERIZATIONS

Dimethyl terephthalate + 1,2-Ethanediol (Ethylene glycol) → Poly(ethylene terephthalate) (Dacron, Mylar), Remove CH$_3$OH, Heat −CH$_3$OH

The crude polyester can be melted, extruded, and then drawn to form the textile fiber Dacron polyester. Dacron's outstanding features include its stiffness (about four times that of nylon-66), very high tensile strength, and remarkable resistance to creasing and wrinkling. Because the early Dacron polyester fibers were harsh to the touch due to their stiffness, they were usually blended with cotton or wool to make acceptable textile fibers. Newly developed fabrication techniques now produce less harsh Dacron polyester textile fibers. PET is also fabricated into Mylar films and recyclable plastic beverage containers.

C Polycarbonates

A **polycarbonate**, the most familiar of which is Lexan, forms from the reaction of the disodium salt of bisphenol A and phosgene. Phosgene is a derivative of carbonic acid, H$_2$CO$_3$, in which both —OH groups have been replaced by chlorine atoms. An ester of carbonic acid is called a carbonate.

Carbonic acid (H$_2$CO$_3$)   Phosgene   Diethyl carbonate (a carbonate ester)

■ Mylar can be made into extremely strong films. Because the film has very tiny pores, it is used for balloons that can be inflated with helium; the helium atoms diffuse only slowly through the pores of the film.

Polycarbonate A polyester in which the carboxyl groups are derived from carbonic acid

---

CHEMICAL CONNECTIONS 18F

Stitches That Dissolve

As the technological capabilities of medicine have grown, the demand for synthetic materials that can be used inside the body has increased as well. Polymers already have many of the characteristics of an ideal biomaterial: They are lightweight and strong, are inert or biodegradable depending on their chemical structure, and have physical properties (softness, rigidity, elasticity) that are easily tailored to match those of natural tissues.

Even though most medical uses of polymeric materials require biostability, some applications require them to be biodegradable. An example is the polymer of glycolic acid and lactic acid used in absorbable sutures, which are marketed under the trade name of Lactomer.

Traditional suture materials such as catgut must be removed by a health care specialist after they have served their purpose. Stitches of Lactomer, however, are hydrolyzed slowly over a period of approximately two weeks. By the time the torn tissues have healed, the stitches have hydrolyzed, and no suture removal is necessary. The body metabolizes and excretes the glycolic and lactic acids formed during this hydrolysis.

Glycolic acid + Lactic acid →(Polymerization, −nH$_2$O) A polymer of glycolic acid and lactic acid, Remove H$_2$O

■ A polycarbonate hockey mask.

In forming a polycarbonate, each mole of phosgene reacts with two moles of the sodium salt of a phenol called bisphenol A:

Disodium salt of Bisphenol A + Phosgene → Lexan (a polycarbonate) + NaCl

Lexan is a tough, transparent polymer that has high impact and tensile strengths and retains its properties over a wide temperature range. It is used in sporting equipment (helmets and face masks); to make light, impact-resistant housings for household appliances; and in the manufacture of safety glass and unbreakable windows.

SUMMARY

The functional group of a **carboxylic acid** is the **carboxyl group**, —**COOH** (Section 18.1). IUPAC names of carboxylic acids are derived from the name of the parent alkane by dropping the suffix *-e* and adding *-oic acid*. Dicarboxylic acids are named as *-dioic acids* (Section 18.2A). Common names for many carboxylic and dicarboxylic acids remain in wide use.

Carboxylic acids are polar compounds; consequently, they have higher boiling points and are more soluble in water than alcohols, aldehydes, ketones, and ethers of comparable molecular weight (Section 18.2C). Carboxylic acids are weak acids that react with strong bases to form salts.

An **anhydride** contains two carbonyl groups bonded to the same oxygen (Section 18.3A). An **ester** contains a carbonyl group bonded to an —OR group derived from an alcohol or phenol (Section 18.3B). An **amide** contains a carbonyl group bonded to a nitrogen atom derived from an amine (Section 18.3C).

Step-growth polymerizations involve the stepwise reaction of difunctional monomers (Section 18.8). Important commercial polymers synthesized through step-growth processes include polyamides, polyesters, and polycarbonates.

KEY REACTIONS

1. **Acidity of Carboxylic Acids (Section 18.2D)** Values of pK_a for most unsubstituted aliphatic and aromatic carboxylic acids are within the range of 4.0 to 5.0.

$$CH_3COOH + H_2O \rightleftharpoons CH_3CO^- + H_3O^+ \quad pK_a = 4.76$$

2. **Reaction of Carboxylic Acids with Bases (Section 18.2E)** Carboxylic acids form water-soluble salts with alkali metal hydroxides, carbonates, bicarbonates, ammonia, and amines.

$$PhCOOH + NaOH \xrightarrow{H_2O} PhCO^-Na^+ + H_2O$$

3. **Fischer Esterification (Section 18.4)** Fischer esterification is reversible. To achieve high yields of ester, it is necessary to force the equilibrium to the right. One way to accomplish this goal is to use an excess of the alcohol. Another way is to remove water as it is formed.

$$CH_3COOH + CH_3CH_2CH_2OH \xrightleftharpoons{H_2SO_4} CH_3COOCH_2CH_2CH_3 + H_2O$$

4. **Preparation of an Amide (Section 18.5)** Reaction of an amine with an anhydride gives an amide.

$$CH_3C-O-CCH_3 + H_2NCH_2CH_3$$
Acetic anhydride

$$\longrightarrow CH_3C-NHCH_2CH_3 + CH_3COOH$$
An amide

5. **Hydrolysis of an Ester (Section 18.6B)** Esters are hydrolyzed rapidly only in the presence of acid or base. Acid is a catalyst. Base is required in an equimolar amount.

$$CH_3CO-\text{(cyclohexyl)} + NaOH$$

$$\xrightarrow{H_2O} CH_3CO^-Na^+ + HO-\text{(cyclohexyl)}$$

6. **Hydrolysis of an Amide (Section 18.5C)** Either acid or base is required in an amount equivalent to that of the amide.

$$CH_3CH_2CH_2CNH_2 + H_2O + HCl$$

$$\xrightarrow[heat]{H_2O} CH_3CH_2CH_2COOH + NH_4^+ Cl^-$$

$$CH_3CH_2CH_2CNH_2 + NaOH$$

$$\xrightarrow[heat]{H_2O} CH_3CH_2CH_2CO^- Na^+ + NH_3$$

PROBLEMS

Numbers that appear in color indicate difficult problems.
▲ designates problems requiring application of principles.

Structure and Nomenclature

18.8 Name and draw structural formulas for the four carboxylic acids with molecular formula $C_5H_{10}O_2$. Which of these carboxylic acids are chiral?

18.9 Write the IUPAC name for each carboxylic acid.

(a) 3,4-dimethylpentanoic acid structure with OH

(b) 2-aminobutanoic acid structure with NH_2 and OH

(c) pentanoic acid structure with COOH

18.10 Write the IUPAC name for each carboxylic acid.

(a) HOOC–CH(OH)–COOH

(b) [benzene ring with COOH and OH in ortho positions]

(c) CCl_3COOH

18.11 Draw a structural formula for each carboxylic acid.
(a) 4-Nitrophenylacetic acid
(b) 4-Aminobutanoic acid
(c) 4-Phenylbutanoic acid
(d) cis-3-Hexenedioic acid

18.12 Draw a structural formula for each carboxylic acid.
(a) 2-Aminopropanoic acid
(b) 3,5-Dinitrobenzoic acid
(c) Dichloroacetic acid
(d) o-Aminobenzoic acid

18.13 Draw a structural formula for each salt.
(a) Sodium benzoate
(b) Lithium acetate
(c) Ammonium acetate
(d) Disodium adipate
(e) Sodium salicylate
(f) Calcium butanoate

18.14 Calcium oxalate is a major component of kidney stones. Draw a structural formula for this compound.

18.15 The monopotassium salt of oxalic acid is present in certain leafy vegetables, including rhubarb. Both oxalic acid and its salts are poisonous in high concentrations. Draw a structural formula for monopotassium oxalate.

Physical Properties

18.16 Draw a structural formula for the dimer formed when two molecules of formic acid form hydrogen bonds with each other.

18.17 Propanedioic (malonic) acid forms an internal hydrogen bond in which the H atom of one COOH group forms a hydrogen bond with an O atom of the other COOH group. Draw a structural formula to show this internal hydrogen bonding. (There are two possible answers.)

18.18 Hexanoic (caproic) acid has a solubility in water of about 1 g/100 mL of water. Which part of the molecule contributes to water solubility, and which part prevents solubility?

18.19 The following compounds all have approximately the same molecular weight: hexanoic acid, heptanal, and 1-heptanol. Arrange them in order of increasing boiling point.

18.20 The following compounds all have approximately the same molecular weight: propanoic acid, 1-butanol, and diethyl ether. Arrange them in order of increasing boiling point.

18.21 Arrange these compounds in order of increasing solubility in water: acetic acid, pentanoic acid, and decanoic acid.

Preparation of Carboxylic Acids

18.22 Complete the equations for these oxidations.

(a) $CH_3(CH_2)_4CH_2OH \xrightarrow[H_2SO_4]{K_2Cr_2O_7}$

(b) [benzene ring with CHO, OCH₃, OH substituents] + $Ag(NH_3)_2^+ \longrightarrow$

(c) HO–[cyclohexane ring]–$CH_2OH \xrightarrow[H_2SO_4]{K_2Cr_2O_7}$

18.23 Draw the structural formula for a compound with the given molecular formula that, on oxidation by potassium dichromate, gives the carboxylic acid or dicarboxylic acid shown.

(a) $C_6H_{14}O \xrightarrow{[O]} CH_3(CH_2)_4COH$ (with C=O)

(b) $C_6H_{12}O \xrightarrow{[O]} CH_3(CH_2)_4COH$ (with C=O)

(c) $C_6H_{14}O_2 \xrightarrow{[O]} HOC(CH_2)_4COH$ (with both C=O)

Acidity of Carboxylic Acids

18.24 Alcohols, phenols, and carboxylic acids all contain an —OH group. Which are the strongest acids? Which are the weakest acids?

18.25 Arrange these compounds in order of increasing acidity: benzoic acid, benzyl alcohol, and phenol.

18.26 Complete the equations for these acid–base reactions.

(a) [phenyl]–$CH_2COOH + NaOH \longrightarrow$

(b) [CH=CH chain]–$COOH + NaHCO_3 \longrightarrow$

(c) [benzene ring with COOH and OCH₃ in ortho positions] + $NaHCO_3 \longrightarrow$

(d) $CH_3\overset{\overset{OH}{|}}{C}HCOOH + H_2NCH_2CH_2OH \longrightarrow$

(e) $\diagup\!\!\!\diagdown\!\!\!\diagup COO^-Na^+ + HCl \longrightarrow$

18.27 Complete the equations for these acid-base reactions.

(a) [2-methylphenol (o-cresol)] + NaOH ⟶

(b) [2-hydroxybenzoate sodium salt] + HCl ⟶

(c) [2-methoxybenzoic acid] + $H_2NCH_2CH_2OH \longrightarrow$

(d) [cyclohexyl]—COOH + $NaHCO_3 \longrightarrow$

18.28 Formic acid is one of the components responsible for the sting of biting ants and is injected under the skin by bee and wasp stings. The pain can be relieved by rubbing the area of the sting with a paste of baking soda ($NaHCO_3$) and water that neutralizes the acid. Write an equation for this reaction.

18.29 The pK_a of acetic acid is 4.76. What form(s) of acetic acid are present at pH 2.0 or lower? At pH 4.76? At pH 8.0 or higher?

18.30 The normal pH range for blood plasma is 7.35 to 7.45. Under these conditions, would you expect the carboxyl group of lactic acid (pK_a 4.07) to exist primarily as a carboxyl group or as a carboxylic anion? Explain.

18.31 The pK_a of ascorbic acid (Chemical Connections 19B) is 4.10. Would you expect ascorbic acid dissolved in blood plasma, pH 7.35 to 7.45, to exist primarily as ascorbic acid or as ascorbate anion? Explain.

18.32 In Chapter 21, we discuss a class of compounds called amino acids, so named because they contain both an amino group and a carboxyl group. Following is a structural formula for the amino acid alanine.

$$CH_3\overset{\overset{}{|}}{\underset{\underset{NH_3^+}{|}}{C}}HCO^- \quad \overset{\overset{O}{\|}}{}$$

Alanine

What would you expect to be the major form of alanine present in aqueous solution at (a) pH 2.0, (b) at pH 5–6, and (c) at pH 11.0? Explain.

18.33 Complete the equations for the following acid–base reactions. Assume one mole of NaOH per mole of amino acid.

(a) $CH_3\overset{\overset{}{|}}{\underset{\underset{NH_3^+}{|}}{C}}HCOOH + NaOH \longrightarrow$

(b) $CH_3\overset{\overset{}{|}}{\underset{\underset{NH_3^+}{|}}{C}}HCOO^-Na^+ + NaOH \longrightarrow$

18.34 Which is the stronger base: $CH_3CH_2NH_2$ or $CH_3CH_2COO^-$? Explain.

18.35 Complete the equations for the following acid–base reactions. Assume one mole of HCl per mole of amino acid.

(a) $CH_3\overset{\overset{}{|}}{\underset{\underset{NH_2}{|}}{C}}HCOO^-Na^+ + HCl \longrightarrow$

(b) $CH_3\overset{\overset{}{|}}{\underset{\underset{NH_3^+}{|}}{C}}HCOO^-Na^+ + HCl \longrightarrow$

Reactions of Carboxylic Acids

18.36 Define and give an example of Fischer esterification.

18.37 Complete these examples of Fischer esterification. In each case, assume an excess of alcohol.

(a) $CH_3COOH + HO\text{—}[isobutyl] \overset{H^+}{\rightleftharpoons}$

(b) $CH_3COOH + HO\text{—}[cyclohexyl] \overset{H^+}{\rightleftharpoons}$

(c) [phthalic acid, benzene-1,2-dicarboxylic acid] + $CH_3CH_2OH \overset{H^+}{\rightleftharpoons}$

18.38 Complete these examples of Fischer esterification. In each case, assume an excess of the alcohol.

(a) [2-methoxybenzoic acid] + $HOCH_2CH_2NH_2 \overset{H^+}{\rightleftharpoons}$

(b) [cyclopentyl]—COOH + $(CH_3)_2CHOH \overset{H^+}{\rightleftharpoons}$

(c) $HO\overset{\overset{O}{\|}}{C}\text{—}CH_2CH_2CH_2\text{—}\overset{\overset{O}{\|}}{C}OH + CH_3OH \overset{H^+}{\rightleftharpoons}$

462 | CHAPTER 18 CARBOXYLIC ACIDS, ANHYDRIDES, ESTERS, AND AMIDES

18.39 From what carboxylic acid and alcohol is each ester derived?

(a) $CH_3CO-\text{(cyclohexane-1,4-diyl)}-OCCH_3$ (both with C=O)

(b) cyclohexyl-$COCH_3$ (with C=O)

(c) $CH_3OCCH_2CH_2COCH_3$ (both with C=O)

(d) (trans-2-pentenoate ester of isobutyl alcohol)

18.40 Methyl 2-hydroxybenzoate (methyl salicylate) has the odor of oil of wintergreen. This compound is prepared by Fischer esterification of 2-hydroxybenzoic acid (salicylic acid) with methanol. Draw a structural formula for methyl 2-hydroxybenzoate.

18.41 Methyl 4-hydroxybenzoate (methylparaben) is used as a preservative in foods, beverages, and cosmetics. This compound is prepared by Fischer esterification of 4-hydroxybenzoic acid with methanol. Draw a structural formula for methylparaben.

18.42 When 5-hydroxypentanoic acid is treated with an acid catalyst, it forms a lactone (a cyclic ester). Draw a structural formula for this lactone.

18.43 From what carboxylic acid and amine or ammonia can each amide be synthesized?

(a) cyclohexyl-$NHC(=O)(CH_2)_4CH_3$

(b) $(CH_3)_2CHCN(CH_3)_2$ (with C=O)

(c) $H_2NC(CH_2)_4CNH_2$ (both with C=O)

18.44 *N,N*-Diethyl *m*-toluamide (DEET) is the active ingredient in several common insect repellents. From what acid and amine can DEET by synthesized?

N,N-Diethyl *m*-toluamide (DEET)

18.45 Following are structural formulas for two local anesthetics. Lidocaine was introduced in 1948 and is now the most widely used local anesthetic for infiltration and regional anesthesia. Its hydrochloride salt is marketed under the name Xylocaine. Mepivacaine is faster acting and somewhat longer in duration than lidocaine. Its hydrochloride salt is marketed under the name Carbocaine.

Lidocaine (Xylocaine)   Mepivacaine (Carbocaine)

(a) Name the functional groups in each anesthetic.
(b) What similarities in structure do you find between these compounds?

Structure and Nomenclature of Anhydrides, Esters, and Amides

18.46 Draw a structural formula for each compound.
(a) Diethyl carbonate
(b) *p*-Nitrobenzamide
(c) Ethyl 3-hydroxybutanoate
(d) Diethyl oxalate
(e) Ethyl *trans*-2-pentenoate
(f) Butanoic anhydride

18.47 Flavoring agents are the largest class of food additives. Each ester listed here is a synthetic flavor additive, with a flavor very close to the natural flavor. While these esters are the major components of the natural flavors, each natural flavor contains many more components, including some in only trace amounts. Draw a structural formula for each ester. (Isopentane is the common name for 2-methylbutane.)
(a) Ethyl formate (rum)
(b) Isopentyl acetate (banana)
(c) Octyl acetate (orange)
(d) Methyl butanoate (apple)
(e) Ethyl butanoate (pineapple)
(f) Methyl 2-aminobenzoate (grape)

18.48 Write the IUPAC name for each compound.

(a) $C_6H_5-COC-C_6H_5$ (anhydride, both C=O)

(b) $CH_3(CH_2)_8COCH_3$ (with C=O)

(c) $CH_3(CH_2)_4CNHCH_3$ (with C=O)

(d) H₂N—C₆H₄—C(=O)NH₂

(e) CH₃C(=O)O—cyclopentyl

(f) CH₃CH(OH)CH₂C(=O)OCH₂CH₃

Reactions of Anhydrides, Esters, and Amides

18.49 Define and give an example of saponification.

18.50 What product forms when ethyl benzoate is treated with each reagent?
(a) H₂O, NaOH, heat
(b) H₂O, HCl, heat

18.51 What product forms when benzamide is treated with each reagent?
(a) H₂O, NaOH, heat
(b) H₂O, HCl, heat

18.52 Which of these types of compounds will produce bubbles of CO₂ when added to an aqueous solution of sodium bicarbonate?
(a) A carboxylic acid
(b) A carboxylic ester
(c) The sodium salt of a carboxylic acid

18.53 Complete the equations for these reactions.

(a) CH₃O—C₆H₄—NH₂ + CH₃C(=O)OC(=O)CH₃ ⟶

(b) piperidine-NH + CH₃C(=O)OC(=O)CH₃ ⟶

18.54 The analgesic phenacetin is synthesized by treating 4-ethoxyaniline with acetic anhydride. Draw a structural formula for phenacetin.

CH₃CH₂O—C₆H₄—NH₂

4-Ethoxyaniline

18.55 Phenobarbital is a long-acting sedative, hypnotic, and anticonvulsant.

Phenobarbital

(a) Name all functional groups in this compound.
(b) Draw structural formulas for the products from complete hydrolysis of all amide groups in aqueous NaOH.

18.56 Following is a structural formula for aspartame, an artificial sweetener about 180 times as sweet as sucrose (table sugar):

Aspartame

(a) Name all functional groups in aspartame.
(b) Is aspartame chiral? If so, how many stereoisomers are possible for it?
(c) Draw structural formulas for the products formed on hydrolysis of all ester and amide bonds in aspartame.

18.57 Why are nylon-66 and Kevlar referred to as polyamides?

18.58 Draw short sections of two parallel chains of nylon-66 (each chain running in the same direction) and show how it is possible to align them such that there is hydrogen bonding between the N—H groups of one chain and C=O groups of the parallel chain.

18.59 Why are Dacron and Mylar referred to as polyesters?

Anhydrides and Esters of Phosphoric Acid

18.60 What type of structural feature do the anhydrides of phosphoric acid and carboxylic acids have in common?

18.61 Draw structural formulas for the mono-, di-, and triethyl esters of phosphoric acid.

18.62 1,3-Dihydroxy-2-propanone (dihydroxyacetone) and phosphoric acid form a monoester called dihydroxyacetone phosphate, which is an intermediate in glycolysis (Section 27.2). Draw a structural formula for this monophosphate ester.

18.63 Show how triphosphoric acid can form from three molecules of phosphoric acid. How many H₂O molecules are split out?

18.64 Write an equation for the hydrolysis of trimethyl phosphate to dimethyl phosphate and methanol.

Chemical Connections

18.65 (Chemical Connections 18A) Locate the ester group in pyrethrin I and draw a structural formula for chrysanthemic acid, the carboxylic acid from which this ester is derived.

18.66 (Chemical Connections 18A) What structural features do pyrethrin I (a natural insecticide) and permethrin (a synthetic pyrethrenoid) have in common?

18.67 (Chemical Connections 18A) A commercial Clothing & Gear Insect Repellant gives the following information about permethrin, its active ingredient:

Cis/trans ratio: Minimum 35% (+/−) cis and maximum 65% (+/−) trans

(a) To what does the cis/trans ratio refer?
(b) To what does the designation "(+/−)" refer?

18.68 (Chemical Connections 18B) Identify the β-lactam portion of amoxicillin and cephalexin.

18.69 (Chemical Connections 18C) What is the compound in willow bark that is responsible for its ability to relieve pain? How is this compound related to salicylic acid?

18.70 (Chemical Connections 18C) Name the two functional groups in aspirin.

18.71 (Chemical Connections 18C) Once it has been opened, and particularly if it has been left open to the air, a bottle of aspirin may develop a vinegar-like odor. Explain how this might happen.

18.72 (Chemical Connections 18C) What is the structural relationship between aspirin and ibuprofen? Between aspirin and naproxen?

18.73 (Chemical Connections 18D) What is the difference in meaning between *sunblock* and *sunscreen*?

18.74 (Chemical Connections 18D) How do sunscreens prevent UV radiation from reaching the skin?

18.75 (Chemical Connections 18D) What structural features do the three sunscreens given in this box have in common?

18.76 (Chemical Connections 18E) Barbiturates are derived from urea. Identify the portion of the structure of pentobarbital and phenobarbital that is derived from urea.

18.77 (Chemical Connections 18F) Why do Lactomer stitches dissolve within two to three weeks following surgery?

Additional Problems

18.78 Potassium sorbate is added as a preservative to certain foods to prevent bacteria and molds from causing food spoilage and to extend the foods' shelf lives. The IUPAC name of potassium sorbate is potassium (*trans,trans*-2,4-hexadienoate). Draw a structural formula of potassium sorbate.

18.79 Propanoic acid and methyl acetate are constitutional isomers, and both are liquids at room temperature. One of these compounds has a boiling point of 141°C; the other has a boiling point of 57°C. Which compound has which boiling point? Explain.

18.80 Excess ascorbic acid is excreted in the urine, the pH of which is normally in the range 4.8–8.4. What form of ascorbic acid (pK_a 4.10) would you expect to be present in urine of pH 8.4: ascorbic acid or ascorbate anion? Explain.

18.81 The pH of human gastric juice is normally in the range 1.0–3.0. What form of lactic acid (pK_a 4.07), would you expect to be present in the stomach: lactic acid or its anion? Explain.

18.82 Benzocaine, a topical anesthetic, is prepared by treatment of 4-aminobenzoic acid with ethanol in the presence of an acid catalyst followed by neutralization. Draw a structural formula for benzocaine.

18.83 The analgesic acetaminophen is synthesized by treating 4-aminophenol with one equivalent of acetic anhydride. Write an equation for the formation of acetaminophen. (*Hint:* The —NH_2 group is more reactive with acetic anhydride than the —OH group.)

18.84 Procaine (its hydrochloride is marketed as Novocain) was one of the first local anesthetics developed for infiltration and regional anesthesia. It is synthesized by the following Fischer esterification. Draw a structural formula for procaine.

H_2N—⟨benzene⟩—C(=O)OH + HO—CH_2CH_2—N(C_2H_5)$_2$

p-Aminobenzoic acid   2-Diethylaminoethanol

$\xrightarrow{\text{Fischer esterification}}$ Procaine

18.85 1,3-Diphosphoglycerate, an intermediate in glycolysis (Section 27.2), contains a mixed anhydride (an anhydride of a carboxylic acid and phosphoric acid) and a phosphoric ester. Draw structural formulas for the products formed by hydrolysis of the anhydride and ester bonds. Show each product as it would exist in solution at pH 7.4.

$$\begin{array}{c} \text{O} \quad\quad \text{O} \\ \overset{1}{\text{C}}-\text{O}-\text{P}-\text{O}^- \\ | \quad\quad\quad | \\ \quad\quad\quad \text{O}^- \\ \overset{2}{\text{HOCH}} \\ | \quad\quad\quad \text{O} \\ \overset{3}{\text{CH}_2}-\text{O}-\text{P}-\text{O}^- \\ \quad\quad\quad | \\ \quad\quad\quad \text{O}^- \end{array}$$

1,3-Diphosphoglycerate

InfoTrac College Edition

For additional readings, go to InfoTrac College Edition, your online research library, at

http://infotrac.thomsonlearning.com

PART 3 BIOCHEMISTRY

Chapter 19
Carbohydrates

Chapter 20
Lipids

Chapter 21
Proteins

Chapter 22
Enzymes

Chapter 23
Chemical Communications:
Neurotransmitters and Hormones

Chapter 24
Nucleotides, Nucleic Acids,
and Heredity

Chapter 25
Gene Expression
and Protein Synthesis

Chapter 26
Bioenergetics. How the Body
Converts Food to Energy

Chapter 27
Specific Catabolic Pathways:
Carbohydrate, Lipid, and
Protein Metabolism

Chapter 28
Biosynthetic Pathways

Chapter 29
Nutrition

Chapter 30
Immunochemistry

Chapter 31
Body Fluids

CHAPTER 19

19.1 Introduction
19.2 Monosaccharides
19.3 The Cyclic Structure of Monosaccharides
19.4 Reactions of Monosaccharides
19.5 Disaccharides and Oligosaccharides
19.6 Polysaccharides
19.7 Acidic Polysaccharides

Breads, grains, and pasta are sources of carbohydrates.

Carbohydrates

19.1 Introduction

Carbohydrates are the most abundant organic compounds in the plant world. They act as storehouses of chemical energy (glucose, starch, glycogen); are components of supportive structures in plants (cellulose), crustacean shells (chitin), and connective tissues in animals (acidic polysaccharides); and are essential components of nucleic acids (D-ribose and 2-deoxy-D-ribose). Carbohydrates account for approximately three-fourths of the dry weight of plants. Animals (including humans) get their carbohydrates by eating plants, but they do not store much of what they consume. In fact, less than 1% of the body weight of animals is made up of carbohydrates.

The word *carbohydrate,* means "hydrate of carbon" and derives from the formula $C_n(H_2O)_m$. Two examples of carbohydrates with molecular formulas that can be written alternatively as hydrates of carbon are

- Glucose (blood sugar): $C_6H_{12}O_6$, which can be written as $C_6(H_2O)_6$
- Sucrose (table sugar): $C_{12}H_{22}O_{11}$, which can be written as $C_{12}(H_2O)_{11}$

Not all carbohydrates, however, have this general formula. Some contain too few oxygen atoms to fit it; others contain too many oxygens. Some also contain nitrogen. The term *carbohydrate* has become so firmly rooted in the chemical nomenclature that, although not completely accurate, it persists as the name for this class of compounds.

At the molecular level, most **carbohydrates** are polyhydroxyaldehydes, polyhydroxyketones, or compounds that yield them after hydrolysis. The simpler members of the carbohydrate family are often referred to as **saccharides** because of their sweet taste (Latin: *saccharum,* sugar). Carbohydrates are classified as monosaccharides, oligosaccharides, or polysaccharides depending on the number of simple sugars they contain.

> **Carbohydrate** A polyhydroxyaldehyde or polyhydroxyketone, or a substance that gives these compounds on hydrolysis

19.2 Monosaccharides

A Structure and Nomenclature

Monosaccharides have the general formula $C_nH_{2n}O_n$, with one of the carbons being the carbonyl group of either an aldehyde or a ketone. The most common monosaccharides have three to nine carbon atoms. The suffix **-ose** indicates that a molecule is a carbohydrate, and the prefixes **tri-, tetr-, pent-,** and so forth indicate the number of carbon atoms in the chain. Monosaccharides containing an aldehyde group are classified as **aldoses;** those containing a ketone group are classified as **ketoses.**

There are only two trioses: the aldotriose glyceraldehyde and the ketotriose dihydroxyacetone.

> **Monosaccharide** A carbohydrate that cannot be hydrolyzed to a simpler compound

> **Aldose** A monosaccharide containing an aldehyde group

```
      CHO              CH₂OH
       |                |
      CHOH             C=O
       |                |
      CH₂OH            CH₂OH
   Glyceraldehyde   Dihydroxyacetone
   (an aldotriose)   (a ketotriose)
```

> **Ketose** A monosaccharide containing a ketone group

Often the designations *aldo-* and *keto-* are omitted, and these molecules are referred to simply as trioses, tetroses, and the like.

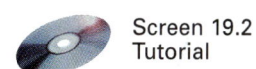

Screen 19.2 Tutorial

B Fischer Projection Formulas

Glyceraldehyde contains a stereocenter and therefore exists as a pair of enantiomers (Figure 19.1).

Chemists commonly use two-dimensional representations called **Fischer projections** to show the configuration of carbohydrates. To draw a Fischer projection, you draw a three-dimensional representation of the molecule oriented so that the vertical bonds from the stereocenter are directed away from you and the horizontal bonds from it are directed toward you.

> **Fischer projection** A two-dimensional representation for showing the configuration of a stereocenter; horizontal lines represent bonds projecting forward from the stereocenter, and vertical lines represent bonds projecting toward the rear

Figure 19.1 The enantiomers of glyceraldehyde.

You then write the molecule as a cross, with the stereocenter indicated by the point at which the bonds cross.

$$\underset{\substack{\text{(R)-Glyceraldehyde}\\\text{(three-dimensional}\\\text{representation)}}}{\overset{\text{CHO}}{\underset{\text{CH}_2\text{OH}}{\text{H}-\text{C}-\text{OH}}}} \xrightarrow{\text{convert to a Fischer projection}} \underset{\substack{\text{(R)-Glyceraldehyde}\\\text{(Fischer projection)}}}{\overset{\text{CHO}}{\underset{\text{CH}_2\text{OH}}{\text{H}-\!\!\!\!\!\!+\!\!\!\!\!\!-\text{OH}}}}$$

The horizontal segments of this Fischer projection represent bonds directed toward you, and the vertical segments represent bonds directed away from you. The only atom in the plane of the paper is the stereocenter.

C D- and L-Monosaccharides

Even though the R,S system is widely accepted today as a standard for designating configuration, the configuration of carbohydrates is commonly designated using the D,L system proposed by Emil Fischer in 1891. At that time, it was known that one enantiomer of glyceraldehyde has a specific rotation (Section 15.5B) of +13.5°; the other has a specific rotation of −13.5°. Fischer proposed that these enantiomers be designated D and L, but he had no experimental way to determine which enantiomer has which specific rotation. Fischer, therefore, did the only possible thing—he made an arbitrary

Emil Fischer, who in 1902 became the second Nobel Prize winner in chemistry, made many fundamental discoveries in the chemistry of carbohydrates, proteins, and other areas of organic and biochemistry.

Table 19.1 Configurational Relationships among the Isomeric D-Aldotetroses, D-Aldopentoses, and D-Aldohexoses

*The configuration of the reference —OH group on the penultimate carbon is shown in color.

assignment. He assigned the dextrorotatory enantiomer the following configuration and named it D-glyceraldehyde. He named its enantiomer L-glyceraldehyde. Fischer could have been wrong, but by a stroke of good fortune, he wasn't. In 1952, x-ray crystallography proved his assignment of configuration to the enantiomers of glyceraldehyde to be correct.

$$
\begin{array}{cc}
\text{CHO} & \text{CHO} \\
\text{H}\!\!-\!\!\text{OH} & \text{HO}\!\!-\!\!\text{H} \\
\text{CH}_2\text{OH} & \text{CH}_2\text{OH} \\
\text{D-Glyceraldehyde} & \text{L-Glyceraldehyde} \\
[\alpha]_D^{25} = +13.5° & [\alpha]_D^{25} = -13.5°
\end{array}
$$

D-glyceraldehyde and L-glyceraldehyde serve as reference points for the assignment of relative configurations to all other aldoses and ketoses. The reference point is the penultimate carbon—that is, the next-to-last carbon on the chain. A **D-monosaccharide** has the same configuration at its penultimate carbon as D-glyceraldehyde (its —OH group is on the right); an **L-monosaccharide** has the same configuration at its penultimate carbon as L-glyceraldehyde (its —OH group is on the left).

Tables 19.1 and 19.2 show names and Fischer projections for all D-aldo- and D-2-ketotetroses, pentoses, and hexoses. Each name consists of three parts. The D specifies the configuration at the stereocenter farthest from the

> **D-Monosaccharide** A monosaccharide that, when written as a Fischer projection, has the —OH group on its penultimate carbon to the right

> **L-Monosaccharide** A monosaccharide that, when written as a Fischer projection, has the —OH group on its penultimate carbon to the left

Table 19.2 Configurational Relationships among the D-2-Ketopentoses and D-2-Ketohexoses

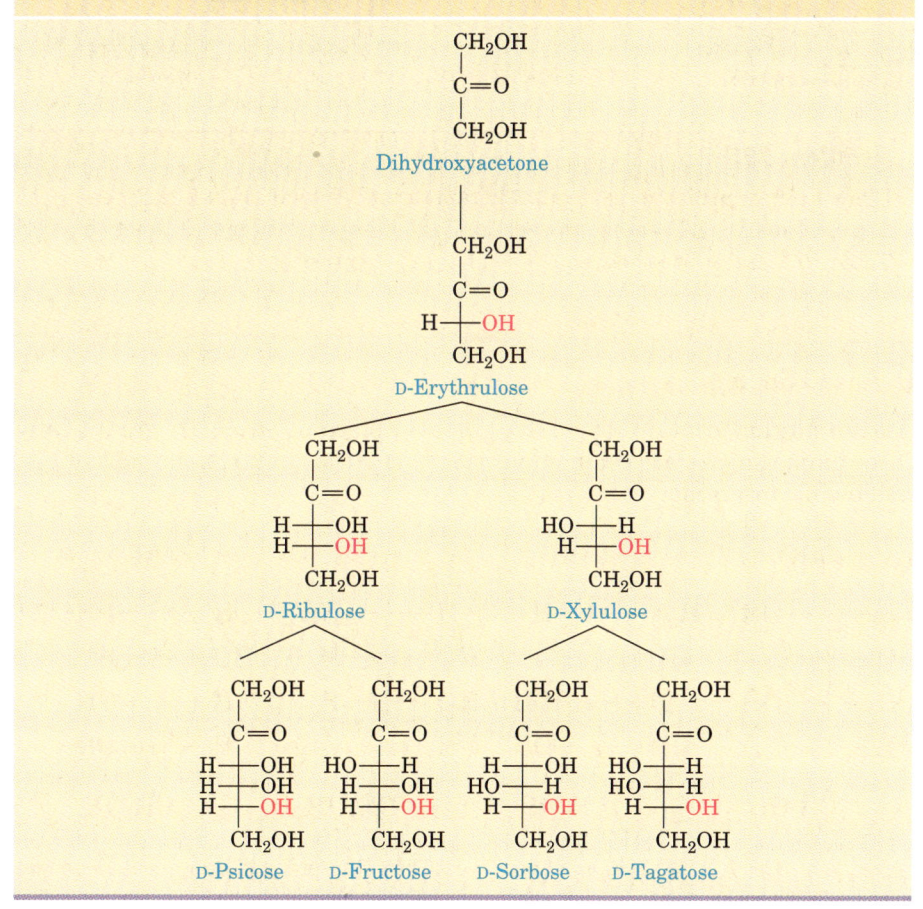

carbonyl group. Prefixes such as *rib-*, *arabin-*, and *gluc-* specify the configuration of all other stereocenters in the monosaccharide relative to one another. The suffix *-ose* indicates that the compound is a carbohydrate.

The three most abundant hexoses in the biological world are D-glucose, D-galactose, and D-fructose. The first two are D-aldohexoses; the third is a D-2-ketohexose. Glucose, by far the most abundant of the three, is also known as dextrose because it is dextrorotatory. Other names for this monosaccharide include grape sugar and blood sugar. Human blood normally contains 65–110 mg of glucose/100 mL of blood.

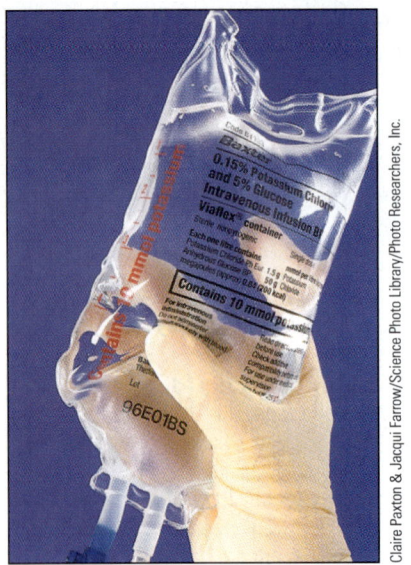

■ Gloved hand holding an intravenous (i.v.) drip bag containing 0.15% potassium chloride (saline) and 5% glucose.

EXAMPLE 19.1

Draw Fischer projections for the four aldotetroses. Which are D-monosaccharides, which are L-monosaccharides, and which are enantiomers? Refer to Table 19.1 and write the name of each aldotetrose.

Solution
Following are Fischer projections for the four aldotetroses. The D- and L- refer to the configuration of the penultimate carbon, which, in the case of aldotetroses, is carbon 3. In the Fischer projection of a D-aldotetrose, the —OH on carbon 3 is on the right; in an L-aldotetrose, it is on the left.

Problem 19.1
Draw Fischer projections for all 2-ketopentoses. Which are D-2-ketopentoses, which are L-2-ketopentoses, and which are enantiomers? Refer to Table 19.2 and write the name of each 2-ketopentose.

D Amino Sugars

Amino sugar A monosaccharide in which an —OH group is replaced by an —NH$_2$ group

Amino sugars contain an —NH$_2$ group in place of an —OH group. Only three amino sugars are common in nature: D-glucosamine, D-mannosamine, and D-galactosamine.

CHEMICAL CONNECTIONS 19A

Galactosemia

One out of every 18,000 infants is born with a genetic defect that renders the child unable to utilize the monosaccharide galactose. Galactose is part of lactose (milk sugar, Section 19.5B). When the body cannot utilize galactose, it accumulates in the blood and in the urine. This buildup in the blood is harmful because it can lead to mental retardation, failure to grow, cataract formation in the eye and, in severe cases, death due to liver damage. When galactose accumulation results from a deficiency of the enzyme galactokinase, the disorder, known as galactosuria, has only mild symptoms. When the enzyme galactose-1-phosphate uridinyltransferase is deficient, however, the disorder is called galactosemia, and its symptoms are severe.

The deleterious effects of galactosemia can be avoided by giving the infant a milk formula in which sucrose is substituted for lactose. Because sucrose contains no galactose, the infant thus consumes a galactose-free diet. A galactose-free diet is critical only in infancy. With maturation, most children develop another enzyme capable of metabolizing galactose. As a consequence, they are able to tolerate galactose as they mature.

N-Acetyl-D-glucosamine, a derivative of D-glucosamine, is a component of connective tissue such as cartilage. It is also a component of chitin, the hard shell-like exoskeleton of lobsters, crabs, shrimp, and other shellfish. Several other amino sugars are components of naturally occurring antibiotics.

E Physical Properties

Monosaccharides are colorless, crystalline solids. Because hydrogen bonding is possible between their polar —OH groups and water, all monosaccharides are very soluble in water. They are only slightly soluble in ethanol and are insoluble in nonpolar solvents such as diethyl ether, dichloromethane, and benzene.

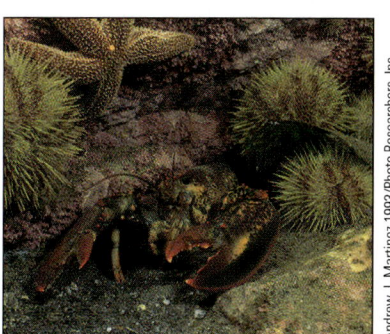

■ **Northern lobster.**

19.3 The Cyclic Structure of Monosaccharides

In Section 17.7, we saw that aldehydes and ketones react with alcohols to form **hemiacetals.** We also saw that cyclic hemiacetals form very readily when hydroxyl and carbonyl groups are part of the same molecule and that their interaction can produce a ring. For example, 4-hydroxypentanal forms a five-membered cyclic hemiacetal. Note that 4-hydroxypentanal contains one stereocenter and that hemiacetal formation generates a second stereocenter at carbon 1.

Screen 19.3
Tutorial

Monosaccharides have hydroxyl and carbonyl groups in the same molecule. As a result, they also exist almost exclusively as five- and six-membered cyclic hemiacetals.

A Haworth Projections

A common way of representing the cyclic structure of monosaccharides is the **Haworth projection,** named after the English chemist Sir Walter N.

Haworth projection A way to view furanose and pyranose forms of monosaccharides; the ring is drawn flat and viewed through its edge with the anomeric carbon on the right and the oxygen atom to the rear

Figure 19.2 Haworth projections for α-D-glucopyranose; and β-D-glucopyranose.

Haworth (Nobel Prize for chemistry, 1937). In a Haworth projection, a five- or six-membered cyclic hemiacetal is represented as a planar pentagon or hexagon, respectively, lying roughly perpendicular to the plane of the paper. Groups attached to the carbons of the ring then lie either above or below the plane of the ring. The new carbon stereocenter created in forming the cyclic structure is called an **anomeric carbon.** Stereoisomers that differ in configuration only at the anomeric carbon are called **anomers.** The anomeric carbon of an aldose is carbon 1; that of the most common ketoses is carbon 2.

Typically, Haworth projections are drawn with the anomeric carbon to the right and the hemiacetal oxygen to the back (Figure 19.2).

In the terminology of carbohydrate chemistry, the designation β means that the —OH on the anomeric carbon of the cyclic hemiacetal lies on the same side of the ring as the terminal —CH$_2$OH. Conversely, the designation α means that the —OH on the anomeric carbon of the cyclic hemiacetal lies on the side of the ring opposite from the terminal —CH$_2$OH.

A six-membered hemiacetal ring is indicated by **-pyran-**, and a five-membered hemiacetal ring is indicated by **-furan-**. The terms **furanose** and **pyranose** are used because monosaccharide five- and six-membered rings correspond to the heterocyclic compounds furan and pyran.

Anomeric carbon The hemiacetal carbon of the cyclic form of a monosaccharide

Anomers Monosaccharides that differ in configuration only at their anomeric carbons

Furanose A five-membered cyclic hemiacetal form of a monosaccharide

Pyranose A six-membered cyclic hemiacetal form of a monosaccharide

Furan   Pyran

Because the α and β forms of glucose are six-membered cyclic hemiacetals, they are named α-D-glucopyranose and β-D-glucopyranose, respectively. The designations *-furan-* and *-pyran-* are not always used in monosaccharide names, however. Thus, the glucopyranoses, for example, are often named simply α-D-glucose and β-D-glucose.

CHEMICAL CONNECTIONS 19B

L-Ascorbic Acid (Vitamin C)

The structure of L-ascorbic acid (vitamin C) resembles that of a monosaccharide. In fact, this vitamin is synthesized both biochemically by plants and some animals and commercially from D-glucose. Humans do not have the enzymes required to carry out this synthesis. For this reason, we must obtain vitamin C in the food we eat or as a vitamin supplement. Approximately 66 million kg of vitamin C is synthesized annually in the United States.

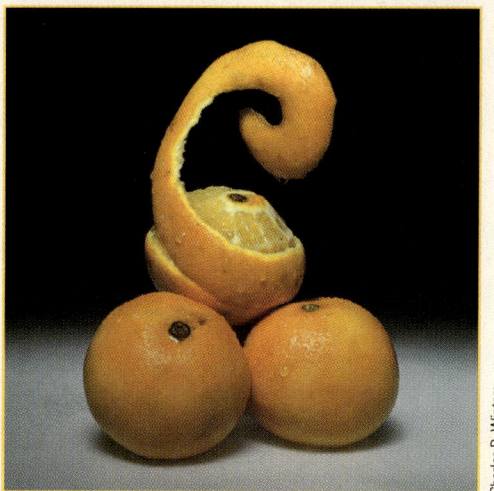

■ Oranges are a major source of vitamin C.

You would do well to remember the configuration of groups on the Haworth projections of α-D-glucopyranose and β-D-glucopyranose as reference structures. Knowing how the open-chain configuration of any other aldohexose differs from that of D-glucose, you can then construct its Haworth projection by referring to the Haworth projection of D-glucose.

EXAMPLE 19.2

Draw Haworth projections for the α and β anomers of D-galactopyranose.

Solution
One way to arrive at these projections is to use the α and β forms of D-glucopyranose as references and to remember (or discover by looking at Table 19.1) that D-galactose differs from D-glucose only in the configu-

ration at carbon 4. Thus, you can begin with the Haworth projections shown in Figure 19.2 and then invert the configuration at carbon 4.

Configuration differs from that of D-glucose at C-4

α-D-Galactopyranose
(α-D-Galactose)

β-D-Galactopyranose
(β-D-Galactose)

Problem 19.2

D-Mannose exists in aqueous solution as a mixture of α-D-mannopyranose and β-D-mannopyranose. Draw Haworth projections for these molecules.

Aldopentoses also form cyclic hemiacetals. The most prevalent forms of D-ribose and other pentoses in the biological world are furanoses. Following are Haworth projections for α-D-ribofuranose (α-D-ribose) and β-2-deoxy-D-ribofuranose (β-2-deoxy-D-ribose). The prefix *2-deoxy* indicates the absence of oxygen at carbon 2. Units of D-ribose and 2-deoxy-D-ribose in nucleic acids and most other biological molecules are found almost exclusively in the β configuration.

α-D-Ribofuranose
(α-D-Ribose)

β-2-Deoxy-D-ribofuranose
(β-2-Deoxy-D-ribose)

Fructose also forms five-membered cyclic hemiacetals. β-D-Fructofuranose, for example, is found in the disaccharide sucrose (Section 19.5A).

α-D-Fructofuranose
(α-D-Fructose)

D-Fructose

β-D-Fructofuranose
(β-D-Fructose)

Anomeric carbon

B Conformation Representations

A five-membered ring is so close to being planar that Haworth projections provide adequate representations of furanoses. For pyranoses, however, the six-membered ring is more accurately represented as a **chair conformation**

(Section 11.6B). Following are structural formulas for α-D-glucopyranose and β-D-glucopyranose, both drawn as chair conformations. Also shown is the open-chain or free aldehyde form with which the cyclic hemiacetal forms are in equilibrium in aqueous solution. Notice that each group, including the anomeric —OH, on the chair conformation of β-D-glucopyranose is equatorial. Notice also that the —OH group on the anomeric carbon is axial in α-D-glucopyranose. Because the —OH on the anomeric carbon of β-D-glucopyranose is in the more stable equatorial position (Section 11.6B), the β anomer predominates in aqueous solution.

β-D-Glucopyranose
$[α]_D^{25} + 18.7°$

D-Glucose

α-D-Glucopyranose
$[α]_D^{25} + 112°$

At this point, you should compare the relative orientations of groups on the D-glucopyranose ring in the Haworth projection and chair conformation. The orientations of groups on carbons 1 through 5 of β-D-glucopyranose, for example, are up, down, up, down, and up in both representations.

β-D-Glucopyranose
(Haworth projection)

β-D-Glucopyranose
(chair conformation)

We do not show hydrogen atoms bonded to the ring in chair conformations. We often show them, however, in Haworth projections.

EXAMPLE 19.3

Draw chair conformations for α-D-galactopyranose and β-D-galactopyranose. Label the anomeric carbon in each.

Solution
The configuration of D-galactose differs from that of D-glucose only at carbon 4. Therefore, draw the α and β forms of D-glucopyranose and then interchange the positions of the —OH and —H groups on carbon 4.

β-D-Galactopyranose
(β-D-Galactose)

D-Galactose

α-D-Galactopyranose
(α-D-Galactose)

Problem 19.3
Draw chair conformations for α-D-mannopyranose and β-D-mannopyranose. Label the anomeric carbon in each.

C Mutarotation

Mutarotation The change in specific rotation that occurs when an α or β form of a carbohydrate is converted to an equilibrium mixture of the two forms

Mutarotation is the change in specific rotation that accompanies the equilibration of α- and β-anomers in aqueous solution. For example, a solution prepared by dissolving crystalline α-D-glucopyranose in water has a specific rotation of +112°, which gradually decreases to an equilibrium value of +52.7° as α-D-glucopyranose reaches equilibrium with β-D-glucopyranose. A solution of β-D-glucopyranose also undergoes mutarotation, during which the specific rotation changes from +18.7° to the same equilibrium value of +52.7°. The equilibrium mixture consists of 64% β-D-glucopyranose and 36% α-D-glucopyranose, with only a trace (0.003 percent) of the open-chain form. Mutarotation is common to all carbohydrates that exist in hemiacetal forms.

β-D-Glucopyranose $[\alpha]_D^{25} = +18.7°$ ⇌ Open-chain form ⇌ α-D-Glucopyranose $[\alpha]_D^{25} = +112°$

19.4 Reactions of Monosaccharides

Screen 19.5 Simulation Tutorial

A Formation of Glycosides (Acetals)

As we saw in Section 17.7, treatment of an aldehyde or ketone with one molecule of alcohol yields a hemiacetal, and treatment of the hemiacetal with a molecule of alcohol yields an acetal. Treatment of a monosaccharide, all forms of which exist almost exclusively as cyclic hemiacetals, with an alcohol also yields an acetal, as illustrated by the reaction of β-D-glucopyranose with methanol.

β-D-Glucopyranose (β-D-Glucose) + CH₃OH →[H⁺, −H₂O] Methyl β-D-glucopyranoside (Methyl β-D-glucoside) + Methyl α-D-glucopyranoside (Methyl α-D-glucoside)

Glycoside A carbohydrate in which the —OH group on its anomeric carbon is replaced by an —OR group

Glycosidic bond The bond from the anomeric carbon of a glycoside to an —OR group

A cyclic acetal derived from a monosaccharide is called a **glycoside**, and the bond from the anomeric carbon to the —OR group is called a **glycosidic bond**. Mutarotation is not possible in a glycoside because an acetal—unlike a hemiacetal—is no longer in equilibrium with the open-chain, carbonyl-containing compound. Glycosides are stable in water and aqueous base, like other acetals (Section 17.7), however, they are hydrolyzed in aqueous acid to an alcohol and a monosaccharide.

We name glycosides by listing the alkyl or aryl group bonded to oxygen, followed by the name of the carbohydrate in which the ending **-e** is replaced by **-ide**. For example, the methyl glycoside derived from β-D-glucopyranose is named methyl β-D-glucopyranoside; that derived from β-D-ribofuranose is named methyl β-D-ribofuranoside.

EXAMPLE 19.4

Draw a structural formula for methyl β-D-ribofuranoside (methyl β-D-riboside). Label the anomeric carbon and the glycosidic bond.

Solution

Methyl β-D-ribofuranoside
(Methyl β-D-riboside)

Problem 19.4

Draw a Haworth projection and a chair conformation for methyl α-D-mannopyranoside (methyl α-D-mannoside). Label the anomeric carbon and the glycosidic bond.

B Reduction to Alditols

The carbonyl group of a monosaccharide can be reduced to an hydroxyl group by a variety of reducing agents, including hydrogen in the presence of a transition metal catalyst and sodium borohydride (Section 17.6). The reduction products are known as **alditols.** Reduction of D-glucose gives D-glucitol, more commonly known as D-sorbitol. Here, D-glucose is shown in the open-chain form. Only a small amount of this form is present in solution but, as it is reduced, the equilibrium between cyclic hemiacetal forms (only the β form is shown here) and the open-chain form shifts to replace it.

Alditol The product formed when the CHO group of a monosaccharide is reduced to a CH_2OH group

β-D-Glucopyranose ⇌ D-Glucose →[NaBH₄] D-Glucitol (D-Sorbitol)

■ Many "sugar free" products contain sugar alcohols, such as D-sorbitol and xylitol.

We name alditols by dropping the **-ose** from the name of the monosaccharide and adding **-itol.** Sorbitol is found in the plant world in many berries and in cherries, plums, pears, apples, seaweed, and algae. It is about 60% as sweet as sucrose (table sugar) and is used in the manufacture of candies and as a sugar substitute for diabetics. Other alditols commonly found in the biological world include erythritol, D-mannitol, and xylitol. Xylitol is used as a sweetening agent in "sugarless" gum, candy, and sweet cereals.

Erythritol, D-Mannitol, Xylitol (Fischer projection structures)

C Oxidation to Aldonic Acids (Reducing Sugars)

Oxidations and reductions of monosaccharides in nature are catalyzed by specific enzymes, as for example glucose oxidase.

As we saw in Section 17.5, aldehydes (RCHO) are oxidized to carboxylic acids (RCOOH) by several agents, including oxygen, O_2. Similarly, the aldehyde group of an aldose can be oxidized, under basic conditions, to a carboxylate group. Under these conditions, the cyclic form of the aldose is in equilibrium with the open-chain form, which is then oxidized by the mild oxidizing agent. D-Glucose, for example, is oxidized to D-gluconate (the anion of D-gluconic acid).

β-D-Glucopyranose (β-D-Glucose) ⇌ D-Glucose $\xrightarrow[\text{Basic solution}]{\text{Oxidizing agent}}$ D-Gluconate

Reducing sugar A carbohydrate that reacts with a mild oxidizing agent under basic conditions to give an aldonic acid; the carbohydrate reduces the oxidizing agent

Any carbohydrate that reacts with an oxidizing agent to form an aldonic acid is classified as a **reducing sugar** (it reduces the oxidizing agent).

Surprisingly, 2-ketoses are also reducing sugars. Carbon 1 (a CH_2OH group) of a ketose is not oxidized directly. Rather, under the basic conditions of this oxidation, a 2-ketose exists in equilibrium with an aldose by way of an enediol intermediate. The aldose is then oxidized by the mild oxidizing agent.

A 2-ketose $\xrightleftharpoons{OH^-}$ An enediol $\xrightleftharpoons{OH^-}$ An aldose $\xrightarrow{\text{Oxidizing agent}}$ An aldonate

CHEMICAL CONNECTIONS 19C

Testing for Glucose

The analytical procedure most often performed in a clinical chemistry laboratory is the determination of glucose in blood, urine, or other biological fluids. The high frequency with which this test is performed reflects the high incidence of diabetes mellitus. Approximately 2 million known diabetics live in the United States, and it is estimated that another 1 million remain undiagnosed.

Diabetes mellitus (Chemical Connections 23G) is characterized by insufficient blood levels of the hormone insulin. If the blood concentration of insulin is too low, muscle and liver cells do not absorb glucose from the blood; this problem, in turn, leads to increased levels of blood glucose (hyperglycemia), impaired metabolism of fats and proteins, ketosis, and possible diabetic coma. A rapid test for blood glucose levels is critical for early diagnosis and effective management of this disease. In addition to giving results quickly, a test must be specific for D-glucose; that is, it must give a positive test for glucose but not react with any other substances normally present in biological fluids.

Today blood glucose levels are measured by an enzyme-based procedure using the enzyme glucose oxidase. This enzyme catalyzes the oxidation of β-D-glucose to D-gluconic acid.

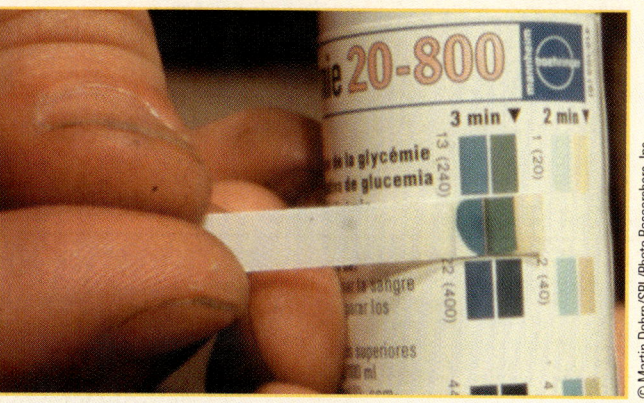

■ Chemstrip kit for blood glucose.

Glucose oxidase is specific for β-D-glucose. Therefore, complete oxidation of any sample containing both β-D-glucose and α-D-glucose requires conversion of the α form to the β form. Fortunately, this interconversion is rapid and complete in the short time required for the test.

Molecular oxygen, O_2, is the oxidizing agent in this reaction and is reduced to hydrogen peroxide, H_2O_2. In one procedure, hydrogen peroxide formed in the glucose oxidase-catalyzed reaction oxidizes colorless o-toluidine to a colored product in a reaction catalyzed by the enzyme peroxidase. The concentration of the colored oxidation product is determined spectrophotometrically and is proportional to the concentration of glucose in the test solution.

Several commercially available test kits use the glucose oxidase reaction for qualitative determination of glucose in urine.

D Oxidation to Uronic Acids

Enzyme-catalyzed oxidation of the primary alcohol at carbon 6 of a hexose yields a uronic acid. Enzyme-catalyzed oxidation of D-glucose, for example, yields D-glucuronic acid, shown here in both its open-chain and cyclic hemiacetal forms.

[D-Glucose] →(Enzyme-catalyzed oxidation) D-Glucuronic acid (a uronic acid) — shown as Fischer projection and chair conformation

D-Glucuronic acid is widely distributed in both the plant and animal worlds. In humans, it serves as an important component of the acidic polysaccharides of connective tissues (Section 19.7). The body also uses it to detoxify foreign phenols and alcohols. In the liver, these compounds are converted to glycosides of glucuronic acid (glucuronides) and then excreted in urine. The intravenous anesthetic propfol, for example, is converted to the following glucuronide and excreted in urine:

Propofol   A urine-soluble glucuronide

E Phosphoric Esters

Mono- and diphosphoric esters are important intermediates in the metabolism of monosaccharides. For example, the first step in glycolysis (Section 27.2) involves conversion of glucose to glucose 6-phosphate. Phosphoric acid is a strong enough acid so that at the pH of cellular and intercellular fluids, both acidic protons of a phosphoric ester are ionized, giving the ester a charge of -2.

D-Glucose →(Enzyme-catalyzed phosphorylation) D-Glucose 6-phosphate α-D-Glucose 6-phosphate

19.5 Disaccharides and Oligosaccharides

Most carbohydrates in nature contain more than one monosaccharide unit. Those that contain two units are called **disaccharides,** those that contain three units are called **trisaccharides,** and so forth. We use the general term **oligosaccharide** to describe any of the carbohydrates that contain from six to ten monosaccharide units. Carbohydrates containing larger numbers of monosaccharide units are called **polysaccharides.**

In a disaccharide, two monosaccharide units are joined by a glycosidic bond between the anomeric carbon of one unit and an —OH group of the other unit. Three important disaccharides are sucrose, lactose, and maltose.

> **Disaccharide** A carbohydrate containing two monosaccharide units joined by a glycosidic bond
>
> **Oligosaccharide** A carbohydrate containing from 6 to 10 monosaccharide units, each joined to the next by a glycosidic bond
>
> **Polysaccharide** A carbohydrate containing a large number of monosaccharide units, each joined to the next by one or more glycosidic bonds

A Sucrose

Sucrose (table sugar) is the most abundant disaccharide in the biological world. It is obtained principally from the juice of sugar cane and sugar beets. In sucrose, carbon 1 of α-D-glucopyranose bonds to carbon 2 of D-fructofuranose by an α-1,2-glycosidic bond. Because the anomeric carbons of both the glucopyranose and fructofuranose units are involved in formation of the glycosidic bond, neither monosaccharide unit is in equilibrium with its open-chain form. Thus sucrose is a nonreducing sugar.

In the production of sucrose, sugar cane or sugar beet is boiled with water, and the resulting solution is cooled. Sucrose crystals separate and are collected. Subsequent boiling to concentrate the solution followed by cooling yields a dark, thick syrup known as molasses.

Sucrose

B Lactose

Lactose is the principal sugar present in milk. It accounts for 5 to 8% of human milk and 4 to 6% of cow's milk. This disaccharide consists of D-galactopyranose bonded by a β-1,4-glycosidic bond to carbon 4 of D-glucopyranose. Lactose is a reducing sugar, because the cyclic hemiacetal of the D-glucopyranose unit is in equilibrium with its open-chain form and can be oxidized to a carboxyl group.

Lactose

CHEMICAL CONNECTIONS 19D

A, B, AB, and O Blood Types

Membranes of animal plasma cells have large numbers of relatively small carbohydrates bound to them. In fact, the outsides of most plasma cell membranes are literally "sugar-coated." These membrane-bound carbohydrates are part of the mechanism by which cell types recognize one another and, in effect, act as biochemical markers. Typically, they contain from 4 to 17 monosaccharide units consisting primarily of relatively few monosaccharides, the most common of which are D-galactose, D-mannose, L-fucose, N-acetyl-D-glucosamine, and N-acetyl-D-galactosamine. L-Fucose is a 6-deoxyaldohexose.

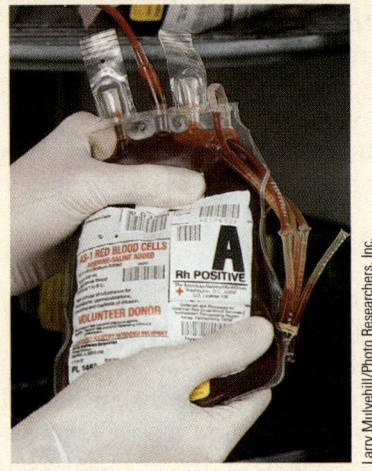

■ Bag of blood showing blood type.

To see the importance of these membrane-bound carbohydrates, consider the ABO blood group system, discovered in 1900 by Karl Landsteiner (1868–1943). Whether an individual belongs to type A, B, AB, or O is genetically determined and depends on the type of trisaccharide or tetrasaccharide bound to the surface of the red blood cells. These surface-bound carbohydrates, designated as A, B, and O, act as antigens. The type of glycosidic bond joining each monosaccharide is shown in the figure.

The blood carries antibodies against foreign substances. When a person receives a blood transfusion, the antibodies clump (aggregate) the foreign blood cells. Type A blood, for example, has A antigens (N-acetyl-D-galactosamine) on the surfaces of its red blood cells, and carries anti-B antibodies (against B antigen). B-type blood carries B antigen (D-galactose) and has anti-A antibodies (against A antigens). Transfusion of type A blood into a person with type B blood can be fatal, and vice versa. The relationships between blood type and donor/receiver relationships are summarized in the figure.

(continued on next page)

CHEMICAL CONNECTIONS 19D

A, B, AB, and O Blood Types, (continued)

Type O
Sugar on cell surface: O
Has antibodies against: A and B
Can receive blood from: O
Can donate blood to: O, A, B, and AB

Type A
Sugar on cell surface: A
Has antibodies against: B
Can receive blood from: A or O
Can donate blood to: A and AB

Type B
Sugar on cell surface: B
Has antibodies against: A
Can receive blood from: B and O
Can donate blood to: B and AB

Type AB
Sugars on cell surface: A and B
Has antibodies against: None
Can receive blood from: O, A, B, and AB
Can donate blood to: AB

People with type O blood are universal donors, and those with type AB blood are universal acceptors. People with A-type blood can accept from type A or type O donors only. Those with type B blood can accept from type B or type O donors only, and type O donors can accept only from type O donors.

C Maltose

Maltose derives its name from its presence in malt, the juice from sprouted barley and other cereal grains. It consists of two units of D-glucopyranose joined by a glycosidic bond between carbon 1 (the anomeric carbon) of one unit and carbon 4 of the other unit. Because the oxygen atom on the anomeric carbon of the first glucopyranose unit is alpha, the bond joining the two units is called an α-1,4-glycosidic bond. Following are a Haworth projection and a chair conformation for β-maltose, so named because the —OH group on the anomeric carbon of the glucose unit on the right is beta.

Maltose is an ingredient in most syrups.

Maltose

Maltose is a reducing sugar: The hemiacetal group on the right unit of D-glucopyranose is in equilibrium with the free aldehyde and can be oxidized to a carboxylic acid.

EXAMPLE 19.5

Draw a chair conformation for the β anomer of a disaccharide in which two units of D-glucopyranose are joined by an α-1,6-glycosidic bond.

Solution
First draw a chair conformation of α-D-glucopyranose. Then connect the anomeric carbon of this monosaccharide to carbon 6 of a second D-glucopyranose unit by an α-1,6 glycosidic bond. The resulting molecule is either α or β depending on the orientation of the —OH group on the reducing end of the disaccharide. The disaccharide shown here is the β form.

Problem 19.5
Draw a chair conformation for the α form of a disaccharide in which two units of D-glucopyranose are joined by a β-1,3-glycosidic bond.

D Relative Sweetness

Among the saccharide sweetening agents, D-fructose tastes the sweetest—even sweeter than sucrose (Table 19.3). The sweet taste of honey is due largely to D-fructose and D-glucose. Lactose has almost no sweetness and is sometimes added to foods as a filler. Some people cannot tolerate lactose well, however, and should avoid these foods.

We have no mechanical way to measure sweetness. Such testing is done by having a group of people taste solutions of varying sweetness.

Table 19.3 Relative Sweetness of Some Carbohydrate and Artificial Sweetening Agents

Carbohydrate	Sweetness Relative to Sucrose	Artificial Sweetener	Sweetness Relative to Sucrose
fructose	1.74	saccharine	450
sucrose (table sugar)	1.00	acesulfame-K	200
honey	0.97	aspartame	180
glucose	0.74		
maltose	0.33		
galactose	0.32		
lactose (milk sugar)	0.16		

CHEMICAL CONNECTIONS 19E

High-Fructose Corn Syrup

If you read the labels of soft drinks and other artificially sweetened food products, you will find that many of them contain high-fructose corn syrup. Looking at Table 19.3, you will see one reason for its use: Fructose is more than 70% sweeter than sucrose (table sugar).

The production of high-fructose corn syrup begins with the partial hydrolysis of corn starch catalyzed by the enzyme α-amylase. This enzyme catalyzes the hydrolysis of α-glucosidic bonds and breaks corn starch into small polysaccharides called dextrins. The enzyme glucoamylase then catalyzes the hydrolysis of the dextrins to D-glucose. Finally, enzyme-catalyzed isomerization of D-glucose gives D-fructose. Several billion pounds of high-fructose corn syrup are produced each year in this way for use by the food processing industry.

$$\text{Corn starch} \xrightarrow{\alpha\text{-Amylase}} \text{Dextrins} \xrightarrow{\text{Glucoamylase}} \text{D-Glucose} \xrightarrow{\text{Glucose isomerase}} \text{D-Fructose}$$

(a polysaccharide of α-D-glucose)   Oligosaccharides of 6–10 α-D-glucose units)

19.6 Polysaccharides

Polysaccharides consist of large numbers of monosaccharide units bonded together by glycosidic bonds. Three important polysaccharides, all made up of glucose units, are starch, glycogen, and cellulose.

A Starch: Amylose and Amylopectin

Starch is used for energy storage in plants. It is found in all plant seeds and tubers and is the form in which glucose is stored for later use. Starch can be separated into two principal polysaccharides: amylose and amylopectin. Although the starch from each plant is unique, most starches contain 20 to 25% amylose and 75 to 80% amylopectin.

Complete hydrolysis of both amylose and amylopectin yields only D-glucose. Amylose is composed of continuous, unbranched chains of as many as 4000 D-glucose units joined by α-1,4-glycosidic bonds. Amylopectin contains chains of as many as 10,000 D-glucose units, also joined by α-1,4-glycosidic bonds. In addition, there is considerable branching from this linear network. At branch points, new chains of 24 to 30 units are started by α-1,6-glycosidic bonds (Figure 19.3).

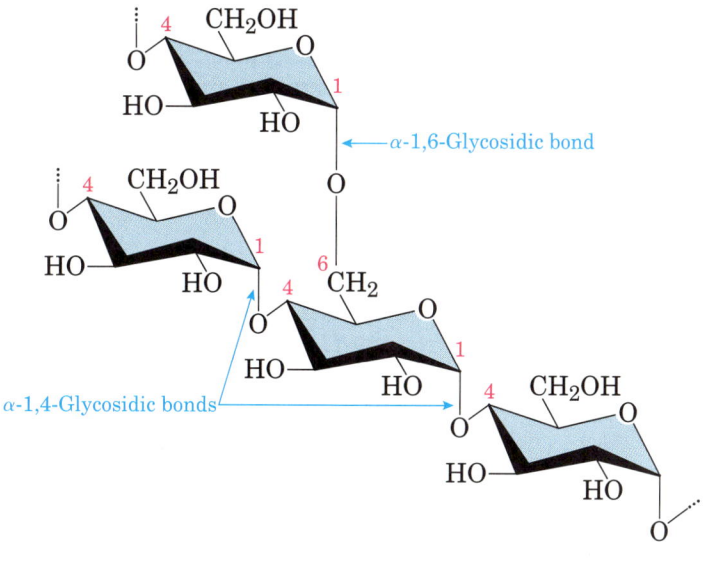

Figure 19.3 Amylopectin is a highly branched polymer of D-glucose. Chains consist of 24–30 units of D-glucose joined by α-1,4-glycosidic bonds and branches created by α-1,6-glycosidic bonds.

Figure 19.4 Cellulose is a linear polymer containing as many as 3000 units of D-glucose joined by β-1,4-glycosidic bonds.

B Glycogen

Glycogen acts as the energy-reserve carbohydrate for animals. Like amylopectin, it is a branched polysaccharide containing approximately 10^6 glucose units joined by α-1,4- and α-1,6-glycosidic bonds. The total amount of glycogen in the body of a well-nourished adult human is about 350 g, divided almost equally between liver and muscle.

C Cellulose

Cellulose, the most widely distributed plant skeletal polysaccharide, constitutes almost half of the cell-wall material of wood. Cotton is almost pure cellulose.

Cellulose is a linear polysaccharide of D-glucose units joined by β-1,4-glycosidic bonds (Figure 19.4). It has an average molecular weight of 400,000 g/mol, corresponding to approximately 2200 glucose units per molecule.

Cellulose molecules act much like stiff rods, a feature that enables them to align themselves side by side into well-organized, water-insoluble fibers in which the OH groups form numerous intermolecular hydrogen bonds. This arrangement of parallel chains in bundles gives cellulose fibers their high mechanical strength. It also explains why cellulose is insoluble in water. When a piece of cellulose-containing material is placed in water, there are not enough —OH groups on the surface of the fiber to pull individual cellulose molecules away from the strongly hydrogen-bonded fiber.

Humans and other animals cannot use cellulose as food because our digestive systems do not contain β-glucosidases, enzymes that catalyze the hydrolysis of β-glucosidic bonds. Instead, we have only α-glucosidases; hence, we use the polysaccharides starch and glycogen as sources of glucose. On the other hand, many bacteria and microorganisms do contain β-glucosidases and so can digest cellulose. Termites (much to our regret) have such bacteria in their intestines and can use wood as their principal food. Ruminants (cud-chewing animals) and horses can also digest grasses and hay because β-glucosidase-containing microorganisms are present in their alimentary systems.

19.7 Acidic Polysaccharides

Acidic polysaccharides—polysaccharides that contain carboxyl groups and/or sulfuric ester groups—play important roles in the structure and function of connective tissues. Because acidic polysaccharides contain amino sugars, a more current name for these substances is glycosaminoglycans. There is no single general type of connective tissue. Rather, there are a large number of highly specialized forms, such as cartilage, bone, synovial fluid, skin, tendons, blood vessels, intervertebral disks, and cornea. Most connective tissues consist of collagen, a structural protein, combined with a variety of acidic polysaccharides (glycosaminoglycans) that interact with collagen to form tight or loose networks.

A Hyaluronic Acid

Hyaluronic acid is the simplest acidic polysaccharide present in connective tissue. It has a molecular weight of between 10^5 and 10^7 g/mol and contains from 300 to 100,000 repeating units, depending on the organ in which it occurs. It is most abundant in embryonic tissues and in specialized connective tissues such as synovial fluid, the lubricant of joints in the body, and the vitreous of the eye, where it provides a clear, elastic gel that holds the retina in its proper position.

In rheumatoid arthritis, inflammation of the synovial tissue results in swelling of the joints.

Hyaluronic acid is composed of D-glucuronic acid linked by a β-1,3-glycosidic bond to N-acetyl-D-glucosamine, which is in turn linked to D-glucuronic acid by a β-1,4-glycosidic bond.

The repeating unit of hyaluronic acid

B Heparin

Heparin is a heterogeneous mixture of variably sulfonated polysaccharide chains, ranging in molecular weight from 6000 to 30,000 g/mol. This acidic polysaccharide is synthesized and stored in mast cells of various tissues—particularly the liver, lungs, and gut. Heparin has many biological functions, the best known and most fully understood of which is its anticoagulant activity. It binds strongly to antithrombin III, a plasma protein involved in terminating the clotting process. A heparin preparation with good anticoagulant activity contians a minimum of eight repeating units (Figure 19.5). The larger the molecule, the better its anticoagulant activity.

Figure 19.5 A pentasaccharide unit of heparin.

SUMMARY

Monosaccharides are polyhydroxyaldehydes or polyhydroxyketones (Section 19.2A). The most common have the general formula $C_nH_{2n}O_n$ where n varies from 3 to 8. Their names contain the suffix *-ose,* and prefixes *tri-, tetr-,* and so on indicate the number of carbon atoms in the chain. The prefix *aldo-* indicates an aldehyde and the prefix *keto-* indicates a ketone.

In a **Fischer projection** of a carbohydrate, we write the carbon chain vertically with the most highly oxidized carbon toward the top (Section 19.2B). Horizontal lines represent groups projecting above the plane of the page; vertical lines represent groups projecting behind the plane of the page.

The penultimate carbon of a monosaccharide is the next-to-last carbon of a Fischer projection (Section 19.2C). A monosaccharide that has the same configuration at the penultimate carbon as D-glyceraldehyde is called a **D-monosaccharide;** one that has the same configuration at the penultimate carbon as L-glyceraldehyde is called an **L-monosaccharide.**

Monosaccharides exist primarily as cyclic hemiacetals (Section 19.3). A six-membered cyclic hemiacetal is a **pyranose;** a five-membered cyclic hemiacetal is a **furanose.** The new stereocenter resulting from hemiacetal formation is called an **anomeric carbon,** and the stereoisomers formed in this way are called **anomers.** The symbol *β*-indicates that the —OH group on the anomeric carbon lies on the same side of the ring as the terminal —CH$_2$OH. The symbol *α*-indicates that the —OH group on the anomeric carbon lies on the opposite side from the terminal —CH$_2$OH. Furanoses and pyranoses can be drawn as **Haworth projections** (Section 19.3A). Pyranoses can also be drawn as **chair conformations** (Section 19.3B). **Mutarotation** is the change in specific rotation that accompanies formation of an equilibrium mixture of α and β anomers in aqueous solution (Section 19.3C).

A **glycoside** is a cyclic acetal derived from a monosaccharide (Section 19.4A). An **alditol** is a polyhydroxy compound formed when the carbonyl group of a monosaccharide is reduced to a hydroxyl group (Section 19.4B). An **aldonic acid** is a carboxylic acid formed when the aldehyde group of an aldose is oxidized to a carboxyl group (Section 19.4C). Any carbohydrate that reacts with an oxidizing agent to form an aldonic acid is classified as a **reducing sugar** (it reduces the oxidizing agent).

A **disaccharide** contains two monosaccharide units joined by a glycosidic bond (Section 19.5). Terms applied to carbohydrates containing larger numbers of monosaccharides are **trisaccharide, tetrasaccharide, oligosaccharide,** and **polysaccharide. Sucrose** is a disaccharide consisting of D-glucose joined to D-fructose by an α-1,2-glycosidic bond (Section 19.5A). **Lactose** is a disaccharide consisting of D-galactose joined to D-glucose by a β-1,4-glycosidic bond (Section 19.5B). **Maltose** is a disaccharide of two molecules of D-glucose joined by an α-1,4-glycosidic bond (Section 19.5C).

Starch can be separated into two fractions, amylose and amylopectin (Section 19.6A). **Amylose** is a linear polysaccharide of as many as 4000 units of D-glucopyranose joined by α-1,4-glycosidic bonds. **Amylopectin** is a highly branched polysaccharide of D-glucose joined by α-1,4-glycosidic bonds and, at branch points, by α-1,6-glycosidic bonds. **Glycogen,** the reserve carbohydrate of animals, is a highly branched polysaccharide of D-glucopyranose joined by α-1,4-glycosidic bonds and, at branch points, by α-1,6-glycosidic bonds (Section 19.6B). **Cellulose,** the skeletal polysaccharide of plants, is a linear polysaccharide of D-glucopyranose joined by β-1,4-glycosidic bonds (Section 19.6C).

The carboxyl and sulfate groups of **acidic polysaccharides** are ionized to —COO$^-$ and —SO$_3^-$ at the pH of body fluids, which gives these polysaccharides net negative charges (Section 19.7).

KEY REACTIONS

1. **Formation of Cyclic Hemiacetals (Section 19.3A)** A monosaccharide existing as a five-membered ring is a furanose; one existing as a six-membered ring is a pyranose. A pyranose is most commonly drawn as either a Haworth projection or a chair conformation.

2. **Mutarotation (Section 19.3C)** Anomeric forms of a monosaccharide are in equilibrium in aqueous solution. Mutarotation is the change in specific rotation that accompanies this equilibration.

3. **Formation of Glycosides (Section 19.4A)** Treatment of a monosaccharide with an alcohol in the presence of an acid catalyst forms a cyclic acetal called a glycoside. The bond to the new —OR group is called a glycosidic bond.

4. **Reduction to Alditols (Section 19.4B)** Reduction of the carbonyl group of an aldose or ketose to a hydroxyl group yields a polyhydroxy compound called an alditol.

5. **Oxidation to an Aldonic Acid (Section 19.4C)** Oxidation of the aldehyde group of an aldose to a carboxyl group by a mild oxidizing agent gives a polyhydroxycarboxylic acid called an aldonic acid.

PROBLEMS

Numbers that appear in color indicate difficult problems. ▶ designates problems requiring application of principles.

19.6 Define carbohydrate.

19.7 What percentage of the dry weight of plants consists of carbohydrates? How does this percentage compare with that in humans?

Monosaccharides

19.8 What is the difference in structure between an aldose and a ketose? Between an aldopentose and a ketopentose?

19.9 Of the eight D-aldohexoses, which is the most abundant in the biological world?

19.10 Name the three most abundant hexoses in the biological world. Which are aldohexoses, and which are ketohexoses?

19.11 Which hexose is also known as "dextrose"?

19.12 What does it mean to say that D- and L-glyceraldehyde are enantiomers?

19.13 Explain the meaning of the designations D and L as used to specify the configuration of a monosaccharide.

19.14 Which carbon of an aldopentose determines whether the pentose has a D or L configuration?

19.15 How many stereocenters are present in D-glucose? In D-ribose? How many stereoisomers are possible for each monosaccharide?

19.16 Which of the following compounds are D-monosaccharides, and which are L-monosaccharides?

19.17 Draw Fischer projections for L-ribose and L-arabinose.

19.18 Draw a Fischer projection for a D-2-ketoheptose.

19.19 Explain why all mono- and disaccharides are soluble in water.

19.20 What is an amino sugar? Name the three amino sugars most commonly found in nature.

The Cyclic Structure of Monosaccharides

19.21 Define the term *anomeric carbon*. Which carbon is the anomeric carbon in glucose? In fructose?

19.22 Define:
(a) Pyranose
(b) Furanose

19.23 Explain the conventions for using α and β to designate the configurations of the cyclic forms of monosaccharides.

19.24 Are α-D-glucose and β-D-glucose anomers? Explain. Are they enantiomers? Explain.

19.25 Are the hydroxyl groups on carbons 1, 2, 3, and 4 of α-D-glucose all in equatorial positions?

19.26 In what way are chair conformations of aldohexoses a more accurate representation of the molecular shape than Haworth projections?

19.27 Convert each of the following Haworth projections to an open-chain form and to a Fischer projection. Name the monosaccharides you have drawn.

19.28 Convert each of the following chair conformations to an open-chain form and to a Fischer projection. Name the monosaccharides you have drawn.

19.29 Explain the phenomenon of mutarotation. How is it detected?

19.30 The specific rotation of α-D-glucose is $+112.2°$. What is the specific rotation of α-L-glucose?

19.31 When α-D-glucose is dissolved in water, the specific rotation of the solution changes from $+112.2°$ to $+52.7°$. Does the specific rotation of α-L-glucose also change when it is dissolved in water? If so, to what value does it change?

Reactions of Monosaccharides

19.32 Define the terms *glycoside* and *glycosidic bond*.

19.33 What is the difference in meaning between the terms *glycosidic bond* and *glucosidic bond*?

19.34 Do glycosides undergo mutarotation?

19.35 Draw Fischer projections for the product formed by treating each of the following monosaccharides with sodium borohydride, NaBH$_4$, in water.
(a) D-Galactose
(b) D-Ribose

19.36 Reduction of D-glucose by NaBH$_4$ gives D-sorbitol, a compound used in the manufacture of sugar-free gums and candies. Draw a structural formula for D-sorbitol.

19.37 Reduction of D-fructose by NaBH$_4$ gives two alditols, one of which is D-sorbitol. Name and draw a structural formula for the other alditol.

19.38 Ribitol and β-D-ribose 1-phosphate are both derivatives of D-ribose. Draw a structural formula for each compound.

19.39 One pathway for the metabolism of D-glucose 6-phosphate is its enzyme-catalyzed conversion to D-fructose 6-phosphate. Show that this transformation can be regarded as two enzyme-catalyzed keto-enol tautomerizations (Section 17.8).

```
       CHO                              CH₂OH
   H—— OH                                C=O
   HO ——H          enzyme          HO ——H
   H —— OH        catalysis         H —— OH
   H —— OH        ⇌                 H —— OH
       CH₂OPO₃²⁻                        CH₂OPO₃²⁻
   D-Glucose 6-phosphate        D-Fructose 6-phosphate
```

19.40 One step in glycolysis, the pathway that converts glucose to pyruvate (Section 27.2), involves an enzyme-catalyzed conversion of dihydroxyacetone phosphate to D-glyceraldehyde 3-phosphate. Show that this transformation can be regarded as two enzyme-catalyzed keto-enol tautomerizations (Section 17.8).

```
   CH₂OH        Enzyme         CHO
   C=O        catalysis     H —— OH
   CH₂OPO₃²⁻     ⇌           CH₂OPO₃²⁻
   Dihydroxyacetone        D-Glyceraldehyde
      phosphate              3-phosphate
```

Disaccharides and Oligosaccharides

19.41 Name three important disaccharides. From which monosaccharides is each derived?

19.42 What does it mean to describe a glycosidic bond as β-1,4-? To describe it as α-1,6-?

19.43 Both maltose and lactose are reducing sugars, but sucrose is not. Explain why.

19.44 Following is a structural formula for a disaccharide.

(a) Name the two monosaccharide units in the disaccharide.
(b) Describe the glycosidic bond.
(c) Is this disaccharide a reducing sugar?
(d) Will this disaccharide undergo mutarotation?

19.45 The disaccharide trehalose is found in young mushrooms and is the chief carbohydrate in the blood of certain insects.

Trehalose

(a) Name the two monosaccharide units in trehalose.
(b) Describe the glycosidic bond in trehalose.
(c) Is trehalose a reducing sugar?
(d) Will trehalose undergo mutarotation?

Polysaccharides

19.46 What is the difference in structure between oligosaccharides and polysaccharides?

19.47 Name three polysaccharides that are composed of units of D-glucose. In which are the glucose units joined by α-glycosidic bonds? In which are they joined by β-glycosidic bonds?

19.48 Starch can be separated into two principal polysaccharides, amylose and amylopectin. What is the major difference in structure between the two?

19.49 Where is glycogen stored in the human body?

19.50 Why is cellulose insoluble in water?

19.51 How is it possible that cows can digest grass but humans cannot?

19.52 A Fischer projection of N-acetyl-D-glucosamine is given in Section 19.2.

(a) Draw a Haworth projection and a chair conformation for the β-pyranose form of this monosaccharide.

(b) Draw a Haworth projection and a chair conformation for the disaccharide formed by joining two units of the pyranose form of N-acetyl-D-glucosamine by a β-1,4-glycosidic bond. If you draw them correctly, you will have a structural formula for the repeating dimer of chitin, the structural polysaccharide component of the shell of lobsters and other crustaceans.

19.53 Propose structural formulas for the repeating disaccharide unit in these polysaccharides.

(a) Alginic acid, isolated from seaweed, is used as a thickening agent in ice cream and other foods. Alginic acid is a polymer of D-mannuronic acid in the pyranose form joined by β-1,4-glycosidic bonds.

(b) Pectic acid is the main component of pectin, which is responsible for the formation of jellies from fruits and berries. Pectic acid is a polymer of D-galacturonic acid in the pyranose form joined by α-1,4-glycosidic bonds.

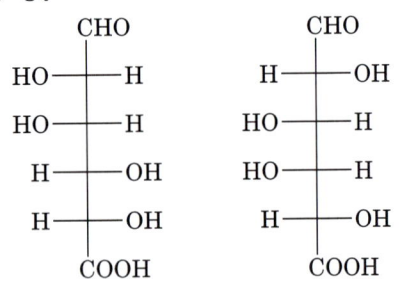

D-Mannuronic acid   D-Galacturonic acid

19.54 Hyaluronic acid acts as a lubricant in the synovial fluid of joints. In rheumatoid arthritis, inflammation breaks hyaluronic acid down to smaller molecules. Under these conditions, what happens to the lubricating power of the synovial fluid?

19.55 The anticlotting property of heparin is partly due to the negative charges it carries.

(a) Identify the functional groups that provide the negative charges.

(b) Which type of heparin is a better anticoagulant, one with a high or a low degree of polymerization?

Chemical Connections

19.56 (Chemical Connections 19A) Why does congenital galactosemia appear only in infants? Why can galactosemia be relieved by feeding an infant a formula containing sucrose as the only carbohydrate?

19.57 (Chemical Connections 19B) What is the difference in structure between L-ascorbic acid and L-dehydroascorbic acid? What does the designation L indicate in these molecules?

19.58 (Chemical Connections 19B) When L-ascorbic acid participates in a redox reaction, it is converted to L-dehydroascorbic acid. In this reaction, is L-ascorbic acid oxidized or reduced? Is L-ascorbic acid a biological oxidizing agent or a biological reducing agent?

19.59 (Chemical Connections 19C) Why is the glucose assay one of the most common analytical tests performed in clinical chemistry laboratories?

19.60 (Chemical Connections 19D) What monosaccharides do types A, B, and O blood have in common? In which monosaccharides do they differ?

19.61 (Chemical Connections 19D) L-Fucose is a monosaccharide unit common to A, B, AB, and O blood types.

(a) Is this monosaccharide an aldose or a ketose? A hexose or a pentose?

(b) What is unusual about this monosaccharide?

(c) If L-fucose were to undergo a reaction in which its terminal —CH_3 group was converted to a —CH_2OH group, which monosaccharide would result?

19.62 (Chemical Connections 19D) Why can't a person with A-type blood donate to a person with B-type blood?

19.63 (Chemical Connections 19E) What is high-fructose corn syrup, and how is it produced?

Additional Problems

19.64 2,6-Dideoxy-D-altrose, also known as D-digitoxose, is a monosaccharide obtained from the hydrolysis of digitoxin, a natural product extracted from purple foxglove (*Digitalis purpurea*). Digitoxin has found wide use in cardiology because it reduces pulse rate, regularizes heart rhythm, and strengthens heart beat. Draw a structural formula for the open-chain form of 2,6-dideoxy-D-altrose.

19.65 In making candy or sugar syrups, sucrose is boiled in water with a little acid, such as lemon juice. Why does the product mixture taste sweeter than the initial sucrose solution?

19.66 Hot-water extracts of ground willow bark are an effective pain reliever (Chemical Connections 18C). Unfortunately, the liquid is so bitter that most persons refuse it. The pain reliever in these infusions is salicin. Name the monosaccharide unit in salicin.

Salicin

19.67 Following is a Haworth projection and a chair conformation of the repeating disaccharide unit in chondroitin 6-sulfate. This biopolymer acts as the flexible connecting matrix between the tough protein filaments in cartilage. It is available as a dietary supplement, often combined with D-glucosamine sulfate. Some believe this combination can strengthen and improve joint flexibility.

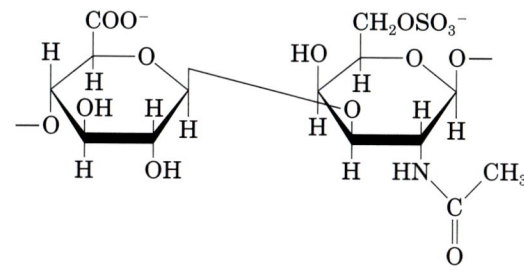

(a) From what two monosaccharide units is the repeating disaccharide unit of chondroitin 6-sulfate derived?

(b) Describe the glycosidic bond between the two units.

InfoTrac College Edition

For additional readings, go to InfoTrac College Edition, your online research library, at

http://infotrac.thomsonlearning.com

CHAPTER 20

- 20.1 Introduction
- 20.2 The Structure of Triglycerides
- 20.3 Properties of Triglycerides
- 20.4 Complex Lipids
- 20.5 Membranes
- 20.6 Glycerophospholipids
- 20.7 Sphingolipids
- 20.8 Glycolipids
- 20.9 Steroids
- 20.10 Steroid Hormones
- 20.11 Bile Salts
- 20.12 Prostaglandins, Thromboxanes, and Leukotrienes

Sea lions are marine mammals that require a heavy layer of fat to survive in cold waters.

Lipids

20.1 Introduction

Found in living organisms, **lipids** are a family of substances that are insoluble in water but soluble in nonpolar solvents and solvents of low polarity, such as diethyl ether.

Unlike in the case of carbohydrates, we define lipids in terms of a property and not in terms of their structure.

A Classification by Function

Lipids play three major roles in human biochemistry: (1) to store energy within fat cells, (2) as parts of membranes to separate compartments of aqueous solutions from each other, and (3) to serve as chemical messengers.

Storage An important use for lipids, especially in animals, is the storage of energy. As we saw in Section 19.6, plants store energy in the form of starch. Animals (including humans) find it more economical to use fats instead. Although our bodies do store some carbohydrates in the form of glycogen for quick energy when we need it, energy stored in the form of fats has much greater importance for us. The reason is simple: The burning of fats produces more than twice as much energy (about 9 kcal/g) as the burning of an equal weight of carbohydrates (about 4 kcal/g).

Membrane Components The lack of water solubility of lipids is an important property because our body chemistry is so firmly based on water. Most body constituents, including carbohydrates and proteins, are soluble in water. But the body also needs insoluble compounds for the membranes that separate compartments containing aqueous solutions from each other, whether they are cells or organelles within the cells. Lipids provide these membranes. Their water insolubility derives from the fact that the polar groups they contain are much smaller than their alkane-like (nonpolar) portions. These nonpolar portions provide the water-repellent, or *hydrophobic,* property.

Messengers Lipids also serve as chemical messengers. Primary messengers, such as steroid hormones, deliver signals from one part of the body to another part. Secondary messengers, such as prostaglandins and thromboxanes, mediate the hormonal response.

Classification by Structure

For purposes of study, we can classify lipids into four groups: (1) simple lipids, such as fats and waxes; (2) complex lipids; (3) steroids; and (4) prostaglandins, thromboxanes, and leukotrienes.

20.2 The Structure of Triglycerides

Animal fats and vegetable oils are triglycerides. **Triglycerides** are triesters of glycerol and long-chain carboxylic acids called fatty acids. In Section 18.3, we saw that esters are made up of an alcohol part and an acid part. As the name indicates, the alcohol of triglycerides is always glycerol.

$$\begin{array}{c} CH_2-OH \\ | \\ CH-OH \\ | \\ CH_2-OH \end{array}$$
Glycerol

In contrast to the alcohol part, the acid component of fats may be any number of acids, which do, however, have certain things in common:

1. Fatty acids are practically all unbranched carboxylic acids.
2. They range in size from about 10 to 20 carbons.
3. They contain an even number of carbon atoms.
4. Apart from the —COOH group, they have no functional groups except that some do have double bonds.
5. In most fatty acids that have double bonds, the cis isomers predominate.

Only even-numbered acids are found in fats because the body builds these acids entirely from acetate units and therefore puts the carbons in two at a time (Section 28.3). Table 20.1 shows the most important acids found in fats.

The fatty acids can be divided into two groups: saturated and unsaturated. **Saturated fatty acids** have only single bonds in the hydrocarbon chain. **Unsaturated fatty acids** have at least one $C=C$ double bond in the chain. All the unsaturated fatty acids listed in Table 20.1 are cis isomers. This fact explains their physical properties, as reflected in their melting points. Saturated fatty acids are solids at room temperature because the

Table 20.1 The Most Important Fatty Acids in Triglycerides

Carbon Atoms: Double Bonds*	Structure	Common Name	Melting Point (°C)
Saturated Fatty Acids			
12 : 0	$CH_3(CH_2)_{10}COOH$	lauric acid	44
14 : 0	$CH_3(CH_2)_{12}COOH$	myristic acid	58
16 : 0	$CH_3(CH_2)_{14}COOH$	palmitic acid	63
18 : 0	$CH_3(CH_2)_{16}COOH$	stearic acid	70
20 : 0	$CH_3(CH_2)_{18}COOH$	arachidic acid	77
Unsaturated Fatty Acids			
16 : 1	$CH_3(CH_2)_5CH=CH(CH_2)_7COOH$	palmitoleic acid	1
18 : 1	$CH_3(CH_2)_7CH=CH(CH_2)_7COOH$	oleic acid	16
18 : 2	$CH_3(CH_2)_4(CH=CHCH_2)_2(CH_2)_6COOH$	linoleic acid	−5
18 : 3	$CH_3CH_2(CH=CHCH_2)_3(CH_2)_6COOH$	linolenic acid	−11
20 : 4	$CH_3(CH_2)_4(CH=CHCH_2)_4(CH_2)_2COOH$	arachidonic acid	−49

*The first number is the number of carbons in the fatty acid; the second number is the number of carbon–carbon double bonds in its hydrocarbon chain.

regular nature of their aliphatic chains allows the molecules to be packed in a close, parallel alignment:

The longer the aliphatic chain, the higher the melting point.

The interactions (London dispersion forces) between neighboring chains are weak. Nevertheless, the regular packing allows these forces to operate over a large portion of the chain, ensuring that a considerable amount of energy is needed to melt them.

Unsaturated fatty acids, in contrast, are all liquids at room temperature because the cis double bonds interrupt the regular packing of the chains:

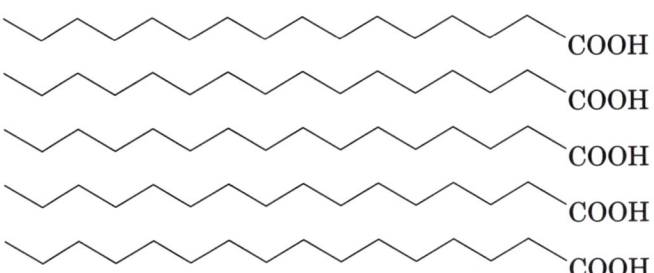

Here the London dispersion forces act over shorter segments of the chain, so much less energy is required to melt them. The greater the degree of unsaturation, the lower the melting point because each double bond introduces more disorder into the packing of the molecules (Table 20.1).

In **triglycerides** (also called **triacylglycerols**) all three —OH groups of glycerol are esterified. Thus a typical fat molecule might be

$$\begin{array}{c} \text{Oleate (18:1)} \qquad\qquad \text{Palmitate (16:0)} \\ \text{CH}_3(\text{CH}_2)_7\text{CH}=\text{CH}(\text{CH}_2)_7\text{COCH} \\ | \\ \text{CH}_2\text{OC}(\text{CH}_2)_{14}\text{CH}_3 \\ | \\ \text{CH}_2\text{OC}(\text{CH}_2)_{16}\text{CH}_3 \qquad \text{Stearate (18:0)} \\ \text{A triglyceride} \end{array}$$

Triglycerides are the most common lipid materials, although **mono-** and **diglycerides** are not infrequent. In the latter two types, only one or two —OH groups of the glycerol are esterified by fatty acids.

Triglycerides are complex mixtures. Although some of the molecules contain three identical fatty acids, in most cases two or three different acids are present. The hydrophobic character of fats is caused by the long hydrocarbon chains. The ester groups (—C(=O)—O—) although polar themselves, are buried in a nonpolar environment, which makes the fats insoluble in water.

■ **Some vegetable oils.**

20.3 Properties of Triglycerides

A Physical State

With some exceptions, **fats** that come from animals are generally solids at room temperature, and those from plants or fish are usually liquids. Liquid fats are often called **oils,** even though they are esters of glycerol just like solid fats and should not be confused with petroleum, which is mostly alkanes.

What is the structural difference between solid fats and liquid oils? In most cases, it is the degree of unsaturation. The physical properties of the fatty acids are carried over to the physical properties of the triglycerides. Solid animal fats contain mainly saturated fatty acids, whereas vegetable

> **Fat** A mixture of triglycerides containing a high proportion of long-chain, saturated fatty acids

> **Oil** A mixture of triglycerides containing a high proportion of long-chain, unsaturated fatty acids or short-chain, saturated fatty acids

CHEMICAL CONNECTIONS 20A

Rancidity

The double bonds in fats and oils are subject to oxidation by the air (see Section 13.5C). When a fat or oil is allowed to stand out in the open, this reaction slowly turns some of the molecules into aldehydes and other compounds with bad tastes and odors. We then say that the fat or oil has become rancid and is no longer edible. Vegetable oils, which generally contain more double bonds, are more susceptible to this than solid fats, but even fats contain some double bonds, so rancidity can be a problem here, too.

Another cause of unpleasant taste is hydrolysis. The hydrolysis of triglycerides may produce short-chain fatty acids, such as butanoic acid (butyric acid), which have bad odors. To prevent rancidity, fats and oils should be kept refrigerated (these reactions are slower at low temperatures) and in dark bottles (the oxidation is catalyzed by ultraviolet light). In addition, antioxidants are often added to fats and oils to prevent rancidity.

CHEMICAL CONNECTIONS 20B

Waxes

Plant and animal waxes are simple esters. They are solids because of their high molecular weights. As in fats, the acid portions of the esters consist of a mixture of fatty acids, but the alcohol portions are not glycerol but simple long-chain alcohols. For example, a major component of beeswax is 1-triacontyl palmitate:

Palmitic acid portion → $CH_3(CH_2)_{13}CH_2\overset{O}{\underset{\|}{C}}-OCH_2(CH_2)_{28}CH_3$ ← 1-Triacontanol portion

Triacontyl palmitate

Waxes generally have higher melting points than fats (60 to 100° C) and are harder. Animals and plants often use them for protective coatings. For example, the leaves of most plants are coated with wax, which helps to prevent microorganisms from attacking them and allows the plants to conserve water. The feathers of birds and the fur of animals are also coated with wax.

Important waxes include carnauba wax (from a Brazilian palm tree), lanolin (from lamb's wool), beeswax, and spermaceti (from whales). These substances are used to make cosmetics, polishes, candles, and ointments. Paraffin waxes are not esters. They are mixtures of high-molecular-weight alkanes. Neither is ear wax a simple ester. It is a gland secretion and contains a mixture of fats (triglycerides), phospholipids, and esters of cholesterol.

---

oils contain high amounts of unsaturated fatty acids. Table 20.2 shows the average fatty acid content of some common fats and oils. Note that even solid fats contain some unsaturated acids and that liquid fats contain some saturated acids. Some unsaturated fatty acids (linoleic and linolenic acids) are called *essential fatty acids* because the body cannot synthesize them from precursors; they must therefore be included in the diet.

Although most vegetable oils contain high amounts of unsaturated fatty acids, there are exceptions. Coconut oil, for example, has only a small amount of unsaturated acids. This oil is a liquid not because it contains

Table 20.2 Average Percentage of Fatty Acids of Some Common Fats and Oils

	Saturated				Unsaturated			
	Lauric	Myristic	Palmitic	Stearic	Oleic	Linoleic	Linolenic	Other
Animal Fats								
Beef tallow	—	6.3	27.4	14.1	49.6	2.5	—	0.1
Butter	2.5	11.1	29.0	9.2	26.7	3.6	—	17.9
Human	—	2.7	24.0	8.4	46.9	10.2	—	7.8
Lard	—	1.3	28.3	11.9	47.5	6.0	—	5.0
Vegetable Oils								
Coconut	45.4	18.0	10.5	2.3	7.5	—	—	16.3
Corn	—	1.4	10.2	3.0	49.6	34.3	—	1.5
Cottonseed	—	1.4	23.4	1.1	22.9	47.8	—	3.4
Linseed	—	—	6.3	2.5	19.0	24.1	47.4	0.7
Olive	—	—	6.9	2.3	84.4	4.6	—	1.8
Peanut	—	—	8.3	3.1	56.0	26.0	—	6.6
Safflower	—	—	6.8	—	18.6	70.1	3.4	1.1
Soybean	0.2	0.1	9.8	2.4	28.9	52.3	3.6	2.7
Sunflower	—	—	5.6	2.2	25.1	66.2	—	—

many double bonds, but rather because it is rich in low-molecular-weight fatty acids (chiefly lauric acid).

Oils with an average of more than one double bond per fatty acid chain are called *polyunsaturated*. Their role in the human diet is discussed in Section 29.4.

Pure fats and oils are colorless, odorless, and tasteless. This statement may seem surprising because we all know the tastes and colors of such fats and oils as butter and olive oil. The tastes, odors, and colors are caused by small amounts of other substances dissolved in the fat or oil.

■ Many common products contain hydrogenated vegetable oils.

B Hydrogenation

In Section 12.6D, we learned that we can reduce carbon–carbon double bonds to single bonds by treating them with hydrogen (H_2) and a catalyst. It is, therefore, not difficult to convert unsaturated liquid oils to solids. For example:

$$\begin{array}{l}
CH_2-O-\overset{O}{\underset{\|}{C}}-(CH_2)_7-CH=CH-(CH_2)_7-CH_3 \quad \text{(Oleic acid)} \\
CH-O-\overset{O}{\underset{\|}{C}}-(CH_2)_7-CH=CHCH_2CH=CH-(CH_2)_4-CH_3 \quad \text{(Linoleic acid)} \\
CH_2-O-\overset{O}{\underset{\|}{C}}-(CH_2)_7-CH=CHCH_2CH=CH-(CH_2)_4-CH_3 \quad \text{(Linoleic acid)}
\end{array} + 5H_2 \xrightarrow{Pt} \begin{array}{l}
CH_2-O-\overset{O}{\underset{\|}{C}}-(CH_2)_{16}-CH_3 \quad \text{(Stearic acid)} \\
CH-O-\overset{O}{\underset{\|}{C}}-(CH_2)_{16}-CH_3 \\
CH_2-O-\overset{O}{\underset{\|}{C}}-(CH_2)_{16}-CH_3
\end{array}$$

This hydrogenation is carried out on a large scale to produce the solid shortening sold in stores under such brand names as Crisco, Spry, and Dexo. In making such products, manufacturers must be careful not to hydrogenate all the double bonds, because a fat with no double bonds would be *too* solid. Partial, but not complete, hydrogenation results in a product with just the right consistency for cooking.

Margarine is also made by partial hydrogenation of vegetable oils.

Because less hydrogen is used, margarine contains more unsaturation than hydrogenated shortenings.

C Saponification

Glycerides, being esters, are subject to hydrolysis, which can be carried out with either acids or bases. As we saw in Section 18.6, the use of bases is more practical. An example of the saponification of a typical fat is

$$\underset{\text{A triglyceride}}{\begin{array}{l} O \\ \| \\ RCOCH \\ | \\ CH_2OCR \\ \| \\ O \end{array} \begin{array}{l} CH_2OCR \\ \| \\ O \end{array}} + 3NaOH \xrightarrow{\text{saponification}} \underset{\substack{\text{1,2,3-Propanetriol} \\ \text{(Glycerol; Glycerin)}}}{\begin{array}{l} CH_2OH \\ | \\ CHOH \\ | \\ CH_2OH \end{array}} + \underset{\text{Sodium soaps}}{3RCO^-Na^+}$$

Thus saponification is the base-catalyzed hydrolysis of fats and oils producing glycerol and a mixture of fatty acid salts called soaps. **Soap** has been used for thousands of years, and saponification is one of the oldest known chemical reactions (see Chemical Connections 20C).

CHEMICAL CONNECTIONS 20C

Soaps and Detergents

Soaps clean because each soap molecule has a hydrophilic head and a hydrophobic tail. The —COO⁻ end of the molecule (the hydrophilic head), being ionic, is soluble in water. The other end (the hydrophobic tail) is a long-chain alkane-like portion, which is insoluble in water but soluble in organic compounds. The soap is needed to remove the water-insoluble portion of dirt. The hydrophilic head of the soap molecule dissolves in the water; the hydrophobic tail dissolves in the dirt. This causes an emulsifying action in which the soap molecules surround the dirt particles: The hydrophobic tail interacts with the hydrophobic dirt particle and the hydrophilic head provides the attraction for water molecules (figure, part c). The cluster of soap molecules in which the dirt is embedded is called a **micelle** (figure, part d).

The solid soaps and soap flakes with which we are all familiar are sodium salts of fatty acids. Liquid soaps are generally potassium salts of the same acids. Soap has the disadvantage of forming precipitates in "hard water," which is water that contains relatively high amounts of Ca^{2+} and Mg^{2+} ions. Most community water supplies in the United States are hard water. Calcium and magnesium salts of fatty acids are insoluble in water. Therefore, when soaps are used in hard water, the Ca^{2+} and Mg^{2+} ions precipitate the fatty acid anions, which deposit on clothes, dishes, sinks, and bathtubs (forming the "ring" around the bathtub).

Because of these difficulties, much of the soap used in the United States has been replaced in the last 50 years by **detergents** for both household and industrial cleaning. Two typical detergent molecules are

$$CH_3(CH_2)_{11}-O-\overset{\overset{O}{\|}}{\underset{\underset{O}{\|}}{S}}-O^-\ Na^+$$

Sodium dodecyl sulfate
(Sodium lauryl sulfate)

$$CH_3(CH_2)_{10}CH_2-\bigcirc-\overset{\overset{O}{\|}}{\underset{\underset{O}{\|}}{S}}-O^-\ Na^+$$

Sodium dodecylbenzenesulfonate

These molecules also have hydrophobic tails and hydrophilic heads and so can function in the same way as soap molecules. However, because they do not precipitate with Ca^{2+} and Mg^{2+}, they work very well in hard water.

■ The cleaning action of soap. (*a*) Without soap, water molecules cannot penetrate through molecules of hydrophobic dirt. (*b* and *c*) Soap molecules line up at the interface between dirt and water. (*d*) Soap molecules carry dirt particles away.

20.4 Complex Lipids

The triglycerides discussed in the previous sections are significant components of fat storage cells. Other kinds of lipids, called complex lipids, are important in a different way. They constitute the main components of membranes (Section 20.5). Complex lipids can be divided into two groups: phospholipids and glycolipids.

Phospholipids contain an alcohol, two fatty acids, and a phosphate group. There are two types: **glycerophospholipids** and **sphingolipids**. In glycerophospholipids, the alcohol is glycerol (Section 20.6). In sphingolipids, the alcohol is sphingosine (Section 20.7).

Glycolipids are complex lipids that contain carbohydrates (Section 20.8). Figure 20.1 shows schematic structures for all of these lipids.

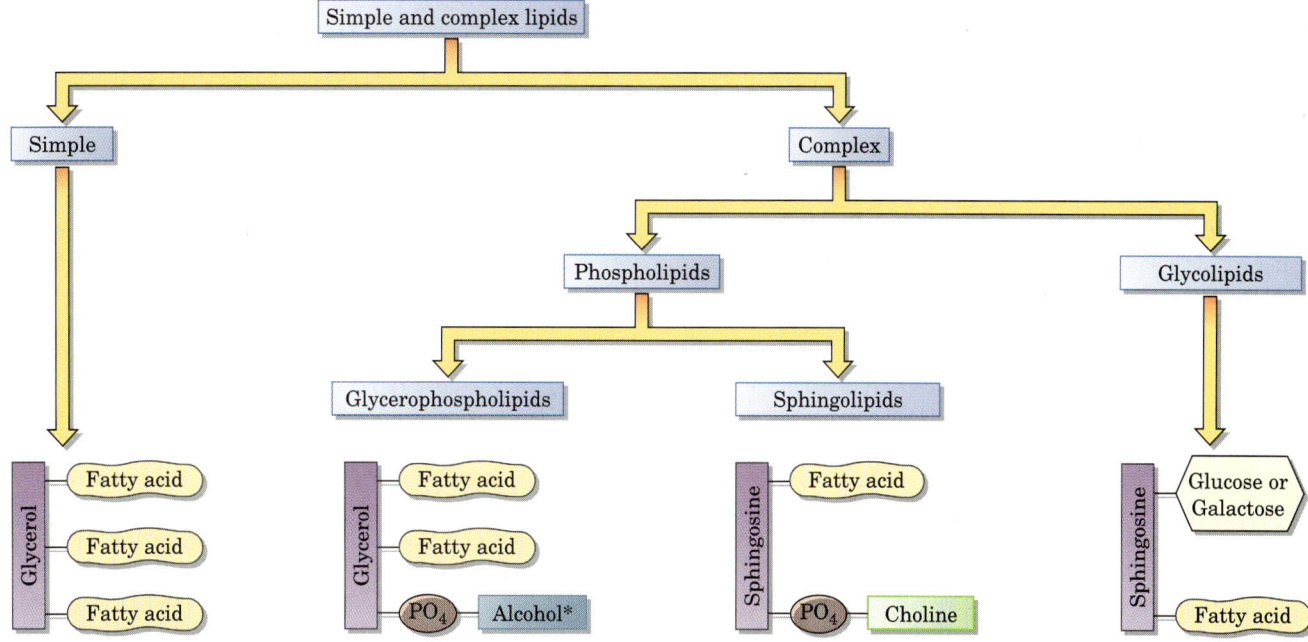

Figure 20.1 Schematic diagram of simple and complex lipids.
*The alcohol can be choline, serine, ethanolamine, inositol, or certain others.

20.5 Membranes

The complex lipids mentioned in Section 20.4 form the **membranes** around body cells and around small structures inside the cells. Unsaturated fatty acids are important components of these lipids. The cell membranes separate cells from the external environment and provide selective transport for nutrients and waste products into and out of cells.

These membranes are made of **lipid bilayers** (Figure 20.2). In a lipid bilayer, two rows (layers) of complex lipid molecules are arranged tail to tail. The hydrophobic tails point toward each other, which enables them to get as far away from the water as possible. This arrangement leaves the hydrophilic heads projecting to the inner and outer surfaces of the membrane. Cholesterol (Section 20.9), another membrane component, also positions the hydrophilic portion of the molecule on the surface of the membranes and the hydrophobic portion inside the bilayer.

The unsaturated fatty acids prevent the tight packing of the hydrophobic chains in the lipid bilayer, thereby providing a liquid-like character to the membranes. This property is of extreme importance because many products of the body's biochemical processes must cross the membrane, and the liquid nature of the lipid bilayer allows such transport.

The lipid part of the membrane serves as a barrier against any movement of ions or polar compounds into and out of the cells. In the lipid bilayer, protein molecules are either suspended on the surface (peripheral proteins) or partly or fully embedded in the bilayer (integral proteins). These proteins stick out either on the inside or on the outside of the membrane; others are thoroughly embedded, going through the bilayer and projecting from both sides. The model shown in Figure 20.2, called the **fluid mosaic model** of membranes, allows the passage of nonpolar compounds by diffusion, as these compounds are soluble in the lipid membranes. In

These small structures inside the cell are called organelles.

Most lipid molecules in the bilayer contain at least one unsaturated fatty acid.

This effect is similar to the one that causes unsaturated fatty acids to have lower melting points than saturated fatty acids.

The term mosaic refers to the topography of the bilayers: protein molecules dispersed in the lipid. The term fluid is used because the free lateral motion in the bilayers makes it liquid-like.

CHEMICAL CONNECTIONS 20D

Transport Across Cell Membranes

Membranes are not just random assemblies of complex lipids that provide a nondescript barrier. In human red blood cells, for example, the outer part of the bilayer is made largely of phosphatidylcholine and sphingomyelin, while the inner part consists mostly of phosphatidylethanolamine and phosphatidyl serine (Sections 20.6 and 20.7). In another example, in the membrane called sarcoplasmic reticulum in the heart muscles, phosphatidylethanolamine is found in the outer part of the membrane, phosphatidylserine is found in the inner part, and phosphatidyl choline is equally distributed in the two layers of the membrane.

Membranes are not static structures, either. In many processes, they fuse with one another; in others, they disintegrate and their building blocks are used elsewhere. When membranes fuse in vacuole fusions inside of cells, for example, certain restrictions prevent incompatible membranes from intermixing.

The protein molecules are not dispersed randomly in the bilayer. Sometimes they cluster in patches, but they also appear in regular geometric patterns. An example of the latter are **gap junctions**, channels made of six proteins that create a central pore. These channels allow neighboring cells to communicate. Gap junctions are an example of **passive transport**. Small polar molecules—which include such essential nutrients as inorganic ions, sugars, amino acids, and nucleotides—can readily pass through gap junctions. Large molecules, such as proteins, polysaccharides, and nucleic acids, cannot.

In **facilitated transport**, a specific interaction takes place between the transporter and the transported molecule. Consider the **anion transporter** of the red blood cells, through which chloride and bicarbonate ions are exchanged in a one-for-one ratio. The transporter is a protein with 14 helical structures that span the membrane. One side of the helices contains the hydrophobic parts of the protein, which can interact with the lipid membrane. The other side of the helices forms a channel. This channel contains the hydrophilic portions of the protein, which can interact with the hydrated ions. In this manner, the anion passes through the erythrocyte membrane.

Active transport involves the passage of ions through a concentration gradient. For example, a higher concentration of K^+ is found inside cells than outside in the surrounding environment. Nevertheless, potassium ions can be transported from the outside into a cell, but at the expense of energy. The transporter, a membrane protein called Na^+, K^+ ATPase, uses the energy from the hydrolysis of ATP molecules to change the conformation of the transporter, which brings in K^+ and exports Na^+. Detailed studies of K^+ ion channels have revealed that K^+ ions enter the channel in pairs. Each of these hydrated ions carries eight water molecules in its solvation layers, with the negative pole of the water molecules (the oxygen atoms) surrounding the positive ion. Deeper in the channel, K^+ ions encounter a constriction, called a selectivity filter. To pass through, the K^+ ions must shed their water molecules. The pairing of the naked K^+ ions generates enough electrostatic repulsion to keep the ions moving through the channel. The channel itself is lined with oxygen atoms to provide a similar attractive environment like that offered by the stable hydrated form before entry.

Polar compounds, in general, are transported through specific **transmembrane channels**.

■ Gap junctions are made of six cylindrical protein subunits. They line up in two plasma membranes parallel to each other, forming a pore. The pores of gap junctions are closed by a sliding and twisting motion of the cylindrical subunits.

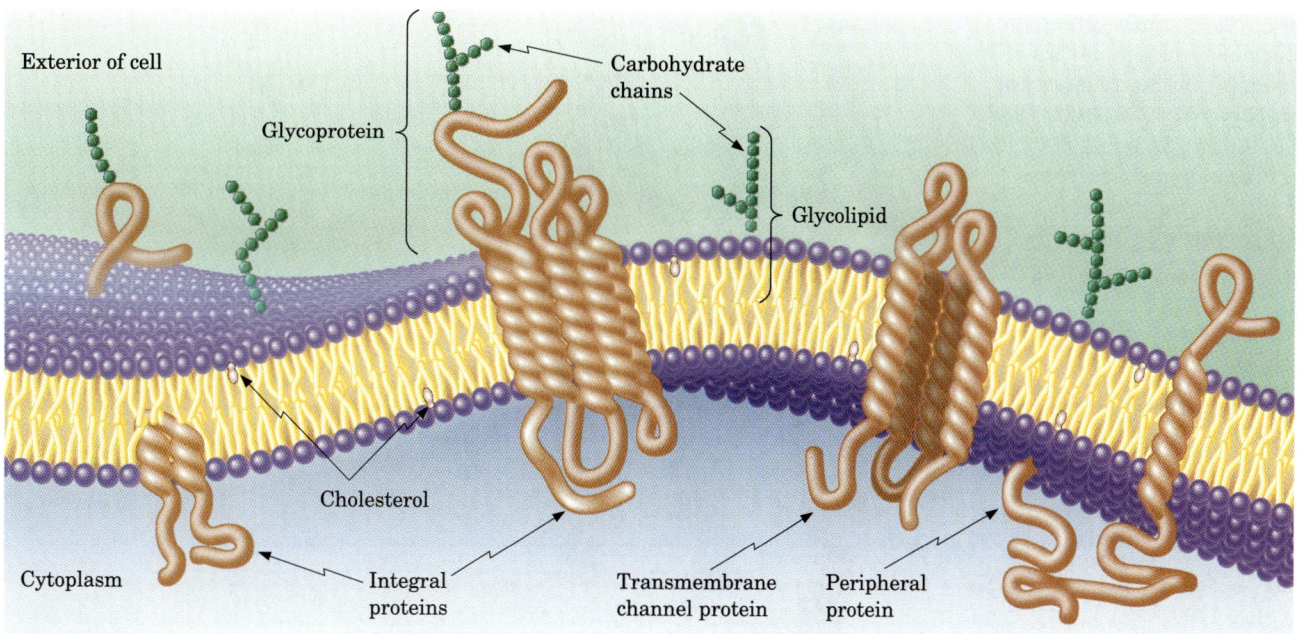

Figure 20.2 The fluid mosaic model of membranes. Note that proteins are embedded in the lipid matrix.

contrast, polar compounds are transported either through specific channels through the protein regions or by a mechanism called active transport (See Chemical Connections 20D). For any transport process, the membrane must behave like a nonrigid liquid so that the proteins can move sideways within the membrane.

20.6 Glycerophospholipids

The structure of glycerophospholipids (also called phosphoglycerides) is very similar to that of fats. The alcohol is glycerol. Two of the three —OH groups are esterified by fatty acids. As with the simple fats, these fatty acids may be any long-chain carboxylic acids, with or without double bonds. The third —OH group is esterified not by a fatty acid, but rather by a phosphate group, which is also esterified to another alcohol. If the other alcohol is choline, a quaternary ammonium compound, the glycerophospholipids are called **phosphatidylcholines** (common name **lecithin**):

Glycerophospholipids are membrane components of cells throughout the body.

In all glycerophospholipids, lecithins, cephalins, and phosphatidylinositols, the fatty acid on carbon 2 of glycerol is always unsaturated.

This typical lecithin molecule has stearic acid on one end and linoleic acid in the middle. Other lecithin molecules contain other fatty acids, but the one on the end is always saturated and the one in the middle is always unsaturated. Lecithin is a major component of egg yolk. Having both polar and nonpolar portions within one molecule, it is an excellent emulsifier (see Chemical Connections 6D) and is used in mayonnaise.

Figure 20.3 Space-filling molecular models of complex lipids in a bilayer. *(From* Biochemistry *by Lubert Stryer © 1975, 1981, 1988, 1995 by Lubert Stryer; © 2002 by W. H. Freeman and Company. Used with the permission of W. H. Freeman and Company.)*

Note that lecithin has a negatively charged phosphate group and a positively charged quaternary nitrogen from the choline. These charged parts of the molecule provide a strongly hydrophilic head, whereas the rest of the molecule is hydrophobic. Thus, when a phospholipid such as lecithin is part of a lipid bilayer, the hydrophobic tail points toward the middle of the bilayer, and the hydrophilic heads line both the inner and outer surfaces of the membranes (Figures 20.2 and 20.3).

Lecithins are just one example of glycerophospholipids. Another is the **cephalins,** which are similar to the lecithins in every way except that, instead of choline, they contain other alcohols, such as ethanolamine or serine:

R = fatty acid portions.

A phosphatidylethanolamine (a cephalin)

A phosphatidylserine (a cephalin)

Another important group of glycerophospholipids is the **phosphatidylinositols (PI).** In PI, the alcohol inositol is linked to the rest of the molecule by a phosphate ester linkage. Such compounds not only are integral structural parts of the biological membranes, but also, in their higher phosphorylated form, such as **phosphatidylinositol 4,5-bisphosphates (PIP2),** serve as signaling molecules in chemical communication (see Chapter 23).

Phosphatidylinositols, PI

20.7 Sphingolipids

Myelin, the coating of nerve axons, contains a different kind of complex lipid: **sphingolipids.** In sphingolipids, the alcohol portion is sphingosine:

$$CH_3(CH_2)_{12}CH=CH-CH(OH)-CH(NH_2)-CH_2OH$$

Sphingosine

A long-chain fatty acid is connected to the —NH_2 group by an amide bond, and the —OH group at the end of the chain is esterified by phosphorylcholine:

A sphingomyelin
(a sphingolipid)

Ceramide portion

Sphingomyelin
(schematic diagram)

Johann Thudichum, who discovered sphingolipids in 1874, named these brain lipids after the monster of Greek mythology, the sphinx. Part woman and part winged lion, the sphinx devoured all who could not provide the correct answer to her riddles. Sphingolipids appeared to Thudichum as part of a dangerous riddle of the brain.

The combination of a fatty acid and sphingosine (shown in color) is called the **ceramide** portion of the molecule, because many of these compounds are also found in cerebrosides (Section 20.8). The ceramide part of complex lipids may contain different fatty acids; stearic acid, for example, occurs mainly in sphingomyelin.

CHEMICAL CONNECTIONS 20E

The Myelin Sheath and Multiple Sclerosis

The human brain and spinal cord can be divided into gray and white regions. Forty % of the human brain is white matter. Microscopic examination reveals that the white matter consists of nerve axons wrapped in a white lipid coating, called the **myelin sheath,** which provides insulation and thereby allows the rapid conduction of electrical signals. The myelin sheath consists of 70% lipids and 30% proteins in the usual lipid bilayer structure.

Specialized cells, called **Schwann cells,** wrap themselves around the peripheral nerve axons to form numerous concentric layers. In the brain, other cells do the wrapping in a similar manner.

Multiple sclerosis affects 250,000 people in the United States. In this disease, the myelin sheath gradually deteriorates. Symptoms, which include muscle weariness, lack of coordination, and loss of vision, may vanish for a time but later return with greater severity. Autopsy of multiple sclerotic brains shows scar-like plaques of white matter, with bare axons not covered by myelin sheaths. The symptoms occur because the demyelinated axons cannot conduct nerve impulses. A secondary effect of the demyelination is damage to the axon itself.

Similar demyelination occurs in the Guillain-Barré syndrome that follows certain viral infections. In 1976, fears of a "swine flu" epidemic prompted a vaccination program that precipitated a number of cases of Guillain-Barré syndrome. This disease can lead to a paralysis that can cause death unless artificial breathing is supplied. In the 1976 cases, the U.S. government assumed legal responsibility for the few bad vaccines and paid compensation to the victims and their families.

■ **Myelination of a nerve axon outside the brain by a Schwann cell. The myelin sheath is produced by the Schwann cell and is rolled around the nerve axon for insulation.**

Sphingomyelins are the most important lipids of myelin sheaths of nerve cells and are associated with diseases such as multiple sclerosis (Chemical Connections 20E). Sphingolipids are not randomly distributed in membranes. In viral membranes, for example, most of the sphingomyelin appears on the inside of the membrane.

20.8 Glycolipids

Glycolipids are complex lipids that contain carbohydrates and ceramides. One group, the **cerebrosides,** are ceramide mono- or oligosaccharides. Other groups, such as the **gangliosides,** contain a more complex carbohydrate structure (see Chemical Connections 20F). In cerebrosides, the fatty acid of the ceramide part may contain either 18-carbon or 24-carbon chains; the latter is found only in these complex lipids. A glucose or galactose carbohydrate unit forms a beta glycosidic bond with the ceramide portion of the molecule. The cerebrosides occur primarily in the brain (7% of the dry weight) and at nerve synapses.

20.8 GLYCOLIPIDS

Glucocerebroside structure showing β-D-Glucose linked to Ceramide.

EXAMPLE 20.1

A lipid isolated from the membrane of red blood cells had the following structure:

$$\begin{array}{l} CH_2-O-\overset{O}{\underset{\|}{C}}-(CH_2)_{14}CH_3 \\ CH-O-\overset{O}{\underset{\|}{C}}-(CH_2)_7CH=CH(CH_2)_7CH_3 \\ CH_2-O-\overset{O}{\underset{\underset{O^-}{|}}{P}}-O-CH_2CH_2NH_3^+ \end{array}$$

(a) To what group of complex lipids does this compound belong?
(b) What are the components?

Solution
(a) The molecule is a triester of glycerol and contains a phosphate group; therefore, it is a glycerophospholipid.
(b) Besides glycerol and phosphate, it has a palmitic acid and an oleic acid component. The other alcohol is ethanolamine. Therefore, it belongs to the subgroup of cephalins.

Problem 20.1

A complex lipid had the following structure:

$$\begin{array}{l} CH_2-O-\overset{O}{\underset{\|}{C}}-(CH_2)_{12}CH_3 \\ CH-O-\overset{O}{\underset{\|}{C}}-(CH_2)_7CH=CHCH_2CH=CH(CH_2)_4CH_3 \\ CH_2-O-\overset{O}{\underset{\underset{O^-}{|}}{P}}-O-CH_2\underset{\underset{NH_3^+}{|}}{C}HCOO^- \end{array}$$

(a) To what group of complex lipids does this compound belong?
(b) What are the components?

CHEMICAL CONNECTIONS 20F

Lipid Storage Diseases

Complex lipids are constantly being synthesized and decomposed in the body. In several genetic diseases classified as lipid storage diseases, some of the enzymes needed to decompose the complex lipids are defective or missing. As a consequence, the complex lipids accumulate and cause enlarged liver and spleen, mental retardation, blindness, and, in certain cases, early death. Table 20F summarizes some of these diseases and indicates the missing enzyme and the accumulating complex lipid.

At present there is no treatment for these diseases. The best way to prevent them is by genetic counseling. Some of the diseases can be diagnosed during fetal development. For example, Tay-Sachs disease, which affects about 1 in every 30 Jewish Americans (versus 1 in 300 in the non-Jewish population), can be diagnosed from amniotic fluid obtained by amniocentesis.

■ The accumulation of glucocerebrosides in the cell of a patient with Gaucher's disease. These Gaucher cells infiltrate the bone marrow.

Table 20F Lipid Storage Diseases

Name	Accumulating Lipid	Missing or Defective Enzyme
Gaucher's disease	Glucocerebroside	β-Glucosidase
Krabbe's leukodystrophy	Galactocerebroside	β-Galactosidase
Fabry's disease	Ceramide trihexoside	α-Galactosidase

CHEMICAL CONNECTIONS 20F

Lipid Storage Diseases, (continued)

Table 20F Lipid Storage Diseases, (continued)

20.9 Steroids

The third major class of lipids is the **steroids,** which are compounds containing this ring system:

In this structure, three cyclohexane rings (A, B, and C) are connected in the same way as in phenanthrene (Section 13.3D); a fused cyclopentane ring (D) is also present. Steroids are thus completely different in structure from the lipids already discussed. Note that they are not necessarily esters, although some of them are.

A Cholesterol

The most abundant steroid in the human body, and the most important, is **cholesterol:**

Cholesterol

Cholesterol serves as a plasma membrane component in all animal cells—for example, in red blood cells. Its second important function is to serve as a raw material for the synthesis of other steroids, such as the sex and adrenocorticoid hormones (Section 20.10) and bile salts (Section 20.11).

Cholesterol exists both in the free form and esterified with fatty acids. Gallstones contain free cholesterol (Figure 20.4).

Because the correlation between high serum cholesterol levels and such diseases as atherosclerosis has received so much publicity, many people are afraid of cholesterol and regard it as some kind of poison. Far from being poisonous, cholesterol is, in fact, necessary for human life. In essence, our livers manufacture cholesterol that satisfies our needs even without dietary intake.

Cholesterol in the body is in a dynamic state. It constantly circulates in the blood. Cholesterol and esters of cholesterol, being hydrophobic, need a water-soluble carrier to circulate in the aqueous medium of blood.

B Lipoproteins—Carriers of Cholesterol

Cholesterol, along with fat, is transported by **lipoproteins.** Most lipoproteins contain a core of hydrophobic lipid molecules surrounded by a shell of hydrophilic molecules such as proteins and phospholipids (Figure 20.5). As summarized in Table 20.3, there are four kinds of lipoproteins:

- **High-density lipoprotein (HDL),** which has a protein content of about 33% and about 30% cholesterol
- **Low-density lipoprotein (LDL),** in which the protein content is only about 25%, but which contains about 50% cholesterol
- **Very-low-density lipoprotein (VLDL),** which mostly carries triglycerides (fats) synthesized by the liver
- **Chylomicrons,** which carry dietary lipids synthesized in the intestines.

C Transport of Cholesterol in LDL

The transport of cholesterol from the liver starts out as a large VLDL particle (55 nanometers in diameter). The core of VLDL contains triglycerides and cholesteryl esters, mainly cholesteryl linoleate. It is surrounded by a polar coat of phospholipids and proteins (Figure 20.5). The VLDL is carried

---

When the cholesterol level goes above 150 mg/100 mL, cholesterol synthesis in the liver is reduced to half the normal rate of production.

Lipoproteins Spherically shaped clusters containing both lipid molecules and protein molecules

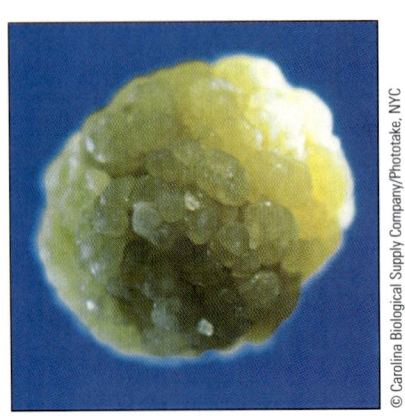

Figure 20.4 A human gallstone is almost pure cholesterol. This gallstone measures 5 mm in diameter.

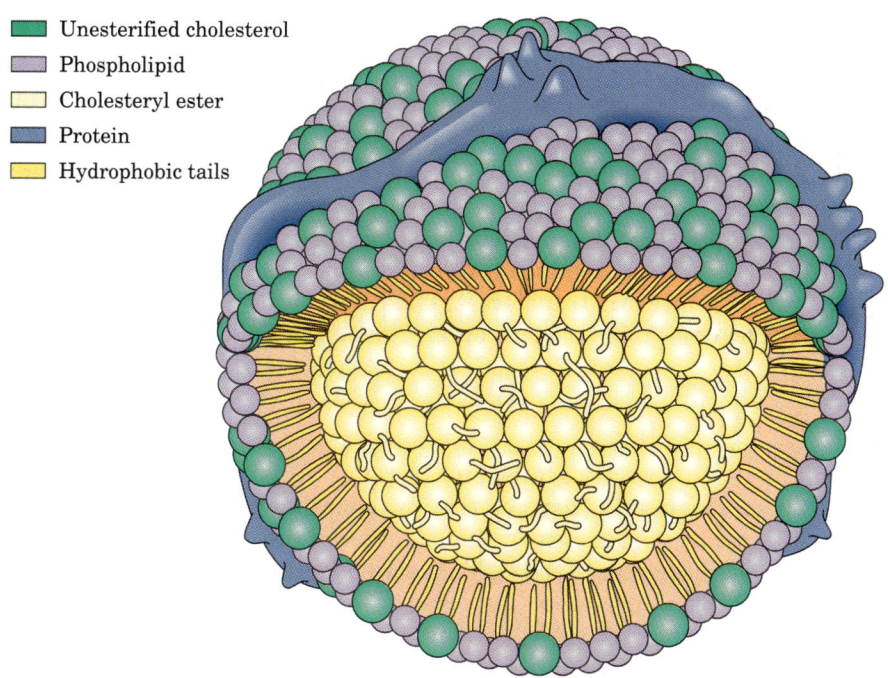

Figure 20.5 Low-density lipoprotein shown schematically.

in the serum. When the capillaries reach muscle or fat tissues, the triglycerides and all proteins except apoB-100 are removed from the VLDL. At this point, the diameter of the lipoprotein shrinks to 22 nanometers and its core contains only cholesteryl esters. Because of the removal of fat, its density increases and it becomes LDL. Low-density lipoprotein stays in the plasma for about 2.5 days.

The LDL carries cholesterol to the cells, where specific LDL-receptor molecules line the cell surface in certain concentrated areas called **coated pits**. The apoB-100 protein on the surface of the LDL binds specifically to the LDL-receptor molecules in the coated pits. After such binding, the LDL is taken inside the cell (endocytosis), where enzymes break down the lipoprotein. In the process, they liberate free cholesterol from cholesteryl esters. In this manner, the cell can, for example, use cholesterol as a component of a membrane. This is the normal fate of LDL and the normal course of cholesterol transport.

Michael Brown and Joseph Goldstein of the University of Texas shared the Nobel Prize in medicine in 1986 for the discovery of the LDL-receptor–mediated pathway.

Table 20.3 Compositions and Properties of Human Lipoproteins

Property	HDL	LDL	VLDL	Chylomicrons
Core				
Cholesterol and Cholesteryl esters (%)	30	50	22	8
Triglycerides (%)	8	4	50	84
Surface				
Phospholipids (%)	29	21	18	7
Proteins (%)	33	25	10	1–2
Density (g/mL)	1.05–1.21	1.02–1.06	0.95–1.00	<0.95
Diameter (nm)	5–15	18–28	30–80	100–500

Percentages are given as % dry weight.

D Transport of Cholesterol in HDL

High-density lipoprotein transports cholesterol from peripheral tissues to the liver and transfers cholesterol to LDL. While in the serum, the free cholesterols in HDL are converted to cholesteryl esters. These esterified cholesterols are delivered to the liver for synthesis of bile acids and steroid hormones. The cholesterol uptake from HDL differs from that noted with LDL. This process does not involve endocytosis and degradation of the lipoprotein particle. Instead, in a selective lipid uptake, the HDL binds to the liver cell surface and transfers its cholesteryl ester to the cell. After that the HDL, depleted from its lipid content, re-enters the circulation.

E Levels of LDL and HDL

Like all lipids, cholesterol is insoluble in water. If its level is elevated in the blood serum, plaque-like deposits may form on the inner surfaces of the arteries. The resulting decrease in the diameter of the blood vessels may lead to a decrease in the flow of blood. This **atherosclerosis,** along with accompanying high blood pressure, may lead to heart attack, stroke, or kidney dysfunction.

Atherosclerosis may exacerbate the blockage of some arteries by a clot at the point where the arteries are constricted by plaque. Furthermore, blockage may deprive cells of oxygen, causing them to cease to function. The death of heart muscles due to lack of oxygen is called *myocardial infarction.*

Most cholesterol is transported by low-density lipoproteins. If a sufficient number of LDL receptors are found on the surface of the cells, LDL is effectively removed from the circulation and its concentration in the blood plasma drops. The number of LDL receptors on the surface of cells is controlled by a feedback mechanism (see Section 22.6). That is, when the concentration of cholesterol molecules inside the cells is high, the synthesis of the LDL receptor is suppressed. Thus less LDL is taken into the cells from the plasma and the LDL concentration in the plasma rises. On the other hand, when the cholesterol level inside the cells is low, the synthesis of the LDL receptor increases. As a consequence, the LDL is taken up more rapidly and its level in the plasma falls.

In certain cases, however, there are not enough LDL receptors. In the disease called *familial hypercholesterolemia,* the cholesterol level in the plasma may be as high as 680 mg/100 mL, as opposed to 175 mg/100 mL in normal subjects. These high levels of cholesterol can lead to premature atherosclerosis and heart attack. The high plasma cholesterol levels in affected patients are caused because the body lacks enough functional LDL receptors or, if enough are produced, they are not concentrated in the coated pits.

In general, high LDL content means high cholesterol content in the plasma because LDL cannot enter the cells and be metabolized. Therefore, a high LDL level combined with a low HDL level is a symptom of faulty cholesterol transport and a warning for possible atherosclerosis.

The serum cholesterol level controls the amount of cholesterol synthesized by the liver. When serum cholesterol is high, the synthesis is at a low level. Conversely, when the serum cholesterol level is low, the synthesis of cholesterol increases.

Diets low in cholesterol and saturated fatty acids usually reduce the serum cholesterol level, and a number of drugs can inhibit the synthesis of cholesterol in the liver. The commonly used statin drugs, such as atorvastatin (Lipicor) or Simvastatin (Zocor) inhibit one of the key enzymes (biocatalyst) in cholesterol synthesis, HMG-CoA reductase (Section 28.4). In this way, they block the synthesis of cholesterol inside the cells and stimulate the synthesis of LDL-receptor proteins. More LDL then enters the cells, diminishing the amount of cholesterol that will be deposited on the inner walls of arteries.

It is generally considered desirable to have high levels of HDL and low levels of LDL in the bloodstream. HDL carries cholesterol from plaques deposited in the arteries to the liver, thereby reducing the risk of atherosclerosis. Premenopausal women have more HDL than men, which is why women have a lower risk of coronary heart disease. HDL levels can be increased by exercise and weight loss.

20.10 Steroid Hormones

Cholesterol is the starting material for the synthesis of steroid hormones. In this process, the aliphatic side chain on the D ring is shortened by the removal of a six-carbon unit, and the secondary alcohol group on carbon 3 is oxidized to a ketone. The resulting molecule, *progesterone,* serves as the starting compound for both the sex hormones and the adrenocorticoid hormones (Figure 20.6).

Figure 20.6 The biosynthesis of hormones from progesterone.

CHEMICAL CONNECTIONS 20G

Anabolic Steroids

Testosterone, the principal male hormone, is responsible for the buildup of muscles in men. Recognizing this fact, many athletes have taken this drug in an effort to increase their muscular development. This practice is especially common among athletes in sports in which strength and muscle mass are important, including weight lifting, shot put, and hammer throw. Participants in other sports, such as running, swimming, and cycling, would also like larger and stronger muscles.

Although used by many athletes, testosterone has two disadvantages:

1. Besides its effect on muscles, it affects secondary sexual characteristics, and too much of it can result in undesired side effects.
2. It is not very effective when taken orally and must be injected to achieve the best results.

For these reasons, a large number of other anabolic steroids, all of them synthetic, have been developed. Examples include the following compounds:

Methandienone

Methenolone

Nandrolone decanoate

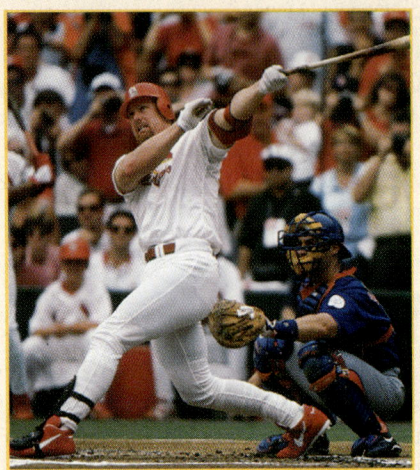

■ Home run champion Mark McGwire used androstenedione, a muscle-building dietary supplement that is allowed in baseball, but is banned in professional football, college athletics, and the Olympic Games.

4-Androstene-3,17-dione

The use of anabolic steroids is forbidden in many sporting events, especially in international competition, largely for two reasons: (1) It gives some competitors an unfair advantage, and (2) these drugs can have many unwanted and even dangerous side effects, ranging from acne to liver tumors. Side effects can be especially disadvantageous for women; they can include growth of facial hair, baldness, deepening of the voice, and menstrual irregularities.

All athletes participating in the Olympic Games are required to pass a urine test for anabolic steroids. A number of medal-winning athletes have had their victories taken away because they tested positive for steroid use. For example, the Canadian Ben Johnson, a world-class sprinter, was stripped of both his world record and his gold medal in the 1988 Olympiad. A similar failed test for andro resulted in the U.S. shot put champion Randy Barnes being banned from competition in 1998. Prohormones such as andro are not listed under the Anabolic Steroid Act of 1990; hence, their nonmedical use is not a federal offense, as is the case with anabolic steroids. Mark McGwire hit his record-breaking home runs in 1998 while taking andro, as baseball rules did not prohibit its usage. Even so, the International Olympic Committee has banned the use of both prohormones and anabolic steroids.

Some women athletes use anabolic steroids, just as their male counterparts do. Because their bodies produce only small amounts of testosterone, women actually have much more to gain from anabolic steroids than men.

Another way to increase testosterone concentration is to use prohormones, which the body converts to testosterone. One such prohormone is 4-androstenedione, or "andro." Some athletes have used it to enhance performance.

A Adrenocorticoid Hormones

The adrenocorticoid hormones (Figure 20.6) are products of the adrenal glands. We divide them into two groups according to function: *Mineralocorticoids* regulate the concentrations of ions (mainly Na^+ and K^+), and *glucocorticoids* control carbohydrate metabolism.

Aldosterone is one of the most important mineralocorticoids. Increased secretion of aldosterone enhances the reabsorption of Na^+ and Cl^- ions in the kidney tubules and increases the loss of K^+. Because Na^+ concentration controls water retention in the tissues, aldosterone also controls tissue swelling.

Cortisol is the major glucocorticoid. Its function is to increase the glucose and glycogen concentrations in the body. This is done at the expense of other nutrients. Fatty acids from fat storage cells and amino acids from body proteins are transported to the liver, which, under the influence of cortisol, manufactures glucose and glycogen from these sources.

Cortisol and its ketone derivative, *cortisone,* have remarkable anti-inflammatory effects. These or similar synthetic derivatives, such as prednisolone, are used to treat inflammatory diseases of many organs, rheumatoid arthritis, and bronchial asthma.

> The term *adrenal* means "adjacent to the renal" (which refers to the kidney).
>
> The term *corticoid* indicates that the site of the secretion is the cortex (outer part) of the gland.

B Sex Hormones

The most important male sex hormone is testosterone (Figure 20.6). This hormone, which promotes the normal growth of the male genital organs, is synthesized in the testes from cholesterol. During puberty, increased testosterone production leads to such secondary male sexual characteristics as deep voice and facial and body hair.

Female sex hormones, the most important of which is estradiol (Figure 20.6), are synthesized from the corresponding male hormone (testosterone) by aromatization of the A ring:

Testosterone (Partial Structure) → (loss of —H and CH_3) → → (keto-enol tautomerism) → Estradiol (Partial Structure)

Estradiol, together with its precursor progesterone, regulates the cyclic changes occurring in the uterus and ovaries known as the *menstrual cycle.* As the cycle begins, the level of estradiol in the body rises, which in turn causes the lining of the uterus to thicken. Then another hormone, called luteinizing hormone (LH), triggers ovulation. If the ovum is fertilized, increased progesterone levels will inhibit any further ovulation. Both estradiol and progesterone then promote further preparation of the uterine lining to receive the fertilized ovum. If no fertilization takes place, progesterone production stops altogether, and estradiol production decreases. This halt decreases the thickening of the uterine lining, which is then sloughed off with accompanying bleeding during menstruation (Figure 20.7).

Because progesterone is essential for the implantation of the fertilized ovum, blocking its action leads to termination of pregnancy (see Chemical Connections 20H). Progesterone interacts with a receptor (a protein molecule) in the nucleus of cells. The receptor changes its shape when progesterone binds to it (see Section 23.7).

Figure 20.7 Events of the menstrual cycle. (*a*) Levels of sex hormones in the bloodstream during the phases of one menstrual cycle in which pregnancy does not occur. (*b*) Development of an ovarian follicle during the cycle. (*c*) Phases of development of the endometrium, the lining of the uterus. The endometrium thickens during the proliferative phase. In the secretory phase, which follows ovulation, the endometrium continues to thicken and the glands secrete a glycogen-rich nutritive material in preparation to receive an embryo. When no embryo is implanted, the new outer layers of the endometrium disintegrate and the blood vessels rupture, producing the menstrual flow.

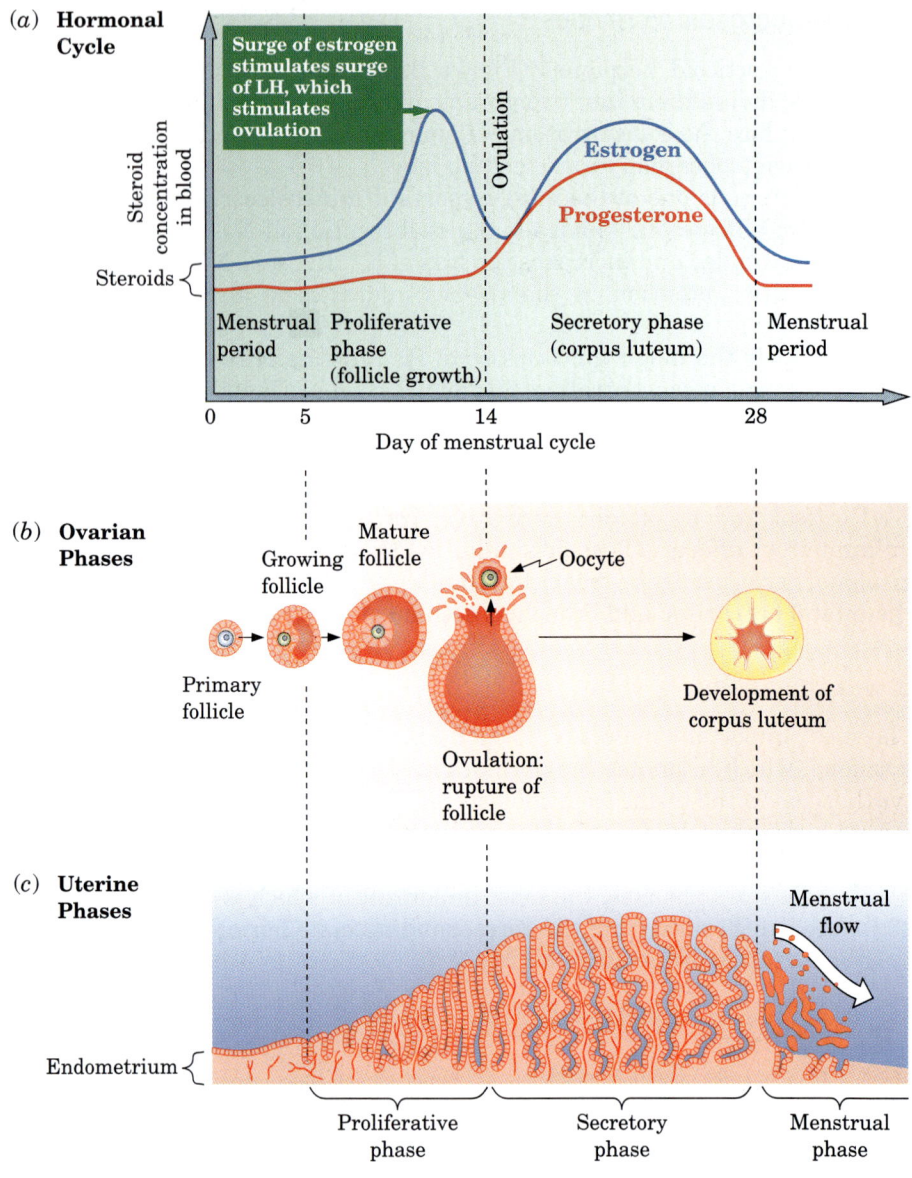

RU486 binds to the receptors of glucocorticoid hormones as well. Its use as an antiglucocorticoid is also recommended to alleviate a disease known as Cushing syndrome, involving the overproduction of cortisone.

A drug, now widely used in France and China, called mifepristone or RU486 acts as a competitor to progesterone:

Mifepristone (RU486)

CHEMICAL CONNECTIONS 20H

Oral Contraception

Because progesterone prevents ovulation during pregnancy, it occurred to investigators that progesterone-like compounds might be used for birth control. Synthetic analogs of progesterone proved to be more effective than the natural compound. In "the Pill," a synthetic progesterone-like compound is supplied together with an estradiol-like compound (the latter prevents irregular menstrual flow). Triple-bond derivatives of testosterone, such as norethindrone, norethynodrel, and ethynodiol diacetate, are used most often in birth-control pills:

Norethindone   Norethynodrel   Ethynodiol diacetate

Mifepristone binds to the same receptor site as progesterone. Because the progesterone molecule cannot reach the receptor molecule, the uterus is not prepared for the implantation of the fertilized ovum, and the ovum is aborted. Once pregnancy has been established, RU486 can be taken up through 49 days of gestation. This chemical form of abortion has been approved by the Food and Drug Administration and in recent years has been gaining clinical use as a supplement to surgical abortion.

A completely different approach is the "morning after pill" that can be taken orally up to 72 hours after an unprotected intercourse. The "morning after pill" is not an abortion pill because it acts before pregnancy takes place. Actually, the components of the pill are regular contraceptives. Two kinds are on the market as prescription drugs: One is a progesterone-like compound, levonogestrel, the other is a combination of levonogestrel and ethynil estradiol marketed as Preven.

Estradiol and progesterone also regulate secondary female sex characteristics, such as the growth of breasts. Thanks to this property, RU486, as an anti-progesterone, has also been reported to be effective against certain types of breast cancer.

Testosterone and estradiol are not exclusive to either males or females. A small amount of estradiol production occurs in males, and a small amount of testosterone production is normal in females. Only when the proportion of these two hormones (hormonal balance) becomes upset can one observe symptoms of abnormal sexual differentiation.

20.11 Bile Salts

Bile salts are oxidation products of cholesterol. First the cholesterol is converted to the trihydroxy derivative, and the end of the aliphatic chain is oxidized to the carboxylic acid. The latter, in turn, forms an amide linkage with an amino acid, either glycine or taurine:

The dispersion of dietary lipids by bile salts is similar to the action of soap on dirt.

Glycocholate

Taurocholate

Bile salts are powerful detergents. One end of the molecule is strongly hydrophilic because of the negative charge, and the rest of the molecule is largely hydrophobic. As a consequence, bile salts can disperse dietary lipids in the small intestine into fine emulsions and thereby help digestion.

Because they are eliminated in the feces, bile salts also remove excess cholesterol in two ways: (1) They are themselves breakdown products of cholesterol (so cholesterol is eliminated via bile salts), and (2) they solubilize deposited cholesterol in the form of bile salt–cholesterol particles.

20.12 Prostaglandins, Thromboxanes, and Leukotrienes

Ulf S. von Euler won the Nobel Prize in physiology and medicine in 1970.

Prostaglandins, a group of fatty-acid-like substances, were discovered by Kurzrok and Leib in the 1930s, when they demonstrated that seminal fluid caused a hysterectomized uterus to contract. Ulf von Euler of Sweden iso-

CHEMICAL CONNECTIONS 20I

Action of Anti-inflammatory Drugs

Anti-inflammatory steroids (such as cortisone; Section 20.10) exert their function by inhibiting phospholipase A_2, the enzyme that releases unsaturated fatty acids from complex lipids in the membranes. For example, arachidonic acid, one of the components of membranes, is made available to the cell through this process. Because arachidonic acid is the precursor of prostaglandins, thromboxanes, and leukotrienes, inhibiting its release stops the synthesis of these compounds and prevents inflammation.

Steroids such as cortisone have many undesirable side effects (duodenal ulcer and cataract formation, among others). Therefore, their use must be controlled. A variety of nonsteroidal anti-inflammatory agents, including aspirin, ibuprofen, ketoprofen, and indomethacin, are available to serve this function.

Aspirin and other NSAIDs (see Chemical Connections 18C) inhibit the cyclooxygenase enzymes, which synthesize prostaglandins and thromboxanes. Aspirin (acetyl salicylic acid), for example, acetylates the enzymes, thereby blocking the entrance of arachidonic acid to the active site. This inhibition of both COX-1 and COX-2 explains why aspirin and the other anti-inflammatory agents have undesirable side effects. NSAIDs also interfere with the COX-1 isoform of the enzyme, which is needed for normal physiological function. Their side effects include stomach and duodenal ulceration and renal (kidney) toxicity.

Obviously it would be desirable to have an anti-inflammatory agent without such side effects, and one that inhibits only the COX-2 isoform. To date, the Food and Drug Administration (FDA) has approved two COX-2 inhibitor drugs: Celebrex, which quickly became the most frequently prescribed drug, and VIOXX, a more recent introduction. Despite their selective inhibition of COX-2, however, these drugs also have ulcer-causing side effects. Many other COX-2 inhibitors are in the clinical trial stage.

The use of COX-2 inhibitors is not limited to rheumatoid arthritis and osteoarthritis. Celebrex has been approved by FDA to treat a type of colon cancer, familial adenomateous polyposis, in an approach called chemoprevention. All anti-inflammatory agents reduce pain and relieve fever and swelling by reducing the prostaglandin production; however, they do not affect the leukotriene production. Thus, asthmatic patients must beware of using these anti-inflammatory agents. Even though they inhibit the prostaglandin synthesis, they may shift the available arachidonic acid to leukotriene production, which could precipitate a severe asthma reaction.

lated these compounds from human semen and, thinking that they had come from the prostate gland, named them **prostaglandin.** Even though the seminal gland secretes 0.1 mg of prostaglandin per day in mature males, small amounts of prostaglandins are present throughout the body in both sexes.

Prostaglandins are synthesized in the body from arachidonic acid by a ring closure at carbons 8 and 12. The enzyme catalyzing this reaction is called **cyclooxygenase** (COX, for short). The product, known as PGG_2, is the common precursor of other prostaglandins, including PGE and PGF. The prostaglandin E group (PGE) has a carbonyl group at carbon 9; the subscript indicates the number of double bonds in the hydrocarbon chain. The prostaglandin F group (PGF) has two hydroxyl groups on the ring at carbons 9 and 11.

Other prostaglandins (PGAs and PGBs) are derived from PGE.

The COX enzyme comes in two forms in the body: COX-1 and COX-2. COX-1 catalyzes the normal physiological production of prostaglandins, which are always present in the body. For example, PGE_2 and $PGF_{2\alpha}$ stimulate uterine contractions and induce labor. PGE_2 lowers blood pressure by relaxing the muscles around blood vessels. In aerosol form, this prostaglandin is used to treat asthma; it opens up the bronchial tubes by relaxing the surrounding muscles. PGE_1 is used as a decongestant; it opens up nasal passages by constricting blood vessels.

COX-2, on the other hand, is responsible for the production of prostaglandins in inflammation. When a tissue is injured or damaged, special inflammatory cells invade the injured tissue and interact with resident cells—for example, smooth muscle cells. This interaction activates the COX-2 enzyme, and prostaglandins are synthesized. Such tissue injury may occur in heart attack (myocardial infarction), rheumatic arthritis, and ulcerative colitis. Nonsteroidal anti-inflammatory drugs (NSAIDs), such as aspirin, inhibit both COX enzymes (see Chemical Connections 20I).

PGH₂   Thromboxane A₂

Another class of arachidonic acid derivatives is the **thromboxanes.** Their synthesis also includes a ring closure. These substances are derived from PGH_2, but their ring is a cyclic acetal. Thromboxane A_2 is known to induce platelet aggregation. When a blood vessel is ruptured, the first line of defense is the platelets circulating in the blood, which form an incipient clot (see Chemical Connections 31B). Thromboxane A_2 causes other platelets to clump, thereby increasing the size of the blood clot. Aspirin and similar anti-inflammatory agents inhibit the COX enzyme. Consequently, PGH_2 and thromboxane A_2 synthesis is inhibited, and blood clotting is impaired. This is the reason that physicians recommend a daily dose of 81 mg aspirin for people at risk for heart attack or stroke. On the other hand, it also explains why physicians forbid patients to use aspirin and other anti-inflammatory agents for a week before a planned surgery because aspirin causes excessive bleeding.

A variety of NSAIDs inhibit COX enzymes. Ibuprofen and indomethacin, both powerful NSAID painkillers, can block the inhibitory effect of aspirin and thus eliminate its anticlotting benefits. Therefore, the use of these NSAIDs together with aspirin is not recommended. Other painkillers, such as acetaminophen and diclofenac, do not interfere with aspirin's anticlotting ability and therefore can be taken together.

Another group of substances that acts to mediate hormonal responses are the **leukotrienes.** Like prostaglandins, leukotrienes are derived from arachidonic acid by an oxidative mechanism. However, in this case, there is no ring closure.

Arachidonic acid   Leukotriene B4

Leukotrienes occur mainly in white blood cells (leukocytes) but are also found in other tissues of the body. They produce long-lasting muscle contractions, especially in the lungs, and can cause asthma-like attacks. In fact, they are 100 times more potent than histamines. Both prostaglandins and leukotrienes cause inflammation and fever, so the inhibition of their production in the body is a major pharmacological concern. One way to counteract the effects of leukotrienes is to inhibit their uptake by leukotriene receptors (LTRs) in the body. A new antagonist of LTRs, zafirlukast, marketed under the brand name Accolate, is used to treat and control chronic asthma. Another antiasthmatic drug, zileuton, inhibits 5-lipoxygenase, which is the initial enzyme in leukotriene biosynthesis from arachidonic acid.

SUMMARY

Lipids are water-insoluble substances (Section 20.1). They are classified into four groups: fats (triglycerides); complex lipids; steroids; and prostaglandins, thromboxanes, and leukotrienes. **Fats** consist of fatty acids and glycerol (Section 20.2). In saturated fatty acids, the hydrocarbon chains have only single bonds; unsaturated fatty acids have hydrocarbon chains with one or more double bonds, all in the cis configuration. Solid fats contain mostly saturated fatty acids, whereas **oils** contain substantial amounts of unsaturated fatty acids (Section 20.3). The alkali salts of fatty acids are called **soaps**.

Complex lipids can be divided into two groups: phospholipids and glycolipids (Section 20.4). **Phospholipids** are made of a central alcohol (glycerol or sphingosine), fatty acids, and a nitrogen-containing phosphate ester, such as phosphorylcholine or inositol phosphate (Section 20.6). **Glycolipids** contain sphingosine and a fatty acid, collectively known as the ceramide portion of the molecule, and a carbohydrate portion (Section 20.8).

Many phospholipids and glycolipids are important constituents of cell **membranes** (Section 20.5). Membranes are made of a **lipid bilayer** in which the hydrophobic parts of phospholipids (fatty acid residues) point toward the middle of the bilayer, and the hydrophilic parts point toward the inner and outer surfaces of the membrane (Section 20.5).

The third major group of lipids comprises the **steroids** (Section 20.9). The characteristic feature of the steroid structure is a fused four-ring nucleus. The most common steroid, **cholesterol,** serves as a starting material for the synthesis of other steroids, such as bile salts and sex and other hormones (Section 20.9A). Cholesterol is also an integral part of membranes, occupying the hydrophobic region of the lipid bilayer. Because of its low solubility in water, cholesterol deposits are implicated in the formation of gallstones and the plaque-like deposits of atherosclerosis.

Cholesterol is transported in the blood plasma mainly by two kinds of lipoproteins: **HDL** and **LDL** (Section 20.9B). LDL delivers cholesterol to the cells to be used mostly as a membrane component. HDL delivers cholesteryl esters mainly to the liver to be used in the synthesis of bile acids and steroid hormones (Section 20.9C). High levels of LDL and low levels of HDL are symptoms of faulty cholesterol transport, indicating greater risk of atherosclerosis (Section 20.9E).

An oxidation product of cholesterol is progesterone, a **sex hormone** (Section 20.10). It also gives rise to the synthesis of other sex hormones, such as testosterone and estradiol, as well as to the **adrenocorticoid hormones**. Among the latter, cortisol and cortisone are best known for their anti-inflammatory action. **Bile salts** are also oxidation products of cholesterol (Section 20.11). They emulsify all kinds of lipids, including cholesterol, and are essential in the digestion of fats.

Prostaglandins, thromboxanes, and **leukotrienes** are derived from arachidonic acid (Section 20.12). They have a wide variety of effects on body chemistry; among other things, they can lower or raise blood pressure, cause inflammation and blood clotting, and induce labor. In general, they mediate hormone action.

PROBLEMS

Numbers that appear in color indicate difficult problems. ▸ designates problems requiring application of principles.

Structure and Properties of Fats

20.2 Why are fats a good source of energy for storage in the body?

20.3 What is the meaning of the term *hydrophobic?* Why is the hydrophobic nature of lipids important?

20.4 Draw the structural formula of a fat molecule (triglyceride) made of myristic acid, oleic acid, palmitic acid, and glycerol.

20.5 Oleic acid has a melting point of 16°C. If you converted the cis double bond into a trans double bond, what would happen to the melting point? Explain.

20.6 Draw schematic formulas for all possible 1,3-diglycerides made up of glycerol, oleic acid, or stearic acid. How many are there? Draw the structure of one of the diglycerides.

20.7 For the diglycerides in Problem 20.6, predict which two will have the highest melting points and which two will have the lowest melting points.

20.8 Predict which acid in each pair has the higher melting point and explain why.
(a) Palmitic acid or stearic acid
(b) Arachidonic acid or arachidic acid

20.9 Which has the higher melting point: (a) a triglyceride containing only lauric acid and glycerol or (b) a triglyceride containing only stearic acid and glycerol?

20.10 Explain why the melting points of the saturated fatty acids increase as we move from lauric acid to stearic acid in Table 20.1.

20.11 Predict the order of the melting points of triglycerides containing fatty acids, as follows.
 (a) Palmitic, palmitic, stearic
 (b) Oleic, stearic, palmitic
 (c) Oleic, linoleic, oleic

20.12 Look at Table 20.2. Which animal fat has the highest percentage of unsaturated fatty acids?

20.13 Rank the following in order of increasing solubility in water (assuming that all are made with the same fatty acids): (a) triglycerides, (b) diglycerides, and (c) monoglycerides. Explain your answer.

20.14 How many moles of H_2 are used up in the catalytic hydrogenation of 1 mol of a triglyceride containing glycerol, palmitic acid, oleic acid, and linoleic acid?

20.15 Name the products of the saponification of this triglyceride:

$$\begin{array}{l} CH_2-O-\overset{O}{\underset{\|}{C}}-(CH_2)_{14}CH_3 \\ CH-O-\overset{O}{\underset{\|}{C}}-(CH_2)_{16}CH_3 \\ CH_2-O-\overset{O}{\underset{\|}{C}}-(CH_2)_7(CH=CHCH_2)_3CH_3 \end{array}$$

20.16 Using the equation in Section 20.3 as a guideline for stoichiometry, calculate the number of moles of NaOH needed to saponify 5 mol of (a) triglycerides, (b) diglycerides, and (c) monoglycerides.

Membranes

20.17 Which portion of the phosphatidylinositol molecule contributes to (a) the fluidity of the bilayer and (b) the surface polarity of the bilayer?

20.18 How do the unsaturated fatty acids of the complex lipids contribute to the fluidity of a membrane?

20.19 Which type of lipid molecules is most likely to be present in membranes?

20.20 What is the difference between an integral and a peripheral membrane protein?

Complex Lipids

20.21 Which glycerophospholipid has the most polar groups capable of forming hydrogen bonds with water?

20.22 Draw the structure of a phosphatidylinositol that contains oleic and arachidonic acid.

20.23 Among the glycerophospholipids containing palmitic acid and linolenic acid, which will have the greatest solubility in water: (a) phosphatidylcholine, (b) phosphatidylethanolamine, or (c) phosphatidylserine? Explain.

20.24 Name all the groups of complex lipids that contain ceramides.

20.25 Are the various phospholipids randomly distributed in membranes? Give an example.

20.26 Enumerate the functional groups that contribute to the hydrophilic character of (a) glucocerebroside and (b) sphingomyelin.

Steroids

20.27 Cholesterol has a fused four-ring steroid nucleus and is a part of body membranes. The —OH group on carbon 3 is the polar head, and the rest of the molecule provides the hydrophobic tail that does not fit into the zig-zag packing of the hydrocarbon portion of the saturated fatty acids. Considering this structure, tell whether small amounts of cholesterol well dispersed in the membrane contribute to the stiffening (rigidity) or the fluidity of a membrane. Explain.

20.28 Where can pure cholesterol crystals be found in the body?

20.29 (a) Find all the carbon stereocenters in a cholesterol molecule.
 (b) How many total stereoisomers are possible?
 (c) How many of these stereoisomers do you think are found in nature?

20.30 Look at the structures of cholesterol and the hormones shown in Figure 20.6. Which ring of the steroid structure undergoes the most substitution?

20.31 What makes LDL soluble in blood plasma?

20.32 How does LDL deliver its cholesterol to the cells?

20.33 How does lovastatin reduce the severity of atherosclerosis?

20.34 How does VLDL become LDL?

20.35 How does HDL deliver its cholesteryl esters to liver cells?

20.36 How does the serum cholesterol level control both cholesterol synthesis in the liver and LDL uptake?

Steroid Hormones and Bile Salts

20.37 What physiological functions are associated with cortisol?

20.38 Estradiol in the body is synthesized starting from progesterone. What chemical modifications occur when estradiol is synthesized?

20.39 Describe the difference in structure between the male hormone testosterone and the female hormone estradiol.

20.40 Considering that RU486 can bind to the receptors of progesterone as well as those of cortisone and cortisol, what can you say regarding the importance of the functional group on carbon 11 of the steroid ring in drug and receptor binding?

20.41 (a) How does the structure of RU486 resemble that of progesterone?
(b) How do the two structures differ?

20.42 What are the common structural features of the oral contraceptive pills, including mifepristone?

20.43 List all of the functional groups that make taurocholate water-soluble.

20.44 Explain how the constant elimination of bile salts through the feces can reduce the danger of plaque formation in atherosclerosis.

Prostaglandins and Leukotrienes

20.45 What is the basic structural difference between:
(a) Arachidonic acid and prostaglandin PGE_2?
(b) PGE_2 and $PGF_{2\alpha}$?

20.46 Find and name all the functional groups in (a) glycocholate, (b) cortisone, (c) prostaglandin PGE_2, and (d) leukotriene B4.

20.47 What are the chemical and physiological functions of the COX-2 enzyme?

20.48 How does aspirin, an anti-inflammatory drug, prevent strokes caused by blood clots in the brain?

Chemical Connections

20.49 (Chemical Connections 20A) What causes rancidity? How can it be prevented?

20.50 (Chemical Connections 20B) What makes waxes harder and more difficult to melt than fats?

20.51 (Chemical Connections 20C) What is the major structural difference between soaps and detergents?

20.52 (Chemical Connections 20D) How do the gap junctions prevent the passage of proteins from cell to cell?

20.53 (Chemical Connections 20D) How does the anion transporter provide a suitable environment for the passage of hydrated chloride ions?

20.54 (Chemical Connections 20D) In what sense is the active transport of K^+ selective? How does K^+ pass through the transporter?

20.55 (Chemical Connections 20E) (a) What role does sphingomyelin play in the conductance of nerve signals?
(b) What happens to this process in multiple sclerosis?

20.56 (Chemical Connections 20F) Compare the complex lipid structures listed for the lipid storage diseases (Table 20F) with the missing or defective enzymes. Explain why the missing enzyme in Fabry's disease is α-galactosidase and not β-galactosidase.

20.57 (Chemical Connections 20F) Identify the monosaccharides in the accumulating glycolipid of Fabry's disease.

20.58 (Chemical Connections 20G) How does the oral anabolic steroid methenolone differ structurally from testosterone?

20.59 (Chemical Connections 20H) What is the role of progesterone and similar compounds in contraceptive pills?

20.60 (Chemical Connections 20I) How does cortisone prevent inflammation?

20.61 (Chemical Connections 20I) How does indomethacin act in the body to reduce inflammation?

20.62 (Chemical Connections 20I) What kind of prostaglandins are synthesized by COX-1 and COX-2 enzymes?

20.63 (Chemical Connections 20I) Steroids prevent asthma-causing leukotriene synthesis as well as inflammation-causing prostaglandin synthesis. Nonsteroidal anti-inflammatory agents (NSAIDs) such as aspirin reduce only prostaglandin production. Why do NSAIDs not affect leukotriene production?

Additional Problems

20.64 What is the role of taurine in lipid digestion?

20.65 Draw a schematic diagram of a lipid bilayer. Show how the bilayer prevents the passage by diffusion of a polar molecule such as glucose. Show why nonpolar molecules, such as $CH_3CH_2-O-CH_2CH_3$, can diffuse through the membrane.

20.66 How many different triglycerides can you create using three different fatty acids (A, B, and C) in each case?

20.67 Prostaglandins have a five-membered ring closure; thromboxanes have a six-membered ring closure. The synthesis of both groups of compounds are prevented by COX inhibitors; the COX enzymes catalyze ring closure. Explain this fact.

20.68 Which lipoprotein is instrumental in removing the cholesterol deposited in the plaques on arteries?

20.69 What are coated pits? What is their function?

20.70 What are the constituents of sphingomyelin?

20.71 What structural feature do detergents (Chemical Connections 20C) and the bile salt, taurocholate, share?

20.72 (Chemical Connections 20D) What is the difference between a facilitated transporter and an active transporter?

20.73 What part of LDL interacts with the LDL receptor?

20.74 What is the major difference between aldosterone and all the other hormones listed in Figure 20.6?

20.75 (Chemical Connections 20I) The new anti-inflammatory drug Celebrex does not have the usual side effect of stomach upset or ulceration commonly observed with the other NSAIDs. Why?

20.76 How many grams of H_2 are needed to saturate 100.0 g of a triglyceride made of glycerol and one unit each of lauric, oleic, and linoleic acids?

20.77 Prednisolone is the synthetic glucocorticoid medicine most frequently prescribed to combat autoimmune diseases. Compare its structure to the natural glucocorticoid hormone, cortisone. What are the similarities and differences in structure?

InfoTrac College Edition

For additional readings, go to InfoTrac College Edition, your online research library, at

http://infotrac.thomsonlearning.com

CHAPTER 21

- 21.1 Introduction
- 21.2 Amino Acids
- 21.3 Zwitterions
- 21.4 Cysteine: A Special Amino Acid
- 21.5 Peptides and Proteins
- 21.6 Some Properties of Peptides and Proteins
- 21.7 The Primary Structure of Proteins
- 21.8 The Secondary Structure of Proteins
- 21.9 The Tertiary and Quaternary Structures of Proteins
- 21.10 Glycoproteins
- 21.11 Denaturation

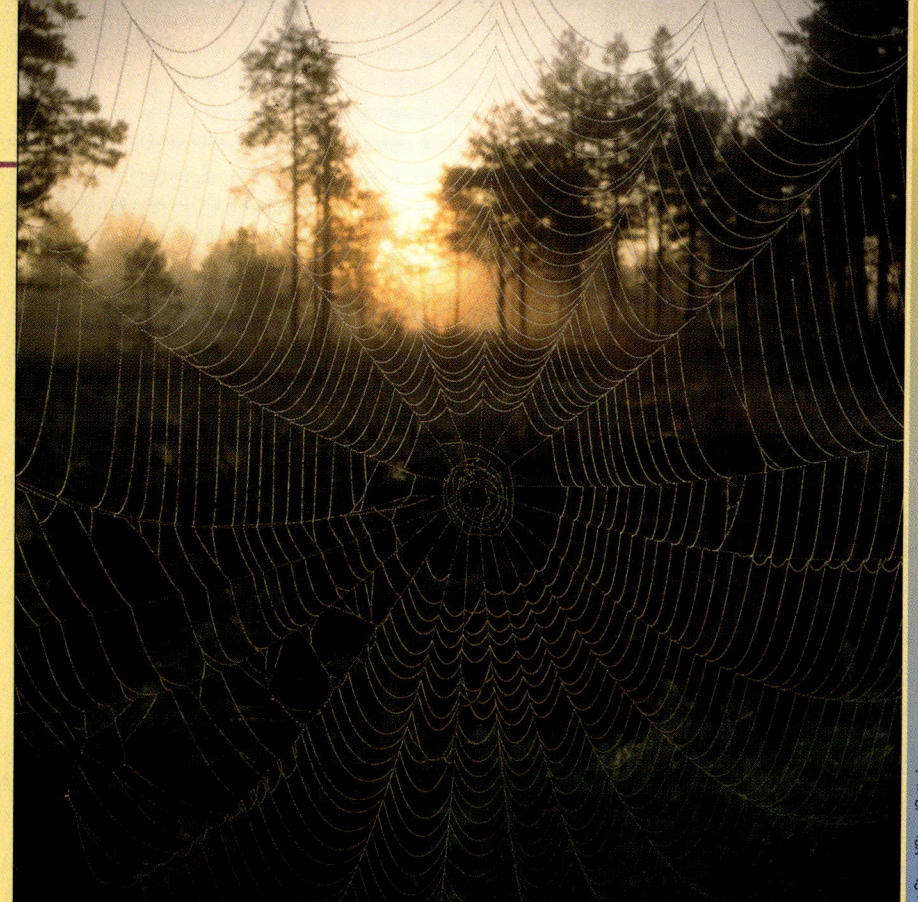

Spider silk is a fibrous protein that exhibits unmatched strength and toughness.

Proteins

21.1 Introduction

Protein A large biological molecule made of numerous amino acids linked together by amide bonds

The heart itself is a muscle, expanding and contracting about 70 to 80 times per minute.

Proteins are by far the most important of all biological compounds. The very word "protein" is derived from the Greek *proteios,* meaning "of first importance," and the scientists who named these compounds more than 100 years ago chose an appropriate term. There are many types of proteins, and they perform a variety of functions, including the following roles:

1. **Structure** In Section 19.6, we saw that the main structural material for plants is cellulose. For animals, it is structural proteins, which are the chief constituents of skin, bones, hair, and nails. Two important structural proteins are collagen and keratin.

2. **Catalysis** Virtually all the reactions that take place in living organisms are catalyzed by proteins called enzymes. Without enzymes, the reactions would take place so slowly as to be useless. We will discuss enzymes in depth in Chapter 22.

3. **Movement** Every time we crook a finger, climb stairs, or blink an eye, we use our muscles. Muscle expansion and contraction are involved in every movement we make. Muscles are made up of protein molecules called myosin and actin.

Antibodies are present in the gamma globulin fraction of blood plasma.

Collagen, actin, and keratin are fibrous proteins. Albumin, hemoglobin, and immunoglobulins are globular proteins.

Screen 21.2 Tutorial

Alpha (α-) amino acid An amino acid in which the amino group is linked to the carbon atom next to the —COOH carbon

The one-letter abbreviations are more recent inventions, but the three-letter abbreviations are still frequently used.

4. **Transport** A large number of proteins perform transportation duties. For example, hemoglobin, a protein in the blood, carries oxygen from the lungs to the cells in which it is used and carbon dioxide from the cells to the lungs. Other proteins transport molecules across cell membranes.

5. **Hormones** Many hormones are proteins, including insulin, oxytocin, and human growth hormone.

6. **Protection** When a protein from an outside source or some other foreign substance (called an antigen) enters the body, the body makes its own proteins (called antibodies) to counteract the foreign protein. This antibody production is the major mechanism that the body uses to fight disease. Blood clotting is another protective function carried out by a protein, called fibrinogen. Without blood clotting, we would bleed to death from any small wound.

7. **Storage** Some proteins store materials in the way that starch and glycogen store energy. For example, casein in milk and ovalbumin in eggs store nutrients for newborn mammals and birds. Ferritin, a protein in the liver, stores iron.

8. **Regulation** Some proteins not only control the expression of genes, thereby regulating the kind of proteins synthesized in a particular cell, but also dictate when such manufacture takes place.

These are not the only functions of proteins, but they are among the most important. Clearly, any individual needs a great many proteins to carry out all these varied functions. A typical cell contains about 9000 different proteins; the entire human body has about 100,000 different proteins.

We can classify proteins into two major types: **fibrous proteins,** which are insoluble in water and are used mainly for structural purposes, and **globular proteins,** which are more or less soluble in water and are used mainly for nonstructural purposes.

21.2 Amino Acids

Although a wide variety of proteins exist, they all have basically the same structure: They are chains of amino acids. As its name implies, an **amino acid** is an organic compound containing an amino group and a carboxyl group. Organic chemists can synthesize many thousands of amino acids, but nature is much more restrictive and uses only 20 different amino acids to make up proteins. Furthermore, all but one of the 20 fit the formula

$$R-\underset{\underset{NH_3^+}{|}}{\overset{\overset{H}{|}}{C}}-COO^-$$

Even the one that doesn't fit this formula (proline) comes fairly close: It differs only in that it has a bond between the R and the N. The 20 amino acids found in proteins are called **alpha amino acids.** They are listed in Table 21.1, which also shows the one- and three-letter abbreviations that chemists and biochemists use for them.

The most important aspect of the R groups is their polarity. On that basis we can classify amino acids into the four groups shown in Table 21.1: nonpolar, polar but neutral, acidic, and basic. Note that the nonpolar side

Table 21.1 The 20 Amino Acids Commonly Found in Proteins (both the three-letter and one-letter abbreviations are shown) and Their Isoelectric Points (pI)

Nonpolar Side Chains

Amino Acid	Structure	Amino Acid	Structure
alanine (Ala, A) pI = 6.01	$CH_3CH(NH_3^+)COO^-$	phenylalanine (Phe, F) pI = 5.48	$C_6H_5CH_2CH(NH_3^+)COO^-$
glycine (Gly, G) pI = 5.97	$HCH(NH_3^+)COO^-$	proline (Pro, P) pI = 6.48	(pyrrolidine ring with COO^-)
isoleucine (Ile, I) pI = 6.02	$CH_3CH_2CH(CH_3)CH(NH_3^+)COO^-$	tryptophan (Trp, W) pI = 5.88	(indole-$CH_2CH(NH_3^+)COO^-$)
leucine (Leu, L) pI = 5.98	$(CH_3)_2CHCH_2CH(NH_3^+)COO^-$	valine (Val, V) pI = 5.97	$(CH_3)_2CHCH(NH_3^+)COO^-$
methionine (Met, M) pI = 5.74	$CH_3SCH_2CH_2CH(NH_3^+)COO^-$		

Polar but Neutral Side Chains

Amino Acid	Structure	Amino Acid	Structure
asparagine (Asn, N) pI = 5.41	$H_2NC(O)CH_2CH(NH_3^+)COO^-$	serine (Ser, S) pI = 5.68	$HOCH_2CH(NH_3^+)COO^-$
glutamine (Gln, Q) pI = 5.65	$H_2NC(O)CH_2CH_2CH(NH_3^+)COO^-$	threonine (Thr, T) pI = 5.87	$CH_3CH(OH)CH(NH_3^+)COO^-$
cysteine (Cys, C) pI = 5.07	$HSCH_2CH(NH_3^+)COO^-$		
tyrosine (Tyr, Y) pI = 5.66	$HO\text{-}C_6H_4\text{-}CH_2CH(NH_3^+)COO^-$		

Acidic Side Chains

Amino Acid	Structure
aspartic acid (Asp, D) pI = 2.77	$^-OOCCH_2CH(NH_3^+)COO^-$
glutamic acid (Glu, E) pI = 3.22	$^-OOCCH_2CH_2CH(NH_3^+)COO^-$

Basic Side Chains

Amino Acid	Structure
arginine (Arg, R) pI = 10.76	$H_2N\text{-}C(=NH_2^+)\text{-}NHCH_2CH_2CH_2CH(NH_3^+)COO^-$
histidine (His, H) pI = 7.59	(imidazole-$CH_2CH(NH_3^+)COO^-$)
lysine (Lys, K) pI = 9.74	$H_3N^+CH_2CH_2CH_2CH_2CH(NH_3^+)COO^-$

*Each ionizable group is shown in the form present in highest concentration at pH 7.0.

Two of the amino acids in Table 21.1 have a second carbon stereocenter and exist as two pairs of enantiomers, but only one of these four possible stereoisomers exists in proteins. Can you find them?

chains are *hydrophobic* (they repel water), whereas polar but neutral, acidic, and basic side chains are *hydrophilic* (attracted to water). This aspect of the R groups is very important in determining both the structure and the function of each protein molecule.

When we look at the general formula for the 20 amino acids, we see at once that all of them (except glycine, in which R = H) are chiral with (carbon) stereocenters, since R, H, COOH, and NH_2 are four different groups. This means that each of the amino acids with one stereocenter exists as two enantiomers. As is the case for most examples of this kind, nature makes only one of the two possible enantiomers for each amino acid, and it is virtually always the L form. Except for glycine, which exists in only one form, all the amino acids in all the proteins in your body are the L form. D amino acids are extremely rare in nature; some are found, for example, in the cell walls of a few types of bacteria.

In Section 19.2, we learned about the systematic use of the D,L system. There we used glyceraldehyde as a reference point for the assignment of relative configuration. Here again, we can use glyceraldehyde as a reference point with amino acids.

L-Glyceraldehyde   L-Alanine

The spatial relationship of the functional groups around the carbon stereocenter in L-amino acids, as in L-alanine, can be compared to that of L-glyceraldehyde. When we put the carbonyl groups of both compounds in the same position (top), the —OH of L-glyceraldehyde and the NH_3^+ of L-alanine lie to the left of the carbon stereocenter.

21.3 Zwitterions

Screen 21.3 Simulation

In Section 18.2 we learned that carboxylic acids, RCOOH, cannot exist in the presence of a moderately weak base (such as NH_3). They donate a proton to become carboxylate ions, $RCOO^-$. Likewise, amines, RNH_2 (Section 16.6), cannot exist as such in the presence of a moderately weak acid (such as acetic acid). They gain a proton to become substituted ammonium ions, RNH_3^+.

An amino acid has —COOH and $—NH_2$ groups in the same molecule. Therefore, in water solution, the —COOH donates a proton to the $—NH_2$, so that an amino acid actually has the structure

$$R-\underset{\underset{NH_3^+}{|}}{\overset{\overset{H}{|}}{C}}-COO^-$$

Zwitterion comes from the German *zwitter*, meaning "hybrid."

Compounds that have a positive charge on one atom and a negative charge on another are called **zwitterions.** Amino acids are zwitterions, not only in water solution but also in the solid state. They are therefore ionic compounds—that is, internal salts. *Un-ionized $RCH(NH_2)COOH$ molecules do not actually exist, in any form.*

The fact that amino acids are zwitterions explains their physical properties. All of them are solids with high melting points (for example, glycine melts at 262°C), just as we would expect for ionic compounds. The 20 amino acids are also fairly soluble in water, as ionic compounds generally are; if they had no charges, we would expect only the smaller ones to be soluble.

If we add an amino acid to water, it dissolves and then has the same zwitterionic structure that it has in the solid state. Let us see what happens if we change the pH of the solution, as we can easily do by adding a source of H_3O^+, such as HCl solution (to lower the pH), or a strong base, such as NaOH (to raise the pH). Because H_3O^+ is a stronger acid than a typical carboxylic acid (Section 18.2), it donates a proton to the $—COO^-$ group, turning the zwitterion into a positive ion. This happens to all amino acids if the pH is sufficiently lowered—let's say, to pH 2.

$$R-\underset{\underset{NH_3^+}{|}}{\overset{\overset{H}{|}}{C}}-COO^- + H_3O^+ \longrightarrow R-\underset{\underset{NH_3^+}{|}}{\overset{\overset{H}{|}}{C}}-COOH + H_2O$$

Addition of OH^- to the zwitterion causes the $—NH_3^+$ to donate its proton to OH^-, turning the zwitterion into a negative ion. This happens to all amino acids if the pH is sufficiently raised—let's say, to pH 10.

$$R-\underset{\underset{NH_3^+}{|}}{\overset{\overset{H}{|}}{C}}-COO^- + OH^- \longrightarrow R-\underset{\underset{NH_2}{|}}{\overset{\overset{H}{|}}{C}}-COO^- + H_2O$$

In both cases, the amino acid is still an ion so it is still soluble in water. There is no pH at which an amino acid has no ionic character at all. If the amino acid is a positive ion at low pH and a negative ion at high pH, there must be some pH at which all the molecules have equal positive and negative charges. This pH is called the **isoelectric point (pI)**.

Every amino acid has a different isoelectric point, although most of them are not very far apart (see the values in Table 21.1). Fifteen of the 20 have isoelectric points near 6. However, the three basic amino acids have higher isoelectric points, and the two acidic amino acids have lower values.

At or near the isoelectric point, amino acids exist in aqueous solution largely or entirely as zwitterions. As we have seen, they react with either a strong acid, by taking a proton (the $—COO^-$ becomes $—COOH$), or a strong base, by giving a proton (the $—NH_3^+$ becomes $—NH_2$). To summarize:

$$R-\underset{\underset{NH_3^+}{|}}{\overset{\overset{H}{|}}{C}}-COOH \underset{H_3O^+}{\overset{OH^-}{\rightleftharpoons}} R-\underset{\underset{NH_3^+}{|}}{\overset{\overset{H}{|}}{C}}-COO^- \underset{H_3O^+}{\overset{OH^-}{\rightleftharpoons}} R-\underset{\underset{NH_2}{|}}{\overset{\overset{H}{|}}{C}}-COO^-$$

In Section 8.3, we learned that a compound that is both an acid and a base is called amphiprotic. We also learned, in Section 8.11, that a solution that neutralizes both acid and base is a buffer solution. Amino acids are therefore *amphiprotic* compounds, and aqueous solutions of them are *buffers*.

Proteins also have isoelectric points and act as buffers. This property will be discussed in Section 21.6.

> **Isoelectric point (pI)** A pH at which a sample of amino acids or protein has an equal number of positive and negative charges

Acidic amino acids have two $—COO^-$ groups, one on the α-carbon and one on the side chain. At the isoelectric point, one of these must be in the un-ionized form ($—COOH$). To achieve this, H_3O^+ must be added to lower the pH. Similar considerations explain why the isoelectric points of basic amino acids are higher than pH 6.

21.4 Cysteine: A Special Amino Acid

One of the 20 amino acids in Table 21.1 has a chemical property not shared by any of the others. This amino acid, cysteine, can easily be dimerized by many mild oxidizing agents:

We met these reactions in Section 14.5D.

[O] symbolizes oxidation, and [H] stands for reduction.

$$2\,HS-CH_2-\underset{NH_3^+}{CH}-COO^- \underset{[H]}{\overset{[O]}{\rightleftharpoons}} {}^-OOC-\underset{NH_3^+}{CH}-CH_2-\overset{\text{A disulfide bond}}{S-S}-CH_2-\underset{NH_3^+}{CH}-COO^-$$

Cysteine $\qquad\qquad\qquad\qquad\qquad\qquad$ Cystine

A dimer is a molecule made up of two units.

The dimer of cysteine, which is called **cystine,** can in turn be fairly easily reduced to give two molecules of cysteine. As we shall see, the presence of cystine has important consequences for the chemical structure and shape of protein molecules of which it is part. The **S—S bond** (shown in color) is also called a **disulfide bond** (Section 14.5D).

21.5 Peptides and Proteins

Each amino acid has a carboxyl group and an amino group. In Section 18.5, we saw that a carboxylic acid and an amine could be combined to form an amide:

$$R-\overset{O}{\underset{\|}{C}}-O^- + R'-NH_3^+ \longrightarrow R-\overset{O}{\underset{\|}{C}}-NH-R' + H_2O$$

In the same way, the —COO⁻ group of one amino acid molecule, say glycine, can combine with the —NH₃⁺ group of a second molecule, say alanine:

$$H_3\overset{+}{N}-CH_2-\overset{O}{\underset{\|}{C}}-O^- + H_3\overset{+}{N}-\underset{CH_3}{CH}-\overset{O}{\underset{\|}{C}}-O^- \longrightarrow H_3\overset{+}{N}-CH_2-\overset{O}{\underset{\|}{C}}-NH-\underset{CH_3}{CH}-\overset{O}{\underset{\|}{C}}-O^- + H_2O$$

Glycine $\qquad\qquad$ Alanine $\qquad\qquad\qquad\qquad$ Glycylalanine (Gly—Ala)

Peptide bond An amide bond that links two amino acids

The naming of the peptides begins with the N-terminal amino acids.

This reaction takes place in the cells by a mechanism that we will examine in Section 25.5. The product is an amide. The two amino acids are joined together by a **peptide bond** (also called a **peptide linkage**). The product is a **dipeptide.**

It is important to realize that glycine and alanine could also be linked the other way:

$$H_3\overset{+}{N}-\underset{CH_3}{CH}-\overset{O}{\underset{\|}{C}}-O^- + H_3\overset{+}{N}-CH_2-\overset{O}{\underset{\|}{C}}-O^- \longrightarrow H_3\overset{+}{N}-\underset{CH_3}{CH}-\overset{O}{\underset{\|}{C}}-NH-CH_2-\overset{O}{\underset{\|}{C}}-O^- + H_2O$$

Alanine $\qquad\qquad$ Glycine $\qquad\qquad\qquad\qquad$ Alanylglycine (Ala—Gly)

In Ala—Gly, the —NH₃⁺ is connected to a —CHCH₃; in Gly—Ala, the —NH₃⁺ is connected to a —CH₂.

In this case we get a *different* dipeptide. The two dipeptides are constitutional isomers, of course; they are different compounds in all respects, with different properties.

EXAMPLE 21.1

Show how to form the dipeptide aspartylserine (Asp—Ser).

Solution

The name implies that this dipeptide is made of two amino acids, aspartic acid (Asp) and serine (Ser), with the amide being formed between the α-carboxyl group of aspartic acid and the amino group of serine. Therefore, we write the formula of aspartic acid with its amino group on the left side. Next we place the formula of serine to the right, with its amino group facing the α-carboxyl group of aspartic acid. Finally, we eliminate a water molecule between the —COO$^-$ and —NH$_3^+$ groups next to each other, forming the peptide bond:

$$H_3\overset{+}{N}-CH-\overset{O}{\underset{\|}{C}}-O^- + H_3\overset{+}{N}-CH-\overset{O}{\underset{\|}{C}}-O^-$$
$$\underset{Asp}{\underset{|}{\underset{COO^-}{\underset{|}{CH_2}}}} \qquad \underset{Ser}{\underset{|}{CH_2OH}}$$

$$\longrightarrow H_3\overset{+}{N}-CH-\overset{O}{\underset{\|}{C}}-NH-CH-\overset{O}{\underset{\|}{C}}-O^- + H_2O$$
$$\underset{|}{\underset{COO^-}{\underset{|}{CH_2}}} \qquad \underset{|}{CH_2OH}$$
$$\text{Asp—Ser}$$

Problem 21.1

Show how to form the dipeptide valylphenylalanine (Val—Phe).

Any two amino acids, whether the same or different, can be linked together to form dipeptides in a similar manner. But the possibilities do not end there. Each dipeptide still contains a —COO$^-$ and an —NH$_3^+$ group. We can therefore add a third amino acid to alanylglycine, say lysine:

$$H_3\overset{+}{N}-CH-\overset{O}{\underset{\|}{C}}-NH-CH_2-\overset{O}{\underset{\|}{C}}-O^- + H_3\overset{+}{N}-CH-\overset{O}{\underset{\|}{C}}-O^-$$
$$\underset{Ala—Gly}{\underset{|}{CH_3}} \qquad\qquad \underset{Lys}{\underset{\underset{NH_3^+}{|}}{\underset{|}{(CH_2)_4}}}$$

$$\xrightarrow{-H_2O} H_3\overset{+}{N}-CH-\overset{O}{\underset{\|}{C}}-NH-CH_2-\overset{O}{\underset{\|}{C}}-NH-CH-\overset{O}{\underset{\|}{C}}-O^-$$
$$\underset{|}{CH_3} \qquad\qquad\qquad\qquad \underset{\underset{NH_3^+}{|}}{\underset{|}{(CH_2)_4}}$$
$$\text{Ala—Gly—Lys}$$
$$\text{A tripeptide}$$

The product is a **tripeptide.** Because it also contains a —COO$^-$ and an —NH$_3^+$ group, we can continue the process to get a tetrapeptide, a

CHEMICAL CONNECTIONS 21A

Glutathione

A very important tripeptide present in high concentrations in all tissues is glutathione. Glutathione contains L-glutamic acid, L-cysteine, and glycine. Its structure is unusual because the glutamic acid is linked to the cysteine by its γ-carboxyl group rather than by the α-carboxyl group, as is usual in most peptides and proteins:

Glutathione functions in the cells as a general protective agent. Oxidizing agents that would damage the cells, such as peroxides, will oxidize glutathione instead (the cysteine portion; see also Section 21.4), thereby protecting the proteins and nucleic acids. Many foreign chemicals also get attached to glutathione, so in a sense it acts as a detoxifying agent.

$$H_3\overset{+}{N}-\underset{\underset{COO^-}{|}}{CH}-CH_2-CH_2-\underset{\underset{O}{\|}}{C}-NH-\underset{\underset{\underset{SH}{|}}{\underset{CH_2}{|}}}{CH}-\underset{\underset{O}{\|}}{C}-NH-CH_2-COO^-$$

Glutathione
(Glu—Cys—Gly)

---

pentapeptide, and so on, until we have a chain containing hundreds or even thousands of amino acids. These chains of amino acids are the proteins that serve so many important functions in living organisms.

A word must be said about the terms used to describe these compounds. The shortest chains are often simply called **peptides,** longer ones are **polypeptides,** and still longer ones are **proteins,** but chemists differ about where to draw the line. Many chemists use the terms "polypeptide" and "protein" almost interchangeably. In this book, we will consider a protein to be a polypeptide chain that contains a minimum of 30 to 50 amino acids.

The amino acids in a chain are often called **residues.** It is customary to use either the one-letter or the three-letter abbreviations shown in Table 21.1 to represent peptides and proteins. For example, the tripeptide shown on the previous page, alanylglycyllysine, is AGK or Ala—Gly—Lys. **C-terminal amino acid** is the residue with the free COO^- group (lysine in Ala—Gly—Lys), and **N-terminal amino acid** is the residue with the free NH_3^+ group (alanine in Ala—Gly—Lys). It is the universal custom to write peptide and protein chains with the N-terminal residue on the left. No matter how long a protein chain gets—hundreds or thousands of units—it always has just two ends: one C-terminal and one N-terminal.

21.6 Some Properties of Peptides and Proteins

The continuing pattern of peptide bonds is the backbone of the peptide or protein molecule; the R groups are called the **side chains.** The six atoms of the peptide backbone lie in the same plane (Figure 21.1), and two adjacent peptide bonds can rotate relative to one another about the C—N and C—C bonds.

$$\sim\sim\sim\overset{H}{\underset{|}{N}}-\underset{\underset{R_1}{|}}{CH}-\overset{O}{\underset{\|}{C}}-\overset{H}{\underset{|}{N}}-\underset{\underset{R_2}{|}}{CH}-\overset{O}{\underset{\|}{C}}\sim\sim\sim$$

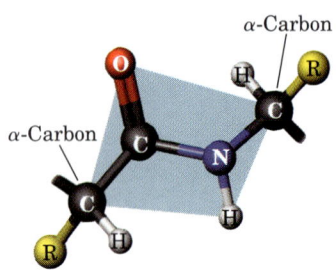

Figure 21.1 Six atoms of the peptide backbone lie in an imaginary (shaded) plane.

The 20 different amino acid side chains supply variety and determine the physical and chemical properties of proteins. Among these properties,

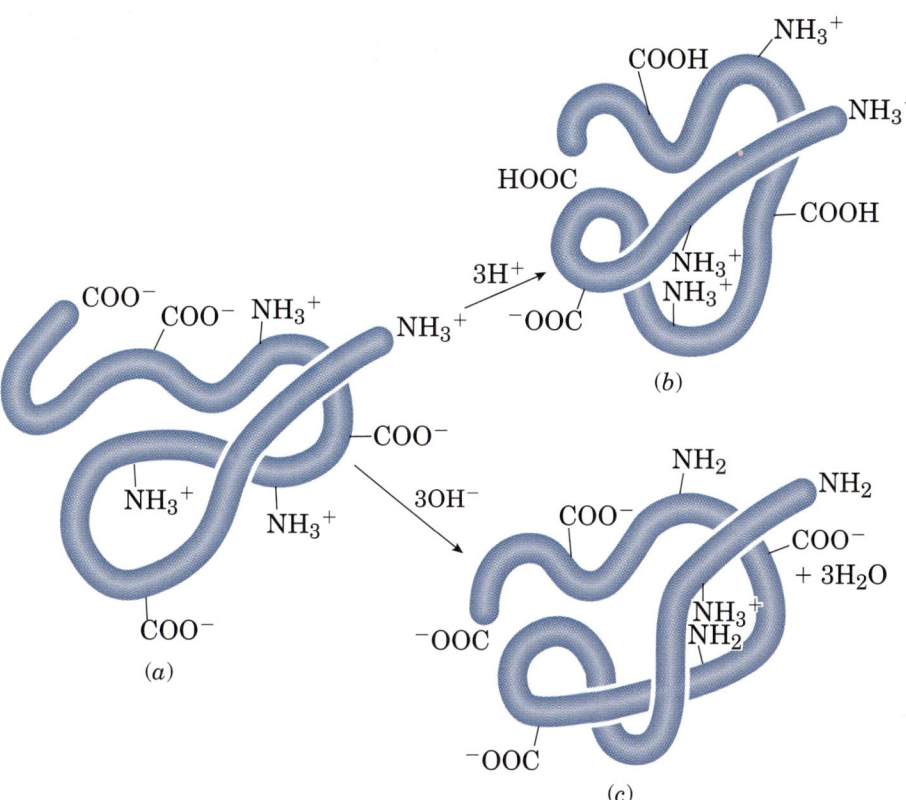

Figure 21.2 Schematic diagram of a protein (a) at its isoelectric point and its buffering action when (b) H⁺ or (c) OH⁻ ions are added.

acid–base behavior is one of the most important. Like amino acids (Section 21.3), proteins behave as zwitterions. The side chains of glutamic and aspartic acids provide acidic groups, whereas lysine and arginine provide basic groups (histidine does, too, but this side chain is less basic than the other two). (See the structures of these amino acids in Table 21.1.)

The isoelectric point of a protein occurs at the pH at which there are an equal number of positive and negative charges (the protein has no *net* charge). At any pH above the isoelectric point, the protein molecules have a net negative charge; at any pH below the isoelectric point, they have a net positive charge. Some proteins, such as hemoglobin, have an almost equal number of acidic and basic groups; the isoelectric point of hemoglobin is at pH 6.8. Others, such as serum albumin, have more acidic groups than basic groups; the isoelectric point of this protein is 4.9. In each case, however, because proteins behave like zwitterions, they act as buffers, for example, in the blood (Figure 21.2).

The water solubility of large molecules such as proteins often depends on the repulsive forces between like charges on their surfaces. When protein molecules are at a pH at which they have a net positive or negative charge, the presence of these like charges causes the protein molecules to repel each other. These repulsive forces are smallest at the isoelectric point, when the net charges are close to zero. When there are no repulsive forces, the protein molecules tend to clump together to form aggregates of two or more molecules, reducing their solubility. Therefore, *proteins are least soluble in water at their isoelectric points and can be precipitated from their solutions.*

As we pointed out in Section 21.1, proteins have many functions. To understand these functions, we must look at four levels of organization in their structures. The *primary structure* describes the linear sequence of amino acids in the polypeptide chain. The *secondary structure* refers to certain re-

The terminal —COOH and —NH₂ groups also ionize, but they are only 2 out of 50 or more residues.

Carbonates and phosphates are the other blood buffers (see Section 8.11).

CHEMICAL CONNECTIONS 21B

AGE and Aging

A reaction can take place between a primary amine and an aldehyde or a ketone, linking the two molecules (shown here for an aldehyde):

$$R-\underset{\underset{H}{\|}}{\overset{O}{C}}-H + H_2N-R' \longrightarrow R-CH=N-R' + H_2O$$
An imine

Because proteins have NH_2 groups and carbohydrates have aldehyde or keto groups, they can undergo this reaction, establishing a link between a sugar and a protein molecule. When this reaction is not catalyzed by enzymes, it is called *glycation* of proteins. The process, however, does not stop there. When these linked products are heated in a test tube, high-molecular-weight, water-insoluble, brownish complexes are formed. These complexes are called **advanced glycation end-products (AGE)**. In the body, they cannot be heated, but the same result happens over long periods of time.

The longer we live, and the higher the blood sugar concentration becomes, the more AGE products accumulate in the body. AGE products show up in all the afflicted organs of diabetic patients: in the lens of the eye (cataracts), in the capillary blood vessels of the retina (diabetic retinopathy), and in the glomeruli of the kidneys (kidney failure). For people who do not have diabetes, these harmful protein modifications become disturbing only in advanced years. In a young person, metabolism functions properly and the AGE products decompose and are eliminated from the body. In an older person, metabolism slows and the AGE products accumulate. AGE products themselves are thought to enhance oxidative damages.

Figure 21.3 The "structure" of a telephone cord: (a) primary, (b) secondary, (c) tertiary.

Screen 21.5 Tutorial

peating patterns, such as the α-helix conformation or the pleated sheet (Section 21.8), or the absence of a repeating pattern, as with the random coil (Section 21.8). The *tertiary structure* describes the overall conformation of the polypeptide chain.

A good analogy for these three types of structure is a coiled telephone cord (Figure 21.3). The primary structure is the stretched-out cord. The secondary structure is the coil in the form of a helix. We can also take the entire coil and twist it into various shapes, thereby creating a tertiary structure. As we shall see, protein molecules twist and curl in a very similar manner.

The quaternary structure (Section 21.9) applies mainly to proteins containing more than one polypeptide chain (subunit) and deals with how the different chains are spatially related to each other. We also call a structure quaternary when a single polypeptide chain interacts with lipid bilayers and that interaction determines how the different parts of the polypeptide chains are related to each other.

21.7 The Primary Structure of Proteins

Very simply, the primary structure of a protein consists of the sequence of amino acids that makes up the chain. Each of the very large number of peptide and protein molecules in biological organisms has a different sequence of amino acids—and that sequence allows the protein to carry out its function, whatever it may be.

How can so many different proteins arise from different sequences of only 20 amino acids? Let us look at a little arithmetic, starting with a dipeptide. How many different dipeptides can be made from 20 amino acids? There are 20 possibilities for the N-terminal amino acid, and for each of these 20 there are 20 possibilities for the C-terminal amino acid. This means that there are $20 \times 20 = 400$ different dipeptides possible from the 20 amino acids. What about tripeptides? We can form a tripeptide by taking any of the 400 dipeptides and adding any of the 20 amino acids. Thus, there are 20 × 20 × 20 = 8000 tripeptides, all different. It is easy to see that we can calculate the total number of possible peptides or proteins for a chain of n amino acids simply by raising 20 to the nth power (20^n).

Taking a typical small protein to be one with 60 amino acid residues, the number of proteins that can be made from the 20 amino acids is $20^{60} = 10^{78}$. This is an enormous number, possibly greater than the total number of atoms in the universe. Clearly, only a tiny fraction of all possible protein molecules has ever been made by biological organisms.

Each peptide or protein in the body has its own unique sequence of amino acids. As with naming of peptides, the *assignment of positions of the amino acids in the sequence starts at the N-terminal end.* Thus in Figure 21.4 glycine is in the number 1 position on the A chain and phenylalanine is number 1 on the B chain. We mentioned that proteins also have secondary, tertiary, and, in some cases, quaternary structures. We will deal with these in Sections 21.8 and 21.9, but here we can say that *the primary structure of a protein determines to a large extent the native* (most frequently occurring) *secondary and tertiary structures.* That is, the particular sequence of amino acids on the chain enables the whole chain to fold and curl in such a way as to assume its final shape. As we will see in Section 21.11, without its particular three-dimensional shape, a protein cannot function.

Just how important is the exact amino acid sequence to the function of a protein? Can a protein perform the same function if its sequence is a little different? The answer to this question is that a change in amino acid sequence may or may not matter, depending on what kind of a change it is. Consider, as an example, cytochrome c, which is a protein of terrestrial vertebrates. Its chain consists of 104 amino acid residues. It performs the same function (electron transport) in humans, chimpanzees, sheep, and other animals. While humans and chimpanzees have exactly the same amino acid sequence of this protein, sheep cytochrome c differs in 10 positions out of the 104. (You can find more about biochemical evolution in Chemical Connections 25E).

Another example is the hormone insulin. Human insulin consists of two chains having a total of 51 amino acids. The two chains are connected by disulfide bonds. Figure 21.4 shows the sequence of amino acids. Insulin is necessary for proper utilization of carbohydrates (Section 27.2), and people with severe diabetes (Chemical Connections 23G) must take insulin injections. The amount of human insulin available is far too small to meet the need for it, so bovine insulin (from cattle) or insulin from hogs or sheep is used instead. Insulin from these sources is similar, but not identical, to human insulin. The differences are entirely in the 8, 9, and 10 positions of the A chain and the C-terminal position (30) of the B chain:

	A Chain			B Chain
	8	9	10	30
Human	—Thr—	Ser—	Ile—	—Thr
Bovine	—Ala—	Ser—	Val—	—Ala
Hog	—Thr—	Ser—	Ile—	—Ala
Sheep	—Ala—	Gly—	Val—	—Ala

The remainder of the molecule is the same in all four varieties of insulin. Despite the slight differences in structure, all these insulins perform the same function and even can be used by humans. However, none of the other three is quite as effective in humans as human insulin.

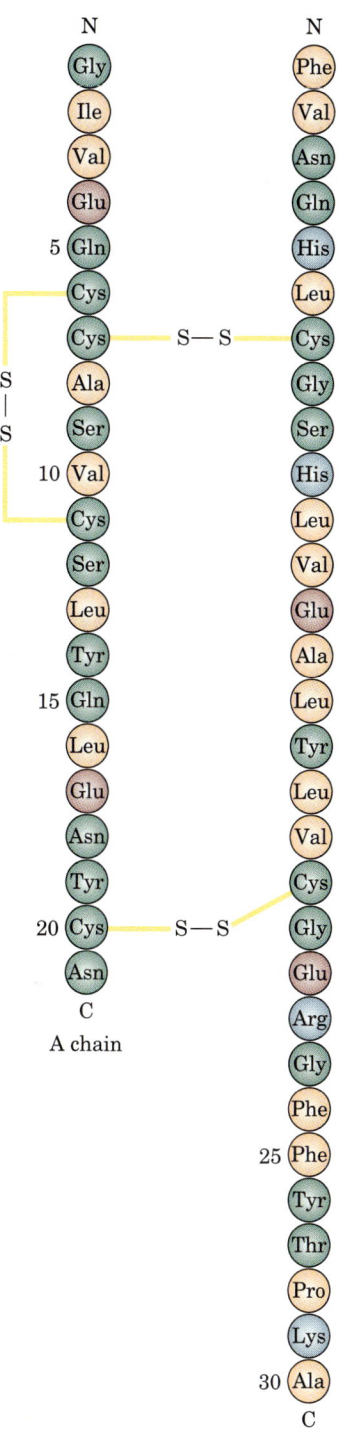

Figure 21.4 The hormone insulin consists of two polypeptide chains, A and B, held together by two disulfide cross-bridges (S—S). The sequence shown is for bovine insulin.

CHEMICAL CONNECTIONS 21C

The Use of Human Insulin

Although human insulin, as manufactured by recombinant DNA techniques (see Section 25.8), is on the market, many diabetic patients continue to use hog or sheep insulin because it is cheaper. Changing from animal to human insulin creates an occasional problem for diabetics. All diabetics experience an insulin reaction (hypoglycemia) when the insulin level in the blood is too high relative to the blood sugar level. Hypoglycemia is preceded by symptoms of hunger, sweating, and poor coordination. These symptoms, called **hypoglycemic awareness**, signal the patient that hypoglycemia is coming and that it must be reversed, which the patient can do by eating sugar.

Some diabetics who changed from animal to human insulin reported that the hypoglycemic awareness from recombinant DNA human insulin is not as strong as that from animal insulin. This lack of recognition can create some hazards, and the effect is probably due to different rates of absorption in the body. The literature supplied with human insulin now incorporates a warning that hypoglycemic awareness may be altered.

The first sequence of an important protein, insulin, was obtained by Frederick Sanger (1918–) in England, for which he received the Nobel Prize in 1958.

Both oxytocin and vasopressin are secreted by the pituitary gland.

Both hormones are used as drugs—vasopressin to combat loss of blood pressure after surgery and oxytocin to induce labor.

Another factor showing the effect of substituting one amino acid for another is that sometimes patients become allergic to, say, bovine insulin but can switch to hog or sheep insulin without experiencing an allergic reaction.

In contrast to the previous examples, there are small changes in amino acid sequence that make a great deal of difference. Consider two peptide hormones, oxytocin and vasopressin (Figure 21.5). These nonapeptides have identical structures, including a disulfide bond, except for different amino acids in positions 2 and 7. Yet their biological functions are quite different. Vasopressin is an antidiuretic hormone. It increases the amount of water reabsorbed by the kidneys and raises blood pressure. Oxytocin has no effect on water in the kidneys and slightly lowers blood pressure. It affects contractions of the uterus in childbirth and the muscles in the breast that aid in the secretion of milk. Vasopressin also stimulates uterine contractions, but to a much lesser extent than oxytocin.

Another instance where a minor change makes a major difference is in the blood protein hemoglobin. A change in only one amino acid in a chain of 146 is enough to cause a fatal disease—sickle cell anemia (Chemical Connections 21D).

In some cases, slight changes in amino acid sequence make little or no difference to the functioning of peptides and proteins, but most times, the sequence is highly important. The sequences of 10,000 protein and peptide molecules have now been determined. The methods for doing so are complicated and will not be discussed in this book.

Figure 21.5 The structures of vasopressin and oxytocin. Differences are shown in color.

```
         9           4    3    2    1
        Cys—S—S—Cys—Pro—Arg—Gly—NH₂
         |             |
       8 Tyr          Asn 5
         |             |
        Phe—Gln
         7    6
```
Vasopressin

```
         9           4    3    2    1
        Cys—S—S—Cys—Pro—Leu—Gly—NH₂
         |             |
       8 Tyr          Asn 5
         |             |
         Ile—Gln
         7    6
```
Oxytocin

CHEMICAL CONNECTIONS 21D

Sickle Cell Anemia

Normal adult human hemoglobin has two alpha chains and two beta chains (see Figure 21.13). Some people, however, have a slightly different kind of hemoglobin in their blood. This hemoglobin (called HbS) differs from the normal type only in the beta chains and only in one position on these two chains: The glutamic acid in the sixth position of normal Hb is replaced by a valine residue in HbS.

	4	5	6	7	8	9
Normal Hb—	Thr—	Pro—	Glu—	Glu—	Lys—	Ala—
Sickle cell Hb—	Thr—	Pro—	Val—	Glu—	Lys—	Ala—

This change affects only two positions in a molecule containing 574 amino acid residues. Yet, it is enough to produce a very serious disease, **sickle cell anemia.**

Red blood cells carrying HbS behave normally when there is an ample oxygen supply. When the oxygen pressure decreases, however, the red blood cells become sickle-shaped. This malformation occurs in the capillaries. As a result of this change in shape, the cells may clog the capillaries. The body's defenses destroy the clogging cells, and the loss of the blood cells causes anemia.

This change at only a single position of a chain consisting of 146 amino acids is severe enough to cause a high death rate. A child who inherits two genes programmed to produce sickle cell hemoglobin (a homozygote) has an 80% smaller chance of surviving to adulthood than a child with only one such gene (a heterozygote) or a child with two normal genes. Despite the high mortality of homozygotes, the genetic trait survives. In central Africa, 40% of the population in malaria-ridden areas carry the sickle cell gene, and 4% are homozygotes. It seems that the sickle cell genes help to acquire immunity against malaria in early childhood so that in malaria-ridden areas the transmission of these genes is advantageous.

Figure 21D Blood cells from a patient with sickle cell anemia. Both normal cells (round) and sickle cells (shriveled) are visible.

There is no known cure for sickle cell anemia. Recently the Food and Drug Administration approved hydroxyurea, (sold under the name Droxia) to treat and control the symptoms of the disease. Hydroxyurea prompts the bone marrow to manufacture fetal hemoglobin (HbF), which does not have beta chains where the mutation occurs. Thus red blood cells containing HbF do not sickle and do not clog the capillaries. With hydroxyurea therapy, the bone marrow still manufactures mutated HbS, but the presence of cells with fetal hemoglobin dilutes the concentration of the sickling cells, thereby relieving the symptoms of the disease.

21.8 The Secondary Structure of Proteins

Proteins can fold or align themselves in such a manner that certain patterns repeat themselves. These repeating patterns are referred to as **secondary structures.** The two most common secondary structures encountered in proteins are the α-helix and the β-pleated sheet (Figure 21.6), which were proposed by Linus Pauling and Robert Corey in the 1940s. In contrast, those protein conformations that do not exhibit a repeated pattern are called random coils (Figure 21.7).

In the α-helix form, a single protein chain twists in such a manner that its shape resembles a right-handed coiled spring—that is, a helix. The shape of the helix is maintained by numerous **intramolecular hydrogen bonds** that exist between the backbone —C=O and H—N— groups. As shown in Figure 21.6, there is a hydrogen bond between the —C=O oxygen atom of each peptide linkage and the —N—H hydrogen atom of another peptide linkage four amino acid residues farther along the chain.

Linus Pauling (1901–1994), who won the 1954 Nobel Prize in chemistry, identified these structures. He also discovered or contributed much to our understanding of certain fundamental concepts, including chemical bonds and electronegativity (Section 3.4B).

● Carbon atom ● Oxygen atom ● Nitrogen atom ● Hydrogen atom ● R group

Figure 21.6 (a) The α-helix. (b) The β-pleated sheet structure.

These hydrogen bonds are in just the right position to cause the molecule (or a portion of it) to maintain a helical shape. Each —N—H points upward and each C=O points downward, roughly parallel to the axis of the helix. All the R— groups (the amino acid side chains) point outward from the helix.

The other important orderly structure in proteins is the **β-pleated sheet.** In this case, the orderly alignment of protein chains is maintained by **intermolecular or intramolecular hydrogen bonds.** The β-pleated sheet structure can occur between molecules when polypeptide chains run parallel (all the N-terminal ends on one side) or antiparallel (neighboring N-terminal ends on opposite sides). β-Pleated sheets can also occur intramolecularly, when the polypeptide chain makes a U-turn, forming a hairpin structure, and the pleated sheet is antiparallel (Figure 21.6).

In all secondary structures, the hydrogen bonding is between backbone —C=O and H—N— groups. This is the distinction between secondary and tertiary structures. In the latter, as we shall see, the hydrogen bonding takes place between R groups on the side chains.

Few proteins have predominantly α-helix or β-pleated sheet structures. Most proteins, especially globular ones, have only certain portions of their molecules in these conformations. The rest of the molecule consists of **random**

> **Beta (β)-pleated sheet** A secondary protein structure in which the backbone of two protein chains in the same or different molecules is held together by hydrogen bonds

An intramolecular hydrogen bond goes from a hydrogen atom in a molecule to an O, N, or F atom in the same molecule.

Figure 21.7 A random coil.

Figure 21.8 Schematic structure of the enzyme carboxypeptidase. The β-pleated sheet portions are shown in blue, the green structures are the α-helix portions, and the orange strings are the random coil areas.

CHEMICAL CONNECTIONS 21E

Protein/Peptide Conformation–Dependent Diseases

In a number of diseases, a normal protein or peptide becomes pathological when its conformation changes. A common feature of these proteins is the property to self-assemble into β-sheet–forming amyloid (starch-like) plaques. These amyloid structures appear in several diseases.

One example of this process is the prion protein, the discovery of which brought Stanley Prusiner of the University of California, San Francisco, the Nobel Prize in 1997. When prion undergoes conformational change, it can cause the so-called mad cow disease. During the conformational change, the α-helical content of the normal prion protein unfolds and re-assembles in the β-pleated sheet form. This new form supposedly has the potential to cause more normal prion proteins to undergo conformational change. In humans, it causes spongiform encephalitis; Creutzfeld-Jakob disease is one variant that mainly afflicts the elderly. Although the transmission of this infection from diseased cows to humans is rare, fear of it caused the wholesale slaughter of British cattle in 1998 and, for a while, an embargo was placed on the importation of such meat in most of Europe and America.

Another such conformational change is found in the protein transthyretin. This protein normally forms a tetramer that carries vitamin A and thyroid hormone in the serum. When an individual inherits a mutant form of this protein from one of his or her parents, the tetramer becomes unstable and dissociates. These monomers, in turn, partially unfold and aggregate to form amyloid fibers that interfere with nerve and muscle function. The resulting disease is called senile systematic amyloidoses. β-amyloid plaques also appear in Alzheimer's disease brains (see Chemical Connections 23D).

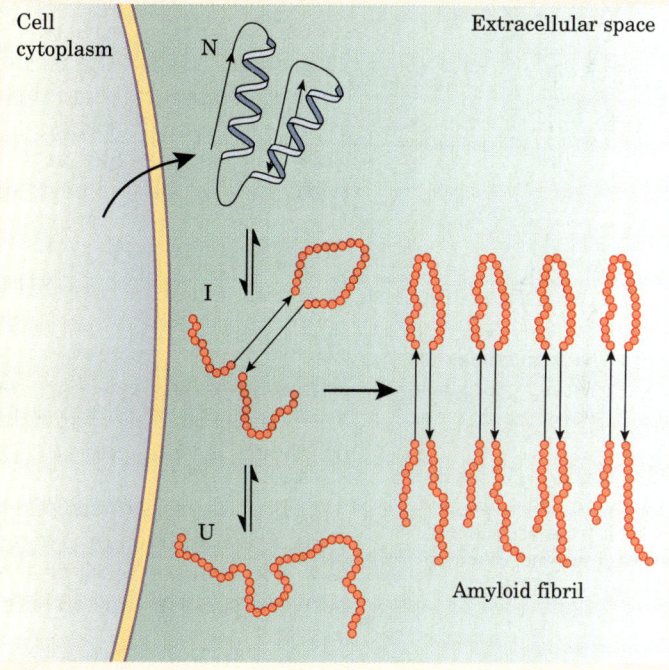

■ **Figure 21E** Schematic representation of a possible mechanism of amyloid fibril formation. After synthesis, the protein is assumed to fold in native (N) secondary structure aided by chaperones. Under certain conditions, the native structure can partially unfold (I) and form β-pleated sheets of amyloid fibrils or even completely unfold (U) as a random coil.

coil. Many globular proteins contain all three kinds of secondary structures in different parts of their molecules: α-helix, β-pleated sheet, and random coil. Figure 21.8 shows a schematic representation of such a structure.

Keratin, a fibrous protein of hair, fingernails, horns, and wool, is one protein that does have a predominantly α-helix structure. Silk is made of fibroin, another fibrous protein, which exists mainly in the β-pleated sheet form. Silkworm silk and especially spider silk exhibit a combination of strength and toughness unmatched by high-performance synthetic fibers. In its primary structure, silk contains sections that consist of only alanine (25%) and glycine (42%). The formation of β-pleated sheets, largely by the alanine sections, allows microcrystals to orient themselves along the fiber axis, which accounts for the material's superior mechanical strength.

Figure 21.9 The triple helix of collagen.

Hydroxyproline is not one of the 20 amino acids listed in Table 21.1, but the body makes it from proline and uses it in certain proteins, of which collagen is but one example.

Another repeating pattern classified as a secondary structure is the **extended helix** of collagen (Figure 21.9). It is quite different from the α-helix. Collagen is the structural protein of connective tissues (bone, cartilage, tendon, blood vessels, skin), where it provides strength and elasticity. The most abundant protein in humans, it makes up about 30% by weight of all the body's protein. The extended helix structure is made possible by the primary structure of collagen. Each strand of collagen consists of repetitive units that can be symbolized as Gly—X—Y; that is, every third amino acid in the chain is glycine. Glycine, of course, has the shortest side chain (—H) of all amino acids. About one-third of the X amino acid is proline, and the Y is often hydroxyproline.

21.9 The Tertiary and Quaternary Structures of Proteins

A Tertiary Structure

In general, tertiary structures are stabilized four ways:

1. Covalent Bonds The covalent bond most often involved in stabilization of the tertiary structure of proteins is the disulfide bond. In Section 21.4, we noted that the amino acid cysteine is easily converted to the dimer cystine. When a cysteine residue is in one chain and another cysteine residue is in another chain (or in another part of the same chain), formation of a disulfide bond provides a covalent linkage that binds together the two chains or the two parts of the same chain:

$$\text{—SH HS—} \xrightarrow{[O]} \text{—S—S—}$$

Examples of both types are found in the structure of insulin (Figure 21.4).

Besides covalent bonds, three other interactions (Figure 21.10) can stabilize tertiary structures.

2. Hydrogen Bonding In Section 21.8, we saw that secondary structures are stabilized by hydrogen bonding between backbone —C=O and H—N— groups. Tertiary structures are stabilized by hydrogen bonding between polar groups on side chains.

3. Salt Bridges Salt bridges occur only between two amino acids with ionized side chains—that is, between an acidic amino acid (—COO$^-$) and a basic amino acid, (—NH$_3^+$ or =NH$_2^+$) side chain. The two are held together by simple ion–ion attraction.

4. Hydrophobic Interactions In aqueous solution, globular proteins usually turn their polar groups outward, toward the aqueous solvent, and their nonpolar groups inward, away from the water molecules. The nonpolar groups prefer to interact with each other, excluding water from these regions. The result is hydrophobic interactions (see Section 20.1). Although this type of interaction is weaker than hydrogen bonding or salt bridges, it usually acts over large surface areas so that cooperatively the interactions are strong enough to stabilize a loop or some other tertiary structure formation.

Figure 21.11 illustrates the four types of interactions that stabilize the tertiary structures of proteins.

Figure 21.10 Noncovalent interactions that stabilize the tertiary and quaternary structures of proteins: (a) hydrogen bonding, (b) salt bridge, (c) hydrophobic interaction.

EXAMPLE 21.2

What kind of noncovalent interaction occurs between the side chains of serine and glutamine?

Solution
The side chain of serine ends in an —OH group; that of glutamine ends in an amide, CO—NH$_2$ group. The two groups can form hydrogen bonds.

Problem 21.2
What kind of noncovalent interaction occurs between the side chains of arginine and glutamic acid?

In Section 21.7, we pointed out that the primary structure of a protein largely determines its secondary and tertiary structures. We can now see the reason for this. When the particular R groups are in the proper positions, all the hydrogen bonds, salt bridges, disulfide linkages, and hydrophobic interactions that stabilize the three-dimensional structure of that molecule can form.

Figure 21.11 The four types of interactions that stabilize the tertiary structure of proteins, shown schematically.

Figure 21.12 Primary, secondary, tertiary, and quaternary structures of protein.

CHEMICAL CONNECTIONS 21F

Proteomics, Ahoy!

Proteins in the body are in a state of dynamic flux. Their multiple functions necessitate that they change constantly: Some are rapidly synthesized, others have their synthesis inhibited; some are degraded, others are modified. There is a new concerted effort being made to catalogue all the proteins in their various forms in a particular cell or a tissue. The name of this venture is *proteomics,* a term coined as an analogy to *genomics* (see Chemical Conenctions 24D), in which all the genes of an organism and their locations in the chromosomes were determined. In proteomics, all the proteins and peptides of a cell or tissue are separated, and their masses are then characterized and compared with a computerized database to obtain and identify a comprehensive protein profile. In this manner, the dynamic states of a large number of proteins are obtained simultaneously and the status of a cell or tissue can be ascertained whether it is healthy or pathological.

The side chains of some proteins allow them to fold (form a tertiary structure) in only one way; other proteins, especially those with long polypeptide chains, can fold in a number of possible ways. Certain proteins in living cells, called **chaperones,** help a newly synthesized polypeptide chain assume the proper secondary and tertiary structures that are necessary for the functioning of that molecule and prevent foldings that would yield biologically inactive molecules.

Chaperone A protein molecule that helps other proteins to fold into the biologically active conformation and enables partially denatured proteins to regain their biologically active conformation

B Quaternary Structure

The highest level of protein organization is the quaternary structure, which applies mainly to proteins with more than one polypeptide chain. Figure 21.12 summarizes schematically the four levels of protein structure. Quaternary structure determines how the different subunits of the protein fit into an organized whole. The subunits are packed and held together by hydrogen bonds, salt bridges, and hydrophobic interactions—the same forces that operate within tertiary structures. The quaternary structure also determines how the different segments of a single polypeptide chain fit into an organized whole when they traverse and interact with lipid bilayers of membranes.

1. Hemoglobin Hemoglobin in adult humans is made of four chains (called globins): two identical chains (alpha) of 141 amino acid residues each and two identical chains (beta) of 146 residues each. Figure 21.13 shows how the four chains fit together.

In hemoglobin, each globin chain surrounds an iron-containing heme unit, the structure of which is shown in Figure 21.14. Proteins that contain

When a protein consists of more than one polypeptide chain, each is called a *subunit.*

Figure 21.13 The quaternary structure of hemoglobin.

Figure 21.14 The structure of heme.

CHEMICAL CONNECTIONS 21G

The Role of Quaternary Structures in Mechanical Stress and Strain

Collagen is the main component of the extracellular matrix that exists between cells of higher organisms. The triple-helix units of collagen (Figure 21.9) are further organized into a pattern called the quarter-stagger arrangement, forming fibrils (Figure 21G.1). Fibers made of these fibrils provide elasticity to such tissues as skin and cornea of the eye.

In the **quarter-stagger arrangement,** the units along a row are not spaced end to end. Rather, there is a gap between the end of one unit and the beginning of another unit. These gaps play an important role in bone formation. Collagen is one of the main constituents of bones and teeth, and the gaps in the quarter-stagger arrangement are essential for the deposition of inorganic crystals of calcium hydroxyapatite, $Ca_5(PO_4)_3OH$. The gaps serve as nucleation sites for the growth of these crystals. The combination of hydroxyapatite crystals and collagen creates a hard material that still has some springiness, owing to the presence of collagen. Dentine, the main constituent of the internal part of a tooth, contains a higher percentage (about 75%) of inorganic crystals than does bone and is therefore harder. Enamel, the outer part of the tooth, has a still higher mineral content (about 95%) and is even harder.

The quaternary structures of other protein molecules residing inside cells provide both shape and strength. These fibrous filaments, commonly called cytoskeleton, come in different sizes.

The microfilaments are 7 nm in diameter and are made of globular G-actin molecules (Figure 21G.2), which under physiological conditions form fibrous F-actin. In F-actin, the single globes are polymerized to form a double-helical structure, which provides the elasticity required when cells move, change their shape, or divide. Some of these actin filaments become anchored to cell membranes by other proteins. Through these proteins, they are connected to the collagen in the extracellular matrix. In this fashion, stresses and strains inside and outside the cells are coordinated. This same F-actin, in the specialized muscle cells, interacts with myosin (see Figure 26.11) to provide the contraction of muscles.

The cytoskeleton with the largest diameter is called a microtubule. These are 30-nm-wide cylindrical structures are formed by assembling two similar proteins, α- and β-tubulin. In the presence of Ca^{2+} ions, the dimer is quite stable. Some of these dimers form a helical structure in which 13 units form a turn. The microtubules are not static, however. At the expense of energy, they constantly shed dimeric units at the "minus end" and add new dimeric units at the "plus end" at about the same rate (Figure 21G.3). Microtubules provide the internal scaffolding of cells in much the same way that bones are the scaffolding in the human body.

(continued on next page)

■ **Figure 21G.1** Under the electron microscope, collagen fibers exhibit alternating light and dark bands. The dark bands correspond to the 40-nm gaps or "holes" between pairs of aligned collagen triple helices.

■ **Figure 21G.2** The three-dimensional structure of an actin monomer from skeletal muscle. This view shows the two domains (left and right) of actin.

CHEMICAL CONNECTIONS 21G

The Role of Quaternary Structures in Mechanical Stress and Strain (continued)

Microtubules also act as "railroads" along which molecules and organelles are transported from one part of the cell to another. For example, in nerve cells organelles and vesicles move at a fast rate, 2–5 μm/s along microtubules. This movement is unidirectional from the minus end to the plus end of microtubules, resulting in the transport from the cell body to the peripheral axons. A 360-kD protein, called kinesin, provides the walking leg of this motion. Kinesin molecules attach themselves with their tails to the organelle or vesicle to be transported. In this way, the heads of kinesin molecules "walk" along the microtubule in 8 nm small steps. At each step, the kinesin hydrolyzes an ATP molecule to supply the energy of the motion.

■ **Figure 21G.3**
A model treadmilling of microtubules.

---

non-amino acid portions are called **conjugated proteins.** The non-amino acid portion of a conjugated protein is called a **prosthetic group.** In hemoglobin, the globins are the amino acid portions and the heme units are the prosthetic groups.

Hemoglobin containing two alpha and two beta chains is not the only kind existing in the human body. In the early development stage of the fetus, the hemoglobin contains two alpha and two gamma chains. The fetal hemoglobin has greater affinity for oxygen than does the adult hemoglobin. In this way, the mother's red blood cells carrying oxygen can pass it to the fetus for its own use. Fetal hemoglobin also alleviates some of the symptoms of sickle cell anemia (see Chemical Connections 21D).

2. Collagen Another example of quaternary structure and higher organizations of subunits can be seen in collagen. The triple helix units, called *tropocollagen,* constitute the soluble form of collagen; they are stabilized by hydrogen bonding between the backbones of the three chains. Collagen is made of many tropocollagen units. Tropocollagen is found only in fetal or young connective tissues. With aging, the triple helixes (Figure 21.9) that organize themselves into fibrils cross-link and form insoluble collagen. In collagen, the **cross-linking** consists of covalent bonds that link together two lysine residues on adjacent chains of the helix. This cross-linking of collagen is an example of the tertiary structures that stabilize the three-dimensional conformations of protein molecules. The quarter-stagger alignment of tropocollagen (Figure 21G.1) forms collagen fibrils, and the twisting of these fibrils into a five- or six-strand helix builds the collagen fibers. These higher organizations provide the superior strength that these fibers exhibit in tissues (Chemical Connections 21G).

The oxygen- and carbon dioxide-carrying functions of hemoglobin are discussed in Sections 31.3 and 31.4.

Figure 21.15 Integral membrane protein of rhodopsin made of α-helices.

Figure 21.16 A β-barrel of integral membrane protein of the outer membrane of mitochondrion made of eight β-pleated sheets.

O-linked saccharide.

α-N-Acetylgalactosyl-serine

N-linked saccharide.

β-N-Acetylglucosyl-aspargine

3. Integral Membrane Proteins Integral membrane proteins traverse partly or completely a membrane bilayer. (See Figure 20.2.) It is estimated that about one-third of all proteins are integral membrane proteins. To keep the protein stable in the nonpolar environment of a lipid bilayer, it must form quaternary structures in which the outer surface is largely nonpolar and interacts with the lipid bilayer. Thus most of the polar groups of the protein must turn inward. Two such quaternary structures exist in integral membrane proteins: (1) often 6 to 10 α-helices crossing the membrane and (2) β-barrels made of 8, 12, 16, or 18 anti-parallel β-sheets (Figures 21.15 and 21.16).

21.10 Glycoproteins

Although many proteins, such as serum albumin, consist exclusively of amino acids, others also contain covalently linked carbohydrates and are therefore classified as **glycoproteins.** These include most of the plasma proteins (for example, fibrinogen), enzymes such as ribonuclease, hormones such as thyroglobulin, storage proteins such as casein and ovalbumins, and protective proteins such as immunoglobulins and interferon. The carbohydrate content of these proteins may vary from a few percent (immunoglobulins) up to 85% (blood group substances, see Chemical Connections 19D). Most of the proteins in membranes (lipid bilayers; see Section 20.5) are glycoproteins.

Glycoproteins have predominantly two kinds of linkages between the protein and the carbohydrate parts. The **O-linked saccharides** are bonded to an —OH group on the serine or threonine side chain of the protein. The linkage itself is a glycosidic bond. For example, mucins, which coat and protect the mucous membranes, are O-linked saccharides of glycoproteins. They line the respiratory and gastrointestinal tracts as well as the cervix. These highly viscous glycoproteins help to protect the underlying cells from harmful environmental agents.

The second kind of linkage between the carbohydrate and the protein is the N-glycosidic bond (—C—N—). This bond exists between the N of an aspargine residue of the protein chain and the C-1 (anomeric carbon) of N-acetylglucosamine (Section 19.2). This kind of glycoprotein is called an **N-linked saccharide.** N-Acetylglucosamine is the first unit in the forklike branched oligosaccharidic chain. Immunoglobulins are an example of N-linked saccharides. In many cases, the function involved in adding N-linked saccharides to the protein core is to make them more water-soluble and to facilitate their transport within the cell.

A special class of glycoproteins are the proteoglycans. These compounds have a protein core, and the side chains are long polysaccharide chains made of acidic polysaccharides (also called glycosaminoglycans), as discussed in Section 19.7. Proteoglycans are essential parts of the extracellular matrix that provide strength, flexibility, and elasticity in tissues such as skin, cartilage, and the cornea of the eye.

21.11 Denaturation

Protein conformations are stabilized in their native states by secondary and tertiary structures and through the aggregation of subunits in quaternary structure. Any physical or chemical agent that destroys these stabilizing structures changes the conformation of the protein (Table 21.2). We call this process **denaturation.**

Table 21.2	Modes of Protein Denaturation (Destruction of secondary and higher structures)
Denaturing Agent	**Affected Regions**
Heat	H bonds
6 M urea	H bonds
Detergents	Hydrophobic regions
Acids, bases	Salt bridges, H bonds
Salts	Salt bridges
Reducing agents	Disulfide bonds
Heavy metals	Disulfide bonds
Alcohol	Hydration layers

For example, heat cleaves hydrogen bonds, so boiling a protein solution destroys the α-helical and β-pleated sheet structure (compare Figures 21.6 and 21.7). In collagen, the triple helixes disappear upon boiling, and the molecules have a largely random-coil conformation in the denatured state, which is gelatin. In other proteins, especially globular proteins, heat causes the unfolding of the polypeptide chains; because of subsequent intermolecular protein–protein interactions, precipitation or coagulation then takes place. That is what happens when we boil an egg.

Similar conformational changes can be brought about by the addition of denaturing chemicals. Solutions such as 6 M aqueous urea, $H_2N-CO-NH_2$, break hydrogen bonds and cause the unfolding of globular proteins. Surface-active agents (detergents) change protein conformation by opening up the hydrophobic regions, whereas acids, bases, and salts affect both salt bridges and hydrogen bonds.

Reducing agents, such as 2-mercaptoethanol ($HOCH_2CH_2SH$), can break the —S—S— disulfide bonds, reducing them to —SH groups. The processes of permanent waving and straightening of curly hair are examples of the latter (Figure 21.17). The protein keratin, which makes up human hair, contains a high percentage of disulfide bonds. These bonds are primarily responsible for the shape of the hair, whether straight or curly. In either permanent waving or straightening, the hair is first treated with a reducing agent that cleaves some of the —S—S— bonds. This treatment allows the molecules to lose their rigid orientations and become more flexi-

Denaturation The loss of the secondary, tertiary, and quaternary structures of a protein by a chemical or physical agent that leaves the primary structure intact

■ **Reduction of disulfide bond.**

Figure 21.17 Permanent wave alters the shape of hair through reduction and oxidation of disulfides and thiols.

CHEMICAL CONNECTIONS 21H

Laser Surgery and Protein Denaturation

Proteins can be denatured by physical means, most notably by heat. For instance, bacteria are killed and surgical instruments are sterilized by heat. A special method of heat denaturation that is seeing increasing use in medicine relies on lasers. A laser beam (a highly coherent light beam of a single wavelength) is absorbed by tissues, and its energy is converted to heat energy. This process can be used to cauterize incisions so that a minimal amount of blood is lost during the operation.

Laser beams can be delivered by an instrument called a **fiberscope**. The laser beam is guided through tiny fibers, thousands of which are fitted into a tube only 1 mm in diameter. In this way, the laser delivers the energy for denaturation only where it is needed. It can, for example, seal wounds or join blood vessels without the necessity of cutting through healthy tissues. Fiberscopes have been used successfully to diagnose and treat many bleeding ulcers in the stomach, intestines, and colon.

A novel use of the laser fiberscope is in treating tumors that cannot be reached for surgical removal. A drug called Photofrin, which is activated by light, is given to patients intravenously. The drug in this form is inactive and harmless. The patient then waits 24 to 48 h, during which time the drug accumulates in the tumor but is removed and excreted from healthy tissues. A laser fiberscope with 630 nm of red light is then directed toward the tumor. An exposure between 10 and 30 min is applied. The energy of the laser beam activates the Photofrin, which destroys the tumor.

This technique does not offer a complete cure because the tumor may grow back, or it may have spread before the treatment. The treatment has only one side effect: The patient remains sensitive to exposure to strong light for approximately 30 days (so sunlight must be avoided). But this inconvenience is minor compared to the pain, nausea, hair loss, and other side effects that accompany radiation or chemotherapy of tumors.

In the United States, Photofrin is approved only to treat esophageal cancer. In Europe, Japan, and Canada, it is also

■ **Figure 21H** Argon krypton laser surgery.

used to treat lung, bladder, gastric, and cervical cancers. The light that activates Photofrin penetrates only a few millimeters, but the new drugs under development may use radiation in the near infrared spectrum that can penetrate tumors up to a few centimeters.

The most common use of laser technology in surgery is its application to correct near-sightedness and astigmatism. In a computer-assisted laser surgery process, the curvature of the cornea is changed. Using the energy of the laser beam, physicians remove part of the cornea. In the procedure called photorefractive keratectomy (PRK), the outer layers of the cornea are denatured—that is, burned off. In the LASIK (*laser in situ keratomileusis*) procedure, the surgeon creates a flap or a hinge of the outer layers of the cornea and then with the laser beam burns off a computer-programmed amount under the flap to change the shape of the cornea. After the 5- to 10-minute procedure is complete, the flap is put back, and it heals without stitches. In successful surgeries, the patient regains good vision one day after the surgery and no longer needs prescription lenses.

■ **Raw egg whites are an antidote to heavy metal poisoning.**

ble. The hair is then set into the desired shape, using curlers or rollers, and an oxidizing agent is applied. The oxidizing agent reverses the preceding reaction, forming new disulfide bonds, which now hold the molecules together in the desired positions.

Heavy metal ions (for example, Pb^{2+}, Hg^{2+}, and Cd^{2+}) also denature protein by attacking the —SH groups. They form salt bridges, as in —$S^-Hg^{2+-}S$—. This very feature is taken advantage of in the antidote for heavy metal poisoning: raw egg whites and milk. The egg and milk proteins are denatured by the metal ions, forming insoluble precipitates in the stomach. These must be pumped out or removed by inducing vomiting. In this way, the poisonous metal ions are removed from the body. If the antidote is not pumped out of the stomach, the digestive enzymes would degrade the proteins and release the poisonous heavy metal ions, which would then be absorbed in the bloodstream.

Other chemical agents such as alcohol also denature proteins, coagulating them. This process is used in sterilizing the skin before injections.

At a concentration of 70%, ethanol penetrates bacteria and kills them by coagulating their proteins, whereas 95% alcohol denatures only surface proteins.

Denaturation changes secondary, tertiary, and quaternary structures. It does not affect primary structures (that is, the sequence of amino acids that make up the chain). If these changes occur to a small extent, denaturation can be reversed. For example, when we remove a denatured protein from a urea solution and put it back into water, it often reassumes its secondary and tertiary structure. This is reversible denaturation. In living cells, some denaturation caused by heat can be reversed by **chaperones.** These proteins help a partially heat-denatured protein to regain its native secondary, tertiary, and quaternary structures. Some denaturation, however, is irreversible. We cannot unboil a hard-boiled egg, for example.

SUMMARY

Proteins are giant molecules made of amino acids linked together by **peptide bonds.** They have many functions (Section 21.1): structural (collagen), enzymatic (ribonuclease), carrier (hemoglobin), storage (casein), protective (immunoglobulin), and hormonal (insulin). **Amino acids** are organic compounds containing an amino (—NH$_2$) and a carboxylic (—COOH) acid group (Section 21.2). The 20 amino acids found in proteins are classified by their side chains: nonpolar, polar but neutral, acidic, and basic. All amino acids in human tissues are L amino acids. Amino acids in the solid state, as well as in water, carry both positive and negative charges; they are called **zwitterions** (Section 21.3). The pH at which the number of positive charges equals the number of negative charges is the **isoelectric point** of an amino acid or protein.

When the amino group of one amino acid condenses with the carboxyl group of another amino acid, an amide (peptide) linkage is formed, with the elimination of water. The two amino acids form a dipeptide (Section 21.5). Three amino acids form a tripeptide. Many amino acids form a **polypeptide chain.** Proteins are made of one or more polypeptide chains.

The linear sequence of amino acids is the **primary structure** of proteins (Section 21.7). The repeating short-range conformations (α-**helix,** β-**pleated sheet, extended helix of collagen,** and **random coil**) are the **secondary structures** (Section 21.8). The **tertiary structure** is the three dimensional conformation of the protein molecule (Section 21.9). Tertiary structures are maintained by covalent cross-links such as **disulfide bonds** and by **salt bridges, hydrogen bonds,** and **hydrophobic interactions** between the side chains. The precise fit of polypeptide subunits into an aggregated whole is called the **quaternary structure** (Section 21.9).

Many proteins are classified as **glycoproteins** because they contain carbohydrate units (Section 21.10).

Secondary and tertiary structures stabilize the native conformation of proteins; physical and chemical agents, such as heat or urea, destroy these structures and **denature** the protein (Section 21.11). Protein functions depend on native conformation; when a protein is denatured, it can no longer carry out its function. Some (but not all) denaturation is reversible; in some cases, **chaperone** molecules reverse denaturation.

PROBLEMS

Numbers that appear in color indicate difficult problems.
▶ designates problems requiring application of principles.

21.3 What are the functions of (a) ovalbumin and (b) myosin?

21.4 The members of which class of proteins are insoluble in water and can serve as structural materials?

Amino Acids

21.5 What is the difference in structure between tyrosine and phenylalanine?

21.6 Classify the following amino acids as nonpolar, polar but neutral, acidic, or basic.
(a) Arginine (b) Leucine
(c) Glutamic acid (d) Asparagine
(e) Tyrosine (f) Phenylalanine
(g) Glycine

21.7 ▶ Which amino acid has the highest percent nitrogen (g N/100 g amino acid)?

21.8 Why does glycine have no D or L form?

21.9 Draw the structure of proline. To which class of heterocyclic compounds does this molecule belong? (See Section 16.2.)

21.10 Which amino acid is also a thiol?

21.11 Why is it necessary to have proteins in our diets?

21.12 Which amino acids in Table 21.1 have more than one stereocenter?

21.13 What are the similarities and differences in the structures of alanine and phenylalanine?

21.14 Two of the 20 amino acids can be obtained by hydroxylation of other amino acids listed in Table 21.1. What are those two, and what are the precursor amino acids?

21.15 Draw the structures of L- and D-valine.

Zwitterions

21.16 Why are all amino acids solids at room temperature?

21.17 Show how alanine, in solution at its isoelectric point, acts as a buffer (write equations to show why the pH does not change much if we add an acid or a base).

21.18 Explain why an amino acid cannot exist in an un-ionized form [RCH(NH$_2$)COOH] at any pH.

21.19 Draw the structure of valine at pH 1 and at pH 12.

Peptides and Proteins

21.20 A tetrapeptide is abbreviated as DPKH. Which amino acid is at the N-terminal and which is at the C-terminal?

21.21 Draw the structure of a tripeptide made of threonine, arginine, and methionine.

21.22 (a) Use the three-letter abbreviations to write the representation of the following tripeptide.

$$H_3\overset{+}{N}-CH-\overset{O}{\overset{\|}{C}}-NH-CH-\overset{O}{\overset{\|}{C}}-NH-CH-\overset{O}{\overset{\|}{C}}-O^-$$

with side chains: CH$_2$–CH$_2$–S–CH$_3$; CH$_2$–CH–CH$_3$ with CH$_3$; CH$_2$–COOH

(b) Which amino acid is the C-terminal end, and which is the N-terminal end?

21.23 Draw the structure of the dipeptide leucylproline.

21.24 A polypeptide chain is made of alternating valine and phenylalanine. Which part of the polypeptide is polar (hydrophilic)?

21.25 Show by chemical equations how alanine and glutamine can be combined to give two different dipeptides.

Properties of Peptides and Proteins

21.26 (a) How many atoms of the peptide bond lie in the same plane?
(b) Which atoms are they?

21.27 (a) Draw the structural formula of the tripeptide Met—Ser—Cys.
(b) Draw the different ionic structures of this tripeptide at pH 2.0, 7.0, and 10.0.

21.28 How can a protein act as a buffer?

21.29 Proteins are least soluble at their isoelectric points. What would happen to a protein precipitated at its isoelectric point if a few drops of dilute HCl were added?

Primary Structure of Proteins

21.30 How many different tripeptides can be made (a) using one, two, or three residues each of leucine, threonine, and valine and (b) using all 20 amino acids?

21.31 How many different tetrapeptides can be made (a) if the peptides contain the residues of asparagine, proline, serine, and methionine and (b) if all 20 amino acids can be used?

21.32 How many amino acid residues in the A chain of insulin are the same in insulin from humans, cattle (bovine), hogs, and sheep?

21.33 Based on your knowledge of the chemical properties of amino acid side chains, suggest a substitution for leucine in the primary structure of a protein that would probably not change the character of the protein very much.

Secondary Structure of Proteins

21.34 Is a random coil a (a) primary, (b) secondary, (c) tertiary, or (d) quaternary structure? Explain.

21.35 Decide whether the following structures that exist in collagen are primary, secondary, tertiary, or quaternary.
(a) Tropocollagen
(b) Collagen fibril
(c) Collagen fiber
(d) The proline-hydroxyproline-glycine repeating sequence

21.36 Proline is often called an α-helix terminator; that is, it is usually in the random-coil secondary structure following an α-helix portion of a protein chain. Why does proline not fit easily into an α-helix structure?

Tertiary and Quaternary Structures of Proteins

21.37 Polyglutamic acid (a polypeptide chain made only of glutamic acid residues) has an α-helix conformation below pH 6.0 and a random-coil conformation above pH 6.0. What is the reason for this conformational change?

21.38 Distinguish between intermolecular and intramolecular hydrogen bonding between backbone groups. Where in protein structures do you find one and where do you find the other?

21.45 *(continued)*
two entities form? To what group of proteins does cytochrome c belong?

21.39 Identify the different primary, secondary, and tertiary structures in the numbered boxes:

21.40 If both cysteine residues on the B chain of insulin were changed to alanine residues, how would it affect the quaternary structure of insulin?

21.41 In a 6 M urea solution, a protein that contained mostly anti-parallel β-pleated sheet became a random coil. Which groups and bonds were affected by urea?

21.42 (a) What is the difference in the quaternary structure between fetal hemoglobin and adult hemoglobin?
(b) Which can carry more oxygen?

21.43 Where are the nonpolar side chains of proteins located in the β-barrel of an integral membrane protein?

21.44 What kind of changes are necessary to transform a protein having a predominantly α-helical structure into one having a β-pleated sheet structure?

21.45 The cytochrome c protein is important in producing energy from food. It contains a heme surrounded by a polypeptide chain. What kind of structure do these

Glycoproteins

21.46 In many glycoproteins, the protein carbohydrate linkage occurs on a serine side chain between the —OH group of the protein and the hemiacetal, of the monosaccharide—for example, D-glucose. Give the structure of such a protein–carbohydrate linkage in which a β-glycosidic bond has been established (see Section 19.5).

21.47 Draw the structure of an N-linked saccharide where the amide nitrogen group on the side chain of asparagine is linked to C-1 of N-acetylglucosamine by a β-N-glycosidic bond.

21.48 What is the difference between a glycoprotein and a proteoglycan?

21.49 Collagen is the most abundant protein in the human body. This fibrous protein contains short, mainly disaccharidic side chains, attached to hydroxylysyl residues. How would you classify collagen?

Protein Denaturation

21.50 Which amino acid side chain is most frequently involved in denaturation by reduction?

21.51 What does the reducing agent do in straightening curly hair?

21.52 Silver nitrate, $AgNO_3$, is sometimes put into the eyes of newborn infants as a disinfectant against gonorrhea. Silver is a heavy metal. Explain how this treatment may work against bacteria.

21.53 Why do nurses and physicians use 70% alcohol to wipe the skin before giving injections?

Chemical Connections

21.54 (Chemical Connections 21A) Write the structure of the oxidation product of glutathione.

21.55 (Chemical Connections 21B) AGE products become disturbing only in elderly people, even though they also form in younger people. Why don't they harm younger people?

21.56 (Chemical Connections 21C) Define *hypoglycemic awareness*.

21.57 (Chemical Connections 21D) How does hydroxyurea therapy alleviate the symptoms of sickle cell anemia?

21.58 (Chemical Connections 21E) What is the difference in the conformation between normal prion protein and the amyloid prion that causes mad cow disease?

21.59 (Chemical Connections 21F) What is the aim of proteomics?

21.60 (Chemical Connections 21G) What constituent of bone determines its hardness?

21.61 (Chemical Connections 21H) How does the fiberscope help to heal bleeding ulcers?

Additional Problems

21.62 Which diseases are associated with amyloid plaques?

21.63 How many different dipeptides can be made (a) using only alanine, tryptophan, glutamic acid, and arginine and (b) using all 20 amino acids?

21.64 Denaturation is usually associated with transitions from helical structures to random coils. If an imaginary process were to transform the keratin in your hair from an α-helix to a β-pleated sheet structure, would you call the process denaturation? Explain.

21.65 What are β-barrels of integral membrane proteins made of?

21.66 Draw the structure of lysine (a) above, (b) below, and (c) at its isoelectric point.

21.67 In collagen, some of the chains of the triple helices in tropocollagen are cross-linked by covalent bonds between two lysyl residues, $-(CH_2)_4-NH-(CH_2)_4-$. What kind of structure is formed by these cross-links? Explain.

21.68 Considering the vast number of animal and plant species on Earth (including those now extinct) and the large variety of protein molecules in each organism, have all possible protein molecules been used already by some species or other? Explain.

21.69 What kind of noncovalent interaction occurs between the following amino acids?
(a) Valine and isoleucine
(b) Glutamic acid and lysine
(c) Tyrosine and threonine
(d) Alanine and alanine

21.70 How many different decapeptides (peptides containing 10 amino acids each) can be made from the 20 amino acids?

21.71 Which amino acid does not rotate the plane of polarized light?

21.72 Write the expected products of the acid hydrolysis of the following tetrapeptide:

$$H_3\overset{+}{N}-CH-\underset{O}{\overset{\parallel}{C}}-NH-CH-\underset{O}{\overset{\parallel}{C}}-NH-CH-\underset{O}{\overset{\parallel}{C}}-NH-CH-\underset{O}{\overset{\parallel}{C}}-O^-$$

with side chains: $H_3C-CH-CH_3$; $CH_2-C_6H_5$; CH_2-indole ; $CH_2-C(=O)-NH_2$

21.73 What charges are on aspartic acid at pH 2.0?

21.74 How many ways can you link the two amino acids in the dipeptide Val—Lys? Which of these peptide bonds will you find in proteins?

21.75 Carbohydrates are integral parts of glycoproteins. Does the addition of carbohydrate to the protein core make the protein more or less soluble in water? Explain.

InfoTrac College Edition

For additional reading, go to InfoTrac College Edition, your online research library, at

http://infotrac.thomsonlearning.com

CHAPTER 22

22.1 Introduction

22.2 Naming and Classifying Enzymes

22.3 Common Terms in Enzyme Chemistry

22.4 Factors Affecting Enzyme Activity

22.5 Mechanism of Enzyme Action

22.6 Enzyme Regulation

22.7 Enzymes in Medical Diagnosis and Treatment

A hydrothermal vent on the ocean floor along with marine life, including tube worms.

Enzymes

22.1 Introduction

The cells in your body are chemical factories. Only a few of the thousands of compounds necessary for the operation of the human organism are obtained from the diet. Instead, most of them are synthesized within the cells, which means that hundreds of chemical reactions take place in your cells every second of your life.

Nearly all of these reactions are catalyzed by **enzymes,** which are large protein molecules that increase the rates of chemical reactions without themselves undergoing any change. Without enzymes, life as we know it would not be possible.

Proteins are not the only biological catalysts, however. **Ribozymes** are enzymes made of ribonucleic acids. They catalyze the self-cleavage of certain portions of their own molecules. Many biochemists believe that during evolution RNA catalysts emerged first, with protein enzymes arriving on the scene later. (We will learn more about RNA catalysts in Section 24.4.)

Like all catalysts, enzymes do not change the position of equilibrium. That is, enzymes cannot make a reaction take place that would not occur without them. What they do is increase the rate; they cause reactions to

For the discovery of ribozymes, Sidney Altman of Yale University and Thomas R. Cech of the University of Colorado were awarded the 1989 Nobel Prize in chemistry.

Enzyme specificity The limitation of an enzyme to catalyze one specific reaction with one specific substrate

take place faster by lowering the activation energy. As catalysts, enzymes are remarkable in two respects:

1. They are extremely effective, increasing reaction rates by anywhere from 10^9 to 10^{20} times.
2. Most of them are extremely **specific.**

As an example of their effectiveness, consider the oxidation of glucose. A lump of glucose or even a glucose solution exposed to oxygen under sterile conditions would show no appreciable change for months. In the human body, however, the same glucose is oxidized within seconds.

Every organism has many enzymes—many more than 3000 in a single cell. Most enzymes are very specific, each of them speeding up only one particular reaction or class of reactions. For example, the enzyme urease catalyzes only the hydrolysis of urea and not that of other amides, even closely related ones.

$$(NH_2)_2C=O + H_2O \xrightarrow{urease} 2\ NH_3 + CO_2$$
Urea

Another type of specificity can be seen with trypsin, an enzyme that cleaves the peptide bonds of protein molecules—but not every peptide bond, only those on the carboxyl side of lysine and arginine residues (Figure 22.1). The enzyme carboxypeptidase specifically catalyzes the hydrolysis on only the last amino acid on a protein chain—the one at the C-terminal end. Lipases are less specific: They cleave any triglyceride, but they still don't affect carbohydrates or proteins.

The specificity of enzymes also extends to stereospecificity. The enzyme arginase hydrolyzes the amino acid L-arginine (the naturally occurring form) to a compound called L-ornithine and urea (Section 27.8) but has no effect on its mirror image, D-arginine.

Enzymes are distributed according to the body's need to catalyze specific reactions. A large number of protein-splitting enzymes are in the blood, ready to promote clotting. Digestive enzymes, which also catalyze the hydrolysis of proteins, are located in the secretions of the stomach and pancreas. Even within the cells themselves, some enzymes are localized according to the need for specific reactions. The enzymes that catalyze the oxidation of compounds that are part of the citric acid cycle (Section 26.4) are located in the mitochondria, and special organelles such as lysosomes contain an enzyme (lysozyme) that catalyzes the dissolution of bacterial cell walls.

The enzyme glucose oxidase that oxidizes D-glucose does not work on L-glucose. It is specific for β-D-glucose.

Figure 22.1 A typical amino acid sequence. The enzyme trypsin catalyzes the hydrolysis of this chain only at the points marked with an arrow (the —COOH side of lysine and arginine).

Gly   Lys   Ala   Ser   Arg   Gly   Asp   Gly

CHEMICAL CONNECTIONS 22A

Muscle Relaxants and Enzyme Specificity

In the body, nerves transmit signals to the muscles. Acetylcholine is a neurotransmitter (Section 23.1) that operates between the nerve endings and muscles. It attaches itself to a specific receptor in the muscle end plate. This attachment transmits a signal to the muscle to contract; shortly thereafter, the muscles relax. A specific enzyme, acetylcholinesterase, then catalyzes the hydrolysis of the acetylcholine, removing it from the receptor site and thus preparing it for the next signal transmission—that is, the next contraction.

Succinylcholine is sufficiently similar to acetylcholine so that it, too, attaches itself to the receptor of the muscle end plate.

However, acetylcholinesterase can hydrolyze succinylcholine only very slowly. While it remains attached to the receptor, no new signal can reach the muscle to allow it to contract again. Thus the muscle stays relaxed for a long time.

This feature makes succinylcholine a good muscle relaxant during minor surgery, especially when a tube must be inserted into the bronchus (bronchoscopy). For example, after intravenous administration of 50 mg of succinylcholine, paralysis and respiratory arrest are observed within 30 s. While respiration is carried on artificially, the bronchoscopy can be performed within minutes.

$$CH_3-\overset{O}{\overset{\|}{C}}-O-CH_2-CH_2-\overset{CH_3}{\overset{|}{\underset{|}{\overset{+}{N}}}}-CH_3 + H_2O \xrightarrow{\text{acetylcholinesterase}} CH_3-\overset{O}{\overset{\|}{C}}-OH + HO-CH_2-CH_2-\overset{CH_3}{\overset{|}{\underset{|}{\overset{+}{N}}}}-CH_3$$
$$\qquad\qquad\qquad\text{Acetylcholine}\qquad\qquad\qquad\qquad\qquad\qquad\text{Acetic acid}\qquad\qquad\text{Choline}$$

$$CH_3-\overset{CH_3}{\overset{|}{\underset{|}{\overset{+}{N}}}}-CH_2-CH_2-O-\overset{O}{\overset{\|}{C}}-CH_2-CH_2-\overset{O}{\overset{\|}{C}}-O-CH_2-CH_2-\overset{CH_3}{\overset{|}{\underset{|}{\overset{+}{N}}}}-CH_3$$
$$\text{Succinylcholine}$$

22.2 Naming and Classifying Enzymes

Enzymes are commonly given names derived from the reaction that they catalyze and/or the compound or type of compound on which they act. For example, lactate dehydrogenase speeds up the removal of hydrogen from lactate (a redox reaction). Acid phosphatase catalyzes the hydrolysis of phosphate ester bonds under acidic conditions. As can be seen from these examples, the names of most enzymes end in "-ase." Some enzymes, however, have older names, which were assigned before their actions were clearly understood. Among these are pepsin, trypsin, and chymotrypsin—all enzymes of the digestive tract.

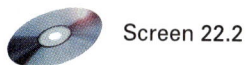

Screen 22.2

Enzymes can be classified into six major groups according to the type of reaction they catalyze (see also Table 22.1):

1. **Oxidoreductases** catalyze oxidations and reductions.
2. **Transferases** catalyze the transfer of a group of atoms, such as CH_3, CH_3CO, or NH_2, from one molecule to another.
3. **Hydrolases** catalyze hydrolysis reactions.
4. **Lyases** catalyze the addition of a group to a double bond or the removal of two groups from adjacent atoms to create a double bond.
5. **Isomerases** catalyze isomerization reactions.
6. **Ligases,** or synthetases, catalyze the joining of two molecules.

Table 22.1 Classification of Enzymes

Class	Typical Example	Reaction Catalyzed	Section Number in This Book
1. Oxidoreductases	Lactate dehydrogenase	$CH_3-\underset{\underset{O}{\|\|}}{C}-COO^- \longrightarrow CH_3-\underset{\underset{OH}{\|}}{CH}-COO^-$ Pyruvate → L-(+)-Lactate	27.2
2. Transferases	Aspartate amino transferase or Aspartate transaminase	Aspartate + α-Ketoglutarate → Oxaloacetate + Glutamate	27.8
3. Hydrolases	Acetylcholinesterase	$CH_3-\underset{\underset{O}{\|\|}}{C}-OCH_2CH_2\overset{+}{N}(CH_3)_3 + H_2O$ Acetylcholine $\longrightarrow CH_3COOH + HOCH_2CH_2\overset{+}{N}(CH_3)_3$ Acetic acid Choline	23.3
4. Lyases	Aconitase	cis-Aconitate + H_2O → Isocitrate	26.4
5. Isomerases	Phosphohexose isomerase	Glucose 6-phosphate → Fructose 6-phosphate	27.2
6. Ligases	Tyrosine-tRNA synthetase	ATP + L-tyrosine + tRNA $\longrightarrow$ L-tyrosyltRNA + AMP + PP$_i$	25.6

22.3 Common Terms in Enzyme Chemistry

Cofactor The nonprotein part of an enzyme necessary for its catalytic function

Some enzymes, such as pepsin and trypsin, consist of polypeptide chains only. Other enzymes contain nonprotein portions called **cofactors**. The protein (polypeptide) portion of the enzyme is called an **apoenzyme**.

The cofactors may be metallic ions, such as Zn^{2+} or Mg^{2+}, or organic compounds. Organic cofactors are called **coenzymes.** An important group of coenzymes are the B vitamins, which are essential to the activity of many enzymes (Section 26.3). Another important coenzyme is heme (Figure 21.14), which is part of several oxidoreductases as well as of hemoglobin. In any case, an apoenzyme cannot catalyze a reaction without its cofactor, nor can the cofactor function without the apoenzyme. When a metal ion is a cofactor, it can bind directly to the protein or to the coenzyme, if the enzyme contains one.

The compound on which the enzyme works, and whose reaction it speeds up, is called the **substrate.** The substrate usually binds to the enzyme surface while it undergoes the reaction. The substrate binds to a specific portion of the enzyme during the reaction, called the **active site.** If the enzyme has coenzymes, they are located at the active site. Therefore, the substrate is simultaneously surrounded by parts of the apoenzyme, coenzyme, and metal ion cofactor (if any), as shown in Figure 22.2.

Activation is any process that initiates or increases the action of an enzyme. It can be the simple addition of a cofactor to an apoenzyme or the cleavage of a polypeptide chain of a proenzyme (Section 22.6).

Inhibition is the opposite—any process that makes an active enzyme less active or inactive (Section 22.5). Inhibitors are compounds that accomplish this. **Competitive inhibitors** bind to the active site of the enzyme surface, thereby preventing the binding of substrate. **Noncompetitive inhibitors,** which bind to some other portion of the enzyme surface, may sufficiently alter the tertiary structure of the enzyme so that its catalytic effectiveness is slowed down. Both competitive and noncompetitive inhibition are *reversible,* but some compounds alter the structure of the enzyme *permanently* and thus make it *irreversibly* inactive.

> **Coenzyme** A nonprotein organic molecule, frequently a B vitamin, that acts as a cofactor

> Some enzymes have two kinds of cofactors: a coenzyme and a metallic ion.

> **Active site** A three-dimensional cavity of the enzyme with specific chemical properties that enable it to accommodate the substrate

> **Competitive inhibition** An enzyme regulation in which an inhibitor competes with the substrate for the active site

> **Noncompetitive inhibition** An enzyme regulation in which an inhibitor binds to the enzyme outside of the active site, thereby changing the shape of the active site and reducing its catalytic activity

22.4 Factors Affecting Enzyme Activity

Enzyme activity is a measure of how much reaction rates are increased. In this section, we examine the effects of concentration, temperature, and pH on enzyme activity.

A Enzyme and Substrate Concentration

If we keep the concentration of substrate constant and increase the concentration of enzyme, the rate increases linearly (Figure 22.3). That is, if the enzyme concentration doubles, the rate doubles as well; if the enzyme concentration triples, the rate also triples. This is the case in practically all enzyme reactions, because the molar concentration of enzyme is almost always much lower than that of substrate (that is, many more molecules of substrate are typically present than molecules of enzyme).

Conversely, if we keep the concentration of enzyme constant and increase the concentration of substrate, we get an entirely different type of curve, called a saturation curve (Figure 22.4). In this case, the rate does not increase continuously. Instead, a point is reached after which the rate stays the same even if we increase the substrate concentration further. This happens because, at the saturation point, substrate molecules are bound to all available active sites of the enzymes. Because the reactions take place at the active sites, once they are all occupied the reaction is proceeding at its maximum rate. Increasing the substrate concentration can no longer increase the rate because the excess substrate cannot find any active sites to which to attach.

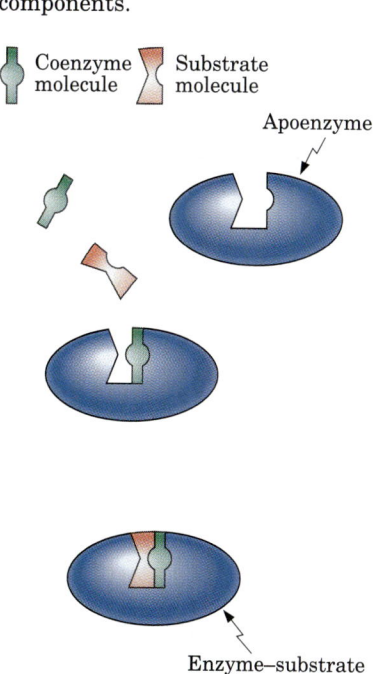

Figure 22.2 Schematic diagram of the active site of an enzyme and the participating components.

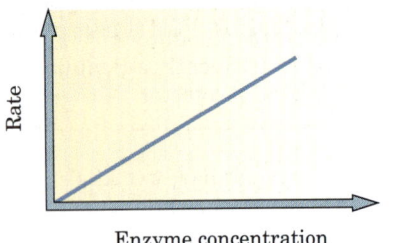

Figure 22.3 The effect of enzyme concentration on the rate of an enzyme-catalyzed reaction. Substrate concentration, temperature, and pH are constant.

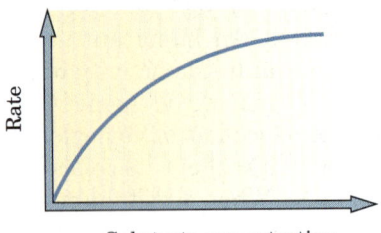

Figure 22.4 The effect of substrate concentration on the rate of an enzyme-catalyzed reaction. Enzyme concentration, temperature, and pH are constant.

Figure 22.5 The effect of temperature on the rate of an enzyme-catalyzed reaction. Substrate and enzyme concentrations and pH are constant.

B Temperature

Temperature affects enzyme activity because it changes the three-dimensional structure of the enzyme. In uncatalyzed reactions, the rate usually increases as the temperature increases (Section 7.4). The effect of temperature on enzyme-catalyzed reactions is different. When we start at a low temperature (Figure 22.5), an increase in temperature first causes an increase in rate. However, protein conformations are very sensitive to temperature changes. Once the optimal temperature is reached, any further increase in temperature causes changes in enzyme conformation. The substrate may then not fit properly onto the changed enzyme surface, so the rate of reaction actually *decreases*.

After a *small* temperature increase above the optimum, the decreased rate could still be increased again by lowering the temperature because, over a narrow temperature range, changes in conformation are reversible. However, at some higher temperature above the optimum, we reach a point where the protein denatures (Section 21.11); the conformation is then altered irreversibly, and the polypeptide chain cannot refold. At this point, the enzyme is completely inactivated. The inactivation of enzymes at low temperatures is used in the preservation of food by refrigeration.

Most enzymes from bacteria and higher organisms have an optimal temperature around 37°C. However, the enzymes of organisms that live at the ocean floor at 2°C have an optimal temperature in that range. Other organisms live in ocean vents under extreme conditions, and their enzymes have optimal conditions at ranges from 90 to 105°C. The enzymes of these hyperthermophile organisms also have other extreme requirements, such as pressures up to 100 atm, and some of them have an optimal pH in the range of 1 to 4. Enzymes from these hyperthermophiles, especially polymerases that catalyze the polymerization of DNA (Section 24.6), have gained commercial importance.

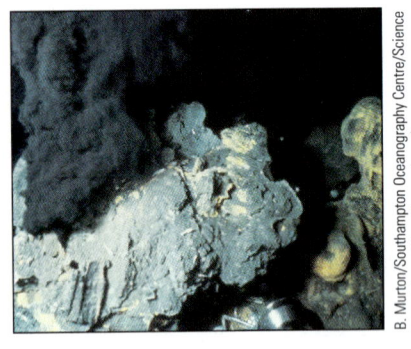

■ A "black smoker" vent pours out a sulfurous (as indicated by the yellow) mineral-rich fluid from its mound or chimney.

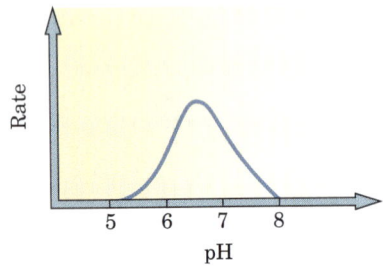

Figure 22.6 The effect of pH on the rate of an enzyme-catalyzed reaction. Substrate and enzyme concentrations and temperature are constant.

C pH

As the pH of its environment changes the conformation of a protein (Section 21.11), we would expect pH-related effects to resemble those observed when the temperature changes. Each enzyme operates best at a certain pH (Figure 22.6). Once again, within a narrow pH range, changes in enzyme activity are reversible. However, at extreme pH values (either acidic or basic), enzymes are denatured irreversibly, and enzyme activity cannot be restored by changing back to the optimal pH.

CHEMICAL CONNECTIONS 22B

Acidic Environment and *Helicobacter*

Certain organisms can live in an extreme pH environment. An example is the bacterium *Helicobacter pylori*, which survives happily in the acidic medium of the stomach wall. These bacteria live under the mucus coating of the stomach and release a toxin that causes ulcers. The role of *H. pylori* in ulceration was discovered only in the last decade, because people did not believe that enzymes of a bacterium could possibly function in such a harsh acidic environment with pH 1.4. In reality, the enzymes in the cytoplasm of *H. pylori* are not really exposed to low pH. The bacteria contain a special enzyme, **urease**, which converts urea to the basic ammonia; therefore, despite the acidic environment, *H. pylori* lives in a protective, neutralizing basic cloud. The urease itself, which is found on the surface of the bacteria, is protected against acid denaturation by its quaternary structure. The subunits of urease form a central cavity that contains the active sites. This cavity is filled with a buffer that protects the active sites before the actual urea hydrolysis starts.

After the destructive role played by *H. pylori* in ulceration was established, new treatments for gastric disorders were designed. Antibiotics such as amoxicillin or tetracyclin can eradicate the bacteria; the ulcers can then heal, and their recurrence is prevented.

22.5 Mechanism of Enzyme Action

We have seen that the action of enzymes is highly specific for a substrate. What kind of mechanism can account for such specificity? About 100 years ago, Arrhenius suggested that catalysts speed up reactions by combining with the substrate to form some kind of intermediate compound. In an enzyme-catalyzed reaction, the intermediate is called the **enzyme–substrate complex.**

> **Lock-and-key model** A model explaining the specificity of enzyme action by comparing the active site to a lock and the substrate to a key

A Lock-and-Key Model

To account for the high specificity of most enzyme-catalyzed reactions, a number of models have been proposed. The simplest and most frequently quoted is the **lock-and-key model** (Figure 22.7). This model assumes that the enzyme is a rigid, three-dimensional body. The surface that contains the active site has a restricted opening into which only one kind of substrate can fit, just as only the proper key can fit exactly into a lock and turn it open.

According to the lock-and-key mechanism, an enzyme molecule has its particular shape because that shape is necessary to maintain the active site in exactly the geometric alignment required for that particular reaction. An enzyme molecule is very large (typically consisting of 100 to 200 amino acid residues), but the active site is usually composed of only two or a few amino acid residues, which may well be located at different places in the chain. The other amino acids—those not part of the active site—are located in the sequence in which we find them because that sequence causes the whole molecule to fold up in exactly the required way. This arrangement emphasizes that the shape and the functional groups on the surface of the active site are of utmost importance in recognizing a substrate.

The lock-and-key model was the first to explain the action of enzymes. For most enzymes, there is evidence that this model is too restrictive. Enzyme molecules are in a dynamic state, not a static one. There are constant motions within them, so that the active site has some flexibility.

Figure 22.7 The lock-and-key model of the enzyme mechanism.

B Induced-Fit Model

From X-ray diffraction, we know that the size and shape of the active site cavity change when the substrate enters. To explain this phenomenon, the American biochemist Daniel Koshland introduced the **induced-fit model**

> **Induced-fit model** A model explaining the specificity of enzyme action by comparing the active site to a glove and the substrate to a hand

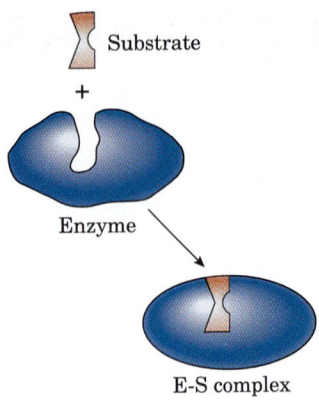

Figure 22.8 The induced-fit model of the enzyme mechanism.

Figure 22.9 The mechanism of competitive inhibition. When a competitive inhibitor enters the active site, the substrate cannot enter.

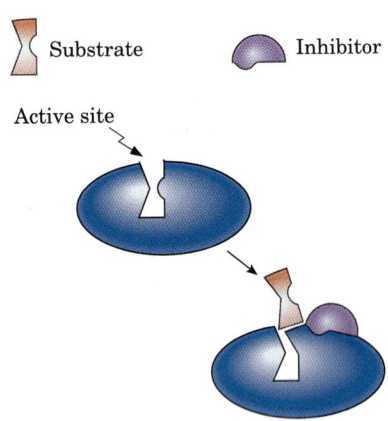

Figure 22.10 Mechanism of noncompetitive inhibition. The inhibitor attaches itself to a site other than the active site (allosterism), thereby changing the conformation of the active site.

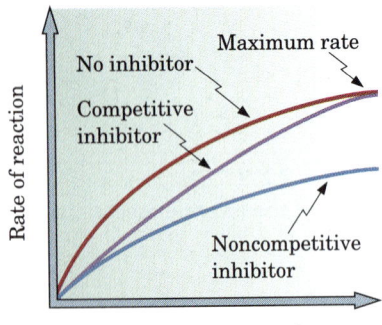

Figure 22.11 Enzyme kinetics in the presence and the absence of inhibitors.

(Figure 22.8), in which he compared the changes occurring in the shape of the cavity upon substrate binding to the changes in the shape of a glove when a hand is inserted. That is, the enzyme modifies the shape of the active site to accommodate the substrate. Recent experiments during actual catalysis have demonstrated that not only does the shape of the active site change with the binding of substrate, but even in the bound state both the backbone and the side chains of the enzyme are in constant motion. Furthermore, as a key residue—for example, the arginyl residue—pulsates, its motion correlates with the catalytic activity.

Both the lock-and-key and the induced-fit model explain the phenomenon of competitive inhibition (Section 22.3). The inhibitor molecule fits into the active site cavity in the same way the substrate does (Figure 22.9), thereby preventing the substrate from entering. The result is that whatever reaction is supposed to take place on the substrate does not occur.

Both the lock-and-key model and the induced-fit model emphasize the shape of the active site. However, the chemistry at the active site is actually the most important factor. A survey of known active sites of enzymes shows that five amino acids participate in the active sites in more than 65% of all cases. They are, in order of their dominance, His > Cys > Asp > Arg > Glu. A quick glance at Table 21.1 reveals that most of these amino acids have either acidic or basic side chains. Thus acid–base chemistry often underlies the mode of catalysis. The example given in Chemical Connections 22C confirms this relationship. Out of the eight amino acids in the catalytic site, two are Arg, one is Glu, one is Asp, and two are the related Asn.

Many cases of noncompetitive inhibition can also be explained by the induced-fit model. In this case, the inhibitor does not bind to the active site but rather to a different part of the enzyme. Nevertheless, the binding causes a change in the three-dimensional shape of the enzyme molecule, which so alters the shape of the active site that the substrate can no longer fit (Figure 22.10).

If we compare enzyme activity in the presence and the absence of an inhibitor, we can tell whether competitive or noncompetitive inhibition is taking place (Figure 22.11). The maximum reaction rate is the same without an inhibitor and in the presence of a competitive inhibitor. The only difference

CHEMICAL CONNECTIONS 22C

Active Sites

The perception of the active site as either a rigid cavity (lock-and-key model) or a partly flexible template (induced-fit model) is an oversimplification. Not only is the geometry of the active site important, but so are the specific interactions that take place between the enzyme surface and the substrate. To illustrate, we take a closer look at the active site of the enzyme pyruvate kinase. This enzyme catalyzes the transfer of the phosphate group from phosphoenol pyruvate (PEP) to ADP, an important step in glycolysis (Section 27.2).

$$CH_2=C(OPO_3^{2-})-COO^- + R-O-P(O)(O^-)-O-P(O)(O^-)-O^- \longrightarrow CH_3-C(O)-COO^- + R-O-P(O)(O^-)-O-P(O)(O^-)-O-P(O)(O^-)-O^-$$

Phosphoenol pyruvate + ADP → Pyruvate + ATP

The active site of the enzyme binds both substrates, PEP and ADP (Figure 22C.1). The rabbit muscle pyruvate kinase has two cofactors, K^+ and either Mn^{2+} or Mg^{2+}. The divalent cation is coordinated to the carbonyl and carboxylate oxygen of the pyruvate substrate and to a glutamyl 271 and aspartyl 295 residue of the enzyme. (The numbers indicate the position of the amino acid in the sequence.) The nonpolar $=CH_2$ group lies in a hydrophobic pocket formed by an alanyl 292, glycyl 294, and threonyl 327 residue. The K^+ on the other side of the active site is coordinated with the phosphate of the substrate and the seryl 76 and aspargyl 74 residue of the enzyme. Lysyl 269 and arginyl 72 are also part of the catalytic apparatus anchoring the ADP. This arrangement of the active site illustrates that specific folding into secondary and tertiary structures is required to bring important functional groups together. The residues of amino acids participating in the active sites are sometimes close in the sequence (aspargyl 74 and seryl 76), but mostly remain far apart (glutamyl 271 and aspartyl 295). Figure 22C.2 illustrates the secondary and tertiary structures providing such a stable active site.

Figure 22C.1 The active site and substrates of pyruvate kinase.

Figure 22C.2 Ribbon cartoon of pyruvate kinase. Pyruvate, Mg^{2+}, K^+ are depicted as space-filling models.

CHEMICAL CONNECTIONS 22D

Sulfa Drugs as Competitive Inhibitors

The vitamin folic acid (see Table 29.2), in its reduced form as tetrahydrofolate, acts as a coenzyme in a number of biosynthetic processes, including the synthesis of amino acids and of nucleotides. Humans obtain folic acid from the diet or from microorganisms in the intestinal tract. These microorganisms can synthesize folic acid if *para*-aminobenzoic acid is available to them.

In the 1930s, researchers discovered that the compound sulfanilamide could kill many types of harmful bacteria, thereby curing several diseases. Because sulfanilamide itself has an unacceptably high toxicity to humans, a number of derivatives of this compound, such as sulfapyridine and sulfathiazole, which act in a similar way, are used instead. These drugs work by "tricking" the bacteria, which normally use *para*-aminobenzoic acid as a raw material in the synthesis of folic acid. When the bacteria get a sulfa drug instead, they cannot tell the difference and use it to make a molecule that also has a folic acid type of structure but is not exactly the same. When they try to use this fake folic acid as a coenzyme, not only doesn't it work, but it is also now a competitive inhibitor of the enzyme's action. Consequently, many of the bacteria's amino acids and nucleotides cannot be synthesized, and the bacteria die.

is that this maximum rate is achieved at a low substrate concentration with no inhibitor but at a high substrate concentration when an inhibitor is present. This is the true sign of competitive inhibition because in this scenario the substrate and the inhibitor are competing for the same active site. If the substrate concentration is sufficiently increased, the inhibitor will be displaced from the active site by Le Chatelier's principle.

If, on the other hand, the inhibitor is noncompetitive, it cannot be displaced by addition of excess substrate because it is bound to a different site. In this case, the enzyme cannot be restored to its maximum activity, and the maximum rate of the reaction is lower than it would be in the absence of the inhibitor. An example of noncompetitive inhibition is heavy metal poisoning (Chemical Connections 22E).

Enzymes can also be inhibited irreversibly if a compound is bound covalently and permanently to or near the active site. Such inhibition occurs with penicillin, which inhibits the enzyme transpeptidase; this enzyme is necessary for cross-linking bacterial cell walls. Without cross-linking, the bacterial cytoplasm spills out, and the bacteria die (Chemical Connections 18B).

CHEMICAL CONNECTIONS 22E

Noncompetitive Inhibition and Heavy Metal Poisoning

Many enzymes contain a number of —SH groups (cysteine residues) that are easily cross-linked by heavy metal ions (Section 21.11):

$$\underset{\substack{|\\ \text{SH}\\ \\ \text{SH}\\ |}}{} + Pb^{2+} \longrightarrow \underset{\substack{|\\ S^-\\ \\ S^-\\ |}}{} Pb^{2+} + 2H^+$$

If these cross-linked cysteine residues are found at or near active sites, the cross-linking can noncompetitively inhibit enzyme activity. This inhibition is the basis of lead and mercury poisoning.

In the seventeenth century, felt hats were made of rabbit hair. The workers treated the hair with mercury salt to enhance the felting process. This step contaminated the air with mercury vapors, which the hat makers inhaled. As a consequence they developed twitches, a symptom of nervous disorders. The term "mad hatter" (as in *Alice in Wonderland*) originates from this industrial disease.

Mercury poisoning has become an acute problem in recent years because the mercury and mercury compounds that factories dumped into streams and lakes for decades have entered the food chain. Mercury is also released into the environment following its use in mining. Certain microorganisms convert mercury metal to organic mercury compounds, mainly dimethylmercury, $(CH_3)_2Hg$. This compound then enters the food chain when it is absorbed by algae. Later, it becomes concentrated in fish, and people who eat the contaminated fish can get mercury poisoning.

In Japan, where mercury was dumped in the Bay of Minamata in the 1950s, more than 100 people suffered mercury poisoning from a diet high in locally caught seafood, and 44 people died. Once such contaminated food was removed from the markets, the number of mercury-poisoning cases reported dropped drastically.

A current campaign focuses on another potential source of contamination: mercury-containing medical thermometers in the home. Advocates are now seeking to replace them with non-mercury thermometers.

22.6 Enzyme Regulation

A Feedback Control

Enzymes are often regulated by environmental conditions. **Feedback control** is an enzyme regulation process in which formation of a product inhibits an earlier reaction in the sequence. The reaction product of one enzyme may control the activity of another, especially in a complex system in which enzymes work cooperatively. For example, in such a system, each step is catalyzed by a different enzyme:

$$A \xrightarrow{E_1} B \xrightarrow{E_2} C \xrightarrow{E_3} D$$

Screen 22.5
Exercise Tutorial

The last product in the chain, D, may inhibit the activity of enzyme E_1 (by competitive or noncompetitive inhibition). When the concentration of D is low, all three reactions proceed rapidly. As the concentration of D increases, however, the action of E_1 becomes inhibited and eventually stops. In this manner, the accumulation of D serves as a message that tells enzyme E_1 to shut down because the cell has enough D for its present needs. Shutting down E_1 stops the entire process.

B Proenzymes

Some enzymes are manufactured by the body in an inactive form. To make them active, a small part of their polypeptide chain must be removed. These inactive forms of enzymes are called **proenzymes** or **zymogens.** After the excess polypeptide chain is removed, the enzyme becomes active. For example, trypsin is manufactured as the inactive molecule trypsinogen (a

Proenzyme (zymogen) A protein that becomes an active enzyme after undergoing a chemical change

CHEMICAL CONNECTIONS 22 F

A Zymogen and Hypertension

Many high blood pressure (hypertension) medications come under the heading of ACE inhibitors. ACE stands for *angiotensin-converting enzyme*, a key player in controlling blood pressure. It catalyzes the activation of angiotensin, a vasoconstrictive hormone, from its inactive zymogen, angiotensinogen.

$$\text{Angiotensinogen} \xrightarrow{\text{ACE}} \text{Angiotensin}$$

This process involves the cleaving of the last two amino acids from the carboxyl terminal of the decapeptide: Asp—Arg—Val—Tyr—Ile—His—Pro—Phe—His—Leu. The octapeptide, angiotensin, is a potent constrictor of the blood vessels and, therefore, increases blood pressure. When a competitive inhibitor occupies the active site of ACE, however, the hormone stays in its inactive zymogen state. As a consequence, blood pressure is lowered.

A large number of ACE inhibitors are currently on the market. Some are sulfur-containing compounds, such as Captopril; others are non–sulfur-containing compounds, such as Accupril and Altace.

The active site of the enzyme, ACE, contains several key chemical entities: arginyl, aspartyl residues, and a Zn^{2+} ion. Each of these interacts with a group on the inhibitor. The arginyl residue attracts the —COO^- group; the aspartyl amide attracts the —C=O group; and the Zn^{2+} forms a salt with the negatively charged sulfide ion or —COO^- group. The key element is the proper separation of these functional groups on the inhibitor molecule.

Figure 22F Structures of the active forms of four ACE inhibitors.

The removal of the six-amino-acid fragment is, of course, also catalyzed by an enzyme.

zymogen). When a fragment containing six-amino-acid residues is removed from the N-terminal end, the molecule becomes a fully active trypsin molecule. Removal of the fragment not only shortens the chain but also changes the three-dimensional structure (the tertiary structure), thereby allowing the molecule to achieve its active form.

Why does the body go to so much trouble? Why not just make the fully active trypsin to begin with? The reason is very simple. Trypsin is a protease—it catalyzes the hydrolysis of proteins (Figure 22.1)—and is, therefore, an important catalyst for the digestion of the proteins we eat. But it would not be good if it cleaved the proteins of which our own bodies are made! Therefore, the body makes trypsin in an inactive form; only after it enters the digestive tract is it allowed to become active.

C Allosterism

Sometimes regulation takes place by means of an event that occurs at a site other than the active site but that eventually affects the active site. This type of interaction is called **allosterism,** and any enzyme regulated by this mechanism is called an **allosteric enzyme.** If a substance binds noncovalently and reversibly to a site *other than the active site,* it may affect the enzyme in either of two ways: It may inhibit enzyme action (**negative modulation**) or it may stimulate enzyme action (**positive modulation**).

The substance that binds to the allosteric enzyme is called a **regulator,** and the site to which it attaches is called a **regulatory site.** In most cases, allosteric enzymes contain more than one polypeptide chain (subunits); the regulatory site is on one polypeptide chain and the active site on another.

Specific regulators can bind reversibly to the regulatory sites. For example, a protein kinase is an allosteric enzyme. In this case, the enzyme has only one polypeptide chain, so it carries both the active site and the regulatory site at different parts of this chain (Figure 22.12). The regulator—another protein molecule—binds reversibly to the regulatory site. As long as the regulator remains bound to the regulatory site, the total enzyme–regulator complex will be inactive. When the regulator is removed from the regulatory site, the protein kinase becomes active. In this way, the regulator controls the allosteric enzyme action.

Allosteric regulation may occur with proteins other than enzymes. Section 31.3 explains how modulators affect the oxygen-carrying ability of hemoglobin, an allosteric protein.

> **Allosterism** An enzyme regulation in which the binding of a regulator on one site on the enzyme modifies the enzyme's ability to bind the substrate in the active site

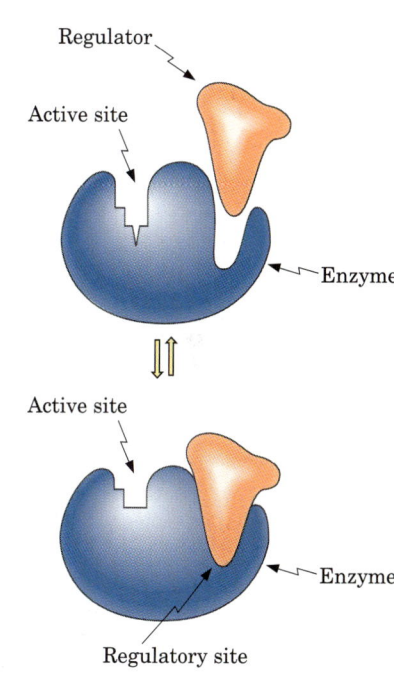

Figure 22.12 The allosteric effect. Binding of a regulator to a site other than the active site changes the shape of the active site.

D Protein Modification

The activity of an enzyme may also be controlled by **protein modification.** The modification is usually a change in the primary structure, typically by addition of a functional group covalently bound to the apoenzyme.

The best-known example of protein modification is the activation/inhibition of enzymes by phosphorylation. A phosphate group is often bonded to a serine or tyrosine residue. In some enzymes, such as glycogen phosphorylase (Section 28.1), the phosphorylated form is the active form of the enzyme. Without it, the enzyme would be less active or inactive.

The opposite example is the enzyme pyruvate kinase, PK (discussed in Chemical Connections 22C). Pyruvate kinase from the liver is inactive when it is phosphorylated. Enzymes that catalyze such phosphorylation go under the common name of *kinases.* When the activity of PK is not needed, it is phosphorylated (to PKP) by a protein kinase using ATP as a substrate as well as a source of energy (Section 26.3). When the system wants to turn on PK activity, the phosphate group, P_i, is removed by another enzyme, phosphatase, which renders PK active.

Phosphoserine

$$H_3N^+ - CH - COO^-$$
$$| $$
$$CH_2$$
$$|$$
$$O - PO_3^{2-}$$

Phosphoserine

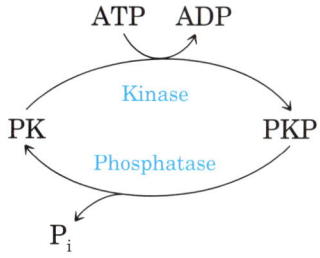

CHEMICAL CONNECTIONS 22G

One Enzyme, Two Functions

The enzyme called prostaglandin enderoperoxide synthase (PGHS) catalyzes the conversion of arachidonic acid to prostaglandin PGH_2 (Section 20.12) in two steps:

In the process, PGHS inserts two molecules of oxygen into arachidonic acid.

The enzyme itself is a single protein molecule. It is associated with the cell membrane and has a heme coenzyme. It has two distinct functions. The enzyme's first activity is to close a substituted cyclopentane ring; this is the *cyclooxygenase activity*. The *peroxidase activity* of the same enzyme yields the 15-hydroxy derivative prostaglandin, PGH_2.

Commercial painkillers act two ways inhibiting the formation of prostaglandin (PGH_2): (1) PGHS's cyclooxygenase activity is inhibited by aspirin and related NSAIDs (Chemical Connections 20I). Aspirin inhibits PGHS by acetylating serine in the active site; as a result, the active site is no longer able to accommodate arachidonic acid. Other NSAIDs also inhibit the cyclooxygenase activity by competitive inhibition, but they do not inhibit the peroxidase activity. (2) Another class of inhibitor has antioxidant activity. For example, acetaminophen (Tylenol) acts as a pain killer by inhibiting the peroxidase activity of PGHS.

E Isoenzymes

Another form of regulation of enzyme activity occurs when the same enzyme appears in different forms in different tissues. Lactate dehydrogenase catalyzes the oxidation of lactate to pyruvate, and vice versa (Figure 27.3, step 11). The enzyme has four subunits (tetramer). Two kinds of subunits, called H and M, exist. The enzyme that dominates in the heart is an H_4 enzyme, meaning that all four subunits are of the H type, although some M-type subunits are present as well. In the liver and skeletal muscles, the M type dominates. Other types of tetramer combinations exist in different tissues: H_3M, H_2M_2, and HM_3. These different forms of the same enzyme are called **isozymes** or **isoenzymes.**

The H_4 enzyme is allosterically inhibited by high levels of pyruvate, while the M_4 enzyme is not. In diagnosing the severity of heart attacks (Section 22.7), the release of H_4 isoenzyme is monitored in the serum.

Isozymes Enzymes that perform the same function but have different combinations of subunits and thus different quaternary structures

22.7 Enzymes in Medical Diagnosis and Treatment

Most enzymes are confined within the cells of the body. However, small amounts of them can also be found in body fluids such as blood, urine, and cerebrospinal fluid. The level of enzyme activity in these fluids can easily be monitored. This information can prove extremely useful: Abnormal activity (either high or low) of particular enzymes in various body fluids signals either the onset of certain diseases or their progression. Table 22.2 lists some enzymes used in medical diagnosis and their activities in normal body fluids.

Table 22.2 Enzyme Assays Useful in Medical Diagnosis

Enzyme	Normal Activity	Body Fluid	Disease Diagnosed
Alanine aminotransferase (ALT)	3–17 U/L*	Serum	Hepatitis
Acid phosphatase	2.5–12 U/L	Serum	Prostate cancer
Alkaline phosphatase (ALP)	13–38 U/L	Serum	Liver or bone disease
Amylase	19–80 U/L	Serum	Pancreatic disease or mumps
Aspartate aminotransferase (AST)	7–19 U/L	Serum	Heart attack or hepatitis
	7–49 U/L	Cerebrospinal fluid	
Lactate dehydrogenase (LDH)	100–350 WU/mL	Serum	Heart attack
Creatine phosphokinase (CPK)	7–60 U/L	Serum	
Phosphohexose isomerase (PHI)	15–75 U/L	Serum	

*U/L = International units per liter; WU/mL = Wrobleski units per milliliter.

For example, a number of enzymes are assayed (measured) during myocardial infarction to diagnose the severity of the heart attack. Dead heart muscle cells spill their enzyme contents into the serum. As a consequence, the level of creatine phosphokinase (CPK) in the serum rises rapidly, reaching a maximum within two days. This increase is followed by a rise in aspartate aminotransferase (AST; formerly called glutamate-oxaloacetate transaminase, or GOT). This second enzyme reaches a maximum two to three days after the heart attack. In addition to CPK and AST, lactate dehydrogenase (LDH) levels are monitored; they peak after five to six days. In infectious hepatitis, the alanine aminotransferase (ALT; formerly called glutamate-pyruvate transaminase, or GPT) level in the serum can rise to ten times normal. There is also a concurrent increase in AST activity in the serum.

In some cases, administration of an enzyme is part of therapy. After duodenal or stomach ulcer operations, for instance, patients are advised to take tablets containing digestive enzymes that are in short supply in the stomach after surgery. Such enzyme preparations contain lipases, either alone or combined with proteolytic enzymes. They are sold under such names as Pancreatin, Acro-lase, and Ku-zyme.

SUMMARY

Enzymes are proteins that catalyze chemical reactions in the body. Most enzymes are very specific—they catalyze only one particular reaction (Section 22.1). The compound whose reaction is catalyzed by an enzyme is called the **substrate**. Enzymes are classified into six major groups according to the type of reaction they catalyze (Section 22.2). They are typically named after the substrate and the type of reaction they catalyze, by adding the ending "-ase."

Some enzymes are made of polypeptide chains only. Others have, besides the polypeptide chain (the **apoenzyme**), nonprotein **cofactors**, which are either organic compounds (**coenzymes**) or inorganic ions (Section 22.3). Only a small part of the enzyme surface, called the **active site,** participates in the actual catalysis of chemical reactions. Cofactors, if any, are part of the active site.

Compounds that slow enzyme action are called **inhibitors.** A **competitive inhibitor** attaches itself to the active site; a **noncompetitive inhibitor** binds to other parts of the enzyme surface (Section 22.3). The higher the enzyme and substrate concentrations, the higher the enzyme activity, except that, at sufficiently high substrate concentrations, a saturation point is reached. After this point, increasing substrate concentration no longer increases the rate (Section 22.4). Each enzyme has an optimal temperature and pH at which it has its greatest activity (Section 22.4).

Two closely related mechanisms by which enzyme activity and specificity are explained are the **lock-and-key model** and the **induced-fit model** (Section 22.5).

Enzyme activity is regulated by five mechanisms. In **feedback control,** the concentration of products influences the rate of the reaction (Section 22.6A). In **allosterism,** an interaction takes place at a position other than the active site but affects the active site, either positively or negatively (Section 22.6C). Enzymes can be activated or inhibited by **protein modification** (Section 22.6D). Some enzymes, called **proenzymes** or **zymogens,** must be activated by removing a small portion of the polypeptide chain (Section 22.6B). Finally, enzyme activity is also regulated by **isozymes,** which are different forms of the same enzyme (Section 22.6E).

Abnormal enzyme activity can be used to diagnose certain diseases (Section 22.7).

PROBLEMS

Numbers that appear in color indicate difficult problems.
▶ designates problems requiring application of principles.

Enzymes

22.1 What is the difference between the terms *catalyst* and *enzyme*?

22.2 What are ribozymes made of?

22.3 Would a lipase hydrolyze two triglycerides, one containing only oleic acid and the other containing only palmitic acid, with equal ease?

22.4 Compare the activation energy in uncatalyzed reactions and in enzyme-catalyzed reactions.

22.5 Why does the body need so many different enzymes?

22.6 Trypsin cleaves polypeptide chains at the carboxyl side of a lysine or arginine residue (Figure 22.1). Chymotrypsin cleaves polypeptide chains on the carboxyl side of an aromatic amino acid residue or any other nonpolar, bulky side chain. Which enzyme is more specific? Explain.

Naming and Classifying Enzymes

22.7 Both lyases and hydrolases catalyze reactions involving water molecules. What is the difference in the types of reactions that these two enzymes catalyze?

22.8 Monoamine oxidases are important enzymes in brain chemistry. Judging from the name, which of the following would be a suitable substrate for this class of enzymes:

(a) HO—⟨⟩—CH(OH)—CH$_2$NH$_2$

(b) CH$_3$—C(=O)—N(CH$_3$)$_2$

(c) ⟨⟩—NO$_2$

22.9 On the basis of the classification given in Section 22.2, decide to which group each of the following enzymes belongs:

(a) Phosphoglyceromutase

$^-$OOC—CH(OH)—CH$_2$—OPO$_3^{2-}$
3-Phosphoglycerate

⇌ $^-$OOC—CH(OPO$_3^{2-}$)—CH$_2$—OH
2-Phosphoglycerate

(b) Urease

H$_2$N—C(=O)—NH$_2$ + H$_2$O ⇌ 2NH$_3$ + CO$_2$
Urea

(c) Succinate dehydrogenase

$^-$OOC—CH$_2$—CH$_2$—COO$^-$ + FAD
Succinate   Coenzyme (oxidized form)

⇌ $^-$OOC—CH=CH—COO$^-$ + FADH$_2$
Fumarate   Coenzyme (reduced form)

(d) Aspartase

$$\begin{array}{c}\text{H}\text{COO}^-\\ \diagdown\diagup\\ \text{C}{=}\text{C}+\text{NH}_4^+\\ \diagup\diagdown\\ ^-\text{OOC}\text{H}\end{array}$$

Fumarate

$$\rightleftharpoons \text{ }^-\text{OOC}-\text{CH}_2-\text{CH}-\text{COO}^-$$
$$\mid$$
$$\text{NH}_3^+$$

L-Asparate

22.10 What kind of reaction does each of the following enzymes catalyze?
(a) Deaminases (b) Hydrolases
(c) Dehydrogenases (d) Isomerases

Common Terms in Enzyme Chemistry

22.11 What is the difference between a *coenzyme* and a *cofactor*?

22.12 In the citric acid cycle, an enzyme converts succinate to fumarate (see the reaction in Problem 22.9c). The enzyme consists of a protein portion and an organic molecule portion called FAD. What term do we use to refer to (a) the protein portion and (b) the organic molecule portion?

22.13 What is the difference between reversible and irreversible noncompetitive inhibition?

Factors Affecting Enzyme Activity

22.14 In most enzyme-catalyzed reactions, the rate of reaction reaches a constant value with increasing substrate concentration. This relationship is described in a saturation curve diagram (Figure 22.4). If the enzyme concentration, on a molar basis, is twice the maximum substrate concentration, would you obtain a saturation curve?

22.15 At a very low concentration of a certain substrate, we find that, when the substrate concentration doubles, the rate of the enzyme-catalyzed reaction also doubles. Would you expect the same finding at a very high substrate concentration? Explain.

22.16 If we wish to double the rate of an enzyme-catalyzed reaction, can we do so by increasing the temperature 10°C? Explain.

22.17 A bacterial enzyme has the following temperature-dependent activity.

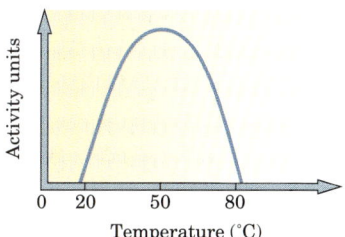

(a) Is this enzyme more or less active at normal body temperature than when a person has a fever?
(b) What happens to the enzyme activity if the patient's temperature is lowered to 35°C?

22.18 The optimal temperature for the action of lactate dehydrogenase is 36°C. It is irreversibly inactivated at 85°C, but a yeast containing this enzyme can survive for months at −10°C. Explain how this can happen.

22.19 The activity of pepsin was measured at various pH values. When the temperature and the concentrations of pepsin and substrate were held constant, the following activities were obtained:

pH	Activity
1.0	0.5
1.5	2.6
2.0	4.8
3.0	2.0
4.0	0.4
5.0	0.0

(a) Plot the pH dependence of pepsin activity.
(b) What is the optimal pH?
(c) Predict the activity of pepsin in the blood at pH 7.4.

Mechanism of Enzyme Action

22.20 Urease can catalyze the hydrolysis of urea but not the hydrolysis of diethylurea. Explain why diethylurea is not hydrolyzed.

Urea   Diethylurea

22.21 The following reaction may be represented by the cartoon figures:

Glucose + ATP $\rightleftharpoons$ glucose 6-phosphate + ADP

In this enzyme-catalyzed reaction, Mg^{2+} is a cofactor, fluoroglucose is a competitive inhibitor, and Cd^{2+} is a noncompetitive inhibitor. Identify each component of the reaction by a cartoon figure and assemble them to show (a) the normal enzyme reaction, (b) a competitive inhibition, and (c) a noncompetitive inhibition.

22.22 Which amino acids appear most frequently in the active sites of enzymes?

22.23 What kind of chemical reaction occurs most frequently at the active site?

22.24 Which of the following is a correct statement describing the induced-fit model of enzyme action? Substrates fit into the active site
  (a) Because both are exactly the same size and shape.
  (b) By changing their size and shape to match those of the active site.
  (c) By changing the size and shape of the active site upon binding.

22.25 What is the maximum rate that can be achieved in competitive inhibition compared with noncompetitive inhibition?

22.26 Enzymes are long protein chains, usually containing more than 100 amino acid residues. Yet the active site contains only a few amino acids. Explain why the other amino acids of the chain are present and what would happen to the enzyme activity if significant changes were made in the enzyme's structure.

Enzyme Regulation

22.27 The hydrolysis of glycogen to yield glucose is catalyzed by the enzyme phosphorylase. Caffeine, which is not a carbohydrate and not a substrate for the enzyme, inhibits phosphorylase. What kind of regulatory mechanism is at work?

22.28 Can the product of a reaction that is part of a sequence act as an inhibitor for another reaction in the sequence? Explain.

22.29 What is the difference between a zymogen and a proenzyme?

22.30 The enzyme trypsin is synthesized by the body in the form of a long polypeptide chain containing 235 amino acids (trypsinogen) from which a piece must be cut before the trypsin can be active. Why does the body not synthesize trypsin directly?

22.31 Give the structure of a tyrosyl residue of an enzyme modified by a protein kinase.

22.32 What is an isozyme?

22.33 The enzyme glycogen phosphorylase initiates the hydrolysis of glycogen to glucose 1-phosphate. It comes in two forms: Phosphorylase b is less active, and phosphorylase a is more active. The difference between b and a is the modification of the apoenzyme. Phosphorylase a has two phosphate groups added to the polypeptide chain. In analogy with the pyruvate kinase discussed in the text, give a scheme indicating the transition between b and a. What enzymes and what cofactors control this reaction?

Enzymes in Medical Diagnosis and Treatment

22.34 After a heart attack, the levels of certain enzymes rise in the serum. Which enzyme would you monitor within 24 hours following a suspected heart attack?

22.35 The enzyme formerly known as GPT (glutamate-pyruvate transaminase) has a new name: ALT (alanine aminotransferase). Looking at the equation in Section 27.9, which is catalyzed by this enzyme, what prompted this change of name?

22.36 If an examination of a patient indicated elevated levels of AST but normal levels of ALT, what would be your tentative diagnosis?

22.37 Which LDH isozyme is monitored in the case of a heart attack?

22.38 Chemists who have been exposed for years to organic vapors usually show higher-than-normal activity when given the alkaline phosphatase test. What organ in the body do organic vapors affect?

22.39 What enzyme preparation is given to patients after duodenal ulcer surgery?

22.40 Chymotrypsin is secreted by the pancreas and passed into the intestine. The optimal pH for this enzyme is 7.8. If a patient's pancreas cannot manufacture chymotrypsin, would it be possible to supply it orally? What happens to chymotrypsin's activity during its passage through the gastrointestinal tract?

Chemical Connections

22.41 (Chemical Connections 22A) Acetylcholine causes muscles to contract. Succinylcholine, a close relative, is a muscle relaxant. Explain the different effects of these related compounds.

22.42 (Chemical Connections 22A) An operating team usually administers succinylcholine before bronchoscopy. What is achieved by this procedure?

22.43 (Chemical Connections 22B) How does *Helicobacter pylori* protect its enzymes against stomach acid?

22.44 (Chemical Connections 22B) How does the quaternary structure of urease of *H. pylori* protect this enzyme against acid denaturation?

22.45 (Chemical Connections 22B) How do we treat ulcers caused by infection with *H. pylori*?

22.46 (Chemical Connections 22C) What role does Mn^{2+} play in anchoring the substrate in the active site of protein kinase?

22.47 (Chemical Connections 22C) Which amino acids of the active site interact with the $=CH_2$ group of the phosphoenol pyruvate? Do both provide the same surface environment? What is the nature of the interaction?

22.48 (Chemical Connections 22D) What part of folic acid is the key structure mimicked by sulfa drugs?

22.49 (Chemical Connections 22D) Sulfa drugs kill certain bacteria by preventing the synthesis of folic acid, a vitamin. Why don't these drugs kill people?

22.50 (Chemical Connections 22E) How does a heavy metal ion such as Hg^{2+} inactivate an enzyme?

22.51 (Chemical Connections 22F) Identify the functional groups that appear in all ACE inhibitors presented in Chemical Connections 22F.

22.52 (Chemical Connections 22G) Which activity of prostaglandin enderoperoxide synthase (PGHS) is inhibited by Tylenol, and which is inhibited by aspirin?

Additional Problems

22.53 Where can one find enzymes that are both stable and active at 90°C?

22.54 What kind of bonds does ACE (angiotensin-converting enzyme) hydrolyze?

22.55 Food can be preserved by inactivation of enzymes that would cause spoilage—for example, by refrigeration. Give an example of food preservation in which the enzymes are inactivated (a) by heat and (b) by lowering the pH.

22.56 Why is enzyme activity during myocardial infarction measured in the patients' serum rather than in their urine?

22.57 In activating angiotensinogen, which two amino acids are cleaved from the decapeptide?

22.58 What is the common characteristic of the amino acids of which the carboxyl groups of the peptide bonds can be hydrolyzed by trypsin?

22.59 Many enzymes are active only in the presence of Zn^{2+}. What common term is used for ions such as Zn^{2+} when discussing enzyme activity?

22.60 An enzyme has the following pH dependence:

At what pH do you think this enzyme works best?

22.61 What enzyme is monitored in the diagnosis of infectious hepatitis?

22.62 The enzyme chymotrypsin catalyzes the following type of reaction:

$$R-\underset{\underset{\text{\Large{C_6H_5}}}{CH_2}}{CH}-\overset{O}{\overset{\|}{C}}-NH-\underset{CH_3}{CH}-R + H_2O$$

$$\longrightarrow R-\underset{\underset{\text{\Large{C_6H_5}}}{CH_2}}{CH}-\overset{O}{\overset{\|}{C}}-O^- + H_3\overset{+}{N}-\underset{CH_3}{CH}-R$$

On the basis of the classification given in Section 22.2, to which group of enzymes does chymotrypsin belong?

22.63 Nerve gases operate by forming covalent bonds at the active site of cholinesterase. Is this an example of competitive inhibition? Can the nerve gas molecules be removed by simply adding more substrate (acetylcholine) to the enzyme?

22.64 What would be the appropriate name for an enzyme that catalyzes each of the following reactions:

(a) $CH_3CH_2OH \longrightarrow CH_3\overset{O}{\overset{\|}{C}}-H$

(b) $CH_3\overset{O}{\overset{\|}{C}}-O-CH_2CH_3 + H_2O$

$\longrightarrow CH_3\overset{O}{\overset{\|}{C}}-OH + CH_3CH_2OH$

22.65 In Section 28.5, a reaction between pyruvate and glutamate, to form alanine and α-ketoglutarate, is given. How would you classify the enzyme that catalyzes this reaction?

22.66 A liver enzyme is made of four subunits: 2A and 2B. The same enzyme, when isolated from the brain, has the following subunits: 3A and 1B. What would you call these two enzymes?

22.67 What is the function of a ribozyme?

InfoTrac College Edition

For additional reading, go to InfoTrac College Edition, your online research library, at

http://infotrac.thomsonlearning.com

CHAPTER 23

23.1 Introduction

23.2 Chemical Messengers: Neurotransmitters and Hormones

23.3 Cholinergic Messenger: Acetylcholine

23.4 Amino Acids as Messengers: Neurotransmitters

23.5 Adrenergic Messengers: Neurotransmitters/Hormones

23.6 Peptidergic Messengers: Neurotransmitters/Hormones

23.7 Steroid Messengers: Hormones

Scanning electron micrograph of neuron touching a muscle cell.

Chemical Communications: Neurotransmitters and Hormones

23.1 Introduction

The importance of chemical communications in health can be ascertained by a quick look at Table 23.1. The list contains only a small sample of drugs pertaining to this chapter and it is of utmost interest to health care providers. In fact, a vast number of drugs in the pharmacopoeia act, in one way or another, to influence chemical communications.

Each cell in the body is an isolated entity enclosed in its own membrane. Furthermore, within each cell of higher organisms, organelles, such as the nucleus or the mitochondrion, are also enclosed by membranes separating them from the rest of the cell. If cells could not communicate with one another, the thousands of reactions in each cell would be uncoordinated. The same is true for organelles within a cell. Such communication allows the activity of a cell in one part of the body to be coordinated with the activity of cells in a different part of the body. There are three principal types of molecules for communications:

- **Receptors** are protein molecules on the surface of cells embedded in the membrane.
- **Chemical messengers** also called ligands, interact with the receptors.
- **Secondary messengers** in many cases carry the message from the receptor to the inside of the cell and amplify the message.

If your house is on fire and the fire threatens your life, external signals—light, smoke, and heat—register alarm at specific receptors in your eyes, nose, and skin. From there the signals are transmitted by specific compounds to nerve cells, or **neurons.** In the neurons, the signals travel as electric impulses along the axons (Figure 23.1). When they reach the end of the neuron, the signals are transmitted to adjacent neurons by specific compounds called **neurotransmitters.** Communication between the eyes and the brain, for example, is by neural transmission.

As soon as the danger signals are processed in the brain, other neurons carry messages to the muscles and to the endocrine glands. The message to the muscles is to run away or to take some other action in response to the fire (save the baby or run to the fire extinguisher, for example). To do so, the muscles must be activated. Again, neurotransmitters carry the necessary messages from the neurons to the muscle cells and the endocrine glands. The endocrine glands are stimulated, and a different chemical signal, called a **hormone,** is secreted into your bloodstream. "The adrenaline begins to flow." The danger signal carried by adrenaline makes quick energy available so that the muscles can contract and relax rapidly, allowing your body to take rapid action to avoid the danger.

Without these chemical communicators, the whole organism—you—would not survive because there is a constant need for coordinated efforts to face a complex outside world. The chemical communications between different cells and different organs play a role in the proper functioning of our bodies. Its significance is illustrated by the fact that *a large percentage of the drugs we encounter in medical practice try to influence this communication.* The scope of these drugs covers all fields—from prescriptions against hypertension, heart disease, through antidepressants, to pain killers, just to mention a few. Principally, there are four ways these drugs act in the body. A drug may affect either the **receptor** or the **messenger.**

1. An **antagonist** drug blocks the receptor and prevents its stimulation.

2. An **agonist** drug competes with the natural messenger for the receptor site. Once there, it stimulates the receptor.

3. Other drugs may decrease the concentration of the messenger by controlling the release of messengers from their storage.

4. Still other drugs may increase the concentration of the messenger by inhibiting its removal from the receptors.

Table 23.1 presents selected drugs and their modes of action that affect neurotransmission.

Chemical messengers fit into the receptor sites in a manner reminiscent of the lock-and-key model mentioned in Section 22.5.

Nerve cells are present throughout the body and, together with the brain, constitute the nervous system.

Neurotransmitter A chemical messenger between a neuron and another target cell: neuron, muscle cell, or cell of a gland

Hormone A chemical messenger released by an endocrine gland into the bloodstream and transported there to reach its target cell

Figure 23.1 Neuron and synapse.

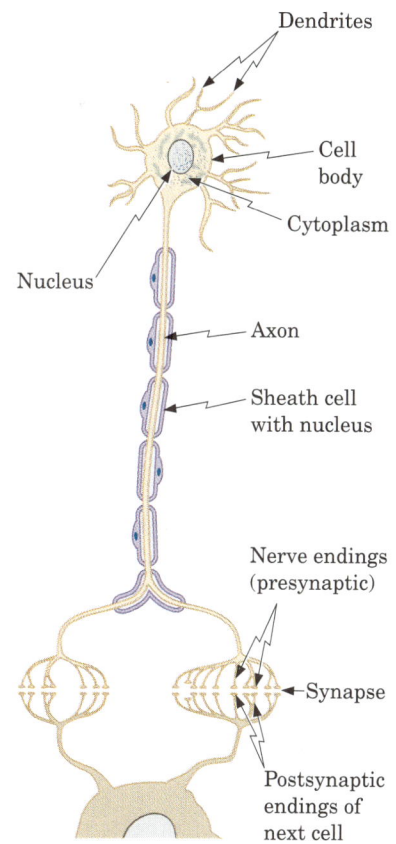

Table 23.1 Drugs That Affect Nerve Transmission

Messenger	Drugs That Affect Receptor Sites		Drugs That Affect Available Concentration of Neurotransmitter or Removal from Receptor Sites	
	Agonists (Activate Receptor Sites)	**Antagonists (Block Receptor Sites)**	**Increase Concentration**	**Decrease Concentration**
Acetylcholine (cholinergic)	Nicotine Pilocarpine Carbachol Succinylcholine Tacrine (Cognex)	Curare Atropine Propantheline (Pro-Banthine)	Malathion Nerve gases Succinylcholine Donepezil (Aricept)	*Clostridium botulinum* toxin
Calcium ion	Nifedipine (Adalat) Diltiazem (Cardizem)		Digitoxin (Lanoxicaps)	
Epinephrine (α-adrenergic)	Terazosin (Hytrin) Prazosin (Minipress)	Clonidine (Catapres)		
Norepinephrine (β-adrenergic)	Phenylephrine (Neo-Synephrine) Epinephrine (Adrenalin) Sotalol (Betaplace)	Propranolol (Inderal) Metoprolol (Lopressor) Selegiline (Deprenyl)	Amphetamines Iproniazide Imirapin (Tofranil) Amitriptyline (Elavil)	Reserpine Methyldopa (Aldomet) Metyrosine (Demser)
Dopamine (adrenergic)	Methylphenidate (Ritalin)	Clozapine (Clorazil) Metoclopramide (Reglan) Promethazine (Phenergan)	(Deprenyl) Entacapon (Comtan)	
Serotonin (adrenergic)		Ondansetron (Zofran)	Antidepressant Fluoxetine (Prozac)	
Histamine (adrenergic)	2-Methylhistamine Betazole Pentagastin	Fexofenadine (Allegra) Promethazine (Phenergan) Diphenhydramine (Benadryl) Ranitidine (Zantac) Cimetidine (Tagamet)	Histidine	Hydrazino-histidine
Glutamic acid (amino acid)	N-Methyl-D-Aspartate	Phencyclidine		
Enkephalin (peptidergic)	Opiate	Morphine Heroin Meperidine (Demerol)	Naloxone (Narcan)	

23.2 Chemical Messengers: Neurotransmitters and Hormones

As mentioned earlier, **neurotransmitters** are compounds that communicate between two nerve cells or between a nerve cell and another cell (such as a muscle cell). A nerve cell (Figure 23.1) consists of a main cell body from which projects a long, fiber-like part called an **axon**. Coming off the other side of the main body are hair-like structures called **dendrites**.

Table 23.2 The Principal Hormones and Their Actions

Gland	Hormone	Action	Structures Shown in
Parathyroid	Parathyroid hormone	Increases blood calcium Excretion of phosphate by kidney	
Thyroid	Thyroxine (T_4) Triiodothyronine (T_3)	Growth, maturation, and metabolic rate Metamorphosis	Chemical Connections 13C
Pancreatic islets			
Beta cells	Insulin	Hypoglycemic factor Regulation of carbohydrates, fats, and proteins	Section 21.7 Chemical Connections 23G
Alpha cells	Glucagon	Liver glycogenolysis	
Adrenal medulla	Epinephrine Norepinephrine	Liver and muscle glycogenolysis	Section 23.5
Adrenal cortex	Cortisol Aldosterone Adrenal androgens	Carbohydrate metabolism Mineral metabolism Androgenic activity (especially females)	Section 20.10 Section 20.10
Kidney	Renin	Hydrolysis of blood precursor protein to yield angiotensin	
Anterior pituitary	Luteinizing hormone Interstitial cell-stimulating hormone Prolactin Mammotropin	Causes ovulation Formation of testosterone and progesterone in interstitial cells Growth of mammary gland Lactation Corpus luteum function	
Posterior pituitary	Vasopressin Oxytocin	Contraction of blood vessels Kidney reabsorption of water Stimulates uterine contraction and milk ejection	Section 21.7 Section 21.7
Ovaries	Estradiol Progesterone	Estrous cycle Female sex characteristics	Section 20.10 Section 20.10
Testes	Testosterone Androgens	Male sex characteristics Spermatogenesis	Section 20.10

Typically, neurons do not touch each other. Between the axon end of one neuron and the cell body or dendrite end of the next neuron is a space filled with an aqueous fluid, called a **synapse.** If the chemical signal travels, say, from axon to dendrite, we call the nerve ends on the axon the **presynaptic** site. The neurotransmitters are stored at the presynaptic site in **vesicles,** which are small, membrane-enclosed packages. Receptors are located on the **postsynaptic** site of the cell body or the dendrite.

Hormones are diverse compounds secreted by specific tissues (the endocrine glands), released into the bloodstream, and then adsorbed onto specific receptor sites, usually relatively far from their source. (This is the physiological definition of a hormone.) Table 23.2 lists some of the principal hormones. Figure 23.2 shows the target organs of hormones secreted by the pituitary gland.

The distinction between hormones and neurotransmitters is physiological, not chemical. Whether a certain compound is considered to be a neurotransmitter or a hormone depends on whether it acts over a short distance across a synapse (2×10^{-6} cm), in which case it is a neurotransmitter, or over a long distance (20 cm) from the secretory gland through the blood-

> **Synapse** An aqueous small space between the tip of a neuron and its target cell

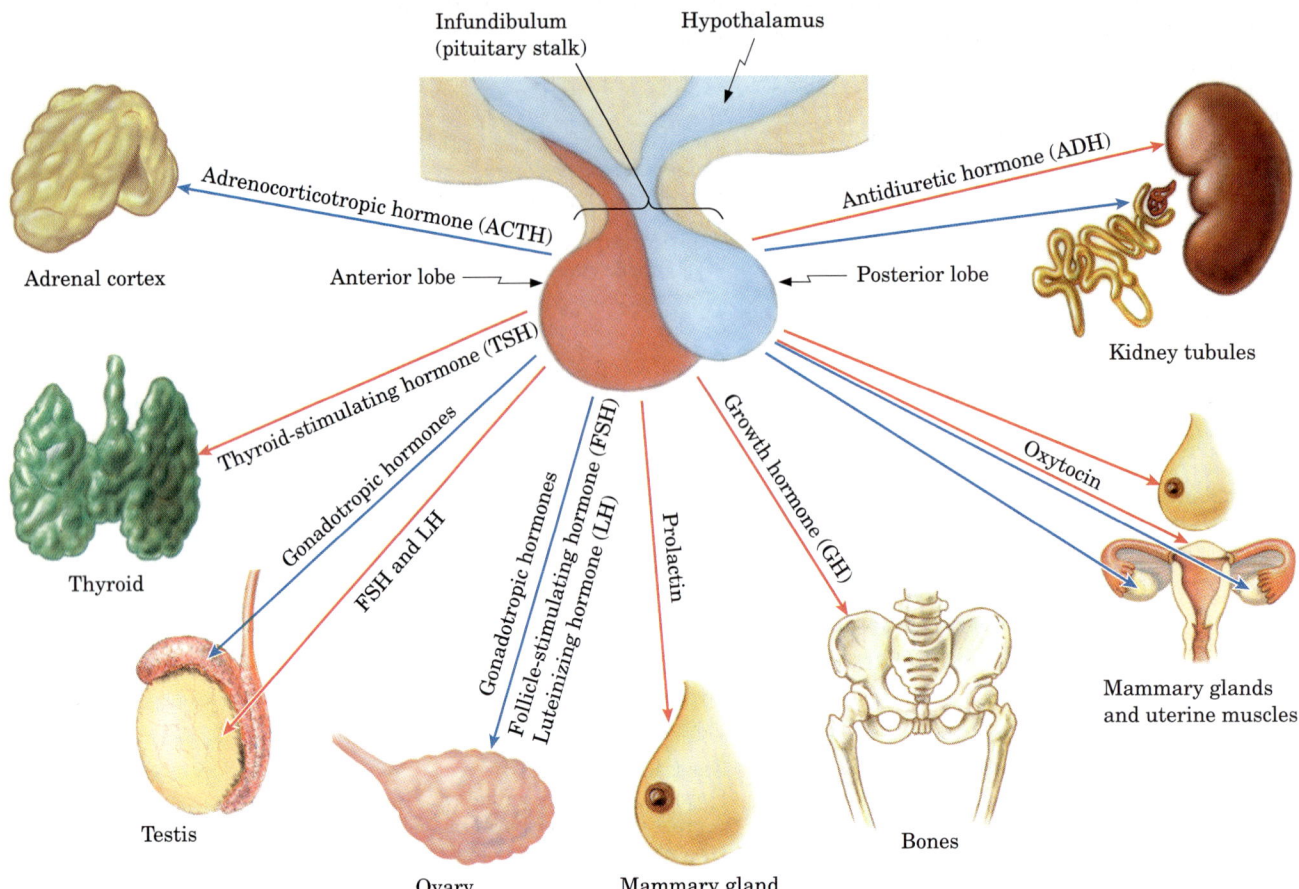

Figure 23.2 The pituitary gland is suspended from the hypothalamus by a stalk of neural tissue. The hormones secreted by the anterior and posterior lobes of the pituitary gland and the target tissues they act upon are shown. *(Modified from P. W. Davis and E. P. Solomon,* The World of Biology. *Philadelphia: Saunders College Publishing, 1986)*

stream to the target cell, in which case it is a hormone. For example, epinephrine and norepinephrine are both neurotransmitters and hormones.

There are, broadly speaking, five classes of chemical messengers: *cholinergic, amino acid, adrenergic, peptidergic,* and *steroid* messengers. This classification is based on the chemical nature of the important messenger (ligand) in each group. Neurotransmitters can belong to all five classes, and hormones can belong to the last three classes.

Messengers can also be classified according to how they work. Some of them—epinephrine, for example—*activate enzymes.* Others affect the *synthesis of enzymes and proteins* by working on the transcription of genes (Section 25.2); steroid hormones (Section 20.10) work in this manner. Finally, some affect the *permeability of membranes;* acetylcholine, insulin, and glucagon belong to this class.

Still another way of classifying messengers is according to their potential to *act directly* or through a *secondary messenger.* The steroid hormones act directly. They can penetrate the cell membrane and also pass through the membrane of the nucleus. For example, estradiol stimulates uterine growth.

Other chemical messengers act through secondary messengers. For example, epinephrine, glucagon, luteinizing hormone, norepinephrine, and vasopressin use cAMP as a secondary messenger (see details in Section 23.5C).

23.3 Cholinergic Messenger: Acetylcholine

The main **cholinergic neurotransmitter** is acetylcholine:

$$CH_3-\overset{O}{\underset{\|}{C}}-O-CH_2-CH_2-\overset{CH_3}{\underset{CH_3}{\overset{|}{\underset{|}{N^+}}}}-CH_3$$

Acetylcholine

A Cholinergic Receptors

There are two kinds of receptors for this messenger. We will look at one that exists on the motor end plates of skeletal muscles or in the sympathetic ganglia. The nerve cells that bring messages contain stored acetylcholine in the vesicles in their axons. The receptor on the muscle cells or neurons is also known as nicotinic receptor because nicotine (see Chemical Connections 16B) inhibits the neurotransmission of these nerves. The receptor itself is a *transmembrane protein* (Figure 20.2) made of five different subunits. The central core of the receptor is an ion channel through which, when open, Na^+ and K^+ ions can pass (Figure 23.3). When the ion channels are closed, the K^+ ion concentration is higher inside the cell than outside; the reverse is true for the Na^+ ion concentration.

B Storage of Messengers

Events begin when a message is transmitted from one neuron to the next by neurotransmitters. The message is initiated by calcium ions (see Chemical Connections 23A). When the Ca^{2+} concentration in a neuron reaches a certain level (more than 0.1 μM), the vesicles containing acetylcholine fuse

Figure 23.3 Acetylcholine in action. The receptor protein has five subunits. When two molecules of acetylcholine bind to the two α subunits, a channel opens to allow the passage of Na^+ and K^+ ions. *(Courtesy of Anthony Tu, Colorado State University)*

The acetylcholine receptor (AChR) binds two molecules of acetylcholine (ACh), one to each α subunit.

After acetylcholine binds to AChR the channel is opened.

CHEMICAL CONNECTIONS 23A

Calcium as a Signaling Agent (Secondary Messenger)

The message delivered to the receptors on the cell membranes by neurotransmitters or hormones must be delivered intracellularly to the various locations within the cell. Figure 23.4 depicts one mode of intracellular signaling, in which cAMP acts as a second messenger or signaling agent. The most universal and, at the same time, most versatile signaling agent is the cation Ca^{2+}.

Calcium ions in the cells come from either extracellular sources or intracellular stores, such as the endoplasmic reticulum. If the ions come from the outside, they enter the cells through specific calcium channels. Calcium ions control our heart beats, our movements through the action of skeletal muscles, and through the release of neurotransmitters in our neurons, learning and memory. They are also involved in signaling the beginning of life at fertilization and its end at death. Calcium ion signaling controls these different functions by two mechanisms: (1) increased concentration (forming sparks and puffs) and (2) duration of the signals.

In the resting state of the neuron, the Ca^{2+} concentration is about 0.1 μM. When neurons are stimulated, this level may increase to 0.5 μM. However, to elicit a fusion between the synaptic vesicles and the plasma membrane of the neuron, much higher concentrations are needed (10–25 μM).

An increase in calcium ion concentration may take the form of sparks or puffs. The source of calcium ions may be external (calcium influx caused by the electric signal of nerve transmission) or internal (calcium released from the stores of the endoplasmic reticulum). Upon receiving the signal of calcium puffs, the vesicles storing acetylcholine travel to the membrane of the presynaptic cleft. There they fuse with the membrane and empty their contents into the synapse.

Calcium ions can also control signaling by the duration of the signal. The signal in arterial smooth muscle lasts for 0.1 to 0.5 s. The wave of Ca^{2+} in the liver lasts 10 to 60 s. The calcium wave in the human egg lasts 1 to 35 min after fertilization. Thus, combining the concentration, localization, and duration of the signal, calcium ions can deliver messages to perform a variety of functions.

The effects of Ca^{2+} are modulated through specific calcium-binding proteins. In all nonmuscle cells and in smooth muscles, *calmodulin* is the calcium-binding protein. Calmodulin-bound calcium activates an enzyme, protein kinase II, which then phosphorylates an appropriate protein substrate. In this way, the signal is translated into metabolic activity.

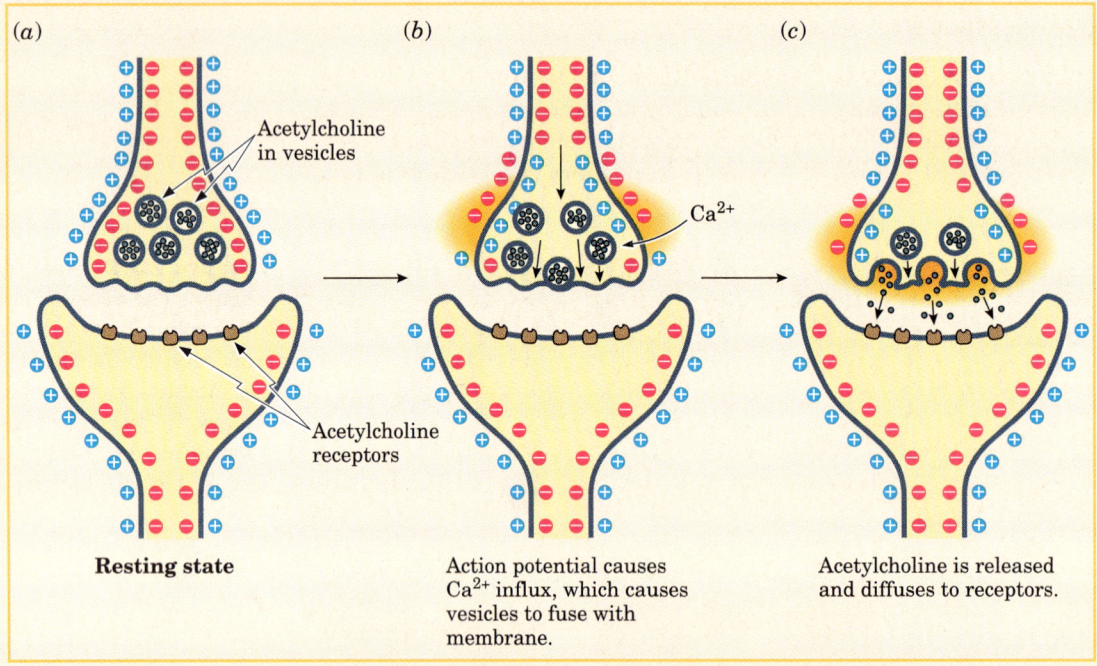

■ **Figure 23A** Calcium signaling to release acetylcholine from its vesicles.

CHEMICAL CONNECTIONS 23B

Nerve Gases and Antidotes

Most nerve gases in the military arsenal exert their lethal effect by binding to acetylcholinesterase. Under normal conditions, this enzyme hydrolyzes the synaptic neurotransmitter acetylcholine within a few milliseconds after it is released at the nerve endings. Nerve gases such as Sarin (agent GB, also called Tabun), Soman (agent GD), and agent VX are organic phosphonates related to such pesticides as parathion (the latter being much less lethal, of course):

$$CH_3-CH(CH_3)-O-P(=O)(CH_3)-F$$
Sarin (Agent GB)

$$CH_3-C(CH_3)(H_3C)-CH(CH_3)-O-P(=O)(CH_3)-F$$
Soman (Agent GD)

$$(CH_3)_2CH-N(CH(CH_3)_2)-CH_2CH_2-S-P(=O)(CH_3)-O-CH_2CH_3$$
Agent VX

$$CH_3CH_2-O-P(=S)(CH_2CH_3)-O-C_6H_4-NO_2$$
Parathion

Phosphonates, $X-\overset{O}{\underset{R}{P}}-O-$, are related to phosphates, $-O-\overset{O}{\underset{O}{P}}-O-$, where R is an alkyl group.

If any of these phosphonates binds to acetylcholinesterase, the enzyme is irreversibly inactivated, and the transmission of nerve signals stops. The result is a cascade of symptoms: sweating, bronchial constriction due to mucus buildup, dimming of vision, vomiting, choking, convulsions, paralysis, and respiratory failure. Direct inhalation of as little as 0.5 mg can cause death within a few minutes. If the dosage is less or if the nerve gas is absorbed through the skin, the lethal effect may take several hours.

In warfare, protective clothing and gas masks are effective countermeasures. An enzyme of the bacteria *Pseudomonas diminuta*, organophosphorous hydroxylase, is capable of detoxifying nerve gases. Their inclusion in cotton towelettes may help to remove nerve gases from contaminated skin. Also, first-aid kits containing injectable antidotes are available. These antidotes contain the alkaloid atropine and pralidoxime chloride. The two substances must be used in tandem—and very quickly—after exposure to nerve gases.

The nations of the world, with the exception of a few Arab countries, signed a treaty in 1993 that obliges all signatory governments to destroy their stockpiles of chemical weapons. In addition, the signees pledged to dismantle the plants that manufacture them within 10 years.

with the presynaptic membrane of the nerve cells. Then they empty the neurotransmitters into the synapse. The messenger molecules travel across the synapse and are adsorbed onto specific receptor sites.

C The Action of Messengers

The presence of acetylcholine molecules at the postsynaptic receptor site triggers a conformational change (Section 21.9A) in the receptor protein. This opens the *ion channel* and allows ions to cross membranes freely. Na^+ ions have higher concentration outside the neuron than do K^+ ions; thus, more Na^+ enters the cell than K^+ leaves. Because it involves ions, which carry electric charges, this process is translated into an electric signal. After a few milliseconds, the channel closes again. The acetylcholine still occupies the receptor. In order that the channel should reopen and transmit a new signal, the acetylcholine must be removed, and the neuron must be reactivated.

CHEMICAL CONNECTIONS 23C

Botulism and Acetylcholine Release

When meat or fish is improperly cooked or preserved, a deadly food poisoning, called botulism, may result. The culprit is the bacterium *Clostridium botulinum,* whose toxin prevents the release of acetylcholine from the presynaptic vesicles. Therefore, no neurotransmitter reaches the receptors on the surface of muscle cells, and the muscles no longer contract. Left untreated, the affected person may die.

Surprisingly, the botulin toxin has a valuable medical use. It is used in the treatment of involuntary muscle spasms—for example, in facial tics. These tics are caused by the uncontrolled release of acetylcholine. Controlled administration of the toxin, when applied locally to the facial muscles, stops the uncontrolled contractions and relieves the facial distortions.

"Facial distortion" has multiple meanings in the cosmetics industry. "Frown lines" and other wrinkles can also be removed by temporarily paralyzing facial muscles. The Food and Drug Administration has recently approved Botox (botulin toxin) for cosmetic use, meaning that its manufacturer can now advertise its rejuvenating appeal. Its use is fast spreading. Botox has been used on an off-label basis for a number of years, especially in Hollywood. Indeed, several movie directors have complained that some actors have used so much Botox that they can no longer show a variety of facial expressions.

■ Figure 23C   *Clostridium botulinum* is a food-poisoning bacterium.

D The Removal of Messengers

Acetylcholine is removed rapidly from the receptor site by the enzyme *acetylcholinesterase,* which hydrolyzes it.

$$CH_3-\overset{O}{\underset{\|}{C}}-O-CH_2-CH_2-\overset{CH_3}{\underset{CH_3}{\overset{|}{N^+}}}-CH_3 + H_2O$$

Acetylcholine

$$\xrightarrow{\text{Acetylcholin-esterase}} CH_3-\overset{O}{\underset{\|}{C}}-O^- + HO-CH_2-CH_2-\overset{CH_3}{\underset{CH_3}{\overset{|}{N^+}}}-CH_3$$

Acetate   Choline

This rapid removal enables the nerves to transmit more than 100 signals per second. By this means, the message moves from neuron to neuron until finally it gets transmitted, again by acetylcholine molecules, to the muscles or endocrine glands that are the ultimate target of the message.

Obviously, the action of the acetylcholinesterase enzyme is essential to the entire process. When this enzyme is inhibited, the removal of acetylcholine is incomplete, and nerve transmission ceases.

CHEMICAL CONNECTIONS 23D

Alzheimer's Disease and Acetylcholine Transferase

Alzheimer's disease is the name given to the symptoms of severe memory loss and other senile behavior that afflict about 1.5 million people in the United States. Postmortem identification of this disease focuses on two pathological hallmarks in the brain: (1) buildup of protein deposits known as β-amyloid plaques outside the nerve cells and (2) neurofibrillar tangles composed of tau proteins. Controversy exists as to which of the two is the primary cause of the neurodegeneration in Alzheimer's disease. Each has its advocates. Those who favor the primacy of the plaque are nicknamed β-aptists, and the other camp sports the name of tauists.

Tau protein binds to microtubules, one of the major cytoskeletons (see Chemical Connections 21G and Figure 21.G3). Genetic mutation of tau protein or environmental factors such as hyperphosphorylation may alter the ability of tau to bind to microtubules. These altered tau proteins form tangles in the cytoplasm of neurons. Neurofibrillar tangles have been found in Alzheimer's brains in the absence of plaque as well, which suggests that tau abnormality may be sufficient to cause neurodegeneration.

In most Alzheimer's brains, the dominant feature is plaques. The plaques consist of protein fibers, some 7–10 nm thick, that are mixed with small peptides called β-amyloid peptides. These peptides originate from a water-soluble precursor, amyloid precursor protein (APP); this transmembrane protein has an unknown function. Certain enzymes, called secretases, cut peptides containing 38, 40, and 42 amino acids from the transmembrane region of APP. Mutations of APP proteins in Alzheimer's disease patients cause preferential accumulation of the 42 amino acid peptide forming β-pleated sheets, which precipitates creating plaques.

In Alzheimer's disease, nerve cells in the cerebral cortex die, the brain becomes smaller, and part of the cortex atrophies. The depression of the folds on the brain surface become deeper.

People with Alzheimer's disease are forgetful, especially about recent events. As the disease advances, they become confused and, in severe cases, lose their ability to speak; at that point, they need total care. As yet, there is no cure for this disease.

Patients with Alzheimer's disease have significantly diminished acetylcholine transferase activity in their brains. This enzyme synthesizes acetylcholine by transferring the acetyl group from acetyl-CoA to choline:

$$CH_3\overset{O}{\overset{\|}{C}}-S-CoA + HO-CH_2CH_2\overset{CH_3}{\overset{|}{\underset{CH_3}{\underset{|}{N}}}}CH_3$$

Acetyl-CoA   Choline

$$\longrightarrow CH_3\overset{O}{\overset{\|}{C}}-O-CH_2CH_2\overset{CH_3}{\overset{|}{\underset{CH_3}{\underset{|}{N}}}}CH_3 + CoA-SH$$

Acetylcholine   Coenzyme A

The diminished concentration of acetylcholine can be partially compensated for by inhibiting the enzyme acetylcholinesterase, which decomposes acetylcholine. Certain drugs, which act as acetylcholinesterase inhibitors, have been shown to improve memory and other cognitive functions in some people with the disease. Drugs such as tacrine (Cognex), donepezil (Aricept), rivastigmine (Exelon), and galantamine (Reminyl) belong to this category; they all moderate the symptoms of Alzheimer's disease. The alkaloid huperzine A, an active ingredient of Chinese herb tea that has been used for centuries to improve memory, is also a potent inhibitor of acetylcholinesterase.

(a)

(b)

(c)

Figure 23D (a) Normal brain. (b) Brain of a person with Alzheimer's disease. (c) PET scan comparing a normal brain with the brain of a person with Alzheimer's disease.

E Control of Neurotransmission

Acetylcholinesterase is inhibited irreversibly by phosphonates in nerve gases and pesticides (Chemical Connections 23B) or reversibly by succinylcholine (Chemical Connections 22A) and decamethonium bromide.

$$\text{Br}^- \quad \text{CH}_3\text{—N}^+(\text{CH}_3)_2\text{—CH}_2(\text{CH}_2)_8\text{CH}_2\text{—N}^+(\text{CH}_3)_2\text{—CH}_3 \quad \text{Br}^-$$

Decamethonium bromide

Succinylcholine and decamethonium bromide resemble the choline end of acetylcholine and, therefore, act as competitive inhibitors of acetylcholinesterase. In small doses, these reversible inhibitors relax the muscles temporarily, making them useful as muscle relaxants in surgery. In large doses, they are just as deadly as the irreversible inhibitors.

The inhibition of acetylcholinesterase is but one way in which cholinergic neurotransmission is controlled. Another way is to modulate the action of the receptor. Because acetylcholine enables the ion channels to open and propagate signals, this mode of action is called *ligand-gated ion channels*. The attachment of the ligand to the receptor is critical in signaling. Nicotine in low doses is a stimulant; it is an agonist because it prolongs the receptor's biochemical response. In large doses, however, nicotine becomes an antagonist and blocks the action on the receptor. As such, it may cause convulsions and respiratory paralysis. Succinylcholine, besides being a reversible inhibitor of acetylcholinesterase, also has this concentration-dependent agonist/antagonist effect on the receptor. A strong antagonist, which blocks the receptor completely, can interrupt the communication between neuron and muscle cell. The venom of a number of snakes, especially that of the cobra, cobratoxin, exerts its deadly influence in this manner. The plant extract curare, which was used in poisoned arrows by the Amazon Indians, works in the same way. In small doses, curare is also used as a muscle relaxant.

Finally, the supply of the acetylcholine messenger can influence the proper nerve transmission. If the acetylcholine messenger is not released from its storage as in botulism (Chemical Connections 23C) or if its synthesis is impaired as in Alzheimer's disease (Chemical Connections 23D), the concentration of acetylcholine is reduced, and nerve transmission is impaired.

23.4 Amino Acids as Messengers: Neurotransmitters

A Messengers

Amino acids are distributed throughout the neurons individually or as parts of peptides and proteins. They can also act as neurotransmitters. Some of them, such as glutamic acid, aspartic acid, and cysteine, act as **excitatory neurotransmitters** similar to acetylcholine and norepinephrine. Others, such as glycine, β-alanine, taurine, and mainly γ-aminobutyric acid (GABA), are **inhibitory neurotransmitters;** they reduce neurotransmission. Note that some of these neurotransmitter amino acids are not found in proteins.

$$^+H_3NCH_2CH_2SO_3^-$$
Taurine

$$^+H_3NCH_2CH_2COO^-$$
β-Alanine

$$^+H_3NCH_2CH_2CH_2COO^-$$
γ-Aminobutyric acid
(GABA)
(IUPAC name:
4-aminobutanoic acid)

B Receptors

Each of these amino acids has its own receptor; in fact, glutamic acid has at least five subclasses of receptors. The best known among them is the *N*-Methyl-D-aspartate (NMDA) receptor. It is a ligand-gated ion channel similar to the nicotinic cholinergic receptor discussed in Section 23.3:

$$\begin{array}{c} CH_3 \\ | \\ NH_2^+ \\ | \\ CHCH_2-COO^- \\ | \\ COO^- \end{array}$$

N-methyl-D-aspartate

When glutamic acid binds to this receptor, the ion channel opens, Na^+ and Ca^{2+} flow into the neuron, and K^+ flows out of the neuron. The same thing happens when NMDA, being an agonist, stimulates the receptor. The gate of this channel is closed by a Mg^{2+} ion.

Phencyclidine (PCP), an antagonist of this receptor, induces hallucination. PCP, known by the street name of "angel dust," is a controlled substance; it causes bizarre psychotic behavior and long-term psychological problems.

C Removal of Messengers

In contrast to acetylcholine, there is no enzyme that would degrade glutamic acid and thereby remove it from its receptor once the signaling has been done. Glutamic acid is removed by **transporter** molecules, which bring it back through the presynaptic membrane into the neuron. This process is called **reuptake**.

> **Transporter** A protein molecule that carries small molecules, such as glucose or glutamic acid, across a membrane

23.5 Adrenergic Messengers: Neurotransmitters/Hormones

A Monoamine Messengers

The third class of neurotransmitters/hormones, the adrenergic messengers, includes such monoamines as epinephrine, serotonin, dopamine, and histamine. (Structures of these compounds can be found later in this section and in Chemical Connections 23E.) These monoamines transmit signals by a mechanism whose beginning is similar to the action of acetylcholine. That is, they are adsorbed on a receptor.

Martin Rodbell (1925–1998) of the National Institutes of Health was one of the pioneers in discovering and naming signal transduction. For this discovery, he was awarded the Nobel Prize in physiology and medicine in 1994.

Signal transduction A cascade of events through which the signal of a neurotransmitter or hormone delivered to its receptor is carried inside the target cell and amplified into many signals that can cause protein modifications, enzyme activation, and the opening of membrane channels

G-protein also participates in another signal transduction cascade, which involves inositol-based compounds (Section 20.6) as signaling molecules. Phosphoinositoldiphosphate (PIP2) mediates the action of hormones and neurotransmitters. These messengers can stimulate the phosphorylation of enzymes, in a manner similar to the cAMP cascade. They also play an important role in the release of calcium ions from their storage areas in the endoplasmic reticulum (ER) or sarcoplasmic reticulum (SR).

B Action of Messengers

Once the monoamine neurotransmitter/hormone (for example, norepinephrine) is adsorbed onto the receptor site, the signal will be amplified inside the cell. In the example shown in Figure 23.4, the receptor has an associated protein called G-protein. This protein is the key to the cascade that produces many signals inside the cell (amplification). The entire process that occurs after the messenger binds to the receptor is called **signal transduction**. The active G-protein has an associated nucleotide, guanosine triphosphate (GTP). It is an analog of adenosine triphosphate (ATP), in which the aromatic base adenine is substituted by guanine (Section 24.2). The G-protein becomes inactive when its associated nucleotide is hydrolyzed to guanosine diphosphate (GDP). Signal transduction starts with the active G-protein, which activates the enzyme adenylate cyclase.

C Secondary Messengers

Adenylate cyclase produces a secondary messenger inside the cell, called cyclic AMP (cAMP). The manufacture of cAMP activates processes that result in the transmission of an electrical signal. The cAMP is manufactured by adenylate cyclase from ATP:

Adenosine triphosphate (ATP)

$$\xrightarrow{\text{adenylate cyclase}}$$

Cyclic-Adenosine monophosphate (cAMP) + Pyrophosphate

The activation of adenylate cyclase accomplishes two important goals:

1. It converts an event occurring at the outer surface of the target cell (adsorption onto receptor site) to a change inside the target cell (formation of cAMP). Thus the primary messenger (neurotransmitter or hormone) does not have to cross the membrane.

2. It amplifies the signal. One molecule adsorbed on the receptor triggers the adenylate cyclase to make many cAMP molecules. Thus the signal is amplified many thousands of times.

Figure 23.4 The sequence of events in the postsynaptic membrane when norepinephrine is absorbed onto the receptor site. (a) The active G-protein hydrolyzes GTP. The energy of hydrolysis of GTP to GDP activates adenylate cyclase. A molecule of cAMP forms when adenylate cyclase cleaves ATP into cAMP and pyrophosphate. (b) Cyclic AMP activates a protein kinase by dissociating the regulatory (R) unit from the catalytic unit (C). A second molecule of ATP, shown in (b), has phosphorylated the catalytic unit and been converted to ADP. (c) The catalytic unit phosphorylates the ion-translocating protein that blocked the channel for ion flow. The phosphorylated ion-translocating protein changes its shape and position and opens the ion gates.

D Removal of Signal

How does this signal amplification stop? When the neurotransmitter or hormone dissociates from the receptor, the adenylate cyclase halts the manufacture of cAMP. The cAMP already produced is destroyed by the enzyme phosphodiesterase, which catalyzes the hydrolysis of the phosphoric ester bond, yielding AMP.

The amplification through the secondary messenger (cAMP) is a relatively slow process. It may take from 0.1 s to a few minutes. Therefore, in cases where the transmission of signals must be fast (ms or μs), a neurotransmitter such as acetylcholine acts on membrane permeability directly, without the mediation of a secondary messenger.

E Control of Neurotransmission

The G-protein–adenylate cyclase cascade in transduction signaling is not limited to monoamine messengers. A wide variety of peptide hormones and neurotransmitters (Section 23.6) use this signaling pathway. Included among them are glucagon, vasopressin, luteinizing hormone, enkephalins, and P-protein. Neither is the opening of ion channels, depicted in Figure 23.4, the only target of this signaling. A number of enzymes can be phosphorylated by protein kinases, and the phosphorylation controls whether these enzymes will be active or inactive (Section 22.6).

The fine control of the G-protein–adenylate cyclase cascade is essential for health. Consider the toxin of the bacteria *Vibrio cholerae,* which perma-

CHEMICAL CONNECTIONS 23E

Parkinson's Disease: Depletion of Dopamine

Parkinson's disease is characterized by spastic motion of the eyelids as well as rhythmic tremors of the hands and other parts of the body, often when the patient is at rest. The posture of the patient changes to a forward, bent-over position; walking becomes slow, with shuffling footsteps. The cause of this degenerative nerve disease is unknown, but genetic factors and environmental effects, such as exposure to pesticides and/or high concentrations of metals such as Mn^{2+} ion, have been implicated.

The neurons affected contain, under normal conditions, mostly dopamine as a neurotransmitter. People with Parkinson's disease have depleted amounts of dopamine in their brains, but the dopamine receptors are not affected. Thus the first line of remedy is to *increase the concentration of dopamine*. Dopamine cannot be administered directly because it cannot penetrate the blood–brain barrier and therefore does not reach the tissue where its action is needed. L-Dopa, on the other hand, is transported through the arterial wall and converted to dopamine in the brain:

[Structure of (S)-3,4-Dihydroxyphenylalanine (L-Dopa)]

enzyme-catalyzed decarboxylation → [Structure of Dopamine] + CO_2

When L-dopa is administered, many Parkinson's disease patients are able to synthesize dopamine and resume normal nerve transmission. In these individuals, L-dopa reverses the symptoms of Parkinson's disease, although the respite is only temporary. In other patients, the L-dopa regimen provides little benefit.

Another way to increase dopamine concentration is to *prevent its metabolic elimination*. The drug entacapon (Comtan) inhibits an enzyme that is instrumental in clearing dopamine from the brain. The enzyme (catechol-O-methyl transferase, COMT) converts dopamine to 3-methoxy-4-hydroxy-L-phenylalanine, which is then eliminated. Entacapon is usually administered together with L-dopa. Another drug, (R)-selegiline (L-Deprenyl), is a monoamine oxidase (MAO) inhibitor. L-Deprenyl, which is also given together with L-dopa, can reduce the symptoms of Parkinson's disease and even increase the life span of patients. It increases the level of dopamine by *preventing its oxidation by MAOs*.

Other drugs may treat only the symptoms of Parkinson's disease: the spastic motions and the tremors. These drugs, such as benzotropin (Cogentin), are similar to atropine (Problem 16.45) and act on cholinergic receptors, thereby preventing muscle spasms.

The real cure for the disease may lie in transplanting human embryonic dopamine neurons. In preliminary work, such grafts have been functionally integrated in the patient's brain, producing dopamine. In the most successful cases, patients have been able to resume normal, independent life following the transplant.

Certain drugs designed to affect one neurotransmitter may also affect another. An example is the drug methylphenidate (Ritalin). In higher doses, this drug promotes the dopamine concentration in the brain and acts as a stimulant. In small doses, it is prescribed to calm hyperactive children or to minimize ADD (attention deficit disorder). It seems that in smaller doses Ritalin raises the concentration of serotonin. This neurotransmitter decreases hyperactivity without affecting the dopamine levels of the brain.

[Structure of Serotonin]

The close connection between two monoamine neurotransmitters, dopamine and serotonin, is also evident in their roles in controlling the nausea and vomiting that often follow general anesthesia and chemotherapy. Blockers of dopamine receptors in the brain, such as metoclopramide (Reglan) or promethazine (Phenergan), can alleviate the symptoms after anesthesia. However, a blocker of serotonin receptors in the brain as well as on the terminals of the vagus nerve in the stomach, such as ondansetrone (Zofran), is the drug of choice for preventing chemotherapy-induced vomiting.

Synthesis and degradation of dopamine are not the only way that the brain keeps its concentration at a steady state. The concentration is also controlled by specific proteins, called *transporters*, that ferry the used dopamine from the receptor back across the synapse into the original neuron for reuptake. Cocaine addiction works through such a transporter. Cocaine binds to the dopamine transporter, like a reversible inhibitor, thereby preventing the reuptake of dopamine. As a consequence, dopamine is not transported back to the original neuron and stays in the synapse, increasing the continuous firing of signals, which is the psychostimulatory effect associated with a cocaine "high."

nently activates G-protein. The result is the symptoms of cholera—namely, severe dehydration as a result of diarrhea. This problem arises because the activated G-proteins overproduce cAMP. This excess, in turn, opens the ion channels, which produces a large outflow of Na^+ ions and accompanying water from the epithelial cells to the intestines. Therefore, the first measure taken in treating cholera victims is to replace the lost water and salt.

F Removal of Neurotransmitters

The inactivation of the adrenergic neurotransmitters differs somewhat from the invactivation of the cholinergic transmitters. While acetylcholine is decomposed by acetylcholinesterase, most of the adrenergic neurotransmitters are inactivated in a different way. *The body inactivates monoamines by oxidizing them to aldehydes.* Enzymes that catalyze these reactions, called monoamine oxidases (MAOs), are very common in the body. For example, one MAO converts both epinephrine and norepinephrine to the corresponding aldehyde:

Many drugs that are used as antidepressants or antihypertensive agents are inhibitors of monoamine oxidases—for example, Tofranil and Elavil. These inhibitors prevent MAOs from converting monoamines to aldehydes, thereby increasing the concentration of the active adrenergic neurotransmitters.

There is also an alternative way to remove adrenergic neurotransmitters. Shortly after adsorption onto the postsynaptic membrane, the neurotransmitter comes off the receptor site and is reabsorbed through the presynaptic membrane and stored again in the vesicles.

G Histamines

The neurotransmitter histamine is present in mammalian brains. It is synthesized from the amino acid, histidine, by decarboxylation:

Histamine cannot readily pass through the blood–brain barrier (Section 31.1) and must be synthesized in the brain neurons.

CHEMICAL CONNECTIONS 23F

Nitric Oxide as a Secondary Messenger

The toxic effects of the simple gaseous molecule NO have long been recognized (see Chemical Connections 3C). Therefore, it came as a surprise to find that this compound plays a major role in chemical communications. This simple molecule is synthesized in the cells when arginine is converted to citrulline. (These two compounds appear in the urea cycle; see Section 27.8). Nitric oxide is a relatively nonpolar molecule, and shortly after it has been produced in the nerve cell, it quickly diffuses across the lipid bilayer membrane. During its short half-life (4–6 s), it can reach a neighboring cell. Because NO passes through membranes, it does not need extracellular receptors to deliver its message. NO is very unstable, so there is no need for a special mechanism for its destruction.

NO acts as an intercellular messenger between the endothelial cells surrounding the blood vessels and the smooth muscles encompassing these cells. It relaxes the muscle cells, thereby dilating the blood vessels. The outcome is less-restricted blood flow and a drop in blood pressure. This reason also explains why nitroglycerin (Chemical Connections 13D) is effective against angina: It produces NO in the body.

Another role of NO in dilating blood vessels lies in remedying impotence. The new impotence-relieving drug, Viagra, enhances the activity of NO by inhibiting an enzyme (phosphodiesterase) that otherwise would reduce NO's effect on smooth muscles. When the NO concentration is sufficiently high, the blood vessels dilate, allowing enough blood to flow to provide an erection. In most cases, this happens within an hour after taking the pill.

Sometimes the dilation of blood vessels is not so beneficial. Headaches are caused by dilated arteries in the head. NO-producing compounds in food—nitrites in smoked and cured meats and sodium glutamate in seasoning (Chemical Connections 27D)—can cause such headaches. Nitroglycerin itself often induces headaches.

Nitric oxide is toxic, as discussed in Chemical Connections 3C. This toxicity is used by our immune system (Section 30.2B) to fight infections by viruses.

The toxic effect of NO is also evident in strokes. A blocked artery restricts the blood flow to certain parts of the brain; the oxygen-starved neurons then die. Next, neurons in the surrounding area, 10 times larger than the place of the initial attack, release glutamic acid, which stimulates other cells. They, in turn, release NO, which kills all cells in the area. Thus the damage to the brain is spread tenfold. A concentrated effort is under way to find inhibitors of the NO-producing enzyme, nitric oxide synthase, that can be used as antistroke drugs. For the discovery of NO and its role in blood pressure control, three pharmacologists—Robert Furchgott, Louis Ignarro, and Ferid Murad—received the 1998 Nobel Prize in physiology.

■ Antihistamines block the H_1 receptor for histamine.

Sir James W. Black of England received the 1988 Nobel Prize in medicine for the invention of cimetidine and such other drugs as propranolol (Table 23.1).

The action of histamine as a neurotransmitter is very similar to that of other monoamines. There are two kinds of receptors for histamine. One receptor, H_1, can be blocked by antihistamines such as dimenhydrinate (Dramamine) and diphenhydramine (Benadryl). The other receptor, H_2, can be blocked by ranitidine (Zantac) and cimetidine (Tagamet).

H_1 receptors are found in the respiratory tract. They affect the vascular, muscular, and secretory changes associated with hay fever and asthma. Therefore, antihistamines that block H_1 receptors relieve these symptoms. The H_2 receptors are found mainly in the stomach and affect the secretion of HCl. Cimetidine and ranitidine, both H_2 blockers, reduce acid secretion and, therefore, are effective drugs for ulcer patients. However, the real cure for ulcers is to kill the bacteria *Helicobacter pylori* (Chemical Connections 22B) by administering either antibiotics or a combination of antibiotics and H_2 blockers.

EXAMPLE 23.1

Three enzymes in the adrenergic neurotransmission pathway affect the transduction of the signals. Identify them and describe how they affect the neurotransmission.

> **Solution**
> The enzyme adenylate cyclase amplifies the signal by producing cAMP secondary messengers. The enzyme phosphatase terminates the signal by hydrolyzing cAMP. The enzyme monoamine oxidase (MAO) reduces the frequency of signals by oxidizing the monoamine neurotransmitters to the corresponding aldehydes.
>
> **Problem 23.1**
> What is the functional difference between G-protein and GTP?

23.6 Peptidergic Messengers: Neurotransmitters/Hormones

A Messengers

Many of the most important hormones affecting metabolism belong to the peptidergic messengers group. Among them are insulin (Section 21.7 and Chemical Connections 23G) and glucagons, hormones of the pancreatic islets, and vasopressin and oxytocin (Section 21.7) products of the posterior pituitary gland.

In the last few years, scientists have isolated a number of brain peptides that have affinity for certain receptors and, therefore, act as if they were neurotransmitters. Some 25 or 30 such peptides are now known.

The first brain peptides isolated were the **enkephalins.** These pentapeptides are present in certain nerve cell terminals. They bind to specific pain receptors and seem to control pain perception. Because they bind to the receptor site that also binds the pain-killing alkaloid morphine, it is assumed that the N-terminal end of the pentapeptide fits the receptor (Figure 23.5).

Even though morphine remains the most effective agent in reducing pain, its clinical use is limited because of its side effects. These include respiratory depression, constipation, and, most significantly, addiction. The

Figure 23.5 Similarities between the structure of morphine and that of the brain's own pain regulators, the enkephalins.

Morphine

Methionine enkephalin

CHEMICAL CONNECTIONS 23G

Diabetes

The disease diabetes mellitus affects approximately 16 million people in the United States. In a normal person, the pancreas, a large gland behind the stomach, secretes insulin as well as several other hormones. Diabetes usually results from low insulin secretion. Insulin is necessary for glucose molecules to penetrate such cells as brain, muscle, and fat cells, where they can be used. It accomplishes this task by being adsorbed onto the receptors in the target cells. This adsorption triggers the manufacture of cyclic GMP (not cAMP), and this secondary messenger in turn increases the transport of glucose molecules into the target cells.

In the resulting cascade of events, the first step is the self-(auto)phosphorylation of the receptor molecule itself, on the cytoplasmic side. The phosphorylated insulin receptor activates enzymes and regulatory proteins by phosphorylating them. As a consequence, glucose transporter molecules (GLUT4) that are stored inside the cells migrate to the plasma membrane. Once there, they facilitate the movement of glucose across the membrane. This transport relieves the accumulation of glucose in blood serum and makes it available for metabolic activity inside the cells. Thus glucose can be used as an energy source, stored as glycogen, or even diverted to enter fat and other molecular biosynthetic pathways.

In diabetic patients, the glucose level rises to 600 mg/100 mL of blood or higher (normal is 80 to 100 mg/100 mL). There are two kinds of diabetes. In insulin-dependent diabetes, patients do not manufacture enough of this hormone in the pancreas. This Type I disease develops early, before the age of 20, and must be treated with daily injections of insulin. Even with daily injections of insulin, the blood sugar level fluctuates, which may cause other disorders, such as cataracts, retinal dystrophy leading to blindness, kidney disease, heart attack, and nervous disorders.

One way to counteract these fluctuations is to monitor the blood sugar and, as the glucose level rises, to administer insulin. Such monitoring requires pricking the finger six times per day, an invasive regimen that few diabetics follow faithfully. Recently, noninvasive techniques have been developed. One of the most promising employs contact lenses. The blood sugar's fall and rise is mimicked by the glucose content of tears. A fluorescent sensor in the contact lens monitors these glucose fluctuations; its data can be read via a hand-held fluorimeter. Thus the patterns of glucose fluctuation can be obtained noninvasively and, if it rises to the danger zone, it can be counteracted by insulin intake.

The delivery of insulin also has undergone a revolution. The tried-and-true methods of injections and insulin pumps are still widely used, but new delivery is available orally or by nasal sniffing.

In Type II (non–insulin-dependent) diabetes, the patient has enough insulin in the blood but cannot utilize it properly because the target cells have an insufficient number of receptors. Such patients typically develop the disease after age 40 and are likely to be obese. Overweight people usually have a lower-than-normal number of insulin receptors in their adipose (fat) cells.

Oral drugs can help this second type of diabetic patient in several ways. For example, sulfonyl urea compounds, such as tolbutamide (Orinase), increase insulin secretion. In addition, insulin concentration in the blood can be increased by enhancing its release from the β-cells of pancreatic islets. The drug repaglinide (Prandin) blocks the K^+-ATP channels of the β-cells, allowing Ca^{2+} influx, which induces the release of insulin from the cells.

These drugs seem to control the symptoms of diabetes, but fluctuations in insulin levels may turn high blood sugar (hyperglycemia) into low blood sugar (hypoglycemia) which is just as dangerous. Other drugs for Type II diabetes that do not elicit hypoglycemia attempt to control the glucose level at its source. Migitol (Glyset), an anti-α-glucosidase drug, inhibits the enzyme that converts glycogen or dietary starch into glucose. The drug metformin (Glucophage) decreases glucose production in the liver, carbohydrate absorption in the intestines, and glucose uptake by fat cells.

Tolbutamide (Orinase)

Metformin (Glucophage)

clinical use of enkephalins yielded only modest relief. The challenge is to develop analgesic drugs that do not involve the opiate receptors in the brain.

Another brain peptide, **neuropeptide Y,** affects the hypothalamus, a region that integrates the body's hormonal and nervous systems. Neuropeptide Y is a potent orexic (appetite-stimulating) agent. When its receptors are blocked (for example, by leptin, the "thin" protein) appetite is suppressed. Leptin is an anorexic agent.

Still another peptidergic neurotransmitter is **substance P** (*P* for "pain"). This 11-amino-acid peptide is involved in transmission of pain signals. In injury or inflammation, sensory nerve fibers transmit signals from the peripheral nervous system (where the injury occurred) to the spinal cord, which processes the pain. The peripheral neurons synthesize and release substance P, which bonds to receptors on the surface of the spinal cord. The substance P, in its turn, removes the magnesium block at the *N*-methyl-D-aspartate (NMDA) receptor. Glutamic acid, an excitatory amino acid, can then bind to this receptor. In doing so, it amplifies the pain signal going to the brain.

B Secondary Messengers

All peptidergic messengers, hormones, and neurotransmitters act through secondary messengers. Glucagon, luteinizing hormone, antidiuretic hormone, angiotensin, enkaphalin, and substance P use the G-protein–adenylate cyclase cascade. Others, such as vasopressin, use membrane-derived phosphatidylinositol (PI) derivates (Section 20.6). As with the cAMP cascade, control of PI secondary messengers is done by phosphorylation. Activation may be attained via phosphorylation, and inactivation by removing a phosphate. The transition between phosphatidylinositol and its various phosphates is exercised by a kinase/phosphatase mechanism:

Still other substances use calcium (Chemical Connections 23A) as secondary messengers.

CHEMICAL CONNECTIONS 23H

Tamoxifen and Breast Cancer

Large-scale clinical studies have shown that tamoxifen (Nolvadex) can reduce the recurrence of breast cancer in women who had been successfully treated for the disease. Even more importantly, a study involving 13,000 women showed that tamoxifen reduced the occurrence of breast cancer by 45% in women who were at high risk because of family history, previous breast abnormality, or age. Tamoxifen is a nonsteroidal compound that binds to the estradiol receptors situated on the nuclei of cells. By binding to the receptor, it causes conformational changes. This then prevents estradiol's binding to its receptor and thus hampers the stimulation of the growth of breast cancer. In addition, tamoxifen has shown promise in treating refractory ovarian cancer.

Although it reduces the risk of breast cancer, tamoxifen increases the risk of both uterine cancer and clot formation. The latter condition can be life-threatening if clots migrate to the lungs, causing pulmonary embolism. However, these last two diseases are much rarer than breast cancer.

Estradiol

Tamoxifen

23.7 Steroid Messengers: Hormones

In Section 20.10, we saw that a large number of hormones possess steroid ring structures. These hormones, the sex hormones among them, are hydrophobic; therefore, they can cross plasma membranes of the cell by diffusion.

Progesterone

There is no need for special receptors embedded in the membrane for these hormones. It has been shown, however, that **steroid hormones** interact inside the cell with protein receptors. Most of these receptors are localized in the nucleus of the cell, but small amounts also exist in the cytoplasm. When they interact with steroids, they facilitate their migration through the aqueous cytoplasm; the proteins themselves are hydrophilic.

Once inside the nucleus, the steroid–receptor complex can either bind directly to the DNA or combine with a transcription factor (Section 25.2), influencing the synthesis of a certain key protein. Thyroid hormones, which also have large hydrophobic domains, have protein receptors as well, which facilitates their transport through aqueous media.

The steroid hormonal response through protein synthesis is not fast. In fact, it takes hours to occur. Steroids can also act at the cell membrane, influencing ligand-gate ion channels. Such a response would take only seconds. An example of such a fast response occurs in fertilization. The sperm head contains proteolytic enzymes, which act on the egg to facilitate its penetration. These enzymes are stored in acrosomes, organelles found on the sperm head. During fertilization, progesterone originating from the follicle cells surrounding the egg acts on the acrosome outer membrane, which disintegrates within seconds and releases the proteolytic enzymes.

The same steroid hormones depicted in Figure 20.6 act as neurotransmitters, too. These neurosteroids are synthesized in the brain cells in both neurons and glia, and they affect receptors, mainly the NMDA and GABA receptors (Section 23.4). Progesterone and progesterone metabolites in brain cells can induce sleep, have analgesic and anticonvulsive effects, and can even serve as natural anesthetics.

SUMMARY

Cell-to-cell communications are carried out by three kinds of molecules (Section 23.1). **Receptors** are protein molecules embedded in the membranes of cells. **Chemical messengers,** or ligands, interact with receptors. **Secondary messengers** carry and amplify the signals from the receptor to inside the cell.

Neurotransmitters send chemical messengers across a short distance—the **synapse** between two neurons or between a neuron and a muscle or endocrine gland cell (Section 23.2). This communication occurs in milliseconds. **Hormones** transmit their signals more slowly and over a longer distance, from the source of their secretion (endocrine gland), through the bloodstream, into target cells (Section 23.2). Many drugs affect these chemical communications: **Antagonists** block receptors; **agonists** stimulate receptors.

Five kinds of chemical messengers exist: **cholinergic, amino acid, adrenergic, peptidergic,** and **steroid.** Neurotransmitters may belong to the first four classes, hormones to the last three classes. Acetylcholine is cholinergic, glutamic acid is an amino acid, epinephrine (adrenaline) and norepinephrine are adrenergic, enkephalins are peptidergic and progesterone is a steroid.

Nerve transmission starts with the neurotransmitters packaged in **vesicles** in the **presynaptic end** of neurons. When these neurotransmitters are released, they cross the membrane and the synapse and are adsorbed onto receptor sites on the **postsynaptic** membranes. This adsorption triggers an electrical response. Some neurotransmitters act directly, whereas others act through a secondary messenger, **cyclic AMP.** After the electrical signal is triggered, the neurotransmitter molecules must be removed from the postsynaptic end. In the case of acetylcholine (Section 23.3), this removal is done by an enzyme called acetylcholinesterase; in the case of monoamines (Section 23.5), enzymes (MAOs) oxidize them to aldehydes.

Amino acids (Section 23.4) bind to their receptors, which are ligand-gated ion channels. Peptides and proteins (Sections 23.6) bind to receptors on the target cell membrane and use secondary messengers to exert their influence. **Signal transduction** is the process that occurs after a ligand binds to its receptor: The signal is carried inside the cell and is amplified. Steroids (Section 23.7) penetrate the cell membrane, and their receptors are found in the cytoplasm. Together with their receptors, they penetrate the cell nucleus. These hormones can act in three ways: (1) They activate enzymes, (2) they affect the gene transcription of an enzyme or protein, and (3) they change membrane permeability. The same steroids can also act as neurotransmitters, when synthesized in neurons.

PROBLEMS

Numbers that appear in color indicate difficult problems. ▶ designates problems requiring application of principles.

Chemical Communications

23.2 What kind of signal travels along the axon of a neuron?

23.3 What is the difference between a *chemical messenger* and a *secondary messenger*?

Neurotransmitters and Hormones

23.4 Define the following:
(a) Synapse (b) Receptor
(c) Presynaptic (d) Postsynaptic
(e) Vesicle

23.5 What is the role of Ca^{2+} in releasing neurotransmitters into the synapse?

23.6 Which signal takes longer: (a) neurotransmitter or (b) hormone? Explain.

23.7 Which gland controls lactation?

23.8 To which of the three groups of chemical messengers do these hormones belong?
(a) Norepinephrine
(b) Thyroxine
(c) Oxytocin
(d) Progesterone

Cholinergic Neurotransmitters

23.9 How does acetylcholine transmit an electric signal from neuron to neuron?

23.10 Which end of the acetylcholine molecule fits into the receptor site?

23.11 Cobra venom and botulin are both deadly toxins, but they affect cholinergic neurotransmission differently. How does each cause paralysis?

23.12 Different ion concentrations across a membrane generate a potential (voltage). We call such a membrane *polarized*. What happens when acetylcholine is adsorbed on its receptor?

Amino Acid Neurotransmitters

23.13 List two features by which taurine differs from the amino acids found in proteins.

23.14 How is glutamic acid removed from its receptor?

23.15 What is unique in the structure of GABA that distinguishes it from all the amino acids that are present in proteins?

23.16 What is the structural difference between NMDA, an agonist of a glutamic acid receptor, and L-aspartic acid?

Adrenergic Neurotransmitters

23.17 (a) Identify two monoamine neurotransmitters in Table 23.1.
(b) Explain how they act.
(c) What medication controls the particular diseases caused by the lack of monoamine neurotransmitters?

23.18 What bond is hydrolyzed and what bond is formed in the synthesis of cAMP?

23.19 How is the catalytic unit of protein kinase activated in adrenergic neurotransmission?

23.20 The formation of cyclic AMP is described in Section 23.5. Show by analogy how cyclic GMP is formed from GTP.

23.21 By analogy to the action of MAO on epinephrine, write the structural formula of the product of the corresponding oxidation of dopamine.

23.22 The action of protein kinase is the next-to-last step in the G-protein–adenylate cyclase cascade signal transduction. What kind of effects can elicit the phosphorylation by this enzyme?

23.23 Explain how adrenergic neurotransmission is affected by (a) amphetamines and (b) reserpine. (See Table 23.1.)

23.24 Which step in the events depicted in Figure 23.3 provides an electrical signal?

23.25 What kind of product is the MAO-catalyzed oxidation of epinephrine?

23.26 How is histamine removed from the receptor site?

23.27 Cyclic AMP affects the permeability of membranes for ion flow.
(a) What blocks the ion channel?
(b) How is this blockage removed?
(c) What is the direct role of cAMP in this process?

23.28 Dramamine and cimetidine are both antihistamines. Would you expect Dramamine to cure ulcers and cimetidine to relieve the symptoms of asthma? Explain.

Peptidergic Neurotransmitters

23.29 What is the chemical nature of enkephalins?

23.30 What is the mode of action of Demerol as a pain killer? (See Table 23.1.)

23.31 What enzyme catalyzes the formation of inositol-1,4,5-triphosphate from inositol-1,4-diphosphate? Give the structures of the reactant and the product.

23.32 How is the secondary messenger inositol-1,4,5-triphosphate inactivated?

Chemical Connections

23.33 (Chemical Connections 23A) What is the difference between calcium sparks (puffs) and calcium waves?

23.34 (Chemical Connections 23A) What is the role of calmodulin in signaling by Ca^{2+} ions?

23.35 (Chemical Connections 23A) To enable a fusion between the synaptic vessel and the plasma membrane, the calcium concentration is increased. How many-fold of an increase in Ca^{2+} concentration is needed?

23.36 (Chemical Connections 23B) (a) What is the effect of nerve gases?
(b) What molecular substance do they affect?

23.37 (Chemical Connections 23C) What is the mode of action of botulinum toxin?

23.38 (Chemical Connections 23C) How can a deadly botulinum toxin contribute to the facial beauty of Hollywood actresses?

23.39 (Chemical Connections 23D) What are the neurofibrillar tangles in the brains of Alzheimer's disease patients made of? How do they affect the cell structure?

23.40 (Chemical Connections 23D) What are the plaques in the brains of Alzheimer's disease patients made of?

23.41 (Chemical Connections 23D) Alzheimer's disease causes loss of memory. What kind of drugs may provide some relief for, if not cure, this disease? How do they act?

23.42 (Chemical Connections 23E) Why would a dopamine pill be ineffective in treating Parkinson's disease?

23.43 (Chemical Connections 23E) What is the mechanism by which cocaine stimulates the continuous firing of signals between neurons?

23.44 (Chemical Connections 23E) Parkinson's disease is due to a paucity of dopamine neurons, yet its symptoms are relieved by drugs that block cholinergic receptors. Explain.

23.45 (Chemical Connections 23E) In certain cases, embryonic dopamine neurons transplanted into the brains of patients with advanced Parkinson's disease resulted in complete remission. How was this result possible?

23.46 (Chemical Connections 23F) How can NO cause headaches?

23.47 (Chemical Connections 23F) How is nitric oxide synthesized in the cells?

23.48 (Chemical Connections 23F) How is the toxicity of NO detrimental in strokes?

23.49 (Chemical Connections 23G) Tolbutamide is called a sulfonyl urea compound. Identify the sulfonyl urea moiety in the structure of the drug.

23.50 (Chemical Connections 23G) What is the difference between insulin-dependent and non–insulin-dependent diabetes?

23.51 (Chemical Connections 23G) How does insulin facilitate the absorption of glucose from blood serum into adipocytes (fat cells)?

23.52 (Chemical Connections 23G) Diabetic patients must frequently monitor the fluctuation of glucose levels in their blood. What is the advantage of the latest technique of monitoring the glucose content in tears over the older technique of obtaining frequent blood samples?

23.53 (Chemical Connections 23H) Tamoxifen is an antagonist. How does it work in reducing the reoccurrence of breast cancer?

Additional Problems

23.54 Considering its chemical nature, how does aldosterone (Section 20.10) affect mineral metabolism (Table 23.2)?

23.55 What is the function of the ion-translocating protein in adrenergic neurotransmission?

23.56 Decamethonium acts as a muscle relaxant. If an overdose of decamethonium occurs, can paralysis be prevented by administering large doses of acetylcholine? Explain.

23.57 Endorphin, a potent pain killer, is a peptide containing 22 amino acids; among them are the same five N-terminal amino acids found in the enkephalins. Does this explain endorphin's pain-killing action?

23.58 How do alanine and beta alanine differ in structure?

23.59 Where is a G-protein located in adrenergic neurotransmission?

23.60 (Chemical Connections 23F) List a number of effects that are caused when NO, acting as a secondary messenger, relaxes smooth muscles.

23.61 (a) In terms of their action, what do the hormone vasopressin and the neurotransmitter dopamine have in common?
(b) What is the difference in their mode's of action?

23.62 What is the difference in the models of action between acetylcholinesterase and acetylcholine transferase?

23.63 How does cholera toxin exert its effect?

23.64 Give the formulas for the following reaction:
$GTP + H_2O \rightleftharpoons GDP + P_i$.

23.65 Insulin is a hormone that, when it binds to a receptor, enables glucose molecules to enter the cell and be metabolized. If you have a drug that is an agonist, how would the glucose level in the serum change upon administering the drug?

23.66 (Chemical Connections 23E) Ritalin is used to alleviate hyperactivity in attention deficit disorder of children. How does this drug work?

23.67 The pituitary gland releases luteinizing hormone (LH), which enhances the production of progesterone in the uterus. Classify these two messengers and discuss how each delivers its message.

InfoTrac College Edition

For additional reading, go to InfoTrac College Edition, your online research library, at

http://infotrac.thomsonlearning.com

CHAPTER 24

24.1 Introduction
24.2 Components of Nucleic Acids
24.3 Structure of DNA and RNA
24.4 RNA
24.5 Genes, Exons, and Introns
24.6 DNA Replication
24.7 DNA Repair
24.8 Cloning

The world's first sheep clones. These Welsh Mountain sheep are the product of research by Dr. Ian Wilmut and colleagues at the Roslin Institute in Edinburgh, Scotland.

Nucleotides, Nucleic Acids, and Heredity

24.1 Introduction

Each cell of our bodies contains thousands of different protein molecules. Recall from Chapter 21 that all of these molecules are made up of the same 20 amino acids, just arranged in different sequences. That is, the hormone insulin has a different amino acid sequence than the globin of the red blood cell. Even the same protein—for example, insulin—has different sequence in different species (Section 21.7). Within the same species, individuals may have some differences in their proteins, although the differences are much less dramatic than those seen between species. This variation is most obvious in cases where people have such conditions as hemophilia, albinism, or color-blindness because they lack certain proteins that normal people have or because the sequence of their amino acids is somewhat different (see Chemical Connections 21D).

After scientists became aware of the differences in amino acid sequences, their next question was, how do the cells know which proteins to synthesize out of the extremely large number of possible amino acid se-

■ Human chromosomes magnified about 8000 times.

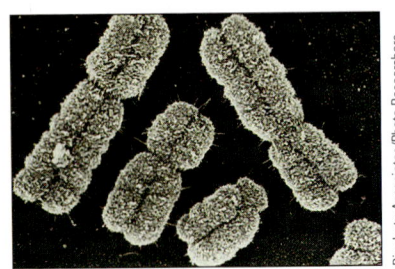

quences? The answer to this question is that an individual gets the information from its parents through *heredity*. Heredity is the transfer of characteristics, anatomical as well as biochemical, from generation to generation. We all know that a pig gives birth to a pig and a mouse gives birth to a mouse.

It was easy to determine that the information is obtained from the parent or parents, but what form does this information take? During the last 60 years, revolutionary developments have enabled us to answer this question—the transmission of heredity on the molecular level.

From about the end of the nineteenth century, biologists suspected that the transmission of hereditary information from one generation to another took place in the nucleus of the cell. More precisely, they believed that structures within the nucleus, called **chromosomes,** have something to do with heredity. Different species have different numbers of chromosomes in the nucleus. The information that determines external characteristics (red hair, blue eyes) and internal characteristics (blood group, hereditary diseases) was thought to reside in **genes** located inside the chromosomes.

Chemical analysis of nuclei showed that they are largely made up of special basic proteins called *histones* and a type of compound called *nucleic acids*. By 1940, it became clear through the work of Oswald Avery (1877–1955) that, of all the material in the nucleus, only a nucleic acid called deoxyribonucleic acid (DNA) carries the hereditary information. That is, the genes are located in the DNA. Other work in the 1940s by George Beadle (1903–1989) and Edward Tatum (1909–1975) demonstrated that each gene controls the manufacture of one protein, and that external and internal characteristics are expressed through this gene. Thus the expression of the gene (DNA) in terms of an enzyme (protein) led to the study of protein synthesis and its control. *The information that tells the cell which proteins to manufacture is carried in the molecules of DNA.*

24.2 Components of Nucleic Acids

Two kinds of nucleic acids are found in cells: **ribonucleic acid (RNA)** and **deoxyribonucleic acid (DNA).** Each has its own role in the transmission of hereditary information. As we just saw, DNA is present in the chromosomes of the nucleus. RNA is not found in the chromosomes, but rather is located elsewhere in the nucleus and even outside the nucleus, in the cytoplasm. As we will see in Section 24.4, there are four types of RNA, all with similar structures.

Both DNA and RNA are polymers. Just as proteins consist of chains of amino acids, and polysaccharides consist of chains of monosaccharides, so nucleic acids are also chains. The building blocks (monomers) of nucleic acid chains are *nucleotides*. Nucleotides themselves, however, are composed of three simpler units: a base, a monosaccharide, and a phosphate. We will look at each of these components in turn.

A Bases

The **bases** found in DNA and RNA are chiefly those shown in Figure 24.1. All of them are basic because they are heterocyclic aromatic amines (Section 16.2). Two of these bases—adenine (A) and guanine (G)—are purines, and the other three—cytosine (C), thymine (T), and uracil (U)—are pyrimidines. The two purines (A and G) and one of the pyrimidines (C) are found in both DNA and RNA, but uracil (U) is found only in RNA, and thymine (T) is found only in DNA. Note that thymine differs from uracil only in the

---

There are two parents if reproduction is sexual but one parent if it is asexual.

Some simple organisms, such as bacteria, do not have a nucleus. Their chromosomes are condensed in a central region.

Gene The unit of heredity; a DNA segment that codes for one protein

Beadle and Tatum shared the 1958 Nobel Prize in physiology with Joshua Lederberg.

The early hypothesis of "one gene, one protein" has undergone modifications. We now know that one gene can participate in the manufacturing of many proteins through processes such as recombination.

Bases Purines and pyrimidines, which are components of nucleotides, DNA, and RNA

The three pyrimidines and guanine are in their keto forms rather than their enol forms (see Section 17.8).

The initial letter of each base is used as an abbreviation for that base.

Figure 24.1 The five principal bases of DNA and RNA. Note how the rings are numbered. The hydrogens shown in blue are lost when the bases are bonded to monosaccharides.

methyl group in the 5 position. Thus both DNA and RNA contain four bases: two pyrimidines and two purines. For DNA, the bases are A, G, C, and T; for RNA, the bases are A, G, C, and U.

B Sugars

The sugar component of RNA is D-ribose (Section 19.2). In DNA, it is 2-deoxy-D-ribose (hence the name deoxyribonucleic acid):

The only difference between these molecules is that ribose has an OH group in the 2 position not found in deoxyribose.

The full name of β-D-ribose is β-D-ribofuranose and that of β-2-deoxy-D-ribose is β-2-deoxy-D-ribofuranose (see Section 19.3).

Nucleoside A compound composed of ribose or deoxyribose and a base

The combination of sugar and base is known as a **nucleoside.** The purine bases are linked to C-1 of the monosaccharide through N-9 (the nitrogen at position 9 of the five-membered ring) by a β-N-glycosidic bond:

The nucleoside made of guanine and ribose is called **guanosine.** Table 24.1 gives the names of the other nucleosides.

The pyrimidine bases are linked to C-1 of the monosaccharide through their N-1 by a β-N-glycosidic bond.

Uridine

C Phosphate

The third component of nucleic acids is phosphoric acid. When this group forms a phosphate ester (Section 18.7) bond with the —CH$_2$OH group of a nucleoside, the result is a compound known as a **nucleotide.** For example, adenosine combines with phosphate to form the nucleotide adenosine 5′-monophosphate (AMP):

> **Nucleotide** A nucleoside bonded to one, two, or three phosphate groups

Table 24.1 gives the names of the other nucleotides. Some of these nucleotides play an important role in metabolism. They are part of the structure of key coenzymes, cofactors, and activators (Section 26.3 and 28.2). Most notably, adenosine 5′-triphosphate (ATP) serves as a common currency into which the energy gained from food is converted and stored. In ATP, two more phosphate groups are joined to AMP in phosphate anhydride bonds (Section 18.7). In adenosine 5′-diphosphate (ADP), only one phosphate group is bonded to the AMP. All other nucleotides have important

CHAPTER 24 NUCLEOTIDES, NUCLEIC ACIDS, AND HEREDITY

Primed numbers are used for the ribose and deoxyribose portions of nucleosides, nucleotides, and nucleic acids. Unprimed numbers are used for the bases.

Table 24.1 The Eight Nucleosides and Eight Nucleotides in DNA and RNA

Base	Nucleoside	Nucleotide
DNA		
Adenine (A)	Deoxyadenosine	Deoxyadenosine 5′-monophosphate (dAMP)*
Guanine (G)	Deoxyguanosine	Deoxyguanosine 5′-monophosphate (dGMP)*
Thymine (T)	Deoxythymidine	Deoxythymidine 5′-monophosphate (dTMP)*
Cytosine (C)	Deoxycytidine	Deoxycytidine 5′-monophosphate (dCMP)*
RNA		
Adenine (A)	Adenosine	Adenosine 5′-monophosphate (AMP)
Guanine (G)	Guanosine	Guanosine 5′-monophosphate (GMP)
Uracil (U)	Uridine	Uridine 5′-monophosphate (UMP)
Cytosine (C)	Cytidine	Cytidine 5′-monophosphate (CMP)

*The *d* indicates that the sugar is deoxyribose.

multiphosphorylated forms. For example, guanosine exists as GMP, GDP, and GTP.

In Section 24.3, we will see how DNA and RNA are chains of nucleotides. In summary:

A nucleoside = Base + Sugar

A nucleotide = Base + Sugar + Phosphoric acid

A nucleic acid = A chain of nucleotides

EXAMPLE 24.1

GTP is an important store of energy. Draw the structure of guanosine triphosphate.

Solution
The base guanine is linked to a ribose unit by β-*N*-glycosidic linkage. The triphosphate is linked to C-5′ of the ribose by an ester bond.

Problem 24.1
Draw the structure of UMP.

CHEMICAL CONNECTIONS 24A

Anticancer Drugs

A major difference between cancer cells and most normal cells is that the cancer cells divide much more rapidly. Rapidly dividing cells require a constant new supply of DNA. One component of DNA is the nucleoside deoxythymidine, which is synthesized in the cell by the methylation of the uracil base.

Fluorouracil

If fluorouracil is administered to a cancer patient as part of chemotherapy, the body converts it to fluorouridine, a compound that irreversibly inhibits the enzyme that manufactures thymidine from uridine, greatly decreasing DNA synthesis. Because this inhibition affects the rapidly dividing cancer cells more than the healthy cells, the growth of the tumor and the spread of the cancer are arrested. Unfortunately, chemotherapy with fluorouracil or other anticancer drugs also weakens the body, because it interferes with DNA synthesis in normal cells.

Chemotherapy is used intermittently to give the body time to recover from the side effects of the drug. During the period after chemotherapy, special precautions must be taken so that bacterial infections do not debilitate the already-weakened body.

24.3 Structure of DNA and RNA

In Chapter 21, we saw that proteins have primary, secondary, and higher structures. Nucleic acids, which are chains of monomers, also have primary, secondary, and higher structures.

A Primary Structure

Nucleic acids are polymers of nucleotides, as shown schematically in Figure 24.2. Primary structure is the sequence of nucleotides. Note that it can be divided into two parts: (1) the backbone of the molecule and (2) the bases that are the side-chain groups. The backbone in DNA consists of alternating deoxyribose and phosphate groups. Each phosphate group is linked to the 3′ carbon of one deoxyribose unit and simultaneously to the 5′ carbon of the next deoxyribose unit (Figure 24.3). Similarly, each monosaccharide unit forms a phosphate ester at the 3′ position and another at the 5′ position. The primary structure of RNA is the same except that each sugar is ribose (so there is an —OH group in the 2′ position) rather than deoxyribose and U is present instead of T.

Thus the backbone of the DNA and RNA chains has two ends: a 3′—OH end and a 5′—OH end. These two ends have roles similar to those of the C-terminal and N-terminal ends in proteins. The backbone provides structural stability for the DNA and RNA molecules.

The bases that are linked, one to each sugar unit, are the side chains, and they carry all the information necessary for protein synthesis. Analysis of the base composition of DNA molecules from many different species was done by Erwin Chargaff (1905–), who showed that in DNA taken from many different species, the quantity of adenine (in moles) is always approximately equal to that of thymine, and the quantity of guanine is always approximately equal to that of cytosine, although the adenine/guanine ratio varies widely from species to species (see Table 24.2). This important information helped to establish the secondary structure of DNA, as we will soon see.

Just as the order of the amino acid residues of protein side chains determines the primary structure of the protein (for example,

Screen 23.3 Models

Nucleic acid A polymer composed of nucleotides

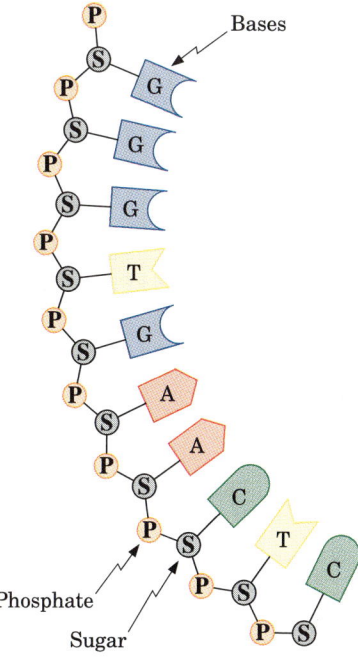

Figure 24.2 Schematic diagram of a nucleic acid molecule. The four bases of each nucleic acid are arranged in various specific sequences.

Figure 24.3 Primary structure of the DNA backbone. The hydrogens shown in blue cause the acidity of nucleic acids. In the body, at neutral pH, the phosphate groups carry a charge of -1, and the hydrogens are replaced by Na^+ and K^+.

The sequence TGA is not the same as AGT, but rather its opposite.

—Ala—Gly—Glu—Met—), the order of the bases (for example, —ATTGAC—) provides the primary structure of DNA. As with proteins, we need a convention to tell us which end to start with when we write the sequence of bases. For nucleic acids, the convention is to begin the sequence with the nucleotide that has the free 5′ terminal. Thus the sequence AGT means that adenine is the base at the 5′ terminal and thymine is the base at the 3′ terminal.

Table 24.2 Base Composition and Base Ratio in Two Species

Organism	Base Composition (mol %)				Base Ratio	
	A	G	C	T	A/T	G/C
Human	30.9	19.9	19.8	29.4	1.05	1.00
Wheat germ	27.3	22.7	22.8	27.1	1.01	1.00

B Secondary Structure of DNA

In 1953, James Watson (1928–) and Francis Crick (1916–) established the three-dimensional structure of DNA. Their work is a cornerstone in the history of biochemistry. The model of DNA established by Watson and Crick was based on two important pieces of information obtained by other workers: (1) the Chargaff rule that (A and T) and (G and C) are present in equimolar quantities and (2) X-ray diffraction photographs obtained by Rosalind Franklin (1920–1958) and Maurice Wilkins (1916–). By the clever use of these facts, Watson and Crick concluded that DNA is composed of two strands entwined around each other in a **double helix,** as shown in Figure 24.4.

In the DNA double helix, the two polynucleotide chains run in opposite directions. Thus, at each end of the double helix, there is one 5′—OH and one 3′—OH terminal. The sugar–phosphate backbone is on the outside, exposed to the aqueous environment, and the bases point inward. The bases are hydrophobic, so they try to avoid contact with water. Through their hydrophobic interactions, they stabilize the double helix. The bases are paired according to Chargaff's rule: For each adenine on one chain, a thymine is aligned opposite it on the other chain; each guanine on one chain has a cytosine aligned with it on the other chain. *The bases so paired form hydrogen bonds with each other, two for A—T and three for G—C, thereby stabilizing the double helix* (Figure 24.5). A—T and G—C are **complementary base pairs.**

Watson, Crick, and Wilkins were awarded the 1962 Nobel Prize in medicine for their discovery. Franklin died in 1958. The Nobel Committee does not award the Nobel Prize posthumously.

Double helix The arrangement in which two strands of DNA are coiled around each other in a screw-like fashion

Recall from Section 21.8 that a helix has a shape like a coiled spring or a spiral staircase. It was Pauling's discovery that human hair protein is helical that led Watson and Crick to look for helixes in DNA.

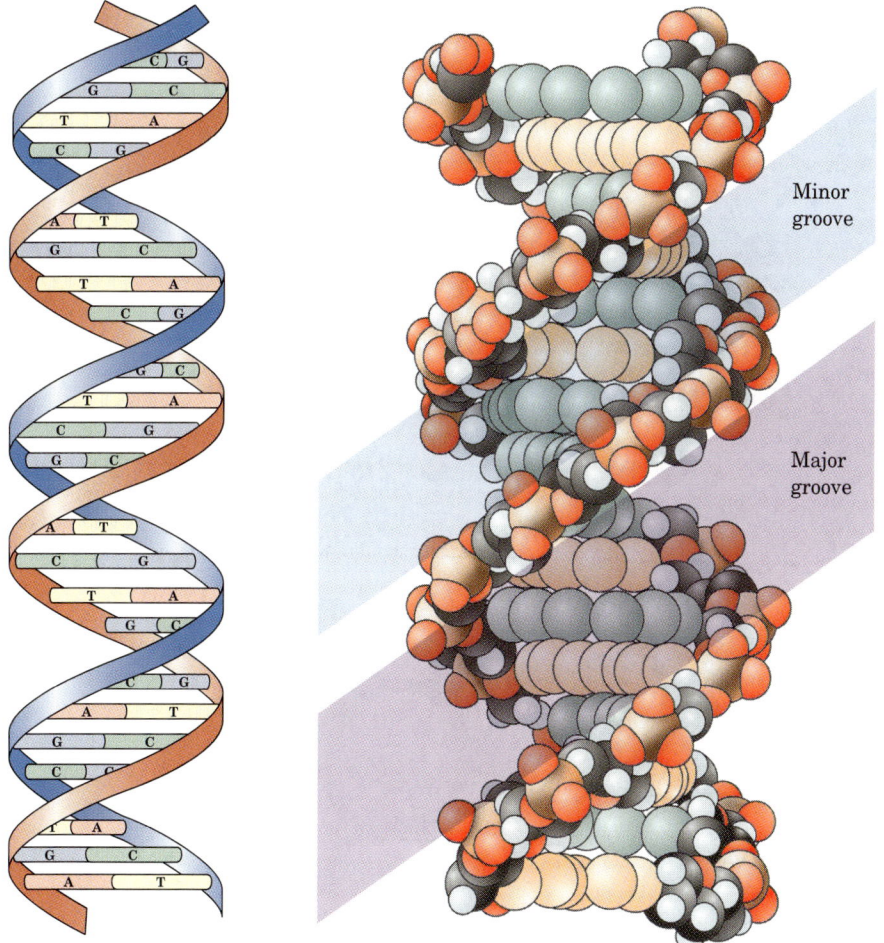

Figure 24.4 Three-dimensional structure of the DNA double helix.

604 | CHAPTER 24 NUCLEOTIDES, NUCLEIC ACIDS, AND HEREDITY

■ Rosalind Franklin (1920–1958).

■ Watson and Crick with their model of the DNA molecule.

The important thing, as Watson and Crick realized, is that only adenine could fit with thymine and only guanine could fit with cytosine. Let us consider the other possibilities. Can two purines (AA, GG, or AG) fit opposite each other? Figure 24.6 shows that they would overlap. How about two pyrimidines (TT, CC, or CT)? As shown in Figure 24.6, they would be too far

Figure 24.5 A and T pair up by forming two hydrogen bonds; G and C pair up by forming three hydrogen bonds.

Thymine — Adenine
AT pair

Cytosine — Guanine
GC pair

apart. *There must be a pyrimidine opposite a purine.* But could A fit opposite C, or G opposite T? Figure 24.7 shows that the hydrogen bonding would be much weaker.

The entire action of DNA—and of the heredity mechanism—depends on the fact that, *wherever there is an adenine on one strand of the helix, there must be a thymine on the other strand because that is the only base that fits and forms strong hydrogen bonds, and similarly for G and C.* The entire heredity mechanism rests on these slender hydrogen bonds (Figure 24.5), as we will see in Section 24.6.

C Higher Structures of DNA

If a human DNA molecule were fully stretched out, its length would be perhaps 1 m. However, the DNA molecules in the nuclei are not stretched out, but rather coiled around basic protein molecules called **histones.** The acidic DNA and the basic histones attract each other by electrostatic (ionic) forces, combining to form units called **nucleosomes.** In a nucleosome, eight histone molecules form a core, around which a 147-base-pair DNA double helix is wound. Nucleosomes are further condensed into chromatin when a 30-nm-wide fiber forms in which nucleosomes are wound in a **solenoid** fashion, with six nucleosomes forming a repeating unit (Figure 24.8). Chromatin fibers are organized still further into loops, and loops are arranged into bands to provide the superstructure of chromosomes.

The beauty of establishing the three-dimensional structure of the DNA molecule was that the knowledge of this structure immediately led to the

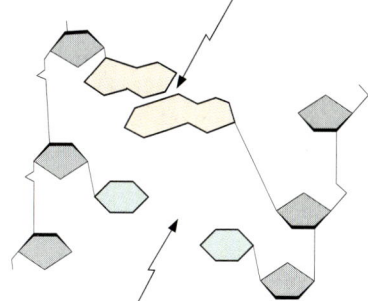

Figure 24.6 The bases of DNA cannot stack properly in the double helix if a purine is opposite a purine or a pyrimidine is opposite a pyrimidine.

Solenoid A coil wound in the form of a helix

Figure 24.7 Only one hydrogen bond is possible for GT or CA. These combinations are not found in DNA. Compare this figure with Figure 24.5.

Figure 24.8 Superstructure of chromosomes. In nucleosomes, the bandlike DNA double helix winds around cores consisting of eight histones. Solenoids of nucleosomes form 30 nm filament. Loops and minibands are other substructures.

explanation for the transmission of heredity—how the genes transmit traits from one generation to another. Before we look at the mechanism of DNA replication (in Section 24.6), let us summarize the three differences in structure between DNA and RNA:

1. DNA has four bases: A, G, C, and T. RNA has three of these bases—A, G, and C—but its fourth base is U, not T.
2. In DNA, the sugar is 2-deoxy-D-ribose. In RNA, it is D-ribose.

3. DNA is almost always double-stranded, with the helical structure shown in Figure 24.4.

There are several kinds of RNA (as we will see in Section 24.4); all of them are single-stranded, although base-pairing can occur within a chain (see, for example, Figure 24.9). When it does, adenine pairs with uracil because thymine is not present.

In DNA–RNA interactions, the complementary bases are

DNA	RNA
A	U
G	C
C	G
T	A

24.4 RNA

We previously noted that there are four types of RNA.

1. **Messenger RNA (mRNA)** mRNA molecules carry the genetic information from the DNA in the nucleus directly to the cytoplasm, where the protein is synthesized. Messenger RNA consists of a chain of nucleotides whose sequence is exactly complementary to that of one of the strands of the DNA. This type of RNA is not long-lived, however. It is synthesized as needed and then degraded, so its concentration at any given time is rather low. The size of mRNA varies widely, with the average unit containing perhaps 750 nucleotides.

   Messenger RNA (mRNA) The RNA that carries genetic information from DNA to the ribosome and acts as a template for protein synthesis

2. **Transfer RNA (tRNA)** Containing from 73 to 93 nucleotides per chain, tRNAs are relatively small molecules. There is at least one different tRNA molecule for each of the 20 amino acids from which the body makes its proteins. The three-dimensional tRNA molecules are L-shaped, but they are conventionally represented as a cloverleaf in two dimensions. Figure 24.9 shows a typical structure. Transfer RNA mole-

   Transfer RNA (tRNA) The RNA that transports amino acids to the site of protein synthesis in ribosomes

Figure 24.9 Structure of tRNA. (a) Two-dimensional simplified cloverleaf structure. (b) Three-dimensional structure. (*From* Biochemistry, *2nd ed., by Lubert Stryer. Copyright 1981 by W. H. Freeman and Co. All rights reserved. Used with permission. Part (b) also courtesy of Dr. Sung-Hou Kim.*)

(a)

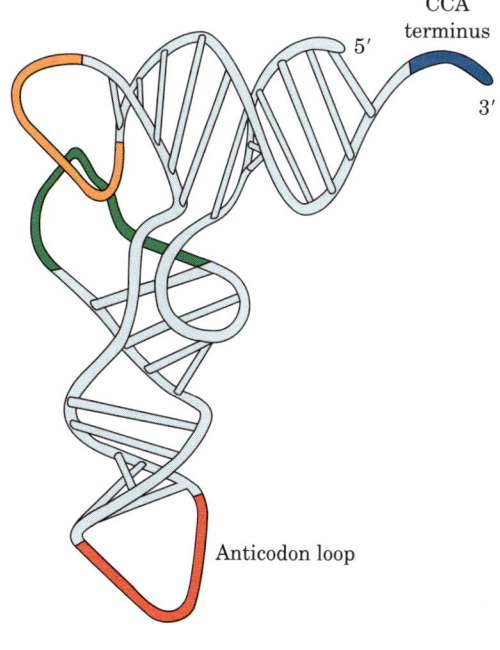

(b)

1-Methylguanosine

Ribosomal RNA (rRNA) The RNA complexed with proteins in ribosomes

Ribosome Small spherical bodies in the cell made of protein and RNA; the site of protein synthesis

Splicing The removal of an internal RNA segment and the joining of the remaining ends of the RNA molecule

cules contain not only cytosine, guanine, adenine, and uracil, but also several other modified nucleotides, such as 1-methylguanosine.

3. **Ribosomal RNA (rRNA)** **Ribosomes,** which are small spherical bodies located in the cells but outside the nuclei, contain rRNA. They consist of about 35% protein and 65% ribosomal RNA (rRNA). These large molecules have molecular weights up to 1 million. As discussed in Section 25.5, protein synthesis takes place on the ribosomes.

4. **Ribozymes** Ribozymes (catalytic RNA) are RNAs with special enzyme functions. In messenger RNA, the precursor molecule is shortened by **splicing** to mature mRNA. The splicing is catalyzed by ribozymes, which are segments of the mRNAs themselves. Similar splicing is catalyzed by ribozymes in forming mature tRNA from its precursor. This ribozyme, however, has both RNA and protein subunits. Both are necessary for the catalytic activity in vivo, although the RNA subunit contains the active site. In higher organisms, ribozymes reside in the ribosome (Section 25.5); hence, they are made of rRNA.

Even though ribozymes have simpler primary structures than proteins (only 4 building blocks in ribozymes versus 20 in proteins), they form complicated secondary structures by hydrogen bonding. These structures are, in turn, twisted into different tertiary structures. Ribozymes also have an active site just like protein enzymes where the substrate binds.

24.5 Genes, Exons, and Introns

A gene is a stretch of DNA, containing a few hundred nucleotides, that carries one particular message—for example, "make a globin molecule." One DNA molecule may have between 1 million and 100 million bases. Therefore, there are many genes in one DNA molecule. In bacteria, this message is continuous; in higher organisms, it is not. That is, stretches of DNA that spell out (code for) the amino acid sequence to be assembled are interrupted by long stretches that seemingly do not code for anything. The coding sequences are called **exons,** short for "expressed sequences," and the noncoding sequences are called **introns,** short for "intervening sequences."

Exon Nucleotide sequence in mRNA that codes for a protein

Intron A nucleotide sequence in mRNA that does not code for a protein

For example, the globin gene has three exons broken up by two introns. Because DNA contains exons and introns, the mRNA transcribed from it also contains exons and introns. The introns are cut out by ribozymes, and the exons are spliced together before the mRNA is used to synthesize a protein (Figure 24.10). In other words, the introns function as spacers and, in rare instances, as enzymes, catalyzing the splicing of exons into mature mRNA.

Philip A. Sharp (1944–) and Richard J. Roberts (1943–) were awarded the 1993 Nobel Prize in medicine for their discovery of the noncoding nature of introns.

In humans, only 3% of the DNA codes for proteins or RNA with clear functions. Introns are not the only noncoding DNA sequences, however. **Satellites** are DNA molecules in which short nucleotide sequences are repeated hundreds or thousands of times. Large satellite stretches appear at

Figure 24.10 Introns are cut out of mRNA before the protein is synthesized.

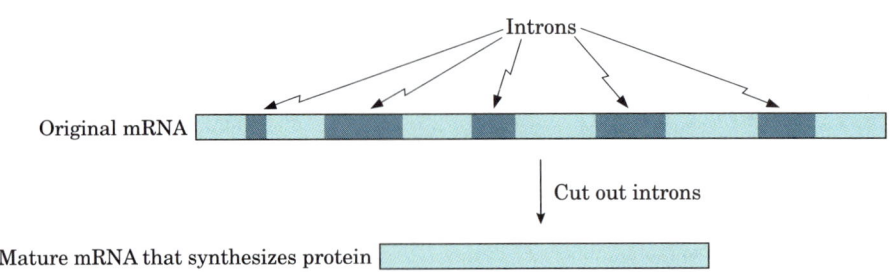

the ends and centers of chromosomes and provide stability for the chromosomes. Smaller repetitive sequences, called **mini-** or **microsatellites,** are associated with cancer when they mutate.

24.6 DNA Replication

The DNA in the chromosomes carries out two functions: (1) It reproduces itself, and (2) it supplies the information necessary to make all the proteins in the body, including enzymes. The second function is covered in Chapter 25. Here we are concerned with the first, **replication.**

Each gene is a section of a DNA molecule that contains a specific sequence of the four bases A, G, T, and C, typically comprising about 1000 to 2000 nucleotides. The base sequence of the gene carries the information necessary to produce one protein molecule. If the sequence is changed (for example, if one A is replaced by a G, or if an extra T is inserted), a different protein is produced, which might have an impaired function as in sickle cell anemia (Chemical Connections 21D).

But consider the task that must be accomplished by the organism. When an individual is conceived, the egg and sperm cells unite to form the zygote. This cell contains only a small amount of DNA, but it nevertheless provides all the genetic information that the individual will ever have. In a human cell, some 6 billion base pairs must be duplicated at each cell cycle, and a fully grown human being may contain more than 1 trillion cells. Each cell contains the same amount of DNA as the original single cell. Furthermore, cells are constantly dying and being replaced. Thus, there must be a mechanism by which DNA molecules can be copied over and over again without error. In Section 25.7, we will see that such errors sometimes do happen and can have serious consequences. Here, however, we want to examine this remarkable mechanism that takes place every day in billions of organisms, from microbes to whales, and has been taking place for billions of years—with only a tiny percentage of errors.

In human cells, the average chromosome has several hundred origins of replication where the copying occurs simultaneously. The DNA double helix has two strands running in opposite directions. The point on the DNA where replication takes place is called the **replication fork.** (See Figure 24.11.)

Screen 23.5
Tutorial

Replication The process by which copies of DNA are made during cell division

The pigments responsible for the blue or brown color of eyes are synthesized with the help of specific enzymes. If one of these enzymes is lacking, the eye color may be different.

Figure 24.11 General features of the replication of DNA. The two strands of the DNA double helix are shown separating at the replication fork.

CHEMICAL CONNECTIONS 24B

Telomers, Telomerase, and Immortality

Every person has a genetic makeup consisting of about 3 billion pairs of nucleotides, distributed over 46 chromosomes. Telomers are specialized structures at the ends of chromosomes. In vertebrates, telomers are TTAGGG sequences that are repeated hundreds to thousands of times. In normal **somatic cells** that divide in a cyclic fashion throughout the life of the organism (via mitosis), chromosomes lose about 50 to 200 nucleotides from their telomers at each cell division.

DNA polymerase, the enzyme that links the fragments, does not work at the end of linear DNA. This fact results in the shortening of the telomers at each replication. The telomer shortening acts as a clock by which the cells count the number of times they have divided. After a certain number of divisions, the cells stop dividing, having reached the limit of the aging process.

In contrast to somatic cells, all immortal cells (germ cells in proliferative stem cells, normal fetal cells, and cancer cells) possess an enzyme, telomerase, that can extend the shortened telomers by synthesizing new chromosomal ends. Telomerase is a ribonucleoprotein; that is, it is made of RNA and protein. The activity of this enzyme seems to confer immortality to the cells.

If the unwinding begins in the middle, the synthesis of new DNA molecules on the old templates continues in both directions until the whole molecule is duplicated. Alternatively, the unwinding can start at one end and proceed in one direction until the whole double helix is unwound.

Replication is bidirectional and it is done with the same speed in both directions. An interesting detail of DNA replication is that the two daughter strands are synthesized in different ways. One of the syntheses is continuous in the $5' \longrightarrow 3'$ direction (see Section 24.3). It is called the **leading strand.** Along the other strand that runs $3' \longrightarrow 5'$ direction, the synthesis is discontinuous. It is called the **lagging strand.**

The newly synthesized strands are called daughter strands.

The replication process is called **semiconservative** because each daughter molecule has one parental strand (conserved) and one newly synthesized one.

The base sequence of each newly synthesized DNA chain is complementary to the chain already there.

Replication is a very complex process involving a number of enzymes and binding proteins. A growing body of evidence indicates that these enzymes assemble in "factories" through which the DNA moves. Such factories may be bound to membranes in bacteria. In higher organisms, the replication factories are not permanent structures. Instead, they may be disassembled and their parts reassembled in ever-larger factories. These assemblies of enzyme "factories" go by the name of **replisomes,** and they contain key enzymes such as polymerases, helicases, and primases (Table 24.3). The primases can shuttle in and out of the replisomes. Other proteins, such as clamp loaders and clamp proteins, through which the newly synthesized primer is threaded, are also parts of the replisomes.

Table 24.3 Components of Replisomes and Their Functions

Component	Function
Helicase	Unwinds the DNA double helix
Primase	Synthesizes short oligonucleotides (primers)
Clamp protein	Allows the leading strand to be threaded through
DNA polymerase	Joins the assembled nucleotides
Ligase	Joins Okazaki fragments in the lagging strand

CHEMICAL CONNECTIONS 24C

DNA Fingerprinting

The base sequence in the nucleus of every one of our billions of cells is identical. However, except for people who have an identical twin, the base sequence in the total DNA of one person is different from that of every other person. This uniqueness makes it possible to identify suspects in criminal cases from a bit of skin or a trace of blood left at the scene of the crime and to prove the identity of a child's father in paternity cases.

To do so, the nuclei of the cells of the criminal evidence are extracted. Their DNA is amplified by PCR techniques (Section 24.8). With the aid of restriction enzymes, the DNA molecules are cut at specific points. The resulting DNA fragments are put on a gel and subjected to **electrophoresis.** In this process, the DNA fragments move with different velocities; the smaller fragments move faster and the larger fragments move slower. After a sufficient amount of time, the fragments separate. When they are made visible in the form of an autoradiogram, one can discern bands in a lane. This is called a **DNA fingerprint.**

When the DNA fingerprint made from a sample taken from a suspect matches that from a sample obtained at the scene of the crime, the police have a positive identification. The accompanying figure shows DNA fingerprints derived by using one particular restriction enzyme. Here, a total of nine lanes can be seen. Three (numbers 1, 5, and 9) are control lanes. They contain the DNA fingerprint of a virus, using one particular restriction enzyme.

Three other lanes (2, 3, and 4) were used in a paternity suit: They contain the DNA fingerprints of the mother, the child, and the alleged father. The child's DNA fingerprint (lane 3) contains six bands. The mother's DNA fingerprint (lane 4) has five bands, all of which match those of the child. The alleged father's DNA fingerprint (lane 2) also contains six bands, of which three match those of the child. This is a positive identification. In such cases, one cannot expect a perfect match even if the man is really the father because the child has inherited only half of its genes from the father. Thus, of six bands, only three are expected to match.

■ **Figure 24C** DNA fingerprint.

In the case just described, the paternity suit was won on the basis of the DNA fingerprint matching.

In the left area of the radiogram are three more lanes (6, 7, and 8). These DNA fingerprints were used in an attempt to identify a rapist. In lanes 7 and 8 are the DNA fingerprints of semen obtained from the rape victim. In lane 6 is the DNA fingerprint of the suspect. The DNA fingerprints of the semen do not match those of the suspect. This is a negative identification and excluded the suspect from the case. When positive identification occurs, the probability that a positive match is due to chance is 1 in 100 billion.

DNA fingerprints are now routinely accepted in court cases. Many convictions are based on such evidence and, just as importantly, many jailed suspects have been released when DNA fingerprinting proved them innocent.

The replication of DNA occurs in a number of distinct steps. A few of the salient features are enumerated below:

1. **Opening up the superstructure.** During replication, the very condensed superstructure of chromosomes must be opened up to be accessible to enzymes and other proteins. A complicated signal transduction mechanism accomplishes this feat. One notable step of the signal transduction is the acetylation and deacetylation of key lysine residues of histones. When histone acetylase, an enzyme, puts acetyl groups on key lysyl residues, some positive charges are eliminated and the strength of the DNA–histone interaction is weakened:

$$\text{Histone}-(CH_2)_4-NH_3^+ + CH_3-COO^- \underset{\text{deacetylation}}{\overset{\text{acetylation}}{\rightleftharpoons}} \text{Histone}-(CH_2)_4-NH-\overset{\overset{\displaystyle O}{\|}}{C}-CH_3$$

Chromatin The DNA complexed with histone and nonhistone proteins that exists in eukaryotic cells between cell divisions

This process allows the opening up of key regions on the DNA molecule. When another enzyme, histone deacetylase, removes these acetyl groups, the positive charges are reestablished. That, in turn, facilitates regaining the highly condensed structure of **chromatin.**

2. **Relaxation of higher structures of DNA.** Topoisomerases (also called gyrases) are enzymes that facilitate the relaxation of supercoiling in DNA. They do so during replication by temporarily introducing either single or double strand breaks in the DNA. The transient break forms a phosphodiester linkage between a tyrosyl residue of the enzyme and either the 5' or 3' end of a phosphate on the DNA. Once the supercoiling is relaxed, the broken strands are joined together, and the topoisomerase diffuses from the location of the replicating fork. Topoisomerases are also involved in the untangling of the replicated chromosomes, before cell division can occur.

3. **Unwinding the DNA double helix.** The replication of DNA molecules starts with the unwinding of the double helix, which can occur at either end or in the middle. Special unwinding protein molecules, called **helicases,** attach themselves to one DNA strand (Figure 24.11) and cause the separation of the double helix. Helicases of eukaryotes are made of six different protein subunits. The subunits form a ring with a hollow core, where the single-stranded DNA sits. The helicases hydrolyze ATP as the DNA strand moves through. The energy of the hydrolysis (Section 24.2C) promotes this movement.

4. **Primer/Primases.** Primers are short—4 to 15 nucleotides long—RNA oligonucleotides synthesized from ribonucleoside triphosphates. They are needed to start the synthesis of both daughter strands. The enzyme catalyzing this synthesis is called primase. Primases form complexes with DNA polymerase in eukaryotes. Primers are placed about every 50 nucleotides in the lagging-strand synthesis.

5. **DNA Polymerase.** The key enzymes in replication are the DNA polymerases. Once the two strands are separated at the replication fork, the DNA nucleotides must be lined up. All four kinds of free DNA nucleotide molecules are present in the vicinity of the replication fork. These nucleotides constantly move into the area and try to fit themselves into new chains. The key to the process is that, as we saw in Section 24.3, *only thymine can fit opposite adenine, and only cytosine can fit opposite guanine.* Wherever a cytosine, for example, is present on one of the strands of an unwound portion of the helix, all four nucleotides may approach, but three of them will be turned away because they do not fit. Only the nucleotide of guanine fits.

In the absence of an enzyme, this alignment is extremely slow. The speed and specificity are provided by DNA polymerase. The active site of this enzyme is quite snug. It surrounds the end of the DNA template–primer complex, creating a specifically shaped pocket for the incoming nucleotide. With such a close contact, the activation energy is lowered and the polymerase enables complementary base pairing with high specificity at a rate of 100 times per second. While the bases of the newly arrived nucleotides are being hydrogen-bonded to their partners, polymerases join the nucleotide backbones.

Along the lagging strand 3' ⟶ 5', the enzymes can synthesize only short fragments because the only way they can work is from 5' to 3'. These short fragments consist of about 200 nucleotides each, named **Okazaki fragments** after their discoverer.

Okazaki fragment A short DNA segment made of about 200 nucleotides in higher organisms (eukaryotes) and 2000 nucleotides in prokaryotes

6. **Ligation.** The Okazaki fragments and any nicks remaining are eventually joined together by another enzyme, DNA ligase. At the end of the process, there are two double-stranded DNA molecules, each exactly the same as the original molecule because only T fits opposite A and only G fits against C in the active site of the polymerase.

Tuneko and Reiji Okazaki provided biochemical verification for the semi-continuous replication of DNA in 1968.

24.7 DNA Repair

The viability of cells depends on DNA repair enzymes that can detect, recognize, and remove mutations from DNA. Such mutations may arise from external or internal sources. Externally, UV radiation or highly reactive oxidizing agents, such as superoxide, O_2^-, may damage a base. Errors in copying or internal chemical reactions—for example, deamidation of a base—can create damage internally. Deamidation of the base cytosine turns it into uracil (Figure 24.1), which creates a mismatch. The former C-G pair becomes a U-G mispair that must be removed.

The same deamidation reaction that can initiate BER plays an important role in providing the enormous variability of immunoglobulins (see Section 30.4).

CHEMICAL CONNECTIONS 24D

Pharmacogenomics: Tailoring Medication to an Individual's Predisposition

A complete DNA sequence of an organism is called a **genome**. The genome of an average human contains approximately 3 billion base pairs, distributed among 22 pairs of chromosomes plus 2 sex chromosomes. Each chromosome consists of a single DNA molecule. Among the 3 billion base pairs are some 90 million that represent 30,000 human genes. It was the task of the Human Genome Project to determine the complete sequence of the genome and, in the process, to identify the sequence and location of the genes. The Human Genome Project was completed in the year 2000. Many other genomes have also been established, ranging from the lowly bacteria *Escherischia coli* (3 million base pairs) to the mouse (3 billion base pairs).

It has always been known that an individual's genetic inheritance plays a role in disease and drug effectiveness. As early as 510 B.C., Phytagoras wrote that some individuals developed hemolytic anemia when eating fava beans, while others thrived on it. Adverse drug reactions are the sixth leading cause of death in the United States (more than 100,000 deaths per year). Problems caused by ineffective drugs are even more numerous. With the new knowledge of an individual's genetic makeup it is now within our power to prescribe drugs in the dosage that best suits the individual and that minimizes adverse reactions or ineffectiveness. **Pharmacogenomics** is the study of how genetic variation influences individual responses to a given drug or class of drugs.

A case in point is CYP2D6, the gene of a member of the cytochrome P-450 enzyme group. This enzyme detoxifies drugs by adding an —OH group, making them more water-soluble and thus able to be excreted in the urine. The normal or wild type of this enzyme is correlated with *extensive metabolism* (EM) of drugs. A mutation, in which a guanine (G) becomes adenine (A) in the CYP2D6 gene, exists approximately in 25% of the population. Its presence makes the individual a *poor metabolizer* (PM). Thus a prescribed drug dosage will stay much longer than normal in the body of the individual having this genetic makeup, possibly leding to toxic effects. Another mutation, in which an adenine (A) base was deleted on the CYP2D6 gene, exists in approximately 3% of the population. It is associated with very fast, *ultra-extensive metabolism* (UEM). In these individuals the drug may be cleared from the body before it can be effective.

These three classes (EM, PM, and UEM) have been known for some time and could be monitored by frequent blood sample analyses during a six-week course of therapy with a drug. With the advances of the Human Genome Project, an individual can now be tested *before* and not during the administration of a drug therapy. This strategy means that the dosage can be adjusted to the individual's particular needs. Drug companies have developed DNA chips that can read a patient's predisposition to a drug using the DNA of the individual taken in one simple blood test. The genetic predisposition is merely one factor among many determining the body's total response to a drug, of course. Nevertheless, eliminating errors based on known genetic makeup can only minimize adverse effects or ineffectiveness.

614 / CHAPTER 24 NUCLEOTIDES, NUCLEIC ACIDS, AND HEREDITY

Figure 24.12 Base excision repair (BER) pathway.

The repair can be effected in a number of ways. One of the most common is called BER, *b*ase *e*xcision *r*epair (Figure 24.12). This pathway contains two steps:

1. A specific DNA glycosylase recognizes the damaged base. It hydrolyzes the N—C′ β-glycosidic bond between the uracil base and the deoxyribose, then flips the damaged base, completing the excision. The sugar–phosphate backbone is still intact. At the **AP site** (*ap*urinic or *ap*yrimidinic site) created in this way, the backbone is cleaved by a second enzyme, endonuclease. A third enzyme, exonuclease, liberates the sugar–phosphate unit of the damaged site.

2. In the synthesis step, the enzyme DNA polymerase inserts the correct nucleotide, cytidine, and the enzyme DNA ligase seals the backbone to complete the repair.

CHEMICAL CONNECTIONS 24E

Apoptosis: Programmed Cell Death

In necrosis, cells of an organism die in large clusters when exposed to toxins or deprived of oxygen. Such a typical necrosis occurs in heart attack and stroke. *Apoptosis*, or programmed cell death, is clearly distinguishable from necrosis. Only scattered cells die at a time, and no inflammation or scar results. In apoptosis, the cells cleave into apoptopic bodies containing intact organelles and large nuclear fragments. These bodies are then swallowed up by roving scavenger cells.

At the heart of this self-immolation process is a family of protein-cleaving enzymes, called **caspases**. The name is an acronym for "cysteine-containing aspartate-specific proteases," which describes the mode of action: Cysteine is in the active site of the enzyme, and the target protein is cleaved by hydrolyzing the amide bond formed by the α-carboxyl group of the amino acid **aspartate**. Caspases are zymogens, and they are activated by self-clipping. Once active, they attack targeted proteins. For example, they cleave gelsolin, a protein that normally binds to actin filaments and that maintains the shape of the cell. As a result of actin degradation, the apoptotic cells acquire one of their signature features: They lose their normal shape, become more rounded, and develop blister-like bumps on their surfaces.

Caspases also attack another protein, a companion of the enzyme, **endonuclease**. When this enzyme is paired with its companion, it is unable to enter the cell nucleus. After it is freed from the companion, however, endonuclease can enter the nucleus and cleave DNA. The result is the second distinguishing fingerprint of death by apoptosis: the appearance of DNA fragments in multiples of 180 base pairs (180, 360, or 540 base pairs). These appear in the membrane-enclosed apoptotic bodies, the remnants of the dying cells.

(*a*)

(*b*)

■ **Figure 24E** Signs of apoptosis. (*a*) Blisters. (*b*) Apoptotic bodies.

A second repair mechanism removes not a single mismatched base but rather whole nucleotides—as many as 24 to 32 residue oligonucleotides. Known as NER, for *n*ucleotide *e*xcision *r*epair, it similarly involves a number of repair enzymes.

Any defect in the repair mechanisms may lead to harmful or even lethal mutations. For example, individuals with inherited XP (xeroderma pigmentosa), a condition in which one or another enzyme in the NER repair pathway is missing or defective, have 1000 times greater risk of skin cancer than normal individuals.

24.8 Cloning

The term **clone** refers to a genetically identical population, be that a sheep, a cell, a virus, or a DNA molecule. Molecular **cloning** is the exact copying of segments of DNA. Millions of copies of selected DNA fragments can be made within a few hours with high precision by a technique called **polymerase chain reaction (PCR),** discovered by Kary B. Mullis (1945–), who shared the 1993 Nobel Prize in chemistry for this achievement.

PCR techniques can be used only if the sequence of a gene to be copied is known. In such a case, one can synthesize two primers that are complementary to the 3' ends of the gene. The primers are polynucleotides consisting of 12 to 16 nucleotides. When added to a target DNA segment, they hybridize with the 3' end of each strand of the gene.

$$5'\text{CATAGGACAGC—OH} \qquad \text{Primer}$$
$$3'\text{TACGTATCCTGTCGTAGG—} \qquad \text{Gene}$$

Figure 24.13 Schematic diagram of PCR. *(Adapted from* The Unusual Origin of the Polymerase Chain Reaction, *by Kary B. Mullis, illustrated by Michael Goodman. Scientific American,* April 1990. Reprinted with the permission of Michael Goodman.)

In cycle 1 (Figure 24.13), the polymerase extends the primers in each direction as individual nucleotides are assembled and connected on the template DNA. In this way, two new copies are created. The two-step process is repeated (cycle 2) when the primers are **hybridized** with the new strands, and the primers are extended again. At that point, four new copies have been created. The process continues, and in 25 cycles, 2^{25} or some 33 million copies can be made. In practice, only a few million are produced, which is sufficient for the isolation of a gene.

This fast process is practical because of the discovery of heat-resistant polymerases isolated from bacteria that live in hot thermal vents on the sea floor (Section 22.4). A temperature of 95°C is needed because the double helix must be unwound to hybridize the primer to the target DNA. Once single strands of DNA have been exposed, the mixture is cooled to 70°C. The primers are hybridized and subsequent extensions take place. The 95°C and 70°C cycles are repeated over and over. No new enzyme is required because the polymerase is stable at both temperatures.

PCR techniques are routinely used when a gene or a segment of DNA must be amplified from a few molecules. It is used in genome study (Chemical Connections 24D), in obtaining evidence from a crime scene (Chemical Connections 24C), and even in obtaining the genes of long-extinct species found fossilized in amber.

> **Hybridization** The process in which two strands of nucleic acids or segments thereof form a double-stranded structure through hydrogen-bonding of complementary base pairs

The temperature at which the hydrogen bonds of double-stranded DNA break and the two strands separate is referred to as the *annealing* temperature.

SUMMARY

Nucleic acids are composed of sugars, phosphates, and organic bases (Section 24.2). Two kinds exist: **ribonucleic acid (RNA)** and **deoxyribonucleic acid (DNA)**. In DNA, the sugar is the monosaccharide 2-deoxy-D-ribose; in RNA, it is D-ribose. In DNA, the heterocyclic amine bases are adenine (A), guanine (G), cytosine (C), and thymine (T). In RNA, they are A, G, C, and uracil (U). Nucleic acids are giant molecules with backbones made of alternating units of sugar and phosphate. The bases are side chains joined by β-N-glycosidic bonds to the sugar units (Section 24.2).

DNA is made of two strands that form a double helix (Section 24.3). The sugar–phosphate backbone runs on the outside of the double helix, and the hydrophobic bases point inward. **Complementary pairing** of the bases occurs in the double helix, such that each A on one strand is hydrogen-bonded to a T on the other, and each G is hydrogen-bonded to a C. No other pairs fit. DNA is coiled around basic protein molecules called **histones**. Together they form **nucleosomes**, which are further condensed into the chromatins of chromosome.

The DNA molecule carries, in the sequence of its bases, all the information necessary to maintain life. When cell division occurs and this information is passed from parent cell to daughter cells, the sequence of the parent DNA is copied.

A **gene** is a segment of a DNA molecule that carries the sequence of bases that directs the synthesis of one particular protein (Section 24.5). There are four kinds of RNA (Section 24.4): **messenger RNA (mRNA), transfer RNA (tRNA), ribosomal RNA (rRNA),** and **ribozyme (catalytic RNA)**. DNA in higher organisms contains sequences, called **introns**, that do not code for proteins (Section 24.5). The sequences that do code for proteins are called **exons.**

DNA replication occurs in several distinct steps (Section 24.6). First, the superstructures of chromosomes are loosened by acetylation of histones. **Topoisomerases** relax the higher structures. **Helicases** at the replication fork separate the two strands of DNA. RNA primers and primerase are needed to start the synthesis of daughter strands. The **leading strand** is synthesized continuously by **DNA polymerase**. The **lagging strand** is synthesized discontinuously as **Okazaki fragments. DNA ligase** seals the nicks and the Okazaki fragments.

An important **DNA repair** mechanism is BER, or single base excision repair (Section 24.7).

Cloning is copying genes or segments of DNA (Section 24.8). The **polymerase chain reaction (PCR)** technique can make millions of copies with high precision in a few hours.

PROBLEMS

Numbers that appear in color indicate difficult problems. ▲ designates problems requiring application of principles.

Nucleic Acids and Heredity

24.2 What structures of the cell, visible in a microscope, contain hereditary information?

24.3 Name one hereditary disease.

Components of Nucleic Acids

24.4 (a) Where in a cell is the DNA located?
(b) Where in a cell is the RNA located?

24.5 What are the components of (a) a nucleotide and (b) a nucleoside?

24.6 Draw the structures of ADP and GDP. Are these structures parts of nucleic acids?

24.7 What is the difference in structure between thymine and uracil?

24.8 Which DNA and RNA bases contain a carbonyl group?

24.9 Draw the structures of (a) cytidine and (b) deoxycytidine.

24.10 Which DNA and RNA bases are primary amines?

24.11 What is the difference in structure between D-ribose and 2-deoxy-D-ribose?

24.12 What is the difference between a nucleoside and a nucleotide?

Structure of DNA and RNA

24.13 In RNA, which carbons of the ribose are linked to the phosphate group and which are linked to the base?

24.14 What constitutes the backbone of DNA?

24.15 Draw the structures of (a) UDP and (b) dAMP.

24.16 In DNA, which carbon atoms of 2-deoxy-D-ribose are bonded to the phosphate groups?

24.17 The sequence of a short DNA segment is ATGGCAATAC.
(a) What name do we give to the two ends (terminals) of a DNA molecule?
(b) In this segment, which end is which?

24.18 Chargaff showed that, in samples of DNA taken from many different species, the molar quantity of A was always approximately equal to the molar quantity of T; the same is true for C and G. How did this information help to establish the structure of DNA?

24.19 How many hydrogen bonds can be formed between uracil and adenine?

24.20 How many histones are in a nucleosome?

24.21 What is the nature of the interaction between histones and DNA in nucleosomes?

24.22 What are chromatin fibers made of?

24.23 What constitutes the superstructure of chromosomes?

RNA

24.24 Which type of RNA has enzyme activity? Where does it function mostly?

24.25 Which has the longest chains: tRNA, mRNA, or rRNA?

24.26 Which type of RNA contains modified nucleotides?

24.27 Which type of RNA has a sequence exactly complementary to that of DNA?

24.28 Where is rRNA located in the cell?

24.29 What kind of functions do ribozymes, in general, perform?

Genes, Exons, and Introns

24.30 Define:
(a) Intron (b) Exon

24.31 Does mRNA also have introns and exons? Explain.

24.32 (a) What percentage of human DNA codes for proteins?
(b) What is the function of the rest of the DNA?

24.33 Do satellites code for a particular protein?

DNA Replication

24.34 A DNA molecule normally replicates itself millions of times, with almost no errors. What single fact about the structure is most responsible for this fidelity of replication?

24.35 What functional groups on the bases form hydrogen bonds in the DNA double helix?

24.36 Draw the structures of adenine and thymine, and show with a diagram the two hydrogen bonds that stabilize A-T pairing in DNA.

24.37 Draw the structures of cytosine and guanine, and show with a diagram the three hydrogen bonds that stabilize C-G pairing in nucleic acids.

24.38 How many different bases are in a DNA double helix?

24.39 What is a replication fork? How many replication forks may exist simultaneously on an average human chromosome?

24.40 Why is the replication called semiconservative?

24.41 How does the removal of some positive charges from histones enable the opening of the chromosomal superstructure?

24.42 Write the chemical reaction for the deacetylation of acetyl-histone.

24.43 What is the quaternary structure of helicases in eukaryotes?

24.44 What are helicases? What is their function?

24.45 Can dATP serve as a source for a primer?

24.46 What are the side products of the action of primase in forming primers?

24.47 What do we call the enzymes that join nucleotides into a DNA strand?

24.48 In which direction is the DNA molecule synthesized continuously?

24.49 What kind of bond formation do polymerases catalyze?

24.50 Which enzyme catalyzes the joining of Okazaki fragments?

DNA Repair

24.51 As a result of damage, a few of the G residues in a gene are methylated. What kind of mechanism could the cell use to repair the damage?

24.52 What is the function of endonuclease in the BER repair mechanism?

24.53 When cytosine is deaminated, uracil is formed. Uracil is a naturally occurring base. Why would the cell use base excision repair to remove it?

24.54 What bonds are cleaved by glycosylase?

24.55 What are AP sites? What enzyme creates them?

24.56 Why are xeroderma pigmentosa patients 1000 times more prone to develop skin cancer than normal individuals are?

Cloning

24.57 What is the advantage of using polymerase from thermophile bacteria that live in hot thermal vents in cloning?

24.58 What 12-nucleotide primer would you use in the PCR technique when you want to clone a gene whose 3′ end is as follows: 3′TACCGTCATCCGGTG—?

Chemical Connections

24.59 (Chemical Connections 24A) Draw the structure of the fluorouridine nucleoside that inhibits DNA synthesis.

24.60 (Chemical Connections 24A) Give an example of how anticancer drugs work in chemotherapy.

24.61 (Chemical Connections 24B) What sequence of nucleotides is repeated many times in telomers?

24.62 (Chemical Connections 24B) Why are up to 200 nucleotides lost at each replication?

24.63 (Chemical Connections 24B) How does telomerase make a cancer cell immortal?

24.64 (Chemical Connections 24C) After having been cut by restriction enzymes, how are DNA fragments separated from each other?

24.65 (Chemical Connections 24C) How is DNA fingerprinting used in paternity suits?

24.66 (Chemical Connections 24D) What is the function of cytochrome P-450?

24.67 (Chemical Connections 24D) How does knowledge of the human genome enable patients to be screened for individual drug tolerance?

24.68 (Chemical Connections 24E) How do caspase enzymes, which operate in the cytoplasm of the cell, cause the cleavage of DNA in the nucleus?

24.69 (Chemical Connections 24E) What are the distinguishing fingerprints of cell death by apoptosis?

Additional Problems

24.70 What is the active site of a ribozyme?

24.71 Why is it important that a DNA molecule be able to replicate itself millions of times without error?

24.72 Draw the structures of (a) uracil and (b) uridine.

24.73 How would you classify the functional groups that bond together the three different components of a nucleotide?

24.74 Why do we call DNA or RNA *nucleic acids*?

24.75 Which nucleic acid molecule is the largest?

24.76 What kind of bonds are broken during replication? Does the primary structure of DNA change during replication?

24.77 In sheep DNA, the mol % of adenine (A) was found to be 29.3. Based on Chargaff's rule, what would be the approximate mol % of G , C, and T?

InfoTrac College Edition

For additional reading, go to InfoTrac College Edition, your online research library, at

http://infotrac.thomsonlearning.com

CHAPTER 25

25.1 Introduction
25.2 Transcription
25.3 The Role of RNA in Translation
25.4 The Genetic Code
25.5 Translation and Protein Synthesis
25.6 Gene Regulation
25.7 Mutations, Mutagens, and Genetic Diseases
25.8 Recombinant DNA

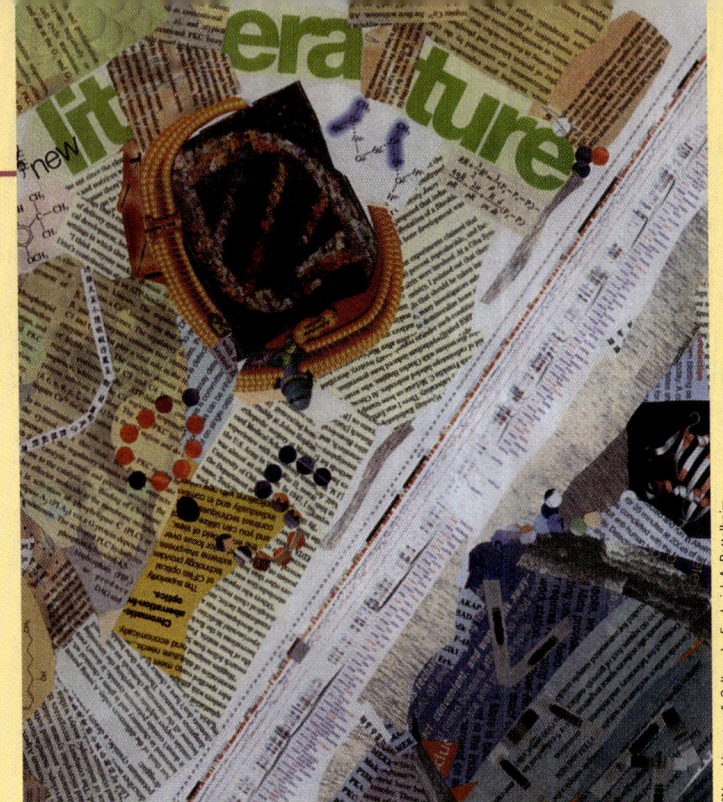

The diagonal that divides the panel is the decoded sequence of chromosome X; hence it represents half of a woman. The upper-left panel is assembled from the molecular fragments of life. Its dominant motif is genomics, written in the language of nucleotides. The lower-right panel depicts cellular components. Its dominant motif is proteomics, written in the language of amino acids.

Gene Expression and Protein Synthesis

25.1 Introduction

We have seen that the DNA molecule is a storehouse of information. We can compare it to a loose-leaf cookbook, each page of which contains one recipe. The pages are the genes. To prepare a meal, we use a number of recipes. Similarly, to provide a certain inheritable trait, a number of genes—segments of DNA—are needed.

Of course, the recipe itself is not the meal. The information in the recipe must be expressed in the proper combination of food ingredients. Similarly, the information stored in DNA must be expressed in the proper combination of amino acids representing a particular protein. The way this expression works is now so well established that it is called the **central dogma of molecular biology.** The dogma states that *the information contained in DNA molecules is transferred to RNA molecules, and then from the RNA molecules the information is expressed in the structure of proteins.* **Gene expression** is the turning on or activation of a gene. Transmission of information occurs in two steps: transcription and translation.

Although this statement is correct in the vast majority of cases, in certain viruses the flow of information goes from RNA to DNA.

Gene expression The activation of a gene to produce a specific protein; it involves both transcription and translation

Transcription Because the information (that is, the DNA) is in the nucleus of the cell and the amino acids are assembled outside the nucleus, the information must first be carried out of the nucleus. This step is analogous to copying a recipe from a cookbook. All the necessary information is copied, albeit in a slightly different format, as if we were converting the printed page into handwriting. On the molecular level, this task is accomplished by transcribing the information from the DNA molecule onto a molecule of messenger RNA (Section 24.4), so named because it carries the message from the nucleus to the site of protein synthesis. Other RNAs are similarly transcribed. rRNA is needed to form ribosomes, and tRNA is required to carry out the translation into protein language. The transcribed information on the different RNA molecules is then carried out of the nucleus.

Transcription The process in which information encoded in a DNA molecule is copied into an mRNA molecule

Translation The mRNA serves as a template on which the amino acids are assembled in the proper sequence. To complete the assembly, the information that is written in the language of nucleotides must be translated into the language of amino acids. The translation is done by the second type of RNA, transfer RNA (Section 24.4). An exact, word-to-word translation occurs. Each amino acid in the protein language has a corresponding word in the RNA language. Each word in the RNA language is a sequence of three bases. This correspondence between three bases and one amino acid is called the genetic code (we will discuss the code in Section 25.4).

Translation The process in which information encoded in an mRNA molecule is used to assemble a specific protein

A summary of the process follows:

$$\text{DNA} \xrightarrow{\text{replication}} \text{DNA} \xrightarrow{\text{transcription}} \text{mRNA} \xrightarrow{\text{translation}} \text{protein}$$

In higher organisms (eukaryotes), transcription and translation occur sequentially. The transcription takes place in the nucleus. After RNA leaves the nucleus and enters the cytoplasm, the translation takes place there. In lower organisms (prokaryotes), there is no nucleus and thus transcription and translation occur simultaneously in the cytoplasm. This extended form of the "central dogma" was challenged in 2001, when it was found that even in eukaryotes about 15% of the proteins are produced in the nucleus itself. Thus some simultaneous transcription and translation do occur even in higher organisms.

We know more about bacterial transcription and translation because they are simpler than the processes operating in higher organisms. Nevertheless, we will concentrate on studying gene expression and protein synthesis in the eukaryotic systems because they are more relevant to human health care.

25.2 Transcription

Transcription starts when the DNA double helix begins to unwind at a point near the gene that is to be transcribed (Figure 25.1). As we saw in Section 24.3C, nucleosomes form chromatin and higher condensed structures in the chromosomes. To make the DNA available for transcription, these superstructures change constantly. There are specific **binding proteins** that bind to the nucleosomes, making the DNA become less dense and more accessible. Only then can the enzyme **helicase,** a ring-shaped complex of six protein subunits, unwind the double helix.

Screen 24.2 Animation

622 / CHAPTER 25 GENE EXPRESSION AND PROTEIN SYNTHESIS

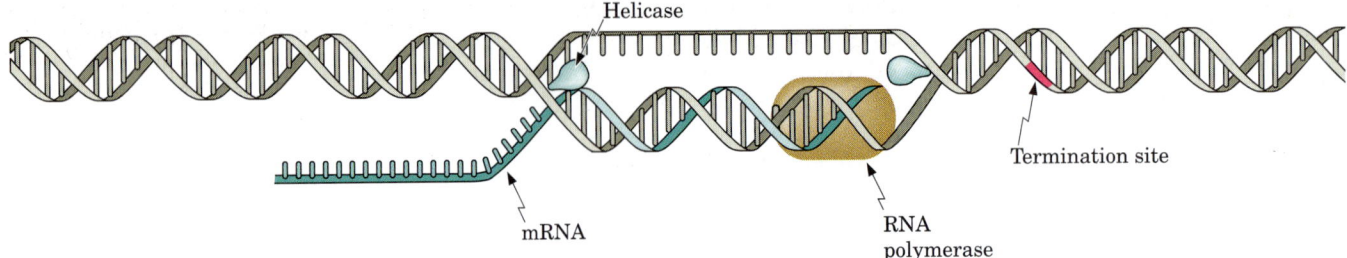

Figure 25.1 Transcription of a gene. The information in one DNA strand is transcribed to a strand of RNA. The termination site is the locus of termination of transcription.

Only one strand of the DNA molecule is transcribed. Ribonucleotides assemble along the unwound DNA strand in the complementary sequence. Opposite each C on the DNA is a G on the growing mRNA; the other complementary bases follow the patterns G ⟶ C, A ⟶ U, and T ⟶ A. The ribonucleotides, thus aligned, are linked to form the appropriate RNA.

Three kinds of **polymerases** catalyze the transcription. RNA polymerase I (poly I) catalyzes the formation of rRNA, poly II catalyzes mRNA formation, and poly III catalyzes tRNA formation. Each enzyme is a complex of 10 or more subunits. Some subunits are unique to each kind of polymerase, whereas other subunits appear in all three polymerases.

The eukaryotic gene has two major parts: the **structural gene** that is transcribed into RNA and the **regulatory gene** that controls the transcription. The structural gene is made of exons and introns (Figure 25.2). The regulatory gene is not transcribed, but rather has control elements. One such control element is the **promoter.** On the DNA strand, there is always a sequence of bases that the polymerase recognizes as an **initiation signal,** saying in essence, "Start here." A promoter is unique to each gene. Besides unique nucleotide sequences, promoters contain **consensus sequences,** such as the TATA box. In TATA boxes the two nucleotides, T and A, are repeated many times. A TATA box lies approximately 25 base pairs upstream—

Note again that RNA contains no thymine but has uracil instead.

Figure 25.2 Organization and transcription of a split eukaryote gene.

CHEMICAL CONNECTIONS 25A

"Antisense" Makes Sense

Many diseases are caused by faulty or excessive production of certain proteins. As we have seen, the message to make proteins is in the mRNA. Antisense drugs are chemically modified strands of short stretches of DNA that are exactly opposite—hence "anti"—the coding, or "sense" of a mRNA. Thus, if a stretch of mRNA is AACGUU, its antisense will be TTGCAA. When such an antisense drug is delivered to a cell, it strongly binds to the mRNA and prevents the translation of the unwanted protein.

The trick is to prevent the degradation of the antisense drug by the cell's enzymes before or after it attaches to the mRNA. One way to protect the antisense drug is to modify the backbone by substituting a sulfur atom for an oxygen atom in the phosphate group of the backbone, thereby forming phosphorothioate moieties.

There is only one antisense drug on the market today. Vitravene is used in treatment of cytomegalovirus-induced retinitis in AIDS patients (see Chemical Connections 25B). Vitravene is a stretch of 21 nucleotides. The structure of the first four of these nucleotides is shown in Figure 25A, spelling out GCGT. Currently, many other antisense oligonucleotide drugs are in clinical trials as potential treatments for various cancers, such as lymphoma.

■ **Figure 25A** First four of the 21 segments of oligonucleotide drug Vitravene.

that is, before the start of the transcription process (see Figure 25.2). TATA boxes are common to all eukaryotes. All three RNA polymerases interact with their promoter regions via **transcription factors** that are binding proteins.

After initiation, the RNA polymerase zips up the complementary bases by forming a phosphate ester bond (Section 18.7) between each ribose and the next phosphate group. This process is called **elongation.**

At the end of the gene is a **termination sequence** that tells the enzyme, "Stop the transcription."

The enzyme poly II has two different forms. At the C-terminal domain, poly II has serine and threonine repeats that can be phosphorylated. When poly II starts the initiation, the enzyme is in its unphosphorylated form. Upon phosphorylation, it performs the elongation process. After termination of the transcription, poly II is dephosphorylated by a phosphatase. In this manner, poly II is constantly recycled between its initiation and elongation roles.

The enzyme synthesizes the mRNA molecule from the 5' to the 3' end (the zipper can move in only one direction). Because the complementary chains (RNA and DNA) run in opposite directions, however, the enzyme must move along the DNA template in the 3' ⟶ 5' direction of the DNA

624 | CHAPTER 25 GENE EXPRESSION AND PROTEIN SYNTHESIS

After RNA molecules are synthesized, they move out of the nucleus and into the cytoplasm.

Guanine methylated at the 7′ position

(Figure 25.1). As the mRNA is synthesized, it moves away from the DNA template, which then rewinds to the original double-helix form. Transfer RNA and ribosomal RNA are also synthesized on DNA templates in this manner.

The RNA products of transcription are not necessarily the functional RNAs. Previously, we have seen that in higher organisms mRNA contains exons and introns (Section 24.5). To ensure that mRNA is functional, the transcribed product is capped at both ends. The 5′ end acquires a methylated guanine (7-mG cap) and a polyA tail that may contain 100 to 200 adenine residue caps at the 3′ end. Once the two ends are capped, the introns are spliced out. This is a **post-transcription process** (Figure 25.2). Similarly, a transcribed tRNA must be trimmed and capped, and some of its nucleotides methylated before becoming functional tRNA. Functional rRNA also undergoes post-transcriptional methylation.

EXAMPLE 25.1

Polymerase II both initiates the transcription and performs the elongation. What are the two forms of the enzyme in these processes? What chemical bond formation occurs in the conversion of these two forms one into the other?

Solution
The phosphorylated form of poly II performs the elongation and the unphosphorylated form initiates the transcription. The chemical bond formed in the phosphorylation is the phosphoric ester between the —OH of serine and threonine residues of the enzyme and phosphoric acid.

Problem 25.1
DNA is highly condensed in the chromosomes. What is the sequence of events that enables the transcription of a gene to begin?

25.3 The Role of RNA in Translation

Translation is the process by which the genetic information preserved in the DNA and transcribed into the mRNA is converted to the language of proteins—that is, the amino acid sequence. Three of the four types of RNA (mRNA, rRNA, and tRNA) participate in the process.

The synthesis of proteins takes place on the ribosomes (Section 24.4). These spheres dissociate into two parts—a larger and a smaller body. Each of these bodies contains rRNA and some polypeptide chains that act as enzymes, speeding up the synthesis. In higher organisms, including humans, the larger ribosomal body is called the 60S ribosome, and the smaller one is called the 40S ribosome. The mature mRNA is transported into the cytoplasm, where its 5′ end is attached to the smaller ribosomal body and later joined by the larger body. Together they form a unit on which the mRNA is stretched out. Triplets of bases on the mRNA are called **codons**. After the mRNA is attached to the ribosome in this way, the 20 amino acids are brought to the site, each carried by its own particular tRNA molecule.

The most important segments of the tRNA molecule are (1) the site to which enzymes attach the amino acids and (2) the recognition site. Figure 25.3 shows that the 3′ terminal of the tRNA molecule is single-stranded; this end carries the amino acid.

As we have said, each tRNA is specific for one amino acid only. How does the body make sure that alanine, for example, attaches only to the one

The S, or Svedberg unit, is a measure of the size of these bodies.

Codon The sequence of three nucleotides in messenger RNA that codes for a specific amino acid

The 20 amino acids are always available in the cytoplasm, near the site of protein synthesis.

tRNA molecule that is specific for alanine? The answer is that each cell carries at least 20 specific enzymes for this purpose. Each of these enzymes recognizes only one amino acid and only one tRNA. The enzyme attaches the activated amino acid to the 3' terminal —OH group of the tRNA, forming an ester bond. The second important segment of the tRNA molecule carries the **codon recognition site,** which is a sequence of three bases called an **anticodon** located at the opposite end of the molecule in the three-dimensional structure of tRNA (see Figure 25.3). This triplet of bases can align itself in a complementary fashion to the codon triplet on mRNA.

25.4 The Genetic Code

By 1961, it was apparent that the order of bases in a DNA molecule corresponds to the order of amino acids in a particular protein. But the code was unknown. Obviously, it could not be a one-to-one code. There are only four bases, so if A coded for glycine, G for alanine, C for valine, and T for serine, there would be 16 amino acids that could not be coded.

In 1961, Marshall Nirenberg (1927–) and his co-workers attempted to break the code in a very ingenious way. They made a synthetic molecule of mRNA consisting of uracil bases only. They put this molecule into a cell-free system that synthesized proteins and then supplied the system with all 20 amino acids. The only polypeptide produced was a chain consisting solely of the amino acid phenylalanine. This experiment showed that the code for phenylalanine must be UUU or some other multiple of U.

A series of similar experiments by Nirenberg and other workers followed, and by 1967 the entire genetic code had been broken. *Each amino acid is coded for by a sequence of three bases,* called a *codon.* Table 25.1 shows the complete code.

The first important aspect of the **genetic code** is that it is almost universal. In virtually every organism, from a bacterium to an elephant to a human, the same sequence of three bases codes for the same amino acid.

Figure 25.3 Three-dimensional structure of tRNA.

Anticodon A sequence of three nucleotides on tRNA complementary to the codon in mRNA

Genetic code The sequence of triplets of nucleotides (codons) that determines the sequence of amino acids in a protein

Table 25.1 The Genetic Code

First Position (5'-end)	Second Position								Third Position (3'-end)
	U		C		A		G		
U	UUU	Phe	UCU	Ser	UAU	Tyr	UGU	Cys	U
	UCC	Phe	UCC	Ser	UAC	Tyr	UGC	Cys	C
	UUA	Leu	UCA	Ser	UAA	Stop	UGA	Stop	A
	UUG	Leu	UCG	Ser	UAG	Stop	UGG	Trp	G
C	CUU	Leu	CCU	Pro	CAU	His	CGU	Arg	U
	CUC	Leu	CCC	Pro	CAC	His	CGC	Arg	C
	CUA	Leu	CCA	Pro	CAA	Gln	CGA	Arg	A
	CUG	Leu	CCG	Pro	CAG	Gln	CGG	Arg	G
A	AUU	Ile	ACU	Thr	AAU	Asn	AGU	Ser	U
	AUC	Ile	ACC	Thr	AAC	Asn	AGC	Ser	C
	AUA	Ile	ACA	Thr	AAA	Lys	AGA	Arg	A
	AUG*	Met	ACG	Thr	AAG	Lys	AGG	Arg	G
G	GUU	Val	GCU	Ala	GAU	Asp	GGU	Gly	U
	GUC	Val	GCC	Ala	GAC	Asp	GGC	Gly	C
	GUA	Val	GCA	Ala	GAA	Glu	GGA	Gly	A
	GUG	Val	GCG	Ala	GAG	Glu	GGG	Gly	G

*AUG also serves as the principal initiation codon.

Some exceptions to the genetic code in Table 25.1 occur in mitochondrial RNA. Because of that and other evidence, it is thought that the mitochondrion may have been an ancient entity. During evolution it developed a symbiotic relationship with eukaryotic cells. For example, some of the respiratory enzymes located on the cristae of the mitochondrion (see Section 26.2) are encoded in the mitochondrial DNA, and other members of the same respiratory chain are encoded in the nucleus of the eukaryotic cell.

Because of this multiple coding, the genetic code is called a *multiple code* or a *degenerate code*.

The universality of the genetic code implies that all living matter on earth arose from the same primordial organisms. This finding is perhaps the strongest evidence for Darwin's theory of evolution.

There are 20 amino acids in proteins, but 64 possible combinations of four bases into triplets. All 64 codons (triplets) have been deciphered. Three of them—UAA, UAG, and UGA—are "stop signs." They terminate protein synthesis. The remaining 61 codons code for amino acids. Since there are only 20 amino acids, there must be more than one codon for each amino acid. Indeed, some amino acids have as many as six codons. Leucine, for example, is coded by UUA, UUG, CUU, CUC, CUA, and CUG.

Just as there are three stop signs in the code, there is also an initiation sign. The initiation sign is AUG, which is also the codon for the amino acid methionine. This means that, in all protein synthesis, the first amino acid is always methionine. Methionine can also be put into the middle of the chain because there are two kinds of tRNA for it.

Although all protein synthesis starts with methionine, most proteins in the body do not have a methionine residue at the N-terminal of the chain. In most cases, the initial methionine is removed by an enzyme before the polypeptide chain is completed. The code on the mRNA is always read in the $5' \longrightarrow 3'$ direction, and the first amino acid to be linked to the initial methionine is the N-terminal end of the translated polypeptide chain.

EXAMPLE 25.2

Which amino acid is represented by the codon CGU? What is its anticodon?

Solution
Looking at Table 25.1, we find that CGU corresponds to arginine; the anticodon is GCA.

Problem 25.2
What are the codons for histidine? What are the anticodons?

Screen 24.3
Tutorial Exercise

All protein synthesis takes place outside the nucleus, in the cytosol.

25.5 Translation and Protein Synthesis

So far we have met the molecules that participate in protein synthesis (Section 25.3) and the dictionary of the translation, the genetic code. Now let us look at the actual mechanism by which the polypeptide chain is assembled.

There are four major stages in protein synthesis: activation, initiation, elongation, and termination. At each stage, a number of molecular entities participate in the process (Table 25.2).

Table 25.2	Molecular Components of Reactions at Four Stages of Protein Synthesis
Stage	**Molecular Components**
Activation	Amino acid, ATP, tRNA, aminoacyl-tRNA synthetase
Initiation	$tRNA_i^{Met}$, 40S ribosome, initiation factor proteins, mRNA, 60S ribosome
Elongation	40S and 60S ribosomes, tRNA, elongation factor proteins, mRNA, peptidyl transferase
Termination	Releasing factor proteins, tRNA, mRNA, 40S and 60S ribosomes

A Activation

Each amino acid is first activated by reacting with a molecule of ATP:

$$\text{Adenosine}-O-\underset{\underset{O^-}{|}}{\overset{\overset{O}{\|}}{P}}-O-\underset{\underset{O^-}{|}}{\overset{\overset{O}{\|}}{P}}-O-\underset{\underset{O^-}{|}}{\overset{\overset{O}{\|}}{P}}-O^- \;+\; {}^-O-\overset{\overset{O}{\|}}{C}-\underset{\underset{R}{|}}{CH}-NH_3{}^+ \;\longrightarrow$$

ATP　　　　　　　　　　　　　　An amino acid

$$\text{Adenosine}-O-\underset{\underset{O^-}{|}}{\overset{\overset{O}{\|}}{P}}-O-\overset{\overset{O}{\|}}{C}-\underset{\underset{R}{|}}{CH}-NH_3{}^+ \;+\; {}^-O-\underset{\underset{O^-}{|}}{\overset{\overset{O}{\|}}{P}}-O-\underset{\underset{O^-}{|}}{\overset{\overset{O}{\|}}{P}}-O^-$$

An amino acid–AMP　　　　　　　　Pyrophosphate

The activated amino acid is then bonded to its own particular tRNA molecule with the aid of an enzyme (a synthetase) that is specific for that particular amino acid and that particular tRNA molecule:

> In the activation step, energy is used up. Two high-energy phosphate bonds are broken for each amino acid added to the chain:
> ATP $\longrightarrow$ AMP + PPi and
> PPi $\longrightarrow$ 2Pi

The specific recognition of an enzyme, aminoacyl-tRNA synthetase (AARS), *of its own tRNA and amino acid is often referred to as the second genetic code.* The 20 different synthetases recognize their substrates by stretches of nucleotide sequences on the tRNA.

B Initiation

The initiation stage in eukaryotes consists of three steps:

1. **Forming the pre-initiation complex** To initiate the protein synthesis a unique tRNA is used, designated as **tRNA$_i{}^{Met}$**. This unique tRNA carries a methionine residue, but it is used solely for the initiation step. It is attached to the 40S ribosomal body and forms the pre-initiation complex (Figure 25.4a). Just as in transcription, each step in translation is aided by a number of factors; these proteins are called **initiation factors**.

2. **Migration to mRNA** Next, the pre-initiation complex binds to the mRNA near the 5′ end that has been capped by a methyl GTP. The complex then migrates to the initiation codon of mRNA, AUG. The anticodon of the tRNA, UAC, lines up against the codon (Figure 25.4b). A number of factors aid this step, too.

> In prokaryotes, the corresponding ribosomes are smaller, 30S and 50S.

Figure 25.4 The initiation of protein synthesis. (*a*) A unique tRNA molecule, methionine tRNA, attaches itself to the 40S ribosome. (*b*) The 40S ribosomal body is attached to the mRNA. It stays there during the entire translational process. (See Figure 25.5.) (*c*) The 60S ribosomal body joins the unit.

Figure 25.5 Alanine tRNA aligning on the ribosome with its complementary codon.

This enzyme, which is part of the 60S ribosome unit, is not a protein but a ribozyme.

3. Forming the full ribosomal complex The 60S ribosomal body joins the complex formed in step 2. Both 40S and 60S ribosomes carry two binding sites. The one shown on the left in Figure 25.4(c) is called the **P site,** because it is where the growing peptide chain will bind. The one next to it is called the **A (acceptor) site,** because that site accepts the incoming tRNA bringing the next amino acid.

C Elongation

1. Binding to the A site At this point, the A site is vacant, and each of the 20 tRNA molecules can come in and try to fit itself in. Only one of the 20 tRNAs carries exactly the right anticodon that corresponds to the next codon on the mRNA. (This is an alanine tRNA in Figures 25.5 and 25.6a). The binding of this tRNA to the A site takes place with the aid of proteins called **elongation factors** (Figure 25.6b).

Figure 25.6 Phases of elongation in protein synthesis. (*a*) After initiation, the tRNA of the second amino acid (alanine in this case) approaches the ribosome. (*b*) The tRNA binds to the A site. (*c*) The enzyme peptidyl transferase connects methionine to alanine, forming a peptide bond. (*d*) The peptide tRNA is moved (translocated) from the A to the P site while the ribosome moves to the right and simultaneously releases the empty tRNA. In this diagram, the alanine tRNA is shown in blue.

Figure 25.7 Ribosome in action. The lower yellow half represents the 30S ribosome; the upper blue half represents the 50S ribosome. The yellow and green twisted cones are tRNAs, and the chain of beads stand for mRNA. The elongation factors are in dark blue. *(Courtesy of Dr. J. Frank, Wadsworth Center, Albany, New York)*

2. **Forming the first peptide bond** At the A site, the new amino acid, alanine (Ala), is linked to the Met in a peptide bond by the enzyme **peptidyl transferase.** The empty tRNA remains on the P site (Figure 25.6c).

3. **Translocation** In the next phase of elongation, the whole ribosome moves one codon along the mRNA. Simultaneously with this move, the dipeptide is **translocated** from the A site to the P site, as shown in Figure 25.6(d); the empty tRNA dissociates and goes back to the tRNA pool to pick up another amino acid.

4. **Forming the second peptide bond** After the translocation, the A site is associated with the next codon on the mRNA, which is UUG in Figure 25.6(d). Once again, each tRNA can try to fit itself in, but only the one whose anticodon is AAC can align itself with UUG. This tRNA, which carries leucine (Leu), now comes in. The transferase establishes a new peptide bond between Leu and Ala, moving the dipeptide from the P site to the A site and forming a tripeptide.

These elongation steps are repeated until the last amino acid is attached.

Figure 25.7 shows a three-dimensional model of the translational process, which has been constructed on the basis of recent cryoelectron microscopy and X ray diffraction studies. This model clearly shows how the elongation factor proteins (in dark blue) are fitting in a cleft between the 50S (blue) and the 30S (pale yellow) bodies of prokaryotic ribosomes. The tRNAs on the P site (green) and on the A site (yellow) occupy a central cavity in the ribosomal complex. The orange beads represent the mRNA.

A similar arrangement exists in eukaryotes, which have 60S and 40S ribosomes.

More recently, a third site was identified on the ribosome. The site from which the tRNA exits when it is no longer needed, is called the E site.

D Termination

After the last translocation, the next codon reads "stop" (UAA, UGA, or UAG). At this point, no more amino acids can be added. Releasing factors then cleave the polypeptide chain from the last tRNA via a mechanism not yet fully understood. The tRNA itself is released from the P site. At the end, the whole mRNA is released from the ribosome. While the mRNA is attached to the ribosomes, many polypeptide chains are synthesized on it simultaneously.

Their work done, the two parts of the ribosome separate.

CHEMICAL CONNECTIONS 25B

Viruses

Nucleic acids are essential for life as we know it. No living thing can exist without them because they carry the information necessary to make protein molecules. The smallest forms of life, the viruses, consist only of a molecule of nucleic acid surrounded by a "coat" of protein molecules. In some viruses, the nucleic acid is DNA; in others, it is RNA. No virus has both. Whether viruses can be considered a true form of life became a topic of debate recently. In the summer of 2002, a group of scientists from the State University of New York, Stony Brook, reported that they had synthesized the polymyelitis virus in the laboratory from fragments of DNA. This new "synthetic" virus caused the same polio symptoms and death as the wild virus.

The shapes and sizes of viruses vary greatly, as shown in Figure 25B.1. Because their structures are so simple, viruses cannot reproduce themselves in the absence of other organisms. They carry DNA or RNA but do not have the nucleotides, enzymes, amino acids, and other molecules necessary to replicate their nucleic acid (Section 24.6) or to synthesize proteins (Section 25.5). Instead, viruses invade the cells of other organisms and cause those cells (the hosts) to do these tasks for them. Typically, the protein coat of a virus remains outside the host cell, attached to the cell wall, while the DNA or RNA is pushed inside.

Once the viral nucleic acid is inside the cell, the cell stops replicating its own DNA and making its own proteins. Instead, it replicates the viral nucleic acid and synthesizes the viral protein, according to the instructions on the viral nucleic acid. One host cell can make many copies of the virus.

In many cases, the cell bursts when a large number of new viruses have been synthesized, sending the new viruses out into the intercellular material, where they can infect other cells. This kind of process causes the host organism to get sick, and perhaps to die. Among the many human diseases caused by viruses are measles, hepatitis, mumps, influenza, the common cold, rabies, and smallpox. There is no cure for most viral diseases. Antibiotics, which can kill bacteria, have no effect on viruses. So far, the best defense against these diseases has been immunization (Chemical Connections 30B), which under the proper circumstances can work spectacularly well. Smallpox, once one of the most dreaded diseases, has been eradicated from this planet by many years of vaccination, and comprehensive programs of vaccination against such diseases as polio and measles have greatly reduced the incidence of these diseases.

Lately, a number of antiviral agents have been developed. They completely stop the reproduction of viral nucleic acids (DNA or RNA) inside infected cells without preventing the DNA of normal cells from replicating. One such drug is called vidarabine, or Ara-A, and is sold under the trade name Vira-A.

Antiviral agents often act like anticancer drugs (Chemical Connections 24A), in that they have structures similar to one of the nucleotides necessary for the synthesis of nucleic acids. Vidarabine is the same as adenosine, except that the sugar is arabinose instead of ribose (Figure 25B.2). Vidarabine is used to fight a life-threatening viral illness, herpes encephalitis. It is also effective in neonatal herpes infection and chickenpox. However, like many other anticancer and antiviral drugs, vidarabine is toxic, causing nausea and diarrhea. In some cases, it has caused chromosomal damage.

■ **Figure 25B.1** The shape of three viruses. (Illustration by Irving Geis, from Scientific American, Jan. 1963. Rights owned by Howard Hughes Medical Institute. Not to be reproduced without permission.)

■ **Figure 25B.2** Vidarabine.

CHEMICAL CONNECTIONS 25C

AIDS

In recent years a deadly virus, commonly called the AIDS virus (for acquired immune deficiency syndrome) or HIV (for human immunodeficiency virus), has spread with alarming speed around the world. Perhaps 100 million people are infected, and this number is increasing daily, especially in underdeveloped countries. The AIDS virus invades the human immune system, and especially enters the T (lymphocyte) cells and kills them (Figure 25C.1). This RNA-containing virus, or **retrovirus**, decreases the population of T cells (blood cells that fight invading foreign cells; see Section 30.2) and thus allows other opportunistic invaders, such as the protozoan *Pneumocystis carinii*, to proliferate, causing pneumonia and eventual death.

At the time of the writing, there is no cure for AIDS. Drug treatment can slow down and even seemingly arrest the development of the disease, however. There are four different types of drugs.

The first type is a **nucleoside analog** that competitively inhibits the enzyme **reverse transcriptase**, which is necessary for the multiplication of the virus. Among the most frequently used agents are azidothymidine (AZT), dideoxycytidine (ddCyd), dideoxyinosine (ddI), and (-)-2-deoxy-3'-thiacytidine (3TC). As illustrated with ddCyd in Figure 25C.2, these drugs resemble nucleosides—in this case, cytidine (Section 24.2), but the two —OH groups on the ribose have been removed. Thus they can enter the active site of the enzyme transcriptase and competitively inhibit its action.

A second type of anti-AIDS drugs is the **non-nucleoside inhibitors** of the enzyme reverse transcriptase. Among these agents are efavirenze and nevirapine. They inhibit the enzyme by noncompetitive binding.

A third group of drugs aims to inhibit a different enzyme, **protease**, that is also necessary for the propagation of the virus. Abacavir, indinavir (Figure 25C.2), nelfinavir, ritonavir, and saquinavir are the most frequently prescribed drugs in this category.

Because the HIV virus quickly mutates and thus becomes resistant to an inhibitor, the present practice is to prescribe a cocktail of drugs called HAART, or *highly active antiretroviral treatment*. HAART may contain, for example, ritonavir, ddCyd, and 3TC, or any other combination of the antitranscriptase and antiprotease drugs. Such a two- or three-drug cocktail greatly reduces the presence of HIV in the patient's blood and decreases the death rate. It even allows, if only temporarily, more or less normal functions for patients.

A fourth group of anti-AIDS drugs, now in development, aims to inhibit the fusion of the virus to the T-cell membrane, thereby preventing its entrance into the cell. HIV particles carry densely arranged spikes on their surfaces. These spikes are made of two glycoproteins: **gp120**, a surface glycoprotein, and **gp41**, a transmembrane glycoprotein. The latter mediates the entry of the

Figure 25C.1 HIV particles (pink) on the surface of a lymphocyte (green).

virus into the target cell. Drugs such as T20 and T1249 are synthetic polypeptides made of 36 and 39 amino acids, respectively, that interact with gp41 (Figure 25C.3).

A **vaccine** against AIDS would be the ideal remedy. People who are ill with AIDS or who have been exposed to the virus carry antibodies against the virus in their blood. That is, the immune system fights back against the infection. If the immune system could be bolstered either by vaccination with "dead" virus or by cytokines such as interleukines (see Section 30.6), AIDS would be curable. Efforts in the direction of vaccination have been hampered by the deadly nature of the virus, however. Even if only one out of a million "dead" viruses survived, it could still kill the vaccinated person.

In the beginning of the AIDS epidemic, some people were infected by blood transfusion. About two to three of every 10,000 blood samples in the United States carry antibodies against the virus. All such blood samples are now rejected; patented screening tests are performed on every blood donor, thereby making blood transfusion a safe procedure.

The strategy of treating a patient with a battery of drugs developed in fighting AIDS paid unexpected dividends in treating other diseases. Hepatitis viruses B and C are present in 1.5% of the U.S. population. A combination of interferon, which bolsters the immune system, and ribavirin, another viral protease inhibitor, is now the cocktail treatment of choice for hepatitis C.

(continued on next page)

CHEMICAL CONNECTIONS 25C

AIDS, (continued)

Figure 25C.2 Structural formulas for some AIDS drugs.

Reverse transcriptase The enzyme that enables a virus to copy the information encoded in an RNA molecule into a DNA molecule

(a)

(b)

Figure 25C.3 AIDS drug preventing the fusion of HIV virus to the T-cell membrane. (a) T-20 (yellow coils), a fusion inhibitor, binds to the matching coil segments that are part of HIV's gp41 (blue threads and red coils), preventing the viral fusion with the cell membrane of the T-cell above. (b) Without the fusion inhibitor, gp41 binds tightly to receptors on the T-cell membrane.

EXAMPLE 25.3

A tRNA has an anticodon, GAA. What amino acid will this tRNA carry? What are the steps necessary for the amino acid to bind to the tRNA?

Solution
The amino acid is leucine, as the codon is CUU. Leucine has to be activated by ATP. A specific enzyme, leucine-tRNA synthetase, catalyzes the carboxyl ester bond formation between the carboxyl group of leucine and the —OH group of tRNA.

Problem 25.3
What are the reactants in the reaction forming valine-tRNA?

25.6 Gene Regulation

Every embryo that is formed by sexual reproduction inherits its genes from both the parent sperm and the egg cells. But the genes in its chromosomal DNA are not active all the time. Rather, they are switched on and off during development and growth of the organism. Soon after formation of the embryo, the cells begin to differentiate. Some cells become neurons, some become muscle cells, some become liver cells, and so on. Each cell is a specialized unit that uses only some of the many genes it carries in its DNA. Thus each cell must switch some of its genes on and off—either permanently or temporarily. How this is done is the subject of **gene regulation**.

We know less about gene regulation in eukaryotes than in the simpler prokaryotes. Even with our limited knowledge, however, we can state that organisms do not have a single, unique way of controlling genes. Many gene regulations occur at the **transcriptional level** (DNA $\longrightarrow$ RNA). Others operate at the **translational level** (mRNA $\longrightarrow$ protein). A few of these processes are listed here as examples.

A Control on the Transcriptional Level

In eukaryotes, transcription is regulated by three entities: promoters, enhancers, and response elements. A fourth way of regulation is the phosphorylation of transcription factors.

1. Promoters of a gene are located adjacent to the transcription site. They are defined by an initiator and conserved sequences, such as a TATA box (see also Section 25.2 and Figure 25.2) or one or more copies of a GGGCGG sequence called a GC box. In eukaryotes, the enzyme RNA-polymerase has little affinity for binding to DNA. Instead, different transcription factors, or binding proteins, bind to the different modules of the promoter. With the aid of these transcription factors, the promoter functions to control the transcription on a steady, normal level. The binding, one after another, of these factors to the promoter site controls the rate of transcription initiation.

Transcription factors may allow the synthesis of mRNA (and from there the target protein) to vary by a factor of 1 million. This wide variation in eukaryotic cells is exemplified by the α-A-crystallin gene, which can be expressed in the lens of the eye, at a rate a-millionfold higher than that in the liver cell of the same organism.

> **Gene regulation** The control process by which the expression of a gene is turned on or off. Since RNA synthesis proceeds in one direction (5' $\longrightarrow$ 3') (see Figure 25.2), the gene (DNA) to be transcribed runs from 3' to 5'. Thus the control sites are in front of, or upstream of, the 3' end of the structural gene.

CHEMICAL CONNECTIONS 25D

The Promoter: A Case of Targeted Expression

Although the major part of a gene contains the region that codes for a particular protein, a portion of the gene does not code for any part of the protein. Instead, it acts as a switch that turns the transcription process on or off. This regulatory part of the gene is called the **promoter.** By using the gene insertion technique discussed in Section 25.8, one can combine the coding sequence of one gene with the promoter of another gene, thereby targeting the expression of a protein to a certain organ.

A recent development in AIDS research (Chemical Connections 25C) involved an enzyme called HIV-1 protease. This enzyme is essential if the virus that causes AIDS is to multiply inside the host cell. It has been argued that if suitable drugs that can inhibit HIV-1 protease could be found, the development of AIDS could be arrested, and perhaps the disease cured. But how can we test the hundreds of organic compounds that show some promise of being HIV-1 protease inhibitors? Here is where the concept of the promoter comes in. For example, if the coding region of the HIV-1 protease is linked to the promoter of α-A-crystallin, a protein molecule that appears mostly in the lens of the eye, the expression of the HIV-1 protease could be targeted to the eye only. True to the design, a DNA sequence containing the gene for HIV-1 protease and the promoter of α-A-crystallin were transfected into a mouse. The result was transgenic mice that became blind 24 days after birth. Otherwise, the mice were not affected at all—they ate, ran, and bred as normal mice do. The blindness was the result of the action of HIV-1 protease in the lens, causing cataract formation.

These transgenic mice could be used to test drugs to determine whether they would be good inhibitors of HIV-1 protease. If injection of a potential drug delayed cataract formation, that would be proof that the drug truly acts as an inhibitor of HIV-1 protease in a living organism and not just in test tubes. Many of the new drugs now used in treatment of AIDS patients (see Chemical Connections 25C), such as indinavir (Figure 25C.2), have first been tested in this manner. This strategy is clearly more efficient than testing all such potential inhibitors directly in humans.

■ **Eye with cataractous lens of a transgenic mouse containing the HIV-1 protease linked to α-A-crystallin promoter, 27 days after birth.**

How do these transcription factors find the specific gene control sequences into which they fit, and how do they bind to them? The interaction between the protein and DNA involves nonspecific electrostatic interactions (positive ions attracting negative ions and repelling other positive ions) as well as more specific hydrogen bonding. The transcription factors find their targeted sites by twisting their protein chains so that a certain amino acid sequence is present at the surface. One such conformational twist is provided by **metal-binding fingers** (Figure 25.8). These finger shapes are created by Zn^{2+} ions, which form covalent bonds with the amino acid side chains of the protein.

The zinc fingers interact with specific DNA (or sometimes RNA) sequences. The recognition comes by hydrogen bonding between a nucleotide (for example, guanine) and the side chain of a specific amino acid (for example, arginine). Besides metal-binding fingers, at least two other prominent transcription factors exist: **helix-turn-helix** and **leucine zipper.** Transcription factors can also be repressors, reducing the rate of transcription. Some can serve as both activators and repressors.

2. A second group of transcription factors, known as **enhancers,** also speeds up the transcription process. These factors may be located several thousand nucleotides away from the transcription site. To stimulate transcription, an enhancer must be brought in the vicinity of the promoter. With the aid of special proteins, called **sensors,** the DNA forms a loop. The enhancer with the sensor interacts with a trancription factor. This complex then allows the RNA polymerase II to speed up the transcription when higher-than-normal production of proteins is needed (Figure 25.9).

3. The third group of transcription factors, **response elements,** are activated by their transcription factors in response to an outside stimulus. The stimulus may be heat shock, heavy metal toxicity, or simply a hormonal signal, such as binding a steroid hormone to its receptor. The response element of steroids is in front of and 250 base pairs upstream from the starting point of transcription. Only the receptor with the bound steroid hormone can interact with its response element, thereby initiating transcription.

4. Transcription does not occur at the same rate throughout the cell life cycle. It is accelerated or slowed down as the need arises. The signal to speed up transcription may originate from outside of the cell. One such signal, in the GTP-adenylate cyclase-cAMP pathway (Section 23.5), produces **phosphorylated protein kinase.** This enzyme enters the nucleus, where it phosphorylates transcription factors, which aid in the transcription cascade.

Figure 25.8 Schematic view of a metal-binding finger (a common motif in many transcription factors). The zinc ion forms covalent bonds to two cysteines and two histidines.

B Control on the Translational Level

During translation, there are also a number of mechanisms that ensure quality control.

1. The Specificity of a tRNA for Its Unique Amino Acid First, the attachment of the proper amino acid to the proper tRNA must be achieved. The enzyme that catalyzes this reaction, aminoacyl-tRNA synthetase (AARS), is specific for each tRNA and amino acid combination. The AARS enzymes recognize their tRNA by specific nucleotide sequences. Furthermore, the active site of the enzyme has two **sieving portions.** For example, in isoleucyl-tRNA synthetase the first sieve excludes any amino acids that are larger than isoleucine. If a similar amino acid such as valine, which is smaller than isoleucine, arrives at the active site, the second sieve eliminates it. The second sieving site is thus a "proofreading" site.

2. Recognition of the Stop Codon Another quality-control measure occurs at the termination. The stop codons must be recognized by release factors, leading to the release of the polypeptide chain and allowing the recycling of the ribosomes. Otherwise, a longer polypeptide chain may be toxic. The release factor combines with GTP and binds to the ribosomal A

Figure 25.9 The enhancer process. A loop in DNA brings the enhancer to the initiation site.

site when that site is occupied by the termination codon. Both the GTP and the peptidyl-tRNA ester bond are hydrolyzed. This hydrolysis releases the polypeptide chain and the deacylated tRNA. Finally, the ribosome dissociates from the mRNA.

3. Post-translational Controls

(a) *Removal of methionine.* In most proteins, the methionine residue at the N-terminal, which was added in the initiation step, is removed. A special enzyme, metaminopeptidase, cleaves the peptide bond.

(b) *Chaperoning.* The tertiary structure of a protein is largely determined by the amino acid sequence (primary structure). Proteins begin to fold even as they are synthesized on ribosomes. Nevertheless, misfolding may occur due to mutation in a gene, lack of fidelity in transcription, or translational errors. All of these errors may lead to aggregation of misfolded proteins that can be detrimental to the cell—for example, causing amyloid diseases. Certain proteins in living cells, called **chaperones,** help the newly synthesized polypeptide chains to fold properly. They recognize hydrophobic regions exposed on unfolded proteins and bind to them. Chaperones then shepherd the proteins to the biologically desirable folding as well as to their local destinations within the cell.

(c) *Degradation of misfolded proteins.* A third post-translational control exists in the form of **proteasomes.** These are cylindrical assemblies of a number of protein subunits with proteolytic activity in the core of the cylinder. If the rescue by chaperones fails, proteases may degrade the misfolded protein first by targeting it with ubiquitination (see Chemical Connections 27E) and finally by proteolysis.

25.7 Mutations, Mutagens, and Genetic Diseases

In Section 24.6, we saw that the base-pairing mechanism provides an almost perfect way to copy a DNA molecule during replication. The key word here is "almost." No machine, not even the copying mechanism of DNA replication, is totally without error. It has been estimated that, on average, one error occurs for every 10^{10} bases (that is, one in 10 billion). An error in the copying of a sequence of bases is called a **mutation.** Mutations can occur during replication. Base errors can also occur during transcription in protein synthesis (a nonheritable error).

These errors may have widely varying consequences. For example, the codon for valine in mRNA can be GUA, GUG, GUC, or GUU. In DNA, these codons correspond to CAT, CAC, CAG, and CAA, respectively. Assume that the original codon in the DNA is CAT. If a mistake is made during replication and CAT is spelled as CAG in the copy, there will be no harmful mutation. Instead, when a protein is synthesized, the CAG will be transcribed onto the mRNA as GUC, which also codes for valine. Therefore, although a mutation has occurred, the same protein is manufactured.

Now assume that the original sequence in the gene's DNA is CTT, which transcribes onto mRNA as GAA and codes for glutamic acid. If a mutation occurs during replication and CTT becomes ATT, the new cells will probably die. The reason is that ATT transcribes to UAA, which does not code for any amino acid but rather is a stop signal. Thus, instead of continuing to build a protein chain with glutamic acid, the synthesis will stop altogether. An important protein will not be manufactured, and the organism may die. In this way, very harmful mutations are not carried over from one generation to the next.

Sickle cell anemia (Chemical Connections 21D) is caused by a single amino acid change: valine for glutamic acid. In this case, CTT is mutated to CAT.

CHEMICAL CONNECTIONS 25E

Mutations and Biochemical Evolution

We can trace the genetic relationship of different species through the variability of their amino acid sequences in different proteins. For example, the blood of all mammals contains hemoglobin, but the amino acid sequences of the hemoglobins are not identical. In the table, we see that the first 10 amino acids in the β-globin of humans and gorillas are exactly the same. As a matter of fact, there is only one amino acid difference, at position 104, between us and apes. The β-globin of the pig differs from ours at 10 positions, of which 2 are in the N-terminal decapeptide. That of the horse differs from ours in 26 positions, of which 4 are in this decapeptide. β-Globin seems to have gone through many mutations during the evolutionary process, because only 26 of the 146 sites are invariant—that is, exactly the same in all species studied so far.

The relationship between different species can also be established by finding similarities in their mRNA primary structures. As the mutations actually occurred on the original DNA molecule, and then were perpetuated in the progeny by the mutant DNA, it is instructive to learn how a point mutation may occur in different species. Looking at position 4 of the β-globin molecule, we see a change from serine to threonine. The code for serine is AGU or AGC, whereas that for threonine is ACU or ACC (Table 25.1). Thus a change from G to C in the second position of the codon brought the divergence between the β-globins of humans and horses. The genes of closely related species, such as humans and apes, have very similar primary structures, presumably because these two species diverged on the evolutionary tree only recently. On the other hand, species far removed from each other diverged long ago and have undergone more mutations, which show up in differences in the primary structures of their DNA, mRNA, and consequently proteins.

The *number* of amino acid substitutions is significant in the evolutionary process caused by mutation, but the *kind* of substitution is even more important. If the substitution involves an amino acid with physicochemical properties similar to those of the amino acid in the ancestor protein, the mutation is most probably viable. For example, in human and gorilla β-globin, position 4 is occupied by threonine, but it is occupied by serine in the pig and horse. Both amino acids provide an —OH-carrying side chain.

Amino Acid Sequence of the N-Terminal Decapeptides of β-Globin in Different Species

Species	1	2	3	4	5	6	7	8	9	10
Human	Val	His	Leu	Thr	Pro	Glu	Glu	Lys	Ser	Ala
Gorilla	Val	His	Leu	Thr	Pro	Glu	Glu	Lys	Ser	Ala
Pig	Val	His	Leu	Ser	Ala	Glu	Glu	Lys	Ser	Ala
Horse	Val	Glu	Leu	Ser	Gly	Glu	Glu	Lys	Ala	Ala

Ionizing radiation (X rays, ultraviolet light, gamma rays) can cause mutations. Furthermore, a large number of organic compounds can induce mutations by reacting with DNA. Such compounds are called **mutagens.** Many changes caused by radiation and mutagens do not become mutations because the cell has its own repair mechanism, called nucleotide excision repair (NER), which can prevent mutations by cutting out damaged areas and resynthesizing them (Section 24.7). In spite of this defense mechanism, certain errors in copying that result in mutations do slip by. Many compounds (both synthetic and natural) are mutagens, and some can cause cancer when introduced into the body. These substances are called **carcinogens** (Chemical Connections 13B). One of the main tasks of the U.S. Food and Drug Administration and the Environmental Protection Agency is to identify carcinogens and eliminate them from our food, drugs, and environment. Even though most carcinogens are mutagens, the reverse is not true.

Not all mutations are harmful. Some are beneficial because they enhance the survival rate of the species. For example, mutation is used to develop new strains of plants that can withstand pests.

Examples of mutagens are benzene, carbon tetrachloride, and vinyl chloride.

CHEMICAL CONNECTIONS 25F

Oncogenes

An **oncogene** is a gene that in some way or other participates in the development of cancer. Cancer cells differ from normal cells in a variety of structural and metabolic ways; however, the most important difference is their uncontrolled proliferation. Most ordinary cells are quiescent. When they lose their controls, they give rise to tumors—benign or malignant. The uncontrolled proliferation allows these cells to spread, invade other tissues, and colonize them, in a process called *metastasis*.

Malignant tumors can be caused by a variety of agents: chemical carcinogens such as benzo(a)pyrene (Chemical Connections 13B), ionizing radiation such as X rays, and viruses. However, for a normal cell to become a cancer cell, certain transformations must occur. Many of these transformations take place on the level of the genes. Certain genes in normal cells have counterparts in viruses, especially in retroviruses (which carry their genes in the form of RNA rather than DNA). Thus normal cells in a human and in a virus both have genes that code for similar proteins. Usually these genes code for proteins that control cell growth.

One such protein is epidermal growth factor (EGF). When a retrovirus gets into the cell, it takes control of the cell's own EGF gene. At that point, the EGF gene becomes an oncogene, a cancer-causing gene, because it produces large amounts of the EGF protein. This excess production, in turn, allows the epidermal cells to grow in an uncontrolled manner.

Because the normal EGF gene in the normal cell was turned into an oncogene, we call it a **proto-oncogene.** Such conversions of a proto-oncogene to an oncogene can occur not just by viral invasion, but also by a mutation of the gene. Although most cancers are caused by multiple mutations, one form of cancer, called human bladder carcinoma, is caused by a single mutation of a gene—a change from guanine (G) to thymine (T). The resulting protein has a valine in place of a glycine. This change is sufficient to transform the cell into a cancerous state because the protein for which the gene codes amplifies signals (Section 23.5). Thus, because of the mutation that caused the proto-oncogene to become an oncogene, the cell stays "on"—its metabolic process is not shut off—and it is now a cancer cell that proliferates without control.

There are a number of other mechanisms by which a proto-oncogene is transformed into an oncogene, but in each case the gene product has something to do with growth, hormonal action, or some other kind of cell regulation. About 100 viral and cellular oncogenes have been identified to date.

CHEMICAL CONNECTIONS 25G

p53: A Central Tumor Suppressor Protein

Not all cancer-causing gene mutations have their origin in oncogenes. There are some 36 known **tumor suppressor genes,** the products of which are proteins controlling cell growth. None of them is more important than a protein with a molar mass of 53,000 simply named **p53.** This protein responds to a variety of cellular stresses including DNA damage, lack of oxygen (hypoxia) and aberrantly activated oncogenes (Chemical Connections 25F). In about 40% of all cancer cases, the tumor contains p53 that underwent mutation. Mutated p53 protein can be found in 55% of lung cancers, about half of all colon and rectal cancers, and some 40% of lymphomas, stomach cancers, and pancreatic cancers. In addition, in one-third of all soft tissue sarcomas, p53 is inactive, even though it did not undergo mutation.

These statistics indicate that the normal function of p53 protein is to suppress tumor growth. When it is mutated or otherwise not present in sufficient or active form, it is unable to perform this protective function and cancer spreads. p53 protein binds to specific sequences of double-stranded DNA. When X rays or γ-rays damage DNA, an increase in p53 protein concentration is observed. The increased binding of p53 controls the cell cycle; it holds it between cell division and DNA replication. The time gained in this arrested cell cycle allows the DNA to repair its damage. If that fails, the p53 protein triggers apoptosis, the programmed cell death of the injured cell (see also Chemical Connections 24E).

Recently, it was reported that p53 has a finely tuned function in cells. It suppresses tumor growth, but if p53 is overly expressed (that is, its concentration is high), it contributes to the premature aging of the organism. In those conditions, p53 arrests the cell cycle not only of damaged cells but also of stem cells. These cells normally differentiate into various types (muscle cells, nerve cells, and so forth) and replace those cells that die with aging. Excess p53 slows down this differentiation. In mice, it was found that the abundance of p53 protein made the animal cancer-free but at a cost: The mice lost weight and muscle, their bones became brittle, and their wounds took longer to heal. All in all, their average life span was 20% shorter than that of normal mice.

If a mutation is harmful, it results in an inborn genetic disease. This condition may be carried as a recessive gene from generation to generation with no individual demonstrating the symptoms of the disease. When both parents carry recessive genes, however, an offspring has a 25% chance of inheriting the disease. If the defective gene is dominant, on the other hand, every carrier will develop symptoms.

Specific diseases that result from defective or missing genes are discussed in Chemical Connections 20F, 21D, 27B, and 27F.

A recessive gene expresses its message only if both parents transmit it to the offspring.

25.8 Recombinant DNA

There are no cures for the inborn genetic diseases that we discussed in the preceding section. The best we can do is detect the carriers and, through genetic counseling of prospective parents, try not to perpetuate the defective genes. However, **recombinant DNA techniques** give us some hope for the future. At this time, recombinant DNA techniques are used mostly in bacteria, plants, and test animals (such as mice), but they could someday be successfully extended to human cells as well.

One example of recombinant DNA techniques begins with certain circular DNA molecules found in the cells of the bacteria *Escherichia coli*. These molecules, called **plasmids** (Figure 25.10), consist of double-stranded DNA arranged in a ring. Certain highly specific enzymes called **restriction endonucleases** cleave DNA molecules at specific locations (a different location for each enzyme). For example, one of these enzymes may split a double-stranded DNA as follows:

B～～GAATTC～～B　　　　B～～G　　　　　AATTC～～B
　　　　　　　　　　enzyme
B～～CTTAAG～～B　　　　B～～CTTAA　　　　　　G～～B

We use "B" to indicate a bacterial plasmid.

The enzyme is so programmed that whenever it finds this specific sequence of bases in a DNA molecule, it cleaves it as shown. Since a plasmid is circular, cleaving it in this way produces a double-stranded chain with two ends (Figure 25.11). These are called "sticky ends" because on one strand each has several free bases that are ready to pair up with a complementary section if they can find one.

Figure 25.10 Plasmids from a bacterium used in the recombinant DNA technique.

We use "H" to indicate a human gene.

Figure 25.11 The recombinant DNA technique can be used to turn a bacterium into an insulin "factory."

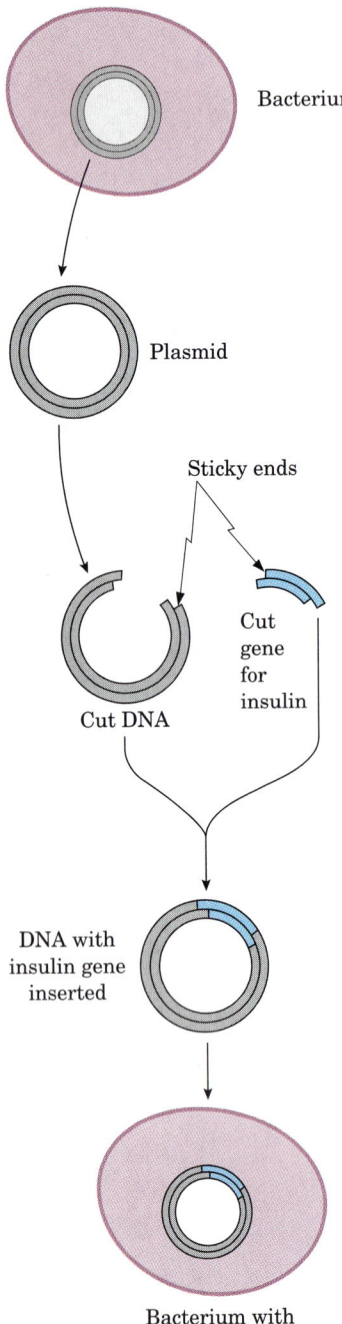

The next step is to give them such a section. This is done by adding a gene from some other species. The gene is a strip of double-stranded DNA that has the necessary base sequence. For example, we can put in the human gene that manufactures insulin, which we can get in two ways:

1. It can be made in a laboratory by chemical synthesis; that is, chemists can combine the nucleotides in the proper sequence to make the gene.
2. We can cut a human chromosome with the same restriction enzyme.

Since it is the same enzyme, it cuts the human gene so as to leave the same sticky ends:

$$\begin{array}{c}\text{H—GAATTC—H} \\ \text{H—CTTAAG—H}\end{array} \xrightarrow{\text{enzyme}} \begin{array}{c}\text{H—G} \\ \text{H—CTTAA}\end{array} + \begin{array}{c}\text{AATTC—H} \\ \text{G—H}\end{array}$$

The human gene must be cut at two places so that a piece of DNA that carries two sticky ends is freed. To splice the human gene into the plasmid, the two are mixed in the presence of DNA ligase, and the sticky ends come together:

$$\begin{array}{c}\text{H—G} \\ \text{H—CTTAA}\end{array} + \begin{array}{c}\text{AATTC—B} \\ \text{G—B}\end{array} \xrightarrow{\text{enzyme}} \begin{array}{c}\text{H—GAATTC—B} \\ \text{H—CTTAAG—B}\end{array}$$

This reaction takes place at both ends of the human gene, turning the plasmid into a circle once again (Figure 25.11).

The modified plasmid is then put back into a bacterial cell, where it replicates naturally every time the cell divides. Bacteria multiply quickly, so soon we have a large number of bacteria, all containing the modified plasmid. All these cells now manufacture human insulin by transcription and translation. Thus, we can use bacteria as a factory to manufacture specific proteins. This new industry has tremendous potential for lowering the price of drugs that are now manufactured by isolation from human or animal tissues (for example, human interferon, a molecule that fights infection). Not only bacteria but also plant cells can be used (Figure 25.12).

Ultimately, if recombinant DNA techniques can be applied to humans and not just to bacteria, genetic diseases might someday be cured by this powerful technology. For instance, an infant or fetus who is missing a gene might be given that gene. Once in the cells, the gene would reproduce itself for the individual's entire lifetime.

EXAMPLE 25.4

Two different restriction endonucleases act on the following sequence of a double-stranded DNA:

$$\text{\textasciitilde\textasciitilde\textasciitilde AATGAATTCGAGGC\textasciitilde\textasciitilde\textasciitilde} \\ \text{\textasciitilde\textasciitilde\textasciitilde TTACTTAAGCTCCG\textasciitilde\textasciitilde\textasciitilde}$$

The endonuclease EcoRI recognizes the sequence GAATTC and cuts the sequence between G and A. The other endonuclease, TaqI, recognizes the sequence TCGA and cuts the sequence between T and C. What are the sticky ends that each of these endonucleases will create?

(a)

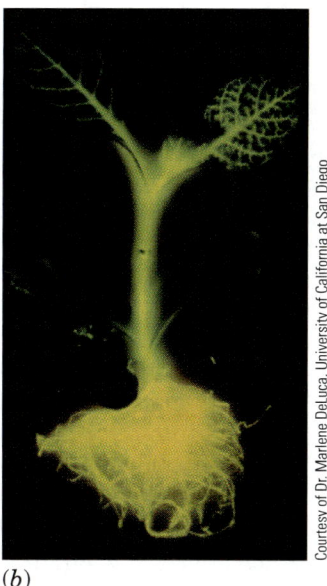

(b)

Figure 25.12 (a) Injection of an aqueous DNA solution into the nucleus of a protoplast. (b) A luminescent tobacco plant. The gene of the enzyme luciferase (from a firefly) has been incorporated into the genetic material of the tobacco plant.

Solution

EcoRI ~~~AATG   AATTCGAGGC~~~
     ~~~TTACTTAA   GCTCCG~~~

TaqI ~~~AATGAATT   CGAGGC~~~
      ~~~TTACTTAAGC   TCCG~~~

Problem 25.4
Show the sticky end of the following double-stranded DNA sequence that is cut by TaqI:

~~~CCTCGATTG~~~
~~~GGAGCTAAC~~~

Human insulin is now marketed by the Lilly Corporation in two forms, called Humulin R and Humulin N (Chemical Connections 21C). The R form has a faster onset of action. Both versions are manufactured by the recombinant DNA technique.

As of mid-2002, no such gene therapy had been successfully achieved in humans. This failure was not for lack of trying. A tremendous amount of research has been devoted to not only trying to cure genetic diseases but also to delivering a fatal bullet to cancer cells by recombinant DNA techniques. The difficulties are numerous, but finding the proper vector (a harmless virus, for example) is one of the bigger challenges. The virus must be proficient enough to deliver the selected gene without eliciting an immune reaction from the organism itself. The hope is that, in the future, genes can be inserted in human organs and they will express themselves in sufficient quantity. Even then the product of the healthy gene will compete with the product of the faulty gene, ideally alleviating the symptoms of the disease or curing it altogether.

By far, the greatest potential of **genetic engineering**—that is, inserting new genes into cells—is in the field of agriculture. Genetic engineering has produced tomatoes that can be picked in their naturally ripened states yet do not spoil quickly on the shelves of supermarkets. Other genetically engineered plants—corn and cotton, for example—can resist damage by insects, fungi, or herbicides. Most of the soybeans grown in the United States today are genetically engineered. In addition to creating herbicide- and insect-resistant plants, genetic engineering offers opportunities to improve crop yields and resist freezing temperatures. All these factors will contribute to the increase of the food supply as is necessary to feed the ever-growing world population.

Some people believe that genetically engineered organisms may create havoc with the ecology of the planet by introducing new, dominant species. Others express concern that genetically engineered organisms may enter the human food chain in ways for which the product has not been approved. This concern is legitimate. For example, genetically engineered corn that has been approved only for animal feed has already found its way in trace amounts into corn chips intended for human consumption. This happened at a single manufacturer. No adverse medical effect has been reported, although some people may prove to be allergic to the introduced gene. Such concerns have restricted the export to European markets of beef that was fed on genetically engineered corn.

Plant breeding has been going on for centuries, with the goal of achieving exactly the same goals as genetic engineering does. Obviously, care must be taken and controls exercised so as not to release harmful newly engineered life-forms into the environment.

SUMMARY

A **gene** is a segment of a DNA molecule that carries the sequence of bases that directs the synthesis of one particular protein (Section 25.1). The information stored in the DNA is transcribed onto RNA and then expressed in the synthesis of a protein molecule. This is done in two steps: **transcription** and **translation.** In transcription (Section 25.2), the information is copied from DNA onto mRNA by complementary base pairing (A $\longrightarrow$ U, T $\longrightarrow$ A, G $\longrightarrow$ C, C $\longrightarrow$ G). There are also start and stop signals.

The mRNA is strung out along the ribosomes (Section 25.3). Transfer RNA carries the individual amino acids, with each tRNA going to a specific site on the mRNA. A sequence of three bases (a triplet) on mRNA constitutes a **codon.** It spells out the particular amino acid that the tRNA brings to this site. Each tRNA has a recognition site, the **anticodon,** that pairs up with the codon. When two tRNA molecules are aligned at adjacent sites, the amino acids that they carry are linked by an enzyme forming a peptide bond. The translation process continues until the whole protein is synthesized (Section 25.5). The **genetic code** (Section 25.4) provides the correspondence between a codon and an amino acid. In most cases, there is more than one codon for each amino acid. Most of human DNA (96 to 98%) does not code for proteins.

There are a number of mechanisms for gene regulation on both the transcriptional level and the translational level (Section 25.6). **Promoters** have initiator and conserved sequences. **Transcription factors** bind to the promoter, thereby regulating the rate of transcription. **Enhancers** are nucleotide sequences far removed from the transcription sites. **Sensors** are proteins that bring the enhancers in the vicinity of the promoter and in this way control the rate of transcription. Some translational controls, such as **release factors,** act during translation, others, such as **chaperones,** act after translation is completed.

A change in the sequence of bases is called a **mutation** (Section 25.7). Mutations can be caused by an internal mistake or induced by chemicals or radiation. In fact, a change in just one base can cause a mutation. This may be harmful or beneficial, or it may cause no change whatsoever in the amino acid sequence. If a mutation is very harmful, the organism may die. Chemicals that cause mutations are called **mutagens** (Section 25.7). Chemicals that cause cancer are called **carcinogens.** Many carcinogens are mutagens, but the reverse is not true.

With the discovery of restriction enzymes that can cut DNA molecules at specific points, scientists have found ways to splice DNA segments together. In this manner, a human gene (for example, the one that codes for insulin) can be spliced into a bacterial plasmid. The bacteria, when multiplied, can then transmit this new information to the daughter cells so that the ensuing generations of bacteria can manufacture human insulin. This powerful method is called the **recombinant DNA technique** (Section 25.8). Genetic engineering is the process by which genes are inserted into cells.

PROBLEMS

Numbers that appear in color indicate difficult problems. ▸ designates problems requiring application of principles.

25.5 Does the term *gene expression* refer to (a) transcription, (b) translation, or (c) transcription plus translation?

25.6 In what part of the eukaryote cell does transcription occur?

25.7 Where does most of the translation occur in a eukaryote cell?

Transcription

25.8 What is the function of RNA polymerase?

25.9 What is the role of helicase in transcription?

25.10 Where is an initiation signal located?

25.11 Which end of the DNA contains the termination signal?

25.12 What would happen to the transcription process if a drug added to a eukaryote cell inhibits the phosphatase?

25.13 Where is the methyl group located in the guanine cap?

25.14 How are the adenine nucleotides linked together in the polyA tails?

The Role of RNA in Translation

25.15 Where are the codons located?

25.16 What are the two most important sites on tRNA molecules?

The Genetic Code

25.17 (a) If a codon is GCU, what is the anticodon?
(b) For what amino acid does this codon code?

25.18 If a segment of DNA is 981 units long, how many amino acids appear in the protein for which this DNA segment codes? (Assume that the entire segment is used to code for the protein and that there is no methionine at the N-terminal end of the protein.)

25.19 In what sense does the universality of the genetic code support the theory of evolution?

Translation and Protein Synthesis

25.20 To which end of the tRNA is the amino acid bonded? Where does the energy come from to form the tRNA–amino acid bond?

25.21 There are three sites on the ribosome, each participating in the translation. Identify them and describe what is happening at each site.

25.22 What is the main role of (a) the 40S ribosome and (b) the 60S ribosome?

25.23 What is the function of elongation proteins?

Gene Regulation

25.24 Which molecules are involved in gene regulation at the transcriptional level?

25.25 Where are enhancers located? How do they work?

25.26 Where are the sieving portions of AARS enzymes located? What is their function?

25.27 What is the function of proteosomes in quality control?

25.28 What kind of interactions exist between metal-binding fingers and DNA?

Mutations

25.29 Using Table 25.1, give an example of a mutation that (a) does not change anything in a protein molecule and (b) might cause fatal changes in a protein.

25.30 How do cells repair mutations caused by X rays?

25.31 Can a harmful mutation-causing genetic disease exist from generation to generation without exhibiting the symptoms of the disease? Explain.

25.32 Are all mutagens also carcinogens?

Recombinant DNA

25.33 How do restriction endonucleases operate?

25.34 What are sticky ends?

25.35 A new genetically engineered corn has been approved by the Food and Drug Administration. This new corn shows increased resistance to a destructive insect called a corn borer. What is the difference, in principle, between this genetically engineered corn and one that developed insect resistance by mutation (natural selection)?

25.36 EcoRI restriction endonuclease recognizes the sequence GAATTC and cuts it between G and A. What will be the sticky ends of the following double-helical sequence when EcoRI acts on it?

C A A A G A A T T C G
G T T T C T T A A G C

Chemical Connections

25.37 (Chemical Connections 25A) What is unique in the backbone of the antisense drug Vitravene?

25.38 (Chemical Connections 25B) What is a viral "coat"? Where do the ingredients—amino acids, enzymes, and so forth—necessary to synthesize the coat come from?

25.39 (Chemical Connections 25C) What are the target cells for the AIDS virus?

25.40 (Chemical Connections 25C) What HIV-1 viral enzymes are inhibited by (a) AZT and (b) indinavir?

25.41 (Chemical Connections 25D) What was the role of the α-A-crystallin promoter in the development of protease-inhibitor drugs as a treatment for AIDS?

25.42 (Chemical Connections 25E) What is an invariant site?

25.43 (Chemical Connections 25F) What kind of mutation transforms the EGF proto-oncogene into an oncogene in human bladder carcinoma?

25.44 (Chemical Connections 25G) What is p53? Why is its mutated form associated with cancer?

25.45 (Chemical Connections 25G) How does p53 promote DNA repair?

Additional Problems

25.46 In both the transcription and the translation steps of protein synthesis, a number of different molecules come together to act as a factor unit. What are these units of (a) transcription and (b) translation?

25.47 In the tRNA structure, there are stretches where complementary base pairing is necessary and other areas where it is absent. Describe two functionally critical areas (a) where base pairing is mandatory and (b) where it is absent.

25.48 Is there any way to prevent a hereditary disease? Explain.

25.49 How does the cell make sure that a specific amino acid (say, valine) attaches itself only to the one tRNA molecule that is specific for valine?

25.50 (a) What is a plasmid?
(b) How does it differ from a gene?

25.51 Why do we call the genetic code *degenerate*?

25.52 Glycine, alanine, and valine are classified as nonpolar amino acids. Compare their codons. What similarities do you find? What differences do you find?

25.53 Looking at the multiplicity (degeneracy) of the genetic code, you may get the impression that the third base of the codon is irrelevant. Point out where this is not the case. Out of the 16 possible combinations of the first and second bases, in how many cases is the third base irrelevant?

25.54 What polypeptide is coded for by the mRNA sequence 5'-GCU-GAA-GUC-GAG-GUG-UGG-3'?

25.55 A new endonuclease is found. It cleaves double-stranded DNA at every location where C and G are paired on opposite strands. Could this enzyme be used in producing human insulin by the recombinant DNA technique? Explain.

InfoTrac College Edition

For additional readings, go to InfoTrac College Edition, your online research library, at

http://infotrac.thomsonlearning.com

CHAPTER 26

26.1 Introduction

26.2 Cells and Mitochondria

26.3 The Principal Compounds of the Common Catabolic Pathway

26.4 The Citric Acid Cycle

26.5 Electron and H^+ Transport

26.6 Phosphorylation and the Chemiosmotic Pump

26.7 The Energy Yield

26.8 Conversion of Chemical Energy to Other Forms of Energy

Wailua Falls, Hawaii, is a natural demonstration of two pathways ending in a common pool.

Bioenergetics: How the Body Converts Food to Energy

26.1 Introduction

The same compounds may be synthesized in one part of a cell and broken down in a different part of the cell.

Living cells are in a dynamic state, which means that compounds are constantly being synthesized and then broken down into smaller fragments. Thousands of different reactions take place at the same time. **Metabolism** is the sum total of all the chemical reactions involved in maintaining the dynamic state of the cell.

In general, we can divide metabolic reactions into two broad groups: (1) those in which molecules are broken down to provide the energy needed by cells and (2) those that synthesize the compounds needed by cells—both simple and complex. **Catabolism** is the process of breaking down molecules to supply energy. The process of synthesizing (building up) molecules is **anabolism.**

In spite of the large number of chemical reactions, a mere handful dominate cell metabolism. In this chapter and Chapter 27, we focus our attention on the catabolic pathways that yield energy. A **biochemical pathway** is a series of consecutive biochemical reactions. We will see the actual reactions that enable the chemical energy stored in our food to be converted to

646 | CHAPTER 26 BIOENERGETICS: HOW THE BODY CONVERTS FOOD TO ENERGY

Figure 26.1 In this simplified schematic diagram of the common catabolic pathway, an imaginary funnel represents what happens in the cell. The diverse catabolic pathways drop their products into the funnel of the common catabolic pathway, mostly in the form of C_2 fragments (Section 26.4). (The source of the C_4 fragments will be shown in Section 27.9.) The spinning wheel of the citric acid cycle breaks these molecules down further. The carbon atoms are released in the form of CO_2, and the hydrogen atoms and electrons are picked up by special compounds such as NAD^+ and FAD. Then the reduced NADH and $FADH_2$ cascade down into the stem of the funnel, where the electrons are transported inside the walls of the stem and the H^+ ions are expelled to the outside. In their drive to get back, the H^+ ions form the energy carrier ATP. Once back inside, they combine with the oxygen that picked up the electrons and produce water.

Common catabolic pathway
A series of chemical reactions through which most degraded food goes to yield energy in the form of ATP; the common catabolic pathway consists of (1) the citric acid cycle (Section 26.4) and (2) oxidative phosphorylation (Sections 26.5 and 26.6)

the energy we use every minute of our lives—to think, to breathe, and to use our muscles to walk, write, eat, and everything else. In Chapter 28, we will look at some synthetic (anabolic) pathways.

The food we eat consists of many types of compounds, largely the ones we discussed in earlier chapters: carbohydrates, lipids, and proteins. All of them can serve as fuel, and we derive our energy from them. To convert those compounds to energy, the body uses a different pathway for each type of compound. *All these diverse pathways converge to one* **common catabolic pathway,** which is illustrated in Figure 26.1. In the figure, the diverse pathways are shown as different food streams. The small C_2 and C_4 molecules produced from the original large molecules in food drop into an imaginary collecting funnel that represents the common catabolic pathway. At the end of the funnel appears the energy carrier molecule adenosine triphosphate (ATP).

The purpose of catabolic pathways is to convert the chemical energy in foods to molecules of ATP. In the process, foods also yield metabolic intermediates, which the body can use for synthesis. In this chapter, we deal with the common catabolic pathway. In Chapter 27, we discuss the ways in which the different types of food (carbohydrates, lipids, and proteins) feed molecules into the common catabolic pathway.

26.2 Cells and Mitochondria

A typical animal cell has many components, as shown in Figure 26.2. Each serves a different function. For example, the replication of DNA (Section 24.6) takes place in the **nucleus; lysosomes** remove damaged cellular components and some unwanted foreign materials; and **Golgi bodies** package and process proteins for secretion and delivery to other cellular compartments. The specialized structures within cells are called **organelles.**

An animal cell

Figure 26.2 Diagram of a rat liver cell, a typical higher animal cell.

Figure 26.3 (a) Schematic of a mitochondrion cut to reveal the internal organization. (b) Colorized transmission electron micrograph of a mitochondrion in a heart muscle cell.

Mitochondria is the plural form; *mitochondrion* is the singular form.

The **mitochondria,** which possess two membranes (Figure 26.3), are the organelles in which the common catabolic pathway takes place in higher organisms. The enzymes that catalyze the common pathway are all located in these organelles. Because these enzymes are synthesized in the cytosol they must be imported through the two membranes. They cross the outer membrane through *t*ranslocator *o*uter *m*embrane (TOM) channels and are accepted in the intermembrane space by chaperone-like *t*ranslocator *i*nner *m*embrane (TIM) complexes, which also insert them into the inner membrane.

Because the enzymes are located inside the inner membrane of mitochondria, the starting materials of the reactions in the common pathway must pass through the two membranes to enter the mitochondria. Products must leave the same way.

The inner membrane of a mitochondrion is quite resistant to the penetration of any ions and of most uncharged molecules. However, ions and molecules can still get through the membrane—they are transported across it by the numerous protein molecules embedded in it (Figure 20.2). The outer membrane, on the other hand, is quite permeable to small molecules and ions and does not have transporting membrane proteins.

The **matrix** is the inner nonmembranous portion of a mitochondrion (Figure 26.3). The inner membrane is highly corrugated and folded. On the basis of electron microscopic studies, Paladin proposed his baffle model of the mitochondrion in 1952. The baffles, which are called **cristae,** project into the matrix like the bellows of an accordion. The enzymes of the oxidative phosphorylation cycle are localized on the cristae. The space between the inner and outer membranes is the **intermembrane space.**

This classical baffle model of mitochondria underwent some changes in the late 1990s after three-dimensional pictures were obtained by the new technique electron-microscope tomography. The 3-D images indicate that the cristae have narrow tubular connections to the inner membrane. These tubular connections may control the diffusion of metabolites from the inside to the intermembrane space. Furthermore, the space between the outer and the inner membrane varies during metabolism, possibly controlling the rate of reactions.

The enzymes of the citric acid cycle are located in the matrix. We will soon see in detail how the specific sequence of these enzymes causes the chain of events in the common catabolic pathway. Furthermore, we will discuss the ways in which nutrients and reaction products move into and out of the mitochondria.

Figure 26.4 Adenosine 5′-monophosphate (AMP).

26.3 The Principal Compounds of the Common Catabolic Pathway

The common catabolic pathway has two parts: the **citric acid cycle** (also called the tricarboxylic acid cycle or the Krebs cycle) and the **electron transport chain** and **phosphorylation,** together called the **oxidative phosphorylation pathway.** To understand what actually happens in these reactions, we must first introduce the principal compounds participating in the common catabolic pathway.

A Agents for Storage of Energy and Transfer of Phosphate Groups

The most important of these agents are three rather complex compounds: **adenosine monophosphate (AMP), adenosine diphosphate (ADP),** and **adenosine triphosphate (ATP)** (Figures 26.4 and 26.5). All three of these molecules contain the heterocyclic amine adenine (Section 24.2) and the sugar D-ribose (Section 19.3A) joined together by a β-N-glycosidic bond, forming adenosine (Section 24.2).

AMP, ADP, and ATP all contain adenosine connected to phosphate groups. The only difference between the three molecules is the number of phosphate groups. As you can see from Figure 26.5, each phosphate is attached to the next by an anhydride bond (Section 18.7A). ATP contains three phosphates—one phosphoric ester and two phosphoric anhydride bonds. In all three molecules, the first phosphate is attached to the ribose by a phosphoric ester bond (Section 18.7B).

A phosphoric anhydride bond, P—O—P, contains more chemical energy (7.3 kcal/mol) than a phosphoric–ester linkage, C—O—P (3.4 kcal/mol). Thus, when ATP and ADP are hydrolyzed to yield phosphate (Figure 26.5), they release more energy per phosphate group than does AMP.

Figure 26.5 Hydrolysis of ATP produces ADP plus dihydrogen phosphate ion plus energy.

650 | CHAPTER 26 BIOENERGETICS: HOW THE BODY CONVERTS FOOD TO ENERGY

The PO_4^{3-} ion is generally called inorganic phosphate.

When one phosphate group is hydrolyzed from each, the following energy yields are obtained:
ATP = 7.3 kcal/mol
ADP = 7.3 kcal/mol
AMP = 3.4 kcal/mol

The body is able to extract only 40 to 60% of the total caloric content of food.

Conversely, when inorganic phosphate bonds to AMP or ADP, greater amounts of energy are added to the chemical bond than when it bonds to adenosine. ADP and ATP contain *high-energy* phosphoric anhydride bonds.

ATP releases the most energy and AMP releases the least energy when each gives up one phosphate group. This property makes ATP a very useful compound for energy storage and release. The energy gained in the oxidation of food is stored in the form of ATP, albeit only for a short while. ATP molecules in the cells normally do not last longer than about 1 min. They are hydrolyzed to ADP and inorganic phosphate to yield energy that drives other processes, such as muscle contraction, nerve signal conduction, and biosynthesis. As a consequence, ATP is constantly being formed and decomposed. Its turnover rate is very high. Estimates suggest that the human body manufactures and degrades as much as 40 kg (approximately 88 lb) of ATP every day.

B Agents for Transfer of Electrons in Biological Oxidation–Reduction Reactions

Two other actors in this drama are the coenzymes (Section 22.3), NAD^+ (nicotinamide adenine dinucleotide) and FAD (flavin adenine dinucleotide), both of which contain an ADP core (Figure 26.6). In NAD^+, the operative part of the coenzyme is the nicotinamide part. In FAD, the operative part is the flavin portion. In both molecules, the ADP is the handle by which the apoenzyme holds onto the coenzyme; the other end of the molecule carries out the actual chemical reaction. For example, when NAD^+ is reduced, the nicotinamide part of the molecule gets reduced:

The $^+$ in NAD^+ refers to the positive charge on the nitrogen.

Nicotinamide and riboflavin are both members of the vitamin B group (see Table 29.1).

The R stands for the rest of the NAD^+ molecule; R′ stands for the rest of the FAD molecule.

$$NAD^+ + H^+ + 2e^- \rightleftharpoons NADH$$

The reduced form of NAD^+ is called NADH. The same reduction happens on two nitrogens of the flavin portion of FAD:

$$FAD + 2H^+ + 2e^- \rightleftharpoons FADH_2$$

The reduced form of FAD is called $FADH_2$. We view NAD^+ and FAD coenzymes as the **hydrogen ion** and **electron-transporting molecules**.

C Agent for Transfer of Acetyl Groups

Acetyl group The CH_3CO- group

The final principal compound in the common catabolic pathway is **coenzyme A** (CoA; Figure 26.7), which is the **acetyl (CCH_3CO-)-transporting molecule**. Coenzyme A also contains ADP, but here the next structural

26.3 THE PRINCIPAL COMPOUNDS OF THE COMMON CATABOLIC PATHWAY

Figure 26.6 The structures of NAD⁺ and FAD.

Figure 26.7 The structure of coenzyme A.

The hydrolysis of acetyl coenzyme A yields 7.51 kcal/mol.

unit is pantothenic acid, another B vitamin. Just as ATP can be viewed as an ADP molecule to which a $-PO_3^{2-}$ group is attached by a high-energy bond, so **acetyl coenzyme A** can be considered a CoA molecule linked to an acetyl group by a high-energy thioester bond (7.51 kcal/mol). The active part of coenzyme A is the mercaptoethylamine. The acetyl group of acetyl coenzyme A is attached to the SH group:

$$CH_3-\overset{\overset{O}{\|}}{C}-S-CoA$$
Acetyl coenzyme A

Hans Krebs (1900–1981), Nobel laureate in 1953, established the relationships among the different components of the cycle.

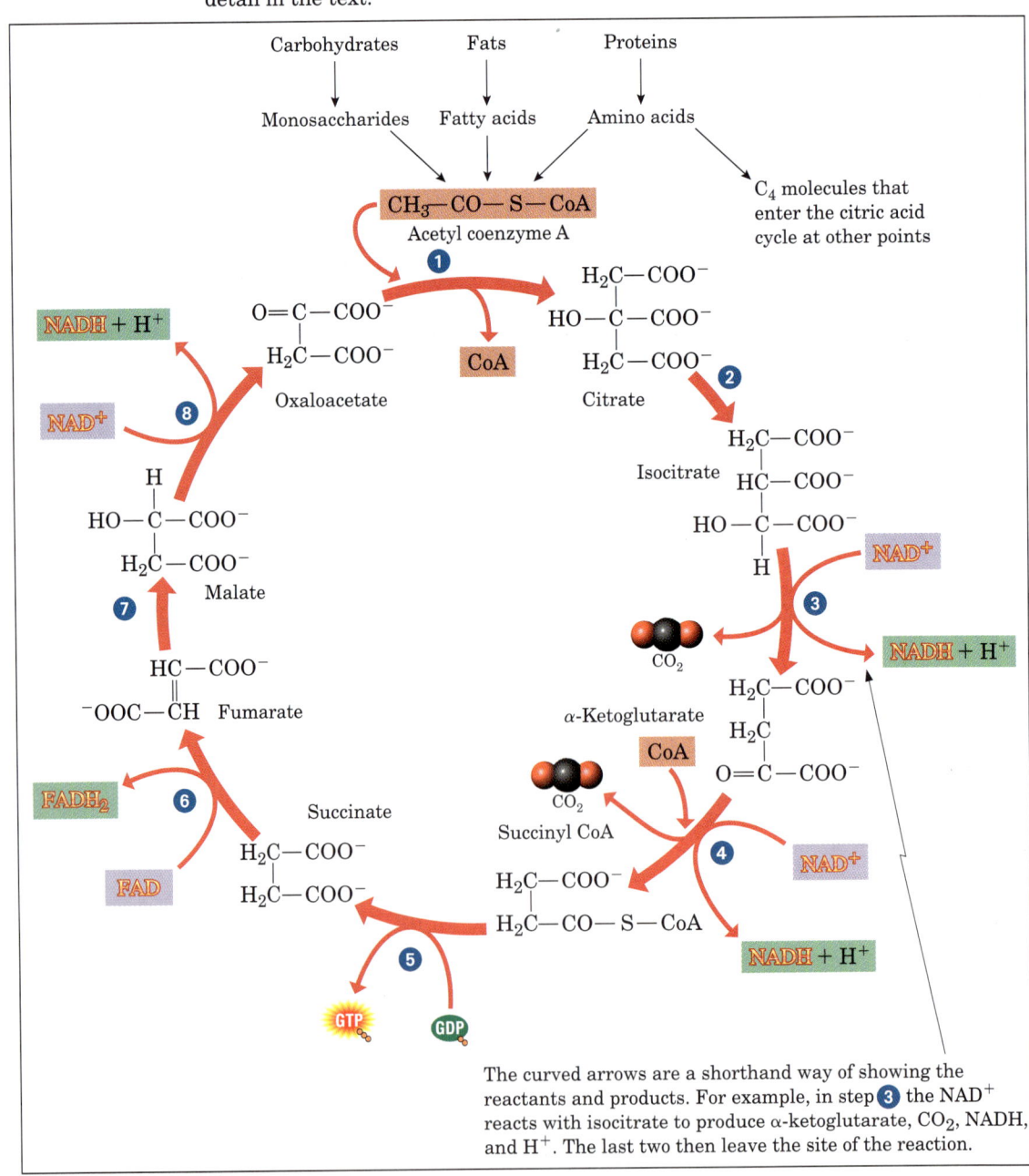

Figure 26.8 The citric acid (Krebs) cycle. The numbered steps are explained in detail in the text.

The curved arrows are a shorthand way of showing the reactants and products. For example, in step ❸ the NAD^+ reacts with isocitrate to produce α-ketoglutarate, CO_2, NADH, and H^+. The last two then leave the site of the reaction.

26.4 The Citric Acid Cycle

The common catabolism of carbohydrates and lipids begins when they have been broken down into pieces of two-carbon atoms each. The two-carbon fragments are the acetyl portions of acetyl coenzyme A. The acetyl is now fragmented further in the citric acid cycle.

Figure 26.8 gives the details of the citric acid cycle. A good way to gain an insight into this cycle is to use Figure 26.8 in conjunction with the simplified schematic diagram shown in Figure 26.9, which shows only the carbon balance.

We will now follow the two carbons of the acetyl group (C_2) through each step in the citric acid cycle. The circled numbers correspond to those in Figure 26.8.

Step 1 Acetyl coenzyme A enters the cycle by combining with a C_4 compound called oxaloacetate:

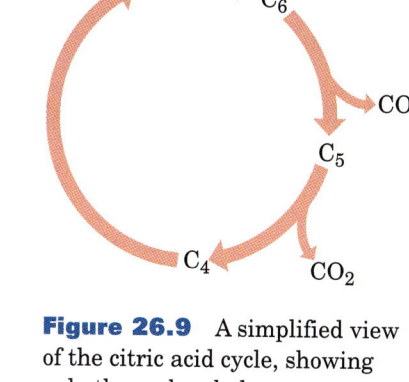

Figure 26.9 A simplified view of the citric acid cycle, showing only the carbon balance.

$$\underset{\text{Oxaloacetate}}{\begin{array}{c}O=C-COO^-\\|\\H_2C-COO^-\end{array}} + \underset{\text{Acetyl CoA}}{CH_3-\overset{O}{\overset{\|}{C}}-S-CoA} \xrightarrow{\text{Citrate synthase}} \underset{\text{Citryl CoA}}{\begin{array}{c}H_2C-\overset{O}{\overset{\|}{C}}-S-CoA\\HO-C-COO^-\\|\\H_2C-COO^-\end{array}} \xrightarrow{H_2O} \underset{\text{Citrate}}{\begin{array}{c}H_2C-COO^-\\|\\HO-C-COO^-\\|\\H_2C-COO^-\end{array}} + HS-CoA$$

In Step ⑧ we will see where the oxaloacetate comes from.

The first thing that happens is the addition of the $-CH_3$ group of the acetyl CoA to the $C=O$ of the oxaloacetate, catalyzed by the enzyme citrate synthase. This event is followed by hydrolysis of the thioester to produce the C_6 compound citrate ion and CoA. Therefore, Step ① is a building-up rather than a breaking-down process.

Step 2 The citrate ion is dehydrated to *cis*-aconitate, after which the *cis*-aconitate is hydrated, but this time to isocitrate instead of citrate:

$$\underset{\text{Citrate}}{\begin{array}{c}H_2C-COO^-\\|\\HO-C-COO^-\\|\\H_2C-COO^-\end{array}} \xrightarrow{-H_2O} \underset{\text{cis-Aconitate}}{\begin{array}{c}H_2C-COO^-\\|\\HC-COO^-\\\|\\C-COO^-\\|\\H\end{array}} \xrightarrow[\text{aconitase}]{H_2O} \underset{\text{Isocitrate}}{\begin{array}{c}H_2C-COO^-\\|\\HC-COO^-\\|\\HO-C-COO^-\\|\\H\end{array}}$$

Screen 25.2 Animation

In citrate, the alcohol is a tertiary alcohol. We learned in Section 14.3 that tertiary alcohols cannot be oxidized. The alcohol in the isocitrate is a secondary alcohol, which upon oxidation yields a ketone.

Step 3 The isocitrate is now oxidized and **decarboxylated** at the same time:

Decarboxylation The process that leads to the loss of CO_2 from a $-COOH$ group

Oxidation:

$$\underset{\text{Isocitrate}}{\begin{array}{c}H_2C-COO^-\\|\\HC-COO^-\\|\\HO-C-COO^-\\|\\H\end{array}} + NAD^+ \xrightarrow{\text{Isocitrate dehydrogenase}} \underset{\text{Oxalosuccinate}}{\begin{array}{c}H_2C-COO^-\\|\\HC-COO^-\\|\\O=C-COO^-\end{array}} + NADH + H^+$$

Decarboxylation:

$$\underset{\text{Oxalosuccinate}}{\begin{array}{c}H_2C-COO^-\\|\\HC-COO^-\\|\\O=C-COO^-\end{array}} + H^+ \longrightarrow \begin{array}{c}H_2C-COO^-\\|\\HC-COOH\\|\\O=C-COO^-\end{array} \longrightarrow \underset{\alpha\text{-Ketoglutarate}}{\begin{array}{c}H_2C-COO^-\\|\\CH_2\\|\\O=C-COO^-\end{array}} + CO_2$$

In oxidizing the secondary alcohol to a ketone, the oxidizing agent NAD^+ removes two hydrogens. One of them is used to reduce NAD^+ to NADH. The other replaces the COO^- that goes into making CO_2. Note that the CO_2 given off comes from the original oxaloacetate and not from the two carbons of the acetyl CoA. Both of these carbons are still present in the α-ketoglutarate. Also note that we are now down to a C_5 compound, α-ketoglutarate.

Steps 4 and 5 Next, a complex system removes another CO_2, once again from the original oxaloacetate portion rather than from the acetyl CoA portion:

> The CO_2 molecules given off in Steps ③ and ④ are the ones we exhale.
>
> The P_i is inorganic phosphate.

$$\underset{\alpha\text{-Ketoglutarate}}{\begin{array}{c}H_2C-COO^-\\|\\H_2C\\|\\O=C-COO^-\end{array}} + NAD^+ + GDP + P_i + H_2O \xrightarrow[\text{system}]{\text{Complex enzyme}} \underset{\text{Succinate}}{\begin{array}{c}H_2C-COO^-\\|\\H_2C-COO^-\end{array}} + CO_2 + NADH + H^+ + GTP$$

We are now down to a C_4 compound, succinate. This oxidative decarboxylation is more complex than the first. It occurs in many steps and requires a number of cofactors. For our purpose, it is sufficient to know that, during this second oxidative decarboxylation, a high-energy compound called **guanosine triphosphate (GTP)** is also formed.

GTP is similar to ATP, except that guanine replaces adenine. Otherwise, the linkages of the base to ribose and the phosphates are exactly the same as in ATP. The function of GTP is also similar to that of ATP—namely, it stores energy in the form of high-energy phosphoric anhydride bonds (chemical energy). The energy from the hydrolysis of GTP drives many important biochemical reactions—for example, the signal transduction in neurotransmission (Section 23.5).

Step 6 In this step, the succinate is oxidized by FAD, which removes two hydrogens to give fumarate (the double bond in this molecule is trans):

$$\underset{\text{Succinate}}{\begin{array}{c}H_2C-COO^-\\|\\H_2C-COO^-\end{array}} + FAD \xrightarrow{\text{Succinate dehydrogenase}} \underset{\text{Fumarate}}{\begin{array}{c}HC-COO^-\\||\\{}^-OOC-CH\end{array}} + FADH_2$$

This reaction cannot be carried out in the laboratory, but with the aid of an enzyme catalyst, the body does it easily.

Step 7 The fumarate is now hydrated to give the malate ion:

$$\underset{\text{Fumarate}}{\begin{array}{c}HC-COO^-\\||\\{}^-OOC-CH\end{array}} + H_2O \xrightarrow{\text{Fumarase}} \underset{\text{Malate}}{\begin{array}{c}H\\|\\HO-C-COO^-\\|\\H_2C-COO^-\end{array}}$$

Step 8 In the final step of the cycle, malate is oxidized by NAD^+ to give oxaloacetate:

$$\underset{\text{Malate}}{\underset{|}{\overset{H}{\underset{H_2C-COO^-}{HO-C-COO^-}}}} + NAD^+ \xrightarrow{\text{Malate dehydrogenase}} \underset{\text{Oxaloacetate}}{\underset{|}{\overset{}{\underset{H_2C-COO^-}{O=C-COO^-}}}} + NADH + H^+$$

Thus, the final product of the Krebs cycle is oxaloacetate, which is the compound with which we started in Step ①.

What has happened in the entire process is that the original two acetyl carbons of acetyl CoA were added to the C_4 oxaloacetate to produce a C_6 unit, which then lost two carbons in the form of CO_2 to produce, at the end of the process, the C_4 unit oxaloacetate. The net effect is that one two-carbon acetyl group enters the cycle and two carbon dioxides are given off.

How does the citric acid cycle produce energy? We have already learned that one step in the process produces a high-energy molecule of GTP. However, most of the energy is produced in the other steps that convert NAD^+ to NADH and FAD to $FADH_2$. These reduced coenzymes carry the H^+ and electrons that eventually will provide the energy for the synthesis of ATP (discussed in detail in Sections 26.5 and 26.6).

This stepwise degradation and oxidation of acetate in the citric acid cycle results in the most efficient extraction of energy. Rather than in one burst, the energy is released in small packets carried away step by step in the form of NADH and $FADH_2$.

The cyclic nature of this acetate degradation has other advantages besides maximizing energy yield:

1. The citric acid cycle components provide raw materials for amino acid synthesis as the need arises (Chapter 28).
2. The many-component cycle provides an excellent method for regulating the speed of catabolic reactions.

For example, α-ketoglutaric acid is used to synthesize glutamic acid.

The regulation can occur at many different parts of the cycle, so that feedback information can be used at many points to speed up or slow down the process as necessary.

The following equation represents the overall reactions in the citric acid cycle:

$$GDP + P_i + CH_3-CO-S-CoA + 2H_2O + 3NAD^+ + FAD$$
$$\longrightarrow CoA + GTP + 2CO_2 + 3NADH + FADH_2 + 3H^+ \quad \text{(Eq. 26.1)}$$

The citric acid cycle is controlled by feedback mechanisms. When the essential product of this cycle, $NADH + H^+$, and the end product of the common catabolic pathway, ATP, accumulate, they inhibit some of the enzymes in the cycle. Citrate synthase (Step ①), isocitrate dehydrogenase (Step ③), and α-ketoglutarate dehydrogenase (part of the complex enzyme system in Step ④) are inhibited by ATP and/or by $NADH + H^+$. This inhibition slows down or shuts off the cycle. On the other hand, when the feed material, acetyl CoA, is in abundance, the cycle accelerates. The enzyme isocitrate dehydrogenase (Step ③) is stimulated by ADP and NAD^+, which are the essential reactants from which the end products of the cycle are derived.

26.5 Electron and H⁺ Transport

The reduced coenzymes NADH and FADH$_2$ are end products of the citric acid cycle. They carry hydrogen ions and electrons and, therefore, have the potential to yield energy when these combine with oxygen to form water:

$$4H^+ + 4e^- + O_2 \longrightarrow 2H_2O + \text{energy}$$

The oxygen in this reaction is the oxygen we breathe.

This simple exothermic reaction is carried out in many steps.

A number of enzymes are involved in this reaction, all of which are embedded in the inner membrane of the mitochondria. These enzymes are situated in a particular *sequence* in the membrane so that the product from one enzyme can be passed on to the next enzyme, in a kind of assembly line. The enzymes are arranged in order of increasing affinity for electrons, so electrons flow through the enzyme system (Figure 26.10).

The sequence of the electron-carrying enzyme systems starts with complex I. The largest complex, it contains some 40 subunits, among them a flavoprotein and several FeS clusters. **Coenzyme Q** (CoQ; also called ubiquinone) is associated with complex I, which oxidizes the NADH produced in the citric acid cycle and reduces the CoQ:

$$NADH + H^+ + CoQ \longrightarrow NAD^+ + CoQH_2$$

Some of the energy released in this reaction is used to move 2H⁺ across the membrane, from the matrix to the intermembrane space. The CoQ is soluble in lipids and can move laterally within the membrane.

Screen 25.3 Tutorial

Figure 26.10 Schematic diagram of the electron and H⁺ transport chain and subsequent phosphorylation. The combined processes are also known as oxidative phosphorylation.

CHEMICAL CONNECTIONS 26A

The Importance of Iron–Sulfur Cluster Assembly

As we have seen, iron–sulfur clusters play an important role in the electron transport system. We have also learned that metals play a critical role in redox reactions (Section 4.7). A less appreciated fact is that sulfur can play an even more active role.

Sulfur has many oxidation states and thus can attract or donate electrons quickly with ease. The careful packaging of the resulting sulfur–iron clusters in surrounding proteins, as in complexes I and II, is necessary because the naked iron and sulfur ions are even more toxic to a cell than is cyanide. A large number of proteins, enzymes, and chaperones facilitate these assemblies. The iron–sulfur clusters themselves are formed inside the mitochondrion. For their role inside the mitochondrion, they are inserted into protein molecules. To perform their role on the inner membrane of mitochondrion, they first must be transported through the membrane by special iron–sulfur cluster transporters. Only then can they be packaged into complexes of the electron transport chain.

In an inherited disease called sideroblastic anemia, the iron–sulfur cluster transporter is inactive in cells that should eventually become red blood cells (RBC). As a consequence, iron accumulates inside the mitochondrion of these RBC precursor cells, causing anemia.

Complex II also catalyzes the transfer of electrons to CoQ. The source of the electrons is the oxidation of succinate in the citric acid cycle, producing $FADH_2$. The final reaction is

$$FADH_2 + CoQ \longrightarrow FAD + CoQH_2$$

The energy of this reaction is not sufficient to pump two protons across the membrane, nor is there an appropriate channel for such a transfer.

Complex III delivers the electrons from $CoQH_2$ to **cytochrome c.** This integral membrane complex contains 11 subunits, including cytochrome b, cytochrome c_1, and FeS clusters. Each cytochrome has an iron-ion-containing heme center (Section 21.9) embedded in its own protein. Complex III has two channels through which two H^+ ions are pumped from the $CoQH_2$ into the intermembrane space. The process is very complicated. To simplify matters, we can imagine that it occurs in two distinct steps, as does the electron transfer. Since each cytochrome c can pick up only one electron, two cytochrome c's are needed:

$CoQH_2$ + 2 cytochrome c (reduced)

$\longrightarrow$ CoQ + $2H^+$ + 2 cytochrome c (oxidized)

Cytochrome c is also a mobile carrier of electrons—it can move laterally in the intermembrane space.

Complex IV, known as cytochrome oxidase, contains 13 subunits—most importantly, cytochrome a_3, a heme that has an associated copper center. Complex IV is an integral membrane protein complex. The electron movement flows from cytochrome c to cytochrome a to cytochrome a_3. There, the electrons are transferred to the oxygen molecule, and the O—O bond is cleaved. The oxidized form of the enzyme takes up two H^+ ions from the matrix for each oxygen atom. The water molecule formed in this way is released into the matrix:

$$\tfrac{1}{2}O_2 + 2H^+ + 2e^- \longrightarrow H_2O$$

The letters used to designate the cytochromes were given in order of their discovery.

CHEMICAL CONNECTIONS 26B

2,4-Dinitrophenol as an Uncoupling Agent

Polynitrated organic compounds are highly explosive (see Chemical Connections 13D). Trinitrotoluene (TNT) is the best known of these agents. During World War I, many ammunition workers were exposed to 2,4-dinitrophenol (DNP), a compound used to prepare the explosive picric acid. After it was observed that these workers lost weight, DNP was used as a weight-reducing drug during the 1920s. Unfortunately, DNP eliminated not only the fat but sometimes also the patient, and its use as a diet pill was discontinued after 1929.

Today we know why DNP works as a weight-reducing drug. It is an effective protonophore, which is a compound that transports H^+ ions through cell membranes passively, without the expenditure of energy. As noted earlier, H^+ ions accumulate in the intermembrane space of mitochondria and, under normal conditions, drive the synthesis of ATP while they are going back to the inside. This is Mitchell's chemiosmotic principle in action. When DNP is ingested, it transfers the H^+ back to the mitochondrion easily, and no ATP is manufactured. The energy of the electron separation is dissipated as heat and is not built in as chemical energy in ATP. The loss of this energy-storing compound makes the utilization of food much less efficient, resulting in weight loss.

The medical literature includes a case history of a woman whose muscles contained mitochondria in which electron transport was not coupled to oxidative phosphorylation. This unfortunate woman could not utilize electron transport to generate ATP. As a consequence, she was severely incapacitated and bedridden. When the uncoupling worsened, the heat generated by the uncontrolled oxidation (no ATP production) was so severe that the patient required continuous cooling.

A similar mechanism provides heat in hibernating bears. The bears have brown fat; its color is derived from the numerous mitochondria in the tissue. The brown fat also contains an uncoupling protein, a protonophore that allows the H^+ ions to stream back into the mitochondrial matrix without manufacturing ATP. The heat generated in this manner keeps the animal alive during cold winter days.

During this process, two more H^+ ions are pumped out of the matrix and into the intermembrane space. Although the mechanism of pumping out protons from the matrix is not known, the energy driving this process is derived from the energy of water formation. This final pumping into the intermembrane space makes a total of six H^+ ions per NADH + H^+ and four H^+ ions per $FADH_2$ molecules.

26.6 Phosphorylation and the Chemiosmotic Pump

How do the electron and H^+ transports produce the chemical energy of ATP? In 1961, Peter Mitchell (1920–1992), an English chemist, proposed the **chemiosmotic theory** to answer this question: The energy in the electron transfer chain creates a proton gradient. A **proton gradient** is a continuous variation in the H^+ concentration along a given region. In this case, there is a higher concentration of H^+ in the intermembrane space than inside the mitochondrion. The driving force, which is the result of the spontaneous flow of ions from a region of high concentration to a region of low concentration, propels the protons back to the mitochondrion through a complex known as **proton translocating ATPase**. This compound is located on the inner membrane of the mitochondrion (Figure 26.10) and is the active enzyme that catalyzes the conversion of ADP and inorganic phosphate to ATP (the reverse of the reaction shown in Figure 26.5):

$$ADP + P_i \xrightleftharpoons{ATPase} ATP + H_2O$$

> **Chemiosmotic theory** Mitchell proposed that the electron transport is accompanied by an accumulation of protons in the intermembrane space of the mitochondrion, which in turn creates an osmotic pressure; the protons driven back to the mitochondrion under this pressure generate ATP

Mitchell received the Nobel Prize in chemistry in 1978.

CHEMICAL CONNECTIONS 26C

Protection Against Oxidative Damage

As we have seen, most of the oxygen we breathe in is used in the final step of the common pathway—namely, in the conversion of our food supply to ATP. This reaction occurs in the mitochondria of cells. However, some 10% of the oxygen we consume is used elsewhere, in specialized oxidation reactions. One such reaction is the hydroxylation of compounds (for example, steroids). It is catalyzed by the enzyme cytochrome P-450, which usually resides outside the mitochondria in the endoplasmic reticulum of cells. P-450 catalyzes the following reaction, where RH represents a steroid or other molecule:

$$RH + O_2 + \boxed{NADPH} + H^+$$
$$\longrightarrow ROH + H_2O + \boxed{NADP^+}$$

Our bodies use this reaction to detoxify foreign substances (for example, barbiturates). The action of P-450 is not always beneficial, however. Some of the most powerful carcinogens are converted to their active forms by P-450 (see Chemical Connections 13B).

Another use of oxygen for detoxification is in the fight against infection by bacteria. Specialized white blood cells called leukocytes (Section 31.2B) engulf bacteria and kill them by generating superoxide, O_2^-, a highly reactive form of oxygen. But superoxide and other highly reactive forms of oxygen, such as ·OH radicals and hydrogen peroxide (H_2O_2), not only destroy foreign cells but also damage our own cells. They especially attack unsaturated fatty acids, thereby damaging cell membranes (Section 20.5). Our bodies have their own defenses against these very reactive compounds. One example is glutathione (Chemical Connections 21A), which protects us against oxidative damage. Furthermore, our cells contain two powerful enzymes that decompose these highly reactive oxidizing agents:

$$2O_2^- + 2H^+ \xrightarrow{\text{superoxide dismutase}} H_2O_2 + O_2$$

$$2H_2O_2 \xrightarrow{\text{catalase}} 2H_2O + O_2$$

These enzymes also play an important defensive role in heart attacks and strokes. When oxygen deprivation of the heart muscles or the brain occurs because arteries are blocked, a condition known as ischemia reperfusion may arise. This means that when these tissues suddenly become reoxygenated, superoxide (O_2^-) can form. This molecule, and the ·OH radicals derived from it, can further damage the tissues. Superoxide dismutase and catalase prevent reperfusion injuries.

Furthermore, the presence of superoxide activates uncoupling proteins (Chemical Connections 26B) and thus allows the leakage of H^+ ions across the mitochondrial membrane. These H^+ ions then serve as substrates for superoxide dismutase in forming hydrogen peroxide from superoxide.

One theory that has been advanced to explain why humans and animals show symptoms of old age suggests that these oxidizing agents damage healthy cells over the years. All the enzymes involved in reactions with highly reactive forms of oxygen contain a heavy metal: iron in the case of P-450 and catalase, and copper and/or zinc in the case of superoxide dismutase. If the theory is correct, then these protecting agents keep us alive for many years, but the oxidizing processes do so much damage that we eventually exhibit symptoms of aging and die.

The proton translocating ATPase is a complex "rotor engine" made of 16 different proteins. The F_0 sector, which is embedded in the membrane, contains the **proton channel** (Figure 26.10). The 12 subunits that form this channel rotate every time a proton passes from the cytoplasmic side (intermembrane) to the matrix side of the mitochondrion. This rotation is transmitted to a "rotor" in the F_1 sector. F_1 contains five kinds of polypeptides. The rotor (γ and ϵ subunits) is surrounded by the catalytic unit (made of α and β subunits) that synthesizes the ATP. The F_1 catalytic unit converts the mechanical energy of the rotor into chemical energy of the ATP molecule. The last unit, the "stator," containing the δ subunit, stabilizes the whole complex.

The proton translocating ATPase can catalyze the reaction in both directions. When protons that have accumulated on the outer surface of the mitochondrion stream inward, the enzyme manufactures ATP and stores the electrical energy (due to the flow of charges) in the form of chemical energy. In the reverse reaction, the enzyme hydrolyzes ATP and, as a consequence, pumps out H^+ from the mitochondrion. Each pair of protons that is translocated gives rise to the formation of one ATP molecule. Only when the two parts of the proton translocating ATPase F_1 and F_0 are linked is energy production possible. When the interaction between F_1 and F_0 is disrupted, the energy transduction is lost.

The protons that enter a mitochondrion combine with the electrons transported through the electron transport chain and with oxygen to form water. The net result of the two processes (electron/H^+ transport and ATP formation) is that each oxygen molecule we breathe in combines with four H^+ ions and four electrons to give two water molecules. The four H^+ ions and four electrons come from the NADH and $FADH_2$ molecules produced in the citric acid cycle. The oxygen, therefore, has two functions:

- To oxidize NADH to NAD^+ and $FADH_2$ to FAD so that these molecules can go back and participate in the citric acid cycle
- To provide energy for the conversion of ADP to ATP

The latter function is accomplished indirectly, not through the reduction of O_2 to H_2O. The entrance of the H^+ ions into the mitochondrion drives the ATP formation, but the H^+ ions enter the mitochondrion because the O_2 depleted the H^+ ion concentration when water was formed. This rather complex process involves the transport of electrons along a whole series of enzyme molecules (which catalyze all these reactions); however, the cell cannot utilize the O_2 molecules without it and eventually will die. The following equations represent the overall reactions in oxidative phosphorylation:

The electron and H^+ transport chain and subsequent phosphorylation process are collectively known as oxidative phosphorylation.

$$\text{NADH} + 3\,\text{ADP} + \tfrac{1}{2}O_2 + 3P_i + H^+ \longrightarrow \text{NAD}^+ + 3\,\text{ATP} + H_2O \qquad \text{(Eq. 26.2)}$$

$$\text{FADH}_2 + 2\,\text{ADP} + \tfrac{1}{2}O_2 + 2P_i \longrightarrow \text{FAD} + 2\,\text{ATP} + H_2O \qquad \text{(Eq. 26.3)}$$

26.7 The Energy Yield

The energy released during electron transport is now finally built into the ATP molecule. Therefore, it is instructive to look at the energy yield in the universal biochemical currency: the number of ATP molecules.

Each pair of protons entering a mitochondrion results in the production of one ATP molecule. For each NADH molecule, three pairs of protons are pumped into the intermembrane space in the electron transport process. Therefore, for each NADH molecule, we get three ATP molecules, as can be seen in Equation 26.2. For each $FADH_2$ molecule, only four protons are pumped out of the mitochondrion. Therefore, only two ATP molecules are produced for each $FADH_2$, as seen in Equation 26.3.

Now we can produce the energy balance for the entire common catabolic pathway (citric acid cycle and oxidative phosphorylation combined). For each C_2 fragment entering the citric acid cycle, we obtain three NADH and one $FADH_2$ (Equation 26.1) plus one GTP, which is equivalent in energy to one ATP. Thus, the total number of ATP molecules produced per C_2 fragment is

$$
\begin{aligned}
3\ \text{NADH} \times 3\ \text{ATP/NADH} &= 9\ \text{ATP} \\
1\ \text{FADH}_2 \times 2\ \text{ATP/FADH}_2 &= 2\ \text{ATP} \\
1\ \text{GTP} &= \underline{1\ \text{ATP}} \\
&= 12\ \text{ATP}
\end{aligned}
$$

Each C_2 fragment that enters the cycle produces 12 ATP molecules and uses up two O_2 molecules. The total effect of the energy-production chain of reactions discussed in this chapter (the common catabolic pathway) is to oxidize one C_2 fragment with two molecules of O_2 to produce two molecules of CO_2 and 12 molecules of ATP:

$$C_2 + 2O_2 + 12 \text{ ADP} + 12 P_i \longrightarrow 12 \text{ ATP} + 2CO_2$$

The important thing is not the waste product, CO_2, but rather the 12 ATP molecules. These molecules will now release their energy when they are converted to ADP.

26.8 Conversion of Chemical Energy to Other Forms of Energy

As mentioned in Section 26.3, the storage of chemical energy in the form of ATP lasts only a short time. Usually, within a minute, the ATP is hydrolyzed (an exothermic reaction) and releases its chemical energy. How does the body utilize this chemical energy? To answer this question, let us look at the different forms in which energy is needed in the body.

A Conversion to Other Forms of Chemical Energy

The activity of many enzymes is controlled and regulated by phosphorylation. For example, the enzyme phosphorylase, which catalyzes the breakdown of glycogen (Chemical Connections 27B), occurs in an inactive form, phosphorylase b. When ATP transfers a phosphate group to a serine residue, the enzyme becomes active. Thus, the chemical energy of ATP is used in the form of chemical energy to activate phosphorylase b so that glycogen can be utilized. We will see several other examples of this energy conversion in Chapters 27 and 28.

B Electrical Energy

The body maintains a high concentration of K^+ ions inside the cells despite the fact that the K^+ concentration is low outside the cells. The reverse is true for Na^+. So that K^+ does not diffuse out of the cells and Na^+ does not enter them, special transport proteins in the cell membranes constantly pump K^+ into and Na^+ out of the cells. This pumping requires energy, which is supplied by the hydrolysis of ATP to ADP. With this pumping, the charges inside and outside the cell are unequal, which generates an electric potential. Thus, the chemical energy of ATP is transformed into electrical energy, which operates in neurotransmission (Section 23.2).

C Mechanical Energy

ATP is the immediate source of energy in muscle contraction. In essence, muscle contraction takes place when thick and thin filaments slide past each other (Figure 26.11). The thick filament is myosin, an ATPase enzyme (that is, one that hydrolyzes ATP). The thin filament, actin, binds strongly to myosin in the contracted state. However, when ATP binds to myosin, the actin–myosin complex dissociates, and the muscle relaxes. When myosin hydrolyzes ATP, it interacts with actin once more, and a new contraction occurs. In this way, the hydrolysis of ATP drives the alternating association and dissociation of actin and myosin and, consequently, the contraction and relaxation of the muscle.

Figure 26.11 Schematic diagram of muscle contraction.

(a) Relaxed muscle

(b) Contracted muscle

D Heat Energy

One molecule of ATP upon hydrolysis to ADP yields 7.3 kcal/mol. Some of this energy is released as heat and used to maintain body temperature. If we estimate that the specific heat of the body is about the same as that of water, a person weighing 60 kg would need to hydrolyze approximately 99 moles (approximately 50 kg) of ATP to raise the temperature of the body from room temperature, 25°C, to 37°C. Not all body heat is derived from ATP hydrolysis; some other exothermic reactions in the body also make heat contributions.

SUMMARY

The sum total of all the chemical reactions involved in maintaining the dynamic state of cells is called **metabolism** (Section 26.1). The breaking down of molecules is **catabolism**; the building up of molecules is **anabolism**. The **common metabolic pathway** uses a two-carbon C_2 fragment (acetyl) from different foods. Through the **citric acid cycle** and **oxidative phosphorylation (electron transport chain),** the C_2 fragment is oxidized. The products formed are water and carbon dioxide. The energy from oxidation is built into the high-chemical-energy-storing molecule ATP.

Both the citric acid cycle and oxidative phosphorylation take place in the **mitochondria** (Section 26.2). The enzymes of the citric acid cycle are located in the matrix, whereas the enzymes of the oxidative phosphorylation chain are on the inner mitochondrial membrane. Some of them project into the intermembrane space.

The principal carriers in the common catabolic pathway (Section 26.3) are as follows: **ATP** is the phosphate carrier; **CoA** is the C_2 fragment carrier; and **NAD⁺** and **FAD** carry the hydrogen ions and electrons. The unit common to all these carriers is **ADP**. The nonactive end of the carriers acts as a handle that fits into the active sites of the enzymes.

In the citric acid cycle (Section 26.4), the C_2 fragment first combines with a C_4 fragment (oxaloacetate) to yield a C_6 fragment (citrate). An oxidative decarboxylation yields a C_5 fragment. One CO_2 is released, and one NADH + H⁺ is passed to the electron transport chain. Another oxidative decarboxylation provides a C_4 fragment. Once again, a CO_2 is released, and another NADH + H⁺ is passed to the electron transport chain. Subsequently, two dehydrogenation (oxidation) steps yield one $FADH_2$ and one additional NADH + H⁺, along with an analog of ATP called GTP. This cycle is controlled by a feedback mechanism.

The electrons of NADH enter the electron transport chain (Section 26.5) at the complex I stage. The coenzyme Q (CoQ) of this complex picks up the electrons and the H⁺ and becomes $CoQH_2$. The energy of this reduction reaction is used to expel two H⁺ ions from the matrix into the intermembrane space. Complex II also has CoQ. Electrons and H⁺ are passed to this complex, and complex II catalyzes the transfer of electrons from $FADH_2$. However, no H⁺ ions are pumped into the intermembrane space at this point. The electrons are passed along by CoQH$_2$ to complex III of the electron transport chain. At complex III, the two H$^+$ ions from the CoQH$_2$ are expelled into the intermembrane space. Cytochrome c of complex III transfers electrons to complex IV through redox reactions. As the electrons are transported from cytochrome c to complex IV, an additional two H$^+$ ions are expelled from the matrix of the mitochondrion to the intermembrane space. For each NADH, six H$^+$ ions are expelled. For each FADH$_2$, four H$^+$ ions are expelled. Finally, the electrons passed to complex IV return to the matrix, where they combine with oxygen and H$^+$ to form water.

When the expelled H$^+$ ions stream back into the mitochondrion, they drive a complex enzyme called **proton translocating ATPase,** which makes one ATP molecule for each two H$^+$ ions that enter the mitochondrion (Section 26.6). Therefore, for each NADH + H$^+$ coming from the citric acid cycle, three ATP molecules are formed. For each FADH$_2$, two ATP molecules are formed. The overall result: For each C$_2$ fragment that enters the citric acid cycle, 12 ATP molecules are produced (Section 26.7).

The proton translocating ATPase is a complex "rotor engine" (Section 26.6). The proton channel part (F$_0$) is embedded in the membrane, and the catalytic unit (F$_1$) converts mechanical energy to chemical energy of the ATP molecule. The chemical energy is stored in ATP only for a short time—ATP is quickly hydrolyzed, usually within a minute. This chemical energy is used to do chemical, mechanical, and electrical work in the body and to maintain body temperature (Section 26.8).

PROBLEMS

Numbers that appear in color indicate difficult problems. ▶ designates problems requiring application of principles.

Metabolism

26.1 To what end product is the energy of foods converted in the catabolic pathways?

26.2 (a) How many sequences are in the common catabolic pathway?
(b) Name these sequences.

Cells and Mitochondria

26.3 (a) How many membranes do mitochondria have?
(b) Which membrane is permeable to ions and small molecules?

26.4 How do the enzymes of the common pathway find their way into the mitochondria?

26.5 What are cristae, and how are they related to the inner membrane of mitochondria?

26.6 (a) Where are the enzymes of the citric acid cycle located?
(b) Where are the enzymes of oxidative phosphorylation located?

Principal Compounds of the Common Catabolic Pathway

26.7 How many high-energy phosphate bonds are in the ATP molecule?

26.8 What are the products of the following reaction?

$$AMP + H_2O \xrightarrow{H^+}$$

26.9 Which yields more energy, the hydrolysis (a) of ATP to ADP or (b) of ADP to AMP?

26.10 How much ATP is needed for normal daily activity in humans?

26.11 What kind of chemical bond exists between the ribitol and the flavin in FAD?

26.12 When NAD$^+$ is reduced, two electrons enter the molecule, together with one H$^+$ ion. Where in the product will the two electrons be located?

26.13 Which atoms in the flavin portion of FAD are reduced to yield FADH$_2$?

26.14 NAD$^+$ has two ribose units in its structure; FAD has a ribose and ribitol. What is the relationship between these molecules?

26.15 In the common catabolic pathway, a number of important molecules act as carriers (transfer agents).
(a) Which is the carrier of phosphate groups?
(b) Which are the coenzymes transferring hydrogen ions and electrons?
(c) What kind of groups does coenzyme A carry?

26.16 The ribitol in FAD is bound to phosphate. What is nature of this bond? On the basis of the energies of the different bonds in ATP, estimate how much energy (in kcal/mol) would be obtained from the hydrolysis of this bond.

26.17 What kind of chemical bond exists between the pantothenic acid and mercaptoethylamine in the structure of CoA?

26.18 Name the vitamin B molecules that are a part of the structure of (a) NAD$^+$, (b) FAD, and (c) coenzyme A.

26.19 In both NAD$^+$ and FAD, the vitamin B portion of the molecule is the active part. Is this also true for CoA?

26.20 What type of compound is formed when coenzyme A reacts with acetate?

26.21 The fats and carbohydrates metabolized by our bodies are eventually converted to a single compound. What is it?

The Citric Acid Cycle

26.22 The first step in the citric acid cycle is abbreviated as

$$C_2 + C_4 = C_6$$

   (a) What do these symbols stand for?
   (b) What are the common names of the three compounds involved in this reaction?

26.23 What is the only C_5 compound in the citric acid cycle?

26.24 Identify by number those steps of the citric acid cycle that are not redox reactions.

26.25 Which substrate in the citric acid cycle is oxidized by FAD? What is the oxidation product?

26.26 In Steps ③ and ⑤ of the citric acid cycle, the compounds are shortened by one carbon each time. What is the form of this one-carbon compound? What happens to it in the body?

26.27 According to Table 22.1, to what class of enzymes does fumarase belong?

26.28 List all the enzymes or enzyme systems of the citric acid cycle that could be classified as oxidoreductases.

26.29 Is ATP directly produced during any step of the citric acid cycle? Explain.

26.30 There are four dicarboxylic acid compounds, each containing four carbons, in the citric acid cycle. Which is (a) the least oxidized and (b) the most oxidized?

26.31 Why is a many-step cyclic process more efficient in utilizing energy from food than a single-step combustion?

26.32 Did the two CO_2 molecules given off in one turn of the citric acid cycle originate from the entering acetyl group?

26.33 Which intermediates of the citric acid cycle contain C=C double bonds?

26.34 The citric acid cycle can be regulated by the body. It can be slowed down or speeded up. What mechanism controls this process?

26.35 Oxidation is defined as loss of electrons. When oxidative decarboxylation occurs, as in Step ④, where do the electrons of the α-ketoglutarate go?

Oxidative Phosphorylation

26.36 What is the main function of oxidative phosphorylation (the electron transport chain)?

26.37 What are the mobile electron carriers of the oxidative phosphorylation?

26.38 In each complex of the electron transport system, the redox reaction occurs around Fe ions.
   (a) Identify the compounds that contain such Fe centers.
   (b) Identify the compounds that contain ion centers other than iron.

26.39 What kind of motion is set up in the proton translocating ATPase by the passage of H$^+$ from the intermembrane space into the matrix?

26.40 The following reaction is a reversible reaction:

$$NADH \rightleftharpoons NAD^+ + H^+ + 2e^-$$

   (a) Where does the forward reaction occur in the common catabolic pathway?
   (b) Where does the reverse reaction occur?

26.41 In oxidative phosphorylation, water is formed from H$^+$, e$^-$, and O_2. Where does this take place?

26.42 At what points in oxidative phosphorylation are the H$^+$ ions and the electrons separated from each other?

26.43 How many ATP molecules are generated (a) for each H$^+$ translocated through the ATPase complex and (b) for each C_2 fragment that goes through the complete common catabolic pathway?

26.44 When H$^+$ is pumped out into the intermembrane space, is the pH there increased, decreased, or unchanged compared with that in the matrix?

The Chemiosmotic Pump

26.45 What is the channel through which H$^+$ ions reenter the matrix of mitochondria?

26.46 The proton gradient accumulated in the intermembrane area of a mitochondrion drives the ATP manufacturing enzyme, ATPase. Why do you think Mitchell called this the "chemiosmotic theory"?

26.47 Which part of the proton translocating ATPase machinery is the catalytic unit? What chemical reaction does it catalyze?

26.48 When the interaction between the two parts of proton translocating ATPase, F_0 and F_1, are disrupted, no energy production is possible. Which subunits maintain connections between F_0 and F_1, and what names are designated for these subunits?

The Energy Yield

26.49 If each mole of ATP yields 7.3 kcal of energy upon hydrolysis, how many kilocalories of energy would you get from 1 g of CH_3COO^- (C_2) entering the citric acid cycle?

26.50 A hexose (C_6) enters the common metabolic pathway in the form of two C_2 fragments.
 (a) How many molecules of ATP are produced from one hexose molecule?
 (b) How many O_2 molecules are used up in the process?

Conversion of Chemical Energy to Other Forms

26.51 (a) How do muscles contract?
 (b) Where does the energy in muscle contraction come from?

26.52 Give an example of the conversion of the chemical energy of ATP to electrical energy.

26.53 How is the enzyme phosphorylase activated?

Chemical Connections

26.54 (Chemical Connections 26A) How can sulfur act as a redox center?

26.55 (Chemical Connections 26A) Sulfur usually is poisonous to cells. How can Fe-S clusters exist and not kill the cells?

26.56 (Chemical Connections 26B) What is a protonophore?

26.57 (Chemical Connections 26B) Oligomycin is an antibiotic that allows electron transport to continue but stops phosphorylation in both bacteria and humans. Would you use it as an antibacterial drug for people? Explain.

26.58 (Chemical Connections 26C) How does superoxide dismutase prevent damage to the brain during a stroke?

26.59 (Chemical Connections 26C) (a) What kind of reaction is catalyzed by cytochrome P-450?
 (b) Where does the oxygen reactant come from?

Additional Problems

26.60 (a) What is the difference in structure between ATP and GTP?
 (b) Compared with ATP, would you expect GTP to carry more, less, or about the same amount of energy?

26.61 How many grams of CH_3COOH (from acetyl CoA) molecules must be metabolized in the common metabolic pathway to yield 87.6 kcal of energy?

26.62 What is the basic difference in the functional groups between citrate and isocitrate?

26.63 The passage of H^+ ions from the cytoplasmic side into the matrix generates mechanical energy. Where is this energy of motion exhibited first?

26.64 What kind of reaction occurs in the citric acid cycle when a C_6 compound is converted to a C_5 compound?

26.65 What structural characteristics do citric acid and malic acid have in common?

26.66 Two keto acids are important in the citric acid cycle. Identify them, and tell how they are manufactured.

26.67 Which filament of muscles is an enzyme, catalyzing the reaction that converts ATP to ADP?

26.68 One of the end products of food metabolism is water. How many molecules of H_2O are formed from the entry of each molecule of (a) NADH + H^+ and (b) $FADH_2$? (*Hint:* Use Figure 26.10.)

26.69 How many stereocenters are in isocitrate?

26.70 Acetyl CoA is labeled with radioactive carbon as shown: $CH_3{}^*CO—S—CoA$. This compound enters the citric acid cycle. If the cycle is allowed to progress to only the α-ketoglutarate level, will the CO_2 expelled by the cell be radioactive?

26.71 Where is the H^+ ion channel located in the proton translocating ATPase complex?

26.72 Is the passage of H^+ ion through the channel converted directly into chemical energy?

26.73 Does all the energy used in ATP synthesis come from the mechanical energy of rotation?

26.74 (a) In the citric acid cycle, how many steps can be classified as decarboxylation reactions?
 (b) In each case, what is the concurrent oxidizing agent? (*Hint:* See Table 22.1.)

26.75 What is the role of succinate dehydrogenase in the citric acid cycle?

26.76 How many stereocenters are in malate?

InfoTrac College Edition

For additional reading, go to InfoTrac College Edition, your online research library, at

http://infotrac.thomsonlearning.com

CHAPTER 27

27.1 Introduction
27.2 Glycolysis
27.3 The Energy Yield from Glucose
27.4 Glycerol Catabolism
27.5 β-Oxidation of Fatty Acids
27.6 The Energy Yield from Stearic Acid
27.7 Ketone Bodies
27.8 Catabolism of the Nitrogen of Amino Acids
27.9 Catabolism of the Carbon Skeleton of Amino Acids
27.10 Catabolism of Heme

Ballet dancer leaping.

Specific Catabolic Pathways: Carbohydrate, Lipid, and Protein Metabolism

27.1 Introduction

The food we eat serves two main purposes: (1) It fulfills our energy needs, and (2) it provides the raw materials to build the compounds our bodies need. Before either of these processes can take place, food—carbohydrates, fats, and proteins—must be broken down into small molecules that can be absorbed through the intestinal walls.

A Carbohydrates

Complex carbohydrates (di- and polysaccharides) in the diet are broken down by enzymes and stomach acid to produce monosaccharides, the most important of which is glucose (Section 29.4). Glucose also comes from the enzymatic breakdown of glycogen that is stored in the liver and muscles until needed. Once monosaccharides are produced, they can be used either to build new oligo- and polysaccharides or to provide energy. The specific pathway by which energy is extracted from monosaccharides is called glycolysis (Sections 27.2 and 27.3).

B Lipids

Ingested fats are broken down by lipases to glycerol and fatty acids or to monoglycerides, which are absorbed through the intestine (Section 29.5). In a similar way, complex lipids are hydrolyzed to smaller units before their absorption. As with carbohydrates, these smaller molecules (fatty acids, glycerol, and so on) can be used to build the complex molecules needed in membranes, or they can be oxidized to provide energy, or they can be stored in **fat storage depots** (Figure 27.1). The stored fats can later be hydrolyzed to glycerol and fatty acids whenever they are needed as fuel.

The specific pathway by which energy is extracted from glycerol involves the same glycolysis pathway as that used for carbohydrates (Section 27.4). The specific pathway used by the cells to obtain energy from fatty acids is called β-oxidation (Section 27.5).

C Proteins

As you might expect from a knowledge of their structures, proteins are broken down by HCl in the stomach and by digestive enzymes in the stomach (pepsin) and intestines (trypsin, chymotrypsin, and carboxypeptidases) to produce their constituent amino acids. The amino acids absorbed through the intestinal wall enter the **amino acid pool.** They serve as building blocks for proteins as needed and, to a smaller extent (especially during starvation), as a fuel for energy. In the latter case, the nitrogen of the amino acids is catabolized through oxidative deamination and the urea cycle and is expelled from the body as urea in the urine (Section 27.8). The carbon skeletons of the amino acids enter the common catabolic pathway (Chapter 26) as either α-ketoacids (pyruvic, oxaloacetic, and α-ketoglutaric acids) or acetyl coenzyme A (Section 27.9).

In all cases, *the specific pathways of carbohydrate, triglyceride (fat), and protein catabolism converge to the common catabolic pathway* (Figure 27.2, next page). In this way, the body needs fewer enzymes to get energy from diverse food materials. Efficiency is achieved because a minimal number of chemical steps are required and because the energy-producing factories of the body are localized in the mitochondria.

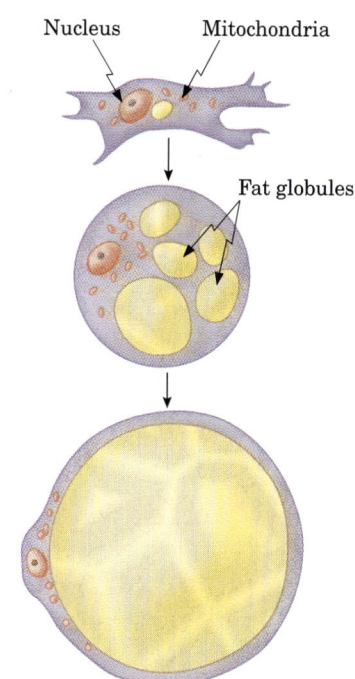

Figure 27.1 Storage of fat in a fat cell. As more and more fat droplets accumulate in the cytoplasm, they coalesce to form a very large globule of fat. Such a fat globule may occupy most of the cell, pushing the cytoplasm and the organelles to the periphery. (Modified from C. A. Villee, E. P. Solomon, and P. W. Davis, Biology, Philadelphia, Saunders College Publishing, 1985.)

Amino acid pool The free amino acids found both inside and outside cells throughout the body

27.2 Glycolysis

Glycolysis is the specific pathway by which the body gets energy from monosaccharides. The detailed steps in glycolysis are shown in Figure 27.3, and the most important features are shown schematically in Figure 27.4.

A Glycolysis of glucose

In the first steps of glucose metabolism, energy is consumed rather than released. At the expense of two molecules of ATP (which are converted to ADP), glucose (C_6) is phosphorylated. First, glucose 6-phosphate is formed in Step ①, then, after isomerization to fructose 6-phosphate in Step ②, a

Glycolysis The biochemical pathway that breaks down glucose to pyruvate, which yields chemical energy in the form of ATP and reduced coenzymes

Screen 26.2 Animation

Figure 27.2 The convergence of the specific pathways of carbohydrate, fat, and protein catabolism into the common catabolic pathway, which is made up of the citric acid cycle and oxidative phosphorylation.

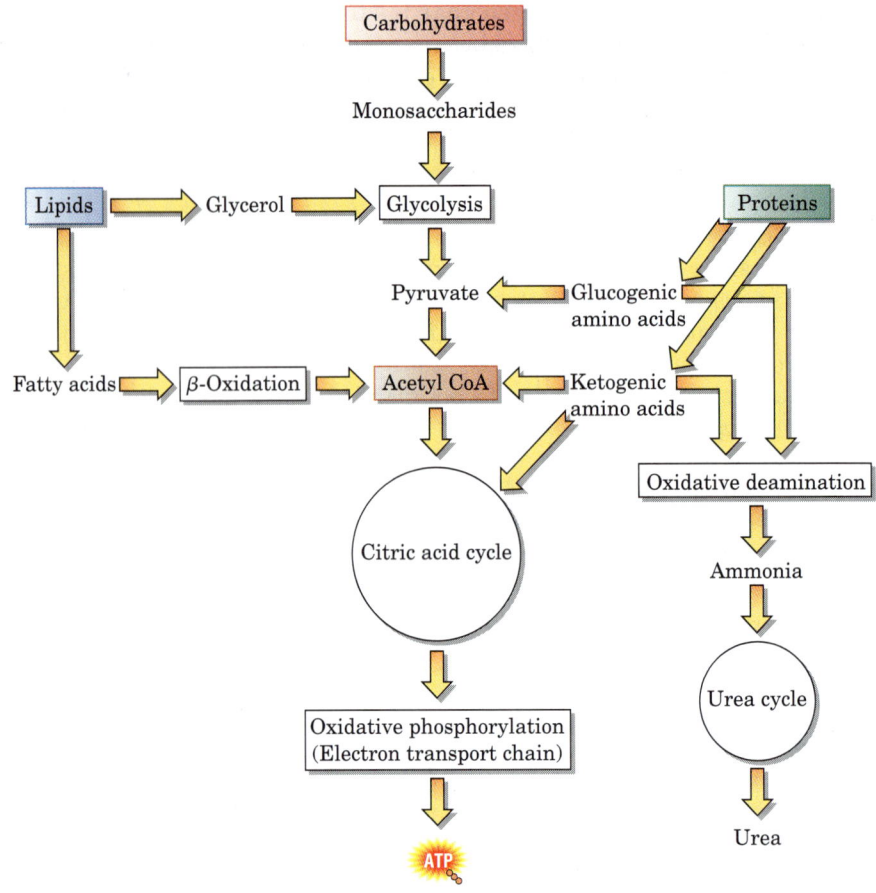

second phosphate group is attached to yield fructose 1,6-bisphosphate in Step ③. We can consider these steps to be the activation process.

In the second stage, the C_6 compound, fructose 1,6-bisphosphate, is broken into two C_3 fragments in Step ④. The two C_3 fragments, glyceraldehyde 3-phosphate and dihydroxyacetone phosphate, are in equilibrium (they can be converted to each other). Only glyceraldehyde 3-phosphate is oxidized in glycolysis, but as this species is removed from the equilibrium mixture, the equilibrium shifts (see the discussion of Le Chatelier's principle in Section 7.8) and dihydroxyacetone phosphate is converted to glyceraldehyde 3-phosphate.

In the third stage, glyceraldehyde 3-phosphate is oxidized to 1,3-bisphosphoglycerate in Step ⑤. The hydrogen of the aldehyde group is removed by the NAD^+ coenzyme. In Step ⑥, the phosphate from the carboxyl group is transferred to ADP, yielding ATP and 3-phosphoglycerate. The latter compound, after isomerization in Step ⑦ and dehydration in Step ⑧, is converted to phosphoenolpyruvate, which loses its remaining phosphate in Step ⑨ and yields pyruvate and another ATP molecule. This last Step (⑨) is also the "payoff" step, as the two ATP molecules produced here (one for each C_3 fragment) represent the net yield of ATPs in glycolysis. Step ⑨ is catalyzed by an enzyme, pyruvate kinase, whose active site was depicted in Chemical Connections 22C. This enzyme plays a key role in the regulation of glycolysis. For example, pyruvate kinase is inhibited by ATP and activated by AMP. Thus, when plenty of ATP is available, glycolysis is shut down; when ATP is scarce and AMP levels are high, the glycolytic pathway is speeded up.

In Step ⑨, after hydrolysis of the phosphate, the resulting enol of pyruvic acid tautomerizes to the more stable keto form (Section 17.8).

Figure 27.3 Glycolysis, the pathway of glucose metabolism. (Steps ⑩, ⑫, and ⑬ are shown in Figure 27.4.) Some of the steps are reversible, but equilibrium arrows are not shown (they appear in Figure 27.4).

670 | CHAPTER 27 SPECIFIC CATABOLIC PATHWAYS: CARBOHYDRATE, LIPID, AND PROTEIN METABOLISM

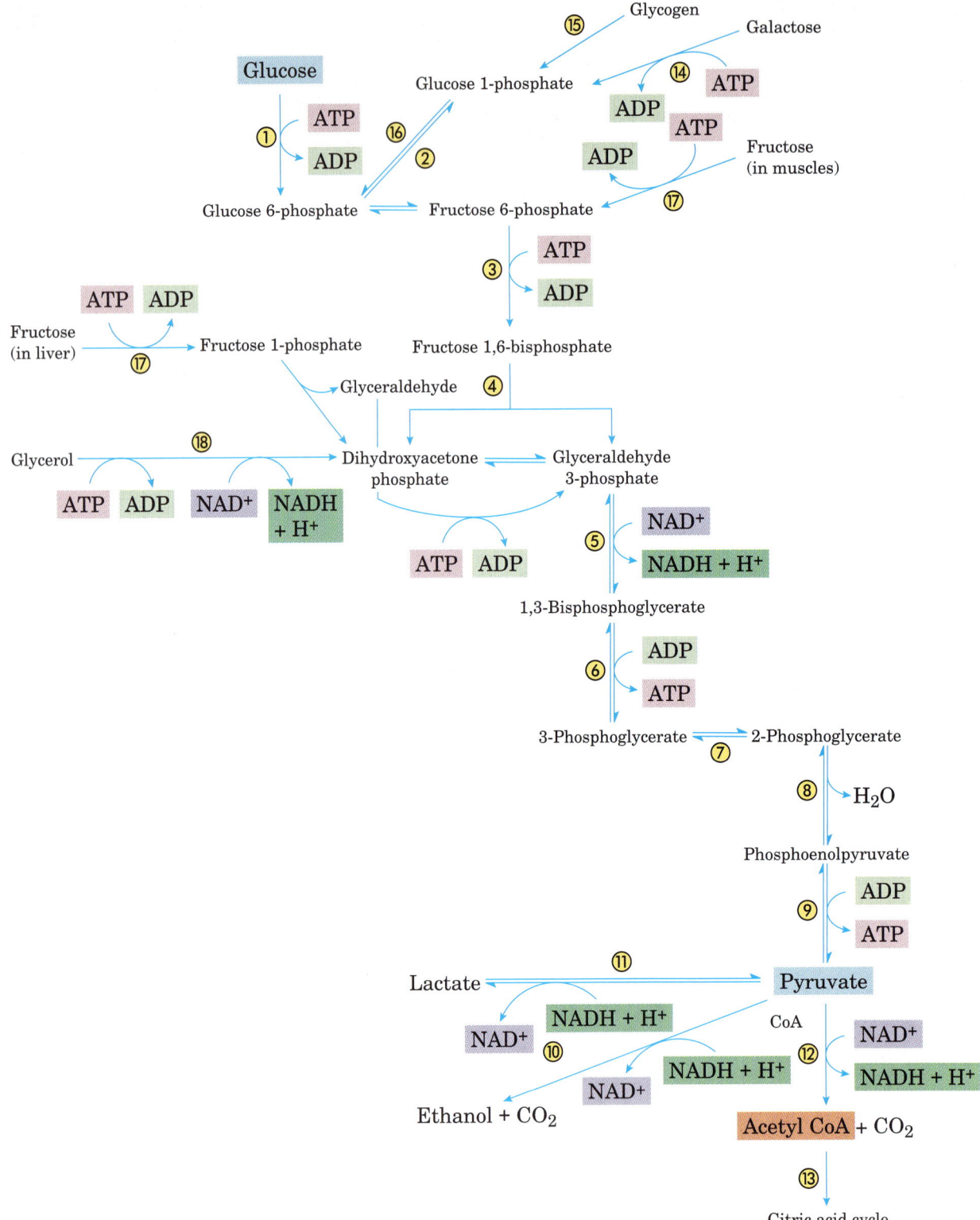

Figure 27.4 An overview of glycolysis and the entries to it and exits from it.

CHEMICAL CONNECTIONS 27A

Lactate Accumulation

Many athletes suffer muscle cramps when they engage in strenuous exercise (see Chapter 8 opener). This problem results from a shift from normal glucose catabolism (glycolysis : citric acid cycle : oxidative phosphorylation) to that of lactate production (see Step ⑪ in Figure 27.4). During exercise, oxygen is used up rapidly, which slows down the rate of the common catabolic pathway. The demand for energy makes anaerobic glycolysis proceed at a high rate, but because the aerobic (oxygen-demanding) pathways are slowed down, not all the pyruvate produced in glycolysis can enter the citric acid cycle. The excess pyruvate ends up as lactate, which causes painful muscle contractions.

The same shift in catabolism also occurs in heart muscle when coronary thrombosis leads to cardiac arrest. The blockage of the artery to the heart muscles cuts off the oxygen supply. The common catabolic pathway and its ATP production are consequently shut off. Glycolysis proceeds at an accelerated rate, causing lactate to accumulate. The heart muscle contracts, producing a cramp. Just as in skeletal muscle, massage of heart muscles can relieve the cramp and start the heart beating. Even if heartbeat is restored within 3 min (the amount of time the brain can survive without being damaged), acidosis may develop as a result of the cardiac arrest. Therefore, at the same time that efforts are under way to start the heart beating by chemical, physical, or electrical means, an intravenous infusion of 8.4% bicarbonate solution is given to combat acidosis.

All of these glycolysis reactions occur in the cytoplasm outside the mitochondria. Because they occur in the absence of O_2, they are also called reactions of the **anaerobic pathway.** As indicated in Figure 27.4, the end product of glycolysis, pyruvate, does not accumulate in the body. In certain bacteria and yeast, pyruvate undergoes reductive decarboxylation in Step ⑩ to produce ethanol. In some bacteria, and in mammals in the absence of oxygen, pyruvate is reduced to lactate in Step ⑪.

B Entrance to the Citric Acid Cycle

Pyruvate is not the end product of glucose metabolism. Most importantly, pyruvate goes through an oxidative decarboxylation in the presence of coenzyme A in Step ⑫ to produce acetyl CoA:

$$NAD^+ + CH_3-\overset{\overset{O}{\|}}{C}-COO^- + CoA-SH \longrightarrow CH_3-\overset{\overset{O}{\|}}{C}-S-CoA + CO_2 + NADH + H^+$$

Pyruvate   Acetyl coenzyme A

This reaction is catalyzed by a complex enzyme system, pyruvate dehydrogenase, that sits on the inner membrane of the mitochondrion. The reaction produces acetyl CoA, CO_2, and $NADH + H^+$. The acetyl CoA then enters the citric acid cycle in Step ⑬ and goes through the common catabolic pathway.

In summary, after converting complex carbohydrates to glucose, the body gets energy from glucose by converting it to acetyl CoA (by way of pyruvate) and then using the acetyl CoA as a starting material for the common catabolic pathway.

C Pentose Phosphate Pathway

As we saw in Figure 27.4, glucose 6-phosphate plays a central role in different entries into the glycolytic pathway. However, glucose 6-phosphate can also be utilized by the body for other purposes, not just for the production of α-D-Glucose 6-phosphate   Ribulose 5-phosphate   Ribose 5-phosphate

Glucose 6-phosphate + 2NADP$^+$ —⑲→ Ribulose 5-phosphate + 2NADPH + CO_2

↓ ⑳

Ribose 5-phosphate

↓ ㉑

Glyceraldehyde 3-phosphate

Figure 27.5 Simplified schematic representation of the pentose phosphate pathway, also called a shunt.

Pentose phosphate pathway
The biochemical pathway that produces ribose and NADPH from glucose 6-phosphate or, alternatively, energy

NADPH is badly needed in red blood cells (RBC) for defense against oxidative damages. Glutathione (Chemical Connections 21A) is the main agent used to keep the hemoglobin in its reduced form. It is regenerated by NADPH, so an insufficient supply of NADPH leads to the destruction of RBC, causing severe anemia.

energy in the form of ATP. Most importantly, glucose 6-phosphate can be shunted to the **pentose phosphate pathway** in Step ⑲ (Figure 27.5). This pathway has the capacity to produce NADPH and ribose in Step ⑳ as well as energy.

NADPH is needed in many biosynthetic processes, including synthesis of unsaturated fatty acids (Section 28.3), cholesterol, and amino acids as well as photosynthesis (Chemical Connections 28A) and the reduction of ribose to deoxyribose for DNA. Ribose is needed for the synthesis of RNA (Section 24.3). Therefore, when the body needs these synthetic ingredients more than energy, glucose 6-phosphate is utilized through the pentose phosphate pathway. When energy is needed, glucose 6-phosphate remains in the glycolytic pathway and even ribose 5-phosphate can be channeled back to glycolysis through glyceraldehyde 3-phosphate. Through this reversible reaction, the cells can also obtain ribose directly from the glycolytic intermediates.

27.3 The Energy Yield from Glucose

Using Figure 27.4, let us sum up the energy derived from glucose catabolism in terms of ATP production. First, however, we must take into account the fact that glycolysis takes place in the cytoplasm, whereas oxidative phosphorylation occurs in the mitochondria. Therefore, the NADH + H$^+$ produced in glycolysis must penetrate the mitochondrial membrane to be utilized in oxidative phosphorylation.

NADH is too large to cross the mitochondrial membrane. Two routes are available to get the electrons of NADH + H$^+$ into the mitochondria; they have different efficiencies. In glycerol 3-phosphate transport, which operates in muscle and nerve cells, only two ATP molecules are produced for each NADH + H⁺. In the other transport route, which operates in the heart and the liver, three ATP molecules are produced for each NADH + H⁺ produced in the cytoplasm, just as is the case in the mitochondria (Section 26.7). Since most energy production takes place in skeletal muscle cells, when we construct the energy balance sheet, we use two ATP molecules for each NADH + H⁺ produced in the cytoplasm.

Armed with this knowledge, we are ready to calculate the energy yield of glucose in terms of ATP molecules produced in skeletal muscles. Table 27.1 shows this calculation. In the first stage of glycolysis (Steps ①, ②, and ③), two ATP molecules are used up, but this loss is more than compensated for by the production of 14 ATP molecules in Steps ⑤, ⑥, ⑨ and ⑫, and in the conversion of pyruvate to acetyl CoA. The net yield of these steps is 12 ATP molecules. As we saw in Section 26.7, the oxidation of one acetyl CoA produces 12 ATP molecules, and one glucose molecule provides two acetyl CoA molecules. Therefore, the total net yield from metabolism of one glucose molecule in skeletal muscle is 36 molecules of ATP, or 6 ATP molecules per carbon atom.

$$C_6H_{12}O_6 + 6O_2 \longrightarrow 6CO_2 + 6H_2O$$

If the same glucose is metabolized in the heart or liver, the electrons of the two NADH produced in the glycolysis are transported into the mitochondrion by the malate-aspartate shuttle. Through this shuttle, the two NADH yield a total of 6 ATP molecules, so that, in this case, 38 ATP molecules are produced for each glucose molecule. It is instructive to note that most of the energy (in the form of ATP) from the glucose is produced in the common metabolic pathway.

Glucose is not the only monosaccharide that can be used as an energy source. Other hexoses, such as galactose (Step ⑭) and fructose (Step ⑰), enter the glycolysis pathway at the stages indicated in Figure 27.4. They also yield 36 molecules of ATP per hexose molecule. Furthermore, the glycogen stored in the liver and muscle cells and elsewhere can be converted by

Nicotinamide adenine dinucleotide phosphate (NADP⁺)

Muscles attached to bones are called skeletal muscles.

Recent investigations suggest that only about 30 to 32 ATP molecules are produced per glucose molecule (2.5 ATP/NADH and 1.5 ATP/FADH₂). Futher research into the complexity of the oxidative phosphorylation pathway is needed to verify these numbers.

Table 27.1 ATP Yield from Complete Glucose Metabolism

Step Numbers in Figure 27.4	Chemical Steps	Number of ATP Molecules Produced
① ② ③	Activation (glucose ⟶ 1,6-fructose bisphosphate)	−2
⑤	Oxidative phosphorylation 2(glyceraldehyde 3-phosphate ⟶ 1,3-bisphosphoglycerate), producing 2(NADH + H⁺) in cytosol	4
⑥ ⑨	Dephosphorylation 2(1,3-bisphosphoglycerate ⟶ pyruvate)	4
⑫	Oxidative decarboxylation 2(pyruvate ⟶ acetyl CoA), producing 2(NADH + H⁺) in mitochondrion	6
⑬	Oxidation of two C₂ fragments in citric acid cycle and oxidative phosphorylation common pathways, producing 12 ATP for each C₂ fragment	24
	Total	36

CHEMICAL CONNECTIONS 27B

Effects of Signal Transduction on Metabolism

The ligand-binding/G-protein/adenylate cyclase cascade, which activates proteins by phosphorylation, has a wide range of effects other than opening or closing ion-gated channels (Figure 23.4). An example of such a target is the glycogen phosphorylase enzyme that participates in the breakdown of glycogen stored in muscles. The enzyme cleaves units of glucose 1-phosphate from glycogen, which enters into the glycolytic pathway and yields quick energy (Section 27.3). The active form of the enzyme, phosphorylase a, is phosphorylated. When it is dephosphorylated (phosphorylase b), it is inactive. When epinephrine signals danger to a muscle cell, phosphorylase is activated through the cascade, and quick energy is produced. Thus, the signal is converted to a metabolic event, allowing the muscles to contract rapidly and enabling the person in danger to fight or run away.

Not all phosphorylation of enzymes results in activation. Consider glycogen synthase. In this case, the phosphorylated form of the enzyme is inactive, and the dephosphorylated form is active. This enzyme participates in glycogenesis, the conversion of glucose to, and storage in the form of, glycogen. The action of glycogen synthase is the opposite of phosphorylase. Nature provides a beautiful balance, inasmuch as the danger signal of the epinephrine hormone has a dual target: It activates phosphorylase to get quick energy, but simultaneously it inactivates glycogen synthase so that the available glucose will be used solely for energy and will not be stored away.

Glycogenolysis The biochemical pathway for the breakdown of glycogen to glucose

enzymatic breakdown and phosphorylation to glucose 1-phosphate (Step ⑮). This compound in turn isomerizes to glucose 6-phosphate, providing an entry to the glycolytic pathway (Step ⑯). The pathway in which glycogen breaks down to glucose is called **glycogenolysis.**

27.4 Glycerol Catabolism

Glycerol 1-phosphate is the same as glycerol 3-phosphate.

The glycerol hydrolyzed from fats or complex lipids (Chapter 20) can also be a rich energy source. The first step in glycerol utilization is an activation step. The body uses one ATP molecule to form glycerol 1-phosphate:

$$\begin{array}{c} CH_2OH \\ | \\ CHOH \\ | \\ CH_2OH \end{array} \xrightarrow{ATP \quad ADP} \begin{array}{c} CH_2O-\text{\textcircled{P}} \\ | \\ CHOH \\ | \\ CH_2OH \end{array} \xrightarrow{NAD^+ \quad NADH + H^+} \begin{array}{c} CH_2O-\text{\textcircled{P}} \\ | \\ C=O \\ | \\ CH_2OH \end{array}$$

Glycerol   Glycerol 1-phosphate   Dihydroxyacetone phosphate

The glycerol phosphate is oxidized by NAD^+ to dihydroxyacetone phosphate, yielding $NADH + H^+$ in the process. Dihydroxyacetone phosphate then enters the glycolysis pathway (Step ⑱ in Figure 27.4) and is isomerized to glyceraldehyde 3-phosphate. A net yield of 20 ATP molecules is produced from each glycerol molecule, or 6.7 ATP molecules per carbon atom.

27.5 β-Oxidation of Fatty Acids

The β-carbon is the second carbon from the COOH group.

As early as 1904, Franz Knoop, working in Germany, proposed that the body utilizes fatty acids as an energy source by breaking them down into C_2 fragments. Prior to fragmentation, the β-carbon is oxidized:

$$-C-C-C-\overset{\beta}{C}-\overset{\alpha}{C}-COOH$$

The name *β-oxidation* has its origin in Knoop's prediction. It took about 50 years to establish the mechanism by which fatty acids are utilized as an energy source.

Figure 27.6 (next page) depicts the overall process of fatty acid metabolism. As is the case with the other foods we have seen, the first step involves activation. This activation occurs in the cytosol, where the fat was previously hydrolyzed to glycerol and fatty acids. It converts ATP to AMP and inorganic phosphate (Step ①), which is equivalent to the cleavage of two high-energy phosphate bonds. The chemical energy derived from the splitting of ATP is built into the compound acyl CoA, which is formed when the fatty acid combines with coenzyme A. The fatty acid oxidation occurs inside the mitochondrion, so the acyl group of acyl CoA must pass through the mitochondrial membrane. Carnitine is the acyl group transporter.

Once the fatty acid in the form of acyl CoA is inside the mitochondrion, the β-oxidation starts. In the first oxidation (dehydrogenation; Step ②), two hydrogens are removed, creating a trans double bond between the alpha and beta carbons of the acyl chain. The hydrogens and electrons are picked up by FAD.

In Step ③, the double bond is hydrated. An enzyme specifically places the hydroxy group on C-3, the β-carbon. The second oxidation (dehydrogenation; Step ④) requires NAD^+ as a coenzyme. The two hydrogens and electrons removed are transferred to the NAD^+ to form $NADH + H^+$. In the process, a secondary alcohol group is oxidized to a keto group at the beta carbon. In Step ⑤, the enzyme thiolase cleaves the terminal C_2 fragment (an acetyl CoA) from the chain, and the rest of the molecule is attached to a new molecule of coenzyme A.

The cycle then starts again with the remaining acyl CoA, which is now two carbon atoms shorter. At each turn of the cycle, one acetyl CoA is produced. Most fatty acids contain an even number of carbon atoms. The cyclic spiral continues until we reach the last four carbon atoms. When this fragment enters the cycle, two acetyl CoA molecules are produced in the fragmentation step.

The β-oxidation of unsaturated fatty acids proceeds in the same way. An extra step is involved, in which the cis double bond is isomerized to a trans bond, but otherwise the spiral is the same.

> **β-oxidation** The biochemical pathway that degrades fatty acids to acetyl CoA by removing two carbons at a time and yielding energy

Transfer into the mitochondrion is accomplished by an enzyme system called carnitine acyltransferase.

Screen 26.3 Animation

27.6 The Energy Yield from Stearic Acid

To compare the energy yield from fatty acids with that of other foods, let us select a typical and quite abundant fatty acid—stearic acid, the C_{18} saturated fatty acid.

We start with the initial step, in which energy is used up rather than produced. The reaction breaks two high-energy phosphoric anhydride bonds:

$$\text{ATP} \longrightarrow \text{AMP} + 2\,P_i$$

This reaction is equivalent to hydrolyzing two molecules of ATP to ADP. In each cycle of the spiral, we obtain one $FADH_2$, one $NADH^+ + H^+$, and one acetyl CoA. Stearic acid (C_{18}) goes through seven cycles in the spiral before it reaches the final C_4 stage. In the last (eighth) cycle, one $FADH_2$, one $NADH^+ + H^+$, and two acetyl CoA molecules are produced. Now we can add up the energy. Table 27.2 shows that, for a C_{18} compound, we obtain a total of 146 ATP molecules.

Figure 27.6 The β-oxidation spiral of fatty acids. Each loop in the spiral contains two dehydrogenations, one hydration, and one fragmentation. At the end of each loop, one acetyl CoA is released.

Table 27.2 ATP Yield from Complete Stearic Acid Metabolism

Step Number in Figure 27.6	Chemical Steps	Happens	Number of ATP Molecules Produced
①	Activation (stearic acid ⟶ stearyl CoA)	Once	−2
②	Dehydrogenation (acyl CoA ⟶ trans-enoyl CoA), producing $FADH_2$	8 times	16
④	Dehydrogenation (hydroxyacyl CoA ⟶ ketoacyl CoA), producing $NADH + H^+$	8 times	24
	C_2 fragment (acetyl CoA ⟶ common catabolic pathway), producing 12 ATP for each C_2 fragment	9 times	108
		Total	146

It is instructive to compare the energy yield from fats with that from carbohydrates, since both are important constituents of the diet. In Section 27.3 we saw that glucose, $C_6H_{12}O_6$, produces 36 ATP molecules—that is, 6 ATPs for each carbon atom. For stearic acid, there are 146 ATP molecules and 18 carbons, or 146/18 = 8.1 ATP molecules per carbon atom. Since additional ATP molecules are produced from the glycerol portion of fats (Section 27.4), fats have a higher caloric value than carbohydrates.

EXAMPLE 27.1

Ketone bodies are a source of energy, especially during dieting and starvation. If acetoacetate (see Section 27.7) is metabolized through β-oxidation, how many ATPs would be produced?

Solution
In Step ①, activation of acetoacetate to acetoacetyl-CoA requires 2 ATPs. Step ⑤ yields two acetyl CoA, which enter the common catabolic pathway, yielding 12 ATPs for each acetyl CoA, for a total of 24 ATPs. The net yield, therefore, is 22 ATP molecules.

Problem 27.1
Which fatty acid yields more ATP molecules per carbon atom: (a) stearic acid or (b) lauric acid?

27.7 Ketone Bodies

In spite of the high caloric value of fats, the body preferentially uses glucose as an energy supply. When an animal is well fed (plenty of sugar intake), fatty acid oxidation is inhibited, and fatty acids are stored in the form of neutral fat in fat depots. When physical exercise demands energy, when the glucose supply dwindles (as in fasting or starvation), or when glucose cannot be utilized (as in the case of diabetes), the β-oxidation pathway of fatty acid metabolism is mobilized.

In some pathological conditions, glucose may not be available at all.

■ Test kit for the presence of ketone bodies in the urine.

Unfortunately, low glucose supply also slows down the citric acid cycle. This lag happens because some oxaloacetate is essential for the continuous operation of the citric acid cycle (Figure 26.8). Oxaloacetate is produced from malate, but it is also produced by the carboxylation of phosphoenolpyruvate (PEP):

$$CO_2 + \underset{\text{PEP}}{\begin{array}{c} O^- \\ | \\ O-P=O \\ | \\ O^- \\ | \\ C-COO^- \\ || \\ CH_2 \end{array}} \underset{\text{GDP} \quad \text{GTP}}{\rightleftharpoons} \underset{\text{Oxaloacetate}}{\begin{array}{c} O \\ || \\ C-COO^- \\ | \\ H_2C-COO^- \end{array}}$$

If there is no glucose, there will be no glycolysis, no PEP formation, and, therefore, greatly reduced oxaloacetate production.

Thus, even though the fatty acids are oxidized, not all of the resulting C_2 fragments (acetyl CoA) can enter the citric acid cycle because not enough oxaloacetate is present. As a result, acetyl CoA builds up in the body, with the following consequences.

The liver is able to condense two acetyl CoA molecules to produce acetoacetyl CoA:

$$2\underset{\text{Acetyl CoA}}{CH_3-\overset{O}{\underset{||}{C}}-SCoA} \longrightarrow \underset{\text{Acetoacetyl CoA}}{CH_3-\overset{O}{\underset{||}{C}}-CH_2-\overset{O}{\underset{||}{C}}-SCoA} + CoASH$$

When the acetoacetyl CoA is hydrolyzed, it yields acetoacetate, which can be reduced to form β-hydroxybutyrate:

$$\underset{\text{Acetoacetyl CoA}}{CH_3-\overset{O}{\underset{||}{C}}-CH_2-\overset{O}{\underset{||}{C}}-SCoA} \xrightarrow{H_2O} \underset{\text{Acetoacetate}}{CH_3-\overset{O}{\underset{||}{C}}-CH_2-\overset{O}{\underset{||}{C}}-O^-} + CoASH + H^+$$

$$\underset{\text{β-Hydroxybutyrate}}{CH_3-\overset{H}{\underset{|}{\underset{OH}{C}}}-CH_2-\overset{O}{\underset{||}{C}}-O^-} \xleftarrow{\text{NADH + H}^+ / \text{NAD}^+} \quad \xrightarrow{\text{H}^+ / CO_2} \underset{\text{Acetone}}{CH_3-\overset{O}{\underset{||}{C}}-CH_3}$$

Ketone bodies A collective name for acetone, acetoacetate and β-hydroxybutyrate; compounds produced from acetyl CoA in the liver that are used as a fuel for energy production by muscle cells and neurons

These two compounds, together with smaller amounts of acetone, are collectively called **ketone bodies.** Under normal conditions, the liver sends these compounds into the bloodstream to be carried to tissues and utilized there as a source of energy via the common catabolic pathway. The brain, for example, normally uses glucose as an energy source. During periods of starvation, however, ketone bodies may be the major energy source for the brain. Normally, the concentration of ketone bodies in the blood is low. But during

CHEMICAL CONNECTIONS 27C

Ketoacidosis in Diabetes

In untreated diabetes, the glucose concentration in the blood is high because the lack of insulin prevents utilization of glucose by the cells. Regular injections of insulin can remedy this situation. However, in some stressful conditions, **ketoacidosis** can still develop.

A typical case was a diabetic patient admitted to the hospital in a semicomatous state. He showed signs of dehydration, his skin was inelastic and wrinkled, his urine showed high concentrations of glucose and ketone bodies, and his blood contained excess glucose and had a pH of 7.0, a drop of 0.4 pH unit from normal, which is an indication of severe acidosis. The patient's urine also contained the bacterium *Escherichia coli*. This indication of urinary tract infection explained why the normal doses of insulin were insufficient to prevent ketoacidosis.

The stress of infection can upset the normal control of diabetes by changing the balance between administered insulin and other hormones produced in the body. This happened during the patient's infection, and his body started to produce ketone bodies in large quantities. Both glucose and ketone bodies appear in the blood before they show up in the urine.

The acidic nature of ketone bodies (acetoacetic acid and β-hydroxybutyric acid) lowers the blood pH. A large drop in pH is prevented by the bicarbonate/carbonic acid buffer (Section 8.11), but even a drop of 0.3 to 0.5 pH unit is sufficient to decrease the Na^+ concentration. Such a decrease of Na^+ ions in the interstitial fluids draws out K^+ ions from the cells. This, in turn, impairs brain function and leads to coma. During the secretion of ketone bodies and glucose in the urine, a lot of water is lost, the body becomes dehydrated, and the blood volume shrinks. As a consequence, the blood pressure drops, and the pulse rate increases to compensate for it. Smaller quantities of nutrients reach the brain cells, which can also cause coma.

The patient mentioned here was infused with physiological saline solution to remedy his dehydration. Extra doses of insulin restored his glucose level to normal, and antibiotics cured the urinary infection.

starvation and in untreated diabetes mellitus, ketone bodies accumulate in the blood and can reach high concentrations. When this occurs, the excess is secreted in the urine. A check of urine for ketone bodies is used in the diagnosis of diabetes.

27.8 Catabolism of the Nitrogen of Amino Acids

The proteins of our foods are hydrolyzed to amino acids in digestion. These amino acids are primarily used to synthesize new proteins. However, unlike carbohydrates and fats, they cannot be stored, so excess amino acids are catabolized for energy production. Section 27.9 explains what happens to the carbon skeleton of the amino acids. Here we discuss the catabolic fate of the nitrogen. Figure 27.7 (next page) gives an overview of the whole process of protein catabolism.

In the tissues, amino groups ($-NH_2$) freely move from one amino acid to another. The enzymes that catalyze these reactions are the transaminases. In essence, there are three stages in nitrogen catabolism in the liver: transamination, oxidative deamination, and the urea cycle.

CHEMICAL CONNECTIONS 27D

Lipomics

Lipomics is the comprehensive analyses of lipid profiles in a cell or a tissue. It is a new approach modeled on the success of genomics (Chemical Connections 24D). It can characterize a sample as small as 50 mg by separating out all the lipid components it contains. By comparing the data with a computer database, lipomics can provide information on the metabolic status of as many as 300 components in one analysis.

Figure 27.7 The overview of pathways in protein catabolism.

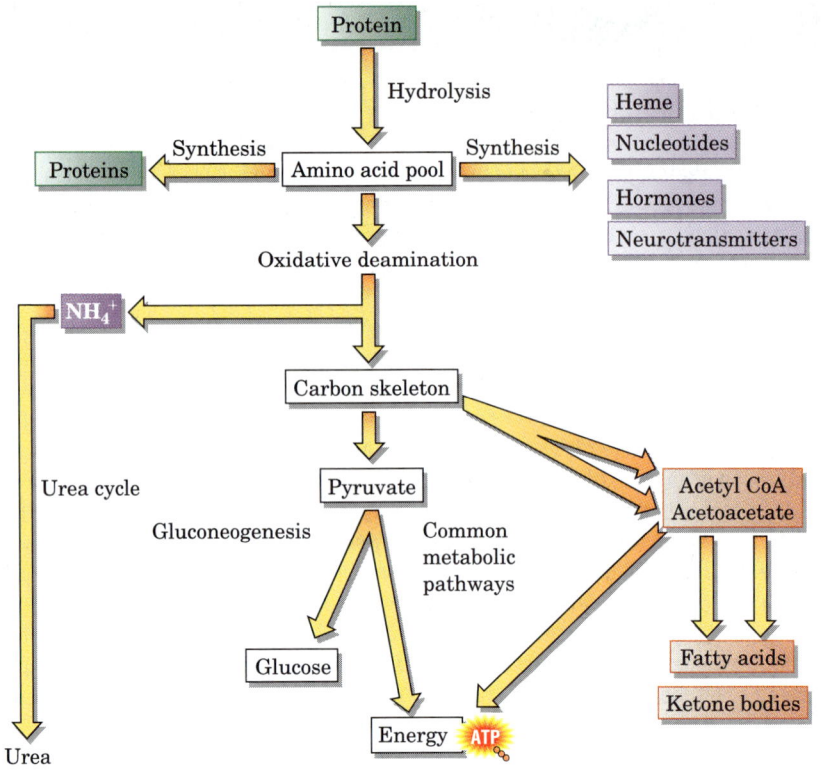

A Transamination

Transamination The exchange of the amino group of an amino acid and a keto group of an α-ketoacid

In the first stage, **transamination,** amino acids transfer their amino groups to α-ketoglutarate:

$$\underset{\substack{\alpha\text{-Amino acid}\\(\text{zwitterion form})}}{\text{R}-\overset{\overset{\displaystyle\text{NH}_3^+}{|}}{\underset{|}{\text{CH}}}-\text{COO}^-} + \underset{\alpha\text{-Ketoglutarate}}{\overset{\text{COO}^-}{\underset{\text{COO}^-}{\overset{|}{\underset{|}{\text{C}=\text{O}}}\,\overset{|}{\underset{|}{\text{CH}_2}}\,\overset{|}{\underset{|}{\text{CH}_2}}}}} \xrightarrow{\text{transaminase}} \underset{\alpha\text{-Ketoacid}}{\text{R}-\overset{\text{O}}{\underset{\|}{\text{C}}}-\text{COO}^-} + \underset{\text{Glutamate}}{\overset{\text{COO}^-}{\underset{\text{COO}^-}{\overset{|}{\underset{|}{\text{CH}-\text{NH}_3^+}}\,\overset{|}{\underset{|}{\text{CH}_2}}\,\overset{|}{\underset{|}{\text{CH}_2}}}}}$$

The catabolism of the α-ketoacid is discussed in the next section.

The carbon skeleton of the amino acid remains behind as an α-ketoacid.

B Oxidative Deamination

Oxidative deamination The reaction in which the amino group of an amino acid is removed and an α-ketoacid is formed

The second stage of the nitrogen catabolism is the **oxidative deamination** of glutamate, which occurs in the mitochondrion:

$$\underset{\text{Glutamate}}{\overset{\text{COO}^-}{\underset{\text{COO}^-}{\overset{|}{\underset{|}{\text{H}-\text{C}-\text{NH}_3^+}}\,\overset{|}{\underset{|}{\text{CH}_2}}\,\overset{|}{\underset{|}{\text{CH}_2}}}}} + \text{NAD}^+ + \text{H}_2\text{O} \rightleftharpoons \text{NH}_4^+ + \underset{\alpha\text{-Ketoglutarate}}{\overset{\text{COO}^-}{\underset{\text{COO}^-}{\overset{|}{\underset{|}{\text{C}=\text{O}}}\,\overset{|}{\underset{|}{\text{CH}_2}}\,\overset{|}{\underset{|}{\text{CH}_2}}}}} + \text{NADH} + \text{H}^+$$

27.8 CATABOLISM OF THE NITROGEN OF AMINO ACIDS / **681**

The oxidative deamination yields NH_4^+ and regenerates α-ketoglutarate, which can again participate in the first stage (transamination). The NADH + H$^+$ produced in the second stage enters the oxidative phosphorylation pathway and eventually produces three ATP molecules.

The body must get rid of NH_4^+ because both it and NH_3 are toxic.

C Urea Cycle

In the third stage, the NH_4^+ is converted to urea through the **urea cycle** (Figure 27.8, next page). In Step ①, NH_4^+ is condensed with CO_2 in the mitochondrion to form an unstable compound, carbamoyl phosphate. This condensation occurs at the expense of two ATP molecules. In Step ②, carbamoyl phosphate is condensed with ornithine, a basic amino acid similar in structure to lysine:

Urea cycle A cyclic pathway that produces urea from ammonia and carbon dioxide

Not all organisms dispose of the metabolic nitrogen in the form of urea. Bacteria and fish release ammonia in the surrounding water as such. Birds and reptiles secrete nitrogen in the form of uric acid, the concentrated white solid so familiar in bird droppings.

Ornithine does not occur in proteins.

Ornithine + Carbamoyl phosphate → Citrulline + PO_4^{3-}

The resulting citrulline diffuses out of the mitochondrion into the cytoplasm.

A second condensation reaction in the cytoplasm takes place between citrulline and aspartate, forming argininosuccinate (Step ③):

ATP + Citrulline + Aspartate → Argininosuccinate + AMP + PP$_i$

The energy for this reaction comes from the hydrolysis of ATP to AMP and pyrophosphate (PP$_i$).

PP$_i$ stands for pyrophosphate.

CHEMICAL CONNECTIONS 27E

Ubiquitin and Protein Targeting

In previous sections, we dealt with the catabolism of dietary proteins, especially with their use as an energy source. The body's own cellular proteins are also broken down and degraded. This occurs sometimes in response to a stress such as starvation, but more often it is done to maintain a steady-state level of regulatory proteins or to eliminate damaged proteins. But how does the cell know which proteins to degrade and which to leave alone?

One pathway that targets proteins for destruction depends on an ancient protein, called **ubiquitin**. (As the name implies, ubiquitin is present in all cells belonging to higher organisms.) Its sequence of 76 amino acids is essentially identical in yeast and in humans. The C-terminal amino acid of ubiquitin is glycine, the carboxyl group of which forms an amide linkage with the side chain of a lysine residue in the protein that is targeted for destruction.

This process, known as **ubiquitinylation**, requires three enzymes and costs energy by using up one ATP molecule per ubiquitin molecule. In many cases, a polymer of ubiquitin molecules (polyubiquitin) is attached to the target protein molecule. In attaching one ubiquitin molecule to the next, the same linkage is used as with the targeted protein: The glycine at the C-terminal of one ubiquitin forms an amide linkage with the amino group on the side chain of a lysyl residue of a second ubiquitin. Once a protein molecule becomes "flagged" by several ubiquitin molecules, it is delivered to an organelle containing the machinery for proteolysis. Part of the proteolytic system is proteasome, a large, water-soluble complex assembly of proteases and regulatory proteins. In this assembly, the polyubiquitin is removed, and the protein is fully or partially degraded. Cleavage of a protein is called **proteolysis**.

One role of the ubiquitin targeting is that of "garbage collector." Damaged proteins that chaperones (Section 21.9) cannot salvage by proper refolding must be removed. For example, oxidation of hemoglobin causes unfolding, and rapid degradation by ubiquitin targeting follows. A second role is the control of the concentration of some regulatory proteins. The protein products of some oncogenes (Chemical Connections 25F) are also removed by ubiquitin targeting.

Both of these functions result in complete degradation of the targeted proteins. A third role involves only a partial degradation of the target protein. As an example, suppose a virus invades a cell. The infected cell recognizes a foreign protein from the virus and targets it for destruction through the ubiquitin system. The targeted protein is partially degraded. A peptide segment of the viral protein is incorporated into a system that rises to the surface of the infected cell and displays the viral peptide as an **antigen**. An antigen is a foreign body recognized by the immune system (Section 30.3). The T cells (Section 30.3) can now recognize the infected cell as a foreign cell; they attack and kill it.

The excessive zeal of ubiquitinylation as a garbage collector lies at the heart of cystic fibrosis, a genetic disease. Cystic fibrosis causes severe bronchopulmonary disorders and pancreatic insufficiency. The molecular culprit is a protein in the membranes that serves as a chloride channel, providing for the transport of chloride ions in and out of cells. In cystic fibrosis, this protein has a mutation that does not allow it to fold properly. The system perceives these chloride channel proteins as faulty and targets them for destruction by ubiquitinylation. The result is an insufficient number of functioning chloride channels. This shortage produces obstructions in the respiratory and intestinal tracts by clogging up the secretion pathways with highly viscous mucins.

In Step ④, the argininosuccinate is split into arginine and fumarate:

$$\text{Argininosuccinate} \longrightarrow \text{Arginine} + \text{Fumarate}$$

27.8 CATABOLISM OF THE NITROGEN OF AMINO ACIDS | 683

Figure 27.8 The urea cycle.

$CO_2 + NH_4^+ + H_2O + 2$ ATP

(1)

$$H_2N-\overset{O}{\underset{\|}{C}}-O-\overset{O}{\underset{\|}{P}}-O^- + 2\,ADP + P_i$$
$\quad\quad\quad\quad\quad\quad\quad\;\; O^-$

Carbamoyl phosphate

$H_2N-\overset{O}{\underset{\|}{C}}-NH_2$
Urea

Ornithine

(2) → P_i

H_2O

(5)

Arginine

Citrulline

ATP

PP$_i$ + AMP

(3) ← Aspartate

Fumarate

(4)

Argininosuccinate

Screen 26.4 Animation

In Step ⑤, the final step, arginine is hydrolyzed to urea and ornithine:

$$\underset{\text{Arginine}}{\begin{array}{c}NH_2^+\\\|\\C-NH_2\\|\\NH\\|\\CH_2\\|\\CH_2\\|\\CH_2\\|\\CH-NH_3^+\\|\\COO^-\end{array}} \xrightarrow{H_2O} \underset{\text{Ornithine}}{\begin{array}{c}NH_3^+\\|\\CH_2\\|\\CH_2\\|\\CH_2\\|\\CH-NH_3^+\\|\\COO^-\end{array}} + \underset{\text{Urea}}{H_2N-\overset{O}{\underset{\|}{C}}-NH_2}$$

The final product of the three stages is urea, which is excreted in the urine of mammals. The ornithine reenters the mitochondrion, completing the cycle. It is then ready to pick up another carbamoyl phosphate.

An important aspect of carbamoyl phosphate's role as an intermediate is that it can be used for synthesis of nucleotide bases (Chapter 24). Furthermore, the urea cycle is linked to the citric acid cycle in that both involve fumarate.

Hans Krebs, who elucidated the citric acid cycle, was also instrumental in establishing the urea cycle.

D Other Pathways of Nitrogen Catabolism

The urea cycle is not the only way that the body can dispose of the toxic NH_4^+ ions. The oxidative deamination process, which produced the NH_4^+ in the first place, is reversible. Therefore, the buildup of glutamate from α-ketoglutarate and NH_4^+ is always possible. A third possibility for disposing of NH_4^+ is the ATP-dependent amidation of glutamate to yield glutamine:

$$NH_4^+ + \text{Glutamate} + ATP \xrightarrow{Mg^{2+}} ADP + P_i + \text{Glutamine}$$

27.9 Catabolism of the Carbon Skeleton of Amino Acids

After transamination of amino acids (Section 27.8) to glutamate, the alpha amino group is removed from glutamate by oxidative deamination (Section 27.8). The remaining carbon skeleton is used as an energy source (Figure 27.9). We will not study the pathways involved except to point out the eventual fate of the skeleton. Not all the carbon skeletons of amino acids are used as fuel. Some may be degraded up to a certain point, and the resulting intermediate then used as a building block to construct another needed molecule.

For example, if the carbon skeleton of an amino acid is catabolized to pyruvate, the body has two possible choices: (1) use the pyruvate as an energy supply via the common catabolic pathway or (2) use it as a building block to synthesize glucose (Section 28.2). Those amino acids that yield a carbon skeleton that is degraded to pyruvate or another intermediate capable of conversion to glucose (such as oxaloacetate) are called **glucogenic**. One example is alanine (Figure 27.9). When alanine reacts with α-ketoglutaric acid, the transamination produces pyruvate directly:

Alanine + α-Ketoglutarate ⟶ Pyruvate + Glutamate

On the other hand, many amino acids are degraded to acetyl CoA and acetoacetic acid. These compounds cannot form glucose but are capable of yielding ketone bodies; they are called **ketogenic**. Leucine is an example of a ketogenic amino acid. Some amino acids are both glucogenic and ketogenic—for example, phenylalanine.

Figure 27.9 Catabolism of the carbon skeletons of amino acids. The glucogenic amino acids are in the purple boxes; the ketogenic ones in the gold boxes.

CHEMICAL CONNECTIONS 27F

Hereditary Defects in Amino Acid Catabolism: PKU

Many hereditary diseases involve missing or malfunctioning enzymes that catalyze the breakdown of amino acids. The oldest known of such diseases is cystinuria, which was described as early as 1810. In this disease, cystine shows up as flat hexagonal crystals in the urine. Stones form because of cystine's low solubility in water. This problem leads to blockage in the kidneys or the ureters and requires surgery. One way to reduce the amount of cystine secreted is to remove as much methionine as possible from the diet. Beyond that, an increased fluid intake increases the volume of the urine, reducing the solubility problem. In addition, penicillamine can prevent cystinuria.

An even more important genetic defect is the absence of the enzyme phenylalanine hydroxylase, which causes a disease called phenylketonuria (PKU). In normal catabolism, this enzyme helps degrade phenylalanine by converting it to tyrosine. If the enzyme is defective, phenylalanine is converted to phenylpyruvate (see the discussion of the conversion of alanine to pyruvate in Section 27.9). Phenylpyruvate (a ketoacid) accumulates in the body and inhibits the conversion of pyruvate to acetyl CoA, thus depriving the cells of energy via the common catabolic pathway. This effect is most important in the brain, which gets its energy from the utilization of glucose. PKU results in mental retardation. This genetic defect can be detected early because phenylpyruvic acid appears in the urine. A federal regulation requires that all infants be tested for this disease. When PKU is detected, mental retardation can be prevented by restricting the intake of phenylalanine in the diet. In particular, patients with PKU should avoid the artificial sweetener aspartame (Chemical Connections 29C) because it yields phenylalanine when hydrolyzed in the stomach.

CHEMICAL CONNECTIONS 27G

Jaundice

In normal individuals, there is a balance between the destruction of heme in the spleen and the removal of bilirubin from the blood by the liver. When this balance is upset, the bilirubin concentration in the blood increases. If this condition continues, the skin and whites of the eyes both become yellow—a condition known as **jaundice**. Jaundice can signal a malfunctioning of the liver, the spleen, or the gallbladder, where bilirubin is stored before excretion.

Hemolytic jaundice is the acceleration of the destruction of red blood cells in the spleen to such a level that the liver cannot cope with the excess bilirubin production. When gallstones obstruct the bile ducts and the bilirubin cannot be excreted in the feces, the backup of bilirubin causes jaundice. In infectious hepatitis, the liver is incapacitated, and bilirubin is not removed from the blood, causing jaundice.

■ Yellowing of the sclera (white outer coat of eyeball) in jaundice. The yellowing is caused by an excess of bilirubin, a bile pigment, in the blood.

Both glucogenic and ketogenic amino acids, when used as an energy supply, enter the citric acid cycle at some point (Figure 27.9) and are eventually oxidized to CO_2 and H_2O. The oxaloacetate (a C_4 compound) produced in this manner enters the citric acid cycle, adding to the oxaloacetate produced from PEP and in the cycle itself.

27.10 Catabolism of Heme

Red blood cells are continuously being manufactured in the bone marrow. Their life span is relatively short—about four months. Aged red blood cells are destroyed in the phagocytic cells.

Phagocytes are specialized blood cells that destroy foreign bodies.

When a red blood cell is destroyed, its hemoglobin is metabolized: The globin (Section 21.9) is hydrolyzed to amino acids, and the heme is first oxidized to biliverdin and finally reduced to bilirubin (Figure 27.10). The color

Figure 27.10 Heme degradation from heme to biliverdine to bilirubin.

Heme → Biliverdine → Bilirubin

changes observed in bruises signal the redox reactions occuring in heme catabolism: black and blue are due to the congealed blood, green to the biliverdin, and yellow to the bilirubin. The iron is preserved in ferritin, an iron-carrying protein, and reused. The bilirubin enters the liver via the blood and is then transferred to the gallbladder, where it is stored in the bile and finally excreted via the small intestine. The color of feces is provided by urobilin, an oxidation product of bilirubin.

SUMMARY

The foods we eat are broken down into small molecules in the stomach and intestines before being absorbed. These small molecules serve two purposes: (1) They can be the building blocks of new materials that the body needs to synthesize (anabolism) or (2) they can be used for energy supply (catabolism). Each group of compounds—carbohydrates, fats, and proteins—has its own catabolic pathway. All the different catabolic pathways converge to the common catabolic pathway (Section 27.1).

The specific pathway of carbohydrate catabolism is **glycolysis** (Section 27.2A). In this process, hexose monosaccharides are activated by ATP and eventually converted to two C_3 fragments, dihydroxyacetone phosphate and glyceraldehyde phosphate. The glyceraldehyde phosphate is further oxidized and eventually ends up as pyruvate. All of these reactions occur in the cytosol. Pyruvate is converted to acetyl CoA, which is further catabolized in the common pathway. When completely metabolized, a hexose yields the energy of 36 ATP molecules (Section 27.3).

When the body needs intermediates for synthesis rather than energy, the glycolytic pathway can be shunted to the **pentose phosphate pathway** (Section 27.2C). In this way, NADPH is obtained, which is necessary for reduction. The pentose phosphate pathway also yields ribose, which is necessary for synthesis of RNA.

Fats are broken down to glycerol and fatty acids. Glycerol is catabolized in the glycolysis pathway and yields 20 ATP molecules (Section 27.4).

Fatty acids are broken down into C_2 fragments in the **β-oxidation** spiral (Section 27.5). At each turn of the spiral, one acetyl CoA is released along with one $FADH_2$ and one $NADH^+ + H^+$. These products go through the common catabolic pathway. Stearic acid, a C_{18} compound, yields 146 molecules of ATP (Section 27.6). In starvation and under certain pathological conditions, not all the acetyl CoA produced in the β-oxidation of fatty acids enters the common catabolic pathway. Instead, some of it forms acetoacetate, β-hydroxybutyrate, and acetone, commonly called **ketone bodies** (Section 27.7). Excess ketone bodies in the blood are secreted in the urine.

Proteins are broken down to amino acids. The nitrogen of the amino acids is first transferred to glutamate (Section 27.8A). This, in turn, is **oxidatively deaminated** to yield ammonia (Section 27.8B). Mammals get rid of the toxic ammonia by converting it to urea in the **urea cycle** (Section 27.8C); urea is secreted in the urine. The carbon skeletons of amino acids are catabolized via the citric acid cycle (Section 27.9). Some of them enter as pyruvate or other intermediates of the citric acid cycle; these are **glucogenic amino acids.** Others are incorporated into acetyl CoA or ketone bodies and are called **ketogenic amino acids.** Heme is catabolized to bilirubin, which is excreted in the feces (Section 27.10).

PROBLEMS

Numbers that appear in color indicate difficult problems. ▸ designates problems requiring application of principles.

Specific Pathways

27.2 What are the products of the lipase-catalyzed hydrolysis of fats?

27.3 What is the main use of amino acids in the body?

Glycolysis

27.4 Although catabolism of a glucose molecule eventually produces a lot of energy, the first step uses up energy. Explain why this step is necessary.

27.5 In one step of the glycolysis pathway, a C_6 chain is broken into two C_3 fragments, only one of which can be further degraded in the glycolysis pathway. What happens to the other C_3 fragment?

27.6 Kinases are enzymes that catalyze the addition (or removal) of a phosphate group to (or from) a substance. ATP is also involved. How many kinases are in glycolysis? Name them.

27.7 (a) Which steps in glycolysis of glucose need ATP?
(b) Which steps in glycolysis yield ATP directly?

27.8 At which intermediate of the glycolytic pathway does the oxidation—and hence the energy production—begin? In what form is the energy produced?

27.9 At what point in glycolysis can ATP act as an inhibitor? What kind of enzyme regulation occurs in this inhibition?

27.10 The end product of glycolysis, pyruvate, cannot enter as such into the citric acid cycle. What is the name of the process that converts this C_3 compound to a C_2 compound?

27.11 What essential compound is produced in the pentose phosphate pathway that is needed for synthesis as well as for defense against oxidative damages?

27.12 Which of the following steps yields energy and which consumes energy?
(a) Pyruvate $\longrightarrow$ lactate
(b) Pyruvate $\longrightarrow$ acetyl CoA + CO_2

27.13 How many moles of lactate are produced from 3 moles of glucose?

27.14 How many moles of net $NADH^+ + H^+$ are produced from 1 mol of glucose going to
(a) Acetyl CoA? (b) Lactate?

Energy Yield from Glucose

27.15 Of the 36 molecules of ATP produced by the complete metabolism of glucose, how many are produced in glycolysis alone—that is, before the common pathway?

27.16 How many net ATP molecules are produced in the skeletal muscles for each glucose molecule
(a) In glycolysis alone (up to pyruvate)?
(b) In converting pyruvate to acetyl CoA?
(c) In the total oxidation of glucose to CO_2 and H_2O?

27.17 (a) If fructose is metabolized in the liver, how many moles of net ATP are produced from each mole during glycolysis?
(b) How many moles are produced if the same thing occurs in a muscle cell?

27.18 In Figure 27.3, Step ⑤ yields one NADH. Yet in Table 27.1, the same step indicates the yield of 2 $NADH + H^+$. Is there a discrepancy between these two statements? Explain.

Glycerol Catabolism

27.19 Based on the names of the enzymes participating in glycolysis, what would be the name of the enzyme catalyzing the activation of glycerol?

27.20 Which yields more energy upon hydrolysis, ATP or glycerol 1-phosphate? Why?

β-Oxidation of Fatty Acids

27.21 Two enzymes participating in β-oxidation have the word "thio" in their names.
(a) Name the two enzymes.
(b) To what chemical group does this name refer?
(c) What is the common feature in the action of these two enzymes?

27.22 (a) Which part of the cells contains the enzymes needed for the β-oxidation of fatty acids?
(b) How does the activated fatty acid get there?

27.23 Assume that lauric acid (C_{12}) is metabolized through β-oxidation. What are the products of the reaction after three turns of the spiral?

27.24 Is the β-oxidation of fatty acid (without the subsequent metabolism of C_2 fragments via the common metabolic pathway) more efficient with a short-chain fatty acid than with a long-chain fatty acid? Is more ATP produced per C atom in a short-chain fatty acid than in a long-chain fatty acid during β-oxidation?

Energy Yield from Fatty Acids

27.25 Calculate the number of ATP molecules obtained in the β-oxidation of myristic acid, $CH_3(CH_2)_{12}COOH$.

27.26 Assume that the cis-trans isomerization in the β-oxidation of unsaturated fatty acids does not require energy. Which C_{18} fatty acid yields the greater amount of energy, saturated (stearic acid) or monounsaturated (oleic acid)? Explain.

27.27 Assuming that both fats and carbohydrates are available, which does the body preferentially use as an energy source?

27.28 If equal weights of fats and carbohydrates are eaten, which will give more calories? Explain.

Ketone Bodies

27.29 Acetoacetate is the common source of acetone and β-hydroxybutyrate. Name the type of reactions that yield these ketone bodies from acetoacetate.

27.30 Do ketone bodies have nutritional value?

27.31 What happens to the oxaloacetate produced from carboxylation of phosphoenolpyruvate?

$$\begin{array}{c} H_3C \\ CH-\underset{\underset{NH_3^+}{|}}{\overset{\overset{H}{|}}{C}}-COO^- \\ H_3C \end{array} + \begin{array}{c} COO^- \\ | \\ C=O \\ | \\ CH_2 \\ | \\ CH_2 \\ | \\ COO^- \end{array} \longrightarrow \begin{array}{c} H_3C \\ CH-C-COO^- \\ H_3C \underset{O}{\overset{\|}{}} \end{array} + \begin{array}{c} COO^- \\ | \\ H-C-NH_3^+ \\ | \\ CH_2 \\ | \\ CH_2 \\ | \\ COO^- \end{array}$$

Catabolism of Amino Acids

27.32 What kind of reaction is the following, and what is its function in the body?

27.33 Write an equation for the oxidative deamination of alanine.

27.34 Ammonia, NH_3, and ammonium ion, NH_4^+, are both soluble in water and could easily be excreted in the urine. Why does the body convert them to urea rather than excreting them directly?

27.35 The metabolism of the carbon skeleton of tyrosine yields pyruvate. Why is tyrosine a glucogenic amino acid?

27.36 What are the sources of the nitrogen in urea?

27.37 What compound is common to both the urea and citric acid cycles?

27.38 (a) What is the toxic product of the oxidative deamination of glutamate?
(b) How does the body get rid of it?

27.39 If the urea cycle is inhibited, in what other ways can the body get rid of NH_4^+ ions?

Catabolism of Heme

27.40 Why is a high bilirubin content in the blood an indication of liver disease?

27.41 When hemoglobin is fully metabolized, what happens to the iron in it?

27.42 Describe which groups on the biliverdin are the oxidation products and which are the reduction products in the degradation of heme.

Chemical Connections

27.43 (Chemical Connections 27A) What causes cramps of the muscles when a person is fatigued?

27.44 (Chemical Connections 27B) How does the signal of epinephrine result in depletion of glycogen in the muscle?

27.45 (Chemical Connections 27C) What system counteracts the acidic effect of ketone bodies in the blood?

27.46 (Chemical Connections 27C) The patient whose condition is described in Chemical Connections 27C was transferred to a hospital in an ambulance. Could a nurse in the ambulance tentatively diagnose his diabetic condition without running blood and urine tests? Explain.

27.47 (Chemical Connections 27D) Assume that you have a 100-mg sample of a diethyl ether extract of a gallstone. What kind of information can you obtain through lipomics?

27.48 (Chemical Connections 27E) What is meant by the following expression: "the amino acid sequence of ubiquitin is highly conserved"?

27.49 (Chemical Connections 27E) How does ubiquitin attach itself to a target protein?

27.50 (Chemical Connections 27E) How does the body eliminate damaged (oxidized) hemoglobin from the blood?

27.51 (Chemical Connections 27F) Draw structural formulas for each reaction component, and complete the following equation:

Phenylalanine $\longrightarrow$ Phenylpyruvate + ?

27.52 (Chemical Connections 27G) What compound causes the yellow coloration of jaundice?

Additional Problems

27.53 If you receive a laboratory report showing the presence of a high concentration of ketone bodies in the urine of a patient, what disease would you suspect?

27.54 What compounds give the sequence in coloration of bruises from black and blue to green and yellow?

27.55 (a) At which step of the glycolysis pathway does NAD^+ participate (see Figures 27.3 and 27.4)?
(b) At which step does $NADH + H^+$ participate?
(c) As a result of the overall pathway, is there a net increase of NAD^+, of $NADH + H^+$, or of neither?

27.56 What is the net energy yield in moles of ATP produced when yeast converts one mole of glucose to ethanol?

27.57 Can the intake of alanine, glycine, and serine relieve hypoglycemia caused by starvation? Explain.

27.58 How can glucose be utilized to produce ribose for RNA synthesis?

27.59 Write the products of the transamination reaction between alanine and oxaloacetate:

$$\underset{\text{CH}_3}{\overset{\text{COO}^-}{\underset{|}{\text{CH}-\text{NH}_3^+}}} + \underset{\underset{\text{COO}^-}{\overset{|}{\text{CH}_2}}}{\overset{\text{COO}^-}{\underset{|}{\overset{|}{\text{C}=\text{O}}}}} \longrightarrow$$

27.60 Phosphoenolpyruvate (PEP) has a high-energy phosphate bond that has more energy than the anhydride bonds in ATP. What step in glycolysis suggests that this is so?

27.61 Suppose that a fatty acid labeled with radioactive carbon-14 is fed to an experimental animal. Where would you look for the radioactivity?

27.62 What functional groups are in carbamoyl phosphate?

27.63 Is the urea cycle an energy-producing or energy-consuming pathway?

27.64 What intermediate of the glycolytic pathway can replenish oxaloacetate in the citric acid cycle?

27.65 How many turns of the spiral are there in the β-oxidation of (a) lauric acid and (b) palmitic acid?

InfoTrac College Edition

For additional readings, go to InfoTrac College Edition, your online research library, at

http://infotrac.thomsonlearning.com

CHAPTER 28

28.1 Introduction

28.2 Biosynthesis of Carbohydrates

28.3 Biosynthesis of Fatty Acids

28.4 Biosynthesis of Membrane Lipids

28.5 Biosynthesis of Amino Acids

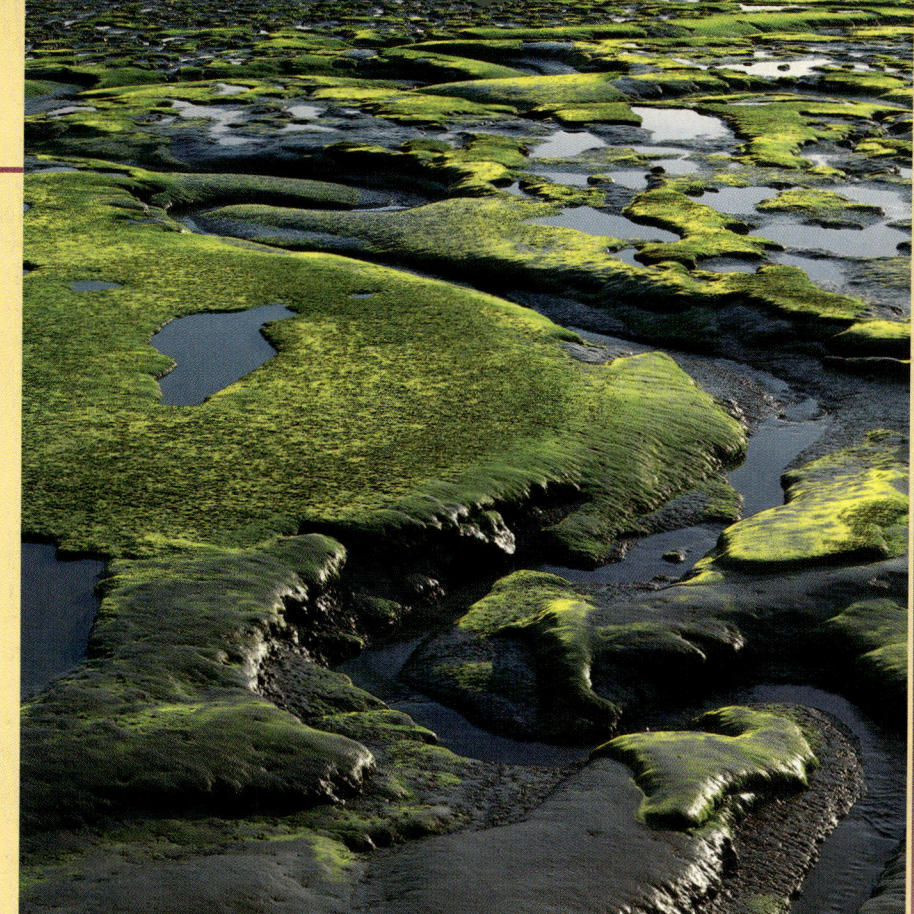

Algae on mudflats.

Biosynthetic Pathways

28.1 Introduction

Anabolic pathways are also called biosynthetic pathways.

In the human body, and in most other living tissues, the pathways by which a compound is synthesized (anabolism) are usually different from the pathways by which it is degraded (catabolism). There are several reasons why it is biologically advantageous for anabolic and catabolic pathways to be different. We will give two of them here:

1. **Flexibility** If the normal **biosynthetic pathway** is blocked, the body can often use the reverse of the degradation pathway instead (recall that most steps in degradation are reversible), thus providing another way to make the necessary compounds.

2. **Overcoming the effect of Le Chatelier's principle** This can be illustrated by the cleavage of a glucose unit from a glycogen molecule, an equilibrium process:

$$(\text{Glucose})_n + P_i \xrightleftharpoons{\text{phosphorylase}} (\text{Glucose})_{n-1} + \text{Glucose 1-phosphate} \quad (28.1)$$

Glycogen   Glycogen (one unit smaller)

Phosphorylase catalyzes not only glycogen degradation (the forward reaction), but also glycogen synthesis (the reverse reaction). However, the body contains a large excess of inorganic phosphate, P_i. This excess would drive the reaction, on the basis of Le Chatelier's principle, to the right, which represents glycogen degradation. To provide a method for the synthesis of glycogen even in the presence of excess inorganic phosphate, a different pathway is needed in which P_i is not a reactant. Thus, the body uses the following synthetic pathway:

$$\underset{\text{Glycogen}}{(\text{Glucose})_{n-1}} + \text{UDP-glucose} \longrightarrow \underset{\substack{\text{Glycogen} \\ \text{(one unit larger)}}}{(\text{Glucose})_n} + \text{UDP} \quad (28.2)$$

Not only are the synthetic pathways different from the catabolic pathways, but the energy requirements are also different, as are the pathways' locations. Most catabolic reactions occur in the mitochondria, whereas anabolic reactions generally take place in the cytoplasm. We will not describe the energy balances of the biosynthetic processes in detail as we did for catabolism. However, keep in mind that, while energy (in the form of ATP) is *obtained* in the degradative processes, biosynthetic processes *consume* energy.

28.2 Biosynthesis of Carbohydrates

We discuss the biosynthesis of carbohydrates by looking at three examples:

- Conversion of atmospheric CO_2 to glucose in plants
- Synthesis of glucose in animals and humans
- Conversion of glucose to other carbohydrate molecules in animals and humans

A Conversion of Atmospheric CO_2 to Glucose in Plants

The most important biosynthesis of carbohydrates takes place in plants, green algae, and cyanobacteria, with the last two representing an important part of the marine food web. This is **photosynthesis.** In this process, the energy of the sun is built into chemical bonds of carbohydrates; the overall reaction is

$$6H_2O + 6CO_2 \xrightarrow[\text{chlorophyll}]{\substack{\text{energy in} \\ \text{the form of} \\ \text{sunlight}}} \underset{\text{Glucose}}{C_6H_{12}O_6} + 6O_2 \quad (28.3)$$

> **Photosynthesis** The process in which plants synthesize carbohydrates from CO_2 and H_2O with the help of sunlight and chlorophyll

Although the primary product of photosynthesis is glucose, it is largely converted to other carbohydrates, mainly cellulose and starch.

This very complicated process takes place in large protein-cofactor complexes (Chemical Connections 28A). We will not discuss it further here except to note that the carbohydrates of plants—starch, cellulose, and other mono- and polysaccharides—serve as the basic carbohydrate supply of all animals, including humans.

B Synthesis of Glucose in Animals

In Chapter 27, we saw that when the body needs energy, carbohydrates are broken down via glycolysis. When energy is not needed, glucose can be syn-

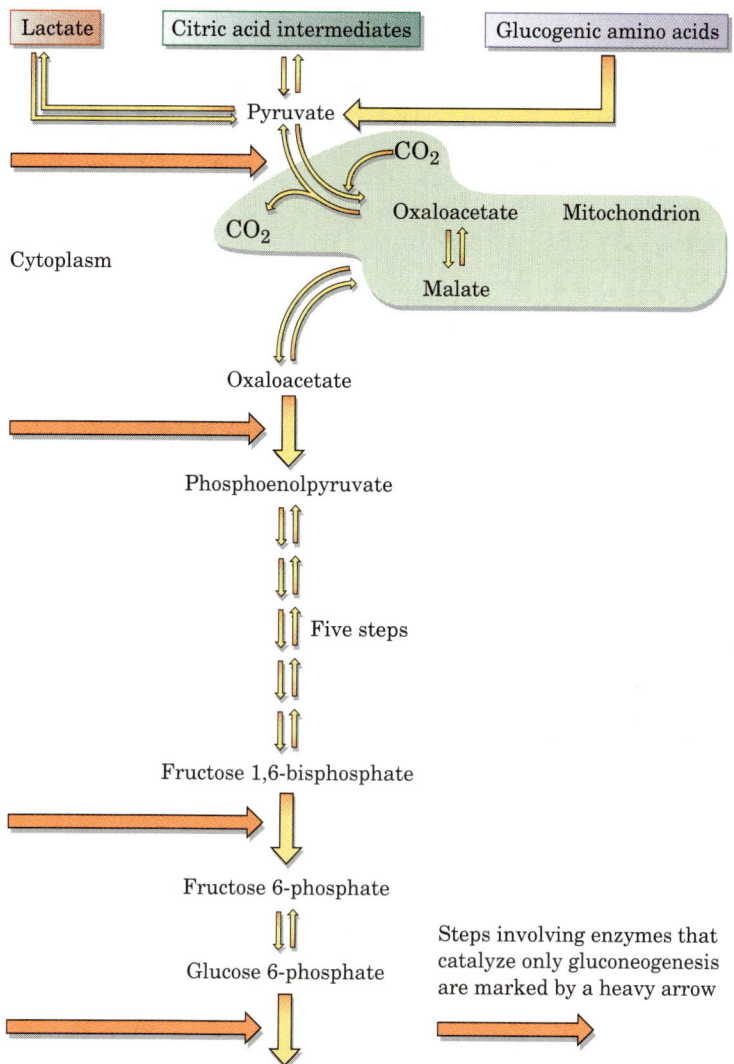

Figure 28.1 Gluconeogenesis. All reactions take place in cytosol, except for those shown in the mitochondria.

thesized from the intermediates of the glycolytic and citric acid pathways. This process is called **gluconeogenesis**. As shown in Figure 28.1, a large number of intermediates—pyruvate, lactate, oxaloacetate, malate, and several amino acids (the glucogenic amino acids we met in Section 27.9)—can serve as starting compounds. Gluconeogenesis proceeds in the reverse order from glycolysis, and many of the enzymes of glycolysis also catalyze gluconeogenesis. At four points, however, unique enzymes (marked in Figure 28.1) catalyze only gluconeogenesis and not the breakdown reactions. These four enzymes make *gluconeogenesis a pathway that is distinct from glycolysis*.

Gluconeogenesis The process by which glucose is synthesized in the body.

ATP is used up in gluconeogenesis and produced in glycolysis.

C Conversion of Glucose to Other Carbohydrates in Animals

The third important biosynthetic pathway for carbohydrates is the conversion of glucose to other hexoses and hexose derivatives and the synthesis of di-, oligo-, and polysaccharides. The common step in all of these processes is the activation of glucose by uridine triphosphate (UTP) to form UDP-glucose:

CHEMICAL CONNECTIONS 28A

Photosynthesis

Photosynthesis requires sunlight, water, CO_2, and pigments found in plants, mainly chlorophyll. The overall reaction shown in Equation 28.3 actually occurs in two distinct steps. First, the light interacts with the pigments that are located in highly membranous organelles of plants, called **chloroplasts** (Figure 28A.1). Chloroplasts resemble mitochondria (Section 26.2) in many respects: They contain a whole chain of oxidation–reduction enzymes similar to the cytochrome and iron–sulfur complexes of mitochondrial membranes, and they contain a proton translocating ATPase. In a manner similar to mitochondria, the proton gradient accumulated in the intermembrane region drives the synthesis of ATP in chloroplasts (see the discussion of the chemiosmotic pump in Section 26.6).

The chlorophyll is the central part of a complex machinery called photosystem I and II. The detailed structure of photosystem I was elucidated in 2001. The photosystem consists of three monomeric entities. These monomeric entities are designated by I, II and III (see Biochemistry Part Opener on page 465). Each monomer contains 12 different proteins, 96 chlorophyll molecules, and 30 cofactors that include iron clusters, lipids, and Ca^{2+} ions. Its most important feature shows that the central Mg^{2+} is bound to the sulfur of a methionine residue of a surrounding protein. This Mg-S linkage makes this whole assembly a powerful oxidizing agent, meaning that it can readily accept electrons.

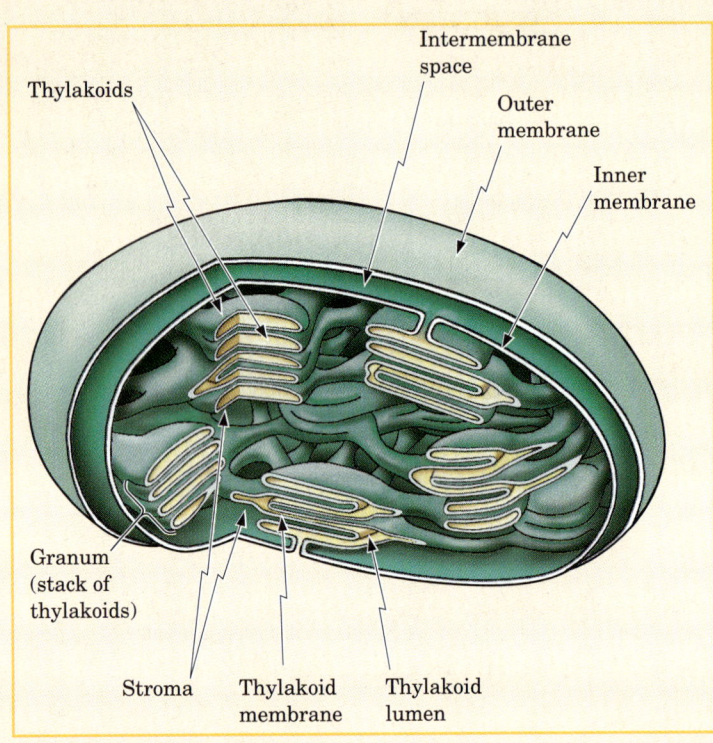

Figure 28.A1 Inside of a chloroplast.

Figure 28.A2 Side view of monomer III in photosystem I.

CHEMICAL CONNECTIONS 28A

Photosynthesis, (continued)

The chlorophyll itself, buried in a complex protein that traverses the chloroplast membranes, is a molecule similar to the heme we have already encountered in hemoglobin (Figure 21.12). In contrast to heme, however, chlorophyll contains Mg^{2+} instead of Fe^{2+}:

Chlorophyll a

The first reaction that takes place in photosynthesis is called the light reaction because chlorophyll captures the energy of sunlight and, with its aid, strips the electrons and protons from water to form oxygen, ATP, and $NADPH + H^+$ (see Section 27.2C):

$$H_2O + ADP + P_i + NADP^+ + \text{sunlight}$$
$$\longrightarrow O_2 + ATP + NADPH + H^+$$

The second reaction, called the dark reaction because it does not need light, in essence converts CO_2 to carbohydrates:

$$CO_2 + ATP + NADPH + H^+$$
$$\longrightarrow (CH_2O)_n + ADP + P_i + NADP^+$$
$$\text{Carbohydrates}$$

In this step, the energy, now in the form of ATP, is used to help $NADPH + H^+$ reduce carbon dioxide to carbohydrates. Thus, the protons and electrons stripped in the light reaction are added to the carbon dioxide in the dark reaction. This reduction takes place in a multistep cyclic process called the **Calvin cycle**, named after its discoverer, Melvin Calvin (1911–1997), who was awarded the 1961 Nobel Prize in chemistry for his work. In this cycle, CO_2 is first attached to a C_5 fragment and breaks down to two C_3 fragments (triose phosphates). Through a complex series of steps, these fragments are converted to a C_6 compound and, eventually, to glucose.

$$CO_2 + C_5 = 2C_3 = C_6$$

The critical step in the dark reaction (Calvin cycle) is the attachment of CO_2 to ribulose 1,5-bisphosphate, a compound derived from ribulose (Table 19.2). The enzyme that catalyzes this reaction, ribulose-1,5-bisphosphate carboxylase-oxygenase, nicknamed RuBisCO, is one of the slowest in nature. As in traffic, the slowest-moving vehicle determines the overall flow, so RuBisCO is the main factor in the low efficiency of the Calvin cycle. Because of this enzyme, most plants convert less than 1% of the absorbed radiant energy into carbohydrates. To overcome its inefficiency, plants must synthesize large quantities of this enzyme. More than half of the soluble proteins in plant leaves are RuBisCO enzymes, whose synthesis requires a lot of energy expenditure.

Uridine diphosphate (UDP)
Uridine diphosphate glucose (UDP-glucose)

UDP is similar to ADP except that the base is uracil instead of adenine. UTP, an analog of ATP, contains two high-energy phosphate anhydride

Glycogenesis The conversion of glucose to glycogen

bonds. For example, when the body has excess glucose and wants to store it as glycogen (a process called **glycogenesis**), the glucose is first converted to glucose 1-phosphate, but then a special enzyme catalyzes the reaction:

$$\text{glucose 1-phosphate} + \boxed{\text{UTP}} \longrightarrow \text{UDP-glucose} + {}^-\text{O}-\underset{\underset{\text{O}^-}{|}}{\overset{\overset{\text{O}}{\|}}{\text{P}}}-\text{O}-\underset{\underset{\text{O}^-}{|}}{\overset{\overset{\text{O}}{\|}}{\text{P}}}-\text{O}^-$$

$$\text{UDP-glucose} + (\text{glucose})_n \longrightarrow \boxed{\text{UDP}} + (\text{glucose})_{n+1}$$
$$\phantom{\text{UDP-glucose} + (\,}\text{Glycogen}\text{Glycogen}$$
$$\phantom{\text{UDP-glucose} + (\text{glucose})_n \longrightarrow \text{UDP} + (\text{gluc}}\text{(one unit larger)}$$

The biosynthesis of many other di- and polysaccharides and their derivatives also uses the common activation step: forming the appropriate UDP compound.

28.3 Biosynthesis of Fatty Acids

The body can synthesize all the fatty acids it needs except for linoleic and linolenic acids (essential fatty acids; see Section 20.2). The source of carbon in this synthesis is acetyl CoA. Since acetyl CoA is also a degradation product of the β-oxidation spiral of fatty acids (Section 27.5), we might expect the synthesis to be the reverse of the degradation. This is not the case. For one thing, the majority of fatty acid synthesis occurs in the cytoplasm, while degradation takes place in the mitochondria. Fatty acid synthesis is catalyzed by a multienzyme system.

However, one aspect of fatty acid synthesis is the same as in fatty acid degradation: Both processes involve acetyl CoA, so both proceed in units of two carbons. Fatty acids are built up two carbons at a time, just as they are broken down two carbons at a time (Section 27.5).

Most of the time, fatty acids are synthesized when excess food is available. That is, when we eat more food than we need for energy, our bodies turn the excess acetyl CoA (produced by catabolism of carbohydrates; see Section 27.2) into fatty acids and then to fats. The fats are stored in the fat depots, which are specialized fat-carrying cells (see Figure 27.1).

The key to fatty acid synthesis is an **acyl carrier protein (ACP)**. It can be looked upon as a merry-go-round—a rotating protein molecule to which the growing chain of fatty acids is bonded. As the growing chain rotates with the ACP, it sweeps over the multienzyme complex; at each enzyme, one reaction of the chain is catalyzed (Figure 28.2).

At the beginning of this cycle, the ACP picks up an acetyl group from acetyl CoA and delivers it to the first enzyme, fatty acid synthase, here called synthase for short:

$$\boxed{\text{CH}_3\overset{\overset{\text{O}}{\|}}{\text{C}}-\text{S}-\text{CoA}} + \text{HS}-\text{ACP} \longrightarrow \text{HS}-\text{CoA} + \text{CH}_3\overset{\overset{\text{O}}{\|}}{\text{C}}-\text{S}-\text{ACP}$$

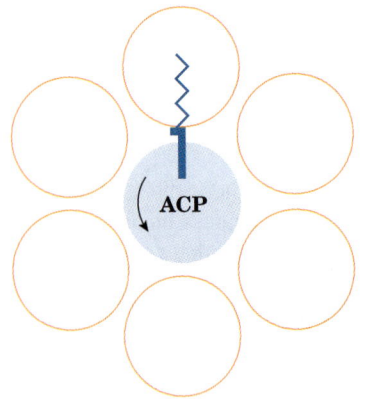

Figure 28.2 The biosynthesis of fatty acids. The ACP (central blue sphere) has a long side chain (↘) that carries the growing fatty acid (∿). The ACP rotates counterclockwise, and its side chain sweeps over a multienzyme system (empty spheres). As each cycle is completed, a C_2 fragment is added to the growing fatty acid chain.

$$\underset{\text{O}}{\overset{\text{O}}{\text{CH}_3\text{C}}}-\text{S}-\text{ACP} + \text{synthase}-\text{SH}$$

$$\longrightarrow \text{CH}_3\overset{\overset{\text{O}}{\|}}{\text{C}}-\text{S}-\text{synthase} + \text{HS}-\text{ACP}$$

The C_2 fragment on the synthase is condensed with a C_3 fragment attached to the ACP in a process in which CO_2 is given off:

$$\text{CH}_3\overset{\overset{\text{O}}{\|}}{\text{C}}-\text{S}-\text{synthase} + \underset{\underset{\text{Malonyl-ACP}}{\text{COO}^-}}{\text{CH}_2}-\overset{\overset{\text{O}}{\|}}{\text{C}}-\text{S}-\text{ACP}$$

$$\longrightarrow \underset{\text{Acetoacetyl-ACP}}{\text{CH}_3\overset{\overset{\text{O}}{\|}}{\text{C}}-\text{CH}_2-\overset{\overset{\text{O}}{\|}}{\text{C}}-\text{S}-\text{ACP}} + \text{CO}_2 + \text{synthase}-\text{SH}$$

The —SH group is the site of acyl binding as a thioester.

The result is a C_4 fragment, which is reduced twice and dehydrated before it becomes a fully saturated C_4 group. This event marks the end of one cycle of the merry-go-round.

In the next cycle, the C_4 fragment is transferred to the synthase, and another malonyl-ACP (C_3 fragment) is added. When CO_2 splits out, a C_6 fragment is obtained. The merry-go-round continues to turn. At each turn, another C_2 fragment is added to the growing chain. Chains up to C_{16} (palmitic acid) can be obtained in this process. If the body needs longer fatty acids—for example, stearic (C_{18})—another C_2 fragment is added to palmitic acid by a different enzyme system.

Unsaturated fatty acids are obtained from saturated fatty acids by an oxidation step in which hydrogen is removed and combined with O_2 to form water:

The three steps are the reverse of what we saw in the β-oxidation of fatty acids in Section 27.5.

$$\text{R}-\text{CH}_2-\text{CH}_2-(\text{CH}_2)_n\text{COOH} + \text{O}_2 + \text{NADPH} + \text{H}^+$$

$$\xrightarrow{\text{enzyme}} \underset{\text{R}}{\overset{\text{H}}{}}\text{C}=\text{C}\underset{(\text{CH}_2)_n\text{COOH}}{\overset{\text{H}}{}} + 2\text{H}_2\text{O} + \text{NADP}^+$$

An example of such elongation and unsaturation is the fatty acid, docosahexenoic acid that as its name indicates is a 22-carbon fatty acid with 6 cis double bonds (22:6). Docosahexenoic acid is part of the glycerophospholipids prevalent in the membranes where the visual pigment rhodopsin resides. Its presence is necessary to provide fluidity to the membrane to process the light signals reaching our retina.

28.4 Biosynthesis of Membrane Lipids

The various membrane lipids (Sections 20.6 to 20.8) are assembled from their constituents. We just saw how fatty acids are synthesized in the body. These fatty acids are activated by CoA, forming acyl CoA. Glycerol 3-

We encountered glycerol 3-phosphate as a vehicle for transporting electrons in and out of mitochondria (Section 27.3).

phosphate, which is obtained from the reduction of dihydroxyacetone phosphate (a C_3 fragment of glycolysis; see Figure 27.4), is the second building block of glycerophospholipids. This compound combines with two acyl CoA molecules, which can be the same or different:

$$\begin{array}{c} CH_2-OH \\ | \\ CH-OH \\ | \\ CH_2-O-\overset{O}{\underset{O^-}{P}}-O^- \end{array} + \begin{array}{c} O \\ \| \\ R-C-S-CoA \\ O \\ \| \\ R'-C-S-CoA \end{array} \longrightarrow \begin{array}{c} CH_2-O-\overset{O}{\overset{\|}{C}}-R \\ | \\ CH-O-\overset{O}{\overset{\|}{C}}-R' \\ | \\ CH_2-O-\overset{O}{\underset{O^-}{P}}-O^- \end{array} + 2CoA-SH$$

Glycerol 3-phosphate   Acyl CoA   A phosphatidate

To complete the molecule, an activated serine or choline or ethanolamine is added to the $-OPO_3^{2-}$ group, forming phosphate esters (see the structures in Section 20.6; Figure 28.3 shows a model of phosphatidylcholine). Choline is activated by cytidine triphosphate (CTP), yielding CDP-choline. This process is similar to the activation of glucose by UTP (Section 28.2) except that the base is cytosine rather than uracil (Section 24.2).

Sphingolipids (Section 20.7) are similarly built up from smaller molecules. An activated phosphocholine is added to the sphingosine part of ceramide (Section 20.7) to make sphingomyelin.

The glycolipids are constructed in a similar fashion. Ceramide is assembled as described above, and the carbohydrate is added one unit at a time in the form of activated monosaccharides (UDP glucose and so on).

Cholesterol, the molecule that controls the fluidity of membranes and is a precursor of all steroid hormones and bile salts, is also synthesized by the human body. It is assembled in the liver from C_2 fragments that come from the acetyl group of acetyl CoA. All the carbon atoms of cholesterol come from the carbons of acetyl CoA molecules (Figure 28.4). Cholesterol in the brain is synthesized in nerve cells themselves; its presence is necessary to form synapses. Cholesterol from our diets and cholesterol that is synthesized in the liver and circulates in the plasma as LDL (see Section 20.9) are not available for synapse formation because LDL cannot cross the blood–brain barrier.

Figure 28.3 A model of phosphatidylcholine, commonly called lecithin.

28.4 BIOSYNTHESIS OF MEMBRANE LIPIDS | 699

Figure 28.4 Biosynthesis of cholesterol. The circled carbon atoms come from the —CH_3 group, and the others come from the —CO— group of the acetyl group of acetyl CoA.

Cholesterol synthesis starts with the sequential condensation of three acetyl CoA molecules to form a C_6 compound, hydroxymethylglutaryl CoA (HMGCoA):

$$3\,CH_3CSCoA \xrightarrow{2CoA\text{-}SH} \text{Hydroxymethylglutaryl CoA} \xrightarrow[CoA\text{-}SH]{HMG\ reductase} \text{Mevalonate}$$

A key enzyme, HMG reductase, controls the rate of cholesterol synthesis. It reduces HMGCoA to a primary alcohol, yielding CoA in the process. The resulting compound, mevalonate, undergoes phosphorylation and decarboxylation to yield a C_5 compound, isopentenyl pyrophosphate:

$$\text{Mevalonate} \xrightarrow[CO_2]{ATP\ AMP} \text{Isopentenyl pyrophosphate}$$

From this basic C_5 unit, the other multiple C_5 compounds are formed that eventually lead to cholesterol synthesis. These intermediates are geranyl, C_{10}, and farnesyl, C_{15}, pyrophosphates:

Geranyl pyrophosphate   Farnesyl pyrophosphate

Finally, cholesterol is synthesized from the condensation of two farnesyl pyrophosphate molecules.

The statin drugs, such as lovastatin, competitively inhibit the key enzyme HMG reductase and thereby the biosynthesis of cholesterol. They are frequently prescribed to control the cholesterol level in the blood so as to prevent atherosclerosis (Section 20.9E).

Note that the intermediates in cholesterol synthesis, the geranyl and farnesyl pyrophosphates, are made of isoprene units; we discussed these C_5 compounds in Section 12.5. These C_{10} and C_{15} compounds are also used to enable protein molecules to be dispersed in the lipid bilayers of membranes. When these multiple isoprene units are attached to a protein in a process called **prenylation,** the protein becomes more hydrophobic and is able to move laterally within the bilayer with greater ease. Prenylation marks proteins to be associated with membranes and to perform other cellular functions, such as signal transduction of G protein (Chemical Connections 28B).

The name *prenylation* originates from isoprene, the five-carbon unit from which the cholesterol intermediates C_{10}, C_{15}, and C_{30} are made.

CHEMICAL CONNECTIONS 28B

Prenylation of Ras Protein and Cancer

Prenylation, or the attachment of multiple isoprene units to a protein, enables the protein to be part of a membrane, an integral protein. The C_{10} and C_{15} units are attached to the protein via a thioether linkage, —C—S—C—. For example, the farnesyl pyrophosphate interacts with a cysteine residue that is located near the carboxyl terminal of the target protein. As a thioether linkage is formed, the pyrophosphate is liberated simultaneously. The resultant modified protein acquires a hydrophobic tail that makes its association with membranes feasible.

A protein named Ras is a GTP-binding protein and plays a key role in the signal transduction (Section 23.5) that regulates cell growth. It is an oncogene (Chemical Connections 25E) in that a Ras-related mutation may cause cancer. The signaling activity of Ras depends on its prenylation. Mutations that inhibit prenylation of Ras cause the cell to die. Other mutations of prenylated Ras may cause uncontrolled growth; indeed, one-third of all human cancers have Ras mutations.

Farnesyl (or geranyl) — Cysteine

28.5 Biosynthesis of Amino Acids

Table 21.1 shows the essential and nonessential amino acids.

The human body needs 20 different amino acids to make its protein chains—all 20 are found in a normal diet. Some of the amino acids can be synthesized from other compounds; these are the nonessential amino acids. Others cannot be synthesized by the human body and must be supplied in the diet; these are the **essential amino acids** (see Section 29.6). Most nonessential amino acids are synthesized from some intermediate of either glycolysis (Section 27.2) or the citric acid cycle (Section 26.4). Glutamate plays a central role in the synthesis of five nonessential amino acids. Glutamate itself is synthesized from α-ketoglutarate, one of the intermediates in the citric acid cycle:

$$\text{NADH} + \text{H}^+ + \text{NH}_4^+ + \underset{\alpha\text{-Ketoglutarate}}{\begin{matrix} \text{COO}^- \\ | \\ \text{C}=\text{O} \\ | \\ \text{CH}_2 \\ | \\ \text{CH}_2 \\ | \\ \text{COO}^- \end{matrix}} \rightleftharpoons \underset{\text{Glutamate}}{\begin{matrix} \text{COO}^- \\ | \\ \text{CH}-\text{NH}_3^+ \\ | \\ \text{CH}_2 \\ | \\ \text{CH}_2 \\ | \\ \text{COO}^- \end{matrix}} + \text{NAD}^+ + \text{H}_2\text{O}$$

The forward reaction is the synthesis, and the reverse reaction is the oxidative deamination (degradation) reaction we encountered in the catabolism of amino acids (Section 27.8). In this case, the synthetic and degradative pathways are exactly the reverse of each other.

Glutamate can serve as an intermediate in the synthesis of alanine, serine, aspartate, asparagine, and glutamine. For example, the transamination reaction we saw in Section 27.8 leads to alanine formation:

CHEMICAL CONNECTIONS 28C

Amino Acid Transport and Blue Diaper Syndrome

Amino acids, both essential and nonessential, are usually more concentrated inside the cells than in the cells' surroundings. In the body, special transport mechanisms are available to carry the amino acids into the cells. Sometimes these transport mechanisms are faulty, however, and the cells lack an amino acid despite the fact that the diet provides the particular amino acid.

An interesting example of this phenomenon is blue diaper syndrome, in which infants excrete blue urine. Indigo blue, a dye, is an oxidation product of the amino acid tryptophan. But how does this oxidation product get into the system to be excreted in the urine? The patient's tryptophan transport mechanism is faulty. Although enough tryptophan is being supplied by the diet, most of it is not absorbed through the intestine. Instead, it accumulates there and is oxidized by bacteria in the gut. The oxidation product is moved into the cells, but because the cells cannot use it, it is excreted in the urine.

Much of the blue diaper syndrome can be eliminated with antibiotics. This treatment kills many of the gut's bacteria, and the nonabsorbed tryptophan is then excreted in the feces because the bacteria cannot now oxidize it to indigo blue.

■ Structures of tryptophan and indigo

Besides being the building blocks of proteins, amino acids serve as intermediates for a large number of biological molecules. We have already seen that serine is needed in the synthesis of membrane lipids (Section 28.4). Certain amino acids are also intermediates in the synthesis of heme and of the purines and pyrimidines that are the raw materials for DNA and RNA (Chapter 24).

SUMMARY

For most biochemical compounds, the biosynthetic pathways are different from the degradation pathways. In **photosynthesis,** carbohydrates are synthesized in plants from CO_2 and H_2O, using sunlight as an energy source (Section 28.2). Glucose can be synthesized by animals from the intermediates of glycolysis, from those of the citric acid cycle, and from glucogenic amino acids. This process is called **gluconeogenesis** (Section 28.2). When glucose or other monosaccharides are built into di-, oligo-, and polysaccharides, each monosaccharide unit in its activated form is added to a growing chain.

Fatty acid biosynthesis is accomplished by a multienzyme system (Section 28.3). The key to this process is the **acyl carrier protein, ACP,** which acts as a merry-go-round transport system: It carries the growing fatty acid chain over a number of enzymes, each of which catalyzes a specific reaction. With each complete turn of the merry-go-round, a C_2 fragment is added to the growing fatty acid chain. The source of the C_2 fragment is malonyl ACP, a C_3 compound attached to the ACP. It becomes C_2 with the loss of CO_2.

Glycerophospholipids (Section 28.4) are synthesized from glycerol 3-phosphate, fatty acids that are activated by conversion to acyl CoA, and activated alcohols such as choline. Cholesterol (Section 28.4) is synthesized from acetyl CoA. Three C_2 fragments are condensed to form C_6, hydroxymethylglutaryl CoA. After reduction and decarboxylation, C_5 isoprene units are formed that condense to the C_{10} and C_{15} intermediates from which cholesterol is built.

Many nonessential amino acids are synthesized in the body from the intermediates of glycolysis or of the citric acid cycle (Section 28.5). In half of these cases, glutamate is the donor of the amino group in transamination. Amino acids serve as building blocks for proteins.

PROBLEMS

Numbers that appear in color indicate difficult problems. ▶ designates problems requiring applications of principles.

Biosynthesis

28.1 Why are the pathways that the body uses for anabolism and catabolism mostly different?

28.2 How does the large excess of inorganic phosphate in a cell affect the amount of glycogen present? Explain.

28.3 Glycogen can be synthesized in the body by the same enzymes that degrade it. Why is this process utilized in glycogen synthesis only to a small extent, while most glycogen biosynthesis occurs via a different synthetic pathway?

28.4 Do most anabolic and catabolic reactions take place in the same location?

Carbohydrate Biosynthesis

28.5 What is the difference in the overall chemical equations between photosynthesis and respiration?

28.6 In photosynthesis, what are the sources of (a) carbon, (b) hydrogen, and (c) energy?

28.7 Name a compound that can serve as a raw material for gluconeogenesis and is (a) from the glycolytic pathway, (b) from the citric acid cycle, and (c) an amino acid.

28.8 How is glucose activated for glycogen synthesis?

28.9 Glucose is the only carbohydrate compound that the brain can use for energy. Which pathway is mobilized to supply the need of the brain during starvation: (a) glycolysis, (b) gluconeogenesis, or (c) glycogenesis? Explain.

28.10 Are the enzymes that combine two C_3 compounds into a C_6 compound in gluconeogenesis the same as or different from those that cleave the C_6 compound into two C_3 compounds in glycolysis?

28.11 Devise a scheme in which maltose is formed starting with UDP-glucose.

28.12 Glycogen is written as (glucose)$_n$.
(a) What does the n stand for?
(b) What is the approximate value of n?

28.13 What are the constituents of UTP?

Fatty Acid Biosynthesis

28.14 What is the source of carbon in fatty acid synthesis?

28.15 (a) Where in the body does fatty acid synthesis occur?
(b) Does fatty acid degradation occur in the same location?

28.16 Is ACP an enzyme?

28.17 In fatty acid biosynthesis, which compound is added repeatedly to the synthase?

28.18 (a) What is the name of the first enzyme in fatty acid synthesis?
(b) What does it do?

28.19 From which compound is CO_2 released in fatty acid synthesis?

28.20 What are the common functional groups in CoA, ACP, and synthase?

28.21 In the synthesis of unsaturated fatty acids, NADPH + H$^+$ is converted to NADP$^+$. Yet this is an oxidation step and not a reduction step. Explain.

28.22 Which of these fatty acids can be synthesized by the multienzyme fatty acid synthesis complex alone?
(a) Oleic (b) Stearic
(c) Myristic (d) Arachidonic
(e) Lauric

28.23 Some enzymes can use NADH as well as NADPH as a coenzyme. Other enzymes use one or the other exclusively. Which features would prevent NADPH from fitting into the active site of an enzyme that otherwise can accommodate NADH?

28.24 Are fatty acids for energy, in the form of fat, synthesized in the same way as fatty acids for the lipid bilayer of membrane?

28.25 Linoleic and linolenic acids cannot be synthesized in the human body. Does this mean that the human body cannot make an unsaturated fatty acid from a saturated one?

Biosynthesis of Membrane Lipids

28.26 When the body synthesizes the following membrane lipid, from which building blocks is it assembled?

$$\begin{array}{l} CH_2-O-\overset{O}{\overset{\|}{C}}-(CH_2)_{14}CH_3 \\ CH-O-\overset{O}{\overset{\|}{C}}-(CH_2)_{10}CH_3 \\ CH_2-O-\overset{O}{\overset{\|}{P}}-O-CH_2-CH-COO^- \\ \quad\quad\quad\;\; O^- \quad\quad\quad\quad NH_3^+ \end{array}$$

28.27 Name the activated constituents necessary to form the glycolipid glucoceramide.

28.28 Why is HMG reductase a key enzyme in cholesterol synthesis?

28.29 Describe by carbon skeleton designation how a C_2 compound ends up as a C_5 compound.

Biosynthesis of Amino Acids

28.30 Which reaction is the reverse of the synthesis of glutamate from α-ketoglutarate, ammonia, and NADH + H$^+$?

28.31 Which amino acid will be synthesized by the following process?

$$\begin{array}{l} COO^- \\ | \\ C=O \\ | \\ CH_2 \\ | \\ COO^- \end{array} + NADH + H^+ + NH_4^+ \longrightarrow$$

28.32 Draw the structure of the compound needed to synthesize asparagine from glutamate by transamination.

28.33 Name the products of the following transamination reaction:

$$(CH_3)_2CH-\overset{O}{\overset{\|}{C}}-COO^-$$

$$+ \;^-OOC-CH_2-CH_2-\underset{\underset{NH_3^+}{|}}{CH}-COO^- \longrightarrow$$

Chemical Connections

28.34 (Chemical Connections 28A) Photosystem I and II are complex factories of proteins, chlorophyll, and many cofactors. Where are these photosystems located in the plant and in which reaction of photosynthesis do they participate?

28.35 (Chemical Connections 28A) Which coenzyme reduces CO_2 in the Calvin cycle?

28.36 (Chemical Connections 28B) Why would an unprenylated Ras protein fail to transduct signals across a membrane?

28.37 (Chemical Connections 28C) (a) What compound produces the colored urine of blue diaper syndrome?
(b) What is the source of this compound?

Additional Problems

28.38 In the structure of NADP$^+$ what are the bonds that connect nicotinamide and adenine to the ribose units?

28.39 Which C_3 fragment carried by ACP is used in fatty acid synthesis?

28.40 When glutamate transaminates phenylpyruvate, what amino acid is produced?

$$C_6H_5-CH_2-\overset{O}{\overset{\|}{C}}-COO^-$$

$$+ \;^-OOC-CH_2-CH_2-\underset{\underset{NH_3^+}{|}}{CH}-COO^- \longrightarrow$$

28.41 Does the prenylation of Ras protein itself cause cancer?

28.42 Each activation step in the synthesis of complex lipids occurs at the expense of one ATP molecule. How many ATP molecules are used in the synthesis of one molecule of lecithin?

28.43 Consider the fact that the deamination of glutamic acid and its synthesis from α-ketoglutaric acid are equilibrium reactions. Which way will the equilibrium shift when the human body is exposed to cold temperature?

28.44 Which compound reacts with glutamate in a transamination process to yield serine?

28.45 What are the names of the C_{10} and C_{15} intermediates in cholesterol biosynthesis?

28.46 Which is carbon number 1 in HMGCoA (3-hydroxy-3-methylglutaryl CoA)?

28.47 In most biosynthetic processes, the reactant is reduced to obtain the desired product. Verify whether this statement holds for the overall reaction of photosynthesis.

28.48 What is the major difference in structure between chlorophyll and heme?

28.49 Can the complex enzyme system participating in every fatty acid synthesis manufacture fatty acids of any length?

InfoTrac College Edition

For additional reading, go to InfoTrac College Edition, your online research library, at

http://infotrac.thomsonlearning.com

CHAPTER 29

- 29.1 Introduction
- 29.2 Nutrition
- 29.3 Calories
- 29.4 Carbohydrates in the Diet and Their Digestion
- 29.5 Fats in the Diet and Their Digestion
- 29.6 Proteins in the Diet and Their Digestion
- 29.7 Vitamins, Minerals, and Water

Foods high in fiber include whole grains, legumes, fruits, and vegetables.

Nutrition

29.1 Introduction

In Chapters 26 and 27, we saw what happens to the food that we eat in its final stages—after the proteins, lipids, and carbohydrates have been broken down into their components. In this chapter, we discuss the earlier stages—nutrition and diet—and then the digestive processes that break down the large molecules to the small ones that undergo metabolism. The purpose of food is to provide energy and new molecules to replace those that the body uses. This synthesis of new molecules is especially important for the period during which a child becomes an adult.

In the days of sailing ships, many sailors contracted scurvy because they had no source of fresh fruit or vegetables on voyages that could last many weeks away from land. To counter this problem, British ships carried limes, which could be stored for long periods. Limes are an excellent source of vitamin C. British sailors acquired the nickname "limey" from this practice, a name that was often extended to any Englishman.

29.2 Nutrition

The components of food and drink that provide growth, replacement, and energy are called **nutrients.** Not all components of food are nutrients. Some components of food and drinks, such as those that provide flavor, color, or aroma, enhance our pleasure in the food but are not themselves nutrients.

CHEMICAL CONNECTIONS 29A

Parenteral Nutrition

In clinical practice, it occasionally happens that a patient cannot be fed through the gastrointestinal tract. Under these conditions, intravenous feeding or parenteral nutrition is required.

In the past, patients could be maintained on parenteral nutrition for only short periods. Calories were provided in the form of glucose, and amino acids were supplied in the form of protein hydrolysates. Over long periods, however, the use of parenteral nutrition resulted in deterioration and progressive emaciation of the patient.

More recently, the solutions used in parenteral nutrition have been manufactured to contain both essential and nonessential amino acids in fixed proportions. Calories now do not come only from glucose. Glucose supplies only about 20% of the caloric needs, fat emulsions provide the rest of the caloric requirements and the essential fatty acids. The parenteral solution also supplies the full vitamin and mineral requirements.

Nutritionists classify nutrients into six groups:

1. Carbohydrates
2. Lipids
3. Proteins
4. Vitamins
5. Minerals
6. Water

For food to be used in our bodies, it must be absorbed through the intestinal walls into the bloodstream or lymph system. Some nutrients, such as vitamins, minerals, glucose, and amino acids, can be absorbed directly. Others, such as starch, fats, and proteins, must first be broken down into smaller components before they can be absorbed. This breakdown process is called **digestion.**

A healthy body needs the proper intake of all nutrients. However, nutrient requirements vary from one person to another. For example, more energy is needed to maintain the body temperature of an adult than that of a child. For this reason, nutritional requirements are usually given per kilogram of body weight. Furthermore, the energy requirements of a physically active body are greater than those of a person in a sedentary occupation. Therefore, when average values are given, as in **Dietary Reference Intakes (DRI)** and in the former guidelines of **Recommended Daily Allowances (RDA),** one should be aware of the wide range that these average values represent.

The public interest in nutrition and diet changes with time and geography. Seventy or eighty years ago, the main nutritional interest of most Americans was getting enough food to eat and avoiding diseases caused by vitamin deficiencies, such as scurvy or beriberi. This issue is still the main concern of the large majority of the world's population. In affluent societies such as ours, however, today's nutritional message is no longer "eat more" but rather "eat less and discriminate more in your selection of food." Dieting to reduce body weight is a constant effort in a sizable percentage of the American population. Many people discriminate in their selection of food to avoid cholesterol (Section 20.9E) and saturated fatty acids to reduce the risk of heart attacks.

Along with such **discriminatory curtailment diets** came many different faddish diets. **Diet faddism** is an exaggerated belief in the effects of

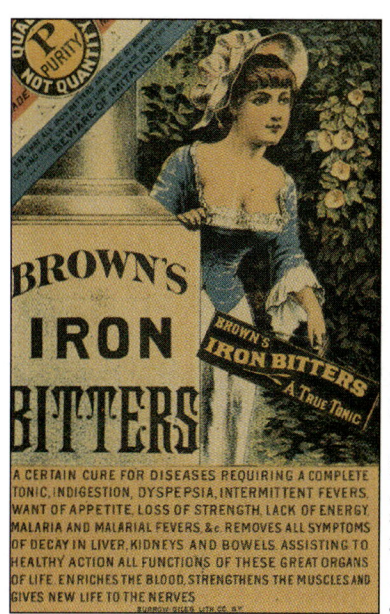

■ A late-nineteenth-century advertisement for a cure-all tonic.

Digestion The process in which the body breaks down large molecules into smaller ones that can then be absorbed and metabolized

Discriminatory curtailment diet A diet that avoids certain food ingredients that are considered harmful to the health of an individual—for example, low-sodium diets for people with high blood pressure

Nutrition Facts
Serving size 1 Bar (28g)
Servings per Container 6

Amount Per Serving	
Calories 120	Calories from Fat 35

	% Daily Value*
Total Fat 4g	6%
Saturated Fat 2g	10%
Cholesterol 0mg	0%
Sodium 45mg	2%
Potassium 100mg	3%
Total Carbohydrate 19g	6%
Dietary fiber 2g	8%
Sugars 13g	
Protein 2g	

Vitamin A 15%	•	Vitamin C 15%	
Calcium 15%	•	Iron 15%	
Vitamin D 15%	•	Vitamin E 15%	
Thiamin 15%	•	Riboflavin 15%	
Niacin 15%	•	Vitamin B_6 15%	
Folate 15%	•	Vitamin B_{12} 15%	
Biotin 10%	•	Pantothenic Acid 10%	
Phosphorus 15%	•	Iodine 2%	
Magnesium 4%	•	Zinc 4%	

*Percent Daily Values are based on a 2,000 calorie diet. Your daily values may be higher or lower depending on your calorie needs.

	Calories:	2,000	2,500
Total Fat	Less than	65g	80g
Sat Fat	Less than	20g	25g
Cholesterol	Less than	300mg	300mg
Sodium	Less than	2,400mg	2,400mg
Potassium		3,500mg	3,500mg
Total Carbohydrate		300g	375g
Dietary Fiber		25g	30g

Figure 29.1 A food label for a peanut butter crunch bar. The portion at the bottom (following the asterisk) is the same on all labels that carry it.

Fiber The cellulosic non-nutrient component in our food

nutrition upon health and disease. This phenomenon is not new; it has been prevalent for many years. Many times it is driven by visionary beliefs, but backed by little science. In the nineteenth century, Dr. Kellogg (of the cornflakes fame) recommended a largely vegetarian diet based on his belief that meat produces sexual excess. Eventually his religious fervor withered and his brother made a commercial success of his inventions of grain-based food. The latest fad is raw food, which bans any application of heat higher than 118°F to food. Heat, these faddists maintain, depletes the nutritional value of proteins and vitamins and concentrates pesticides in food. Obviously raw food diet is vegetarian, as it excludes meat and meat products.

A recommended food is rarely as good, and a condemned food is never as bad, as faddists claim. Scientific studies have proved, for instance, that food grown with chemical fertilizers is just as healthy as food grown with organic (natural) fertilizers.

Each food contains a large variety of nutrients. For example, a typical breakfast cereal lists the following items as its ingredients: milled corn, sugar, salt, malt flavoring, and vitamins A, B, C, and D, plus flavorings and preservatives. U.S. consumer laws require that most packaged food be labeled in a uniform manner to show the nutritional values of the food. Figure 29.1 shows a typical label of the type found on almost every can, bottle, or box of food that we buy.

Such labels must list the percentages of **Daily Values** for four key vitamins and minerals: vitamins A and C, calcium, and iron. If other vitamins or minerals have been added, or if the product makes a nutritional claim about other nutrients, their values must be shown as well. The percent daily values on the labels are based on a daily intake of 2000 Cal. For anyone who eats more than that amount, the actual percentage figures would be lower (and higher for those who eat less). Note that each label specifies the serving size; the percentages are based on that portion, not on the entire contents of the package. The section at the bottom of the label is exactly the same on all labels, no matter what the food, and shows the daily amounts of nutrients recommended by the government, based on a consumption of either 2000 or 2500 Cal. Some food packages are allowed to carry shorter labels, either because they have only a few nutrients or because the package has limited label space. The uniform labels make it much easier for the consumer to know exactly what he or she is eating.

The U.S. Department of Agriculture issues occasional guidelines as to what constitutes a healthy diet, depicted in the form of a pyramid (Figure 29.2). These guidelines consider the basis of a healthy diet to be the foods richest in starch (bread, rice, and so on), plus lots of fruits and vegetables (which are rich in vitamins and minerals). Protein-rich foods (meat, fish, dairy products) are to be consumed more sparingly, and fats, oils, and sweets are not considered necessary at all.

An important non-nutrient in some foods is **fiber,** which generally consists of the indigestible portion of vegetables and grains. Lettuce, cabbage, celery, whole wheat, brown rice, peas, and beans are all high in fiber. Chemically, fiber is made of cellulose, which, we saw in Section 19.7C, cannot be digested by humans. But though we cannot digest it, fiber is necessary for proper operation of the digestive system; without it constipation may result. In more serious cases, a diet lacking sufficient fiber may lead to colon cancer. The DRI recommendation is to ingest 35 g/day for men 50 and younger, and 25 g/day for women of the same age.

Figure 29.2 The Food Guide Pyramid developed by the U.S. Department of Agriculture as a general guide to a healthful diet.

Food Guide Pyramid
A Guide to Daily Food Choices

- Fats, Oils, & Sweets **Use sparingly**
- Milk, Yogurt, & Cheese Group **2-3 Servings**
- Meat, Poultry, Fish, Dry Beans, Eggs, & Nuts Group **2-3 Servings**
- Vegetable Group **3-5 Servings**
- Fruit Group **2-4 Servings**
- Bread, Cereal, Rice, & Pasta Group **6-11 Servings**

Key
• Fat (naturally occurring and added) ▼ Sugars (added)
These symbols show fats, oils and added sugars in foods.

USDA, 1992

FOOD GROUP SERVING SIZES

Bread, Cereal, Rice, and Pasta
- 1/2 cup cooked cereal, rice, pasta
- 1 ounce dry cereal
- 1 slice bread
- 2 cookies
- 1/2 medium doughnut

Vegetables
- 1/2 cup cooked or raw chopped vegetables
- 1 cup raw leafy vegetables
- 3/4 cup vegetable juice
- 10 french fries

Fruit
- 1 medium apple, banana, or orange
- 1/2 cup chopped, cooked, or canned fruit
- 3/4 cup fruit juice
- 1/4 cup dried fruit

Milk, Yogurt, and Cheese
- 1 cup milk or yogurt
- 1½ ounces natural cheese
- 2 ounces process cheese
- 2 cups cottage cheese
- 1½ cups ice cream
- 1 cup frozen yogurt

Meat, Poultry, Fish, Dry Beans, Eggs, and Nuts
- 2–3 ounces cooked lean meat, fish, or poultry
- 2–3 eggs
- 4–6 tablespoons peanut butter
- 1½ cups cooked dry beans
- 1 cup nuts

Fats, Oils, and Sweets
- Butter, mayonnaise, salad dressing, cream cheese, sour cream, jam, jelly

29.3 Calories

The largest part of our food supply goes to provide energy for our bodies. As we saw in Chapters 26 and 27, this energy comes from the oxidation of carbohydrates, fats, and proteins. The energy derived from food is usually measured in calories. One nutritional calorie (Cal) equals 1000 cal or 1 kcal. Thus, when we say that the average daily nutritional requirement for a young adult male is 2900 Cal, we mean the same amount of energy needed

One calorie is the amount of heat necessary to raise the temperature of 1 g of liquid water by 1°C (Section 1.9).

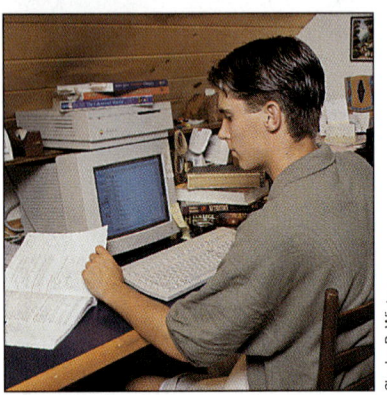

■ Different activity levels are associated with different caloric needs.

Basal caloric requirement The caloric requirement for a resting body

Excessive caloric intake results from all sources of food—carbohydrates and proteins as well as fats.

Obesity is defined by the National Institutes of Health as being a person who has a body mass index (BMI) of 30 or greater. The BMI is a measure of body fat based on height and weight that applies to both adult men and women. For example, a person 70 inches high is *normal* (BMI less than 25) if he weighs 174 lb or less. A person of the same height is *overweight* if he weighs more than 174 lb but less than 209 lb; he is *obese* if he weighs more than 209 lb. More than 97 million Americans are overweight or obese.

to raise the temperature of 2900 kg of water by 1°C or of 29 kg (64 lb) of water by 100°C. A young adult female needs 2100 Cal/day. These are peak requirements—children and older people, on average, require less energy. Keep in mind that these energy requirements apply to active people. For bodies completely at rest, the corresponding energy requirement for young adult males is 1800 Cal/day, and that for females is 1300 Cal/day. The requirement for a resting body is called the **basal caloric requirement.**

An imbalance between the caloric requirement of the body and the caloric intake creates health problems. Chronic caloric starvation exists in many parts of the world where people simply do not have enough food to eat because of prolonged drought, the devastation of war, natural disasters, or overpopulation. Famine particularly affects infants and children. Chronic starvation, called **marasmus,** increases infant mortality by as much as 50%. It results in arrested growth, muscle wasting, anemia, and general weakness. Even if starvation is later alleviated, it leaves permanent damage, insufficient body growth, and lowered resistance to disease.

At the other end of the caloric spectrum is excessive caloric intake. It results in obesity, or the accumulation of body fat. **Obesity** is becoming an epidemic in the U.S. population, with important consequences: It increases the risk of hypertension, cardiovascular disease, and diabetes. Reducing diets aim at decreasing caloric intake without sacrificing any essential nutrients. A combination of exercise and lower caloric intake can eliminate obesity, but usually these diets must achieve their goal over an extended period.

Crash diets give the illusion of quick weight loss, but most of this decrease is due to loss of water, which can be regained very quickly. To reduce obesity, we must lose body fat and not water; achieving this goal takes a lot of effort, because fats contain so much energy. A pound of body fat is equivalent to 3500 Cal. Thus, to lose 10 pounds, it is necessary either to consume 35,000 fewer Cal, which can be achieved if one reduces caloric intake by 350 Cal every day for 100 days (or by 700 Cal daily for 50 days), or to use up, through exercise, the same number of food calories.

29.4 Carbohydrates in the Diet and Their Digestion

Carbohydrates are the major source of energy in the diet. They also furnish important compounds for the synthesis of cell components (Chapter 28). The main dietary carbohydrates are the polysaccharide starch, the disaccharides lactose and sucrose, and the monosaccharides glucose and fructose. Before the body can absorb carbohydrates, it must break down di-, oligo-, and polysaccharides into monosaccharides, because only monosaccharides can pass into the bloodstream.

The monosaccharide units are connected to each other by glycosidic bonds. Glycosidic bonds are cleaved by hydrolysis. In the body, this hydrolysis is catalyzed by acids and by enzymes. When a metabolic need arises, storage polysaccharides—amylose, amylopectin, and glycogen—are hydrolyzed to yield glucose and maltose.

This hydrolysis is aided by a number of enzymes:

- **α-Amylase** attacks all three storage polysaccharides at random, hydrolyzing the α-1,4 glycosidic bonds,
- **β-Amylase** also hydrolyzes the α-1,4 glycosidic bonds but in an orderly fashion, cutting disaccharidic maltose units one by one from the nonreducing end of a chain.
- The **debranching enzyme** hydrolyzes the α-1,6 glycosidic bonds (Figure 29.3).

In acid-catalyzed hydrolysis, storage polysaccharides are cut at random points. At body temperature, acid catalysis is slower than the enzyme-catalyzed hydrolysis.

The digestion (hydrolysis) of starch and glycogen in our food supply starts in the mouth, where α-amylase is one of the main components of saliva. Hydrochloric acid in the stomach and other hydrolytic enzymes in the intestinal tract decompose starch and glycogen to produce mono- and disaccharides (D-glucose and maltose).

D-glucose enters the bloodstream and is carried to the cells to be utilized (Section 27.2). For this reason, D-glucose is often called blood sugar. In healthy people, little or none of this sugar ends up in the urine except for short periods of time (binge eating).

In diabetes mellitus, however, glucose is not completely metabolized and does appear in the urine (Section 31.5B). Because of this, it is necessary to test the urine of diabetic patients for the presence of glucose (Chemical Connections 19C).

The latest DRI guideline, issued by the National Academy of Sciences in 2002, recommends a minimum carbohydrate intake of 130 g/day. Most people exceed this value. Artificial sweeteners (Chemical Connections 29C) can be used to reduce mono- and disaccharide intake.

Not everyone considers this recommendation to be the final word. Certain diets, such as the Atkins diet, recommend reducing carbohydrate intake to force the body to burn stored fat for energy supply. For example, during the introductory two-week period in the Atkins diet, only 20 g of carbohydrates per day is recommended in the form of salad and vegetables; no fruits or starchy vegetables are allowed. For the longer term, an additional 5 g of carbohydrates per day is added in the form of fruits. This restriction induces ketosis, the production of ketone bodies (Section 27.7) that may create muscle weakness and kidney problems.

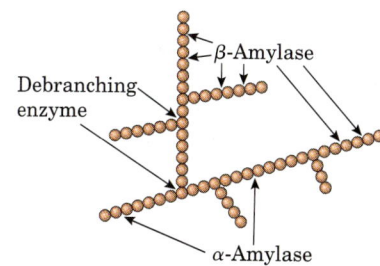

Figure 29.3 The action of different enzymes on glycogen and starch.

29.5 Fats in the Diet and Their Digestion

Fats are the most concentrated source of energy. About 98% of the lipids in our diet are fats and oils (triglycerides); the remaining 2% consist of complex lipids and cholesterol.

The lipids in the food we eat must be hydrolyzed into smaller components before they can be absorbed into the blood or lymph system through the intestinal walls. The enzymes that promote this hydrolysis are located in the small intestine and are called *lipases*. However, since lipids are insoluble in the aqueous environment of the gastrointestinal tract, they must be dispersed into fine colloidal particles before the enzymes can act on them.

Bile salts (Section 20.11) perform this important function. Bile salts are manufactured in the liver from cholesterol and stored in the gallbladder. From there, they are secreted through the bile ducts into the intestine. Lipases act on the emulsion produced by bile salts and dietary fats, breaking the fats into glycerol and fatty acids and complex lipids into fatty acids, alcohols (glycerol, choline, ethanolamine, sphingosine), and carbohydrates. These hydrolysis products are then absorbed through the intestinal walls.

Only two fatty acids are essential in higher animals, including humans: linolenic and linoleic acids (Section 20.3). Nutritionists occasionally list arachidonic acid as an **essential fatty acid.** However, our bodies can synthesize arachidonic acid from linoleic acid.

Figure 29.4 Different enzymes hydrolyze peptide chains in different but specific ways.

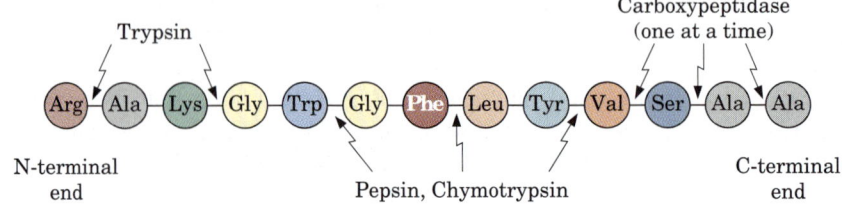

29.6 Proteins in the Diet and Their Digestion

Although the proteins in our diet can be used for energy (Section 27.9), their main use is to furnish amino acids from which the body synthesizes its own proteins (Section 25.5).

The digestion of dietary proteins begins with cooking, which denatures proteins. (Denatured proteins are hydrolyzed more easily by hydrochloric acid in the stomach and by digestive enzymes than are native proteins.) *Stomach acid* contains about 0.5% HCl. This HCl both denatures the proteins and hydrolyzes the peptide bonds randomly. *Pepsin,* the proteolytic enzyme of stomach juice, breaks peptide bonds on the C=O side of three amino acids only: tryptophan, phenylalanine, and tyrosine (Figure 29.4).

Most protein digestion occurs in the small intestine. There, the enzyme *chymotrypsin* breaks internal peptide bonds at the same positions as does pepsin. Another enzyme, *trypsin,* hydrolyzes them only on the C=O side of arginine and lysine. Other enzymes, such as *carboxypeptidase,* hydrolyze amino acids one by one from the C-terminal end of the protein. The amino acids and small peptides are then absorbed through the intestinal walls.

The human body is incapable of synthesizing ten of the amino acids needed to make proteins. These ten (the **essential amino acids**) must be obtained from our food; they are shown in Table 21.1. The body hydrolyzes food proteins into their amino acid constituents and then puts the amino acids together again to make body proteins. For proper nutrition, the human diet should contain about 20% protein.

A dietary protein that contains all of the essential amino acids is called a **complete protein.** Casein, the protein of milk, is a complete protein, as are most other animal proteins—those found in meat, fish, and eggs. People who eat adequate quantities of meat, fish, eggs, and dairy products get all the amino acids they need to keep healthy. About 50 g/day of complete proteins constitutes an adequate quantity.

An important animal protein that is not complete is gelatin, which is made by denaturing collagen (Section 21.11). Gelatin lacks tryptophan and is low in several other amino acids, including isoleucine and methionine. Many people on quick-reducing diets consume "liquid protein." This substance is simply denatured and partially hydrolyzed collagen (gelatin). Therefore, if it is the only protein source in the diet, some essential amino acids will be lacking.

Most plant proteins are incomplete. For example, corn protein lacks lysine and tryptophan; rice protein lacks lysine and threonine; wheat protein lacks lysine; and legumes are low in methionine and cystein. Even soy protein, one of the best plant proteins, is very low in methionine. Adequate amino acid nutrition is possible with a vegetarian diet, but only if a wide range of vegetables is eaten. **Protein complementation** is one such diet. In protein complementation, two or more foods complement the others' deficiencies. For example, grains and legumes complement each other, with grains being low in lysine but high in methionine. Over time, such protein

Essential amino acid An amino acid that the body cannot synthesize in the required amounts and so must be obtained in the diet

Table 29.1 Vitamins and Trace Elements as Coenzymes and Cofactors

Vitamin/Trace Element	Form of Coenzyme	Representative Enzyme	Reference
B_1, thiamine	Thiamine pyrophosphate, TPP	Pyruvate dehydrogenase	Step 12, Section 27.2
B_2, riboflavin	Flavin adenine dinucleotide, FAD	Succinate dehydrogenase	Step 6, Section 26.4
Niacin	Nicotinamide adenine dinucleotide, NAD^+	D-Glyceraldehyde-3-phosphate dehydrogenase	Step 5, Section 27.2
Panthotenic acid	Coenzyme A, CoA	Fatty acid synthase	Step 1, Section 28.3
B_6, pyridoxal	Pyridoxal phosphate, PLP	Aspartate amino transferase	Class 2, Section 22.2
B_{12}, cobalamine		Ribose reductase	Step 1, Section 24.2
Biotin	N-carboxybiotin	Acetyl CoA carboxylase	Malonyl CoA, Section 28.3
Folic acid		Purin biosynthesis	Section 24.2
Mg		Pyruvate kinase	Chemical Connections 22C
Fe		Cytochrome oxidase	Section 26.5
Cu		Cytochrome oxidase	Section 26.5
Zn		DNA polymerase	Section 24.6
Mn		Arginase	Step 5, Section 27.8
K		Pyruvate kinase	Chemical Connections 22C
Ni		Urease	Section 22.1
Mo		Nitrate reductase	
Se		Glutathione peroxidase	Chemical Connections 21A

complementation in vegetarian diets became the staple in many parts of the world—corn tortillas and beans in Central and South America, rice and lentils in India, and rice and tofu in China and Japan.

In many developing countries, protein deficiency diseases are widespread because the people get their protein mostly from plants. Among these diseases is **kwashiorkor,** whose symptoms include a swollen stomach, skin discoloration, and retarded growth.

29.7 Vitamins, Minerals, and Water

Vitamins and **minerals** are essential for good nutrition. Animals maintained on diets that contain sufficient carbohydrates, fats, and proteins and provided with an ample water supply cannot survive on these alone; they also need the essential organic components called vitamins and the inorganic ions called minerals. Many vitamins, especially those in the B group, act as coenzymes and inorganic ions as cofactors in enzyme-catalyzed reactions (Table 29.1). Table 29.2 lists the structures, dietary sources and functions of the vitamins and minerals. Deficiencies in vitamins and minerals lead to many nutritionally controllable diseases (one example is shown in Figure 29.5); these conditions are also listed in Table 29.2.

The recent trend in vitamin appreciation is connected to their general role rather than any specific action against a particular disease. For example, today the role of vitamin C in prevention of scurvy is barely mentioned but it is hailed as an important antioxidant. Similarly, other antioxidant vitamins or vitamin precursors dominate the medical literature. As an example, it has been shown that consumption of carotenoids (other than β-carotene), as well as vitamin E and C contributes significantly to respiratory health. The most important of the three is vitamin E. Furthermore, the loss of vitamin C during hemodialysis contributes significantly to oxidative damage in patients, leading to accelerated atherosclerosis.

Some vitamins have unusual effects besides their normal participation as coenzymes in many metabolic pathways or action as antioxidants in the body. Among these are the well-known effect of riboflavin, vitamin B_2, as a

Figure 29.5 Symptoms of ricketse, a vitamin D deficiency in children. The nonmineralization of the bones of the radius and the ulna results in prominence of the wrist.

The name "vitamin" comes from a mistaken generalization. Casimir Funk (1884–1967) discovered that certain diseases such as beriberi, scurvy, and pellagra are caused by lack of certain nutrients. He found that the "antiberiberi factor" compound is an amine and mistakenly thought that all such nutrients are amines. Hence, Funk coined the name "Vitamine."

Table 29.2 Vitamins and Minerals: Sources, Functions, Deficiency Diseases, and Daily Requirements

Name	Structure	Best Food Source	Function	Deficiency Symptoms and Diseases	Recommended Dietary Allowance[a]
Fat-Soluble Vitamins					
A		Liver, butter, egg yolk, carrots, spinach, sweet potatoes	Vision; to heal eye and skin injuries	Night blindness; blindness; keratinization of epithelium and cornea	800 μg (1500 μg)[b]
D		Salmon, sardines, cod liver oil, cheese, eggs, milk	Promotes calcium and phosphate absorption and mobilization	Rickets (in children): pliable bones; osteomalacia (in adults); fragile bones	5–10 μg; exposure to sunlight
E		Vegetable oils, nuts, potato chips, spinach	Antioxidant	In cases of malabsorption such as in cystic fibrosis: anemia. In premature infants: anemia	8–10 mg
K		Spinach, potatoes, cauliflower, beef liver	Blood clotting	Uncontrolled bleeding (mostly in newborn infants)	65–80 μg
Water-Soluble Vitamins					
B_1 (thiamine)		Beans, soybeans, cereals, ham, liver	Coenzyme in oxidative decarboxylation and in pentose phosphate shunt	Beriberi. In alcoholics: heart failure; pulmonary congestion	1.1 mg
B_2 (riboflavin)		Kidney, liver, yeast, almonds, mushrooms, beans	Coenzyme of oxidative processes	Invasion of cornea by capillaries; cheilosis; dermatitis	1.4 mg

Vitamin	Structure	Source	Function	Deficiency	RDA
Nicotinic acid (niacin)	(pyridine-3-COOH)	Chickpeas, lentils, prunes, peaches, avocados, figs, fish, meat, mushrooms, peanuts, bread, rice, beans, berries	Coenzyme of oxidative processes	Pellagra	15–18 mg
B$_6$ (pyridoxal)	(pyridoxal structure)	Meat, fish, nuts, oats, wheat germ, potato chips	Coenzyme in transamination; heme synthesis	Convulsions; chronic anemia; peripheral neuropathy	1.6–2.2 mg
Folic acid	(folic acid structure)	Liver, kidney, eggs, spinach, beets, orange juice, avocados, cantaloupe	Coenzyme in methylation and in DNA synthesis	Anemia	400 μg
B$_{12}$	(cobalamin structure)	Oysters, salmon, liver, kidney	Part of methyl-removing enzyme in folate metabolism	Patchy demyelination; degradation of nerves, spinal cord, and brain	1–3 μg

Table 29.2 Vitamins and Minerals: Sources, Functions, Deficiency Diseases, and Daily Requirements *(continued)*

Name	Structure	Best Food Source	Function	Deficiency Symptoms and Diseases	Recommended Dietary Allowance[a]
Pantothenic acid	HOCH$_2$C(CH$_3$)(CH$_3$)—CHCNHCH$_2$CH$_2$COOH, OH, O	Peanuts, buckwheat, soybeans, broccoli, lima beans, liver, kidney, brain, heart	Part of CoA; fat and carbohydrate metabolism	Gastrointestinal disturbances; depression	4–7 mg
Biotin	(biotin structure with (CH$_2$)$_4$COOH)	Yeast, liver, kidney, nuts, egg yolk	Synthesis of fatty acids	Dermatitis; nausea; depression	30–100 μg
C (ascorbic acid)	(ascorbic acid structure)	Citrus fruit, berries, broccoli, cabbage, peppers, tomatoes	Hydroxylation of collagen, wound healing; bond formation; antioxidant	Scurvy; capillary fragility	60 mg
Minerals					
Potassium		Apricots, bananas, dates, figs, nuts, raisins, beans, chickpeas, cress, lentils	Provides membrane potential	Muscle weakness	3500 mg
Sodium		Meat, cheese, cold cuts, smoked fish, table salt	Osmotic pressure	None	2000–2400 mg
Calcium		Milk, cheese, sardines, caviar	Bone formation; hormonal function; blood coagulation; muscle contraction	Muscle cramps; osteoporosis; fragile bones	800–1200 mg
Chloride		Meat, cheese, cold cuts, smoked fish, table salt	Osmotic pressure	None	1700–5100 mg
Phosphorus		Lentils, nuts, oats, grain flours, cocoa, egg yolk, cheese, meat (brain, sweetbreads)	Balancing calcium in diet	Excess causes structural weakness in bones	800–1200 mg

Magnesium	Cheeses, cocoa, chocolate, nuts, soybeans, beans	Cofactor in enzymes	Hypocalcemia	280–350 mg
Iron	Raisins, beans, chickpeas, parsley, smoked fish, liver, kidney, spleen, heart, clams, oysters	Oxidative phosphorylation; hemoglobin	Anemia	15 mg
Zinc	Yeast, soybeans, nuts, corn, cheese, meat, poultry	Cofactor in enzymes, insulin	Retarded growth; enlarged liver	12–15 mg
Copper	Oysters, sardines, lamb, liver	Oxidative enzymes cofactor	Loss of hair pigmentation, anemia	1.5–3 mg
Manganese	Nuts, fruits, vegetables, whole-grain cereals	Bone formation	Low serum cholesterol levels; retarded growth of hair and nails	2.0–5.0 mg
Chromium	Meat, beer, whole wheat and rye flours	Glucose metabolism	Glucose not available to cells	0.05–0.2 mg
Molybdenum	Liver, kidney, spinach, beans, peas	Protein synthesis	Retarded growth	0.075–0.250 mg
Cobalt	Meat, dairy products	Component of vitamin B_{12}	Pernicious anemia	0.05 mg (20–30 mg)[b]
Selenium	Meat, seafood	Fat metabolism	Muscular disorders	0.05–0.07 mg (2.4–3.0 mg)[b]
Iodine	Seafood, vegetables, meat	Thyroid glands	Goiter	150–170 μg (1000 μg)[b]
Fluorine	Fluoridated water; fluoridated toothpaste	Enamel formation	Tooth decay	1.5–4.0 mg (8–20 mg)[b]

[a] The U.S. RDAs are set by the Food and Nutrition Board of the National Research Council. The numbers given here are based on the latest recommendations (*National Research Council Recommended Dietary Allowances*, 10th ed., 1989, National Academy Press, Washington, D.C.). The RDA varies with age, sex, and level of activity; the numbers given are average values for both sexes between the ages of 18 and 54.
[b] Toxic if doses above the level shown in parentheses are taken.

CHEMICAL CONNECTIONS 29B

Too Much Vitamin Intake?

One might suppose that, since vitamins are essential nutrients, the more you take, the better off you are. Using this reasoning, some people consider that the RDAs listed for each vitamin and mineral in Table 29.2 are merely a minimum, and that larger quantities will be even more beneficial. It has become fashionable to take large doses of vitamins, especially those with antioxidant properties—namely, vitamins A, C, and E. No scientific proof exists for such claims. Larger doses are usually harmless because the body excretes the excess, but not in all cases.

High dosages can do unexpected harm in certain instances, and complying with the RDA can prevent harm not usually associated with the vitamin. The case of folic acid illustrates this point. Table 29.2 shows an RDA of 400 μg for this vitamin, as the amount necessary to prevent anemia. It has also been shown that this amount of folic acid is especially important during pregnancy. Consumption of this vitamin can prevent birth defects that result in spina bifida, the birth of a child with an open spine, and anencephaly, the birth of a child without most of the brain. Since B vitamins are not stored in the body, it is important that a pregnant woman take the RDA every day—but not more. An intake of 1000 μg or more of folic acid daily is harmful because it can hide the symptoms of certain anemias without actually preventing these diseases. The U.S. Food and Drug Administration now requires that food labeled as "enriched" include certain amounts of folic acid. The amount for a slice of enriched bread is 40 to 70 μg, and that for enriched cereals is 100 μg per serving.

Similarly, high dosages for niacin and of vitamins A and D have been shown to cause health problems.

CHEMICAL CONNECTIONS 29C

Dieting and Artificial Sweeteners

Obesity is a serious problem in the United States. Many people would like to lose weight but are unable to control their appetites enough to do so, which is why artificial sweeteners are popular. These substances have a sweet taste but do not add calories. Many people restrict their sugar intake. Some are forced to do so by diseases such as diabetes; others, by the desire to lose weight. Since most of us like to eat sweet foods, artificial sweeteners are added to many foods and drinks for those who must (or want to) restrict their sugar intake.

Four noncaloric artificial sweeteners are now approved by the U.S. Food and Drug Administration (FDA). The oldest of these, saccharin, is 450 times sweeter than sucrose. Saccharin has been in use for about 100 years. Unfortunately, some tests have shown that this sweetener, when fed in massive quantities to rats, caused some of these rats to develop cancer. Other tests have given negative results. Recent research has shown that this cancer in rats is not relevant to human consumption. Saccharin continues to be sold on the market.

A newer artificial sweetener, aspartame, does not have the slight aftertaste that saccharin does. Aspartame is the methyl ester of a dipeptide, Asp-Phe (aspartylphenylalanine). Its sweetness was discovered in 1969. After extensive biological testing, it was approved by the FDA in 1981 for use in cold cereals, drink mixes, and gelatins and as tablets or powder to be used as a sugar substitute. Aspartame is 100 to 150 times sweeter than sucrose. It is made from natural amino acids, so both the aspartic acid and the phenylalanine have the L configuration. Other possibilities have also been synthesized: the L-D, D-L and D-D configurations. However, they are all bitter rather than sweet.

(continued on next page)

Saccharin   Aspartame   Acesulfame-K

CHEMICAL CONNECTIONS 29C

Dieting and Artificial Sweeteners (continued)

Aspartame is sold under such brand names as Equal and NutraSweet. A newer version of aspartame, neotame, was approved by the FDA in 2002. Twenty-five times sweeter than aspartame, it is essentially the same compound but with one —H substituted by a —CH_2—CH_2—$C(CH_3)_3$ group at the amino terminal.

A third artificial sweetener, acesulfame-K, is 200 times sweeter than sucrose and is used (under the name Sunette) mostly in dry mixes. The fourth option, Sucralose, is used in sodas, baked goods, and tabletop packets. This trichloro derivative of a disaccharide is 600 times sweeter than sucrose and has no aftertaste. Neither Sucralose nor acesulfame-K is metabolized in the body; that is, they pass through unchanged.

Olestra

Of course, calories in the diet do not come only from sugar. Dietary fat is an even more important source (Section 29.5). It has long been hoped that some kind of artificial fat, which would taste the same but have no (or only a few) calories, would help in losing weight. Procter & Gamble Company has developed such a product, called Olestra. Like natural fats, this molecule is a carboxylic ester; but instead of glycerol, the alcohol component is sucrose (Section 19.5). All eight of sucrose's OH groups are converted to ester groups; the carboxylic acids are the fatty acids containing eight to ten carbons. The one illustrated here is made with the C_{10} decanoic (capric) acid.

Although Olestra has a chemical structure similar to a fat, the human body cannot digest it because the enzymes that digest ordinary fats are not designed for the particular size and shape of this molecule. Therefore, it passes through the digestive system unchanged, and we derive no calories from it. Olestra can be used instead of an ordinary fat in cooking such items as cookies and potato chips. People who eat these will then be consuming fewer calories.

On the downside, Olestra can cause diarrhea, cramps, and nausea in some people, effects that increase with the amount consumed. Individuals who are susceptible to such side effects will have to decide if the potential to lose weight is worth the discomfort involved. Furthermore, Olestra dissolves and sweeps away some vitamins and nutrients in other foods being digested at the same time. To counter this effect, food manufacturers must add some of these nutrients (vitamins A, D, E, and K) to the product. FDA requires that all packages of Olestra-containing foods carry a warning label explaining these side effects.

---

photosensitizer. Niacinamide, the amide form of vitamin B (niacin), is used in megadoses (2 g/day) to treat an autoimmune disease called bullus pemphigoid, which causes blisters on the skin. Conversely, the same megadoses in healthy patients may cause harm (Chemical Connections 29B).

Although the concept of RDA has been used since 1940 and periodically updated as new knowledge dictated, a new concept is being developed in the field of nutrition. The DRI, currently in the process of completion, is designed to replace the RDA and is tailor-made to different ages and genders. It gives a set of two to four values for a particular nutrient in RDI:

- The estimated average requirement
- The recommended dietary allowance

PABA or p-aminobenzoic acid, a precursor of folic acid (Chemical Connections 22D), actually protects the skin against sunburn. As such, it is an integral part of many sunscreen lotions.

CHEMICAL CONNECTIONS 29D

Food for Performance Enhancement

Athletes do whatever they can legally to enhance their performance. After vigorous exercise lasting 15 minutes, however, performance typically declines because the glycogen stores in the muscle are depleted. If athletes can maintain their blood glucose levels, their performance might improve. This fact explains why many athletes load up on carbohydrates in the form of pasta meals, energy bars, and sport drinks.

In recent years, a new performance-enhancer food has appeared on the market and quickly become a best-seller: creatine. It is sold over the counter. Creatine is a naturally occurring amino acid in the muscles, which store energy in the form of high-energy phosphocreatine. During a short and strenuous bout of exercise, such as a 100-meter sprint, the muscles first use up the ATP obtainable from the reaction of phosphocreatine with ADP; only then do they rely on the glycogen stores.

Phosphocreatine   Creatine

Both creatine and carbohydrates are natural food and body components and, therefore, cannot be considered equivalent to banned performance enhancers such as anabolic steroids or "andro" (Chemical Connections 20G). Creatine also has no known hazards, even in long-term use. It is beneficial in improving performance in sports where short bursts of energy are needed, such as weight lifting, jumping, and sprinting. Lately, it has been used experimentally to preserve muscle neurons in degenerative diseases such as Parkinson's disease, Huntington's disease, and muscular dystrophy.

■ **Figure 29D** Creatine is sold over the counter.

- The adequate intake level
- The tolerable upper intake level

For example, the RDA for vitamin D is 5-10 μg, the adequate intake listed in the DRI for vitamin D for a person between age 9 and 50 years is 5 μg, and the tolerable upper intake level for the same age group is 50 μg.

A third set of standards appears on food labels in the United States, the Daily Values discussed in Section 29.2. Each give a single value for each nutrient and reflects the need of an average healthy person eating 2000 to 2500 calories per day. The Daily Value for vitamin D, as appears on vitamin bottles, is 400 International Units, which is 10 μg, the same as the RDA.

Water makes up 60% of our body weight. Most of the compounds in our body are dissolved in water, which also serves as a transporting medium to carry nutrients and waste materials. These functions are discussed in Chapter 31. We must maintain a proper balance between water intake and water excretion via urine, feces, sweat, and exhalation of breath. A normal diet requires about 1200 to 1500 mL of water per day, in addition to the water consumed as part of our foods.

Public drinking water systems are regulated by the Environmental Protection Agency (EPA) which sets minimum standards for protection of pub-

SUMMARY

Nutrients are components of foods that provide growth, replacement, and energy (Section 29.2). Nutrients are classified into six groups: carbohydrates, lipids, proteins, vitamins, minerals, and water. Each food contains a variety of nutrients. The largest part of our food intake is used to provide energy for our bodies. A typical young adult needs 2900 Cal (male) or 2100 Cal (female) as an average daily caloric intake (Section 29.3). **Basal caloric requirements,** the energy needs when the body is completely at rest, are less than the normal requirements. An Imbalance between need and caloric intake may create health problems, for example, chronic starvation increases infant mortality, whereas obesity leads to hypertension, cardiovascular disease, and diabetes.

Carbohydrates are the major source of energy in our diet (Section 29.4). Monosaccharides are directly absorbed in the intestines, while oligo- and polysaccharides, such as starch, are digested with the aid of stomach acid, α- and β-amylases, and debranching enzymes.

Fats are the most concentrated source of energy (Section 29.5). They are emulsified by bile salts and digested by lipase before being absorbed as fatty acids and glycerol in the intestines. Essential fatty and amino acids are needed as building blocks because our bodies cannot synthesize them.

Proteins are hydrolyzed by stomach acid and further digested by enzymes such as pepsin and trypsin before being absorbed as amino acids (Section 29.6).

Vitamins and minerals are essential constituents of the diet that are needed in small quantities (Section 29.7). The fat-soluble vitamins are A, D, E, and K. Vitamins C and the B group are water-soluble vitamins. Most of the B vitamins are essential coenzymes. The most important dietary minerals are Na^+, Cl^-, K^+, PO_4^{3-}, Ca^{2+}, Fe^{2+}, and Mg^{2+}, but trace minerals are also necessary.

Water makes up 60% of body weight.

PROBLEMS

Numbers that appear in color indicate difficult problems. ▶ designates problems requiring application of principles.

Nutrition

29.1 Are nutrient requirements uniform for everyone?

29.2 Is banana flavoring, isopentyl acetate, a nutrient?

29.3 If sodium benzoate, a food preservative, is excreted as such, and if calcium propionate, another food preservative, is metabolized to CO_2 and H_2O, would you consider either of these preservatives to be a nutrient? If so, why?

29.4 Is corn grown solely with organic fertilizers more nutritious than that grown with chemical fertilizers?

29.5 Which part of the Nutrition Facts label found on food packages is the same for all labels carrying it?

29.6 Of which kinds of food does the government recommend that we have the most servings each day?

29.7 What is the importance of fiber in the diet?

29.8 ▶ Can a chemical that, in essence, goes through the body unchanged be an essential nutrient? Explain.

Calories

29.9 A young adult female needs a caloric intake of 2100 Cal/day. Her basal caloric requirement is only 1300 Cal/day. Why is the extra 800 Cal needed?

29.10 What ill effects may obesity bring?

29.11 Assume that you want to lose 20 lb of body fat in 60 days. Your present dietary intake is 3000 Cal/day. What should your caloric intake be, in Cal/day, to achieve this goal, assuming no change in exercise habits?

29.12 What is marasmus?

29.13 Diuretics help to secrete water from the body. Would diuretic pills be a good way to reduce body weight?

Carbohydrates in the Diet and Their Digestion

29.14 Humans cannot digest wood; termites do so with the aid of bacteria in their digestive tract. Is there a basic difference in the digestive enzymes present in humans and termites?

29.15 What is the product of the reaction when α-amylase acts on amylose?

29.16 Does HCl in the stomach hydrolyze both the 1,4- and 1,6-glycosidic bonds?

29.17 Beer contains maltose. Can beer consumption be detected by analyzing the maltose content of a blood sample?

Fats in the Diet and Their Digestion

29.18 Which nutrient provides energy in its most concentrated form?

29.19 What is the precursor of arachidonic acid in the body?

29.20 How many (a) essential fatty acids and (b) essential amino acids do humans need in their diets?

29.21 Do lipases degrade (a) cholesterol or (b) fatty acids?

29.22 What is the function of bile salts in digestion of fats?

Proteins in the Diet and Their Digestion

29.23 Is it possible to get a sufficient supply of nutritionally adequate proteins by eating only vegetables?

29.24 Suggest a way to cure kwashiorkor.

29.25 What is the difference between protein digestion by trypsin and by HCl?

29.26 Which one will be digested faster: (a) a raw egg or (b) a hard-boiled egg? Explain.

Vitamins, Minerals, and Water

29.27 In a prison camp during a war, the prisoners are fed plenty of rice and water but nothing else. What would be the result of such a diet in the long run?

29.28 (a) How many milliliters of water per day does a normal diet require?
(b) How many calories does this amount of water contribute?

29.29 Why did British sailing ships carry a supply of limes?

29.30 What are the symptoms of vitamin A deficiency?

29.31 What is the function of vitamin K?

29.32 (a) Which vitamin contains cobalt?
(b) What is the function of this vitamin?

29.33 Vitamin C is recommended in megadoses by some people for prevention of all kinds of diseases, ranging from colds to cancer. What disease has been scientifically proven to be prevented when sufficient daily doses of vitamin C are in the diet?

29.34 Why is the Recommended Dietary Allowances (RDA) being phased out in preference of Daily Reference Intake (DRI)?

29.35 What are the nonspecific effects of vitamin E, C, and carotenoids?

29.36 What are the best dietary sources of calcium, phosphorus, and cobalt?

29.37 Which vitamin contains a sulfur atom?

29.38 What are the symptoms of vitamin B_{12} deficiency?

Chemical Connections

29.39 (Chemical Connections 29A) Why cannot glucose supply all the caloric needs in parenteral nutrition over a long period?

29.40 (Chemical Connections 29B) The RDA for niacin is 18 mg. Would it be still better to take a 100-mg tablet of this vitamin every day?

29.41 (Chemical Connections 29C) Describe the difference between the structure of aspartame and the methyl ester of phenylalanylaspartic acid.

29.42 (Chemical Connections 29C) (a) Which artificial sweeteners are not metabolized in the body at all?
(b) What could be products of digested aspartame?

29.43 (Chemical Connections 29C) What is common in the structures of Olestra and Sucralose?

29.44 (Chemical Connections 29D) Looking in Table 21.1, which lists the common amino acids found in proteins, which amino acid most resembles creatine?

Additional Problems

29.45 Which two chemicals are used most frequently to disinfect public water supply?

29.46 Which vitamin is part of coenzyme A (CoA)? List a step (or the enzyme) that has CoA as a coenzyme in (a) glycolysis and (b) fatty acid synthesis.

29.47 What vitamin in megadoses is prescribed to combat autoimmune blisters?

29.48 Why is it necessary to have proteins in our diets?

29.49 What chemical processes take place during digestion?

29.50 According to the government's food pyramid, are there any foods that we can completely omit from our diets and still be healthy?

29.51 Does the debranching enzyme help in digesting amylose?

29.52 As an employee of a company that markets walnuts, you are asked to provide information for an ad that would stress the nutritional value of walnuts. What information would you provide?

29.53 In diabetes, insulin is administered intravenously. Explain why this hormone protein cannot be taken orally.

29.54 Egg yolk contains a lot of lecithin (a phosphoglyceride). After ingesting a hard-boiled egg, would you find an increase in the lecithin level of your blood? Explain

29.55 What would you call a diet that scrupulously avoids phenylalanine-containing compounds? Could aspartame be used in such a diet?

29.56 What kind of supplemental enzyme would you recommend for a patient after a peptic ulcer operation?

29.57 In a trial, a woman was accused of poisoning her husband by adding arsenic to his meals. Her attorney stated that this was done to promote her husband's health, as arsenic is an essential nutrient. Would you accept this argument?

InfoTrac College Edition

For additional reading, go to InfoTrac College Edition, your online research library, at

http://infotrac.thomsonlearning.com

CHAPTER 30

- *30.1* Introduction
- *30.2* Organs and Cells of the Immune System
- *30.3* Antigens and Their Presentation by Major Histocompatibility Complex
- *30.4* Immunoglobulin
- *30.5* T Cells and Their Receptors
- *30.6* Control of Immune Response: Cytokines
- *30.7* How to Recognize "Self"

Two natural killer (NK) cells, shown in yellow-orange, attacking a leukemia cell, shown in red.

Immunochemistry

30.1 Introduction

When you were in elementary school, you may have had chickenpox along with many other children. The viral disease passed from one person to the next and ran its course, but after the children recovered, they never had chickenpox again. Those who were infected became *immune* to this disease. Humans and other vertebrates possess a highly developed, complex immune system that defends the body against foreign invaders. Figure 30.1, which presents an overview of this complex system, could serve as a road map as you read the different sections in this chapter. Referring back to Figure 30.1, you will find the precise location of the topic under discussion and its relationship to the immune system as a whole.

A Innate Immunity

The first line of defense is the natural resistance of the body: the **innate immunity.** It comes in two forms, external and internal. The salient characteristics of innate immunity are that it is nonspecific and without memory.

One example of **external innate immunity** is the skin, which provides a barrier against penetration of pathogens. The skin also secretes lactic acid

Innate immunity The natural nonspecific resistance of the body against foreign invaders, which has no memory

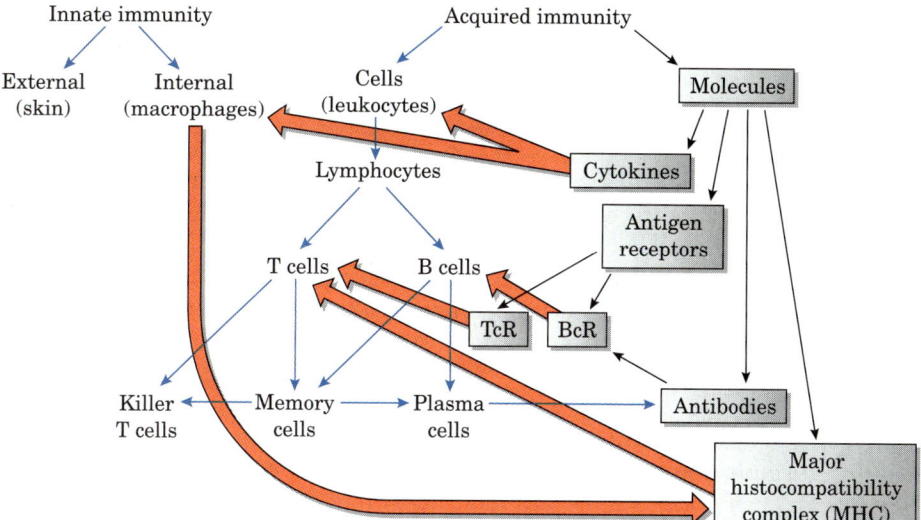

Figure 30.1 Overview of the immune system: its components and their interactions.

and fatty acids, both of which create a low pH, uncomfortable environment for bacteria. Tears in the eyes and mucus in the respiratory and gastrointestinal tracts perform the same function.

After a pathogen penetrates a tissue, **internal innate immunity** takes over. This system creates physiological barriers such as high oxygen pressures that kill anaerobic bacteria. Natural killer cells (NK) function as policemen of internal innate immunity. When they encounter cancerous cells, cells infected by a virus, or any other suspicious cells, they attach themselves to those abnormal cells (see the chapter opening photo). Other nonspecific responses include the proliferation of macrophages, which engulf and digest bacteria and reduce inflammation. In the inflammatory response to injury or infection, the capillaries dilate to allow greater flow of blood to the site of the injury, enabling agents of the internal innate immunity system to congregate there.

Acute inflammation and its symptoms were described two millennia ago by one of the founding fathers of systematic medicine, the Roman physician Celsus. These symptoms are redness, swelling, heat, and pain.

B Adaptive Immunity

Vertebrates have a second line of defense, called the **adaptive or acquired immunity.** We refer to this type of immunity when we colloquially talk about the **immune system.** The key features of the immune system are *specificity* and *memory*. The immune system uses antibodies designed specifically for each type of invader. In a second encounter with the same invader, the response is more rapid, more vigorous, and more prolonged than it was in the first case, because the immune system remembers the nature of the invader from the first encounter.

This is why, for example, if we have measles once, we will not get this disease a second time.

The invaders may be bacteria, viruses, molds, or pollen grains. A body with no defense against such invaders could not survive. In a rare genetic disease, a person is born without a functioning immune system. Attempts have been made to bring up such children in an enclosure totally sealed from the environment. While in this environment, they can survive, but when the environment is removed, such people always die within a short time. The disease AIDS (Chemical Connections 25C) slowly destroys the immune system, leaving its victims to die from some invading organism that a person without AIDS would easily be able to fight off.

As we shall see, the beauty of the body's immune system is that it is flexible. The system is capable of making millions of potential defenders, so

C Components of the Immune System

Foreign substances that invade the body are called **antigens.** The immune system is made of both cells and molecules. Two types of **white blood cells,** called lymphocytes, fight against the invaders: (1) T cells kill the invader by contact and (2) B cells manufacture **antibodies,** which are soluble immunoglobulin molecules that immobilize antigens.

The basic molecules of the immune system belong to the **immunoglobulin superfamily.** All molecules of this class have a certain portion of the molecule that can interact with antigens, and all are glycoproteins. The polypeptide chains in this superfamily have two domains: a constant region and a variable region. The constant region has the same amino acid sequence in each of the same class molecules. In contrast, the variable region is antigen-specific, which means that the amino acid sequence in this region is unique for each antigen. The variable regions are designed to recognize only one specific antigen.

There are three representatives of the immunoglobulin superfamily in the immune system:

1. Antibodies are soluble immunoglobulins secreted by the plasma cells.
2. Receptors on the surface of T cells (TcR) recognize and bind antigens presented to them.
3. Molecules that present antigens also belong to this superfamily. They reside inside the cells. These protein molecules are known as major histocompatibility complex (MHC).

When a cell is infected by an antigen, MHC molecules interact with it and bring a characteristic portion of the antigen to the surface of the cell. Such a surface presentation then marks the diseased cell for destruction. This can happen in a cell that was infected by a virus, and it can happen in macrophages that engulf and digest bacteria and virus.

Antigen A substance foreign to the body that triggers an immune response

Antibody A glycoprotein molecule that interacts with an antigen

Immunoglobulin superfamily Glycoproteins that are composed of constant and variable protein segments having significant homologies to suggest that they evolved from a common ancestry

Other glycoproteins belonging to the immunoglobulin superfamily are the cell-adhesion molecules that operate among cells and cause loose cells to adhere to each other to form layers and then tissues such as epithelium. They also play a significant part in the development of embryos.

D The Speed of the Immune Response

The process of finding and making just the right immunoglobulin to fight a particular invader is relatively slow, compared to the actions of the chemical messengers discussed in Chapter 23. While neurotransmitters act within a millisecond and hormones within seconds, minutes, or hours, immunoglobulins respond to an antigen over a longer span of time—weeks and months.

Although the immune system can be considered another form of chemical communication (Chapter 23), it is much more complex than neurotransmission because it involves molecular signals and interplay between various cells. In these constant interactions, the major elements are as follows: (1) the cells of the immune system, (2) the antigens and their perception by the immune system, (3) the antibodies (immunoglobulin molecules) that are designed to immobilize antigens, (4) the receptor molecules on the surface of the cells that recognize antigens, and (5) the cytokine molecules that control these interactions. Because the immune system is the cornerstone of the body's defenses, its importance for students in health-related sciences is undeniable.

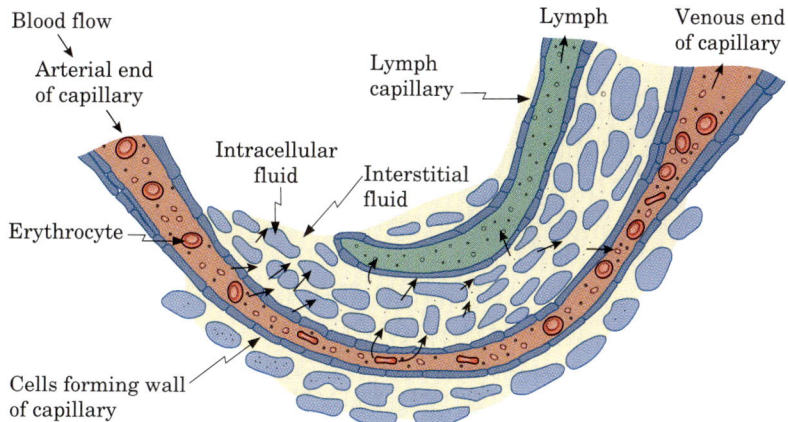

Figure 30.2 Exchange of compounds among three body fluids: blood, interstitial fluid, and lymph. *(After J. R. Holum, Fundamentals of General, Organic and Biological Chemistry. New York: John Wiley & Sons, 1978, p. 569)*

30.2 Organs and Cells of the Immune System

The blood plasma circulates in the body and comes in contact with the other body fluids through the semipermeable membranes of the blood vessels. Therefore, blood can exchange chemical compounds with other body fluids and, through them, with the cells and organs of the body (Figure 30.2).

A Lymphoid Organs

The lymphatic capillary vessels drain the fluids that bathe the cells of the body. The fluid within these vessels is the **lymph**. Lymphatic vessels circulate throughout the body and enter certain organs, called **lymphoid organs**, such as the thymus, spleen, tonsils, and lymph nodes (Figure 30.3). The cells primarily responsible for the functioning of the immune system are the specialized white blood cells called **lymphocytes**. As their name implies, these cells are mostly found in the lymphoid organs. Lymphocytes may be either specific for a given antigen or nonspecific.

T cells are lymphocytes that originate in the bone marrow but mature in the thymus gland. **B cells** are lymphocytes that originate and mature in the bone marrow. Both B and T cells are found mostly in the lymph, where they circulate looking for invaders. Small numbers of lymphocytes are also found in the blood. To get there, they must squeeze through tiny openings between the endothelial cells. This process is aided by signaling molecules called cytokines (Section 30.6). The sequence of the response of the body to a foreign invader is depicted schematically in Table 30.1.

B Cells of the Internal Innate Immunity

The two most important cells of innate immunity are macrophages and natural killer cells. **Macrophages** are the first cells in the blood that encounter an antigen. They belong to the internal innate immune system; inasmuch as they are nonspecific, macrophages attack virtually anything that is not recognized as part of the body, including pathogens, cancer cells, and damaged tissues. Macrophages engulf an invading bacterium or virus and kill it. The "magic bullet" in this case is the NO molecule, which we have seen is both toxic (Chemical Connections 3C) and can act as a secondary messenger (Chemical Connections 23F).

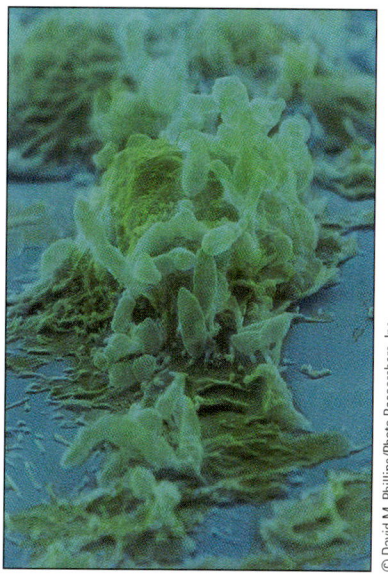

■ Macrophage-ingesting bacteria (the rod-shaped structures). The bacteria will be pulled inside the cell within a membrane-bound vesicle and quickly killed.

Figure 30.3 The lymphatic system is a web of lymphatic vessels containing a clear fluid, called lymph, and various lymphatic tissues and organs located throughout the body. Lymph nodes are masses of lymphatic tissue covered with a fibrous capsule. Lymph nodes filter the lymph. In addition, they are packed with macrophages and lymphocytes.

Table 30.1 Interactions Among the Different Cells of the Immune System

Nonspecific:
Infection → Interferon → NO synthase → Macrophage engulfing bacteria, or virus digests them → Antigen (digested) presented on the surface of macrophage → T cell, B cell

Specific: Killer T cell, Memory cell, Plasma cell

The NO molecule is short-lived and must be constantly manufactured anew. When an infection begins, the immune system manufactures the protein interferon. It, in turn, activates a gene that produces an enzyme, nitric oxide synthase. With the aid of this enzyme, the macrophages, endowed with NO, kill the invading organisms. Macrophages then digest the engulfed antigen and display a small portion of it on their surface.

Natural killer (NK) cells target abnormal cells. Once in physical contact with such cells, NK cells release proteins, aptly called perforins, that perforate the target cell membranes and create pores. The membrane of the target cell becomes leaky, allowing hypotonic (Section 6.8) liquid from the surroundings to enter the cell, which swells and eventually bursts.

The interferon-induced, enzyme-manufactured NO is critical for the immune system. This enzyme is also implicated in the overproduction of NO, however. Such an overproduction may result in pathologies, such as septic shock, multiple sclerosis, rheumatoid arthritis, and cancer formation.

C Cells of Adaptive Immunity: T and B Cells

T cells interact with the antigen presented by the macrophage and produce other T cells that are highly specific to the antigen. When these T cells differentiate, some of them become **killer T cells,** which kill the invading foreign cells by cell-to-cell contact. Killer T cells, like the natural killer cells, act through perforins, which attach themselves to the target cell, in effect punching holes in its membranes. Through these holes water rushes into the target cell; it swells and eventually bursts.

Both the natural killer cells and the T killer cells act the same way, through perforin. The T killer cells attack specific targets; the natural killer cells attack all suspicious targets.

Other T cells will become **memory cells.** They will remain in the bloodstream, so that if the same antigen enters the body again, even years after the primary infection, the body will not need to build up its defenses anew but is ready to kill the invader instantly.

The production of antibodies is the task of **plasma cells.** These cells are derived from B cells after the B cells have been exposed to an antigen.

The lymphatic vessels, in which most of the antigen attacking takes place, flow through a number of lymph nodes (Figure 30.3). These nodes are essentially filters. Most plasma cells reside in lymph nodes, so most antibodies are produced there. Each lymph node is also packed with millions of other lymphocytes. More than 99% of all invading bacteria and foreign particles are filtered out in the lymph nodes. As a consequence, the outflowing lymph is almost free of invaders and is packed with antibodies produced by the plasma cells.

30.3 Antigens and Their Presentation by Major Histocompatibility Complex

A Antigens

Antigens are any foreign substances that elicit an immune response; for this reason, they are also called immunogens. Three features characterize an antigen. The first condition is foreignness—molecules of your own body will not elicit an immune response. The second condition is that the antigen must be of molecular weight greater than 6000. The third condition is that the molecule must have sufficient complexity. A polypeptide made of lysine only, for example, is not immunogenic.

There are rare exceptions: In an autoimmune disease, the body mistakes its own protein as foreign (see Chemical Connections 30D).

Antigens can be proteins, polysaccharides, or nucleic acids, as all of these substances are large molecules. Antigens may be soluble in the cytoplasm, or they may be found at the surface of cells, either embedded in the membrane or just absorbed on it. An example of polysaccharidic antigenicity is ABO blood groups (Chemical Connections 19D).

In protein antigens, only part of the primary structure is needed to cause an immune response. Between 5 and 7 amino acids are needed to interact with an antibody, and between 10 and 15 amino acids are necessary

Epitope The smallest number of amino acids on an antigen that elicits an immune response

to bind to a receptor on a T cell. The smallest unit of an antigen capable of binding with an antibody is called the **epitope.** The amino acids in an epitope do not have to be in sequence in the primary structure, as folding and secondary structures may bring amino acids that are not in sequence into each other's proximity. For example, the amino acids in positions 20 and 28 may form part of an epitope. *Antibodies can recognize all types of antigens, but T cell receptors recognize only peptide antigens.*

As noted earlier, antigens may be in the interior of an infected cell or on the surface of a virus or bacteria that penetrated the cell. To elicit an immune reaction, the antigen or its epitope must be brought to the surface of the infected cell. Similarly, after a macrophage swallows up and partially digests an antigen, the macrophage must bring the epitope back to its surface to elicit an immune response from T cells (Table 30.1).

B Major Histocompatibility Complexes

The task of bringing an antigen's epitope to the infected cell's surface is performed by a protein complex called **major histocompatibility complex (MHC).** As the name implies, its role in immune response was first discovered in organ transplants. MHC molecules are transmembrane proteins belonging to the immunoglobulin superfamily. The two classes of MHC molecules (Figure 30.4) both have peptide-binding variable domains. Class I MHC is made of a single polypeptide chain, whereas class II MHC is a dimer. Class I MHC seeks out antigen molecules that have been *synthesized*

Figure 30.4 Differential processing of antigens in the MHC class II pathway (left) or MHC class I pathway (right). Cluster determinants (CD) are parts of the T cell receptor complex (see Section 30.5B).

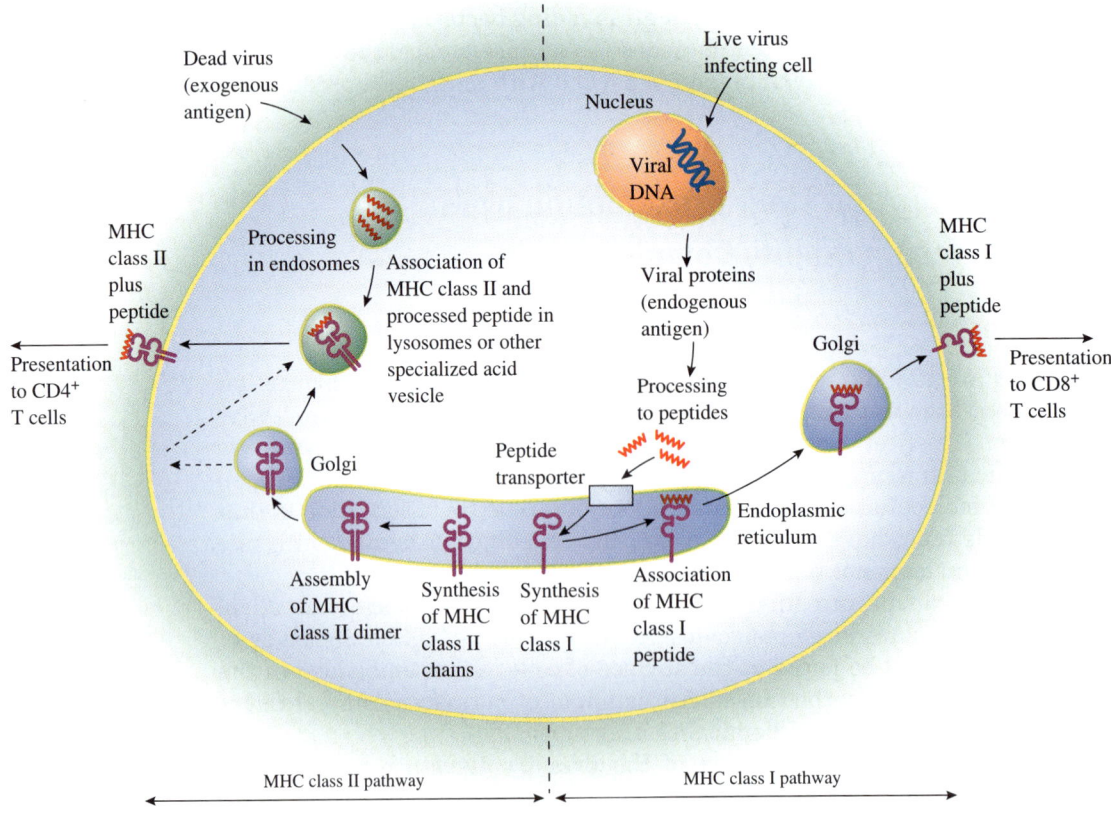

Table 30.2	Immunoglobulin Classes		
Class	Molecular Weight (MW)	Carbohydrate Content (%)	Concentration in Serum (mg/100 mL)
IgA	200 000–700 000	7–12	90–420
IgD	160 000	<1	1–40
IgE	190 000	10–12	0.01–0.1
IgG	150 000	2–3	600–1800
IgM	950 000	10–12	50–190

inside a virus-infected cell. Class II MHC molecules pick up exogeneous "dead" antigens. In each case, the epitope attached to the MHC is brought to the cell surface to be presented to T cells.

For example, if a macrophage engulfed and digested a virus, the result would be dead antigens. The digestion occurs in several steps. First, the antigen is processed in lysosomes, special organelles of cells that contain proteolytic enzymes. A special enzyme called GILT (gamma-interferon inducible lysosomal thiol reductase) breaks the disulfide bridges of the antigen by reduction. The reduced peptide antigen unfolds and is exposed to proteolytic enzymes that hydrolyze it to smaller peptides. These peptides serve as epitopes that are recognized by class II MHC.

30.4 Immunoglobulin

A Classes of Immunoglobulins

Immunoglobulins are glycoproteins—that is, carbohydrate-carrying protein molecules. Not only do the different classes of immunoglobulins vary in molecular weight and carbohydrate content, but their concentration in the blood differs dramatically as well (Table 30.2). The IgG and IgM antibodies are the most important antibodies in the blood. They interact with antigens and trigger the swallowing up (phagocytosis) of these cells by such specialized cells as macrophages. The IgA molecules are found mostly in secretions: tears, milk, and mucus. Therefore, these immunoglobulins attack the invading material before it gets into the bloodstream. IgE plays a part in such allergic reactions as asthma and hay fever.

In immunoglobulin IgG, the light chains have a molecular weight of 25,000 and the heavy chains have a molecular weight of 50,000.

B Structure of Immunoglobulins

Each immunoglobulin molecule is made of four polypeptide chains: two identical light chains and two identical heavy chains. The four polypeptide chains are arranged symmetrically, forming a Y shape (Figure 30.5). Four disulfide bridges link the four chains into a unit. Both light and heavy chains have constant and variable regions. The constant regions have the same amino acid sequences in different antibodies, and the variable regions have different amino acid sequences in different antibodies.

The variable regions of the antibody recognize the foreign substance (the antigen) and bind to it (Figure 30.6). Since each antibody contains two variable regions, it can bind two antigens, thereby forming a large aggregate.

The binding of the antigen to the variable region of the antibody is not by covalent bonds but by much weaker intermolecular forces such as London dispersion forces, dipole–dipole interactions, and hydrogen bonds

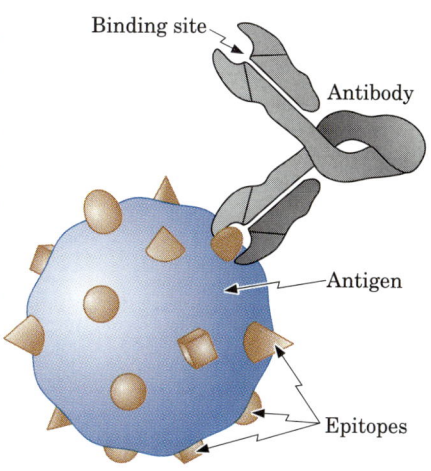

■ **Antibody binding to the epitope of an antigen.**

Figure 30.5 (a) Schematic diagram of an IgG-type antibody consisting of two heavy chains and two light chains connected by disulfide linkages. The amino terminal end of each chain has the variable portion. (b) The same in the ribbon model. (c) Model showing how an antibody bonds to an antigen.

Figure 30.6 (a) An antigen–antibody complex. The antigen (shown in green) is lysozyme. The heavy chain of the antibody is shown in blue; the light chain in yellow. The most important amino acid residue (glutamine in the 121 position) on the antigen is the one that fits into the antibody groove (shown in red). (b) The antigen–antibody complex has been pulled apart. Note how they fit each other. *(Courtesy of Dr. R. J. Poijak, Pasteur Institute, Paris, France)*

(Section 5.7). This binding is similar to the way in which substrates bind to enzymes or hormones and neurotransmitters bind to a receptor site. That is, the antigen must fit into the antibody surface. Since there are a large variety of antigens against which the human body must fight, our systems contain more than 10,000 different antibodies.

Studies have shown that many other molecules that function as communication signals between cells also have the same basic structural design as the immunoglobulins.

C B Cells and Antibodies

Each B cell synthesizes only one unique immunoglobulin antibody, and that antibody contains a unique antigen-binding site to one epitope. Before encountering an antigen, these antibodies are inserted in the plasma membrane of the B cells, where they serve as receptors. When an antigen interacts with its receptor, it stimulates the B cell to divide and differentiate into plasma cells. These daughter cells secrete soluble antibodies that have the same antigen-binding sites as the original antibody/receptor. The soluble secreted antibodies appear in the serum (the noncellular part of blood) and can react with the antigen. Thus, an immunoglobulin produced in B cells acts both as a receptor to be stimulated by the antigen and as a secreted messenger ready to neutralize and eventually destroy the antigen.

D How Does the Body Acquire the Diversity Needed to React to Different Antigens?

During B cell development, the variable regions of the H chains are assembled by a process called V(J)D recombination. There are a number of exons in each of three different areas—V, J, and D—of the immunoglobin gene. Combining one exon from each area creates a *new V(J)D gene*. This process creates a great diversity because of the large number of ways that this combination can be performed (Figure 30.7). But that is merely the first step. When the body encounters a new antigen, mutations are induced in the new V(J)D genes; these mutations then create even greater diversity.

There are three ways to create mutations, two of which affect the variable regions and one that alters the constant region:

1. Somatic hypermutation (SHM) creates a point mutation (only one nucleotide). The resultant protein of the mutation is able to bind more strongly with the antigen.

2. In a gene conversion (GC) mutation, stretches of nucleotide sequences are copied from a pseudogene V and entered into the V(J)D. This also enables the proteins synthesized from the GC mutation to make stronger contact with the antigen than without the mutation.

3. Mutations on the constant region of the chain can be achieved by class switch recombination (CSR). In this case, exons of the constant regions are swapped between highly repetitive regions.

The diversity of antibodies created by the V(J)D recombination is greatly amplified and finely tuned by the mutations on these genes. Since the response to the antigen occurs on the gene level, it is easily preserved and transmitted from one generation of cells to the next.

Figure 30.7 Diversification of immunoglobulins by V(J)D recombination. Exons of three genes—the V(ariable) (A, B, C, D), J(oin) (1, 2, 3, 4), and D(iverse) (a, b, c, d) genes—combine to form new V(J)D genes that are transcribed to the corresponding mRNAs. The expressions of these new genes result in a wide variety of immunoglobulins having different variable regions on their heavy chains.

CHEMICAL CONNECTIONS 30A

Antibodies and Cancer Therapy

Antibodies come in a variety of forms. As a response to a specific antigen, the body may develop a number of closely related antibodies, the so-called **multiclonal antibodies**. Monoclonal antibodies, in contrast, are identical copies of a single antibody molecule that bind to a specific antigen. Georg Kohler and Cesar Milstein (1927–2002) developed techniques for producing monoclonal antibodies in large quantities. For their pioneering work, they received the Nobel Prize in 1984.

Monoclonal antibody therapy is the latest advance in fighting cancer. For instance, the drug Rituxan is a monoclonal antibody against the CD20 antigen, which appears in high concentration on the surface of B cells in patients suffering from non-Hodgkins lymphoma. Rituxan kills tumor cells in two ways. First, by binding to the diseased B cells, it enlists the whole immune system to attack the marked cells and to destroy them. Second, the monoclonal antibody causes the tumor cells to stop growing and die through apoptosis, programmed cell death (Chemical Connections 24E). Another monoclonal antibody is Herceptin, which is recommended as a treatment for metastatic breast cancers that have a specific antigen, c-erb-B2, on the surface of cancer cells.

One advantage that monoclonal antibody treatment has over chemotherapy is its relatively minimal toxicity. Even when the monoclonal antibody is employed to deliver damaging radiation, as in the case of the antibody against the prostate-specific antigen to which actinium-225 is attached (Section 9.6B), the antibody delivers the radiation to only the specific cancer cells, killing them without significantly harming healthy tissues.

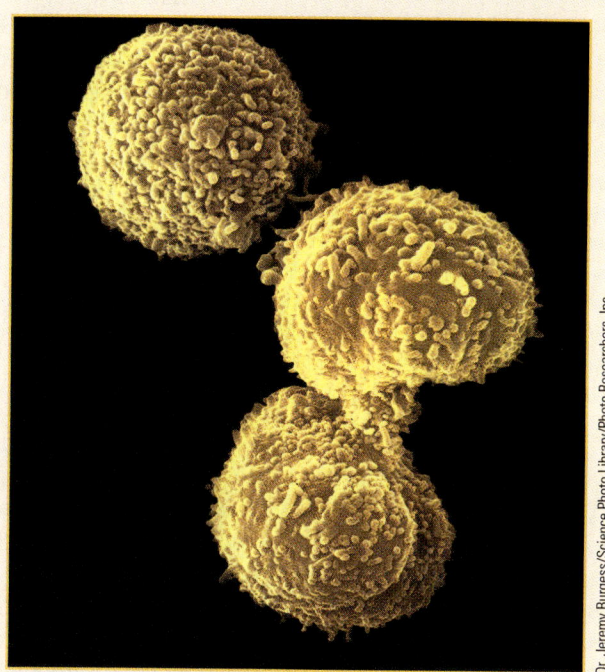

■ False-color scanning electron micrograph of hybridoma cells producing a monoclonal antibody to cytoskeleton protein.

E Monoclonal Antibodies

When an antigen is injected into an organism (for example, human lysozyme into a rabbit), the initial response is quite slow. It may take from one to two weeks before the antilysozyme antibody shows up in the rabbit's serum. Those antibodies, however, are not uniform. The antigen may have many epitopes, and the antisera contains a mixture of immunoglobulins with varying specificity for all the epitopes. Even antibodies to a single epitope usually have a variety of specificities. **Monoclonal antibodies,** in contrast, are the products of cells cloned from a single B cell and have a single specificity.

F Roles of Antibodies That Are Not Antigen-Specific

Recent discoveries indicate that antibodies are versatile and may play other roles besides binding to an antigen and marking it for destruction. Antibodies are able to form hydrogen peroxide, H_2O_2, from water and singlet oxygen. The latter is a free radical and very toxic to cells. Thus, when singlet oxygen is formed in a process, antibodies are able to detoxify it and convert it to the less toxic H_2O_2. Furthermore, the hydrogen peroxide formed in this way may be used to kill pathogens directly.

30.5 T Cells and Their Receptors

A T Cell Receptors

Like B cells, T cells carry on their surface unique receptors that interact with antigens. We noted earlier that T cells respond only to protein antigens. An individual has millions of different T cells, each of which carries on its surface a unique T cell receptor (TcR) that is specific for one antigen only. The TcR is a glycoprotein made of two subunits cross-linked by disulfide bridges. Like immunoglobulins, TcRs have constant (C) and variable (V) regions. The antigen binding occurs on the variable region. The similarity in amino acid sequence between immunoglobulins (Ig) and TcR, as well as the organization of the polypeptide chains, makes TcR molecules members of the immunoglobulin superfamily.

There are, however, basic differences between immunoglobulins and TcRs. For instance, immunoglobulins have four polypeptide chains, whereas TcRs contain only two subunits. Immunoglobulins can interact directly with antigens, but TcRs can interact with them only when the epitope of an antigen is presented by an MHC molecule. Lastly, immunoglobulins can undergo somatic mutation. This kind of mutation can occur in all body cells except the ones involved in sexual reproduction. Thus, Ig molecules can increase their diversity by somatic mutation; TcRs cannot.

B T Cell Receptor Complex

A TcR is anchored in the membrane by hydrophobic transmembrane segments (Figure 30.8). TcR alone, however, is not sufficient for antigen binding. Also needed are other protein molecules that act as coreceptors and/or

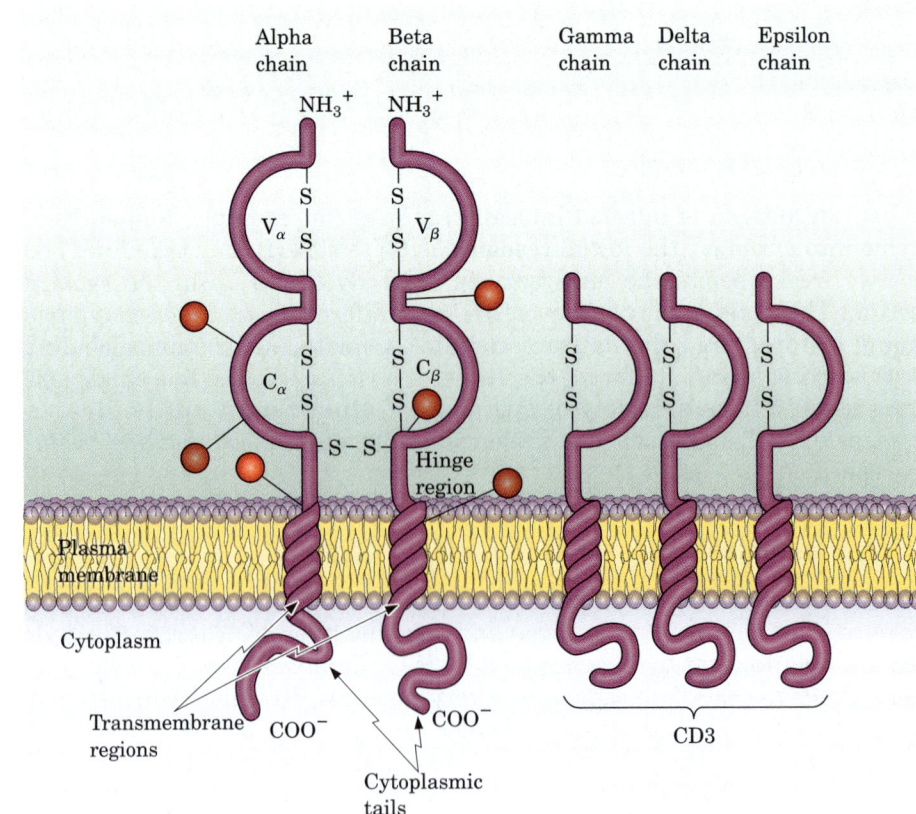

Figure 30.8 Schematic structure of TcR complex. TcR consists of two chains, α and β. Each has two extracellular domains, an amino-terminal V domain and a carboxyl-terminal C domain. Domains are stabilized by intrachain disulfide bonds between cysteine residues. The α and β chains are linked by an interchain disulfide bridge near the cell membrane (hinge region). Each chain is anchored on the membrane by a hydrophobic transmembrane segment and ends in the cytoplasm with a carboxyl-terminal segment rich in cationic residues. Both chains are glycosylated (red spheres). The cluster determinant (CD) coreceptor consists of three chains, γ, δ, and ϵ. Each is anchored in the plasma membrane by a hydrophobic transmembrane segment. Each is also cross-linked by a disulfide bridge, and the carboxylic terminal is located in the cytoplasm.

CHEMICAL CONNECTIONS 30B

Immunization

Smallpox was a scourge over the centuries, each outbreak leaving many dead and others maimed by deep pits on the face and body. A form of immunization was practiced in ancient China and the Middle East by intentionally exposing people to scabs and fluids of lesions of victims of smallpox. This practice was known as variolation in the Western world, where the disease was called variola. Variolation was introduced to England and the American Colonies in 1721.

Edward Jenner, an English physician, noted that milkmaids who had contracted cowpox from infected cows seemed to be immune to smallpox. Cowpox was a mild disease, whereas smallpox could be lethal. In 1796, Jenner performed a daring experiment: He dipped a needle into the pus of a cowpox-infected milkmaid and then scratched a boy's hand with the needle. Two months later, Jenner injected the boy with a lethal dose of smallpox-carrying agent. The boy survived, and did not develop any symptoms of the disease. The word spread, and Jenner was soon established in the immunization business. When the news reached France, the skeptics there coined a derogatory term, *vaccination,* which literally means "encowment." The derision did not last long, and the practice was soon adopted worldwide.

A century later, in 1879, Louis Pasteur found that tissue infected with rabies has much weakened virus in it. When injected into patients, it elicits an immune response that protects against rabies. Pasteur named these attenuated, protective antigens *vaccines* in honor of Jenner's work. Today, immunization and vaccination are synonymous.

Vaccines are available for several diseases, including polio, measles, and smallpox, to name a few. A vaccine may be made of either dead viruses or bacteria or weakened ones. For example, the Salk polio vaccine is a polio virus that has been made harmless by treatment with formaldehyde; it is given by intramuscular injection. In contrast, the Sabin polio vaccine is a mutated form of the wild virus; the mutation makes the virus sensitive to temperature. The mutated live virus is taken orally. The body temperature and the gastric juices render it harmless before it penetrates the bloodstream.

Many cancers have specific carbohydrate cell surface tumor markers. If such antigens could be introduced by injection without endangering the individual, they could provide an ideal vaccine. Obviously, one cannot use live or even attenuated cancer cells for vaccination. However, the expectation is that a synthetic analog of a natural tumor marker will elicit the same immune reaction that the original cancer surface makers do. Thus injecting such innocuous synthetic compounds may prompt the body to produce immunoglobulins that can provide a cure for the cancer—or at least prevent the occurrence of it. A compound called 12:13 dEpoB, a derivative of the macrolid epothilon B, is currently being explored for its potential to become the first anticancer vaccine.

Vaccines change lymphocytes into plasma cells that produce large quantities of antibodies to fight any invading antigens. This is only the immediate, short-term response, however. Some lymphocytes become memory cells rather than plasma cells. These memory cells do not secrete antibodies, but instead store them to serve as a detecting device for future invasion of the same foreign cells. In this way, long-term immunity is conferred. If a second invasion occurs, these memory cells divide directly into both antibody-secreting plasma cells and more memory cells. This time the response is faster because it does not have to go through the process of activation and differentiation into plasma cells, which usually takes two weeks.

Smallpox, which was once one of humanity's worst scourges, has been totally wiped out, and smallpox vaccination is no longer required. Because smallpox is a potential threat of bioterrorism, the U.S. government has recently begun gearing up its smallpox vaccine production.

---

signal transducers. These molecules go under the name of CD3, CD4, and CD8, with "CD" standing for **cluster determinant.** TcR and CD together form the **T cell receptor complex.**

The CD3 molecule adheres to the TcR in the complex not through covalent bonds but by intermolecular forces (Section 5.7). It is a signal transducer because, upon antigen binding, CD3 becomes phosphorylated. This event sets off a signaling cascade inside the cell, which is carried out by different kinases. We saw a similar signaling cascade in neurotransmission (Section 23.5).

The CD4 and CD8 molecules act as **adhesion molecules** as well as signal transducers. A T cell has either a CD4 or a CD8 molecule to help bind the antigen to the receptor (Figure 30.4). A unique characteristic of the CD4 molecule is that it binds strongly to a special glycoprotein that has a molecular weight of 120,000 (gp120). This glycoprotein exists on the surface of the human immunodeficiency virus (HIV). Through this binding to CD4, the HIV virus can enter and infect T cells and cause AIDS (Chemical Connec-

Adhesion molecules Various protein molecules that help to bind an antigen to the T cell receptor tions 25C). T cells die as a result of HIV infection, which depletes the T cell population so drastically that the immune system can no longer function. As a consequence, the body succumbs to opportunistic pathogen infections.

30.6 Control of Immune Response: Cytokines

A Nature of Cytokines

Cytokine A glycoprotein that traffics between cells and alters the function of a target cell

Cytokines are glycoprotein molecules that are produced by one cell but alter the function of another cell. They have no antigen specificity. Cytokines transmit intercellular communications between different types of cells at diverse sites in the body. They are short-lived and are not stored in cells.

Cytokines facilitate a coordinated and appropriate inflammatory response by controlling many aspects of the immune reaction. They are released in bursts, in response to all manner of insult or injury (real or perceived). They travel and bind to specific cytokine receptors on the surface of macrophages and B and T cells and induce cell proliferation.

One set of cytokines are called **interleukins** (ILs) because they communicate between and coordinate the actions of leukocytes (all kinds of white blood cells). Macrophages secrete IL-1 upon a bacterial infection. The presence of IL-1 then induces shivering and fever. The elevated body temperature both reduces the bacterial growth and speeds up the mobilization of the immune system (Chemical Connections 7A). One leukocyte may make many different cytokines, and one cell may be the target of many cytokines.

B Classes of Cytokines

Cytokines can be classified according to their mode of action, origin, or target. The best way to classify them is by their structure—namely, the secondary structure of their polypeptide chains.

1. One class of cytokines is made of four α-helical segments. A typical example is interleukin-2 (IL-2), which is a 15,000 MW polypeptide chain. A prominent source of IL-2 is T cells. IL-2 activates other B and T cells

CHEMICAL CONNECTIONS 30C

Mobilization of Leukocytes: Tight Squeeze

When injury occurs, leukocytes in the bloodstream must migrate into the lymph or inflammatory sites to provide protection. A number of adhesion and signaling molecules guide this migration, and two major protein molecules on the surface of leukocytes facilitate their movement.

Selectins are protein molecules belonging to the CD (cluster determinant) class. When activated, they interact with the carbohydrate portions of glycoproteins on the surface of endothelial cells. These endothelial cells line the inner wall of blood vessels and, with tight junctions, seal the vessels against leakiness. Endothelial cells have their own selectin molecules that interact with the carbohydrates of the leukocytes. These alternate interactions create a rolling motion of leukocytes along the endothelial wall. When a leukocyte arrives at a point near the injury or infection, another surface molecule is activated. This molecule, called integrin, exists in almost every cell type of the animal kingdom. Activated integrin on endothelial cells strongly interacts with surface adhesion molecules of the leukocyte, which stops rolling. This attachment and the force of the streaming blood make the leukocyte flatten out.

Chemokines, such as IL-8, also play a role in this leukocyte trafficking. When a chemokine is bound to its receptor on the leukocyte, the cell undergoes a remarkable transformation. Inside the cell, the cytoskeleton—actin—is first broken down and then repolymerized, resulting in armlike and leglike projections (lamellipodia) that enable the cells to migrate. In addition, chemokines signal the endothelial cells to change their shape. This results in the opening of gaps between the cells. The flattened cells then manage to squeeze through the gaps between the tight junctions and enter the tissue or lymph to migrate to the point of injury. After accumulating in the inflamed tissue, the leukocytes may kill the invading bacteria by engulfing and digesting them.

and macrophages. Its action is to enhance proliferation and differentiation of the target cells.

2. Another class of cytokines has only β-pleated sheets in its secondary structure. Tumor necrosis factor (TNF), for example, is produced mostly by T cells and macrophages. Its name comes from its ability to destroy susceptible tumor cells through lysis, after it binds to receptors on the tumor cell.

3. A third class of cytokines has both α-helix and β-sheet secondary structures. A representative of this class is epidermal growth factor (EGF), which is a cysteine-rich protein. As its name indicates, EGF stimulates the growth of epidermal cells and its main function is in wound healing.

4. A subgroup of cytokines, the chemotactic cytokines, are also called **chemokines.** In humans, there are some 40 chemokines, all low-molecular-weight proteins with distinct structural characteristics. They attract leukocytes to the site of infection or inflammation. All chemokines have four cysteine residues, which form two disulfide bridges: Cys1-Cys3 and Cys2-Cys4. Chemokines have a variety of names, such as interleukin-8 (IL-8) and monocyte chemotactic proteins (MCP-1 to MCP-4). Chemokines interact with specific receptors, which consist of seven helical segments coupled to GTP-binding proteins.

TNF targets other cells as well. For example, it plays a role in rheumatoid arthritis. When the immune system goes awry, the TNF attacks the patient's own cartilage. It binds to the receptors of the cartilage cells, which lyse. In this manner, it eats away even the bones. A new drug, called Enebrel, interacts with TNF in the blood. It immobilizes the cytokine and prevents its migration to the joints, where it would cause damage.

Chemokine A low-molecular-weight polypeptide that interacts with special receptors on target cells and alters their functions

Two subgroups or chemokines exist in which the first two Cys residues (1) are adjacent (CC) or (2) are separated by one amino acid (CXC).

Special receptors for the CC and CXC chemokines are found on the surface of leukocytes and lymphocytes.

The crucial parts of the chemokines in these interactions are the N-terminal domain and the loop between Cys2 and Cys4.

C Mode of Action of Cytokines

When a tissue is injured, leukocytes are rushed to the inflamed area. Chemokines help leukocytes migrate out of the blood vessels to the site of injury (Chemical Connections 30C). There leukocytes, in all their forms—neutrophils, monocytes, lymphocytes—accumulate and attack the invaders by engulfing them (phagocytosis) and later killing them. Other phagocytic cells, the macrophages, which reside in the tissues and thus do not have to migrate, do the same. These activated phagocytic cells destroy their prey by releasing endotoxins that kill bacteria and/or by producing such highly reactive oxygen intermediates as superoxide, singlet oxygen, hydrogen peroxide, and hydroxyl radicals (Chemical Connections 26C).

Chemokines are also major players in chronic inflammation, autoimmune diseases, asthma, and other forms of allergic inflammation, and even in transplant rejection.

30.7 How to Recognize "Self"

One major problem facing body defenses is how to recognize the foreign body as "not self" and thereby avoid attacking the "self"—that is, the healthy cells of the organism.

A Selection of T and B Cells

The members of the adaptive immunity system, B cells and T cells, are all specific and have memory, so they target only truly foreign invaders. The T cells mature in the thymus gland. During the maturation process, those T cells that fail to interact with MHC and thus cannot respond to foreign antigens are eliminated through a selection process. T cells that express receptors (TcR) that are prone to interact with normal self antigens are also eliminated through the selection process (Figure 30.9). Thus the activated T cells leaving the thymus gland carry TcRs that can respond to foreign antigens. Even if some T cells prone to react with self antigen escape the selection detection, they can be deactivated through the signal transduction

Figure 30.9 Maturation and activation of T cells. (*a*) T cells in the absence of an antigen do not proliferate. (*b*) In the presence of an antigen, the TcR binds to the antigen presented by MHC. No proliferation occurs because a second signal is needed. In this manner the body can avoid responding to a "self" antigen. (*c*) A second signal is provided when the infected cell produces a stress signal—for example, a "heat shock" protein that normally has a chaperoning function (Section 21.11). The primary MHC-antigen-TcR signal and the second signal are enhanced by the synthesis of cytokines, such as interleukins. Under such triple stimulation, T cells mature and proliferate.

system that, among other functions, performs tyrosine kinase activation and phosphatase deactivation similar to those processes seen in adrenergic neurotransmitter signaling (Section 23.5).

Similarly, the maturation of B cells in the bone marrow depends on the engagement of their receptors, BcR, with antigen. Those B cells that are prone to interact with self antigen are also eliminated before they leave the bone marrow. Again, as with the T cells, many signaling pathways control the proliferation of B cells. Among them, the activation by tyrosine kinase and the deactivation by phosphatase provide a secondary control.

B Discrimination of the Cells of the Innate Immunity System

The first line of defense is the innate immunity system, in which cells such as natural killer cells or macrophages have no specific targets and no memory of which epitope signals danger. Still, these cells must somehow dis-

CHEMICAL CONNECTIONS 30D

Myasthenia Gravis: An Autoimmune Disease

Myasthenia gravis manifests itself as muscular weakness and excessive fatigue. In its severe form, the patient has difficulty chewing, swallowing, and even breathing. Death follows from respiratory failure. Patients with myasthenia gravis develop antibodies against their own acetylcholine receptors (Figure 23.3). This condition is an example of an autoimmune disease, in which the body mistakes its own protein for a foreign body. The antibodies produced interact with the acetylcholine receptor in the neuromuscular junctions. This interaction blocks both the reception of acetylcholine molecules and the signaling nerve impulses. As a result, the muscles do not respond, and paralysis sets in. In addition, the interaction leads to the degradation and lysis of the blocked receptors.

criminate between normal and abnormal cells to identify targets. The mechanism by which this identification is accomplished has only recently been explored and is not yet fully understood. The main point is that the cells of the innate immunity have two kinds of receptors on their surface: an **activating receptor** and an **inhibitory receptor.** When a healthy cell of the body encounters a macrophage or a natural killer cell, the inhibitory receptor on the latter's surface recognizes the epitope of the normal cell, binds to it, and prevents the activation of the killer cell or macrophage. On the other hand, when a macrophage encounters a bacterium with a foreign antigen on its surface, the antigen binds to the activating receptor of the macrophage. This ligand binding prompts the macrophage to engulf the bacterium by phagocytosis. Such foreign antigens may be the lipopolysaccharides of gram-negative bacteria or the peptidoglycans (Chemical Connections 18B) of gram-positive bacteria.

When a cell is infected, damaged, or transformed into a malignant cell, the epitopes that signaled healthy cells diminish greatly, and unusual epitopes are presented on the surface of these cancer cells. The effect is that fewer inhibitory receptors of macrophages or natural killer cells can bind to the surface of the target cell and more activating receptors find inviting ligands. As a consequence, the balance shifts in favor of activation and the macrophages or killer cells will do their job.

C Autoimmune Diseases

In spite of the safeguards in the body intended to prevent acting against "self," in the form of healthy cells, many diseases exist in which some part of a pathway in the immune system goes awry. The skin disease psoriasis is thought to be a T cell-mediated disease in which cytokines and chemokines play an essential role. Other autoimmune diseases, such as myasthenia gravis (Chemical Connections 30D), rheumatoid arthritis, multiple sclerosis (Chemical Connections 20E), and insulin-dependent diabetes (Chemical Connections 23G) also involve cytokines and chemokines. Allergies are another example of malfunctioning of the immune system. Pollens and animal furs are allergens that can provoke asthma attacks. Some people are so sensitive to certain food allergens that even a knife used in smearing peanut butter may prove fatal in a person known to be allergic to peanuts.

The major drug treatment for autoimmune diseases involves the glucocorticoids, the most important of which is cortisol (Section 20.10). It is a standard therapy for rheumatoid arthritis, asthma, inflammatory bone diseases (Crohn's disease), psoriasis, and eczema. The wide beneficial effects of glucocorticoids are overshadowed, however, by their undesirable side effects, which include osteoporosis, skin atrophy, and diabetes. Glucocorti-

CHEMICAL CONNECTIONS 30E

Glycomics

You may have noticed that in this chapter we frequently called the participant molecules *glycoproteins*. Immunoglobulins, cytokines, and T cell receptors are all glycoproteins. As we saw in Section 21.10, a glycoprotein is a complex molecule with a protein core and carbohydrate side chains. In essence, the carbohydrate portions of glycoproteins as well as those of glycolipids constitute a third language of chemical communication.

The language of DNA was deciphered through genomics (Chemical Connections 24D), which in essence elucidated all the information coded in the human chromosomes as well as in the chromosomes of other species. In the wake of the success of genomics, a similar effort was launched under the name of proteomics (Chemical Connections 21F) to find and describe all the proteins of an organism and their functions. Besides the language of the nucleotides and the language written in amino acids, there is a third language of communication written in monosaccharides. This language is more variable and specific to individuals. For example, the ABO blood group to which an individual belongs is written in the carbohydrate language (Chemical Connections 19D). Another example of information written in the carbohydrate language is the coat around the *Pneumococcus* bacterium. Different carbohydrates make the same bacterium either highly infectious or dormant. Cancer cells often display abnormal sugars on their surface, marking their malignancy.

Because of the importance of deciphering this information, an effort was launched recently to elucidate the structures of all the carbohydrates (including those that are parts of the glycoproteins and glycolipids) that an organism possesses and, if possible, to describe their functions. This effort is called **glycomics** in analogy with genomics and proteomics.

But consider the complexity of this third language. Genomics has an alphabet of 4 letters (the nucleotides); the alphabet of proteomics consists of 20 letters (amino acids). If we include only the D-sugars of pentoses and hexoses in Table 19.1, we have 12 original letters in the glycomics alphabet. While nucleotides and amino acids can form only one kind of bond, each of these sugars can form a minimum of four kinds of bonds (1,2; 1,4; and so on) and in the α and β configurations. Thus the carbohydrate alphabet has $12 \times 4 \times 2 = 96$ letters at minimum. Now consider just the two-letter words. There are $4^2 = 16$ such words in the nucleotide language, $20^2 = 400$ words in the amino acid language, and $96^2 = 9216$ such words in the monosaccharide language. Clearly, the task of glycomics is much more daunting than that of the genomics and even that of the proteomics.

coids regulate the synthesis of cytokines either directly, by interacting with their genes, or indirectly, through transcription factors.

Macrolide drugs are used to suppress the immune system during tissue transplantation or in the case of certain autoimmune diseases. Drugs such as cyclosporin A or rapamycin bind to receptors in the cytosol and, through secondary messengers, inhibit the entrance of nuclear factors into the nucleus. Normally those nuclear factors signal a need for transcription, so their absence prevents the transcription of cytokines—for example, interleukin-2.

SUMMARY

The human immune system protects us against foreign invaders. It consists of two parts: (1) the natural resistance of the body, called innate immunity, and (2) adaptive or acquired immunity (Section 30.1). **Innate immunity** is nonspecific. **Macrophages** and **natural killer (NK) cells** are cells of innate immunity that function as policemen. **Adaptive or acquired immunity** is highly specific, being directed against one particular invader. Acquired immunity (known as the immune system) also has memory, unlike innate immunity.

The principal cellular components of the immune system are the white blood cells, or **leukocytes** (Section 30.2). The specialized leukocytes in the lymph system are called **lymphocytes** (Section 30.2). They circulate mostly in the **lymphoid organs**. The **lymph** is a collection of vessels extending throughout the body and connected to the interstitial fluid on the one hand, and to the blood vessels on the other hand. Lymphocytes that mature in the bone marrow and produce soluble immunoglobulins are **B cells**. Lymphocytes that mature in the thymus gland are **T cells.**

Antigens are large, complex molecules of foreign origin (Section 30.3). An antigen may be a bacterium, a virus, or a toxin. Such an invader may interact with antibodies, T cell receptors (TcR), or major histocompatibility complex (MHC) molecules. All three types of molecules belong to the **immunoglobulin superfamily.** An **epitope** is the smallest part of an antigen that binds to antibodies, TcRs, and MHCs. Antibodies are **immunoglobulins** (Section 30.4). These water-soluble glycoproteins are made of two heavy chains and two light chains. All four chains are linked together by disulfide bridges. Immunoglobulins contain variable regions in which the amino acid composition of each antibody is different. These regions interact with antigens to form insoluble large aggregates.

A large variety of antibodies are synthesized by a number of processes in the body. During B cell development, the **variable regions** of the H chains are assembled by a process called V(J)D recombination. Mutations on these new genes then create even greater diversity. Somatic hypermutation (SHM) that creates a point mutation (only one nucleotide) is one way; patches of mutation introduced into the V(J)D gene are the gene conversion (GC) mutation. Immunoglobulins respond to an antigen over a long span of time, lasting for weeks and even months.

All antigens—whether proteins, polysaccharides, or nucleic acids—interact with soluble immunoglobulins produced by B cells. Protein antigens also interact with T cells. The binding of the epitope to the TcR is facilitated by MHC, which carries the epitope to the T cell surface where it is presented to the receptor (Section 30.5).

Upon epitope binding to the receptor, the T cell is stimulated. It proliferates and can differentiate into (1) killer T cells or (2) memory cells. The T cell receptor has a number of helper molecules, such as CD4 or CD8, that enable it to bind the epitope tightly. CD **(cluster determinant)** molecules also belong to the immunoglobulin superfamily. Antibodies can recognize all antigens, but T cell receptors recognize only peptide antigens.

The control and coordination of the immune response are handled by **cytokines,** which are small protein molecules (Section 30.6). Chemotactic cytokines, the **chemokines,** such as interleukin-8, facilitate the migration of leukocytes from blood vessels into the site of injury or inflammation. Other cytokines activate B and T cells and macrophages, enabling them to engulf the foreign body, digest it, or destroy it by releasing special toxins. Some cytokines, such as tumor necrosis factor (TNF), can lyse tumor cells.

A number of mechanisms in the body ensure that the body recognizes "self" (Section 30.8). In adaptive immunity, the selection of T and B cells that are prone to interact with "self" antigens are eliminated. In innate immunity, two kinds of receptors exist on their surface: an **activating receptor** and an **inhibitory receptor.** The inhibitory receptor on their surface recognizes the epitope of a normal cell, binds to it, and prevents the activation of the killer cell or macrophage. Many autoimmune diseases are T cell-mediated diseases, in which cytokines and chemokines play an essential role. The standard drug treatment for autoimmune diseases involves glucocorticoids, which prevent the transcription and hence the synthesis of cytokines.

PROBLEMS

Numbers that appear in color indicate difficult problems.
▶ designates problems requiring application of principles.

30.1 Give two examples of external innate immunity in humans.

30.2 Which form of immunity is characteristic of vertebrates only?

30.3 How does the skin fight bacterial invasion?

30.4 ▶ T cell receptors and MHC molecules both interact with an antigen. What is the difference in the mode of interaction between the two?

30.5 What differentiates innate immunity from adaptive (acquired) immunity?

Cells and Organs of the Immune System

30.6 Where in the body do you find the largest concentration of antibodies as well as T cells?

30.7 Where do T and B cells mature and differentiate?

30.8 What are memory cells? What is their function?

30.9 What are the favorite targets of macrophages? How do they kill the target cells?

Antigens and Their Presentation

30.10 Would a foreign substance, such as aspirin (MW 180), be considered an antigen by the body?

30.11 What kind of antigen does a T cell recognize?

30.12 What is the smallest unit of an antigen that is capable of binding to an antibody?

30.13 ▶ How does the body process antigens to be recognized by class II MHC?

30.14 What role does the MHC play in the immune response of the ABO blood groups?

30.15 To what class of compounds do MHCs belong? Where would you find them?

Immunoglobulins

30.16 When a foreign substance is injected in a rabbit, how long does it take to find antibodies against the foreign substance in the rabbit serum?

30.17 Distinguish among the roles of the IgA, IgE, and IgG immunoglobulins.

30.18 (a) Which immunoglobulin has the highest carbohydrate content and the lowest concentration in the serum?
(b) What is its main function?

30.19 Chemical Connections 19D states that the antigen in the red blood cells of a person with B-type blood is a galactose unit. Show schematically how the antibody of a person with A-type blood would aggregate the red blood cells of a B-type person if such a transfusion were made by mistake.

30.20 In the immunoglobulin structure, the "hinge region" joins the stem of the Y to the arms. The hinge region can be cleaved by a specific enzyme to yield one Fc fragment (the stem of the Y) and two Fab fragments (the two arms). Which of these two kinds of fragments can interact with an antigen? Explain.

30.21 How are the light and heavy chains of an antibody held together?

30.22 What do we mean by the term *immunoglobulin superfamily*?

30.23 If you could isolate two monoclonal antibodies from a certain population of lymphocytes, in what sense would they be similar to each other and in what sense would they differ?

30.24 What kind of interaction takes place between an antigen and an antibody?

30.25 How is a new protein created on the variable H chain by V(J)D recombination?

T Cells and Their Receptors

30.26 T cell receptor molecules are made of two polypeptide chains. Which part of the chain acts as a binding site and what binds to it?

30.27 What is the difference between a T cell receptor (TcR) and a TcR complex?

30.28 What kind of tertiary structure characterizes the TcR?

30.29 What are the components of the T cell receptor complex?

30.30 By what chemical process does CD3 transduct signals inside the cell?

30.31 Which adhesion molecule in the TcR complex helps the HIV virus infect a leukocyte?

30.32 Three kinds of molecules in the T cell belong to the immunoglobulin superfamily. List them and indicate briefly their function.

Control of Immune Response: Cytokines

30.33 What kind of molecules are cytokines?

30.34 With what do cytokines interact? Do they bind to antigens?

30.35 As in most biochemistry literature, one encounters a veritable alphabet soup when reading about cytokines. Identify by their full name (a) TNF, (b) IL, and (c) EGF.

30.36 What are chemokines? How do they deliver their message?

30.37 What is the characteristic chemical signature in the structure of chemokines?

30.38 What are the chemical characteristics of cytokines that allow their classification?

30.39 What specific amino acid appears in all chemokines?

How to Recognize "Self"

30.40 How does the body prevent T cells from being active against a "self" antigen?

30.41 What makes a tumor cell different from a normal cell?

30.42 Name a signaling pathway that controls the maturation of B cells and prevents those with affinity toward self antigen from becoming active.

30.43 How does the inhibitory receptor on a macrophage prevent an attack on normal cells?

30.44 Which components of the immune system are principally involved in autoimmune diseases?

30.45 How do glucocorticoids make autoimmune disease sufferers feel more comfortable?

Chemical Connections

30.46 (Chemical Connections 30A) How does Rituxan, a monoclonal antibody, find its way to the surface of non-Hodgkins lymphoma cells?

30.47 (Chemical Connections 30A) Why is monoclonal antibody treatment better than chemotherapy?

30.48 (Chemical Connections 30B) What made Edward Jenner the father of immunization? In your opinion, could one do such an experiment today legally?

30.49 (Chemical Connections 30B) What is the difference between memory cells and plasma cells?

30.50 (Chemical Connections 30C) In leukocyte trafficking, what keeps the leukocytes rolling along the endothelial wall of the blood vessels?

30.51 (Chemical Connections 30C) What is the role of a chemokine in facilitating the passage of leukocytes from a blood vessel to the site of injury?

30.52 (Chemical Connections 30D) What causes the weakening and eventual paralysis of the muscles in myasthenia gravis?

30.53 (Chemical Connections 30D) One finds elevated antibodies in the serum of myasthenia gravis patients. What is the source of the antibodies? Which cells of the immune system manufacture them?

Additional Problems

30.54 Which immunoglobulins form the first line of defense against invading bacteria?

30.55 Where do all the lymphatic vessels come together?

30.56 (Chemical Connections 30A) Why is a monoclonal antibody treatment preferable to a multiclonal antibody treatment?

30.57 Which compound or complex of compounds of the immune system is mostly responsible for the proliferation of leukocytes?

30.58 Name a process beside the V(J)D recombination that can enhance immunoglobulin diversity in the variable region.

30.59 Name a tumor cell marker, a synthetic analogue of which may be the first anticancer vaccine.

30.60 Is the light chain of an immunoglobulin the same as the V region?

30.61 Where are TNF receptors located?

30.62 The variable regions of immunoglobulins bind the antigens. How many polypeptide chains carry variable regions in one immunoglobulin molecule?

InfoTrac College Edition

For additional readings, go to InfoTrac College Edition, your online research library, at

http://infotrac.thomsonlearning.com

CHAPTER 31

31.1 Introduction

31.2 Functions and Composition of Blood

31.3 Blood as a Carrier of Oxygen

31.4 Transport of Carbon Dioxide in the Blood

31.5 Blood Cleansing: A Kidney Function

31.6 Buffer Production: Another Kidney Function

31.7 Water and Salt Balance in Blood and Kidneys

31.8 Blood Pressure

Human circulatory cells: red blood cells, platelets, and white blood cells.

Body Fluids

31.1 Introduction

Single-cell organisms receive their nutrients directly from the environment and discard waste products directly into it. In multicellular organisms, the situation is not so simple. There, too, each cell needs nutrients and produces wastes, but most of the cells are not directly in contact with the environment. Body fluids serve as a medium for carrying nutrients to and waste products from the cells as well as a means for carrying the chemical communicators (Chapter 23) that coordinate activities among cells.

The body fluids that are not inside the cells are collectively known as **extracellular fluids.** These fluids make up about one-fourth of a person's body weight. The most abundant is **interstitial fluid,** which directly surrounds most cells and fills the spaces between them. It makes up about 17% of body weight. Another body fluid is **blood plasma,** which flows in the arteries and veins. It makes up about 5% of body weight. Other body fluids that occur in lesser amounts are urine, lymph, cerebrospinal fluid, aqueous humor, and synovial fluid. All body fluids are aqueous solutions—water is the only solvent in the body.

The blood plasma circulates in the body and comes in contact with the other body fluids through the semipermeable membranes of the blood ves-

---

The fluid inside the cells is called **intracellular fluid.**

Interstitial fluid The fluid that surrounds the cells and fills the spaces between them

Blood plasma The noncellular portion of blood

CHEMICAL CONNECTIONS 31A

Using the Blood–Brain Barrier to Eliminate Undesirable Side Effects of Drugs

Many drugs have undesirable side effects. For example, many antihistamines, such as Dramamine and Benadryl (Section 23.5), cause drowsiness. These antihistamines are supposed to act on the peripheral H_1 histamine receptors to relieve seasickness, hay fever, or asthma. Because they penetrate the blood–brain barrier, they also act as antagonists to the H_1 receptors in the brain, causing sleepiness.

One drug that also acts on the peripheral H_1 receptors, fexofenadine (sold under the trade name Allegra), cannot penetrate the blood–brain barrier. This antihistamine alleviates seasickness and asthma in the same way as the old antihistamines, but it does not cause drowsiness as a side effect.

---

sels (Figure 30.2). In this way, blood can exchange chemical compounds with other body fluids, such as lymph and interstitial fluid, and, through them, with the cells and organs of the body.

There is only a limited exchange between blood and cerebrospinal fluid, on the one hand, and between blood and the interstitial fluid of the brain, on the other hand. The limited exchange between blood and interstitial fluid of the brain is referred to as the **blood–brain barrier.** This barrier is permeable to water, oxygen, carbon dioxide, glucose, alcohols, and most anesthetics, but only slightly permeable to electrolytes such as Na^+, K^+, and Cl^- ions. Many higher-molecular-weight compounds are also excluded.

The blood–brain barrier protects the cerebral tissue from detrimental substances in the blood and allows it to maintain a low K^+ concentration, which is needed to generate the high electrical potential essential for neurotransmission. It is vital that the body maintain a proper balance of levels of salts, proteins, and all other components in the blood. **Homeostasis** is the process of maintaining the levels of nutrients in the blood as well as the body's temperature.

Body fluids have special importance for the health care professions. Samples of body fluid can be taken with relative ease. The chemical analysis of blood plasma, blood serum, urine, and occasionally cerebrospinal fluid is of major importance in diagnosing disease.

> **Blood–brain barrier** A barrier limiting the exchange of blood components between blood and the cerebrospinal and brain interstitial fluids to water, carbon dioxide, glucose, and other small molecules but excluding electrolytes and large molecules

31.2 Functions and Composition of Blood

It has been known for centuries that "life's blood" is essential to human life. Blood has many functions, including the following:

1. It carries O_2 from the lungs to the tissues.
2. It carries CO_2 from the tissues to the lungs.
3. It carries nutrients from the digestive system to the tissues.
4. It carries waste products from the tissues to the excretory organs.
5. With its buffer systems, it maintains the pH of the body (with the help of the kidneys).
6. It maintains a constant body temperature.
7. It carries hormones from the endocrine glands to wherever they are needed.
8. It transports infection-fighting blood cells (leukocytes) and antibodies.

Figure 31.1 Nutrients, oxygen, and other materials diffuse out of the blood and through the tissue fluid that bathes the cells. Carbon dioxide and other waste products diffuse out of the cells and enter the blood through the capillary wall.

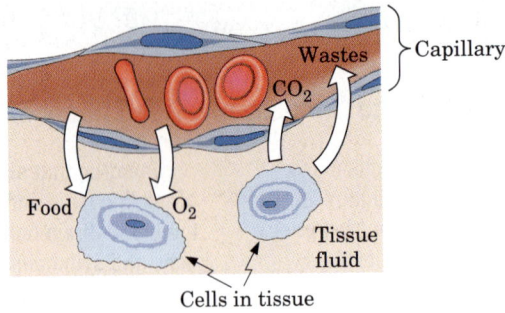

Figure 31.1 shows some of these functions. The rest of this chapter describes how the blood carries out some of these functions.

Whole blood is a complicated mixture. It contains several types of cells (Figure 31.2) plus a liquid, noncellular portion called **plasma,** in which many substances are dissolved (Table 31.1). There are three main types of cellular elements of blood: erythrocytes, leukocytes, and platelets.

A 150-lb (68-kg) man has about 6 L of whole blood, 50 to 60% of which is plasma.

A Erythrocytes

The most prevalent cells of blood are the red blood cells, which are called **erythrocytes.** There are about 5 million red blood cells in every cubic millimeter of blood, or roughly 100 million in every drop. Erythrocytes are very specialized cells. They have no nuclei and, hence, no DNA. Their main function is to carry oxygen to the cells and carbon dioxide away from the cells.

Erythrocytes are formed in the bone marrow, and they stay in the bloodstream for about 120 days. Old erythrocytes are removed by the liver and

Erythrocytes Red blood cells; transporters of blood gases

In an adult male, there are approximately 30 trillion erythrocytes.

Figure 31.2 Some cellular elements of blood. The dimensions are shown in microns (μm; also called micrometers).

Table 31.1 Blood Components and Some Diseases Associated with Their Abnormal Presence in the Blood

Cellular Elements:
- Erythrocytes (high: polycythemia; low: anemia)
- Leukocytes (high: leukemia; low: typhoid fever)
- Platelets (low: thrombocytopenia)

Serum:
- Water (high: edema; low: dehydration)
- Albumin (low: edema)
- Globulins (high: transplant rejection; low: infection)
- Clotting factors (low: hemophilia)
- Glucose (high: diabetes; low: hypoglycemia)
- Cholesterol (high: gallstones, atherosclerosis)
- Urea
- Inorganic salts
- Gases (N_2, O_2, CO_2)
- Enzymes, hormones, vitamins

spleen and destroyed. The constant formation and destruction of red blood cells maintain a steady number of erythrocytes in the body.

B Leukocytes

Leukocytes (white blood cells) are relatively minor cellular components (minor in numbers, not in function). For each 1000 red blood cells, there are only one or two white blood cells. Most of the different leukocytes destroy invading bacteria or other foreign substances by devouring them **(phagocytosis)**.

Like erythrocytes, leukocytes are manufactured in the bone marrow. Specialized white blood cells formed in the lymph nodes and in the spleen are called lymphocytes. They synthesize immunoglobulins (antibodies; see Section 30.4) and store them.

> **Leukocytes** White blood cells; part of the immune system

C Platelets

When a blood vessel is cut or injured, the bleeding is controlled by a third type of cellular element: the **platelets** (also called **thrombocytes**). They are formed in the bone marrow and spleen and are more numerous than leukocytes but less numerous than erythrocytes.

> **Platelets** Cellular element of blood; essential to clot formation

There are about 300,000 platelets in each mm^3 of blood, or one for every 10 or 20 red blood cells.

D Plasma

If all the cellular elements from whole blood are removed by centrifugation, the resulting liquid is the plasma. The cellular elements—mainly red blood cells, which settle at the bottom of the centrifuge tube—account for between 40 and 50% of the blood volume.

Blood plasma is 92% water. The dissolved solids in the plasma are mainly proteins (7%). The remaining 1% contains glucose, lipids, enzymes, vitamins, hormones, and waste products such as urea and CO_2. Of the plasma proteins, 55% is albumin, 38.5% is globulin, and 6.5% is fibrinogen. If plasma is allowed to stand, it forms a clot, a gel-like substance. We

A dried clot becomes a scab.

CHEMICAL CONNECTIONS 31B

Blood Clotting

When body tissues are damaged, the blood flow must be stopped or else enough of it will pour out to cause death. The mechanism used by the body to stem leaks in the blood vessels is clotting. This complicated process involves many factors. Here we mention only a few important steps.

When a blood vessel is injured, the first line of defense is the platelets, which circulate constantly in the blood. They rush to the site of injury, adhere to the collagen molecules in the capillary wall that are exposed by the cut, and form a gel-like plug. At the signal of thromboxane A_2, more platelets enlarge the size of the clot (Section 20.12). This plug is porous, however, and a firmer gel (a clot) is needed to seal the site. A clot is a three-dimensional network of fibrin molecules that also contains platelets. The fibrin network is formed from the blood fibrinogen by the enzyme thrombin. Together with the embedded platelets, it constitutes the blood clot.

Why doesn't the blood clot in the blood vessels under normal conditions (with no injury or disease)? The reason is that the enzyme that starts clot formation, thrombin, exists in the blood only in its inactive form, called prothrombin. Prothrombin itself is manufactured in the liver, and vitamin K is needed for its production. Even when prothrombin is in sufficient supply, a number of proteins are needed to change it to thrombin. These proteins are given the collective name thromboplastin. Any thromboplastic substance can activate prothrombin in the presence of Ca^{2+} ions. Thromboplastic substances exist in the platelets, in the plasma, and in the injured tissue itself.

Clotting is nature's way of protecting us from loss of blood. However, we don't want the blood to clot during blood transfusions because this would stop the flow. To prevent this problem, we add sodium citrate to the blood. Sodium citrate interacts with Ca^{2+} ions and removes them from the solution, so the thromboplastic substances cannot activate prothrombin and no clot forms.

A clot is not dangerous if it stays near the injury because, once the body repairs the tissue, the clot is digested and removed. However, a clot formed in one part of the body may break loose and travel to other parts, where it may lodge in an artery. This condition is called **thrombosis**. If the clot then blocks the oxygen and nutrient supply to the heart and brain, it can result in paralysis and death. After surgery, **anticoagulant drugs** are occasionally administered to prevent clot formation.

The most common anticoagulants are heparin and dicumarol. Heparin enhances the inhibition of thrombin by antithrombin (Section 19.7B), and dicumarol blocks the transport of vitamin K to the liver, preventing prothrombin formation.

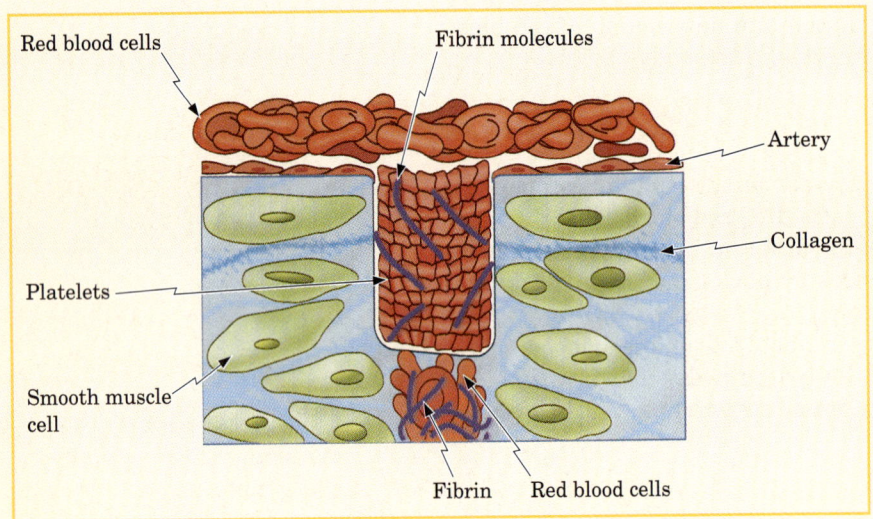

■ **Components of a blood clot.** (From *The Functioning of Blood Platelets* by M. B. Zucker. Copyright 1980 by Scientific American, Inc. All rights reserved.)

Serum Blood plasma from which fibrinogen has been removed

See the discussion of osmotic pressure in Section 6.8.

can squeeze out a clear liquid from the clot, called **serum**. It contains all the components of the plasma except the fibrinogen. This protein is involved in the complicated process of clot formation (Chemical Connections 31B).

As for the other plasma proteins, most of the globulins take part in the immune reactions (Section 30.4), and the albumin provides proper osmotic pressure. If the albumin concentration drops (from malnutrition or from kidney disease, for example), the water from the blood oozes into the interstitial fluid and creates the swelling of tissues called **edema**.

31.3 Blood as a Carrier of Oxygen

One of the most important functions of blood is to carry oxygen from the lungs to the tissues. This task is accomplished by hemoglobin molecules located inside the erythrocytes. As we saw in Section 21.9, hemoglobin is made up of two alpha and two beta protein chains, each attached to a molecule of heme.

The active sites are the hemes, and at the center of each heme is an iron(II) ion. The heme with its central Fe^{2+} ion forms a plane. Because each hemoglobin molecule has four hemes, it can hold a total of four O_2 molecules. In reality, the ability of the hemoglobin molecule to hold O_2 depends on how much oxygen is in the environment. Consider Figure 31.3, which shows how the oxygen-carrying ability of hemoglobin depends on oxygen pressure. When oxygen enters the lungs, the pressure is high (100 mm Hg). At this pressure, all the Fe^{2+} ions of the hemes bind oxygen molecules; they are fully saturated. By the time the blood reaches the muscles through the capillaries, the oxygen pressure in the muscle is only 20 mm Hg (Chemical Connections 31C). At this pressure, only 30% of the binding sites carry oxygen. Thus, 70% of the oxygen carried will be released to the tissues. The S shape of the binding (dissociation) curve (Figure 31.3) implies that the reaction is not a simple equilibrium reaction.

$$HbO_2 \rightleftharpoons Hb + O_2$$

In hemoglobin, each heme has a cooperative effect on the other hemes. This cooperative action allows hemoglobin to deliver twice as much oxygen to the tissues as it would if each heme acted independently. The reason for this is as follows: When a hemoglobin molecule is carrying no oxygen, the four globin units coil into a certain shape (Section 21.9). When the first oxygen molecule attaches to one of the heme subunits, it changes the shape not only of that subunit but also of a second subunit, making it easier for the second subunit to bind an oxygen. When the second heme binds, the shape

■ A model of heme with a central Fe^{2+} ion.

The O_2 picked up by the hemoglobin binds to the Fe^{2+}.

This allosteric effect (Section 22.6) explains the S-shaped curve shown in Figure 31.3.

Figure 31.3 An oxygen dissociation curve. Saturation (%) means the percentage of Fe^{2+} ions that carry O_2 molecules.

CHEMICAL CONNECTIONS 31C

Breathing and Dalton's Law

A gas naturally flows from an area of higher pressure to one of lower pressure. This fact is responsible for respiration, though we must look not at the total pressure of air but only at the partial pressure of O_2 and CO_2 (see the discussion of Dalton's law in Section 5.5). Respiration is the process by which blood carries O_2 from the lungs to tissues, collects the CO_2 manufactured by cells, and brings it to the lungs, where it is breathed out.

The air we breathe contains about 21% O_2 at a partial pressure of approximately 159 mm Hg. The partial pressure of O_2 in the alveoli (tiny air sacs) in the lungs is about 100 mm Hg. Because the partial pressure is higher in the inhaled air, O_2 flows to the alveoli. At this point, venous blood flows past the alveoli. The partial pressure of O_2 in the venous blood is only 40 mm Hg, so O_2 flows from the alveoli to the venous blood. This new supply of O_2 raises the partial pressure to about 100 mm Hg, and the blood becomes arterial blood and flows to the body tissues. There, the partial pressure of O_2 is 30 mm Hg or less (because the metabolic activity of the tissue cells has used up most of their oxygen to provide energy). Oxygen now flows from the arterial blood to the body tissues. The partial pressure of O_2 in the blood decreases (to about 40 mm Hg), and the blood becomes venous blood again and returns to the lungs for a fresh supply of O_2. At each step of the cycle, O_2 flows from a region of higher partial pressure to one of lower partial pressure.

Meanwhile, CO_2 goes the opposite way for the same reason. The partial pressure of CO_2 as a result of metabolic activity is about 60 mm Hg in tissues. CO_2 flows from tissues to arterial blood, changing it to venous blood and reaching a CO_2 pressure of 46 mm Hg. The blood then flows to the lungs. The CO_2 pressure in the alveoli is 40 mm Hg, and so the CO_2 flows to the alveoli. Because the partial pressure of CO_2 in air is only 0.3 mm Hg, CO_2 flows from the alveoli and is exhaled.

■ Blood circulation showing partial pressures of O_2 and CO_2 in the different parts of the system.

Lowering the pH decreases the oxygen-binding capacity of hemoglobin.

Bohr effect The effect caused by a change in the pH on the oxygen-carrying capacity of hemoglobin

This is how the body delivers more oxygen to those tissues that need it.

changes once again, and the oxygen-binding ability of the two remaining subunits increases still more.

The oxygen-delivering capacity of hemoglobin is also affected by its environment. A slight change in the pH of the environment changes the oxygen-binding capacity, a phenomenon called the **Bohr effect.** An increase in CO_2 pressure also decreases the oxygen-binding capacity of hemoglobin. When a muscle contracts, H^+ ions and CO_2 are produced. The newly produced CO_2 enhances the release of oxygen. At the same time, H^+ ions lower the pH of the muscle. Thus, at the same pressure (20 mm Hg), more oxygen is released for an active muscle than for a muscle at rest.

31.4 Transport of Carbon Dioxide in the Blood

What happens to the CO_2 and H^+ produced in metabolically active cells? They bind to the hemoglobin of erythrocytes, and carbaminohemoglobin is formed. This is an equilibrium reaction.

$$Hb-NH_2 + CO_2 \rightleftharpoons \underset{\text{Carbaminohemoglobin}}{Hb-NH-COO^-} + H^+$$

The CO_2 is bound to the terminal α-NH_2 groups of the four polypeptide chains, so each hemoglobin can carry a maximum of four CO_2 molecules, one for each chain. How much CO_2 each hemoglobin actually carries depends on the pressure of CO_2. The higher the CO_2 pressure, the more carbaminohemoglobin formed.

Only 25% of the total CO_2 produced by the cells is transported to the lungs in the form of carbaminohemoglobin. Another 70% is converted in the red blood cells to carbonic acid by the enzyme carbonic anhydrase. A large part of this H_2CO_3 is carried as such to the lungs, where it is converted back to CO_2 by carbonic anhydrase and released. The remaining 5% of the total CO_2 is carried in the plasma as dissolved gas.

Peripheral tissues

$2CO_2 + 2H_2O$
↕ Carbonic anhydrase
$2H_2CO_3$
↕
$2HCO_3^- + 2H^+$

Blood

$4O_2$
$Hb \cdot 4O_2$
$Hb \cdot 2H^+$

Lungs

$2CO_2 + 2H_2O$
↕ Carbonic anhydrase
$2H_2CO_3$
↕
$2H^+ + 2HCO_3^-$

$4O_2$

The reaction proceeds to the CO_2 production (top) because loss of CO_2 from the lungs causes this equilibrium to shift to replace it (Le Chatelier's principle).

31.5 Blood Cleansing: A Kidney Function

We have seen that one of blood's functions is to maintain the pH at 7.4 (Section 8.11). Another is to carry waste products away from the cells. CO_2, one of the principal waste products of respiration, is carried by the blood to the lungs and exhaled. The other waste products are filtered out by the kidneys and eliminated in the urine.

The kidney is a superfiltration machine. In the event of kidney failure, filtration can be done by hemodialysis (Figure 31.4). About 100 L of blood passes through a normal human kidney daily. Of this amount, only about 1.5 L of liquid is excreted as urine. Obviously, we do not have here just a simple filtration system in which small molecules are lost and large ones are retained. The kidneys also reabsorb from the urine those small molecules that are not waste products.

Hemodialysis is discussed in Chemical Connections 6F.

The balance between filtration and reabsorption is controlled by a number of hormones.

A Kidney

The biological units inside the kidneys that perform these functions are called **nephrons,** and each kidney contains about a million of them. A nephron is made up of a filtration head called a **Bowman's capsule** con-

Figure 31.4 A patient undergoing hemodialysis for kidney disease.

Arteries are blood vessels that carry oxygenated blood away from the heart.

Glomeruli Part of the kidney filtration cell, or nephron; a tangle of capillaries surrounded by a fluid-filled space

The singular is *glomerulus*.

nected to a tiny, U-shaped tube called a tubule. The part of the tubule close to the Bowman's capsule is the **proximal tubule,** the U-shaped twist is called **Henle's loop,** and the part of the tubule farthest from the capsule is the **distal tubule** (Figure 31.5).

Blood vessels penetrate the kidney throughout. The arteries branch into capillaries, and one tiny capillary enters each Bowman's capsule. Inside the capsule, the capillary first branches into even smaller vessels, called **glomeruli,** and then leaves the capsule. The blood enters the glomeruli with every heartbeat, and the pressure forces the water, ions, and all small molecules (urea, sugars, salts, amino acids) through the walls of the glomeruli and the Bowman's capsule. These molecules and ions enter the proximal tubule. Blood cells and large molecules (proteins) are retained in the capillary and exit with the blood.

As shown in Figure 31.5, the tubules and Henle's loop are surrounded by blood vessels; these blood vessels reabsorb vital nutrients. Eighty percent of the water is reabsorbed in the proximal tubule. Almost all the glucose and amino acids are reabsorbed here. When excess sugar is present in the blood (diabetes; see Chemical Connections 23G), some of it passes into the urine. As a consequence, the measurement of glucose concentration in the urine is used in diagnosing diabetes.

By the time the glomerular filtrate reaches Henle's loop, the solids and most of the water have been reabsorbed. Only wastes (such as urea, creatinine, uric acid, ammonia, and some salts) pass into the collecting tubules that bring the urine to the ureter, from which it passes to the bladder, as shown in Figure 31.5(a).

B Urine

Normal urine contains about 4% dissolved waste products; the rest is water. Although the daily amount of urine varies greatly, it averages about 1.5 L per day. The pH varies from 5.5 to 7.5. The main solute is urea, the end product of protein metabolism (Section 27.8). Other nitrogenous waste

Figure 31.5 Excretion through the kidneys. (a) The human urinary system. (b) A kidney nephron and its components, with the surrounding circulatory system.

products, such as creatine, creatinine, ammonia, and hippuric acid, are also present, albeit in much smaller amounts.

Screen 27.4
Animation

Creatine   Creatinine   Hippuric acid

In addition, normal urine contains inorganic ions such as Na^+, Ca^{2+}, Mg^{2+}, Cl^-, PO_4^{3-}, SO_4^{2-}, and HCO_3^-.

Under certain pathological conditions, other substances can appear in the urine. This possibility makes urine analysis a highly important part of diagnostic medicine. We have already noted that the presence of glucose and ketone bodies is an indication of diabetes. Other abnormal constituents may include proteins, which can indicate kidney diseases such as nephritis.

31.6 Buffer Production: Another Kidney Function

Among the waste products sent into the blood by the tissues are H^+ ions. They are neutralized by the HCO_3^- ions that are a part of the blood's buffer system:

$$H^+ + HCO_3^- \rightleftharpoons H_2CO_3$$

When the blood reaches the lungs, as we saw in Section 31.4, H_2CO_3 is decomposed by carbonic anhydrase, and CO_2 is exhaled. If the body had no mechanism for replacing HCO_3^-, we would lose most of the bicarbonate buffer from the blood, and the blood pH would eventually drop (acidosis). In reality, the lost HCO_3^- ions are continuously replaced by the kidneys—another principal kidney function. This replacement takes place in the distal tubules. The cells lining the walls of the distal tubules reabsorb the CO_2 that was lost in the glomeruli. With the aid of carbonic anhydrase, carbonic acid forms quickly and then dissociates to HCO_3^- and H^+ ions:

$$CO_2 + H_2O \rightleftharpoons H_2CO_3 \rightleftharpoons H^+ + HCO_3^-$$

The H^+ ions move from the cells into the urine in the tubule, where they are partially neutralized by a phosphate buffer. To compensate for the lost positive ions, Na^+ ions from the tubule enter the cells. When this happens, Na^+ and HCO_3^- ions move from the cells into the capillaries. In this way, the H^+ ions picked up at the tissues and temporarily neutralized in the blood by HCO_3^- are finally pumped out into the urine. At the same time, the HCO_3^- ions lost in the lungs are regained by the blood in the distal tubules.

31.7 Water and Salt Balance in Blood and Kidneys

In Section 31.5, we mentioned that there is a balance in the kidneys between filtration and reabsorption. This balance is under hormonal control. The reabsorption of water is promoted by vasopressin, a small peptide hormone manufactured in the pituitary gland (Figure 23.2). In the absence of this hormone, only the proximal tubules—not the distal tubules or the collecting tubules—reabsorb water. As a consequence, too much water passes into the urine without vasopressin. In the presence of vasopressin, water is reabsorbed in these parts of the nephrons; thus, vasopressin causes the blood to retain more water and produces a more concentrated urine. The production of urine is called **diuresis**. Any agent that reduces the volume of urine is called an **antidiuretic**.

Usually, the vasopressin level in the body is sufficient to maintain the proper amount of water in tissues under various levels of water intake. However, when severe dehydration occurs (as a result of diarrhea, excessive sweating, or insufficient water intake), another hormone helps to maintain proper fluid level. This hormone, aldosterone (Section 20.10), controls the Na^+ ion concentration in the blood. In the presence of aldosterone, the reabsorption of Na^+ ions increases. When more Na^+ ions enter the blood, more Cl^- ions follow (to maintain electroneutrality) as well as more water to solvate these ions. Thus, increased aldosterone production enables the body to retain more water. When the Na^+ ion and water levels in the blood return to normal, aldosterone production stops.

A high concentration of solids in the urine is a symptom of diabetes insipidus.

Diuresis Enhanced excretion of water in urine

For this reason, vasopressin is also called antidiuretic hormone (ADH).

CHEMICAL CONNECTIONS 31D

Sex Hormones and Old Age

The same sex hormones that provide a healthy reproductive life in young people can and do cause problems in the aging body. For male fertility, the prostate gland and testis provide semen. The prostate gland is situated at the exit of the bladder, surrounding the ureter. Its growth is promoted by the conversion of the male sex hormone, testosterone, to 5-α-dihydrotestosterone by the enzyme 5-α-reductase. The majority of men over the age of 50 have enlarged prostate glands, a condition known as benign prostatic hyperplasia. In fact, this condition affects more than 90% of men older than age 70. The enlarged prostate gland constricts the ureter and causes a restricted flow of urine.

This symptom can be partially relieved by an oral drug called terazosin (trade name Hytrin). This drug was originally approved for treatment of high blood pressure, where its action results from its ability to block the α-1-adrenoreceptors, thereby reducing the constriction around peripheral blood vessels. In a similar way, it reduces constriction around the ureter (the duct that conveys urine from the kidney to the bladder), allowing an unrestricted flow of urine. Terazosin, if taken as the sole antihypertensive, antiprostate drug, has come under scrutiny lately, however. A study released in 2000 concluded that this drug may contribute to the occurrence of heart attacks.

One measure, the concentration of prostate-specific antigen (PSA) in the serum, may indicate the possibility of developing prostate cancer. A PSA of 4.0 ng/mL or less is considered to be normal or an indicator of benign prostatic hyperplasia. A PSA concentration of more than 10 ng/mL is an indicator for prostate cancer. The area between 4.0 and 10 ng/mL is a gray area that deserves to be closely monitored.

A more accurate measure of prostate cancer is the percentage of free (not bound) PSA. When this level is higher than 15%, the likelihood of cancer diminishes with increasing percentage. When it is 9% or lower, the prostate cancer has probably already spread. The ultimate confirmation of cancer is microscopic examination of tissue taken by biopsy.

Estradiol and progesterone control sexual characteristics in the female body (Section 20.10). With the onset of menopause, smaller amounts of these and other sex hormones are secreted. This drop-off not only stops fertility but occasionally creates medical problems. The most serious complication is osteoporosis, which is the loss of bone tissue through absorption. It results in brittle and occasionally deformed bone structure. Estrogens (a general name for female sex hormones other than progesterone) in the proper amount can help to absorb calcium from the diet and thus prevent osteoporosis. The lack of enough female sex hormones can also result in atrophic vaginitis and atrophic urethritis.

A large number of postmenopausal women take pills that contain a synthetic progesterone analog and natural estrogens. However, continuous medication with these pills—for example, Premarin (mixed estrogens) and Provera (a progesterone analog)—is not without danger. A slight but statistically significant increase in cancer of the uterus has been reported in postmenopausal women who use such medication. In the latest study released in 2002, a slight risk of heart problems was also noticed in women taking such drugs. A physician's advice is needed in weighing the benefits against the risks.

31.8 Blood Pressure

Blood pressure is produced by the pumping action of the heart. The blood pressure at the arterial capillary end is about 32 mm Hg, a level higher than the osmotic pressure of the blood (18 mm Hg). The osmotic pressure of the blood is caused by the fact that more solutes are dissolved in the blood than in the interstitial fluid. At the capillary end, therefore, nutrient solutes flow from the capillaries into the interstitial fluid and from there into the cells (see Figure 30.2). On the other hand, the blood pressure in the venous capillaries is about 12 mm Hg. This level is less than the osmotic pressure, so solutes (waste products) from the interstitial fluid flow into the capillaries.

The blood pressure is maintained by the total volume of blood, by the pumping of the heart, and by the muscles that surround the blood vessels and provide the proper resistance to blood flow (see Chemical Connections 5D). Blood pressure is controlled by several very complex systems—some that act within seconds and some that take days to respond after a change in blood pressure occurs. For example, if a patient hemorrhages, three nervous control systems begin to function within seconds. First, the **baroreceptors** in the neck detect the consequent drop in pressure and send appropriate signals to the heart to pump harder and to the muscles sur-

> **Baroreceptors** A feedback mechanism in the neck that detects changes in the blood pressure and sends signals to the heart to counteract those changes by pumping harder or slower as the need arises

CHEMICAL CONNECTIONS 31E

Hypertension and Its Control

Nearly one-third of the U.S. population has high blood pressure (**hypertension**). Epidemiological studies have shown that even mild hypertension (diastolic pressure between 90 and 104 mm Hg) brings the risk of cardiovascular disease. It can lead to heart attack, stroke, or kidney failure. Hypertension can be managed effectively with diet and drugs. As noted in the text, blood pressure is under the control of a complex system, so hypertension must be managed on more than one level. Recommended dietary practices are low Na^+ ion intake and abstention from caffeine and alcohol.

The most commonly used drugs for lowering blood pressure are **diuretics.** A number of synthetic organic compounds (mostly thiazides such as Diuril and Enduron) increase urine excretion. By doing so, they decrease the blood volume as well as the Na^+ ion concentration, thereby lowering blood pressure.

Other antihypertensive drugs affect the nerve control of blood pressure. Propranolol hydrochloride blocks the adrenoreceptor sites of beta-adrenergic neurotransmitters (Section 23.5). By doing so, it reduces the flow of nerve signals as well as the blood output of the heart; it affects the responses of baroreceptors and chemoreceptors in the central nervous system and the adrenergic receptors in the smooth muscles surrounding the blood vessels. Propranolol and other beta blockers (such as metoprolol) not only lower blood pressure, but also reduce the risk of myocardial infarction. Similar lowering of blood pressure is achieved by drugs that block calcium channels (for example, nifedipine/Adalat or diltiazem hydrochloride/Cardizem). For proper muscle contraction, calcium ions must pass through muscle cell membranes, an activity facilitated by calcium channels. If a drug blocks these channels, the heart muscles will contract less frequently, pumping less blood and reducing blood pressure.

Other drugs that reduce hypertension include the vasodilators that relax the vascular smooth muscles. However, many of these agents have side effects such as causing fast heartbeat (tachycardia) and promoting Na^+ and water retention. Some drugs on the market act on one specific site rather than on many sites as the blockers do, which obviously reduces the possible side effects. An example is an enzyme inhibitor that prevents the production of angiotensin. Captopril (Chemical Connections 22F) and Accupril are both ACE (angiotensin-converting enzyme) inhibitors that lower blood pressure only if the sole cause of hypertension is angiotensin production. Prostaglandins PGE and PGA (Section 20.12) lower the blood pressure and, when taken orally, have prolonged duration of action. Since hypertension is such a common problem, and it is manageable by drugs, there is great activity in drug research to come up with even more effective and safer antihypertension drugs.

rounding the blood vessels to contract and thus restore pressure. Next, chemical receptors on the cells that detect less O_2 delivery or CO_2 removal send nerve signals. Finally, the central nervous system reacts to oxygen deficiency with a feedback mechanism.

Hormonal controls act somewhat more slowly, taking minutes or even days. The kidneys secrete an enzyme called renin, which acts on an inactive blood protein called angiotensinogen by converting it to **angiotensin,** a potent vasoconstrictor. The action of this peptide increases blood pressure. Aldosterone (Section 31.7) also increases blood pressure by increasing Na^+ ion levels and water reabsorption in the kidneys.

Finally, there is a long-term renal blood control for volume and pressure. When blood pressure falls, the kidneys retain more water and salt, thereby increasing blood volume and pressure.

SUMMARY

The most important body fluid is **blood plasma,** or whole blood from which cellular elements have been removed (Section 31.1). Other important body fluids are urine and the **interstitial fluids** that directly surround the cells of the tissues. The cellular elements of the blood are the red blood cells (**erythrocytes**), which carry O_2 to the tissues and CO_2 away from the tissues; the white blood cells (**leukocytes**), which fight infection; and the **platelets,** which control bleeding (Section 31.2). The plasma contains fibrinogen, which is necessary for blood clot formation. When fibrinogen is removed from the plasma, what remains is the **serum.** The serum contains albumins, globulins, nutrients, waste products, inorganic salts, enzymes, hormones, and vitamins dissolved in water. The body uses blood to maintain **homeostasis.**

The red blood cells carry O_2 (Section 31.3). The Fe(II) in the heme portion of hemoglobin binds the oxygen, which is then released at the tissues by a combination of factors: low O_2 pressure in the tissue cells, high H⁺ ion concentration **(Bohr effect)**, and high CO_2 concentration. Of the CO_2 in the venous blood, 25% binds to the terminal —NH_2 of the polypeptide chains of hemoglobin to form carbaminohemoglobin, 70% is carried in the plasma as H_2CO_3, and 5% is carried as dissolved gas (Section 31.4).

Waste products are removed from the blood in the kidneys (Section 31.5). The filtration units—**nephrons**—contain an entering blood vessel that branches out into fine vessels called **glomeruli**. Water and all the small molecules in the blood diffuse out of the glomeruli, enter the **Bowman's capsules** of the nephrons, and from there go into the **tubules**. The blood cells and large molecules remain in the blood vessels. The water, inorganic salts, and organic nutrients are then reabsorbed into the blood vessels. The waste products plus water form the urine, which goes from the tubule into the ureter and finally into the bladder.

The kidneys not only filter out waste products, but also produce HCO_3^- ions for buffering, replacing those lost through the lungs (Section 31.6). The water and salt balance of the blood and urine are under hormonal control (Section 31.7). Vasopressin helps to retain water, and aldosterone increases the Na^+ ion concentration in the blood by reabsorption.

Blood pressure is controlled by a large number of factors (Section 31.8).

PROBLEMS

Numbers that appear in color indicate difficult problems. ▸ designates problems requiring application of principles.

Body Fluids

31.1 What is the blood–brain barrier?

31.2 What percentage of our body weight is blood plasma?

31.3 (a) Is the blood plasma in contact with the other body fluids?
(b) If so, in what manner?

Functions and Composition of Blood

31.4 Which blood cell is the most abundant?

31.5 What is the function of erythrocytes?

31.6 Which blood cells are the largest?

31.7 Define:
(a) Interstitial fluid (b) Plasma
(c) Serum (d) Cellular elements

31.8 Where are the following manufactured?
(a) Erythrocytes (b) Leukocytes
(c) Lymphocytes (d) Platelets

31.9 How is plasma prepared from whole blood?

31.10 State the function of the following:
(a) Albumin (b) Globulin
(c) Fibrinogen

31.11 Which blood plasma component protects against infections?

31.12 How is serum prepared from blood plasma?

Blood as a Carrier of Oxygen

31.13 How many molecules of O_2 are bound to a hemoglobin molecule at full saturation?

31.14 A chemist analyzes contaminated air from an industrial accident and finds that the partial pressure of oxygen is only 38 mm Hg. How many O_2 molecules, on average, will be transported by each hemoglobin molecule when someone breathes such air? (Consult Figure 31.3.)

31.15 Which inorganic ion is involved in the definition of percent saturation? Explain.

31.16 From Figure 31.3, predict the O_2 pressure at which the hemoglobin molecule binds four, three, two, one, and zero O_2 molecules to its heme.

31.17 Explain how the absorption of O_2 on one binding site in the hemoglobin molecule increases the absorption capacity at other sites.

31.18 (a) What happens to the oxygen-carrying capacity of hemoglobin when the pH is lowered?
(b) What is the name of this effect?

Transport of CO_2 in the Blood

31.19 (a) Where does CO_2 bind to hemoglobin?
(b) What is the name of the complex formed?

31.20 In carbaminohemoglobin, the N-terminal ends of globin chains carry a carbamate group, R—NH—COO⁻. The anionic site can form a salt bridge with a cationic amino group on a side chain of the globin molecule and stabilize it. Draw the formula of the salt bridge.

31.21 If the plasma carries 70% of its carbon dioxide in the form of H_2CO_3 and 5% in the form of dissolved CO_2 gas, calculate the equilibrium constant of the reaction catalyzed by carbonic anhydrase:

$$H_2O + CO_2 \rightleftharpoons H_2CO_3$$

(Assume that the H_2O concentration is included in the equilibrium constant.)

31.22 How is the equilibrium reaction of carbaminohemoglobin formation influenced by the CO_2 concentration of the surroundings? Apply Le Chatelier's principle.

Kidney Functions and Urine

31.23 Where is most of the water reabsorbed in the kidney?

31.24 What nitrogenous waste products are eliminated in the urine?

31.25 Besides nitrogenous waste products, what other components are normal constituents of urine?

31.26 What is the difference between glomeruli and nephrons?

Buffer Production

31.27 What happens to the H^+ ions sent into the blood as waste products by the cells of the tissues?

31.28 The pH of blood is maintained at 7.4. What would be the pH of an active muscle cell relative to that of blood? What component and process generates this pH?

Water and Salt Balance in the Blood and Kidneys

31.29 In which part of the kidneys does vasopressin change the permeability of membranes?

31.30 How does aldosterone production counteract the excessive sweating that usually accompanies a high fever?

31.31 Caffeine is a diuretic. What effect will drinking a lot of coffee have on the specific gravity of your urine?

Blood Pressure

31.32 What is the function of baroreceptors?

31.33 What would happen to the glucose in an arterial capillary if the blood pressure at the end of the capillary were 12 mm Hg rather than the normal 32 mm Hg?

31.34 Some drugs used as antihypertensive agents—for example, Captopril (Chemical Connections 31E)—inhibit the enzyme that activates angiotensinogen. What is the mode of action of such drugs in lowering blood pressure?

31.35 A number of drugs combatting high blood pressure go under the name ACE (angiotensin-converting enzyme) inhibitors. A patient takes diuretics and ACE inhibitors. How do these drugs lower blood pressure?

Chemical Connections

31.36 (Chemical Connections 31A) What side effect of antihistamines is avoided by using the drug fexofenadine?

31.37 (Chemical Connections 31B) Which molecules help to anchor the platelets when a blood vessel is cut?

31.38 (Chemical Connections 31B) What are the functions of (a) thrombin, (b) vitamin K, and (c) thromboplastin in clot formation?

31.39 (Chemical Connections 31B) What is the first line of defense when a blood vessel is cut?

31.40 (Chemical Connections 31C) In circulating blood, where is the partial pressure of CO_2 the highest? The lowest?

31.41 (Chemical Connections 31D) What are the symptoms of an enlarged prostate gland (benign prostatic hyperplasia)?

31.42 (Chemical Connections 31D) What causes osteoporosis in postmenopausal women?

31.43 (Chemical Connections 31E) How do calcium-channel blockers act in lowering blood pressure?

31.44 (Chemical Connections 31E) People with high blood pressure are advised to follow a low-sodium diet. How does such a diet lower blood pressure?

Additional Problems

31.45 Albumin makes up 55% of the protein content of plasma. Is the albumin concentration of the serum higher, lower, or the same as that of the plasma? Explain.

31.46 Which nitrogenous waste product of the body can be classified as an amino acid?

31.47 What happens to the oxygen-carrying ability of hemoglobin when acidosis occurs in the blood?

31.48 Does the heme of hemoglobin participate in CO_2 transport as well as in O_2 transport?

31.49 Is the blood–brain barrier permeable to insulin? Explain.

31.50 When kidneys fail, the body swells, a condition called edema. Why and where does the water accumulate?

31.51 Which hormone controls water retention in the body?

31.52 Describe the action of renin.

31.53 The oxygen dissociation curve of fetal hemoglobin is higher than that of adult hemoglobin. How does the fetus obtain its oxygen supply from the mother's blood when the two circulations meet?

InfoTrac College Edition

For additional readings, go to InfoTrac College Edition, your online research library, at

http://infotrac.thomsonlearning.com

Exponential Notation

APPENDIX I

The **exponential notation** system is based on powers of 10 (see table). For example, if we multiply $10 \times 10 \times 10 = 1000$, we express this as 10^3. The 3 in this expression is called the **exponent** or the **power**, and it indicates how many times we multiplied 10 by itself and how many zeros follow the 1.

There are also negative powers of 10. For example, 10^{-3} means 1 divided by 10^3:

$$10^{-3} = \frac{1}{10^3} = \frac{1}{1000} = 0.001$$

Numbers are frequently expressed like this: 6.4×10^3. In a number of this type, 6.4 is the **coefficient** and 3 is the exponent, or power of 10. This number means exactly what it says:

$$6.4 \times 10^3 = 6.4 \times 1000 = 6400$$

Similarly, we can have coefficients with negative exponents:

$$2.7 \times 10^{-5} = 2.7 \times \frac{1}{10^5} = 2.7 \times 0.00001 = 0.000027$$

For numbers greater than 10 in exponential notation, we proceed as follows: *Move the decimal point to the left,* to just after the first digit. The (positive) exponent is equal to the number of places we moved the decimal point.

Exponential notation is also called scientific notation.

For example, 10^6 means a one followed by six zeros, or 1,000,000, and 10^2 means 100.

APP. I.1 Examples of Exponential Notation

10,000	$= 10^4$
1000	$= 10^3$
100	$= 10^2$
10	$= 10^1$
1	$= 10^0$
0.1	$= 10^{-1}$
0.01	$= 10^{-2}$
0.001	$= 10^{-3}$

EXAMPLE

$37500 = 3.75 \times 10^4$ — 4 because we went four places to the left

Four places to the left — Coefficient

$628 = 6.28 \times 10^2$

Two places to the left — Coefficient

$859{,}600{,}000{,}000 = 8.596 \times 10^{11}$

Eleven places to the left — Coefficient

We don't really have to place the decimal point after the first digit, but by doing so we get a coefficient between 1 and 10, and that is the custom.

Using exponential notation, we can say that there are 2.95×10^{22} copper atoms in a copper penny. For large numbers, the exponent is always *positive*. Note that we do not usually write out the zeros at the end of the number.

For small numbers (less than 1), we move the decimal point *to the right*, to just after the first nonzero digit, and use a *negative exponent*.

EXAMPLE

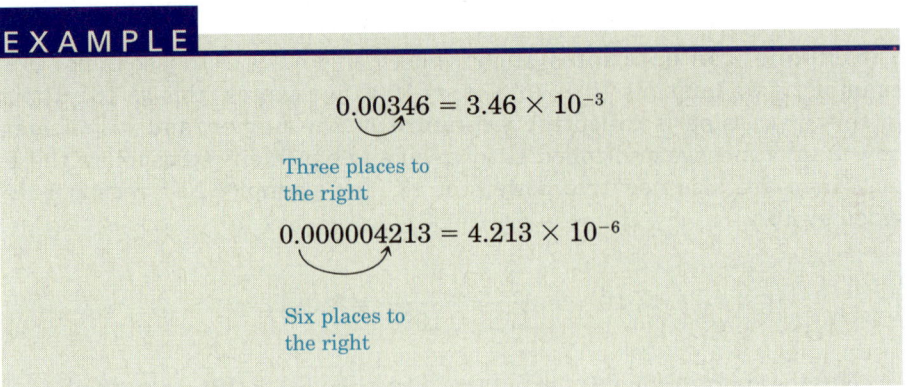

$0.00346 = 3.46 \times 10^{-3}$

Three places to the right

$0.000004213 = 4.213 \times 10^{-6}$

Six places to the right

In exponential notation, a copper atom weighs 2.3×10^{-25} pounds.

To convert exponential notation into fully written-out numbers, we do the same thing backward.

EXAMPLE

Write out in full: (a) 8.16×10^7 (b) 3.44×10^{-4}.

Solution

(a) $8.16 \times 10^7 = 81,600,000$

Seven places to the right (add enough zeros)

(b) $3.44 \times 10^{-4} = 0.000344$

Four places to the left

When scientists add, subtract, multiply, and divide, they are always careful to express their answers with the proper number of digits, called significant figures. This method is described in Appendix II.

Adding and Subtracting Numbers in Exponential Notation

We are allowed to add or subtract numbers expressed in exponential notation *only if they have the same exponent*. All we do is add or subtract the coefficients and leave the exponent as it is.

EXAMPLE

Add 3.6×10^{-3} and 9.1×10^{-3}.

Solution

$$\begin{array}{r} 3.6 \times 10^{-3} \\ + 9.1 \times 10^{-3} \\ \hline 12.7 \times 10^{-3} \end{array}$$

The answer could also be written in other, equally valid ways:

$$12.7 \times 10^{-3} = 0.0127 = 1.27 \times 10^{-2}$$

When it is necessary to add or subtract two numbers that have different exponents, we first must change them so that the exponents are the same.

A calculator with exponential notation changes the exponent automatically.

EXAMPLE

Add 1.95×10^{-2} and 2.8×10^{-3}.

Solution
To add these two numbers, we make both exponents -2. Thus, $2.8 \times 10^{-3} = 0.28 \times 10^{-2}$. Now we can add:

$$\begin{aligned} & 1.95 \times 10^{-2} \\ +\ & 0.28 \times 10^{-2} \\ \hline & 2.23 \times 10^{-2} \end{aligned}$$

Multiplying and Dividing Numbers in Exponential Notation

To multiply numbers in exponential notation, we first multiply the coefficients in the usual way and then algebraically *add* the exponents.

EXAMPLE

Multiply 7.40×10^5 by 3.12×10^9.

Solution

$$7.40 \times 3.12 = 23.1$$

Add exponents:

$$10^5 \times 10^9 = 10^{5+9} = 10^{14}$$

Answer:

$$23.1 \times 10^{14} = 2.31 \times 10^{15}$$

EXAMPLE

Multiply 4.6×10^{-7} by 9.2×10^4

Solution

$$4.6 \times 9.2 = 42$$

Add exponents:

$$10^{-7} \times 10^4 = 10^{-7+4} = 10^{-3}$$

Answer:

$$42 \times 10^{-3} = 4.2 \times 10^{-2}$$

To divide numbers expressed in exponential notation, the process is reversed. We first divide the coefficients and then algebraically *subtract* the exponents.

EXAMPLE

Divide: $\dfrac{6.4 \times 10^8}{2.57 \times 10^{10}}$

Solution

$$6.4 \div 2.57 = 2.5$$

Subtract exponents:

$$10^8 \div 10^{10} = 10^{8-10} = 10^{-2}$$

Answer:

$$2.5 \times 10^{-2}$$

EXAMPLE

Divide: $\dfrac{1.62 \times 10^{-4}}{7.94 \times 10^7}$

Solution

$$1.62 \div 7.94 = 0.204$$

Subtract exponents:

$$10^{-4} \div 10^7 = 10^{-4-7} = 10^{-11}$$

Answer:

$$0.204 \times 10^{-11} = 2.04 \times 10^{-12}$$

Scientific calculators do these calculations automatically. All that is necessary is to enter the first number, press $+$, $-$, $\times$, or $\div$, enter the second number, and press $=$. (The method for entering numbers of this form varies; consult the instructions that come with the calculator.) Many scientific calculators also have a key that will automatically convert a number such as 0.00047 to its scientific notation form (4.7×10^{-4}), and vice versa. For problems relating to exponential notation, see Chapter 1, Problems 1.16 through 1.23.

Significant Figures

APPENDIX II

If you measure the volume of a liquid in a graduated cylinder, you might find that it is 36 mL, to the nearest milliliter, but you cannot tell if it is 36.2, or 35.6, or 36.0 mL because this measuring instrument does not give the last digit with any certainty. A buret gives more digits, and if you use one you should be able to say, for instance, that the volume is 36.3 mL and not 36.4 mL. But even with a buret, you could not say whether the volume is 36.32 or 36.33 mL. For that, you would need an instrument that gives still more digits. This example should show you that *no measured number can ever be known exactly*. No matter how good the measuring instrument, there is always a limit to the number of digits it can measure with certainty.

We define the number of **significant figures** as the number of digits of a measured number that have uncertainty only in the last digit.

What do we mean by this definition? Assume that you are weighing a small object on a laboratory balance that can weigh to the nearest 0.1 g, and you find that the object weighs 16 g. Because the balance weighs to the nearest 0.1 g, you can be sure that the object does not weigh 16.1 g or 15.9 g. In this case, you would write the weight as 16.0 g. To a scientist, there is a difference between 16 g and 16.0 g. Writing 16 g says that you don't know the digit after the 6. Writing 16.0 g says that you do know it: It is 0. However, you don't know the digit after that. Several rules govern the use of significant figures in reporting measured numbers.

A. Determining the Number of Significant Figures

1. Nonzero digits are always significant.

For example, 233.1 m has four significant figures; 2.3 g has two significant figures.

2. Zeros at the beginning of a number are never significant.

For example, 0.0055 L has two significant figures; 0.3456 g has four significant figures.

3. Zeros between nonzero digits are always significant.

For example, 2.045 kcal has four significant figures; 8.0506 g has five significant figures.

4. Zeros at the end of a number that contains a decimal point are always significant.

For example, 3.00 L has three significant figures; 0.0450 mm has three significant figures.

5. Zeros at the end of a number that contains no decimal point may or may not be significant.

We cannot tell whether they are significant without knowing something about the number. This is the ambiguous case. If you know that a certain small business made a profit of $36,000 last year, you can be sure that the 3 and 6 are significant, but what about the rest? The profit might have been $36,126 or $35,786.53, or maybe even exactly $36,000. We just don't know because it is customary to round off such numbers. On the other hand, if the profit were reported as $36,000.00, then all seven digits would be significant.

In science, to get around the ambiguous case we use exponential notation (Section 1.3). Suppose a measurement comes out to be 2500 g. If we made the measurement, we of course know whether the two zeros are significant, but we need to tell others. If these digits are *not* significant, we write our number as 2.5×10^3. If one zero is significant, we write 2.50×10^3. If both zeros are significant, we write 2.500×10^3. Since we now have a decimal point, all the digits shown are significant.

B. Multiplying and Dividing

The rule in multiplication and division is that the final answer should have the *same* number of significant figures as there are in the number with the *fewest* significant figures.

EXAMPLE

Do the following multiplications and divisions:
(a) 3.6×4.27
(b) 0.004×217.38
(c) $\dfrac{42.1}{3.695}$
(d) $\dfrac{0.30652 \times 138}{2.1}$

Solution
(a) 15 (3.6 has two significant figures)
(b) 0.9 (0.004 has one significant figure)
(c) 11.4 (42.1 has three significant figures)
(d) 2.0×10^1 (2.1 has two significant figures)

C. Adding and Subtracting

In addition and subtraction, the rule is completely different. The number of significant figures in each number doesn't matter. The answer is given to the *same number of decimal places* as the term with the fewest decimal places.

EXAMPLE

Add or subtract:

(a) 320.084
    80.47
  200.23
    20.0
  620.8

(b) 61.4532
  13.7
  22
    0.003
  97

(c) 14.26
  −1.05041
  13.21

Solution
In each case, we add or subtract in the normal way but then round off so that the only digits that appear in the answer are those in the columns in which every digit is significant.

D. Rounding Off

When we have too many significant figures in our answer, it is necessary to round off. In this book we have used the rule that if *the first digit dropped* is 5, 6, 7, 8, or 9, we raise *the last digit kept* to the next number; otherwise, we do not.

EXAMPLE

In each case, drop the last two digits:
(a) 33.679   (b) 2.4715   (c) 1.1145   (d) 0.001309   (e) 3.52

Solution
(a)   33.679 = 33.7
(b)   2.4715 = 2.47
(c)   1.1145 = 1.11
(d) 0.001309 = 0.0013
(e)   3.52 = 4

E. Counted or Defined Numbers

All of the preceding rules apply to *measured* numbers and **not** to any numbers that are *counted* or *defined*. Counted and defined numbers are known exactly. For example, a triangle is defined as having 3 sides, not 3.1 or 2.9. Here, we treat the number 3 as if it has an infinite number of zeros following the decimal point.

EXAMPLE

Multiply 53.692 (a measured number) × 6 (a counted number).

Solution

$$322.15$$

Because 6 is a counted number, we know it exactly, and 53.692 is the number with the fewest significant figures. All we really are doing is adding 53.692 six times.

For problems relating to significant figures, see Chapter 1, Problems 1.24 to 1.29.

ANSWERS*

Chapter 1 Matter, Energy, and Measurement

1.1 (a) 147°F   (b) 8.3°C
1.2 109 kg
1.3 13.8 km
1.4 743 mph
1.5 78.5 g
1.6 2.43 g/mL
1.7 1.016 g/mL
1.8 48×10^3 cal = 48 kcal
1.9 46°C
1.10 0.0430 cal/g·deg
1.11 The major contributor is the advance in medical sciences, among them chemistry. Advances in both the diagnosis and treatment of diseases, especially in drug development, contribute to longer life expectancy.
1.13 The hypothesis is "Organic compounds have higher specific heats than metals."
1.15 Chemical changes: (a) (e) (f) (g)
Physical changes: (b) (c) (d)
1.17 (a) 403,000   (b) 3200   (c) 0.0000713
(d) 0.000000000555
1.19 (a) 2.00×10^{18} (b) 1.37×10^5
(c) 2.36×10^{-10} (d) 8.8×10^1
(e) 1.56×10^{-8}
1.21 (a) 7.74×10^3 (b) 8.81×10^{-2} (c) 1.3×10^2
1.23 3.25×10^{-5}
1.25 (a) 3   (b) 2   (c) 1   (d) 4   (e) 5
1.27 (a) 25,000   (b) 4.1   (c) 15.5
1.29 (a) 10963.1   (b) 244   (c) 172.34
1.31 (a) 1000 g   (b) 0.001 g
1.33 (a) 100 cm   (b) 230 mL   (c) 75 kg   (d) 15 mL
(e) 50 mg   (f) 100 mm   (g) 40 g
1.35 (a) 160°C, 433 K   (b) 100°C, 373 K
(c) −18°C, 255 K   (d) −157°C, 116 K
1.37 (a) 93.9 lb   (b) 735 g   (c) 86 cm
(d) 23.1 miles   (e) 10.3 L   (f) 2.2 oz (avdp)
(g) 31.80 L   (h) 11.5 gal   (i) 1.8 km
(j) 1.18 fl oz
1.39 You will arrive within an hour. The difference between kilometers and miles is much greater (1.6 km equals 1 mile) than the difference between 76 and 80.
1.41 10.6 km/L
1.43 bottom: manganese; top: sodium acetate; middle: calcium chloride
1.45 1.023 g/cc
1.47 water
1.49 One should raise the temperature of water to 4°C. During this temperature change the density of the crystals decreases, while the density of water increases. This brings the less dense crystals to the surface of the more dense water.
1.51 The motion of the wheels of the car generates kinetic energy, which is stored in your battery as potential energy.
1.53 0.34 cal/g·C°
1.55 334 mg
1.57 The body shivers. Further temperature lowering results in unconsciousness and then death.
1.59 Methanol, because its higher specific heat allows it to retain the heat longer.
1.61 0.732
1.63 Kinetic: (b), (d), (e)
Potential: (a), (c)
1.65 the European car
1.67 kinetic energy
1.69 The largest is 41 g. The smallest is 4.1×10^{-8} kg.
1.71 10.9 h
1.73 The heavy water. When converting the specific heat given in J/g·°C to cal/g·°C, one finds that the specific heat of heavy water is 1.008 cal/g·°C, which is somewhat greater than that of ordinary water.
1.75 (a) 1.57 g/mL   (b) 1.25 g/mL
1.77 two
1.79 60 J would raise the temperature by 4.5°C; thus the final temperature will be 24.5°C.

Chapter 2 Atoms

2.1 (a) $NaClO_3$ (b) AlF_3
2.2 (a) The mass number is 31.
(b) The mass number is 222.
2.3 (a) The element is phosphorus (P); its symbol is $^{31}_{15}P$.
(b) The element is radon (Rn); its symbol is $^{222}_{86}Rn$.
2.4 (a) The atomic number of mercury (Hg) is 80; that of lead (Pb) is 82.
(b) An atom of Hg has 80 protons; an atom of Pb has 82 protons.
(c) The mass number of this isotope of Hg is 200; the mass number of this isotope of Pb is 202.
(d) The symbols of these isotopes are $^{200}_{80}Hg$ and $^{202}_{82}Pb$.
2.5 The atomic number of iodine (I) is 53. The number of neutrons in each isotope is 72 for iodine-125 and 78 for iodine-131. The symbols for these two isotopes are $^{125}_{53}I$ and $^{131}_{53}I$, respectively.
2.6 Lithium-7 is the more abundant isotope.
2.7 The element is aluminum (Al).
2.9 The ring is made of a mixture.
2.11 (a) sulfur   (b) iron   (c) hydrogen   (d) potassium
(e) silver   (f) gold
2.13 Given here is the element, its symbol, and its atomic number:
(a) Americium (Am, 95)   (b) Berkelium (Bk, 97)
(c) Californium (Cf, 98)   (d) Dubnium (Db, 105)
(e) Europium (Eu, 63)   (f) Francium (Fr, 87)
(g) Gallium (Ga, 31)   (h) Germanium (Ge, 32)
(i) Hafnium (Hf, 72)   (j) Hassium (Hs, 108)
(k) Lutetium (Lu, 71)   (l) Holmium (Ho, 67)
(m) Magnesium (Mg, 12)   (n) Polonium (Po, 84)
(o) Rhenium (Re, 75)   (p) Ruthenium (Ru, 44)
(q) Scandium (Sc, 21)   (r) Strontium (Sr, 38)

---

*Answers to in-text and odd-numbered end-of-chapter problems.

(s) Ytterbium (Yb, 70), terbium (Tb, 65), and yttrium (Y, 39) (t) Thulium (Tm, 69)

2.15 (a) K_2O (b) Na_3PO_4 (c) $LiNO_3$

2.17 (a) The law of conservation of mass states that matter can be neither created nor destroyed. According to Dalton's theory, if all matter is made up of indestructible atoms, then any chemical reaction just changes the attachments between atoms and does not destroy the atoms themselves.
(b) The law of constant composition states that any compound is always made up of elements in the same proportion by mass. According to Dalton's theory, molecules consist of tightly bound groups of atoms, each of which has a particular mass. Therefore, each element in a compound always constitutes a fixed proportion of the total mass.

2.19 No. CO and CO_2 are different compounds, and each obeys the law of constant composition.

2.21 The statement is true in the sense that the number of protons (the atomic number) determines the identity of the atom.

2.23 (a) The element with 22 protons is titanium (Ti).
(b) The element with 76 protons is osmium (Os).
(c) The element with 34 protons is selenium (Se).
(d) The element with 94 protons is plutonium (Pu).

2.25 Each would still be the same element because the number of protons has not changed.

2.27 Radon (Rn) has an atomic number of 86, so each isotope has 86 protons. The number of neutrons is mass number − atomic number.
(a) Radon-210 has $210 - 86 = 124$ neutrons.
(b) Radon-218 has $218 - 86 = 132$ neutrons.
(c) Radon-222 has $222 - 86 = 136$ neutrons.

2.29 Two more neutrons: tin-120
Three more neutrons: tin-121
Six more neutrons: tin-124

2.31 (a) An ion is an atom with an unequal number of protons and electrons.
(b) Isotopes are atoms with the same number of protons in their nuclei but different numbers of neutrons.

2.33 Rounded to three significant figures, the calculated value is 12.0 amu. The value given in the Periodic Table is 12.011 amu.

$$\left(\frac{98.90}{100} \times 12.000\right) + \left(\frac{1.10}{100} \times 13.000\right)$$
$$= 11.868 + 0.1430 = 12.011$$

2.35 Carbon-11 has 6 protons, 6 electrons, and 5 neutrons.

2.37 Americium-241 (Am) has atomic number 95. This isotope has 91 protons, 91 electrons, and $241 - 95 = 146$ neutrons.

2.39 (a) Groups 2A, 3B, 4B, 5B, 6B, 7B, 8B, 1B, and 2B contain only metals. Note that Group 1A contains one nonmetal, hydrogen.
(b) No group contains only metalloids.
(c) Only Groups 7A and 8A contain only nonmetals.

2.41 Elements in the same group in the Periodic Table should have similar properties: As, N, and P; I and F; Ne and He; Mg, Ca, and Ba; K and Li.

2.43 (a) Aluminum > silicon (b) Arsenic > phosphorus (c) Gallium > germanium (d) Gallium > aluminum

2.45 These numbers arise because (1) an orbital can hold two electrons, (2) s orbitals occur singly and can hold two electrons, (3) p orbitals occur in sets of three and can hold 6 electrons, (4) d orbitals occur in sets of five and can hold ten electrons, and (5) f orbitals come in sets of seven and can hold 14 electrons. The first shell has only an s orbital and, therefore, can hold only two electrons. The second shell has $2s$ and $2p$ orbitals and can hold 8 electrons. The third shell has $3s$, $3p$, and $3d$ orbitals and can hold 18 electrons. The fourth shell has $4s$, $4p$, $4d$, and $4f$ orbitals and can hold 32 electrons.

2.47 The group number tells the number of electrons in the valence shell of the element in the group.

2.49 (a) Li(3): $1s^2 2s^1$ (b) Ne(10): $1s^2 2s^2 2p^6$
(c) Be(4): $1s^2 2s^2$ (d) C(6): $1s^2 2s^2 2p^2$
(e) Mg(12): $1s^2 2s^2 2p^6 3s^2$

2.51 (a) He(2): $1s^2$ (b) Na(11): $1s^2 2s^2 2p^6 3s^1$
(c) Cl(17): $1s^2 2s^2 2p^6 3s^2 3p^5$ (d) P(15): $1s^2 2s^2 2p^6 3s^2 3p^3$
(e) H(1): $1s^1$

2.53 In all three parts of this problem, the outer-shell electron configurations are the same. The only difference is the number of the valence shell being filled.

2.55 The element might be in Group 2A, all of whose members have two valence electrons. It might also be helium.

2.57 The total is $(6s = 2) + (6p = 6) + (6d = 10) + (6f = 14) = 32$ electrons.

2.59 The properties are similar because all of them have the same outer-shell electron configuration. They are not identical because each has a different number of filled inner shells.

2.61 Sulfur and iron are essential components of proteins, and calcium is a major component of bones and teeth.

2.63 Because the $^2H/^1H$ ratio on Mars is five times larger than that on Earth, the atomic weight of hydrogen on Mars would be greater than that on Earth.

2.65 Bronze is an alloy of copper and tin.

2.67 (a) $1s$ (b) $2s, 2p$ (c) $3s, 3p, 3d$ (d) $4s, 4p, 4d, 4f$

2.69 (a) s^2p^1 (b) s^2p^5 (c) s^2p^3

2.71 (a) Phosphorus-32 has 15 protons, 15 electrons, and $32 - 15 = 17$ neutrons.
(b) Molybdenum-98 has 42 protons, 42 electrons, and $98 - 42 = 56$ neutrons.
(c) Calcium-44 has 20 protons, 20 electrons, and $44 - 20 = 24$ neutrons.
(d) Hydrogen-3 has 1 proton, 1 electron, and $3 - 1 = 2$ neutrons.
(e) Gadolinium-158 has 64 protons, 64 electrons, and $158 - 64 = 94$ neutrons.
(f) Bismuth-212 has 83 protons, 83 electrons, and $212 - 83 = 129$ neutrons.

2.73 (a) P (b) K (c) Na (d) N (e) Br
(f) Ag (g) Ca (h) C (i) Sn (j) Zn

2.75 (a) Silicon is in Group 4A. It has four outer-shell electrons.
(b) Bromine is in Group 7A. It has seven outer-shell electrons.
(c) Phosphorus is in Group 5A. It has five outer-shell electrons.
(d) Potassium is in Group 1A. It has one outer-shell electron.
(e) Helium is in Group 8A. It has two outer-shell electrons.
(f) Calcium is in Group 2A. It has two outer-shell electrons.

(g) Krypton is in Group 8A. It has eight outer-shell electrons.
(h) Lead is in Group 4A. It has four outer-shell electrons.
(i) Selenium is in Group 6A. It has six outer-shell electrons.
(j) Oxygen is in Group 6A. It has six outer-shell electrons.

2.77 For elements with atomic numbers less than that of iron (Fe), the number of neutrons is close to the number of protons. Elements with atomic numbers greater than that of Fe have more neutrons than protons. Therefore, most elements have more neutrons than protons.

2.79 (a) An electron has a charge of -1, a proton has a charge of $+1$, and a neutron has no charge.
(b) An electron has a mass of 0.0005 amu; a proton and a neutron each have a mass of 1 amu.

2.81 This isotope of xenon has 77 neutrons.

Chapter 3   Chemical Bonds

3.1 By losing two electrons, Mg acquires a complete octet. By gaining two electrons, sulfur acquires a complete octet.
(a) Mg (12 electrons): $1s^22s^22p^63s^2 \longrightarrow$ Mg^{2+} (10 electrons): $1s^22s^22p^6$
(b) S (16 electrons): $1s^22s^22p^63s^23p^4 \longrightarrow$ S^{2-} (18 electrons): $1s^22s^22p^63s^23p^6$

3.2 Each pair of elements is in the same column of the Periodic Table, and electronegativity increases from bottom to top within a column. Therefore,
(a) Li > K   (b) N > P   (c) C > Si

3.3 (a) KCl   (b) CaF$_2$ (c) Fe$_2$O$_3$

3.4 (a) magnesium oxide   (b) barium iodide
(c) potassium chloride

3.5 (a) MgCl$_2$ (b) Al$_2$O$_3$ (c) LiI

3.6 (a) iron(II) oxide, ferrous oxide
(b) iron(III) oxide, ferric oxide

3.7 (a) potassium hydrogen phosphate
(b) aluminum sulfate
(c) iron(II) carbonate, ferrous carbonate

3.8 (a) S—H (2.5 − 2.1 = 0.4); nonpolar covalent
(b) P—H (2.1 − 2.1 = 0.0); nonpolar covalent
(c) C—F (4.0 − 2.5 = 1.5); polar covalent
(d) C—Cl (3.0 − 2.5 = 0.5); polar covalent

3.9 (a) $\overset{\delta+}{C}-\overset{\delta-}{N}$ (b) $\overset{\delta+}{N}-\overset{\delta-}{O}$ (c) $\overset{\delta+}{C}-\overset{\delta-}{Cl}$

3.10 (a) H—C(H)(H)—C(H)(H)—H   (b) H—C(H)(H)—Cl̈:
(c) H—C≡N:

3.11 (a) nitrogen dioxide   (b) phosphorus tribromide
(c) sulfur dichloride

3.12 Given are three-dimensional structures showing all unshared electron pairs.

(a) CH$_3$OH with 109.5° angles
(b) CH$_2$Cl$_2$ with 109.5° angles
(c) HCOOH with 109.5° and 120° angles

3.13 (a) CH$_3$Cl contains a polar C—Cl bond and is a polar molecule.
(b) HCN contains a polar C—N bond and is a polar molecule.
(c) C$_2$H$_6$ contains no polar bonds and is not a polar molecule.

3.15 (a) A lithium atom has the electron configuration $1s^22s^1$. When Li loses its single $2s$ electron, it forms Li$^+$, which has the electron configuration $1s^2$. This configuration is the same as that of helium, the noble gas nearest Li in atomic number.
(b) An oxygen atom has the electron configuration $1s^22s^22p^4$. When O gains two electrons, it forms O^{2-}, which has the electron configuration $1s^22s^22p^6$. This configuration is the same as that of neon, the noble gas nearest oxygen in atomic number.

3.17 (a) Mg^{2+} (b) F$^-$ (c) Al^{3+} (d) S^{2-} (e) K$^+$ (f) Br$^-$

3.19 Li$^-$ would have the electron configuration $1s^22s^2$, which places two electrons in the valence shell of lithium. This electron configuration is not the same as that of helium, the noble gas nearest lithium in atomic number.

3.21 No. Copper is a transition metal, almost all of which form two or more different positively charged ions.

3.23 Each sodium ion is surrounded by six chloride ions, and each chloride ion is surrounded by six sodium ions.

3.25 In completing this table, be certain to balance charges in each ionic compound.

	Br$^-$	ClO$_4^-$	O^{2-}	NO$_3^-$	SO$_4^{2-}$	PO$_4^{3-}$	OH$^-$
Li$^+$	LiBr	LiClO$_4$	Li$_2$O	LiNO$_3$	Li$_2$SO$_4$	Li$_3$PO$_4$	LiOH
Ca^{2+}	CaBr$_2$	Ca(ClO$_4$)$_2$	CaO	Ca(NO$_3$)$_2$	CaSO$_4$	Ca$_3$(PO$_4$)$_2$	Ca(OH)$_2$
Co^{3+}	CoBr$_3$	Co(ClO$_4$)$_3$	Co$_2$O$_3$	Co(NO$_3$)$_3$	Co$_2$(SO$_4$)$_3$	CoPO$_4$	Co(OH)$_3$
K$^+$	KBr	KClO$_4$	K$_2$O	KNO$_3$	K$_2$SO$_4$	K$_3$PO$_4$	KOH
Cu^{2+}	CuBr$_2$	Cu(ClO$_4$)$_2$	CuO	Cu(NO$_3$)$_2$	CuSO$_4$	Cu$_3$(PO$_4$)$_2$	Cu(OH)$_2$

3.27 (a) hydrogen carbonate ion (bicarbonate ion)
(b) nitrite ion (c) sulfate ion (d) hydrogen sulfate ion (bisulfate ion) (e) dihydrogen phosphate ion
3.29 KNO_2
3.31 (a) Na^+ and Br^- (b) Fe^{2+} and SO_3^{2-}
(c) Mg^{2+} and PO_4^{3-}
(d) K^+ and $H_2PO_4^-$ (e) Na^+ and HCO_3^-
(f) Ba^{2+} and NO_3^-
3.33 (a) $Fe(OH)_3$ (b) $BaCl_2$ (c) $Ca_3(PO_4)_2$ (d) $NaMnO_4$
3.35 Parts (a) and (b) are incorrect. The correct formulas are (a) $(NH_4)_3PO_4$ and (b) $BaCO_3$.
3.37 (a) KBr (b) CaO (c) HgO (d) $Cu_3(PO_4)_2$
(e) Li_2SO_4 (f) Fe_2S_3
3.39 (a) sodium fluoride (b) magnesium sulfide
(c) aluminum oxide (d) barium chloride
(e) calcium hydrogen sulfite (calcium bisulfite)
(f) potassium iodide (g) strontium phosphate
(h) iron(II) hydroxide (ferrous hydroxide)
(i) sodium dihydrogen phosphate
(j) lead(II) acetate (plumbous acetate)
(k) barium hydride (l) ammonium hydrogen phosphate
3.41 (a) N normally forms three covalent bonds and has one unshared pair of electrons.
(b) F normally forms one covalent bond and has three unshared pairs of electrons.
(c) C normally forms four covalent bonds and has no unshared pairs of electrons.
(d) Br normally forms one covalent bond and has three unshared pairs of electrons.
(e) O normally forms two covalent bonds and has two unshared pairs of electrons.
3.43 (a) Following are Lewis structures for each diatomic element.
(b) At room temperature, H_2, O_2, N_2, F_2, and Cl_2 are gases. Br_2 is a liquid, and I_2 is a solid.

H—H   Ö=Ö   :N≡N: :F̈—F̈:
Hydrogen Oxygen Nitrogen Fluorine

:C̈l—C̈l: :B̈r—B̈r: :Ï—Ï:
Chlorine   Bromine   Iodine

3.45 A molecular formula shows the number of each atom present in the simplest unit of a compound. A structural formula shows how atoms are bonded in a compound. A Lewis structure shows how atoms are bonded in a compound as well as the location of all unshared pairs of electrons.
3.47 (a) NH_3 has 8 (b) C_3H_6 has 18 (c) $C_2H_4O_2$ has 24
(d) C_2H_6O has 20 (e) CCl_4 has 32 (f) HNO_2 has 18
(g) CCl_2F_3 has 32 (h) O_2 has 12
3.49 A bromine atom contains seven electrons in its valence shell. A bromine molecule contains two bromine atoms bonded by a single bond. A bromide ion is a bromine atom that has gained one electron in its valence shell; it has a complete octet and a charge of −1.
(a) :B̈r· (b) :B̈r—B̈r: (c) :B̈r:$^-$
3.51 (a) Incorrect. The left carbon has five bonds.
(b) Incorrect. The carbon in the middle has only three bonds.
(c) Incorrect. The second carbon from the right has only three bonds and the oxygen on the right has only one bond.

(d) Incorrect. Fluorine has two bonds.
(e) Correct.
(f) Incorrect. The second carbon from the left has five bonds.
3.53 (a) H_2O has 8 valence electrons, and H_2O_2 has 14 valence electrons. (c) Predict bond angles of 109.5° about each oxygen atom.

H—Ö—H   H—Ö—Ö—H
Water   Hydrogen peroxide

3.55 Under each Lewis structure is the number of valence electrons in the molecule and the predicted bond angles about nitrogen.

H—N̈—H   H—N̈—N̈—H   H—N̈=N̈—H
   | | |
   H   H   H
Ammonia   Hydrazine   Diimide
8 valence electrons 14 valence electrons 12 valence electrons
109.5°   109.5°   120°

3.57 Given is the shape of the molecule and approximate bond angles about its central atom.
(a) tetrahedral, 109.5° (b) pyramidal, 109.5°
(c) tetrahedral, 109.5° (d) bent, 120°
(e) trigonal planar, 120° (f) tetrahedral, 109.5°
(g) bent, 120° (h) pyramidal, 109.5°
(i) tetrahedral, 109.5° (j) pyramidal, 109.5°
3.59 Electronegativity generally increases in going from left to right across a row because the number of positive charges in the nucleus of each element in a row increases by one unit. The increasing nuclear charge exerts a stronger and stronger pull on the valence electrons.
3.61 Electrons are shifted toward the more electronegative atom.
(a) Cl (b) O (c) O (d) Cl (e) C (f) S (g) O
3.63 The differences are in their shapes. Because CO_2 is a linear molecule, it is nonpolar. Because SO_2 is a bent molecule, it is polar.
(a) Ö=C=Ö

(b) :Ö=S̈=Ö:

3.65 No.
3.67 For parts in which a bond will form, use differences in electronegativity to predict the polarity of the bond.
(a) Cl—I (3.0 − 2.5 = 0.5); a polar covalent bond.
(b) No bond. Each loses electrons; neither gains electrons to become a negative ion.
(c) K—O (3.5 − 0.8 = 2.7); an ionic bond.
(d) Al—N (3.0 − 1.5 = 1.5); a polar covalent bond.
(e) No bond. Each loses electrons; neither gains electrons to become a negative ion.
(f) No bond. Neon is a noble gas and neither gains, loses, nor shares electrons.
(g) No bond. Each loses electrons; neither gains electrons to become a negative ion.
(h) No bond. Helium is a noble gas and neither gains, loses, nor shares electrons.
3.69 titanium dioxide, TiO_2
3.71 lead(IV) oxide, PbO_2, and lead(IV) carbonate, $Pb(CO_3)_2$

3.73 Fe^{2+}
3.75 (a) $CaSO_3$ (b) $Ca(HSO_3)_2$ (c) $Ca(OH)_2$ (d) $CaHPO_4$
3.77 calcium dihydrogen phosphate, calcium phosphate, and calcium carbonate
3.79 To visualize the gastrointestinal tract by X ray examination.
3.81 calcium ion, Ca^{2+}
3.83 Hydrogen has the electron configuration $1s^1$. Hydrogen's valence shell has only a $1s$ orbital, which can hold two electrons.
3.85 When carbon has four bonds, it is surrounded by eight valence electrons and has a complete octet.
3.87 Nitrogen has five valence electrons. By sharing three more electrons with another atom or atoms, it can achieve an outer-shell electron configuration like that of neon, the noble gas nearest it in atomic number. The three shared pairs of electrons may be in the form of three single bonds, one double bond and one single bond, or one triple bond. With any of these combinations, there is one unshared pair on nitrogen.
3.89 Oxygen has six valence electrons. By sharing two electrons with another atom or atoms, it can achieve an outer-shell electron configuration like that of neon, the noble gas nearest it in atomic number. The two shared electrons may be in the form of two single bonds or one double bond. With either of these possibilities, there are two unshared pairs of electrons on oxygen.
3.91 O^{6+} has a charge too concentrated for a small ion.
3.93 Use the difference in electronegativity to determine the character of the bond.
(a) C—Cl $(3.0 - 2.5 = 0.5)$; polar covalent
(b) C—Li $(2.5 - 1.0 = 1.5)$; polar covalent
(c) C—N $(3.0 - 2.5 = 0.5)$; polar covalent
3.95 The problem is not with carbon forming a double bond to chlorine, but with chlorine forming a double bond to anything. Because chlorine needs only one electron to complete its valence shell, it forms only a single covalent bond with another element.
3.97 (a) BF_3 (b) CF_2 (e) CH_3 (g) NO
3.99 Predict a shape like that created by putting together the bases of two square-based pyramids. This shape is called octahedral, because it has eight sides.

3.101 (a) $SiCl_4$ (b) PH_3 (c) H_2S
3.103 (a) Following is a Lewis structure for vinyl chloride.

(b) Predict that all bond angles are 120°. (c) Vinyl chloride has a polar C—Cl bond, is a polar molecule, and has a dipole.
3.105 ClO_2 has 19 valence electrons. In a Lewis structure, either chlorine or one of the oxygens must have only seven valence electrons. In the following structure, the odd electron is placed on chlorine, the less electronegative element.

$$\ddot{O}=\ddot{Cl}=\ddot{O}$$

Chapter 4 Chemical Reactions

4.1 (a) $C_9H_8O_4 = 180.0$ amu (b) $Ba_3(PO_4)_2 = 601$ amu
4.2 1500 g water is 83.3 mol water.
4.3 2.84 mol Na_2S is 222 g Na_2S.
4.4 In 2.5 mol of glucose, there are 15 mol C, 30 mol H, and 15 mol O.
4.5 0.062 g $CuNO_3$ contains 4.9×10^{-4} mol Cu^+.
4.6 235 g H_2O contains 7.86×10^{24} molecules H_2O.
4.7 The balanced equation is

$$6CO_2(g) + 6H_2O(\ell) \xrightarrow{\text{Photosynthesis}} C_6H_{12}O_6(aq) + 6O_2(g)$$

4.8 The balanced equation is
$2C_6H_{14}(g) + 19O_2(g) \longrightarrow 12CO_2(g) + 14H_2O(g)$
4.9 The balanced equation is
$3K_2C_2O_4(aq) + Ca_3(AsO_4)_2(s) \longrightarrow 2K_3AsO_4(aq) + 3CaC_2O_4(s)$
4.10 (a) The balanced equation is

$$2Al_2O_3(s) \xrightarrow{\text{Electrolysis}} 4Al(s) + 3O_2(g)$$

(b) It requires 51 g of alumina to prepare 27 g of aluminum.
4.11 It requires 16.6 mol CO to produce 16.6 mol acetic acid.
4.12 334 g of ethanol is produced.
4.13 (a) H_2 is in excess and C is the limiting reagent.
(b) 8.0 g CH_4 is produced.
4.14 The percent yield is 80.7%.
4.15 The net ionic equation is
$Cu^{2+}(aq) + S^{2-}(aq) \longrightarrow CuS(s)$.
4.16 (a) Ni^{2+} has gained two electrons so it has been reduced. Cr has lost two electrons so it has been oxidized. Ni^{2+} is the oxidizing agent and Cr is the reducing agent.
(b) CH_2O has gained hydrogens so it has been reduced. H_2 gains oxygens in being converted to CH_3OH so it is reduced. CH_2O is the oxidizing agent and H_2 is the reducing agent.
4.17 (a) $Cl_2 = 71.0$ amu (b) Ar = 39.9 amu
(c) $P_4 = 124.0$ amu (d) $N_2 = 28.0$ amu
(e) He = 4.0 amu
4.19 (a) 2.0 mol CH_4 (b) 11.52 mol NO (c) 2.754 mol ClO_2 (d) 7.82 mol $C_3H_8O_3$
4.21 (a) 18.1 mol O atoms (b) 1.2 mol Br atoms
(c) 4.2×10^4 O atoms (d) 0.40 mol Hg atoms
4.23 The same—that is, 2 : 1
4.25 1.0×10^{21} molecules aspirin
4.27 6.1×10^{18} molecules cholesterol
4.29 (a) HI + NaOH $\longrightarrow$ NaI + H_2O
(b) $Ba(NO_3)_2 + H_2S \longrightarrow BaS + 2HNO_3$
(c) $CH_4 + 2O_2 \longrightarrow CO_2 + 2H_2O$
(d) $2C_4H_{10} + 13O_2 \longrightarrow 8CO_2 + 10H_2O$
(e) $2Fe + 3CO_2 \longrightarrow Fe_2O_3 + 3CO$
4.31 $CO_2(g) + Ca(OH)_2(aq) \longrightarrow CaCO_3(s) + H2O(\ell)$
4.33 $2Mg(s) + O_2(g) \longrightarrow 2MgO(s)$
4.35 $N_2O_5(g) + H_2O(\ell) \longrightarrow 2HNO_3(aq)$

4.37 $(NH_4)_2CO_3(s) \longrightarrow 2NH_3(g) + CO_2(g) + H_2O(\ell)$
4.39 $2Al(s) + 3HCl(aq) \longrightarrow 2AlCl_3(aq) + 3H_2(g)$
4.41 (a) 0.67 mol N_2 (b) 0.67 mol N_2O_3 (c) 12 mol O_2
4.43 319 g Cl_2
4.45 (a) The balanced equation is

$$2NaClO_2(aq) + Cl_2(g) \longrightarrow 2ClO_2(g) + 2NaCl(aq)$$

(b) 4.1 kg of ClO_2
4.47 7.5 g CO_2
4.49 0.13 g C
4.51 (a) The reactants are 0.769 mol C_6H_6 and 0.845 mol Br_2. Therefore, according to the balanced equation, Br_2 is in excess and C_6H_6 is the limiting reagent.
(b) 121 g bromobenzene
4.53 68%
4.55 (a) Spectator ion: an ion that does not take part in a chemical reaction.
(b) Net ionic equation: a balanced equation showing only the ions that react.
(c) Aqueous solution: a solution in which water is the solvent.
4.57 The spectator ions are Na^+ and Cl^-. The balanced net ionic equation is

$$CO_3^{2-}(aq) + Sr^{2+}(aq) \longrightarrow SrCO_3(s)$$

4.59 $Pb^{2+}(aq) + 2Cl^-(aq) \longrightarrow PbCl_2(s)$
4.61 $NH_4^+(aq) + OH^-(aq) \longrightarrow NH_3(g) + H_2O(\ell)$
4.63 (a) $MgCl_2$ is soluble; most compounds containing Cl^- are soluble.
(b) $CaCO_3$ is insoluble; most compounds containing CO_3^{2-} are insoluble.
(c) Na_2SO_4 is soluble; all compounds containing Na^+ are soluble.
(d) NH_4NO_3 is soluble; all compounds containing NH_4^+ and NO_3^- are soluble.
(e) $Pb(OH)_2$ is insoluble; most compounds containing OH^- are insoluble.
4.65 No, one species gains electrons and the other loses electrons. Electrons are not destroyed, but rather transferred from one chemical species to another.
4.67 (a) C_7H_{12} is oxidized and O_2 is reduced.
(b) O_2 is the oxidizing agent and C_7H_{12} is the reducing agent.
4.69 *Exothermic* means that a chemical reaction gives off heat; *endothermic* means that it absorbs heat.
4.71 19.6 kcal is given off.
4.73 55.8 kcal is given off.
4.75 (a) The synthesis of starch is endothermic.
(b) 26.4 kcal
4.77 88.68 g Fe
4.79 Calcium ions, phosphate ions, and hydroxide ions
4.81 C_2H_4O is oxidized and H_2O_2 is reduced. H_2O_2 is the oxidizing agent and C_2H_4O is the reducing agent.
4.83 (a) Fe_2O_3 loses oxygen; it is reduced. CO gains oxygen; it is oxidized.
(b) 19.2 mol Fe_2O_3
(c) 1.61×10^3 g CO
4.85 The spectator ions are Na^+ and NO_3^-. The net ionic equation is

$$3Cd^{2+}(aq) + 2PO_4^{3-} \longrightarrow Cd_3(PO_4)_2(s)$$

4.87 893 amu

4.89 (a) The reactants are 0.0232 mol $Pb(NO_3)_2$ and 0.0200 mol $AlCl_3$. 0.0200 mol $AlCl_3$ requires 0.300 mol $Pb(NO_3)_2$ for complete reaction, so $Pb(NO_3)_2$ is the limiting reagent and $AlCl_3$ is in excess.
(b) The percent yield is 82.5%.
4.91 (a) $C_5H_{12}(g) + 8O_2(g) \longrightarrow 5CO_2(g) + 6H_2O(g)$
(b) Pentane is oxidized and oxygen is reduced.
(c) Oxygen is the oxidizing agent and pentane is the reducing agent.

Chapter 5 Gases, Liquids, and Solids

5.1 0.41 atm
5.2 16.4 atm
5.3 0.053 atm
5.4 4.84 atm
5.5 0.422 mol Ne
5.6 9.91 g He
5.7 0.107 atm water vapor
5.8 (a) yes (b) no
5.9 83.3 g H_2O
5.10 230 cal
5.11 The water vapor will condense to form solid water (ice).
5.13 1.5 atm CH_4
5.15 2.5 atm
5.17 1.87 L ethane
5.19 475°C
5.21 1.17 atm
5.23

V_1	T_1	P_1	V_2	T_2	P_2
546 L	43°C	6.5 atm	1198 L	65°C	1.9 atm
43 mL	−56°C	865 torr	48 mL	43°C	1.5 atm
4.2 L	234 K	0.87 atm	3.2 L	29°C	1.5 atm
1.3 L	25°C	740 mm Hg	1.2 L	0°C	1.0 atm

5.25 0.31 L
5.27 4.35 atm
5.29 61.0 mL
5.31 (a) 2.33 moles (b) No, it is not necessary to know the identity of the gas.
5.33 4.00 g/mol
5.35 As the pressure increases, the density increases. As the temperature increases at constant pressure, the volume increases and, therefore, the density decreases.
5.37 (a) 24.7 mol O_2 (b) 790 g O_2
5.39 1.4×10^{23} molecules O_2
5.41 (a) The mass of one mole of air is 28.95 g. (b) 1.29 g/L
5.43 Molecules in the gas phase are in constant, random motion. Over time, it mixes thoroughly with air and, in so doing, finds it way to the stratosphere.
5.45 2.69×10^{24} molecules CO
5.47 56.9 L N_2
5.49 O_2 = 1.475 mm Hg (0.194 atm)
CO_2 = 3.0 mm Hg (0.004 atm)
H_2O = 47.1 mm Hg (0.062 atm)
N_2 = 562.4 mm Hg
(0.740 atm)
5.51 Intramolecular covalent bonds are stronger.

5.53 Yes. Oxygen of S=O is a hydrogen bond acceptor, and H of —OH is a hydrogen bond donor.
5.55 Ethanol is a polar molecule with polar C—O and O—H bonds. Ethanol molecules are held in the liquid state by hydrogen bonding. Carbon dioxide is a nonpolar molecule and the only attractive forces between its molecules are London dispersion forces.
5.57 Hexane, C_6H_{14}, has the higher boiling point.
5.59 $0.232\ cal \cdot g^{-1} \cdot °C^{-1}$
5.61 240 cal
5.63 CH_4 has the higher vapor pressure.
5.65 1860 cal
5.67 (a) 5400 cal + 630 cal = 6030 cal = 6.0 kcal
(b) 630 cal
(c) The heat of vaporization of liquid water (540 cal/g) is much greater than the heat capacity of liquid water ($1\ cal \cdot g^{-1} \cdot °C^{-1}$). More heat is withdrawn from the skin by condensing steam than by cooling liquid water from 100°C to 20°C.
5.69 89.1 L
5.71 The ice will sublime and become water vapor.
5.73 Soot has the most disorder.
5.75 CO is bound to hemoglobin and does not allow it to carry oxygen. Oxygen under pressure in the hyperbaric chamber dissolves in the plasma and is carried to the tissues without the aid of hemoglobin.
5.77 Frostbite occurs when tissue freezes. The frozen water in the cells expands and ruptures the cells. The resulting damage is sometimes irreversible.
5.79 Supercritical carbon dioxide has the density of a liquid but maintains its gas-like property of being able to flow with little viscosity or surface tension.
5.81 (a) Pressure results from the collisions of gas molecules with the walls of the container.
(b) Temperature is a measure of the average kinetic energy of the gas particles.
5.83 6.82 atm
5.85 The barometric pressure drops because the air column above us weighs less than before.
5.87 CO, because it is a polar molecule, whereas CO_2 is a nonpolar molecule.
5.89 You must add 0.34 mol N_2 to the cylinder.
5.91 Gases are transparent because there is a lot of empty space between gas particles. Visible light passes through a gas sample without much interaction with the gas particles.
5.93 Pentane has the higher vapor pressure at 20°C.
5.95 H_2O molecules are held in the liquid stable by hydrogen bonding. H_2S molecules are held in the liquid state only weak dipole–dipole interactions.
5.97 11.4 L air
5.99 water vapor

Chapter 6 Solutions and Colloids

6.1 To 11 g KBr add enough water to make 250 mL of solution.
6.2 1.9% w/v
6.3 To 158.2 g KCl (1.06 moles × 2) add enough water to make 2.0 L of solution.
6.4 0.013 M
6.5 185 mL
6.6 The concentration is known to four significant figures. The molecular weight of $NaHSO_3$ is 104.1 g/mol; thus 0.010 mol = 1.041 g/L. Employing the factor label technique:

$$100\ \text{gal} \times \frac{3.785\ L}{1\ \text{gal}} \times \frac{1.041\ g}{1\ L} = 393.6\ g\ NaHSO_3$$

6.7 Take 16.5 mL of concentrated HCl solution and add enough water to make 300 mL solution.
6.8 To 0.13 mL of the concentrated solution add enough water to make 20.0 mL solution.
6.9 $-12.5°C$
6.10 (c)
6.11 0.80 osmol
6.12 (a)
6.13 water
6.15 (a) solid in solid (b) solid in liquid (c) gas in gas (d) gas (carbon dioxide) and liquid (ethyl alcohol) in liquid (water)
6.17 True solutions. They mix in all proportions.
6.19 unsaturated
6.21 (c)
6.23 (a) and (c)
6.25 The warm water contains less oxygen than cold water. The fish die because of lack of oxygen.
6.27 (a)
6.29 (a) 76 mL ethyl alcohol (b) 7.8 mL ethyl acetate (c) 132 mL benzene
6.31 (a) 4.15% w/v casein (b) 0.030% w/v vitamin C (c) 1.75% w/v sucrose
6.33 In each case add enough water to the given weight of solute to make the total volume specified:
(a) 19.5 g of NH_4Br—175 mL of solution
(b) 167 g NaI—1.35 L of solution
(c) 2.4 g C_2H_5OH—330 mL of solution
6.35 0.17 M
6.37 (a) (22.0 g $C_6H_{12}O_6$/240 mL) × (1000 mL/1 L)
        × (1 mol/180 g $C_6H_{12}O_6$) = $5.08 \times 10^{-1}\ M$
(b) (0.190 g K^+/240 mL) × (1000 mL/1 L)
        × (1 mol/39.0 g K^+) = $2.02 \times 10^{-2}\ M$
(c) (0.00400 g Na^+/240 mL) × (1000 mL/1 L)
        × (1 mol/23.0 g Na^+) = $7.25 \times 10^{-4}\ M$
6.39 (13 g sucrose/15 mL) × (1000 mL/1 L)
        × (1 mol/342 g sucrose) = 2.53 M
6.41 25.0 mL
6.43 To 2.08 mL of the stock solution add enough water to make 250 mL of solution.
6.45 (a) 3.85×10^4 ppm captopril
(b) 6.77×10^4 ppm Mg^{2+} and (c) 831 ppm Ca^{2+}
6.47 No, the concentration is 0.01 ppb.
6.49 (a) (15.0×10^{-3} g Fe^{2+}/dose RDA) × (2 RDA/100)
        × (1 cheese/28.0 g) × (1×10^6 g cheese/million g)
        = 10.7 ppm
(b)   (1.200 g Ca^{2+}/dose RDA) × (6 RDA/100)
        × (1 cheese/28 g) × (1×10^6 g cheese/million g)
        = 2570 ppm
(c) 800×10^{-6} g A/dose RDA) × (6 RDA/100)
        × (1 cheese/28 g) × (1×10^6 g cheese/million g)
        = 1.71 ppm
6.51 (b)

6.53 soluble: (b), (c), (d)
6.55 homogeneous: (a); heterogeneous: (b), (d); colloid: (c), (e), (f)
6.57 In cooling, the protein molecules must have been aggregated to large colloidal size particles, which scattered light, causing a turbid appearance.
6.59 (a) $-3.72°C$ (b) $-5.58°C$ (c) $-5.58°C$ (d) $-7.44°C$
6.61 344 g methanol
6.63 The acetic acid did not dissociate to any extent, while the KF dissociated and yielded two ions.
6.65 (a) B (b) B (c) A (d) B (e) A (f) No rise. In each case, the side with the higher osmolarity rises.
6.67 (a) 1.2 (b) 2.5 (c) 8.4 (d) 0.036 osmol
6.69 (b)
6.71 dissolved CO_2
6.73 Nitrogen under high pressure creates a feeling similar to alcohol intoxication.
6.75 $CaCO_3$
6.77 the pressure that is larger than the osmotic pressure of the sea water
6.79 Using the same percentage but of a salt with higher molecular weight ($KHCO_3$) makes the solution lower in osmolarity and thus hypotonic. This error would also create an electrolyte imbalance in the blood, increasing the K^+ concentration and decreasing the Na^+ concentration.
6.81 The ethanol removed the water from the solvation layer of the hyaluronic acid and thus allowed the molecules to stick together upon collision and aggregate.
6.83 4 ppm
6.85 Methanol is more efficient. It has a lower molecular weight; hence there are more moles in the same mass.
6.87 $CO_2(g) + H_2O(\ell) \longrightarrow H_2CO_3(aq)$,
$SO_2(g) + H_2O(\ell) \longrightarrow H_2SO_3(aq)$

6.89 Take 38.6 mL of the concentrated solution, add enough water to bring to 300-mL volume.
6.91 0.006 g
6.93 354 ppm; 7.09 kg

Chapter 7 Reaction Rates and Chemical Equilibria

7.1 0.022 L O_2/min
7.2 -3.6×10^{-3} mol H_2O_2/L min
7.3 $K = \dfrac{[H_2SO_4]}{[SO_3][H_2O]}$
7.4 $K = \dfrac{[N_2][H_2]^3}{[NH_3]^2}$
7.5 $K = 0.602$
7.6 (a) $K = \dfrac{[CH_3COOC_2H_5][H_2O]}{[CH_3COOH][C_2H_5OH]}$
(b) $K = \dfrac{[CH_3COOC_2H_5]}{[CH_3COOH][C_2H_5OH]}$
7.7 The equilibrium shifts to the left.
7.9 endothermic
7.10 The equilibrium shifts to the right.
7.11 7.25×10^{-3} mol/L/min
7.13 The activation energy is low because the breaking of covalent bonds is not required.
7.15 It is unlikely that five molecules will collide simultaneously.
7.17 50°C
7.19 (1) increase the concentrations (2) increase the temperature (3) add a catalyst
7.21 It provides an alternate pathway of lower activation energy.
7.23 digesting a piece of candy; explosion of TNT; rusting of iron; addition of Na or K metal to water
7.25 (a) $K = [H_2O]^2[O_2]/[H_2O_2]^2$
(b) $K = [N_2O_4]^2[O_2]/[N_2O_5]^2$
(c) $K = [C_6H_{12}O_6][O_2]^6/[H_2O]^6[CO_2]^6$
7.27 $K = 0.67$
7.29 $K = 9.9 \times 10^{-2} M$
7.31 products: (a), (b), (c); reactants; (d), (e)
7.33 (a) No, the speed of reaction depends on the energy of activation, not on the exothermic character.
(b) Yes, endothermic reactions always have high energies of activation (Fig. 7.5) and so must be very slow.
7.35 (a) shift to the right (b) shift to the right (c) shift to the left (d) shift to the left (e) no shift
7.37 (a) The equilibrium will shift right. (b) The equilibrium constant will stay the same.
7.39 (a) no change (b) no change (c) K smaller
7.41 The higher temperature causes chemical reactions in the body to go faster. If the reactions go too fast, it is dangerous to the health of the patient. At low temperature, the reactions are too slow.
7.43 The capsule, because the beads have much larger surface area than the solid pill.
7.45 because the reaction is too slow at low temperature
7.47

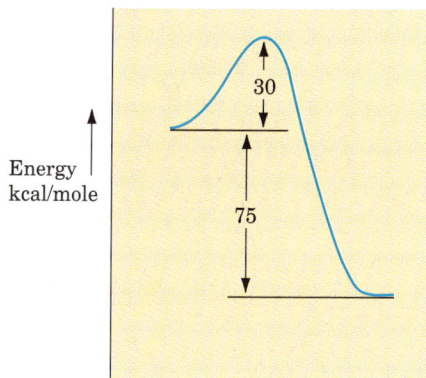

7.49 -0.37/h
7.51 -2.0×10^{-3} mole of N_2O_4 per L per s
7.53 $0.4 M$
7.55 $4NH_3(g) + 7O_2(g) \rightleftharpoons 4NO_2(g) + 6H_2O(g)$
7.57 A, because the orientation does not matter, but rod-like molecules need special orientations for effective collisions to occur.
7.59 0.030 mol/L per min
7.61 (a) The second step is the slowest, and hence the rate-determining step. (b) The third step is the fastest, and hence has the lowest energy of activation.

Chapter 8 Acids and Bases

8.1 (a) toward the left (b) toward the left

(a)
H_3O^+ + I^- ⇌ H_2O + HI
Weaker acid — Weaker base — Stronger base — Stronger acid

(b) CH_3COO^- + H_2S ⇌ CH_3COOH + HS^-
Weaker acid — Weaker acid — Stronger acid — Stronger base

8.2 $pK_a = 9.31$
8.3 (a) ascorbic acid (b) aspirin
8.4 $[H_3O^+] = 1.0 \times 10^{-2}$
8.5 (a) pH = 2.46 (b) $[H_3O^+] = 7.9 \times 10^{-5}$
8.6 pOH = 4 and pH = 10
8.7 0.0960 M
8.8 (a) pH = 9.25 (b) pH = 4.75
8.9 pH = 9.44
8.10 (a) An acid is a substance that produces H_3O^+ in aqueous solution. (b) A base is a substance that produces OH^- in aqueous solution.
8.11 (a) $HNO_3(aq) + H_2O(\ell) \longrightarrow NO_3^-(aq) + H_3O^+(aq)$
(b) $HBr(aq) + H_2O(\ell) \longrightarrow Br^-(aq) + H_3O^+(aq)$
(c) $H_2SO_3(aq) + H_2O(\ell) \longrightarrow HSO_3^-(aq) + H_3O^+(aq)$
(d) $H_2SO_4(aq) + H_2O(\ell) \longrightarrow HSO_4^-(aq) + H_3O^+(aq)$
(e) $HCO_3^-(aq) + H_2O(\ell) \longrightarrow CO_3^{2-}(aq) + H_3O^+(aq)$
(f) $H_3BO_3(aq) + H_2O(\ell) \longrightarrow H_2BO_3^-(aq) + H_3O^+(aq)$
8.13 (a) $LiOH(s) \xrightarrow{H_2O} Li^+(aq) + OH^-(aq)$
(b) $(CH_3)_2NH(aq) + H_2O(\ell) \longrightarrow (CH_3)_2NH_2^+(aq) + OH^-(aq)$
8.15 (a) HSO_4^- (b) $H_2BO_3^-$ (c) I^- (d) H_2O (e) NH_3 (f) PO_4^{3-}
8.17 (a) H_2O (b) H_2S (c) NH_4^+ (d) C_6H_5OH (e) HCO_3^- (f) H_2CO_3
8.19
(a) H_3PO_4 + OH^- ⇌ $H_2PO_4^-$ + H_2O
Stronger acid — Stronger base — Weaker base — Weaker acid

(b) H_2O + Cl^- ⇌ HCl + OH^-
Weaker acid — Weaker base — Stronger acid — Stronger base

(c) HCO_3^- + OH^- ⇌ CO_3^{2-} + H_2O
Stronger acid — Stronger base — Weaker base — Weaker acid

8.21 $H_2CO_3(aq) \longrightarrow CO_2(g) + H_2O(\ell)$
8.23 (a) the pK_a of the weak acid (b) the K_a of the strong acid
8.25 (a) 0.10 M HCl (b) 0.10 M H_3PO_4 (c) 0.010 M H_2CO_3 (d) 0.10 M NaH_2PO_4 (e) 0.10 M aspirin
8.27 Only (b) is a redox reaction. All others are acid–base reactions.
(a) $Na_2CO_3 + 2HCl \longrightarrow 2NaCl + CO_2 + H_2O$
(b) $Mg + 2HCl \longrightarrow MgCl_2 + H_2$
(c) $NaOH + HCl \longrightarrow NaCl + H_2O$
(d) $Fe_2O_3 + 6HCl \longrightarrow 2FeCl_3 + 3H_2O$
(e) $NH_3 + HCl \longrightarrow NH_4Cl$
(f) $CH_3NH_2 + HCl \longrightarrow CH_3NH_3Cl$
(g) $NaHCO_3 + HCl \longrightarrow NaCl + H_2O + CO_2$
8.29 (a) 10^{-3} M (b) 10^{-10} M (c) 10^{-7} M (d) 10^{-15} M

8.31 (a) pH 8; basic (b) pH = 10; basic (c) pH = 2; acidic (d) pH 0; acidic (e) pH = 7; neutral
8.33 pH = 8.5; basic (b) pH = 1.2; acidic (c) pH = 11.1; basic (d) pH = 6.3; acidic
8.35 (a) $[OH^-] = 0.10$ M; pOH = 1
(b) pOH = 2.4; $[OH^-] = 4.0 \times 10^{-3}$
(c) pOH = 2.0; $[OH^-] = 10^{-2}$
(d) pOH = 5.6; $[OH^-] = 2.5 \times 10^{-6}$
8.37 (a) $Ba(OH)_2 + 2HCl \longrightarrow BaCl_2 + 2H_2O$
(b) $NaOH + HNO_3 \longrightarrow NaNO_3 + H_2O$
(c) $HCOOH + NH_3 \longrightarrow HCOONH_4$
(d) $Ca(OH)_2 + H_2SO_4 \longrightarrow CaSO_4 + 2H_2O$
(e) $Mg(OH)_2 + H_2SO_4 \longrightarrow MgSO_4 + 2H_2O$
(f) $2NaOH + H_3PO_4 \longrightarrow Na_2HPO_4 + 2H_2O$
or $NaOH + NaH_2PO_4 \longrightarrow Na_2HPO_4 + H_2O$
(g) $NH_3 + HCl \longrightarrow NH_4Cl$
8.39 Sodium carbonate is a basic salt. The carbonate ion reacts with water to produce bicarbonate ion and hydroxide ion. Even though the position of this acid–base equilibrium lies far to the left, there is a sufficient concentration of hydroxide ion to make the solution basic.
8.41 0.348 M
8.43 (a) To 12 g NaOH, add enough water to make a volume of 400 mL.
(b) To 12 g of $Ba(OH)_2$, add enough water to make a volume of 1.0 L.
8.45 5.66 mL of 0.740 M H_2SO_4
8.47 3.30×10^{-3} mol of unknown base
8.49 (a) $CH_3COO^- + H_3O^+ \longrightarrow CH_3COOH + H_2O$
(b) $CH_3COOH + OH^- \longrightarrow CH_3COO^- + H_2O$
8.51 The pH of the buffer is 3.75.
8.53 Buffer capacity is the ability of a buffer to neutralize added amounts of H_3O^+ or OH^-. The greater its ability to neutralize these species, the greater its buffer capacity.
8.55 (a) The pH of the buffer would be the same. (b) The buffer containing 1.0 mol of the weak acid and its conjugate base has the greater buffer capacity.
8.57 When $[HA] = [A^-]$, then $\log [A^-]/[HA] = 0$, and pH = pK_a.
8.59 The $HPO_4^{2-}/H_2PO_4^-$ ratio is 1.55/1.
8.61 The 0.10 mol of sodium acetate reacts completely with the 0.10 mol of HCl to form 0.10 mol of acetic acid. The pH of 0.10 M acetic acid is 2.9.
8.63 Rinse the eyes immediately and thoroughly with a stream of water.
8.65 The bases that do not contain hydroxide ion are $CaCO_3$, $MgCO_3$, and $NaHCO_3$.
8.67 (a) Will dissolve (b) Will not dissolve (c) Will not dissolve
8.69 The 0.10 M HCl is more acidic. Because HCl is completely ionized in aqueous solution, the concentration of H_3O^+ is 0.10 M. Acetic acid is a weak acid and the extent of its ionization in aqueous solution is very small.
8.71 3.70×10^{-3} M
8.73 0.90 M
8.75 Yes. If the concentration of HCl is 1.0 M, $-\log(1.0) = 0$.
8.77 No. The pH of 1.0 M HCl is smaller (it is a more acidic solution) than the pH of 1.0 M acetic acid.
8.79 Dissolve 0.182 mol NaH_2BO_3 and 1.00 mol of H_3BO_3 in enough water to make 1 L of solution.

8.81 The acid with the higher pK_a is the weaker acid. The equilibrium for an acid–base reaction will always lie on the side of the weaker acid and the weaker base.
8.83 (a) $HCOO^- + H_3O^+ \longrightarrow HCOOH + H_2O$
(b) $HCOOH + OH^- \longrightarrow HCOO^- + H_2O$
8.85 (a) 0.050 mol Na_2HPO_4 (b) 0.0050 mol Na_2HPO_4
(c) 0.500 mol Na_2HPO_4
8.87 Decrease (become more acidic). You are adding more of the weak acid to the solution, so it becomes more acidic and its pH decreases.

Chapter 9 Nuclear Chemistry

9.1 xenon-139
9.2 radium-219
9.3 germanium-74
9.4 0.31 g
9.5 3.3×10^{-3} mCi
9.6 1.5 mL
9.7 Yes; all electromagnetic radiations have the same speed in vacuum. Gamma rays are one form of electromagnetic radiation.
9.9 5.2×10^9 s^{-1}
9.11 4.6×10^{14}/s
9.13 (a) $^{19}_{9}F$ (b) $^{32}_{15}P$ (c) $^{87}_{37}Rb$
9.15 boron-10
9.17 (a) $^{159}_{64}Gd$ (b) $^{141}_{57}La$ (c) $^{242}_{96}Cm$
9.19 (a) $^{206}_{81}Tl$ (b) $^{234}_{92}U$ (c) $^{170}_{70}Yb$
9.21 $^{29}_{15}P \longrightarrow ^{29}_{14}Si + ^{0}_{+1}e$
9.23 In the first step, alpha emission occurs; in the second and third steps, beta emissions occur.
9.25 (a) $^{28}_{14}Si$ (b) $^{143}_{54}Xe$ (c) $^{1}_{0}n$ (d) $^{210}_{84}Po$ (e) $^{13}_{7}N$
9.27 12.5%
9.29 No, the conversion $Ra \longrightarrow Ra^{2+}$ involves electrons from the electron cloud. Radioactivity is a nuclear phenomenon.
9.31 15 min
9.33 No; scintillation counters count each particle one by one.
9.35 (a)
9.37 (a), (e) the amount of radiation absorbed by tissues
(b), (g) the effective dose absorbed by humans
(c) the effective dose delivered
(d), (f) intensity of radiation
9.39 The penetrating power of alpha particles is so low that most of them are stopped by the outer skin. But when they get into the lung, the membranes offer little resistance so the particles can damage the cells of the lung.
9.41 alpha particles
9.43 (a) cobalt-60 (b) selenium-75 (c) strontium-85
(d) carbon-11, technetium-99m (e) mercury-197, technetium-99m
9.45 The half-life is too short. It would not last long enough to reach a hospital.
9.47 $^{1}_{1}H$
9.49 curium-248
9.51 $^{290}_{118}X$
9.53 because gamma rays carry no electrical charges
9.55 one-tenth of a half-life
9.57 It gives off an alpha particle.
9.59 See Chemical Connections 9E.
9.61 because they are destroyed by collisions with electrons, and electrons are in all matter
9.63 If ingested, iodine concentrates in the thyroid gland where radioactive iodine can cause cancer.
9.65 fluorine-19 and neon-20
9.67 the same
9.69 (a) 82% (b) 11% (c) 0.1%
9.71 X rays
9.73 Neptunium-237. Approximately 25% of the original americium-241 will be around after 1000 years.
9.75 One sievert is 100 rem. This is sufficient to cause radiation sickness but not certain death.
9.77 (a) because a radioactive sample decays, and the decay products are always mixed with the starting isotope
(b) because a neutron in the nucleus is converted to a proton
9.79 Oxygen-16 is stable because it has an equal number of protons and neutrons. The others are unstable because the numbers are unequal; in this case the less equal, the faster the isotope decays.

Chapter 10 Organic Chemistry

10.1 Following are Lewis structures showing all bond angles.

10.2 Of the four alcohols with the molecular formula $C_4H_{10}O$, two are 1°, one is 2°, and one is 3°. For the Lewis structures the 3° alcohol and two of the 1° alcohols, three $C-CH_3$ bonds are drawn longer to avoid crowding in the formulas.

(a) $CH_3CH_2CH_2CH_2OH$ Primary (1°)

(b) $CH_3CH_2CHCH_3$ Secondary (2°)
      |
      OH

(c) CH_3CHCH_2OH Primary (1°)
       |
       CH_3

(d)
```
        H
        |
      H-C-H
        |           CH₃
        H           |
    H-C-C-ÖH    CH₃COH
        |           |
        H           CH₃
      H-C-H      Tertiary (3°)
        |
        H
```

10.3 The three secondary (2°) amines with molecular formula $C_4H_{11}N$ are

$$CH_3CH_2CH_2NHCH_3 \quad CH_3\overset{\overset{\displaystyle CH_3}{|}}{C}HNHCH_3 \quad CH_3CH_2NHCH_2CH_3$$

10.4 The three ketones with the molecular formula $C_5H_{10}O$ are

$$CH_3CH_2CH_2\overset{\overset{\displaystyle O}{\|}}{C}CH_3 \quad CH_3CH_2\overset{\overset{\displaystyle O}{\|}}{C}CH_2CH_3 \quad CH_3\overset{\overset{\displaystyle O}{\|}}{C}\underset{\underset{\displaystyle CH_3}{|}}{C}HCH_3$$

10.5 The two carboxylic acids with the molecular formula $C_4H_8O_2$ are

$$CH_3CH_2CH_2\overset{\overset{\displaystyle O}{\|}}{C}OH \quad CH_3\underset{\underset{\displaystyle CH_3}{|}}{C}H\overset{\overset{\displaystyle O}{\|}}{C}OH$$

10.7 For hydrogen, the number of valence electrons plus the number of bonds equals 2. For carbon, nitrogen, and oxygen, the number of valence electrons plus the number of bonds equals 8. Carbon, with four valence electrons, forms four bonds. Nitrogen, with five valence electrons, forms three bonds. Oxygen, with six valence electrons, forms two bonds.

10.9 Following are the Lewis structures for each compound:

(a)
```
    H       H
    |       |
  H-C-Ö-C-H
    |       |
    H       H
```

(b)
```
    H  H
    |  |
  H-C-C-H
    |  |
    H  H
```

(c)
```
    H      H
     \    /
      C=C
     /    \
    H      H
```

(d) $H-C\equiv C-H$

(e) $\ddot{\ddot{O}}=C=\ddot{\ddot{O}}$

(f)
```
        H
         \
          C=Ö
         /
        H
```

(g)
```
         :O:
         ‖
    H-Ö-C-Ö-H
```

(h)
```
       H  :O:
       |   ‖
    H-C-C-Ö-H
       |
       H
```

10.11 In stable organic compounds, carbon must have four covalent bonds to other atoms. In (a), carbon has five bonds to other atoms. In (b), one carbon has four bonds to other atoms, but the second carbon has five bonds to other atoms.

10.13 You would predict 120° for the H—N—H bond angles.

10.15 (a) 109.5° about C and O (b) 120° about C (c) 180° about C

10.17 (a) Predict 120° for all H—C—C and C—C—C bond angles. (b) planar

10.19 Living organisms use the substances they take in from their environment to produce organic compounds. They also use them to produce inorganic compounds—for example, the calcium-containing compounds of bones, teeth, and shells.

10.21 Carbon, which forms four bonds; hydrogen, which forms one bond; nitrogen, which forms three bonds and has one unshared pair of electrons; and oxygen, which forms two bonds and has two unshared pairs of electrons

10.23 A functional group is a part of an organic molecule that undergoes chemical reactions.

10.25

(a)
```
      :Ö:
      ‖
    —C—
```

(b)
```
      :Ö:
      ‖
    —C—Ö—H
```

(c) $-\ddot{\ddot{O}}-H$

(d)
```
    —N̈—H
      |
      H
```

10.27 (a) Incorrect. The carbon on the left has five bonds.
(b) Incorrect. The carbon bearing the Cl has five bonds.
(c) Correct.
(d) Incorrect. Oxygen has three bonds and one carbon has five bonds.
(e) Correct.
(f) Incorrect. Carbon on the right has five bonds.

10.29 The one tertiary alcohol with molecular formula $C_4H_{10}O$ is

$$CH_3\overset{\overset{\displaystyle CH_3}{|}}{\underset{\underset{\displaystyle CH_3}{|}}{C}}OH$$

10.31 The one tertiary amine with molecular formula $C_4H_{11}N$ is

$$CH_3CH_2\overset{\overset{\displaystyle CH_3}{|}}{N}CH_3$$

10.33 (a) a ketone (b) two carboxylic acids
(c) a primary amine, a secondary amine, and a carboxylic acid (d) two primary alcohols and one ketone

10.35 (a) The four primary alcohols with the molecular formula $C_5H_{12}O$ are

$$CH_3CH_2CH_2CH_2CH_2OH \quad CH_3\underset{\underset{\displaystyle CH_3}{|}}{C}HCH_2CH_2OH$$

CH₃CH₂CHCH₂OH CH₃CCH₂OH
        | |
        CH₃   CH₃
                       (CH₃ on top)

(b) The three secondary alcohols with the molecular formula C₅H₁₂O are

           OH   OH   OH
           | | |
CH₃CHCH₂CH₂CH₃   CH₃CH₂CHCH₂CH₃   CH₃CHCHCH₃
                                              |
                                              CH₃

(c) The tertiary alcohol with the molecular formula C₅H₁₂O is

         CH₃
         |
CH₃CH₂COH
         |
         CH₃

10.37 The eight carboxylic acids with the molecular formula C₆H₁₂O₂ are

CH₃CH₂CH₂CH₂CH₂COOH   CH₃CHCH₂CH₂COOH   CH₃CH₂CHCH₂COOH
                               | |
                               CH₃   CH₃

                    CH₃   CH₃
                    | |
CH₃CH₂CH₂CHCOOH   CH₃CCH₂COOH   CH₃CHCHCOOH
         | | |
         CH₃   CH₃   CH₃

    CH₃
    |
CH₃CH₂CCOOH   CH₃CH₂CHCOOH
    | |
    CH₃   CH₂CH₃

10.39 Taxol was discovered by a survey of indigenous plants sponsored by the National Cancer Institute with the goal of discovering new chemicals for fighting cancer.
10.41 The goal of combinatorial chemistry is to synthesize large numbers of closely related compounds in one reaction mixture.
10.43 Predict 109.5° for each C—Si—C bond angle.
10.45

(a) CH₃CH₂OH   (b) CH₃CH₂CH
                         ‖
                         O

(c) CH₃CCH₃   (d) CH₃CH₂COH
       ‖   ‖
       O   O

10.47 The three tertiary amines with the molecular formula C₅H₁₃N are

                         CH₃
                         |
CH₃NCH₂CH₂CH₃   CH₃NCHCH₃   CH₃CH₂NCH₂CH₃
    | | |
    CH₃   CH₃   CH₃

10.49 (a) O—H is the most polar bond. (b) C—C and C=C are the least polar bonds.

Chapter 11 Alkanes and Cycloalkanes

11.1 The alkane is octane, and its molecular formula is C₈H₁₈.
11.2 (a) constitutional isomers   (b) the same compound
11.3 The three constitutional isomers with the molecular formula C₅H₁₂ are

11.4 (a) 5-isopropyl-2-methyloctane, C₁₂H₂₆
(b) 4-isopropyl-4-propyloctane, C₁₄H₃₀
11.5 (a) isobutylcyclopentane, C₉H₁₈
(b) sec-butylcycloheptane, C₁₁H₂₂
(c) 1-ethyl-1-methylcyclopropane, C₆H₁₂
11.6 The structure with the three methyl groups equatorial is

H₃C — [cyclohexane with positions 4, 2, 1 labeled] — CH₃
                              CH₃

11.7 Cycloalkanes (a) and (c) show cis-trans isomerism.

(a)

cis-1,3-Dimethylcyclopentane

trans-1,3-Dimethylcyclopentane

(c)

cis-1,3-Dimethylcyclohexane

trans-1,3-Dimethylcyclohexane

11.8 In order of increasing boiling point, they are:
(a) 2,2-dimethylpropane (9.5°C), 2-methylbutane (27.8°C), pentane (36.1°C)
(b) 2,2,4-trimethylhexane, 3,3-dimethylheptane, nonane
11.9 The two products with the formula C₃H₇Cl are

CH₃CH₂CH₂Cl   CH₃CHClCH₃
1-Chloropropane   2-Chloropropane
(Propyl chloride)   (Isopropyl chloride)

11.11 The carbon chain of an alkane is not straight; it is bent with all C—C—C bond angles approximately 109.5°.

11.13 Line-angle formulas are

(a), (b), (c), (d), (e), (f) [line-angle structural formulas]

11.15 Statements (a) and (b) are true. Statements (c) and (d) are false.

11.17 Structures (a) and (g) represent the same compound. Structures (a, g), (c), (d), (e), and (f) represent constitutional isomers.

11.19 The structural formulas in parts (b), (c), (e), and (f) represent constitutional isomers.

11.21 (a) ethyl (b) isopropyl (c) isobutyl (d) *tert*-butyl

11.23 (a) 2-methylpentane (b) 2,5-dimethylhexane (c) 3-ethyloctane (d) 1-isopropyl-2-methylcyclohexane (e) isobutylcyclopentane (f) 1-ethyl-2,4-dimethylcyclohexane

11.25 A conformation is any three-dimensional arrangement of the atoms in a molecule that results from rotation about a single bond.

11.27 Here are representations of two conformations of ethane. In the staggered conformation, the hydrogen atoms on one carbon are as far apart as possible from the hydrogens on the other carbon. In the eclipsed conformation, they are as close as possible.

[Staggered conformation diagram] — Hydrogens are as far apart as possible

[Eclipsed conformation diagram] — Hydrogens are as close as possible

11.29 no

11.31 Structural formulas for the six cycloalkanes with the molecular formula C_5H_{10} are

Cyclopentane   Methylcyclobutane   1,1-Dimethylcyclopropane

trans-1,2-Dimethylcyclopropane   *cis*-1,2-Dimethylcyclopropane   Ethylcyclopropane

11.33 The first two representations of menthol show the cyclohexane ring as a planar hexagon. The third shows it as a chair conformation. Notice that in the chair conformation, all three substituents are equatorial.

11.35 The alternative representation is

[structure of furanose-like ring with OH groups]

11.37 Heptane, C_7H_{16}, has a boiling point of 98°C and a molecular weight of 100. Its molecular weight is approximately 5.5 times that of water. Although considerably smaller, water molecules associate in the liquid by relatively strong hydrogen bonding, whereas the much larger heptane molecules associate only by relatively weak London dispersion forces.

11.39 Alkanes are insoluble in water.

11.41 Boiling points of unbranched alkanes are related to their surface area; the larger the surface area, the higher the boiling point. The relative increase in size per CH_2 group is greatest between CH_4 and CH_3CH_3 and becomes progressively smaller as the molecular weight increases. Therefore, the increase in boiling point per added CH_2 group is greatest between CH_4 and CH_3CH_3, and becomes progressively smaller for higher alkanes.

11.43 Balanced equations for the complete combustion of each are

(a) $2CH_3(CH_2)_4CH_3 + 19O_2 \longrightarrow 12CO_2 + 14H_2O$
Hexane

(b) ⬡ + 9 O$_2$ ⟶ 6 CO$_2$ + 6 H$_2$O
   Cyclohexane

(c) 2 CH$_3$CH(CH$_3$)CH$_2$CH$_2$CH$_3$ + 19 O$_2$ ⟶ 12 CO$_2$ + 14 H$_2$O
   2-Methylpentane

11.45 Structural formulas for each part are
(a) CH$_3$Br
(b) ⬡—Cl
(c) BrCH$_2$CH$_2$Br
(d) CH$_3$C(CH$_3$)(Cl)CH$_3$
(e) CCl$_2$F$_2$

11.47 (a) Only one ring contains just carbon atoms.
(b) One ring contains two nitrogen atoms. (c) One ring contains two oxygen atoms.

11.49 It has been demonstrated that their presence in the stratosphere results in destruction of the stratospheric ozone layer.

11.51 An octane rating indicates the relative smoothness with which a gasoline blend burns in an automobile engine. The higher the octane rating, the less the engine knocks. The reference hydrocarbons are 2,2,4-trimethylpentane (isooctane), which is assigned an octane rating of 100, and heptane, which is assigned an octane rating of 0.

11.53 2,2,4-Trimethylpentane has a higher heat of combustion both per mole and per gram.

11.55 (a) The longest chain is pentane. Its IUPAC name is 2-methylpentane.
(b) The pentane chain is numbered incorrectly. Its IUPAC name is 2-methylpentane.
(c) The longest chain is pentane. Its IUPAC name is 3-ethyl-3-methylpentane.
(d) The longest chain is hexane. Its IUPAC name is 3,4-dimethylhexane.
(e) The longest chain is heptane. Its IUPAC name is 4-methylheptane.
(f) The longest chain is octane. Its IUPAC name is 3-ethyl-3-methyloctane.
(g) The ring is numbered incorrectly. Its IUPAC name is 1,1-dimethylcyclopropane.
(h) The ring is numbered incorrectly. Its IUPAC name is 1-ethyl-3-methylcyclohexane.

11.57 Tetradecane is a liquid at room temperature.

11.59 Water, a polar molecule, cannot penetrate the surface layer created by this nonpolar hydrocarbon.

Chapter 12 Alkenes and Alkynes

12.1 (a) 3,3-dimethyl-1-pentene (b) 2,3-dimethyl-2-butene
(c) 3,3-dimethyl-1-butyne

12.2 (a) *trans*-3,4-dimethyl-2-pentene
(b) *cis*-4-ethyl-3-heptene

12.3 (a) 1-isopropyl-4-methylcyclohexene (b) cyclooctene
(c) 4-*tert*-butylcyclohexene

12.4 Line-angle formulas for the other two dienes are:

cis,trans-2,4-Heptadiene   *cis,cis*-2,4-Heptadiene

12.5 Four cis-trans isomers are possible.

12.6
(a) CH$_3$CH(Br)CH$_3$ (b) ⬡(Br)(CH$_3$)

12.7 Propose a two-step mechanism similar to that proposed for the addition of HCl to propene.
Step 1: Reaction of H$^+$ with the carbon–carbon double bond gives a 3° carbocation intermediate.

⬡=CH$_3$ + H$^+$ ⟶ ⬡$^+$—CH$_3$
A 3° carbocation intermediate

Step 2: Reaction of the 3° carbocation intermediate with bromide ion completes the valence shell of carbon and gives the product.

⬡$^+$—CH$_3$ + :Br:$^-$ ⟶ ⬡(Br)(CH$_3$)

12.8 The product from each acid-catalyzed hydration is the same alcohol.

CH$_3$C(CH$_3$)(OH)CH$_2$CH$_3$

12.9 Propose a three-step mechanism similar to that for the acid-catalyzed hydration of propene.
Step 1: Reaction of the carbon–carbon double bond with H$^+$ gives a 3° carbocation intermediate.

⬡=CH$_3$ + H$^+$ ⟶ ⬡$^+$—CH$_3$
A 3° carbocation intermediate

Step 2: Reaction of the 3° carbocation intermediate with water completes the valence shell of carbon and gives an oxonium ion.

⬡$^+$—CH$_3$ + :Ö—H ⟶ ⬡(CH$_3$)(Ö$^+$(H)—H)
                    H
An oxonium ion

Step 3: Loss of H⁺ from the oxonium ion completes the reaction and generates a new H⁺ catalyst.

12.10 (a) $CH_3-\underset{\underset{H_3C}{|}}{\overset{\overset{CH_3}{|}}{C}}-\underset{\underset{Br}{|}}{CH}-\underset{\underset{Br}{|}}{CH_2}$

(b)

12.11 A saturated hydrocarbon contains only carbon–carbon single bonds. An unsaturated hydrocarbon contains one or more carbon–carbon double or triple bonds.

12.13

(a)   (b)

(c) $HC\equiv C-CH=CH_2$ (180°, 120°)   (d) (120°)

12.15 Line-angle formulas for each compound are

(a), (b), (c), (d), (e), (f)

12.17 There are six parts to this problem. The solutions are:
(a) 1-heptene   (b) 1,4,4-trimethylcyclopentene
(c) 1,3-dimethylcyclohexene   (d) 2,4-dimethyl-2-pentene
(e) 1-octyne   (f) 2,2-dimethyl-3-hexyne

12.19 (a) The longest chain is four carbons. The correct name is 2-butene.
(b) The chain is not numbered correctly. The correct name is 2-pentene.
(c) The ring is not numbered correctly. The correct name is 1-methylcyclohexene.
(d) The double bond must be located. The correct name is 3,3-dimethyl-1-pentene.
(e) The chain is not numbered correctly. The correct name is 2-hexyne.
(f) The longest chain is five carbon atoms. The correct name is 3,4-dimethyl-2-pentene.

12.21 Parts (b), (c), and (e) show cis-trans isomerism. Following are line-angle formulas for the cis and trans isomers of each.

(b) cis-2-Hexene, trans-2-Hexene
(c) cis-3-Hexene, trans-3-Hexene
(e) cis-3-Methyl-2-hexene, trans-3-Methyl-2-hexene

12.23 There are six alkenes of molecular formula C_5H_{10}.

1-Pentene, 2-Methyl-1-butene, 3-Methyl-1-butene, cis-2-Pentene, trans-2-Pentene, 2-Methyl-2-butene

12.25 Line-angle formulas for these three unsaturated fatty acids are:

Oleic acid

Linoleic acid

Linolenic acid

12.27

Arachidonic acid

12.29 Only compounds (b) and (d) show cis-trans isomerism.

(b)

(d)

12.31 The structural formula of β-ocimene is drawn on the left showing all atoms and on the right as a line-angle formula.

$$CH_2=CH-\underset{3}{\underset{|}{C}}=\underset{H}{\underset{|}{C}}-CH_2CH=CCH_3$$ with labels 1,2,3,6,7 and CH₃ groups

12.33 The four isoprene units in vitamin A are shown in bold.

12.35 (a) HBr (b) H₂O/H₂SO₄ (c) HI (d) Br₂

12.37

(a) $CH_3CH_2\overset{CH_3}{\underset{+}{C}}-CH_2CH_3$ and $CH_3CH_2\overset{CH_3}{\underset{+}{C}H}-CHCH_3$

    Tertiary   Secondary

(b) $CH_3CH_2\overset{+}{C}H-CH_2CH_3$ and $CH_3CH_2CH_2-\overset{+}{C}HCH_3$

    Secondary   Secondary

(c) cyclopentyl–CH₃ (Tertiary, + on ring carbon) and cyclopentyl–CH₃ (Secondary)

(d) cyclohexyl–CH₃ (Tertiary, + on ring) and cyclohexyl–CH₂⁺ (Primary)

12.39

(a) 1-methyl-1,2-dibromocyclohexane (CH₃, Br, Br)

(b) 1-methyl-1,2-dichlorocyclopentane (CH₃, Cl, Cl, CH₃)

12.41

(a) $CH_3\overset{CH_3}{\underset{|}{C}}=CHCH_3$ (b) $CH_2=\overset{CH_3}{\underset{|}{C}}CH_2CH_3$

(c) $CH_2=CHCH_2CH_2CH_3$

12.43

(a) $CH_3CH_2CH=CHCH_2CH_3$

(b) cyclobutene with CH₃ substituent or methylenecyclobutane (CH₂)

(c) $CH_2=\overset{CH_3}{\underset{|}{C}}CH_2CH_3$ or $CH_3\overset{CH_3}{\underset{|}{C}}=CHCH_3$

(d) $CH_3CH=CH_2$

12.45

(a) $CH_3CH_2CH_2CH_2CH_3$ (b) $CH_3CH_2CH_2CH_2CH_3$

(c) cyclopentane (b) methylcyclopentane

12.47 Reagents are shown over each arrow.

12.49 Ethylene is a natural ripening agent for fruits.

12.51 Its molecular formula is $C_{16}H_{32}O_2$, its molecular weight is 254 amu, and its molar mass is 254 g/mol.

12.53 Rods are primarily responsible for black-and-white vision. Cones are responsible for color vision.

12.55 The most common consumer items made of high-density polyethylene (HDPE) are milk and water jugs, grocery bags, and squeeze bottles. The most common consumer items made of low-density polyethylene (LDPE) are packaging for baked goods, vegetables, and other produce, as well as trash bags. Currently only HDPE materials are recyclable.

12.57 There are five compounds with the molecular formula C_4H_8. All are constitutional isomers. The only cis-trans isomers are *cis*-2-butene and *trans*-2-butene.

Cyclobutane   Methyl-cyclopropane   1-Butene   *cis*-2-Butene   *trans*-2-Butene

12.59 (a) There are five alkenes with this carbon skeleton.

2-Methyl-1-pentene   2-Methyl-2-pentene   *trans*-4-Methyl-2-pentene

cis-4-Methyl-2-pentene   4-Methyl-1-pentene

(b) There are two alkenes with this carbon skeleton.

2,3-Dimethyl-1-butene   2,3-Dimethyl-2-butene

(c) There is one alkene with this carbon skeleton.

3,3-Dimethyl-1-butene

12.61 The carbon skeletons of lycopene and β-carotene are almost identical. The difference is that, at each end, the second carbon of β-carotene is bonded to the seventh carbon to form the two six-membered rings of lycopene.

12.63 Each alkene hydration reaction follows Markovnikov's rule. In (a), —H adds preferentially to carbon-1 and —OH to carbon-2 to give 2-hexanol. In (b), each carbon of the double bond has the same pattern of substitution, so 2-hexanol and 3-hexanol are formed in approximately equal amounts. In (c), each carbon of the double bond again has the same pattern of substitution, but no matter which way H—OH adds, the only product possible is 3-hexanol.

(a) 2-Hexanol

(b) 2-Hexanol + 3-Hexanol

(c) 3-Hexanol

12.65 (a) cyclohexene

(b) 1-methylcyclohexene or methylenecyclohexane

(c) 1,2-dimethylcyclopentene

(d) 2-methyl-1-butene or 2-methyl-2-butene

Chapter 13 Benzene and Its Derivatives

13.1 (a) 2,4,6-tri-*tert*-butylphenol (b) 2,6-dichloroaniline (c) 3-nitrobenzoic acid

13.2 A saturated compound contains only single covalent bonds. An unsaturated compound contains one or more double or triple bonds. The most common double bonds are C=C, C=O, and C=N. The only triple bond is C≡C.

13.3 An aromatic compound is one that contains one or more benzene rings.

13.5 Yes, they have double bonds, at least in the contributing structures we normally use to represent them. Yes, they are unsaturated, because they have fewer hydrogens than a cycloalkane with the same number of carbons.

13.7 (a) An alkene of six carbons has the molecular formula C_6H_{12} and contains one carbon–carbon double bond. Three examples are

1-Hexene, trans-3-Hexene, cis-3-Hexene

(b) A cycloalkene of six carbons has the molecular formula C_6H_{10} and contains one ring and one carbon–carbon double bond. Three examples are

Cyclohexene, 4-Methylcyclopentene, 1-Methylcyclopentene

(c) An alkyne of six carbons has the molecular formula C_6H_{10} and contains one carbon–carbon triple bond. Three examples are

1-Hexyne, 2-Hexyne, 4-Methyl-2-pentyne

(d) An aromatic hydrocarbon of eight carbons has the molecular formula C_8H_{10} and contains one benzene ring. Three examples are

Ethylbenzene, 1,3-Dimethylbenzene (*m*-Xylene), 1,4-Dimethylbenzene (*p*-Xylene)

13.9 Benzene consists of carbons, each surrounded by three regions of electron density, which gives 120° for all bond angles. Bond angles of 120° in benzene can be maintained only if the molecule is planar. Cyclohexane, on the other hand, consists of carbons, each surrounded by four regions of electron density, which gives 109.5° for all bond angles. Angles of 109.5° in cyclohexane can be maintained only if the molecule is nonplanar.

13.11 Neither a unicorn, which has a horn like a rhinoceros, nor a dragon, which has a tough, leathery hide like a rhinoceros, exists. If they did and you made a hybrid of them, you would have a rhinoceros. Furthermore, a rhinoceros is not a dragon part of the time and a unicorn the rest of the time; a rhinoceros is a rhinoceros all of the time. To carry this analogy to aromatic compounds, resonance-contributing structures for them do not exist; they are imaginary. If they did exist and you made a hybrid of them, you would have the real aromatic compound.

13.13

(a) benzene with Br, Cl, and CH_2CH_3 substituents

(b) benzene with two CH_3 groups and Br

(c) benzene with CH_3, two NO_2 groups (from O_2N) and NO_2

(d) benzene with a CH_3-bearing chain with a terminal alkene

(e) 4-methylphenol structure with OH and CH₃

(f) 2,4-dichlorophenol structure

13.15 Only cyclohexene will react with a solution of bromine in dichloromethane. A solution of Br_2/CH_2Cl_2 is red, whereas a dibromocycloalkane is colorless. To tell which bottle contains which compound, place a small quantity of each compound in a test tube and to each add a few drops of Br_2/CH_2Cl_2 solution. If the red color disappears, the compound is cyclohexene. If it remains, the compound is benzene.

13.17

2-Bromotoluene (o-Bromotoluene)   3-Bromotoluene (m-Bromotoluene)   4-Bromotoluene (p-Bromotoluene)

13.19 (a) Nitration using HNO_3/H_2SO_4, followed by sulfonation using H_2SO_4. The order of the steps may also be reversed.
(b) Bromination using $Br_2/FeCl_3$, followed by chlorination using $Cl_2/FeCl_3$. The order of the steps may also be reversed.

13.21 Phenol is a sufficiently strong acid that reacts with strong bases such as sodium hydroxide to form sodium phenoxide, a water-soluble salt. Cyclohexanol has no comparable acidity and does not react with sodium hydroxide.

13.23 *Radical* indicates a molecule or ion with an unpaired electron. *Chain* means a cycle of two or more steps, called propagation steps, that repeat over and over. The net effect of a radical chain reaction is the conversion of starting materials to products. *Chain length* refers to the number of times the cycle of chain propagation steps repeats.

13.25 Vitamin E participates in one or the other of the chain propagation steps and forms a stable radical, which breaks the cycle of propagation steps.

13.27 The abbreviation DDT is derived from **D**ichloro**D**iphenyl**T**richloroethane.

13.29 Insoluble in water. It is entirely a hydrophobic molecule.

13.31 Biodegradable means that a substance can be broken down into one or more harmless compounds by living organisms in the environment.

13.33 Iodine is an element that is found primarily in sea water and, therefore, seafoods are rich sources of it. Individuals in inland areas where seafood is only a limited part of the diet are the most susceptible to developing goiter.

13.35 One of the aromatic rings in Allura Red has a —CH_3 group and an —OCH_3 group. These groups are not present in Sunset Yellow.

13.37 orange

13.39 Capsaicin is isolated from the fruit of various species of *Capsicum,* otherwise known as chili peppers.

13.41 Two cis-trans isomers are possible for capsaicin.

13.43 Following are the three contributing structures for naphthalene.

(1)   (2)   (3)

13.45 BHT participates in one of the chain propagation steps of autoxidation, forms a stable radical, and thus terminates autoxidation.

13.47

phenyl-CHBr-CH₂Br structure

Chapter 14 Alcohols, Ethers, and Thiols

14.1 (a) 2-heptanol (b) 2,2-dimethyl-1-propanol
(c) *trans*-3-isopropylcyclohexanol

14.2 (a) primary (b) secondary (c) primary
(d) tertiary

14.3 In each case, the major product (circled) contains the more substituted double bond.

(a) [CH₃C=CHCH₃ with CH₃] + CH₂=CCH₂CH₃ with CH₃

(b) [cyclopentene with CH₃] + methylenecyclopentane

14.4

2-Methyl cyclohexanol →(H_2SO_4)→ 1-Methyl cyclohexene (C) →(H_2O, H_2SO_4)→ 1-Methyl cyclohexanol (D)

14.5 Each secondary alcohol is oxidized to a ketone.

(a) cyclohexanone (b) $CH_3CCH_2CH_2CH_3$ (with C=O)

14.6 (a) ethyl isobutyl ether (b) cyclopentyl methyl ether

14.7 (a) 3-methyl-1-butanethiol (b) 3-methyl-2-butanethiol

14.9 Only (c) and (d) are secondary alcohols.

14.11 (a) 1-pentanol (b) 1,3-propanediol
(c) 1,2-butanediol (d) 3-methyl-1-butanol
(e) *cis*-1,2-cyclohexanediol (f) 2,6-dimethylcyclohexanol

14.13

(a) isobutanol structure   (b) HO-CH₂CH₂CH₂CH₂-OH

(c) HO-CH₂CH₂CH₂CH(CH₃)CH₂CH₃ structure (6-methyl-1-hexanol type) (d) HO-(CH₂)₅-OH
(e) trans-1,4-cyclohexanediol (HO⋯⋯ ⬡ ⋯OH) (f) CH₃CH(OH)CH₂Cl

14.15 (a) ethyl isopropyl ether (b) dibutyl ether
(c) diphenyl ether
14.17 (a) *sec*-butyl mercaptan (b) butyl mercaptan
(c) cyclohexyl mercaptan
14.19 Prednisone contains three ketones, one primary alcohol, one tertiary alcohol, one disubstituted carbon–carbon double bond, and one trisubstituted carbon–carbon double bond. Estradiol contains one secondary alcohol and one disubstituted phenol.
14.21 Low-molecular-weight alcohols form hydrogen bonds with water molecules through both the oxygen and hydrogen atoms of their —OH groups. Low-molecular-weight ethers form hydrogen bonds with water molecules only through the oxygen atom of their —O— group. The greater extent of hydrogen bonding between alcohol and water molecules makes the low-molecular-weight alcohols more soluble in water than the low-molecular-weight ethers.
14.23 Both types of hydrogen bonding are shown on the same illustration.

$$CH_3-\overset{\delta-}{O}\cdots H-\overset{\delta+}{}\quad \overset{\delta+}{H}-\overset{\delta-}{O}-CH_3$$
(illustration showing hydrogen bonding between CH₃OH, H₂O, and CH₃OCH₃)

14.25 In order of increasing boiling point, they are

$$\underset{-42°C}{CH_3CH_2CH_3} \quad \underset{78°C}{CH_3CH_2OH}$$

$$\underset{117°C}{CH_3CH_2CH_2CH_2OH} \quad \underset{198°C}{HOCH_2CH_2OH}$$

14.27 Because 1-butanol molecules associate in the liquid state by hydrogen bonding, it has the higher boiling point (117°C). There is very little polarity to an S—H bond. The only interactions among 1-butanethiol molecules in the liquid state are the considerably weaker London dispersion forces. For this reason, 1-butanethiol has the lower boiling point (98°C).
14.29 Evaporation of a liquid from the surface of the skin cools because heat is absorbed from the skin in converting molecules from the liquid state to the gaseous state. 2-Propanol (isopropyl alcohol), which has a boiling point of 82°C, absorbs heat from the skin, evaporates rapidly, and has a cooling effect. 2-Hexanol, which has a boiling point of 140°C, also absorbs heat from the surface of the skin but, because of its higher boiling point, evaporates much more slowly and does not have the same cooling effect as 2-propanol.
14.31 The more water-soluble compound is circled.
(a) $\boxed{CH_3OH}$ or CH_3OCH_3

(b) $\boxed{CH_3CH(OH)CH_3}$ or $CH_3C(=CH_2)CH_3$

(c) $CH_3CH_2CH_2SH$ or $\boxed{CH_3CH_2CH_2OH}$

14.33 For three parts, two constitutional isomers will give the desired alcohol. For two parts, only one alkene will give the desired alcohol.
(a) $CH_2=CHCH_2CH_3$ or $CH_3CH=CHCH_3$
(b) methylcyclohexene (with CH₃ on ring double bond) or methylenecyclohexane (=CH₂)
(c) $CH_3CH_2CH=CHCH_2CH_3$
(d) $CH_2=C(CH_3)CH_2CH_2CH_3$ or $CH_3C(CH_3)=CHCH_2CH_3$
(e) cyclopentene

14.35 Phenols are weak acids, with pK_a values approximately equal to 10. Alcohols, considerably weaker acids, have about the same acidity as water.
14.37 The first reaction is an acid-catalyzed dehydration; the second is an oxidation.
(a) $CH_3CH_2CH_2CH_2OH \xrightarrow[\text{heat}]{H_2SO_4} CH_3CH_2CH=CH_2 + H_2O$

(b) $CH_3CH_2CH_2CH_2OH \xrightarrow[H_2SO_4]{K_2Cr_2O_7} CH_3CH_2CH_2COOH$

14.39 Oxidation of a primary alcohol by $K_2Cr_2O_7/H_2SO_4$ gives a carboxylic acid.
(a) $CH_3(CH_2)_6COOH$ (b) $HOOCCH_2CH_2COOH$

14.41 Each can be prepared from 1-propanol (circled) as shown in this flow chart.

(flow chart showing conversions from CH₃CH₂CH₂OH:
(a) CH₃CH₂CHO via K₂Cr₂O₇/H₂SO₄
(b) CH₃CH₂COOH via K₂Cr₂O₇/H₂SO₄
(c) CH₃CH=CH₂ via H₂SO₄
(d) CH₃CHOHCH₃ via H₂O/H₂SO₄
(e) CH₃COCH₃ via K₂Cr₂O₇/H₂SO₄ from OH
(f) CH₃CH₂CH₃ via H₂/M
(g) CH₃CHBrCH₃ via HBr
(h) CH₃CHBrCH₂Br via Br₂)

14.43 Ethanol and ethylene glycol are derived from ethylene. Ethanol is a solvent and is the starting material for the synthesis of diethyl ether, also an important solvent. Ethylene glycol is used in automotive antifreezes and is one of the two starting materials required for the synthesis of poly(ethylene terephthalate), better known as PET (Section 18.8B).
14.45 (a) The three functional groups are a thiol, a primary amine, and a carboxylic acid. (b) Oxidation of the thiol group gives a disulfide (—S—S—).

(a)

```
      O                    O
      ‖                    ‖
HOCCHCH₂S—SCH₂CHCOH
      |                    |
      NH₂                NH₂
```

14.47 Nitroglycerin was discovered in 1847. It is a pale yellow, oily liquid.

14.49 One of the products the body derives from metabolism of nitroglycerin is nitric oxide, which causes the coronary artery to dilate, thus relieving angina.

14.51 The relationship is that 2100 mL of breath contains the same amount of ethanol as 1.00 mL of blood.

14.53 Diethyl ether is easy to use and causes excellent muscle relaxation. Blood pressure, pulse rate, and respiration are usually only slightly affected. Diethyl ether's chief drawbacks are its irritating effect on the respiratory passages and its after-effect of nausea.

14.55 Enflurane and isoflurane are insoluble in water but soluble in hexane.

14.57 $2CH_3OH + 3O_2 \longrightarrow 2CO_2 + 4H_2O$

14.59 The eight isomeric alcohols with the molecular formula C_5H_{12} are

1-Pentanol, 2-Pentanol, 3-Pentanol, 2-Methyl-1-butanol, 2-Methyl-2-butanol, 3-Methyl-2-butanol, 3-Methyl-1-butanol, 2,2-Dimethyl-1-propanol

14.61 Ethylene glycol has two —OH groups by which each molecule participates in hydrogen bonding, whereas 1-propanol has only one. The stronger intermolecular forces of attraction between molecules of ethylene glycol give it the higher boiling point.

14.63 Arranged in order of increasing solubility in water, they are

CH₃CH₂CH₂CH₂CH₂CH₃   CH₃CH₂CH₂CH₂CH₂OH
Hexane   1-Pentanol
(Insoluble)   (2.3 g/mL water)

HOCH₂CH₂CH₂CH₂OH
1,4-Butanediol
(Infinitely soluble)

14.65 Each can be prepared from 2-methyl-1-propanol (circled) as shown in this flow chart.

Chapter 15 Chirality

15.1 The enantiomers of each part are drawn with two groups in the plane of the paper, a third group toward you in front of the plane, and the fourth group away from you behind the plane.

15.2 The group of higher priority in each set is circled.

(a) ⎡—CH₂OH⎤ and —CH₂CH₂COOH

(b) ⎡—CH₂NH₂⎤ and —CH₂CH₂COOH

15.3 The order of priorities and the configuration of each is shown in the drawings.

15.4 (a) Compounds 1 and 3 are one pair of enantiomers. Compounds 2 and 4 are a second pair of enantiomers. (b) Compounds 1 and 2, 1 and 4, 2 and 3, and 3 and 4 are diastereomers.

15.5 Four stereoisomers are possible for 3-methylcyclohexanol. The cis isomer is one pair of enantiomers; the trans isomer is a second pair of enantiomers.

15.6 Stereocenters are marked by an asterisk, and the number of stereoisomers possible is shown under the structural formula.

(a) $2^1 = 2$

(b) $CH_2=CHCHCH_2CH_3$ $2^1 = 2$

(c) $2^2 = 4$

15.7 Chirality is a property of an object. The object is not superposable on its mirror image; that is, the object has handedness. 2-Butanol is a chiral molecule.

15.9 Stereoisomers are isomers that have the same molecular formula and the same connectivity, but a different orientation of their atoms in space. Three examples are cis-trans isomers (in cycloalkanes and alkenes), enantiomers, and diastereomers.

15.11 (a) Chiral. (b) Achiral. (c) Achiral. (d) Achiral. (e) Chiral, unless you are on the equator, in which case it goes straight down and has no chirality.

15.13 Neither cis-2-butene nor trans-2-butene is chiral. Each is superposable on its mirror image.

15.15 Compounds (a), (c), and (d) contain stereocenters, here marked by asterisks.

(a) CH$_3$CHCH$_2$CH$_2$CH$_3$ with Cl (starred)
(b) CH$_2$=CHCHCH$_3$ with Cl (starred)
(c) CH$_2$CHCH$_3$ with Cl, Cl (starred)

15.17 The stereocenter in each chiral compound is marked by an asterisk.

(a), (b), (c), (d), (e), (f) structures shown.

15.19 Following are mirror images of each.

(a) H$_2$N—C—H / H—C—NH$_2$ with COOH / HOOC and CH$_3$

(b) cyclohexane structures with H and OH

15.21 Parts (b) and (c) contain stereocenters.

(b) HO—*—OH structure
(c) OH—*— with vinyl

15.23 Each stereocenter is marked by an asterisk. Under each structural formula is the number of stereoisomers possible. Compound (b) has no stereocenter.

(a) cyclopentane with OH, two * ($2^2 = 4$)
(c) tetrahydrofuran with OH, * ($2^1 = 2$)
(d) decalin structure with O, two * ($2^2 = 4$)

15.25 Molecules (b) and (d) have R configurations. Molecules (a) and (c) have S configurations.

(a) Br(1), CH$_3$, H, CH$_2$OH(2), S
(b) CH$_3$, H, HOCH$_2$(2), Br(1), R
(c) CH$_2$OH(2), H$_3$C, H, Br(1), S
(d) R—CH$_2$OH(2), H, Br(1), CH$_3$

15.27 The optical rotation of its enantiomer is +41°.

15.29 To say that a drug is *chiral* means that it has handedness—that it has one or more stereocenters and the possibility for two or more stereoisomers. Just because a compound is chiral does not mean that it will be optically active. It may be chiral and present as a racemic mixture, in which case it will have no effect on the plane of polarized light. If, however, it is present as a single enantiomer, it will rotate the plane of polarized light.

15.31 The two stereocenters are marked by asterisks.

OH with two * structure

3-Methyl-2-pentanol

15.33 Each stereocenter is marked by an asterisk. Under the name of each compound is the number of stereoisomers possible for it.

(a) Fluoxetine (Prozac) ($2^1 = 2$)

(b) Sertraline (Zoloft) ($2^2 = 4$)

(c) Paroxetine (Paxil) ($2^2 = 4$)

15.35 (a) In this chair conformation of glucose, carbons 1, 2, 3, 4, and 5 are stereocenters. (b) There are $2^5 = 32$ stereoisomers possible. (c) Because enantiomers always occur in pairs, there are 16 pairs of enantiomers possible.

15.37 (a) The eight stereocenters are marked by asterisks. (b) There are are $2^8 = 256$ stereoisomers possible.

Triamcinolone acetonide

15.39 The majority have a right-handed twist because the machines that make them all impart the same twist.

Chapter 16 Amines

16.1 Pyrollidine has nine hydrogens; its molecular formula is C_4H_9N. Purine has four hydrogens; its molecular formula is $C_5H_4N_4$.

16.2 Following is a line-angle formula for each compound.

16.3 Following is a line-angle formula for each compound.

16.4 The stronger base is circled.

16.5 The product of each reaction is an amine salt.
(a) $(CH_3CH_2)_3\overset{+}{N}H\ Cl^-$

(b)

16.7 In an aliphatic amine, all carbon groups bonded to nitrogen are alkyl groups. In an aromatic amine, one or more of the carbon groups bonded to nitrogen are aryl (aromatic) groups.

16.9 Following is a structural formula for each amine.

(a)

(b) $CH_3(CH_2)_6CH_2NH_2$

(c)

(d) $H_2N(CH_2)_5NH_2$

(e)

(f) $(CH_3CH_2CH_2CH_2)_3N$

16.11 Each amine is classified by type.

(a) 5-hydroxytryptamine structure with labels:
- NH₂ group: 1° aliphatic amine
- Indole N–H: Heterocyclic aromatic amine

(b) Ethyl 4-aminobenzoate (benzocaine) structure with label:
- H₂N–: 1° aromatic amine

(c) Structure with labels:
- N(CH₃)₂ group: 3° aliphatic amine

16.13 There are four primary amines of this molecular formula, three secondary amines, and one tertiary amine. Only 2-butanamine is chiral.

1° amines:

- 1-Butanamine (Butylamine): CH₃CH₂CH₂CH₂NH₂
- 2-Butanamine (sec-Butylamine): CH₃CH₂CH*(NH₂)CH₃
- 2-Methyl-1-propanamine (Isobutylamine): (CH₃)₂CHCH₂NH₂
- 1,1-Dimethylethanamine (tert-Butylamine): (CH₃)₃CNH₂

2° amines:

- Methylpropylamine: CH₃CH₂CH₂NHCH₃
- Methylisopropylamine: (CH₃)₂CHNHCH₃
- Diethylamine: CH₃CH₂NHCH₂CH₃

3° amines:

- Dimethylethylamine: CH₃CH₂N(CH₃)₂

16.15 Both propylamine (a 1° amine) and ethylmethylamine (a 2° amine) have an N—H group and show hydrogen bonding between their molecules in the liquid state. Because of this intermolecular force of attraction, these two amines have higher boiling points than trimethylamine, which has no N—H bond and, therefore, cannot participate in intermolecular hydrogen bonding.

16.17 2-Methylpropane is a nonpolar hydrocarbon and the only attractive forces between its molecules in the liquid state are the very weak London dispersion forces. Both 2-propanol and 2-propanamine are polar molecules and associate in the liquid state by hydrogen bonding. Hydrogen bonding is stronger between alcohol molecules than between amine molecules because of the greater strength of an O—H----O hydrogen bond compared to an N—H----N hydrogen bond. It takes more energy (a higher temperature) to separate an alcohol molecule in the liquid state from its neighbors than to separate an amine molecule from its neighbors and, therefore, the alcohol has the higher boiling point.

16.19 Nitrogen is less electronegative than oxygen and, therefore, more willing to donate its unshared pair of electrons to H⁺ in an acid–base reaction to form a salt.

16.21 (a) ethylammonium chloride
(b) diethylammonium chloride
(c) anilinium hydrogen sulfate

16.23 Structural formula A contains both an acid (the carboxyl group) and a base (the 1° amino group). The acid–base reaction between them gives structural formula B, which is the better representation of alanine.

16.25

(a) CH_3COH + pyridine → pyridinium + CH_3CO^-

(b) 1-phenyl-2-propanamine + HCl → 1-phenyl-2-propanaminium chloride ($NH_3^+Cl^-$)

(c) N-methyl-1-phenyl-2-propanamine + H_2SO_4 → corresponding ammonium hydrogen sulfate (HSO_4^-)

16.27 (a) The primary aliphatic amine is the stronger base and forms the salt with HCl. The salt is named pyridoxamine hydrochloride.

Pyridoxamine hydrochloride structure: pyridine ring with HO–, $CH_2NH_3^+$ Cl^-, CH_2OH, and H_3C– substituents.

16.29 Albuterol contains one 1° alcohol, one 2° alcohol, one phenol, and one 2° amine. Albuterol differs from epinephrine in that one phenolic —OH group of epinephrine is converted to a —CH$_2$OH group, and the N-methyl group of epinephrine is converted to a *tert*-butyl group.
16.31 Possible negative effects are long periods of sleeplessness, loss of weight, and paranoia.
16.33 Both coniine and nicotine have one stereocenter; two stereoisomers (one pair of enantiomers) are possible for each.
16.35 The four stereocenters of cocaine are marked with asterisks. Following is the structural formula of the salt formed by the reaction of cocaine with HCl.

[Structure of cocaine·HCl salt with four stereocenters marked by asterisks, showing H$_3$C—N$^+$—H, Cl$^−$, COCH$_3$, and OCC$_6$H$_5$ groups]

16.37 Neither Librium nor Valium is chiral.
16.39 No. No unreacted HCl is present.
16.41

(a) C$_6$H$_5$—NHCH$_3$ (b) C$_6$H$_5$—N(CH$_3$)$_2$

(c) C$_6$H$_5$—CH$_2$NH$_2$ (d) CH$_3$CH(NH$_2$)CH$_2$CH$_3$

(e) N-methylpyrrolidine   (f) 2,6-dimethyl-4-methylaniline (H$_3$C— —NH$_2$ with CH$_3$ groups)

(g) CH$_3$N$^+$(CH$_2$CH$_3$)(CH(CH$_3$)$_2$)(CH$_2$CH$_2$CH$_3$) Cl$^−$

16.43 (a) CH$_3$SH is the stronger acid.
(b) (CH$_3$)$_2$NH is the stronger base.
(c) CH$_3$OH has the highest boiling point.
(d) Molecules of CH$_3$OH form the strongest hydrogen bonds.
16.45 Both alcohols and amines can interact with water by hydrogen bonding. Because amines and alcohols of the same molecular weight have about the same solubility in water, the strength of hydrogen-bonding between their molecules must be comparable.

16.47 (a) Following is its structural formula. (b) It has two stereocenters, marked with asterisks, and $2^2 = 4$ stereoisomers are possible.

C$_6$H$_5$—*CH(OH)*CH(NH$_2$)CH$_3$ + HCl ⟶ C$_6$H$_5$—CH(OH)CH(NH$_3^+$)CH$_3$ Cl$^−$

16.49 Gabapentin is better represented by structural formula A, the internal salt. Structural formula B contains both an acid (—COOH) and a base (—NH$_2$) which will react by an internal proton-transfer reaction to form A.
16.51 The 2° aliphatic amine is more basic than the heterocyclic aromatic amine. The three stereocenters are marked by asterisks.

Epibatidine [structure with Cl-substituted pyridine and bicyclic amine; arrow indicates "This nitrogen is more basic"]

Chapter 17 Aldehydes and Ketones

17.1 (a) 3,3-dimethylbutanal (b) cyclopentanone
(c) (S)-2-phenylpropanal
17.2 Following are line-angle formulas for each aldehyde with the molecular formula C$_6$H$_{12}$O. In the three that are chiral, the stereocenter is marked by an asterisk.

Hexanal   4-Methylpentanal

3-Methylpentanal   2-Methylpentanal

2,3-Dimethylbutanal   3,3-Dimethylbutanal

2,2-Dimethylbutanal

17.3 (a) 2,3-dihydroxypropanal (b) 2-aminobenzaldehyde
(c) 5-amino-2-pentanone
17.4 Each aldehyde is oxidized to a carboxylic acid.

(a) HOOC—(CH$_2$)$_4$—COOH
Hexanedioic acid
(Adipic acid)

(b) 3-Phenylpropanoic acid

17.5 Each primary alcohol comes from reduction of an aldehyde. Each secondary alcohol comes from reduction of a ketone.

(a) cyclohexanone
(b) CH₃O–C₆H₄–CH₂CHO (4-methoxyphenylacetaldehyde)
(c) hexane-2,5-dione

7.6 Shown first is the hemiacetal and then the acetal.

(Hemiacetal: PhCH(OH)(OCH₃)) → CH₃OH → (Acetal: PhCH(OCH₃)₂)

17.7 (a) A hemiacetal formed from 3-pentanone, a ketone, and ethanol.
(b) Neither a hemiacetal nor an acetal. This compound is the dimethyl ether of ethylene glycol.
(c) An acetal derived from 5-hydroxypentanal and methanol.

17.8 Following is the keto form of each enol.

(a) cyclohexanone
(b) 2-hydroxycyclohexanone
(c) cyclohexanecarbaldehyde

17.9 Oxidation of a primary alcohol gives either an aldehyde or a carboxylic acid, depending on the experimental conditions. Oxidation of a secondary alcohol gives a ketone.

(a) cyclooctanone
(b) cyclopentanecarbaldehyde or cyclopentanecarboxylic acid

17.11 The carbonyl carbon of an aldehyde is bonded to at least one hydrogen. The carbonyl carbon of a ketone is bonded to two carbon groups.

17.13 No. To be a carbon stereocenter, the carbon atom must have four different groups bonded to it. The carbon atom of a carbonyl group has only three groups bonded to it.

17.15 (a) Cortisone contains three ketones, one 3° alcohol, one 1° alcohol, and one carbon–carbon double bond.
(b) Aldosterone contains two ketones, one aldehyde, one 1° alcohol, one 2° alcohol, and one carbon–carbon double bond.

17.17 Following are structural formulas for each aldehyde.

(a) HCHO
(b) CH₃CH₂CHO
(c) 4-methylhexanal (structure shown)
(d) CH₃(CH₂)₈CHO
(e) HO–C₆H₄–C(=O)–H
(f) HOCH₂CH(OH)CHO

17.19 (a) 4-heptanone (b) 2-methylcyclopentanone
(c) cis-2-methyl-2-butenal (d) (S)-2-hydroxypropanal
(e) 1-phenyl-2-propanone (f) hexanedial

17.21 (a) The chain is not numbered correctly. Its name is 2-butanone.
(b) The compound is an aldehyde. Its name is butanal.
(c) The longest chain is five carbons. Its name is pentanal.
(d) The location of the ketone takes precedence. Its name is 3,3-dimethyl-2-butanone.

17.23 (a) ethanol (b) 3-pentanone (c) butanal (d) 2-butanol

17.25 Acetone has the higher boiling point because of the intermolecular attraction between the carbonyl groups of acetone molecules.

17.27 Acetaldehyde forms hydrogen bonds with water primarily through its carbonyl oxygen.

(diagram of acetaldehyde hydrogen bonding with water, showing $\delta+$ on C and H, $\delta-$ on O)

17.29 Aldehydes are oxidized by this reagent to carboxylic acids. Ketones are not oxidized under these conditions. Secondary alcohols are oxidized to ketones.

(a) CH₃CH₂CH₂COOH
(b) C₆H₅COOH
(c) no reaction
(d) cyclohexanone

17.31 (a) Treat each with Tollens' reagent. Only pentanal will give a silver mirror.
(b) Treat each with $K_2Cr_2O_7/H_2SO_4$. Only 2-pentanol is oxidized (to 2-pentanone), which causes the red color of $Cr_2O_7^{2-}$ ion to disappear and be replaced by the green color of Cr^{3+} ion.

17.33 The white solid is benzoic acid, formed by air oxidation of benzaldehyde.

17.35 These experimental conditions reduce an aldehyde to a primary alcohol and a ketone to a secondary alcohol.

(a) $CH_3\overset{\underset{|}{OH}}{C}HCH_2CH_3$ (b) $CH_3(CH_2)_4CH_2OH$

(c) cyclopentane with OH and CH₃ substituents (d) benzene ring with CH₂OH and OH substituents

17.37 (a) Following is its structural formula.

$HOCH_2\overset{\overset{O}{\|}}{C}CH_2OH + H_2 \xrightarrow{\text{Metal catalyst}} HOCH_2\overset{\underset{|}{OH}}{C}HCH_2OH$

1,3-Dihydroxy-2-propanone (Dihydroxyacetone) → 1,2,3-Propanetriol (Glycerol, glycerin)

(b) Because it has two hydroxyl groups and one carbonyl group, all of which can interact with water molecules by hydrogen bonding, predict that it is soluble in water.
(c) Its reduction gives 1,2,3-propanetriol, better known as glycerol or glycerin.

17.39 The first two reactions are reduction. There is no reaction in (c) or (d); ketones are not oxidized by these reagents.

(a, b) phenyl-CHCH₃ with OH

17.41 (a) an acetal (b) a hemiacetal (c) an acetal (d) an acetal (e) an acetal (f) neither

17.43 Following are equations for the formation of each hemiacetal and acetal.

(a) $CH_3CH_2\overset{\overset{O}{\|}}{C}-H \xrightarrow{CH_3OH} CH_3CH_2\overset{\underset{|}{OH}}{\underset{\underset{|}{OCH_3}}{C}}-H$

$\xrightarrow{CH_3OH} CH_3CH_2\overset{\underset{|}{OCH_3}}{\underset{\underset{|}{OCH_3}}{C}}-H + H_2O$

(b) cyclopentanone $\xrightarrow{CH_3OH}$ cyclopentane with OH and OCH₃ $\xrightarrow{CH_3OH}$ cyclopentane with two OCH₃ $+ H_2O$

17.45 *Hydration* refers to the addition of one or more molecules of water to a substance. An example of hydration is the acid-catalyzed addition of water to propene to give 2-propanol. *Hydrolysis* refers to the reaction of a substance with water, generally with the breaking (lysis) of one or more bonds in the substance. An example of hydrolysis is the acid-catalyzed reaction of an acetal with a molecule of water to give an aldehyde or ketone and two molecules of alcohol.

17.47 Compounds (a), (b), (d), and (f) undergo keto-enol tautomerism.

17.49 Following are the keto forms of each enol.

(a) cyclopentanone (b) $CH_3\overset{\overset{O}{\|}}{C}CH_2CH_2CH_2CH_3$

(c) phenyl—$CH_2\overset{\overset{O}{\|}}{C}CH_3$

17.51 Compounds (a), (b), and (c) can be formed by reduction of the aldehyde or ketone shown. Compound (d) is a 3° alcohol and cannot be formed in this manner.

(a) $CH_3\overset{\overset{O}{\|}}{C}CH_3$ (b) phenyl—$\overset{\overset{O}{\|}}{C}-H$

(c) $H-\overset{\overset{O}{\|}}{C}-H$

17.53 The alkene of three carbons is propene. Acid-catalyzed hydration of this alkene gives 2-propanol as the major product, not 1-propanol.

17.55 Each conversion can be brought about by acid-catalyzed hydration of the alkene to a secondary alcohol, followed by oxidation of the secondary alcohol to a ketone.

(a) $CH_2=CHCH_2CH_2CH_3 \xrightarrow[H_2SO_4]{H_2O}$

$CH_3\overset{\underset{|}{OH}}{C}HCH_2CH_2CH_3 \xrightarrow[H_2SO_4]{K_2Cr_2O_7} CH_3\overset{\overset{O}{\|}}{C}CH_2CH_2CH_3$

(b) cyclohexene $\xrightarrow[H_2SO_4]{H_2O}$ cyclohexanol $\xrightarrow[H_2SO_4]{K_2Cr_2O_7}$ cyclohexanone

17.57 The aldehyde and ketone functional groups are circled.

(a) $H\overset{\overset{O}{\|}}{C}CH_2CH_2CH_2\overset{\overset{O}{\|}}{C}CH_3$ (b) cyclohexane with =O and —CH=O

(c) $HOCH_2\overset{\underset{|}{HO}}{C}H\overset{\overset{O}{\|}}{C}H$ (d) bicyclic structure with ketone

(e) phenyl—$\overset{\overset{O}{\|}}{C}CH_2CH_3$ (f) HO—benzene with CH₃O and CH=O

17.59 Each compound reduces a carbonyl group of an aldehyde or ketone to an alcohol by delivering a hydride ion, H:⁻, to the carbonyl carbon.

17.61 Formulas for this one ketone and two aldehydes of molecular formula C_4H_8O are

(a) $CH_3\overset{O}{\overset{\|}{C}}CH_2CH_3$

(b) $CH_3CH_2CH_2\overset{O}{\overset{\|}{C}}H$ and $CH_3\overset{O}{\overset{\|}{C}}HCH$
                                                            |
                                                           CH_3

17.63 2-Propanol has the higher boiling point because of the greater attraction between its molecules due to hydrogen bonding through its hydroxyl group.

17.65 (a) Treat each compound with Tollens' reagent. Only benzaldehyde reduces Ag⁺ to give a precipitate of silver metal as a silver mirror.
(b) Treat each compound with Tollens' reagent as in part (a). Only acetaldehyde reduces Ag⁺ to give a precipitate of silver metal as a silver mirror.

17.67 Shown in the equation are structural formulas for the equilibrium products.

```
   HC=O              HC—OH            CH₂OH
   |                 ‖                |
   CH—OH    ⇌       C—OH     ⇌       C=O
   |                 |                |
   CH₃               CH₃              CH₃
An α-hydroxyaldehyde  An enediol   An α-hydroxyketone
```

17.69 (a) Carbon 5 provides the —OH group, and carbon 1 provides the —CHO group. (b) Following is a structural formula for the free aldehyde.

[Structural formula of open-chain sugar with numbered carbons 1–6, showing CH₂OH at C6, OH groups and H atoms]

Chapter 18   Carboxylic Acids, Anhydrides, Esters, and Amides

18.1 (a) 2,3-dihydroxypropanoic acid
(b) 3-aminopropanoic acid
(c) 3,5-dihydroxy-3-methylpentanoic acid

18.2 Each acid is converted to its ammonium salt.

(a) $CH_3(CH_2)_2COOH + NH_3 \longrightarrow CH_3(CH_2)_2COO^- NH_4^+$
   Butanoic acid   Ammonium butanoate
   (Butyric acid)   (Ammonium butyrate)

(b) $CH_3\underset{\underset{OH}{|}}{C}HCOOH + NH_3 \longrightarrow CH_3\underset{\underset{OH}{|}}{C}HCOO^- NH_4^+$
   2-Hydroxypropanoic   Ammonium 2-hydroxypropanoate
   acid   (Ammonium lactate)
   (Lactic acid)

18.3 (a) ethyl benzoate (b) phenyl acetate

18.4 Following is a structural formula for each amide.

(a) $CH_3\overset{O}{\overset{\|}{C}}NH$—⟨cyclohexyl⟩

(b) ⟨phenyl⟩—$\overset{O}{\overset{\|}{C}}NH_2$

8.5 Following is the structural formula of each ester.

(a) [γ-butyrolactone structure]

(b) [cyclohexyl isobutyrate structure]

18.6 Under basic conditions, as in part (a), each carboxyl group is present as a carboxylic anion. Under acidic conditions, as in part (b), each carboxyl group is present in its un-ionized form.

(a) [ortho-diacetoxybenzene + 2NaOH] $\xrightarrow{H_2O}$ [disodium phthalate-like structure with two $CO^- Na^+$ groups] + $2CH_3OH$

(b) [ketoester structure] + H_2O

[then $\xrightarrow{HCl}$ ketoacid + HO—ethyl]

18.7 In aqueous NaOH, each carboxyl group is present as a carboxylic anion, and each amine is present in its unprotonated form.

(a) $CH_3\overset{O}{\overset{\|}{C}}N(CH_3)_2 + NaOH \xrightarrow[\text{heat}]{H_2O} CH_3\overset{O}{\overset{\|}{C}}O^- Na^+ + (CH_3)_2NH$

(b) [piperidinone (δ-valerolactam)] + NaOH $\xrightarrow[\text{heat}]{H_2O}$ H_2N—[chain]—$\overset{O}{\overset{\|}{C}}O^- Na^+$

18.9 (a) 3,4-dimethylpentanoic acid
(b) 2-aminobutanoic acid (c) hexanoic acid

18.11 Following are structural formulas for each carboxylic acid.

(a) O_2N—⟨para-substituted benzene⟩—CH_2COOH

(b) H$_2$N‒‒‒COOH

(c) C$_6$H$_5$‒CH$_2$CH$_2$CH$_2$‒COOH

(d) cis-HOOC‒CH=CH‒COOH

18.13 Following are structural formulas for each salt.

(a) C$_6$H$_5$‒CO$^-$ Na$^+$ (with C=O)

(b) CH$_3$CO$^-$ Li$^+$ (with C=O)

(c) CH$_3$CO$^-$ NH$_4^+$ (with C=O)

(d) Na$^+$ $^-$OC(CH$_2$)$_4$CO$^-$ Na$^+$

(e) 2-hydroxybenzoate: ortho-HO-C$_6$H$_4$-CO$^-$ Na$^+$

(f) (CH$_3$CH$_2$CH$_2$CO$^-$)$_2$Ca^{2+}

18.15 One of the carboxyl groups in this salt is present as —COO$^-$, the other as —COOH.

HOC(=O)‒C(=O)O$^-$ K$^+$

18.17 If you draw this molecule correctly to show the internal hydrogen bonding, you will see that the hydrogen-bonded part of the molecule forms a six-membered ring.

[two structures showing six-membered ring H-bonding with Hδ+ and Oδ−]

18.19 In order of increasing boiling point, they are heptanal (bp 153°C), 1-heptanol (bp 176°C), and heptanoic acid (bp 223°C).

18.21 In order of increasing solubility in water, they are decanoic acid, pentanoic acid, and acetic acid.

18.23 Following are structural formulas for the indicated starting material.

(a) CH$_3$(CH$_2$)$_4$CH$_2$OH
    C$_6$H$_{14}$O

(b) CH$_3$(CH$_2$)$_4$CHO
    C$_6$H$_{12}$O

(c) HOCH$_2$(CH$_2$)$_4$CH$_2$OH
    C$_6$H$_{14}$O$_2$

18.25 In order of increasing acidity, they are benzyl alcohol, phenol, and benzoic acid.

18.27 Following are completed equations for these acid–base reactions.

(a) 2-methylphenol + NaOH ⟶ 2-methylphenoxide Na$^+$ + H$_2$O

(b) sodium salicylate + HCl ⟶ salicylic acid + NaCl

(c) methyl salicylate (2-methoxybenzoic acid) + H$_2$$\overset{+}{\text{N}}CH_2CH_2$OH ⟶ 2-methoxybenzoate$^-$ + H$_3$$\overset{+}{\text{N}}CH_2CH_2$OH

(d) cyclohexyl-COOH + NaHCO$_3$ ⟶ cyclohexyl-COO$^-$ Na$^+$ + H$_2$O + CO$_2$

18.29 CH$_3$COOH at pH 2.0, equal amounts of CH$_3$COOH and CH$_3$COO$^-$ at pH 4.76, and CH$_3$COO$^-$ at pH 8.0 or higher

18.31 The pK_a of ascorbic acid is 4.10. At this pH, ascorbic acid is present 50 percent as ascorbic acid and 50 percent as ascorbate ion. At pH 7.35 to 7.45, which is more basic than pH 4.10, ascorbic acid would be present as ascorbate ion.

18.33 In part (a), the —COOH group is a stronger acid than the —NH$_3^+$ group.

(a) CH$_3$CH(NH$_3^+$)COOH + NaOH ⟶ CH$_3$CH(NH$_3^+$)COO$^-$ + H$_2$O

(b) CH$_3$CH(NH$_3^+$)COO$^-$Na$^+$ + NaOH ⟶ CH$_3$CH(NH$_2$)COO$^-$Na$^+$ + H$_2$O

18.35 In part (a), the amine is the stronger base.

(a) CH$_3$CH(NH$_2$)COO$^-$Na$^+$ + HCl ⟶ CH$_3$CH(NH$_3^+$)COO$^-$Na$^+$ + NaCl

(b) CH$_3$CH(NH$_3^+$)COO$^-$Na$^+$ + HCl ⟶ CH$_3$CH(NH$_3^+$)COOH + NaCl

18.37 Following is a structural formula for the ester formed in each reaction.

(a) CH$_3$C(=O)O‒CH$_2$CH(CH$_3$)$_2$ (isobutyl acetate)

(b) CH$_3$CO‒O‒cyclohexyl

(c) diethyl phthalate: ortho-C$_6$H$_4$(COOCH$_2$CH$_3$)$_2$

18.39 Following is a structural formula for acid and alcohol from which each ester is derived.

(a) 2CH$_3$COOH + HO—⌬—OH

(b) ⌬—COOH + CH$_3$OH

(c) 2CH$_3$OH + HOOCCH$_2$CH$_2$COOH

(d) ⌇COOH + HO—CH(CH$_3$)$_2$

18.41 Following is a structural formula for methyl 4-hydroxybenzoate.

HO—⌬—COOCH$_3$

18.43 Following are structural formulas for the reagents to synthesize each amide.

(a) ⌬—NH$_2$ + HOC(=O)(CH$_2$)$_4$CH$_3$

(b) (CH$_3$)$_2$CHCOOH + HN(CH$_3$)$_2$

(c) 2NH$_3$ + HOC(=O)(CH$_2$)$_4$C(=O)OH

18.45 (a) Both lidocaine and mepivacaine contain an amide and a 3° amine. (b) Both are derived from 2,6-dimethylaniline. In addition, the aliphatic amine nitrogen in each is separated by one carbon from the carbonyl group of the amide.

18.47 Following is a structural formula for each synthetic flavoring agent.

(a) HCOO-ethyl

(b) isoamyl acetate

(c) octyl acetate

(d) propanoate ester

(e) butanoate ester

(f) methyl anthranilate (with NH$_2$)

18.49 Saponification is the hydrolysis of an ester using aqueous NaOH or KOH to give the sodium or potassium salt of the carboxylic acid and an alcohol. Following is a balanced equation for the saponification of methyl acetate.

CH$_3$COCH$_3$ + NaOH $\xrightarrow{H_2O}$ CH$_3$CO$^-$Na$^+$ + CH$_3$OH

Methyl acetate   Sodium acetate   Methanol

18.51 Each reaction brings about hydrolysis of the amide bond. Each product is shown as it would exist under the specified reaction conditions.

(a) Ph—CNH$_2$ + NaOH $\xrightarrow{H_2O}$ Ph—CO$^-$Na$^+$ + NH$_3$

(b) Ph—CNH$_2$ + HCl $\xrightarrow{H_2O}$ Ph—COH + NH$_4^+$Cl$^-$

18.53 The products in each part are an amide and acetic acid.

(a) CH$_3$O—⌬—NH$_2$ + CH$_3$COCCH$_3$ → CH$_3$O—⌬—NH—CCH$_3$ + CH$_3$COH

(b) ⌬NH + CH$_3$COCCH$_3$ → ⌬N—CCH$_3$ + CH$_3$COH

18.55 (a) Phenobarbital contains four amide groups. (b) Complete hydrolysis of all amide bonds gives a dicarboxylic acid dianion, two moles of ammonia, and one mole of sodium carbonate.

phenobarbital $\xrightarrow[H_2O]{NaOH}$ dianion + 2NH$_3$ + Na$^+$ $^-$OCO$^-$ Na$^+$

18.57 In nylon-66 and Kevlar, the monomer units are joined by amide bonds.

18.59 In Dacron and Mylar, the monomer units are joined by ester bonds.

18.61 Following are structural formulas for the mono-, di-, and triethyl esters.

$$HO-\overset{\overset{O}{\|}}{\underset{OH}{P}}-OCH_2CH_3 \qquad HO-\overset{\overset{O}{\|}}{\underset{OCH_2CH_3}{P}}-OCH_2CH_3$$

Ethyl phosphate   Diethyl phosphate

$$CH_3CH_2O-\overset{\overset{O}{\|}}{\underset{OCH_2CH_3}{P}}-OCH_2CH_3$$

Triethyl phosphate

18.63 Two molecules of water are split out.

$$HO-\overset{\overset{O}{\|}}{\underset{OH}{P}}-OH + HO-\overset{\overset{O}{\|}}{\underset{OH}{P}}-OH + HO-\overset{\overset{O}{\|}}{\underset{OH}{P}}-OH$$

$$\longrightarrow HO-\overset{\overset{O}{\|}}{\underset{OH}{P}}-O-\overset{\overset{O}{\|}}{\underset{OH}{P}}-O-\overset{\overset{O}{\|}}{\underset{OH}{P}}-OH + 2H_2O$$

18.65 The arrow points to the ester group. Below is chrysanthemic acid.

Pyrethrin I

Chrysanthemic acid

18.67 (a) The *cis/trans ratio* refers to the cis-trans relationship between the ester group and the carbon–carbon double bond in the three-membered ring. (b) Permethrin has three stereocenters, and eight stereoisomers (four pairs of enantiomers) are possible for it. The designation "(+/−)" refers to the fact that the members of each pair of possible enantiomers are present in equal amounts; that is, each pair of enantiomers is present as a racemic mixture.

18.69 The compound is salicin. Removal of the glucose unit and oxidation of the primary alcohol to a carboxylic acid gives salicylic acid.

18.71 The moisture present in humid air may be sufficient to bring about hydrolysis of the ester to yield salicylic acid and acetic acid. The vinegar-like odor is due to the presence of acetic acid.

18.73 A *sunblock* prevents all ultraviolet radiation from reaching protected skin by reflecting it away from the skin. A *sunscreen* prevents a portion of the ultraviolet radiation from reaching protected skin. Its effectiveness is related to its skin protection factor (SPF).

18.75 They all contain an ester bonded to an alkyl group as well as a benzene ring. The benzene has either a nitrogen atom or an oxygen atom on it.

18.77 Lactomer stitches dissolve as the ester groups in the polymer chain are hydrolyzed until only glycolic acid and lactic acid remain. These small molecules are metabolized and excreted by existing biochemical pathways.

18.79 Propanoic acid has a boiling point of 141°C, and methyl acetate has a boiling point of 57°C. The boiling point of propanoic acid is higher because of the association of adjacent molecules by hydrogen bonding.

18.81 In a solution of pH 4.07, lactic acid is 50 percent in the un-ionized (lactic acid) form and 50 percent in the ionized (lactate anion) form. In gastric juice of pH 1.0 to 3.0, which is more acidic than pH 4.10, lactic acid is present primarily as un-ionized lactic acid molecules.

18.83 Following is an equation for this synthesis of acetaminophen.

$$CH_3\overset{\overset{O}{\|}}{C}O\overset{\overset{O}{\|}}{C}CH_3 + H_2N-\underset{}{\bigcirc}-OH \longrightarrow$$

Acetic anhydride   4-Aminophenol

$$CH_3\overset{\overset{O}{\|}}{C}NH-\underset{}{\bigcirc}-OH + CH_3\overset{\overset{O}{\|}}{C}OH$$

Acetaminophen

18.85 Hydrolysis gives one mole of 2,3-dihydroxypropanoic acid and two moles of phosphoric acid. At pH 7.35–7.45, the carboxyl group is present as its anion, and phosphoric acid is present as its dianion.

$$\begin{array}{c} \overset{O}{\|} \\ CO^- \\ | \\ CHOH \\ | \\ CH_2OH \end{array} + 2HO-\overset{\overset{O}{\|}}{\underset{O^-}{P}}-O^-$$

Chapter 19   Carbohydrates

19.1 Following are Fischer projections for the four 2-ketopentoses. They consist of two pairs of enantiomers.

19.2 D-Mannose differs in configuration from D-glucose only at carbon 2. One way to arrive at the structures of the α and β forms of D-mannopyranose is to draw the corresponding α and β forms of D-glucopyranose, and then invert the configuration in each at carbon 2.

19.3 D-Mannose differs in configuration from D-glucose only at carbon 2.

19.4 Following is a Haworth projection and a chair conformation for this glycoside.

19.5 The β-glycosidic bond is between carbon 1 of the left unit and carbon 3 of the right unit.

19.7 Carbohydrates make up about three-fourths of the dry weight of plants. Less than 1 percent of the body weight of humans is made up of carbohydrates.

19.9 The most abundant D-aldohexose in the biological world is glucose.

19.11 D-glucose

19.13 The designations D and L refer to the configuration of the stereocenter farthest from the aldehyde or ketone group on a monosaccharide chain. When the monosaccharide chain is drawn as a Fischer projection, a D-monosaccharide has the —OH on this carbon on the right; an L monosaccharide has it on the left.

19.15 In D-glucose, carbons 2, 3, 4, and 5 are stereocenters. In D-ribose, carbons 2, 3, and 4 are stereocenters. D-Glucose is one of 16 possible stereoisomers; D-ribose is one of eight possible stereoisomers.

19.17 First draw the D form of each, and then its mirror image.

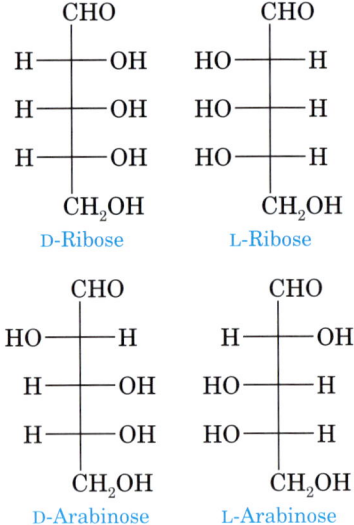

19.19 Each carbon of a monosaccharide and a disaccharide has an oxygen that is able to participate in hydrogen bonding with water molecules.

19.21 An anomeric carbon is the hemiacetal or acetal carbon in the cyclic form of a monosaccharide. Said another way, it is the carbon that was the carbonyl carbon in the open-chain form of the monosaccharide.

19.23 The designation β means that the —OH group on the anomeric carbon of a cyclic hemiacetal is on the same side of the ring as the terminal —CH₂OH group. The designation α means that it is on the opposite side of the ring from the terminal —CH₂OH group.

19.25 No. The hydroxyl groups on carbon 2, 3, and 4 of α-D-glucose are equatorial, but the hydroxyl group on carbon 1 is axial.

19.27 First compare each Haworth projection with that of β-D-glucose. Compound (a) differs in configuration at carbon 3. Compound (b) differs in configuration at carbons 2 and 3.

(a) [Haworth and Fischer projections leading to D-Allose]

(b) [Haworth and Fischer projections leading to D-Altrose]

19.29 During mutarotation, an α or β form of a carbohydrate is converted to an equilibrium mixture of the two. Mutarotation can be detected by observing the change in optical activity over time as the two forms equilibrate.

19.31 The specific rotation of α-L-glucose changes to −52.7°.

19.33 A glycosidic bond is the bond to the —OR group of the cyclic acetal form of any monosaccharide. A glucosidic bond is the bond to the —OR group of a cyclic acetal form of glucose.

19.35 Each aldehyde group is reduced by $NaBH_4$ to a primary alcohol.

(a) [Fischer projection] (b) [Fischer projection]

19.37 In reduction of the ketone group of D-fructose, the —OH group at carbon 2 may be either on the right or the left in the Fischer projection.

D-Fructose $\xrightarrow{NaBH_4}$ D-Mannitol + D-Sorbitol

19.39 Review Section 17.8 on keto-enol tautomerism and your answer to Problem 17.67. The intermediate in this conversion is an enediol; that is, it contains a carbon–carbon double bond with two OH groups on it.

D-Glucose 6-phosphate ⇌ An enediol intermediate ⇌ D-Fructose 6-phosphate

19.41 Sucrose contains one unit of D-glucose and one unit of D-fructose. Lactose contains one unit of D-galactose and one unit of D-glucose. Maltose contains two units of D-glucose.

19.43 The anomeric carbons of both monosaccharide units in sucrose are involved in forming the glycosidic bond. In maltose and lactose, one of the monosaccharide units is in the cyclic hemiacetal form, which is in equilibrium with the open-chain form and, therefore, undergoes oxidation.

19.45 (a) Both monosaccharides are units of D-glucose. (b) The glycosidic bond can be described as α-1,1- or β-1,1- depending on which D-glucose unit you take as the reference. (c) Trehalose is not a reducing sugar, because both anomeric carbons participate in forming the glycosidic bond. (d) Trehalose will not undergo mutarotation for the reason given in (c).

19.47 Cellulose, starch, and glycogen are all composed of units of D-glucose. The glycosidic bonds are alpha in starch and glycogen, and beta in cellulose.

19.49 Glycogen is stored in roughly equal amounts in the liver and muscle tissue.

19.51 Within their digestive systems, cows have enzymes that catalyze the hydrolysis of the glycosidic bonds of cellulose. We do not have these enzymes in our digestive systems.

19.53 One way to construct each repeating disaccharide is to draw a disaccharide of two units of D-glucose, and then modify each glucose unit to make it the appropriate unit in alginic acid or pectic acid.

(a) [Structure with β-1,4-Glycosidic bond]

Units of D-mannuronic acid

19.55 (a) The negative charges are provided by —OSO_3^- and —COO^- groups. (b) The higher the degree of polymerization, the better the anticoagulant activity.
19.57 The difference is two hydrogens. L-Ascorbic acid is the reduced form of this vitamin; L-dehydroascorbic acid is its oxidized form. The L indicates that each has the L configuration at carbon 5.
19.59 It is used to monitor blood glucose levels in diabetics and pre-diabetics.
19.61 (a) L-Fucose is an L-aldohexose. (b) What is unusual about it in human biochemistry is that it belongs to the L series of monosaccharides, and it has no oxygen atom at carbon 6. (c) If its terminal —CH_3 group were converted to a —CH_2OH group, the monosaccharide formed would be L-galactose.
19.63 High-fructose corn syrup, as its name suggests, is produced from corn syrup by partial enzyme-catalyzed hydrolysis of cornstarch to D-glucose and then partial enzyme-catalyzed isomerization of D-glucose to D-fructose.
19.65 During boiling in water with an acid catalyst, some of the sucrose is hydrolyzed to glucose and fructose, and fructose is sweeter than sucrose.
19.67 (a) The left unit is D-glucuronic acid, which is derived from D-glucose. The right unit is a sulfate ester derived from N-acetyl-D-galactosamine, which is in turn derived from D-galactosamine. (b) The two units are joined by a β-1,3-glycosidic bond.

Chapter 20 Lipids

20.1 It is an ester of glycerol and contains a phosphate group; therefore it is a glycerophospholipid. Besides glycerol and phosphate it has a myristic acid and a linoleic acid component. The other alcohol is serine. Therefore, it belongs to the subgroup of cephalins.
20.3 *Hydrophobic* means water hating. It is important because if the body did not have such molecules there could be no structure since the water would dissolve everything.
20.5 The melting point would increase. This happens because trans double bonds would fit more in the packing of the long hydrophobic tails, creating more order and therefore more interaction between chains. This would require more energy to disrupt, and hence a higher melting point.
20.7 highest: A and B (no double bonds); lowest C and D (two double bonds in each diglyceride)
20.9 (b) because its molecular weight is higher
20.11 lowest (c); then (b); highest (a)
20.13 the more long chain groups, the lower the solubility; lowest (a); then (b); highest (c)

20.15 glycerol, sodium palmitate, sodium stearate, and sodium linolenate
20.17 (a) the unsaturated fatty acid (b) inositol and the phosphate group
20.19 complex lipids and cholesterol
20.21 Phosphatidyl inositol; the inositol has five —OH groups, which can form H-bonds with water.
20.23 (c) because it has three charges, while the others have only two
20.25 No. For example, in red blood cells lecithins are on the outside, facing the plasma; cephalins are on the inside.
20.27 Cholesterol contributes to the fluidity of the membrane because it interrupts the alignments of the saturated fatty acids similar to the cis isomers of the unsaturated fatty acids. Thus there is less London force interaction between the fatty acid tails, contributing to more liquid-like behavior.
20.29 (a) Carbon stereocenters are circled.

(b) $2^8 = 256$ (c) only 1
20.31 It contains polar groups of phospholipids, cholesterol, and proteins.
20.33 It blocks cholesterol synthesis, thus helping to remove cholesterol from the blood.
20.35 HDL delivers its content by selective lipid uptake that does not involve endocytosis and degradation of the lipoprotein. The lipoprotein never enters the cell but returns to the circulation.
20.37 It increases the glucose and glycogen concentrations in the body. It is also an anti-inflammatory agent.
20.39 Loss of a methyl group at the junction of rings A and B, together with a hydrogen at position 1, produces a double bond. This causes tautomerization of the keto group at C-3, resulting in an aromatic ring A, with a phenolic group at the C-3 position.
20.41 (a) They both have a steroid ring structure.
(b) RU486 has a *para*-aminophenyl group on ring C and a triple-bond group on ring D.
20.43 —OH groups on the A, B, C rings of steroid and the sulfate on taurine.
20.45 (a) PGE_2 has a ring, a ketone group, and two OH groups. (b) The C=O group on PGE_2 is an OH group on $PGF_{2\alpha}$.
20.47 COX-2 catalyzes the ring closure between C-8 and C-12 in prostaglandin synthesis. COX-2 produces prostaglandin that take part in inflammation.
20.49 (a) oxidation of the double bond to aldehydes and other compounds
(b) exclude oxygen and sunlight; keep refrigerated
20.51 Soaps are salts of carboxylic acids; detergents are salts of sulfonic acids.
20.53 The transporter is a helical transmembrane protein. The hydrophobic groups on the helices are turned outward and interact with the membrane. The hydrophilic groups of the helices are on the inside and interact with the hydrated chloride ions.

20.55 (a) Sphingomyelin acts as an insulator.
(b) The insulator is degraded, impairing nerve conduction.
20.57 α-D-galactose, β-D-glucose, β-D-glucose
20.59 They prevent ovulation.
20.61 It inhibits prostaglandin formation by preventing ring closure.
20.63 NSAIDS inhibit cyclooxygenases (COX enzymes) that are needed for ring closure. Leukotrienes have no ring in their structure; therefore they are not affected by COX inhibitors.
20.65 (See Figure 20.2.) Polar molecules cannot penetrate the bilayer. They are insoluble in lipids. Nonpolar molecules can interact with the interior of the bilayer (like dissolves like).
20.67 Both groups are derived from a common precursor, PGH_2, which is catalyzed by the COX enzymes.
20.69 Coated pits are concentrations of LDL receptors on the surface of cells. They bind LDL and by endocytosis transfer it inside the cell.
20.71 They are both salts of sulfonic acids.
20.73 apoB-100 protein
20.75 Celebrex inhibits only COX-2 enzymes that cause inflammation, but not COX-1, which is needed to manufacture prostaglandins needed for normal physiological functions.
20.77 Both have the steroid ring structure and the same side chain. Ring A contains one more double bond in prednisolone than in cortisone. In ring C, the carbonyl group of cortisone is reduced to —OH.

Chapter 21 Proteins

21.1

$$H_3\overset{+}{N}-CH-\underset{\underset{O^-}{\overset{\overset{O}{\|}}{C}}}{}+H_3\overset{+}{N}-CH-\underset{\underset{O^-}{\overset{\overset{O}{\|}}{C}}}{}$$
$$\underset{CH_3}{\overset{CH-CH_3}{|}} \qquad \underset{}{\overset{CH_2}{|}}$$

Valine (Val)   Phenylalanine (Phe)

↓

$$H_3\overset{+}{N}-CH-\overset{\overset{O}{\|}}{C}-NH-CH-\overset{\overset{O}{\|}}{C}-O^- + H_2O$$
$$\underset{CH_3}{\overset{CH-CH_3}{|}} \qquad \overset{CH_2}{|}$$

Valylphenylalanine (Val-Phe)

21.2 salt bridge
21.3 (a) storage (b) movement
21.5 Tyrosine has an extra —OH group in the side chain. The terminal group of the side chain of tyrosine is a phenol; in phenylalanine, it is a benzene.
21.7 arginine

21.9

pyrrolidines (saturated pyrroles)

21.11 They supply most of the amino acids we need in our bodies.
21.13 Their structures are the same, except that a hydrogen of alanine is replaced by a phenyl group in phenylalanine.
21.15 These Fischer projections show the relationships between an L-amino acid and an L-monosaccharide and between a D-amino acid and a D-monosaccharide.

L-Valine   D-Valine

L-Glyceraldehyde   D-Glyceraldehyde

21.17

$$CH_3-\underset{\underset{NH_3^+}{|}}{CH}-COO^- + H_3O^+ \longrightarrow CH_3-\underset{\underset{NH_3^+}{|}}{CH}-COOH + H_2O$$

$$CH_3-\underset{\underset{NH_3^+}{|}}{CH}-COO^- + OH^- \longrightarrow CH_3-\underset{\underset{NH_2}{|}}{CH}-COO^- + H_2O$$

21.19

at pH1 $CH_3-\underset{\underset{NH_3^+}{|}}{CH}-\underset{\underset{}{}}{\overset{\overset{CH_3}{|}}{CH}}-COOH$

at pH12 $CH_3-\underset{\underset{NH_2}{|}}{CH}-\underset{\underset{}{}}{\overset{\overset{CH_3}{|}}{CH}}-COO^-$

21.21 There are six. One of them is

$$H_3\overset{+}{N}-CH-\overset{\overset{O}{\|}}{C}-NH-CH-\overset{\overset{O}{\|}}{C}-NH-CH-COO^-$$
with side chains:
CH—OH, CH_3 / CH_2CH_2CH_2NH—C(=NH_2^+)—NH_2 / CH_2CH_2—S—CH_3

21.23

$H_3\overset{+}{N}-CH-\underset{\underset{CH_3-CH}{|}}{C}-N$ (pyrrolidine ring with COO⁻)
with CH₃-CH-CH₃ side chain

Leu-Pro

21.25

$H_3\overset{+}{N}-CH(CH_3)-C(=O)-NH-CH-C(=O)-O^-$
side chain: $-CH_2-CH_2-C(=O)-NH_2$

Alanylglutamine (Ala-Gln)

$H_3\overset{+}{N}-CH-C(=O)-NH-CH(CH_3)-C(=O)-O^-$
side chain: $-CH_2-CH_2-C(=O)-NH_2$

Glutaminylalanine (Gln-Ala)

21.27 at pH 2.0

$\overset{+}{NH_3}-CH-C(=O)-NH-CH-C(=O)-NH-CH-COOH$
side chains: $-CH_2-CH_2-S-CH_3$; $-CH_2-OH$; $-CH_2-SH$

at pH 7.0

$\overset{+}{NH_3}-CH-C(=O)-NH-CH-C(=O)-NH-CH-COO^-$
side chains: $-CH_2-CH_2-S-CH_3$; $-CH_2-OH$; $-CH_2-SH$

at pH 10.0

$NH_2-CH-C(=O)-NH-CH-C(=O)-NH-CH-COO^-$
side chains: $-CH_2-CH_2-S-CH_3$; $-CH_2-OH$; $-CH_2-SH$

21.29 It would acquire a net positive charge and become more water-soluble.

21.31 (a) 256 (b) $20^4 = 160\,000$

21.33 valine or isoleucine

21.35 (a) quaternary (b) quaternary (c) quaternary (d) primary

21.37 Above pH 6.0 the COOH groups are converted to COO⁻ groups. The negative charges repel each other, disrupting the compact α-helix and converting it to a random coil.

21.39 (1) C-terminal end (2) N-terminal end (3) pleated sheet (4) random coil (5) hydrophobic interaction (6) disulfide bridge (7) α-helix (8) salt bridge (9) hydrogen bonds

21.41 the intramolecular hydrogen bonds between C=O and —N—H

21.43 on the outside interacting with the lipid bilayer

21.45 The heme and the polypeptide chain form the quaternary structure of cytochrome c. This is a conjugated protein.

21.47

Sugar ring (CH₂OH, OH, HO, H, H, OH) with NH—C(=O)—CH₃ linked; and $NH_2-C(=O)-CH_2-CH(NH_3^+)-COO^-$

21.49 Collagen is a glycoprotein with O-linked saccharide side chains.

21.51 It breaks the disulfide bridges of keratin and allows the hair to be set into a desired shape.

21.53 to destroy bacteria on the surface of the skin

21.55 In younger people, the AGE products are degraded; in older people, the metabolism slows down and the AGE products accumulate.

21.57 Hydroxyurea therapy promotes the manufacture of fetal hemoglobin, which does not carry beta chains and, therefore, the sickle cell mutations. Cells with fetal hemoglobin do not sickle.

21.59 calcium hydroxyapatite

21.61 The fiberscope carries the laser light in optical fibers in a tiny tube that can be inserted into the body and led to organs of interest. It can repair wounds and seal blood vessels in bleeding ulcers.

21.63 (a) $4^2 = 16$ (b) $20^2 = 400$

21.65 They are made of 8 to 18 β-sheets.

21.67 a quaternary structure, because subunits are cross-linked

21.69 (a) hydrophobic (b) salt bridge (c) hydrogen bond (d) hydrophobic

21.71 glycine

21.73 one positive charge: NH_3^+

21.75 more soluble because carbohydrates have many OH groups that form hydrogen bonds with water

Chapter 22 Enzymes

22.1 A catalyst is a substance that speeds up a reaction without being used up itself. An enzyme is a catalyst that is a protein or a nucleic acid.

22.3 Yes; lipases are not very specific.
22.5 because enzymes are highly specific, and there are a great many different reactions that must be catalyzed
22.7 In a lyase-catalyzed reaction, the water is either added to a double bond or, if removed from a compound, generates a double bond. Hydrolases, on the other hand, add water to an ester or an amide bond, thereby splitting the molecule into two constituent parts.
22.9 (a) isomerase (b) hydrolase
(c) oxidoreductase (d) lyase
22.11 A cofactor is a nonprotein portion of an enzyme. A coenzyme is an organic cofactor.
22.13 See Section 22.3.
22.15 No, at high substrate concentration the enzyme surface is saturated, and doubling of the substrate concentration will produce only a slight increase in the rate of the reaction or no increase at all.
22.17 (a) less active at normal body temperature
(b) The activity decreases.
22.19 (a)

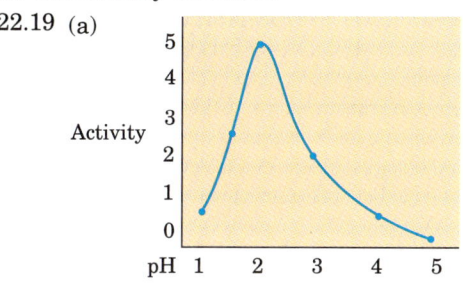

(b) 2 (c) zero activity
22.21 (A) large circle = enzyme
(B) hexagon = active site
(C) triangle = glucose
(D) larger triangle = fluoroglucose
(E) small square = Mg^{2+}
(F) half hexagon = ATP
(G) black circle = Cd^{2+}
(a) assemble ABCEF
(b) ABDEF
(c) ABCEFH
22.23 acid–base reactions
22.25 In competitive inhibition the same maximum rate can be obtained as in the reaction without the inhibitor, although at a higher substrate concentration. In noncompetitive inhibition the maximum rate is always lower than that without the inhibitor.
22.27 an allosteric mechanism
22.29 There is no difference. They are the same.
22.31

$$-NH-CH-\overset{O}{\underset{|}{C}}- $$
$$\underset{CH_2}{|}$$

(phenyl group with -O-P(=O)(-O⁻)-O- substituent)

22.33

Phosphorylase b $\underset{\underset{2P_i}{phosphatase}}{\overset{\overset{2ATP \quad ADP}{kinase}}{\rightleftharpoons}}$ Phosphorylase a

22.35 The enzyme is now named for the compound (alanine) that donates the amino group. The former name refers to the products.
22.37 H_4
22.39 digestive enzymes such as Pancreatin or Acro-lase
22.41 Succinylcholine is a competitive inhibitor of acetylcholinesterase and prevents acetylcholine from triggering muscle contractions.
22.43 *Helicobacter pylori* has an enzyme, urease, that catalyzes the hydolysis of urea to ammonia and carbon dioxide. The ammonia is a base and neutralizes the acid in the stomach. In this way the enzymes of the bacteria experience no acidic environment.
22.45 by administering antibiotics
22.47 Alanyl and threonyl residues both provide the same surface environment. In both, the terminal group of the side chain is a nonpolar methyl group. The nature of the interaction is hydrophobic interaction. Glycyl residue also contributes to the hydrophobic interaction.
22.49 Humans do not synthesize their folic acid; they get it in their diets.
22.51

$$-NH-\underset{|}{\overset{|}{C}}-COO^-$$

22.53 in hyperthermophile organisms that live in ocean vents
22.55 (a) pasteurized milk, canned tomatoes
(b) yogurt, pickles
22.57 His-Leu
22.59 cofactors
22.61 ALT
22.63 No, the nerve gas binds irreversibly by a covalent bond, so it cannot be removed simply by adding substrate.
22.65 a transferase
22.67 They catalyze the self-splicing of introns.

Chapter 23 Chemical Communications: Neurotransmitters and Hormones

23.1 G-protein is an enzyme; it catalyzes the hydrolysis of GTP to GDP. GTP, therefore, is a substrate.
23.3 A chemical messenger operates between cells; secondary messengers signal inside a cell in the cytoplasm.
23.5 The concentration of Ca^{2+} in neurons controls the process. When it reaches 10^{-4} M the vesicles release the neurotransmitters into the synapse.
23.7 anterior pituitary gland
23.9 Upon binding of acetylcholine, the conformation of the proteins in the receptor changes and the central core of ion channel opens.
23.11 When acetyl choline binds to its receptor, it opens up the ion channels. Thus the concentrations of ions on both sides of the membrane become equalized. The potential (voltage) drops, and the membrane becomes depolarized.

23.13 Taurine is a β-amino acid; its acidic group is —SO₂OH instead of —COOH.
23.15 The amino group in GABA is in the gamma position; proteins contain only alpha amino acids.
23.17 (a) norepinephrine and histamine
(b) They activate a secondary messenger, cAMP, inside the cell.
(c) amphetamines and histidine
23.19 It is phosphorylated by an ATP molecule.
23.21

HO—, HO— (benzene ring) —CH₂—C(=O)—H

23.23 (a) Amphetamines increase and (b) reserpine decreases the concentration of the adrenergic neurotransmitter.
23.25 the corresponding aldehyde
23.27 (a) the ion-translocating protein
(b) It gets phosphorylated and changes its shape.
(c) It activates the protein kinase that does the phosphorylation of the ion-translocating protein.
23.29 They are pentapeptides.
23.31 The enzyme is a kinase. The reaction is

I-1, 4-P₂

I-1, 4, 5-P₃

23.33 Calcium sparks are localized at specific parts of the cell; global calcium waves travel to different parts of the cell.
23.35 100 to 250-fold
23.37 Botulinum toxin prevents the release of acetylcholine from the presynaptic vesicles, preventing neurotransmission.
23.39 They are made of tau proteins. A mutation on tau proteins prevents their normal interaction with microtubules, a cytoskeletal component.
23.41 Acetylcholinesterase inhibitors, such as Cognex, inhibit the enzyme that decomposes the neurotransmitter.
23.43 by binding to the dopamine transporter
23.45 The transplanted neurons became integrated into the brain and produce dopamine.
23.47 in the conversion of arginine to citrulline

23.49 —O₂S—NH—CO—NH—
23.51 When insulin is adsorbed on its receptor, it elicits the autophosphorylation of the receptor on the cytosolic site. This starts a phosphorylation cascade of a number of compounds, the net result of which is the mobilization of glucose transporters and their movement to the surface of the cell, where they carry the glucose across the membrane.
23.53 Tamoxifen binds to the estrogen receptor. Therefore, estrogen cannot bind to its receptor and cannot reach the DNA in the nucleus to influence protein synthesis.
23.55 It opens the ion gates and allows the translocation of ions.
23.57 Yes. It fits into the same receptor site as enkephalins.
23.59 It is associated with the receptor.
23.61 (a) Both deliver a message to a receptor.
(b) Neurotransmitters act rapidly over short distances. Hormones act more slowly and over longer distances. They are carried by the blood from the source to the target cell.
23.63 Cholera toxin affects G-protein.
23.65 An agonist stimulates the receptor; therefore, more glucose would enter the cells, and its concentration in the serum would decrease.
23.67 The luteinizing hormone is a peptide. It acts through the G-protein-adenylate cyclase cascade, producing secondary messengers. Progesterone is a steroid hormone; it acts in the nucleus of cells, influencing protein synthesis.

Chapter 24 Nucleotides, Nucleic Acids, and Heredity

24.1 Structure of UMP

24.3 hemophilia, sickle-cell anemia, etc.
24.5 (a) base, sugar, phosphate
(b) base, sugar
24.7 Thymine has a methyl group in the 5 position; uracil has a hydrogen there.
24.9 (a)

(b) the same, but without the —OH in the 2 position.

24.11 Ribose has an —OH in the 2 position; deoxyribose has an H there.
24.13 C-3 and C-5 to the phosphate group; C-1 to the base
24.15 (a) [structure of uridine diphosphate]
(b) [structure of adenosine monophosphate]

4.17 (a) the 3′—OH end and the 5′—OH end
(b) A is the 5′—OH end. C is the 3′—OH end.
24.19 two
24.21 It is an electrostatic attraction.
24.23 The superstructure of chromosome is made of chromatin fibers, in which the fibers are folded into loops and the loops are further folded into bands.
24.25 rRNA
24.27 mRNA
24.29 Ribozymes in general shorten preexisting RNA molecules: from mRNA to mature RNA, from precursor tRNA to mature tRNA.
24.31 mRNA contains exons and introns when it is first synthesized, but before it actually functions the introns are cut out.
24.33 No; satellites are noncoding regions, repeated nucleotide sequences.
24.35 amino and carbonyl groups
24.37 [structure showing base-pair hydrogen bonding between cytosine and guanine]

24.39 The replication fork is where the replication starts. There may be hundreds of replication forks on a human chromosome at any time.
24.41 The DNA is wrapped around the histones in the nucleosome. The negative charges on the DNA interact strongly with the positive charges of the histones, making for tight packing. The removal of some of the positive charges loosens the electrostatic interactions.
24.43 Six subunits form a ring with a hollow core through which a single strand of DNA can move.
24.45 No. The sugar in the primer must be ribose. Only ATP, GTP, CTP, and UTP can serve as source for a primer.
24.47 polymerases
24.49 phosphodiester
24.51 If only a few G residues are damaged, the likely mechanism is nucleotide excision repair (NER).
24.53 Uracil is a naturally occurring base only when its sugar is ribose. At the damaged DNA site, the uracil is linked to a deoxyribose.
24.55 AP sites are apurinic or apyrimidinic sites—that is, the sites from which the purine or pyrimidine bases have been excised. The enzyme glycosylase generates AP sites.
24.57 The polymerases from thermophile bacteria are heat resistant, so they can work at the high temperatures that are needed for the unwinding of the DNA double helix.
24.59 [structure of 5-fluorouracil nucleoside]

24.61 TTAGGG
24.63 Telomerase resynthesizes the shortened DNA end in telomers; thus, the cell has no signal telling it when to stop dividing and therefore never dies.
24.65 The DNA fragments of the child, mother, and alleged father are run on the same gel, and their patterns are compared.
24.67 One can screen patients for their individual genetic make-up with respect to cytochrome P-450. In this way it can be established whether the patient is an extensive, ultraextensive, or poor metabolizer of a drug even before the drug is administered.
24.69 the DNA fragments in multiple of 180 base pairs
24.71 Every time an error is introduced, it may cause a mutation that can be harmful or even detrimental to the propagation of the species.
24.73 the glycosidic bond between the sugar and the base, and the phosphate ester bond between the sugar and the phosphate
24.75 DNA; it contains up to millions of base pairs.
24.77 T = 29.3; G = 20.7; C = 20.7

Chapter 25 Gene Expression and Protein Synthesis

25.1 First, binding proteins must make the portion of the chromosome where the gene is less condensed and more accessible. Second, the helicase enzyme must unwind the double helix near the gene. Third, the polymerase must recognize the initiation signal on the gene.
25.2 (a) CAU and CAC   (b) GCA and GTG
25.3 valine + ATP + tRNA$_{val}$

25.4 CCT   CGATTG
     GGAGC   TAAC
25.5 (c)
25.7 in the cytoplasm
25.9 to unwind the DNA double helix
25.11 5′ end
25.13 on the N atom of guanine
25.15 on the mRNA
25.17 (a) CGA   (b) alanine
25.19 Because all the diverse life forms on Earth share the same genetic code, they must have evolved from a common ancestry.
25.21 The A site is where the incoming tRNA carrying the amino acid binds. The P site is where the growing peptide chain binds. The E site is where the exiting tRNA binds before being recycled.
25.23 They enable the tRNA to bind to the A site of the rRNA.
25.25 Thousands of nucleotides upstream from the transcription site. By interacting with the promoter, they speed up the transcription process.
25.27 to degrade misfolded proteins
25.29 (a) UAU and UAC; both code for tyrosine.
(b) If UAU is changed to UAA, then instead of producing tyrosine, the chain will be terminated. This might well be fatal.
25.31 Yes; if the mutation occurred on a recessive gene.
25.33 Restriction endonucleases cleave DNA at specific sites.
25.35 In genetic engineering, a new gene is inserted into the genome. However, in the corn that acquired insect resistance by natural selection, mutations occurred in the native genes.
25.37 the —S—P—O— bonding
25.39 T (lymphocyte) cells
25.41 The promoter enabled the HIV protease to be expressed in the lens of the mouse. Thus expressed, the lens became cataractous. In this manner, it could serve as a bioessay to prove the efficacy of new HIV protease inhibitor drugs.
25.43 A change from guanine to thymine in the gene results in a valine in place of a glycine.
25.45 When p53 binds to DNA, it arrests the cell cycle; thus it allows more time for the cell defenses to repair damaged DNA.
25.47 (a) The 5′ end and the section just before the 3′ end must be base paired.   (b) There must be no base pairing in the anticodon loop.
25.49 The aminoacyl-tRNA synthtase is specific for one amino acid and all the possible tRNA molecules carrying anticodons. This specificity is called the second genetic code because it ensures the fidelity of translation.
25.51 *Degenerate* means multiple; that is, more than one codon specifies the same amino acid. There is a certain amount of redundancy in the genetic code.
25.53 The third base is not irrelevant for phe, ile, met, tyr, his, gln, asn, lys, asp, glu, cys, trp, ser, and arg. In eight cases, the third base is irrelevant.
25.55 No. It would produce too many small fragments and no sticky ends.

Chapter 26   Bioenergetics: How the Body Converts Food to Energy

26.1 ATP
26.3 (a) 2   (b) the outer membrane
26.5 Christae are folded membranes originating from the inner membrane. They are connected to the inner membrane by tubular channels.
26.7 two
26.9 Neither; they yield the same energy.
26.11 phosphate ester
26.13 the 2 N atoms that are part of C=N bonds
26.15 (a) adenosine   (b) NAD^+ and FAD   (c) acetyl groups
26.17 amide
26.19 No. The pantothenic acid portion is not the active part.
26.21 acetyl coenzyme A
26.23 α-ketoglutarate
26.25 Succinate is oxidized by FAD, and the oxidation product is fumarate.
26.27 lyase
26.29 No, but GTP is produced in step **5**.
26.31 It allows the energy to be released in small packets.
26.33 *cis*-aconitate and fumarate
26.35 to NAD^+, which becomes $NADH + H^+$
26.37 cytochrome C and CoQ
26.39 When H^+ passes through the ion channel, the proteins of the channel rotate. The kinetic energy of this rotatory motion is converted to and stored as the chemical energy in ATP.
26.41 in the inner membranes of the mitochondria
26.43 (a) 0.5   (b) 12
26.45 the proton-translocating ATPase
26.47 F_1; it catalyzes the conversion of ADP to ATP.
26.49 1.5 kcal
26.51 (a) by sliding the thick filaments (myosin) and the thin filaments (actin) past each other   (b) from the hydrolysis of ATP
26.53 ATP transfers a phosphate group to the serine residue at the active site of the enzyme.
26.55 Sulfur, in the Fe-S clusters, is embedded in a protective protein coat.
26.57 No. It would harm humans because they would not synthesize enough ATP molecules.
26.59 (a) hydroxylation   (b) the air we breathe in
26.61 60 g or 1 mol of CH_3COOH
26.63 The energy of motion appears first in the ion channel where, upon passage of H^+, the proteins making this channel rotate.
26.65 They are both hydroxy acids.
26.67 myosin
26.69 two
26.71 F_0; it is made of 12 subunits.
26.73 No, it largely comes from the chemical energy as a result of the breaking of bonds in the O_2 molecule.
26.75 It removes two hydrogens from succinate to produce fumarate.

Chapter 27 Specific Catabolic Pathways: Carbohydrate, Lipid, and Protein Metabolism

27.1 According to Table 27.2 the ATP yield from stearic acid (C_{18}) is 146 ATP. This makes 146/18 = 8.1 ATP/carbon atom. For lauric acid (C_{12}):

Step **1** Activation	−2 ATP
Step **2** Dehydrogenation five times	10 ATP
Step **3** Dehydrogenation five times	15 ATP
Six C_2 fragments in common pathway	<u>72 ATP</u>
Total	95 ATP

95/12 = 7.9 ATP per carbon atom for lauric acid. Thus stearic acid yields more ATP/C atom.

27.3 They serve as building blocks for the synthesis of proteins.

27.5 The two C_3 fragments are in equilibrium. As the glyceraldehyde phosphate is used up, the equilibrium shifts and converts the other C_3 fragment (dihydroxyacetone phosphate) to glyceraldehyde phosphate.

27.7 (a) steps **1** and **3** (b) steps **6** and **9**

27.9 At step **9**; the pyruvate kinase; it is a feedback regulation

27.11 NADPH

27.13 6 moles

27.15 12

27.17 Two net ATP molecules are produced in both cases.

27.19 glycerol kinase

27.21 (a) thiokinase and thiolase (b) —SH (c) Both enzymes insert a CoA—SH into a compound.

27.23 $CH_3(CH_2)_4CO$—CoA, three acetyl CoA, three $FADH_2$, three NADH + H$^+$

27.25 112

27.27 carbohydrates

27.29 (a) reduction (b) decarboxylation

27.31 It enters the citric acid cycle.

27.33

(a) CH_3—CH—COO$^-$ + NAD$^+$ + H_2O ⟶
         |
         NH$_3^+$

CH_3—C—COO$^-$ + NADH + H$^+$ + NH$_4^+$
       ‖
       O

27.35 because pyruvate can be converted to glucose when the body needs it

27.37 fumarate

27.39 (a) by reductive amidation, which is the reverse of oxidative deamidation (b) by converting glutamate to glutamine

27.41 It is stored in ferritin and reused.

27.43 lactic acid accumulation

27.45 the bicarbonate/carbonic acid buffer

27.47 You can find out what kind of lipids are in the gallstone extract and what their metabolic products are in a single analysis.

27.49 Glycine, the C-terminal amino acid of ubiquitin, forms an amide linkage with a lysine residue in the target protein.

27.51

Phenylalanine + α-Ketoglutarate ⟶

Phenylpyruvate + Glutamate

27.53 diabetes

27.55 (a) in steps **5** and **12**
(b) in steps **10** and **11**
(c) If enough O_2 is present in muscles, a net increase of NADH + H$^+$ occurs. If not enough O_2 is present (as in yeast), there is no net increase of either NAD$^+$ or NADH + H$^+$.

27.57 Yes, they are glucogenic and can be converted to glucose.

27.59 CH_3—C—COO$^-$ + COO$^-$
              ‖ |
              O   CH—NH$_3^+$
                           |
                           CH$_2$
                           |
                           COO$^-$

27.61 in the carbon dioxide exhaled by the animal

27.63 It is energy consuming. Three ATP molecules are hydrolyzed per cycle.

27.65 (a) 5 (b) 7

Chapter 28 Biosynthetic Pathways

28.1 for flexibility and to overcome unfavorable equilibria

28.3 because the presence of a large inorganic phosphate pool would shift the reaction to the degradation process such that no substantial amount of glycogen would be synthesized

28.5 Photosynthesis is the reverse of respiration.

28.7 (a) pyruvate (b) oxaloacetate (c) alanine

28.9 Gluconeogenesis, because the other pathways metabolize glucose, and only gluconeogenesis manufactures it.

28.11 UDP-glucose + glucose = maltose

28.13 uracil, ribose, and three phosphates

28.15 (a) the cytoplasm (b) no

28.17 malonyl ACP

28.19 malonyl ACP

28.21 It is an oxidation step because hydrogen is removed from the substrate. The oxidizing agent is O_2. NADPH is also oxidized during this step.

28.23 NADPH is bulkier than NADH; it also has two more negative charges.
28.25 No, the body makes other unsaturated fatty acids such as oleic acid and arachidonic acid.
28.27 palmitoyl CoA, serine, acyl CoA, UDP-glucose
28.29 $3C_2 \longrightarrow C_6 \longrightarrow C_5 + C$
28.31 aspartic acid
28.33 valine and α-ketoglutarate
28.35 NADPH
28.37 Indigo blue, an oxidation product of tryptophan. Indigo blue cannot enter the cells and ends up in the urine.
28.39 malonyl ACP
28.41 No, the cancer is caused when Ras is mutated, but prenylation is a precondition for the activity of RAS.
28.43 When we are exposed to cold, energy is needed in the form of heat for survival. α-ketoglutarate is part of the citric acid cycle, which produces energy. Thus α-ketoglutarate is used up and the equilibrium shifts to the left, deamidating glutamic acid. This produces α-ketoglutaric acid, replacing that used up in the citric acid cycle.
28.45 geranyl- and farnesylpyrophosphate
28.47 The statement is true. Carbohydrate is synthesized by reducing CO_2.
28.49 No, it can add C_2 fragments to a growing chain up to C_{16}. After that, another enzyme takes over.

Chapter 29 Nutrition and Digestion

29.1 No, they differ according to body weight, age, occupation, and sex.
29.3 Only the calcium propionate. When it is metabolized, it yields energy.
29.5 the section at the bottom that gives daily caloric needs
29.7 It is necessary for proper operation of the digestive system, and a lack of it may lead to colon cancer.
29.9 for daily activities involving motion
29.11 1830 Cal/day
29.13 No, the loss of water is only temporary and the weight is regained very quickly.
29.15 different oligosaccharides of α-D-glucose
29.17 No; maltose cannot be absorbed into the blood; it is hydrolyzed to glucose before being absorbed.
29.19 linoleic acid
29.21 neither
29.23 Yes, but one must eat a wide range of vegetables and cereals.
29.25 Both catalyze the hydrolysis of amide bonds. Acid catalyzes random hydrolysis, and trypsin catalyzes the hydrolysis on the C=O side of arginine and lysine.
29.27 dietary deficiency diseases, since rice cannot supply the essential amino acids lysine or threonine or the essential fatty acids
29.29 to prevent scurvy
29.31 It assists in blood clotting.
29.33 scurvy
29.35 They act as antioxidants in the body.
29.37 biotin, thiamine
29.39 If glucose supplied all caloric needs, it would cause deterioration of the patient.

29.41 No, these are two different dipeptides. In aspartame, aspartic acid is at the N-terminal; in phenylalanylaspartic acid, it is at the C-terminal.
29.43 Both are derivatives of disaccharides.
29.45 chlorine and ozone
29.47 niacinamide
29.49 hydrolysis of acetal, ester, and amide linkages
29.51 No, amylase does not have branches with 1,6-glycosidic linkage.
29.53 The body would treat it as a food protein and digest it (hydrolyze it) before it could get into the blood serum.
29.55 Discriminatory curtailment diet; no, aspartame contains phenylalanine.
29.57 No. Arsenic is not an essential nutrient.

Chapter 30 Immunochemistry

30.1 skin and mucin (or tear)
30.3 The skin fights bacteria by secreting lactic and fatty acids, thereby lowering the pH.
30.5 Acquired immunity is selective, and it has memory.
30.7 T cells mature and differentiate in the thymus gland, B cells in the bone marrow.
30.9 Cancer cells and virus-infected cells; NO molecules kill them.
30.11 only peptide antigens
30.13 The polypeptide antigen is digested by proteolytic enzymes and becomes small peptides. These peptides can serve as targets for the class II MHC bringing them to the surface of the cell.
30.15 They belong to the immunoglobulin superfamily. They are found in the membranes of cells.
30.17 IgA are in secretions. They are the largest, with molecular weights of 200,000 to 700,000. IgE are involved in allergic reactions; their molecular weight is 190,000. IgG are the most important antibodies in blood; their molecular weight is 150,000.
30.19

Red blood cell   Anti-A antibody   Red blood cell

30.21 by disulfide bridges
30.23 Two monoclonal antibodies isolated from the same population of lymphocytes would differ in their V region. They would interact with different antigens. They would probably have the same C (constant) region.
30.25 Exons from three regions, V, J, and D, of the variable H gene are selected and combine to form a new gene. The transcription of this new gene yields mRNA that is translated to a new protein.
30.27 TcR receptor complex contains at least two more proteins besides the TcR receptor—for example, the CD3 and CD4 molecules, which are also members of the immunoglobulin superfamily.
30.29 TcR and coreceptors such as CD3 and CD4 or CD8
30.31 CD4 adhesion molecule
30.33 Cytokines are small glycoprotein molecules.

30.35 (a) tumor necrosis factor (b) interleukin (c) epidermal growth factor
30.37 They have four cysteine residues. Cysteine residue number 1 is linked to residue number 3 by disulfide linkage. Cysteine number 2 is linked to cysteine number 4.
30.39 cysteine
30.41 In malignant tumor cells, the epitopes that signal healthy cells diminish greatly, and unusual epitopes are presented on the surface of these cells.
30.43 The inhibitory receptor on their surface recognizes the epitope of a normal cell, binds to it, and prevents the activation of the killer cell or macrophage.
30.45 Glucocorticoids regulate the synthesis of TcR-cytokines directly by interacting with their genes or indirectly through transcription factors.
30.47 Monoclonal antibody treatment is only mildly toxic. It specifically attacks the lymphoma cells. Chemotherapy kills many other cells besides the quickly reproducing cancer cells. Therefore, it is highly toxic.
30.49 Plasma cells are differentiated B cells, and memory cells are differentiated T cells.
30.51 The chemokine, when it interacts with the surface of a leukocyte, leads to alternative polymerization and breakdown of actin inside the cells. This allows the cell to change its shape. The leukocyte can then extend leg-like extrusions for moving along the endothelium and can flatten out. The endothelial cells can change their shape and develop gaps among the tight junctions. This process enables the flattened cells to pass through the gaps.
30.53 Antibody is manufactured and secreted by B cells. They are the only cells that manufacture soluble immunoglobulins.
30.55 in the thoracic duct
30.57 Cytokines carry signals from cell to cell to induce proliferation of leukocytes.
30.59 Carbohydrate cell surface marker; epothilon is a synthetic analogue of such marker.
30.61 on many cells, not just tumor cells

Chapter 31 Body Fluids

31.1 the limited exchange of components between blood and the interstitial fluid of brain
31.3 (a) yes (b) through semipermeable membranes
31.5 They carry oxygen to the cells and carbon dioxide away from the cells.
31.7 See Sections 31.1 and 31.2.
31.9 By centrifugation; the cellular components settle and the supernatant is the plasma.
31.11 globulins
31.13 four
31.15 Fe^{2+}; percent of saturation is the percentage of oxygen carrying Fe^{2+} out of the total iron.
31.17 Each heme has a cooperative effect on the other hemes.
31.19 (a) at the terminal NH_2 groups of the four polypeptide chains (b) carbaminohemoglobin
31.21 $K = 10$ (significant figures!)
31.23 in the proximal tubule
31.25 inorganic ions
31.27 The H^+ ion from cell metabolism goes to the blood, where it reacts with bicarbonate ion to form H_2CO_3. In the lungs, carbonic anhydrase catalyzes its conversion to CO_2 and water. CO_2 is exhaled. To avoid the loss of bicarbonate ion from blood, CO_2 is reabsorbed into the cells of the distal tubule. There it reacts with water to form carbonic acid, which dissociates and gives bicarbonate and H^+ ion. The latter is filtered out into the urine and gives an acidic pH. In this manner, the H^+ of metabolism ends up in the urine.
31.29 the distal tubule and the collecting tubule
31.31 The specific gravity of urine becomes lower because the urine is more diluted with water.
31.33 No glucose reaches the tissues or the cells. Because 12 mm Hg is lower than the osmotic pressure, nutrients flow from the interstitial fluid into the blood.
31.35 The inhibition of ACE stops the production of the vasoconstriction.
31.37 collagen
31.39 platelets
31.41 restricted flow of urine
31.43 By blocking calcium channels, heart muscles contract less frequently, thereby reducing blood pressure.
31.45 The concentration is higher. Because fibrinogen is removed when plasma becomes serum, the albumin is dissolved in a smaller amount of material, and its concentration is higher.
31.47 Less oxygen is carried; more is released to the tissues.
31.49 No; it is too large of a molecule.
31.51 vasopressin
31.53 Where the two circulations meet, the oxygen pressure is the same. However, the saturation of fetal hemoglobin is higher than that of adult hemoglobin. Thus, at the same oxygen pressure, the fetal hemoglobin can carry more oxygen than the adult hemoglobin. Therefore, some of the oxygen will be transferred from the adult hemoglobin to the fetal hemoglobin when the two circulations meet.

GLOSSARY

A site *(Section 25.5)* The site on the large ribosomal subunit where the incoming tRNA molecule binds.

Acetal *(Section 17.7)* A molecule containing two —OR groups bonded to the same carbon.

Acetyl coenzyme A *(Section 26.3)* A biomolecule in which an acetyl group is bonded to the —SH group of coenzyme A by a thioester bond.

Achiral *(Section 15.2)* An object that lacks chirality; an object that is superposable on its mirror image.

Acid *(Section 8.1)* An Arrhenius acid is a substance that ionizes in aqueous solution to give a hydronium ion, H_3O^+, and an anion *(Section 8.3)*. A Brønsted-Lowry acid is a proton donor.

Acid ionization constant (K_a) *(Section 8.5)* An equilibrium constant for the ionization of an acid in aqueous solution to H_3O^+ and its conjugate base. K_a is also called an acid dissociation constant.

Acidic salt *(Section 8.9)* The salt of a strong acid and a weak base; an aqueous solution of an acidic salt is basic.

Acidosis *(Chemical Connections 8D)* A condition in which the pH of blood is lower than 7.35.

ACP *(Section 28.3)* Acyl carrier protein, a large molecule that activates a fatty acid for fatty acid synthesis. The fatty acid is bonded to ACP as a thioester.

Acquired immunity *(Section 30.1)* The second line of defense that vertebrates have against invading organisms.

Activation energy *(Section 7.2)* The minimum energy necessary to cause a chemical reaction.

Activation, of an amino acid *(Section 25.5)* The process by which an amino acid is bonded to an AMP molecule and then bonded to the 3'—OH of a tRNA molecule.

Activation, of enzymes *(Section 22.3)* Any process by which an inactive enzyme is transformed into an active enzyme.

Active site *(Section 22.5; Chemical Connections 22C)* A three-dimensional cavity of the enzyme with specific chemical properties to accommodate the substrate.

Actual yield *(Section 4.5C)* The mass of product formed in a chemical reaction.

Acyl group *(Section 18.3A)* An RCO— group.

Adaptive immunity *(Section 30.1)* Acquired immunity with specificity and memory.

Adhesion molecules *(Section 30.5)* Various protein molecules that help to bind an antigen to the T-cell receptor.

Adrenergic neurotransmitter *(Section 23.4)* A monoamine neurotransmitter/hormone, the most common of which are epinephrine (adrenalin), serotonin, histamine, and dopamine.

Adrenocorticoid hormone *(Section 20.10)* A hormone produced by the adrenal glands. Adrenocorticoid hormones regulate the concentrations of ions and carbohydrates.

Agonist *(Section 23.1)* A molecule that mimics the structure of a natural neurotransmitter/hormone, binds to the same receptor, and elicits the same response.

AIDS *(Section 30.5)* Acquired immune deficiency syndrome. The disease caused by the human immunodeficiency virus, which attacks and depletes T cells.

Alcohol *(Section 10.4A)* A compound containing an —OH (hydroxyl) group bonded to a tetrahedral carbon atom.

Aldehyde *(Section 10.4C)* A compound containing a carbonyl group bonded to a hydrogen; a —CHO group.

Alditol *(Section 19.4B)* The product formed when the CHO group of a monosaccharide is reduced to a CH_2OH group.

Aldose *(Section 19.2A)* A monosaccharide containing an aldehyde group.

Aliphatic amine *(Section 16.2)* An amine in which nitrogen is bonded only to alkyl groups.

Aliphatic hydrocarbon *(Section 11.2)* An alkane.

Alkali *(Chemical Connections 8B)* An older term for a base.

Alkali metal *(Section 2.5A)* An element in Group 1A of the Periodic Table other than hydrogen, H.

Alkaloid *(Chemical Connections 16B)* A basic nitrogen-containing compound of plant origin, many of which have physiological activity when administered to humans.

Alkalosis *(Chemical Connections 8D)* A condition in which the pH of blood is greater than 7.45.

Alkane *(Section 11.2)* A saturated hydrocarbon whose carbon atoms are arranged in an open chain.

Alkene *(Section 12.2)* An unsaturated hydrocarbon that contains a carbon–carbon double bond.

Alkyl group *(Section 11.4A)* A group derived by removing a hydrogen from an alkane; given the symbol R—.

Alkyne *(Section 12.2)* An unsaturated hydrocarbon that contains a carbon–carbon triple bond.

Allosterism *(Section 22.6)* An enzyme regulation in which the binding of a regulator on one site on the enzyme modifies the enzyme's ability to bind the substrate in the active site. Allosteric enzymes often have multiple polypeptide chains with the possibility of chemical communication between the chains.

Alloy *(Section 6.2)* A homogeneous mixture of metals.

Alpha (α-) amino acid *(Section 21.2)* An amino acid in which the amino group is linked to the carbon atom next to the —COOH carbon.

Alpha helix *(Section 21.8)* A repeating secondary structure of a protein where the chain adopts a helical conformation and the structure is held together by hydrogen bonds from the peptide backbone N—H to the backbone C=O four amino acids farther up the chain.

Alpha particle *(Section 9.2)* A helium nucleus ($\alpha = {}^{4}_{2}He$).

Amide *(Section 18.3C)* A compound in which a carbonyl group is bonded to a nitrogen atom, $RCONR'_2$; one or both of the R' groups may be H, alkyl, or aryl groups.

Amino acid *(Section 21.2)* An organic compound containing an amino group and a carboxyl group.

Amino acid neurotransmitter *(Section 23.5)* A class of neurotransmitter/hormones that are amino acids.

Amino acid pool *(Section 27.1)* The name used for the free amino acids found both inside and outside cells throughout the body.

Amino group *(Section 10.4B)* An —NH_2 group.

Amino sugar *(Section 19.2D)* A monosaccharide in which an —OH group is replaced by an —NH_2 group.

Amorphous solid *(Section 5.10)* A solid whose atoms, molecules, or ions are not in an orderly arrangement.

Amphiprotic *(Section 8.3)* A substance that can act as either an acid or a base.

Amphoteric *(Section 8.3)* An alternative for amphiprotic.

Anabolism *(Section 26.1)* The pathways by which biomolecules are synthesized.

Anaerobic pathway *(Section 27.2)* The pathway that generates lactic acid from pyruvate when the oxygen supply is too low for the pyruvate to be oxidized aerobically through the common pathway.

Angiotensin *(Section 31.8)* A hormone that is a vasoconstrictor and raises the blood pressure.

Anhydride *(Section 18.3A)* A compound in which two carbonyl groups are bonded to the same oxygen atom; RCO—O—COR'.

Anhydrous *(Section 6.6)* A term meaning "without water."

Anion *(Section 3.2)* An ion with a negative electric charge.

Anode *(Section 6.6)* An electrode toward which anions move; the electrode at which oxidation occurs.

Anomeric carbon *(Section 19.3A)* The hemiacetal carbon of the cyclic form of a monosaccharide.

Anomers *(Section 19.3A)* Monosaccharides that differ in configuration only at their anomeric carbons.

Antacid *(Chemical Connections 8C)* A substance taken orally to neutralize stomach acidity.

Antagonist *(Section 23.1)* A molecule that binds to a neurotransmitter receptor but does not elicit the natural response.

Antibody *(Section 30.4)* A defense glycoprotein synthesized by the immune system of vertebrates that interacts with an antigen. It is also called an immunoglobulin.

Anticoagulant *(Chemical Connections 31B)* A drug given to prevent blood clot formation.

Anticodon *(Section 25.3)* A sequence of three nucleotides on tRNA complementary to the codon in mRNA.

Antidiuretic *(Section 31.7)* Any agent that reduces the volume of urine produced.

Antigen *(Section 30.1)* A substance foreign to the body that triggers an immune response.

Apoenzyme *(Section 22.3)* The protein portion of an enzyme that has cofactors or prosthetic groups.

Apoptosis *(Chemical Connections 24E)* Planned cell death characterized by small numbers of cells dying at a time and being cleaved into apoptopic bodies containing organelles and large nuclear fragments.

Aqueous solution *(Section 4.6A)* A solution in which the solvent is water.

Ar— *(Section 13.1)* The symbol used for an aryl group.

Arene *(Section 13.1)* A compound containing one or more benzene rings.

Aromatic amine *(Section 16.2)* An amine in which nitrogen is bonded to one or more aromatic rings.

Aromatic compound *(Section 13.1)* A term used to classify benzene and its derivatives.

Aryl group *(Section 13.1)* A group derived from an arene by removal of a hydrogen.

Atom *(Section 2.3)* The smallest particle of an element that retains all of the chemical properties of the element.

Atomic mass unit, amu *(Section 2.4A)* A unit of the scale of relative masses of the elements; 1 amu = 1.6605×10^{-24} g. One amu is defined as $\frac{1}{12}$ the mass of a carbon atom that has six protons and six neutrons.

Atomic number *(Section 2.4C)* The number of protons and neutrons in the nucleus of an atom of an element.

Atomic weight *(Section 2.6E)* The weighted average of the masses of the naturally occurring isotopes of an element. The units of atomic weight are atomic mass units (amu).

ATP *(Section 26.1)* Adenosine triphosphate.

Autoxidation *(Section 13.5C)* The reaction of a C—H group with oxygen, O_2, to form a hydroperoxide, R—OOH.

Avogadro's Law *(Section 5.5)* Equal volumes of gases at the same temperature and pressure contain the same number of molecules.

Avogadro's number *(Section 4.4)* 6.02×10^{23} formula units per mole; the amount of any substance that contains as many formula units as the number of atoms in 12 g of carbon-12.

Axial position *(Section 11.6B)* A position on a chair conformation of a cyclohexane ring that extends from the ring parallel to the imaginary axis of the ring.

Axon *(Section 23.2)* The long part of a nerve cell that comes out of the main cell body and eventually connects with another nerve cell or a tissue cell.

B cell *(Section 30.2)* A type of lymphocyte that is produced in and matures in the bone marrow. B cells produce antibody molecules.

Barometer *(Section 5.2)* An instrument for measuring atmospheric pressure.

Baroreceptors *(Section 31.8)* The receptors in the neck that detect a drop in blood pressure and send a signal to the heart to pump harder and to the muscles around the blood vessels to contract to raise blood pressure.

Basal caloric requirement *(Section 29.3)* The caloric requirement for an individual at rest, usually given in Cal/day.

Base *(Section 8.1)* An Arrhenius base is a substance that ionizes in aqueous solution to give hydroxide (OH^-) ions. *(Section 8.3)* A Brønsted-Lowry base is a proton acceptor.

Bases *(Section 24.2)* Purines and pyrimidines, which are components of nucleotides, DNA, and RNA.

Basic salt *(Section 8.9)* The salt of a weak acid and a strong base. An aqueous solution of a basic salt is acidic.

Becquerel (Bq) *(Section 9.5)* One disintegration (count) per second.

Beta (β) pleated sheet *(Section 21.8)* A secondary protein structure in which the backbone of two protein chains in the same or different molecules is held together by hydrogen bonds.

Beta particle *(Section 9.2)* An electron, $_{-1}^{0}e$.

Beta (β) oxidation *(Section 27.5)* The biochemical pathway that degrades fatty acids to acetyl CoA by removing two carbons at a time and yielding energy.

Bile salts *(Section 20.11)* Oxidation products of cholesterol that are emulsifying agents for fatty acids. They are produced in the liver and stored in the gallbladder.

Binary compound *(Section 3.6A)* A compound containing two elements.

Biochemical pathway *(Section 26.1)* A series of consecutive biochemical reactions that converts one molecule into another molecule; the pathway may be anabolic or catabolic.

Biosynthetic pathway *(Section 28.1)* A pathway that leads to the production of a compound. These pathways often build larger molecules from smaller ones and consume energy.

Blood cellular elements *(Section 31.2)* The cell components that are suspended in blood—namely, the erythrocytes, leukocytes, and platelets.

Blood plasma *(Section 31.1)* The noncellular portion of blood.

Blood–brain barrier *(Section 31.1)* A barrier limiting the exchange of blood components between blood and the cerebrospinal and brain interstitial fluids to water, carbon dioxide, glucose, and other small molecules but excluding electrolytes and large molecules.

Bohr effect *(Section 31.3)* The effect caused by a change in the pH on the oxygen-carrying capacity of hemoglobin.

Boiling point *(Section 5.9B)* The temperature at which the vapor pressure of a liquid is equal to the atmospheric pressure.

Bond angle *(Section 3.9)* The angle between two atoms bonded to a central atom.

Bonding electrons *(Section 3.7C)* Valence electrons involved in forming a covalent bond; that is, shared electrons.

Bowman's capsule *(Section 31.5)* A cup-shaped receptacle in the kidney that is the initial, expanded segment of the nephron where filtrate enters from the blood.

Boyle's Law *(Section 5.3A)* The volume of a gas at constant temperature is inversely proportional to the pressure applied to the gas.

Brackish water *(Chemical Connections 6E)* A mixture of sea water (salt water) and fresh water.

Brownian motion *(Section 6.7)* The random motion of colloidal-size particles.

Buffer capacity *(Section 8.11C)* The extent to which a buffer solution can prevent a significant change in pH of a solution upon addition of an acid or a base.

Buffer solution *(Section 8.11A)* A solution that resists change in pH when limited amounts of acid or base are added to it; a aqueous solution containing a weak acid and its conjugate base.

Calorie *(Section 1.9)* The amount of heat necessary to raise 1 g of liquid water by 1°C.

Calvin cycle *(Chemical Connections 28A)* The dark reactions of photosynthesis that produce glucose from CO_2 and H_2O by a cyclic process.

Carbocation *(Section 12.6A)* A species containing a carbon atom with only three bonds to it and bearing a positive charge.

Carbohydrate *(Section 19.1)* A polyhydroxyaldehyde or polyhydroxyketone, or a substance that gives these compounds on hydrolysis.

Carbonyl group *(Section 10.4C)* A C=O group.

Carboxyl group *(Section 10.4D)* A —COOH group.

Carboxylic acid *(Section 10.4D)* A compound containing a —COOH group.

Carcinogen *(Section 25.7)* A chemical mutagen that can cause cancer.

Catabolism *(Section 26.1)* The biochemical pathways that are involved in generating energy by breaking down large nutrient molecules into smaller molecules with the concurrent production of ATP.

Catalyst *(Section 7.4)* A substance that increases the rate of a chemical reaction by providing an alternative pathway with a lower activation energy.

Cathode *(Section 6.6)* An electrode to which cations are attracted or move; the electrode at which reduction occurs.

Cation *(Section 3.2)* An ion with a positive electric charge.

Celsius scale (°C) *(Section 1.4)* A temperature scale based on 0° as the freezing point of water and 100° as the normal boiling point of water.

Central dogma *(Section 25.1)* Doctrine stating the basic directionality of heredity where DNA leads to RNA, which leads to protein. This is true in almost all life forms, except certain viruses.

Cephalin *(Section 20.6)* A glycerophospholipid similar to phosphatidylcholine with the exception that another alcohol is attached to the phosphate instead of choline.

Ceramide *(Section 20.7)* The combination of a fatty acid linked to sphingosine by an amide bond.

Cerebroside *(Section 20.8)* A glycolipid based on the ceramide backbone where one or more carbohydrates are attached to the ceramide.

Chain reaction, nuclear *(Section 9.7)* A nuclear reaction that results from fission of a nucleus, which produces a particle that causes fission in another nucleus and continues with accelerated speed.

Chair conformation *(Section 11.6B)* The most stable conformation of a cyclohexane ring; all bond angles are approximately 109.5°.

Chaperone *(Section 21.11)* Protein molecules that help other proteins to fold into the biologically active conformation and enable partially denatured proteins to regain their biologically active conformation.

Charles's Law *(Section 5.3B)* The volume of a gas at constant pressure is directly proportional to its temperature in kelvins.

Chemical change *(Section 1.1)* Chemical reaction.

Chemical equation *(Section 4.4A)* A representation using chemical formulas of the process that occurs when reactants are converted to products.

Chemical equilibrium *(Section 7.5)* A state in which the rate of the forward reaction equals the rate of the reverse reaction.

Chemical kinetics *(Section 7.1)* The study of chemical reaction rates.

Chemical messenger *(Section 23.1)* Any chemical that is released from one location and travels to another location before acting. It may be a hormone, neurotransmitter, or simple ion.

Chemical property *(Section 1.1)* The chemical reactions that a substance undergoes.

Chemical reaction *(Section 1.1)* As substances are used up, others appear in their place.

Chemiosmotic hypothesis *(Section 26.6)* The hypothesis proposed by Peter Mitchell to explain how movement of electrons down the electron transport chain creates a proton gradient that, in turn, leads to the production of ATP.

Chemiosmotic theory *(Section 26.6)* Peter Mitchell proposed that electron transport is accompanied by an accumulation of protons in the intermembrane space of the mitochondrion, which in turn creates an osmotic pressure; the protons driven back to the mitochondrion under this pressure generate ATP.

Chemistry *(Section 1.1)* The science that deals with matter.

Chemokine *(Section 30.6)* A chemotactic cytokine that facilitates the migration of leukocytes from the blood vessels to the site of injury or inflammation.

Chiral *(Section 15.2)* From the Greek *cheir*, meaning "hand"; an object that is not superposable on its mirror image.

Chloroplast *(Chemical Connections 22A)* The subcellular organelle in plants that contains chlorophyll and is responsible for photosynthesis.

Cholinergic neurotransmitter *(Section 23.3)* A neurotransmitter/hormone based on acetylcholine.

Chromatin *(Section 24.6)* The DNA complexed with histone and nonhistone proteins that exists in eukaryotic cells between cell divisions.

Chromosomes *(Section 24.1)* Structures within the nucleus of eukaryotes that contain DNA and protein and that are replicated as units during mitosis. Each chromosome is made up of one long DNA molecule that contains many heritable genes.

cis *(Section 11.7)* A prefix meaning "on the same side."

Cis-trans isomers *(Sections 11.7 and 12.2B)* Isomers that have the same order of attachment of their atoms but a different arrangement of their atoms in space due to the presence of either a ring or a carbon–carbon double bond.

Citric acid cycle *(Section 26.4)* The central biochemical pathway in metabolism whereby an acetyl-CoA combines with oxaloacetate to form citrate, which then undergoes a series of reactions to regenerate oxaloacetate while producing energy in the form of GTP, NADH, and $FADH_2$.

Cloning *(Section 24.6)* A process whereby DNA is amplified by inserting it into a host and having the host replicate it along with the host's own DNA.

Cluster determinant *(Section 30.6)* A set of membrane proteins on T cells that help the binding of antigens to the T-cell receptors.

Coated pit *(Section 20.9)* An area of a cell membrane that contains LDL receptors and is involved in bringing LDL into the cell.

Codon *(Section 25.3)* The sequence of three nucleotides in messenger RNA that codes for a specific amino acid.

Coefficient *(Section 4.4B)* A number placed in front of a chemical entity in a balanced chemical equation.

Coenzyme *(Section 22.3)* An organic molecule, frequently a B vitamin, that acts as a cofactor.

Coenzyme A *(Section 26.3)* An important activating group often attached to a fatty acid or an acetate group. Coenzyme A is based on a structure composed of an adenine-based nucleotide, pantothenic acid, and mercaptoethylamine.

Coenzyme Q *(Section 26.5)* An electron carrier found in the inner mitochondrial membrane that acts as an electron receptor from a variety of sources during electron transport; also called ubiquinone.

Cofactor *(Section 22.3)* The nonprotein part of an enzyme necessary for its catalytic function.

Colligative property *(Section 6.8)* A physical property of a solvent that depends on the concentration of solute particles—for example, freezing point, boiling point, vapor pressure, and osmotic pressure.

Colloid *(Section 6.7)* A two-part homogeneous mixture in which the suspended solute particles range from 1 to 1000 nm in size.

Combined gas law *(Section 5.3C)* The pressure, volume, and temperature in kelvins of two samples of the same gas are related by the equation $P_1V_1/T_1 = P_2V_2/T_2$.

Combustion *(Section 4.4A)* Burning in air.

Common catabolic pathway *(Section 26.1)* A series of chemical reactions that most degraded food goes through to yield energy in the form of ATP. The common catabolic pathway consists of (1) the citric acid cycle *(Section 26.4)* and (2) oxidative phosphorylation *(Sections 26.5 and 26.6)*.

Competitive inhibition *(Section 22.5)* An enzyme regulation in which an inhibitor competes with the substrate for the active site.

Complete protein *(Section 29.4)* A protein source that contains sufficient quantities of all amino acids required for normal growth and development.

Complex lipids *(Section 20.4)* Lipids found in membranes. The two main classes are glycolipids and phospholipids.

Compound *(Section 2.2B)* A pure substance composed of more than one type of atom with a fixed ratio of the component atoms.

Computer-assisted tomography (CAT or CT scan) *(Chemical Connections 9E)* Multiple x-rays used to image tissues on the basis of their density.

Condensation *(Section 5.1)* The change of a substance from the vapor or gaseous state to the liquid state.

Conformation *(Section 11.7)* Any three-dimensional arrangement of atoms in a molecule that results from rotation about a single bond.

Conjugate acid *(Section 8.3)* In the Brønsted-Lowry theory, a substance formed when a base accepts a proton.

Conjugate base *(Section 8.3)* In the Brønsted-Lowry theory, a substance formed when an acid donates a proton to another molecule or ion.

Conjugated acid–base pair *(Section 8.3)* A pair of molecules or ions that are related to one another by the gain or loss of a proton.

Conjugated protein *(Section 21.9)* A protein that contains a nonprotein part, such as hemoglobin contains a heme.

Constitutional isomers *(Section 11.3)* Compounds with the same molecular formula but a different order of attachment of their atoms.

Control site *(Section 25.6)* A DNA sequence that is part of a prokaryotic operon. This sequence is upstream of the structural gene DNA and plays a role in controlling whether the structural gene is transcribed.

Cosmic rays *(Section 9.5)* High-energy particles, mainly protons, from outer space bombarding the Earth.

Covalent bond *(Section 3.4A)* A bond resulting from the sharing of electrons between two atoms.

Crenation *(Section 6.8)* The shriveling of red blood cells.

Cristae *(Section 26.2)* The inner mitochondrial membrane.

Crystallization *(Section 5.10)* The formation of a solid from a liquid.

C-terminus *(Section 21.5)* The amino acid at the end of a peptide chain that has a free carboxyl group.

Curie (Ci) *(Section 9.5)* A measure of radioactive decay equal to 3.7×10^{10} disintegrations per second.

Cyclic ether *(Section 14.4A)* An ether in which the ether oxygen is one of the atoms of a ring.

Cycloalkane *(Section 11.5)* A saturated hydrocarbon that contains carbon atoms bonded to form a ring.

Cyclooxygenase *(Section 20.12)* The enzyme that catalyzes the first step in the conversion of arachidonic acid to prostaglandins and thromboxanes.

Cystine *(Section 21.4)* A dimer of cysteine in which the two amino acids are covalently bonded by disulfide bond between their side-chain —SH groups.

Cytochrome *(Section 26.5)* A class of compounds that act as electron carriers in the inner mitochondrial membrane during electron transport. They are based on an iron-containing porphyrin ring similar to the heme group found in hemoglobin.

Cytokine *(Section 30.6)* A glycoprotein that traffics between cells and alters the function of a target cell.

Cytosol *(Section 27.2)* The soluble portion of a cell that is inside the cell membrane but outside any of the subcellular organelles.

Dalton's Law *(Section 5.5)* The pressure of a mixture of gases is equal to the sum of the partial pressure of each gas in the mixture.

Debranching enzyme *(Section 29.7)* The enzyme that catalyzes the hydrolysis of the 1,6-glycosidic bonds in starch and glycogen.

Decarboxylation *(Section 26.4)* The process that leads to the loss of CO_2 from a —COOH group.

Decay *(Section 9.3)* The change of the nucleus of a radioactive atom into a different nucleus.

Degenerate code *(Section 25.4)* The genetic code is said to be degenerate because more than one codon can code for the same amino acid.

Dehydration *(Section 14.3B)* Elimination of a molecule of water from an alcohol. An OH is removed from one carbon, and an H is removed from an adjacent carbon.

Dehydrogenase *(Section 22.6)* A class of enzymes that catalyze oxidation–reduction reactions, often using NAD^+.

Denaturation *(Section 21.11)* The loss of the secondary, tertiary, and quaternary structure of a protein by a chemical or physical agent that leaves the primary structure intact.

Dendrite *(Section 23.2)* The hair-like projections that extend from the cell body of a nerve cell on the opposite side from the axon.

Density *(Section 1.7)* The ratio of mass to volume for a substance.

Deoxyribonucleic acid *(Section 24.2)* The macromolecule of heredity in eukaryotes and prokaryotes. It is composed of chains of nucleotide monomers of a nitrogenous base, deoxyribose, and phosphate.

Dextrorotatory *(Section 15.5B)* Clockwise (to the right) rotation of the plane of polarized light in a polarimeter.

Dialysis *(Section 6.8)* The separation of larger molecules, dissolved substances, or colloidal particles from smaller molecules, substances, or particles by a semipermeable membrane.

Diastereomers *(Section 15.4A)* Stereoisomers that are not mirror images of each other.

Diatomic element *(Section 2.3B)* An element that occurs in nature as two atoms of the element joined by a covalent bond.

Digestion *(Section 29.4)* The process in which the body breaks down large molecules into smaller ones that can then be absorbed and metabolized.

Digestion *(Section 29.6)* The process of hydrolysis of starches, fats, and proteins into smaller units.

Diglyceride *(Section 20.2)* An ester of glycerol and two fatty acids.

Dilute *(Section 6.5)* A qualitative term referring to a solution that contains less solute than does a concentrated solution.

Dilution *(Section 6.5)* The process of reducing the concentration of the solute in a solution.

Diol *(Section 14.2B)* A compound containing two —OH (hydroxyl) groups.

Dipeptide *(Section 21.5)* A peptide with two amino acids.

Dipole *(Section 3.7B)* A chemical species in which there is a separation of charge; there is a positive pole in one part of the species and a negative pole in another part.

Dipole–dipole attraction *(Section 5.7B)* The attraction between the dipole of one molecule and a dipole of opposite charge of the same or another molecule.

Diprotic acid *(Section 8.3)* An acid that can give up two protons.

Disaccharide *(Section 19.5)* A carbohydrate containing two monosaccharide units joined by a glycosidic bond.

Discriminatory curtailment diet *(Section 29.2)* A diet that avoids certain food ingredients that are considered harmful to the health of an individual—for example, low-sodium diets for people with high blood pressure.

Distal tubule *(Section 31.5)* The part of the kidney tubule farthest from the Bowman's capsule.

Disulfide *(Section 14.5D)* A compound containing an —S—S— group.

Diuresis *(Section 31.7)* Enhanced excretion of water in urine.

Diuretic *(Chemical Connections 31E)* A compound that increases the flow of urine from the kidney.

DNA *(Section 24.2)* Deoxyribonucleic acid.

Double bond *(Section 3.7B)* A bond formed by sharing two pairs of electrons; represented by two lines between the two bonded atoms.

Double helix *(Section 24.3)* The arrangement of which two strands of DNA are coiled around each other in a screw-like fashion.

Ductile *(Section 2.5B)* Can be drawn into a wire.

Edema *(Section 31.2)* A swelling of tissues due to water seeping from the blood into the interstitial fluids.

Effective collision *(Section 7.2)* A collision between two molecules that results in chemical reaction.

EGF *(Section 30.6)* Epidermal growth factor, a cytokine that stimulates epidermal cells during healing of wounds.

Electrolyte *(Section 6.6)* A compound that, when dissolved in water, produces a solution that conducts electricity.

Electromagnetic spectrum *(Section 9.1)* The array of electromagnetic phenomena by wavelength.

Electron *(Section 2.4A)* A subatomic particle with a charge of -1 and a mass of approximately 0.0005 amu; it is found in the space surrounding a nucleus.

Electron configuration *(Section 2.6C)* A description of the orbitals of an atom or ion that are occupied by electrons.

Electron transport chain *(Section 26.6)* A series of electron carriers imbedded or attached to the inner mitochondrial membrane that are alternately reduced and oxidized as electrons pass from an initial electron carrier (NADH or $FADH_2$) to oxygen.

Electronegativity *(Section 3.4B)* A measure of an atom's attraction for the electrons it shares in a chemical bond with another atom.

Element *(Section 2.2A)* A pure substance composed of one type of atom.

Elongation *(Section 25.5)* The phase of protein synthesis during which activated tRNA molecules deliver new amino acids to ribosomes where they are joined by peptide bonds to form a polypeptide.

Elongation factor *(Section 25.5)* Small protein molecules that are involved in the process of tRNA binding and movement of the ribosome on the mRNA during elongation.

Emulsion *(Chemical Connections 6D)* A suspension produced from small globules of a liquid in another liquid.

Enantiomers *(Section 15.2)* Stereoisomers that are nonsuperposable mirror images; refers to a relationship between pairs of objects.

End point *(Section 8.10)* The point at which there is an equal amount of acid and base in a neutralization reaction.

Endocrine gland *(Section 23.2)* A gland, such as the pancreas, pituitary, and hypothalamus, that produces hormones involved in the control of the chemical reactions and metabolism.

Endothermic reaction *(Section 4.8)* A chemical reaction that absorbs heat.

Energy *(Section 1.8)* The capacity to do work; the joule is the base SI unit.

Energy yield *(Section 26.7)* The calculation of how many ATP molecules are formed from electrons passing down the electron transport chain. It is usually calculated as a number of ATP per NADH used, $FADH_2$ used, or H^+.

Enkephalins *(Section 23.6)* Pentapeptides found in nerve cells of the brain that act to control pain perception.

Enol *(Section 17.8)* A molecule containing an —OH group bonded to a carbon of a carbon–carbon double bond.

Enzyme *(Section 22.1)* A biological catalyst that increases the rate of a chemical reaction by providing an alternative pathway with a lower activation energy.

Enzyme activity *(Section 22.4)* The rate at which an enzyme-catalyzed reaction proceeds; commonly measured as the amount of product produced per minute.

Enzyme specificity *(Section 22.1)* The limitation of an enzyme to catalyze one specific reaction with one specific substrate.

Enzyme–substrate complex *(Section 22.5)* A part of an enzyme reaction mechanism where the enzyme is bound to the substrate.

Epitope *(Section 30.3)* The smallest number of amino acids on an antigen that elicits an immune response.

Equatorial position *(Section 11.6B)* A position on a chair conformation of a cyclohexane ring that extends from the ring roughly perpendicular to the imaginary axis of the ring.

Equilibrium *(Section 5.9A)* A condition in which two opposing physical forces are equal; a reversible chemical reaction in which the rate of the forward reaction is equal to the rate of the reverse reaction.

Equilibrium constant *(Section 7.6)* The concentrations of the products of a chemical equilibrium divided by the concentration of the reactants, with each term raised to the power equal to its coefficient in the balanced chemical equation.

Equivalent dose *(Section 9.5)* The sum of the doses of different forms of radiation.

Erythrocytes *(Section 31.2)* Red blood cells; transporters of blood gases.

Essential amino acid *(Section 29.4)* An amino acid that the body cannot synthesize in the required amounts and so must be obtained in the diet.

Essential fatty acid *(Section 20.3)* A polyunsaturated fatty acid that cannot be synthesized in the human body in sufficient quantities for normal growth and development.

Ester *(Section 18.3B)* A compound in which the OH of a carboxyl group, RCOOH, is replaced by an —OR′ group; RCOOR′.

Ether *(Section 14.4A)* A compound containing an oxygen atom bonded to two carbon atoms.

Eukaryote *(Section 25.6)* An organism that has a true nucleus and organelles. Eukaryotes include animals, plants, and fungi.

Excitatory neurotransmitter *(Section 23.4)* A neurotransmitter that increases the transmission of nerve impulses.

Exon *(Section 24.5)* A nucleotide sequence in mRNA that codes for a protein.

Exothermic reaction *(Section 4.8)* A chemical reaction that gives off heat.

Exponential notation *(Section 1.3)* The expression of a number in base 10.

External innate immunity *(Section 30.1)* The innate protection against foreign invaders characteristic of the skin barrier, tears, and mucus.

Extracellular fluids *(Section 31.1)* All the body fluids that are not inside the cell.

Fact *(Section 1.2)* A statement based on experience.

FAD/$FADH_2$ *(Section 26.3)* Flavin adenine dinucleotide, a coenzyme involved in metabolic oxidation–reduction reactions.

Fat *(Section 20.2)* A mixture of triglycerides containing a high proportion of long-chain, saturated fatty acids.

Fat depot *(Section 27.1)* The supply of stored fat in the form of triacylglycerols, usually found in fat cells or distributed in other tissues.

Feedback control *(Section 22.6)* A type of enzyme regulation where the product of a series of reactions inhibits the enzyme that catalyzes the first reaction in the series.

Fiber *(Section 29.2)* The cellulosic, non-nutrient component in our food.

Fibrous protein *(Section 21.1)* A protein used for structural purposes. Fibrous proteins are insoluble in water and have a high percentage of secondary structures, such as alpha helices and/or beta-pleated sheets.

Fischer esterification *(Section 18.4)* The process of forming an ester by refluxing a carboxylic acid and an alcohol in the presence of an acid catalyst, commonly sulfuric acid.

Fischer projection *(Section 19.2B)* A two-dimensional representation for showing the configuration of a stereocenter; horizontal lines represent bonds projecting forward from the stereocenter, and vertical lines represent bonds projecting toward the rear.

Fission, nuclear *(Section 9.8)* The fragmentation of a heavy nucleus into smaller nuclei.

Fluid mosaic model *(Section 20.5)* A model for the structure and function of biological membranes.

Formula weight (FW) *(Section 4.2)* The sum of the atomic weights of all atoms is a compound's formula expressed in atomic mass units (amu). Formula weight can be used for both ionic and molecular compounds.

Free radical *(Chemical Connections 9C)* A molecule or ion with one or more unpaired electrons.

Freezing point depression *(Section 6.8)* The decrease in the freezing point of a liquid caused by adding a solute.

Frequency *(Section 9.1)* The number of wave crests passing a point per minute.

Functional group *(Section 10.4)* An atom or group of atoms within a molecule that shows a characteristic set of physical and chemical properties.

Furanose *(Section 19.3A)* A five-membered cyclic hemiacetal form of a monosaccharide.

Fusion, nuclear *(Section 9.7)* A combining of two or more lighter nuclei to form a heavier nucleus.

Gamma ray *(Section 9.1)* A form of electromagnetic radiation characterized by a very short wavelength and very high energy.

Gas *(Section 1.6)* A physical state of matter in which there is no definite volume or shape.

Gay-Lussac's Law *(Section 5.3C)* The pressure of a gas at constant volume is directly proportional to its temperature in kelvins.

Geiger-Müller counter *(Section 9.4)* An instrument for measuring ionizing radiation.

Gene *(Section 24.1)* The unit of heredity; a DNA segment that codes for one protein.

Gene expression *(Section 25.1)* The activation of a gene to produce a specific protein. It involves both transcription and translation.

Gene regulation *(Section 25.6)* The various methods used by organisms to control which genes will be expressed and when.

Genetic code *(Section 25.4)* The sequence of triplets of nucleotides (codons) that determines the sequence of amino acids in a protein.

Genetic engineering *(Section 25.8)* The process by which genes are inserted into cells.

Genome *(Chemical Connections 24D)* A complete DNA sequence of an organism.

Globular protein *(Section 21.1)* Protein that is used mainly for non-structural purposes and is largely soluble in water.

Glomeruli *(Section 31.5)* Part of the kidney filtration cell, nephron; a tangle of capillaries surrounded by a fluid-filled space.

Glucogenic amino acid *(Sections 27.9, 28.2)* An amino acid whose carbon skeleton can be used for the synthesis of glucose.

Gluconeogenesis *(Section 28.2)* The biosynthetic pathway that converts pyruvate, citric acid cycle intermediates, or glucogenic amino acids into glucose.

Glycerophospholipid *(Section 20.4)* A diester of glycerol and two fatty acids. The third —OH group of glycerol is esterified with phosphoric acid, which is, in turn, esterified with a low-molecular-weight alcohol such as choline or ethanolamine.

Glycogenesis *(Section 28.2)* The synthesis of glycogen from glucose.

Glycogenolysis *(Section 27.3)* The biochemical pathway for the breakdown of glycogen to glucose.

Glycol *(Section 14.2B)* A compound with two hydroxyl (—OH) groups on adjacent carbons.

Glycolipid *(Section 20.4)* A complex lipid that contains at least one monosaccharide.

Glycolysis *(Section 27.2)* The biochemical pathway that breaks down glucose to pyruvate, which yields chemical energy in the form of ATP and reduced coenzymes.

Glycomics *(Chemical Connections 30E)* The collective knowledge of all the carbohydrates, including glycoproteins and glycolipids, that a cell or a tissue contains and the determination of their functions.

Glycoprotein *(Section 21.10)* A protein to which one or more carbohydrate molecules are bonded.

Glycoside *(Section 19.4A)* A carbohydrate in which the —OH group on its anomeric carbon is replaced by an —OR group.

Glycosidic bond *(Section 19.4A)* The bond from the anomeric carbon of a glycoside to an —OR group.

Gp120 *(Section 30.5)* A 120,000-molecular-weight glycoprotein on the surface of the human immunodeficiency virus that binds strongly to the CD4 molecules on T-cells.

G-protein *(Section 23.5)* A protein that is either stimulated or inhibited when a hormone binds to a receptor and that subsequently alters the activity of another protein, such as adenylate cyclase.

Gray (Gy) *(Section 9.5)* The SI unit of measure of the rad; 100 rad.

Ground-state electron configuration *(Section 2.6)* The electron configuration of an atom of lowest energy.

Haber process *(Chemical Connections 7E)* The industrial process by which N_2 and H_2 are converted to NH_3.

Half-life, of a radioisotope *(Section 9.3)* The time it takes for one-half of a sample to decay.

Halogen *(Section 2.5A)* An element in Group 7A of the Periodic Table.

Haworth projection *(Section 19.3A)* A way to view furanose and pyranose forms of monosaccharides; the ring is drawn flat and viewed through its edge, with the anomeric carbon on the right and the oxygen atom to the rear.

HDL *(Section 20.9)* High-density lipoprotein; a lipoprotein with high protein content (approximately 50 percent).

Heat of combustion *(Section 4.8)* The heat given off in a combustion reaction.

Heat of reaction *(Section 4.8)* The heat given off or absorbed in a chemical reaction.

Helicase *(Section 24.4)* An unwinding protein that acts at a replication fork to unwind DNA so that DNA polymerase can synthesize a new DNA strand.

Hemiacetal *(Section 17.7)* A molecule containing a carbon bonded to one —OH and one —OR group; the product of adding one molecule of alcohol to the carbonyl group of an aldehyde or ketone.

Hemodialysis *(Chemical Connections 6F)* The dialysis of blood.

Hemolysis *(Section 6.8)* The bursting of red blood cells.

Henderson-Hasselbalch equation *(Section 8.12)* A mathematical relationship between pH, the pK_a of a weak acid, and the concentrations of the weak acid and its conjugate base.

$$pH = pK_a + \log \frac{[A^-]}{[HA]}$$

Henle's loop *(Section 31.5)* A U-shaped twist in a kidney tubule between the proximal tubule and the distal tubule.

Henry's Law *(Section 6.4)* The solubility of a gas in a liquid is directly proportional to the pressure.

Heterocyclic aliphatic amine *(Section 16.2)* A heterocyclic amine in which nitrogen is bonded only to alkyl groups.

Heterocyclic amine *(Section 16.2)* An amine in which nitrogen is one of the atoms of a ring.

Heterocyclic aromatic amine *(Section 16.2)* An amine in which nitrogen is one of the atoms of an aromatic ring.

Heterogeneous catalyst *(Section 7.4)* A catalyst in a separate phase from the reactants.

Histone *(Section 24.1)* A basic protein that is found in complexes with DNA in eukaryotes.

HIV *(Section 30.5)* Human immunodeficiency virus.

Homeostasis *(Section 31.1)* The process of maintaining the proper levels of nutrients and temperature of the blood.

Homogeneous catalyst *(Section 7.4)* A catalyst in the same phase as the reactants—for example, enzymes in body tissues.

Hormone *(Section 23.2)* A chemical messenger released by an endocrine gland into the bloodstream and transported there to reach its target cell.

Hybridization *(Section 24.6)* A process whereby two strands of nucleic acids or segments thereof form a double-stranded structure through H-bonding of complementary base pairs.

Hydrate *(Section 6.6)* Substances that contain water in their crystals.

Hydration *(Section 12.6B)* Addition of water.

Hydrocarbon *(Section 11.2)* A compound that contains only carbon and hydrogen atoms.

Hydrogen bond *(Section 5.7C)* A noncovalent force of attraction between the partial positive charge on a hydrogen bonded to an atom of high electronegativity, most commonly O or N, and the partial negative charge on a nearby O or N.

Hydrolase *(Section 22.2)* An enzyme that catalyzes a hydrolysis reaction.

Hydrolysis *(Section 17.7)* A reaction of a compound with water (*hydro-*) in which one or more bonds are broken (*-lysis*) and the —H and —OH of water add to the ends of the bond or bonds broken.

Hydronium ion *(Section 8.1)* The H_3O^+ ion.

Hydrophobic interaction *(Section 21.9)* Interaction by London dispersion forces between hydrophobic groups.

Hydroxyl group *(Section 10.4A)* An —OH group bonded to a tetrahedral carbon atom.

Hygroscopic *(Section 6.6)* A substance that absorbs water from the air.

Hyperglycemia *(Chemical Connections 29D)* A condition whereby the blood glucose level reaches 140 mg/dL or higher.

Hypertension *(Chemical Connections 31E)* The clinical term for high blood pressure.

Hyperthermophile *(Section 22.4)* An organism that lives at extremely high temperatures.

Hypertonic solution *(Section 6.8)* A solution that has higher osmolarity than another.

Hypoglycemia *(Chemical Connections 29D)* A condition whereby the blood glucose level drops below the normal 65–100 mg/dL.

Hypothermia *(Chemical Connections 1B)* Having a body temperature lower than normal.

Hypothesis *(Section 1.2)* A statement that is proposed, without actual proof, to explain certain facts and their relationship.

Hypotonic solution *(Section 6.8)* A solution that has lower osmolarity than another.

Ideal gas *(Section 5.6)* A gas that obeys the assumptions of the kinetic molecular theory.

Ideal gas constant (R) *(Section 5.4)* 0.0821 L·atm·mol^{-1}·K^{-1}.

Ideal gas law *(Section 5.4)* $PV = nRT$.

Immunogen *(Section 30.3)* Another term for antigen.

Immunoglobulin *(Section 30.4)* An antibody protein generated against and capable of binding specifically to an antigen.

Immunoglobulin superfamily *(Section 30.4)* A family of molecules based on a similar structure that includes the immunoglobulins, T-cell receptors, and other membrane proteins that are involved in cell communications. All molecules in this class have a certain portion that can react with antigens.

Indicator, acid–base *(Section 8.8)* A substance that changes color within a given pH range.

Induced-fit model *(Section 22.5)* A model explaining the specificity of enzyme action by comparing the active site to a glove and the substrate to a hand.

Inhibition, of enzyme activity (Section 22.3) Any reversible or irreversible process that makes an enzyme less active.

Inhibitory neurotransmitters (Section 23.4) Neurotransmitters that decrease the transmission of nerve impulses.

Initiation, of protein synthesis (Section 25.5) The first step in the process whereby the base sequence of an mRNA is translated into the primary structure of a polypeptide.

Initiation signal (Section 25.2) A sequence on DNA that identifies the location where transcription is to begin.

Innate immunity (Section 30.1) The first line of defense against foreign invaders, which includes skin resistance to penetration, tears, mucus, and nonspecific macrophages that engulf bacteria.

Inner transition element (Section 2.5A) One of a group of elements in the Periodic Table characterized by the filling of f orbitals.

Inorganic phosphate (Section 26.3) PO_4^{3-}.

Interleukin (Section 30.6) A cytokine that controls and coordinates the action of leukocytes.

Internal innate immunity (Section 30.1) The type of innate immunity that is used once a pathogen has already penetrated a tissue.

International System of Units (SI) (Section 1.4) A system of units of measurement based in part on the metric system.

Interstitial fluid (Section 30.2) The fluid that surrounds and fills in spaces between cells.

Intron (Section 24.5) A nucleotide sequence in mRNA that does not code for a protein.

Ion (Section 2.8) An atom with an unequal number of protons and electrons.

Ion product of water, K_w (Section 8.7) The concentration of H_3O^+ multiplied by the concentration of OH^-; $[H_3O^+][OH^-] = 1 \times 10^{-14}$.

Ionic bond (Section 3.4A) A chemical bond resulting from the attraction between a positive ion and a negative ion.

Ionic compound (Section 3.5A) A compound formed by the combination of positive and negative ions.

Ionization (Section 6.6) Dissociation of a compound of compounds into ions.

Ionization energy (Section 2.8) The energy required to remove the most loosely held electron from at atom.

Ionizing radiation (Section 9.4) Radiation that causes one or more electrons to be ejected from an atom or molecule, thereby creating a positive ion.

Isoelectric point (Section 21.3) The pH at which a molecule has no net charge; abbreviated pI.

Isoenzyme (Section 22.6) An enzyme that can be found in multiple forms but with each form catalyzing the same reaction; also called an isozyme.

Isomerase (Section 22.2) An enzyme that catalyzes an isomerization reaction.

Isomers (Section 11.3) Different compounds that have the same molecular formula.

Isotopes (Section 2.4D) Atoms that have the same number of protons in their nuclei but different numbers of neutrons.

Isozymes (Section 22.6) Enzymes that perform the same function but have different combinations of subunits—that is, different quaternary structures.

Joule (J) (Section 1.9) SI base unit for heat; 1 J is 4.184 cal.

Kelvin scale (K) (Section 1.4) SI base unit for temperature.

Ketoacidosis (Chemical Connections 27C) A biochemical condition caused by starvation or diabetes where ketone bodies build up in the blood and cause the pH to decrease.

Ketogenic amino acid (Section 27.9) An amino acid that, when catabolized, can only form acetyl-CoA or acetoacetyl-CoA and therefore cannot lead to glucose production.

Ketone (Section 10.4C) A compound containing a carbonyl group bonded to two carbons.

Ketone bodies (Section 27.7) A collective name for acetone, acetoacetate and β-hydroxybutyrate; compounds produced from acetyl-CoA in the liver that are used as a fuel for energy production by muscle cells and neurons.

Ketose (Section 19.2A) A monosaccharide containing a ketone group.

Killer T cell (Section 30.2) A T cell that kills invading foreign cells by cell-to-cell contact.

Kinase (Section 22.6) An enzyme that covalently modifies a protein with a phosphate group, usually through the —OH group on the side chain of a serine, threonine, or tyrosine.

Kinetic energy (Section 1.8) The energy of motion; energy that is in the process of doing work.

Kinetic molecular theory (Section 5.6) A theory advanced to explain the behavior of gases.

Krebs cycle (Section 26.4) Another name for the citric acid cycle, used to honor Hans Krebs, the discoverer of the pathway.

Kwashiorkor (Section 29.4) A disease caused by insufficient protein intake and characterized by a swollen stomach, skin discoloration, and retarded growth.

Lactam (Section 18.3C) A cyclic amide.

Lactone (Section 18.3B) A cyclic ester.

Lagging strand (Section 24.4) A discontinuously synthesized DNA that elongates in a direction away from the replication fork.

Law of conservation of energy (Section 1.8) Energy can be neither created nor destroyed.

LDL (Section 20.9) Low-density lipoprotein; a circulating lipoprotein where the cholesterol content is very high (45 percent) and the protein content is 25 percent or less.

Le Chatelier's principle (Section 7.8) Principle stating that when a stress is applied to a chemical system in equilibrium, the position of the equilibrium shifts in the direction to relieve the applied stress.

Leading strand (Section 24.4) The continuously synthesized DNA strand that elongates toward the replication fork.

Lecithin (Section 20.6) The common name for a phosphatidylcholine.

Leukocytes (Sections 30.2, 31.2) White blood cells, which are the principal parts of the acquired immunity system and act via phagocytosis or antibody production.

Leukotrienes (Section 20.12) A class of acyclic arachidonic acid derivatives that act as mediators of hormonal responses.

Levorotatory (Section 15.5B) Counterclockwise rotation of the plane of polarized light in a polarimeter.

Lewis dot structure (Section 2.6F) The symbol of an element surrounded by a number of dots equal to the number of electrons in the valence shell of an atom of the element.

Lewis structure (Section 3.7) A formula for a molecule or ion showing all pairs of bonding electrons as single, double, or triple lines, and all nonbonding electrons as pairs of Lewis dots.

Ligases (Section 22.2) A class of enzymes that catalyze a reaction joining two molecules. They are often called synthetases or synthases.

Limiting reagent (Section 4.5B) The reactant that is consumed, leaving an excess of another reagent or reagents unreacted.

Line-angle formula (Section 11.2) An abbreviated way to draw structural formulas in which each angle and line terminus represents a carbon atom and each line represents a bond.

Lipase (Section 29.8) An enzyme that catalyzes the hydrolysis of an ester bond between a fatty acid and glycerol.

Lipid bilayer (Section 20.5) A back-to-back arrangement of phospholipids, often forming a closed vesicle or membrane.

Lipids (Section 20.1) A class of compounds found in living organisms that are insoluble in water but soluble in nonpolar solvents.

Lipoproteins (Section 20.9) Spherically shaped clusters containing both lipid molecules and protein molecules.

Liquid state (Section 1.6) A physical state of matter in which there is a definite volume but no specific shape.

Lock-and-key model (Section 22.5) A model for enzyme substrate interaction based on the postulate that the active site of an enzyme is a perfect fit for the substrate.

London dispersion forces (Section 5.7A) Extremely weak attractive forces between atoms or molecules caused by the electrostatic attraction between temporary induced positive and negative dipoles.

Lyase (Section 22.2) An enzyme that catalyzes the addition of two atoms or groups of atoms to a double bond or their removal to form a double bond.

Lymphocyte (Section 30.2) A white blood cell that spends most of its time in the lymphatic tissues. Those that mature in the bone marrow are B cells. Those that mature in the thymus are T cells.

Lymphoid organs *(Section 30.2)* The main organs of the immune system, such as the lymph nodes, spleen, and thymus, which are connected together by lymphatic capillary vessels.

Lysosome *(Section 26.2)* A subcellular organelle that cleanses a cell of damaged cellular components and some foreign material.

Macrophage *(Section 30.2)* An amoeboid white blood cell that moves through tissue fibers, engulfing dead cells and bacteria by phagocytosis, and then displays some of the engulfed antigens on its surface.

Magnetic resonance imaging (MRI) *(Chemical Connections 9E)* The detection of the inversion of spin (flip) of hydrogen nuclei caused by exposure to radiofrequency radiation.

Main-group element *(Section 2.5A)* An element in any of the A groups of the Periodic Table.

Major histocompatibility complex (MHC) *(Section 30.3)* A transmembrane protein complex that brings the epitope of an antigen to the surface of the infected cell to be presented to the T cells.

Manometer *(Section 5.2)* An instrument for measuring the pressure of a gas within a container.

Marasmus *(Section 29.3)* Another term for chronic starvation whereby the individual does not have adequate caloric intake. It is characterized by arrested growth, muscle wasting, anemia, and general weakness.

Markovnikov's rule *(Section 12.6A)* In the addition of HX or H_2O to an alkene, hydrogen adds to the carbon of the double bond having the greater number of hydrogens.

Mass *(Section 1.4)* The quantity of matter in an object; the SI base unit is the kilogram; often referred to as weight.

Mass number *(Section 2.4B)* The sum of the number of protons and neutrons in the nucleus of an atom.

Matter *(Section 1.1)* Anything that has mass and takes up space.

Membrane *(Section 20.5)* A structure composed of a lipid bilayer, protein, and often carbohydrate that separates one cell compartment from another or one cell from another.

Memory *(Section 30.1)* The characteristic of the acquired immunity system seen as a more rapid and vigorous response to a pathogen the second time it is encountered.

Memory cell *(Section 30.2)* A type of T cell that stays in the blood after an infection is over and acts as a quick line of defense if the same antigen is encountered again.

Mercaptan *(Section 14.5A)* A common name for any molecule containing an —SH group.

Messenger RNA (mRNA) *(Section 24.4)* The RNA that carries genetic information from DNA to the ribosome and acts as a template for protein synthesis.

Meta (m) *(Section 13.3B)* Refers to groups occupying the 1 and 3 positions on a benzene ring.

Metabolism *(Section 26.1)* The sum of all the chemical reactions involved in maintaining the dynamic state of a cell or organism.

Metal *(Section 2.5B)* An element that is solid (except for mercury, which is a liquid), shiny, conducts electricity, ductile, and malleable, and that forms alloys.

Metal-binding finger *(Section 25.6)* A type of transcription factor containing heavy metal ions, such as Zn^{2+}, that is involved in helping RNA polymerase bind to the DNA to be transcribed.

Metalloid *(Section 2.5B)* An element that displays some of the characteristic properties of metals and some of nonmetals.

Metastasis *(Chemical Connections 25E)* The process whereby cancerous cells proliferate and spread to other tissues.

Meter *(Section 1.4)* The SI base unit of length.

Metric system *(Section 1.4)* A system in which measurements of parameter are related by powers of 10.

Mineral in diet *(Section 29.5)* An inorganic ion that is required for structure and metabolism, such as Ca^{2+}, Fe^{2+}, and Mg^{2+}.

Mirror image *(Section 15.1)* The reflection of an object in a mirror.

Mitochondrion *(Section 26.2)* A subcellular organelle characterized by an outer membrane and a highly folded inner membrane. Mitochondria are responsible for the generation of most of the energy for cells.

Mixture *(Section 2.2C)* A sample of matter that is composed of two or more pure substances and that can be separated into its components by appropriate physical means.

Molar mass *(Section 4.3)* The mass of one mole of a substance expressed in grams; the formula weight of a compound expressed in grams.

Molarity (M) *(Section 6.5)* Concentration expressed as the number of moles of a solute per liter of solution.

Mole (mol) *(Section 4.3)* The formula weight of a substance expressed in grams.

Molecular weight (MW) *(Section 4.2)* The sum of the atomic weights of all atoms in a molecular compound expressed in atomic mass units (amu).

Molecule *(Section 2.3)* A compound in which all bonds are covalent.

Monatomic element *(Section 2.3B)* An element that occurs in nature as single atoms.

Monoclonal antibody *(Section 30.5)* An antibody produced by clones of a single B cell specific to a single epitope.

Monoglyceride *(Section 20.2)* An ester of glycerol and a single fatty acid.

Monomer *(Section 12.7A)* From the Greek *mono*, "single," and *meros*, "part"; the simplest nonredundant unit from which a polymer is synthesized.

Monoprotic acid *(Section 8.3)* An acid that can give up only one proton.

Monosaccharide *(Section 19.2)* A carbohydrate that cannot be hydrolyzed to a simpler compound.

D-Monosaccharide *(Section 19.2C)* A monosaccharide that, when written as a Fischer projection, has the —OH group on its penultimate carbon to the right.

L-Monosaccharide *(Section 19.2C)* A monosaccharide that, when written as a Fischer projection, has the —OH group on its penultimate carbon to the left.

Mutagen *(Section 25.7)* A chemical substance that induces a base change, or mutation, in DNA.

Mutarotation *(Section 19.3C)* The change in specific rotation that occurs when an α or β form of a carbohydrate is converted to a equilibrium mixture of the two forms.

Mutation *(Section 9.5)* A heritable change in the base sequence of DNA.

NAD⁺/NADH *(Section 26.3)* Nicotinamide adenine dinucleotide, a coenzyme participating in many metabolic oxidation–reduction reactions.

Native structure *(Section 21.10)* The natural or normal conformation of a protein in the biological setting.

Negative modulation *(Section 22.6)* The process whereby an allosteric regulator inhibits enzyme action.

Nephron *(Section 31.5)* The tubular excretory unit of the kidney that filters the blood to remove waste products and to reabsorb nutrients.

Net ionic equation *(Section 4.6)* A chemical equation that does not contain spectator ions.

Network solid *(Section 5.10)* A extremely large solid composed of a single crystal.

Neuron *(Section 23.1)* Another name for a nerve cell.

Neuropeptide Y *(Section 23.6)* A brain peptide that affects the hypothalamus and is an appetite-stimulating agent.

Neurotransmitter *(Section 23.2)* Chemical messengers between a neuron and another target cell: neuron, muscle cell, or cell of a gland.

Neutral salt *(Section 8.9)* A salt whose aqueous solution is neither acidic nor basic.

Neutral solution *(Section 8.7)* An aqueous solution whose pH is 7.0; an aqueous solution that is neither acidic nor basic.

Neutron *(Section 2.4A)* A subatomic particle with a mass of 1 amu and a zero charge; it is found in the nucleus.

N-linked saccharide *(Section 21.10)* A carbohydrate that is bonded by a N-glycosidic bond to a protein, most commonly to the side-chain nitrogen of asparagine.

Noble gas *(Section 2.5C)* An element in Group 8A of the Periodic Table.

Nonbonding electrons *(Section 3.7C)* Valence electrons not involved in forming covalent bonds; that is, unshared electrons.

Noncompetitive inhibition (Section 22.5) An enzyme regulation in which an inhibitor binds to the enzyme outside of the active site, thereby changing the shape of the active site and reducing its catalytic activity.

Nonmetal (Section 2.5B) An element that does not have the characteristic properties of a metal and tends to accept electrons.

Nonpolar covalent bond (Section 3.7B) A covalent bond between two atoms whose difference in electronegativity is less than 0.5.

Normal boiling point (Section 5.9B) The temperature at which a liquid boils under a pressure of 1 atm.

N-terminus (Section 21.5) The amino acid at the end of a peptide chain that has a free amino group.

Nuclear fission (Section 9.8) The process of splitting a nucleus.

Nuclear fusion (Section 9.7) Joining together atomic nuclei in forming nuclides heavier than the starting material.

Nuclear reaction (Section 9.2) A reaction that changes atomic nuclei of elements (usually to those of other elements).

Nucleic acid (Section 24.3) A polymer composed of nucleotides.

Nucleoside (Section 24.2) The combination of a heterocyclic aromatic amine attached via a β-glycosidic bond to either D-ribose or 2-deoxy-D-ribose.

Nucleosome (Section 24.6) Combinations of DNA and histone proteins.

Nucleotide (Section 24.2) A phosphoric ester of a nucleoside.

Nucleus (Section 26.2) A subcellular organelle found in eukaryotes where the DNA is stored and replicated.

Nutrient (Section 29.2) Components of food and drink that provide energy, replacement, and growth.

Obesity (Section 29.3) Accumulation of body fat beyond the norm.

Octet rule (Section 3.2) When undergoing chemical reaction, atoms of Group 1A–7A elements tend to gain, lose, or share electrons to achieve an election configuration having eight valence electrons.

Oil (Section 20.2) A mixture of triglycerides containing a high proportion of long-chain unsaturated fatty acids or short-chain saturated fatty acids.

Okazaki fragment (Section 24.6) A short DNA segment made of about 200 nucleotides in higher organisms (eukaryotes) and 2000 nucleotides in prokaryotes.

Oligosaccharide (Section 19.5) A carbohydrate containing from 4 to 10 monosaccharide units, each joined to the next by a glycosidic bond.

O-linked saccharide (Section 21.10) A carbohydrate that is bonded by a glycosidic bond to a protein, most commonly to the side-chain OH of a serine, threonine, or hydroxylysine.

Oncogene (Chemical Connections 25E) A gene that in some way participates in the development of cancer.

Optically active (Section 15.5B) Showing that a compound rotates the plane of polarized light.

Orbital (Section 2.6A) A region of space around a nucleus that can hold up to two electrons.

Organelle (Section 26.2) A special structure within a cell that carries out a particular part or parts of metabolism.

Organic chemistry (Section 10.1) The study of the compounds of carbon.

Ortho (o) (Section 13.3B) Refers to groups occupying the 1 and 2 positions on a benzene ring.

Osmotic pressure (Section 6.8) The amount of external pressure applied to the more concentrated solution to stop the passage of solvent molecules across a semipermeable membrane.

Oxidation (Section 4.7) The loss of electrons; the gain of oxygen or the loss of hydrogen.

Oxidative deamination (Section 27.8) A reaction in which the amino group of an amino acid is removed in the form of NH_4^+ and an α-ketoacid is formed.

Oxidative phosphorylation (Section 26.5) The process of generating ATPs by producing a hydrogen ion gradient during electron transport and then harnessing the energy of the gradient by letting the ions flow through the ATPase.

Oxidizing agent (Section 4.7) An entity that accepts electrons in an oxidation–reduction reaction.

Oxidoreductase (Section 22.2) An enzyme that catalyzes an oxidation–reduction reaction.

Oxonium ion (Section 12.6B) An ion in which oxygen is bonded to three other atoms and bears a positive charge.

P site (Section 25.5) The site on the large ribosomal subunit where the current peptide is bound before peptidyl transferase links it to the amino acid attached at the A site during elongation.

Para (p) (Section 13.3B) Refers to groups occupying the 1 and 4 positions on a benzene ring.

Parenteral nutrition (Chemical Connections 29A) The technical term for intravenous feeding.

Partial pressure (Section 5.5) The pressure that a gas in a mixture of gases would exert if it were alone in the container.

Pentose phosphate pathway (Section 27.2) A pathway, often called the hexose monophosphate shunt, in which glucose 6-phosphate is decarboxylated to yield ribulose 5-phosphate. When necessary, this pathway can produce ribose for nucleotide synthesis and NADPH, which is required for many synthetic pathways.

Penultimate carbon (Section 19.2B) The stereocenter of a monosaccharide farthest from the carbonyl group—for example, carbon 5 of glucose.

Peptide (Section 21.5) A short chain of amino acids linked via peptide bonds.

Peptide backbone (Section 21.6) The repeating pattern of peptide bonds in a polypeptide or protein.

Peptide bond (Section 21.5) An amide bond that links two amino acids.

Peptidergic neurotransmitter (Section 23.6) A type of neurotransmitter/hormone that is based on a peptide, such as glucagon, insulin, and the enkephalins.

Percent composition (Section 6.5) A concentration expression in which the amount of a substance in a solution is compared to the total amounts of the solution components (weight/volume, weight/weight, volume/volume).

Percent yield (Section 4.5C) The actual yield divided by the theoretical yield times 100.

Perforin (Section 30.2) A protein produced by killer T cells that punches holes in the membrane of target cells.

Period, of the Periodic Table (Section 2.5A) A horizontal row of the Periodic Table.

pH (Section 8.8) The negative logarithm of the hydronium ion concentration; $pH = -\log[H_3O^+]$.

Phagocytosis (Section 30.6) The process by which large particulates, including bacteria, are pulled inside a white cell called a phagocyte.

Phase change (Section 5.11) A change from one physical state (gas, liquid, or solid) to another.

Phenol (Section 13.5) A compound that contains an —OH group bonded to a benzene ring.

Phenyl group (Section 13.3) C_6H_5—, the aryl group derived by removing a hydrogen from benzene.

Pheromone (Chemical Connections 12B) A chemical secreted by an organism to influence the behavior of another member of the same species.

Phosphatidylcholine (Section 20.6) A glycerophospholipid where the substituent attached via a phosphoric ester bond is choline.

Phosphatidylinositol (Section 20.6) A glycerophospholipid where the substituent attached via a phosphoric ester bond is inositol.

Phospholipid (Section 20.4) A complex lipid comprising an alcohol, such as glycerol or sphingosine, a phosphate group, and fatty acids.

Photon (Section 9.1) The smallest unit of electromagnetic radiation.

Photosynthesis (Section 28.2) The process in which plants synthesize carbohydrates from CO_2 and H_2O with the help of sunlight and chlorophyll.

Physical change (Section 1.1) A change in matter in which it does not lose its identity.

Physical property (Section 1.1) Characteristics of a substance that are not chemical properties; those properties that are not a result of a chemical change.

PIP$_2$ (Section 20.6) Phosphatidylinositol 4,5-bisphosphate.

Plane-polarized light (Section 15.5A) Light vibrating in only parallel planes.

Plasma cell *(Section 30.2)* A cell derived from a B cell that has been exposed to an antigen.

Plasmid *(Section 25.8)* A small, circular, double-stranded DNA molecule of bacterial origin. Plasmids were found to naturally confer antibiotic resistance for certain bacteria. Now they are used for genetic engineering purposes.

Plastic *(Chemical Connections 12D)* A polymer that can be molded when hot and that retains its shape when cooled.

Platelets *(Section 31.2)* Special cells in the blood that control bleeding when there is an injury; also called thrombocytes.

pOH *(Section 8.8)* The negative logarithm of the hydroxide ion concentration; pOH = $-\log[OH^-]$.

Polar covalent bond *(Section 3.7B)* A covalent bond between two atoms whose difference in electronegativity is between 0.5 and 1.9.

Polarimeter *(Section 15.5B)* An instrument for measuring the ability of a compound to rotate the plane of polarized light.

Polyamide *(Section 18.8A)* A polymer in which each monomer unit is bonded to the next by an amide bond—or example, nylon-66.

Polyatomic element *(Section 2.3B)* An element that occurs in nature as four or more atoms of the element joined by covalent bonds.

Polyatomic ion *(Section 3.3C)* An ion that contains more than one atom.

Polycarbonate *(Section 18.8C)* A polyester in which the carboxyl groups are derived from carbonic acid.

Polyester *(Section 18.8B)* A polymer in which each monomer unit is bonded to the next by an ester bond—for example, poly(ethylene terephthalate).

Polymer *(Section 12.7A)* From the Greek *poly*, "many," and *meros*, "parts"; any long-chain molecule synthesized by bonding together many single parts called monomers.

Polymerase chain reaction (PCR) *(Sections 22.4, 24.6)* An automated technique for amplifying DNA using a heat-stable DNA polymerase from thermophilic bacteria.

Polynuclear aromatic hydrocarbon *(Section 13.3D)* A hydrocarbon containing two or more benzene rings, each of which shares two carbon atoms with another benzene ring.

Polypeptide *(Section 21.5)* A long chain of amino acids bonded via peptide bonds.

Polysaccharide *(Section 19.6)* A carbohydrate containing a large number of monosaccharide units, each joined to the next by one or more glycosidic bonds.

Positive modulation *(Section 22.6)* The process whereby an allosteric regulator increases enzyme action.

Positron *(Section 9.3)* A particle with the mass of an electron and a positive charge, $(^{0}_{+1}e)$.

Positron emission tomography (PET) *(Chemical Connections 9E)* The detection scan of positrons emitting isotopes in different organs, i.e. brains.

Postsynaptic membrane *(Section 23.2)* The membrane on the side of the synapse nearest the dendrite of the neuron receiving the transmission.

Potential energy *(Section 1.8)* Energy that is being stored; energy that is available for later use.

Pressure *(Section 5.2)* The force per unit area exerted against a surface.

Presynaptic membrane *(Section 23.2)* The membrane on the side of the synapse nearest the axon of the neuron transmitting the signal.

Primary (1°) alcohol *(Section 10.4A)* An alcohol in which the carbon atom bearing the —OH group is bonded to only one other carbon.

Primary (1°) amine *(Section 10.4B)* An amine in which nitrogen is bonded to one carbon and two hydrogens.

Primary structure, of DNA *(Section 24.3)* The order of the bases in DNA.

Primary structure, of proteins *(Section 21.7)* The order of amino acids in a peptide, polypeptide, or protein.

Principal energy level *(Section 2.6A)* An energy level containing orbitals of all the same number (1, 2, 3, 4, and so forth).

Proenzyme *(Section 22.6)* An inactive form of an enzyme that must have part of its polypeptide chain cleaved before it becomes active.

Prokaryote *(Section 25.6)* An organism that has no true nucleus or organelles.

Promoter *(Chemical Connections 25C)* An upstream DNA sequence that is used for RNA polymerase recognition and binding to DNA.

Proportional counter *(Section 9.4)* An instrument that counts the number of He^+ ions caused by radiation.

Prostaglandin *(Section 20.12)* A fatty acid-like substance derived from arachidonic acid that is involved in inflammation and smooth muscle metabolism.

Prosthetic group *(Section 21.9)* The non-amino-acid part of a conjugated protein.

Protein *(Section 21.1)* A long chain of amino acids linked via peptide bonds. There are usually 30 to 50 amino acids in a chain before it is considered a protein.

Protein modification *(Section 22.7)* The process of affecting the enzyme activity by covalently modifying the enzyme, such as phosphorylating a particular amino acid.

Proteoglycans *(Section 21.10)* A special class of glycoproteins where the protein is the core and the carbohydrate part is a long side chains of acidic polysaccharides also called glycosaminoglycans.

Proteomics *(Chemical Connections 21G)* The collective knowledge of all the proteins and peptides of a cell or a tissue and their functions.

Proton *(Section 2.4A)* A subatomic particle with a charge of +1 and a mass of approximately 1 amu; it is found in a nucleus.

Proton channel *(Section 26.6)* An opening in a membrane that will let hydrogen ions pass through it. The ATPase has such a channel that will allow hydrogen ions to pass back into the mitochondria while the energy is used to make ATP.

Proton gradient *(Section 26.6)* A difference in proton (hydrogen ion) concentration across a membrane. During the electron transport process, hydrogen ions are pumped out of the mitochondria, causing a gradient to form, with the outside having a higher $[H^+]$.

Proton-translocating ATPase *(Section 26.6)* The ATPase of the inner mitochondrial membrane allows protons to pass through it and uses the energy to phosphorylate ADP to ATP. The enzyme could also hydrolyze ATP and use this energy to move hydrogen ions out of the matrix.

Proximal tubule *(Section 31.5)* The part of the kidney tubule close to the Bowman's capsule.

Pyranose *(Section 19.3A)* A six-membered cyclic hemiacetal form of a monosaccharide.

Quaternary structure *(Section 21.9)* The organization of a protein that has multiple polypeptide chains, or subunits; Refers principally to the way the multiple chains interact.

R *(Section 15.3)* From the Latin *rectus*, meaning "straight, correct"; used in the *R,S* system to show that, when the lowest-priority group is away from you, the order of priority of groups on a stereocenter is clockwise.

R,S system *(Section 15.3)* A set of rules for specifying configuration about a stereocenter.

R— *(Section 11.4)* A symbol used to represent an alkyl group.

Racemic mixture *(Section 15.2)* A mixture of equal amounts of two enantiomers.

Rad *(Section 9.5)* Radiation absorbed dose. The SI unit is the gray (Gy).

Radiation, nuclear *(Section 9.1)* Radiation emitted from a nucleus during nuclear decay; includes alpha particles, beta particles, gamma rays, and positrons.

Radioactive *(Section 9.1)* Refers to a substance that emits radiation during nuclear decay.

Radioactive dating *(Chemical Connections 9B)* The process of establishing the age of a substance by analyzing radioisotope abundance as compared with the current amount.

Radioactive isotope *(Section 9.2)* A radiation-emitting isotope of an element.

Radioactivity *(Section 9.1)* The process by which radiation is emitted from a nucleus and/or the nuclear components change and/or the nucleus splits.

Random coils *(Section 21.8)* Proteins that do not exhibit any repeated pattern.

Rate constant *(Section 7.4)* A proportionality constant between the molar concentration of reactant(s) and the rate of reaction (rate = k[compound]).

RDA *(Section 29.2)* Recommended Dietary Allowance; an average daily requirement for nutrients published by the U.S. Food and Drug Administration.

Reaction mechanism *(Section 12.6)* A step-by-step description of how a chemical reaction occurs.

Receptor *(Section 23.1)* A membrane protein that can bind a chemical messenger and then perform a function, such as synthesizing a second messenger or opening an ion channel.

Recognition site *(Section 25.3)* The area of the tRNA molecule that recognizes the mRNA codon.

Recombinant DNA *(Section 25.8)* DNAs from two sources that have been combined into one molecule.

Redox reaction *(Section 4.7)* An oxidation–reduction reaction.

Reducing agent *(Section 4.7)* An entity that donates electrons in an oxidation–reduction reaction.

Reducing sugar *(Section 19.4C)* A carbohydrate that reacts with a mild oxidizing agent under basic conditions to give an aldonic acid; the carbohydrate reduces the oxidizing agent.

Reduction *(Section 4.7)* The gain of electrons; the loss of oxygen or the gain of hydrogen.

Regioselective reaction *(Section 12.6)* A reaction in which one direction of bond forming or bond breaking occurs in preference to all other directions.

Regulator *(Section 22.6)* A molecule that binds to an allosteric enzyme and changes its activity. This change could be positive or negative.

Regulatory gene *(Section 25.6)* A gene that produces a protein that controls the transcription of a structural gene.

Regulatory site *(Section 22.6)* A site, other than the active site, where a regulator binds to an allosteric enzyme and affects the rate of reaction.

Rem Roentgen equivalent man: a biological measure of radiation.

Replication *(Section 24.4)* The process whereby DNA is duplicated to form two exact replicas from the original DNA molecule.

Residue *(Section 21.5)* Another term for an amino acid in a peptide chain.

Respiration *(Section 4.8)* Oxidation of substances occurring within living organisms.

Respiratory chain *(Section 26.3)* Another term for the electron transport chain.

Response element *(Section 25.6)* A sequence of DNA upstream from a promoter that interacts with a transcription factor to initiate transcription in eukaryotes.

Restriction endonuclease *(Section 25.8)* An enzyme, usually purified from bacteria, that cuts DNA at a specific base sequence. Restriction endonucleases are the tools that allow the creation of recombinant DNA.

Retrovirus *(Chemical Connections 25B)* A virus, such as HIV, that has an RNA genome.

Reuptake *(Section 23.4)* The transport of a neurotransmitter from its receptor back through the presynaptic membrane into the neuron.

Reverse transcriptase *(Chemical Connections 25C)* The enzyme that enables a virus to copy the information encoded in an RNA molecule into a DNA molecule.

Ribonucleic acid *(Section 24.2)* A type of nucleic acid consisting of nucleotide monomers of a nitrogenous base, ribose, and phosphate.

Ribosomal RNA (rRNA) *(Section 24.5)* The type of RNA that is complexed with proteins and makes up the ribosomes used in translation of mRNA into protein.

Ribosome *(Section 24.4)* Small spherical bodies in the cell made of protein and RNA; the site of protein synthesis.

Ribozyme *(Sections 22.1, 24.5)* An enzyme that is made up of ribonucleic acid. The currently recognized ribozymes catalyze cleavage of part of their own sequences in mRNA and tRNA.

RNA *(Section 24.2)* Ribonucleic acid.

Roentgen (R) *(Section 9.5)* The amount of radiation that produces ions having 2.58×10^{-4} coulombs per kilogram.

s *(Section 15.3)* From the Latin *sinister*, meaning "left"; used in the *R,S* system to show that, when the lowest-priority group is away from you, the order of priority of groups on a stereocenter is counterclockwise.

Saccharide *(Section 19.1)* A simpler member of the carbohydrate family, such as glucose.

Saponification *(Section 18.6B)* Hydrolysis of an ester in aqueous NaOH or KOH to an alcohol and the sodium or potassium salt of a carboxylic acid.

Satellites *(Section 24.6)* Short sequences of DNA that are repeated hundreds or thousands of times but that do not code for any protein or RNA.

Saturated fatty acid *(Section 20.2)* A fatty acid that has no carbon–carbon double bonds.

Saturated hydrocarbon *(Section 11.2)* A hydrocarbon that contains only carbon–carbon single bonds.

Saturation curve *(Section 22.4)* A graph of enzyme rate versus substrate concentration. At high levels of substrate, the enzyme becomes saturated, and the velocity does not increase linearly with increasing substrate.

Scientific method *(Section 1.2)* A method of acquiring knowledge by testing theories.

Scintillation counter *(Section 9.4)* A device containing a phosphor that emits light on exposure to ionizing radiation that can be sensed.

Secondary (2°) alcohol *(Section 10.4A)* An alcohol in which the carbon atom bearing the —OH group is bonded to two other carbons.

Secondary (2°) amine *(Section 10.4B)* An amine in which nitrogen is bonded to two carbons and one hydrogen.

Secondary messenger *(Section 23.1)* A molecule that is created/released due to the binding of a hormone or neurotransmitter, which then proceeds to carry and amplify the signal inside the cell.

Secondary structure, of DNA *(Section 24.3)* Specific forms taken by DNA due to pairing of complementary bases.

Secondary structure, of proteins *(Section 21.8)* Repeating structures within polypeptides that are based solely on interactions of the peptide backbone. Examples are the alpha helix and the beta-pleated sheet.

Semiconservative replication *(Section 24.4)* Replication of DNA strands whereby each daughter molecule has one parental strand and one newly synthesized strand.

Semipermeable membrane *(Section 6.8)* A membrane that allows only certain substances to pass through.

Serendipity *(Section 1.2)* A chance observation.

Serum *(Section 31.2)* A clear liquid that can be extracted from blood plasma. It has all the soluble components of plasma but lacks the fibrinogen that makes blood clots.

Shell *(Section 2.6A)* All of the orbitals of a principal energy level of an atom.

SI *(Section 1.4)* International System of Units.

Side chains *(Section 21.6)* The part of an amino acid that varies one from the other. The side chain is attached to the alpha carbon and the nature of the side chain determines the characteristics of the amino acid.

Sievert (Sv) *(Section 9.5)* A biological measure of radiation. One sievert is the value of 100 rem.

Signal transduction *(Section 23.5)* A cascade of events through which the signal of a neurotransmitter or hormone delivered to its receptor is carried inside the target cell and amplified into many signals that can cause protein modifications, enzyme activation, and the opening of membrane channels.

Significant figures *(Appendix II)* The total number of digits in a number with the exception of placeholders; the digits that express the certainty of a number.

Single bond *(Section 3.7C)* A bond formed by sharing one pair of electrons; represented by a single line between two bonded atoms.

Solenoid *(Section 24.3)* A coil wound in the form of a helix.

Solid *(Section 1.6)* A physical state of matter in which there is a definite volume and shape.

Solubility *(Section 6.4)* The maximum amount of a substance that can be dissolved in a solute at a specific temperature and pressure.

Soluble immunoglobulins *(Section 30.1)* The antibodies secreted by B cells, which surround and immobilize antigens.

Solute *(Section 6.2)* The substance or substances that are dissolved in a solvent to produce a solution.

Solution *(Section 6.1)* A homogeneous mixture composed of a solvent and solute or solutes (solute particle size less than 1 nm).

Solvent *(Section 6.2)* The fraction of a solution in which the other components are dissolved.

Specific gravity *(Section 1.7)* The density of a substance compared to water as a standard.

Specific heat *(Section 1.9)* The amount of heat (calories) necessary to raise the temperature of 1 g of a substance by 1°C.

Specificity *(Section 30.1)* A characteristic of acquired immunity based on the fact that cells make specific antibodies to a wide range of pathogens.

Spectator ion *(Section 4.6)* An ion that appears unchanged on both sides of a chemical equation.

Sphingolipid *(Section 20.4)* A phospholipid with sphingosine as the alcohol backbone.

Splicing *(Section 24.4)* The removal of an internal RNA segment and the joining of the remaining ends of the RNA molecule.

Standard temperature and pressure (STP) *(Section 5.4)* One atmosphere pressure and 0°C (273 K).

Step-growth polymerization *(Section 18.8)* A polymerization in which chain growth occurs in a stepwise manner between difunctional monomers, as, for example, between adipic acid and hexamethylenediamine to form nylon-66.

Stereocenter *(Section 15.2)* A tetrahedral carbon atom that has four different groups bonded to it.

Stereoisomers *(Section 11.7)* Isomers that have the same connectivity (the same order of attachment of their atoms) but different orientations of their atoms in space.

Steroid *(Section 20.9)* A molecule that contains the characteristic four-ring steroid nucleus.

Stoichiometry *(Section 4.5A)* The mass relationships in a chemical reaction.

Strong acid *(Section 8.2)* An acid that ionizes completely in aqueous solution.

Strong base *(Section 8.2)* A base that ionizes completely in aqueous solution.

Structural formula *(Section 3.7)* A formula showing how atoms in a molecule or ion are bonded to each other. Similar to a Lewis structure except that a structural formula shows only bonding pairs of electrons.

Structural genes *(Section 25.6)* Genes that code for the product proteins.

Subshell *(Section 2.6A)* All of the orbitals of an atom having the same principal energy level and the same letter designation (either s, p, d, or f).

Substance P *(Section 23.6)* An 11-amino-acid peptidergic neurotransmitter involved in the transmission of pain signals.

Substrate *(Section 22.3)* The compound or compounds whose reaction an enzyme catalyzes.

Subunit *(Section 22.6)* An individual polypeptide chain of an enzyme that has multiple chains.

Supersaturated *(Section 6.4)* A solution in which the solvent has dissolved an amount of solute beyond the maximum amount at a specific temperature and pressure.

Surface presentation *(Section 30.1)* The process whereby a portion of an antigen from a foreign pathogen that infected a cell is brought to the surface of the cell.

Surface tension *(Section 5.8)* The layer on the surface of a liquid produced by uneven intermolecular attractions at its surface.

Suspension *(Section 6.7)* An unstable system in which the solute particles are larger than 1000 nm and that separates with time.

Synapse *(Section 23.2)* An aqueous small space between the tip of a neuron and its target cell.

T cell *(Section 30.2)* A type of lymphoid cell that matures in the thymus and that reacts with antigens via bound receptors on its cell surface. T cells can differentiate into memory T cells or killer T cells.

Tautomers *(Section 17.8)* Constitutional isomers that differ in the location of an H atom and a double bond relative to an O or N atom.

T-cell receptor *(Section 30.5)* Glycoprotein of the immunoglobulin superfamily on the surface of T cells that interacts with the epitope presented by MHC.

T-cell receptor complex *(Section 30.5)* The combination of T-cell receptors, antigens, and cluster determinants (CD) that are all involved in the T cell's ability to bind antigen.

Temperature *(Section 1.9)* The measure of the amount of heat in a substance. Measurement is made in a temperature scale (°F, °C, or K); K is the SI base unit.

Termination *(Section 25.5)* The final stage of translation during which a termination sequence on mRNA tells the ribosomes to dissociate and release the newly synthesized peptide.

Termination sequence *(Section 25.2)* A sequence of DNA that tells the RNA polymerase to terminate synthesis.

Terpene *(Section 12.5)* A compound whose carbon skeleton can be divided into two or more units identical to the carbon skeleton of isoprene.

Tertiary (3°) alcohol *(Section 10.4A)* An alcohol in which the carbon atom bearing the —OH group is bonded to three other carbons.

Tertiary (3°) amine *(Section 10.4B)* An amine in which nitrogen is bonded to three carbons.

Tertiary structure *(Section 21.9)* The overall conformation of a polypeptide chain, including the interactions of the side chains and the position of every atom in the polypeptide.

Theoretical yield *(Section 4.5C)* The amount of product that should be formed in a chemical reaction according to the stoichiometry of the balanced equation.

Theory *(Section 1.2)* A hypothesis that is supported by evidence; a hypothesis that has passed tests.

Thiol *(Section 14.5A)* A compound containing an —SH (sulfhydryl) group bonded to a tetrahedral carbon atom.

Thrombocyte *(Section 31.2)* A cell whose function is to control bleeding after an injury; also called a platelet.

Thrombosis *(Chemical Connections 31B)* The process whereby a blood clot breaks loose from one part of the body and lodges in an artery in another part of the body.

Thromboxane *(Section 20.12)* An arachidonic acid derivative involved in blood clotting.

Titration *(Section 8.10)* An analytical procedure whereby we react a known volume of a solution of known concentration with a known volume of a solution of unknown concentration.

TNF *(Section 30.6)* Tumor necrosis factor; a type of cytokine produced by T cells and macrophages that has the ability to lyse susceptible tumor cells.

trans *(Section 11.7)* A prefix meaning "across from."

Transamination *(Section 27.8)* A reaction that involves the interchange of the alpha amino group from the amino acid and a keto group from an alpha ketoacid.

Transcription *(Section 25.1)* The process in which information encoded in a DNA molecule is copied into an mRNA molecule.

Transcription factors *(Section 25.6)* Binding proteins that facilitate the binding of RNA polymerase to the DNA to be transcribed.

Transfection *(Chemical Connections 25C)* The process wherein a virus infects another organism and the viral genome is then expressed in the host cells.

Transfer RNA (tRNA) *(Section 24.4)* The RNA that transports amino acids to the site of protein synthesis in ribosomes.

Transferases *(Section 22.2)* A class of enzymes that catalyze a reaction where a group of atoms, such as an acetyl group or amino group, is transferred from one molecule to another. *(Section 25.5)* Peptidyl transferase; the enzyme activity of the ribosomal complex that is responsible for formation of peptide bonds between the amino acids of the growing peptide.

Transgenic *(Chemical Connections 25C)* An organism that expresses foreign proteins due to transfection.

Transition element *(Section 2.5A)* An element of the Periodic Table characterized by filling d orbitals.

Translation *(Section 25.1)* The process in which information encoded in an mRNA molecule is used to assemble a specific protein.

Translocation *(Section 25.5)* The part of translation where the ribosome moves down the mRNA a distance of three bases, so that the new codon is on the A site.

Transporter *(Section 23.5)* A protein molecule carrying small molecules, such as glucose or glutamic acid, across a membrane.

Transuranium element *(Section 9.7)* An element with an atomic number greater than 92.

Triacylglycerol *(Section 20.2)* Triglyceride.

Tricarboxylic acid cycle *(Section 26.4)* The citric acid cycle.

Triglyceride *(Section 20.2)* A lipid composed of glycerol esterified to a fatty acid at each of its three hydroxyls.

Triple bond *(Section 3.7C)* A bond formed by sharing three pairs of electrons; represented by three lines between the two bonded atoms.

Triple helix *(Section 21.8)* The collagen triple helix is composed of three peptide chains. Each chain is itself a left-handed helix. These chains are twisted around each other in a right-handed helix.

Triprotic acid *(Section 8.3)* An acid that can give up three protons.

Trisaccharide *(Section 19.7)* A combination of three monosaccharides linked by glycosidic bonds.

Tumor suppression factor *(Chemical Connections 25F)* A protein that controls replication of DNA so that cells do not divide constantly. Many cancers are caused by mutated tumor suppression factors.

Tyndall effect *(Section 6.7)* Light passing through and scattered by a colloid viewed at a right angle.

Ubiquitinylation *(Chemical Connections 27E)* The process of attaching one or more units of the protein called ubiquitin to an aging protein that is ready for degradation.

Unsaturated *(Section 6.4)* A solution in which the solvent has not dissolved the maximum amount of solute at a specific temperature and pressure.

Unsaturated fatty acid *(Section 20.2)* A fatty acid containing one or more carbon–carbon double bonds.

Unwinding proteins *(Section 24.4)* Special proteins that help unwind DNA so that it can be replicated.

Urea cycle *(Section 27.8)* A cyclic pathway that produces urea from ammonia and carbon dioxide.

Valence electron *(Section 2.6F)* An electron in the outermost occupied (valence) shell of an atom.

Valence shell *(Section 2.6F)* The outermost occupied shell of an atom.

Vapor pressure *(Section 5.9A)* The pressure of a gas in equilibrium with its liquid form in a closed container.

Vapor *(Section 5.9A)* A gas.

Vesicle, synaptic *(Section 23.2)* A compartment containing a neurotransmitter that fuses with a presynaptic membrane and releases its contents when a nerve impulse arrives.

Vitamin *(Section 29.5)* An organic substance required in small quantities in the diet of most species, which generally functions as a cofactor in important metabolic reactions.

Volume *(Section 1.4)* The space that a substance occupies; the cubic meter is the base SI unit.

VSEPR model *(Section 3.9)* Valence-shell electron-pair repulsion model.

Wavelength *(Section 9.1)* The distance from the top of one wave to the top of the next wave.

Weak acid *(Section 8.2)* An acid that is only partially ionized in aqueous solution.

Weak base *(Section 8.2)* A base that is only partially ionized in aqueous solution.

Weight *(Section 1.4)* The result of a mass acted upon by gravity; the base unit of measure is a gram (g).

X-ray *(Section 9.1)* A type of electromagnetic radiation with a wavelength shorter than that of ultraviolet light and longer than that of gamma rays.

Zwitterion *(Section 21.3)* A molecule that has equal numbers of positive and negative charges, giving it a net charge of zero.

Zymogen *(Section 22.6)* An inactive form of an enzyme that must have part of its polypeptide chain cleaved before it becomes active; a proenzyme.

CREDITS

Text and Illustrations

Chapter 20 **Fig. 20.3:** From *Biochemistry* by Lubert Stryer © 1975, 1981, 1988, 1995 by Lubert Stryer; © 2002 by W. H. Freeman and Company. Used with permission of W. H. Freeman and Company.
Chapter 23 **Fig. 23.3:** Courtesy of Anthony Tu, Colorado State University.
Chapter 24 **Fig. 24.9a:** From *Biochemistry*, 2nd ed., by Lubert Stryer. Copyright 1981 by W. H. Freeman and Company. All rights reserved. Used with permission of W. H. Freeman and Company. **Fig. 24.9b:** Courtesy of Dr. Sung-Hou Kim. **Fig. 24.13:** Adapted from "The Unusual Origin of the Polymerase Reaction," by Kary B. Mullis, illustrated by Michael Goodman. *Scientific American,* April 1990. Reprinted with permission of Michael Goodman.
Chapter 25 **Figure, p. 627:** Illustration by Irving Geis, from *Scientific American,* Jan. 1963. Rights owned by Howard Hughes Medical Institute. Not to be reproduced without permission. **Fig. 25.7:** Courtesy of Dr. J. Frank, Wadsworth Center, Albany, New York.
Chapter 29 **Fig. 29.3:** Courtesy of Drs. P. G. Bullough and V. J. Vigorita and the Gower Medical Publishing Co., New York.
Chapter 31 **Figure, p. 748:** From *The Functioning of Blood Platelets* by M. B. Zucker. Copyright 1980 by Scientific American, Inc. All rights reserved. Reprinted by permission of Patricia J. Wynne.

Photographs

Pages xii, xvi, xvii, xviii, 4, 15, 31, 40, 43, 58, 60, 61, 62, 69 (top), 76, 77, 86, 87, 91, 95, 106 (left), 108, 112 (bottom), 113, 127, 137, 145, 153, 154, 158, 159, 164, 167, 171, 173, 175, 177 (top), 184, 191, 193 (center), 202, 204, 213, 216, 217, 220, 221, 222, 225, 226, 227, 228, 229, 232, 235, 251, 255, 278, 281, 288, 302, 308, 314, 334, 336, 345, 352 (top), 361, 362 (left), 373, 384, 385, 394, 408, 419, 425, 438, 441, 443, 445, 450, 455, 456, 457, 458, 466, 473, 497, 499, 547, 548, 678, 704, 708: Charles D. Winters.
Pages 7, 10, 105, 106 (top), 107, 141, 154 (center), 194, 258 (bottom), 260, 331, 334, 412, 752: Beverly March.
Contents **p. v:** IBMRL/Visuals Unlimited, Inc. **p. vi:** Tom and Pat Leeson/Photo Researchers, Inc. **p. vii:** Joseph Nettis/Photo Researchers, Inc. **p. viii:** Lester Lefkowitz/Stone/Getty Images. **p. ix:** NSF Oasis Project/Norbert Wu Photography. **p. x:** Douglas Brown. **p. xi** A & L Sinibaldi/Stone/Getty Images.
Preface **p. xiii:** Christi Carter/Grant Heilman Photography, Inc. **p. xiv:** Doug Perrine/TCL/Masterfile. **p. xv:** NASA. **Summary p. xix:** Ken Eward/Biografx/Photo Researchers, Inc. **p. xx:** Bruce M. Herman/Photo Researchers, Inc. **p. xxi:** Meckes/Ottawa/Photo Researchers, Inc. **p. 1:** Tom Till/Stone/Getty Images.
Chapter 1 **p. 2:** James Balog/Stone/Getty Images. **pp. 3, 5:** National Library of Medicine. **p. 7 (right):** George Semple. **p. 9:** Brinkmann Instruments, Co. **p. 16:** Ken Graham/Stone/Getty Images. **p. 18:** Eric Lars Baleke/Black Star. **p. 19:** Courtesy of Siemen Solar Industries/Shell Solar.
Chapter 2 **p. 27:** IBMRL/Visuals Unlimited, Inc. **p. 30 (top):** Martin Dohrn/Science Photo Library/Photo Researchers, Inc. **p. 30 (bottom):** Ken Eward/Science Photo Library/Photo Researchers, Inc. **p. 38 (top):** Courtesy of Roger Freedman and Paul Hansma, University of CA, Santa Barbara. **p. 38 (bottom):** NASA. **p. 39:** E.F. Smith Collection, University of Pennsylvania. **p. 41:** Werner Forman/Art Resource, NY. **p. 44:** AIP, Neils Bohr Library.
Chapter 3 **p. 59:** Francis Simon/AIP, Neils Bohr Library. **p. 63:** Nordian Corporation. **p. 69:** Harold Levin. **p. 79:** Arthur C. Smith, III/Grant Heilman Photography, Inc.
Chapter 4 **p. 89:** Joseph Nettis/Photo Researchers, Inc. **p. 96:** Jeff Daly/Visuals Unlimited, Inc. **p. 100:** AT&T Bell Labs. **p. 105 (top):** Charles Steele. **p. 111 (top):** Yoav Levy/Phototake, NYC. **p. 111 (bottom):** Russell Curtis/Photo Researchers, Inc. **p. 112 (center):** Jeff Greenberg/Visuals Unlimited, Inc. **p. 115:** John Sohlden/Visuals Unlimited, Inc.
Chapter 5 **p. 119:** Vince Streano/Corbis. **p. 124 (top):** Vandystadt/Photo Researchers, Inc. **p. 124 (bottom):** Custom Medical Stock Photo. **p. 129:** Gregory G. Dimijian/Photo Researchers, Inc. **p. 135:** Hermann Eisenbeiss/Photo Researchers, Inc. **p. 140:** Mehau Kulyk/Science Photo Library/Photo Researchers, Inc. **p. 147:** George Semple.
Chapter 6 **p. 153:** Peter Arnold, Inc. **p. 168 (center):** Courtesy of P. A. Dieppe, P. A. Bacon, A.N. Bamji, I. Watt, and the Gower Medical Publishing Co, Ltd., London. **p. 168 (right):** National Library of Medicine. **p. 176:** Science Source/Photo Researchers, Inc. **p. 177 (center):** SIU/Photo Researchers, Inc.
Chapter 7 **p. 191:** Leon Lewandowski. **p. 193:** Michael Engleish, M.D./Custom Medical Stock Photo. **p. 201:** John D. Cunningham/Visuals Unlimited. **p. 205:** C. P. Vance/Visuals Unlimited, Inc.
Chapter 8 **p. 210:** RB-GM-J.O. Atlanta/Liason Agency/Getty Images.

Chapter 9 **p. 242:** NASA. **p. 247:** USDA. **p. 249 (top):** Sci-Vu/Visuals Unlimited, Inc. **p. 249 (bottom):** Santa Visalli/The Image Bank. **p. 254:** Klaus Guldbrandsen/Science Photo Library/Photo Researchers, Inc. **p. 256:** CNRI/SPL/Photo Researchers, Inc. **p. 257:** CRC Handbook in Clinical Lab Science. **p. 258 (bottom):** Lakeville-Nassau CT Center, NY. **p. 258 (top):** Debbie Breen/Phototake. **p. 259 (top):** Radiology Department, Suburban Hospital, Bethesda, MD. **p. 259 (bottom):** Dr. Peter T. Fox, Washingtion University, School of Medicine. **p. 263:** PSE&G. **p. 264:** U.S. Department of Energy.
Chapter 10 **p. 269:** Christi Carter/Grant Heilman Photography, Inc. **p. 270:** Pat and Tom Leeson/Photo Researchers, Inc. **p. 272:** D.E. Cox/Stone/Getty Images. **p. 273:** Pete K. Ziminski/Visuals Unlimited. **p. 274:** George Semple.
Chapter 11 **p. 286:** J.L. Bohin/Photo Researchers, Inc. **p. 299:** Tim Rock/Animals/Animals. **p. 307:** Ashland Oil. Co.
Chapter 12 **p. 323:** David Sieren/Visuals Unlimited, Inc. **p. 324:** DonSuzio. **p. 334 (bottom left):** The Stock Market.
Chapter 13 **p. 343:** Douglas Brown. **p. 352 (center):** John D. Cunningham/Visuals Unlimited, Inc. **p. 353:** Chuck Pefley/Stone/Getty Images.
Chapter 14 **p. 359:** Earl Robber/Photo Researchers, Inc. **p. 362:** The Bettmann Archive/Corbis **p. 368:** Kathy Merrifield/Photo Researchers, Inc. **p. 372:** Steven J. Krasemann/Photo Researchers, Inc. **p. 373 (top):** Boston Medical Library in the Francis A. Countway Library. **p. 375:** D. Young/Tom Stack & Associates.
Chapter 15 **p. 382:** Lester Lefkowitz/Stone/Getty Images. **p. 397:** William Brown.
Chapter 16 **p. 403:** Jack Ballard/Visuals Unlimited. **p. 405:** Inga Spence/Visuals Unlimited. **p. 418:** Tom McHugh/Photo Researchers, Inc.
Chapter 18 **p. 434:** George Semple.
Chapter 18 **p. 439:** Hans Reinhard/OKAPIA/Photo Researchers, Inc. **p. 440:** Ted Nelson/Dembinsky Photo Associates. **p. 447:** Andrew McClenaghan/Science Photo Library/Photo Researchers, Inc.
Chapter 19 **p. 465:** Courtesy of Dr. Petra Fromme. **p. 470:** Claire Paxton and Jacqui Farrow/Science Photo Library/Photo Researchers, Inc. **p. 471:** Andrew Martinez 1993/Photo Researchers, Inc. **p. 477:** Gregory Smolin. **p. 479:** Martin Dohrn/SPL/Photo Researchers, Inc. **p. 482:** Larry Mulvehill/Photo Researchers, Inc.
Chapter 20 **p. 494:** Doug Perrine/TCL/Masterfile. **p. 508:** Drs. P. G. Bullogh and V. J. Vigorita and the Gower Medical Publishing Co. Ltd. **p. 510:** Carolina Biological Supply/Phototake, NYC. **p. 514:** Jed Jacobson/AllSport USA/Getty Images.
Chapter 21 **p. 525:** Hans Strand/Stone/Getty Images. **p. 537:** G. W. Willis/Biological Photo Service. **p. 544:** J. Gross/Biozentrum/Science Photo Library/Photo Researchers, Inc. **p. 548 (bottom):** Larry Mulvehill/Photo Researchers, Inc.
Chapter 22 **p. 553:** NSF Oasis Project/Norbert Wu Photography. **p. 558:** B. Murton/Southampton Oceanography Centre/SPL/Photo Researchers, Inc.
Chapter 23 **p. 572:** Don Fawcett/Science Source/Photo Researchers, Inc. **p. 580:** Alfred Pasieka/SPL/Photo Researchers, Inc. **p. 581:** Courtesy of Anatomical Collection, National Museum of Health and Medicine. **p. 588:** George Semple.
Chapter 24 **p. 596:** James King-Holmes/SPL/Photo Researchers, Inc. **p. 596 (bottom):** Biophoto Associates/Photo Researchers, Inc. **p. 604 (left):** Chemical Heritage Foundation. **p. 604 (right):** Barrington Brown/Science Source/Photo Researchers, Inc. **p. 611:** Dr. Lawrence Koblinsky. **p. 615:** Larry Schwartz, University of Massachusetts, Amherst.
Chapter 25 **p. 620:** Fredrick A. Bettelheim. **p. 629:** Dr. J. Frank, Wadsworth Center, Albany, NY. **p. 631:** NIBSC/Science Photo Library/Photo Researchers, Inc. **p. 632:** 2001 Trimeris, Inc. and Hoffman-Laroche, Ltd. **p. 634:** Dr. Paul Russell, National Eye Institute, NIH. **p. 639:** Thomas Broker/Cold Spring Harbor Laboratory. **p. 641: (left)** Courtesy of Dr. Anne Crossway, Calgene, Inc., Biotechniques, and Eaton Publishing. **p. 641 (right):** Dr. Marlene DeLuca, University of CA, San Diego.
Chapter 26 **p. 645:** A & L Sinibaldi/Stone/Getty Images. **p. 648:** Dennis Kunkel/Phototake, NYC.
Chapter 27 **p. 666:** Dennis Degnan/Corbis. **p. 686:** Science Photo Library/Photo Researchers, Inc.
Chapter 28 **p. 691:** Bruce M. Herman/Photo Researchers, Inc. **p. 694:** Dr. Petra Fromme.
Chapter 29 **p. 705:** National Library of Medicine. **p. 711:** Courtesy of Drs. P. G. Bullogh and V. J. Vigorita and the Gower Medical Publishing Co. Ltd. **p. 718:** George Semple.
Chapter 30 **p. 722:** Meckes/Ottawa/Photo Researchers, Inc. **p. 725:** David M. Phillips/Photo Researchers, Inc. **p. 731:** Courtesy of Dr. R.J. Poijak, Pasteur Institute, Paris, France. **p. 733:** Dr. Jeremy Burgess/Science Photo Library/Photo Researchers, Inc.
Chapter 31 **p. 744:** Ken Eward/Biografx/Photo Researchers, Inc.

INDEX

Key: **Boldface** indicates glossary entry; *italics* indicates figure; 30*t* indicates table.

A (acceptor) site, *628*–*629*
Abacavir, in AIDS treatment, 631
Absolute zero, 11, 120
Accolate, 520
Accupril, 564, 756
ACE (angiotensin-converting enzyme), 564, 756
Acesulfame-K, 484*t*, 716, 717
Acetaldehyde, 280, 420, 423
Acetals, **428**–430, 476–477. *See also* Hemiacetals
Acetamide, 446
Acetaminophen, 520, 566
Acetanilide, hydrolysis of, 454
Acetate ion, 212–213, 214, 216–217, 227, 230–231, 233, 495, 580
Acetic acid, 74, 100, 103, *138*, 139*t*, 160, 164, 170, 200, 201, 202, 214, 216–217, 227, 230–231, 233, 275–276, 281, 423, 440*t*, 441–442, 442*t*, 443, 448, 449, 451, 454, 555, 556*t*
Acetic anhydride, 444, 451
Acetoacetate, 677, 678–679, *680*
Acetoacetyl-ACP, in fatty acid biosynthesis, 697
Acetoacyl-CoA, 677, 678
Acetone, 133, 185, 186, 280, 420, *430*, 678–679
Acetylation, during DNA replication, 611–612
Acetylcholine, 574*t*, 555, 556*t*, 576, 577–582
Acetylcholine receptor (AChR), 577, *578*, 739
Acetylcholinesterase, 555, 556*t*, 579, 580, 582
Acetylcholine transferase, 581
Acetyl-CoA, 581, *652*, 653–655, 667, *668*, 671, 673*t*, 675–677, *676*, 678, 679, *680*, 685, 696, 699
Acetyl-CoA carboxylase, as cofactor, 711*t*
Acetylene, 73, 78, 79, 80*t*, *82*, 275*t*, 287, 314, 318
N-Acetyl-D-galactosamine, 482
N-Acetyl-D-glucosamine, 470, 482, 487, 546
Acetyl group, *650*–652, 653–655
Acetylsalicylic acid, 518
Acetyl-transporting molecules, 650–652
Achiral objects, **386**
Acid–base indicators, 225–226, 226*t*, *227*, 228–*229*, 232
Acid–base reactions, 105, 214–216, 216–218
Acid–base titration, **228**–230, *229*
Acid dissociation constant, 219
Acidic amino acids, 526, 527*t*, 529
Acidic polysaccharides, 466, 486–487, 546
Acidic salts, 226–228, **227**
Acidic solutions, 224–228, 230–236
Acid ionization constants (K_a), 218–220, **219**
Acidity, 107, 364–365, 374–375. *See also* Acids
Acidosis, 235, 671, 754
Acid phosphatase, 555, 567*t*
Acid rain, 76, 155, 167, *201*, 203
Acids, 210–236, *213*, 352, 411–412, 442, 476, 558, 559
  amphiprotic, monoprotic, diprotic, and triprotic, **215**, 216, 529
  strong and weak, **212**–213, 216–218, 219
Aconitase, 556*t*, 653
cis-Aconitate, 556*t*, 653
Acquired immunity, *723*–724
Acrolein, molecular structure of, 420
Acrylonitrile, polymers of, 333*t*
ACTH. *See* Adrenocorticotropic hormone (ACTH)
Actin, 525, 526, 544, 661, *662*, 736
Actinide series, 51
Actinium-225, 260–261, 733
Activation
  of enzymes, 557
  of glycolysis, 669–670, 673*t*, 677*t*
  of protein synthesis, 626, 626*t*, 627
Activation energy, 187, *188*–190, 194, *195*
Active sites, **557**, 559–562, 565

Active transport, across cell membranes, 502, 503
Actual yield, from chemical reactions, 103
Acyl carrier protein (ACP), *696*–697
Acyl-CoA, 675, *676*, 677*t*, 697–698
Acyl-CoA dehydrogenase, *676*, 677*t*
Acyl group, **444**, 675
Adalat, 574*t*, 756
Adaptive immunity, 723–724, 727, 737–738
Addition reactions, of alkenes, 325–332
Adenine, 597–598, 600*t*, 601–*602*, 603–605, *604*, 649
Adenosine, 598, 600*t*, 649–650
Adenosine diphosphate (ADP), 561, *585*, 649–650, *651*, 658–660, 661–662, 667–668, *669*–*670*, 682, 695
Adenosine monophosphate (AMP), 556*t*, 627, 649–650, 675–677, *676*, 682. *See also* Cyclic AMP (cAMP)
Adenosine triphosphate (ATP), 502, 561, 584, *585*, 627, *646*, 647, 649–650, 654, 658–661, 667–668, *669*–*670*, 673, 673*t*, 675–677, *676*, 680, 681, *682*, 693, 695
Adenylate cyclase, 584, *585*–587, 588–589
Adipamide, 447–448
Adipic acid, 439, 456
Adipocytes, 667
ADP. *See* Adenosine diphosphate (ADP)
Adrenal cortex, 575*t*, 576
Adrenal glands, 413, 515
Adrenaline, *403*, 413, 573, 574*t*. *See also* Epinephrine
Adrenal medulla, hormone from, 575*t*
Adrenergic neurotransmitters, 574*t*, 576, 583–589
Adrenocorticoid hormones, 513, 515
Adrenocorticotropic hormone (ACTH), action of, *576*
Advanced glycation end-products (AGE), 534
Advil, 450
Aerobic pathways, lactate accumulation and, 671
Agents GB, GD, and VX, 579
Aging, 534, 638, 755
Agonist drugs, 573, 582
AIDS (acquired immunodeficiency syndrome), 623, 631–632, 634, 723, 735–736. *See also* Human immunodeficiency virus (HIV)
Air pollution, 76, *111*, 167, 307. *See also* Acid rain
Alanine, Ala, A, 527*t*, 528, 530, *542*, 625*t*, 684, 685, 700–701
β-Alanine, 441, 582–583
Alanine aminotransferase (ALT), 567, 567*t*
Albinism, 596
Albumin, 526, 747, 747*t*, 748
Albuterol, *403*, 413, 414
Alcohols, **277**–279, 323, 359–369, 364*t*, 375–376, 425–427, 428–430, 448–449, 453, *501*, 505, 548–549. *See also* Breath-alcohol screening
Aldehyde group, 277*t*, 467
Aldehydes, **280**–281, 367–369, 419–431, 424–*425*, 534, 587
Alditols, **477**–478
Aldohexoses, **472**–474, 474–475, 476
Aldolase, in glycolysis, *669*
Aldonic acids, 478, 479
Aldopentoses, furanose form of, **472**, 474
Aldoses, **467**, 469–470, 478
Aldosterone, 513, 515, 575*t*, 754
Aleve, 450
Aliphatic amines, **403**–405, 406–407, 409–411
Aliphatic hydrocarbons, **286**
Alkadienes, 321
Alkali metals, **42**, 43, 43*t*, 51*t*
Alkaline phosphatase (ALP), 567*t*
Alkalis. *See* Bases

Alkaloids, 405, 581
Alkalosis, of blood, 235
Alkanes, **286**–308, 303*t*, 332, 360, 364*t*
  cis-trans isomerism in, **300**–302
  cyclic, **295**–296, 296–300, 300–302
  as hydrocarbons, 286–287
Alkenes, 286, 287, **314**–336, 325*t*, 344
  cis-trans isomerism in, **316**, 318–320
  converting alcohols to, 365–367
  hydration of, 325*t*, **328**–330, 366–367
Alkylbenzenes, 345–348
Alkylbenzesulfonate detergents, 350–351
Alkyl groups, **292**, 293*t*, 423, 444–446
N-Alkylpyridinium chloride, 408
Alkyl substitutions, 345–348
Alkynes, 286, 287, **314**, 317, 318, 323
Allegra, 574*t*, 745
Allergens, 739
Allergies, 536, 739. *See also* Side effects
D-Allose, 468*t*
Allosteric effect, 749
Allosteric enzymes, 565
Allosterism, enzyme regulation via, **565**
Alloys, **154**
α-1-Adrenoreceptors, 755
Alpha amino acids, **526**, 527*t*
Alpha particles, 243, 244, 246–247, 252, 254
Altace, molecular structure of, 564
Altman, Sidney, 553
Aluminum (Al), *4*, 47, 110, *271*
Aluminum bromide, *4*
Alzheimer's disease, 539, 581, 582
Amidation, of glutamate, 684
Amides, **445**–448, 449–451, 454–455
Amine group, 277*t*, **279**–280, 422, 470–471
Amines, 221–222, **279**–280, *403*–414, 449–451
  aliphatic and aromatic, **403**–405, 409–411
  heterocyclic, **404**, 410, 411
Amino acid-AMP molecule, 627
Amino acid pool, **667**, *680*
Amino acids, **526**–528, 527*t*, 528–540, 554, 559, 561, 574*t*, 581, 596–597, 607, 627–629, 635–636, 637, *652*, 667, *668*, 672, 679–686, *680*, 685, 705, 729, *730*
  antigenicity of, 727–728
  biosynthesis of, 700–701
  chemical communication via, 576, 582–583, 740
  essential and nonessential, 700, **710**
  genetic code for, 621, 624–625, 625*t*, **625**–626
  peptide bonds between, **530**–532, 554
Amino acid substitutions, in biochemical evolution, 637
Amino acid transport, 701
Amino acid-tRNA molecule, 627, 633
Amino acid-tRNA synthetase (AARS), 627, 633, 635
p-Aminobenzoic acid (PABA), 350, 439, 562, 718,
4-Aminobutanoic acid, 440, 582–583
γ-Aminobutyric acid (GABA), 440, 582–583
Amino group, 350
Amino sugars, **470**–471
Ammonia, 73, 77–78, 79, 80*t*, *81*–82, 99, *120*, 188–189, 199–200, 205, 211–212, 213, 214, 216–217, 221–222, 409, 425, 438, 443, 559, *668*, 681, 684, 753
Ammonium benzoate, 443
Ammonium chloride, 271
Ammonium hydroxide, 211–212
Ammonium ion, 199–200, 214, 216–217, 221–222, 454, *680*, 681, 684
Ammonium nitrate, 193
Ammonium salts, 408, 449

INDEX / I-2

Amniocentesis, Tay-Sachs disease and, 508
Amorphous solids, 142, 143t
Amoxicillin, 447, 559
AMP. *See* Adenosine monophosphate (AMP)
Amphetamine, 404, 406–407, 574t
Amphiprotic acids, **215**, 529
Ampicillin, 447
Amylase, 567t
α-Amylase, 485, 708, 709
β-Amylase, in carbohydrate digestion, 708, 709
Amyloid fibril formation, 539
Amyloid plaques, 539
Amyloid precursor protein (APP), in Alzheimer's disease, 581
Amylopectin, 485, 708
Amylose, 485, 708
Anabolic steroids, 514, 718
Anabolism, 645, 691–692
Anaerobic pathway, reactions of, 671
Analgesics, 450
Androgens, action of, 575t
4-Androstene-3,17-dione ("Andro"), 514, 718
Anemia, 536, 537, 609, 613, 636, 657, 671, 747t
Anesthetics, ethers as, 359, *373*
"Angel dust," 583
Angiotensin, 591, 756
Angiotensinogen, 564, 756
Anhydrides, 438–439, **444**, 451, 455, 599, 649–650
Anhydrous substances, 167
Aniline, 345, 404, 406, 454
Anilinium, 408
Anions, *59*, 61, 62, 62t, 67, 165–*166*, 167–168
Anion transporter, of red blood cells, 502
Anisole, 345
Anode, 109, *112*, 167–*168*
Anomeric carbon, 471, **472**, 474, 475, 476, 477, 481, 483
Anomers, of monosaccharides, **472**, 473–474, 475, 476, 477
Anorexic agents, 591
Antacids, treating heartburn with, 221
Antagonist drugs, 573, 582
Anterior pituitary gland, hormones secreted by, 575t, *576*
Anthracene, 348
Antibiotics, 394, 446, 447, 471, 630, 701
Antibodies, 482–483, 526, 631, *723*, **724**, 727, 728, *729–733*, 746
Anticancer drugs, 601, 630
Anticoagulants, 487, 748
Anticodon loops, in tRNA, *607*
Anticodons, **625**, 626, 627, 628–629, 633
Antidepressants, 574t
Antidiuretic hormone (ADH), 536, 576, 591, 754
Antidiuretics, 754
Antifreeze, 173, 363
Antigen–antibody complex, 729, *731*
Antigen receptors, in immune system, *723*, 724
Antigens, 482–483, 526, 683, **724**, 726t, 727–729, *728, 730, 732*, 734
Antihistamines, 588, 745
Anti-inflammatory drugs, 515, 518
Antiknock ratings, of gasoline, 308
Antimony (Sb), 41, 42
Antioxidants, 352–355, 497, 711
Antisense drugs, 623
Antiseptics, 111, 213
Antithrombin, 487, 748
Antiviral agents, 630
AP (apurinic/apyrimidinic) site, 615
apoB-100 protein, in cholesterol transport, 511
Apoenzyme, 556, *557*
Apoptosis, 615, 638
APP. *See* Amyloid precursor protein (APP)
Appetite-stimulating agents, 591
Aqueous body fluids, 170
Aqueous humor, 744

Aqueous solution, **104**–107, 224, 226–228, 363–365, 409, 454, 475, 476, 528–529
Ara-A, molecular structure of, 630
D-Arabinose, 468t
Arachidic acid, 440t, 496t
Arachidonic acid, 450, 496t, 518, 519, 520, 566, 709
Arenes, 286, 287, **343**–355
Arginase, 554, 711t
Arginine (Arg, R), 527t, 554, 588, 625t, *682–683*, 685
D-Arginine, 554
L-Arginine (Arg, R), 554
Argininosuccinate, 681–682
Argon (Ar), 47
Argon krypton laser surgery, 548
Aricept, 574t, 581
Aromatic amines, **403**–405, 406–407, 409–411
heterocyclic, **404**, 410, 411
Aromatic compounds, **343**–355
Aromatic sextet, 345
Aromatic substitutions, 349–351
Arrhenius, Svante, 210, 559
Arsenic (As), as metalloid, 42
Arterial blood, *750*
Arteries, 512, 513, *750*, 752, *753*
Arthritis, 518, 737, 739
Artificial heavy elements, 250, 261–262
Artificial isotopes, 245
Artificial kidney machines, *178*
Artificial sweeteners, 484t, 716–717
Aryl groups, 343, 423, 445–446
Ascorbic acid, 274, 714t. *See also* Vitamin C
Asparagine (Asn, N), 527t, 546, 625t, *685*, 700
Aspartame, 484t, 685, 716–717
Aspartate, 556t, 681, *682*, 700. *See also* Aspartic acid
Aspartate aminotransferase (AST), 556t, 567, 567t, 711t
Aspartic acid (Asp, D), 527t, 531, 582, 625t, *685*. *See also* Aspartate
Aspirin, 94, 184, 274, 450, 518, 520, 566
Astatine (At), 42
Asthma, 413, 518, 520, 745
Astigmatism, laser surgery to correct, 548
Atherosclerosis, 510, 512, 699, 711, 747t
Atkins diet, 709
Atmosphere, water vapor in, 165
Atmosphere (atm), 122, 127, 138–139
Atmospheric pressure, 121–122
Atomic energy, 262–264
Atomic mass unit (amu), 33–**34**, 89–90
Atomic nuclei, 33–39, *34*, 119, 244–251
Atomic number, 35–36, *52*, 246, 247, 388–391, 389t
Atomic weight, 35, **37**–38, 89–90, 91
Atoms, 27–53, **31**, *34*, 119–121, 245, 246
chemical bonds between, 58–82
electrons in, 44–50, **46**, 50–52
formula weight and, 89–90
ions and, **52**, 58–61
mole and, 90–**91**
ATP. *See* Adenosine triphosphate (ATP)
ATPase, *646*, 658–660, 694
Atropine, 574t, 579
Attention deficit disorder (ADD), treatment of, 586
Attractive forces
between molecules, 131–135
in matter, 119–121
of immunoglobulins, 729–731
Autoclave, *124–125*, 125–126
Autoimmune diseases, 739–740. *See also* AIDS
vitamin treatment of, 717
Autoxidation, 352–355
Avery, Oswald, 597
Avogadro, Amadeo, 91
Avogadro's law, **127**–130
Avogadro's number, **91**
Axial bonds, **298**–300

Axons, 574–575
electrical signal transmission by, *573*
myelin sheaths on, 505–506
Azidothymidine (AZT), 631
Background radiation, 254, 256
Bacteria, 111, 205, 254, 350, 486, 558, 559, 562, 563, 597, 617, 621, 625–626, 630, *639–640*, 659, 681, 701, 723, *725*, 726t, 728, 736, 737, 739
Baking soda, 221. *See also* Sodium bicarbonate
Balancing chemical equations, 95–98, 105, 197
Balancing nuclear equations, 246, 247
Ball-and-stick models, 77–80
Balloon models, 77
Barbiturates, 453, 659
Barium chloride, *106*
Barium hydroxide, 212t
Barium sulfate, 69, *106*
Barium-122, 250
Barnes, Randy, 514
Barometer, *121–122*
Baroreceptors, 755–**756**
Basal caloric requirement, **708**
Base excision repair (BER), 613, *614*, 615
Base pairs, in DNA molecule, *603–605*
Bases, 210–236, *213*, 374, 403, 409–411, 443–444, 454, 558
strong, **212**–213, 216–218
weak, **212**–213, 216–218
Bases (of nucleic acids), **597**–598, 601–602, *603–605*, 606, 609, 612, 613
Basic amino acids, 526, 527t, 529
Basic salts, 226–228, **227**
Basic solutions, 224–226, 226–228, 230–236
B cell receptors (BcR), in immune system, *723*, 738
B cells, *723*, 724, 725, 726t, 727, 731, *732*, 736, 737–738
Beadle, George, 597
Becquerel, Henri, 242
Becquerel (Bq), 253, 254t
Beeswax, 498
Benadryl, 574t, 588, 745
Bends, the, 158
Benzaldehyde, 345, *419*, 421, 422, 424, 428
Benzene, 157, 287, 343, 344–345, 349–351, 637
1-4-Benzenediamine, 456
1,4-Benzenedicarboxylic acid, 456–457
Benzenediol, 352
Benzene rings, 286, 287, 344–345, 509–510
Benzenesulfonic acid, 350
Benzodiazepine, 408
Benzoic acid, 170, 219–220, 345, 424, 443, 449, 451–452
Benzo[a]pyrene, as carcinogen, 348, 638
Benzylamine, 411
Benzyldimethylamine, 404
Beriberi, 705, 711
Berkelium (Bk), 261
Beryllium (Be), 47
Beta blockers, in hypertension treatment, 756
Beta emission, 246, 247
Beta particles, 243, 244, 246, 247, 252, 254
Betazole, 574t
Bicarbonate ion, *105*, 215, *216*, 234, 754
Bicarbonates, reactions with acids, 221–222
Bile, heme catabolism and, 687
Bile salts, 517–518, 709
Bilirubin, 686, 687
Biliverdin, 686, 687
Binary ionic compounds, 67–68
Binary molecular compounds, 75–76
Binding curve, 749
Binding energy, neutrons and, 245
Binding proteins, in transcription, 621
Binding sites, 397–398, 729–731, *730*
Biochemical evolution, mutations and, 637
Biochemical pathways, 645–647

Biodegradable sutures, 457
Bioenergetics, 645–662
Biosynthesis, 513, 691–701. *See also* Anabolism
Bioterrorism, vaccines and, 735
Biotin, 711*t*, 714*t*
Birth-control pills, 517
Bisphenol A, 457–458
1,3-Bisphosphoglycerate, 668, *669–670*, 673*t*
Black, James W., 588
Bleaching, as redox reaction, 112
Blood, 30, 170, 410, 745–751
  bilirubin in, 687
  buffers in, 234, 754
  checking alcohol level in, 369
  chemical balance of, 754
  cholesterol in, 510, 512
  clotting of, 487, 520, 747–748
  deep-sea diving and, 158
  dialysis of, 177, 178
  enzymes in, 554, 566–567
  glucose in, 470, 479
  hormones in, 575–576
  immune system and, 725
  ketone bodies in, 679
  kidneys and, 746, 751–753, 754
  lactate in, 427
  nitric oxide and, 588
  pH of, 225, 234, 235, 746
  salt balance in, 754
  solution tonicity and, 175–177
  thomboxane A_2 and, 520
Blood–brain barrier, 698, **745**
Blood circulation, 4–5, *750*
Blood plasma, **744**–745
Blood pressure, 136, 536, 755–756
Blood serum, 162–163, *163*
Blood sugar, 466, 470, 479, 590, 709
Blood transfusion, 631, 748
Blood types, 482–483, 727, 740
Blood vessels, 588, 752, *753*
Blood volume, in blood pressure regulation, 756
Blow-molding techniques, 334, *335*
Blue diaper syndrome, 701
Blue litmus paper, *213*, 225
Body fluids, 170, 410, 412, 566–567, *725*, 744–756
Body mass, drug dosage and, 10
Body mass index (BMI), 708
Body temperature, 20, 192, 193, 746
Bohr, Niels, *44*
Bohr effect, **750**
Boiling point, 121, 135, **138**–141, 146
  of alcohols, 363, 364*t*, 442*t*
  of aldehydes, 423–424, 423*t*, 442*t*
  of alkali metals, 43*t*
  of alkanes, 302–303, 304
  of amines, 408–409
  of carboxylic acids, 441, 442*t*
  of ethers, 371–372
  of halogens, 42*t*
  of ketones, 423–424, 423*t*
  of noble gases, 43*t*
  of thiols, 374, 375*t*
Bond angles, **77**–80, 274–276, 287–288
  in alcohols, *360*
  in alkenes, 315–316
  in ethers, *370*
  in thiols, 372
Bond formation, 63–65, 188–190
Bonding electrons, **72**–74
Bone marrow, 537, 738, 746, 747
Boric acid, 213
Boron (B), 42, 47, 248
Botox (botulin toxin), 580
Botulism, 580, 582
Bowman's capsule, 751–752, *753*
Boyle, Robert, 122
Boyle's law, 122–123, 125

Brain damage, 193
Brain peptides, as neurotransmitters, 589–591
Brain scans, 257
Brass, 41, *154*
Bread dough, rising of, 221
Breast cancer, 517, 592, 733
Breath-alcohol screening, 369
Breathing, 123, 750, 751. *See also* Respiration
Brethaire, 413
Bright's disease, 18
Bromination, 325*t*, 330–331, 349
Bromine (Br_2), 4, **15**, 33, 39, 42, 202, 270, 275, 305–306, 330–*331*, 344
Bromobenzene, 349
p-Bromobenzoic acid, 346
Bromocresol green, as acid–base indicator, 226*t*, 232
Bronchodilators, *403*, 413–414
Bronchoscopy, muscle relaxants during, 555
Brϕnsted, Johannes, 214
Brϕnsted-Lowry acids and bases, 214–216
Bronze Age, 41
Brown, Michael, 511
Brown, Robert, 171
Brown fat, hibernation and, 658
Brownian motion, *171*–172
Bruises, heme catabolism and, 687
Buckminsterfullerene (C_{60}, "buckyballs"), 142, *142*
Buffer capacity, **233**
Buffers, **230**–236, 529, 754
Bullus pemphigoid, vitamin treatment of, 717
Burning, as chemical change, 3
Butanal, 423*t*, 424
Butanamide, 447–448, 454
2-Butanamine, 406–407
Butane, 96–97, 133, 288, *288*, 289, 295, *297*, 303*t*, 304, 364*t*
Butanedioic acid, 439, 449
Butanethiol, 373–374, 375*t*
Butanoic acid, 440*t*, 442, 442*t*, 443, 454, 497
1-Butanol, 360, 364*t*, 375*t*, 423–424, 423*t*
2-Butanol, 360, 365, 382–386
(*R*)-2-Butanol, 390
2-Butanone, 423*t*, 424, 429
1-Butanol, from butanol dehydration, 365
2-Butene, 327, 330, 365, *316*
Butyl alcohol, 360
sec-Butyl alcohol, 360
tert-Butyl alcohol, 360, 365
sec-Butylamine, 407
Butylated hydroxyanisole (BHA), 355
Butylated hydroxytoluene (BHA), 355
tert-Butyl ethyl ether, 370
Butyl group, 293*t*
sec-Butyl group, 293*t*
tert-Butyl group, 293*t*
sec-Butyl mercaptan, 373–374
2-Butyne, 318
Butyramide, 447–448
Butyric acid, 440*t*, 497
Cadaverine, 406–407
Caffeine, 147
Calcium (Ca), 29*t*, 40, 271, 714*t*
Calcium carbonate, 69, 106, 167, 203
Calcium channel blockers, 756
Calcium channels, 578
Calcium chloride, 96, 173
Calcium hydroxide, 96
Calcium hydroxyapatite, 63, 107, 544
Calcium ions, 500, 574*t*, 577–579, 694, 753, 756
Calcium oxide, 221
Calcium propanoate, *443*
Calcium sulfate, 203
Californium (Cf), 261
Calmodulin, 578
Calories, 7*t*, 19, 705, 707–708
Calvin, Melvin, 695
Calvin cycle, in photosynthesis, 695

Camphor, 157
Cancer, 260–261, 273, 348, 548, 601, 609, 700, 735, 755
Cancer cells, 610, 725, 739, 740
Capillaries, 537, 723, *725*, *745*, *750*, *753*, 755
Capillary dialyzer, *178*
Capping, in transcription, 622
Capric acid, 440*t*
Caproic acid, 439, 440*t*
Caprylic acid, 440*t*
Capsaicin, 353
Captopril, 394, 564, 756
Carbachol, 574*t*
Carbaminohemoglobin, 751
Carbamoyl phosphate, 681, *682*, 683–684
Carbocations, **327**, 329, 330
Carbohydrate intake, minimum daily, 709
Carbohydrates, 375, 391–392, 466–487, **467**, 534, 546, 646, *652*, 653–655, 666–667, *668*, 692–696, 705, 708–709, 740
  disaccharides as, **481**–484
  monosaccharides as, **467**–480
  oligosaccharides as, **481**–484
  polysaccharides as, **481**, 485–487
Carbon (C), solid forms of, 141–142
α-Carbon, 430–431, **526**, 528, *532*, 675
Carbon-11, 247
Carbon-14, 246, 249
Carbonate buffer system, 234
Carbon dioxide, **31**, 78–79, 80*t*, 81, 94–96, *95*, 99, 102, *105*, 128, 130, *145*, 147, 155, 158–159, 194, 195–196, 197, 203, 221–222, *646*, *652*, 654, 660–661, 681, 692, *693*, 694–695, *745*, 746, 747*t*, 750, 751, 756
Carbonic acid, 73, 155, 215, 221–222, 224, 234, 457–458, 751
Carbonic anhydrase, 751, 754
Carbon monoxide, **31**, 100, 103, 111, 129, 194–196, 197
Carbon nanotubes, 142
Carbon radical, in autoxidation, 353–354
Carbon tetrachloride, **15**, 157, 305, 306–307, 330–*331*, 637
Carbonyl group, **280**–281, 419, 420, 423–424, 430–431, 438–439, 445–446, 471. *See also* Ketone group
N-Carboxybiotin, as coenzyme, 711*t*
Carboxylation, of PEP, 678
Carboxyl group, **281**–282, 438–439, 444–445, 486
Carboxylic acids, 277, 277*t*, **281**–282, 367–369, 423, 424, 425, 438–449, 451, 454. *See also* Fatty acids
Carboxypeptidases, *538*, 554, 667, 710
Carcinogens, 348, 637, 638
Cardiac arrest, lactate accumulation and, 671
Cardiovascular disease, obesity and, 708
Carnauba wax, 498
Carnitine, in fatty acid β-oxidation, 675
Carnitine acyltransferase, 675
Carotene, *314*
Carotenoids, 711
Carson, Rachel, 347
Cartilage, 471, 486–487, 737
Casein, 526, 710
Caspases, in apoptosis, 615
Catabolism, 645, 674, 679–684, 686–687, 691–692
Catalase, in detoxification, 659
Catalysts, 193–194, 206, 332, 365–367, 426–427, 525, 553–554
Cataracts, 534, *634*
Catechol, 352, 413
Catecholamines, 413
Catechol-*o*-methyl transferase (COMT), 586
Cathode, 109, *112*, 167–168
Cations, **59**, 61–62, 61*t*, 62*t*, 67, 165–*166*, 167–168
CAT scan, 258, 259
CDP-choline, in lipid biosynthesis, 698
CDs. *See* Cluster determinants (CDs)
Cech, Thomas R., 553
Cefalexin, 447

Celebrex, 518
Cell-adhesion molecules. *See* Adhesion molecules
Cell membranes, 482–483, 494, 495, 501–503, 518, 572–573, *585*, 588, 592–593, *647*, 683, *694*, 697–700, *734–735*, 744–745. *See also* Thylakoid membrane
Cell nucleus, 572, *573*, 597, 615, 621, 624, 626, *647*, *667*, 746
Cellophane, 174, 177, 178
Cells, 544–545, 572–573, 574, 645–662, *647*
Cellulose, 466, 486, 692, 706
Celsius (C), 7*t*, 10–11, 126
Centigrade scale. *See* Celsius (C)
Centimeter (cm), 8
Central dogma of molecular biology, 620–621
Cephalins, 504
Cephalosporin, 446, 447
Ceramide, 505, 506–507, 698
Ceramide oligosaccharide, in Tay-Sachs disease, 509
Cerebrosides, 506–507
Cerebrospinal fluid, 566–567, 567*t*, 744, **745**
Cesium-137, in nuclear fallout, 263
cGMP. *See* Cyclic GMP (cGMP)
Chain, Ernst, 447
Chain-growth polymers, 332
Chain initiation, in autoxidation, 353
Chain length, in autoxidation, 354
Chain propagation, in autoxidation, 353–354
Chain reactions, *262–263*
Chair conformation, **297**–300, 474–475, 476, 481, 483, 484
Chaperones, 539, **543**, 549, 636, 648, 683
Chargaff, Erwin, 601
Chargaff's rule, 601, 603
Charles, Jacques, 124
Charles's law, 124, 125
Chemical bonds, 58–82, 537
 in acid–base reactions, 214, 216–217
 angles between, **77**–80, 274–276, 287–288, 315–316
 single, double, and triple, **72**–74
Chemical change, 31
Chemical communication, 572–593, 574*t*, **584**, 724, 731, 740
Chemical energy, 18, 19, 661–662
Chemical equations, **94**–98, 105, 197
Chemical equilibrium, 184, 216–218
Chemical kinetics, 185–186
Chemical messengers, 494, 495, 573, 576–577
Chemical properties, 3
Chemical reactions, 3, 4, 89–113
 activation energy for, 187, *188*–190
 heat of, **112**–113
 irreversible, 194
 limiting reagents in, 101–**102**
 molecular collisions and, 186–188, 192–193
 pressure changes and, 204–205
 rates of, 184–206, 557–559
 reversible, 194–197
 stress on, 200–206
 temperature changes and, 203–204
Chemical weapons, 579
Chemiosmotic pump, 658–660, 694
Chemiosmotic theory, **658**
Chemokines, 736, **737**, 739
Chemotactic cytokines, 737
Chemotherapy, 586, 601. *See also* Drugs; Medicine; Therapy
Chemstrip blood sugar test kit, 479
Chickenpox, 630, 722
Chinese herb tea, for Alzheimer's disease, 581
Chiral molecules, 382–398, **385**, **395**, 528
Chitin, 466, 471
Chloride channels, cystic fibrosis and, 683
Chloride ions, 65–66, 168, 211, 214, 216, 714*t*, 753, 754

Chlorinated water, 719
Chlorination, of benzene, 349
Chlorine (Cl$_2$), 29*t*, 33, 37, 38, 39, 42, 47, 59–60, 65–66, 111, 198, 270, 275, 305–306, 325*t*, 326, 327, 330–331, 457–458
Chloroacetic acid, 442
Chloroaniline, 346, 347
Chlorobenzene, 349
2-Chlorobutane, 327
Chlorodiazepoxide, 408
Chlorofluorocarbons (CFCs), 306, 307
Chloroform, 16–17, *138*, 139*t*, 305
Chloromethane, 170, 185, 186, 305
Chlorophyll, 692, 694–695
Chloroplasts, *694*
Cholera, 587
Cholesterol, 395, *503*, **510**–513, *511*, 517–518, 672, 698–699, 705, 709, 747*t*
Cholesteryl linoleate, 510–511
Choline, *501*, 503, 555, 556*t*, 580, 581, 698, 709
Cholinergic neurotransmitters, 574*t*, 576, 577–582
Cholinergic receptors, 577
Chromatin, 605, *606*, **612**, 621
Chromium (Cr), 29*t*, 715*t*
Chromosomes, 597, 605, *606*, 611–612, *613*, 621
 human, *596*, 609, 610
Chylomicrons, 510
Chymotrypsin, 555, 667, 710
Cinnamaldehyde, *419*, 422, 426
Cinnamyl alcohol, 426
Circulation, 4–5, *750*
Cis-trans isomers, **300**–302, *316*, 318–320, 321–323, *383*
Citrate synthase, 653, 655
Citric acid, *438*, 442, *652*, 653, *685*
Citric acid cycle, *646*, 648, 649, *652*, 653–655, *668*, *670*, 671, 673*t*, 678, *685*, 686, *693*, 700
Citronellal, 422
Citrulline, 588, 681, *682*
Citryl-CoA, in citric acid cycle, 653
Clamp loaders, 610
Clamp protein, *609*, 610*t*
Class I MHC, *728*
Class II MHC, *728*–729
Class switch recombination (CSR) mutation of V(J)D genes, 732
Clones, *596*, 616
Clonidine, 574*t*
Cloning, of DNA, 616–617
Clorazil, 574*t*
Clotting, 487, 520, 526, 554, 747–748
Cloverleaf structure, of tRNA, *607*
Clozapine, 574*t*
Cluster determinants (CDs), *728*, *734*, *735*
CMP. *See* Cytidine 5'-monophosphate (CMP)
Coated pits, in cholesterol transport, 511
Cobalamine, 711*t*. *See also* Vitamin B$_{12}$
Cobalt (Co), 29*t*, 715*t*
Cobalt-60, in cancer therapy, 260
Cobra venom, 582
Cocaine, 405
Codon recognition site, 625
Codons, **624**, 625–626, 625*t*, 627, 628–629, 636
Coefficients, in chemical equations, 95–98, 98–104
Coenzyme A, 650–*652*, *651*, 675, *676*, 677*t*, 697–698, 711*t*
Coenzyme Q (CoQ), 656, 657
Coenzymes, 557, 562, 711*t*
Cofactors, **556**, 561, 694–695, 711*t*
Cold compresses, 22
Cold virus, 630
Collagen, 487, 525, 526, 540, 544, 545, 547, 710, 748
Collecting tubule, in kidneys, *753*
Colligative properties, **172**–178
Collins, Robert John, *373*
Collisions, among molecules, 186–188, 192–193
Colloidal dispersions/systems, 153–154, 170, 171, 171*t*
Colloids, **170**–172

Colon cancer, 518, 706
Color-blindness, 596
Combinatorial chemistry, 274
Combined gas law, **125**–126
Combustion, **94**, 111, 113
Common catabolic pathway, **646**–647, 649–652, 653–655, 656–658, 667, *668*
Common cold, 630
Competitive inhibition of enzymes, **557**, 560–562
Complementary base pairs, in DNA molecule, *603*–605
Complete protein, 710
Complexes I–IV, in electron and hydrogen ion transport, 656, 657–658
Complex lipids, 495, 500–*501*, *504*, 507, 508–509, 518. *See also* Glycerophospholipids; Glycolipids; Sphingolipids
Compounds, 28–30, 31, 43
 aromatic, **343**–355
 binary, 67
 inorganic, 271–272, 272*t*
 ionic, 65–69
 molecular, 70–76
 organic, 271–272, 272*t*
 polar, 371–372
Computer-assisted tomography (CT, CAT) scan, 258, 259
Concentration units, for solutions, 159–165
Condensation, **131**, *137*. *See also* Liquefaction
Condensed structural formulas, 278, 279, 288*t*
Conductance of electrolytes, 168–169
Conformations, **297**–300
(*s*)-Coniine, 405
Conjugate acids and bases, **214**–216, 215*t*, 234–236
Conjugated proteins, 545
Connective tissue, acidic polysaccharides in, 486–487
Consensus sequences, in transcription, 622–623
Conservation of energy, law of, 19
Constant composition, law of, 32
Constitutional isomers, **289**–291, 291*t*, 291–294, *383*, 530
Contraceptive drugs, 516–517
Contributing structures, 344–345
Conversion, 11–14
 among forms of energy, 19
 in English and metric systems, 7–8
Conversion factors, 8*t*, **11**
Cooking oils, autoxidation of, 352
Cooling effect, of evaporation, 138
Copper (Cu), 29*t*, *40*, 41, 61, 107–109, 711*t*, 715*t*
Copper(I) oxide, *61*
Copper(II) oxide, *61*
Copper(II) chloride, 106–107
Copper(II) nitrate, 105–*106*
Copper(II) sulfate, 227
Copper(II) sulfide, 106–107
Copper carbonate, *107*
Copper sulfate pentahydrate, *167*
Coral chemistry, 63
Cormack, Allan M., 258
Corn, incomplete proteins in, 710, 711
Corn syrup, high-fructose, 485
Cornea, 217, 548
Coronary thrombosis, lactate accumulation and, 671
Cortisol, 513, 515, 575*t*
Cortisone, 513, 515, 516, 518
Cosmic rays, 244, **256**
Cotton, cellulose in, 486
Covalent bonds, 63–**64**, 70–75, 133–134, 188–189, 272, 274–276, 540. *See also* Polar covalent bonds
Covalent compounds, 90, 92, 169–170. *See also* Molecular compounds
Cowpox, in smallpox immunization, 735
COX-1 and COX-2 enzymes and inhibitors, 518, 519
Cracking, 315
Cramps, lactate accumulation and, 671
Crash diets, 708
Creatine, 718, 753

Creatine phosphokinase (CPK), after heart attack, 567, 567t
Creatinine, in urine, 753
Cresol, 352
Creutzfeld-Jakob disease, 539
Crick, Francis, 603, *604*
Cristae, of mitochondria, 648
Critical temperature, for carbon dioxide, 147
Crohn's disease, treatment of, 739
Cross-linking, 545, 562, 563
Crutzen, Paul, 307
Cryoelectron microscopy, of ribosomes, 629
Crystallization, **141**. *See also* Freezing; Solidification
Crystals, 119, *141*, 166–167
C-terminal amino acid in peptides, 532, 534, *538*
CT scan, 258, 259
Curare, 574t, 582
Curie (Ci), 253, 254t
Curie, Marie, 242, 253
Curie, Pierre, 242
Curium (Cm), 261
Curl, Robert F., Jr., 142
Cushing syndrome, 516
Cyanobacteria, 692
Cyclic amides, 446, 447
Cyclic AMP (cAMP), 576, 578, 584, *585*, 589
Cyclic esters, 445, 446
Cyclic ethers, **371**
Cyclic GMP (cGMP), diabetes and, 590
Cyclic hemiacetals, creation of, 428
Cyclic hydrocarbons, 295
Cycloalkanes, **295**–300, **300**–302
Cycloalkenes, 320–321, 331
Cyclohexane, 295, **297**–300
Cyclonite (RDX), 350
Cyclooxygenase (COX), 450, 518, 519, 520, 566. *See also* COX-1 and COX-2 enzymes and inhibitors
Cyclopentane, 295, *297*
Cyclosporin A, for autoimmune diseases, 740
CYP2D6 gene, mutations of, 613
Cysteine, 527t, 530, 540, 541, 563, 582, 625t, *685*, 700, 710
Cystic fibrosis, ubiquitin and, 683
Cystine, 530, 685
Cystinuria, 685
Cytidine, 600t, 631
Cytidine diphosphate (CDP), 698
Cytidine 5′-monophosphate (CMP), 600t
Cytochrome *c*, 535, 657
Cytochrome oxidase, 657, 711t
Cytochrome P-450 enzyme, 613, 659
Cytochromes, 657, 694
Cytokines, 631, **736**–737, 739–740
Cytoplasm, *502*, *503*, *573*, 621, 624, *647*, 671, 672, *693*, 696
Cytosine (C), 597–598, 600t, 601–*602*, 603–605, *604*, 698
Cytoskeleton, 544, *647*
Cytosol, 626, 675, *676*
Dacron, 346, 457
Daily Values, for nutrients, 706, 718
Dalton, John, 31–32, 36
Dalton's law of partial pressures, 130, 750
Dark reaction, in photosynthesis, 695
Daughter strands, in DNA replication, 610
Davy, Humphry, 373
ddCyd. *See* Dideoxycytidine (ddCyd)
ddI. *See* Dideoxyinosine (ddI)
DDT (dichlorodiphenyltrichloroethane), 347
Deacetylation, during DNA replication, 611–612
Deamidation, DNA mutations from, 613
Debranching enzyme, in carbohydrate digestion, 708, *709*
Decamethonium bromide, 582
Decarboxylation, 586, 587–588, **653**–654, 699
Decay of atomic nuclei, 248
Deep-sea diving, dissolved gases and, 158
Deficiency diseases, 705

Dehydration, **365**–367, 587, 679, 747t
L-Dehydroascorbic acid, 473
Dehydrogenation, of fatty acids, 675, *676*, 677t
Democritus, 27–28, 31
Demyelination, in multiple sclerosis, 506
Denaturation, 546–549, **547**
Denaturing agents, of proteins, 547t
Dendrites, *573*, 574–575
Density, 15–17, 140, 303t, 304
Dentine, composition of, 544
Deoxyadenosine, 600t
Deoxyadenosine 5′-monophosphate (dAMP), 600t
Deoxycytidine, 600t
Deoxycytidine 5′-monophosphate (dCMP), 600t
Deoxyglucose, in PET scans, 259
Deoxyguanosine, 600t
Deoxyguanosine 5′-monophosphate (dGMP), 600t
β-2-Deoxy-D-ribose, 474, 598–599, 606, 672
(−)-2-Deoxy-3′-thiacytidine (3TC), 631, *632*
Deoxythymidine, 600t
Deoxythymidine 5′-monophosphate (dTMP), 600t
Dephosphorylation, ATP yield from, 673t
Deprenyl, 574t, 586
Desalinization, 176
Detergents, 350–351, 500, 517–518
Detoxification, 659
Deuterium, 245, 261
Dextrins, 485
Dextrorotatory compounds, **396**, 468–470
Dextrose, 470
dGMP. *See* Deoxyguanosine 5′-monophosphate (dGMP)
Diabetes, 353, 534, 536, 590, 679, 708, 739, 747t, 753
Diabetes insipidus, 754
Diabetes mellitus, 18, 479, 590, 679, 709
Diabetic retinopathy, 534
Diagnosis, radioactive isotopes for, 256–260
Dialysis, *177*, *177*, 178
Diamond, 33, 141, *142*, 143
Diaphragm, 123
Diastereomers, *383*, **391**–392
Diastolic pressure, 136
Diatomaceous earth, in dynamite, 362
Diatomic elements, 33
Diazepam, molecular structure of, 408
Dicarboxylic acids, 439
Dichloroacetic acid, 442
Dichlorodifluoromethane, 306
1,1-Dichloroethylene, 333t, *334*
Dichloromethane, 82, 305, 307, 330
Dichromate ion, in breath-alcohol screening, 369
Diclofenac, 520
Dicumarol, 748
Dicyclohexyl ether, 370, 423
Dideoxycytidine (ddCyd), 631
Dideoxyinosine (ddI), 631
Dienes, 321
Diet faddism, 705–706
Dietary fat substitutes, 717
Dietary Reference Intakes (DRI), 705, 706, 712–715t, 717–718
Diethylammonium chloride, 412
Diethyl butanedioate, 449
Diethyl carbonate, 457–458
Diethyl ether, **15**, *156*, 370, 376, 423*t*
Diethyl glutarate, 445
Diethyl malonate, 453
Diethyl pentanedioate, 445
Diethyl propanedioate, 453
Diethyl succinate, 449
Dieting, 705–706, 716–717.
Digestion, 3, 486, 679–680, **705**, 708–710
Digestive enzymes, 554, 667
Digitoxin, 574t
Diglycerides, 497
5-α-Dihydroxytestosterone, aging and, 755
Dihydroxyacetone, 469t

Dihydroxyacetone phosphate, 455, 668, *669–670*, 674
1,2-Dihydroxybenzene, 413
(*S*)-3,4-Dihydroxyphenylalanine, 586
Dilation of blood vessels, nitric oxide and, 588
Diltiazem, 574t, 756
Dilute solutions, 159, 199
Dilution, of solutions, 163–165, *164*
Dimenhydrinate, as antihistamine, 588
Dimerization, 441, 530
N,*N*-Dimethylacetamide, 454
Dimethylacetylene, 318
Dimethylamine, 279, 403, 454
1,3-Dimethylbenzene, 346
3,3-Dimethyl-2-butanone, 421
cis-1,4-Dimethylcyclohexane, *301*
trans-1,4-Dimethylcyclohexane, *301*
N,*N*-Dimethylcyclopentanamine, 406
cis-1,2-Dimethylcyclopentane, *301*
trans-1,2-Dimethylcyclopentane, *301*
Dimethyl ether, 371–372
N,*N*-Dimethylformamide, 446
Dimethyl phosphate, 455
Dimethyl sulfide, 374
Dimethyl terephthalate, 456–457
Dinitrogen tetroxide, 202, *204*
2,4-Dinitrophenol, 658
Diols, 363
Dioxin, toxicity of, 165
Dipeptides, amino acids in, 530–531, 534
Diphenhydramine, 574t, 588
Diphosphate ion, 455
Diphosphoric acid, 455
Diphosphoric esters, of monosaccharides, 480
Dipoles, **71**, *132*
Dipole–dipole interactions, 131–132, 132–**133**, *371*, 729–731
Dipropyl ether, 170
Diprotic acids, **215**
Diradicals, in autoxidation, 353
Disaccharides, **481**–484, 666–667, 692, 693–696, 708–709
Discriminatory curtailment diets, **705**–706
Disease
  autoimmune, 717, 739–740
  of brain and nervous system, 506
  deficiency, 596, 705, 711, 712–715t
  diets to prevent, 705–706
  excessive protein production in, 623
  immunity to, 722–724
  kidney, 752, 753
  lipid storage, 508–509
  protein/peptide-conformation-dependent, 539
  sulfa drugs and, 562
Disinfectants, 111
Dissociation, 104
Distal tubules, of kidneys, 752, *753*, 754
Disubstituted alkylbenzenes, 345–346
Disulfide bond, 530, 535, 536, 540, 541, 547–548, 729, *730*, 734, 737
Disulfides, **374**–375
Diuresis, **754**, 756
Diuril, 756
D,L system, 391, 468–471, 528
DNA (deoxyribonucleic acid), 348, 404, 597–600, 600t, 601–607, *602*, 602t, *603*, *604*, *605*, *606*, 609, 616–617, 630, 636–639, 740
  genes and, 608, 620–621, 625, 625t, 633, 634, 635
  NADPH and, 672
  recombinant, 639–642
  repair of, 613–616
  replication of, 609–613
  transcription of, **621**–624
DNA chips, pharmacogenomics and, 613
DNA fingerprinting, 611
DNA fragments, in apoptosis, 615
DNA gyrase, *609*
DNA ligase, *609*, 613, *614*, 615

DNA polymerase, *609*, 610, 610*t*, 612, *614*, 615, 711*t*
DNA polymerase I, *609*
DNA replication, 613–616, 633–636
DNA viruses, 630
DNP. *See* 2,4-Dinitrophenol
Docosahexanoic acid, 697
Dodecanoic acid, 440*t*
Dodecylbenzene, sulfonation of, 350–351
Donepezil, 574*t*, 581
L-Dopa, 586
Dopamine, 410, 574*t*, 583, 586
Dosimetry, 253–256, 254*t*
Double bond, **72–74**, 274, 275, 276, 286, 287, 352, 425–427, 430–431, 495, 496, 496*t*, 497, 499
   in alkenes, 314, 315, 316, 321–323, 325–332
   bond angles for, 77, 78–79
Double helix, DNA molecule as, **603**–605, *603*, *606*, *607*, 612, 621–*622*
Dramamine, 588, 745
DRI. *See* Dietary Reference Intakes (DRI)
Drug dosage, body mass and, 10
Drugs, 257, 413–414, 453, 572–573, 574*t*, 659
   active stereoisomers of, 387–388, 394, 398
   for AIDS, 631–*632*, 634
   for Alzheimer's disease, 581
   anabolic steroids as, 514
   anti-inflammatory, 515, 518
   anticancer, 601
   antihistamine, 588
   antisense, 623
   antistroke, 588
   antiviral, 630
   for autoimmune diseases, 739–740
   blood–brain barrier and, *745*
   cholesterol and, 512–513, 699
   contraceptive, 516–517
   eliminating side effects of, 745
   extensive metabolism of, 613
   for diabetes, 590
   as hydrochlorides, 412
   for hypertension, 756
   monoamine-oxidase-inhibiting, 587
   in monoclonal antibody therapy, 733
   pain-relieving, 450
   for Parkinson's disease, 586
   pharmacogenomics and, 613
   for prostate treatment, 755
   recombinant DNA techniques to produce, 640–641
   solubility in body fluids, 412
   statin, 699
   sulfa, 562
   synthetic, 274
   weight-loss, 658
Dry cell batteries, *112*
Dry ice, 147
dTMP. *See* Deoxythymidine 5′-monophosphate (dTMP)
Dynamic equilibrium, **195**–197, **197**–200
Dynamite, nitroglycerin in, 362
Ear wax, 498
EcoRI restriction endonuclease, 640–641
Edema, 747*t*, 748
E (exit) site, in protein elongation, 629
Efavirenze, in AIDS treatment, 631
Effective collisions, 186–188, 192–193
EGF. *See* Epidermal growth factor (EGF)
Egg, zygote from, 609
Egg whites, as heavy metal poisoning antidote, *548*
Eicosanoic acid, 440*t*
Einstein, Albert, 261
Eka-silicon, 5
Elavil, 574*t*
Electrical energy, metabolic conversion of chemical energy to, 661
Electricity, in chemical reactions, 113. *See also* Battery
Electrodes, 167–168
Electrolytes, 167–169, 213, 223

Electrolytic balance, *169*
Electromagnetic spectrum, *244*
Electron configurations, **46**–52, 58, **59**–61, 63–64
Electron density model, 71
Electronegativity, 64–66, 70–72, 442–443
Electron-microscope tomography, of mitochondria, 648
Electron pairs, in acid–base reactions, 214, 216–217
Electrons, 33–39, **34**, 44–52, 59–60, 243, 244–245, 245*t*, 246
   in bonds, 64, 65–66, 70–72, **72**–74, 77–80
   in ions, 52, 58–59
   in oxidation–reduction reactions, 107–112
Electron shielding, ionization energy and, 53
Electron transport chain, 649, 650, 656–661, *668*
Electrophoresis, in DNA fingerprinting, 611
Elements, 28, 29*t*, 32*t*, 33, 34–36, 39–44, 245, 246–247, 270–*271*
Elongation, 623, 624, 626, 626*t*, *628*–629, 697
Elongation factors, 628, *629*
Emulsions, 171, 172
Enamel, composition of, 544
Enantiomers, 301, *383*, **385**, 387–388, 390–391, 392, 394, 398, 450, *467*, 470, 528
   in racemic mixtures, **386**
   R,S system for naming, 387–391, **388**
Encephalitis, spongiform, 539
Endocrine glands, 573, 575–*576*, 575*t*, 746
Endocytosis, in cholesterol transport, 511, 512
Endometrium, in menstrual cycle, *516*
Endonuclease, *614*, 615
Endoplasmic reticulum (ER), 578, 584, *647*, *728*
Endothelial cells, in leukocyte mobilization, 736
Endothermic reactions, 112–**113**, *189*, 203, 204
End point, of titration, **228**
Enduron, 756
Enebrel, 737
Enediols, in sugar oxidation, 478
Energy, 7*t*, 18–19, 112–113, 649–650, 655, 672–674, 677–678, *680*, 692, 695, 704, 705, 707–708, 709. *See also* Activation energy; Bioenergetics
   atomic, 262
   conversion of mass into, 261
   from electron transport, 660–661
   ionization, **52**–53, 64–65
   radiation, 251, 252
Energy diagrams, for chemical reactions, *188*–190
Energy storage, 494, 649–650
Enflurane, 373
English system, 6–8
Enhancers, in gene regulation, 635
Enkephalins, 574*t*, 585, 589–591
Enolase, in glycolysis, *669*
Enols, **430**–431
trans-Enoyl-CoA hydratase, *676*
Entacapon, 574*t*
Entropy, 120
Envelope conformation, of cyclopentane, *297*
Environmental Protection Agency (EPA), public drinking water standards by, 718–719
Enzyme assays, 567*t*
Enzymes, 5, 478, 479, 508–509, 525, 553–567, 556*t*, 557, 581, 610, 648, 656–658, 696–697, 708–709, 729. *See also* Ribozymes
   active sites of, **557**, 559–562
   in amino acid catabolism, 679–684
   in base excision repair, *614*, 615
   chemical messengers affecting, 576
   chirality of, 397–398
   in cholesterol synthesis, 511, 512–513
   in clotting, 748
   cofactors of, **556**–557
   in DNA fingerprinting, 611
   for eye color, 609
   in gluconeogenesis, *693*
   in immune system, 727
   inhibitors of, **557**, 560–562
   nitric oxide and, 588

   in nucleotide excision repair, 616
   phosphorylation of, 584, 585
   in photosynthesis, 694, 695
   in protein digestion, 710
   in protein metabolism, 667
   in recombinant DNA techniques, 639–640
   in serum, 747*t*
   signal transduction and, 674
   in transcription, 622–624
   for tRNA, 624–625, 627, 633
   ubiquitin and, 683
Enzyme specificity, **554**, 555
Enzyme–substrate complexes, 557, 559–563
EPA. *See* Environmental Protection Agency (EPA)
Epichlorohydrin, 376
Epidermal growth factor (EGF), 638, 737
Epinephrine, *403*, 413–414, 574*t*, 575*t*, 576, 583, 587, 674
Epitope, 728–729, 733, 739
   antigenicity of, **728**, *729*, 731
Epothilon B, 735
Epoxides, as carcinogens, 348
Epoxy glues, production of, 376
Epoxy resins, production of, 376
Equal sweetener, 717
Equatorial bonds, **298**–300
Equilibrium, **137**, 200
   in acid–base reactions, 216–218
   dynamic, **195**–197, **197**–200
   enzyme catalysis and, 553–554
   of glucose anomers in solution, 475
   between keto and enol tautomers, 430–431
   Le Chatelier's principle and, 201–206
   in self-ionization of water, 223
   with solutions, 156
Equilibrium constant (K), **197**–200, 201, 203, 218–220
Equilibrium expressions, 197–200, 218–220
ER. *See* Endoplasmic reticulum (ER)
Errors, in DNA replication, 613–616, 633–636, 636–639
Erythritol, 477–478
Erythrocytes, 725, **746**–747, *746*, 747*t*
Erythromycin, stereoisomers of, 394
Erythrose, 391–392, 470
D-Erythrulose, 469*t*
Escherichia coli, 613, 639, 679
Essential amino acids, 700, **710**
Essential fatty acids, 498, 709
Essential oils, terpenes in, 323–324
Esterification, 448–449, 599
Esters, 438–439, **444**–445, 446, 448–449, 451–453, 455, 457, 480, 486, 495–497, 498, 499, 510, 698
Estradiol, 513, 515, 517, 575*t*, 592, 755
Estrogen, *516*, 755
Ethanal, 420
Ethane, 275*t*, 287, 303*t*, 364*t*, 409*t*
Ethanedioic acid, 439
1,2-Ethanediol, 363, 451–452, 456–457
Ethanethiol, 373, 374, 375*t*
Ethanoic acid, 440*t*, 448
Ethanol, 21, 92–93, 111, *138*, 139*t*, 156, 160, 170, 276, 278, 308, 360, 364*t*, 365, 369, 371–372, 375–376, 375*t*
   in esterification, 448
   from fermentation, 221
   from hydrolysis, 451
   monosaccharide solubility in, 471
   natural versus synthetic, 273
   protein denaturation by, 548–549
   reactions with aldehydes and ketones, 428–429
   in reversible reaction, 200, 201, 202
Ethanolamine, 504, 698, 709
Ethene, 318
Ethephon, 315
Ethers, 359, **370**–372, 373

Ethrane, 373
Ethyl acetate, 200, 201, 202, 445, 448, 451
Ethyl acrylate 333*t*
Ethyl alcohol. *See* Ethanol
Ethylamine, 275–276, 449
Ethylbenzene, 345
Ethyl butanoate, 445
Ethyl butyrate, 445
Ethylene, *73*, 78, *79*, 80*t*, 275*t*, 287, 314, 315, *316*, 318, 325, 326, 328, 365, 376
 polymerization of, 315, 332–336, 333*t*
Ethylene glycol, 173, *361*, 363, 376, 451–452, 456–457
Ethylene oxide, 371, 376
Ethyl ethanoate, 445
Ethyl group, 293*t*
Ethyl isopropyl ketone, 423
Ethyl mercaptan, 373
2-Ethylpentanoic acid, 440–441
Ethyndiol diacetate, 517
Ethyne, 317, 318
Ethynil estradiol, as contraceptive, 517
Eukaryotes, 621, 622–624, 627–628
 genes of, 611–613, 625–626, 633–636
Evaporation, 135–*137*, 137–141
Evolution, 626, 637
Excitatory neurotransmitters, 582–583
Excretion, *753*, 756
Exercise, 678, 708
Exons, **608**, *622*, *732*
Exonuclease, in base excision repair, *614*, 615
Exothermic reactions, 112–**113**, 203, 204, *188*, 189, 325
Explosives, nitro group in, 350
Exponential notation, **6**
Extended helix, *540*
External innate immunity, 722–723
Extracellular fluids, 744
Extracellular water, 165
Eyes, 217, 548, 609, 687, 723
Fabricius, 5
Fabry's disease, characteristics of, 508
Facial tics, botulin treatment of, 580
Facilitated transport, across cell membranes, 502
F-actin, quaternary structure of, 544
Factor-label method, for unit conversions, 11–14
FAD$^+$. *See* Flavin adenine dinucleotide (FAD$^+$)
FADH$_2$. *See* Reduced FAD$^+$ (FADH$_2$)
Fahrenheit scale, Celsius scale and, 10–11
Familial adenomatous polyposis, treatment of, 518
Familial hypercholesterolemia, 512
Famine, 708
Faraday, Michael, 344
Farnesol, 324
Farnesyl pyrophosphate, 699
Fastin, 404
Fasting, fat metabolism and, 678
Fat, lipids in, 494
Fat depots, *667*, 677, 696
Fat globules, *667*
Fats, 495, **497**–500, 498*t*, *652*, 658, 666, 667, *668*, 677, 696, 705, *707*, 708, 709, 717. *See also* Fat; Lipids
Fat-soluble vitamins, 712*t*
Fatty acids, 353–354, 440*t*, 495–497, 496*t*, 498, 498*t*, 500–*501*, *652*, 667, *668*, 677–678, 696–697, 705, 709, *723*
 β-oxidation of, 674–675, *675*–677, *676*
 cholesterol esterified with, 510
 in cooking oils, 352
 in glycerophospholipids, 503–505
 NADPH in synthesis of, 672
 from protein catabolism, *680*
 saturated, 705
 in sphingolipids, 505
 unsaturated, 501–503, 675, 697
Fatty acid synthase, 696–697, 711*t*
Fava beans, hereditary disease involving, 613
FDA. *See* Food and Drug Administration (FDA)

FD & C. *See* Food, drugs, and cosmetics (FD & C)
Feces, 518, 687
Feedback control, 563, 655
Fermentation, 221, 375
Fermi, Enrico, 262
Ferredoxin, in photosynthesis, *694*
Ferric hydroxide, *107*
Ferritin, 526, 687
Fertility, aging and, 755
Fertilization, 515, *516*, 593
Fertilizers, 205, 706
Fetal cells, 610
Fetal hemoglobin (HbF), 537, 545
Fever, 192
Fexofenadine, 574*t*, 745
Fiber, dietary, **706**
Fiberscope, for laser surgery, 548
Fibrils, of collagen, 544, 545
Fibrinogen, 526, 747–748, 747*t*
Fibrous proteins, 526, 539
50S ribosome, 627, *629*
Filamentous cytoskeleton, *647*
Filtration, 155
 as kidney function, 751–753
Fischer, Emil, 448, 468–469
Fischer esterification, **448**–449
Fischer projections, 467–471
Fission, nuclear, 262–264
Flavin, in FAD molecular structure, *651*
Flavin adenine dinucleotide (FAD$^+$), 646, 650, *651*, *652*, 654–655, *656*, 657, 658, 660, 675, *676*, 711*t*
Flavodoxin, in photosynthesis, *694*
Fleming, Alexander, 447
Florey, Howard, 447
Fluid mosaic model, 501–*503*
Fluoridation, 61
Fluoride ions, 61
Fluorine (F, F$_2$), 29*t*, 33, 39, 42, 47, 61, 715*t*
 in organic chemistry, 270, 275
18-Fluorodeoxyglucose, in PET scans, 259
Fluorouracil, 601
Fluorouridine, 601
Fluoxetine, 574*t*
Folic acid, 562, 711*t*, 713*t*, 716
Follicle-stimulating hormone (FSH), *576*
Food allergens, 739
Food colorants, synthetic, 351
Food, drugs, and cosmetics (FD & C), 351
Food guide pyramid, 706, *707*
Forane, 373
f orbitals, 45, 51
Forces, attractive, 119–121, 131–135
Formaldehyde, *73*, 78, *79*, 80*t*, *82*, 134, 275*t*, 280, 375, 419–420, 423
Formica, *440*
Formic acid, 110, *423*, *440*, 440*t*
Formulas. *See also* Structural formulas
 of alkanes, 288*t*
 of compounds, 29–30
 of ionic compounds, 66–67
 water of hydration in, 166
Formula units, 90–91
Formula weight, 89–90, 91
40S ribosome, 624, 626*t*, 627, *628*
Fractional distillation, of petroleum, 307–308
Franklin, Rosalind, 603, *604*
Free cholesterol, 510
Free radicals, from ionizing radiation, 255
Freeze-drying, 145
Freezing point depression, **173**–174
Freons, *137*, 306, 307
Frequency (*f*), of electromagnetic radiation, 243–244
α-D-Fructofuranose, 474
β-D-Fructofuranose, 474, 481
D-Fructose, 469*t*, 470, 474, 484, 485
Fructose 1,6-biphosphate, 668, *669*–670, 673*t*, *693*
Fructose 1-phosphate, *670*, 674

Fructose 6-phosphate, 556*t*, 667–668, *669*–670, *693*
L-Fucose, 482
Fugu fish, 299
Fuller, Buckminster, 142
Fullerenes, 142, *142*
Fumarase, in citric acid cycle, 654
Fumarate, *652*, 654, *682*, 684, *685*
Fumigants, ethylene oxide as, 371
Functional groups, **276**–282, 277*t*, 438–439
Funk, Casimir, 711
Fur, as allergen, 739
Furan, 472
Furanose form, of monosaccharides, 471, **472**–476, 477
Furchgott, Robert, 588
Fusion, nuclear, **261**–262
GABA. *See* γ-Aminobutyric acid (GABA)
G-actin, 544
Galactocerebroside, in Krabbe's leukodystrophy, 508
Galactokinase deficiency, 471
D-Galactopyranose, 473–474, 475, 481
D-Galactosamine, 470
Galactose, 471, 484*t*, *501*, 506–507, *670*, 673–674
D-Galactose, 468*t*, 470, 473–474, 475, 482
Galactosemia, 471
Galactose-1-phosphate uridinyltransferase deficiency, 471
α-Galactosidase, in Fabry's disease, 508
β-Galactosidase, in Krabbe's leukodystrophy, 508
Galactosuria, 471
Galantimine, for Alzheimer's disease, 581
Galen, 4–**5**, *5*
Gallbladder, 687, 709
Gallons, converting to liters, 13–14
Gallstones, *510*, 747*t*
Gamma rays, 243, 244, 245, 252, 254, 260, 637, 638
Gangliosides, 506, 509
Gap junctions, *502*
Garlic, thiols in, 373
Garnet, *141*
Gas laws, 122–126, 127–130
Gas pressure, 121–122
Gases, 14–15, *105*, 119–121, *120*, *127*, 147, 158, 171*t*, 190, 194–197, 204–205, 747*t*, *750*, 751. *See also* Gas laws; Noble gases; Vapor
 condensation of, **131**
 kinetic molecular theory of, 130–**131**
 partial pressure of, **130**
 as solutes and solvents, 154, 154*t*, 155*t*, 156, 158
Gasohol, 359, 375–376
Gasoline, *123*, 157, 308, 375–376
Gastric disorders, 559
Gastric fluid, hydrochloric acid in, 213
Gaucher's disease, characteristics of, 508
Gay-Lussac, Joseph, 124
Gay-Lussac's law, 124–126
GC box, 633
GDP. *See* Guanosine diphosphate (GDP)
Geiger-Müller counter, *251*
Gelatin, as incomplete protein, 710
Gelsolin, in apoptosis, 615
Gene conversion (GC) mutation, of V(J)D genes, 732
Gene expression, **620**–621
Gene insertion technique, 634
Gene therapy, 640–641
Genes, 256, 526, 537, *597*, 608, 609, 613, 616–617, **620**–621, *622*–624, 636, 638, 639, *732*
 regulation of, 526, **633**–636
Genetic code, 621, 624–625, 625*t*, **625**–626
 second, 627
Genetic diseases, 639, 640–641
Genetic engineering, 641–642
Genome, 613
Genomics, 543, *620*, 740
Geraniol, 322–323, 324
Geranyl pyrophosphate, 699
Germ cells, immortality of, 610
Germane, 43–44

Germanium (Ge), 42
GH. *See* Growth hormone (GH)
GILT (gamma-interferon inducible lysosomal thiol) reductase, 729
Glacial acetic acid, 213, *441*
β-Globin, biochemical evolution of, 637
Globin chains, in hemoglobin, 543, 545, 637, 637*t*
Globular proteins, 526, 538–539, 547
Globulins, 747, 747*t*, 748
Glomeruli of kidneys, 177, 752, *753*, 754
Glucagon, 575*t*, 576, 585, 589, 591
D-Glucitol, 477
Glucoamylase, 485
Glucocerebroside, 507, 508
Glucocorticoid hormones, 516, 739–740
Glucogenic amino acids, *668*, 684, *685*, 686, *693*
Glucokinase, in glycolysis, *669*
D-Gluconate anion, 478
Gluconeogenesis, *680*, **693**, *693*
D-Gluconic acid, 479
Glucophage, for diabetes, 590
D-Glucopyranose, 481, 483–484
D-Glucosamine, 470
Glucose, 90, 222, 258, 427, 466, 484*t*, 666–667, 672–674, 677–679, 685, 691–692, 695, 705, 708, 709, 747, 747*t*
  ATP yield from metabolism of, 673*t*
  diabetes and, 590
  in gluconeogenesis, *693*
  glycolipids and, *501*
  glycolysis of, 667–671, *669–670*, 672–674
  oxidation of, 184, 188, 554
  from protein catabolism, *680*, 684
D-Glucose, 468*t*, 470, 472, 473–474, 475, 477, 478–479, 480, 484, 485, 486, 709
L-Glucose, 554
Glucose isomerase, 485
Glucose oxidase, 478, 479, 554
Glucose 1-phosphate, 691, 696
Glucose 6-phosphate, 556*t*, 667, *669–670*, *693*
D-Glucose 6-phosphate, 480
Glucose solution, isotonic, 176–177, 178
Glucose transporter molecules (GLUT4), 590
β-Glucosidase, in Gaucher's disease, 508
α-Glucosidases, in glycogen digestion, 486
β-Glucosidases, in cellulose digestion, 486
D-Glucuronic acid, 479–480, 487
Glucuronides, 480
Glues, 375, 376
Glutamate, 556*t*, 680, 681, 684, *685*. *See also* Glutamic acid (Glu, E)
Glutamate-oxaloacetate transaminase (GOT), 567
Glutamate-pyruvate transaminase (GPT), 567
Glutamic acid, 527*t*, 532, *542*, 574*t*, 582, 583, 591, 625*t*, 636, 655, *685*, 700–701
Glutamine (Gln, Q), 527*t*, 541, 625*t*, 684, *685*, 700
Glutaric acid, 439
Glutathione, 532, 659, 671
Glutathione peroxidase, as cofactor, 711*t*
Glycation, of proteins, 534
Glyceraldehyde, *467–468*, 469, *670*
(*R*)-Glyceraldehyde, 397–398, 468
(*S*)-Glyceraldehyde, 397–398
Glyceraldehyde dehydrogenase, in glycolysis, *669*
Glyceraldehyde 3-phosphate, in glycolysis, 668, *669–670*, 673*t*
D-Glyceraldehyde-3-phosphate dehydrogenase, as cofactor, 711*t*
Glyceric acid, 441
Glycerin, 362, 363, 376, 499
Glycerol, 363, 376, 495, 497, 499, 500–*501*, 667, *668*, *670*, 674, 709. *See also* Glycerin
Glycerol 1-phosphate, 674. *See also* Glycerol 3-phosphate
Glycerol 3-phosphate, 697–698. *See also* Glycerol 1-phosphate
Glycerol 3-phosphate transport, 672–673, 698
Glycerophospholipids, 500–*501*, 503–505, 507, 698

Glycine (Gly, G), *157*, 517, 527*t*, 528, 530, 582–583, 625*t*, 683, *685*
Glycocholate, 518
Glycogen, 466, 486, 661, 666–667, *670*, 674, 691–692, 708, 709
Glycogenesis, 674, **696**
Glycogenolysis, **674**
Glycogen phosphorylase, signal transduction and, 674
Glycogen synthase, signal transduction and, 674
Glycolic acid, in Lactomer polymerization, 457
Glycolipids, 500–*501*, 506–509, 698
Glycolysis, 427, 561, **667**–674, *668*, *669–670*, 693, 700
Glycomics, 740
Glycoproteins, 546, 631, 724, 729, 734, 735, 736, 740
Glycosaminoglycans, 487, 546
Glycosides, formation of, **476**–477
Glycosidic bonds, **476**–477, 481, 483, 484, 485, 486, 546, 598–599
Glycosylase, in base excision repair, 614, 615
Glyset, for diabetes, 590
GMP. *See* Guanosine 5′-monophosphate (GMP)
Goiter, iodine therapy for, 349
Gold (Au), *40*
Goldstein, Joseph, 511
Golgi bodies, *647*, 728
Gonadotropic hormones, action of, 576
GOT. *See* Glutamate-oxaloacetate transaminase (GOT)
Gout, 167, *168*
gp41 glycoprotein, in HIV, 631, *632*
gp120 glycoprotein, in HIV, 631
G-protein, 584, *585*, 674
G-protein–adenylate cyclase cascade, 585–587, 591
GPT. *See* Glutamate-pyruvate transaminase (GPT)
Grain alcohol. *See* Ethanol
Gram (g), 7*t*, 9, 11–12, 91–94, 98–99, 129
Granite, 154
Granulocytes, *746*
Grapes, *359*
Grape sugar, 470
Graphite, *38*, 141–*142*
Gravity, mass and weight and, 9
Gray (Gy), 253, 254*t*
Green algae, photosynthesis in, 692
Ground state, **44**, 46, 248
Groups, in Periodic Table, 40
Growth hormone (GH), 576
Guanine (G), 597–598, 599, 600, 600*t*, 601–*602*, 603–605, *604*, 624
Guanosine, 599, 600*t*
Guanosine diphosphate (GDP), *652*, 654
Guanosine 5′-monophosphate (GMP), 600*t*
Guanosine triphosphate (GTP), 584, *585*, 600, *652*, 654
Guillain-Barré syndrome, 506
Guldberg, Cato, 197
D-Gulose, 468*t*
Gypsum, 167
Gyrase, during DNA replication, 612
H_1 and H_2 histamine receptors, 588, 745
HAART (highly active antiretroviral treatment) drug cocktail, 631
Haber, Fritz, 205
Haber process, 205
Hahn, Otto, 262
Half-life ($t_{1/2}$), 248–251, 250*t*
Haloalkanes, 305–307
Halogenation, 305–306, 349, 373
Halogens, 39, **42**, 42*t*, 43, 270.
Hard water, soaps and, 500
Harvey, William, 5
Haworth, Walter N., 471–472
Haworth projections, 471–476
Hayfever, treatment of, 413
Hb. *See* Hemoglobin (Hb)
HbF and HbS hemoglobin, in sickle cell anemia, 537
HDL. *See* High-density lipoprotein (HDL)
HDPE. *See* High-density polyethylene (HDPE)
Headaches, nitric oxide and, 588

Health
  basal caloric requirement and, 708
  chemical communication and, 572–573
  immune system and, 722–724
  nutrients and, 705
  radiation dosimetry and, 253–256
  vitamins and minerals and, 711–718, 712–715*t*
Heart, 4–5, 136, 512, 520, 525, 708, *726*, 750, 755–756
Heart attack, 567, 659, 671
Heartburn, treating, 221
Heat, 18, 19–22, 143–147, 203–204, 547, 548, 662, 706
Heating curves, 143–**144**
Heat of combustion, 113
Heat of fusion, 144
Heat of reaction, **112**–113
Heat of vaporization, 144
Heat-resistant polymerases, in cloning, 617
"Heat shock" protein, *738*
Heat stroke, 192
Heavy chains, in immunoglobulins, 729, *730*, *731*, *732*
Heavy elements, 246–247, 250
Heavy metal ions, 548, 659
Heavy metal poisoning, *548*, 562, 563
Helicase, 609, 610*t*, 612, 621–*622*
Helicobacter pylori, 559, 588
Helium (He), 47, 48, 132, 158, 261, *457*
Helium nuclei, 243, 245*t*
Helix (helices), *384*. *See also* α-Helix; Double helix; Extended helix; Triple helix
α-Helix
  denaturation of, 547
  of interleukins, 736–737
  in protein secondary structure, 537–540
Helix-turn-helix transcription factor, in gene regulation, 634
Heme, 545, 557, *680*, 686–687, 701, 749
Hemiacetals, **428**–430, 471–477
Hemlock, poison, 405
Hemodialysis, 178, 751, *752*
Hemoglobin (Hb), 129, 526, 533, 536, 537, *542*, *543*–545, 557, 637, 671, 683, 686–687, 749–750, 751
Hemolysis, 176
Hemolytic anemia, 613
Hemophilia, 596, 747*t*
Henderson-Hasselbalch equation, 234–236
Henle's loop, of kidneys, 752, *753*
Henry's law, **158**–159
Heparin, 487, 748
Hepatitis, 567, 630, 631, 687
Heptane, 289, 290, 303*t*, 308
Herb tea, for Alzheimer's disease, 581
Herceptin, 733
Hereditary diseases, 597, 613, 685. *See also* Genetic diseases
Heredity, 597, 605
Heroin, 574*t*
Herpes virus, *630*
Heterocyclic aliphatic amines, **404**
Heterocyclic aromatic amines, **404**, 410, 411
Heterogeneous catalyst, **194**
Heterogeneous matter, *28*
Heterogeneous mixtures, 153–154, 170
Hexadecanoic acid, 440*t*
Hexamethylenediamine, 456
Hexanal, 369, 420, 424
Hexane, 139, 288, 289–291, 303, 303*t*, 304
1,6-Hexanediamine, 406, 456
Hexanediamide, 447–448
Hexanedioic acid, 439, 456
Hexanoic acid, 369, 424, 439, 440*t*
1-Hexanol, 368–369
2-Hexanol, 368–369
2-Hexanone, 369, 421
3-Hexanone, 421
trans-3-Hexene, 319
Hexoaminadase A, in Tay-Sachs disease, 509

Hexokinase, in glycolysis, *669*
Hexoses, 470, 474–475, 476, 673–674, 693–696, 740
2-Hexyne, 317, 318
HFC-134a, 307
HFC-141b, 307
Hibernation, mitochondria during, 658
High blood pressure, diets for, 705
High-density lipoprotein (HDL), 510, 512–513
High-density polyethylene (HDPE), 334–335, 336*t*
High-energy phosphate anhydride bond, 649–650
High fever, dangers of, 192
High-fructose corn syrup, 485
Hinckley, Robert, *373*
Hippuric acid, 753
Histamine, 574*t*, 583, 587–588
Histidine (His, H), 527*t*, 542, 574*t*, 587–588, 625*t*, 685
Histone acetylase, during DNA replication, 611–612
Histones, 597, 605, *606*, 611–612
HIV-1 protease, inhibition of, 634
HIV. *See* Human immunodeficiency virus (HIV)
HMG CoA. *See* Hydroxymethylglutaryl CoA (HMG CoA)
HMG-CoA reductase, in cholesterol synthesis, 512–513, 699
Hog insulin, primary structure of, 535
Hollow-fiber dialyzer, *178*
Homeostasis, of body fluids, 745
Homogeneous catalysts, **194**
Homogeneous matter, *28*, 30
Homogeneous mixtures, 153, 154–155, 170
Homosalate, 452
Homozygotes, sickle cell anemia and, 537
Honey, 484, 484*t*
Hormones, 495, 510, 513, 515–517, 526, 535, 536, **573**, 574–577, 575*t*, 583–589, 589–591, 592–593, *680*, 747*t*, 751, 756
Hot-air balloons, Charles's law and, 124
Hounsfield, Godfrey N., 258
Hour (h), 10
Human chromosomes, *596*
Human Genome Project, 613
Human growth hormone, 526
Human immunodeficiency virus (HIV), *631*, 735–736
Human insulin, 535–536. *See also* Insulin
Humans, 602*t*, 609, 610, 611, 637, 637*t*
Huntington's disease, creatine and, 718
Huperzine A, for Alzheimer's disease, 581
Hyaluronic acid, 487
Hybridization, 616–**617**
Hybridoma cells, *733*
Hydrated ions, 165–166
Hydrates, 166–167
Hydration, 166–167, 325*t*, **328**–330, 366–367, *675*, *676*
Hydration–dehydration reactions, 366–367
Hydrazinohistidine, 574*t*
Hydride ion (H⁻), in aldehyde reduction, 426
Hydrobromic acid, 212*t*
Hydrocarbons, **286**–287, 295, 305, 314–315, 343–355
  aliphatic, **286**
  saturated, **286**
  unsaturated, 286, 314
Hydrochloric acid, 212*t*, 213
Hydrochlorides, of drugs, 412
Hydrochlorination, of alkenes, 325*t*, 326, 327
Hydroelectric plants, 18
Hydrofluorocarbons (HFCs), 307
Hydrogen (H₂), 29*t*, 32, 33, 38, 42, 47, 77, 78, 102, 203, 215–216, 245, 261, 270, 275. *See also* Hydride ion (H⁻); Hydrogen gas (H₂)
  alkali metals and, 42
  in ammonia, 77–78, 188–189, 199
  in autoxidation, 353–354
  in catalyzed reaction with cyclohexanone, 194, 426
  in common catabolic pathway, *646*
  in DNA, *602*
  in Earth's crust, *271*
  liquefaction of, 132
  in proteins, *538*

Hydrogen-2. *See* Deuterium
Hydrogen-3. *See* Tritium
Hydrogenation, 325*t*, 332, 499
Hydrogen bomb, 261
Hydrogen bonds, 131–132, **133**–135, 139, 140, 165, 169–170, 363–*364*, 371–372, 374, *408*–*409*, 423, 424, 441, 471, 537–*538*, 540, *541* 547, 603–605, 608, 634, 729–731
Hydrogen bromide, 326, 327
Hydrogen chloride, *71*, *73*, 163, 169, *187*, 188, 201, 211, 214, 216, *220*–221, 221–222, 231, *232*, 233, 326, 327, 454, 709
Hydrogen cyanide, 275*t*
Hydrogen halides, 326–328
Hydrogen iodide, 326, 327, 328
Hydrogen ions (H⁺), 210–211, *646*, 650, 653–655, 656–661, 750, 754. *See also* pH
Hydrogen peroxide, 73, 111, 189–190, 190–191, 479, 659, 733
Hydrogen sulfide, 75
Hydroiodic acid, 212*t*
Hydrolases, action of, 555, 556*t*
Hydrolysis, 450–455, 499, 554, *649*–650, 661–662, 682–683
  of acetyl-CoA, 652, *668*
  of carbohydrates, 467, 485, 708–709
  in fat metabolism, 497, 678, 709
  of glycosides, 476
  in protein catabolism, *680*
Hydrometer, 17
Hydronium ion, 169, **210**–211, 212, 214, 216, 219, 221–222, 223–226, 228–229, 230–234
Hydroperoxide group, in autoxidation, 353–355
Hydroperoxy radical, in autoxidation, 353–354
Hydrophilic substances, 500, 501–503, 518, 526–528
Hydrophobic interactions, protein tertiary structure and, 540, *541*
Hydrophobic pocket, in pyruvate kinase, 561
Hydrophobic substances, 495, 500, 501–503, 510, 518, 526–528, 603, *734*–735
Hydroquinone, 352
Hydrothermal vent, *553*, 558
Hydroxide ion, 199, 210, 211–212, 223–224, 227, 228–229, 230–234, 451, 529
Hydroxyacyl CoA, in stearic acid metabolism, 677*t*
Hydroxyacyldehydrogenase, in fatty acid β-oxidation, *676*
Hydroxyapatite. *See* Calcium hydroxyapatite
4-Hydroxybenzoic acid, 425, 440–441
β-Hydroxybutyrate, 678–679
Hydroxyl group, **277**–279, 277*t*, 328–330, 422, 471
4-Hydroxy-3-methoxybenzaldehyde, 352
Hydroxymethylglutaryl CoA (HMG CoA), 699
Hydroxyproline, 540
2-Hydroxypropanoic acid, 386, 440. *See also* Lactic acid
(*R*)-2-Hydroxypropanoic acid, 440–441
Hydroxyurea, 537
Hygroscopic substances, **167**
Hyperbaric chamber, *129*, 158
Hyperbaric medicine, 129
Hypercholesterolemia, familial, 512
Hyperglycemia, diabetes and, 590
Hyper-osmotic solutions, 176
Hyperphosphorylation, in Alzheimer's disease, 581
Hypertension, 564, 708, 756
Hyperthermia, 20, 192
Hyperthermophile bacteria, 558, 617
Hyperthyroidism, 257
Hypertonic solutions, 176–177
Hypoglycemia, 536, 590, 747*t*
Hypo-osmotic solutions, 176
Hypothalamus, *576*, 591
Hypothermia, 20
Hypothesis, **4**
Hypothyroidism, 257, 349
Hypotonic solutions, 176–177

Hypoxia, 129, 638
Hytrin, 574*t*, 755
Ibuprofen, 387–388, 390–391, 394, 398, 450, 518, 520
Ice, 140, *144*, 144*t*, 145
Ideal gas constant (*R*), **127**–130
Ideal gases, 127–128, 131
Ideal gas law, **127**–130
D-Idose, 468*t*
IgA, IgD, IgE, IgG, and IgM antibodies, 729, 729*t*, *730*
Ignarro, Louis, 588
Imidazole, molecular structure of, 404
Imines, formation of, 534
Imirapin, 574*t*
Immune system, 631, 722, *723*, 724, 725–727, 737–740
Immunity, 722–724, 727, 735, 737–738
  innate, **722**–723, 738–739
Immunization, 630, 735
Immunochemistry, 722–740
Immunogens, antigens as, 727
Immunoglobulins, 526, 546, 724, 729–733, 729*t*, *730*, *732*, 734, 740, 747
Impotence, nitric oxide treatment of, 588
Impurities, in organic compounds, 273–274
Inderal, 574*t*
Indigo, in blue diaper syndrome, 701
Indinavir, in AIDS treatment, 631, *632*, 634
Indomethacin, 518, 520
Induced-fit model, of enzyme action, **559**–562
Infection, 679, 736, 737, 747*t*
Infectious hepatitis, 567, 687
Inflammation, as response to injury, 723
Inflammatory bowel disease, treatment of, 739
Influenza, 630
Information transfer, 597, 605, 609, 620–621
Infrared (IR) radiation, *244*
Inhibition, of enzymes, **557**, 560–562, 563
Inhibitors, **557**, 560–562, 588, 634
Inhibitory neurotransmitters, 582–583
Inhibitory receptors, innate immunity and, 739
Initiation, of protein synthesis, 626, 626*t*, 627–628
Initiation codon, 625*t*, 626
Initiation factors, in protein synthesis, 627, *628*
Initiation signal, in transcription, 622, 624
Initiator, in transcription, *622*–623
Injury, responses to, 736, 737, 747, 748
Innate immunity, **722**–*723*, 738–739
Inner transition elements, 40, 51
Inorganic chemistry, 270
Inorganic compounds, 271–272, 272*t*, 747*t*
Inositol-1,4-diphosphate (I-1,4-P), 591
Inositol-1-phosphate (I-1-P), 591
Insecticides, pyrethrins as, 446
Insoluble solids, 156
Insulin, 479, 526, *535*–536, 575*t*, 576, 589, 590, 596, 640, 679
Insulin-dependent diabetes, 590, 739
Integral membrane proteins, 546
Integral proteins, 501, *503*
Integrin, in leukocyte mobilization, 736
Intensity, of radiation, 251–252
Interferon, 631, 726*t*, 727
Interleukins (ILs), 631, 736–737
Intermembrane space, of mitochondria, 648
Intermolecular forces, boiling point and, 139
Internal innate immunity, 722, *723*, 725–727, 726*t*
International System of Units (SI), 7–11
International Union of Pure and Applied Chemistry (IUPAC), 61
Interstitial fluid, **744**, 745, 755
Intestine, 667, 687
Intracellular fluid, 165, 725, 744
Intramolecular hydrogen bonds, in proteins, 537–*538*
Introns, **608**, *622*
Iodide ion, goiter and, 349

Iodine, 29t, 33, 37, 39, 42, 111, 189–190, 198, 199, 270, 275, 715t
Iodine-125, in cancer therapy, 260
Iodine-131, 248, 250, 256, 257, 263
Iodomethane, 185, 186
Ion channels, 579, 582, 583. *See also* Calcium channels; Sodium ion channels; Transmembrane channels
Ionic bonds, **63**–64, 65–66, 70–71
Ionic compounds, 65–69, 90, 92, 106–107, 157, 165–*166*, 528–529
Ionic solids, 141, 143t
Ionization energy, **52**–53, 64–65
Ionizing radiation, 251, 252, 255–256, 260, 637, 638
β-Ionone, 422
Ion product of water (K_w), 223, 225
Ions, **52**, 58–61, 61–63, 104–107, 141, 142, 165–166, 167–169, 502, 634, 753
Iproniazide, 574t
Iron (Fe), 16, 29t, **31**, 110, 112, *271*, 526, 687, 711t, 715t
Iron Age, 41
Iron(II) ion, in heme, 543, 686, 749
Iron–sulfur clusters, 656, 657, 694
Irreversible inhibition, 557, 562, 582
Irreversible reactions, 194
Ischemia reperfusion, 659
Isobutane, 295
Isobutyl alcohol, 360
Isobutylene, 318, 365
Isobutyl group, 293t
Isobutyl mercaptan, 373
Isocitrate, 556t, *652*, 653
Isocitrate dehydrogenase, 653, 655
Isoelectric point (pI), 527t, **529**, 533
Isoenzymes, enzyme regulation via, **566**
Isoflurane, 373
Isoleucine (Ile, I), 527t, 625t, 685
Isomerases, action of, 555, 556t
Isomers, 289, 291–294, 303, 304, 345–346, *383*
  chirality in, 382–398, **385**
  cis-trans, **300**–302, **316**, 318–320, 321–323, *383*
  constitutional, **289**–291, *383*
Isooctane, in gasoline, 308
Iso-osmotic solutions, 176
Isopentane, 295
Isopentenyl pyrophosphate, 699
Iso prefix, in alkane names, 295
Isoprene, 323–324, 699, 700
Isoprene units, 324
Isopropyl alcohol, 278–279, 360, 376, 451–452
Isopropylamine, 407
Isopropyl group, 293t
2-Isopropyl-5-methylcyclohexanol, 368, 395
2-Isopropyl-5-methylcyclohexanone, 368
2-Isopropyl-5-methylphenol, 352
(R)-Isoproterenol, 413
Isotonic solution, *153*, 175–177, *177*, 178
Isotopes, 36–38, 245, 246–247, 248–251
Isotopic abundances, 38
Isovaleric acid, 439
Isozymes, enzyme regulation via, **566**
IUPAC. *See* International Union of Pure and Applied Chemistry (IUPAC)
Jackson, Charles, 373
Jaundice, 687
Jenner, Edward, 735
Joule (J), 19
Keflex, 447
Kekulé, August, 271, 344
Kekulé structures for benzene, 344–345
Kelvin (K), 11, 120, 124–126, 127
Keratin, 525, 526, 539, 547–548
Ketoacidosis, in diabetes, 679
Ketoacids, in amino acid catabolism, 667, 680
Ketoacyl CoA, in stearic acid metabolism, 677t
Keto-enol tautomerism, 430–431, 515
Ketogenic amino acids, 668, 685, 686

α-Ketoglutarate, 556t, *652*, 654, 655, 680, 681, 684, *685*, 700
α-Ketoglutarate dehydrogenase, 655
α-Ketoglutaric acid, 667, 680, *685*, 701
D-2-Ketohexoses, 469t
Ketone bodies, 677–**679**, *680*, 686, 709, 753
Ketone group, 467
Ketones, 277t, **280**–281, 419–431, 534
Ketoprofen, anti-inflammatory drugs and, 518
Ketoses, **467**, 469–470
2-Ketoses, as reducing sugars, 478
Ketosis, 709. *See also* Ketone bodies
D-2-Ketotetroses, 469t
β-Ketothiolase, in fatty acid β-oxidation, *676*
Ketotrioses, 467
Kevlar, 456
Kidney failure, 512, 534
Kidney machines, *178*
Kidneys, 177, 178, *259*, 536, 575t, 685, 746, 751–*753*, 754, 756
Kidney tubules, *576*, 753
Killer T cells, in immune system, *723*, 726t, 727
Kilocalorie (kcal), 19, 707–708
Kilogram (kg), 9
Kilometer (km), 8, 12–13
Kinases, 565, 591
Kinesin, 545
Kinetic energy (KE), 18, 44, 119–121, 131, 135–137, 137–138, 187, *192*, *193*
Kinetic molecular theory, 130–**131**
Kinetics, chemical, 185–186
Knocking, in engines, 308
Knoop, Franz, 674–675
Kohler, Georg, 733
Koshland, Daniel, 559
Krabbe's leukodystrophy, 508
Krebs cycle, 649, *652*. *See also* Citric acid cycle
Krebs, Hans, 652, 684
Kroto, Harold, 142
Kwashiorkor, 711
β-Lactam, in penicillin, 447
Lactams, **446**, 447, 454
Lactate, 235, 427, *669–670*, 671, *693*. *See also* Lactic acid
Lactate dehydrogenase (LDH), 555, 556t, 566, 567, 567t, *669*
Lactic acid, *386*, 397, 440, 443–444, 457, 722–723. *See also* Lactate
Lactomer, 457
Lactones, **445**, 446
Lactose, 484, 484t, 471, 481, 708
Lagging strand, of DNA, *609*, 610
L-Amino acids, 528
Landsteiner, Karl, 482
Lanolin, 498
Lanoxicaps, 574t
Lanthanide series, 51
Laser surgery, protein denaturation and, 548
LASIK (laser in situ keratomileusis), 548
"Laughing gas," 373
Lauric acid, 440t, 496t, 498t, 499
Lavoisier, Antoine Laurent, 31
Law of conservation of energy, 19
Law of conservation of mass, 31, 95
Law of constant composition, 32
Lawrencium (Lr), creation of, 261
LDL. *See* Low-density lipoprotein (LDL)
LDL-receptor–mediated pathway, 510–511
LDPE. *See* Low-density polyethylene (LDPE)
Le Chatelier, Henri, 201
Le Chatelier's principle, 200–206, 366–367, 691–692
Lead (Pb), **31**, 38–39, 165, 245
Lead-206, 247, 249
Leading strand, of DNA, *609*, 610
Lead oxide, molecule of, **31**
Lead poisoning, 563

Lead storage battery, *112*
Lecithins, 503–504, *698*
Lemon grass, *323*
Length, 7–8, 7t, 8t
Leptin, 591
Lethal mutations, 636
Leucine (Leu,L), 527t, 542, 625t, 633, *685*, 686
Leucine-tRNA synthetase, 631
Leucine zipper, in gene regulation, 634
Leukemia, 255, *722*, 747t
Leukocytes, 520, 659, *723*, 736, *746*, **747**
Leukotriene B4, 520
Leukotriene receptors (LTRs), 520
Leukotrienes, 518, 520
Levonogestrel, as contraceptive, 517
Levorotary compounds, **396**
Lewis dot structures, **49**–50, 50t, 51, 72–74, **73**, 211, 274, 275–276, 278–279, 344–345. *See also* Lewis model of chemical bonding
Lewis, Gilbert N., 49, 58, *59*
Lewis model of chemical bonding, 58–61, 63–66, 70–75, 77–80. *See also* Lewis dot structures
Lexan, synthesis of, 458
Librium, molecular structure of, 408
Ligand-binding G-protein/adenylate cyclase cascade, signal transduction and, 674
Ligand-gated ion channels, 582, 583
Ligands, 576
Ligases, 555, 556t, 610t, 613, **614**, 615
Ligation step, in DNA replication, 613
Light, 7, 18, 19, 113, *244*, 395–397
Light chains, in immunoglobulins, 729, *730*
Light reaction, in photosynthesis, 695
Limestone, acid rain and, 167
Limiting reagents, 101–**102**
Limonene, 324
Line-angle formulas, *287*–288
Linear molecules, 77, 80t
Linoleic acid, 496t, 498, 498t, 499, 696, 709
Linolenic acid, 496t, 498, 498t, 499, 503, 696, 709
Lipases, 554, 567, 667, *668*, 709
Lipicor, 512
Lipid bilayers, 501–503, *504*, 534, 588, 699
Lipids, 494–520, *501*, *504*, **510**, 646, 653–655, *668*, 679, 697–700, 705, 709, 747. *See also* Fats
Lipid storage diseases, 508–509
Lipomics, 679
Lipopolysaccharides, innate immunity and, 739
Lipoproteins, 354, **510**–513, *511*, 511t
5-Lipoxygenase, inhibition of, 520
Liquefaction, 137
"Liquid protein" diets, 710
Liquids, 14–17, 119–121, *120*, 131, 135, *137*, **138**–141, 147, 154, 154t, 156, 171t
  vapor pressure of, 135–**138**
Liquid soaps, 500
Liquid state, 121
Liter (L), 7t, 8, 13–14
Lithium (Li), electron configuration of, 47, 48
Lithium carbonate, 61
Lithium hydroxide, 212t
Lithium iodide, 160
Lithium ions, manic depression and, 61
Lithium oxide, 162
Lithium sulfate, *157*
Litmus paper, *213*, 225, 226t
Liver, 480, 510–511, 512, 515, 578, *647*, 678–679, 687, 746–747
Lock-and-key model, of enzyme action, **559**
London, Fritz, 132
London dispersion forces, 131–**132**, 139–141, 363, 409, 496, 729–731
Loops, DNA molecules arranged in, *606*, 635
Lopressor, 574t
Low body temperature, dangers of, 193
Low-density lipoprotein (LDL), 510–*511*, 512–513, 698
Low-density polyethylene (LDPE), 334, *335*, 336t

Lowry, Thomas, 214
Low-sodium diets, 705
Luciferase, in luminescent tobacco plant, *641*
Luminescent tobacco plant, *641*
Lung cancer, from radon radiation, 255
Lungs, 123, 745, 749, *750*, 751
Luteinizing hormone (LH), 515, 575t, 576, 585, 591
Lyases, action of, 555, 556t
Lye, 213
Lymph, *725*, *726*, 744, 745
Lymphatic system, *726*
Lymphatic vessels, *726*, 727
Lymph capillaries, *725*, *726*
Lymph nodes, in immune system, 725, *726*, 727
Lymphocytes, *631*, *632*, *723*, *726*, 735–736, *737*, 746. *See also* T cell entries
Lymphoid organs, 725, *726*
Lymphoma, antisense drugs for, 623
Lysine (Lys, K), 527t, 531, *542*, 554, 625t, *685*, 710
Lysosomes, 554, *647*, 729
Lysozyme, 554, *731*
D-Lyxose, 468t
Macrolide drugs, for autoimmune diseases, 740
Macrophages, *723*, *724*, *725*, *726*, 726t, 727, *728–729*, 736–737, *738–739*
Magnesium (Mg), 29t, 37–38, *40*, 47, *184*, 220, *271*, 711t, 715t
Magnesium hydroxide, 213
Magnesium ions, 500, 583, *669*, *676*, 694, 695, 753
Magnetic resonance imaging (MRI), 258
Main-group elements, **40**, 50–51
Major groove, in DNA molecule, *603*
Major histocompatibility complex (MHC), *723*, *724*, 727, *728–729*, 734, *738–739*
Malachite, copper from, 41
Malaria, sickle cell anemia and, 537
Malate, *652*, 654–655, 677, *693*
Malate dehydrogenase, 655
Malathion, 574t
Malignant tumors, oncogenes and, 638
Malonic acid, 439
Malonyl-ACP, in fatty acid biosynthesis, 697
Maltose, 483–484, 484t, 708, 709
Mammotropin, action of, 575t
Manganese (Mn), 29t, 711t, 715t
Manic depression, lithium ions and, 61
D-Mannitol, 477–478
D-Mannosamine, 470
D-Mannose, 468t, 482
Manometer, *122*
MAOs. *See* Monoamine oxidases (MAOs)
Marasmus, 708
Marble, acid rain destruction of, 203
Margarine, production of, 499
Markovnikov, Vladimir, 326
Markovnikov's rule, **326**, 327, 328, 329, 367
Mars, isotopic abundances on, 38
Mass, 7t, 8t, 9–10, 15, 18, 33, 38–39, 98, 261
 law of conservation of, 31, 95
Mass number, 34–35, 246, 247
Matrix, of mitochondria, 648
Matter, 3, 9, 19, 27–31
 attractive forces within, 119–121
 phase changes of, 14–15, 119–121, *120*, **143**–147
Measles, 630, 723
Measurements, 6–11
Mechanical energy, 18, 19, 661–662
Medicine, 2–3, 69, 129, 256–261, 257t, 362, 382, 387–388, 566–567, 573, 723, 753. *See also* Disease; Drugs; Therapy
Meitner, Lisa, 262
Melting, *120*
Melting point, 121, 142, 143, 146
 of alkali metals, 43t
 of alkanes, 302–303
 of amino acids, 529
 of fatty acids, 495–496, 496t
 of halogens, 42t
 of network crystals, 143
 of noble gases, 43t
 of solids, 143t
 of waxes, 498
Memory cells, 722, *723*, 726t, 727, 735
Mendeleev, Dmitri, 5, *39*, 50
Menstrual cycle, 515, *516*
Mental retardation, from PKU, 685
Menthol, 324, 368, 394–395
Menthone, 368
Meperidine, 574t
Mercaptans, **372**–374
Mercaptoethylamine, *651*, 652
Mercury (Hg), 40
Mercury barometer, *121–122*
Mercury(II) oxide, decomposition of, *113*
Mercury ion, mercaptans and, 372
Mercury poisoning, 563
Messenger RNA (mRNA), **607**, *608*, 623, 624, 625, 626t, 627, *628*, 629, 633, 637
 in transcription and translation, **621**, *622*–624
Messengers, drugs affecting, 573, 574t
Metabolic acidosis and alkalosis of blood, 235
Metabolism, 259, 534, 645, 666–687, *668*, *669–670*, *676*, 677t
Metal-binding fingers, in gene regulation, 634, *635*
Metal catalysts, for alkene hydrogenation, 332
Metal hydroxides, 211, 212, 212t, 220–221
Metallic ions, as coenzymes, 557
Metalloids, **42**, 64
Metal oxides, reactions with acids, 221
Metals, **40**–41, 64, 65–66, 68, 141, 220
Metaminopeptidase, methionine removal by, 636
Metastable nuclei, 250
Metastasis, 260, 638
Meter (m), 7–8
Metformin, for diabetes, 590
Methadone hydrochloride, 412
(S)-Methamphetamine, 404
Methanal, 419–420. *See also* Formaldehyde
Methanamine, 409t. *See also* Methylamine
Methandienone, 514
Methane, 43–44, 73, 77, 80t, 102, 109–110, 128, 139, 287, 288t, 303t, 305
Methanethiol, *372*, 375t
Methanoic acid, 440t. *See also* Formic acid
Methanol, 73–74, 100, 103, 110, 134, 169, 173, 275t, 364t, 375, 375t, 409t
Methedrine, 404
Methenolone, 514
Methicillin, 447
Methionine (Met, M), 527t, 625t, 626, 636, *685*, 694, 710
Methionine enkephalin, 589
Methoxycyclohexane, 370
Methoxy group, 370
N-Methylacetamide, 446
Methylacetylene, 318
Methylamine, 221–222, 275t, 279, 403, 409. *See also* Methanamine
Methylammonium hydroxide, 409
2-Methylaniline, 479
3-Methylaniline, 406
N-Methylaniline, 404, 406
N-Methyl-D-aspartate, (NMDA), 574t, 583, 591
Methylation, 622, 624
Methyl benzoate, 449
2-Methyl-1,3-butadiene, 324
Methyl chloride, 305
1-Methylcyclohexanol, 329, 330
trans-2-Methylcyclohexanol, 361
Methylcyclohexane, 299–300
Methyldopa, 574t
Methylene chloride, 305
Methylenecyclohexane, 328, 330
Methylene group, 287–288
Methyleneimine, 275t
Methyl ethyl ketone (MEK), 423
Methyl α-D-glucoside, 476–477
Methyl β-D-glucoside, 476–477
Methyl group, 293t
Methyl GTP, in protein synthesis initiation, 627
1-Methylguanosine, 608
2-Methylhistamine, 574t
Methyl isovalerate, 445
Methyl methacrylate, 333t
Methyl 3-methylbutanoate, 445
Methyl orange, as acid–base indicator, 226
Methylphenidate, 574t, 586
3-Methylphenol, 352
2-Methylpropane, 289, 292
2-Methyl-1-propanethiol, 373
2-Methyl-1-propanol, 360
2-Methyl-2-propanol, 328, 360, 365
2-Methylpropene, 318, 328, 365
Methyl red, as acid–base indicator, 226t
Methyl β-D-riboside, 477
Metoclopramide, 574t, 586
Metoprolol, 574t
Metric system, **6**–11, 7t, 8t
Metyrosine, 574t
Mevalonate, 699
Mevalonic acid, 441
MHC. *See* Major histocompatibility complex (MHC)
Micelles, of soap and dirt, 500
Microfilaments, in cytoskeletons, 544
Microsatellites, in DNA, 609
Microtubules, 273, 544–545, *647*
Microwave radiation, *244*
Mifepristone, 516, 517
Migitol, for diabetes, 590
Milk of magnesia, 213. *See also* Magnesium hydroxide
Milk sugar, 484t. *See also* Lactose
Millimeters of mercury (mm Hg), *121–122*
Millirem (mrem), 254
Milstein, Cesar, 733
Mineralocorticoid hormones, 515
Minerals, as nutrients, 705, 711–718, 711t, 712–715t
Minibands, DNA molecules arranged in, *606*
Minipress, 574t
Minisatellites, in DNA, 609
Minor groove, in DNA molecule, *603*
Minute (min), 10
Mirror-image molecules, **383**–387
Mirror images, *385*
Mirrors, silvering, 425
Miscible liquids, 156
Mitchell, Peter, 658
Mitochondria, 554, 572, 625t, *647*–648, 657–660, *667* 671, 672, 675, *676*, 680–681, *693*, 694, 696
Mitochondrial DNA and RNA, 625t
Mitochondrion membrane protein, *546*
Mixed anhydrides, 444
Mixed oxide (MOX) fuel, from nuclear waste, 264
Mixtures, 28, 30–31, *31*, 37, 130, 153–154, *155*, 170, 171t
 racemic, **386**
Modulation, enzyme regulation via, 565
Molarity (M), 161–163, 175
Molar mass, **91**–94
Molar volume, *127*
Mole (mol), 7t, 90–94, **91**, 98–99, 127, 128, 129, 161–165, 173–174
Molecular cloning, 616–617
Molecular collisions, reaction rate and, 189–190, 192–193
Molecular compounds, 70–76.
Molecular shape, boiling point and, 139–141
Molecular solids, 143t
Molecular weight (MW), 90, 409t, 423–424, 423t, 441–442, 442t, 498, 729, 729t

Molecules, *30*, 31, 130–131, 135–143, 153, 186–188, 192–193, *723*, 724
  achiral, **386**
  chiral, 382–398, **385**, 528
  functional groups in, **276**–282, 277*t*, 438–439
  mirror-image, **383**–387
  mole and, 90–**91**
  with multiple stereocenters, 391–395
  osmotic pressure and, 174–175
  polar and nonpolar, 81–82
  shapes of, *77*, *78*, *79*–80, 80*t*
  states of matter and, 119–121, 147
  surface tension and, *135*
Molina, Mario, 307
Molybdenum (Mo), 29*t*, 711*t*, 715*t*
Monatomic anions, 62, 62*t*
Monatomic elements, 33
Monoamine messengers, 583, 584
Monoamine oxidases (MAOs), 586, 587, 588–589
Monoamines, oxidation of, 587
Monoclonal antibodies, *733*
Monocyte chemotactic proteins (MCPs), 737
Monocytes, 737, *746*
Monoglycerides. 497
Monomers, **332**, 333, 333*t*, 694
Monophosphoric esters, of monosaccharides, 480
Monoprotic acids, **215**, 216
Monosaccharides, **467**–480, 482–483, 597, 598–599, *652*, 667–671, *668*, 692, 708, 709
  mutarotation of, **476**
"Morning after pill," 517
Morphine, 574*t*, 589–591
Morpholine, 411
Morton, William T. G., *373*
MRI. *See* Magnetic resonance imaging (MRI)
Mucins, cystic fibrosis and, 683
Mucus, external innate immunity of, 723
Mullis, Kary B., 616
Multiple sclerosis, 506, 739
Mumps, 630
Murad, Ferid, 588
Muscle, 514, 525, 555, *572*–574, 578, 580, 588, 661, *662*, 673, 674, 739, 749, 750, 755
Muscle relaxants, enzyme specificity and, 555
Muscone, molecular structure of, 422
Muscular dystrophy, creatine and, 718
Mushrooms, thiols in, *373*
Mutagens, 637
Mutarotation, **476**
Mutations, 609, 613–616, 636–639, 700, *732*, 735
Myasthenia gravis, 739
Myelin, composition of, 505
Myelin sheaths, 505–506
Mylar, 346, 457
Myocardial infarction, 512, 567
Myosin, 525, 661, *662*
Myrcene, 324
Myristic acid, 440*t*, 496*t*, 498*t*
NAD⁺. *See* Nicotinamide adenine dinucleotide (NAD⁺)
NADH. *See* Reduced NAD⁺ (NADH)
NADP⁺. *See* Nicotinamide adenine dinucleotide phosphate (NADP⁺)
NADPH. *See* Reduced NADP⁺ (NADPH)
Na⁺, K⁺ ATPase, 502
Naloxone, 574*t*
Nandrolone decanoate, 514
Nanometer (nm), 8
Nanoparticles, in colloids, 170
Nanopores, 172
Nanotubes, *142*, 170, 172
Nanowires, 172
Naphthalene, 348
Naproxen, 394, 398, 450
Narcan, 574*t*
Natural gas, *286*, 307, 359, 375.

Natural killer (NK) cells, *722*, 723, 727, 738–739
Naturally occurring radiation, 256*t*
Near-sightedness, laser surgery to correct, 548
Necrosis, 615
Negative modulation, enzyme regulation via, 565
Nelfinavir, in AIDS treatment, 631
Nembutal, 453
Neon (Ne), 47, 48, 132, 245
Neonatal herpes, 630
Neo-Synephrine, 574*t*
Neotame, 717
Nephritis, 753
Nephrons, 177, 751, *753*
Nerve axons, 505–506
Nerve gases, 574*t*, 579
Nerves, 555, 580
Nerve transmission. *See* Neurotransmitters
Nervous system, in blood pressure regulation, 756
Net ionic equation, **105**
Network crystals, 143
Network solids, 142–143, 143*t*
Neuralgia, capsaicin and, 353
Neurofibrillar tangles, in Alzheimer's disease, 581
Neurons, 545, *572*, 573, 574–575, 578, 579, 581, 586
Neuropeptide Y, 591
Neurotransmitters, 555, **573**, 574–577, 582–591, 592–593, *680*
  adrenergic, 574*t*, 576, 583–589
  cholinergic, 574*t*, 576, 577–582
  peptidergic, 574*t*, 576, 589–591
Neutral salts, **227**
Neutral solutions, 220, 224–226
Neutralization, acid-base, 220
Neutrons, 33–39, 33*t*, **34**, 245, 245*t*, 246, 247, 262–263
Neutrophils, 737, *746*
Nevirapine, in AIDS treatment, 631
Niacin, 711*t*, 713*t*, 716, 717
Niacinamide, as nutrient, 717
Nickel (Ni), 332, 711*t*
Nicotinamide, 650, *651*
Nicotinamide adenine dinucleotide (NAD⁺), 424, 426–427, *646*, 650, *651*, *652*, 653–655, 656, 658, 660, *670*, 674, *675*, *676*, **676**, 678, 711*t*
Nicotinamide adenine dinucleotide phosphate (NADP⁺), 672, 695, 697
Nicotine, 574*t*, 577, 582
Nicotinic acid, 713*t*
Nicotinic receptor, 577
Niemann-Pick disease, 509
Nifedipine, 574*t*, 756
Nirenberg, Marshall, 625
Nitrate reductase, 711*t*
Nitration, of arenes, 349–350
Nitric acid, 76, 212*t*, 213, 349
Nitric acid test, for proteins, 213
Nitric oxide, *76*, 202, 588, 725–727, 726*t*
Nitric oxide synthase, 588, 726*t*, 727
p-Nitroaniline, 406
p-Nitrobenzoic acid, 350
Nitrobenzene, 349
Nitrogen (N₂), 29*t*, 33, 47, 77–78, 93, 130, 133–134, 155, 158, 188–189, 199, 205, 270, 275, 279–280, 403, 405, 445–446, *538*, 667, 679–684, 747*t*
Nitrogen-13, 247
Nitrogen dioxide, 76, 202, *204*
Nitrogen-fixing bacteria, 205
Nitrogenous waste, in urine, 752–753
Nitroglycerin, 350, 362, 376
Nitro group, 350
Nitrosyl bromide, 202
Nitrous oxide, as anesthetic, 373
N-linked saccharides, 546
Nobel, Alfred, *362*
Noble gases, 42, 43*t*, **43**, 49, 58
Nolvadex, treating breast cancer with, 592

Nomenclature.
  of alcohols, 360–363
  of aldehydes and ketones, 420–423
  of alditols, 477
  of alkanes, 291–295
  of amides, 445–448
  of amines, 406–408
  of amino acids, 526, 527*t*
  of anhydrides, 444
  of arenes, 345–348
  of carboxylic acids, 439–441
  of carboxylic acid salts, 443–444
  of cycloalkanes, 295–296
  of enzymes, 555, 556*t*
  of esters, 444–445
  of ethers, 370
  of glycosides, 476–477
  of halogenated alkanes, 306
  of monosaccharides, 467, 469–470, 474
  of nucleic acid bases, 597–598
  of nucleotides, 599–600, 600*t*
  of peptides, 530, 532
  of phenols, 351–352
  of polymers, 333, 333*t*
  of proteins, 532, 535
  of thiols, 372–374
Non-insulin-dependent diabetes, 590
Non-nucleoside inhibitor drugs, for AIDS, 631
Nonane, physical properties of, 303*t*
Nonapeptides, 536
Nonbonding electrons, **72**
Noncompetitive inhibition, **557**, 560–562, 563
Nonelectrolytes, 168
Nonessential amino acids, biosynthesis of, 700
Nonheritable errors, in protein transcription, 636
Nonmetals, **42**, 64, 65–66
Nonpolar amino acids, 526, 527*t*
Nonpolar compounds, solubility of, 156–157
Nonpolar covalent bonds, **70**–72
Nonpolar molecules, 81–82
Nonsteroidal anti-inflammatory drugs (NSAIDs), 450, 518, 519, 520, 566
Nonsuperposable mirror images, 384
Norepinephrine, 411, 574*t*, 575*t*, 576, 584, *585*
Norethindone, 517
Norethynodrel, 517
Normal boiling point, **138**, 146
Normal freezing point, 146
Novocain, 412
NSAIDs. *See* Nonsteroidal anti-inflammatory drugs (NSAIDs)
N-terminal amino acid in peptides, 530, 532, 534, 535, *538*
Nuclear chemistry, 242–264
Nuclear energy, 18–19
Nuclear equations, balancing, 246, 247
Nuclear fission, 262–264
Nuclear fusion, **261**–262
Nuclear membrane, *647*
Nuclear power plants, 250–251, *263*
Nuclear reactions, **246**
Nuclear testing, strontium-90 and, 40
Nuclear waste disposal, 263–264
Nucleic acids, 466, 474, 597–600, **601**, 630, 727. *See also* DNA; RNA
Nucleoside analog drugs, for AIDS, 631
Nucleosides, **598**–599
Nucleosomes, DNA molecules arranged in, 605, *606*
Nucleotide excision repair (NER), 616, 637
Nucleotides, 584, 597, 599–600, 600*t*, 612, 616–617, *680*, 740
Nucleus, atomic, 33–39, *34*, 119
Nutrasweet, 717
Nutrients, 648, 704–707, 718, *745*, 746
Nutrition, 704–719, 712–715*t*
Nylon-66, 457

Obesity, 590, 705–706, 708, 716
O blood type, 482–483
Octadecanoic acid, 440t
Octanal, 368
Octane, 303t, 304
Octane antiknock ratings, 308
Octanoic acid, 368, 440t
1-Octanol, 368
1-Octene, 317
Octet rule, **59**–61, 74–75, 76
Octyl p-methoxycinnamate, 452
Odors
    of aldehydes and ketones, 424
    of amines, 408
    of carboxylic acids, 442
    of thiols, 372
Oil of almonds, 422
Oil of cinnamon, 422
Oils, **497**–500, 498t, 707, 709. See also Essential oils
Okazaki, Reiji, 613
Okazaki, Tuneko, 613
Okazaki fragments, 609, **612**, 613
Oleic acid, 496t, 498t, 499
Olestra, 717
Oligonucleotide drugs, 623
Oligosaccharides, 467, **481**–484, 509, 693–696, 708–709
O-linked saccharides, 546
Oncogenes, 638, 683, 700
Ondansetrone, 574t, 586
Onions, thiols in, 373
Oocytes, in menstrual cycle, 516
Opening step, in DNA replication, 611–612
Opiate, 574t
Opsin, in retina, 322
Optical activity, of chiral molecules, **395**–397
Oral contraceptives, 517
Orbital box diagrams, 46–49
Orbitals, **45**, **46**–50, 50–52
Orexic agents, 591
Organelles, 501, 545, 554, 572–573, 647, 648, 667
Organic chemistry, 270–282
Organic compounds, 271–276, 272t
Organic fertilizers, 706
Organisms, chemical communication within, 573
Organophosphorous hydroxylase, nerve gas detoxification via, 579
Orientation of molecules, in molecular collisions, 187
Origins of replication, 609
Orinase, for diabetes, 590
Ornithine, 681, 682, 683
L-Ornithine, 554
Osmolarity (osmol), 175–177
Osmosis, 174–177
Osmotic pressure, 174–177, **175**, 748, 755
Osteoarthritis, treatment of, 518
Osteoporosis, aging and, 755
Ovalbumin, 526
Ovarian follicle, 516
Ovarian phases, in menstrual cycle, 516
Ovulation, hormones in, 515, 516
Oxalic acid, 439, 440
Oxaloacetate, 556t, 652, 653, 655, 678, 685, 686, 693
Oxaloacetic acid, 667, 685
Oxalosuccinate, 653–654
Oxidation, 105, **107**–112, 367–369, 374–375, 412, 424–425, 478–480, 497, 530, 554, 587, 653, 673t
β-Oxidation, 667, 668,
    fatty acids and, 674–675, 675–677, 676, 696–697
Oxidation–reduction reactions, **107**–112, 694, 695
Oxidative damage, 659, 671
Oxidative deamination, 667, **680**–**681**, 684, 700
Oxidative decarboxylation, 671, 673t
Oxidative phosphorylation, 672, 673t, 646, 649–650, 656, 660, 668. See also Electron transport chain
Oxidizing agents, **108**, 532
Oxidizing antiseptics, 111

Oxidoreductases, 555, 556t, 557
Oxonium ion, **329**, 330
Oxygen, 29t, 32, 33, 47, 48, 78, 93, 94–96, 95, 99, 102, 109–110, 112–113, 155, 158, 191, 193, 203, 270, 271, 275, 374–375, 412, 512, 538, 545, 656, 657, 660–661, 723, 747t
    aldehyde reactions with, 424
    autoxidation and, 352–355
    blood and, 479, 745, 749–750, 756
    in detoxification, 659
    glucose oxidation by, 184, 188
    in hydrogen bonding, 133–135
    in hyperbaric chambers, 129
    lactate accumulation and, 671
    reactions of alkanes with, 305
Oxygen deprivation, 659
Oxygen dissociation curve, 749
Oxytocin, 526, 536, 575t, 576, 589
Ozone, 33, 111, 719
Ozone layer, CFCs and, 307
p53 protein, as tumor suppressor, 638
PABA. See p-Aminobenzoic acid (PABA)
Paclitaxel, 270, 273
Padimate A, 452
PAHs. See Polynuclear aromatic hydrocarbons (PAHs)
Pain relievers, 450, 566
Pain transmission, 591
Palladium (Pd), as hydrogenation catalyst, 332
Palmitic acid, 440t, 496t, 498t
Palmitoleic acid, 496t
Pancreas, in diabetes, 590
Pancreatic islets, 575t, 589, 590
Pantothenic acid, 651, 652, 711t, 714t
Paraffin waxes, 302, 498
Parathion, 579
Parathyroid gland, 575t
Parathyroid hormone, 575t
Parenteral nutrition, 705
Parkinson's disease, 586, 718
Partial pressure, **130**, 750, 751
Parts per billion (ppb), 165
Parts per million (ppm), 165
Pascal, 122
Passive transport, across cell membranes, 502
Pasteur, Louis, 735
Pathogens, immunity against, 722–723
Pauling, Linus, 64, 344, 537, 603
PCR. See Polymerase chain reaction (PCR)
PCR techniques, 611, 616–617
Pellagra, 711
Penicillamine, in treating cystinuria, 685
Penicillin, 394, 446, 447, 562
Penicillin G, 447
Pentaerythritol tetranitrate (PETN), 350
Pentagastin, 574t
Pentanal, 426, 442t
Pentane, 140–141, 288, 295, 303t, 364t, 423t
1,5-Pentanediamine, 406–407
Pentanedioic acid, 439
1-Pentanethiol, 373–374
Pentanoic acid, 440t, 425, 442
2-Pentanone, 429–430
Pentapeptides, 532
Pentobarbital, 453
Pentose phosphate pathway, **672**
Pentyl mercaptan, 373–374
PEP. See Phosphoenol pyruvate (PEP)
Peppers, capsaicin in, 353
Pep pills, 404
Pepsin, 555, 565, 667, 710
Peptide backbone, 532, 538
Peptide bond, **530**–532, 629–630
Peptidergic neurotransmitters, 574t, 576, 589–591
Peptides, 530–534, 581, 710, 728
Peptidoglycans, innate immunity and, 739
Peptidyl transferase, 628, 629

Percent concentration, **159**–160
Percent yield, 103–104
Perchloric acid, 212t
Perforins, macrophages and, 727
Periodic Table, 5, 35, 39–44, 41, 50–53, 64–65, 64
Periods, in Periodic Table, **39**
Peripheral proteins, 501, 503
Permanent wave, protein denaturation and, 547–548
Permethrin, 446
Peroxidase, 479, 566
Peroxidase activity, 566
Peroxides, 334, 532
PET. See Poly(ethylene terephthalate) (PET)
PET scan, 258–259, 581
Petrolatum, 302
Petroleum, 286, 307–308, 315
Petroleum refinery, 286, 307, 308
Petroleum spills, **16**
Pewter, 41
PGHS. See Prostaglandin endoperoxide synthase (PGHS)
pH, 224–228, 226t, 230–236, 364, 444, 455, 529, 533, 557, 558, 559, 602, 723, 746, 750, 752, 754. See also Hydrogen ions (H^+)
pH indicators, 225–226
pH meter, 226, 227, 228
Phagocytes, 686–687, 737
Phagocytosis, 729, 737, 747
Pharmacogenomics, 613
Pharmacology, radioactive isotopes in, 257–260
Phase changes, **143**–147
Phase diagrams, 145–147, 146
Phases, 143
Phenanthrene, 348, 509–510
Phencyclidine (PCP), 574t, 583
Phenergan, 574t, 586
Phenobarbital, 453
Phenol, 216, 345, 352
Phenolic antiseptics, 111
Phenolphthalein, 226t, 229
Phenols, **351**–355, 364–365
Phenoxide ion, 216
Phenteramine, 404
N-Phenylacetamide, 454
Phenylalanine, 527t, 625, 625t, 685, 686
Phenylalanine hydroxylase, 685
1-Phenylcyclohexene, 345
p-Phenylenediamine, 456
Phenylephrine, 574t
Phenyl group, **345**
Phenylketonuria (PKU), 685
(S)-1-Phenyl-2-propanamine, 406–407
Phenylpyruvate, hereditary diseases involving, 685
Pheromones, of European corn borer, 320
PHI. See Phosphohexose isomerase (PHI)
Phosgene, 457–458
Phosphatase, 565, 588–589, 591, 738
Phosphate buffer solution, 231, 232, 234, 235–236
Phosphate esters, in lipid biosynthesis, 698
Phosphate group, 503–504, 561, 565, 599–600, 601–602, 649–650
Phosphate ion, in urine, 753
Phosphates, 579, 597, 599–600
Phosphatidates, in lipid biosynthesis, 698
Phosphatidylcholines, 503–504, 698
Phosphatidylethanolamine, 502, 504–505
Phosphatidyl inositol (PI), 591
Phosphatidylinositol-4,5-biphosphates (PIP2), 504
Phosphatidylserine, 502, 504
Phosphatylcholine, 502
Phosphine, 75
Phosphocreatine, 718
Phosphodiesterase, nitric oxide and, 588
Phosphoenol pyruvate (PEP), 561, 668, 669–670, 678, 685, 686, 693
Phosphofructokinase, 669

2-Phosphoglycerate, *669–670*
3-Phosphoglycerate, 668, *669–670*
Phosphoglycerate kinase, *669*
Phosphoglycerate mutase, *669*
Phosphohexose isomerase (PHI), 556*t*, 567*t*, *669*
Phosphoinositoldiphosphate (PIP2), 584
Phospholipase, 518, 591
Phospholipids, 500–*501*, 510–511
Phosphonates, 579, 582
Phosphoric acid, 75, 213, 365, 455, 480
Phosphoric anhydride bond, 649–650
Phosphoric anhydrides, 455
Phosphoric esters, 455, 480
Phosphorothioate moieties, in antisense drugs, 623
Phosphorus (P), 29*t*, 33, 42, 47, 48, 75, 270, 714*t*
Phosphorus-32, 246
Phosphorus pentachloride, 75
Phosphoryl group, 455
Phosphorylase, 661, 674, 691–692
Phosphorylated ADP, in coenzyme A, *651*
Phosphorylated protein kinase, in gene regulation, 635
Phosphorylation, 480, 565, 584, 585, 590, 591, 623, 624, 633, 635, *646*, 649–650, *656*, 658–660, *667–668*, *669–670*, 674, 699
Phosphoserine, 565
Phosphotriose isomerase, in glycolysis, *669*
Photofrin, 548
Photons, in electromagnetic radiation, 244
Photorefractive keratectomy (PRK), 548
Photosynthesis, 96, 113, 672, **692**, 694–695
Photosystem I, 694–695
Photosystem II, 694–695
Physical changes, 3
Physical properties
  of alcohols, 363–364
  of alkanes, 302–304
  of alkenes and alkynes, 323
  of amines, 408–409
  of amino acids, 529
  of carboxylic acids, 441–442
  of cellulose, 486
  of fatty acids, 495–496, 496*t*
  of lipoproteins, 511*t*
  of monosaccharides, 471
Picric acid, 658
Pigs, hemoglobin of, 637, 637*t*
Pilocarpine, 574*t*
Pinpointing technique, in cancer therapy, 260
Piperidine, 404, 405
Pituitary gland, 575*t*, *576*
pK_a, 219–220, 232–233, 234–236, 364, 374, 409–410, 409*t*, 442–443, 444
PKU. *See* Phenylketonuria (PKU)
Plane-polarized light, 395–**396**
Plant proteins, 710–711
Plants, *384*, 446, 641–642, 692, 694–695
Plant waxes, 498
Plaque, cholesterol and, 512
Plasma, 176, 512, 746, 747–748, 747*t*, 751. *See also* Blood plasma
Plasma cells, 482, *723*, 726*t*, 727, 731, 735
Plasma membrane, 510, *647*
Plasmids, in recombinant DNA techniques, *639*–640
Plaster of Paris, 167
Plastics, recycling of, 336
Platelets, 520, *744*, 746, **747**, 747*t*, 748
Platinum (Pt), as hydrogenation catalyst, 332
β-Pleated sheet, 537–540, **538**, 547, 581, 737
Plutonium (Pu), in nuclear waste, 264
PM mutant. *See* Poor metabolism (PM) mutant
Pneumocystis carinii, AIDS and, 631
pOH, 224, 225
Poison hemlock, 405
Poison ivy, *352*
Polar amino acids, 526, 527*t*, 528–529
Polar compounds, 156–157, 371–372, 408–409, 502, 503

Polar covalent bonds, **70**–72, 363
Polarimeter, 396–397, *396*
Polarity
  of aldehydes and ketones, 423
  of amino acids, 526–528
  of carboxylic acids, 441–442
  of cell-membrane lipids, 501–503
  of esters, 445
  of lipids, 495, 497
  of methanol bonds, *363*
  of monosaccharides, 471
  of solvent and solute, 156–157
  of water, 165
Polar molecules, 81–82
Poliomyelitis virus, 630
Polio vaccines, 735
Pollen, as allergen, 723, 739
Polonium-210, -214, and -218, 247, 255
Poly A addition, in transcription, *622*
Polyamides, 455, **456**
Polyatomic elements, 33
Polyatomic ions, 62–63, 62*t*, 69, 97–98
Polycarbonates, 455, **457**–458
Polycythemia, 747*t*
Polyenes, 321–323
Polyesters, 455, **456**–457
Polyethylene, 332–333, 334–335
Polyethylene films, *334*
Poly(ethylene terephthalate) (PET), 336, 346, 376, 456–457
Poly II. *See* RNA polymerase II
Polymerase chain reaction (PCR), *616*–617
Polymerases, *614*, 615, 616–617, 622–624
Polymeric solids, 143*t*
Polymerization, 315, 332–336, 455–458
Polymers, *332*, 333, 333*t*. *See also* Polymerization DNA and RNA as, 597, 601
Polynuclear aromatic hydrocarbons (PAHs), **348**
Polynucleotide chains, in DNA molecule, *603*–605, *606*
Polypeptides, 532, 631, 727, 728
Polypropylene (PP), *334*, 336*t*
Polysaccharides, 466, 467, **481**, 485–487, 666–667, 693–696, 708–709, 727
Polystyrene (PS), *334*, 336*t*
Polysubstituted alkylbenzenes, 346–348
Polyubiquitin, 683
Poly(vinyl chloride) (PVC), 333, 336*t*
Poor metabolism (PM) mutant, of cytochrome P-450 enzyme, 613
p orbitals, 45–46, *46*, 47–49, 50–51, 74–75
Portable dialysis unit, *177*
Positive modulation, enzyme regulation via, 565
Positron emission, *245*, 247
Positron emission tomography (PET) scan, 258–259
Posterior pituitary gland, hormones secreted by, 575*t*, *576*
Postsynaptic membrane, *585*
Postsynaptic nerve endings, *573*, 575
Postsynaptic receptor site, acetylcholine at, 579
Post-translational controls, in gene regulation, 636
Potassium (K), 271, 711*t*, 714*t*
Potassium chloride, 177, 220–221
Potassium dichromate, *164*, 423, 424
Potassium dihydrogen phosphate, 221
Potassium hydroxide, 212*t*, 213, 220–221, 443
Potassium iodide, nuclear fallout and, 263
Potassium ions, 502, 577, 579, 583, 661, 679
Potassium permanganate, solution of, *159*
Potassium sulfate, *106*, 174
Potassium sulfide, in aqueous solution, 106–107
Potential energy, 18
Powders, 190, 446
P (peptide) site, *628*–629
P-protein, action of, 585
Pralidoxime chloride, nerve gas detoxification via, 579

Prandin, for diabetes, 590
Prazosin, 574*t*
Precipitates, 104–*105*, *106*, 106–107, 533
Pre-initiation complex, in protein synthesis, *627*–628
Premarin, 755
Prenylation, 699, 700
Preservatives, carboxylic acid salts as, *443*
Pressure (*P*), 121, 122–123, 124–126, 127–130, 131, 146
  liquids under, 135, 136
  osmotic, 174–177, **175**
  reaction rate and, 190, 204–205
  solubility and, 158–159
Pressure cookers, 138–139
Presynaptic cleft, 578–579
Presynaptic nerve endings, *573*, 575
Preven, as contraceptive, 517
Priestly, Joseph, 373
Primary (1°) alcohols, 278, 361–362, 365, 367–369
Primary (1°) amides, 446
Primary (1°) amines, 279–280, 403, 404, 406
Primary (1°) carbocations, 327
Primary structure
  of DNA, *601–602*
  of nucleic acids, 601–*602*
  of proteins, 533, *534–537*, 542
Primases, *609*, 610*t*, 612
Primatine Mist, 413
Primers, *609*, 612, 616–617
Principal energy level, **44**–45
Prion protein, 539
Priorities, in R,S system, 388–391, 389*t*
Procaine hydrochloride, 412
Products, of chemical reactions, 89
Proenzymes, enzyme regulation via, **563**–564
Progesterone, 513, 515, *516*–517, 575*t*, 592, 593, 755
Prohormones, testosterone and, 514
Prokaryotes, 612, 621, 627, 629, 633
Prolactin, action of, 575*t*, *576*
Proline (Pro, P), 526, 527*t*, 625*t*, 685
Promethazine, 574*t*, 586
Promoters, 622, 633, 634, *635*
Proof, ethanol and, 160
Pro Osteon, 63
Propanal, 442*t*
2-Propanamine, 406
Propane, 94–96, *95*, 99, 288, 303*t*, 305, 364*t*
Propanedioic acid, 439
1,2-Propanediol, 363
1,2,3-Propanetriol, 362, 363, 499
1,2,3-Propanetriol trinitrate, 362. *See also* Nitroglycerin
Propanoic acid, 423*t*, 424, 440*t*
1-Propanol, 360, 364*t*, 442*t*
2-Propanol, 278–279, 328, 360, 386, 451–452
Propanone, 420. *See also* Acetone
2-Propanone, 420
Propantheline, 574*t*
2-Propenal, 420
Propene, 276, *316*, 318, 326, 328, 329, 376
Properties, 3. *See also* Physical properties
  colligative, **172**–178
  of metalloids, 42
  of metals, 40–41
  of mixtures, 171*t*
  of nonmetals, 42
  of subatomic particles, 33*t*
Propionic acid, 440*t*
Propofol, 480
Proportional counter, 251
Propranolol, 574*c*, 588
Propyl alcohol, 360
Propylamine, 407
Propylene, 318, 333*t*
Propylene glycol, 363
Propyl group, 293*t*
Propyne, 317, 318

Prostaglandin A (PGA), 519, 756
Prostaglandin B (PGB), 519
Prostaglandin E (PGE), 519, 756
Prostaglandin endoperoxide synthase (PGHS), 566
Prostaglandin F (PGF), 519
Prostaglandins, 450, 495, 518–519, 566
Prostate cancer, 260–261, 755
Prostate gland, 519, 755
Prostate-specific antigen (PSA), 260, 755
Prosthetic group, in proteins, 545
Protease inhibitor drugs, 631
Proteases, 564, 636, 683
Proteasomes, 636, 683
Protein coats, of viruses, 630
Protein complementation, 710–711
Protein deficiency diseases, 596, 711
Protein kinase, *585*
Protein kinase II, 578
Protein modification, enzyme regulation via, 565
Protein/peptide-conformation-dependent diseases, 539
Proteins, 5, 213, 501, *502, 503,* **525**–549, *534, 538, 542,* 553, 559, 576, 578, 581, 596–597, 615, 636, 666, 667, *668,* 727–728
   amino acids in, **510**, **526**–528, 527*t*, 528–537, 554, 596–597
   chirality of, 397–398
   in cholesterol transport, 510–511
   in citric acid cycle, *652*
   complete, 710
   degradation of misfolded, 636, 683
   digestion of, 679–*680*, 710
   diseases caused by production of excess, 623
   genes and, 597, 620–621
   genetic code and, 621
   incomplete, 710, 711
   in leukocyte mobilization, 736
   as nutrients, 646, 705, 710–711
   peptide bonds in, **530**–532
   in photosynthesis, 694–695
   in plasma, 747
   as receptors, 573
   in ribosomes, 608
   in serum, 747*t*
   steroid hormones and, 592–593
   in T cell receptor complex, 734–735
   transmembrane, 577
   as transporters, **583**
   tRNA and, 607
   ubiquitin and, 683
   in urine, 753
Protein synthesis, *608*, 624–626, **625**, 625*t*, 626–633, 635–636
Proteoglycans, as glycoproteins, 546
Proteolysis, 683
Proteolytic enzymes, of macrophages, 729
Proteomics, 543, *620*, 740
Prothrombin, in clotting, 748
Proton beams, in cancer therapy, 260
Proton channel, in electron and hydrogen ion transport, 659–660
Proton donors and acceptors, 214–216, 528–529
Proton gradient, 658–660, 694
Proton transfer, in acids and bases, 211
Proton translocating ATPase, 658–660
Protonophores, 658
Protons, 33–39, **33**, 33*t*, 36–38, 52, 245, 245*t*, 246, 247
Proto-oncogenes, 638
Protozoa, AIDS and, 631
Proust, Joseph, 32
Proventil, *403*, 413
Provera, 755
Proximal tubules, of kidneys, 752, *753*
Prozac, 574*t*
Prusiner, Stanley, 539
D-Psicose, 469*t*
Psoriasis, as autoimmune disease, 739

Puffer fish, poison of, 299
Pure substances, *28*
Purine, molecular structure of, 404
Purines, as nucleic acid bases, 597–599, 603–605, *604,* 701
Pyramidal molecules, 78
Pyran, 472
Pyranose form, of monosaccharides, 471, **472**–477
Pyrethrin I and II, 446
Pyrethrum powder, 446
Pyridine, 404, 405, 411
Pyridinium acetate, 412
Pyridoxal, *455,* 711*t,* 713*t*
Pyridoxal phosphate (PLP), 455, 711*t*
Pyrimidine, 404
Pyrimidines, as nucleic acid bases, 597–599, 603–605, *604,* 701
Pyrite, *141*
Pyrophosphate ion, 455, 584, *585,* 627, 682
Pyrophosphoric acid, 455
Pyrrolidine, 404
Pyruvate, 427, 556*t,* 561, 668, 669–670, 671, 673*t,* 680, 684, *685, 693,* 701
Pyruvate dehydrogenase, 671, 711*t*
Pyruvate kinase (PK), 561, 565, 668, *669,* 711*t*
Pyruvic acid, 667, *685*. See also Pyruvate
Quantization, 44, *45*
Quarter-stagger arrangement, of collagen fibers, 544, 545
Quartz, *141*
Quaternary molecular structure, of proteins, 534, 540–549, *542,* 559
Rabies, 630, 735
Racemic mixtures, **386**, 394
Rad (radiation absorbed dose), 253–254, 254*t*
Radiation, 242, 245*t,* 251–252, 256*t,* 637
   dosimetry of, 253–256, 254*t*
Radiation badge, *254*
Radiation sickness, 255, 260
Radicals, in autoxidation, 353–355
Radio waves, *244*
Radioactive dating, 249
Radioactive decay curve, *248*
Radioactive fallout, 263
Radioactive isotopes, **245**, 256–261, 257*t*
Radioactive materials, 244–245, 248–251, 250*t,* 256–261
Radioactive nuclei, 245, 250*t*
Radioactivity, 242–243, 244–245, 245*t,* 251–252
Radium-230, 249
Radon (Rn), as health hazard, 255
Radon-222, radiation from, 255
Rancidity, 354, 497
Random coils, in protein secondary structure, *538*–539
Ranitidine, 574*t,* 588
Rapamycin, for autoimmune diseases, 740
Ras protein, prenylation of, 700
Rate constant, 191–192
Rat liver cell, *647*
RDA. *See* Recommended Daily Allowances (RDA)
RDX. *See* Cyclonite (RDX)
Reactants, in chemical reactions, 89, 190
Reaction mechanisms, **327**–328
Reaction rates, 184–206
   of alkene addition reactions, 325–326
   of enzyme activity, 557–559
Receptors, 573, 574*t,* 577, 583, 588, 590, *723,* 724, 739
Recessive genes, 639
Recombinant DNA, 639–642
Recommended Daily Allowances (RDA), 705, 712–715*t,* 716, 717–718
Recycling, of plastics, 336
Red blood cells (RBC), *153,* 175–177, 482–483, 502, 510, *537,* 545, 657, 671, 686–687, *744,* 747, 748. *See also* Erythrocytes

Red cabbage juice, as acid–base indicator, 227
Red litmus paper, *213,* 225
Red No. 40, 351
Redox reactions, **107**–112, 220
Reduced FAD$^+$ (FADH$_2$), *646,* 650, *652,* 654–655, *656,* 657, 658, 660, 675, *676,* 677*t*
Reduced NAD$^+$ (NADH), *646,* 650, *652,* 653–655, *656,* 658, 660, 671, 672–673, 673*t,* 674, 675–677, *676,* 678, 681
Reduced NADP$^+$ (NADPH), 671, 672, 695, 697
Reducing agents, **108**
Reducing sugars, **478**, 479, 481, 484
5-α-Reductase, aging and, 755
Reduction, **107**–112
   of aldehydes and ketones, 425–427
   of alkenes, 332
   of cystine, 530
   of monosaccharides, 477–478
Refining of petroleum, 307–308
Regioselective reactions, **326**
Reglan, 574*t,* 586
Regulation
   of acetylcholinesterase, 582
   of B cells and T cells, 737–738
   of blood pressure, 755–756
   of enzymes, 563–566
   of genes, **633**–636
   as protein function, 526
Regulators, allosteric enzymes and, 565
Regulatory genes, in transcription, 622
Regulatory sites, allosteric enzymes and, 565
Relaxation step, in DNA replication, 612
Release factors, 629, 635–636
Rem (roentgen equivalent for man), 254–256
Reminyl, for Alzheimer's disease, 581
Renin, action of, 575*t*
Repaglinide, for diabetes, 590
Repair, of DNA, 613–616
Replication, **609**, *609*–613
Replication fork, in DNA, *609*–610, 612
Replication origins, on DNA molecule, 609
Replisomes, 610
Reprocessing, of nuclear waste, 264
Reserpine, 574*t*
Resins, 375, 376
Resonance, 344
Resonance hybrid, **344**–345
Resorcinol, 352
Respiration, as redox reaction, 112
Respiratory acidosis, 235
Respiratory alkalosis, 235
Response elements, in gene regulation, 635
Restriction endonucleases, 639–640
Restriction enzymes, in DNA fingerprinting, 611
Retina, 322, 697
Retinal, 322
Retinitis, cytomegalovirus-induced, 623
Retinol, 322, 323
Retinopathy, diabetic, 534
Retroviruses, 631, 638
Reuptake, of glutamic acid, 583
Reverse osmosis, 176
Reverse transcriptase, in AIDS, 631, **632**
Reversibility, of phase changes, 144
Reversible denaturation, 549
Reversible inhibition, 557, 582
Reversible reactions, 194–197
Rheumatoid arthritis, 518, 737, 739
Rhodopsin, 322, *546*
Ribavirin, in hepatitis treatment, 631
Ribitol, in FAD molecular structure, *651*
Riboflavin, 650, *651,* 711–717, 711*t,* 712*t*
Ribonucleoproteins, telomerase as, 610
Ribonucleoside triphosphates, during DNA replication, 612
Ribose, *651, 672*

D-Ribose, 466, 468t, 598–599, 600, 649
α-D-Ribose, 474
β-D-Ribose, 598–599
Ribose 5-phosphate, 672
Ribose reductase, as cofactor, 711t
Ribosomal complex, in protein synthesis initiation, 628
Ribosomal RNA (rRNA), **608**, **621**, 622, 624
Ribosomes, 607, **608**, 624–625, 626–629
Ribozymes, 553, 608, 628
D-Ribulose, 469t
Ribulose 1,5-bisphosphate carboxylase-oxygenase (RuBisCO), 695
Ribulose 5-phosphate, 672
Rice, incomplete proteins in, 710, 711
Ritalin, 574t, 586
Ritonavir, in AIDS treatment, 631
Rituxan, 733
Rivastigmine, for Alzheimer's disease, 581
RNA (ribonucleic acid), 177, 597–600, 600t, **621**–624, 625t, 633, 634, 635, 636
    heterocyclic aromatic amines in, 404
    molecular structure of, 601, 606–608
    NADPH and, 672
    in ribozymes, 553
    translation from, **621**, 624–625, 626–629
    types of, **607–608**
    in viruses, 630, 631
RNA polymerase I, 622
RNA polymerase II, 622, 623–624, **635**
RNA polymerase III, 622
RNA polymerases, 622–624, 633
RNA viruses, 630, 631
Roberts, Richard J., 608
Rock samples, radioactive dating of, 249
Rod cells, in retina, 322
Rodbell, Martin, 584
Roentgen, Wilhelm, 253
Roentgen (R), 253, 254, 254t
"Rotor engine," proton translocating ATPase as, 659
Rough endoplasmic reticulum, 647
Rowland, Sherwood, 307
R (rectus) configuration, in R,S system, 388–391, 389
R,S system, 387–391, **388**, 468
RU486, 516, 517
Rubbing alcohol, 279. See also Isopropyl alcohol
RuBisCO. See Ribulose 1,5-bisphosphate carboxylase-oxygenase (RuBisCO)
Rusting, as redox reaction, 112
Ruthenium (Ru), as hydrogenation catalyst, 332
Ruthenium-103, in nuclear fallout, 263
Rutherford, Ernest, 242, 243
Sabin polio vaccine, 735
Saccharides, 467, 546
Saccharin, 484t, 716
Salicin, 450
Salicylic acid, aspirin and, 450
Saline solution, 177, 178, 470
Salk polio vaccine, 735
Salmeterol, 414
Salt balance, in blood and kidneys, 754
Salt bridge
    in protein tertiary structure, 540, 541
    of a battery, 109
Salting, 173
Salts, 220–222, 747t
    acidic and basic, 226–228, **227**
    amino acids as, 528–529
    of carboxylic acids, 443–444
    neutral, **227**
Sanger, Frederick, 536
Saponification, **451**, 499–500
Saquinavir, in AIDS treatment, 631
Saran Wrap, 334
Sarcoplasmic reticulum (SR), 502, 584
Sarin, 579

Satellites, in DNA, 608–609
Saturated fatty acids, 495–496, 496t, 705
Saturated hydrocarbons, **286**
Saturated solutions, 156
Scanning tunneling microscope, 38
Schwann cells, 506
Science, 4, 5, 6
Scientific method, 4
Scientific notation, 6
Scintillation counters, 251
Scurvy, 705, 711
SDS. See Sodium 4-dodecylbenzenesulfonate (SDS)
Seaborg, Glenn, 261
Sea water, 153, 176
Second (s), 7t, 10
Secondary (2°) alcohols, 278, 279, 361–362, 365, 367–369
Secondary (2°) amides, 446
Secondary (2°) amines, 279, 403, 404, 406
Secondary (2°) carbocations, 327
Secondary messengers, 573, 576, 578, 584, 585, 588–589, 591, 725–727
Secondary structure
    of DNA, 603–605, 606
    of nucleic acids, 601
    of proteins, 533–534, 537–540, 542, 546–549
    of RNA, 607
Secondary sexual characteristics, 515
Second genetic code, 627
Secretases, in Alzheimer's disease, 581
Sedatives, 408, 453
Seeding, 158
Selectins, in leukocyte mobilization, 736
Selegiline, 574t
Selenium (Se), 711t, 715t
Self-ionization of water, 223–224
Self recognition, 737–740
Semen, 518–519, 611, 755
Semiconservative replication, 610
Semipermeable membranes, 174–175, 175–177, 178
Sensor proteins, in gene regulation, 635
Sequences
    of amino acids in proteins, 535–536, 537, 554, 596–597
    of bases in DNA fingerprinting, 611
    in DNA transcription, 622–623
    genome as, 613
    of nucleotides in nucleic acids, 597–598, 601–602, 603–605, 608–609
Serendipity, in science, 5
Serevent, 414
Serine (Ser, S), 504, 527t, 531, 541, 546, 625t, 685
Serotonin, 574t, 583, 586
Serum, 170, 512, 566–567, 567t, 731, 733, 747t, **748**, 755. See also Blood serum
Sex hormones, 513, 515–517, 516, 755
Sexual reproduction, 597
Sharp, Philip A., 608
Sheath cells, 573. See also Myelin sheaths; Schwann cells
Sheep clones, 596
Sheep insulin, primary structure of, 535
Shells
    of atomic electrons, 44–45
    monosaccharides in, 471
Shortenings, production of, 499
Shroud of Turin, radioactive dating of, 249
Shunt, pentose, 672
SI. See International System of Units (SI)
Sickle cell anemia, 536, 537, 609, 636
Side chains
    of nucleic acids, 601–602
    of proteins, 532–534, 540, 541, 543
Side reactions, 103
Sideroblastic anemia, 657
Sievert (Sv), 254, 254t

Sieving portions of AARS, in gene regulation, 635
Signal transduction, **584**, 674
    during DNA replication, 611–612
    in immune system, 734–735, 737–738
    in self recognition, 738–739
Significant figures, 6
Silane, 43–44
Silicon (Si), 42, 47, 100, 271
Silicon tetrachloride, 100
Silk, secondary structure of, 539. See also Spider silk
Silver acetate, 202
Silver chloride, 104–105, 185, 204, 271
Silver cyanate, 271
Silvering of mirrors, 425
Silver-mirror test, for aldehydes, 425
Silver nitrate, 104–105, 184–185
    in Tollens' reagent, 425
Simple lipids, 495–501
Simvastatin, 512
Sinus, of lymph node, 726
60S ribosome, 624, 626t, 628
Size, of atoms, 38–39
Skin, external innate immunity of, 722–723
Skin cancer, xeroderma pigmentosa and, 616
Skunk, thiols in scent of, 372
Smalley, Richard E., 142
Smallpox, 630, 735
Smelting, origin of, 41
Smoke, as colloid, 170
Smooth endoplasmic reticulum, 647
Snake venom, effects on acetylcholinesterase regulation, 582
Soaps, 451, 499, 500
Sobrero, Ascanio, 362
Sodium (Na), 29t, 40, 43, 47, 48, 59–60, 65–66, 271, 457–458, 714t
Sodium acetate, 158, 226–227, 230–231, 233, 443, 451–452, 454
Sodium benzoate, 443, 451–452
Sodium bicarbonate, 216, 221, 352, 443, 444
Sodium borohydride, 426, 477
Sodium butanoate, 443
Sodium carbonate, 362, 443
Sodium chloride, 58, 59–60, 65–66, 156, 157, 160, 172. See also Table salt
    in aqueous solution, 104–105
    in blood serum, 162–163
    as electrolyte, 168
    freezing point depression using, 173
    in isotonic solution, 175
    reaction with silver nitrate, 184–185
Sodium citrate, 748
Sodium dihydrogen phosphate, 221
Sodium D line, in polarimetry, 396–397
Sodium dodecylbenzenesulfonate, 500
Sodium 4-dodecylbenzenesulfonate (SDS), 350–351
Sodium dodecyl sulfate, 500
Sodium ethoxide, 453
Sodium fluoride, 92
Sodium hydroxide, 161–162, 164–165, 212t, 213, 217, 228–230, 231, 232, 352, 374, 443, 451, 452, 454, 455
Sodium ion channels, 299, 577, 579, 583
Sodium ions, 65–66, 93–94, 168, 216, 577, 579, 661, 679, 753, 754, 756
Sodium lactate, 443–444
Sodium lauryl sulfate, 500
Sodium pentobarbital, 453
Sodium phenoxide, 352, 365
Sodium propanoate, 443
Sodium soaps, 499, 500
Sodium sulfate, 93–94, 106
Sodium urate monohydrate, 167, 168
Solar system, isotopic abundances in, 38
Solatol, 574t
Solenoid, DNA molecules arranged in, **605**, 606

Solid hydrates, 166–167
Solidification, 131
Solids, 14–15, 119–121, *120*, 141–147, 143*t*, 154, 154*t*
  from aqueous solution, *105*
  in colloidal systems, 171*t*
  density of, 15–17
  insoluble, 156
  reaction rates in, 190
  solidification to, 131
  soluble, 156
Solid shortenings, production of, 499
Solid soaps, 500
Solid solutions, 154
Solid state, 121
Solubility, 156–159, 155*t*
  of alcohols, 364*t*
  of aldehydes and ketones, 424
  of alkanes, 303–304
  of amine salts, 411
  of carboxylic acids, 441–442
  of ethers, 372
  of ionic compounds in water, 106–107
  of lipids, 494, 495
  of monosaccharides, 471
  of nonpolar compounds, 156–157
  of polar compounds, 156–157
  of proteins, 526, 533
  of soaps, 500
  tooth decay and, 107
Solubility limit, 156
Solutes, 154, 154*t*, 156–157
  colligative properties and, **172**–178
  solution concentration and, 159–165
Solutions, 153–*154*, 154–155, 154*t*, 156, 159–165, *164*
  acidic and basic, 224–228, 230–236
  buffering of, 230–234, 234–236
  colligative properties of, **172**–178
  dilute, 159, 199
  hypertonic and hypotonic, 176–177
  isotonic, *153*, 175–177, 178
  neutral, 220, 224–226
  percent concentration of, **159**–160
Solvated ions, *166*
Solvation layer, 166, 172
Solvents, 154, 154*t*, 156–157, 165–170, 177, 306–307
  colligative properties and, **172**–178
  osmotic pressure and, 174–175
  solution concentration and, 159–165
Soman, 579
Somatic cells, telomers in, 610
Somatic hypermutation (SHM), of V(J)D genes, 732
Sonograph, 258, *259*
Soot, *142*
s orbitals, 45–46, *46*, 47–49, 50–51, 74–75
D-Sorbitol, 477
D-Sorbose, 469*t*
Sour taste, of acids, 220, 442
Soybeans, incomplete proteins in, 710, 711
Specific gravity, 15, 17–18
Specific heat (SH), 20–21, 20*t*, 144
Specificity
  of enzyme action, **554**, 555
  of immune system, 723
Specific rotation, 396–397, 468
Spectator ions, in aqueous solution, *105*
Spectrum, visible, *244*
Speed of light (*c*), 7, 244, 261
Sperm, zygote from, 609
Spermaceti, 498
Sphingolipids, 500–*501*, 505–506, 698
Sphingomyelinase, in Niemann-Pick disease, 509
Sphingomyelins, 502, 505–506, 509, 698
Sphingosine, 500–*501*, 505, 709
Sphygmomanometer, 136
Spider silk, *525*, 539
Spin, of electrons, 46, 47–49

Spiral, left-handed, *382*
Spleen, 687, 725, *726*, 746–747
Splicing, of RNA molecules, 608, *622*
Spongiform encephalitis, 539
Spotted skunk, thiols in scent of, 372
S (*sinister*) configuration, in R,S system, 388–391, *389*
Stable nuclei, 245, 246–247
Stainless steel, as solid solution, 154
Standard temperature and pressure (STP), **127**–128, 131
Starch, 466, 485, 692, 705, 708–709
Starch-like plaques, 539
Start codon, 625*t*, 626
Starvation, 678, 683, 708
States of matter, 14–15, 119–121, *120*
Statin drugs, cholesterol and, 512–513, 699
Steam, *124*–126, 144
Stearic acid, 440*t*, 496*t*, 498*t*, 499, 503, 505, 675–677, *676*, 677*t*
Stearyl-CoA, in stearic acid metabolism, 677*t*
Steel, 112, 154
Stem cells, 610
Step-growth polymerization, 455–458
Stereocenters, **386**, 387–395, 467–471, 528
Stereoisomers, **301**, 382–398
  of amino acids, 528
  with anomeric carbons, 471, *472*, 474, 475, 476, 477
  of monosaccharides, 467–471
Stereospecificity, of enzymes, 554
Steroid hormones, 513–517, 592–593
Steroids, 509–518 576, 592–593
  anabolic, 514, 718
  as lipids, 495
Stoichiometry, **98**–104, 228, 230
Stomach acid, in protein digestion, 710
Stomach fluid, excess acids in, 221
Stomach inflammation, by aspirin, 450
Stomach ulcers, treatment of, 221
Stone Age, 41
Stop codons, 625*t*, 626, 629, 635–636
Strassman, Fritz, 262
Stress, on chemical reactions, 200–206
Stroke, 512, 588, 659
Stroma, *694*
Strong acids, **212**–213, 216–218, 219, 221–222, 231, *232*, 233, 529
Strong bases, **212**–213, 216–218, 231, *232*, 233, 374, 443–444, 529
Strong electrolytes, 169
Strontium-90, 40, 248
Structural formulas, **73**, 274–276, 315–316
  condensed, 278, 279, 288*t*
Structural genes, in transcription, 622
Structural isomers, 289
Structural proteins, 525, 526
Styrene, 333*t*, 345
Subatomic particles, 33–39, 245*t*
Sublimation, 145
Subshells, of atomic electrons, **45**
Substance(s), 7*t*, 28
  anhydrous, 167
  hydrophilic, 500, 501–503, 518, 526–528
  hydrophobic, 495, 500, 501–503, 518, 526–528
  hygroscopic, 167
  pH of common, 226*t*
  radioactive, 244–245
  specific heats of common, 20*t*
  temperature and states of, 15
Substance P, 591
Substitutions
  alkyl, 345–348
  aromatic, 349–351
Substrate
  of an enzyme, 557, 559–562
  of pyruvate kinase, 561
Substrate concentration, 557–558

Subunits
  in collagen, 545
  of proteins, 543
Succinate, *652*, 654
Succinate dehydrogenase, 654, 711*t*
Succinic acid, 439, 449
Succinylcholine, 555, 574*t*, 582
Succinyl-CoA, *652*, 685
Sucralose, 717
Sucrose, 466, 471, 474, 481, 484*t*, 708
Sugar, 121, 157, 536
Sugar beets, sucrose from, 481
Sugar cane, sucrose from, 481
"Sugar-free" products, alditols as, *477*, 478
"Sugarless" candy, xylitol in, 478
Sugars, 484, 484*t*, *707*
  glycation of proteins by, 534
  in glycomics, 740
  in nucleic acids, 598–599, 606
  reducing, **478**, 479, 481, 484
Sulfa drugs, 562
Sulfanilamide, 562
Sulfapyridine, 562
Sulfathiazole, 562
Sulfhydryl group, in thiols, 372–375
Sulfonated polysaccharides, heparin as, 487
Sulfonation, of arenes, 350–351
Sulfonyl urea compounds, for diabetes, 590
Sulfur (S), 29*t*, **31**, 33, 47, 75, *141*, 167, 270, 530, 564, 657, 694
Sulfur-32, 246
Sulfur dioxide, 75, 155
Sulfur trioxide, 155, 169
Sulfuric acid, 75, 155, 203, 212*t*, 213, 228–230, 350–351, 365, 369
Sulfuric esters, acidic polysaccharides as, 486
Sumner, James, 5
Sun, in photosynthesis, 692, 694–695
Sunglasses, adjustable, 204
Sunlight, as health hazard, 255–256
Sun protection factor (SPF), 452
Sunscreens, 718
Sunset Yellow (Yellow No. 6), 351
Supercritical carbon dioxide, 147, 264
Supercritical state, 147
Superimposable molecules. *See* Superposable molecules
Superoxide, 613, 659
Superoxide dismutase, in detoxification, 659
Superposable molecules, 384
Supersaturated solutions, 156, 157–158
Surface tension, *135*
Surgical implants, in bone, 63
Suspensions, 154, 171, 171*t*
Sutures, biodegradable, 457
Svedberg unit (S), 624
Sweet taste, 471, 476, 484*t*
Sweeteners. *See* Artificial sweeteners
Symmetrical anhydrides, 444
Synapses, *573*, **575**, 578–579
Synergistic effects, of alcohol and barbiturates, 453
Synovial fluid, 744
Synthetases, 555, 556*t*, 627
Synthetic fibers, creation of, 457
Synthetic polypeptides, in AIDS treatment, 631
Systéme Internationale (SI), 7
Systolic pressure, 136
Table salt (NaCl), 29
Table sugar, 157, 170, 466, 481, 484*t*
Tacrine, 574*t*, 581
Tagamet, 574*t*, 588
D-Tagatose, 469*t*
Tagging of drugs, 257
D-Talose, 468*t*
Tamoxifen, breast cancer and, 592
TaqI restriction endonuclease, 640–641

TATA box, *622–623*, 633
Tatum, Edward, 597
Tau proteins, in Alzheimer's disease, 581
Taurine, 517, 582–583
Taurocholate, molecular structure of, 518
Tautomers, 430–**431**
Taxol, 273
Tay-Sachs disease, 508, 509
T cell receptor complex, formation of, *734–736*
T cell receptors (TcR), *723*, 724, 728, *734–736*, *738–739*, 740
T cells, *723*, 724, 725, 726*t*, 727, 728, *734–736*, 737–*738*
Technetium-99m, 250, *257*
Teflon, 334
Tellurium (Te), as metalloid, 42
Telomerase and telomers, 610
Temperature (*T*), 7*t*, 9, 10–11, 15, 17
  Charles's law and, 124, 125
  combined gas law and, 125–126
  critical, 147
  effect on enzyme activity, 557, *558*
  entropy and, 120
  Gay-Lussac's law and, 124–126
  of human body, 20, 192, 193
  ideal gas law and, 127–130
  in kinetic molecular theory, 121, 131, 192, *193*
  phase changes and, 143–147
  reaction rate and, 192–193, 203–204
  solubility and, 156, 157–158
  specific heat and, 20–21
  vapor pressure and, 135–*138*
Temperature scales, **11**
Terazosin, 574*t*, 755
Terbutaline, 413, 414
Terephthalic acid, 376, 456–457
Termination, of transcription, 626, 626*t*, 629
Termination sequence, in transcription, 623
Termination site, in transcription, *622*, 623
Terpenes, **323**–325
Tertiary (3°) alcohols, 278, 361–362, 365, 368
Tertiary (3°) amides, 446
Tertiary (3°) amines, 279, 403, 404, 406
Tertiary (3°) carbocations, 327
Tertiary structure
  of proteins, *534*, 540–549, *542*, 636
  of RNA, *607*, 608
Testes, 575*t*, *576*, 755
Testosterone, 513, 514, 515, 517, 575*t*, 755
Tetrachloromethane. *See* Carbon tetrachloride
Tetracycline, 257–260, 559
11-Tetradecenyl acetate, 320
Tetradecanoic acid, 440*t*
Tetrafluoroethylene, polymers of, 333*t*
Tetrahedral molecules, 77–78, 80*t*
Tetrahedron, 77
Tetrahydrofolate, 562
Tetrahydrofuran (THF, C₄H₈O), 371
Tetraodontidae, 299
Tetrapeptides, 531–532
Tetrasaccharides, blood type and, 482
Tetrodotoxin, 299
Tetroses, 467
Theoretical yield, from chemical reactions, 103
Theory, 4, **5**
Therapy. *See also* Chemotherapy; Drugs; Medicine
  anticancer, 601
  gene, 640–641
  iodine, 349
  laser, 548
  monoclonal antibody, 733
  radioisotope, 260–261
Thermal pollution, 158
Thermometers, mercury poisoning from, 563
Thiamine, 711*t*, 712*t*
Thiamine pyrophosphate (TPP), as coenzyme, 711*t*

Thiazides, in hypertension treatment, 756
Thioesters, in fatty acid biosynthesis, 696–697
Thioether linkage, in prenylation, 700
Thiokinase, in fatty acid β-oxidation, *676*
Thiolase, in fatty acid β-oxidation, 675, *676*
Thiolate salts, 374
Thiols, 359, **372**–375
Thiophenol, 372
30S ribosome, 627, *629*
Thorium-234, 249
3TC. *See* (−)-2-Deoxy-3′-thiacytidine (3TC)
Threonine (Thr, T), 527*t*, 625*t*, *685*, 710
Threose, 391–392
D-Threose, 468*t*, 470
L-Threose, 470
Thrombin, in clotting, 748
Thrombocytes, *746*. *See also* Platelets
Thromboplastin, in clotting, 748
Thrombosis, 748
Thromboxanes, 495, 518, 520
Thudichum, Johann, 505
Thylakoids, *694*
Thymine, 597–598, 600*t*, 601–*602*, 603–605, *604*, 622
Thymol, molecular structure of, 352
Thymol blue, as acid–base indicator, 226*t*
Thymus, in immune system, 725, *726*
Thyroid gland, *256*, 257, 575*t*, *576*
Thyroid hormones, action of, 592
Thyroid-stimulating hormone (TSH), *576*
Thyroxine, 349, 575*t*
Time, measuring, 7*t*, 10
Tin (Sn), 41
Tincture, 111, 360
Titanium (Ti), in Earth's crust, *271*
Titanium dioxide, 172, 452
Titration, **228**–230
T lymphocytes, AIDS and, 631, *632*, 735–736. *See also* T cells
TNT (2,4,5-trinitrotoluene), 93, 350, 658
Tofranil, 574*t*
Tolbutamine, for diabetes, 590
Tollens' reagent, aldehydes and, 424–425
Toluene, 345
Toluidine, 406, 411
Tonicity, of solutions, 175–177
Tonsils, 725, *726*
Tooth decay, 107
Tooth enamel, 544
Topoisomerases, during DNA replication, 612
Torr, *121*–122
Torricelli, Evangelista, 121
Toxicity
  of alkaloids, 405
  of ammonia and ammonium ion, 681, 684
  of benzene, 344
  of botulin, 580
  of dioxin, 165
  of ethylene oxide, 371
  of monoclonal antibody therapy, 733
  of nerve gases, 579
  of nitric oxide, 588, 725
Toxins, from *Vibrio cholerae*, 585–587
Trace elements, 29*t*, 711*t*, 714–715*t*
Tranquilizers, amines in, 408
Transaminase, in amino acid catabolism, 680
Transamination, in amino acid catabolism, **680**
Transcription, **621**–624, *622*
Transcriptional level gene regulation, 633–635
Transcription factors, 623, 633–635
Transenoyl CoA, in stearic acid metabolism, 677*t*
Transferases, action of, 555, 556*t*
Transfer RNA (tRNA), 556*t*, **607**–608, *607*, 624–625, 626–633, 635, 636
  in transcription and translation, **621**, 624
Transgenic mice, 634

Transition elements, 40, 51
Transition metal catalysts, in aldehyde reduction, 426
Transition state, in chemical reactions, 188, *189*
Translation, **621**, 624–625, 626–633
Translational level gene regulation, 633, 635–636
Translocation, in protein elongation, 629
Translocator inner membrane (TIM) complexes, 648
Translocator outer membrane (TOM) channels, 658
Transmembrane channel proteins, *503*, 577
Transmembrane channels, 502
Transmutation, of elements, 246–247
Transmutation series, for uranium-238, 249
Transparency
  of colloids, 170–171
  of solutions, 153, *154*, 155
Transpeptidase, 562
Transplant rejection, 747*t*
Transport
  of chemicals across cell membranes, 501–503
  via microtubules, 545
  proteins for, 526
Transporters, **583**, 586
Transthyretin, 539
Transuranium elements, 261–262
2,4,6-Tribromophenol, 346
Tricarboxylic acid cycle, 649. *See also* Citric acid cycle
Trichloroacetic acid, 442
Trichlorofluoromethane, 306
Trichloromethane, 305
Trienes, 321
Triethylamine, 407
Triethylammonium chloride, 408
Triglycerides, 495–500, 510–511, 709
Trigonal planar molecules, 77, *78*, 79, 80*t*
2,3,4-Trihydroxybutanal, 391–392
Triiodothyronine, 575*t*
Trimethylamine, 279, 403
2,2,4-Trimethylpentane, 304, 308
Trinitroglycerin, 350. *See also* Nitroglycerin
Triols, 363
Triose phosphates, in photosynthesis, 695
Trioses, 467
Tripeptides, 531–532, 535
Triphosphate ion, 455
Triple bond, **72–74**, 77, 274, 275, 286, 287, 314
Triple helix, *540*, 544, 545, 547
Triple point, 146
Triprotic acids, **215**
Trisaccharides, 481, 482
Tritium, 245, 257–260, 261
tRNA$_i$^{Met} transfer RNA, 627, *628*
Tropocollagen, 545
Trypsin, 554, 555, 556, 563–564, 667, 710
Trypsinogen, 563–564
Tryptophan (Trp, W), 527*t*, 625*t*, *685*, 701, 710
Tubules, of kidneys, 752, *753*
α-Tubulin, in microtubules, 544
β-Tubulin, in microtubules, 544
Tumor cells, 733, 737
Tumor necrosis factor (TNF), 736
Tumors, oncogenes and, 638
Tumor suppressor genes, 638
12:13 dEpoB, 735
Tylenol, 566
Tyndall effect, **170**–171
Type I diabetes, 590
Type II diabetes, 590
Typhoid fever, 747*t*
Tyrosine, 527*t*, *542*, 625*t*, *685*
L-Tyrosine, 556*t*
Tyrosine kinase, in B cell and T cell regulation, 738
Tyrosine-tRNA synthetase, 556*t*
L-TyrosyltRNA, 556*t*
Ubiquinone, 656

Ubiquitin, 683
Ubiquitinylation, 683
UDP-glucose, 692, 693–696
Ulcers, 559, 588
Ultra-extensive metabolism (UEM) mutant of cytochrome P-450 enzyme, 613
Ultrasound scan, 258, *259*
Ultraviolet (UV) radiation, *244*, 255–256, 307, 452, 497, 613, 637. *See also* UVA and UVB radiation
UMP. *See* Uridine 5′-monophosphate (UMP)
Units of measurement, 6–11, 7t, 8t, 11–14
 in radiation dosimetry, 253–256, 254t
Unsaturated fatty acids, 495–496, 496t, 498–499, 501–503, 675, 697
Unsaturated hydrocarbons, 286, 314
Unsaturated solutions, 156
Unwinding step, in DNA replication, 612
Uracil, 597–598, 600t, 622, 625, 695, 698
Uranium (U), 242, 249, 255, 261, *262–263*, 264
Urea, 90, 101, 271, 273, 554, *668*, *680*, 681, *682*, 683–684
 protein denaturation by, 547
 in serum, 747t
 in urine, 752–753
Urea cycle, 588, 667, *668*, *680*, 681–684, *682*
Urease, 101, 559, 711t
Uric acid, nitrogen excretion via, 681
Uridine, 600t
Uridine diphosphate (UDP), 692, 693–696
Uridine 5′-monophosphate (UMP), 600t
Urine, 170, 177, 744, 751, 752–753
 blue, 701
Urine tests
 for glucose, 479
 for anabolic steroids, 514
Urinometer, **17**
Urobilin, heme catabolism and, 687
Uronic acids, 479–480
Urushiol, 352
UVA and UVB radiation, 452
Vaccination (vaccines), 630, 631, 735
Valence electrons, **49**–50, 72–74, 76, 77–80
Valence-shell electron-pair repulsion (VSEPR) model, 77–80, 274–276, 296, 315–316
Valence shells, **49**–50, 63–64, 64–65, 65–66, 75
Valeric acid, 440t

Valine (Val, V), 527t, 625t, 636, *685*
Valium, 408
Vanilla beans, *352*, 422
Vanillin, 352, 422
Vapor pressure, 135–**138**
Variola, 735. *See also* Smallpox
Variolation, 735
Vaseline, 302
Vasopressin, 536, 575t, 576, 585, 589, 591, 754
V domains, in T cell receptor complex, *734*
Vegetable oils, 495, *497*, 498t
Vegetarian diets, 706, 710–711
Velocity, kinetic energy and, 18
Venom, effects on acetylcholinesterase regulation, 582
Venous blood, *750*
Very-low-density lipoprotein (VLDL), 510–511
Viagra, nitric oxide and, 588
Vibrio cholerae, G-protein and, 585–587
Vidarabine, 630
Vinegar, acetic acid in, 160, 213
Vinyl chloride, 333, 333t, *334*, 637
VIOXX, 518
Vira-A, molecular structure of, 630
Viruses, 254, 630, 638, 641, 683, 722, 723, 724, 725, 726t, *728–729*
 AIDS, 631–632, 634
Vision, *cis-trans* isomerism in, 322
Visual purple, 322
Vitamin A, 322, 323, 325, 539, 712t, 716
Vitamin A aldehyde, 322
Vitamin B$_1$, 711t, 712t
Vitamin B$_2$, 711–717, 711t, 712t
Vitamin B$_6$, 455, 711t, 713t
Vitamin B$_{12}$, 711t, 713t
Vitamin B complex, 557
Vitamin C, 274, 473, 704, 711, 714t, 716
Vitamin D, *711*, 712t, 716
Vitamin E, 354–355, 711, 712t, 716
Vitamin K, 712t, 748
Vitamins, 705, 711–718, 711t, 712–715t, 747t
Vitravene, 623
V(J)D (variable-join-diverse) recombination, *732*
VLDL. *See* Very-low-density lipoprotein (VLDL)
Voltaic cells, 109, 112
Volume (V), 7t, 8, 8t, 15, *120*, 122–123, 124, 125–126, 127–130, 131, 135

von Baeyer, Adolph, 453
von Euler, Ulf S., 518
VSEPR model, 77–80, 274–276, 296, 315–316
Waage, Peter, 197
Water, 29, *30*, 32, *73*, *78*, 80t, 81, 138–139, *138*, 139t, 143–145, *144*, 146, 156, 160, 165, 170, 173, 177, *187*, 188, 199–200, 203, 210–212, 219, 226–228, 230–*232*, 409, 466–467, *646*, 657–658, 660, 744
Water balance, in blood and kidneys, 754
Water of hydration, 166–167
Water-soluble vitamins, 712–714t
Watson, James, 603, *604*
Wavelength (λ), 243–244
Waxes, 302, 495, 498
Weak acids, **212**–213, 216–218, 219, 219t, 230–231, 233, 234–236, 352, 364–365, 374–375, 442–443
Weak bases, **212**–213, 216–218, 230–231, 233, 374, 409–411
Weak electrolytes, 169
Weight, mass and, 9
White blood cells, 520, 659, 724, *744*, 747. *See also* B cells; Leukocytes; Lymphocytes; Macrophages; Natural killer (NK) cells; T cells
White matter, in brain and spinal cord, 506
Whole blood, 746, 747t. *See also* Blood
Wilkins, Maurice, 603
Wöhler, Friedrich, 271–283
Wood alcohol, 375. *See also* Methanol
Work, energy and, 18
Xeroderma pigmentosa (XP), 616
X-ray crystallography, of glyceraldehyde, 469
X-ray diffraction studies, 603, 629
X rays, 244, 245, 252, 254, 260, 637, 638
Xylene, 346
Xylitol, 477–478
D-Xylose, 468t
Zafirlukast, 520
Zalcitabone, in AIDS treatment, *632*
Zantac, 574t, 588
Zeno of Elea, 27–28
Ziegler, Karl, 334
Zileuton, 520
Zinc (Zn), 29t, 41, 107–109, 634, *635*, 711t, 715t
Zinc fingers, in gene regulation, 634, *635*
Zwitterions, 528–529, 533, 680
Zymogens, **563**–564, 615

Standard Atomic Weights of the Elements 1997
Based on Relative Atomic Mass of $^{12}C = 12$

Name	Symbol	Atomic Number	Atomic Weight
Actinium*	Ac	89	(227)
Aluminum	Al	13	26.981538
Americium*	Am	95	(243)
Antimony	Sb	51	121.760
Argon	Ar	18	39.948
Arsenic	As	33	74.92160
Astatine*	At	85	(210)
Barium	Ba	56	137.327
Berkelium*	Bk	97	(247)
Beryllium	Be	4	9.012182
Bismuth	Bi	83	208.98038
Bohrium*	Bh	107	(264)
Boron	B	5	10.811
Bromine	Br	35	79.904
Cadmium	Cd	48	112.411
Calcium	Ca	20	40.078
Californium*	Cf	98	(251)
Carbon	C	6	12.0107
Cerium	Ce	58	140.116
Cesium	Cs	55	132.90545
Chlorine	Cl	17	35.4527
Chromium	Cr	24	51.9961
Cobalt	Co	27	58.933200
Copper	Cu	29	63.546
Curium*	Cm	96	(247)
Dubnium*	Db	105	(262)
Dysprosium	Dy	66	162.50
Einsteinium*	Es	99	(252)
Erbium	Er	68	167.26
Europium	Eu	63	151.964
Fermium*	Fm	100	(257)
Fluorine	F	9	18.9984032
Francium*	Fr	87	(223)
Gadolinium	Gd	64	157.25
Gallium	Ga	31	69.723
Germanium	Ge	32	72.61
Gold	Au	79	196.96655
Hafnium	Hf	72	178.49
Hassium*	Hs	108	(269)
Helium	He	2	4.002602
Holmium	Ho	67	164.93032
Hydrogen	H	1	1.00794
Indium	In	49	114.818
Iodine	I	53	126.90447
Iridium	Ir	77	192.217
Iron	Fe	26	55.845
Krypton	Kr	36	83.80
Lanthanum	La	57	138.9055
Lawrencium*	Lr	103	(262)
Lead	Pb	82	207.2
Lithium	Li	3	6.941
Lutetium	Lu	71	174.967
Magnesium	Mg	12	24.3050
Manganese	Mn	25	54.938049
Meitnerium*	Mt	109	(268)
Mendelevium*	Md	101	(258)
Mercury	Hg	80	200.59
Molybdenum	Mo	42	95.94
Neodymium	Nd	60	144.24
Neon	Ne	10	20.1797
Neptunium*	Np	93	(237)
Nickel	Ni	28	58.6934
Niobium	Nb	41	92.90638
Nitrogen	N	7	14.00674
Nobelium*	No	102	(259)
Osmium	Os	76	190.23
Oxygen	O	8	15.9994
Palladium	Pd	46	106.42
Phosphorus	P	15	30.973762
Platinum	Pt	78	195.078
Plutonium*	Pu	94	(244)
Polonium*	Po	84	(210)
Potassium	K	19	39.0983
Praseodymium	Pr	59	140.90765
Promethium*	Pm	61	(145)
Protactinium*	Pa	91	231.03588
Radium*	Ra	88	(226)
Radon*	Rn	86	(222)
Rhenium	Re	75	186.207
Rhodium	Rh	45	102.90550
Rubidium	Rb	37	85.4678
Ruthenium	Ru	44	101.07
Rutherfordium*	Rf	104	(261)
Samarium	Sm	62	150.36
Scandium	Sc	21	44.955910
Seaborgium*	Sg	106	(266)
Selenium	Se	34	78.96
Silicon	Si	14	28.0855
Silver	Ag	47	107.8682
Sodium	Na	11	22.989770
Strontium	Sr	38	87.62
Sulfur	S	16	32.066
Tantalum	Ta	73	180.9479
Technetium*	Tc	43	(98)
Tellurium	Te	52	127.60
Terbium	Tb	65	158.92534
Thallium	Tl	81	204.3833
Thorium*	Th	90	232.0381
Thulium	Tm	69	168.93421
Tin	Sn	50	118.710
Titanium	Ti	22	47.867
Tungsten	W	74	183.84
Uranium*	U	92	238.0289
Vanadium	V	23	50.9415
Xenon	Xe	54	131.29
Ytterbium	Yb	70	173.04
Yttrium	Y	39	88.90585
Zinc	Zn	30	65.39
Zirconium	Zr	40	91.224
—†		110	(269)
—†		111	(272)
—†		112	(277)
—†		114	(285)
—†		116	(289)

* Elements with no stable nuclide; the value given in parentheses is the atomic mass number of the isotope of longest known half-life. However, three such elements (Th, Pa, and U) have a characteristic terrestrial isotopic composition, and the atomic weight is tabulated for these.

† Not yet named.